A Brief Guide to **Getting the Most** from This Book

1 Read the Book

Feature	Description	Benefit
Applications Using Real-World Data	From the chapter and section openers through the examples and exercises, interesting applications from nearly every discipline, supported by up-to-date real-world data, are included in every section.	Ever wondered how you'll use algebra? This feature will show you how algebra can solve real problems.
Detailed Worked-Out Examples	Examples are clearly written and provide step-by-step solutions. No steps are omitted, and key steps are thoroughly explained to the right of the mathematics.	The blue annotations will help you to understand the solutions by providing the reason why the algebraic steps are true.
Explanatory Voice Balloons	Voice balloons help to demystify algebra. They translate algebraic language into plain English, clarify problem-solving procedures, and present alternative ways of understanding.	Does math ever look foreign to you? This feature translates math into everyday English.
Great Question!	Answers to students' questions offer suggestions for problem solving, point out common errors to avoid, and provide informal hints and suggestions.	This feature should help you not to feel anxious or threatened when asking questions in class.
Achieving Success	The book's Achieving Success boxes offer strategies for success in learning algebra.	Follow these suggestions to help achieve your full academic potential in mathematics.

2 Work the Problems

Feature	Description	Benefit
Check Point Examples	Each example is followed by a similar problem, called a Check Point, that offers you the opportunity to work a similar exercise. Answers to all Check Points are provided in the answer section.	You learn best by doing. You'll solidify your understanding of worked examples if you try a similar problem right away to be sure you understand what you've just read.
Concept and Vocabulary Checks	These short-answer questions, mainly fill-in-the blank and true/false items, assess your understanding of the definitions and concepts presented in each section.	It is difficult to learn algebra without knowing its special language. These exercises test your understanding of the vocabulary and concepts.
Extensive and Varied Exercise Sets	An abundant collection of exercises is included in an Exercise Set at the end of each section. Exercises are organized within several categories. Practice Exercises follow the same order as the section's worked examples. Practice PLUS Exercises contain more challenging problems that often require you to combine several skills or concepts.	The parallel order of the Practice Exercises lets you refer to the worked examples and use them as models for solving these problems. Practice PLUS provides you with ample opportunity to dig in and develop your problem-solving skills.

3 Review for Quizzes and Tests

Feature	Description	Benefit
Mid-Chapter Check Points	Near the midway point in the chapter, an integrated set of review exercises allows you to review the skills and concepts you learned separately over several sections.	Combining exercises from the first half of the chapter gives you a comprehensive review before you continue on.
Chapter Review Charts	Each chapter contains a review chart that summarizes the definitions and concepts in every section of the chapter, complete with examples.	Review this chart and you'll know the most important material in the chapter.
Chapter Tests	Each chapter contains a practice test with problems that cover the important concepts in the chapter. Take the test, check your answers, and then watch the Chapter Test Prep Videos.	You can use the Chapter Test to determine whether you have mastered the material covered in the chapter.
Chapter Test Prep Videos	These videos contain worked-out solutions to every exercise in each chapter test.	These videos let you review any exercises you miss on the chapter test.
Lecture Series	These interactive lecture videos highlight key examples from every section of the textbook.	These videos let you review each objective from the textbook that you need extra help on.

Get the most out of MyMathLab®

MyMathLab, Pearson's online learning management system, creates personalized experiences for students and provides powerful tools for instructors. With a wealth of tested and proven resources, each course can be tailored to fit your specific needs. Talk to your Pearson Representative about ways to integrate MyMathLab into your course for the best results.

Data-Driven Reporting for Instructors

- MyMathLab's comprehensive online gradebook automatically tracks students' results to tests, quizzes, homework, and work in the study plan.

- The Reporting Dashboard, found under More Gradebook Tools, makes it easier than ever to identify topics where students are struggling, or specific students who may need extra help.

Learning in Any Environment

- Because classroom formats and student needs continually change and evolve, MyMathLab has built-in flexibility to accommodate various course designs and formats.

- With a new, streamlined, mobile-friendly design, students and instructors can access courses from most mobile devices to work on exercises and review completed assignments.

Visit www.mymathlab.com and click Get Trained to make sure you're getting the most out of MyMathLab.

SEVENTH EDITION

Intermediate Algebra
for College Students

Robert Blitzer

Miami Dade College

PEARSON

Boston • Columbus • Indianapolis • New York • San Francisco
Amsterdam • Cape Town • Dubai • London • Madrid • Milan • Munich • Paris • Montréal • Toronto
Delhi • Mexico City • São Paulo • Sydney • Hong Kong • Seoul • Singapore • Taipei • Tokyo

Editorial Director: *Chris Hoag*

Editor-in-Chief: *Michael Hirsch*

Senior Acquisitions Editor: *Dawn Giovanniello*

Editorial Assistant: *Megan Tripp*

Program Manager: *Beth Kaufman*

Project Manager: *Kathleen A. Manley*

Program Management Team Lead: *Karen Wernholm*

Project Management Team Lead: *Christina Lepre*

Media Producer: *Shana Siegmund*

TestGen Content Manager: *Marty Wright*

MathXL Content Developer: *Rebecca Williams and Eric Gregg*

Product Marketing Manager: *Alicia Frankel*

Field Marketing Managers: *Jenny Crum and Lauren Schur*

Marketing Assistant: *Alexandra Habashi*

Senior Author Support/Technology Specialist: *Joe Vetere*

Rights and Permissions Project Manager: *Gina Cheselka*

Procurement Specialist: *Carol Melville*

Associate Director of Design: *Andrea Nix*

Program Design Lead: *Beth Paquin*

Composition: *codeMantra*

Illustrations: *Scientific Illustrators*

Cover Design: *Studio Montage*

Cover Image: brick wall painted white: *Raimund Linke/ Getty;* brick wall with ladder: *nejuras/Shutterstock;* concrete wall: *ilbusca/Getty;* lemons: *RedHelga/Getty*

Library of Congress Cataloging-in-Publication Data

Blitzer, Robert.
Intermediate algebra for college students/Robert F. Blitzer, Miami Dade College. —7th edition.
    pages cm
ISBN 978-0-13-417894-3
1. Algebra—Textbooks.   I. Title.

QA154.3.B584 2017

512.9—dc23

2015031909

2   16

www.pearsonhighered.com

ISBN 13: 978-0-13-417894-3
ISBN 10: 0-13-417894-7

Table of Contents

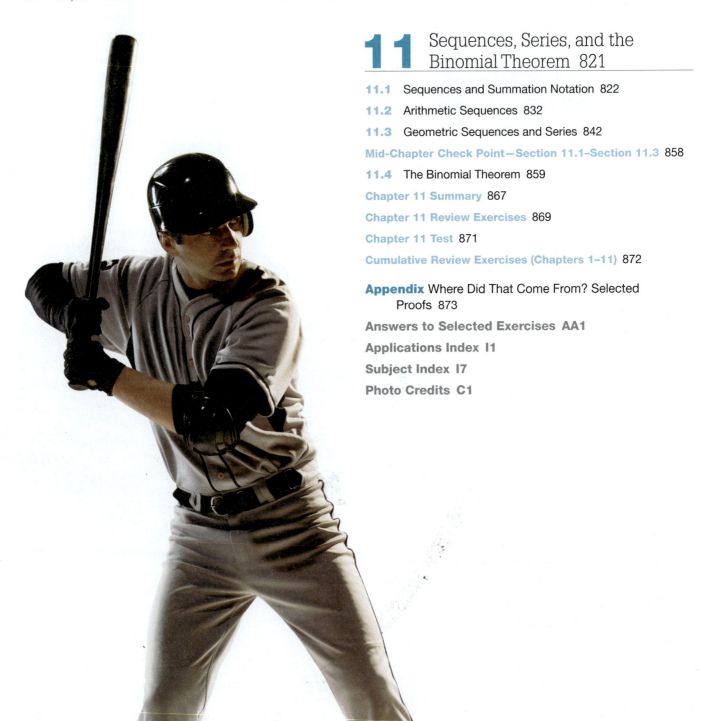

Preface

Intermediate Algebra for College Students, Seventh Edition, provides comprehensive, in-depth coverage of the topics required in a one-term course in intermediate algebra. The book is written for college students who have had a course in introductory algebra or who have demonstrated proficiency in the objectives of such a course. I wrote the book to help diverse students with different backgrounds and career plans succeed in intermediate algebra. *Intermediate Algebra for College Students,* Seventh Edition, has two primary goals:

1. To help students acquire a solid foundation in the skills and concepts of intermediate algebra needed for success in future STEM and non-STEM directed math courses.
2. To show students how algebra can model and solve authentic real-world problems.

One major obstacle in the way of achieving these goals is the fact that very few students actually read their textbook. This has been a regular source of frustration for me and for my colleagues in the classroom. Anecdotal evidence gathered over years highlights two basic reasons students give when asked why they do not take advantage of their textbook:

- "I'll never use this information."
- "I can't follow the explanations."

I've written every page of the Seventh Edition with the intent of eliminating these two objections. The ideas and tools I've used to do so are described in the features that follow. These features and their benefits are highlighted for the student in "A Brief Guide to Getting the Most from This Book," which appears inside the front cover.

What's New in the Seventh Edition?

- **New Applications and Real-World Data.** The Seventh Edition contains 131 new or revised worked-out examples and exercises based on updated and new data sets. Many of these applications involve topics relevant to college students and newsworthy items. Among topics of interest to college students, you'll find new and updated data sets describing student loan debt (Section 1.1, Example 4), tuition costs (Section 1.4, Example 7; Exercise Set 1.4, Exercises 67–68), dormitory charges (Exercise Set 11.2, Exercises 65–66), the number of hours college students study, by major (Exercise Set 3.3, Exercises 39–40), sleep hours of college students (Exercise Set 3.2, Exercise 10), college students and video games (Chapter 6 opener, Section 6.7 opener; Section 6.7, Example 3), and self-assesment of physical health by college freshmen (Chapter 7 Review, Exercise 85). Among newsworthy items, new applications range from the frivolous (Hamachiphobia: correlating those who won't try sushi and those who don't approve of marriage equality: Exercise Set 9.6, Exercise 41) to the weighty (mentally ill adults in the United States: Chapter 9 Review Exercise 90). Other new and updated applications address climate change (Section 2.4, Example 10 and Check Point 10), the war on drugs and nonviolent inmates in federal prisons (Exercise Set 3.1, Exercise 94; Exercise Set 10.5, Exercise 63), and the changing U.S. population by race/ethnicity (Section 11.2, Example 3).

- **New Blitzer Bonus Videos with Assessment.** The Blitzer Bonus features throughout the textbook have been turned into animated videos that are built into the MyMathLab course. These videos help students make visual connections to algebra and the world around them. Assignable exercises have been created within the MyMathLab course to assess conceptual understanding and mastery. These videos and exercises can be turned into a media assignment within the Blitzer MyMathLab course.

- **Updated Learning Guide.** Organized by the textbook's learning objectives, this updated Learning Guide helps students learn how to make the most of their textbook for test preparation. Projects are now included to give students an opportunity to discover and reinforce the concepts in an active learning environment and are ideal for group work in class.

- **Updated Graphing Calculator Screens.** All screens have been updated using the TI-84 Plus C.

What's New in the Blitzer Developmental Mathematics Series?

Two new textbooks and MyMathLab courses have been added to the series:

- *Developmental Mathematics,* First Edition, is intended for a course sequence covering prealgebra, introductory algebra, and intermediate algebra. The text provides a solid foundation in arithmetic and algebra.

- *Pathways to College Mathematics,* First Edition, provides a general survey of topics to prepare STEM and non-STEM students for success in a variety of college math courses, including college algebra, statistics, liberal arts mathematics, quantitative reasoning, finite mathematics, and mathematics for education majors. The prerequisite is basic math or prealgebra.

- *MyMathLab with Integrated Review* courses are also available for select Blitzer titles. These MyMathLab courses provide the full suite of resources for the core textbook, but also add in study aids and skills check assignments keyed to the prerequisite topics that students need to know, helping them quickly get up to speed.

What Familiar Features Have Been Retained in the Seventh Edition of *Intermediate Algebra for College Students?*

- **Learning Objectives.** Learning objectives, framed in the context of a student question (What am I supposed to learn?), are clearly stated at the beginning of each section. These objectives help students recognize and focus on the section's most important ideas. The objectives are restated in the margin at their point of use.

- **Chapter-Opening and Section-Opening Scenarios.** Every chapter and every section open with a scenario presenting a unique application of mathematics in students' lives outside the classroom. These scenarios are revisited in the course of the chapter or section in an example, discussion, or exercise.

- **Innovative Applications.** A wide variety of interesting applications, supported by up-to-date, real-world data, are included in every section.

- **Detailed Worked-Out Examples.** Each example is titled, making the purpose of the example clear. Examples are clearly written and provide students with detailed step-by-step solutions. No steps are omitted and each step is thoroughly explained to the right of the mathematics.

- **Explanatory Voice Balloons.** Voice balloons are used in a variety of ways to demystify mathematics. They translate algebraic ideas into everyday English, help clarify problem-solving procedures, present alternative ways of understanding concepts, and connect problem solving to concepts students have already learned.

- **Check Point Examples.** Each example is followed by a similar matched problem, called a Check Point, offering students the opportunity to test their understanding of the example by working a similar exercise. The answers to the Check Points are provided in the answer section.

- **Concept and Vocabulary Checks.** This feature offers short-answer exercises, mainly fill-in-the-blank and true/false items, that assess students' understanding of the definitions and concepts presented in each section. The Concept and Vocabulary Checks appear as separate features preceding the Exercise Sets.

- **Extensive and Varied Exercise Sets.** An abundant collection of exercises is included in an Exercise Set at the end of each section. Exercises are organized within eight category types: Practice Exercises, Practice Plus Exercises, Application Exercises, Explaining the Concepts, Critical Thinking Exercises, Technology Exercises, Review Exercises, and Preview Exercises. This format makes it easy to create well-rounded homework assignments. The order of the Practice Exercises is exactly the same as the order of the section's worked examples. This parallel order enables students to refer to the titled examples and their detailed explanations to achieve success working the Practice Exercises.

- **Practice Plus Problems.** This category of exercises contains more challenging practice problems that often require students to combine several skills or concepts. With an average of ten Practice Plus problems per Exercise Set, instructors are provided with the option of creating assignments that take Practice Exercises to a more challenging level.

- **Mid-Chapter Check Points.** At approximately the midway point in each chapter, an integrated set of Review Exercises allows students to review and assimilate the skills and concepts they learned separately over several sections.

- **Graphing and Functions.** Graphing is introduced in Chapter 1 and functions are introduced in Chapter 2, with an integrated graphing functional approach emphasized throughout the book. Graphs and functions that model data appear in nearly every section and Exercise Set. Examples and exercises use graphs of functions to explore relationships between data and to provide ways of visualizing a problem's solution. Because functions are the core of this course, students are repeatedly shown how functions relate to equations and graphs.

- **Integration of Technology Using Graphic and Numerical Approaches to Problems.** Side-by-side features in the technology boxes connect algebraic solutions to graphic and numerical approaches to problems. Although the use of graphing utilities is optional, students can use the explanatory voice balloons to understand different approaches to problems even if they are not using a graphing utility in the course.

- **Great Question!** This feature presents a variety of study tips in the context of students' questions. Answers to questions offer suggestions for problem solving, point out common errors to avoid, and provide informal hints and suggestions. As a secondary benefit, this feature should help students not to feel anxious or threatened when asking questions in class.

- **Achieving Success.** The Achieving Success boxes at the end of many sections offer strategies for persistence and success in college mathematics courses.

- **Chapter Review Grids.** Each chapter contains a review chart that summarizes the definitions and concepts in every section of the chapter. Examples that illustrate these key concepts are also included in the chart.

- **End-of-Chapter Materials.** A comprehensive collection of Review Exercises for each of the chapter's sections follows the review grid. This is followed by a Chapter Test that enables students to test their understanding of the material covered in the chapter. Beginning with Chapter 2, each chapter concludes with a comprehensive collection of mixed Cumulative Review Exercises.

- **Blitzer Bonuses.** These enrichment essays provide historical, interdisciplinary, and otherwise interesting connections to the algebra under study, showing students that math is an interesting and dynamic discipline.

- **Discovery.** Discover for Yourself boxes, found throughout the text, encourage students to further explore algebraic concepts. These explorations are optional and their omission does not interfere with the continuity of the topic under consideration.

I hope that my passion for teaching, as well as my respect for the diversity of students I have taught and learned from over the years, is apparent throughout this new edition. By connecting algebra to the whole spectrum of learning, it is my intent to show students that their world is profoundly mathematical, and indeed, π is in the sky.

Robert Blitzer

Resources for Success
MyMathLab for the Blitzer Developmental Algebra Series

MyMathLab is available to accompany Pearson's market-leading text offerings. This text's flavor and approach are tightly integrated throughout the accompanying MyMathLab course, giving students a consistent tone, voice, and teaching method that make learning the material as seamless as possible.

Section Lecture and Chapter Test Prep Videos

An **updated** video program provides a multitude of resources for students. Section Lecture videos walk students through the concepts from every section of the text in a fresh, modern presentation format. Chapter Test Prep videos walk students through the solution of every problem in the text's Chapter Tests, giving students video resources when they might need it most.

Blitzer Bonus Videos

NEW! Animated videos have been created to mirror the Blitzer Bonus features throughout the textbook. Blitzer Bonus features in the text provide interesting real-world connections to the mathematical topics at hand, conveying Bob Blitzer's signature style to engage students. These new Blitzer Bonus videos will help students to connect the topics to the world around them in a visual way. Corresponding assignable exercises in MyMathLab are also available, allowing these new videos to be turned into a media assignment to truly ensure that students have understood what they've watched.

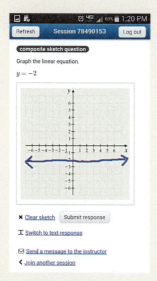

Learning Catalytics

Integrated into MyMathLab, the Learning Catalytics feature uses students' devices in the classroom for an engagement, assessment, and classroom intelligence system that gives instructors real-time feedback on student learning. Learning Catalytics contains Pearson-created content for developmental math topics that allows you to take advantage of this exciting technology immediately.

Student Success Modules

These modules are integrated within the MyMathLab course to help students succeed in college courses and prepare for future professions.

www.mymathlab.com

Resources for Success

Instructor Resources

Annotated Instructor's Edition

This version of the text contains answers to exercises printed on the same page, with graphing answers in a special Graphing Answer Section at the back of the text.

The following resources can be downloaded from www.pearsonhighered.com or in MyMathLab.

PowerPoint® Lecture Slides

Fully editable slides correlated with the textbook include definitions, key concepts, and examples for use in a lecture setting.

Instructor's Solutions Manual

This manual includes fully worked-out solutions to all text exercises.

Instructor's Resource Manual

This manual includes a Mini-Lecture, and Additional Exercises for every section of the text. It also includes Chapter Test forms, as well as Cumulative and Final Exams, with answers.

TestGen®

TestGen® (www.pearsoned.com/testgen) enables instructors to build, edit, print, and administer tests using a computerized bank of questions developed to cover all the objectives of the text.

Student Resources

The following additional resources are available to support student success:

Learning Guide

UPDATED! Organized by learning objectives, the Learning Guide helps students make the most of their textbook and prepare for tests. Now updated to include projects, students will have the opportunity to discover and reinforce the concepts in an active learning environment. These projects are ideal for group work in class. The Learning Guide is available in MyMathLab, and available as a printed supplement.

Video Lecture Series

Available in MyMathLab, the video program covers every section in the text, providing students with a video tutor at home, in lab, or on the go. The program includes Section Lecture Videos and Chapter Test Prep videos.

Student Solutions Manual

This manual provides detailed, worked-out solutions to odd-numbered section exercises, plus all Check Points, Review/Preview Exercises, Mid-Chapter Check Points, Chapter Reviews, Chapter Tests, and Cumulative Reviews.

www.mymathlab.com

Acknowledgments

An enormous benefit of authoring a successful series is the broad-based feedback I receive from the students, dedicated users, and reviewers. Every change to this edition is the result of their thoughtful comments and suggestions. I would like to express my appreciation to all the reviewers, whose collective insights form the backbone of this revision. In particular, I would like to thank the following people for reviewing *Intermediate Algebra for College Students.*

Cindy Adams, *San Jacinto College*

Gwen P. Aldridge, *Northwest Mississippi Community College*

Ronnie Allen, *Central New Mexico Community College*

Dr. Simon Aman, *Harry S. Truman College*

Howard Anderson, *Skagit Valley College*

John Anderson, *Illinois Valley Community College*

Michael H. Andreoli, *Miami Dade College – North Campus*

Michele Bach, *Kansas City Kansas Community College*

Jana Barnard, *Angelo State University*

Rosanne Benn, *Prince George's Community College*

Christine Brady, *Suffolk County Community College*

Gale Brewer, *Amarillo College*

Carmen Buhler, *Minneapolis Community & Technical College*

Warren J. Burch, *Brevard College*

Alice Burstein, *Middlesex Community College*

Edie Carter, *Amarillo College*

Jerry Chen, *Suffolk County Community College*

Sandra Pryor Clarkson, *Hunter College*

Sally Copeland, *Johnson County Community College*

Valerie Cox, *Calhoun Community College*

Carol Curtis, *Fresno City College*

Robert A. Davies, *Cuyahoga Community College*

Deborah Detrick, *Kansas City Kansas Community College*

Jill DeWitt, *Baker College of Muskegon*

Ben Divers, Jr., *Ferrum College*

Irene Doo, *Austin Community College*

Charles C. Edgar, *Onondaga Community College*

Karen Edwards, *Diablo Valley College*

Scott Fallstrom, *MiraCosta College*

Elise Fischer, *Johnson County Community College*

Susan Forman, *Bronx Community College*

Wendy Fresh, *Portland Community College*

Jennifer Garnes, *Cuyahoga Community College*

Gary Glaze, *Eastern Washington University*

Jay Graening, *University of Arkansas*

Robert B. Hafer, *Brevard College*

Andrea Hendricks, *Georgia Perimeter College*

Donald Herrick, *Northern Illinois University*

Beth Hooper, *Golden West College*

Sandee House, *Georgia Perimeter College*

Tracy Hoy, *College of Lake County*

Laura Hoye, *Trident Community College*

Margaret Huddleston, *Schreiner University*

Marcella Jones, *Minneapolis Community & Technical College*

Shelbra B. Jones, *Wake Technical Community College*

Sharon Keenee, *Georgia Perimeter College*

Regina Keller, *Suffolk County Community College*

Gary Kersting, *North Central Michigan College*

Dennis Kimzey, *Rogue Community College*

Kandace Kling, *Portland Community College*

Gray Knippenberg, *Lansing Community College*

Mary Kochler, *Cuyahoga Community College*

Scot Leavitt, *Portland Community College*

Robert Leibman, *Austin Community College*

Jennifer Lempke, *North Central Michigan College*

Ann M. Loving, *J. Sargent Reynolds Community College*

Kent MacDougall, *Temple College*

Jean-Marie Magnier, *Springfield Technical Community College*

Hank Martel, *Broward College*

Kim Martin, *Southeastern Illinois College*

John Robert Martin, *Tarrant County College*

Lisa McMillen, *Baker College of Auburn Hills*

Irwin Metviner, *State University of New York at Old Westbury*

Jean P. Millen, *Georgia Perimeter College*

Lawrence Morales, *Seattle Central Community College*

Morteza Shafii-Mousavi, *Indiana University South Bend*

Lois Jean Nieme, *Minneapolis Community & Technical College*

Allen R. Newhart, *Parkersburg Community College*

Karen Pain, *Palm Beach State College*

Peg Pankowski, *Community College of Allegheny County – South Campus*

Robert Patenaude, *College of the Canyons*

Matthew Peace, *Florida Gateway College*

Dr. Bernard J. Piña, *New Mexico State University – Doña Ana Community College*

Jill Rafael, *Sierra College*

James Razavi, *Sierra College*

Christopher Reisch, *The State University of New York at Buffalo*

Nancy Ressler, *Oakton Community College*

Katalin Rozsa, *Mesa Community College*

Haazim Sabree, *Georgia Perimeter College*

Chris Schultz, *Iowa State University*

Shannon Schumann, *University of Phoenix*

Barbara Sehr, *Indiana University Kokomo*

Brian Smith, *Northwest Shoals Community College*

Gayle Smith, *Lane Community College*

Dick Spangler, *Tacoma Community College*

Janette Summers, *University of Arkansas*

Robert Thornton, *Loyola University*

Lucy C. Thrower, *Francis Marion College*

Mary Thurow, *Minneapolis Community & Tech College*

Richard Townsend, *North Carolina Central University*

Cindie Wade, *St. Clair County Community College*

Andrew Walker, *North Seattle Community College*

Kathryn Wetzel, *Amarillo College*

Additional acknowledgments are extended to Dan Miller and Kelly Barber for preparing the solutions manuals and the new Learning Guide; Brad Davis, for preparing the answer section and serving as accuracy checker; the codeMantra formatting team for the book's brilliant paging; Brian Morris and Kevin Morris at Scientific Illustrators, for superbly illustrating the book; and Francesca Monaco, project manager, and Kathleen Manley, production editor, whose collective talents kept every aspect of this complex project moving through its many stages.

I would like to thank my editors at Pearson, Dawn Giovanniello and Megan Tripp, who guided and coordinated the book from manuscript through production. Thanks to Beth Paquin and Studio Montage for the quirky cover and interior design. Finally, thanks to marketing manager Alicia Frankel for your innovative marketing efforts, and to the entire Pearson sales force, for your confidence and enthusiasm about the book.

To the Student

The bar graph shows some of the qualities that students say make a great teacher.

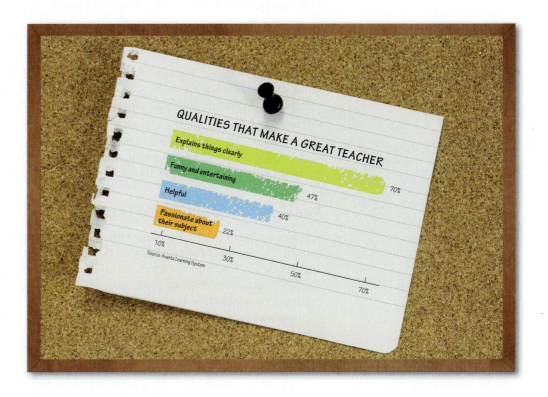

QUALITIES THAT MAKE A GREAT TEACHER

Explains things clearly — 70%
Funny and entertaining — 47%
Helpful — 40%
Passionate about their subject — 22%

10%
30%
50%
70%

Source: Avanta Learning System

It was my goal to incorporate each of the qualities that make a great teacher throughout the pages of this book.

Explains Things Clearly

I understand that your primary purpose in reading *Intermediate Algebra for College Students* is to acquire a solid understanding of the required topics in your algebra course. In order to achieve this goal, I've carefully explained each topic. Important definitions and procedures are set off in boxes, and worked-out examples that present solutions in a step-by-step manner appear in every section. Each example is followed by a similar matched problem, called a Check Point, for you to try so that you can actively participate in the learning process as you read the book. (Answers to all Check Points appear in the back of the book.)

Funny/Entertaining

Who says that an algebra textbook can't be entertaining? From our quirky cover to the photos in the chapter and section openers, prepare to expect the unexpected. I hope some of the book's enrichment essays, called Blitzer Bonuses, will put a smile on your face from time to time.

Helpful

I designed the book's features to help you acquire knowledge of intermediate algebra, as well as to show you how algebra can solve authentic problems that apply to your life. These helpful features include:

- *Explanatory Voice Balloons:* Voice balloons are used in a variety of ways to make math less intimidating. They translate algebraic language into everyday English, help clarify problem-solving procedures, present alternative ways of understanding concepts, and connect new concepts to concepts you have already learned.

- *Great Question!:* The book's Great Question! boxes are based on questions students ask in class. The answers to these questions give suggestions for problem solving, point out common errors to avoid, and provide informal hints and suggestions.

- *Achieving Success:* The book's Achieving Success boxes give you helpful strategies for success in learning algebra, as well as suggestions that can be applied for achieving your full academic potential in future college coursework.

- *Detailed Chapter Review Charts:* Each chapter contains a review chart that summarizes the definitions and concepts in every section of the chapter. Examples that illustrate these key concepts are also included in the chart. Review these summaries and you'll know the most important material in the chapter!

Passionate about Their Subject

I passionately believe that no other discipline comes close to math in offering a more extensive set of tools for application and development of your mind. I wrote the book in Point Reyes National Seashore, 40 miles north of San Francisco. The park consists of 75,000 acres with miles of pristine surf-washed beaches, forested ridges, and bays bordered by white cliffs. It was my hope to convey the beauty and excitement of mathematics using nature's unspoiled beauty as a source of inspiration and creativity. Enjoy the pages that follow as you empower yourself with the algebra needed to succeed in college, your career, and your life.

Regards,
Bob
Robert Blitzer

About the Author

Bob Blitzer is a native of Manhattan and received a Bachelor of Arts degree with dual majors in mathematics and psychology (minor: English literature) from the City College of New York. His unusual combination of academic interests led him toward a Master of Arts in mathematics from the University of Miami and a doctorate in behavioral sciences from Nova University. Bob's love for teaching mathematics was nourished for nearly 30 years at Miami Dade College, where he received numerous teaching awards, including Innovator of the Year from the League for Innovations in the Community College and an endowed chair based on excellence in the classroom. In addition to *Intermediate Algebra for College Students*, Bob has written textbooks covering developmental mathematics, introductory algebra, college algebra, algebra and trigonometry, precalculus, and liberal arts mathematics, all published by Pearson. When not secluded in his Northern California writer's cabin, Bob can be found hiking the beaches and trails of Point Reyes National Seashore, and tending to the chores required by his beloved entourage of horses, chickens, and irritable roosters.

Algebra, Mathematical Models, and Problem Solving

How would your lifestyle change if a gallon of gas cost $9.15? Of if the price of a staple such as milk were $15? That's how much those products would cost if their prices had increased at the same rate as college tuition has increased since 1980. If this trend continues, what can we expect in the 2020s and beyond?

We can answer this question by representing data for tuition and fees at U.S. colleges mathematically. With such representations, called *mathematical models*, we can gain insights and predict what might occur in the future on a variety of issues, ranging from college costs to a possible Social Security doomsday, and even the changes that occur as we age.

Here's where you'll find these applications:

- Mathematical models involving college costs appear as Example 7 in Section 1.4 and Exercises 67–68 in Exercise Set 1.4.

- The insecurities of Social Security are explored in Exercise 78 in the Review Exercises.

- Some surprising changes that occur with aging appear as Example 2 in Section 1.1, Exercises 89–92 in Exercise Set 1.1, and Exercises 53–56 in Exercise Set 1.3.

SECTION

1.1 Algebraic Expressions, Real Numbers, and Interval Notation

What am I supposed to learn?

After studying this section, you should be able to:

1. Translate English phrases into algebraic expressions. ▶

2. Evaluate algebraic expressions. ▶

3. Use mathematical models. ▶

4. Recognize the sets that make up the real numbers. ▶

5. Use set-builder notation. ▶

6. Use the symbols ∈ and ∉ . ▶

7. Use inequality symbols. ▶

8. Use interval notation. ▶

As we get older, do we mellow out or become more neurotic? In this section, you will learn how the special language of algebra describes your world, including our improving emotional health with age.

Algebraic Expressions

Algebra uses letters, such as x and y, to represent numbers. If a letter is used to represent various numbers, it is called a **variable**. For example, imagine that you are basking in the sun on the beach. We can let x represent the number of minutes that you can stay in the sun without burning with no sunscreen. With a number 6 sunscreen, exposure time without burning is six times as long, or 6 times x. This can be written $6 \cdot x$, but it is usually expressed as $6x$. Placing a number and a letter next to one another indicates multiplication.

Notice that $6x$ combines the number 6 and the variable x using the operation of multiplication. A combination of variables and numbers using the operations of addition, subtraction, multiplication, or division, as well as powers or roots, is called an **algebraic expression**. Here are some examples of algebraic expressions:

$$x + 6, \quad x - 6, \quad 6x, \quad \frac{x}{6}, \quad 3x + 5, \quad x^2 - 3, \quad \sqrt{x} + 7.$$

Is every letter in algebra a variable? No. Some letters stand for a particular number. Such a letter is called a **constant**. For example, let $d =$ the number of days in a week. The letter d represents just one number, namely 7, and is a constant.

1. Translate English phrases into algebraic expressions. ▶

Translating English Phrases into Algebraic Expressions

Problem solving in algebra involves translating English phrases into algebraic expressions. Here is a list of words and phrases for the four basic operations:

Addition	Subtraction	Multiplication	Division
sum	difference	product	quotient
plus	minus	times	divide
increased by	decreased by	of (used with fractions)	per
more than	less than	twice	ratio

EXAMPLE 1   Translating English Phrases into Algebraic Expressions

Write each English phrase as an algebraic expression. Let x represent the number.

a. Nine less than six times a number

b. The quotient of five and a number, increased by twice the number

Great Question!

Why is it so important to work each of the book's Check Points?

You learn best by doing. Do not simply look at the worked examples and conclude that you know how to solve them. To be sure that you understand the worked examples, try each Check Point. Check your answer in the answer section before continuing your reading. Expect to read this book with pencil and paper handy to work the Check Points.

2 Evaluate algebraic expressions.

Solution

a.
Nine less than | six times a number
$$6x - 9$$

b.
The quotient of five and a number, | increased by | twice the number
$$\frac{5}{x} \quad + \quad 2x$$ ∎

✓ **CHECK POINT 1**   Write each English phrase as an algebraic expression. Let x represent the number.

a. Five more than 8 times a number

b. The quotient of a number and 7, decreased by twice the number

Evaluating Algebraic Expressions

Evaluating an algebraic expression means to find the value of the expression for a given value of the variable.

EXAMPLE 2   Evaluating an Algebraic Expression

We opened the section with a comment about our improving emotional health with age. A test measuring neurotic traits, such as anxiety and hostility, indicates that people may become less neurotic as they get older. **Figure 1.1** shows the average level of neuroticism, on a scale of 0 to 50, for persons at various ages.

Neurosis and Age

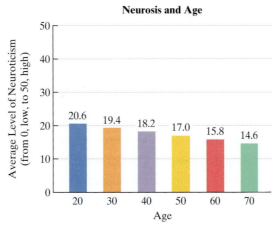

Figure 1.1

Source: L. M. Williams, "The Mellow Years? Neural Basis of Improving Emotional Stability over Age," *The Journal of Neuroscience,* June 14, 2006.

The algebraic expression $23 - 0.12x$ describes the average neurotic level for people who are x years old. Evaluate the expression for $x = 80$. Describe what the answer means in practical terms.

Solution   We begin by substituting 80 for x. Because $x = 80$, we will be finding the average neurotic level at age 80.

$$23 - 0.12x$$

Replace x with 80.

$$= 23 - 0.12(80) = 23 - 9.6 = 13.4$$

Thus, at age 80, the average level of neuroticism on a scale of 0 to 50 is 13.4. ∎

✓ **CHECK POINT 2**   Evaluate the expression from Example 2, $23 - 0.12x$, for $x = 10$. Describe what the answer means in practical terms.

Many algebraic expressions involve *exponents*. For example, the algebraic expression

$$46x^2 + 541x + 17{,}650$$

approximates student-loan debt in the United States, in dollars, x years after 2000. The expression x^2 means $x \cdot x$, and is read "x to the second power" or "x squared." The exponent, 2, indicates that the base, x, appears as a factor two times.

Exponential Notation

If n is a counting number (1, 2, 3, and so on),

Exponent or power

$$b^n = b \cdot b \cdot b \cdot \cdots \cdot b.$$

Base

b appears as a factor n times.

b^n is read "the nth power of b" or "b to the nth power." Thus, the nth power of b is defined as the product of n factors of b. The expression b^n is called an **exponential expression**. Furthermore, $b^1 = b$.

For example,

$$8^2 = 8 \cdot 8 = 64, \quad 5^3 = 5 \cdot 5 \cdot 5 = 125, \quad \text{and} \quad 2^4 = 2 \cdot 2 \cdot 2 \cdot 2 = 16.$$

Many algebraic expressions involve more than one operation. Evaluating an algebraic expression without a calculator involves carefully applying the following order of operations agreement:

The Order of Operations Agreement

1. Perform operations within the innermost parentheses and work outward. If the algebraic expression involves a fraction, treat the numerator and the denominator as if they were each enclosed in parentheses.
2. Evaluate all exponential expressions.
3. Perform multiplications and divisions as they occur, working from left to right.
4. Perform additions and subtractions as they occur, working from left to right.

EXAMPLE 3 Evaluating an Algebraic Expression

Evaluate $7 + 5(x - 4)^3$ for $x = 6$.

Solution

$$
\begin{aligned}
7 + 5(x - 4)^3 &= 7 + 5(6 - 4)^3 && \text{Replace } x \text{ with 6.} \\
&= 7 + 5(2)^3 && \text{First work inside parentheses: } 6 - 4 = 2. \\
&= 7 + 5(8) && \text{Evaluate the exponential expression:} \\
& && 2^3 = 2 \cdot 2 \cdot 2 = 8. \\
&= 7 + 40 && \text{Multiply: } 5(8) = 40. \\
&= 47 && \text{Add.} \quad \blacksquare
\end{aligned}
$$

☑ **CHECK POINT 3** Evaluate $8 + 6(x - 3)^2$ for $x = 13$.

Using Technology

You can use a calculator to evaluate exponential expressions. For example, to evaluate 2^4, press the following keys:

Many Scientific Calculators

2 $\boxed{y^x}$ 4 $\boxed{=}$

Many Graphing Calculators

2 $\boxed{\wedge}$ 4 $\boxed{\text{ENTER}}$

Although calculators have special keys to evaluate powers of ten and to square bases, you can always use one of the sequences shown here.

Great Question!

What am I supposed to do with the worked examples?

Study the step-by-step solutions in the examples. Reading the solutions slowly and with great care will prepare you for success with the Check Points and the exercises in the Exercise Sets.

3 Use mathematical models.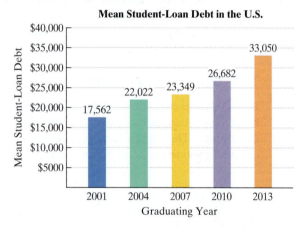

Formulas and Mathematical Models

An **equation** is formed when an equal sign is placed between two algebraic expressions. One aim of algebra is to provide a compact, symbolic description of the world. These descriptions involve the use of *formulas*. A **formula** is an equation that uses variables to express a relationship between two or more quantities. Here is an example of a formula:

$$C = \frac{5}{9}(F - 32).$$

| Celsius temperature | is | $\frac{5}{9}$ of | the difference between Fahrenheit temperature and 32°. |

The process of finding formulas to describe real-world phenomena is called **mathematical modeling**. Such formulas, together with the meaning assigned to the variables, are called **mathematical models**. We often say that these formulas model, or describe, the relationships among the variables.

EXAMPLE 4 Modeling Student-Loan Debt

College students are graduating with the highest debt burden in history. **Figure 1.2** shows the mean, or average, student-loan debt in the United States for five selected graduating years from 2001 through 2013.

Mean Student-Loan Debt in the U.S.

[Bar chart with y-axis "Mean Student-Loan Debt" ranging from $5000 to $40,000 and x-axis "Graduating Year":
2001: 17,562
2004: 22,022
2007: 23,349
2010: 26,682
2013: 33,050]

Figure 1.2
Source: Pew Research Center

The formula

$$D = 46x^2 + 541x + 17{,}650$$

models mean student-loan debt, D, in dollars, x years after 2000.

a. Use the formula to find mean student-loan debt for college students who graduated in 2013.

b. Does the mathematical model underestimate or overestimate mean student-loan debt for 2013 shown in **Figure 1.2**? By how much?

Solution

a. Because 2013 is 13 years after 2000, we substitute 13 for x in the given formula. Then we use the order of operations to find D, the mean student-loan debt for the graduating class of 2013.

$D = 46x^2 + 541x + 17{,}650$	This is the given mathematical model.
$D = 46(13)^2 + 541(13) + 17{,}650$	Replace each occurrence of x with 13.
$D = 46(169) + 541(13) + 17{,}650$	Evaluate the exponential expression: $13^2 = 13 \cdot 13 = 169$.
$D = 7774 + 7033 + 17{,}650$	Multiply from left to right: $46(169) = 7774$ and $541(13) = 7033$.
$D = 32{,}457$	Add.

The formula indicates that the mean student-loan debt for college students who graduated in 2013 was $32,457.

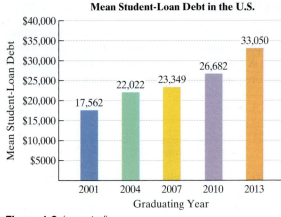

Mean Student-Loan Debt in the U.S.

Figure 1.2 (repeated)

b. The mean student-loan debt for 2013 given in **Figure 1.2** is $33,050. The model value, $32,457, is less than the actual data value, $33,050. Thus, the mathematical model underestimates mean student-loan debt for 2013 by

$$\$33,050 - \$32,457,$$

or by $593. ■

☑ **CHECK POINT 4**

a. Use the formula in Example 4 to find mean student–loan debt for college students who graduated in 2010.

b. Does the mathematical model underestimate or overestimate mean student-loan debt for 2010 shown in **Figure 1.2**? By how much?

Sometimes a mathematical model gives an estimate that is not a good approximation or is extended to include values of the variable that do not make sense. In these cases, we say that **model breakdown** has occurred. Models that accurately describe data for the past ten years might not serve as reliable predictions for what can reasonably be expected to occur in the future. Model breakdown can occur when formulas are extended too far into the future.

4 Recognize the sets that make up the real numbers. ⊙

The Set of Real Numbers

Before we describe the set of real numbers, let's be sure you are familiar with some basic ideas about sets. A **set** is a collection of objects whose contents can be clearly determined. The objects in a set are called the **elements** of the set. For example, the set of numbers used for counting can be represented by

$$\{1, 2, 3, 4, 5, \dots\}.$$

The braces, { }, indicate that we are representing a set. This form of representation, called the **roster method**, uses commas to separate the elements of the set. The three dots after the 5, called an *ellipsis*, indicate that there is no final element and that the listing goes on forever.

Three common sets of numbers are the *natural numbers*, the *whole numbers*, and the *integers*.

Great Question!

Can I use symbols other than braces to indicate sets in the roster method?

No. Grouping symbols such as parentheses, ( ), and square brackets, [ ], are not used to represent sets in the roster method. Furthermore, only commas are used to separate the elements of a set. Separators such as colons or semicolons are not used.

Natural Numbers, Whole Numbers, and Integers

The Set of Natural Numbers

$$\{1, 2, 3, 4, 5, \dots\}$$

These are the numbers that we use for counting.

The Set of Whole Numbers

$$\{0, 1, 2, 3, 4, 5, \dots\}$$

The set of whole numbers includes 0 and the natural numbers.

The Set of Integers

$$\{\dots, -5, -4, -3, -2, -1, 0, 1, 2, 3, 4, 5, \dots\}$$

The set of integers includes the negatives of the natural numbers and the whole numbers.

5 Use set-builder notation. ▶

A set can also be written in **set-builder notation**. In this notation, the elements of the set are described, but not listed. Here is an example:

$$\{x \mid x \text{ is a natural number less than } 6\}.$$

| The set of all x | such that | x is a natural number less than 6. |

The same set is written using the roster method as

$$\{1, 2, 3, 4, 5\}.$$

6 Use the symbols $\in$ and $\notin$. ▶

The symbol $\in$ is used to indicate that a number or object is in a particular set. The symbol $\in$ is read "is an element of." Here is an example:

$$7 \in \{1, 2, 3, 4, 5, \ldots\}.$$

| 7 | is an element of | the set of natural numbers. |

The symbol $\notin$ is used to indicate that a number or object is not in a particular set. The symbol $\notin$ is read "is not an element of." Here is an example:

$$\frac{1}{2} \notin \{1, 2, 3, 4, 5, \ldots\}.$$

| $\frac{1}{2}$ | is not an element of | the set of natural numbers. |

EXAMPLE 5   Using the Symbols $\in$ and $\notin$

Determine whether each statement is true or false:

a. $100 \in \{x \mid x \text{ is an integer}\}$ **b.** $20 \notin \{5, 10, 15\}.$

Solution

a. Because 100 is an integer, the statement

$$100 \in \{x \mid x \text{ is an integer}\}$$

is true. The number 100 is an element of the set of integers.

b. Because 20 is not an element of $\{5, 10, 15\}$, the statement $20 \notin \{5, 10, 15\}$ is true. ∎

✓ **CHECK POINT 5**   Determine whether each statement is true or false:

a. $13 \in \{x \mid x \text{ is an integer}\}$ **b.** $6 \notin \{7, 8, 9, 10\}.$

Another common set is the set of *rational numbers*. Each of these numbers can be expressed as an integer divided by a nonzero integer.

Rational Numbers

The set of **rational numbers** is the set of all numbers that can be expressed as a quotient of two integers, with the denominator not 0.

This means that b is not equal to zero.

$$\left\{ \frac{a}{b} \,\middle|\, a \text{ and } b \text{ are integers and } b \neq 0 \right\}$$

Three examples of rational numbers are

$$\frac{1}{4} \quad (a = 1, \ b = 4), \quad \frac{-2}{3} \quad (a = -2, \ b = 3), \quad \text{and} \quad 5 = \frac{5}{1} \quad (a = 5, \ b = 1).$$

Can you see that integers are also rational numbers because they can be written in terms of division by 1?

Rational numbers can be expressed in fraction or decimal notation. To express the fraction $\frac{a}{b}$ as a decimal, divide the denominator, b, into the numerator, a. In decimal notation, rational numbers either terminate (stop) or have a digit, or block of digits, that repeats. For example,

$$\frac{3}{8} = 3 \div 8 = 0.375 \qquad \text{and} \qquad \frac{7}{11} = 7 \div 11 = 0.6363... = 0.\overline{63}.$$

> The decimal stops: it is a terminating decimal.

> This is a repeating decimal. The bar is written over the repeating part.

Some numbers cannot be expressed as terminating or repeating decimals. An example of such a number is $\sqrt{2}$, the square root of 2. The number $\sqrt{2}$ is a number that can be squared to give 2. No terminating or repeating decimal can be squared to get 2. However, some approximations have squares that come close to 2. We use the symbol $\approx$, which means "is approximately equal to."

- $\sqrt{2} \approx 1.4$ because $(1.4)^2 = (1.4)(1.4) = 1.96$.
- $\sqrt{2} \approx 1.41$ because $(1.41)^2 = (1.41)(1.41) = 1.9881$.
- $\sqrt{2} \approx 1.4142$ because $(1.4142)^2 = (1.4142)(1.4142) = 1.99996164$.

$\sqrt{2}$ is an example of an *irrational number*.

Irrational Numbers

The set of **irrational numbers** is the set of numbers whose decimal representations neither terminate nor repeat. Irrational numbers cannot be expressed as quotients of integers.

Examples of irrational numbers include

$$\sqrt{3} \approx 1.73205 \qquad \text{and} \qquad \pi(\text{pi}) \approx 3.141593.$$

Not all square roots are irrational. For example, $\sqrt{25} = 5$ because $5^2 = 5 \cdot 5 = 25$. Thus, $\sqrt{25}$ is a natural number, a whole number, an integer, and a rational number $\left(\sqrt{25} = \frac{5}{1}\right)$.

The set of *real numbers* is formed by combining the sets of rational numbers and irrational numbers. Thus, every real number is either rational or irrational, as shown in **Figure 1.3**.

Real Numbers

The set of **real numbers** is the set of numbers that are either rational or irrational:

$$\{x \,|\, x \text{ is rational or } x \text{ is irrational}\}.$$

Figure 1.3 Every real number is either rational or irrational.

The Real Number Line

The **real number line** is a graph used to represent the set of real numbers. An arbitrary point, called the **origin**, is labeled 0. Select a point to the right of 0 and label it 1. The distance from 0 to 1 is called the **unit distance**. Numbers to the right of the origin are **positive** and numbers to the left of the origin are **negative**. The real number line is shown in **Figure 1.4**.

Figure 1.4 The real number line

Great Question!

How did you locate $\sqrt{2}$ as a precise point on the number line in Figure 1.5?

We used a right triangle with two legs of length 1. The remaining side has a length measuring $\sqrt{2}$.

We'll have lots more to say about right triangles later in the book.

7 Use inequality symbols.

Real numbers are **graphed** on a number line by placing a dot at the correct location for each number. The integers are easiest to locate. In **Figure 1.5**, we've graphed six rational numbers and three irrational numbers on a real number line.

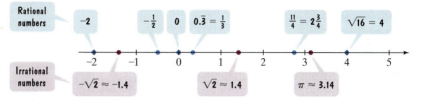

Figure 1.5 Graphing numbers on a real number line

Every real number corresponds to a point on the number line and every point on the number line corresponds to a real number. We say that there is a **one-to-one correspondence** between all the real numbers and all points on a real number line.

Ordering the Real Numbers

On the real number line, the real numbers increase from left to right. The lesser of two real numbers is the one farther to the left on a number line. The greater of two real numbers is the one farther to the right on a number line.

Look at the number line in **Figure 1.6**. The integers -4 and -1 are graphed.

Figure 1.6

Observe that -4 is to the left of -1 on the number line. This means that -4 is less than -1.

$$-4 < -1 \qquad \boxed{\text{-4 is less than -1 because -4 is to the left of -1 on the number line.}}$$

In **Figure 1.6**, we can also observe that -1 is to the right of -4 on the number line. This means that -1 is greater than -4.

$$-1 > -4 \qquad \boxed{\text{-1 is greater than -4 because -1 is to the right of -4 on the number line.}}$$

The symbols $<$ and $>$ are called **inequality symbols**. These symbols always point to the lesser of the two real numbers when the inequality statement is true.

$\boxed{\text{-4 is less than -1.}}$ $\quad -4 < -1 \quad$ The symbol points to -4, the lesser number.

$\boxed{\text{-1 is greater than -4.}}$ $\quad -1 > -4 \quad$ The symbol still points to -4, the lesser number.

The symbols $<$ and $>$ may be combined with an equal sign, as shown in the following table:

	Symbol	Meaning	Examples	Explanation
This inequality is true if either the $<$ part or the $=$ part is true.	$a \le b$	a is less than or equal to b.	$2 \le 9$ $9 \le 9$	Because $2 < 9$ Because $9 = 9$
This inequality is true if either the $>$ part or the $=$ part is true.	$b \ge a$	b is greater than or equal to a.	$9 \ge 2$ $2 \ge 2$	Because $9 > 2$ Because $2 = 2$

EXAMPLE 6   Using Inequality Symbols

Write out the meaning of each inequality. Then determine whether the inequality is true or false.

a. $-5 < -1$ **b.** $6 > -2$ **c.** $-6 \le 3$ **d.** $10 \ge 10$ **e.** $-9 \ge 6$

Solution The solution is illustrated by the number line in **Figure 1.7**.

Figure 1.7

$$-9 \quad -8 \quad -7 \quad -6 \quad -5 \quad -4 \quad -3 \quad -2 \quad -1 \quad 0 \quad 1 \quad 2 \quad 3 \quad 4 \quad 5 \quad 6 \quad 7 \quad 8 \quad 9 \quad 10$$

Inequality	Meaning
a. $-5 < -1$	"-5 is less than -1." Because -5 is to the left of -1 on the number line, the inequality is true.
b. $6 > -2$	"6 is greater than -2." Because 6 is to the right of -2 on the number line, the inequality is true.
c. $-6 \leq 3$	"-6 is less than or equal to 3." Because $-6 < 3$ is true (-6 is to the left of 3 on the number line), the inequality is true.
d. $10 \geq 10$	"10 is greater than or equal to 10." Because $10 = 10$ is true, the inequality is true.
e. $-9 \geq 6$	"-9 is greater than or equal to 6." Because neither $-9 > 6$ nor $-9 = 6$ is true, the inequality is false.

■

✓ **CHECK POINT 6**   Write out the meaning of each inequality. Then determine whether the inequality is true or false.

a. $-8 < -2$ **b.** $7 > -3$

c. $-1 \leq -4$ **d.** $5 \geq 5$

e. $2 \geq -14$

Great Question!

Can similar English phrases have different algebraic representations?

Yes. Here are three similar English phrases that have very different translations:

- 7 minus a number: $7 - x$
- 7 less than a number: $x - 7$
- 7 is less than a number: $7 < x$

Think carefully about what is expressed in English before you translate into the language of algebra.

---

8 Use interval notation. ▶

Interval Notation

Some sets of real numbers can be represented using **interval notation**. Suppose that a and b are two real numbers such that $a < b$.

Interval Notation	Graph
The **open interval** (a, b) represents the set of real numbers between, but not including, a and b. $(a, b) = \{x \mid a < x < b\}$ x is greater than a ($a < x$) and x is less than b ($x < b$).	a (a, b) b → x The parentheses in the graph and in interval notation indicate that a and b, the endpoints, are excluded from the interval.
The **closed interval** $[a, b]$ represents the set of real numbers between, and including, a and b. $[a, b] = \{x \mid a \leq x \leq b\}$ x is greater than or equal to a ($a \leq x$) and x is less than or equal to b ($x \leq b$).	a $[a, b]$ b → x The square brackets in the graph and in interval notation indicate that a and b, the endpoints, are included in the interval.
The **infinite interval** (a, ∞) represents the set of real numbers that are greater than a. $(a, \infty) = \{x \mid x > a\}$ The infinity symbol does not represent a real number. It indicates that the interval extends indefinitely to the right.	a (a, ∞) → x The parenthesis indicates that a is excluded from the interval.
The **infinite interval** $(-\infty, b]$ represents the set of real numbers that are less than or equal to b. $(-\infty, b] = \{x \mid x \leq b\}$ The negative infinity symbol indicates that the interval extends indefinitely to the left.	$(-\infty, b]$ b → x The square bracket indicates that b is included in the interval.

Parentheses and Brackets in Interval Notation

Parentheses indicate endpoints that are not included in an interval. Square brackets indicate endpoints that are included in an interval. Parentheses are always used with ∞ or $-\infty$.

Table 1.1 lists nine possible types of intervals used to describe sets of real numbers.

Table 1.1	Intervals on the Real Number Line	
Let a and b be real numbers such that $a < b$.		
Interval Notation	**Set-Builder Notation**	**Graph**
(a, b)	$\{x \mid a < x < b\}$	
$[a, b]$	$\{x \mid a \leq x \leq b\}$	
$[a, b)$	$\{x \mid a \leq x < b\}$	
$(a, b]$	$\{x \mid a < x \leq b\}$	
(a, ∞)	$\{x \mid x > a\}$	
$[a, \infty)$	$\{x \mid x \geq a\}$	
$(-\infty, b)$	$\{x \mid x < b\}$	
$(-\infty, b]$	$\{x \mid x \leq b\}$	
$(-\infty, \infty)$	$\{x \mid x \text{ is a real number}\}$ or $\mathbb{R}$ (set of real numbers)	

EXAMPLE 7   Interpreting Interval Notation

Express each interval in set-builder notation and graph:

a. $(-1, 4]$ **b.** $[2.5, 4]$ **c.** $(-4, \infty)$.

Solution

a. $(-1, 4] = \{x \mid -1 < x \leq 4\}$

b. $[2.5, 4] = \{x \mid 2.5 \leq x \leq 4\}$

c. $(-4, \infty) = \{x \mid x > -4\}$

☑ **CHECK POINT 7**   Express each interval in set-builder notation and graph:

a. $[-2, 5)$ **b.** $[1, 3.5]$ **c.** $(-\infty, -1)$.

CONCEPT AND VOCABULARY CHECK

Fill in each blank so that the resulting statement is true.

1. A letter that represents a variety of different numbers is called a/an _____.

2. A combination of numbers, letters that represent numbers, and operation symbols is called an algebraic _____.

3. If n is a counting number, b^n, read _____, indicates that there are n factors of b. The number b is called the _____ and the number n is called the _____.

4. A statement that expresses a relationship between two or more variables, such as $C = \frac{5}{9}(F - 32)$, is called a/an _____. The process of finding such statements to describe real-world phenomena is called mathematical _____. Such statements, together with the meaning assigned to the variables, are called mathematical _____.

5. The set $\{1, 2, 3, 4, 5, \dots\}$ is called the set of _____ numbers.

6. The set $\{0, 1, 2, 3, 4, 5, \dots\}$ is called the set of _____ numbers.

7. The set $\{\dots, -4, -3, -2, -1, 0, 1, 2, 3, 4, \dots\}$ is called the set of _____.

8. The set of numbers in the form $\frac{a}{b}$, where a and b belong to the set in statement 7 above and $b \neq 0$, is called the set of _____ numbers.

9. The set of numbers whose decimal representations are neither terminating nor repeating is called the set of _____ numbers.

10. Every real number is either a/an _____ number or a/an _____ number.

11. The notation $2 < 5$ means that 2 is to the _____ of 5 on a number line.

12. In interval notation, $[2, 5)$ represents the set of real numbers between _____ and _____, including _____ but not including _____.

13. In interval notation, $(-2, \infty)$ represents the set of real numbers _____ -2.

14. In interval notation, $(-\infty, -1]$ represents the set of real numbers _____ -1.

1.1 EXERCISE SET   ▶ MyMathLab®

Practice Exercises

In Exercises 1–14, write each English phrase as an algebraic expression. Let x represent the number.

1. Five more than a number

2. A number increased by six

3. Four less than a number

4. Nine less than a number

5. Four times a number

6. Twice a number

7. Ten more than twice a number

8. Four more than five times a number

9. The difference of six and half of a number

10. The difference of three and half of a number

11. Two less than the quotient of four and a number

12. Three less than the quotient of five and a number

13. The quotient of three and the difference of five and a number

14. The quotient of six and the difference of ten and a number

In Exercises 15–26, evaluate each algebraic expression for the given value or values of the variable(s).

15. $7 + 5x$, for $x = 10$

16. $8 + 6x$, for $x = 5$

17. $6x - y$, for $x = 3$ and $y = 8$

18. $8x - y$, for $x = 3$ and $y = 4$

19. $x^2 + 3x$, for $x = \frac{1}{3}$

20. $x^2 + 2x$ for $x = \frac{1}{2}$

21. $x^2 - 6x + 3$, for $x = 7$

22. $x^2 - 7x + 4$, for $x = 8$

23. $4 + 5(x - 7)^3$, for $x = 9$

24. $6 + 5(x - 6)^3$, for $x = 8$

25. $x^2 - 3(x - y)$, for $x = 8$ and $y = 2$

26. $x^2 - 4(x - y)$, for $x = 8$ and $y = 3$

In Exercises 27–34, use the roster method to list the elements in each set.

27. $\{x \mid x$ is a natural number less than 5$\}$

28. $\{x \mid x$ is a natural number less than 4$\}$

29. $\{x \mid x$ is an integer between -8 and $-3\}$

30. $\{x \mid x$ is an integer between -7 and $-2\}$

31. $\{x \mid x$ is a natural number greater than 7$\}$

32. $\{x \mid x$ is a natural number greater than 9$\}$

33. $\{x \mid x$ is an odd whole number less than 11$\}$

34. $\{x \mid x$ is an odd whole number less than 9$\}$

In Exercises 35–48, use the meaning of the symbols $\in$ and $\notin$ to determine whether each statement is true or false.

35. $7 \in \{x \mid x$ is an integer$\}$

36. $9 \in \{x \mid x$ is an integer$\}$

37. $7 \in \{x \mid x$ is a rational number$\}$

38. $9 \in \{x \mid x$ is a rational number$\}$

39. $7 \in \{x \mid x$ is an irrational number$\}$

40. $9 \in \{x \mid x$ is an irrational number$\}$

41. $3 \notin \{x \mid x$ is an irrational number$\}$

42. $5 \notin \{x \mid x$ is an irrational number$\}$

43. $\frac{1}{2} \notin \{x \mid x$ is a rational number$\}$

44. $\frac{1}{4} \notin \{x \mid x$ is a rational number$\}$

45. $\sqrt{2} \notin \{x \mid x$ is a rational number$\}$

46. $\pi \notin \{x \mid x$ is a rational number$\}$

47. $\sqrt{2} \notin \{x \mid x$ is a real number$\}$

48. $\pi \notin \{x \mid x$ is a real number$\}$

In Exercises 49–64, write out the meaning of each inequality. Then determine whether the inequality is true or false.

49. $-6 < -2$

50. $-7 < -3$

51. $5 > -7$

52. $3 > -8$

53. $0 < -4$

54. $0 < -5$

55. $-4 \le 1$

56. $-5 \le 1$

57. $-2 \le -6$

58. $-3 \le -7$

59. $-2 \le -2$

60. $-3 \le -3$

61. $-2 \ge -2$

62. $-3 \ge -3$

63. $2 \le -\frac{1}{2}$

64. $4 \le -\frac{1}{2}$

In Exercises 65–78, express each interval in set-builder notation and graph the interval on a number line.

65. $(1, 6]$

66. $(-2, 4]$

67. $[-5, 2)$

68. $[-4, 3)$

69. $[-3, 1]$

70. $[-2, 5]$

71. $(2, \infty)$

72. $(3, \infty)$

73. $[-3, \infty)$

74. $[-5, \infty)$

75. $(-\infty, 3)$

76. $(-\infty, 2)$

77. $(-\infty, 5.5)$

78. $(-\infty, 3.5]$

Practice PLUS

By definition, an "and" statement is true only when the statements before and after the "and" connective are both true. Use this definition to determine whether each statement in Exercises 79–88 is true or false.

79. $0.\overline{3} > 0.3$ and $-10 < 4 + 6$.

80. $0.6 < 0.\overline{6}$ and $2 \cdot 5 \le 4 + 6$.

81. $12 \in \{1, 2, 3, \ldots\}$ and $\{3\} \in \{1, 2, 3, 4\}$

82. $17 \in \{1, 2, 3, \ldots\}$ and $\{4\} \in \{1, 2, 3, 4, 5\}$

83. $\left(\frac{2}{5} + \frac{3}{5}\right) \in \{x \mid x$ is a natural number$\}$ and the value of $9x^2(x + 11) - 9(x + 11)x^2$, for $x = 100$, is 0.

84. $\left(\frac{14}{19} + \frac{5}{19}\right) \in \{x \mid x$ is a natural number$\}$ and the value of $12x^2(x + 10) - 12(x + 10)x^2$, for $x = 50$, is 0.

85. $\{x \mid x$ is an integer between -3 and $0\} = \{-3, -2, -1, 0\}$ and $-\pi > -3.5$.

86. $\{x \mid x$ is an integer between -4 and $0\} = \{-4, -3, -2, -1, 0\}$ and $-\frac{\pi}{2} > -2.3$.

87. Twice the sum of a number and three is represented by $2x + 3$ and $-1{,}100{,}000 \in \{x \mid x$ is an integer$\}$

88. Three times the sum of a number and five is represented by $3x + 5$ and $-4{,}500{,}000 \in \{x \mid x$ is an integer$\}$

Application Exercises

We opened the section with a comment about our improving emotional health with age. We also saw an example indicating people become less neurotic as they get older. How can this be explained? One theory is that key centers of the brain tend to create less resistance to feelings of happiness as we age. The graph shows the average resistance to happiness, on a scale of 0 (no resistance) to 8 (completely resistant), for persons at various ages.

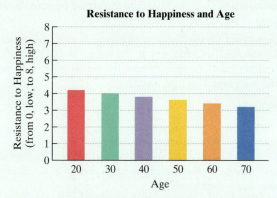

Resistance to Happiness and Age

Source: L. M. Williams, "The Mellow Years? Neural Basis for Improving Stability over Age." THE JOURNAL OF NEUROSCIENCE, June 14, 2006

The data in the graph above can be modeled by the formula

$$R = 4.6 - 0.02x,$$

where R represents the average resistance to happiness, on a scale of 0 to 8, for a person who is x years old. Use this formula to solve Exercises 89–92.

89. According to the formula, what is the average resistance to happiness at age 20?

(In Exercises 90–92, refer to the formula and the graph at the bottom of the previous page.)

90. According to the formula, what is the average resistance to happiness at age 30?

91. What is the difference between the average resistance to happiness at age 30 and at age 50?

92. What is the difference between the average resistance to happiness at age 20 and at age 70?

The majority of American adults use their smartphones to go online. The bar graph shows the percentage of Americans using phones to go online from 2009 through 2013.

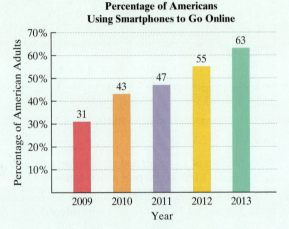

Percentage of Americans Using Smartphones to Go Online

Source: Pew Internet and American Life Project

The data in the graph can be modeled by the formula

$$S = 32 + 8.7x - 0.3x^2,$$

where S represents the percentage of American adults using smartphones to go online x years after 2009. Use this formula to solve Exercises 93–94.

93. According to the formula, what percentage of American adults used smartphones to go online in 2013? Does this underestimate or overestimate the actual percent shown by the bar graph? By how much?

94. According to the formula, what percentage of American adults used smartphones to go online in 2012? Does this underestimate or overestimate the actual percent shown by the bar graph? By how much?

The formula

$$C = \frac{5}{9}(F - 32)$$

expresses the relationship between Fahrenheit temperature, F, and Celsius temperature, C. In Exercises 95–96, use the formula to convert the given Fahrenheit temperature to its equivalent temperature on the Celsius scale.

95. 50°F

96. 86°F

A football was kicked vertically upward from a height of 4 feet with an initial speed of 60 feet per second. The formula

$$h = 4 + 60t - 16t^2$$

describes the ball's height above the ground, h, in feet, t seconds after it was kicked. Use this formula to solve Exercises 97–98.

97. What was the ball's height 2 seconds after it was kicked?

98. What was the ball's height 3 seconds after it was kicked?

Explaining the Concepts

99. What is a variable?

100. What is an algebraic expression? Give an example with your explanation.

101. If n is a natural number, what does b^n mean? Give an example with your explanation.

102. What does it mean when we say that a formula models real-world phenomena?

103. What is model breakdown?

104. What is a set?

105. Describe the roster method for representing a set.

106. What are the natural numbers?

107. What are the whole numbers?

108. What are the integers?

109. Describe the rational numbers.

110. Describe the difference between a rational number and an irrational number.

111. What are the real numbers?

112. What is set-builder notation?

113. Describe the meanings of the symbols $\in$ and $\notin$. Provide an example showing the correct use of each symbol.

114. What is the real number line?

115. If you are given two real numbers, explain how to determine which one is the lesser.

116. In interval notation, what does a parenthesis signify? What does a bracket signify?

Critical Thinking Exercises

Make Sense?   In Exercises 117–120, determine whether each statement makes sense or does not make sense, and explain your reasoning.

117. My mathematical model describes the data for the past ten years extremely well, so it will serve as an accurate prediction for what will occur in 2050.

118. My calculator will not display the value of 13^{1500}, so the algebraic expression $4x^{1500} - 3x + 7$ cannot be evaluated for $x = 13$ even without a calculator.

119. Regardless of what real numbers I substitute for x and y, I will always obtain zero when evaluating $2x^2y - 2yx^2$.

120. A model that describes the number of smartphone users x years after 2007 cannot be used to estimate the number in 2007.

In Exercises 121–124, determine whether each statement is true or false. If the statement is false, make the necessary change(s) to produce a true statement.

121. Every rational number is an integer.

122. Some whole numbers are not integers.

123. Some rational numbers are not positive.

124. Some irrational numbers are negative.

125. A bird lover visited a pet shop where there were twice 4 and 20 parrots. The bird lover purchased $\frac{1}{7}$ of the birds. English, of course, can be ambiguous and "twice 4 and 20" can mean $2(4 + 20)$ or $2 \cdot 4 + 20$. Explain how the conditions of the situation determine if "twice 4 and 20" means $2(4 + 20)$ or $2 \cdot 4 + 20$.

In Exercises 126–127, insert parentheses to make each statement true.

126. $2 \cdot 3 + 3 \cdot 5 = 45$

127. $8 + 2 \cdot 4 - 3 = 10$

128. Between which two consecutive integers is $-\sqrt{26}$? Do not use a calculator.

Preview Exercises

Exercises 129–131 will help you prepare for the material covered in the next section.

129. There are two real numbers whose distance is five units from zero on a real number line. What are these numbers?

130. Simplify: $\dfrac{16 + 3(2)^4}{12 - (10 - 6)}$.

131. Evaluate $2(3x + 5)$ and $6x + 10$ for $x = 4$.

Operations with Real Numbers and Simplifying Algebraic Expressions ▶

What am I supposed to learn?

After studying this section, you should be able to:

1. Find a number's absolute value. ▶
2. Add real numbers. ▶
3. Find opposites. ▶
4. Subtract real numbers. ▶
5. Multiply real numbers. ▶
6. Evaluate exponential expressions. ▶
7. Divide real numbers. ▶
8. Use the order of operations. ▶
9. Use commutative, associative, and distributive properties. ▶
10. Simplify algebraic expressions. ▶

College students have money to spend, often courtesy of Mom and Dad. The most sophisticated college marketers, from American Eagle to Apple to Red Bull, are increasingly turning to social media focused on students' wants and needs. In 2011, there were nearly 10,000 student reps on U.S. campuses facebooking, tweeting, and partying their way to selling things vital and specific to college students. Call it New World College Marketing 101.

How much do college students have to spend? In this section's Exercise Set (see Exercises 147–148), you'll be working with a model that addresses this question. In order to use the model, we need to review operations with real numbers, our focus of this section.

Absolute Value

Absolute value is used to describe how to operate with positive and negative numbers.

Geometric Meaning of Absolute Value

The **absolute value** of a real number a, denoted by $|a|$, is the distance from 0 to a on the number line. This distance is always taken to be nonnegative.

1 Find a number's absolute value.

EXAMPLE 1 Finding Absolute Value

Find the absolute value:

a. $|-4|$ **b.** $|3.5|$ **c.** $|0|$.

Solution The solution is illustrated in **Figure 1.8**.

a. $|-4| = 4$ The absolute value of -4 is 4 because -4 is 4 units from O.

b. $|3.5| = 3.5$ The absolute value of 3.5 is 3.5 because 3.5 is 3.5 units from O.

c. $|0| = 0$ The absolute value of O is O because O is O units from itself.

$$|-4| = 4 \qquad |3.5| = 3.5$$

$$-5 \quad -4 \quad -3 \quad -2 \quad -1 \quad 0 \quad 1 \quad 2 \quad 3 \quad 4 \quad 5$$

$$|0| = 0$$

Figure 1.8 ■

Can you see that the absolute value of a real number is either positive or zero? Zero is the only real number whose absolute value is 0:

$$|0| = 0.$$

The absolute value of a real number is never negative.

☑ **CHECK POINT 1** Find the absolute value:

a. $|-6|$ **b.** $|4.5|$ **c.** $|0|$.

2 Add real numbers.

Adding Real Numbers

Table 1.2 reviews how to add real numbers.

Table 1.2 Adding Real Numbers

Rule	Example
To add two real numbers with the same sign, add their absolute values. Use the common sign as the sign of the sum.	$(-7) + (-4) = -(\lvert -7 \rvert + \lvert -4 \rvert)$ $= -(7 + 4)$ $= -11$
To add two real numbers with different signs, subtract the smaller absolute value from the greater absolute value. Use the sign of the number with the greater absolute value as the sign of the sum.	$7 + (-15) = -(\lvert -15 \rvert - \lvert 7 \rvert)$ $= -(15 - 7)$ $= -8$

EXAMPLE 2 Adding Real Numbers

Add:

a. $-12 + (-14)$ **b.** $-0.3 + 0.7$ **c.** $-\dfrac{3}{4} + \dfrac{1}{2}$.

Solution

a. $-12 + (-14) = -26$ — Add absolute values: $12 + 14 = 26$.

Use the common sign.

b. $-0.3 + 0.7 = 0.4$ — Subtract absolute values: $0.7 - 0.3 = 0.4$.

Use the sign of the number with the greater absolute value. The sign of the sum is assumed to be positive.

c. $-\dfrac{3}{4} + \dfrac{1}{2} = -\dfrac{1}{4}$ — Subtract absolute values: $\frac{3}{4} - \frac{1}{2} = \frac{3}{4} - \frac{2}{4} = \frac{1}{4}$.

Use the sign of the number with the greater absolute value. ■

✓ **CHECK POINT 2**   Add:

a. $-10 + (-18)$ **b.** $-0.2 + 0.9$ **c.** $-\dfrac{3}{5} + \dfrac{1}{2}$.

If one of two numbers being added is zero, the sum is the other number. For example,

$$-3 + 0 = -3 \quad \text{and} \quad 0 + 2 = 2.$$

In general,

$$a + 0 = a \quad \text{and} \quad 0 + a = a.$$

We call 0 the **identity element of addition** or the **additive identity**. Thus, the additive identity can be deleted from a sum.

3 Find opposites. ▶

Numbers with different signs but the same absolute value are called **opposites** or **additive inverses**. For example, 3 and -3 are additive inverses. When additive inverses are added, their sum is 0. For example,

$$3 + (-3) = 0 \quad \text{and} \quad -3 + 3 = 0.$$

Inverse Property of Addition

The sum of a real number and its additive inverse is 0, the additive identity.

$$a + (-a) = 0 \quad \text{and} \quad (-a) + a = 0$$

The symbol "$-$" is used to name the opposite, or additive inverse, of a. When a is a negative number, $-a$, its opposite, is positive. For example, if a is -4, its opposite is 4. Thus,

$$-(-4) = 4.$$

The opposite of -4 is 4.

In general, if a is any real number,

$$-(-a) = a.$$

EXAMPLE 3   Finding Opposites

Find $-x$ if

a. $x = -6$ **b.** $x = \dfrac{1}{2}$.

Solution

a. If $x = -6$, then $-x = -(-6) = 6$.   The opposite of -6 is 6.

b. If $x = \dfrac{1}{2}$, then $-x = -\dfrac{1}{2}$.   The opposite of $\dfrac{1}{2}$ is $-\dfrac{1}{2}$. ∎

✓ **CHECK POINT 3**   Find $-x$ if

  a. $x = -8$ **b.** $x = \dfrac{1}{3}$.

We can define the absolute value of the real number a using opposites, without referring to a number line. The algebraic definition of the absolute value of a is given as follows:

> **Definition of Absolute Value**
>
> $$|a| = \begin{cases} a & \text{if } a \geq 0 \\ -a & \text{if } a < 0 \end{cases}$$

If a is nonnegative (that is, $a \geq 0$), the absolute value of a is the number itself: $|a| = a$. For example,

$$|5| = 5 \qquad |\pi| = \pi \qquad \left|\frac{1}{3}\right| = \frac{1}{3} \qquad |0| = 0$$

Zero is the only number whose absolute value is 0.

If a is a negative number (that is, $a < 0$), the absolute value of a is the opposite of a: $|a| = -a$. This makes the absolute value positive. For example,

$$|-3| = -(-3) = 3 \qquad |-\pi| = -(-\pi) = \pi \qquad \left|-\frac{1}{3}\right| = -\left(-\frac{1}{3}\right) = \frac{1}{3}.$$

This middle step is usually omitted.

4 Subtract real numbers. ▶

Subtracting Real Numbers

Subtraction of real numbers is defined in terms of addition.

> **Definition of Subtraction**
>
> If a and b are real numbers,
>
> $$a - b = a + (-b).$$
>
> To subtract a real number, add its opposite or additive inverse.

Thus, to subtract real numbers,

1. Change the subtraction to addition.
2. Change the sign of the number being subtracted.
3. Add, using one of the rules for adding numbers with the same sign or different signs.

EXAMPLE 4   Subtracting Real Numbers

Subtract:

  a. $6 - 13$ **b.** $5.1 - (-4.2)$ **c.** $-\dfrac{11}{3} - \left(-\dfrac{4}{3}\right)$.

Solution

a. $6 - 13 = 6 + (-13) = -7$

> Change the subtraction to addition.

> Replace 13 with its opposite.

b. $5.1 - (-4.2) = 5.1 + 4.2 = 9.3$

> Change the subtraction to addition.

> Replace -4.2 with its opposite.

c. $-\dfrac{11}{3} - \left(-\dfrac{4}{3}\right) = -\dfrac{11}{3} + \dfrac{4}{3} = -\dfrac{7}{3}$

> Change the subtraction to addition.

> Replace $-\frac{4}{3}$ with its opposite.

∎

✓ **CHECK POINT 4**   Subtract:

a. $7 - 10$ **b.** $4.3 - (-6.2)$ **c.** $-\dfrac{4}{5} - \left(-\dfrac{1}{5}\right).$

5 Multiply real numbers. ▶

Multiplying Real Numbers

You can think of multiplication as repeated addition or subtraction that starts at 0. For example,

$$3(-4) = 0 + (-4) + (-4) + (-4) = -12$$

> The numbers have different signs and the product is negative.

and

$$(-3)(-4) = 0 - (-4) - (-4) - (-4) = 0 + 4 + 4 + 4 = 12$$

> The numbers have the same sign and the product is positive.

Table 1.3 reviews how to multiply real numbers.

Table 1.3   Multiplying Real Numbers

Rule	Example
The product of two real numbers with different signs is found by multiplying their absolute values. The product is negative.	$7(-5) = -35$
The product of two real numbers with the same sign is found by multiplying their absolute values. The product is positive.	$(-6)(-11) = 66$
The product of 0 and any real number is 0: $a \cdot 0 = 0$ and $0 \cdot a = 0$.	$-17(0) = 0$
If no number is 0, a product with an odd number of negative factors is found by multiplying their absolute values. The product is negative.	$-2(-3)(-5) = -30$ Three (odd) negative factors
If no number is 0, a product with an even number of negative factors is found by multiplying absolute values. The product is positive.	$-2(3)(-5) = 30$ Two (even) negative factors

6 Evaluate exponential expressions. ▶

Because exponents indicate repeated multiplication, rules for multiplying real numbers can be used to evaluate exponential expressions.

EXAMPLE 5 Evaluating Exponential Expressions

Evaluate:

a. $(-6)^2$ **b.** -6^2 **c.** $(-5)^3$ **d.** $\left(-\dfrac{2}{3}\right)^4$.

Solution

a. $(-6)^2 = (-6)(-6) = 36$

> Base is −6.   Same signs give positive product.

b. $-6^2 = -(6 \cdot 6) = -36$

> Base is 6. The negative is not inside parentheses and is not taken to the second power.

c. $(-5)^3 = (-5)(-5)(-5) = -125$

> An odd number of negative factors gives a negative product.

d. $\left(-\dfrac{2}{3}\right)^4 = \left(-\dfrac{2}{3}\right)\left(-\dfrac{2}{3}\right)\left(-\dfrac{2}{3}\right)\left(-\dfrac{2}{3}\right) = \dfrac{16}{81}$

> Base is $-\frac{2}{3}$.   An even number of negative factors gives a positive product. ■

✓ **CHECK POINT 5** Evaluate:

a. $(-5)^2$ **b.** -5^2 **c.** $(-4)^3$ **d.** $\left(-\dfrac{3}{5}\right)^4$.

7 Divide real numbers. ▶

Dividing Real Numbers

If a and b are real numbers and b is not 0, then the quotient of a and b is defined as follows:

$$a \div b = a \cdot \frac{1}{b} \quad \text{or} \quad \frac{a}{b} = a \cdot \frac{1}{b}.$$

Thus, to find the quotient of a and b, we can divide by b or multiply by $\frac{1}{b}$. The nonzero real numbers b and $\frac{1}{b}$ are called **reciprocals**, or **multiplicative inverses**, of one another. When reciprocals are multiplied, their product is 1:

$$b \cdot \frac{1}{b} = 1.$$

Because division is defined in terms of multiplication, the sign rules for dividing numbers are the same as the sign rules for multiplying them.

Dividing Real Numbers

The quotient of two numbers with different signs is negative. The quotient of two numbers with the same sign is positive. The quotient is found by dividing absolute values.

EXAMPLE 6   Dividing Real Numbers

Divide:

a. $\dfrac{20}{-5}$ **b.** $-\dfrac{3}{4} \div \left(-\dfrac{5}{9}\right)$.

Solution

a. $\dfrac{20}{-5} = -4$ Divide absolute values: $\frac{20}{5} = 4.$

Different signs: negative quotient

b. $-\dfrac{3}{4} \div \left(-\dfrac{5}{9}\right) = \dfrac{27}{20}$ Divide absolute values: $\frac{3}{4} \div \frac{5}{9} = \frac{3}{4} \cdot \frac{9}{5} = \frac{27}{20}.$

Same sign: positive quotient   ∎

✓ **CHECK POINT 6**   Divide:

a. $\dfrac{32}{-4}$ **b.** $-\dfrac{2}{3} \div \left(-\dfrac{5}{4}\right)$.

We must be careful with division when 0 is involved. Zero divided by any nonzero real number is 0. For example,

$$\frac{0}{-5} = 0.$$

Can you see why $\dfrac{0}{-5}$ must be 0? The definition of division tells us that

$$\frac{0}{-5} = 0 \cdot \left(-\frac{1}{5}\right)$$

and the product of 0 and any real number is 0. By contrast, what happens if we divide −5 by 0. The answer must be a number that, when multiplied by 0, gives −5. However, any number multiplied by 0 is 0. Thus, we cannot divide −5, or any other real number, by 0.

Division by Zero

Division by zero is not allowed; it is undefined. A real number can never have a denominator of 0.

8 Use the order of operations. ▶

Order of Operations

The rules for order of operations can be applied to positive and negative real numbers. Recall that if no grouping symbols are present, we

- Evaluate exponential expressions.
- Multiply and divide, from left to right.
- Add and subtract, from left to right.

EXAMPLE 7   Using the Order of Operations

Simplify: $4 - 7^2 + 8 \div 2(-3)^2.$

Solution

$$4 - 7^2 + 8 \div 2(-3)^2$$

$$= 4 - 49 + 8 \div 2(9)$$ Evaluate exponential expressions:
$7^2 = 7 \cdot 7 = 49$ and $(-3)^2 = (-3)(-3) = 9$.

$$= 4 - 49 + 4(9)$$ Divide: $8 \div 2 = 4$.

$$= 4 - 49 + 36$$ Multiply: $4(9) = 36$.

$$= -45 + 36$$ Subtract: $4 - 49 = 4 + (-49) = -45$.

$$= -9$$ Add. ■

✓ **CHECK POINT 7**   Simplify: $3 - 5^2 + 12 \div 2(-4)^2$.

If an expression contains grouping symbols, we perform operations within these symbols first. Common grouping symbols are parentheses, brackets, and braces. Other grouping symbols include fraction bars, absolute value symbols, and radical symbols, such as square root signs ($\sqrt{\ }$).

EXAMPLE 8   Using the Order of Operations

Simplify: $\dfrac{13 - 3(-2)^4}{3 - (6 - 10)}$.

Solution   Simplify the numerator and the denominator separately. Then divide.

$$\dfrac{13 - 3(-2)^4}{3 - (6 - 10)}$$ Evaluate the exponential expression in the numerator: $(-2)^4 = (-2)(-2)(-2)(-2) = 16$.

$$= \dfrac{13 - 3(16)}{3 - (-4)}$$ Subtract inside parentheses in the denominator: $6 - 10 = 6 + (-10) = -4$.

$$= \dfrac{13 - 48}{7}$$ Multiply in the numerator: $3(16) = 48$. Subtract in the denominator: $3 - (-4) = 3 + 4 = 7$.

$$= \dfrac{-35}{7}$$ Subtract in the numerator: $13 - 48 = 13 + (-48) = -35$.

$$= -5$$ Divide. ■

✓ **CHECK POINT 8**   Simplify: $\dfrac{4 + 3(-2)^3}{2 - (6 - 9)}$.

9 Use commutative, associative, and distributive properties. ▶

The Commutative, Associative, and Distributive Properties

Basic algebraic properties enable us to write *equivalent algebraic expressions*. Two algebraic expressions that have the same value for all replacements are called **equivalent algebraic expressions**. In Section 1.4, you will use such expressions to solve equations.

In arithmetic, when two numbers are added or multiplied, the order in which the numbers are written does not affect the answer. These facts are called **commutative properties**.

The Commutative Properties

Let a and b represent real numbers, variables, or algebraic expressions.

$$\text{Addition:} \quad a + b = b + a$$

$$\text{Multiplication:} \quad ab = ba$$

Changing order when adding or multiplying does not affect a sum or product.

EXAMPLE 9   Using the Commutative Properties

Write an algebraic expression equivalent to $3x + 7$ using each of the commutative properties.

Solution

Commutative of Addition	Commutative of Multiplication
$3x + 7 = 7 + 3x$	$3x + 7 = x \cdot 3 + 7$
Change the order of the addition.	Change the order of the multiplication.

■

✓ **CHECK POINT 9**   Write an algebraic expression equivalent to $4x + 9$ using each of the commutative properties.

The **associative properties** enable us to form equivalent expressions by regrouping.

The Associative Properties

Let a, b, and c represent real numbers, variables, or algebraic expressions.

$$\text{Addition:} \quad (a + b) + c = a + (b + c)$$
$$\text{Multiplication:} \quad (ab)c = a(bc)$$

Changing grouping when adding or multiplying does not affect a sum or product.

EXAMPLE 10   Using the Associative Properties

Use an associative property to write an equivalent expression and simplify:

a. $7 + (3 + x)$ **b.** $-6(5x)$.

Solution

a. $7 + (3 + x) = (7 + 3) + x = 10 + x$
b. $-6(5x) = (-6 \cdot 5)x = -30x$ ■

✓ **CHECK POINT 10**   Use an associative property to write an equivalent expression and simplify:

a. $6 + (12 + x)$ **b.** $-7(4x)$.

The **distributive property** allows us to rewrite the product of a number and a sum as the sum of two products.

The Distributive Property

Let a, b, and c represent real numbers, variables, or algebraic expressions.

$$a(b + c) = ab + ac$$

Multiplication distributes over addition.

Great Question!

What's the most important thing I should keep in mind when using the distributive property?

When using a distributive property to remove parentheses, be sure to multiply *each term* inside the parentheses by the factor outside.

Incorrect

$5(3y + 7) = 5 \cdot 3y + 7$
$= 15y + 7$

EXAMPLE 11   Using the Distributive Property

Use the distributive property to write an equivalent expression:

$$-2(3x + 5).$$

Solution

$$-2(3x + 5) = -2 \cdot 3x + (-2) \cdot 5 = -6x + (-10) = -6x - 10 \quad \blacksquare$$

✓ **CHECK POINT 11**   Use the distributive property to write an equivalent expression: $-4(7x + 2)$.

Table 1.4 shows a number of other forms of the distributive property.

Table 1.4   Other Forms of the Distributive Property

Property	Meaning	Example
$a(b - c) = ab - ac$	Multiplication distributes over subtraction.	$5(4x - 3) = 5 \cdot 4x - 5 \cdot 3$ $= 20x - 15$
$a(b + c + d) = ab + ac + ad$	Multiplication distributes over three or more terms in parentheses.	$4(x + 10 + 3y)$ $= 4x + 4 \cdot 10 + 4 \cdot 3y$ $= 4x + 40 + 12y$
$(b + c)a = ba + ca$	Multiplication on the right distributes over addition (or subtraction).	$(x + 7)9 = x \cdot 9 + 7 \cdot 9$ $= 9x + 63$

10 Simplify algebraic expressions. ▶

Combining Like Terms and Simplifying Algebraic Expressions

The **terms** of an algebraic expression are those parts that are separated by addition. For example, consider the algebraic expression

$$7x - 9y + z - 3,$$

which can be expressed as

$$7x + (-9y) + z + (-3).$$

This expression contains four terms, namely $7x$, $-9y$, z, and -3.

The numerical part of a term is called its **coefficient**. In the term $7x$, the 7 is the coefficient. If a term containing one or more variables is written without a coefficient, the coefficient is understood to be 1. Thus, z means $1z$. If a term is a constant, its coefficient is that constant. Thus, the coefficient of the constant term -3 is -3.

$$7x + (-9y) + z + (-3)$$

Coefficient is 7.   Coefficient is −9.   Coefficient is 1; z means $1z$.   Coefficient is −3.

The parts of each term that are multiplied are called the **factors** of the term. The factors of the term $7x$ are 7 and x.

Like terms are terms that have exactly the same variable factors. For example, $3x$ and $7x$ are like terms. The distributive property in the form

$$ba + ca = (b + c)a$$

enables us to add or subtract like terms. For example,

$$3x + 7x = (3 + 7)x = 10x$$
$$7y^2 - y^2 = 7y^2 - 1y^2 = (7 - 1)y^2 = 6y^2$$

This process is called **combining like terms**.

An algebraic expression is **simplified** when grouping symbols have been removed and like terms have been combined.

Great Question!

What is the bottom line for combining like terms?

To combine like terms mentally, add or subtract the coefficients of the terms. Use this result as the coefficient of the terms' variable factor(s).

EXAMPLE 12   Simplifying an Algebraic Expression

Simplify: $7x + 12x^2 + 3x + x^2$.

Solution

$$7x + 12x^2 + 3x + x^2$$

$$= (7x + 3x) + (12x^2 + x^2)$$ Rearrange terms and group like terms using commutative and associative properties. This step is often done mentally.

$x^2 = 1x^2$

$$= (7 + 3)x + (12 + 1)x^2$$ Apply the distributive property.
$$= 10x + 13x^2$$ Simplify. Because $10x$ and $13x^2$ are not like terms, this is the final answer.

Using the commutative property of addition, we can write this simplified expression as $13x^2 + 10x$. ∎

✓ **CHECK POINT 12**   Simplify: $3x + 14x^2 + 11x + x^2$.

EXAMPLE 13   Simplifying an Algebraic Expression

Simplify: $4(7x - 3) - 10x$.

Solution

$$4(7x - 3) - 10x$$

$$= 4 \cdot 7x - 4 \cdot 3 - 10x$$ Use the distributive property to remove the parentheses.

$$= 28x - 12 - 10x$$ Multiply.
$$= (28x - 10x) - 12$$ Group like terms.
$$= (28 - 10)x - 12$$ Apply the distributive property.
$$= 18x - 12$$ Simplify. ∎

✓ **CHECK POINT 13**   Simplify: $8(2x - 5) - 4x$.

It is not uncommon to see algebraic expressions with parentheses preceded by a negative sign or subtraction. An expression of the form $-(b + c)$ can be simplified as follows:

$$-(b + c) = -1(b + c) = (-1)b + (-1)c = -b + (-c) = -b - c.$$

Do you see a fast way to obtain the simplified expression on the right? **If a negative sign or a subtraction symbol appears outside parentheses, drop the parentheses and change the sign of every term within the parentheses.** For example,

$$-(3x^2 - 7x - 4) = -3x^2 + 7x + 4.$$

EXAMPLE 14 Simplifying an Algebraic Expression

Simplify: $8x + 2[5 - (x - 3)]$.

Solution

$$8x + 2[5 - (x - 3)]$$
$$= 8x + 2[5 - x + 3]$$ Drop parentheses and change the sign of each term in parentheses: $-(x - 3) = -x + 3$.
$$= 8x + 2[8 - x]$$ Simplify inside brackets: $5 + 3 = 8$.
$$= 8x + 16 - 2x$$ Apply the distributive property:
$$2[8 - x] = 2 \cdot 8 - 2x = 16 - 2x.$$
$$= (8x - 2x) + 16$$ Group like terms.
$$= (8 - 2)x + 16$$ Apply the distributive property.
$$= 6x + 16$$ Simplify. ∎

✓ **CHECK POINT 14**   Simplify: $6 + 4[7 - (x - 2)]$.

CONCEPT AND VOCABULARY CHECK

In items 1–7, state whether each result is a positive number, a negative number, or 0. Do not actually perform the computation.

1. $-16 + (-30)$ _____

2. $-16 + 16$ _____

3. $-(-16)$ _____

4. $|-16|$ _____

5. $-16 - (-30)$ _____

6. $(-16)^3$ _____

7. $(-16)(-30)$ _____

In items 8–9, use the choices below to fill in each blank:

   add   subtract   multiply   divide.

8. To simplify $4 - 49 + 8 \div 4 \cdot 2$, first _____.

9. To simplify $15 - [3 - (-1)] + 12 \div 2 \cdot 3$, first _____.

In the remaining items, fill in each blank so that the resulting statement is true.

10. $|a|$, called the _____ of a, represents the distance from _____ to _____ on the number line.

11. If $a \geq 0$, then $|a| = $ _____. If $a < 0$, then $|a| = $ _____.

12. $a + (-a) = $ _____ : The sum of a real number and its additive _____ is _____, the additive _____.

13. If a and b are real numbers, the commutative property of addition states that $a + b = $ _____.

14. If a, b, and c are real numbers, the associative property of multiplication states that _____ $= a(bc)$.

15. If a, b, and c are real numbers, the distributive property states that $a(b + c) = $ _____.

16. An algebraic expression is _____ when parentheses have been removed and like terms have been combined.

1.2 EXERCISE SET   ▶ MyMathLab®

Practice Exercises

In Exercises 1–12, find each absolute value.

1. $|-7|$
2. $|-10|$
3. $|4|$
4. $|13|$
5. $|-7.6|$
6. $|-8.3|$
7. $\left|\dfrac{\pi}{2}\right|$
8. $\left|\dfrac{\pi}{3}\right|$
9. $|-\sqrt{2}|$
10. $|-\sqrt{3}|$
11. $-\left|-\dfrac{2}{5}\right|$
12. $-\left|-\dfrac{7}{10}\right|$

In Exercises 13–28, add as indicated.

13. $-3 + (-8)$
14. $-5 + (-10)$
15. $-14 + 10$
16. $-15 + 6$
17. $-6.8 + 2.3$
18. $-7.9 + 2.4$
19. $\dfrac{11}{15} + \left(-\dfrac{3}{5}\right)$
20. $\dfrac{7}{10} + \left(-\dfrac{4}{5}\right)$
21. $-\dfrac{2}{9} - \dfrac{3}{4}$
22. $-\dfrac{3}{5} - \dfrac{4}{7}$
23. $-3.7 + (-4.5)$
24. $-6.2 + (-5.9)$
25. $0 + (-12.4)$
26. $0 + (-15.3)$
27. $12.4 + (-12.4)$
28. $15.3 + (-15.3)$

In Exercises 29–34, find $-x$ for the given value of x.

29. $x = 11$
30. $x = 13$
31. $x = -5$
32. $x = -9$
33. $x = 0$
34. $x = -\sqrt{2}$

In Exercises 35–46, subtract as indicated.

35. $3 - 15$
36. $4 - 20$
37. $8 - (-10)$
38. $7 - (-13)$
39. $-20 - (-5)$
40. $-30 - (-10)$
41. $\dfrac{1}{4} - \dfrac{1}{2}$
42. $\dfrac{1}{10} - \dfrac{2}{5}$
43. $-2.3 - (-7.8)$
44. $-4.3 - (-8.7)$
45. $0 - (-\sqrt{2})$
46. $0 - (-\sqrt{3})$

In Exercises 47–58, multiply as indicated.

47. $9(-10)$
48. $8(-10)$
49. $(-3)(-11)$
50. $(-7)(-11)$
51. $\dfrac{15}{13}(-1)$
52. $\dfrac{11}{13}(-1)$
53. $-\sqrt{2} \cdot 0$
54. $-\sqrt{3} \cdot 0$
55. $(-4)(-2)(-1)$
56. $(-5)(-3)(-2)$
57. $2(-3)(-1)(-2)(-4)$
58. $3(-2)(-1)(-5)(-3)$

In Exercises 59–70, evaluate each exponential expression.

59. $(-10)^2$
60. $(-8)^2$
61. -10^2
62. -8^2
63. $(-2)^3$
64. $(-3)^3$
65. $(-1)^4$
66. $(-4)^4$
67. $(-1)^{33}$
68. $(-1)^{35}$
69. $-\left(-\dfrac{1}{2}\right)^3$
70. $-\left(-\dfrac{1}{4}\right)^3$

In Exercises 71–82, divide as indicated or state that the division is undefined.

71. $\dfrac{12}{-4}$
72. $\dfrac{30}{-5}$
73. $\dfrac{-90}{-2}$
74. $\dfrac{-55}{-5}$
75. $\dfrac{0}{-4.6}$
76. $\dfrac{0}{-5.3}$
77. $-\dfrac{4.6}{0}$
78. $-\dfrac{5.3}{0}$
79. $-\dfrac{1}{2} \div \left(-\dfrac{7}{9}\right)$
80. $-\dfrac{1}{2} \div \left(-\dfrac{3}{5}\right)$
81. $6 \div \left(-\dfrac{2}{5}\right)$
82. $8 \div \left(-\dfrac{2}{9}\right)$

In Exercises 83–100, use the order of operations to simplify each expression.

83. $4(-5) - 6(-3)$
84. $8(-3) - 5(-6)$
85. $3(-2)^2 - 4(-3)^2$
86. $5(-3)^2 - 2(-2)^2$
87. $8^2 - 16 \div 2^2 \cdot 4 - 3$
88. $10^2 - 100 \div 5^2 \cdot 2 - 3$
89. $\dfrac{5 \cdot 2 - 3^2}{[3^2 - (-2)]^2}$
90. $\dfrac{10 \div 2 + 3 \cdot 4}{(12 - 3 \cdot 2)^2}$
91. $8 - 3[-2(2 - 5) - 4(8 - 6)]$
92. $8 - 3[-2(5 - 7) - 5(4 - 2)]$
93. $\dfrac{2(-2) - 4(-3)}{5 - 8}$
94. $\dfrac{6(-4) - 5(-3)}{9 - 10}$
95. $\dfrac{(5 - 6)^2 - 2|3 - 7|}{89 - 3 \cdot 5^2}$
96. $\dfrac{12 \div 3 \cdot 5 |2^2 + 3^2|}{7 + 3 - 6^2}$
97. $15 - \sqrt{3 - (-1)} + 12 \div 2 \cdot 3$
98. $17 - |5 - (-2)| + 12 \div 2 \cdot 3$
99. $20 + 1 - \sqrt{10^2 - (5 + 1)^2}(-2)$
100. $24 \div \sqrt{3 \cdot (5 - 2)} \div [-1 - (-3)]^2$

In Exercises 101–104, write an algebraic expression equivalent to the given expression using each of the commutative properties.

101. $4x + 10$

102. $5x + 30$

103. $7x - 5$

104. $3x - 7$

In Exercises 105–110, use an associative property to write an algebraic expression equivalent to each expression and simplify.

105. $4 + (6 + x)$

106. $12 + (3 + x)$

107. $-7(3x)$

108. $-10(5x)$

109. $-\dfrac{1}{3}(-3y)$

110. $-\dfrac{1}{4}(-4y)$

In Exercises 111–116, use the distributive property to write an equivalent expression.

111. $3(2x + 5)$ **112.** $5(4x + 7)$

113. $-7(2x + 3)$ **114.** $-9(3x + 2)$

115. $-(3x - 6)$ **116.** $-(6x - 3)$

In Exercises 117–130, simplify each algebraic expression.

117. $7x + 5x$ **118.** $8x + 10x$

119. $6x^2 - x^2$ **120.** $9x^2 - x^2$

121. $6x + 10x^2 + 4x + 2x^2$

122. $9x + 5x^2 + 3x + 4x^2$

123. $8(3x - 5) - 6x$

124. $7(4x - 5) - 8x$

125. $5(3y - 2) - (7y + 2)$

126. $4(5y - 3) - (6y + 3)$

127. $7 - 4[3 - (4y - 5)]$

128. $6 - 5[8 - (2y - 4)]$

129. $18x^2 + 4 - [6(x^2 - 2) + 5]$

130. $14x^2 + 5 - [7(x^2 - 2) + 4]$

Practice PLUS

In Exercises 131–138, write each English phrase as an algebraic expression. Then simplify the expression. Let x represent the number.

131. A number decreased by the sum of the number and four

132. A number decreased by the difference between eight and the number

133. Six times the product of negative five and a number

134. Ten times the product of negative four and a number

135. The difference between the product of five and a number and twice the number

136. The difference between the product of six and a number and negative two times the number

137. The difference between eight times a number and six more than three times the number

138. Eight decreased by three times the sum of a number and six

Application Exercises

The number line shows the approval ratings by Americans of five selected countries. Use this information to solve Exercises 139–146.

**Approval Ratings by Americans of
Selected Countries**

50: Very Favorable
0: Neutral
−50: Very Unfavorable

Source: www.thechicagocouncil.org

139. What is the combined approval rating of the UK and Iran?

140. What is the combined approval rating of Israel and China?

141. What is the difference between the approval rating of the UK and the approval rating of Iran?

142. What is the difference between the approval rating of Israel and the approval rating of China?

143. By how much does the approval rating of France exceed the approval rating of China?

144. By how much does the approval rating of France exceed the approval rating of Iran?

145. What is the average of the approval ratings for China, France, and Israel?

146. What is the average of the approval ratings for Iran, China, and the UK?

We opened the section by noting that college students have money to spend, often courtesy of Mom and Dad. The bar graph shows discretionary spending on non-essentials, excluding food and college costs, by full- and part-time college students in the United States for four selected years from 2007 through 2013.

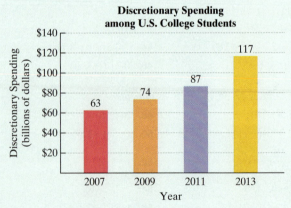

Source: Alloy Media and Marketing

The data in the graph can be modeled by the formula

$$D = 1.2x^2 + 1.6(x + 40),$$

where D represents discretionary spending among U.S. college students, in billions of dollars, x years after 2007. Use this formula to solve Exercises 147–148.

147. According to the formula on the previous page, how much money did college students spend in 2013? Does the model underestimate or overestimate the amount of discretionary spending displayed by the bar graph? By how much?

148. According to the formula on the previous page, how much money did college students spend in 2011? Does the model underestimate or overestimate the amount of discretionary spending displayed by the bar graph? By how much?

149. You had $10,000 to invest. You put x dollars in a safe, government-insured certificate of deposit paying 5% per year. You invested the remainder of the money in noninsured corporate bonds paying 12% per year. Your total interest earned at the end of the year is given by the algebraic expression

$$0.05x + 0.12(10,000 - x).$$

a. Simplify the algebraic expression.

b. Use each form of the algebraic expression to determine your total interest earned at the end of the year if you invested $6000 in the safe, government-insured certificate of deposit.

150. It takes you 50 minutes to get to campus. You spend t minutes walking to the bus stop and the rest of the time riding the bus. Your walking rate is 0.06 mile per minute and the bus travels at a rate of 0.5 mile per minute. The total distance walking and traveling by bus is given by the algebraic expression

$$0.06t + 0.5(50 - t).$$

a. Simplify the algebraic expression.

b. Use each form of the algebraic expression to determine the total distance that you travel if you spend 20 minutes walking to the bus stop.

Explaining the Concepts

151. What is the meaning of $|a|$ in terms of a number line?

152. Explain how to add two numbers with the same sign. Give an example with your explanation.

153. Explain how to add two numbers with different signs. Give an example with your explanation.

154. What are opposites, or additive inverses? What happens when finding the sum of a number and its opposite?

155. Explain how to subtract real numbers.

156. Explain how to multiply two numbers with different signs. Give an example with your explanation.

157. Explain how to multiply two numbers with the same sign. Give an example with your explanation.

158. Explain how to determine the sign of a product that involves more than two numbers.

159. Explain how to divide real numbers.

160. Why is $\frac{0}{4} = 0$, although $\frac{4}{0}$ is undefined?

161. What are equivalent algebraic expressions?

162. State a commutative property and give an example of how it is used to write equivalent algebraic expressions.

163. State an associative property and give an example of how it is used to write equivalent algebraic expressions.

164. State a distributive property and give an example of how it is used to write equivalent algebraic expressions.

165. What are the terms of an algebraic expression? How can you tell if terms are like terms?

166. What does it mean to simplify an algebraic expression?

167. If a negative sign appears outside parentheses, explain how to simplify the expression. Give an example.

Critical Thinking Exercises

Make Sense? In Exercises 168–171, determine whether each statement makes sense or does not make sense, and explain your reasoning.

168. My mathematical model, although it contains an algebraic expression that is not simplified, describes the data perfectly well, so it will describe the data equally well when simplified.

169. Subtraction actually means the addition of an additive inverse.

170. The terms $13x^2$ and $10x$ both contain the variable x, so I can combine them to obtain $23x^3$.

171. There is no number in front of the term x, so this means that the term has no coefficient.

In Exercises 172–176, determine whether each statement is true or false. If the statement is false, make the necessary change(s) to produce a true statement.

172. $16 \div 4 \cdot 2 = 16 \div 8 = 2$

173. $6 - 2(4 + 3) = 4(4 + 3) = 4(7) = 28$

174. $5 + 3(x - 4) = 8(x - 4) = 8x - 32$

175. $-x - x = -x + (-x) = 0$

176. $x - 0.02(x + 200) = 0.98x - 4$

In Exercises 177–178, insert parentheses to make each statement true.

177. $8 - 2 \cdot 3 - 4 = 14$

178. $2 \cdot 5 - \frac{1}{2} \cdot 10 \cdot 9 = 45$

179. Simplify: $\dfrac{9[4 - (1 + 6)] - (3 - 9)^2}{5 + \dfrac{12}{5 - \dfrac{6}{2 + 1}}}$.

Review Exercises

From here on, each Exercise Set will contain three review exercises. It is important to review previously covered topics to improve your understanding of the topics and to help maintain your mastery of the material. If you are not certain how to solve a review exercise, turn to the section and the worked-out example given in parentheses at the end of each exercise.

180. Write the following English phrase as an algebraic expression: "The quotient of ten and a number, decreased by four times the number." Let x represent the number. (Section 1.1, Example 1)

181. Evaluate $10 + 2(x - 5)^4$ for $x = 7$.
(Section 1.1, Example 3)

182. Graph $(-5, \infty)$ on a number line.
(Section 1.1, Example 7)

Preview Exercises

Exercises 183–185 will help you prepare for the material covered in the next section.

183. If $y = 4 - x^2$, find the value of y that corresponds to values of x for each integer starting with -3 and ending with 3.

184. If $y = 1 - x^2$, find the value of y that corresponds to values of x for each integer starting with -3 and ending with 3.

185. If $y = |x + 1|$, find the value of y that corresponds to values of x for each integer starting with -4 and ending with 2.

1.3

Graphing Equations

What am I supposed to learn?

After studying this section, you should be able to:

1 Plot points in the rectangular coordinate system.

2 Graph equations in the rectangular coordinate system.

3 Use the rectangular coordinate system to visualize relationships between variables.

4 Interpret information about a graphing utility's viewing rectangle or table.

The beginning of the seventeenth century was a time of innovative ideas and enormous intellectual progress in Europe. English theatergoers enjoyed a succession of exciting new plays by Shakespeare. William Harvey proposed the radical notion that the heart was a pump for blood rather than the center of emotion. Galileo, with his new-fangled invention called the telescope, supported the theory of Polish astronomer Copernicus that the sun, not the Earth, was the center of the solar system. Monteverdi was writing the world's first grand operas. French mathematicians Pascal and Fermat invented a new field of mathematics called probability theory.

Into this arena of intellectual electricity stepped French aristocrat René Descartes (1596–1650). Descartes (pronounced "day cart"), propelled by the creativity surrounding him, developed a new branch of mathematics that brought together algebra and geometry in a unified way—a way that visualized numbers as points on a graph, equations as geometric figures, and geometric figures as equations. This new branch of mathematics, called *analytic geometry*, established Descartes as one of the founders of modern thought and among the most original mathematicians and philosophers of any age. We begin this section by looking at Descartes's deceptively simple idea, called the **rectangular coordinate system** or (in his honor) the **Cartesian coordinate system.**

1 Plot points in the rectangular coordinate system.

Points and Ordered Pairs

Descartes used two number lines that intersect at right angles at their zero points, as shown in **Figure 1.9**. The horizontal number line is the *x*-axis. The vertical number line is the *y*-axis. The point of intersection of these axes is their zero points, called the **origin**.

Figure 1.9 The rectangular coordinate system

Great Question!

What's the significance of the word "ordered" when describing a pair of real numbers?

The phrase *ordered pair* is used because order is important. The order in which coordinates appear makes a difference in a point's location. This is illustrated in **Figure 1.10**.

Positive numbers are shown to the right and above the origin. Negative numbers are shown to the left and below the origin. The axes divide the plane into four quarters, called **quadrants**. The points located on the axes are not in any quadrant.

Each point in the rectangular coordinate system corresponds to an **ordered pair** of real numbers, (x, y). Examples of such pairs are $(-5, 3)$ and $(3, -5)$. The first number in each pair, called the **x-coordinate**, denotes the distance and direction from the origin along the x-axis. The second number, called the **y-coordinate**, denotes vertical distance and direction along a line parallel to the y-axis or along the y-axis itself.

Figure 1.10 shows how we **plot**, or locate, the points corresponding to the ordered pairs $(-5, 3)$ and $(3, -5)$. We plot $(-5, 3)$ by going 5 units from 0 to the left along the x-axis. Then we go 3 units up parallel to the y-axis. We plot $(3, -5)$ by going 3 units from 0 to the right along the x-axis and 5 units down parallel to the y-axis. The phrase "the points corresponding to the ordered pairs $(-5, 3)$ and $(3, -5)$" is often abbreviated as "the points $(-5, 3)$ and $(3, -5)$."

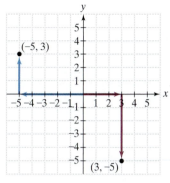

Figure 1.10 Plotting $(-5, 3)$ and $(3, -5)$

Great Question!

Isn't it easy to confuse the notation for an ordered pair and the notation for an open interval? For example, how do I know if $(-4, 5)$ represents a point in the rectangular coordinate system or the set of numbers between, but not including, -4 and 5?

The only way to tell the difference between ordered-pair notation and open-interval notation involves the situation in which the notation is used. If the situation involves plotting points, $(-4, 5)$ is an ordered pair. If the situation involves a set of numbers along a number line, $(-4, 5)$ is an open interval of numbers.

② Graph equations in the rectangular coordinate system. ▶

EXAMPLE 1 Plotting Points in the Rectangular Coordinate System

Plot the points: $A(-4, 5)$, $B(3, -4)$, $C(-5, 0)$, $D(-4, -2)$, $E(0, 3.5)$, and $F(0, 0)$.

Solution See **Figure 1.11**. We move from the origin and **plot** the points in the following way:

$A(-4, 5)$:	4 units left, 5 units up
$B(3, -4)$:	3 units right, 4 units down
$C(-5, 0)$:	5 units left, 0 units up or down
$D(-4, -2)$:	4 units left, 2 units down
$E(0, 3.5)$:	0 units right or left, 3.5 units up
$F(0, 0)$:	0 units right or left, 0 units up or down

Notice that the origin is represented by (0, 0).

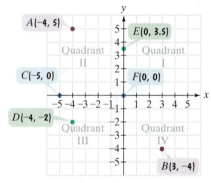

Figure 1.11 Plotting points ■

☑ CHECK POINT 1 Plot the points: $A(2, 5)$, $B(-1, 3)$, $C(-1.5, -4.5)$, and $D(0, -2)$.

Graphs of Equations

A relationship between two quantities can be expressed as an **equation in two variables**, such as

$$y = 4 - x^2.$$

A **solution of an equation in two variables**, x and y, is an ordered pair of real numbers with the following property: When the x-coordinate is substituted for x and the y-coordinate is substituted for y in the equation, we obtain a true statement. For example, consider the equation $y = 4 - x^2$ and the ordered pair $(3, -5)$. When 3 is substituted for x and -5 is substituted for y, we obtain the statement $-5 = 4 - 3^2$, or $-5 = 4 - 9$, or $-5 = -5$. Because this statement is true, the ordered pair $(3, -5)$ is a solution of the equation $y = 4 - x^2$. We also say that $(3, -5)$ **satisfies** the equation.

We can generate as many ordered-pair solutions as desired to $y = 4 - x^2$ by substituting numbers for x and then finding the corresponding values for y. For example, suppose we let $x = 3$:

		Form the ordered pair (x, y).
Start with x.	Compute y.	
x	**$y = 4 - x^2$**	**Ordered Pair (x, y)**
3	$y = 4 - 3^2 = 4 - 9 = -5$	$(3, -5)$
Let $x = 3$.		$(3, -5)$ is a solution of $y = 4 - x^2$.

The **graph of an equation in two variables** is the set of all points whose coordinates satisfy the equation. One method for graphing such equations is the **point-plotting method**. First, we find several ordered pairs that are solutions of the equation. Next, we plot these ordered pairs as points in the rectangular coordinate system. Finally, we connect the points with a smooth curve or line. This often gives us a picture of all ordered pairs that satisfy the equation.

EXAMPLE 2 Graphing an Equation Using the Point-Plotting Method

Graph $y = 4 - x^2$. Select integers for x, starting with -3 and ending with 3.

Solution For each value of x, we find the corresponding value for y.

We selected integers from -3 to 3, inclusive, to include three negative numbers, 0, and three positive numbers. We also wanted to keep the resulting computations for y relatively simple.

		Form the ordered pair (x, y).
Start with x.	Compute y.	
x	**$y = 4 - x^2$**	**Ordered Pair (x, y)**
-3	$y = 4 - (-3)^2 = 4 - 9 = -5$	$(-3, -5)$
-2	$y = 4 - (-2)^2 = 4 - 4 = 0$	$(-2, 0)$
-1	$y = 4 - (-1)^2 = 4 - 1 = 3$	$(-1, 3)$
0	$y = 4 - 0^2 = 4 - 0 = 4$	$(0, 4)$
1	$y = 4 - 1^2 = 4 - 1 = 3$	$(1, 3)$
2	$y = 4 - 2^2 = 4 - 4 = 0$	$(2, 0)$
3	$y = 4 - 3^2 = 4 - 9 = -5$	$(3, -5)$

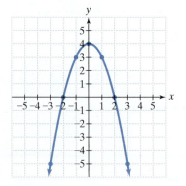

Figure 1.12 The graph of $y = 4 - x^2$

Now we plot the seven points and join them with a smooth curve, as shown in **Figure 1.12.** The graph of $y = 4 - x^2$ is a curve where the part of the graph to the right of the y-axis is a reflection of the part to the left of it and vice versa. The arrows on the left and the right of the curve indicate that it extends indefinitely in both directions. ∎

☑ **CHECK POINT 2** Graph $y = 1 - x^2$. Select integers for x, starting with -3 and ending with 3.

EXAMPLE 3 Graphing an Equation Using the Point-Plotting Method

Graph $y = |x|$. Select integers for x, starting with -3 and ending with 3.

Solution For each value of x, we find the corresponding value for y.

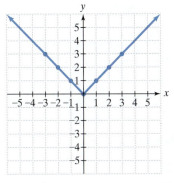

Figure 1.13 The graph of $y = |x|$

x	$y = |x|$	Ordered Pair (x, y)
-3	$y = |-3| = 3$	$(-3, 3)$
-2	$y = |-2| = 2$	$(-2, 2)$
-1	$y = |-1| = 1$	$(-1, 1)$
0	$y = |0| = 0$	$(0, 0)$
1	$y = |1| = 1$	$(1, 1)$
2	$y = |2| = 2$	$(2, 2)$
3	$y = |3| = 3$	$(3, 3)$

We plot the points and connect them, resulting in the graph shown in **Figure 1.13**. The graph is V-shaped and centered at the origin. For every point (x, y) on the graph, the point $(-x, y)$ is also on the graph. This shows that the absolute value of a positive number is the same as the absolute value of its opposite. ∎

✓ **CHECK POINT 3** Graph $y = |x + 1|$. Select integers for x, starting with -4 and ending with 2.

3 Use the rectangular coordinate system to visualize relationships between variables. ⏵

Applications

The rectangular coordinate system allows us to visualize relationships between two variables by associating any equation in two variables with a graph. Graphs in the rectangular coordinate system can also be used to tell a story.

EXAMPLE 4 Telling a Story with a Graph

Too late for that flu shot now! It's only 8 A.M. and you're feeling lousy. Fascinated by the way that algebra models the world (your author is projecting a bit here), you construct a graph showing your body temperature from 8 A.M. through 3 P.M. You decide to let x represent the number of hours after 8 A.M. and y your body temperature at time x. The graph is shown in **Figure 1.14**. The symbol ⌇ on the y-axis shows that there is a break in values between 0 and 98. Thus, the first tick mark on the y-axis represents a temperature of 98°F.

a. What is your temperature at 8 A.M.?

b. During which period of time is your temperature decreasing?

c. Estimate your minimum temperature during the time period shown. How many hours after 8 A.M. does this occur? At what time does this occur?

d. During which period of time is your temperature increasing?

e. Part of the graph is shown as a horizontal line segment. What does this mean about your temperature and when does this occur?

Figure 1.14 Body temperature from 8 A.M. through 3 P.M.

Solution

a. Because x is the number of hours after 8 A.M., your temperature at 8 A.M. corresponds to $x = 0$. Locate 0 on the horizontal axis and look at the point on the graph above 0. **Figure 1.15** shows that your temperature at 8 A.M. is 101°F.

b. Your temperature is decreasing when the graph falls from left to right. This occurs between $x = 0$ and $x = 3$, also shown in **Figure 1.15**. Because x represents the number of hours after 8 A.M., your temperature is decreasing between 8 A.M. and 11 A.M.

Figure 1.15

c. Your minimum temperature can be found by locating the lowest point on the graph. This point lies above 3 on the horizontal axis, shown in **Figure 1.16**. The y-coordinate of this point falls more than midway between 98 and 99, at approximately 98.6. The lowest point on the graph, (3, 98.6), shows that your minimum temperature, 98.6°F, occurs 3 hours after 8 A.M., at 11 A.M.

d. Your temperature is increasing when the graph rises from left to right. This occurs between $x = 3$ and $x = 5$, shown in **Figure 1.16**. Because x represents the number of hours after 8 A.M., your temperature is increasing between 11 A.M. and 1 P.M.

Figure 1.16

e. The horizontal line segment shown in **Figure 1.17** indicates that your temperature is neither increasing nor decreasing. Your temperature remains the same, 100°F, between $x = 5$ and $x = 7$. Thus, your temperature is at a constant 100°F between 1 P.M. and 3 P.M. ■

Figure 1.17

☑ **CHECK POINT 4**   When a physician injects a drug into a patient's muscle, the concentration of the drug in the body, measured in milligrams per 100 milliliters, depends on the time elapsed after the injection, measured in hours. **Figure 1.18** shows the graph of drug concentration over time, where x represents hours after the injection and y represents the drug concentration at time x.

a. During which period of time is the drug concentration increasing?

b. During which period of time is the drug concentration decreasing?

c. What is the drug's maximum concentration and when does this occur?

d. What happens by the end of 13 hours?

Figure 1.18

Graphing Equations and Creating Tables Using a Graphing Utility

4   Interpret information about a graphing utility's viewing rectangle or table. ▶

Graphing calculators and graphing software packages for computers are referred to as **graphing utilities** or graphers. A graphing utility is a powerful tool that quickly generates the graph of an equation in two variables. **Figures 1.19(a)** and **1.19(b)** show two such graphs for the equations in Examples 2 and 3.

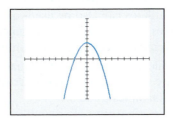

Figure 1.19(a) The graph of $y = 4 - x^2$

Figure 1.19(b) The graph of $y = |x|$

Great Question!

I'm not using a graphing calculator, so should I just skip to the end of this section?

Even if you are not using a graphing utility in the course, read this part of the section. Knowing about viewing rectangles will enable you to understand the graphs that we display in the technology boxes throughout the book.

What differences do you notice between these graphs and the graphs that we drew by hand? They do seem a bit "jittery." Arrows do not appear on the left and right ends of the graphs. Furthermore, numbers are not given along the axes. For both graphs in **Figure 1.19**, the x-axis extends from -10 to 10 and the y-axis also extends from -10 to 10. The distance represented by each consecutive tick mark is one unit. We say that the **viewing rectangle**, or the **viewing window**, is $[-10, 10, 1]$ by $[-10, 10, 1]$.

To graph an equation in x and y using a graphing utility, enter the equation and specify the size of the viewing rectangle. The size of the viewing rectangle sets minimum and maximum values for both the x- and y-axes. Enter these values, as well as the values between consecutive tick marks, on the respective axes. The $[-10, 10, 1]$ by $[-10, 10, 1]$ viewing rectangle used in **Figure 1.19** is called the **standard viewing rectangle**.

EXAMPLE 5   Understanding the Viewing Rectangle

What is the meaning of a $[-2, 3, 0.5]$ by $[-10, 20, 5]$ viewing rectangle?

Solution   We begin with $[-2, 3, 0.5]$, which describes the x-axis. The minimum x-value is -2 and the maximum x-value is 3. The distance between consecutive tick marks is 0.5.

Next, consider $[-10, 20, 5]$, which describes the y-axis. The minimum y-value is -10 and the maximum y-value is 20. The distance between consecutive tick marks is 5.

Figure 1.20 illustrates a $[-2, 3, 0.5]$ by $[-10, 20, 5]$ viewing rectangle. To make things clearer, we've placed numbers by each tick mark. These numbers do not appear on the axes when you use a graphing utility to graph an equation.   ∎

Figure 1.20 A $[-2, 3, 0.5]$ by $[-10, 20, 5]$ viewing rectangle

---

☑ **CHECK POINT 5**   What is the meaning of a $[-100, 100, 50]$ by $[-100, 100, 10]$ viewing rectangle? Create a figure like the one in **Figure 1.20** that illustrates this viewing rectangle.

On many graphing utilities, the display screen is five-eights as high as it is wide. By using a square setting, you can equally space the x and y tick marks. (This does not occur in the standard viewing rectangle.) Graphing utilities can also *zoom in* and *zoom out*. When you zoom in, you see a smaller portion of the graph, but you do so in greater detail. When you zoom out, you see a larger portion of the graph. Thus, zooming out may help you to develop a better understanding of the overall character of the graph. With practice, you will become more comfortable with graphing equations in two variables using your graphing utility. You will also develop a better sense of the size of the viewing rectangle that will reveal needed information about a particular graph.

Graphing utilities can also be used to create tables showing solutions of equations in two variables. Use the Table Setup function to choose the starting value of x and to input the increment, or change, between the consecutive x-values. The corresponding y-values are calculated based on the equation(s) in two variables in the $\boxed{Y=}$ screen. In **Figure 1.21**, we used a TI-84 Plus C to create a table for $y = 4 - x^2$ and $y_2 = |x|$, the equations in Examples 2 and 3.

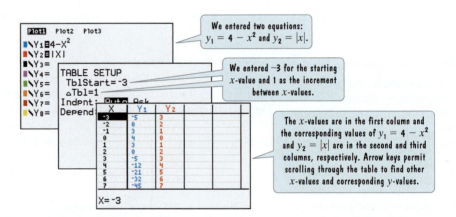

We entered two equations: $y_1 = 4 - x^2$ and $y_2 = |x|$.

We entered -3 for the starting x-value and 1 as the increment between x-values.

The x-values are in the first column and the corresponding values of $y_1 = 4 - x^2$ and $y_2 = |x|$ are in the second and third columns, respectively. Arrow keys permit scrolling through the table to find other x-values and corresponding y-values.

Figure 1.21 Creating a table for $y_1 = 4 - x^2$ and $y_2 = |x|$

Achieving Success

Check! Check! Check!

After completing each Check Point or odd-numbered exercise, compare your answer with the one given in the answer section at the back of the book. To make this process more convenient, place a Post-it® or some other marker at the appropriate page of the answer section. If your answer is different from the one given in the answer section, try to figure out your mistake. Then correct the error. If you cannot determine what went wrong, show your work to your professor. **By recording each step neatly and using as much paper as you need**, your professor will find it easier to determine where you had trouble.

CONCEPT AND VOCABULARY CHECK

Fill in each blank so that the resulting statement is true.

1. In the rectangular coordinate system, the horizontal number line is called the _____.

2. In the rectangular coordinate system, the vertical number line is called the _____.

3. In the rectangular coordinate system, the point of intersection of the horizontal axis and the vertical axis is called the _____.

4. The axes of the rectangular coordinate system divide the plane into regions, called _____. There are _____ of these regions.

5. The first number in an ordered pair such as (8, 3) is called the _____. The second number in such an ordered pair is called the _____.

6. The ordered pair (4, 19) is a/an _____ of the equation $y = x^2 + 3$ because when 4 is substituted for x and 19 is substituted for y, we obtain a true statement. We also say that (4, 19) _____ the equation.

1.3 EXERCISE SET   ▶ MyMathLab®

Practice Exercises

In Exercises 1–10, plot the given point in a rectangular coordinate system.

1. $(1, 4)$ **2.** $(2, 5)$

3. $(-2, 3)$ **4.** $(-1, 4)$

5. $(-3, -5)$ **6.** $(-4, -2)$

7. $(4, -1)$ **8.** $(3, -2)$

9. $(-4, 0)$ **10.** $(0, -3)$

Graph each equation in Exercises 11–26. Let $x = -3, -2, -1, 0,$ 1, 2, and 3.

11. $y = x^2 - 4$ **12.** $y = x^2 - 9$

13. $y = x - 2$ **14.** $y = x + 2$

15. $y = 2x + 1$ **16.** $y = 2x - 4$

17. $y = -\dfrac{1}{2}x$ **18.** $y = -\dfrac{1}{2}x + 2$

19. $y = |x| + 1$ **20.** $y = |x| - 1$

21. $y = 2|x|$ **22.** $y = -2|x|$

23. $y = -x^2$ **24.** $y = -\dfrac{1}{2}x^2$

25. $y = x^3$ **26.** $y = x^3 - 1$

In Exercises 27–30, match the viewing rectangle with the correct figure. Then label the tick marks in the figure to illustrate this viewing rectangle.

27. $[-5, 5, 1]$ by $[-5, 5, 1]$

28. $[-10, 10, 2]$ by $[-4, 4, 2]$

29. $[-20, 80, 10]$ by $[-30, 70, 10]$

30. $[-40, 40, 20]$ by $[-1000, 1000, 100]$

a.   **b.**

c.   **d.**

The table of values was generated by a graphing utility with a TABLE feature. Use the table to solve Exercises 31–38.

X	Y₁	Y₂
-3	9	5
-2	4	4
-1	1	3
0	0	2
1	1	1
2	4	0
3	9	-1
4	16	-2
5	25	-3
6	36	-4
7	49	-5

X= -3

31. Which equation corresponds to Y_2 in the table?

  a. $y_2 = x + 8$ **b.** $y_2 = x - 2$

  c. $y_2 = 2 - x$ **d.** $y_2 = 1 - 2x$

32. Which equation corresponds to Y_1 in the table?

  a. $y_1 = -3x$ **b.** $y_1 = x^2$

  c. $y_1 = -x^2$ **d.** $y_1 = 2 - x$

33. Does the graph of Y_2 pass through the origin?

34. Does the graph of Y_1 pass through the origin?

35. At which point does the graph of Y_2 cross the x-axis?

36. At which point does the graph of Y_2 cross the y-axis?

37. At which points do the graphs of Y_1 and Y_2 intersect?

38. For which values of x is $Y_1 = Y_2$?

Practice PLUS

In Exercises 39–42, write each English sentence as an equation in two variables. Then graph the equation.

39. The y-value is four more than twice the x-value.

40. The y-value is the difference between four and twice the x-value.

41. The y-value is three decreased by the square of the x-value.

42. The y-value is two more than the square of the x-value.

In Exercises 43–46, graph each equation.

43. $y = 5$ (Let $x = -3, -2, -1, 0, 1, 2,$ and 3.)

44. $y = -1$ (Let $x = -3, -2, -1, 0, 1, 2,$ and 3.)

45. $y = \dfrac{1}{x}$ (Let $x = -2, -1, -\dfrac{1}{2}, -\dfrac{1}{3}, \dfrac{1}{3}, \dfrac{1}{2}, 1,$ and 2.)

46. $y = -\dfrac{1}{x}$ (Let $x = -2, -1, -\dfrac{1}{2}, -\dfrac{1}{3}, \dfrac{1}{3}, \dfrac{1}{2}, 1,$ and 2.)

Application Exercises

The line graph shows the percentage of the U.S. population using the Internet from 2007 through 2013. Use the graph to solve Exercises 47–52.

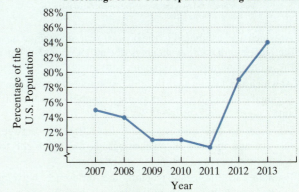

Percentage of the U.S. Population Using the Internet

Source: International Telecommunication Union

(In Exercises 47–52, be sure to refer to the graph at the bottom of the previous page.)

47. For the period shown, in which year did the greatest percentage of the U.S. population use the Internet? What percentage of the population used the Internet in that year?

48. For the period shown, in which year did the least percentage of the U.S. population use the Internet? What percentage of the population used the Internet in that year?

49. For the period shown, between which two years did the percentage of the U.S. population using the Internet remain constant? What percentage of the population used the Internet in those years?

50. For the period shown, between which two years did the percentage of the U.S. population using the Internet increase most rapidly? What was the increase in the percentage using the Internet between those years?

51. For the period shown, between which two years did the percentage of the U.S. population using the Internet decrease most rapidly? What was the decrease in the percentage using the Internet between those years?

52. Between 2007 and 2013, what was the increase in the percentage of the U.S. population using the Internet?

Contrary to popular belief, older people do not need less sleep than younger adults. However, the line graphs show that they awaken more often during the night. The numerous awakenings are one reason why some elderly individuals report that sleep is less restful than it had been in the past. Use the line graphs to solve Exercises 53–56.

Average Number of Awakenings During the Night, by Age and Gender

Source: Stephen Davis and Joseph Palladino, *Psychology*, 5th Edition, Prentice Hall, 2007

53. At which age, estimated to the nearest year, do women have the least number of awakenings during the night? What is the average number of awakenings at that age?

54. At which age do men have the greatest number of awakenings during the night? What is the average number of awakenings at that age?

55. Estimate, to the nearest tenth, the difference between the average number of awakenings during the night for 25-year-old men and for 25-year-old women.

56. Estimate, to the nearest tenth, the difference between the average number of awakenings during the night for 18-year-old men and for 18-year-old women.

In Exercises 57–60, match the story with the correct figure. The figures are labeled (a), (b), (c), and (d).

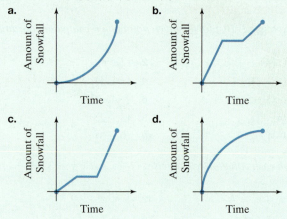

57. As the blizzard got worse, the snow fell harder and harder.

58. The snow fell more and more softly.

59. It snowed hard, but then it stopped. After a short time, the snow started falling softly.

60. It snowed softly, and then it stopped. After a short time, the snow started falling hard.

In Exercises 61–64, select the graph that best illustrates each story.

61. An airplane flew from Miami to San Francisco.

62. At noon, you begin to breathe in.

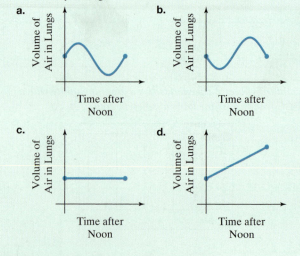

63. Measurements are taken of a person's height from birth to age 100.

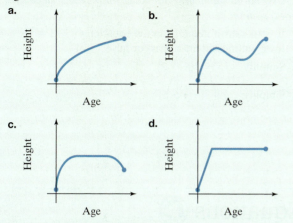

64. You begin your bike ride by riding down a hill. Then you ride up another hill. Finally, you ride along a level surface before coming to a stop.

Explaining the Concepts

65. What is the rectangular coordinate system?

66. Explain how to plot a point in the rectangular coordinate system. Give an example with your explanation.

67. Explain why $(5, -2)$ and $(-2, 5)$ do not represent the same point.

68. Explain how to graph an equation in the rectangular coordinate system.

69. What does a $[-20, 2, 1]$ by $[-4, 5, 0.5]$ viewing rectangle mean?

In Exercises 70–71, write a story, or description, to match each title and graph.

70. Checking Account Balance

71. Hair Length

Technology Exercise

72. Use a graphing utility to verify each of your hand-drawn graphs in Exercises 11–26. Experiment with the viewing rectangle to make the graph displayed by the graphing utility resemble your hand-drawn graph as much as possible.

Critical Thinking Exercises

Make Sense? *In Exercises 73–76, determine whether each statement makes sense or does not make sense, and explain your reasoning.*

73. The rectangular coordinate system provides a geometric picture of what an equation in two variables looks like.

74. There is something wrong with my graphing utility because it is not displaying numbers along the *x*- and *y*-axes.

75. A horizontal line is not a graph that tells the story of the number of calories that I burn throughout the day.

76. I told my story with a graph, so I can be confident that there is a mathematical model that perfectly describes the graph's data.

In Exercises 77–80, determine whether each statement is true or false. If the statement is false, make the necessary change(s) to produce a true statement.

77. If the product of a point's coordinates is positive, the point must be in quadrant I.

78. If a point is on the *x*-axis, it is neither up nor down, so $x = 0$.

79. If a point is on the *y*-axis, its *x*-coordinate must be 0.

80. The ordered pair $(2, 5)$ satisfies $3y - 2x = -4$.

The graph shows the costs at a parking garage that allows cars to be parked for up to ten hours per day. Closed dots indicate that points belong to the graph and open dots indicate that points are not part of the graph. Use the graph to solve Exercises 81–82.

81. You park your car at the garage for four hours on Tuesday and five hours on Wednesday. What are the total parking garage costs for the two days?

82. On Thursday, you paid $12 for parking at the garage. Describe how long your car was parked.

Review Exercises

83. Find the absolute value: $|-14.3|$. (Section 1.2, Example 1)

84. Simplify: $[12 - (13 - 17)] - [9 - (6 - 10)]$. (Section 1.2, Examples 7 and 8)

85. Simplify: $6x - 5(4x + 3) - 10$. (Section 1.2, Example 13)

Preview Exercises

Exercises 86–88 will help you prepare for the material covered in the next section.

86. If -9 is substituted for x in the equation $4x - 3 = 5x + 6$, is the resulting statement true or false?

87. Simplify: $13 - 3(x + 2)$.

88. Simplify: $10\left(\dfrac{3x + 1}{2}\right)$.

SECTION

1.4   Solving Linear Equations

What am I supposed to learn?

After studying this section, you should be able to:

1. Solve linear equations.
2. Recognize identities, conditional equations, and inconsistent equations.
3. Solve applied problems using mathematical models.

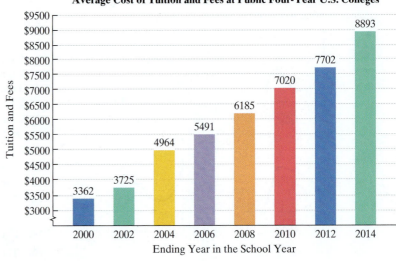

Average Cost of Tuition and Fees at Public Four-Year U.S. Colleges

Figure 1.22

Source: The College Board

"Questions have intensified about whether going to college is worthwhile," says *Education Pays*, released by the College Board Advocacy & Policy Center. "For the typical student, the investment pays off very well over the course of a lifetime, even considering the expense."

Among the findings in *Education Pays*:

- Mean (average) full-time earnings with a bachelor's degree in 2012 were $60,159, which is $27,529 more than high school graduates.

- Compared with a high school graduate, a four-year college graduate who enrolled in a public university at age 18 will break even by age 33. The college graduate will have earned enough by then to compensate for being out of the labor force for four years and for borrowing enough to pay tuition and fees, shown in **Figure 1.22**, adjusted for inflation.

The rising college costs from 2000 through 2014 can be modeled by the formula

$$T = 394x + 3123,$$

where T represents the average cost of tuition and fees at public four-year colleges for the school year ending x years after 2000. So, when will tuition and fees top $10,000 and reach $10,215? Substitute 10,215 for T in the formula $T = 394x + 3123$:

$$10{,}215 = 394x + 3123.$$

Our goal is to determine the value of x, the number of years after 2000, when tuition and fees will reach \$10,215. Notice that the exponent on the variable in $10{,}215 = 394x + 3123$ is 1. In this section, we will study how to determine the value of x in such equations. With this skill, you will be able to use certain mathematical models, such as the model for tuition and fees at public colleges, to project what might occur in the future.

1 Solve linear equations. ▶

Solving Linear Equations in One Variable

We begin with a general definition of a linear equation in one variable.

Definition of a Linear Equation

A **linear equation in one variable** x is an equation that can be written in the form

$$ax + b = 0,$$

where a and b are real numbers, and $a \neq 0$ (a is not equal to 0).

An example of a linear equation in one variable is

$$4x + 12 = 0.$$

Solving an equation in x involves determining all values of x that result in a true statement when substituted into the equation. Such values are **solutions**, or **roots**, of the equation. For example, substitute -3 for x in $4x + 12 = 0$. We obtain

$$4(-3) + 12 = 0, \quad \text{or} \quad -12 + 12 = 0.$$

This simplifies to the true statement $0 = 0$. Thus, -3 is a solution of the equation $4x + 12 = 0$. We also say that -3 **satisfies** the equation $4x + 12 = 0$, because when we substitute -3 for x, a true statement results. The set of all such solutions is called the equation's **solution set**. For example, the solution set of the equation $4x + 12 = 0$ is $\{-3\}$.

Two or more equations that have the same solution set are called **equivalent equations**. For example, the equations

$$4x + 12 = 0 \quad \text{and} \quad 4x = -12 \quad \text{and} \quad x = -3$$

are equivalent equations because the solution set for each is $\{-3\}$. To solve a linear equation in x, we transform the equation into an equivalent equation one or more times. Our final equivalent equation should be of the form

$$x = \text{a number}.$$

The solution set of this equation is the set consisting of the number.

To generate equivalent equations, we will use the following properties:

The Addition and Multiplication Properties of Equality

The Addition Property of Equality

The same real number or algebraic expression may be added to both sides of an equation without changing the equation's solution set.

$$a = b \text{ and } a + c = b + c \text{ are equivalent equations.}$$

The Multiplication Property of Equality

The same nonzero real number may multiply both sides of an equation without changing the equation's solution set.

$$a = b \text{ and } ac = bc \text{ are equivalent equations as long as } c \neq 0.$$

Because subtraction is defined in terms of addition, the addition property also lets us subtract the same number from both sides of an equation without changing the equation's solution set. Similarly, because division is defined in terms of multiplication, the multiplication property of equality can be used to divide both sides of an equation by the same nonzero number to obtain an equivalent equation.

Table 1.5 illustrates how these properties are used to isolate x to obtain an equation of the form $x = $ a number.

Great Question!

Have I solved an equation if I obtain something like $-x = 3$?

No. Your final equivalent equation should not be of the form

$$-x = \text{a number.}$$

We're not finished. A negative sign should not precede the variable.

Isolate x by multiplying or dividing both sides of this equation by -1.

Table 1.5 Using Properties of Equality to Solve Linear Equations

	Equation	How to Isolate x	Solving the Equation	The Equation's Solution Set
These equations are solved using the Addition Property of Equality.	$x - 3 = 8$	Add 3 to both sides.	$x - 3 + 3 = 8 + 3$ $x = 11$	$\{11\}$
	$x + 7 = -15$	Subtract 7 from both sides.	$x + 7 - 7 = -15 - 7$ $x = -22$	$\{-22\}$
These equations are solved using the Multiplication Property of Equality.	$6x = 30$	Divide both sides by 6 (or multiply both sides by $\frac{1}{6}$).	$\frac{6x}{6} = \frac{30}{6}$ $x = 5$	$\{5\}$
	$\frac{x}{5} = 9$	Multiply both sides by 5.	$5 \cdot \frac{x}{5} = 5 \cdot 9$ $x = 45$	$\{45\}$

EXAMPLE 1 Solving a Linear Equation

Solve and check: $2x + 3 = 17$.

Solution Our goal is to obtain an equivalent equation with x isolated on one side and a number on the other side.

$$2x + 3 = 17 \quad \text{This is the given equation.}$$
$$2x + 3 - 3 = 17 - 3 \quad \text{Subtract 3 from both sides.}$$
$$2x = 14 \quad \text{Simplify.}$$
$$\frac{2x}{2} = \frac{14}{2} \quad \text{Divide both sides by 2.}$$
$$x = 7 \quad \text{Simplify.}$$

Now we check the proposed solution, 7, by replacing x with 7 in the original equation.

$$2x + 3 = 17 \quad \text{This is the original equation.}$$
$$2 \cdot 7 + 3 \stackrel{?}{=} 17 \quad \text{Substitute 7 for } x. \text{ The question mark indicates that we do not yet know if the two sides are equal.}$$
$$14 + 3 \stackrel{?}{=} 17 \quad \text{Multiply: } 2 \cdot 7 = 14.$$
$$17 = 17 \quad \text{Add: } 14 + 3 = 17.$$

This statement is true.

Because the check results in a true statement, we conclude that the solution of the given equation is 7, or the solution set is $\{7\}$. ■

✓ **CHECK POINT 1** Solve and check: $4x + 5 = 29$.

Great Question!

What are the differences between what I'm supposed to do with algebraic expressions and algebraic equations?

We simplify algebraic expressions. We solve algebraic equations. Notice the differences between the procedures:

Simplifying an Algebraic Expression

Simplify: $3(x - 7) - (5x - 11)$.

> This is not an equation. There is no equal sign.

Solution $\quad 3(x - 7) - (5x - 11)$
$= 3x - 21 - 5x + 11$
$= (3x - 5x) + (-21 + 11)$
$= -2x + (-10)$
$= -2x - 10$

> Stop! Further simplification is not possible. Avoid the common error of setting $-2x - 10$ equal to 0.

Solving an Algebraic Equation

Solve: $3(x - 7) - (5x - 11) = 14$.

> This is an equation. There is an equal sign.

Solution $3(x - 7) - (5x - 11) = 14$
$3x - 21 - 5x + 11 = 14$
$-2x - 10 = 14$

> Add 10 to both sides.

$-2x - 10 + 10 = 14 + 10$
$-2x = 24$

> Divide both sides by -2.

$\dfrac{-2x}{-2} = \dfrac{24}{-2}$
$x = -12$

The solution set is $\{-12\}$.

Here is a step-by-step procedure for solving a linear equation in one variable. Not all of these steps are necessary to solve every equation.

Solving a Linear Equation

1. Simplify the algebraic expression on each side by removing grouping symbols and combining like terms.
2. Collect all the variable terms on one side and all the numbers, or constant terms, on the other side.
3. Isolate the variable and solve.
4. Check the proposed solution in the original equation.

EXAMPLE 2 Solving a Linear Equation

Solve and check: $\quad 2x - 7 + x = 3x + 1 + 2x$.

Solution

Step 1. Simplify the algebraic expression on each side.

$2x - 7 + x = 3x + 1 + 2x$ This is the given equation.

$3x - 7 = 5x + 1$ Combine like terms: $2x + x = 3x$ and $3x + 2x = 5x$.

Step 2. Collect variable terms on one side and constant terms on the other side. We will collect variable terms on the left by subtracting $5x$ from both sides. We will collect the numbers on the right by adding 7 to both sides.

$3x - 5x - 7 = 5x - 5x + 1$ Subtract $5x$ from both sides.

$-2x - 7 = 1$ Simplify.

$-2x - 7 + 7 = 1 + 7$ Add 7 to both sides.

$-2x = 8$ Simplify.

Discover for Yourself

Solve the equation in Example 2 by collecting terms with the variable on the right and constant terms on the left. What do you observe?

Step 3. Isolate the variable and solve. We isolate x in $-2x = 8$ by dividing both sides by -2.

$$\frac{-2x}{-2} = \frac{8}{-2}$$ Divide both sides by -2.

$$x = -4$$ Simplify.

Step 4. Check the proposed solution in the original equation. Substitute -4 for x in the original equation.

$2x - 7 + x = 3x + 1 + 2x$	This is the original equation.
$2(-4) - 7 + (-4) \overset{?}{=} 3(-4) + 1 + 2(-4)$	Substitute -4 for x.
$-8 - 7 + (-4) \overset{?}{=} -12 + 1 + (-8)$	Multiply: $2(-4) = -8$, $3(-4) = -12$, and $2(-4) = -8$.
$-15 + (-4) \overset{?}{=} -11 + (-8)$	Add or subtract from left to right: $-8 - 7 = -15$ and $-12 + 1 = -11$.
$-19 = -19$	Add.

The true statement $-19 = -19$ verifies that -4 is the solution, or the solution set is $\{-4\}$. ∎

☑ **CHECK POINT 2**   Solve and check: $2x - 12 + x = 6x - 4 + 5x$.

EXAMPLE 3   Solving a Linear Equation

Solve and check: $4(2x + 1) - 29 = 3(2x - 5)$.

Solution

Step 1. Simplify the algebraic expression on each side.

$4(2x + 1) - 29 = 3(2x - 5)$	This is the given equation.
$8x + 4 - 29 = 6x - 15$	Use the distributive property.
$8x - 25 = 6x - 15$	Simplify.

Step 2. Collect variable terms on one side and constant terms on the other side. We will collect the variable terms on the left by subtracting $6x$ from both sides. We will collect the numbers on the right by adding 25 to both sides.

$8x - 6x - 25 = 6x - 6x - 15$	Subtract $6x$ from both sides.
$2x - 25 = -15$	Simplify.
$2x - 25 + 25 = -15 + 25$	Add 25 to both sides.
$2x = 10$	Simplify.

Step 3. Isolate the variable and solve. We isolate x by dividing both sides by 2.

$\dfrac{2x}{2} = \dfrac{10}{2}$	Divide both sides by 2.
$x = 5$	Simplify.

Step 4. Check the proposed solution in the original equation. Substitute 5 for x in the original equation.

$4(2x + 1) - 29 = 3(2x - 5)$	This is the original equation.
$4(2 \cdot 5 + 1) - 29 \overset{?}{=} 3(2 \cdot 5 - 5)$	Substitute 5 for x.
$4(11) - 29 \overset{?}{=} 3(5)$	Simplify inside parentheses: $2 \cdot 5 + 1 = 10 + 1 = 11$ and $2 \cdot 5 - 5 = 10 - 5 = 5$.
$44 - 29 \overset{?}{=} 15$	Multiply: $4(11) = 44$ and $3(5) = 15$.
$15 = 15$	Subtract.

The true statement $15 = 15$ verifies that 5 is the solution, or the solution set is $\{5\}$. ∎

☑ **CHECK POINT 3**   Solve and check: $2(x - 3) - 17 = 13 - 3(x + 2)$.

Using Technology

Numeric and Graphic Connections

In many algebraic situations, technology provides numeric and visual insights into problem solving. For example, you can use a graphing utility to check the solution of a linear equation, giving numeric and geometric meaning to the solution. Enter each side of the equation separately under y_1 and y_2. Then use the table or the graphs to locate the x-value for which the y-values are the same. This x-value is the solution.

Let's verify our work in Example 3 and show that 5 is the solution of

$$4(2x + 1) - 29 = 3(2x - 5).$$

Enter $y_1 = 4(2x + 1) - 29$ in the $\boxed{y=}$ screen.

Enter $y_2 = 3(2x - 5)$ in the $\boxed{y=}$ screen.

Numeric Check

Display a table for y_1 and y_2.

$y_1 = 4(2x + 1) - 29$ $y_2 = 3(2x - 5)$

When $x = 5$, y_1 and y_2 have the same value, namely 15. This verifies 5 is the solution of $4(2x + 1) - 29 = 3(2x - 5)$.

Graphic Check

Display graphs for y_1 and y_2 and use the intersection feature. The solution is the x-coordinate of the intersection point.

Graphs intersect at $(5, 15)$. When $x = 5$, y_1 and y_2 have the same value, namely, 15. This graphically verifies 5 is the solution of $4(2x + 1) - 29 = 3(2x - 5)$.

$y_2 = 3(2x - 5)$

$y_1 = 4(2x + 1) - 29$

$[0, 7, 1]$ by $[-1, 20, 1]$

Choose a large enough viewing rectangle so that you can see the intersection point.

Linear Equations with Fractions

Equations are easier to solve when they do not contain fractions. How do we remove fractions from an equation? We begin by multiplying both sides of the equation by the least common denominator (LCD) of any fractions in the equation. The least common denominator is the smallest number that all denominators will divide into. Multiplying every term on both sides of the equation by the least common denominator will eliminate the fractions in the equation. Example 4 shows how we "clear an equation of fractions."

EXAMPLE 4   Solving a Linear Equation Involving Fractions

Solve: $\dfrac{2x + 5}{5} + \dfrac{x - 7}{2} = \dfrac{3x + 1}{2}$.

Solution   The denominators are 5, 2, and 2. The smallest number that is divisible by 5, 2, and 2 is 10. We begin by multiplying both sides of the equation by 10, the least common denominator.

$$\frac{2x + 5}{5} + \frac{x - 7}{2} = \frac{3x + 1}{2}$$ This is the given equation.

$$10\left(\frac{2x + 5}{5} + \frac{x - 7}{2}\right) = 10\left(\frac{3x + 1}{2}\right)$$ Multiply both sides by 10.

$$\frac{10}{1} \cdot \left(\frac{2x + 5}{5}\right) + \frac{10}{1} \cdot \left(\frac{x - 7}{2}\right) = \frac{10}{1} \cdot \left(\frac{3x + 1}{2}\right)$$ Use the distributive property and multiply each term by 10.

$$\frac{\overset{2}{10}}{1} \cdot \left(\frac{2x + 5}{\underset{1}{5}}\right) + \frac{\overset{5}{10}}{1} \cdot \left(\frac{x - 7}{\underset{1}{2}}\right) = \frac{\overset{5}{10}}{1} \cdot \left(\frac{3x + 1}{\underset{1}{2}}\right)$$ Divide out common factors in each multiplication.

$$2(2x + 5) + 5(x - 7) = 5(3x + 1)$$ The fractions are now cleared.

At this point, we have an equation similar to those we have previously solved. Use the distributive property to begin simplifying each side.

$$4x + 10 + 5x - 35 = 15x + 5$$ Use the distributive property.

$$9x - 25 = 15x + 5$$ Combine like terms on the left side: $4x + 5x = 9x$ and $10 - 35 = -25$.

For variety, let's collect variable terms on the right and constant terms on the left.

$$9x - 9x - 25 = 15x - 9x + 5$$ Subtract $9x$ from both sides.

$$-25 = 6x + 5$$ Simplify.

$$-25 - 5 = 6x + 5 - 5$$ Subtract 5 from both sides.

$$-30 = 6x$$ Simplify.

Isolate x on the right side by dividing both sides by 6.

$$\frac{-30}{6} = \frac{6x}{6}$$ Divide both sides by 6.

$$-5 = x$$ Simplify.

Check the proposed solution in the original equation. Substitute -5 for x in the original equation. You should obtain $-7 = -7$. This true statement verifies that -5 is the solution, or the solution set is $\{-5\}$. ∎

✓ **CHECK POINT 4**   Solve: $\dfrac{x + 5}{7} + \dfrac{x - 3}{4} = \dfrac{5}{14}$.

---

2 Recognize identities, conditional equations, and inconsistent equations.

Types of Equations

Equations can be placed into categories that depend on their solution sets.

An equation that is true for all real numbers for which both sides are defined is called an **identity**. An example of an identity is

$$x + 3 = x + 2 + 1.$$

Every number plus 3 is equal to that number plus 2 plus 1. Therefore, the solution set to this equation is the set of all real numbers. This set is written either as

$$\{x \,|\, x \text{ is a real number}\} \quad \text{or} \quad (-\infty, \infty) \quad \text{or} \quad \mathbb{R}.$$

An equation that is not an identity, but that is true for at least one real number, is called a **conditional equation**. The equation $2x + 3 = 17$ is an example of a conditional equation. The equation is not an identity and is true only if x is 7.

An **inconsistent equation** is an equation that is not true for even one real number. An example of an inconsistent equation is

$$x = x + 7.$$

There is no number that is equal to itself plus 7. The equation $x = x + 7$ has no solution. Its solution set is written either as

$$\{ \ \} \quad \text{or} \quad \varnothing.$$

> These symbols stand for the empty set, a set with no elements.

If you attempt to solve an identity or an inconsistent equation, you will eliminate the variable. A true statement such as $6 = 6$ or a false statement such as $2 = 3$ will be the result. **If a true statement results, the equation is an identity that is true for all real numbers. If a false statement results, the equation is an inconsistent equation with no solution.**

EXAMPLE 5 Categorizing an Equation

Solve and determine whether the equation

$$2(x + 1) = 2x + 3$$

is an identity, a conditional equation, or an inconsistent equation.

Solution Begin by applying the distributive property on the left side. We obtain

$$2x + 2 = 2x + 3.$$

Does something look strange about $2x + 2 = 2x + 3$? Can doubling a number and increasing the product by 2 give the same result as doubling the same number and increasing the product by 3? No. Let's continue solving the equation by subtracting $2x$ from both sides of $2x + 2 = 2x + 3$.

$$2x - 2x + 2 = 2x - 2x + 3$$

> Keep reading. $2 = 3$ is not the solution.

$$2 = 3$$

The original equation is equivalent to the statement $2 = 3$, which is false for every value of x. The equation is inconsistent and has no solution. You can express this by writing "no solution" or using one of the symbols for the empty set, $\{ \ \}$ or $\varnothing$. ∎

Using Technology

Graphic Connections

How can technology visually reinforce the fact that the equation

$$2(x + 1) = 2x + 3$$

has no solution? Enter $y_1 = 2(x + 1)$ and $y_2 = 2x + 3$. The graphs of y_1 and y_2 appear to be parallel lines with no intersection point. This supports our conclusion that $2(x + 1) = 2x + 3$ is an inconsistent equation with no solution.

$[-5, 2, 1]$ by $[-5, 5, 1]$

✓ **CHECK POINT 5** Solve and determine whether the equation

$$4x - 7 = 4(x - 1) + 3$$

is an identity, a conditional equation, or an inconsistent equation.

EXAMPLE 6 Categorizing an Equation

Solve and determine whether the equation

$$4x + 6 = 6(x + 1) - 2x$$

is an identity, a conditional equation, or an inconsistent equation.

Solution

$4x + 6 = 6(x + 1) - 2x$	This is the given equation.
$4x + 6 = 6x + 6 - 2x$	Apply the distributive property on the right side.
$4x + 6 = 4x + 6$	Combine like terms on the right side: $6x - 2x = 4x$.

Can you see that the equation $4x + 6 = 4x + 6$ is true for every value of x? Let's continue solving the equation by subtracting $4x$ from both sides.

$$4x - 4x + 6 = 4x - 4x + 6$$

$$6 = 6$$ Keep reading. $6 = 6$ is not the solution.

The original equation is equivalent to the statement $6 = 6$, which is true for every value of x. The equation is an identity, and all real numbers are solutions. You can express this by writing "all real numbers" or using one of the following notations:

$$\{x \mid x \text{ is a real number}\} \quad \text{or} \quad (-\infty, \infty) \quad \text{or} \quad \mathbb{R}. \blacksquare$$

Using Technology

Numeric Connections

A graphing utility's ⎡TABLE⎤ feature can be used to numerically verify that the solution set of

$$4x + 6 = 6(x + 1) - 2x$$

is the set of all real numbers.

No matter how far up or down you scroll, y_1 and y_2 have the same corresponding value for each x-value. The expressions $4x + 6$ and $6(x + 1) - 2x$ are always equal.

✓ **CHECK POINT 6** Solve and determine whether the equation

$$7x + 9 = 9(x + 1) - 2x$$

is an identity, a conditional equation, or an inconsistent equation.

3 Solve applied problems using mathematical models. ▶

Applications

We opened both the chapter and this section with a discussion of skyrocketing college costs. In our next example, we use a formula that models these costs. The example shows how the procedure for solving linear equations can be used to find the value of a variable in a mathematical model.

EXAMPLE 7   The Cost of Attending a Public College

The formula

$$T = 394x + 3123$$

models the average cost of tuition and fees, T, at public four-year colleges for the school year ending x years after 2000. When will tuition and fees reach $10,215?

Solution   We are interested in when tuition and fees will reach $10,215. We substitute 10,215 for T in the formula and solve for x, the number of years after 2000.

$T = 394x + 3123$	This is the given formula.
$10,215 = 394x + 3123$	Replace T with 10,215.
$10,215 - 3123 = 394x + 3123 - 3123$	Subtract 3123 from both sides.
$7092 = 394x$	Simplify.
$\dfrac{7092}{394} = \dfrac{394x}{394}$	Divide both sides by 394.
$18 = x$	Simplify.

The model indicates that for the school year ending 18 years after 2000, or the school year ending 2018, tuition and fees will reach $10,215. ■

☑ **CHECK POINT 7**   Use the formula in Example 7 to find when tuition and fees will reach $11,397.

Achieving Success

Because concepts in mathematics build on each other, **it is extremely important that you complete all homework assignments.** This requires more than attempting a few of the assigned exercises. When it comes to assigned homework, you need to do four things and to do these things consistently throughout any math course:

1. Attempt to work every assigned problem.
2. Check your answers.
3. Correct your errors.
4. Ask for help with the problems you have attempted, but do not understand.

Having said this, **don't panic at the length of the Exercise Sets.** You are not expected to work all, or even most, of the problems. Your professor will provide guidance on which exercises to work by assigning those problems that are consistent with the goals and objectives of your course.

CONCEPT AND VOCABULARY CHECK

Fill in each blank so that the resulting statement is true.

1. An equation in the form $ax + b = 0$, $a \neq 0$, such as $3x + 17 = 0$, is called a/an _____ equation in one variable.

2. Two or more equations that have the same solution set are called _____ equations.

3. The addition property of equality states that if $a = b$, then $a + c =$ _____.

4. The multiplication property of equality states that if $a = b$ and $c \neq 0$, then $ac =$ _____.

5. The first step in solving $7 + 3(x - 2) = 2x + 10$ is to _____.

6. The equation

$$\frac{x}{4} = 2 + \frac{x-3}{3}$$

can be solved by multiplying both sides by the _____ of the fractions, which is _____.

7. In solving an equation, if you eliminate the variable and obtain a false statement such as $2 = 3$, the equation is a/an _____ equation. The solution set can be expressed using the symbol _____.

8. In solving an equation, if you eliminate the variable and obtain a true statement such as $6 = 6$, the equation is a/an _____. The solution set can be expressed using interval notation as _____.

1.4 EXERCISE SET ▶ MyMathLab®

Practice Exercises

In Exercises 1–24, solve and check each linear equation.

1. $5x + 3 = 18$
2. $3x + 8 = 50$
3. $6x - 3 = 63$
4. $5x - 8 = 72$
5. $14 - 5x = -41$
6. $25 - 6x = -83$
7. $11x - (6x - 5) = 40$
8. $5x - (2x - 8) = 35$
9. $2x - 7 = 6 + x$
10. $3x + 5 = 2x + 13$
11. $7x + 4 = x + 16$
12. $8x + 1 = x + 43$
13. $8y - 3 = 11y + 9$
14. $5y - 2 = 9y + 2$
15. $3(x - 2) + 7 = 2(x + 5)$
16. $2(x - 1) + 3 = x - 3(x + 1)$
17. $3(x - 4) - 4(x - 3) = x + 3 - (x - 2)$
18. $2 - (7x + 5) = 13 - 3x$
19. $16 = 3(x - 1) - (x - 7)$
20. $5x - (2x + 2) = x + (3x - 5)$
21. $7(x + 1) = 4[x - (3 - x)]$
22. $2[3x - (4x - 6)] = 5(x - 6)$
23. $\frac{1}{2}(4z + 8) - 16 = -\frac{2}{3}(9z - 12)$
24. $\frac{3}{4}(24 - 8z) - 16 = -\frac{2}{3}(6z - 9)$

In Exercises 25–38, solve each equation.

25. $\frac{x}{3} = \frac{x}{2} - 2$
26. $\frac{x}{5} = \frac{x}{6} + 1$
27. $20 - \frac{x}{3} = \frac{x}{2}$
28. $\frac{x}{5} - \frac{1}{2} = \frac{x}{6}$
29. $\frac{3x}{5} = \frac{2x}{3} + 1$
30. $\frac{x}{2} = \frac{3x}{4} + 5$
31. $\frac{3x}{5} - x = \frac{x}{10} - \frac{5}{2}$
32. $2x - \frac{2x}{7} = \frac{x}{2} + \frac{17}{2}$
33. $\frac{x + 3}{6} = \frac{2}{3} + \frac{x - 5}{4}$
34. $\frac{x + 1}{4} = \frac{1}{6} + \frac{2 - x}{3}$
35. $\frac{x}{4} = 2 + \frac{x - 3}{3}$
36. $5 + \frac{x - 2}{3} = \frac{x + 3}{8}$
37. $\frac{x + 1}{3} = 5 - \frac{x + 2}{7}$
38. $\frac{3x}{5} - \frac{x - 3}{2} = \frac{x + 2}{3}$

In Exercises 39–50, solve each equation. Then state whether the equation is an identity, a conditional equation, or an inconsistent equation.

39. $5x + 9 = 9(x + 1) - 4x$
40. $4x + 7 = 7(x + 1) - 3x$
41. $3(y + 2) = 7 + 3y$
42. $4(y + 5) = 21 + 4y$
43. $10x + 3 = 8x + 3$
44. $5x + 7 = 2x + 7$
45. $\frac{1}{2}(6z + 20) - 8 = 2(z - 4)$

46. $\frac{1}{3}(6z + 12) = \frac{1}{5}(20z + 30) - 8$

47. $-4x - 3(2 - 2x) = 7 + 2x$

48. $3x - 3(2 - x) = 6(x - 1)$

49. $y + 3(4y + 2) = 6(y + 1) + 5y$

50. $9y - 3(6 - 5y) = y - 2(3y + 9)$

In Exercises 51–54, use the $\boxed{Y=}$ *screen to write the equation being solved. Then use the table to solve the equation.*

51.

52.

53.

54.

Practice PLUS

55. Evaluate $x^2 - x$ for the value of x satisfying
$4(x - 2) + 2 = 4x - 2(2 - x)$.

56. Evaluate $x^2 - x$ for the value of x satisfying
$2(x - 6) = 3x + 2(2x - 1)$.

57. Evaluate $x^2 - (xy - y)$ for x satisfying $\dfrac{3(x + 3)}{5} = 2x + 6$
and y satisfying $-2y - 10 = 5y + 18$.

58. Evaluate $x^2 - (xy - y)$ for x satisfying $\dfrac{13x - 6}{4} = 5x + 2$
and y satisfying $5 - y = 7(y + 4) + 1$.

In Exercises 59–66, solve each equation.

59. $[(3 + 6)^2 \div 3] \cdot 4 = -54x$

60. $2^3 - [4(5 - 3)^3] = -8x$

61. $5 - 12x = 8 - 7x - [6 \div 3(2 + 5^3) + 5x]$

62. $2(5x + 58) = 10x + 4(21 \div 3.5 - 11)$

63. $0.7x + 0.4(20) = 0.5(x + 20)$

64. $0.5(x + 2) = 0.1 + 3(0.1x + 0.3)$

65. $4x + 13 - \{2x - [4(x - 3) - 5]\} = 2(x - 6)$

66. $-2\{7 - [4 - 2(1 - x) + 3]\} = 10 - [4x - 2(x - 3)]$

Application Exercises

The bar graph shows the average cost of tuition and fees at private four-year colleges in the United States.

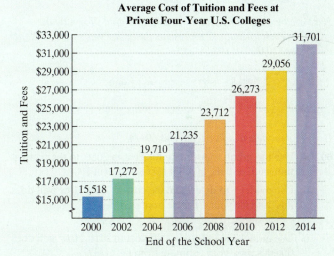

Source: The College Board

Here are two mathematical models for the data shown in the graph. In each formula, T represents the average cost of tuition and fees at private U.S. colleges for the school year ending x years after 2000.

Model 1 ▸ $T = 1157x + 14{,}961$

Model 2 ▸ $T = 21x^2 + 862x + 15{,}552$

Use this information to solve Exercises 67–68.

67. **a.** Use each model to find the average cost of tuition and fees at private U.S. colleges for the school year ending in 2014. By how much does each model underestimate or overestimate the actual cost shown for the school year ending in 2014?

b. Use model 1 to determine when tuition and fees at private four-year colleges will average $36,944.

68. a. Use each model at the bottom of the previous page to find the average cost of tuition and fees at private U.S. colleges for the school year ending in 2012. By how much does each model underestimate or overestimate the actual cost shown for the school year ending in 2012?

   b. Use model 1 to determine when tuition and fees at private four-year colleges will average $39,258.

The line graph shows the cost of inflation. What cost $10,000 in 1984 would cost the amount shown by the graph in subsequent years.

The Cost of Inflation

$13,100

$23,300

What cost you $10,000 in 1984 would cost you this much in subsequent years.

Source: U.S. Bureau of Labor Statistics

Here are two mathematical models for the data shown by the graph. In each formula, C represents the cost x years after 1990 of what cost $10,000 in 1984.

Model 1 $C = 442x + 12{,}969$

Model 2 $C = 2x^2 + 390x + 13{,}126$

Use these models to solve Exercises 69–74.

69. a. Use the graph to estimate the cost in 2010, to the nearest thousand dollars, of what cost $10,000 in 1984.

   b. Use model 1 to determine the cost in 2010. How well does this describe your estimate from part (a)?

   c. Use model 2 to determine the cost in 2010. How well does this describe your estimate from part (a)?

70. a. Use the graph to estimate the cost in 2000, to the nearest thousand dollars, of what cost $10,000 in 1984.

   b. Use model 1 to determine the cost in 2000. How well does this describe your estimate from part (a)?

   c. Use model 2 to determine the cost in 2000. How well does this describe your estimate from part (a)?

71. Which model is a better description for the cost in 1990 of what cost $10,000 in 1984? Does this model underestimate or overestimate the cost shown by the graph? By how much?

72. Which model is a better description for the cost in 2013 of what cost $10,000 in 1984? Does this model underestimate or overestimate the cost shown by the graph? By how much?

73. Use model 1 to determine in which year the cost will be $26,229 for what cost $10,000 in 1984.

74. Use model 1 to determine in which year the cost will be $25,345 for what cost $10,000 in 1984.

Explaining the Concepts

75. What is a linear equation in one variable? Give an example of this type of equation.

76. What does it mean to solve an equation?

77. How do you determine if a number is a solution of an equation?

78. What are equivalent equations? Give an example.

79. What is the addition property of equality?

80. What is the multiplication property of equality?

81. Explain how to clear an equation of fractions.

82. What is an identity? Give an example.

83. What is a conditional equation? Give an example.

84. What is an inconsistent equation? Give an example.

85. Despite low rates of inflation, the cost of a college education continues to skyrocket. This is a departure from the trend during the 1970s: In constant dollars (which negate the effect of inflation), the cost of college actually decreased several times. What explanations can you offer for the increasing cost of a college education?

Technology Exercises

In Exercises 86–89, use your graphing utility to enter each side of the equation separately under y_1 and y_2. Then use the utility's TABLE *or* GRAPH *feature to solve the equation.*

86. $5x + 2(x - 1) = 3x + 10$

87. $2x + 3(x - 4) = 4x - 7$

88. $3(2x + 11) = 3(5 + x)$

89. $\dfrac{2x - 1}{3} - \dfrac{x - 5}{6} = \dfrac{x - 3}{4}$

Critical Thinking Exercises

Make Sense? *In Exercises 90–93, determine whether each statement makes sense or does not make sense, and explain your reasoning.*

90. Because $x = x + 5$ is an inconsistent equation, the graphs of $y = x$ and $y = x + 5$ should not intersect.

91. Because subtraction is defined in terms of addition, it's not necessary to state a separate subtraction property of equality to generate equivalent equations.

92. The number 3 satisfies the equation $7x + 9 = 9(x + 1) - 2x$, so {3} is the equation's solution set.

93. I can solve $-2x = 10$ using the addition property of equality.

In Exercises 94–97, determine whether each statement is true or false. If the statement is false, make the necessary change(s) to produce a true statement.

94. The equation $-7x = x$ has no solution.

95. The equations $\dfrac{x}{x-4} = \dfrac{4}{x-4}$ and $x = 4$ are equivalent.

96. The equations $3y - 1 = 11$ and $3y - 7 = 5$ are equivalent.

97. If a and b are any real numbers, then $ax + b = 0$ always has only one number in its solution set.

98. Solve for x: $ax + b = c$.

99. Write three equations that are equivalent to $x = 5$.

100. If x represents a number, write an English sentence about the number that results in an inconsistent equation.

101. Find b such that $\dfrac{7x+4}{b} + 13 = x$ will have a solution set given by $\{-6\}$.

Review Exercises

In Exercises 102–103, perform the indicated operations.

102. $-\dfrac{1}{5} - \left(-\dfrac{1}{2}\right)$ (Section 1.2, Example 4)

103. $4(-3)(-1)(-5)$ (Section 1.2, Examples in **Table 1.3**)

104. Graph $y = x^2 - 4$. Let $x = -3, -2, -1, 0, 1, 2,$ and 3. (Section 1.3, Example 2)

Preview Exercises

Exercises 105–107 will help you prepare for the material covered in the next section.

105. Let x represent a number.

   a. Write an equation in x that describes the following conditions:

   Four less than three times the number is 32.

   b. Solve the equation and determine the number.

106. Let x represent the number of countries in the world that are not free. The number of free countries exceeds the number of not-free countries by 44. Write an algebraic expression that represents the number of free countries.

107. You purchase a new car for $20,000. Each year the value of the car decreases by $2500. Write an algebraic expression that represents the car's value, in dollars, after x years.

MID-CHAPTER CHECK POINT   Section 1.1 – Section 1.4

 What You Know: We reviewed a number of topics from introductory algebra, including the real numbers and their representations on number lines. We used interval notation to represent sets of real numbers, where parentheses indicate endpoints that are not included and square brackets indicate endpoints that are included. We performed operations with real numbers and applied the order-of-operations agreement to expressions containing more than one operation. We used commutative, associative, and distributive properties to simplify algebraic expressions. We used the rectangular coordinate system to represent ordered pairs of real numbers and graph equations in two variables. Finally, we solved linear equations, including equations with fractions. We saw that some equations have no solution, whereas others have all real numbers as solutions.

In Exercises 1–14, simplify the expression or solve the equation, whichever is appropriate.

1. $-5 + 3(x + 5)$

2. $-5 + 3(x + 5) = 2(3x - 4)$

3. $3[7 - 4(5 - 2)]$

4. $\dfrac{x-3}{5} - 1 = \dfrac{x-5}{4}$

5. $\dfrac{-2^4 + (-2)^2}{-4 - (2-2)}$

6. $7x - [8 - 3(2x - 5)]$

7. $3(2x - 5) - 2(4x + 1) = -5(x + 3) - 2$

8. $3(2x - 5) - 2(4x + 1) - 5(x + 3) - 2$

9. $-4^2 \div 2 + (-3)(-5)$

10. $3x + 1 - (x - 5) = 2x - 4$

11. $\dfrac{3x}{4} - \dfrac{x}{3} + 1 = \dfrac{4x}{5} - \dfrac{3}{20}$

12. $(6 - 9)(8 - 12) \div \dfrac{5^2 + 4 \div 2}{8^2 - 9^2 + 8}$

13. $4x - 2(1 - x) = 3(2x + 1) - 5$

14. $\dfrac{3[4 - 3(-2)^2]}{2^2 - 2^4}$

In Exercises 15–16, express each interval in set-builder notation and graph.

15. $[-2, 0)$ **16.** $(-\infty, 0]$

In Exercises 17–19, graph each equation in a rectangular coordinate system.

17. $y = 2x - 1$ **18.** $y = 1 - |x|$ **19.** $y = x^2 + 2$

In Exercises 20–23, determine whether each statement is true or false.

20. $-\left| -\dfrac{\sqrt{3}}{5} \right| = -\dfrac{\sqrt{3}}{5}$

21. $\{x \mid x \text{ is a negative integer greater than } -4\} = \{-4, -3, -2, -1\}$

22. $-17 \notin \{x \mid x \text{ is a rational number}\}$

23. $-128 \div (2 \cdot 4) > (-128 \div 2) \cdot 4$

1.5

Problem Solving and Using Formulas

What am I supposed to learn?

After studying this section, you should be able to:

1 Solve algebraic word problems using linear equations. ▶

2 Solve a formula for a variable. ▶

1969   2013

Many changes occurred from 1969 to 2013, including the shifting goals of college students. Compared to 1969, college freshmen in 2013 had making money on their minds. In this section, you will learn a problem-solving strategy that uses linear equations to model the changing attitudes of college freshmen.

Problem Solving with Linear Equations

We have seen that a model is a mathematical representation of a real-world situation. In this section, we will be solving problems that are presented in English. This means that we must obtain models by translating from the ordinary language of English into the language of algebraic equations. To translate, however, we must understand the English prose and be familiar with the forms of algebraic language. Here are some general steps we will follow in solving word problems:

1 Solve algebraic word problems using linear equations. ▶

Strategy for Solving Word Problems

Step 1. Read the problem carefully several times until you can state in your own words what is given and what the problem is looking for. Let x (or any variable) represent one of the unknown quantities in the problem.

Step 2. If necessary, write expressions for any other unknown quantities in the problem in terms of x.

Step 3. Write an equation in x that models the verbal conditions of the problem.

Step 4. Solve the equation and answer the problem's question.

Step 5. Check the solution *in the original wording* of the problem, not in the equation obtained from the words.

Great Question!

Why are word problems important?

There is great value in reasoning through the steps for solving a word problem. This value comes from the problem-solving skills that you will attain and is often more important than the specific problem or its solution.

EXAMPLE 1   Education Pays Off

The graph in **Figure 1.23** at the top of the next page shows average yearly earnings in the United States by highest educational attainment.

The average yearly salary of a man with a bachelor's degree exceeds that of a man with an associate's degree by $25 thousand. The average yearly salary of a man with a master's degree exceeds that of a man with an associate's degree by $45 thousand. Combined, three men with each of these degrees earn $214 thousand. Find the average yearly salary of men with each of these levels of education.

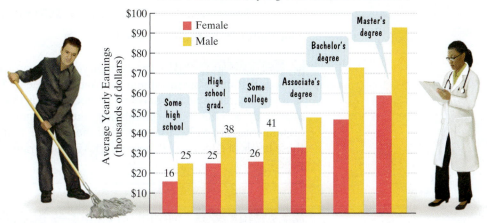

**Average Earnings of Full-Time Workers
in the United States, by Highest Educational Attainment**

Figure 1.23

Source: U.S. Census Bureau (2012 data)

Solution

Step 1. Let x represent one of the unknown quantities. We know something about salaries of men with bachelor's degrees and master's degrees: They exceed the salary of a man with an associate's degree by $25 thousand and $45 thousand, respectively. We will let

x = the average yearly salary of a man with an associate's degree
(in thousands of dollars).

Step 2. Represent other unknown quantities in terms of x. Because a man with a bachelor's degree earns $25 thousand more than a man with an associate's degree, let

$x + 25$ = the average yearly salary of a man with a bachelor's degree.

Because a man with a master's degree earns $45 thousand more than a man with an associate's degree, let

$x + 45$ = the average yearly salary of a man with a master's degree.

Step 3. Write an equation in x that models the conditions. Combined, three men with each of these degrees earn $214 thousand.

Salary: associate's degree	plus	salary: bachelor's degree	plus	salary: master's degree	equals	$214 thousand.
x	$+$	$(x + 25)$	$+$	$(x + 45)$	$=$	214

Step 4. Solve the equation and answer the question.

$$x + (x + 25) + (x + 45) = 214 \quad \text{This is the equation that models the problem's conditions.}$$

$$3x + 70 = 214 \quad \text{Remove parentheses, regroup, and combine like terms.}$$

$$3x = 144 \quad \text{Subtract 70 from both sides.}$$

$$x = 48 \quad \text{Divide both sides by 3.}$$

Because we isolated the variable in the model and obtained $x = 48$,

average salary with an associate's degree = $x = 48$

average salary with a bachelor's degree = $x + 25 = 48 + 25 = 73$

average salary with a master's degree = $x + 45 = 48 + 45 = 93$.

Men with associate's degrees average $48 thousand per year, men with bachelor's degrees average $73 thousand per year, and men with master's degrees average $93 thousand per year.

Great Question!

Example 1 involves using the word "exceeds" to represent two of the unknown quantities. Can you help me write algebraic expressions for quantities described using "exceeds"?

Modeling with the word *exceeds* can be a bit tricky. It's helpful to identify the smaller quantity. Then add to this quantity to represent the larger quantity. For example, suppose that Tim's height exceeds Tom's height by *a* inches. Tom is the shorter person. If Tom's height is represented by *x*, then Tim's height is represented by $x + a$.

Step 5. Check the proposed solution in the original wording of the problem. The problem states that combined, three men with each of these educational attainments earn $214 thousand. Using the salaries we determined in step 4, the sum is

$48 thousand + $73 thousand + $93 thousand, or $214 thousand,

which satisfies the problem's conditions. ∎

✓ **CHECK POINT 1**   The average yearly salary of a woman with a bachelor's degree exceeds that of a woman with an associate's degree by $14 thousand. The average yearly salary of a woman with a master's degree exceeds that of a woman with an associate's degree by $26 thousand. Combined, three women with each of these educational attainments earn $139 thousand. Find the average yearly salary of women with each of these levels of education. (These salaries are illustrated by the bar graph on the previous page.)

EXAMPLE 2   Modeling Attitudes of College Freshmen

Researchers have surveyed college freshmen every year since 1969. **Figure 1.24** shows that attitudes about some life goals have changed dramatically over the years. In particular, the freshmen class of 2013 was more interested in making money than the freshmen of 1969 had been. In 1969, 42% of first-year college students considered "being well-off financially" essential or very important. For the period from 1969 through 2013, this percentage increased by approximately 0.9 each year. If this trend continues, by which year will all college freshmen consider "being well-off financially" essential or very important?

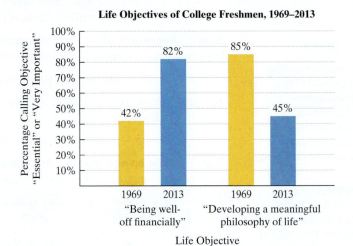

Figure 1.24

Source: Higher Education Research Institute

Solution

Step 1. Let x represent one of the unknown quantities. We are interested in the year when all college freshmen, or 100% of the freshmen, will consider this life objective essential or very important. Let

x = the number of years after 1969 when all freshmen will consider "being well-off financially" essential or very important.

Step 2. Represent other unknown quantities in terms of x. There are no other unknown quantities to find, so we can skip this step.

Step 3. Write an equation in x that models the conditions.

The 1969 percentage	increased by	0.9 each year for x years	equals	100% of the freshmen.
42	+	0.9x	=	100

Your author teaching math in 1969

Step 4. Solve the equation and answer the question.

$$42 + 0.9x = 100$$

This is the equation that models the problem's conditions.

$$42 - 42 + 0.9x = 100 - 42$$

Subtract 42 from both sides.

$$0.9x = 58$$

Simplify.

$$\frac{0.9x}{0.9} = \frac{58}{0.9}$$

Divide both sides by 0.9.

$$x = 64.\overline{4} \approx 64$$

Simplify and round to the nearest whole number.

Using current trends, by approximately 64 years after 1969, or in 2033, all freshmen will consider "being well-off financially" essential or very important.

Step 5. Check the proposed solution in the original wording of the problem. The problem states that all freshmen (100%, represented by 100 using the model) will consider the objective essential or very important. Does this approximately occur if we increase the 1969 percentage, 42%, by 0.9 each year for 64 years, our proposed solution?

$$42 + 0.9(64) = 42 + 57.6 = 99.6 \approx 100$$

This verifies that using trends shown in **Figure 1.24**, all first-year college students will consider the objective essential or very important approximately 64 years after 1969. ∎

Great Question!

I notice that the equation in Example 2, 42 + 0.9x = 100, contains a decimal. Can I clear an equation of decimals much like I cleared equations of fractions?

- You can clear an equation of decimals by multiplying each side by a power of 10. The exponent on 10 will be equal to the greatest number of digits to the right of any decimal point in the equation.

- Multiplying a decimal number by 10^n has the effect of moving the decimal point n places to the right.

Example

$$42 + 0.9x = 100$$

The greatest number of digits to the right of any decimal point in the equation is 1. Multiply each side by 10^1, or 10.

$$10(42 + 0.9x) = 10(100)$$
$$10(42) + 10(0.9x) = 10(100)$$
$$420 + 9x = 1000$$
$$420 - 420 + 9x = 1000 - 420$$
$$9x = 580$$
$$\frac{9x}{9} = \frac{580}{9}$$
$$x = 64.\overline{4} \approx 64$$

It is not a requirement to clear decimals before solving an equation. Compare this solution to the one in step 4 of Example 2. Which method do you prefer?

✓ **CHECK POINT 2**   **Figure 1.24** on the previous page shows that the freshmen class of 2013 was less interested in developing a philosophy of life than the freshmen of 1969 had been. In 1969, 85% of the freshmen considered this objective essential or very important. Since then, this percentage has decreased by approximately 0.9 each year. If this trend continues, by which year will only 25% of college freshmen consider "developing a meaningful philosophy of life" essential or very important?

EXAMPLE 3 Modeling Options for a Toll

The toll to a bridge costs $7. Commuters who use the bridge frequently have the option of purchasing a monthly discount pass for $30. With the discount pass, the toll is reduced to $4. For how many bridge crossings per month will the total monthly cost without the discount pass be the same as the total monthly cost with the discount pass?

Solution

Step 1. Let *x* represent one of the unknown quantities. Let

$$x = \text{the number of bridge crossings per month.}$$

Step 2. Represent other unknown quantities in terms of *x*. There are no other unknown quantities, so we can skip this step.

Step 3. Write an equation in *x* that models the conditions. The monthly cost without the discount pass is the toll, $7, times the number of bridge crossings per month, x. The monthly cost with the discount pass is the cost of the pass, $30, plus the toll, $4, times the number of bridge crossings per month, x.

The monthly cost without the discount pass	must equal	the monthly cost with the discount pass.
$7x$	$=$	$30 + 4x$

Step 4. Solve the equation and answer the question.

$$7x = 30 + 4x \qquad \text{This is the equation that models the problem's conditions.}$$

$$3x = 30 \qquad \text{Subtract } 4x \text{ from both sides.}$$

$$x = 10 \qquad \text{Divide both sides by 3.}$$

Because x represents the number of bridge crossings per month, the total monthly cost without the discount pass will be the same as the total monthly cost with the discount pass for 10 bridge crossings per month.

Step 5. Check the proposed solution in the original wording of the problem. The problem states that the monthly cost without the discount pass should be the same as the monthly cost with the discount pass. Let's see if they are the same with 10 bridge crossings per month.

$$\text{Cost without the discount pass} = \$7(10) = \$70$$

Cost of the pass   Toll

$$\text{Cost with the discount pass} = \$30 + \$4(10) = \$30 + \$40 = \$70$$

With 10 bridge crossings per month, both options cost $70 for the month. Thus the proposed solution, 10 bridge crossings, satisfies the problem's conditions. ■

Using Technology

Numeric and Graphic Connections

We can use a graphing utility to numerically or graphically verify our work in Example 3.

The monthly cost without the discount pass   must equal   the monthly cost with the discount pass.

$$7x = 30 + 4x$$

Enter $y_1 = 7x$.   Enter $y_2 = 30 + 4x$.

Numeric Check

Display a table for y_1 and y_2.

When $x = 10$, y_1 and y_2 have the same value, 70. With 10 bridge crossings, costs are the same, $70, for both options.

Graphic Check

Display graphs for y_1 and y_2. Use the intersection feature.

$y_2 = 30 + 4x$

$y_1 = 7x$

Graphs intersect at (10, 70). With 10 bridge crossings, costs are the same, $70, for both options.

[0, 20, 2] by [0, 150, 5]

✓ **CHECK POINT 3**   The toll to a bridge costs $5. Commuters who use the bridge frequently have the option of purchasing a monthly discount pass for $40. With the discount pass, the toll is reduced to $3. For how many bridge crossings per month will the total monthly cost without the discount pass be the same as the total monthly cost with the discount pass?

EXAMPLE 4   A Price Reduction on a Digital Camera

Your local computer store is having a terrific sale on digital cameras. After a 40% price reduction, you purchase a digital camera for $276. What was the camera's price before the reduction?

Solution

Step 1. Let x represent one of the unknown quantities. We will let

$x = $ the original price of the digital camera prior to the reduction.

Step 2. Represent other unknown quantities in terms of x. There are no other unknown quantities to find, so we can skip this step.

Step 3. Write an equation in x that models the conditions. The camera's original price minus the 40% reduction is the reduced price, $276.

| Original price | minus | the reduction (40% of the original price) | is | the reduced price, $276. |

$$x - 0.4x = 276$$

Great Question!

Why is the 40% reduction written as $0.4x$ in Example 4?

- 40% is written 0.40 or 0.4.
- "Of" represents multiplication, so 40% of the original price is $0.4x$.

Notice that the original price, x, reduced by 40% is $x - 0.4x$ and *not* $x - 0.4$.

Step 4. Solve the equation and answer the question.

$$x - 0.4x = 276$$ This is the equation that models the problem's conditions.

$$0.6x = 276$$ Combine like terms: $x - 0.4x = 1x - 0.4x = (1 - 0.4)x = 0.6x.$

$$\frac{0.6x}{0.6} = \frac{276}{0.6}$$ Divide both sides by 0.6.

$$x = 460$$ Simplify: $0.6\overline{)276.0}$ = 460.

The digital camera's price before the reduction was $460.

Step 5. Check the proposed solution in the original wording of the problem. The price before the reduction, $460, minus the 40% reduction should equal the reduced price given in the original wording, $276:

$$460 - 40\% \text{ of } 460 = 460 - 0.4(460) = 460 - 184 = 276.$$

This verifies that the digital camera's price before the reduction was $460. ∎

✓ **CHECK POINT 4**   After a 30% price reduction, you purchase a new computer for $840. What was the computer's price before the reduction?

Solving geometry problems usually requires a knowledge of basic geometric ideas and formulas. Formulas for area, perimeter, and volume are given in **Table 1.6**.

Table 1.6   Common Formulas for Area, Perimeter, and Volume

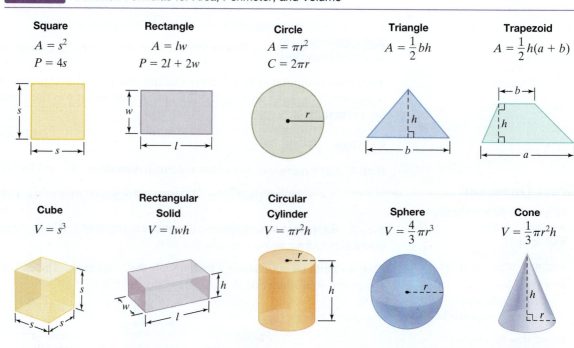

Square	Rectangle	Circle	Triangle	Trapezoid
$A = s^2$	$A = lw$	$A = \pi r^2$	$A = \frac{1}{2}bh$	$A = \frac{1}{2}h(a + b)$
$P = 4s$	$P = 2l + 2w$	$C = 2\pi r$		

Cube	Rectangular Solid	Circular Cylinder	Sphere	Cone
$V = s^3$	$V = lwh$	$V = \pi r^2 h$	$V = \frac{4}{3}\pi r^3$	$V = \frac{1}{3}\pi r^2 h$

We will be using the formula for the perimeter of a rectangle, $P = 2l + 2w$, in our next example. The formula states that a rectangle's perimeter is the sum of twice its length and twice its width.

EXAMPLE 5   Finding the Dimensions of an American Football Field

The length of an American football field is 200 feet more than the width. If the perimeter of the field is 1040 feet, what are its dimensions?

Solution
Step 1. Let x represent one of the unknown quantities. We know something about the length; the length is 200 feet more than the width. We will let

$$x = \text{the width.}$$

Step 2. Represent other unknown quantities in terms of x. Because the length is 200 feet more than the width, let

$$x + 200 = \text{the length.}$$

Figure 1.25 illustrates an American football field and its dimensions.

Step 3. Write an equation in x that models the conditions. Because the perimeter of the field is 1040 feet,

$$2(x + 200) \quad + \quad 2x \quad = \quad 1040.$$

Step 4. Solve the equation and answer the question.

$$2(x + 200) + 2x = 1040 \qquad \text{This is the equation that models the problem's conditions.}$$
$$2x + 400 + 2x = 1040 \qquad \text{Apply the distributive property.}$$
$$4x + 400 = 1040 \qquad \text{Combine like terms: } 2x + 2x = 4x.$$
$$4x = 640 \qquad \text{Subtract 400 from both sides.}$$
$$x = 160 \qquad \text{Divide both sides by 4.}$$

Thus,
$$\text{width} = x = 160.$$
$$\text{length} = x + 200 = 160 + 200 = 360.$$

The dimensions of an American football field are 160 feet by 360 feet. (The 360-foot length is usually described as 120 yards.)

Step 5. Check the proposed solution in the original wording of the problem. The perimeter of the football field using the dimensions that we found is

$$2(360 \text{ feet}) + 2(160 \text{ feet}) = 720 \text{ feet} + 320 \text{ feet} = 1040 \text{ feet.}$$

Because the problem's wording tells us that the perimeter is 1040 feet, our dimensions are correct. ∎

☑ **CHECK POINT 5**   The length of a rectangular basketball court is 44 feet more than the width. If the perimeter of the basketball court is 288 feet, what are its dimensions?

---

2 Solve a formula for a variable.

Solving a Formula for One of Its Variables

We know that solving an equation is the process of finding the number (or numbers) that make the equation a true statement. All of the equations we have solved contained only one letter, x.

Sidebar (left column)

Figure 1.25 An American football field

$x + 200$

Length

x

Width

Great Question!

Should I draw pictures like Figure 1.25 when solving geometry problems?

When solving word problems, particularly problems involving geometric figures, drawing a picture of the situation is often helpful. Label x on your drawing and, where appropriate, label other parts of the drawing in terms of x.

By contrast, formulas contain two or more letters, representing two or more variables. An example is the formula for the perimeter of a rectangle:

$$2l + 2w = P.$$

We say that this formula is solved for the variable P because P is alone on one side of the equation and the other side does not contain a P.

Solving a formula for a variable means using the addition and multiplication properties of equality to rewrite the formula so that the variable is isolated on one side of the equation. It does not mean obtaining a numerical value for that variable.

To solve a formula for one of its variables, treat that variable as if it were the only variable in the equation. Think of the other variables as if they were numbers. Use the addition property of equality to isolate all terms with the specified variable on one side of the equation and all terms without the specified variable on the other side. Then use the multiplication property of equality to get the specified variable alone. The next example shows how to do this.

EXAMPLE 6   Solving a Formula for a Variable

Solve the formula $2l + 2w = P$ for l.

Solution   First, isolate $2l$ on the left by subtracting $2w$ from both sides. Then solve for l by dividing both sides by 2.

> We need to isolate l.

$$2l + 2w = P$$ This is the given formula.

$$2l + 2w - 2w = P - 2w$$ Isolate $2l$ by subtracting $2w$ from both sides.

$$2l = P - 2w$$ Simplify.

$$\frac{2l}{2} = \frac{P - 2w}{2}$$ Solve for l by dividing both sides by 2.

$$l = \frac{P - 2w}{2}$$ Simplify. ∎

✓ **CHECK POINT 6**   Solve the formula $2l + 2w = P$ for w.

EXAMPLE 7   Solving a Formula for a Variable

Circular Cylinder

$V = \pi r^2 h$

Table 1.6 on page 60 shows that the volume of a circular cylinder is given by the formula

$$V = \pi r^2 h,$$

where r is the radius of the circle at either end and h is the height. Solve this formula for h.

Solution   Our goal is to get h by itself on one side of the formula. There is only one term with h, $\pi r^2 h$, and it is already isolated on the right side. We isolate h on the right by dividing both sides by πr^2.

> We need to isolate h.

$$V = \pi r^2 h$$ This is the given formula.

$$\frac{V}{\pi r^2} = \frac{\pi r^2 h}{\pi r^2}$$ Isolate h by dividing both sides by πr^2.

$$\frac{V}{\pi r^2} = h$$ Simplify: $\frac{\pi r^2 h}{\pi r^2} = \frac{\pi r^2}{\pi r^2} \cdot h = 1h = h.$

Equivalently,

$$h = \frac{V}{\pi r^2}. \quad \blacksquare$$

✓ **CHECK POINT 7**   The volume of a rectangular solid is the product of its length, width, and height:

$$V = lwh.$$

Solve this formula for h.

You'll be leaving the cold of winter for a vacation to Hawaii. CNN International reports a temperature in Hawaii of 30°C. Should you pack a winter coat? You can convert from Celsius temperature, C, to Fahrenheit temperature, F, using the formula

$$F = \frac{9}{5}C + 32.$$

A temperature of 30°C corresponds to a Fahrenheit temperature of

$$F = \frac{9}{5} \cdot 30 + 32 = \frac{9}{\underset{1}{5}} \cdot \frac{\overset{6}{30}}{1} + 32 = 54 + 32 = 86,$$

or a balmy 86°F. (Don't pack the coat.)

Visitors to the United States are more likely to be familiar with the Celsius temperature scale. For them, a useful formula is one that can be used to convert from Fahrenheit to Celsius. In Example 8, you will see how to obtain such a formula.

The Celsius scale is on the left and the Fahrenheit scale is on the right.

EXAMPLE 8   Solving a Formula for a Variable

Solve the formula

$$F = \frac{9}{5}C + 32$$

for C.

Solution   We begin by multiplying both sides of the formula by 5 to clear the fraction. Then we isolate the variable C.

$$F = \frac{9}{5}C + 32 \qquad \text{This is the given formula.}$$

$$5F = 5\left(\frac{9}{5}C + 32\right) \qquad \text{Multiply both sides by 5.}$$

$$5F = 5 \cdot \frac{9}{5}C + 5 \cdot 32 \qquad \text{Apply the distributive property.}$$

We need to isolate C.

$$5F = 9C + 160 \qquad \text{Simplify.}$$

$$5F - 160 = 9C + 160 - 160 \qquad \text{Subtract 160 from both sides.}$$

$$5F - 160 = 9C \qquad \text{Simplify.}$$

$$\frac{5F - 160}{9} = \frac{9C}{9} \qquad \text{Divide both sides by 9.}$$

$$\frac{5F - 160}{9} = C \qquad \text{Simplify.}$$

Using the distributive property, we can express $5F - 160$ as $5(F - 32)$. Thus,

$$C = \frac{5F - 160}{9} = \frac{5(F - 32)}{9}.$$

This formula, used to convert from Fahrenheit to Celsius, is usually given as

$$C = \frac{5}{9}(F - 32). \quad \blacksquare$$

✓ **CHECK POINT 8**   The formula

$$\frac{W}{2} - 3H = 53$$

models the recommended weight, W, in pounds, for a male, where H represents his height, in inches, over 5 feet. Solve this formula for W.

EXAMPLE 9   Solving a Formula for a Variable That Occurs Twice

The formula

$$A = P + Prt$$

describes the amount, A, that a principal of P dollars is worth after t years when invested at a simple annual interest rate, r. Solve this formula for P.

Solution   Notice that all the terms with P already occur on the right side of the formula.

We need to isolate P.

$$A = P + Prt$$

We can use the distributive property in the form $ab + ac = a(b + c)$ to convert the two occurrences of P into one.

$$A = P + Prt \qquad \text{This is the given formula.}$$
$$A = P(1 + rt) \qquad \text{Use the distributive property to obtain a single occurrence of } P.$$
$$\frac{A}{1 + rt} = \frac{P(1 + rt)}{1 + rt} \qquad \text{Divide both sides by } 1 + rt.$$
$$\frac{A}{1 + rt} = P \qquad \text{Simplify: } \frac{P(1 + rt)}{1(1 + rt)} = \frac{P}{1} = P.$$

Equivalently,

$$P = \frac{A}{1 + rt}. \quad \blacksquare$$

Great Question!

Can I solve $A = P + Prt$ for P by subtracting Prt from both sides and writing

$$A - Prt = P?$$

No. When a formula is solved for a specified variable, that variable must be isolated on one side. The variable P occurs on both sides of

$$A - Prt = P.$$

✓ **CHECK POINT 9**   Solve the formula $P = C + MC$ for C.

Blitzer Bonus

Einstein's Famous Formula: $E = mc^2$

One of the most famous formulas in the world is $E = mc^2$, formulated by Albert Einstein. Einstein showed that any form of energy has mass and that mass itself has an associated energy that can be released if the matter is destroyed. In this formula, E represents energy, in ergs, m represents mass, in grams, and c represents the speed of light. Because light travels at 30 billion centimeters per second, the formula indicates that 1 gram of mass will produce 900 billion billion ergs of energy.

Einstein's formula implies that the mass of a golf ball could provide the daily energy needs of the metropolitan Boston area. Mass and energy are equivalent, and the transformation of even a tiny amount of mass releases an enormous amount of energy. If this energy is released suddenly, a destructive force is unleashed, as in an atom bomb. When the release is gradual and controlled, the energy can be used to generate power.

The theoretical results implied by Einstein's formula $E = mc^2$ have not been realized because scientists have not yet developed a way of converting a mass completely to energy. Japan's devastating earthquake–tsunami crisis rekindled debate over safety problems associated with nuclear energy.

Achieving Success

Do not expect to solve every word problem immediately. As you read each problem, underline the important parts. It's a good idea to read the problem at least twice. Be persistent, but use the **"Ten Minutes of Frustration" Rule**. If you have exhausted every possible means for solving a problem and you are still bogged down, stop after ten minutes. Put a question mark by the exercise and move on. When you return to class, ask your professor for assistance.

CONCEPT AND VOCABULARY CHECK

Fill in each blank so that the resulting statement is true.

1. According to the U.S. Office of Management and Budget, the 2011 budget for defense exceeded the budget for education by $658.6 billion. If x represents the budget for education, in billions of dollars, the budget for defense can be represented by _____.

2. In 2000, 31% of U.S. adults viewed a college education as essential for success. For the period from 2000 through 2010, this percentage increased by approximately 2.4 each year. The percentage of U.S. adults who viewed a college education as essential for success x years after 2000 can be represented by _____.

3. A text message plan costs $4 per month plus $0.15 per text. The monthly cost for x text messages can be represented by _____.

4. I purchased a computer after a 15% price reduction. If x represents the computer's original price, the reduced price can be represented by _____.

5. Solving a formula for a variable means rewriting the formula so that the variable is _____.

6. The first step in solving $IR + Ir = E$ for I is to obtain a single occurrence of I using the _____ property.

1.5 EXERCISE SET   ▶ MyMathLab®

Practice Exercises

Use the five-step strategy for solving word problems to find the number or numbers described in Exercises 1–10.

1. When five times a number is decreased by 4, the result is 26. What is the number?

2. When two times a number is decreased by 3, the result is 11. What is the number?

3. When a number is decreased by 20% of itself, the result is 20. What is the number?

4. When a number is decreased by 30% of itself, the result is 28. What is the number?

5. When 60% of a number is added to the number, the result is 192. What is the number?

6. When 80% of a number is added to the number, the result is 252. What is the number?

7. 70% of what number is 224?

8. 70% of what number is 252?

9. One number exceeds another by 26. The sum of the numbers is 64. What are the numbers?

10. One number exceeds another by 24. The sum of the numbers is 58. What are the numbers?

Practice PLUS

In Exercises 11–16, find all values of x satisfying the given conditions.

11. $y_1 = 13x - 4$, $y_2 = 5x + 10$, and y_1 exceeds y_2 by 2.

12. $y_1 = 10x + 6$, $y_2 = 12x - 7$, and y_1 exceeds y_2 by 3.

13. $y_1 = 10(2x - 1)$, $y_2 = 2x + 1$, and y_1 is 14 more than 8 times y_2.

14. $y_1 = 9(3x - 5)$, $y_2 = 3x - 1$, and y_1 is 51 less than 12 times y_2.

15. $y_1 = 2x + 6$, $y_2 = x + 8$, $y_3 = x$, and the difference between 3 times y_1 and 5 times y_2 is 22 less than y_3.

16. $y_1 = 2.5$, $y_2 = 2x + 1$, $y_3 = x$, and the difference between 2 times y_1 and 3 times y_2 is 8 less than 4 times y_3.

Application Exercises

In Exercises 17–18, use the five-step strategy for solving word problems.

17. Some animals stick around much longer than their species' life expectancy would predict. The bar graph shows three critters that lived to a ripe old age.

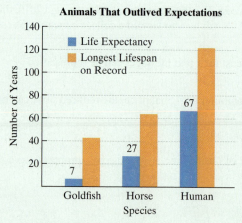

Animals That Outlived Expectations

Source: Listomania, HarperCollins

The longest lifespan of a horse exceeded that of a goldfish by 21 years. The longest lifespan of a human exceeded that of a goldfish by 79 years. Combined, the longest lifespan on record for a goldfish, a horse, and a human was 229 years. Determine the longest lifespan on record for a goldfish, a horse, and a human.

18. How many words are there? The bar graph shows the number of words in English, in thousands, compared with four other languages.

Number of Words in English and Other Major Languages

Source: Global Language Monitor

The number of words in English exceeds the number of words in Japanese by 767 thousand. The number of words in Chinese (various dialects) exceeds the number of words in Japanese by 268 thousand. Combined, these three languages contain 1731 thousand words. Determine the number of words, in thousands, for each of these three languages.

Solve Exercises 19–22 using the fact that the sum of the measures of the three angles of a triangle is 180°.

19. In a triangle, the measure of the first angle is twice the measure of the second angle. The measure of the third angle is 8° less than the measure of the second angle. What is the measure of each angle?

20. In a triangle, the measure of the first angle is three times the measure of the second angle. The measure of the third angle is 35° less than the measure of the second angle. What is the measure of each angle?

21. In a triangle, the measures of the three angles are consecutive integers. What is the measure of each angle?

22. In a triangle, the measures of the three angles are consecutive even integers. What is the measure of each angle?

Even as Americans increasingly view a college education as essential for success, many believe that a college education is becoming less available to qualified students. Exercises 23–24 are based on the data displayed by the graph.

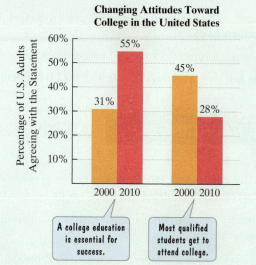

Changing Attitudes Toward College in the United States

Source: Public Agenda

23. In 2000, 31% of U.S. adults viewed a college education as essential for success. For the period 2000 through 2010, the percentage viewing a college education as essential for success increased on average by approximately 2.4 each year. If this trend continues, by which year will 67% of all American adults view a college education as essential for success?

24. In 2000, 45% of U.S. adults believed that most qualified students get to attend college. For the period from 2000 through 2010, the percentage who believed that a college education is available to most qualified students decreased by approximately 1.7 each year. If this trend continues, by which year will only 11% of all American adults believe that most qualified students get to attend college?

The line graph indicates that in 1960, 23% of U.S. taxes came from corporate income tax. For the period from 1960 through 2010, this percentage decreased by approximately 0.28 each year. Use this information to solve Exercises 25–26.

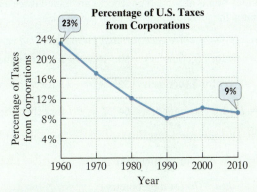

Percentage of U.S. Taxes from Corporations

Source: Office of Management and Budget

25. If this trend continues, by which year will corporations pay zero taxes? Round to the nearest year.

26. If this trend continues, by which year will the percentage of federal tax receipts coming from the corporate income tax drop to 5%? Round to the nearest year.

On average, every minute of every day, 158 babies are born. The bar graph represents the results of a single day of births, deaths, and population increase worldwide. Exercises 27–28 are based on the information displayed by the graph.

Daily Growth of World Population

Source: James Henslin, *Sociology*, Eleventh Edition, Pearson, 2012.

27. Each day, the number of births in the world is 84 thousand less than three times the number of deaths.

   a. If the population increase in a single day is 228 thousand, determine the number of births and deaths per day.

   b. If the population increase in a single day is 228 thousand, by how many millions of people does the worldwide population increase each year? Round to the nearest million.

   c. Based on your answer to part (b), approximately how many years does it take for the population of the world to increase by an amount greater than the entire U.S. population (320 million)?

28. Each day, the number of births in the world exceeds twice the number of deaths by 72 thousand.

   a. If the population increase in a single day is 228 thousand, determine the number of births and deaths per day.

   b. If the population increase in a single day is 228 thousand, by how many millions of people does the worldwide population increase each year? Round to the nearest million.

   c. Based on your answer to part (b), approximately how many years does it take for the population of the world to increase by an amount greater than the entire U.S. population (320 million)?

29. The bus fare in a city is $1.25. People who use the bus have the option of purchasing a monthly discount pass for $15.00. With the discount pass, the fare is reduced to $0.75. Determine the number of times in a month the bus must be used so that the total monthly cost without the discount pass is the same as the total monthly cost with the discount pass.

30. You are choosing between two health clubs. Club A offers membership for a fee of $40 plus a monthly fee of $25. Club B offers membership for a fee of $15 plus a monthly fee of $30. After how many months will the total cost at each health club be the same? What will be the total cost for each club?

31. A discount pass for a bridge costs $30 per month. The toll for the bridge is normally $5.00, but it is reduced to $3.50 for people who have purchased the discount pass. Determine the number of times in a month the bridge must be crossed so that the total monthly cost without the discount pass is the same as the total monthly cost with the discount pass.

32. You are choosing between two cellphone plans. Data Plan A has a monthly fee of $40 with a charge of $15 per gigabyte (GB). Data Plan B has a monthly fee of $30 with a charge of $20 per GB. For how many GB of data will the costs for the two data plans be the same?

33. In 2008, there were 13,300 students at college A, with a projected enrollment increase of 1000 students per year. In the same year, there were 26,800 students at college B, with a projected enrollment decline of 500 students per year.

   a. According to these projections, when will the colleges have the same enrollment? What will be the enrollment in each college at that time?

   b. Use the following table to numerically check your work in part (a). What equations were entered for y_1 and y_2 to obtain this table?

X	Y₁	Y₂		
7	20300	23300		
8	21300	22800		
9	22300	22300		
10	23300	21800		
11	24300	21300		
12	25300	20800		
13	26300	20300		
14	27300	19800		
15	28300	19300		
16	29300	18800		
17	30300	18300		

X=7

34. In 2000, the population of Greece was 10,600,000, with projections of a population decrease of 28,000 people per year. In the same year, the population of Belgium was 10,200,000, with projections of a population decrease of 12,000 people per year. (*Source:* United Nations) According to these projections, when will the two countries have the same population? What will be the population at that time?

35. After a 20% reduction, you purchase a television for $336. What was the television's price before the reduction?

36. After a 30% reduction, you purchase a dictionary for $30.80. What was the dictionary's price before the reduction?

37. Including 8% sales tax, an inn charges $162 per night. Find the inn's nightly cost before the tax is added.

38. Including 5% sales tax, an inn charges $252 per night. Find the inn's nightly cost before the tax is added.

Exercises 39–40 involve markup, the amount added to the dealer's cost of an item to arrive at the selling price of that item.

39. The selling price of a refrigerator is $584. If the markup is 25% of the dealer's cost, what is the dealer's cost of the refrigerator?

40. The selling price of a scientific calculator is $15. If the markup is 25% of the dealer's cost, what is the dealer's cost of the calculator?

41. A rectangular soccer field is twice as long as it is wide. If the perimeter of the soccer field is 300 yards, what are its dimensions?

42. A rectangular swimming pool is three times as long as it is wide. If the perimeter of the pool is 320 feet, what are its dimensions?

43. The length of the rectangular tennis court at Wimbledon is 6 feet longer than twice the width. If the court's perimeter is 228 feet, what are the court's dimensions?

44. The length of a rectangular pool is 6 meters less than twice the width. If the pool's perimeter is 126 meters, what are its dimensions?

45. The rectangular painting in the figure shown measures 12 inches by 16 inches and contains a frame of uniform width around the four edges. The perimeter of the rectangle formed by the painting and its frame is 72 inches. Determine the width of the frame.

46. The rectangular swimming pool in the figure shown measures 40 feet by 60 feet and contains a path of uniform width around the four edges. The perimeter of the rectangle formed by the pool and the surrounding path is 248 feet. Determine the width of the path.

47. For an international telephone call, a telephone company charges $0.43 for the first minute, $0.32 for each additional minute, and a $2.10 service charge. If the cost of a call is $5.73, how long did the person talk?

48. A job pays an annual salary of $33,150, which includes a holiday bonus of $750. If paychecks are issued twice a month, what is the gross amount for each paycheck?

In Exercises 49–74, solve each formula for the specified variable.

49. $A = lw$ for l (geometry) **50.** $A = lw$ for w (geometry)

51. $A = \frac{1}{2}bh$ for b (geometry) **52.** $A = \frac{1}{2}bh$ for h (geometry)

53. $I = Prt$ for P (finance) **54.** $I = Prt$ for t (finance)

55. $T = D + pm$ for p (finance)

56. $P = C + MC$ for M (finance)

57. $A = \frac{1}{2}h(a + b)$ for a (geometry)

58. $A = \frac{1}{2}h(a + b)$ for b (geometry)

59. $V = \frac{1}{3}\pi r^2 h$ for h (geometry)

60. $V = \frac{1}{3}\pi r^2 h$ for r^2 (geometry)

61. $y - y_1 = m(x - x_1)$ for m (algebra)

62. $y_2 - y_1 = m(x_2 - x_1)$ for m (algebra)

63. $V = \frac{d_1 - d_2}{t}$ for d_1 (physics)

64. $z = \frac{x - u}{s}$ for x (statistics)

65. $Ax + By = C$ for x (algebra)

66. $Ax + By = C$ for y (algebra)

67. $s = \frac{1}{2}at^2 + vt$ for v (physics)

68. $s = \frac{1}{2}at^2 + vt$ for a (physics)

69. $L = a + (n - 1)d$ for n (algebra)

70. $L = a + (n - 1)d$ for d (algebra)

71. $A = 2lw + 2lh + 2wh$ for l (geometry)

72. $A = 2lw + 2lh + 2wh$ for h (geometry)

73. $IR + Ir = E$ for I (physics)

74. $A = \frac{x_1 + x_2 + x_3}{n}$ for n (statistics)

Explaining the Concepts

75. In your own words, describe a step-by-step approach for solving algebraic word problems.
76. Write an original word problem that can be solved using a linear equation. Then solve the problem.
77. Explain what it means to solve a formula for a variable.
78. Did you have difficulties solving some of the problems that were assigned in this Exercise Set? Discuss what you did if this happened to you. Did your course of action enhance your ability to solve algebraic word problems?
79. The bar graph in **Figure 1.23** on page 55 indicates a gender gap in pay at all levels of education. What explanations can you offer for this phenomenon?

Technology Exercises

80. Use a graphing utility to numerically or graphically verify your work in any one exercise from Exercises 29–32. For assistance on how to do this, refer to the Using Technology box on page 59.
81. The formula $y = 31 + 2.4x$ models the percentage of U.S. adults, y, viewing a college education as essential for success x years after 2000. Graph the formula in a $[0, 20, 2]$ by $[0, 100, 10]$ viewing rectangle. Then use the $\boxed{\text{TRACE}}$ or $\boxed{\text{ZOOM}}$ feature to verify your answer in Exercise 23.
82. In Exercises 25–26, we saw that in 1960, 23% of U.S. taxes came from corporate income tax, decreasing by approximately 0.28% per year.
    a. Write a formula that models the percentage of U.S. taxes from corporations, y, x years after 1960.
    b. Enter the formula from part (a) as y_1 in your graphing utility. Then use either a table for y_1 or a graph of y_1 to numerically or graphically verify your answer to Exercise 25 or Exercise 26.

Critical Thinking Exercises

Make Sense? *In Exercises 83–86, determine whether each statement makes sense or does not make sense, and explain your reasoning.*

83. I solved the formula for one of its variables, so now I have a numerical value for that variable.
84. Reasoning through a word problem often increases problem-solving skills in general.
85. The hardest part in solving a word problem is writing the equation that models the verbal conditions.
86. When traveling in Europe, the most useful form of the two Celsius-Fahrenheit conversion formulas is the formula used to convert from Fahrenheit to Celsius.

In Exercises 87–90, determine whether each statement is true or false. If the statement is false, make the necessary change(s) to produce a true statement.

87. If $I = prt$, then $t = I - pr$.
88. If $y = \dfrac{kx}{z}$, then $z = \dfrac{kx}{y}$.
89. It is not necessary to use the distributive property to solve $P = C + MC$ for C.
90. An item's price, x, reduced by $\dfrac{1}{3}$ is modeled by $x - \dfrac{1}{3}$.

91. The price of a dress is reduced by 40%. When the dress still does not sell, it is reduced by 40% of the reduced price. If the price of the dress after both reductions is $72, what was the original price?
92. Suppose that we agree to pay you 8¢ for every problem in this chapter that you solve correctly and fine you 5¢ for every problem done incorrectly. If at the end of 26 problems we do not owe each other any money, how many problems did you solve correctly?
93. It was wartime when the Ricardos found out Mrs. Ricardo was pregnant. Ricky Ricardo was drafted and made out a will, deciding that $14,000 in a savings account was to be divided between his wife and his child-to-be. Rather strangely, and certainly with gender bias, Ricky stipulated that if the child were a boy, he would get twice the amount of the mother's portion. If it were a girl, the mother would get twice the amount the girl was to receive. We'll never know what Ricky was thinking, for (as fate would have it) he did not return from war. Mrs. Ricardo gave birth to twins—a boy and a girl. How was the money divided?
94. A thief steals a number of rare plants from a nursery. On the way out, the thief meets three security guards, one after another. To each security guard, the thief is forced to give one-half the plants that he still has, plus 2 more. Finally, the thief leaves the nursery with 1 lone palm. How many plants were originally stolen?
95. Solve for C: $V = C - \dfrac{C - S}{L} N$.

Review Exercises

96. Express in set-builder notation and graph: $(-4, 0]$. (Section 1.1, Example 7)
97. Simplify: $\dfrac{(2 + 4)^2 + (-1)^5}{12 \div 2 \cdot 3 - 3}$. (Section 1.2, Example 8)
98. Solve: $\dfrac{2x}{3} - \dfrac{8}{3} = x$. (Section 1.4, Example 4)

Preview Exercises

Exercises 99–101 will help you prepare for the material covered in the next section.

99. In parts (a) and (b), complete each statement.
    a. $b^4 \cdot b^3 = (b \cdot b \cdot b \cdot b)(b \cdot b \cdot b) = b^?$
    b. $b^5 \cdot b^5 = (b \cdot b \cdot b \cdot b \cdot b)(b \cdot b \cdot b \cdot b \cdot b) = b^?$
    c. Generalizing from parts (a) and (b), what should be done with the exponents when multiplying exponential expressions with the same base?
100. In parts (a) and (b), complete each statement.
    a. $\dfrac{b^7}{b^3} = \dfrac{\not{b} \cdot \not{b} \cdot \not{b} \cdot b \cdot b \cdot b \cdot b}{\not{b} \cdot \not{b} \cdot \not{b}} = b^?$
    b. $\dfrac{b^8}{b^2} = \dfrac{\not{b} \cdot \not{b} \cdot b \cdot b \cdot b \cdot b \cdot b \cdot b}{\not{b} \cdot \not{b}} = b^?$
    c. Generalizing from parts (a) and (b), what should be done with the exponents when dividing exponential expressions with the same base?
101. Simplify: $\dfrac{1}{\left(-\dfrac{1}{2}\right)^3}$.

SECTION

1.6

Properties of Integral Exponents

What am I supposed to learn?

After studying this section, you should be able to:

1 Use the product rule. ▶

2 Use the quotient rule. ▶

3 Use the zero-exponent rule. ▶

4 Use the negative-exponent rule. ▶

5 Use the power rule. ▶

6 Find the power of a product. ▶

7 Find the power of a quotient. ▶

8 Simplify exponential expressions. ▶

Our opening photo shows the head of a fly as seen under an electronic microscope. Some electronic microscopes can view objects that are less than 10^{-4} meter, or 0.0001 meter, in size. In this section, we'll make sense of the negative exponent in 10^{-4}, as we turn to integral exponents and their properties.

The Product and Quotient Rules

We have seen that exponents are used to indicate repeated multiplication. Now consider the multiplication of two exponential expressions, such as $b^4 \cdot b^3$. We are multiplying 4 factors of b and 3 factors of b. We have a total of 7 factors of b:

$$
\underbrace{4 \text{ factors of } b}\qquad \underbrace{3 \text{ factors of } b}
$$

$$
b^4 \cdot b^3 = (b \cdot b \cdot b \cdot b)(b \cdot b \cdot b) = b^7.
$$

Total: 7 factors of b

The product is exactly the same if we add the exponents:

$$
b^4 \cdot b^3 = b^{4+3} = b^7.
$$

This suggests the following rule:

1 Use the product rule. ▶

> **The Product Rule**
>
> $$b^m \cdot b^n = b^{m+n}$$
>
> When multiplying exponential expressions with the same base, add the exponents. Use this sum as the exponent of the common base.

EXAMPLE 1   Using the Product Rule

Multiply each expression using the product rule:

a. $b^8 \cdot b^{10}$ **b.** $(6x^4y^3)(5x^2y^7)$.

Solution

a. $b^8 \cdot b^{10} = b^{8+10} = b^{18}$

b. $(6x^4y^3)(5x^2y^7)$

$$= 6 \cdot 5 \cdot x^4 \cdot x^2 \cdot y^3 \cdot y^7 \qquad \text{Use the associative and commutative properties.}$$
$$\text{This step can be done mentally.}$$
$$= 30x^{4+2}y^{3+7}$$
$$= 30x^6y^{10} \quad \blacksquare$$

☑ **CHECK POINT 1**   Multiply each expression using the product rule:
a. $b^6 \cdot b^5$ **b.** $(4x^3y^4)(10x^2y^6)$.

2 Use the quotient rule. ▶

Now, consider the division of two exponential expressions, such as the quotient of b^7 and b^3. We are dividing 7 factors of b by 3 factors of b.

$$\frac{b^7}{b^3} = \frac{b \cdot b \cdot b \cdot b \cdot b \cdot b \cdot b}{b \cdot b \cdot b} = \boxed{\frac{b \cdot b \cdot b}{b \cdot b \cdot b}} \cdot b \cdot b \cdot b \cdot b = 1 \cdot b \cdot b \cdot b \cdot b = b^4$$

This factor is equal to 1.

The quotient is exactly the same if we subtract the exponents:

$$\frac{b^7}{b^3} = b^{7-3} = b^4.$$

This suggests the following rule:

The Quotient Rule

$$\frac{b^m}{b^n} = b^{m-n}, \quad b \neq 0$$

When dividing exponential expressions with the same nonzero base, subtract the exponent in the denominator from the exponent in the numerator. Use this difference as the exponent of the common base.

EXAMPLE 2   Using the Quotient Rule

Divide each expression using the quotient rule:

a. $\dfrac{(-2)^7}{(-2)^4}$ **b.** $\dfrac{30x^{12}y^9}{5x^3y^7}$.

Solution

a. $\dfrac{(-2)^7}{(-2)^4} = (-2)^{7-4} = (-2)^3$ or -8 $(-2)^3 = (-2)(-2)(-2) = -8$

b. $\dfrac{30x^{12}y^9}{5x^3y^7} = \dfrac{30}{5} \cdot \dfrac{x^{12}}{x^3} \cdot \dfrac{y^9}{y^7} = 6x^{12-3}y^{9-7} = 6x^9y^2$ $\blacksquare$

☑ **CHECK POINT 2**   Divide each expression using the quotient rule:

a. $\dfrac{(-3)^6}{(-3)^3}$ **b.** $\dfrac{27x^{14}y^8}{3x^3y^5}$.

3 Use the zero-exponent rule. ▶

Zero as an Exponent

A nonzero base can be raised to the 0 power. The quotient rule can be used to help determine what zero as an exponent should mean. Consider the quotient of b^4 and b^4, where b is not zero. We can determine this quotient in two ways.

$$\frac{b^4}{b^4} = 1 \qquad\qquad \frac{b^4}{b^4} = b^{4-4} = b^0$$

Any nonzero expression divided by itself is 1.

Use the quotient rule and subtract exponents.

This means that b^0 must equal 1.

The Zero-Exponent Rule

If b is any real number other than 0,

$$b^0 = 1.$$

EXAMPLE 3   Using the Zero-Exponent Rule

Use the zero-exponent rule to simplify each expression:

a. 8^0 **b.** $(-6)^0$ **c.** -6^0 **d.** $5x^0$ **e.** $(5x)^0$.

Solution

a. $8^0 = 1$ — Any nonzero number raised to the 0 power is 1.

b. $(-6)^0 = 1$

c. $-6^0 = -(6^0) = -1$

Only 6 is raised to the 0 power.

d. $5x^0 = 5 \cdot 1 = 5$ — Only x is raised to the 0 power.

e. $(5x)^0 = 1$ — The entire expression, $5x$, is raised to the 0 power. ∎

✓ **CHECK POINT 3**   Use the zero-exponent rule to simplify each expression:

a. 7^0 **b.** $(-5)^0$ **c.** -5^0 **d.** $10x^0$ **e.** $(10x)^0$.

---

4 Use the negative-exponent rule. ▶

Negative Integers as Exponents

A nonzero base can be raised to a negative power. The quotient rule can be used to help determine what a negative integer as an exponent should mean. Consider the quotient of b^3 and b^5, where b is not zero. We can determine this quotient in two ways.

$$\frac{b^3}{b^5} = \frac{b \cdot b \cdot b}{b \cdot b \cdot b \cdot b \cdot b} = \frac{1}{b^2} \qquad\qquad \frac{b^3}{b^5} = b^{3-5} = b^{-2}$$

After dividing common factors, we have two factors of b in the denominator.

Use the quotient rule and subtract exponents.

Notice that $\dfrac{b^3}{b^5}$ equals both b^{-2} and $\dfrac{1}{b^2}$. This means that b^{-2} must equal $\dfrac{1}{b^2}$. This example is a special case of the **negative-exponent rule**.

The Negative-Exponent Rule

If b is any real number other than 0 and n is a natural number, then

$$b^{-n} = \frac{1}{b^n}.$$

EXAMPLE 4   Using the Negative-Exponent Rule

Use the negative-exponent rule to write each expression with a positive exponent. Simplify, if possible:

a. 9^{-2} **b.** $(-2)^{-5}$ **c.** $\dfrac{1}{6^{-2}}$ **d.** $7x^{-5}y^2$.

Solution

a. $9^{-2} = \dfrac{1}{9^2} = \dfrac{1}{81}$

b. $(-2)^{-5} = \dfrac{1}{(-2)^5} = \dfrac{1}{(-2)(-2)(-2)(-2)(-2)} = \dfrac{1}{-32} = -\dfrac{1}{32}$

> Only the sign of the exponent, −5, changes. The base, −2, does not change sign.

c. $\dfrac{1}{6^{-2}} = \dfrac{1}{\frac{1}{6^2}} = 1 \cdot \dfrac{6^2}{1} = 6^2 = 36$

d. $7x^{-5}y^2 = 7 \cdot \dfrac{1}{x^5} \cdot y^2 = \dfrac{7y^2}{x^5}$ ∎

✓ **CHECK POINT 4**   Use the negative-exponent rule to write each expression with a positive exponent. Simplify, if possible:

a. 5^{-2} **b.** $(-3)^{-3}$ **c.** $\dfrac{1}{4^{-2}}$ **d.** $3x^{-6}y^4$.

In Example 4 and Check Point 4, did you notice that

$$\frac{1}{6^{-2}} = 6^2 \quad \text{and} \quad \frac{1}{4^{-2}} = 4^2?$$

In general, if a negative exponent appears in a denominator, an expression can be written with a positive exponent using

$$\frac{1}{b^{-n}} = b^n.$$

Negative Exponents in Numerators and Denominators

If b is any real number other than 0 and n is a natural number, then

$$b^{-n} = \frac{1}{b^n} \quad \text{and} \quad \frac{1}{b^{-n}} = b^n.$$

When a negative number appears as an exponent, switch the position of the base (from numerator to denominator or from denominator to numerator) and make the exponent positive. The sign of the base does not change.

EXAMPLE 5   Using Negative Exponents

Write each expression with positive exponents only. Then simplify, if possible:

a. $\dfrac{5^{-3}}{4^{-2}}$ **b.** $\dfrac{1}{6x^{-4}}$.

Solution

a. $\dfrac{5^{-3}}{4^{-2}} = \dfrac{4^2}{5^3} = \dfrac{4 \cdot 4}{5 \cdot 5 \cdot 5} = \dfrac{16}{125}$ Switch the position of each base to the other side of the fraction bar and change the sign of the exponent.

b. $\dfrac{1}{6x^{-4}} = \dfrac{x^4}{6}$ Switch the position of x to the other side of the fraction bar and change -4 to 4.

Don't switch the position of 6. It is not affected by a negative exponent.

■

✓ **CHECK POINT 5**   Write each expression with positive exponents only. Then simplify, if possible:

a. $\dfrac{7^{-2}}{4^{-3}}$ **b.** $\dfrac{1}{5x^{-2}}$.

---

5 Use the power rule. ▶

The Power Rule for Exponents (Powers to Powers)

The next property of exponents applies when an exponential expression is raised to a power. Here is an example:

$$(b^2)^4.$$

The exponential expression b^2 is raised to the fourth power.

There are 4 factors of b^2. Thus,

$$(b^2)^4 = b^2 \cdot b^2 \cdot b^2 \cdot b^2 = b^{2+2+2+2} = b^8.$$

Add exponents when multiplying with the same base.

We can obtain the answer, b^8, by multiplying the exponents:

$$(b^2)^4 = b^{2 \cdot 4} = b^8.$$

This suggests the following rule:

The Power Rule (Powers to Powers)

$$(b^m)^n = b^{mn}$$

When an exponential expression is raised to a power, multiply the exponents. Place the product of the exponents on the base and remove the parentheses.

EXAMPLE 6   Using the Power Rule (Powers to Powers)

Simplify each expression using the power rule:

a. $(x^6)^4$ **b.** $(y^5)^{-3}$ **c.** $(b^{-4})^{-2}$.

Solution

a. $(x^6)^4 = x^{6 \cdot 4} = x^{24}$

b. $(y^5)^{-3} = y^{5(-3)} = y^{-15} = \dfrac{1}{y^{15}}$

c. $(b^{-4})^{-2} = b^{(-4)(-2)} = b^8$ ■

✓ **CHECK POINT 6**   Simplify each expression using the power rule:

a. $(x^5)^3$ **b.** $(y^7)^{-2}$ **c.** $(b^{-3})^{-4}$.

6 Find the power of a product. ▶

The Products-to-Powers Rule for Exponents

The next property of exponents applies when we are raising a product to a power. Here is an example:

$$(2x)^4.$$

The product $2x$ is raised to the fourth power.

There are four factors of $2x$. Thus,

$$(2x)^4 = 2x \cdot 2x \cdot 2x \cdot 2x = 2 \cdot 2 \cdot 2 \cdot 2 \cdot x \cdot x \cdot x \cdot x = 2^4 x^4.$$

We can obtain the answer, $2^4 x^4$, by raising each factor within the parentheses to the fourth power:

$$(2x)^4 = 2^4 x^4.$$

This suggests the following rule:

Products to Powers

$$(ab)^n = a^n b^n$$

When a product is raised to a power, raise each factor to that power.

EXAMPLE 7 Using the Products-to-Powers Rule

Simplify each expression using the products-to-powers rule:

a. $(6x)^3$ **b.** $(-2y^2)^4$ **c.** $(-3x^{-1}y^3)^{-2}$.

Solution

a. $(6x)^3 = 6^3 x^3$ Raise each factor to the third power.

$= 216x^3$ Simplify: $6^3 = 6 \cdot 6 \cdot 6 = 216$.

b. $(-2y^2)^4 = (-2)^4(y^2)^4$ Raise each factor to the fourth power.

$= (-2)^4 y^{2 \cdot 4}$ To raise an exponential expression to a power, multiply exponents: $(b^m)^n = b^{mn}$.

$= 16y^8$ Simplify: $(-2)^4 = (-2)(-2)(-2)(-2) = 16$.

c. $(-3x^{-1}y^3)^{-2} = (-3)^{-2}(x^{-1})^{-2}(y^3)^{-2}$ Raise each factor to the -2 power.

$= (-3)^{-2} x^{(-1)(-2)} y^{3(-2)}$ Use $(b^m)^n = b^{mn}$ on the second and third factors.

$= (-3)^{-2} x^2 y^{-6}$ Simplify.

$= \dfrac{1}{(-3)^2} \cdot x^2 \cdot \dfrac{1}{y^6}$ Apply $b^{-n} = \dfrac{1}{b^n}$ to the first and last factors.

$= \dfrac{x^2}{9y^6}$ Simplify: $(-3)^2 = (-3)(-3) = 9$. ■

✓ **CHECK POINT 7** Simplify each expression using the products-to-powers rule:

a. $(2x)^4$ **b.** $(-3y^2)^3$ **c.** $(-4x^5y^{-1})^{-2}$.

7 Find the power of a quotient. ▶

The Quotients-to-Powers Rule for Exponents

The following rule is used to raise a quotient to a power:

> ### Quotients to Powers
>
> If b is a nonzero real number, then
>
> $$\left(\frac{a}{b}\right)^n = \frac{a^n}{b^n}.$$
>
> When a quotient is raised to a power, raise the numerator to that power and divide by the denominator to that power.

EXAMPLE 8   Using the Quotients-to-Powers Rule

Simplify each expression using the quotients-to-powers rule:

a. $\left(\dfrac{x^2}{4}\right)^3$ **b.** $\left(\dfrac{2x^3}{y^{-4}}\right)^5$ **c.** $\left(\dfrac{x^3}{y^2}\right)^{-4}$.

Great Question!

In Example 8, why didn't you start by simplifying inside parentheses?

When simplifying exponential expressions, the first step should be to simplify inside parentheses. In Example 8, the expressions inside parentheses have different bases and cannot be simplified. This is why we begin with the quotients-to-powers rule.

Solution

a. $\left(\dfrac{x^2}{4}\right)^3 = \dfrac{(x^2)^3}{4^3} = \dfrac{x^{2\cdot3}}{4\cdot4\cdot4} = \dfrac{x^6}{64}$ Cube the numerator and the denominator.

b. $\left(\dfrac{2x^3}{y^{-4}}\right)^5 = \dfrac{(2x^3)^5}{(y^{-4})^5}$ Raise the numerator and the denominator to the fifth power.

$= \dfrac{2^5(x^3)^5}{(y^{-4})^5}$ Raise each factor in the numerator to the fifth power.

$= \dfrac{2^5 \cdot x^{3\cdot5}}{y^{(-4)(5)}}$ Multiply exponents in both powers-to-powers expressions: $(b^m)^n = b^{mn}$.

$= \dfrac{32x^{15}}{y^{-20}}$ Simplify.

$= 32x^{15}y^{20}$ Move y to the other side of the fraction bar and change -20 to 20: $\dfrac{1}{b^{-n}} = b^n$.

c. $\left(\dfrac{x^3}{y^2}\right)^{-4} = \dfrac{(x^3)^{-4}}{(y^2)^{-4}}$ Raise the numerator and the denominator to the -4 power.

$= \dfrac{x^{3(-4)}}{y^{2(-4)}}$ Multiply exponents in both powers-to-powers expressions: $(b^m)^n = b^{mn}$.

$= \dfrac{x^{-12}}{y^{-8}}$ Simplify.

$= \dfrac{y^8}{x^{12}}$ Move each base to the other side of the fraction bar and make each exponent positive. ∎

✓ **CHECK POINT 8**   Simplify each expression using the quotients-to-powers rule:

a. $\left(\dfrac{x^5}{4}\right)^3$ **b.** $\left(\dfrac{2x^{-3}}{y^2}\right)^4$ **c.** $\left(\dfrac{x^{-3}}{y^4}\right)^{-5}$.

8 Simplify exponential expressions. ▶️

Simplifying Exponential Expressions

Properties of exponents are used to simplify exponential expressions. An exponential expression is **simplified** when

- No parentheses appear.
- No powers are raised to powers.
- Each base occurs only once.
- No negative or zero exponents appear.

Simplifying Exponential Expressions

Example

1. If necessary, remove parentheses by using

$$(ab)^n = a^n b^n \quad \text{or} \quad \left(\frac{a}{b}\right)^n = \frac{a^n}{b^n}.$$

$$(xy)^3 = x^3 y^3$$

2. If necessary, simplify powers to powers by using

$$(b^m)^n = b^{mn}.$$

$$(x^4)^3 = x^{4\cdot3} = x^{12}$$

3. If necessary, be sure that each base appears only once by using

$$b^m \cdot b^n = b^{m+n} \quad \text{or} \quad \frac{b^m}{b^n} = b^{m-n}.$$

$$x^4 \cdot x^3 = x^{4+3} = x^7$$

4. If necessary, rewrite exponential expressions with zero powers as 1 ($b^0 = 1$). Furthermore, write the answer with positive exponents by using

$$b^{-n} = \frac{1}{b^n} \quad \text{or} \quad \frac{1}{b^{-n}} = b^n.$$

$$\frac{x^5}{x^8} = x^{5-8} = x^{-3} = \frac{1}{x^3}$$

The following example shows how to simplify exponential expressions. Throughout the example, assume that no variable in a denominator is equal to zero.

EXAMPLE 9   Simplifying Exponential Expressions

Simplify:

a. $(-2xy^{-14})(-3x^4 y^5)^3$ **b.** $\left(\dfrac{25x^2 y^4}{-5x^6 y^{-8}}\right)^2$ **c.** $\left(\dfrac{x^{-4} y^7}{2}\right)^{-5}.$

Solution

a. $(-2xy^{-14})(-3x^4 y^5)^3$

$$= (-2xy^{-14})(-3)^3(x^4)^3(y^5)^3 \qquad \text{Cube each factor in the second parentheses.}$$

$$= (-2xy^{-14})(-27)x^{12}y^{15} \qquad \text{Multiply the exponents when raising a power to a power:}$$
$$(x^4)^3 = x^{4\cdot3} = x^{12} \text{ and } (y^5)^3 = y^{5\cdot3} = y^{15}.$$

$$= (-2)(-27)x^{1+12}y^{-14+15} \qquad \text{Mentally rearrange factors and multiply like bases by adding the exponents.}$$

$$= 54x^{13}y \qquad \text{Simplify.}$$

b. $\left(\dfrac{25x^2y^4}{-5x^6y^{-8}}\right)^2$

The expression inside parentheses contains more than one occurrence of each base and can be simplified. Begin with this simplification.

$= (-5x^{2-6}y^{4-(-8)})^2$

Simplify inside the parentheses. Subtract the exponents when dividing.

$= (-5x^{-4}y^{12})^2$

Simplify.

$= (-5)^2(x^{-4})^2(y^{12})^2$

Square each factor in parentheses.

$= 25x^{-8}y^{24}$

Multiply the exponents when raising a power to a power: $(x^{-4})^2 = x^{-4(2)} = x^{-8}$ and $(y^{12})^2 = y^{12\cdot2} = y^{24}$.

$= \dfrac{25y^{24}}{x^8}$

Simplify x^{-8} using $b^{-n} = \dfrac{1}{b^n}$.

c. $\left(\dfrac{x^{-4}y^7}{2}\right)^{-5}$

$= \dfrac{(x^{-4}y^7)^{-5}}{2^{-5}}$

Raise the numerator and the denominator to the -5 power.

$= \dfrac{(x^{-4})^{-5}(y^7)^{-5}}{2^{-5}}$

Raise each factor in the numerator to the -5 power.

$= \dfrac{x^{20}y^{-35}}{2^{-5}}$

Multiply the exponents when raising a power to a power: $(x^{-4})^{-5} = x^{-4(-5)} = x^{20}$ and $(y^7)^{-5} = y^{7(-5)} = y^{-35}$.

$= \dfrac{2^5x^{20}}{y^{35}}$

Move each base with a negative exponent to the other side of the fraction bar and make each negative exponent positive.

$= \dfrac{32x^{20}}{y^{35}}$

Simplify: $2^5 = 2\cdot2\cdot2\cdot2\cdot2 = 32$. ■

☑ **CHECK POINT 9** Simplify:

a. $(-3x^{-6}y)(-2x^3y^4)^2$ **b.** $\left(\dfrac{10x^3y^5}{5x^6y^{-2}}\right)^2$ **c.** $\left(\dfrac{x^3y^5}{4}\right)^{-3}$.

Great Question!

Simplifying exponential expressions seems to involve lots of steps. Are there common errors I can avoid along the way?

Yes. Here's a list. The first column has the correct simplification. The second column contains common errors you should try to avoid.

Correct	Incorrect	Description of Error
$b^3 \cdot b^4 = b^7$	$b^3 \cdot b^4 = b^{12}$	The exponents should be added, not multiplied.
$3^2 \cdot 3^4 = 3^6$	$3^2 \cdot 3^4 = 9^6$	The common base should be retained, not multiplied.
$\dfrac{5^{16}}{5^4} = 5^{12}$	$\dfrac{5^{16}}{5^4} = 5^4$	The exponents should be subtracted, not divided.
$(4a)^3 = 64a^3$	$(4a)^3 = 4a^3$	Both factors should be cubed.
$b^{-n} = \dfrac{1}{b^n}$	$b^{-n} = -\dfrac{1}{b^n}$	Only the exponent should change sign.
$(a + b)^{-1} = \dfrac{1}{a + b}$	$(a + b)^{-1} = \dfrac{1}{a} + \dfrac{1}{b}$	The exponent applies to the entire expression $a + b$.

CONCEPT AND VOCABULARY CHECK

Fill in each blank so that the resulting statement is true.

1. The product rule for exponents states that $b^m \cdot b^n =$ _____. When multiplying exponential expressions with the same base, _____ the exponents.

2. The quotient rule for exponents states that $\dfrac{b^m}{b^n} =$ _____, $b \neq 0$. When dividing exponential expressions with the same nonzero base, _____ the exponents.

3. If $b \neq 0$, then $b^0 =$ _____.

4. The negative-exponent rule states that $b^{-n} =$ _____, $b \neq 0$.

5. True or false: $5^{-2} = -5^2$ _____

6. Negative exponents in denominators can be evaluated using $\dfrac{1}{b^{-n}} =$ _____, $b \neq 0$.

7. True or false: $\dfrac{1}{8^{-2}} = 8^2$ _____

1.6 EXERCISE SET ▶ MyMathLab®

Practice Exercises

In Exercises 1–14, multiply using the product rule.

1. $b^4 \cdot b^7$
2. $b^5 \cdot b^9$
3. $x \cdot x^3$
4. $x \cdot x^4$
5. $2^3 \cdot 2^2$
6. $2^4 \cdot 2^2$
7. $3x^4 \cdot 2x^2$
8. $5x^3 \cdot 3x^2$
9. $(-2y^{10})(-10y^2)$
10. $(-4y^8)(-8y^4)$
11. $(5x^3y^4)(20x^7y^8)$
12. $(4x^5y^6)(20x^7y^4)$
13. $(-3x^4y^0z)(-7xyz^3)$
14. $(-9x^3yz^4)(-5xy^0z^2)$

In Exercises 15–24, divide using the quotient rule.

15. $\dfrac{b^{12}}{b^3}$
16. $\dfrac{b^{25}}{b^5}$
17. $\dfrac{15x^9}{3x^4}$
18. $\dfrac{18x^{11}}{3x^4}$
19. $\dfrac{x^9y^7}{x^4y^2}$
20. $\dfrac{x^9y^{12}}{x^2y^6}$
21. $\dfrac{50x^2y^7}{5xy^4}$
22. $\dfrac{36x^{12}y^4}{4xy^2}$
23. $\dfrac{-56a^{12}b^{10}c^8}{7ab^2c^4}$
24. $\dfrac{-66a^9b^7c^6}{6a^3bc^2}$

In Exercises 25–34, use the zero-exponent rule to simplify each expression.

25. 6^0
26. 9^0
27. $(-4)^0$
28. $(-2)^0$
29. -4^0
30. -2^0
31. $13y^0$
32. $17y^0$
33. $(13y)^0$
34. $(17y)^0$

In Exercises 35–52, write each expression with positive exponents only. Then simplify, if possible.

35. 3^{-2}
36. 4^{-2}
37. $(-5)^{-2}$
38. $(-7)^{-2}$
39. -5^{-2}
40. -7^{-2}
41. x^2y^{-3}
42. x^3y^{-4}
43. $8x^{-7}y^3$
44. $9x^{-8}y^4$
45. $\dfrac{1}{5^{-3}}$
46. $\dfrac{1}{2^{-5}}$
47. $\dfrac{1}{(-3)^{-4}}$
48. $\dfrac{1}{(-2)^{-4}}$
49. $\dfrac{x^{-2}}{y^{-5}}$
50. $\dfrac{x^{-3}}{y^{-7}}$
51. $\dfrac{a^{-4}b^7}{c^{-3}}$
52. $\dfrac{a^{-3}b^8}{c^{-2}}$

In Exercises 53–58, simplify each expression using the power rule.

53. $(x^6)^{10}$
54. $(x^3)^2$
55. $(b^4)^{-3}$
56. $(b^8)^{-3}$
57. $(7^{-4})^{-5}$
58. $(9^{-4})^{-5}$

In Exercises 59–72, simplify each expression using the products-to-powers rule.

59. $(4x)^3$
60. $(2x)^5$
61. $(-3x^7)^2$
62. $(-4x^9)^2$
63. $(2xy^2)^3$
64. $(3x^2y)^4$

65. $(-3x^2y^5)^2$

66. $(-3x^4y^6)^2$

67. $(-3x^{-2})^{-3}$

68. $(-2x^{-4})^{-3}$

69. $(5x^3y^{-4})^{-2}$

70. $(7x^2y^{-5})^{-2}$

71. $(-2x^{-5}y^4z^2)^{-4}$

72. $(-2x^{-4}y^5z^3)^{-4}$

In Exercises 73–84, simplify each expression using the quotients-to-powers rule.

73. $\left(\dfrac{2}{x}\right)^4$

74. $\left(\dfrac{y}{2}\right)^5$

75. $\left(\dfrac{x^3}{5}\right)^2$

76. $\left(\dfrac{x^4}{6}\right)^2$

77. $\left(-\dfrac{3x}{y}\right)^4$

78. $\left(-\dfrac{2x}{y}\right)^5$

79. $\left(\dfrac{x^4}{y^2}\right)^6$

80. $\left(\dfrac{x^5}{y^3}\right)^6$

81. $\left(\dfrac{x^3}{y^{-4}}\right)^3$

82. $\left(\dfrac{x^4}{y^{-2}}\right)^3$

83. $\left(\dfrac{a^{-2}}{b^3}\right)^{-4}$

84. $\left(\dfrac{a^{-3}}{b^5}\right)^{-4}$

In Exercises 85–116, simplify each exponential expression.

85. $\dfrac{x^3}{x^9}$

86. $\dfrac{x^6}{x^{10}}$

87. $\dfrac{20x^3}{-5x^4}$

88. $\dfrac{10x^5}{-2x^6}$

89. $\dfrac{16x^3}{8x^{10}}$

90. $\dfrac{15x^2}{3x^{11}}$

91. $\dfrac{20a^3b^8}{2ab^{13}}$

92. $\dfrac{72a^5b^{11}}{9ab^{17}}$

93. $x^3 \cdot x^{-12}$

94. $x^4 \cdot x^{-12}$

95. $(2a^5)(-3a^{-7})$

96. $(4a^2)(-2a^{-5})$

97. $\left(-\dfrac{1}{4}x^{-4}y^5z^{-1}\right)(-12x^{-3}y^{-1}z^4)$

98. $\left(-\dfrac{1}{3}x^{-5}y^4z^6\right)(-18x^{-2}y^{-1}z^{-7})$

99. $\dfrac{6x^2}{2x^{-8}}$

100. $\dfrac{12x^5}{3x^{-10}}$

101. $\dfrac{x^{-7}}{x^3}$

102. $\dfrac{x^{-10}}{x^4}$

103. $\dfrac{30x^2y^5}{-6x^8y^{-3}}$

104. $\dfrac{24x^2y^{13}}{-2x^5y^{-2}}$

105. $\dfrac{-24a^3b^{-5}c^5}{-3a^{-6}b^{-4}c^{-7}}$

106. $\dfrac{-24a^2b^{-2}c^8}{-8a^{-5}b^{-1}c^{-3}}$

107. $\left(\dfrac{x^3}{x^{-5}}\right)^2$

108. $\left(\dfrac{x^4}{x^{-11}}\right)^3$

109. $\left(\dfrac{-15a^4b^2}{5a^{10}b^{-3}}\right)^3$

110. $\left(\dfrac{-30a^{14}b^8}{10a^{17}b^{-2}}\right)^3$

111. $\left(\dfrac{3a^{-5}b^2}{12a^3b^{-4}}\right)^0$

112. $\left(\dfrac{4a^{-5}b^3}{12a^3b^{-5}}\right)^0$

113. $\left(\dfrac{x^{-5}y^8}{3}\right)^{-4}$

114. $\left(\dfrac{x^6y^{-7}}{2}\right)^{-3}$

115. $\left(\dfrac{20a^{-3}b^4c^5}{-2a^{-5}b^{-2}c}\right)^{-2}$

116. $\left(\dfrac{-2a^{-4}b^3c^{-1}}{3a^{-2}b^{-5}c^{-2}}\right)^{-4}$

Practice PLUS

In Exercises 117–124, simplify each exponential expression.

117. $\dfrac{9y^4}{x^{-2}} + \left(\dfrac{x^{-1}}{y^2}\right)^{-2}$

118. $\dfrac{7x^3}{y^{-9}} + \left(\dfrac{x^{-1}}{y^3}\right)^{-3}$

119. $\left(\dfrac{3x^4}{y^{-4}}\right)^{-1}\left(\dfrac{2x}{y^2}\right)^3$

120. $\left(\dfrac{2^{-1}x^{-2}y}{x^4y^{-1}}\right)^{-2}\left(\dfrac{xy^{-3}}{x^{-3}y}\right)^3$

121. $(-4x^3y^{-5})^{-2}(2x^{-8}y^{-5})$

122. $(-4x^{-4}y^5)^{-2}(-2x^5y^{-6})$

123. $\dfrac{(2x^2y^4)^{-1}(4xy^3)^{-3}}{(x^2y)^{-5}(x^3y^2)^4}$

124. $\dfrac{(3x^3y^2)^{-1}(2x^2y)^{-2}}{(xy^2)^{-5}(x^2y^3)^3}$

Application Exercises

The formula

$$A = 1000 \cdot 2^t$$

models the population, A, of aphids in a field of potato plants after t weeks. Use this formula to solve Exercises 125–126.

125. a. What is the present aphid population?

   b. What will the aphid population be in 4 weeks?

   c. What was the aphid population 3 weeks ago?

126. a. What is the present aphid population?

   b. What will the aphid population be in 3 weeks?

   c. What was the aphid population 2 weeks ago?

A rumor about algebra CDs that you can listen to as you sleep, allowing you to awaken refreshed and algebraically empowered, is spreading among the students in your math class. The formula

$$N = \dfrac{25}{1 + 24 \cdot 2^{-t}}$$

models the number of people in the class, N, who have heard the rumor after t minutes. Use this formula to solve Exercises 127–128.

127. a. How many people in the class started the rumor?

   b. How many people in the class have heard the rumor after 4 minutes?

128. a. How many people in the class started the rumor?

   b. How many people in the class, rounded to the nearest whole number, have heard the rumor after 6 minutes?

Use the graph of the rumor model to solve Exercises 129–132.

Modeling a Spreading Rumor

$$N = \frac{25}{1 + 24 \cdot 2^{-t}}$$

(Number of Students Who Have Heard the Rumor — vertical axis; Time (minutes) — horizontal axis)

129. Identify your answers to Exercise 127, parts (a) and (b), as points on the graph.

130. Identify your answers to Exercise 128, parts (a) and (b), as points on the graph.

131. Which one of the following best describes the rate of growth of the rumor as shown by the graph?

  a. The number of people in the class who heard the rumor grew steadily over time.

  b. The number of people in the class who heard the rumor remained constant over time.

  c. The number of people in the class who heard the rumor increased slowly at the beginning, but this rate of increase continued to escalate over time.

  d. The number of people in the class who heard the rumor increased quite rapidly at the beginning, but this rate of increase eventually slowed down, ultimately limited by the number of students in the class.

132. Use the graph to determine how many people in the class eventually heard the rumor.

The astronomical unit (AU) is often used to measure distances within the solar system. One AU is equal to the average distance between Earth and the sun, or 92,955,630 miles. The distance, d, of the nth planet from the sun is modeled by the formula

$$d = \frac{3(2^{n-2}) + 4}{10},$$

where d is measured in astronomical units. Use this formula to solve Exercises 133–136.

Relative Distances of Planets from the Sun

Neptune

Uranus

Saturn

Jupiter

Mars   Earth   Venus   Mercury

Sun

133. Substitute 1 for n and find the distance between Mercury and the sun.

134. Substitute 2 for n and find the distance between Venus and the sun.

135. How much farther from the sun is Jupiter than Earth?

136. How much farther from the sun is Uranus than Earth?

Explaining the Concepts

137. Explain the product rule for exponents. Use $b^2 \cdot b^3$ in your explanation.

138. Explain the quotient rule for exponents. Use $\dfrac{b^8}{b^2}$ in your explanation.

139. Explain how to find any nonzero number to the 0 power.

140. Explain the negative-exponent rule and give an example.

141. Explain the power rule for exponents. Use $(b^2)^3$ in your explanation.

142. Explain how to simplify an expression that involves a product raised to a power. Give an example.

143. Explain how to simplify an expression that involves a quotient raised to a power. Give an example.

144. How do you know if an exponential expression is simplified?

Technology Exercise

145. Enter the rumor formula

$$N = \frac{25}{1 + 24 \cdot 2^{-t}}$$

in your graphing utility as

$y_1 = 25 \boxed{\div} \boxed{(} \boxed{1} \boxed{+} 24 \boxed{\times} 2 \boxed{\wedge} \boxed{(-)} \boxed{x} \boxed{)}$.

Then use a table for y_1 to numerically verify your answers to Exercise 127 or 128.

Critical Thinking Exercises

Make Sense? *In Exercises 146–149, determine whether each statement makes sense or does not make sense, and explain your reasoning.*

146. The properties $(ab)^n = a^n b^n$ and $\left(\dfrac{a}{b}\right)^n = \dfrac{a^n}{b^n}$ are like distributive properties of powers over multiplication and division.

147. If 7^{-2} is raised to the third power, the result is a number between 0 and 1.

148. There are many exponential expressions that are equal to $25x^{12}$, such as $(5x^6)^2$, $(5x^3)(5x^9)$, $25(x^3)^9$, and $5^2(x^2)^6$.

149. The expression $\dfrac{a^n}{b^0}$ is undefined because division by 0 is undefined.

In Exercises 150–157, determine whether each statement is true or false. If the statement is false, make the necessary change(s) to produce a true statement.

150. $2^2 \cdot 2^4 = 2^8$ **151.** $5^6 \cdot 5^2 = 25^8$

152. $2^3 \cdot 3^2 = 6^5$

153. $\dfrac{1}{(-2)^3} = 2^{-3}$

154. $\dfrac{2^8}{2^{-3}} = 2^5$

155. $2^4 + 2^5 = 2^9$

156. $2000.002 = (2 \times 10^3) + (2 \times 10^{-3})$

157. $40{,}000.04 = (4 \times 10^4) + (4 \times 10^{-2})$

In Exercises 158–161, simplify the expression. Assume that all variables used as exponents represent integers and that all other variables represent nonzero real numbers.

158. $x^{n-1} \cdot x^{3n+4}$

159. $(x^{-4n} \cdot x^n)^{-3}$

160. $\left(\dfrac{x^{3-n}}{x^{6-n}} \right)^{-2}$

161. $\left(\dfrac{x^n y^{3n+1}}{y^n} \right)^3$

Review Exercises

162. Graph $y = 2x - 1$ in a rectangular coordinate system. Let $x = -3, -2, -1, 0, 1, 2,$ and 3. (Section 1.3, Example 2)

163. Solve $Ax + By = C$ for y. (Section 1.5, Example 6)

164. The length of a rectangular playing field is 5 meters less than twice its width. If 230 meters of fencing enclose the field, what are its dimensions? (Section 1.5, Example 5)

Preview Exercises

Exercises 165–167 will help you prepare for the material covered in the next section.

165. If 6.2 is multiplied by 10^3, what does this multiplication do to the decimal point in 6.2?

166. If 8.5 is multiplied by 10^{-2}, what does this multiplication do to the decimal point in 8.5?

167. Write each computation as a single power of 10. Then evaluate this exponential expression.

a. $10^9 \times 10^{-4}$

b. $\dfrac{10^4}{10^{-2}}$

SECTION

1.7

Scientific Notation

What am I supposed to learn?

After studying this section, you should be able to:

1 Convert from scientific to decimal notation. ▶

2 Convert from decimal to scientific notation. ▶

3 Perform computations with scientific notation. ▶

4 Use scientific notation to solve problems. ▶

People who complain about paying their income tax can be divided into two types: men and women. Perhaps we can quantify the complaining by examining the data in **Figure 1.26**. The bar graphs show the U.S. population, in millions, and the total amount we paid in federal taxes, in trillions of dollars, from 2004 through 2012.

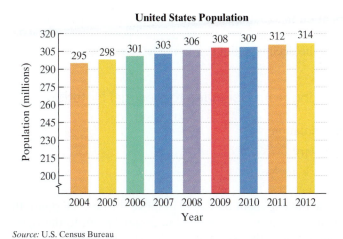

United States Population

Total Tax Collections in the United States

Source: U.S. Census Bureau

Source: Internal Revenue Service

Figure 1.26   Population and total tax collections in the United States

The bar graph on the right shows that in 2012 total tax collections were $2.52 trillion. How can we place this amount in the proper perspective? If the total tax collections were evenly divided among all Americans, how much would each citizen pay in taxes?

In this section, you will learn to use exponents to provide a way of putting large and small numbers in perspective. Using this skill, we will explore the per capita tax for some of the years shown in **Figure 1.26**.

1 Convert from scientific to decimal notation.

Scientific Notation

We have seen that in 2012 total tax collections were $2.52 trillion. Because a trillion is 10^{12} (see **Table 1.7**), this amount can be expressed as

$$2.52 \times 10^{12}.$$

The number 2.52×10^{12} is written in a form called *scientific notation*.

Table 1.7	Names of Large Numbers
10^2	hundred
10^3	thousand
10^6	million
10^9	billion
10^{12}	trillion
10^{15}	quadrillion
10^{18}	quintillion
10^{21}	sextillion
10^{24}	septillion
10^{27}	octillion
10^{30}	nonillion
10^{100}	googol

Scientific Notation

A number is written in **scientific notation** when it is expressed in the form

$$a \times 10^n,$$

where the absolute value of a is greater than or equal to 1 and less than 10 $(1 \le |a| < 10)$, and n is an integer.

It is customary to use the multiplication symbol, $\times$, rather than a dot, when writing a number in scientific notation.

Converting from Scientific to Decimal Notation

Here are two examples of numbers in scientific notation:

$$6.4 \times 10^5 \quad \text{means} \quad 640{,}000.$$
$$2.17 \times 10^{-3} \quad \text{means} \quad 0.00217.$$

Do you see that the number with the positive exponent is relatively large and the number with the negative exponent is relatively small?

We can use n, the exponent on the 10 in $a \times 10^n$, to change a number in scientific notation to decimal notation. If n is **positive**, move the decimal point in a to the **right** n places. If n is **negative**, move the decimal point in a to the **left** $|n|$ places.

EXAMPLE 1   Converting from Scientific to Decimal Notation

Write each number in decimal notation:

a. 6.2×10^7 **b.** -6.2×10^7

c. 2.019×10^{-3} **d.** -2.019×10^{-3}.

Solution   In each case, we use the exponent on the 10 to move the decimal point. In parts (a) and (b), the exponent is positive, so we move the decimal point to the right. In parts (c) and (d), the exponent is negative, so we move the decimal point to the left.

a. $6.2 \times 10^7 = 62{,}000{,}000$

     $n = 7$ Move the decimal point 7 places to the right.

b. $-6.2 \times 10^7 = -62{,}000{,}000$

     $n = 7$ Move the decimal point 7 places to the right.

c. $2.019 \times 10^{-3} = 0.002019$

     $n = -3$ Move the decimal point $|-3|$ places, or 3 places, to the left.

d. $-2.019 \times 10^{-3} = -0.002019$

     $n = -3$ Move the decimal point $|-3|$ places, or 3 places, to the left.

✓ **CHECK POINT 1**   Write each number in decimal notation:

     a. -2.6×10^9 **b.** 3.017×10^{-6}.

---

2 Convert from decimal to scientific notation. ▶

Converting from Decimal to Scientific Notation

To convert from decimal notation to scientific notation, we reverse the procedure of Example 1.

Converting from Decimal to Scientific Notation

Write the number in the form $a \times 10^n$.

- Determine a, the numerical factor. Move the decimal point in the given number to obtain a number whose absolute value is between 1 and 10, including 1.
- Determine n, the exponent on 10^n. The absolute value of n is the number of places the decimal point was moved. The exponent n is positive if the decimal point was moved to the left, negative if the decimal point was moved to the right, and 0 if the decimal point was not moved.

EXAMPLE 2   Converting from Decimal Notation to Scientific Notation

Write each number in scientific notation:

     a. $34{,}970{,}000{,}000{,}000$ **b.** $-34{,}970{,}000{,}000{,}000$

     c. 0.0000000000802 **d.** -0.0000000000802.

Using Technology

You can use your calculator's $\boxed{\text{EE}}$ (enter exponent) or $\boxed{\text{EXP}}$ key to convert from decimal to scientific notation. Here is how it's done for 0.0000000000802.

Many Scientific Calculators

Keystrokes

.0000000000802 $\boxed{\text{EE}}$ $\boxed{=}$

Display

$$8.02 - 11$$

Many Graphing Calculators

Use the mode setting for scientific notation.

Keystrokes

.0000000000802 $\boxed{\text{ENTER}}$

Display

$$8.02\text{E} - 11$$

Solution

a. $34{,}970{,}000{,}000{,}000 = 3.497 \times 10^{13}$

> Move the decimal point to get a number whose absolute value is between 1 and 10.

> The decimal point was moved 13 places to the left, so $n = 13$.

b. $-34{,}970{,}000{,}000{,}000 = -3.497 \times 10^{13}$

c. $0.0000000000802 = 8.02 \times 10^{-11}$

> Move the decimal point to get a number whose absolute value is between 1 and 10.

> The decimal point was moved 11 places to the right, so $n = -11$.

d. $-0.0000000000802 = -8.02 \times 10^{-11}$ ∎

✓ CHECK POINT 2   Write each number in scientific notation:

a. $5{,}210{,}000{,}000$ **b.** -0.00000006893.

Great Question!

In scientific notation, which numbers have positive exponents and which have negative exponents?

If the absolute value of a number is greater than 10, it will have a positive exponent in scientific notation. If the absolute value of a number is less than 1, it will have a negative exponent in scientific notation.

EXAMPLE 3   What a Difference a Day Makes in Scientific Notation

What's happening to the planet's 6.9 billion people today? A very partial answer is provided by the bar graph in **Figure 1.27**.

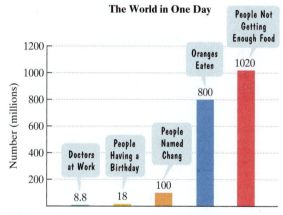

The World in One Day

Figure 1.27

Source: Fullman et al., *Look Now*, DK Publishing, 2010

Express the number of oranges eaten in the world in one day in scientific notation.

Solution   Because a million is 10^6, the number of oranges eaten in a day can be expressed as

$$800 \times 10^6.$$

> This factor is not between 1 and 10, so the number is not in scientific notation.

The voice balloon indicates that we need to convert 800 to scientific notation.

$$800 \times 10^6 = (8 \times 10^2) \times 10^6 = 8 \times (10^2 \times 10^6) = 8 \times 10^{2+6} = 8 \times 10^8$$

There are 8×10^8 oranges eaten in the world in one day. ∎

Figure 1.27 (repeated)

Great Question!

Is there more than one way to describe a large number, such as the number of oranges eaten in a day?

Yes. Many of the large numbers you encounter in newspapers, magazines, and online are expressed in millions (10^6), billions (10^9), and trillions (10^{12}). We can use exponential properties to describe these large numbers, such as the number of oranges eaten in a day.

$$800 \times 10^6 \quad = \quad 8 \times 10^8 \quad = \quad 0.8 \times 10^9$$

There are 800 million oranges eaten.

This expresses the number of oranges in scientific notation.

There are $\frac{4}{5}$ (0.8) of a of a billion oranges eaten.

✓ **CHECK POINT 3** Use **Figure 1.27** to express the number of people in the world who have a birthday today in scientific notation.

---

3 Perform computations with scientific notation.

Computations with Scientific Notation

Properties of exponents are used to perform computations with numbers that are expressed in scientific notation.

Computations with Numbers in Scientific Notation

Multiplication

$$(a \times 10^n)(b \times 10^m) = (a \times b) \times 10^{n+m}$$

Add the exponents on 10 and multiply the other parts of the numbers separately.

Division

$$\frac{a \times 10^n}{b \times 10^m} = \left(\frac{a}{b}\right) \times 10^{n-m}$$

Subtract the exponents on 10 and divide the other parts of the numbers separately.

After the computation is completed, the answer may require an adjustment before it is back in scientific notation.

EXAMPLE 4 Computations with Scientific Notation

Perform the indicated computations, writing the answers in scientific notation:

a. $(6.1 \times 10^5)(4 \times 10^{-9})$

b. $\dfrac{1.8 \times 10^4}{3 \times 10^{-2}}$.

Solution

a. $(6.1 \times 10^5)(4 \times 10^{-9})$

$= (6.1 \times 4) \times (10^5 \times 10^{-9})$ Regroup factors.

$= 24.4 \times 10^{5+(-9)}$ Add the exponents on 10 and multiply the other parts.

$= 24.4 \times 10^{-4}$ Simplify.

$= (2.44 \times 10^1) \times 10^{-4}$ Convert 24.4 to scientific notation: $24.4 = 2.44 \times 10^1$.

$= 2.44 \times 10^{-3}$ $10^1 \times 10^{-4} = 10^{1+(-4)} = 10^{-3}$

b. $\dfrac{1.8 \times 10^4}{3 \times 10^{-2}} = \left(\dfrac{1.8}{3}\right) \times \left(\dfrac{10^4}{10^{-2}}\right)$ Regroup factors.

$= 0.6 \times 10^{4-(-2)}$ Subtract the exponents on 10 and divide the other parts.

$= 0.6 \times 10^6$ Simplify: $4 - (-2) = 4 + 2 = 6$.

$= (6 \times 10^{-1}) \times 10^6$ Convert 0.6 to scientific notation: $0.6 = 6 \times 10^{-1}$.

$= 6 \times 10^5$ $0.6 = 6 \times 10^{-1}$. ■

Using Technology

$(6.1 \times 10^5)(4 \times 10^{-9})$
on a Calculator:

Many Scientific Calculators

6.1 | EE | 5 | × | 4 | EE | 9 | +/− | = |

Display

2.44 − 03

Many Graphing Calculators

6.1 | EE | 5 | × | 4 | EE | (−) | 9 | ENTER |

Display (in scientific notation mode)

2.44ᴇ − 3

☑ **CHECK POINT 4**   Perform the indicated computations, writing the answers in scientific notation:

a. $(7.1 \times 10^5)(5 \times 10^{-7})$

b. $\dfrac{1.2 \times 10^6}{3 \times 10^{-3}}$.

4 Use scientific notation to solve problems. ⊚

Applications: Putting Numbers in Perspective

In the section opener, we saw that in 2012 the U.S. government collected $2.52 trillion in taxes. Example 5 shows how we can use scientific notation to comprehend the meaning of a number such as 2.52 trillion.

EXAMPLE 5   Tax per Capita

In 2012, the U.S. government collected 2.52×10^{12} dollars in taxes. At that time, the U.S. population was approximately 314 million, or 3.14×10^8. If the total tax collections were evenly divided among all Americans, how much would each citizen pay? Express the answer in decimal notation, rounded to the nearest dollar.

Solution The amount that we would each pay, or the tax per capita, is the total amount collected, 2.52×10^{12}, divided by the number of Americans, 3.14×10^8.

$$\frac{2.52 \times 10^{12}}{3.14 \times 10^8} = \left(\frac{2.52}{3.14}\right) \times \left(\frac{10^{12}}{10^8}\right) \approx 0.8025 \times 10^{12-8} = 0.8025 \times 10^4 = 8025$$

To obtain an answer in decimal notation, it is not necessary to express this number in scientific notation.

Move the decimal point 4 places to the right.

If total tax collections were evenly divided, we would each pay approximately $8025 in taxes. ■

☑ **CHECK POINT 5** In 2011, the U.S. government collected 2.42×10^{12} dollars in taxes. At that time, the U.S. population was approximately 312 million or 3.12×10^8. Find the per capita tax, rounded to the nearest dollar, in 2011.

Many problems in algebra involve motion. Suppose that you ride your bike at an average speed of 12 miles per hour. What distance do you cover in 2 hours? Your distance is the product of your speed and the time that you travel:

$$\frac{12 \text{ miles}}{\text{hour}} \times 2 \text{ hours} = 24 \text{ miles}.$$

Your distance is 24 miles. Notice how the hour units cancel. The distance is expressed in miles.

In general, the distance covered by any moving body is the product of its average speed, or rate, and its time in motion.

A Formula for Motion

$$d = rt$$

Distance equals rate times time.

EXAMPLE 6 Using the Motion Formula

Light travels at a rate of approximately 1.86×10^5 miles per second. It takes light 5×10^2 seconds to travel from the sun to Earth. What is the distance between Earth and the sun?

Solution

$d = rt$	Use the motion formula.
$d = (1.86 \times 10^5) \times (5 \times 10^2)$	Substitute the given values.
$d = (1.86 \times 5) \times (10^5 \times 10^2)$	Rearrange factors.
$d = 9.3 \times 10^7$	Add the exponents on 10 and multiply the other parts.

The distance between Earth and the sun is approximately 9.3×10^7 miles, or 93 million miles. ■

☑ **CHECK POINT 6** A futuristic spacecraft traveling at 1.55×10^3 miles per hour takes 20,000 hours (about 833 days) to travel from Venus to Mercury. What is the distance from Venus to Mercury?

Achieving Success

Do not wait until the last minute to study for an exam. Cramming is a high-stress activity that forces your brain to make a lot of weak connections. No wonder crammers tend to forget everything they learned minutes after taking a test.

Preparing for Tests Using the Book

- Study the appropriate sections from the review chart in the Chapter Summary. The chart contains definitions, concepts, procedures, and examples. Review this chart and you'll know the most important material in each section!

- Work the assigned exercises from the Review Exercises. The Review Exercises contain the most significant problems for each of the chapter's sections.

- Find a quiet place to take the Chapter Test. Do not use notes, index cards, or any other resources. Check your answers and ask your professor to review any exercises you missed.

CONCEPT AND VOCABULARY CHECK

Fill in each blank so that the resulting statement is true.

1. A positive number is written in scientific notation when it is expressed in the form $a \times 10^n$, where a is _____ and n is a/an _____.

2. True or false: 7×10^4 is written in scientific notation. _____.

3. True or false: 70×10^3 is written in scientific notation. _____

1.7 EXERCISE SET ▶ MyMathLab®

Practice Exercises

In Exercises 1–14, write each number in decimal notation without the use of exponents.

1. 3.8×10^2
2. 9.2×10^2
3. 6×10^{-4}
4. 7×10^{-5}
5. -7.16×10^6
6. -8.17×10^6
7. 1.4×10^0
8. 2.4×10^0
9. 7.9×10^{-1}
10. 6.8×10^{-1}
11. -4.15×10^{-3}
12. -3.14×10^{-3}
13. -6.00001×10^{10}
14. -7.00001×10^{10}

In Exercises 15–30, write each number in scientific notation.

15. 32,000
16. 64,000
17. 638,000,000,000,000,000
18. 579,000,000,000,000,000
19. -317
20. -326
21. -5716
22. -3829
23. 0.0027
24. 0.0083
25. -0.00000000504
26. -0.00000000405
27. 0.007
28. 0.005
29. 3.14159
30. 2.71828

In Exercises 31–50, perform the indicated computations. Write the answers in scientific notation. If necessary, round the decimal factor in your scientific notation answer to two decimal places.

31. $(3 \times 10^4)(2.1 \times 10^3)$

32. $(2 \times 10^4)(4.1 \times 10^3)$
33. $(1.6 \times 10^{15})(4 \times 10^{-11})$
34. $(1.4 \times 10^{15})(3 \times 10^{-11})$
35. $(6.1 \times 10^{-8})(2 \times 10^{-4})$
36. $(5.1 \times 10^{-8})(3 \times 10^{-4})$
37. $(4.3 \times 10^8)(6.2 \times 10^4)$
38. $(8.2 \times 10^8)(4.6 \times 10^4)$
39. $\dfrac{8.4 \times 10^8}{4 \times 10^5}$
40. $\dfrac{6.9 \times 10^8}{3 \times 10^5}$
41. $\dfrac{3.6 \times 10^4}{9 \times 10^{-2}}$
42. $\dfrac{1.2 \times 10^4}{2 \times 10^{-2}}$
43. $\dfrac{4.8 \times 10^{-2}}{2.4 \times 10^6}$
44. $\dfrac{7.5 \times 10^{-2}}{2.5 \times 10^6}$
45. $\dfrac{2.4 \times 10^{-2}}{4.8 \times 10^{-6}}$
46. $\dfrac{1.5 \times 10^{-2}}{3 \times 10^{-6}}$
47. $\dfrac{480,000,000,000}{0.00012}$
48. $\dfrac{282,000,000,000}{0.00141}$
49. $\dfrac{0.00072 \times 0.003}{0.00024}$
50. $\dfrac{66,000 \times 0.001}{0.003 \times 0.002}$

Practice PLUS

In Exercises 51–58, solve each equation. Express the solution in scientific notation.

51. $(2 \times 10^{-5})x = 1.2 \times 10^9$
52. $(3 \times 10^{-2})x = 1.2 \times 10^4$
53. $\dfrac{x}{2 \times 10^8} = -3.1 \times 10^{-5}$

54. $\dfrac{x}{5 \times 10^{11}} = -2.9 \times 10^{-3}$

55. $x - (7.2 \times 10^{18}) = 9.1 \times 10^{18}$

56. $x - (5.3 \times 10^{-16}) = 8.4 \times 10^{-16}$

57. $(-1.2 \times 10^{-3})x = (1.8 \times 10^{-4})(2.4 \times 10^{6})$

58. $(-7.8 \times 10^{-4})x = (3.9 \times 10^{-7})(6.8 \times 10^{5})$

Application Exercises

The graph shows the net worth, in billions of dollars, of the five richest Americans in 2014. Each bar also indicates the source of each person's wealth. Use 10^9 for one billion and the figures shown to solve Exercises 59–62. Express all answers in scientific notation.

Net Worth of the Five Richest Americans

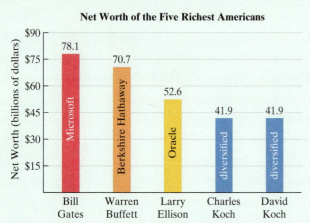

Source: Forbes

59. How much is Bill Gates worth?

60. How much is Warren Buffet worth?

61. By how much does Larry Ellison's worth exceed that of Charles Koch?

62. What is the combined worth of the Koch brothers (Charles and David)?

Our ancient ancestors hunted for their meat and expended a great deal of energy chasing it down. Today, our animal protein is raised in cages and on feedlots, delivered in great abundance nearly to our door. Use the numbers shown below to solve Exercises 63–66. Use 10^6 for one million and 10^9 for one billion.

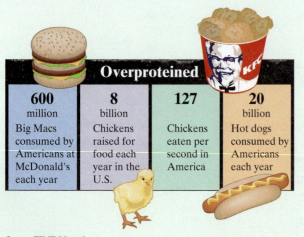

Overproteined

| **600** million Big Macs consumed by Americans at McDonald's each year | **8** billion Chickens raised for food each year in the U.S. | **127** Chickens eaten per second in America | **20** billion Hot dogs consumed by Americans each year |

Source: TIME Magazine

In Exercises 63–64, use 300 million, or 3×10^8, for an approximation of the U.S. population. Express answers in decimal notation, rounded, if necessary, to the nearest whole number.

63. Find the number of hot dogs consumed by each American in a year.

64. If the consumption of Big Macs was divided evenly among all Americans, how many Big Macs would we each consume in a year?

In Exercises 65–66, use the fact that there are approximately 3.2×10^7 seconds in a year.

65. How many chickens are raised for food each second in the United States? Express the answer in scientific and decimal notations.

66. How many chickens are eaten per year in the United States? Express the answer in scientific notation.

The graph shows the cost, in billions of dollars, and the enrollment, in millions of people, for various federal social programs for a recent year. Use the numbers shown to solve Exercises 67–69.

Cost and Enrollment for Federal Social Programs

- Federal Cost
- Enrollment

Program	Federal Cost (billions of dollars)	Enrollment (millions of people)
Medicaid	198	53.4
Social Security	519	48
Medicare	294	42.3
Child Nutrition	12	32.3
Food Stamps	33	25.7

Source: Office of Management and Budget

67. a. What was the average per person benefit for Social Security? Express the answer in scientific notation and in decimal notation, rounded to the nearest dollar.

   b. What was the average monthly per person benefit, rounded to the nearest dollar, for Social Security?

68. a. What was the average per person benefit for the food stamps program? Express the answer in scientific notation and in decimal notation, rounded to the nearest dollar.

   b. What was the average monthly per person benefit, rounded to the nearest dollar, for the food stamps program?

69. Medicaid provides health insurance for the poor. Medicare provides health insurance for people 65 and older, as well as younger people who are disabled. Which program provides the greater per person benefit? By how much, rounded to the nearest dollar?

70. The area of Alaska is approximately 3.66×10^8 acres. The state was purchased in 1867 from Russia for $7.2 million. What price per acre, to the nearest cent, did the United States pay Russia?

71. The mass of one oxygen molecule is 5.3×10^{-23} gram. Find the mass of 20,000 molecules of oxygen. Express the answer in scientific notation.

72. The mass of one hydrogen atom is 1.67×10^{-24} gram. Find the mass of 80,000 hydrogen atoms. Express the answer in scientific notation.

73. In Exercises 65–66, we used 3.2×10^7 as an approximation for the number of seconds in a year. Convert 365 days (one year) to hours, to minutes, and, finally, to seconds, to determine precisely how many seconds there are in a year. Express the answer in scientific notation.

Explaining the Concepts

74. How do you know if a number is written in scientific notation?

75. Explain how to convert from scientific to decimal notation and give an example.

76. Explain how to convert from decimal to scientific notation and give an example.

77. Describe one advantage of expressing a number in scientific notation over decimal notation.

Technology Exercises

78. Use a calculator to check any three of your answers in Exercises 1–14.

79. Use a calculator to check any three of your answers in Exercises 15–30.

80. Use a calculator with an $\boxed{\text{EE}}$ or $\boxed{\text{EXP}}$ key to check any four of your computations in Exercises 31–50. Display the result of the computation in scientific notation.

Critical Thinking Exercises

Make Sense? In Exercises 81–84, determine whether each statement makes sense or does not make sense, and explain your reasoning.

81. For a recent year, total tax collections in the United States were $\$2.02 \times 10^7$.

82. I just finished reading a book that contained approximately 1.04×10^5 words.

83. If numbers in the form $a \times 10^n$ are listed from least to greatest, values of a need not appear from least to greatest.

84. When expressed in scientific notation, 58 million and 58 millionths have exponents on 10 with the same absolute value.

In Exercises 85–89, determine whether each statement is true or false. If the statement is false, make the necessary change(s) to produce a true statement.

85. $534.7 = 5.347 \times 10^3$

86. $\dfrac{8 \times 10^{30}}{4 \times 10^{-5}} = 2 \times 10^{25}$

87. $(7 \times 10^5) + (2 \times 10^{-3}) = 9 \times 10^2$

88. $(4 \times 10^3) + (3 \times 10^2) = 43 \times 10^2$

89. The numbers $8.7 \times 10^{25}, 1.0 \times 10^{26}, 5.7 \times 10^{26}$, and 3.7×10^{27} are listed from least to greatest.

In Exercises 90–91, perform the indicated additions. Write the answers in scientific notation.

90. $5.6 \times 10^{13} + 3.1 \times 10^{13}$

91. $8.2 \times 10^{-16} + 4.3 \times 10^{-16}$

92. Our hearts beat approximately 70 times per minute. Express in scientific notation how many times the heart beats over a lifetime of 80 years. Round the decimal factor in your scientific notation answer to two decimal places.

93. Give an example of a number where there is no advantage in using scientific notation over decimal notation.

Review Exercises

94. Simplify: $9(10x - 4) - (5x - 10)$. (Section 1.2, Example 14)

95. Solve: $\dfrac{4x - 1}{10} = \dfrac{5x + 2}{4} - 4$. (Section 1.4, Example 4)

96. Simplify: $(8x^4 y^{-3})^{-2}$. (Section 1.6, Example 7)

Preview Exercises

Exercises 97–99 will help you prepare for the material covered in the first section of the next chapter.

97. Here are two sets of ordered pairs:

$$\text{set 1: } \{(1, 5), (2, 5)\}$$
$$\text{set 2: } \{(5, 1), (5, 2)\}$$

In which set is each x-coordinate paired with only one y-coordinate?

98. Evaluate $r^3 - 2r^2 + 5$ for $r = -5$.

99. Evaluate $5x + 7$ for $x = a + h$.

Chapter 1 Summary

Definitions and Concepts	Examples

Section 1.1 Algebraic Expressions, Real Numbers, and Interval Notation

Letters that represent numbers are called variables.

An algebraic expression is a combination of variables, numbers, and operation symbols. English phrases can be translated into algebraic expressions:

- Addition: sum, plus, increased by, more than
- Subtraction: difference, minus, decreased by, less than
- Multiplication: product, times, of, twice
- Division: quotient, divide, per, ratio

Translate: Six less than the product of a number and five.

$$5x - 6$$

Many algebraic expressions contain exponents. If b is a natural number, b^n, the nth power of b, is the product of n factors of b.

Furthermore, $b^1 = b$.

Evaluating an algebraic expression means to find the value of the expression for a given value of the variable.

Evaluate $6 + 5(x - 10)^3$ for $x = 12$.

$$6 + 5(12 - 10)^3$$
$$= 6 + 5 \cdot 2^3$$
$$= 6 + 5 \cdot 8$$
$$= 6 + 40 = 46$$

An equation is a statement that two expressions are equal. Formulas are equations that express relationships among two or more variables. Mathematical modeling is the process of finding formulas to describe real-world phenomena. Such formulas, together with the meaning assigned to the variables, are called mathematical models. The formulas are said to model, or describe, the relationships among the variables.

The formula

$$h = -16t^2 + 200t + 4$$

models the height, h, in feet, of fireworks t seconds after launch. What is the height after 2 seconds?

$$h = -16(2)^2 + 200(2) + 4$$
$$= -16(4) + 200(2) + 4$$
$$= -64 + 400 + 4 = 340$$

The height is 340 feet.

A set is a collection of objects, called elements, enclosed in braces. The roster method uses commas to separate the elements of the set. Set-builder notation describes the elements of a set, but does not list them. The symbol $\in$ means that a number or object is in a set; $\notin$ means that a number or object is not in a set. The set of real numbers is the set of all numbers that can be represented by points on the number line. Sets that make up the real numbers include

Natural numbers: $\{1, 2, 3, 4, \dots\}$

Whole numbers: $\{0, 1, 2, 3, 4, \dots\}$

Integers: $\{\dots, -4, -3, -2, -1, 0, 1, 2, 3, 4, \dots\}$

Rational numbers: $\left\{\frac{a}{b} \mid a \text{ and } b \text{ are integers and } b \neq 0\right\}$

Irrational numbers:
$\{x \mid x \text{ is a real number and } x \text{ is not a rational number}\}$.

In decimal form, rational numbers terminate or repeat.

In decimal form, irrational numbers do neither.

- Use the roster method to list the elements of
$$\{x \mid x \text{ is a natural number less than } 6\}.$$

Solution

$$\{1, 2, 3, 4, 5\}$$

- True or false:
$$\sqrt{2} \notin \{x \mid x \text{ is a rational number}\}.$$

Solution

The statement is true:

$$\sqrt{2} \text{ is not a rational number.}$$

The decimal form of $\sqrt{2}$ neither terminates nor repeats. Thus, $\sqrt{2}$ is an irrational number.

Definitions and Concepts	Examples

Section 1.1 Algebraic Expressions, Real Numbers, and Interval Notation (continued)

For any two real numbers, a and b, a is less than b if a is to the left of b on the number line.

Inequality Symbols

$<$: is less than
$>$: is greater than
$\leq$: is less than or equal to
$\geq$: is greater than or equal to

- $-1 < 5$, or -1 is less than 5, is true because -1 is to the left of 5 on a number line.
- $-3 \geq 7$, -3 is greater than or equal to 7, is false. Neither $-3 > 7$ nor $-3 = 7$ is true.

In interval notation, parentheses indicate endpoints that are not included in an interval. Square brackets indicate endpoints that are included in an interval. Parentheses are always used with ∞ or $-\infty$.

- $(-2, 1] = \{x \mid -2 < x \leq 1\}$

  $-4 \ -3 \ -2 \ -1 \ 0 \ 1 \ 2 \ 3 \ 4$

- $[-2, \infty) = \{x \mid x \geq -2\}$

  $-4 \ -3 \ -2 \ -1 \ 0 \ 1 \ 2 \ 3 \ 4$

Section 1.2 Operations with Real Numbers and Simplifying Algebraic Expressions

Absolute Value

$$|a| = \begin{cases} a & \text{if } a \geq 0 \\ -a & \text{if } a < 0 \end{cases}$$

The opposite, or additive inverse, of a is $-a$. When a is a negative number, $-a$ is positive.

- $|6.03| = 6.03$
- $|0| = 0$
- $|-4.9| = -(-4.9) = 4.9$

Adding Real Numbers

To add two numbers with the same sign, add their absolute values and use their common sign. To add two numbers with different signs, subtract the smaller absolute value from the greater absolute value and use the sign of the number with the greater absolute value.

- $-4.1 + (-6.2) = -10.3$
- $-30 + 25 = -5$
- $12 + (-8) = 4$

Subtracting Real Numbers

$$a - b = a + (-b)$$

$$-\frac{3}{4} - \left(-\frac{1}{2}\right) = -\frac{3}{4} + \frac{1}{2} = -\frac{3}{4} + \frac{2}{4} = -\frac{1}{4}$$

Multiplying and Dividing Real Numbers

The product or quotient of two numbers with the same sign is positive and with different signs is negative. If no number is 0, a product with an even number of negative factors is positive and a product with an odd number of negative factors is negative. Division by 0 is undefined.

- $2(-6)(-1)(-5) = -60$

  Three (odd) negative factors give a negative product.

- $(-2)^3 = (-2)(-2)(-2) = -8$

- $-\frac{1}{3}\left(-\frac{2}{5}\right) = \frac{2}{15}$

- $\dfrac{-14}{2} = -7$

Definitions and Concepts	**Examples**

Section 1.2 Operations with Real Numbers and Simplifying Algebraic Expressions (continued)

Order of Operations

1. Perform operations within grouping symbols, starting with the innermost grouping symbols. Grouping symbols include parentheses, brackets, fraction bars, absolute value symbols, and square root signs.
2. Evaluate exponential expressions.
3. Multiply and divide from left to right.
4. Add and subtract from left to right.

Simplify: $\dfrac{6(8-10)^3 + (-2)}{(-5)^2(-2)}$.

$$= \dfrac{6(-2)^3 + (-2)}{(-5)^2(-2)} = \dfrac{6(-8) + (-2)}{25(-2)}$$

$$= \dfrac{-48 + (-2)}{-50} = \dfrac{-50}{-50} = 1$$

Basic Algebraic Properties

Commutative: $a + b = b + a$
$$ab = ba$$

Associative: $(a + b) + c = a + (b + c)$
$$(ab)c = a(bc)$$

Distributive: $a(b + c) = ab + ac$
$$a(b - c) = ab - ac$$
$$(b + c)a = ba + ca$$

- Commutative
$$3x + 5 = 5 + 3x = 5 + x \cdot 3$$
- Associative
$$-4(6x) = (-4 \cdot 6)x = -24x$$
- Distributive

$$-4(9x + 3) = -4(9x) + (-4) \cdot 3$$
$$= -36x + (-12)$$
$$= -36x - 12$$

Simplifying Algebraic Expressions

Terms are separated by addition. Like terms have the same variable factors and are combined using the distributive property. An algebraic expression is simplified when grouping symbols have been removed and like terms have been combined.

Simplify: $7(3x - 4) - (10x - 5)$.
$$= 21x - 28 - 10x + 5$$
$$= 21x - 10x - 28 + 5$$
$$= 11x - 23$$

Section 1.3 Graphing Equations

The rectangular coordinate system consists of a horizontal number line, the x-axis, and a vertical number line, the y-axis, intersecting at their zero points, the origin. Each point in the system corresponds to an ordered pair of real numbers (x, y). The first number in the pair is the x-coordinate; the second number is the y-coordinate.

Plot: $(4, 2)$, $(-3, 4)$, $(-5, -4)$, and $(4, -3)$.

An ordered pair is a solution of an equation in two variables if replacing the variables by the corresponding coordinates results in a true statement. The ordered pair is said to satisfy the equation. The graph of the equation is the set of all points whose coordinates satisfy the equation. One method for graphing an equation is to plot ordered-pair solutions and connect them with a smooth curve or line.

Graph: $y = x^2 - 1$.

x	$y = x^2 - 1$
-2	$(-2)^2 - 1 = 3$
-1	$(-1)^2 - 1 = 0$
0	$0^2 - 1 = -1$
1	$1^2 - 1 = 0$
2	$2^2 - 1 = 3$

Definitions and Concepts	Examples

Section 1.4 Solving Linear Equations

A linear equation in one variable can be written in the form $ax + b = 0, a \neq 0$. A solution is a value of the variable that makes the equation a true statement. The set of all such solutions is the equation's solution set. Equivalent equations have the same solution set. To solve a linear equation,

1. Simplify each side.
2. Collect variable terms on one side and constant terms on the other side.
3. Isolate the variable and solve.
4. Check the proposed solution in the original equation.

Solve: $4(x - 5) = 2x - 14$.

$$4x - 20 = 2x - 14$$
$$4x - 2x - 20 = 2x - 2x - 14$$
$$2x - 20 = -14$$
$$2x - 20 + 20 = -14 + 20$$
$$2x = 6$$
$$\frac{2x}{2} = \frac{6}{2}$$
$$x = 3$$

Checking gives $-8 = -8$, so 3 is the solution, or $\{3\}$ is the solution set.

Equations Containing Fractions

Multiply both sides (all terms) by the least common denominator. This clears the equation of fractions.

Solve: $\dfrac{x - 2}{5} + \dfrac{x + 2}{2} = \dfrac{x + 4}{3}$.

$$30\left(\frac{x - 2}{5} + \frac{x + 2}{2}\right) = 30\left(\frac{x + 4}{3}\right)$$
$$6(x - 2) + 15(x + 2) = 10(x + 4)$$
$$6x - 12 + 15x + 30 = 10x + 40$$
$$21x + 18 = 10x + 40$$
$$11x = 22$$
$$x = 2$$

Checking gives $2 = 2$, so 2 is the solution, or $\{2\}$ is the solution set.

Types of Equations

An equation that is true for all real numbers, $(-\infty, \infty)$, is called an identity. When solving an identity, the variable is eliminated and a true statement, such as $3 = 3$, results. An equation that is not true for even one real number is called an inconsistent equation. A false statement, such as $3 = 7$, results when solving such an equation, whose solution set is $\varnothing$, the empty set. A conditional equation is not an identity, but is true for at least one real number.

Solve: $4x + 5 = 4(x + 2)$.

$$4x + 5 = 4x + 8$$
$$5 = 8, \quad \text{false}$$

The inconsistent equation has no solution: $\varnothing$.

Solve: $5x - 4 = 5(x + 1) - 9$.

$$5x - 4 = 5x + 5 - 9$$
$$5x - 4 = 5x - 4$$
$$-4 = -4, \quad \text{true}$$

All real numbers satisfy the identity: $(-\infty, \infty)$.

Definitions and Concepts	**Examples**

Section 1.5 Problem Solving and Using Formulas

Strategy for Solving Algebraic Word Problems

1. Let x represent one of the unknown quantities.
2. Represent other unknown quantities in terms of x.
3. Write an equation that models the conditions.
4. Solve the equation and answer the question.
5. Check the proposed solution in the original wording of the problem.

After a 60% reduction, a suit sold for $32. What was the original price?

Let x = the original price.

Original price	minus	60% reduction		reduced price
x	$-$	$0.6x$	$=$	32

$$0.4x = 32$$
$$\frac{0.4x}{0.4} = \frac{32}{0.4}$$
$$x = 80$$

The original price was $80. Check this amount using the first sentence in the problem's conditions.

To solve a formula for a variable, use the steps for solving a linear equation and isolate that variable on one side of the equation.

Solve for r: $\qquad E = I(R + r)$.

$$E = IR + Ir \quad \boxed{\text{We need to isolate } r.}$$
$$E - IR = Ir$$
$$\frac{E - IR}{I} = r$$

Section 1.6 Properties of Integral Exponents

The Product Rule

$$b^m \cdot b^n = b^{m+n}$$

$$(-3x^{10})(5x^{20}) = -3 \cdot 5x^{10+20}$$
$$= -15x^{30}$$

The Quotient Rule

$$\frac{b^m}{b^n} = b^{m-n}, b \neq 0$$

$$\frac{5x^{20}}{10x^{10}} = \frac{5}{10} \cdot x^{20-10} = \frac{x^{10}}{2}$$

Zero and Negative Exponents

$$b^0 = 1, b \neq 0$$

$$b^{-n} = \frac{1}{b^n} \quad \text{and} \quad \frac{1}{b^{-n}} = b^n, b \neq 0$$

- $(3x)^0 = 1$
- $3x^0 = 3 \cdot 1 = 3$
- $\dfrac{2^{-3}}{4^{-2}} = \dfrac{4^2}{2^3} = \dfrac{16}{8} = 2$

Power Rule

$$(b^m)^n = b^{mn}$$

$$(x^5)^{-4} = x^{5(-4)} = x^{-20} = \frac{1}{x^{20}}$$

Products to Powers

$$(ab)^n = a^n b^n$$

$$(5x^3 y^{-4})^{-2} = 5^{-2} \cdot (x^3)^{-2} \cdot (y^{-4})^{-2}$$
$$= 5^{-2} x^{-6} y^8$$
$$= \frac{y^8}{5^2 x^6} = \frac{y^8}{25x^6}$$

Definitions and Concepts	**Examples**

Section 1.6 Properties of Integral Exponents (continued)

Quotients to Powers

$$\left(\frac{a}{b}\right)^n = \frac{a^n}{b^n}$$

$$\left(\frac{2}{x^3}\right)^{-4} = \frac{2^{-4}}{(x^3)^{-4}} = \frac{2^{-4}}{x^{-12}}$$

$$= \frac{x^{12}}{2^4} = \frac{x^{12}}{16}$$

An exponential expression is simplified when

- No parentheses appear.
- No powers are raised to powers.
- Each base occurs only once.
- No negative or zero exponents appear.

Simplify: $\dfrac{-5x^{-3}y^2}{-20x^2y^{-6}}.$

$$= \frac{-5}{-20} \cdot x^{-3-2} \cdot y^{2-(-6)}$$

$$= \frac{1}{4}x^{-5}y^8 = \frac{y^8}{4x^5}$$

Section 1.7 Scientific Notation

A number in scientific notation is expressed in the form

$$a \times 10^n,$$

where $|a|$ is greater than or equal to 1 and less than 10, and n is an integer.

Write in decimal notation: 3.8×10^{-3}.

$$3.8 \times 10^{-3} = .0038 = 0.0038$$

Write in scientific notation: 26,000.

$$26{,}000 = 2.6 \times 10^4$$

Computations with Numbers in Scientific Notation

$$(a \times 10^n)(b \times 10^m) = (a \times b) \times 10^{n+m}$$

$$\frac{a \times 10^n}{b \times 10^m} = \left(\frac{a}{b}\right) \times 10^{n-m}$$

$$(8 \times 10^3)(5 \times 10^{-8})$$

$$= 8 \cdot 5 \times 10^{3+(-8)}$$

$$= 40 \times 10^{-5}$$

$$= (4 \times 10^1) \times 10^{-5} = 4 \times 10^{-4}$$

CHAPTER 1 REVIEW EXERCISES

1.1   *In Exercises 1–3, write each English phrase as an algebraic expression. Let x represent the number.*

1. Ten less than twice a number

2. Four more than the product of six and a number

3. The quotient of nine and a number, increased by half of the number

In Exercises 4–6, evaluate each algebraic expression for the given value or values of the variable.

4. $x^2 - 7x + 4$, for $x = 10$

5. $6 + 2(x - 8)^3$, for $x = 11$

6. $x^4 - (x - y)$, for $x = 2$ and $y = 1$

In Exercises 7–8, use the roster method to list the elements in each set.

7. $\{x \mid x \text{ is a natural number less than } 3\}$

8. $\{x \mid x \text{ is an integer greater than } -4 \text{ and less than } 2\}$

In Exercises 9–11, determine whether each statement is true or false.

9. $0 \in \{x \mid x \text{ is a natural number}\}$

10. $-2 \in \{x \mid x \text{ is a rational number}\}$

11. $\frac{1}{3} \notin \{x \mid x \text{ is an irrational number}\}$

In Exercises 12–14, write out the meaning of each inequality. Then determine whether the inequality is true or false.

12. $-5 < 2$

13. $-7 \geq -3$

14. $-7 \leq -7$

15. On average, smartphone users in the United States look at their phones more than 150 times per day, spending approximately 2.5 hours on their phones. The bar graph shows the rise of iPhone sales, in millions, from 2007 through 2013.

iPhone Sales over Time

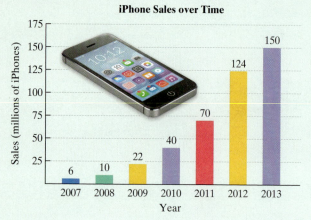

Source: Statista

The data can be modeled by the formula

$$S = 4x^2 + 0.7x + 5,$$

where S represents iPhone sales, in millions of units, x years after 2007. Does the model underestimate or overestimate sales shown for 2013? By how much?

In Exercises 16–18, express each interval in set-builder notation and graph the interval on a number line.

16. $(-2, 3]$

17. $[-1.5, 2]$

18. $(-1, \infty)$

1.2 *In Exercises 19–21, find each absolute value.*

19. $|-9.7|$ **20.** $|5.003|$ **21.** $|0|$

In Exercises 22–33, perform the indicated operation.

22. $-2.4 + (-5.2)$ **23.** $-6.8 + 2.4$

24. $-7 - (-20)$ **25.** $(-3)(-20)$

26. $-\dfrac{3}{5} - \left(-\dfrac{1}{2}\right)$ **27.** $\left(\dfrac{2}{7}\right)\left(-\dfrac{3}{10}\right)$

28. $4(-3)(-2)(-10)$ **29.** $(-2)^4$

30. -2^5 **31.** $-\dfrac{2}{3} \div \dfrac{8}{5}$

32. $\dfrac{-35}{-5}$ **33.** $\dfrac{54.6}{-6}$

34. Find $-x$ if $x = -7$.

In Exercises 35–41, simplify each expression.

35. $-11 - [-17 + (-3)]$ **36.** $\left(-\dfrac{1}{2}\right)^3 \cdot 2^4$

37. $-3[4 - (6 - 8)]$ **38.** $8^2 - 36 \div 3^2 \cdot 4 - (-7)$

39. $\dfrac{(-2)^4 + (-3)^2}{2^2 - (-21)}$ **40.** $\dfrac{(7 - 9)^3 - (-4)^2}{2 + 2(8) \div 4}$

41. $4 - (3 - 8)^2 + 3 \div 6 \cdot 4^2$

In Exercises 42–46, simplify each algebraic expression.

42. $5(2x - 3) + 7x$

43. $5x + 7x^2 - 4x + 2x^2$

44. $3(4y - 5) - (7y + 2)$

45. $8 - 2[3 - (5x - 1)]$

46. $6(2x - 3) - 5(3x - 2)$

1.3 *In Exercises 47–49, plot the given point in a rectangular coordinate system.*

47. $(-1, 3)$ **48.** $(2, -5)$ **49.** $(0, -6)$

In Exercises 50–53, graph each equation. Let $x = -3, -2, -1, 0, 1, 2,$ and 3.

50. $y = 2x - 2$ **51.** $y = x^2 - 3$

52. $y = x$ **53.** $y = |x| - 2$

54. What does a $[-20, 40, 10]$ by $[-5, 5, 1]$ viewing rectangle mean? Draw axes with tick marks and label the tick marks to illustrate this viewing rectangle.

The caseload of Alzheimer's disease in the United States is expected to explode as baby boomers head into their later years. The graph shows the percentage of Americans with the disease, by age. Use the graph to solve Exercises 55–57.

Alzheimer's Prevalence in the United States, by Age

Source: U.S. Centers for Disease Control

55. What percentage of Americans who are 75 have Alzheimer's disease?

56. What age represents 50% prevalence of Alzheimer's disease?

57. Describe the trend shown by the graph.

58. Select the graph that best illustrates the following description: A train pulls into a station and lets off its passengers.

a.   b.

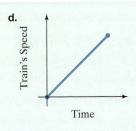

c. Train's Speed vs. Time

d. Train's Speed vs. Time

1.4 *In Exercises 59–64, solve and check each linear equation.*

59. $2x - 5 = 7$

60. $5x + 20 = 3x$

61. $7(x - 4) = x + 2$

62. $1 - 2(6 - x) = 3x + 2$

63. $2(x - 4) + 3(x + 5) = 2x - 2$

64. $2x - 4(5x + 1) = 3x + 17$

In Exercises 65–69, solve each equation.

65. $\dfrac{2x}{3} = \dfrac{x}{6} + 1$

66. $\dfrac{x}{2} - \dfrac{1}{10} = \dfrac{x}{5} + \dfrac{1}{2}$

67. $\dfrac{2x}{3} = 6 - \dfrac{x}{4}$

68. $\dfrac{x}{4} = 2 + \dfrac{x - 3}{3}$

69. $\dfrac{3x + 1}{3} - \dfrac{13}{2} = \dfrac{1 - x}{4}$

In Exercises 70–74, solve each equation. Then state whether the equation is an identity, a conditional equation, or an inconsistent equation.

70. $7x + 5 = 5(x + 3) + 2x$

71. $7x + 13 = 4x - 10 + 3x + 23$

72. $7x + 13 = 3x - 10 + 2x + 23$

73. $4(x - 3) + 5 = x + 5(x - 2)$

74. $(2x - 3)2 - 3(x + 1) = (x - 2)4 - 3(x + 5)$

75. The bar graph shows the changing face of America's federal tax returns.

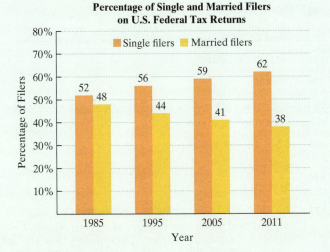

Percentage of Single and Married Filers on U.S. Federal Tax Returns

Source: IRS

The data can be modeled by the formula

$$M = -0.4x + 48,$$

where M represents the percentage of married filers x years after 1985.

a. Does the model underestimate or overestimate the percentage of married filers in 2005? By how much?

b. According to the model, when will 34% of federal tax returns be submitted by married filers?

1.5 *In Exercises 76–82, use the five-step strategy for solving word problems.*

76. Although you want to choose a career that fits your interests and abilities, it is good to have an idea of what jobs pay when looking at career options. The bar graph shows the average yearly earnings of full-time employed college graduates with only a bachelor's degree based on their college major.

Average Earnings, by College Major

Source: Arthur J. Keown, *Personal Finance*, Pearson.

The average yearly earnings of engineering majors exceeds the earnings of marketing majors by $19 thousand. The average yearly earnings of accounting majors exceeds the earnings of marketing majors by $6 thousand. Combined, the average yearly earnings for these three college majors is $196 thousand. Determine the average yearly earnings, in thousands of dollars, for each of these three college majors.

77. One angle of a triangle measures 10° more than the second angle. The measure of the third angle is twice the sum of the measures of the first two angles. Determine the measure of each angle.

78. Without changes, the graphs show projections for the amount being paid in Social Security benefits and the amount going into the system. All data are expressed in billions of dollars.

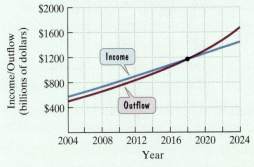

Social Insecurity: Projected Income and Outflow of the Social Security System

Source: 2004 Social Security Trustees Report

(Be sure to refer to the information at the bottom of the previous page.)

a. In 2004, the Social Security system's income was $575 billion, projected to increase at an average rate of $43 billion per year. In which year will the system's income be $1177 billion?

b. The data for the Social Security system's outflow can be modeled by the formula

$$B = 0.07x^2 + 47.4x + 500,$$

where B represents the amount paid in benefits, in billions of dollars, x years after 2004. According to this model, what will be the amount paid in benefits for the year you determined in part (a)? Round to the nearest billion dollars.

c. How are your answers to parts (a) and (b) shown by the graphs at the bottom of the previous page?

79. You are choosing between two cellphone plans. Data Plan A has a monthly fee of $52 with a charge of $18 per gigabyte (GB). Data Plan B has a monthly fee of $32 with a charge of $22 per GB. For how many GB of data will the costs for the two data plans be the same?

80. After a 20% price reduction, a cordless phone sold for $48. What was the phone's price before the reduction?

81. A salesperson earns $300 per week plus 5% commission of sales. How much must be sold to earn $800 in a week?

82. The length of a rectangular field is 6 yards less than triple the width. If the perimeter of the field is 340 yards, what are its dimensions?

83. In 2015, there were 14,100 students at college A, with a projected enrollment increase of 1500 students per year. In the same year, there were 41,700 students at college B, with a projected enrollment decline of 800 students per year.

a. Let x represent the number of years after 2015. Write, but do not solve, an equation that can be used to find how many years after 2015 the colleges will have the same enrollment.

b. The following table is based on your equation in part (a). Y_1 represents one side of the equation and Y_2 represents the other side of the equation. Use the table to answer these questions: In which year will the colleges have the same enrollment? What will be the enrollment in each college at that time?

X	Y₁	Y₂			
7	24600	36100			
8	26100	35300			
9	27600	34500			
10	29100	33700			
11	30600	32900			
12	32100	32100			
13	33600	31300			
14	35100	30500			
15	36600	29700			
16	38100	28900			
17	39600	28100			

X=7

In Exercises 84–89, solve each formula for the specified variable.

84. $V = \frac{1}{3}Bh$ for h

85. $y - y_1 = m(x - x_1)$ for x

86. $E = I(R + r)$ for R

87. $C = \dfrac{5F - 160}{9}$ for F

88. $s = vt + gt^2$ for g

89. $T = gr + gvt$ for g

1.6 *In Exercises 90–104, simplify each exponential expression. Assume that no denominators are 0.*

90. $(-3x^7)(-5x^6)$

91. x^2y^{-5}

92. $\dfrac{3^{-2}x^4}{y^{-7}}$

93. $(x^3)^{-6}$

94. $(7x^3y)^2$

95. $\dfrac{16y^3}{-2y^{10}}$

96. $(-3x^4)(4x^{-11})$

97. $\dfrac{12x^7}{4x^{-3}}$

98. $\dfrac{-10a^5b^6}{20a^{-3}b^{11}}$

99. $(-3xy^4)(2x^2)^3$

100. $2^{-2} + \dfrac{1}{2}x^0$

101. $(5x^2y^{-4})^{-3}$

102. $(3x^4y^{-2})(-2x^5y^{-3})$

103. $\left(\dfrac{3xy^3}{5x^{-3}y^{-4}}\right)^2$

104. $\left(\dfrac{-20x^{-2}y^3}{10x^5y^{-6}}\right)^{-3}$

1.7 *In Exercises 105–106, write each number in decimal notation.*

105. 7.16×10^6

106. 1.07×10^{-4}

In Exercises 107–108, write each number in scientific notation.

107. $-41,000,000,000,000$

108. 0.00809

In Exercises 109–110, perform the indicated computations. Write the answers in scientific notation.

109. $(4.2 \times 10^{13})(3 \times 10^{-6})$

110. $\dfrac{5 \times 10^{-6}}{20 \times 10^{-8}}$

111. The human body contains approximately 3.2×10^4 microliters of blood for every pound of body weight. Each microliter of blood contains approximately 5×10^6 red blood cells. Express in scientific notation the approximate number of red blood cells in the body of a 180-pound person.

CHAPTER 1 TEST

1. Write the following English phrase as an algebraic expression:

   Five less than the product of a number and four.

   Let x represent the number.

2. Evaluate $8 + 2(x - 7)^4$ for $x = 10$.

3. Use the roster method to list the elements in the set:

   $\{x \mid x$ is a negative integer greater than $-5\}$.

4. Determine whether the following statement is true or false:

   $\dfrac{1}{4} \notin \{x \mid x$ is a natural number$\}$.

5. Write out the meaning of the inequality $-3 > -1$. Then determine whether the inequality is true or false.

In Exercises 6–7, express each interval in set-builder notation and graph the interval on a number line.

6. $[-3, 2)$
7. $(-\infty, -1]$

8. One possible reason for the explosion of college tuition involves the decrease in government aid per student. In 2001, higher-education revenues per student averaged $8500. The bar graph shows government aid per U.S. college student from 2005 through 2012. (All figures are adjusted for inflation and expressed in 2012 dollars.)

Higher-Education Government Aid per U.S. College Student

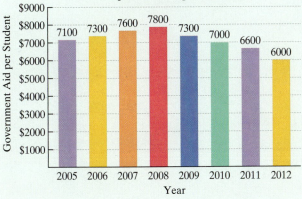

Source: State Higher Education Executive Officers Association

   The mathematical model

   $$G = -82x^2 + 410x + 7079$$

   describes government aid per college student, G, x years after 2005. Does the model underestimate or overestimate aid per student in 2011? By how much?

9. Find the absolute value: $|-17.9|$.

In Exercises 10–14, perform the indicated operation.

10. $-10.8 + 3.2$

11. $-\dfrac{1}{4} - \left(-\dfrac{1}{2}\right)$

12. $2(-3)(-1)(-10)$

13. $-\dfrac{1}{4}\left(-\dfrac{1}{2}\right)$

14. $\dfrac{-27.9}{-9}$

In Exercises 15–20, simplify each expression.

15. $24 - 36 \div 4 \cdot 3$

16. $(5^2 - 2^4) + [9 \div (-3)]$

17. $\dfrac{(8 - 10)^3 - (-4)^2}{2 + 8(2) \div 4}$

18. $7x - 4(3x + 2) - 10$

19. $5(2y - 6) - (4y - 3)$

20. $9x - [10 - 4(2x - 3)]$

21. Plot $(-2, -4)$ in a rectangular coordinate system.

22. Graph $y = x^2 - 4$ in a rectangular coordinate system.

In Exercises 23–25, solve each equation. If the solution set is $\varnothing$ or $(-\infty, \infty)$, classify the equation as an inconsistent equation or an identity.

23. $3(2x - 4) = 9 - 3(x + 1)$

24. $\dfrac{2x - 3}{4} = \dfrac{x - 4}{2} - \dfrac{x + 1}{4}$

25. $3(x - 4) + x = 2(6 + 2x)$

In Exercises 26–30, use the five-step strategy for solving word problems.

26. Find two numbers such that the second number is 3 more than twice the first number and the sum of the two numbers is 72.

27. You bought a new car for $13,805. Its value is decreasing by $1820 per year. After how many years will its value be $4705?

28. The toll to a bridge costs $8. Commuters who use the bridge frequently have the option of purchasing a monthly discount pass for $45. With the discount pass, the toll is reduced to $5. For how many crossings per month will the monthly cost without the discount pass be the same as the monthly cost with the discount pass?

29. After a 60% reduction, a jacket sold for $20. What was the jacket's price before the reduction?

30. The length of a rectangular field exceeds the width by 260 yards. If the perimeter of the field is 1000 yards, what are its dimensions?

In Exercises 31–32, solve each formula for the specified variable.

31. $V = \dfrac{1}{3} lwh$ for h

32. $Ax + By = C$ for y

In Exercises 33–37, simplify each exponential expression.

33. $(-2x^5)(7x^{-10})$

34. $(-8x^{-5}y^{-3})(-5x^2y^{-5})$

35. $\dfrac{-10x^4y^3}{-40x^{-2}y^6}$

36. $(4x^{-5}y^2)^{-3}$

37. $\left(\dfrac{-6x^{-5}y}{2x^3y^{-4}}\right)^{-2}$

38. Write in decimal notation: 3.8×10^{-6}.

39. Write in scientific notation: 407,000,000,000.

40. Divide and write the answer in scientific notation:

$$\frac{4 \times 10^{-3}}{8 \times 10^{-7}}.$$

41. In 2010, world population was approximately 6.9×10^9. By some projections, world population will double by 2080. Express the population at that time in scientific notation.

Functions and Linear Functions

A vast expanse of open water at the top of our world was once covered with ice. The melting of the Arctic ice caps has forced polar bears to swim as far as 40 miles, causing them to drown in significant numbers. Such deaths were rare in the past.

There is strong scientific consensus that human activities are changing the Earth's climate. Scientists now believe that there is a striking correlation between atmospheric carbon dioxide concentration and global temperature. As both of these variables increase at significant rates, there are warnings of a planetary emergency that threatens to condemn coming generations to a catastrophically diminished future.*

In this chapter, you'll learn to approach our climate crisis mathematically by creating formulas, called functions, that model data for average global temperature and carbon dioxide concentration over time. Understanding the concept of a function will give you a new perspective on many situations, ranging from climate change to using mathematics in a way that is similar to making a movie.

*Sources: Al Gore, An Inconvenient Truth, Rodale, 2006; Time, April 3, 2006; Rolling Stone, September 26, 2013

Here's where you'll find these applications:

- Mathematical models involving climate change are developed in Example 10 and Check Point 10 in Section 2.4.
- Using mathematics in a way that is similar to making a movie is discussed in the Blitzer Bonus on page 148.

SECTION

2.1 Introduction to Functions

Relations

What am I supposed to learn?

After studying this section, you should be able to:

1 Find the domain and range of a relation. ▶

2 Determine whether a relation is a function. ▶

3 Evaluate a function. ▶

Forbes magazine published a list of the highest-paid TV celebrities between June 2013 and June 2014. The results are shown in **Figure 2.1**.

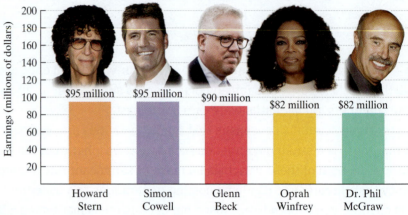

Figure 2.1

Source: Forbes

The graph indicates a correspondence between a TV celebrity and that person's earnings, in millions of dollars. We can write this correspondence using a set of ordered pairs:

{(Stern, 95), (Cowell, 95), (Beck, 90), (Winfrey, 82), (McGraw, 82)}.

> These braces indicate we are representing a set.

The mathematical term for a set of ordered pairs is a *relation*.

> ### Definition of a Relation
>
> A **relation** is any set of ordered pairs. The set of all first components of the ordered pairs is called the **domain** of the relation and the set of all second components is called the **range** of the relation.

1 Find the domain and range of a relation. ▶

EXAMPLE 1 Finding the Domain and Range of a Relation

Find the domain and range of the relation:

{(Stern, 95), (Cowell, 95), (Beck, 90), (Winfrey, 82), (McGraw, 82)}.

Solution The domain is the set of all first components. Thus, the domain is

{Stern, Cowell, Beck, Winfrey, McGraw}.

The range is the set of all second components. Thus, the range is

{95, 90, 82}.

> Although Stern and Cowell both earned $95 million, it is not necessary to list 95 twice.

> Although Winfrey and McGraw both earned $82 million, it is not necessary to list 82 twice.

☑ **CHECK POINT 1** Find the domain and range of the relation:

$$\{(0, 9.1), (10, 6.7), (20, 10.7), (30, 13.2), (40, 21.2)\}.$$

As you worked Check Point 1, did you wonder if there was a rule that assigned the "inputs" in the domain to the "outputs" in the range? For example, for the ordered pair (30, 13.2), how does the output 13.2 depend on the input 30? The ordered pair is based on the data in **Figure 2.2(a)**, which shows the percentage of first-year U.S. college students claiming no religious affiliation.

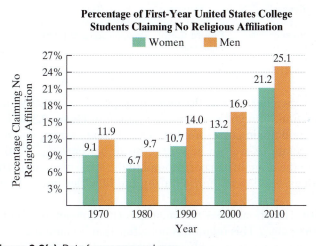

Figure 2.2(a) Data for women and men

Source: John Macionis, *Sociology*, Fourteenth Edition, Pearson, 2012.

Percentage of First-Year United States College Women Claiming No Religious Affiliation

Figure 2.2(b) Visually representing the relation for the women's data

In **Figure 2.2(b)**, we used the data for college women to create the following ordered pairs:

$$\left(\text{years after 1970,} \quad \begin{array}{l} \text{percentage of first-year college} \\ \text{women claiming no religious} \\ \text{affiliation} \end{array} \right).$$

Consider, for example, the ordered pair (30, 13.2).

(30, 13.2)

30 years after 1970, or in 2000,   13.2% of first-year college women claimed no religious affiliation.

The five points in **Figure 2.2(b)** visually represent the relation formed from the women's data. Another way to visually represent the relation is as follows:

2 Determine whether a relation is a function.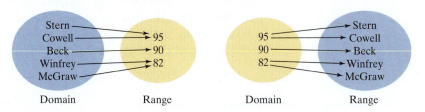

Table 2.1	Highest-Paid TV Celebrities
Celebrity	**Earnings (millions of dollars)**
Stern	95
Cowell	95
Beck	90
Winfrey	82
McGraw	82

Functions

Table 2.1, based on our earlier discussion, shows the highest-paid TV celebrities and their earnings between June 2013 and June 2014, in millions of dollars. We've used this information to define two relations.

Figure 2.3(a) shows a correspondence between celebrities and their earnings. **Figure 2.3(b)** shows a correspondence between earnings and celebrities.

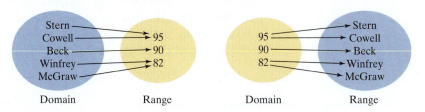

Figure 2.3(a) Celebrities correspond to earnings.

Figure 2.3(b) Earnings correspond to celebrities.

A relation in which each member of the domain corresponds to exactly one member of the range is a **function**. Can you see that the relation in **Figure 2.3(a)** is a function? Each celebrity in the domain corresponds to exactly one earning amount in the range. If we know the celebrity, we can be sure of his or her earnings. Notice that more than one element in the domain can correspond to the same element in the range: Stern and Cowell both earned $95 million. Winfrey and McGraw both earned $82 million.

Is the relation in **Figure 2.3(b)** a function? Does each member of the domain correspond to precisely one member of the range? This relation is not a function because there are members of the domain that correspond to two different members of the range:

(95, Stern) (95, Cowell)   (82, Winfrey) (82, McGraw).

The member of the domain 95 corresponds to both Stern and Cowell. The member of the domain 82 corresponds to both Winfrey and McGraw. If we know that the earnings are $95 million or $82 million, we cannot be sure of the celebrity. Because **a function is a relation in which no two ordered pairs have the same first component and different second components**, the ordered pairs (95, Stern) and (95, Cowell) are not ordered pairs of a function. Similarly, (82, Winfrey) and (82, McGraw) are not ordered pairs of a function.

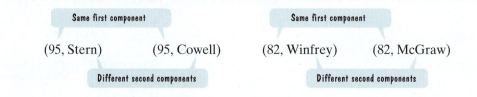

Definition of a Function

A **function** is a correspondence from a first set, called the **domain**, to a second set, called the **range**, such that each element in the domain corresponds to *exactly one* element in the range.

Example 2 illustrates that not every correspondence between sets is a function.

EXAMPLE 2 Determining Whether a Relation Is a Function

Determine whether each relation is a function:

a. {(1, 5), (2, 5), (3, 7), (4, 8)}   **b.** {(5, 1), (5, 2), (7, 3), (8, 4)}.

Solution We begin by making a figure for each relation that shows the domain and the range (**Figure 2.4**).

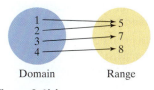

Domain   Range

Figure 2.4(a)

a. Figure 2.4(a) shows that every element in the domain corresponds to exactly one element in the range. The element 1 in the domain corresponds to the element 5 in the range. Furthermore, 2 corresponds to 5, 3 corresponds to 7, and 4 corresponds to 8. No two ordered pairs in the given relation have the same first component and different second components. Thus, the relation is a function.

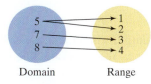

Domain   Range

Figure 2.4(b)

b. Figure 2.4(b) shows that 5 corresponds to both 1 and 2. If any element in the domain corresponds to more than one element in the range, the relation is not a function. This relation is not a function because two ordered pairs have the same first component and different second components.

Look at **Figure 2.4(a)** again. The fact that 1 and 2 in the domain correspond to the same number, 5, in the range does not violate the definition of a function. **A function can have two different first components with the same second component.** By contrast, a relation is not a function when two different ordered pairs have the same first component and different second components. Thus, the relation in Example 2(b) is not a function.

✓ **CHECK POINT 2** Determine whether each relation is a function:

a. {(1, 2), (3, 4), (5, 6), (5, 7)}

b. {(1, 2), (3, 4), (6, 5), (7, 5)}.

Great Question!

If I reverse a function's components, will this new relation be a function?

If a relation is a function, reversing the components in each of its ordered pairs may result in a relation that is not a function.

3 Evaluate a function. ▶

Functions as Equations and Function Notation

Functions are usually given in terms of equations rather than as sets of ordered pairs. For example, here is an equation that models the percentage of first-year college women claiming no religious affiliation as a function of time:

$$y = 0.014x^2 - 0.24x + 8.8.$$

The variable x represents the number of years after 1970. The variable y represents the percentage of first-year college women claiming no religious affiliation. The variable y is a function of the variable x. For each value of x, there is one and only one value of y. The variable x is called the **independent variable** because it can be assigned any value from the domain. Thus, x can be assigned any nonnegative integer representing the number of years after 1970. The variable y is called the **dependent variable** because its value depends on x. The percentage claiming no religious affiliation depends on the number of years after 1970. The value of the dependent variable, y, is calculated after selecting a value for the independent variable, x.

If an equation in x and y gives one and only one value of y for each value of x, then the variable y is a function of the variable x. When an equation represents a function, the function is often named by a letter such as f, g, h, F, G, or H. Any letter can be used to name a function. Suppose that f names a function. Think of the domain as the set of the function's inputs and the range as the set of the function's outputs. As shown in **Figure 2.5**, the input is represented by x and the output by $f(x)$. The special notation $f(x)$, read "f of x" or "f at x," represents the **value of the function at the number x.**

Input x

Output
$f(x)$

Figure 2.5 A "function machine" with inputs and outputs

Let's make this clearer by considering a specific example. We know that the equation

$$y = 0.014x^2 - 0.24x + 8.8$$

defines y as a function of x. We'll name the function f. Now, we can apply our new function notation.

> We read this equation as "f of x equals $0.014x^2 - 0.24x + 8.8$."

Input	Output	Equation
x	$f(x)$	$f(x) = 0.014x^2 - 0.24x + 8.8$

Great Question!

Doesn't $f(x)$ indicate that I need to multiply f and x?

The notation $f(x)$ does *not* mean "f times x." The notation describes the value of the function at x.

Suppose we are interested in finding $f(30)$, the function's output when the input is 30. To find the value of the function at 30, we substitute 30 for x. We are **evaluating the function** at 30.

$f(x) = 0.014x^2 - 0.24x + 8.8$
Input
$x = 30$

$0.014(30)^2 - 0.24(30) + 8.8$

Output
$f(30) = 14.2$

Figure 2.6 A function machine at work

$$f(x) = 0.014x^2 - 0.24x + 8.8 \qquad \text{This is the given function.}$$
$$f(30) = 0.014(30)^2 - 0.24(30) + 8.8 \qquad \text{Replace each occurrence of } x \text{ with 30.}$$
$$= 0.014(900) - 0.24(30) + 8.8 \qquad \text{Evaluate the exponential expression: } 30^2 = 30 \cdot 30 = 900.$$
$$= 12.6 - 7.2 + 8.8 \qquad \text{Perform the multiplications.}$$
$$f(30) = 14.2 \qquad \text{Subtract and add from left to right.}$$

The statement $f(30) = 14.2$, read "f of 30 equals 14.2," tells us that the value of the function at 30 is 14.2. When the function's input is 30, its output is 14.2. **Figure 2.6** illustrates the input and output in terms of a function machine.

$$f(30) = 14.2$$

> 30 years after 1970, or in 2000, 14.2% of first-year college women claimed no religious affiliation.

We have seen that in 2000, 13.2% actually claimed nonaffiliation, so our function that models the data slightly overestimates the percent for 2000.

Using Technology

Graphing utilities can be used to evaluate functions. The screens on the right show the evaluation of

$$f(x) = 0.014x^2 - 0.24x + 8.8$$

at 30 on a TI-84 Plus C graphing calculator. The function f is named Y_1.

We used $f(x) = 0.014x^2 - 0.24x + 8.8$ to find $f(30)$. To find other function values, such as $f(40)$ or $f(55)$, substitute the specified input value, 40 or 55, for x in the function's equation.

If a function is named f and x represents the independent variable, the notation $f(x)$ corresponds to the y-value for a given x. Thus,

$$f(x) = 0.014x^2 - 0.24x + 8.8 \quad \text{and} \quad y = 0.014x^2 - 0.24x + 8.8$$

define the same function. This function may be written as

$$y = f(x) = 0.014x^2 - 0.24x + 8.8.$$

EXAMPLE 3   Using Function Notation

Find the indicated function value:

a. $f(4)$ for $f(x) = 2x + 3$ **b.** $g(-2)$ for $g(x) = 2x^2 - 1$
c. $h(-5)$ for $h(r) = r^3 - 2r^2 + 5$ **d.** $F(a + h)$ for $F(x) = 5x + 7$.

Solution

a. $f(x) = 2x + 3$ This is the given function.
    $f(4) = 2 \cdot 4 + 3$ To find f of 4, replace x with 4.
        $= 8 + 3$ Multiply: $2 \cdot 4 = 8$.
    $f(4) = 11$ *f of 4 is 11.*   Add.

b. $g(x) = 2x^2 - 1$ This is the given function.
    $g(-2) = 2(-2)^2 - 1$ To find g of -2, replace x with -2.
        $= 2(4) - 1$ Evaluate the exponential expression: $(-2)^2 = 4$.
        $= 8 - 1$ Multiply: $2(4) = 8$.
    $g(-2) = 7$ *g of -2 is 7.*   Subtract.

c. $h(r) = r^3 - 2r^2 + 5$ The function's name is h and r represents the independent variable.

    $h(-5) = (-5)^3 - 2(-5)^2 + 5$ To find h of -5, replace each occurrence of r with -5.
        $= -125 - 2(25) + 5$ Evaluate exponential expressions.
        $= -125 - 50 + 5$ Multiply.
    $h(-5) = -170$ *h of -5 is -170.* $-125 - 50 = -175$ and $-175 + 5 = -170$.

d. $F(x) = 5x + 7$ This is the given function.
    $F(a + h) = 5(a + h) + 7$ Replace x with $a + h$.
    $F(a + h) = 5a + 5h + 7$ Apply the distributive property. ∎

F of $a + h$ is $5a + 5h + 7$.

✓ CHECK POINT 3   Find the indicated function value:

a. $f(6)$ for $f(x) = 4x + 5$
b. $g(-5)$ for $g(x) = 3x^2 - 10$
c. $h(-4)$ for $h(r) = r^2 - 7r + 2$
d. $F(a + h)$ for $F(x) = 6x + 9$.

Great Question!

In Example 3 and Check Point 3, finding some of the function values involved evaluating exponential expressions. Can't this be a bit tricky when such functions are evaluated at negative numbers?

Yes. Be particularly careful if there is a term with a coefficient of −1. Notice the following differences:

$$f(x) = -x^2 \qquad\qquad g(x) = (-x)^2$$

Replace x with −4. Replace x with −4.

$$f(-4) = -(-4)^2 \qquad\qquad g(-4) = (-(-4))^2$$
$$= -16 \qquad\qquad\qquad = 4^2 = 16$$

Functions Represented by Tables and Function Notation

Function notation can be applied to functions that are represented by tables.

EXAMPLE 4 Using Function Notation

Function f is defined by the following table:

x	f(x)
−2	5
−1	0
0	3
1	1
2	4

a. Explain why the table defines a function.

b. Find the domain and the range of the function.

Find the indicated function value:

c. $f(-1)$

d. $f(0)$

e. Find x such that $f(x) = 4$.

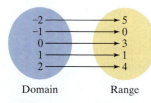

Domain   Range

Figure 2.7

Solution

a. Values in the first column of the table make up the domain, or input values. Values in the second column of the table make up the range, or output values. We see that every element in the domain corresponds to exactly one element in the range, shown in **Figure 2.7**. Therefore, the relation given by the table is a function.

The voice balloons pointing to appropriate parts of the table illustrate the solution to parts (b)–(e).

x	f(x)
−2	5
−1	0
0	3
1	1
2	4

c. $f(-1) = 0$: When the input is −1, the output is 0.

d. $f(0) = 3$: When the input is 0, the output is 3.

e. $f(x) = 4$ when $x = 2$: The output, $f(x)$, is 4 when the input, x, is 2.

b. The domain is the set of inputs: {−2, −1, 0, 1, 2}.

b. The range is the set of outputs: {5, 0, 3, 1, 4}.

| ✓ CHECK POINT 4 | Function g is defined by the following table: |

x	$g(x)$
0	3
1	0
2	1
3	2
4	3

a. Explain why the table defines a function.

b. Find the domain and the range of the function.

Find the indicated function value:

c. $g(1)$

d. $g(3)$

e. Find x such that $g(x) = 3$.

Achieving Success

Check out the *Learning Guide* that accompanies this textbook.

Benefits of using the *Learning Guide* include:

- It will help you become better organized. This includes organizing your class notes, assigned homework, quizzes, and tests.
- It will enable you to use your textbook more efficiently.
- It will bring together the learning tools for this course, including the textbook, the Video Lecture Series, and the PowerPoint Presentation.
- It will help increase your study skills.
- It will help you prepare for the chapter tests.

Ask your professor about the availability of this textbook supplement.

CONCEPT AND VOCABULARY CHECK

Fill in each blank so that the resulting statement is true.

1. Any set of ordered pairs is called a/an _____. The set of all first components of the ordered pairs is called the _____. The set of all second components of the ordered pairs is called the _____.

2. A set of ordered pairs in which each member of the set of first components corresponds to exactly one member of the set of second components is called a/an _____.

3. The notation $f(x)$ describes the value of _____ at _____.

4. If $h(r) = -r^2 + 4r - 7$, we can find $h(-2)$ by replacing each occurrence of _____ by _____.

2.1 EXERCISE SET ▶ MyMathLab®

Practice Exercises

In Exercises 1–8, determine whether each relation is a function. Give the domain and range for each relation.

1. $\{(1, 2), (3, 4), (5, 5)\}$

2. $\{(4, 5), (6, 7), (8, 8)\}$

3. $\{(3, 4), (3, 5), (4, 4), (4, 5)\}$

4. $\{(5, 6), (5, 7), (6, 6), (6, 7)\}$

5. $\{(-3, -3), (-2, -2), (-1, -1), (0, 0)\}$

6. $\{(-7, -7), (-5, -5), (-3, -3), (0, 0)\}$

7. $\{(1, 4), (1, 5), (1, 6)\}$

8. $\{(4, 1), (5, 1), (6, 1)\}$

In Exercises 9–24, find the indicated function values.

9. $f(x) = x + 1$

   a. $f(0)$ **b.** $f(5)$ **c.** $f(-8)$

   d. $f(2a)$ **e.** $f(a + 2)$

10. $f(x) = x + 3$

   a. $f(0)$ **b.** $f(5)$ **c.** $f(-8)$

   d. $f(2a)$ **e.** $f(a + 2)$

11. $g(x) = 3x - 2$

   a. $g(0)$ **b.** $g(-5)$ **c.** $g\left(\dfrac{2}{3}\right)$

   d. $g(4b)$ **e.** $g(b + 4)$

12. $g(x) = 4x - 3$

   a. $g(0)$ **b.** $g(-5)$ **c.** $g\left(\dfrac{3}{4}\right)$

   d. $g(5b)$ **e.** $g(b + 5)$

13. $h(x) = 3x^2 + 5$

   a. $h(0)$ **b.** $h(-1)$ **c.** $h(4)$

   d. $h(-3)$ **e.** $h(4b)$

14. $h(x) = 2x^2 - 4$

   a. $h(0)$ **b.** $h(-1)$ **c.** $h(5)$

   d. $h(-3)$ **e.** $h(5b)$

15. $f(x) = 2x^2 + 3x - 1$

   a. $f(0)$ **b.** $f(3)$ **c.** $f(-4)$

   d. $f(b)$ **e.** $f(5a)$

16. $f(x) = 3x^2 + 4x - 2$

   a. $f(0)$ **b.** $f(3)$ **c.** $f(-5)$

   d. $f(b)$ **e.** $f(5a)$

17. $f(x) = (-x)^3 - x^2 - x + 7$

   a. $f(0)$ **b.** $f(2)$

   c. $f(-2)$ **d.** $f(1) + f(-1)$

18. $f(x) = (-x)^3 - x^2 - x + 10$

   a. $f(0)$ **b.** $f(2)$

   c. $f(-2)$ **d.** $f(1) + f(-1)$

19. $f(x) = \dfrac{2x - 3}{x - 4}$

   a. $f(0)$ **b.** $f(3)$ **c.** $f(-4)$

   d. $f(-5)$ **e.** $f(a + h)$

   f. Why must 4 be excluded from the domain of f?

20. $f(x) = \dfrac{3x - 1}{x - 5}$

   a. $f(0)$ **b.** $f(3)$ **c.** $f(-3)$

   d. $f(10)$ **e.** $f(a + h)$

   f. Why must 5 be excluded from the domain of f?

21.

x	$f(x)$
-4	3
-2	6
0	9
2	12
4	15

   a. $f(-2)$

   b. $f(2)$

   c. For what value of x is $f(x) = 9$?

22.

x	$f(x)$
-5	4
-3	8
0	12
3	16
5	20

   a. $f(-3)$

   b. $f(3)$

   c. For what value of x is $f(x) = 12$?

23.

x	$h(x)$
-2	2
-1	1
0	0
1	1
2	2

   a. $h(-2)$

   b. $h(1)$

   c. For what values of x is $h(x) = 1$?

24.

x	$h(x)$
-2	-2
-1	-1
0	0
1	-1
2	-2

   a. $h(-2)$

   b. $h(1)$

   c. For what values of x is $h(x) = -1$?

Practice PLUS

In Exercises 25–26, let $f(x) = x^2 - x + 4$ and $g(x) = 3x - 5$.

25. Find $g(1)$ and $f(g(1))$.

26. Find $g(-1)$ and $f(g(-1))$.

In Exercises 27–28, let f and g be defined by the following table:

x	$f(x)$	$g(x)$
-2	6	0
-1	3	4
0	-1	1
1	-4	-3
2	0	-6

27. Find $\sqrt{f(-1) - f(0)} - [g(2)]^2 + f(-2) \div g(2) \cdot g(-1)$.

28. Find $|f(1) - f(0)| - [g(1)]^2 + g(1) \div f(-1) \cdot g(2)$.

In Exercises 29–30, find $f(-x) - f(x)$ for the given function f. Then simplify the expression.

29. $f(x) = x^3 + x - 5$

30. $f(x) = x^2 - 3x + 7$

In Exercises 31–32, each function is defined by two equations. The equation in the first row gives the output for negative numbers in the domain. The equation in the second row gives the output for nonnegative numbers in the domain. Find the indicated function values.

31. $f(x) = \begin{cases} 3x + 5 & \text{if } x < 0 \\ 4x + 7 & \text{if } x \ge 0 \end{cases}$

   a. $f(-2)$

   b. $f(0)$

   c. $f(3)$

   d. $f(-100) + f(100)$

32. $f(x) = \begin{cases} 6x - 1 & \text{if } x < 0 \\ 7x + 3 & \text{if } x \geq 0 \end{cases}$

   a. $f(-3)$

   b. $f(0)$

   c. $f(4)$

   d. $f(-100) + f(100)$

Application Exercises

The Corruption Perceptions Index uses perceptions of the general public, business people, and risk analysts to rate countries by how likely they are to accept bribes. The ratings are on a scale from 0 to 10, where higher scores represent less corruption. The graph shows the corruption ratings for the world's least corrupt and most corrupt countries. (The rating for the United States is 7.6.) Use the graph to solve Exercises 33–34.

Top Four Least Corrupt and Most Corrupt Countries

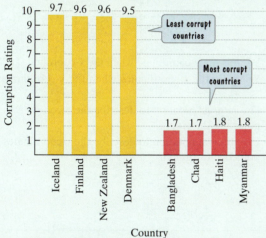

Source: Transparency International, *Corruption Perceptions Index*

33. a. Write a set of four ordered pairs in which each of the least corrupt countries corresponds to a corruption rating. Each ordered pair should be in the form

      (country, corruption rating).

   b. Is the relation in part (a) a function? Explain your answer.

   c. Write a set of four ordered pairs in which corruption ratings for the least corrupt countries correspond to countries. Each ordered pair should be in the form

      (corruption rating, country).

   d. Is the relation in part (c) a function? Explain your answer.

34. a. Write a set of four ordered pairs in which each of the most corrupt countries corresponds to a corruption rating. Each ordered pair should be in the form

      (country, corruption rating).

   b. Is the relation in part (a) a function? Explain your answer.

   c. Write a set of four ordered pairs in which corruption ratings for the most corrupt countries correspond to countries. Each ordered pair should be in the form

      (corruption rating, country).

   d. Is the relation in part (c) a function? Explain your answer.

Explaining the Concepts

35. What is a relation? Describe what is meant by its domain and its range.

36. Explain how to determine whether a relation is a function. What is a function?

37. Does $f(x)$ mean f times x when referring to function f? If not, what does $f(x)$ mean? Provide an example with your explanation.

38. For people filing a single return, federal income tax is a function of adjusted gross income because for each value of adjusted gross income there is a specific tax to be paid. By contrast, the price of a house is not a function of the lot size on which the house sits because houses on same-sized lots can sell for many different prices.

   a. Describe an everyday situation between variables that is a function.

   b. Describe an everyday situation between variables that is not a function.

Critical Thinking Exercises

Make Sense?   In Exercises 39–42, determine whether each statement makes sense or does not make sense, and explain your reasoning.

39. Today's temperature is a function of the time of day.

40. My height is a function of my age.

41. Although I presented my function as a set of ordered pairs, I could have shown the correspondences using a table or using points plotted in a rectangular coordinate system.

42. My function models how the chance of divorce depends on the number of years of marriage, so the range is $\{x \mid x$ is the number of years of marriage$\}$.

In Exercises 43–48, determine whether each statement is true or false. If the statement is false, make the necessary change(s) to produce a true statement.

43. All relations are functions.

44. No two ordered pairs of a function can have the same second components and different first components.

Using the tables that define f and g, determine whether each statement in Exercises 45–48 is true or false.

x	f(x)		x	g(x)
−4	−1		−1	−4
−3	−2		−2	−3
−2	−3		−3	−2
−1	−4		−4	−1

45. The domain of f = the range of f.

46. The range of f = the domain of g.

47. $f(-4) - f(-2) = 2$

48. $g(-4) + f(-4) = 0$

49. If $f(x) = 3x + 7$, find $\dfrac{f(a + h) - f(a)}{h}$.

50. Give an example of a relation with the following characteristics: The relation is a function containing two ordered pairs. Reversing the components in each ordered pair results in a relation that is not a function.

51. If $f(x + y) = f(x) + f(y)$ and $f(1) = 3$, find $f(2)$, $f(3)$, and $f(4)$. Is $f(x + y) = f(x) + f(y)$ for all functions?

Review Exercises

52. Simplify: $24 \div 4[2 - (5 - 2)]^2 - 6$
(Section 1.2, Example 7)

53. Simplify: $\left(\dfrac{3x^2 y^{-2}}{y^3}\right)^{-2}$. (Section 1.6, Example 9)

54. Solve: $\dfrac{x}{3} = \dfrac{3x}{5} + 4$. (Section 1.4, Example 4)

Preview Exercises

Exercises 55–57 will help you prepare for the material covered in the next section.

55. Graph $y = 2x$. Select integers for x, starting with -2 and ending with 2.

56. Graph $y = 2x + 4$. Select integers for x, starting with -2 and ending with 2.

57. Use the following graph to solve this exercise.

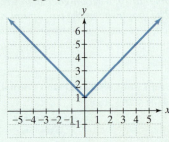

a. What is the y-coordinate when the x-coordinate is 2?

b. What are the x-coordinates when the y-coordinate is 4?

c. Use interval notation to describe the x-coordinates of all points on the graph.

d. Use interval notation to describe the y-coordinates of all points on the graph.

SECTION

2.2

Graphs of Functions ▶

Magnified 6000 times, this color-scanned image shows a T-lymphocyte blood cell (green) infected with the HIV virus (red). Depletion of the number of T cells causes destruction of the immune system.

What am I supposed to learn?

After studying this section, you should be able to:

1 Graph functions by plotting points. ▶

2 Use the vertical line test to identify functions. ▶

3 Obtain information about a function from its graph. ▶

4 Identify the domain and range of a function from its graph. ▶

The number of T cells in a person with HIV is a function of time after infection. In this section, we'll analyze the graph of this function, using the rectangular coordinate system to visualize what functions look like.

Graphs of Functions

The **graph of a function** is the graph of its ordered pairs. For example, the graph of $f(x) = 2x$ is the set of points (x, y) in the rectangular coordinate system satisfying $y = 2x$. Similarly, the graph of $g(x) = 2x + 4$ is the set of points (x, y) in the rectangular coordinate system satisfying the equation $y = 2x + 4$. In the next example, we graph both of these functions in the same rectangular coordinate system.

1 Graph functions by plotting points.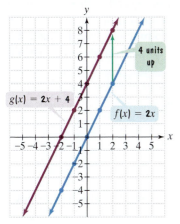

EXAMPLE 1   Graphing Functions

Graph the functions $f(x) = 2x$ and $g(x) = 2x + 4$ in the same rectangular coordinate system. Select integers for x, starting with -2 and ending with 2.

Solution   We begin by setting up a partial table of coordinates for each function. Then, we plot the five points in each table and connect them, as shown in **Figure 2.8**. The graph of each function is a straight line. Do you see a relationship between the two graphs? The graph of g is the graph of f shifted vertically up by 4 units.

Figure 2.8

x	$f(x) = 2x$	(x, y) or $(x, f(x))$	x	$g(x) = 2x + 4$	(x, y) or $(x, g(x))$
-2	$f(-2) = 2(-2) = -4$	$(-2, -4)$	-2	$g(-2) = 2(-2) + 4 = 0$	$(-2, 0)$
-1	$f(-1) = 2(-1) = -2$	$(-1, -2)$	-1	$g(-1) = 2(-1) + 4 = 2$	$(-1, 2)$
0	$f(0) = 2 \cdot 0 = 0$	$(0, 0)$	0	$g(0) = 2 \cdot 0 + 4 = 4$	$(0, 4)$
1	$f(1) = 2 \cdot 1 = 2$	$(1, 2)$	1	$g(1) = 2 \cdot 1 + 4 = 6$	$(1, 6)$
2	$f(2) = 2 \cdot 2 = 4$	$(2, 4)$	2	$g(2) = 2 \cdot 2 + 4 = 8$	$(2, 8)$

Choose x.   Compute $f(x)$ by evaluating f at x.   Form the ordered pair.   Choose x.   Compute $g(x)$ by evaluating g at x.   Form the ordered pair.

The graphs in Example 1 are straight lines. All functions with equations of the form $f(x) = mx + b$ graph as straight lines. Such functions, called **linear functions**, will be discussed in detail in Sections 2.4–2.5.

Using Technology

We can use a graphing utility to check the tables and the graphs in Example 1 for the functions

$$f(x) = 2x \quad \text{and} \quad g(x) = 2x + 4.$$

Enter $y_1 = 2x$ in the $\boxed{y=}$ screen.   Enter $y_2 = 2x + 4$ in the $\boxed{y=}$ screen.

We entered -2 for the starting x-value and 1 as an increment between x-values to check our tables in Example 1.

Checking Tables

Checking Graphs

Use the first five ordered pairs (x, y_1) to check the first table.   Use the first five ordered pairs (x, y_2) to check the second table.

$[-6, 6, 1]$ by $[-6, 9, 1]$

We selected this viewing rectangle, or window, to match **Figure 2.8**.

✓ **CHECK POINT 1**   Graph the functions $f(x) = 2x$ and $g(x) = 2x - 3$ in the same rectangular coordinate system. Select integers for x, starting with -2 and ending with 2. How is the graph of g related to the graph of f?

---

2 Use the vertical line test to identify functions. ▶

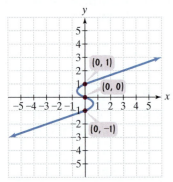

Figure 2.9 y is not a function of x because 0 is paired with three values of y, namely, 1, 0, and -1.

The Vertical Line Test

Not every graph in the rectangular coordinate system is the graph of a function. The definition of a function specifies that no value of x can be paired with two or more different values of y. Consequently, if a graph contains two or more different points with the same first coordinate, the graph cannot represent a function. This is illustrated in **Figure 2.9**. Observe that points sharing a common first coordinate are vertically above or below each other.

This observation is the basis of a useful test for determining whether a graph defines y as a function of x. The test is called the **vertical line test**.

The Vertical Line Test for Functions

If any vertical line intersects a graph in more than one point, the graph does not define y as a function of x.

EXAMPLE 2   Using the Vertical Line Test

Use the vertical line test to identify graphs in which y is a function of x.

a.

b.

c.

d.

Solution y is a function of x for the graphs in (b) and (c).

a.

y **is not a function** of x.
Two values of y correspond to one x-value.

b.

y **is a function** of x.

c.

y **is a function** of x.

d.

y **is not a function** of x.
Two values of y correspond to one x-value.

■

☑ **CHECK POINT 2**   Use the vertical line test to identify graphs in which *y* is a function of *x*.

③ Obtain information about a function from its graph. ▶

Obtaining Information from Graphs

You can obtain information about a function from its graph. At the right or left of a graph, you will often find closed dots, open dots, or arrows.

- A closed dot indicates that the graph does not extend beyond this point and the point belongs to the graph.
- An open dot indicates that the graph does not extend beyond this point and the point does not belong to the graph.
- An arrow indicates that the graph extends indefinitely in the direction in which the arrow points.

EXAMPLE 3   Analyzing the Graph of a Function

The human immunodeficiency virus, or HIV, infects and kills helper T cells. Because T cells stimulate the immune system to produce antibodies, their destruction disables the body's defenses against other pathogens. By counting the number of T cells that remain active in the body, the progression of HIV can be monitored. The fewer helper T cells, the more advanced the disease. Without the drugs that are now used to inhibit the progression of the virus, **Figure 2.10** shows a graph that is used to monitor the average progression of the disease. The number of T cells, $f(x)$, is a function of time after infection, *x*.

Figure 2.10

Source: Human Sexuality by B. Pruitt, et al. © 1997 Prentice Hall

a. Explain why *f* represents the graph of a function.

b. Use the graph to find $f(8)$.

c. For what value of *x* is $f(x) = 350$?

d. Describe the general trend shown by the graph.

Solution

a. No vertical line can be drawn that intersects the graph of *f* in **Figure 2.10** more than once. By the vertical line test, *f* represents the graph of a function.

b. To find $f(8)$, or f of 8, we locate 8 on the x-axis. **Figure 2.11** shows the point on the graph of f for which 8 is the first coordinate. From this point, we look to the y-axis to find the corresponding y-coordinate. We see that the y-coordinate is 200. Thus,

$$f(8) = 200.$$

When the time after infection is 8 years, the T cell count is 200 cells per milliliter of blood. (AIDS clinical diagnosis is given at a T cell count of 200 or below.)

c. To find the value of x for which $f(x) = 350$, we approximately locate 350 on the y-axis. **Figure 2.12** shows that there is one point on the graph of f for which 350 is the second coordinate. From this point, we look to the x-axis to find the corresponding x-coordinate. We see that the x-coordinate is 6. Thus,

$$f(x) = 350 \text{ for } x = 6.$$

A T cell count of 350 occurs 6 years after infection.

d. **Figure 2.13** uses voice balloons to describe the general trend shown by the graph.

Figure 2.11 Finding $f(8)$

Figure 2.12 Finding x for which $f(x) = 350$

Figure 2.13 Describing changing T cell count over time in a person infected with HIV ■

✓ **CHECK POINT 3**

a. Use the graph of f in **Figure 2.10** on page 117 to find $f(5)$.

b. For what value of x is $f(x) = 100$?

c. Estimate the minimum T cell count during the asymptomatic stage.

4 Identify the domain
and range of a function
from its graph.

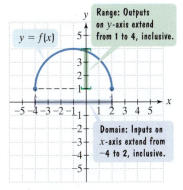

Figure 2.14 Domain and range of *f*

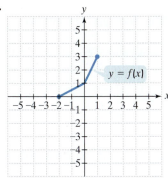

Figure 2.15 Domain and range of *f*

Identifying Domain and Range from a Function's Graph

Figure 2.14 illustrates how the graph of a function is used to determine the function's domain and its range.

Domain: set of inputs

Found on the *x*-axis

Range: set of outputs

Found on the *y*-axis

Let's apply these ideas to the graph of the function shown in **Figure 2.15**. To find the domain, look for all the inputs on the *x*-axis that correspond to points on the graph. Can you see that they extend from −4 to 2, inclusive? Using interval notation, the function's domain can be represented as follows:

$$[-4, 2].$$

The square brackets indicate −4 and 2 are included. Note the square brackets on the *x*-axis in **Figure 2.15**.

To find the range, look for all the outputs on the *y*-axis that correspond to points on the graph. They extend from 1 to 4, inclusive. Using interval notation, the function's range can be represented as follows:

$$[1, 4].$$

The square brackets indicate 1 and 4 are included. Note the square brackets on the *y*-axis in **Figure 2.15**.

EXAMPLE 4   Identifying the Domain and Range of a Function from Its Graph

Use the graph of each function to identify its domain and its range.

a.

b.

c.

d.

e.

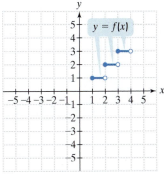

Solution For the graph of each function, the domain is highlighted in purple on the x-axis and the range is highlighted in green on the y-axis.

a.

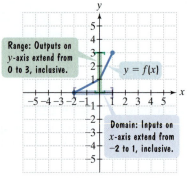

Range: Outputs on y-axis extend from 0 to 3, inclusive.

$y = f(x)$

Domain: Inputs on x-axis extend from −2 to 1, inclusive.

Domain = [−2, 1]; Range = [0, 3]

b.

Range: Outputs on y-axis extend from 1, excluding 1, to 2, including 2.

$y = f(x)$

Domain: Inputs on x-axis extend from −3, excluding −3, to 2, including 2.

Domain = (−3, 2]; Range = (1, 2]

c.

Range: Outputs on y-axis extend from 1 to 5, inclusive.

$y = f(x)$

Domain: Inputs on x-axis extend from −2, including −2, to 1, excluding 1.

Domain = [−2, 1); Range = [1, 5]

d.

Range: Outputs on y-axis include real numbers greater than or equal to 0.

$y = f(x)$

Domain: Inputs on x-axis include real numbers less than or equal to 4.

Domain = (−∞, 4]; Range = [0, ∞)

e.

Range: Outputs on y-axis "step" from 1 to 2 to 3.

$y = f(x)$

Domain: Inputs on x-axis extend from 1, including 1, to 4, excluding 4.

Domain = [1, 4); Range = {1, 2, 3} ■

Great Question!

The range in Example 4(e) was identified as {1, 2, 3}. Why didn't you use interval notation like you did in the other parts of Example 4?

Interval notation is not appropriate for describing a set of distinct numbers such as {1, 2, 3}. Interval notation, [1, 3], would mean that numbers such as 1.5 and 2.99 are in the range, but they are not. That's why we used the roster method.

✓ **CHECK POINT 4** Use the graph of each function to identify its domain and its range.

a.

$y = f(x)$

b.

$y = f(x)$

c.

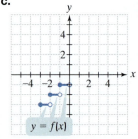

$y = f(x)$

CONCEPT AND VOCABULARY CHECK

Fill in each blank so that the resulting statement is true.

1. The graph of a function is the graph of its _____.

2. If any vertical line intersects a graph _____, the graph does not define y as a/an _____ of x.

3. The shaded set of numbers shown on the x-axis can be expressed in interval notation as _____. This set represents the function's _____.

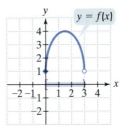

4. The shaded set of numbers shown on the y-axis can be expressed in interval notation as _____. This set represents the function's _____.

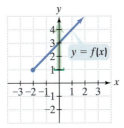

2.2 EXERCISE SET ▶ MyMathLab®

Practice Exercises

In Exercises 1–10, graph the given functions, f and g, in the same rectangular coordinate system. Select integers for x, starting with −2 and ending with 2. Once you have obtained your graphs, describe how the graph of g is related to the graph of f.

1. $f(x) = x, g(x) = x + 3$

2. $f(x) = x, g(x) = x - 4$

3. $f(x) = -2x, g(x) = -2x - 1$

4. $f(x) = -2x, g(x) = -2x + 3$

5. $f(x) = x^2, g(x) = x^2 + 1$

6. $f(x) = x^2, g(x) = x^2 - 2$

7. $f(x) = |x|, g(x) = |x| - 2$

8. $f(x) = |x|, g(x) = |x| + 1$

9. $f(x) = x^3, g(x) = x^3 + 2$

10. $f(x) = x^3, g(x) = x^3 - 1$

In Exercises 11–18, use the vertical line test to identify graphs in which y is a function of x.

11.

12.

13.

14.

15.

16.

17.

18.

In Exercises 19–24, use the graph of f to find each indicated function value.

19. $f(-2)$

20. $f(2)$

21. $f(4)$

22. $f(-4)$

23. $f(-3)$

24. $f(-1)$

Use the graph of g to solve Exercises 25–30.

25. Find $g(-4)$.

26. Find $g(2)$.

27. Find $g(-10)$.

28. Find $g(10)$.

29. For what value of x is $g(x) = 1$?

30. For what value of x is $g(x) = -1$?

In Exercises 31–40, use the graph of each function to identify its domain and its range.

31.

32.

33.

34.

35.

36.

37.

38.

39.

40.

Practice PLUS

41. Use the graph of *f* to determine (a) through (h) at the top of the next column. Where applicable, use interval notation.

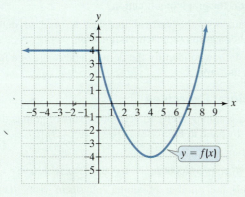

a. the domain of *f*

b. the range of *f*

c. $f(-3)$

d. the values of *x* for which $f(x) = -2$

e. the points where the graph of *f* crosses the *x*-axis

f. the point where the graph of *f* crosses the *y*-axis

g. values of *x* for which $f(x) < 0$

h. Is $f(-8)$ positive or negative?

42. Use the graph of *f* to determine each of the following. Where applicable, use interval notation.

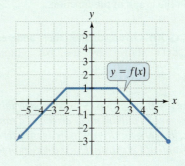

a. the domain of *f*

b. the range of *f*

c. $f(-4)$

d. the values of *x* for which $f(x) = -3$

e. the points where the graph of *f* crosses the *x*-axis

f. the point where the graph of *f* crosses the *y*-axis

g. values of *x* for which $f(x) > 0$

h. Is $f(-2)$ positive or negative?

Application Exercises

The wage gap is used to compare the status of women's earnings relative to men's. The wage gap is expressed as a percent and is calculated by dividing the median, or middlemost, annual earnings for women by the median annual earnings for men. The bar graph shows the wage gap for selected years from 1980 through 2010.

Source: U.S. Bureau of Labor Statistics

(Refer to the graphs at the bottom of the previous page.) The function $G(x) = -0.01x^2 + x + 60$ models the wage gap, as a percent, x years after 1980. The graph of function G is shown to the right of the actual data. Use this information to solve Exercises 43–44.

43. a. Find and interpret $G(30)$. Identify this information as a point on the graph of the function.

   b. Does $G(30)$ overestimate or underestimate the actual data shown by the bar graph? By how much?

44. a. Find and interpret $G(10)$. Identify this information as a point on the graph of the function.

   b. Does $G(10)$ overestimate or underestimate the actual data shown by the bar graph? By how much?

The function $f(x) = 0.4x^2 - 36x + 1000$ models the number of accidents, $f(x)$, per 50 million miles driven as a function of a driver's age, x, in years, for drivers from ages 16 through 74, inclusive. The graph of f is shown. Use the equation for f to solve Exercises 45–48.

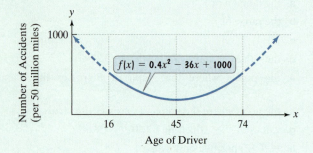

45. Find and interpret $f(20)$. Identify this information as a point on the graph of f.

46. Find and interpret $f(50)$. Identify this information as a point on the graph of f.

47. For what value of x does the graph reach its lowest point? Use the equation for f to find the minimum value of y. Describe the practical significance of this minimum value.

48. Use the graph to identify two different ages for which drivers have the same number of accidents. Use the equation for f to find the number of accidents for drivers at each of these ages.

The figure shows the cost of mailing a first-class letter, $f(x)$, as a function of its weight, x, in ounces, for weights not exceeding 3.5 ounces. Use the graph to solve Exercises 49–52.

49. Find $f(3)$. What does this mean in terms of the variables in this situation?

50. Find $f(3.5)$. What does this mean in terms of the variables in this situation?

51. What is the cost of mailing a letter that weighs 1.5 ounces?

52. What is the cost of mailing a letter that weighs 1.8 ounces?

Explaining the Concepts

53. What is the graph of a function?

54. Explain how the vertical line test is used to determine whether a graph represents a function.

55. Explain how to identify the domain and range of a function from its graph.

Technology Exercises

56. Use a graphing utility to verify the pairs of graphs that you drew by hand in Exercises 1–10.

57. The function
$$f(x) = -0.00002x^3 + 0.008x^2 - 0.3x + 6.95$$
models the number of annual physician visits, $f(x)$, by a person of age x. Graph the function in a $[0, 100, 5]$ by $[0, 40, 2]$ viewing rectangle. What does the shape of the graph indicate about the relationship between one's age and the number of annual physician visits? Use the $\boxed{\text{TRACE}}$ or minimum function capability to find the coordinates of the minimum point on the graph of the function. What does this mean?

Critical Thinking Exercises

Make Sense? *In Exercises 58–61, determine whether each statement makes sense or does not make sense, and explain your reasoning.*

58. I knew how to use point plotting to graph the equation $y = x^2 - 1$, so there was really nothing new to learn when I used the same technique to graph the function $f(x) = x^2 - 1$.

59. The graph of my function revealed aspects of its behavior that were not obvious by just looking at its equation.

60. I graphed a function showing how paid vacation days depend on the number of years a person works for a company. The domain was the number of paid vacation days.

61. I graphed a function showing how the number of annual physician visits depends on a person's age. The domain was the number of annual physician visits.

In Exercises 62–67, determine whether each statement is true or false. If the statement is false, make the necessary change(s) to produce a true statement.

62. The graph of every line is a function.

63. A horizontal line can intersect the graph of a function in more than one point.

Use the graph of f to determine whether each statement in Exercises 64–67 is true or false.

Graph of f

64. The domain of f is [−4, 4].

65. The range of f is [−2, 2].

66. $f(-1) - f(4) = 2$

67. $f(0) = 2.1$

In Exercises 68–69, let f be defined by the following graph.

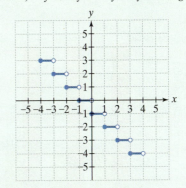

68. Find
$$\sqrt{f(-1.5) + f(-0.9)} - [f(\pi)]^2 + f(-3) \div f(1) \cdot f(-\pi).$$

69. Find
$$\sqrt{f(-2.5) - f(1.9)} - [f(-\pi)]^2 + f(-3) \div f(1) \cdot f(\pi).$$

Review Exercises

70. Is {(1, 1), (2, 2), (3, 3), (4, 4)} a function? (Section 2.1, Example 2)

71. Solve: $12 - 2(3x + 1) = 4x - 5$. (Section 1.4, Example 3)

72. The length of a rectangle exceeds 3 times the width by 8 yards. If the perimeter of the rectangle is 624 yards, what are its dimensions? (Section 1.5, Example 5)

Preview Exercises

Exercises 73–75 will help you prepare for the material covered in the next section.

73. If $f(x) = \dfrac{4}{x - 3}$, why must 3 be excluded from the domain of f?

74. If $f(x) = x^2 + x$ and $g(x) = x - 5$, find $f(4) + g(4)$.

75. Simplify: $-2.6x^2 + 49x + 3994 - (-0.6x^2 + 7x + 2412)$.

SECTION

2.3

The Algebra of Functions

What am I supposed to learn?

After studying this section, you should be able to:

1 Find the domain of a function. ▶

2 Use the algebra of functions to combine functions and determine domains. ▶

We're born. We die. **Figure 2.16** at the top of the next page quantifies these statements by showing the number of births and deaths in the United States from 2000 through 2011.

In this section, we look at these data from the perspective of functions. By considering the yearly change in the U.S. population, you will see that functions can be subtracted using procedures that will remind you of combining algebraic expressions.

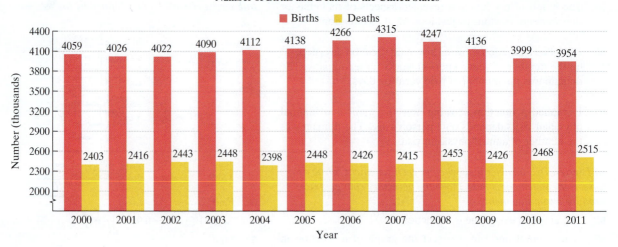

Figure 2.16

Source: U.S. Department of Health and Human Services

1 Find the domain of a function.

The Domain of a Function

We begin with two functions that model the data in **Figure 2.16**.

$$B(x) = -2.6x^2 + 49x + 3994 \qquad D(x) = -0.6x^2 + 7x + 2412.$$

Number of births, $B(x)$, in thousands, x years after **2000**

Number of deaths, $D(x)$, in thousands, x years after **2000**

The years in **Figure 2.16** extend from 2000 through 2011. Because x represents the number of years after 2000,

$$\text{Domain of } B = \{0, 1, 2, 3, \ldots, 11\}$$

and

$$\text{Domain of } D = \{0, 1, 2, 3, \ldots, 11\}.$$

Functions that model data often have their domains explicitly given with the function's equation. However, for most functions, only an equation is given and the domain is not specified. In cases like this, the domain of a function f is the largest set of real numbers for which the value of $f(x)$ is a real number. For example, consider the function

$$f(x) = \frac{1}{x - 3}.$$

Because division by 0 is undefined (and not a real number), the denominator, $x - 3$, cannot be 0. Thus, x cannot equal 3. The domain of the function consists of all real numbers other than 3, represented by

$$\text{Domain of } f = (-\infty, 3) \text{ or } (3, \infty).$$

In Chapter 7, we will be studying square root functions such as

$$g(x) = \sqrt{x}.$$

The equation tells us to take the square root of x. Because only nonnegative numbers have square roots that are real numbers, the expression under the square root sign, x, must be nonnegative. Thus,

$$\text{Domain of } g = [0, \infty).$$

Great Question!

Can you explain how $(-\infty, 3)$ or $(3, \infty)$ represents all real numbers other than 3?

The following voice balloons should help:

$$(-\infty, 3) \quad \text{or} \quad (3, \infty).$$

This consists of all real numbers less than 3, excluding 3.

This consists of all real numbers greater than 3, excluding 3.

Using Technology

Graphic Connections

You can graph a function and visually determine its domain. Look for inputs on the x-axis that correspond to points on the graph. For example, consider

$g(x) = \sqrt{x}$, or $y = \sqrt{x}$.

$y = \sqrt{x}$

Domain: Inputs on x-axis include real numbers greater than or equal to 0.

$[-10, 10, 1]$ by $[-4, 4, 1]$

This verifies that $[0, \infty)$ is the domain.

Finding a Function's Domain

If a function f does not model data or verbal conditions, its domain is the largest set of real numbers for which the value of $f(x)$ is a real number. Exclude from a function's domain real numbers that cause division by zero and real numbers that result in a square root of a negative number.

EXAMPLE 1   Finding the Domain of a Function

Find the domain of each function:

a. $f(x) = 3x + 2$ **b.** $g(x) = \dfrac{3x + 2}{x + 1}$.

Solution

a. The function $f(x) = 3x + 2$ contains neither division nor a square root. For every real number, x, the algebraic expression $3x + 2$ is a real number. Thus, the domain of f is the set of all real numbers.

$$\text{Domain of } f = (-\infty, \infty).$$

b. The function $g(x) = \dfrac{3x + 2}{x + 1}$ contains division. Because division by 0 is undefined, we must exclude from the domain the value of x that causes $x + 1$ to be 0. Thus, x cannot equal -1.

$$\text{Domain of } g = (-\infty, -1) \text{ or } (-1, \infty). \quad \blacksquare$$

☑ **CHECK POINT 1**   Find the domain of each function:

a. $f(x) = \dfrac{1}{2}x + 3$

b. $g(x) = \dfrac{7x + 4}{x + 5}$.

---

② Use the algebra of functions to combine functions and determine domains. ▶

The Algebra of Functions

We can combine functions using addition, subtraction, multiplication, and division by performing operations with the algebraic expressions that appear on the right side of the equations. For example, the functions $f(x) = 2x$ and $g(x) = x - 1$ can be combined to form the sum, difference, product, and quotient of f and g. Here's how it's done:

For each function,
$f(x) = 2x$
and
$g(x) = x - 1$.

Sum: $f + g$
$$(f + g)(x) = f(x) + g(x)$$
$$= 2x + (x - 1) = 3x - 1$$

Difference: $f - g$
$$(f - g)(x) = f(x) - g(x)$$
$$= 2x - (x - 1) = 2x - x + 1 = x + 1$$

Product: fg
$$(fg)(x) = f(x) \cdot g(x)$$
$$= 2x(x - 1) = 2x^2 - 2x$$

Quotient: $\dfrac{f}{g}$
$$\left(\dfrac{f}{g}\right)(x) = \dfrac{f(x)}{g(x)} = \dfrac{2x}{x - 1}, x \neq 1.$$

The domain for each of these functions consists of all real numbers that are common to the domains of f and g. In the case of the quotient function $\dfrac{f(x)}{g(x)}$, we must remember not to divide by 0, so we add the further restriction that $g(x) \neq 0$.

The Algebra of Functions: Sum, Difference, Product, and Quotient of Functions

Let f and g be two functions. The **sum** $f + g$, the **difference** $f - g$, the **product** fg, and the **quotient** $\dfrac{f}{g}$ are functions whose domains are the set of all real numbers common to the domains of f and g, defined as follows:

1. Sum: $(f + g)(x) = f(x) + g(x)$
2. Difference: $(f - g)(x) = f(x) - g(x)$
3. Product: $(fg)(x) = f(x) \cdot g(x)$
4. Quotient: $\left(\dfrac{f}{g}\right)(x) = \dfrac{f(x)}{g(x)}$, provided $g(x) \neq 0$.

EXAMPLE 2　Using the Algebra of Functions

Let $f(x) = x^2 - 3$ and $g(x) = 4x + 5$. Find each of the following:

a. $(f + g)(x)$

b. $(f + g)(3)$.

Solution

a. $(f + g)(x) = f(x) + g(x) = (x^2 - 3) + (4x + 5) = x^2 + 4x + 2$

Thus,

$$(f + g)(x) = x^2 + 4x + 2.$$

b. We find $(f + g)(3)$ by substituting 3 for x in the equation for $f + g$.

$$(f + g)(x) = x^2 + 4x + 2 \qquad \text{This is the equation for } f + g.$$

Substitute 3 for x.

$$(f + g)(3) = 3^2 + 4 \cdot 3 + 2 = 9 + 12 + 2 = 23 \quad \blacksquare$$

✓ **CHECK POINT 2**　Let $f(x) = 3x^2 + 4x - 1$ and $g(x) = 2x + 7$. Find each of the following:

　　a. $(f + g)(x)$　　　　　　　　b. $(f + g)(4)$.

EXAMPLE 3　Using the Algebra of Functions

Let $f(x) = \dfrac{4}{x}$ and $g(x) = \dfrac{3}{x + 2}$. Find each of the following:

a. $(f - g)(x)$

b. the domain of $f - g$.

Solution

a. $(f - g)(x) = f(x) - g(x) = \dfrac{4}{x} - \dfrac{3}{x + 2}$

(In Chapter 6, we will discuss how to perform the subtraction with these algebraic fractions. Perhaps you remember how to do so from your work in introductory algebra. For now, we will leave these fractions in the form shown.)

b. The domain of $f - g$ is the set of all real numbers that are common to the domain of f and the domain of g. Thus, we must find the domains of f and g. We will do so for f first.

Note that $f(x) = \dfrac{4}{x}$ is a function involving division. Because division by 0 is undefined, x cannot equal 0.

The function $g(x) = \dfrac{3}{x + 2}$ is also a function involving division. Because division by 0 is undefined, x cannot equal -2.

To be in the domain of $f - g$, x must be in both the domain of f and the domain of g. This means that $x \neq 0$ and $x \neq -2$. **Figure 2.17** shows these excluded values on a number line.

−2 is excluded from the domain of $f - g$.

0 is excluded from the domain of $f - g$.

Figure 2.17

The number line in **Figure 2.17** is helpful in writing each of the three intervals that describe the domain of $f - g$.

Domain of $f - g = (-\infty, -2)$ or $(-2, 0)$ or $(0, \infty)$

All real numbers less than −2, excluding −2

All real numbers between −2 and 0, excluding −2 and excluding 0

All real numbers greater than 0, excluding 0

∎

Great Question!

I find the interval notation in Example 3(b), where you excluded –2 and 0 from the domain of $f - g$, somewhat confusing. Is there another way to describe such a domain?

You could use set-builder notation:

Domain of $f - g = \{x \mid x$ is a real number and $x \neq -2$ and $x \neq 0\}$.

Ask your professor if set-builder notation is an acceptable alternative for describing domains that exclude two or more numbers. The downside: We'll be using interval notation throughout the book to indicate both domains and ranges.

✓ **CHECK POINT 3**   Let $f(x) = \dfrac{5}{x}$ and $g(x) = \dfrac{7}{x - 8}$. Find each of the following:

a. $(f - g)(x)$ **b.** the domain of $f - g$.

EXAMPLE 4   Using the Algebra of Functions

Let $f(x) = x^2 + x$ and $g(x) = x - 5$. Find each of the following:

a. $(f + g)(4)$ **b.** $(f - g)(x)$ and $(f - g)(-3)$

c. $\left(\dfrac{f}{g}\right)(x)$ and $\left(\dfrac{f}{g}\right)(7)$ **d.** $(fg)(-2)$.

$f(x) = x^2 + x$

$g(x) = x - 5$

The given functions in Example 4 (repeated)

Solution

a. We can find $(f + g)(4)$ using $f(4)$ and $g(4)$.

$$f(x) = x^2 + x \qquad g(x) = x - 5$$
$$f(4) = 4^2 + 4 = 20 \qquad g(4) = 4 - 5 = -1$$

Thus,

$$(f + g)(4) = f(4) + g(4) = 20 + (-1) = 19.$$

We can also find $(f + g)(4)$ by first finding $(f + g)(x)$ and then substituting 4 for x:

$(f + g)(x) = f(x) + g(x)$ This is the definition of the sum $f + g$.

$\qquad\qquad = (x^2 + x) + (x - 5)$ Substitute the given functions.

$\qquad\qquad = x^2 + 2x - 5.$ Simplify.

Using $(f + g)(x) = x^2 + 2x - 5$, we have

$$(f + g)(4) = 4^2 + 2 \cdot 4 - 5 = 16 + 8 - 5 = 19.$$

b. $(f - g)(x) = f(x) - g(x)$ This is the definition of the difference $f - g$.

$\qquad\qquad = (x^2 + x) - (x - 5)$ Substitute the given functions.

$\qquad\qquad = x^2 + x - x + 5$ Remove parentheses and change the sign of each term in the second set of parentheses.

$\qquad\qquad = x^2 + 5$ Simplify.

Using $(f - g)(x) = x^2 + 5$, we have

$$(f - g)(-3) = (-3)^2 + 5 = 9 + 5 = 14.$$

c. $\left(\dfrac{f}{g}\right)(x) = \dfrac{f(x)}{g(x)}$ This is the definition of the quotient $\dfrac{f}{g}$.

$\qquad\qquad = \dfrac{x^2 + x}{x - 5}$ Substitute the given functions.

Using $\left(\dfrac{f}{g}\right)(x) = \dfrac{x^2 + x}{x - 5}$, we have

$$\left(\frac{f}{g}\right)(7) = \frac{7^2 + 7}{7 - 5} = \frac{56}{2} = 28.$$

d. We can find $(fg)(-2)$ using the fact that

$$(fg)(-2) = f(-2) \cdot g(-2).$$

$$f(x) = x^2 + x \qquad\qquad\qquad g(x) = x - 5$$
$$f(-2) = (-2)^2 + (-2) = 4 - 2 = 2 \qquad g(-2) = -2 - 5 = -7$$

Thus,

$$(fg)(-2) = f(-2) \cdot g(-2) = 2(-7) = -14.$$

We could also have found $(fg)(-2)$ by multiplying $f(x) \cdot g(x)$ and then substituting -2 into the product. We will discuss how to multiply expressions such as $x^2 + x$ and $x - 5$ in Chapter 5. ∎

✓ **CHECK POINT 4** Let $f(x) = x^2 - 2x$ and $g(x) = x + 3$. Find each of the following:

a. $(f + g)(5)$ **b.** $(f - g)(x)$ and $(f - g)(-1)$

c. $\left(\dfrac{f}{g}\right)(x)$ and $\left(\dfrac{f}{g}\right)(7)$ **d.** $(fg)(-4)$.

EXAMPLE 5   Applying the Algebra of Functions

We opened the section with functions that model the number of births and deaths in the United States from 2000 through 2011:

$$B(x) = -2.6x^2 + 49x + 3994 \qquad D(x) = -0.6x^2 + 7x + 2412$$

Number of births, $B(x)$, in thousands, x years after **2000**

Number of deaths, $D(x)$, in thousands, x years after **2000**

a. Write a function that models the change in U.S. population for the years from 2000 through 2011.

b. Use the function from part (a) to find the change in U.S. population in 2008.

c. Does the result in part (b) overestimate or underestimate the actual population change in 2008 obtained from the data in **Figure 2.16** on page 126? By how much?

Solution

a. The change in population is the number of births minus the number of deaths. Thus, we will find the difference function, $B - D$.

$(B - D)(x)$
$= B(x) - D(x)$
$= (-2.6x^2 + 49x + 3994) - (-0.6x^2 + 7x + 2412)$ Substitute the given functions.
$= -2.6x^2 + 49x + 3994 + 0.6x^2 - 7x - 2412$ Remove parentheses and change the sign of each term in the second set of parentheses.

$= (-2.6x^2 + 0.6x^2) + (49x - 7x) + (3994 - 2412)$ Group like terms.

$= -2x^2 + 42x + 1582$ Combine like terms.

The function

$$(B - D)(x) = -2x^2 + 42x + 1582$$

models the change in U.S. population, in thousands, x years after 2000.

b. Because 2008 is 8 years after 2000, we substitute 8 for x in the difference function $(B - D)(x)$.

$(B - D)(x) = -2x^2 + 42x + 1582$ Use the difference function $B - D$.
$(B - D)(8) = -2(8)^2 + 42(8) + 1582$ Substitute 8 for x.
$= -2(64) + 42(8) + 1582$ Evaluate the exponential expression: $8^2 = 64$.
$= -128 + 336 + 1582$ Perform the multiplications.
$= 1790$ Add from left to right.

We see that $(B - D)(8) = 1790$. The model indicates that there was a population increase of 1790 thousand, or approximately 1,790,000 people, in 2008.

c. The data for 2008 in **Figure 2.16** on page 126 show 4247 thousand births and 2453 thousand deaths.

$$\text{population change} = \text{births} - \text{deaths}$$
$$= 4247 - 2453 = 1794$$

The actual population increase was 1794 thousand, or 1,794,000. Our model gave us an increase of 1790 thousand. Thus, the model underestimates the actual increase by $1794 - 1790$, or 4 thousand people. ∎

Achieving Success

Ask! Ask! Ask!

Do not be afraid to ask questions in class. Your professor may not realize that a concept is unclear until you raise your hand and ask a question. Other students who have problems asking questions in class will be appreciative that you have spoken up. Be polite and professional, but ask as many questions as required.

✓ **CHECK POINT 5** Use the birth and death models from Example 5.

a. Write a function that models the total number of births and deaths in the United States for the years from 2000 through 2011.

b. Use the function from part (a) to find the total number of births and deaths in the United States in 2003.

c. Does the result in part (b) overestimate or underestimate the actual number of total births and deaths in 2003 obtained from the data in **Figure 2.16** on page 126? By how much?

CONCEPT AND VOCABULARY CHECK

Fill in each blank so that the resulting statement is true.

1. We exclude from a function's domain real numbers that cause division by _____.

2. We exclude from a function's domain real numbers that result in a square root of a/an _____ number.

3. $(f + g)(x) = $ _____

4. $(f - g)(x) = $ _____

5. $(fg)(x) = $ _____

6. $\dfrac{f}{g}(x) = $ _____, provided _____ $\neq 0$

7. The domain of $f(x) = 5x + 7$ consists of all real numbers, represented in interval notation as _____.

8. The domain of $g(x) = \dfrac{3}{x - 2}$ consists of all real numbers except 2, represented in interval notation as $(-\infty, 2)$ or _____.

9. The domain of $h(x) = \dfrac{1}{x} + \dfrac{7}{x - 3}$ consists of all real numbers except 0 and 3, represented in interval notation as $(-\infty, 0)$ or _____ or _____.

2.3 EXERCISE SET ▶ MyMathLab®

Practice Exercises

In Exercises 1–10, find the domain of each function.

1. $f(x) = 3x + 5$

2. $f(x) = 4x + 7$

3. $g(x) = \dfrac{1}{x + 4}$

4. $g(x) = \dfrac{1}{x + 5}$

5. $f(x) = \dfrac{2x}{x - 3}$

6. $f(x) = \dfrac{4x}{x - 2}$

7. $g(x) = x + \dfrac{3}{5 - x}$

8. $g(x) = x + \dfrac{7}{6 - x}$

9. $f(x) = \dfrac{1}{x + 7} + \dfrac{3}{x - 9}$

10. $f(x) = \dfrac{1}{x + 8} + \dfrac{3}{x - 10}$

In Exercises 11–16, find $(f + g)(x)$ and $(f + g)(5)$.

11. $f(x) = 3x + 1, g(x) = 2x - 6$

12. $f(x) = 4x + 2, g(x) = 2x - 9$

13. $f(x) = x - 5, g(x) = 3x^2$

14. $f(x) = x - 6, g(x) = 2x^2$

15. $f(x) = 2x^2 - x - 3, g(x) = x + 1$

16. $f(x) = 4x^2 - x - 3, g(x) = x + 1$

17. Let $f(x) = 5x$ and $g(x) = -2x - 3$. Find $(f + g)(x)$, $(f - g)(x)$, $(fg)(x)$, and $\left(\dfrac{f}{g}\right)(x)$.

18. Let $f(x) = -4x$ and $g(x) = -3x + 5$. Find $(f + g)(x)$, $(f - g)(x)$, $(fg)(x)$, and $\left(\dfrac{f}{g}\right)(x)$.

In Exercises 19–30, for each pair of functions, f and g, determine the domain of f + g.

19. $f(x) = 3x + 7, g(x) = 9x + 10$

20. $f(x) = 7x + 4, g(x) = 5x - 2$

21. $f(x) = 3x + 7, g(x) = \dfrac{2}{x - 5}$

22. $f(x) = 7x + 4, g(x) = \dfrac{2}{x - 6}$

23. $f(x) = \dfrac{1}{x}, g(x) = \dfrac{2}{x - 5}$

24. $f(x) = \dfrac{1}{x}, g(x) = \dfrac{2}{x - 6}$

25. $f(x) = \dfrac{8x}{x - 2}, g(x) = \dfrac{6}{x + 3}$

26. $f(x) = \dfrac{9x}{x - 4}, g(x) = \dfrac{7}{x + 8}$

27. $f(x) = \dfrac{8x}{x - 2}, g(x) = \dfrac{6}{2 - x}$

28. $f(x) = \dfrac{9x}{x - 4}, g(x) = \dfrac{7}{4 - x}$

29. $f(x) = x^2, g(x) = x^3$

30. $f(x) = x^2 + 1, g(x) = x^3 - 1$

In Exercises 31–50, let

$$f(x) = x^2 + 4x \quad \text{and} \quad g(x) = 2 - x.$$

Find each of the following.

31. $(f + g)(x)$ and $(f + g)(3)$

32. $(f + g)(x)$ and $(f + g)(4)$

33. $f(-2) + g(-2)$

34. $f(-3) + g(-3)$

35. $(f - g)(x)$ and $(f - g)(5)$

36. $(f - g)(x)$ and $(f - g)(6)$

37. $f(-2) - g(-2)$ **38.** $f(-3) - g(-3)$

39. $(fg)(-2)$ **40.** $(fg)(-3)$

41. $(fg)(5)$ **42.** $(fg)(6)$

43. $\left(\dfrac{f}{g}\right)(x)$ and $\left(\dfrac{f}{g}\right)(1)$

44. $\left(\dfrac{f}{g}\right)(x)$ and $\left(\dfrac{f}{g}\right)(3)$

45. $\left(\dfrac{f}{g}\right)(-1)$ **46.** $\left(\dfrac{f}{g}\right)(0)$

47. The domain of $f + g$

48. The domain of $f - g$

49. The domain of $\dfrac{f}{g}$

50. The domain of fg

Practice PLUS

Use the graphs of f and g below to solve Exercises 51–58.

51. Find $(f + g)(-3)$.

52. Find $(g - f)(-2)$.

53. Find $(fg)(2)$.

54. Find $\left(\dfrac{g}{f}\right)(3)$.

55. Find the domain of $f + g$.

56. Find the domain of $\dfrac{f}{g}$.

57. Graph $f + g$.

58. Graph $f - g$.

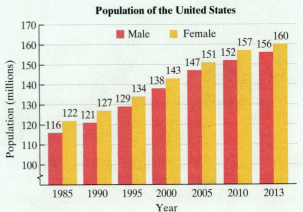

Use the table defining f and g to solve Exercises 59–62.

x	$f(x)$	$g(x)$
-2	5	0
-1	3	-2
0	-2	4
1	-6	-3
2	0	1

59. Find $(f + g)(1) - (g - f)(-1)$.

60. Find $(f + g)(-1) - (g - f)(0)$.

61. Find $(fg)(-2) - \left[\left(\dfrac{f}{g}\right)(1)\right]^2$.

62. Find $(fg)(2) - \left[\left(\dfrac{g}{f}\right)(0)\right]^2$.

Application Exercises

The bar graph shows the population of the United States, in millions, for seven selected years.

Population of the United States

Year	Male	Female
1985	116	122
1990	121	127
1995	129	134
2000	138	143
2005	147	151
2010	152	157
2013	156	160

Source: U.S. Census Bureau

Here are two functions that model the data in the bar graph at the bottom of the previous page:

$M(x) = 1.5x + 115$ ⟵ Male U.S. population, $M(x)$, in millions, x years after 1985

$F(x) = 1.4x + 121.$ ⟵ Female U.S. population, $F(x)$, in millions, x years after 1985

Use the functions above to solve Exercises 63–65.

63. a. Write a function that models the total U.S. population for the years shown in the bar graph.
   b. Use the function from part (a) to find the total U.S. population in 2010.
   c. Does the result in part (b) overestimate or underestimate the actual total U.S. population in 2010 shown by the bar graph? By how much?

64. a. Write a function that models the difference between the female U.S. population and the male U.S. population for the years shown in the bar graph.
   b. Use the function from part (a) to find how many more women than men there were in the U.S. population in 2005.
   c. How well does the result in part (b) describe the actual difference between the female and male population in 2005 shown by the bar graph?

65. a. Write a function that models the ratio of men to women in the U.S. population for the years shown in the bar graph.
   b. Use the function from part (a) to find the ratio of men to women, correct to three decimal places, in 2000.
   c. Does the result in part (b) overestimate or underestimate the actual ratio of men to women in 2000 shown by the bar graph? By how much?

66. A company that sells radios has yearly fixed costs of $600,000. It costs the company $45 to produce each radio. Each radio will sell for $65. The company's costs and revenue are modeled by the following functions:

$C(x) = 600,000 + 45x$ This function models the company's costs.

$R(x) = 65x$ This function models the company's revenue.

Find and interpret $(R - C)(20,000)$, $(R - C)(30,000)$, and $(R - C)(40,000)$.

Explaining the Concepts

67. If a function is defined by an equation, explain how to find its domain.

68. If equations for functions f and g are given, explain how to find $f + g$.

69. If the equations of two functions are given, explain how to obtain the quotient function and its domain.

70. If equations for functions f and g are given, describe two ways to find $(f - g)(3)$.

Technology Exercises

In Exercises 71–74, graph each of the three functions in the same $[-10, 10, 1]$ by $[-10, 10, 1]$ viewing rectangle.

71. $y_1 = 2x + 3$
   $y_2 = 2 - 2x$
   $y_3 = y_1 + y_2$

72. $y_1 = x - 4$
   $y_2 = 2x$
   $y_3 = y_1 - y_2$

73. $y_1 = x$
   $y_2 = x - 4$
   $y_3 = y_1 \cdot y_2$

74. $y_1 = x^2 - 2x$
   $y_2 = x$
   $y_3 = \dfrac{y_1}{y_2}$

75. In Exercise 74, use the ⎢TRACE⎥ feature to trace along y_3. What happens at $x = 0$? Explain why this occurs.

Critical Thinking Exercises

Make Sense? *In Exercises 76–79, determine whether each statement makes sense or does not make sense, and explain your reasoning.*

76. There is an endless list of real numbers that cannot be included in the domain of $f(x) = \sqrt{x}$.

77. I used a function to model data from 1990 through 2015. The independent variable in my model represented the number of years after 1990, so the function's domain was $\{0, 1, 2, 3, \ldots, 25\}$.

78. If I have equations for functions f and g, and 3 is in both domains, then there are always two ways to determine $(f + g)(3)$.

79. I have two functions. Function f models total world population x years after 2000 and function g models population of the world's more-developed regions x years after 2000. I can use $f - g$ to determine the population of the world's less-developed regions for the years in both functions' domains.

In Exercises 80–83, determine whether each statement is true or false. If the statement is false, make the necessary change(s) to produce a true statement.

80. If $(f + g)(a) = 0$, then $f(a)$ and $g(a)$ must be opposites, or additive inverses.

81. If $(f - g)(a) = 0$, then $f(a)$ and $g(a)$ must be equal.

82. If $\left(\dfrac{f}{g}\right)(a) = 0$, then $f(a)$ must be 0.

83. If $(fg)(a) = 0$, then $f(a)$ must be 0.

Review Exercises

84. Solve for b: $R = 3(a + b)$. (Section 1.5, Example 6)

85. Solve: $3(6 - x) = 3 - 2(x - 4)$. (Section 1.4, Example 3)

86. If $f(x) = 6x - 4$, find $f(b + 2)$. (Section 2.1, Example 3)

Preview Exercises

Exercises 87–89 will help you prepare for the material covered in the next section.

87. Consider $4x - 3y = 6$.

   a. What is the value of x when $y = 0$?

   b. What is the value of y when $x = 0$?

88. a. Graph $y = 2x + 4$. Select integers for x from -3 to 1, inclusive.

   b. At what point does the graph cross the x-axis?

   c. At what point does the graph cross the y-axis?

89. Solve for y: $5x + 3y = -12$.

MID-CHAPTER CHECK POINT   Section 2.1–Section 2.3

 What You Know: We learned that a function is a relation in which no two ordered pairs have the same first component and different second components. We represented functions as equations and used function notation. We graphed functions and applied the vertical line test to identify graphs of functions. We determined the domain and range of a function from its graph, using inputs on the x-axis for the domain and outputs on the y-axis for the range. Finally, we developed an algebra of functions to combine functions and determine their domains.

In Exercises 1–6, determine whether each relation is a function. Give the domain and range for each relation.

1. $\{(2, 6), (1, 4), (2, -6)\}$

2. $\{(0, 1), (2, 1), (3, 4)\}$

3.

4.

5.

6.

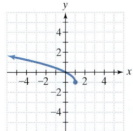

Use the graph of f to solve Exercises 7–12.

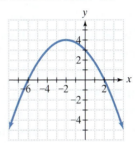

7. Explain why f represents the graph of a function.

8. Use the graph to find $f(-4)$.

9. For what value or values of x is $f(x) = 4$?

10. For what value or values of x is $f(x) = 0$?

11. Find the domain of f.

12. Find the range of f.

In Exercises 13–14, find the domain of each function.

13. $f(x) = (x + 2)(x - 2)$

14. $g(x) = \dfrac{1}{(x + 2)(x - 2)}$

In Exercises 15–22, let

$$f(x) = x^2 - 3x + 8 \text{ and } g(x) = -2x - 5.$$

Find each of the following.

15. $f(0) + g(-10)$

16. $f(-1) - g(3)$

17. $f(a) + g(a + 3)$

18. $(f + g)(x)$ and $(f + g)(-2)$

19. $(f - g)(x)$ and $(f - g)(5)$

20. $(fg)(-1)$

21. $\left(\dfrac{f}{g}\right)(x)$ and $\left(\dfrac{f}{g}\right)(-4)$

22. The domain of $\dfrac{f}{g}$

2.4

Linear Functions and Slope

Is there a relationship between literacy and child mortality? As the percentage of adult females who are literate increases, does the mortality of children under age five decrease? **Figure 2.18** is based on data from the United Nations and indicates that this is, indeed, the case. Each point in the figure represents one country.

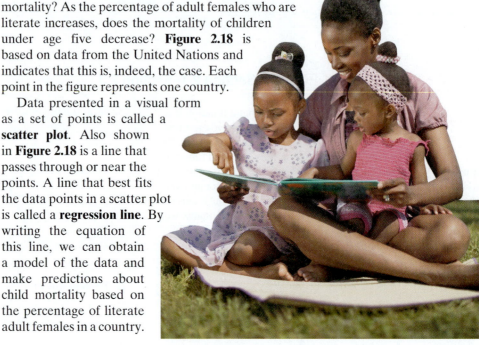

Data presented in a visual form as a set of points is called a **scatter plot**. Also shown in **Figure 2.18** is a line that passes through or near the points. A line that best fits the data points in a scatter plot is called a **regression line**. By writing the equation of this line, we can obtain a model of the data and make predictions about child mortality based on the percentage of literate adult females in a country.

What am I supposed to learn?

After studying this section, you should be able to:

1. Use intercepts to graph a linear function in standard form. ▶

2. Compute a line's slope. ▶

3. Find a line's slope and y-intercept from its equation. ▶

4. Graph linear functions in slope-intercept form. ▶

5. Graph horizontal or vertical lines. ▶

6. Interpret slope as rate of change. ▶

7. Find a function's average rate of change. ▶

8. Use slope and y-intercept to model data. ▶

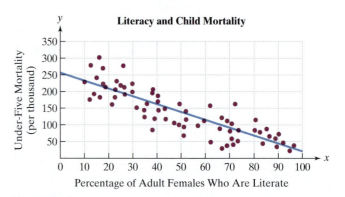

Figure 2.18

Source: United Nations Development Programme

Data often fall on or near a line. In the remainder of this chapter, we will use equations to model such data and make predictions. We begin with a discussion of graphing linear functions using intercepts.

1. Use intercepts to graph a linear function in standard form. ▶

Graphing Using Intercepts

The equation of the regression line in **Figure 2.18** is

$$y = -2.39x + 254.47.$$

The variable x represents the percentage of adult females in a country who are literate. The variable y represents child mortality, per thousand, for children under five in that country. Using function notation, we can rewrite the equation as

$$f(x) = -2.39x + 254.47.$$

A function such as this, whose graph is a straight line, is called a **linear function**. There is another way that we can write the function's equation

$$y = -2.39x + 254.47.$$

We will collect the x- and y-terms on the left side. This is done by adding $2.39x$ to both sides:

$$2.39x + y = 254.47.$$

The form of this equation is $Ax + By = C$.

$$2.39x \quad + \quad y = 254.47$$

| A, the coefficient of x, is **2.39**. | B, the coefficient of y, is **1**. | C, the constant on the right, is **254.47**. |

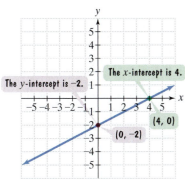

Figure 2.19 The graph of $2x - 4y = 8$

All equations of the form $Ax + By = C$ are straight lines when graphed, as long as A and B are not both zero. Such an equation is called the **standard form of the equation of a line**. To graph equations of this form, we will use two important points: the **intercepts**.

An **x-intercept** of a graph is the x-coordinate of a point where the graph intersects the x-axis. For example, look at the graph of $2x - 4y = 8$ in **Figure 2.19**. The graph crosses the x-axis at $(4, 0)$. Thus, the x-intercept is 4. **The y-coordinate corresponding to an x-intercept is always zero.**

A **y-intercept** of a graph is the y-coordinate of a point where the graph intersects the y-axis. The graph of $2x - 4y = 8$ in **Figure 2.19** shows that the graph crosses the y-axis at $(0, -2)$. Thus, the y-intercept is -2. **The x-coordinate corresponding to a y-intercept is always zero.**

Great Question!

Are single numbers the only way to represent intercepts? Can ordered pairs also be used?

Mathematicians tend to use two ways to describe intercepts. Did you notice that we are using single numbers? If a graph's x-intercept is a, it passes through the point $(a, 0)$. If a graph's y-intercept is b, it passes through the point $(0, b)$.

Some books state that the x-intercept is the *point* $(a, 0)$ and the x-intercept is *at a* on the x-axis. Similarly, the y-intercept is the *point* $(0, b)$ and the y-intercept is *at b* on the y-axis. In these descriptions, the intercepts are the actual points where a graph crosses the axes.

Although we'll describe intercepts as single numbers, we'll immediately state the point on the x- or y-axis that the graph passes through. Here's the important thing to keep in mind:

x-intercept: The corresponding y-coordinate is 0.
y-intercept: The corresponding x-coordinate is 0.

When graphing using intercepts, it is a good idea to use a third point, a checkpoint, before drawing the line. A checkpoint can be obtained by selecting a value for either variable, other than 0, and finding the corresponding value for the other variable. The checkpoint should lie on the same line as the x- and y-intercepts. If it does not, recheck your work and find the error.

Using Intercepts to Graph $Ax + By = C$

1. Find the x-intercept. Let $y = 0$ and solve for x.
2. Find the y-intercept. Let $x = 0$ and solve for y.
3. Find a checkpoint, a third ordered-pair solution.
4. Graph the equation by drawing a line through the three points.

EXAMPLE 1 Using Intercepts to Graph a Linear Equation

Graph: $4x - 3y = 6$.

Solution
Step 1. Find the *x*-intercept. Let *y* = 0 and solve for *x*.

$$4x - 3 \cdot 0 = 6 \qquad \text{Replace } y \text{ with } 0 \text{ in } 4x - 3y = 6.$$
$$4x = 6 \qquad \text{Simplify.}$$
$$x = \frac{6}{4} = \frac{3}{2} \qquad \text{Divide both sides by 4.}$$

The *x*-intercept is $\frac{3}{2}$, so the line passes through $\left(\frac{3}{2}, 0\right)$ or $(1.5, 0)$.

Step 2. Find the *y*-intercept. Let *x* = 0 and solve for *y*.

$$4 \cdot 0 - 3y = 6 \qquad \text{Replace } x \text{ with } 0 \text{ in } 4x - 3y = 6.$$
$$-3y = 6 \qquad \text{Simplify.}$$
$$y = -2 \qquad \text{Divide both sides by } -3.$$

The *y*-intercept is -2, so the line passes through $(0, -2)$.

Step 3. Find a checkpoint, a third ordered-pair solution. For our checkpoint, we will let $x = 1$ and find the corresponding value for *y*.

$$4x - 3y = 6 \qquad \text{This is the given equation.}$$
$$4 \cdot 1 - 3y = 6 \qquad \text{Substitute 1 for } x.$$
$$4 - 3y = 6 \qquad \text{Simplify.}$$
$$-3y = 2 \qquad \text{Subtract 4 from both sides.}$$
$$y = -\frac{2}{3} \qquad \text{Divide both sides by } -3.$$

The checkpoint is the ordered pair $\left(1, -\frac{2}{3}\right)$.

Step 4. Graph the equation by drawing a line through the three points. The three points in **Figure 2.20** lie along the same line. Drawing a line through the three points results in the graph of $4x - 3y = 6$. ∎

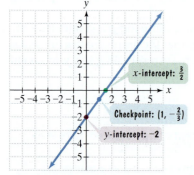

Figure 2.20 The graph of $4x - 3y = 6$

Using Technology

You can use a graphing utility to graph equations of the form $Ax + By = C$. Begin by solving the equation for *y*. For example, to graph $4x - 3y = 6$, solve the equation for *y*.

$$4x - 3y = 6 \qquad \text{This is the equation to be graphed.}$$
$$4x - 4x - 3y = -4x + 6 \qquad \text{Add } -4x \text{ to both sides.}$$
$$-3y = -4x + 6 \qquad \text{Simplify.}$$
$$\frac{-3y}{-3} = \frac{-4x + 6}{-3} \qquad \text{Divide both sides by } -3.$$
$$y = \frac{4}{3}x - 2 \qquad \text{Simplify.}$$

This is the equation to enter in your graphing utility. The graph of $y = \frac{4}{3}x - 2$, or, equivalently, $4x - 3y = 6$ is shown above in a $[-6, 6, 1]$ by $[-6, 6, 1]$ viewing rectangle.

✓ **CHECK POINT 1** Graph: $3x - 2y = 6$.

2 Compute a line's slope.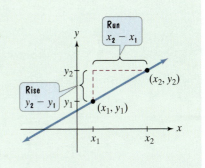

The Slope of a Line

Mathematicians have developed a useful measure of the steepness of a line, called the **slope** of the line. Slope compares the vertical change (the **rise**) to the horizontal change (the **run**) when moving from one fixed point to another along the line. To calculate the slope of a line, we use a ratio that compares the change in y (the rise) to the change in x (the run).

Definition of Slope

The **slope** of the line through the distinct points (x_1, y_1) and (x_2, y_2) is

$$\frac{\text{Change in } y}{\text{Change in } x} = \frac{\text{Rise}}{\text{Run}} \quad \begin{array}{l}\text{Vertical change}\\ \text{Horizontal change}\end{array}$$

$$= \frac{y_2 - y_1}{x_2 - x_1},$$

where $x_2 - x_1 \neq 0$.

It is common notation to let the letter m represent the slope of a line. The letter m is used because it is the first letter of the French verb *monter*, meaning "to rise," or "to ascend."

Run: 2 units
Rise: 10 units

Figure 2.21 Visualizing a slope of 5

EXAMPLE 2 Using the Definition of Slope

Find the slope of the line passing through each pair of points:

a. $(-3, -4)$ and $(-1, 6)$ **b.** $(-1, 3)$ and $(-4, 5)$.

Solution

a. Let $(x_1, y_1) = (-3, -4)$ and $(x_2, y_2) = (-1, 6)$. The slope is obtained as follows:

$$m = \frac{\text{Change in } y}{\text{Change in } x} = \frac{y_2 - y_1}{x_2 - x_1} = \frac{6 - (-4)}{-1 - (-3)} = \frac{6 + 4}{-1 + 3} = \frac{10}{2} = 5.$$

The situation is illustrated in **Figure 2.21**. The slope of the line is 5, or $\frac{10}{2}$. For every vertical change, or rise, of 10 units, there is a corresponding horizontal change, or run, of 2 units. The slope is positive and the line rises from left to right.

Great Question!

When using the definition of slope, how do I know which point to call (x_1, y_1) and which point to call (x_2, y_2)?

When computing slope, it makes no difference which point you call (x_1, y_1) and which point you call (x_2, y_2). If we let $(x_1, y_1) = (-1, 6)$ and $(x_2, y_2) = (-3, -4)$, the slope is still 5:

$$m = \frac{y_2 - y_1}{x_2 - x_1} = \frac{-4 - 6}{-3 - (-1)} = \frac{-10}{-2} = 5.$$

However, you should not subtract in one order in the numerator ($y_2 - y_1$) and then in a different order in the denominator ($x_1 - x_2$).

$$\frac{-4 - 6}{-1 - (-3)} = \frac{-10}{2} = -5. \quad \text{Incorrect! The slope is not } -5.$$

b. To find the slope of the line passing through $(-1, 3)$ and $(-4, 5)$, we can let $(x_1, y_1) = (-1, 3)$ and $(x_2, y_2) = (-4, 5)$. The slope is computed as follows:

$$m = \frac{\text{Change in } y}{\text{Change in } x} = \frac{y_2 - y_1}{x_2 - x_1} = \frac{5 - 3}{-4 - (-1)} = \frac{2}{-3} = -\frac{2}{3}.$$

The situation is illustrated in **Figure 2.22**. The slope of the line is $-\frac{2}{3}$. For every vertical change of -2 units (2 units down), there is a corresponding horizontal change of 3 units. The slope is negative and the line falls from left to right.

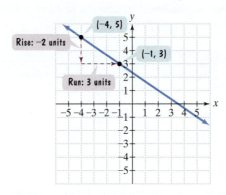

Figure 2.22 Visualizing a slope of $-\dfrac{2}{3}$ ∎

✓ **CHECK POINT 2** Find the slope of the line passing through each pair of points:

a. $(-3, 4)$ and $(-4, -2)$ **b.** $(4, -2)$ and $(-1, 5)$.

Example 2 illustrates that a line with a positive slope is rising from left to right and a line with a negative slope is falling from left to right. By contrast, a horizontal line neither rises nor falls and has a slope of zero. A vertical line has no horizontal change, so $x_2 - x_1 = 0$ in the formula for slope. Because we cannot divide by zero, the slope of a vertical line is undefined. This discussion is summarized in **Table 2.2.**

Table 2.2   Possibilities for a Line's Slope

Positive Slope	Negative Slope	Zero Slope	Undefined Slope
$m > 0$	$m < 0$	$m = 0$	m is undefined.
Line rises from left to right.	Line falls from left to right.	Line is horizontal.	Line is vertical.

Great Question!

Is it OK to say that a vertical line has no slope?

Always be clear in the way you use language, especially in mathematics. For example, it's not a good idea to say that a line has "no slope." This could mean that the slope is zero or that the slope is undefined.

The Slope-Intercept Form of the Equation of a Line

We opened this section with a linear function that models child mortality as a function of literacy. The function's equation can be expressed as

$$y = -2.39x + 254.47 \quad \text{or} \quad f(x) = -2.39x + 254.47.$$

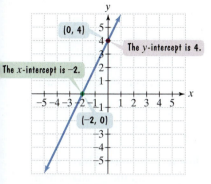

Figure 2.23 The graph of $y = 2x + 4$

What is the significance of -2.39, the x-coefficient, or of 254.47, the constant term? To answer this question, let's look at an equation in the same form with simpler numbers. In particular, consider the equation $y = 2x + 4$.

Figure 2.23 shows the graph of $y = 2x + 4$. Verify that the x-intercept is -2 by setting y equal to 0 and solving for x. Similarly, verify that the y-intercept is 4 by setting x equal to 0 and solving for y.

Now that we have two points on the line, we can calculate the slope of the graph of $y = 2x + 4$.

$$\text{Slope} = \frac{\text{Change in } y}{\text{Change in } x} = \frac{4 - 0}{0 - (-2)} = \frac{4}{2} = 2$$

We see that the slope of the line is 2, the same as the coefficient of x in the equation $y = 2x + 4$. The y-intercept is 4, the same as the constant in the equation $y = 2x + 4$.

It is not merely a coincidence that the x-coefficient is the line's slope and the constant term is the y-intercept. Let's find the equation of any nonvertical line with slope m and y-intercept b. Because the y-intercept is b, the point $(0, b)$ lies on the line. Now, let (x, y) represent any other point on the line, shown in **Figure 2.24**. Keep in mind that the point (x, y) is arbitrary and is not in one fixed position. By contrast, the point $(0, b)$ is fixed.

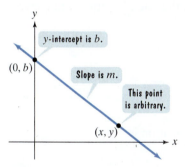

Figure 2.24 A line with slope m and y-intercept b

Regardless of where the point (x, y) is located, the steepness of the line in **Figure 2.24** remains the same. Thus, the ratio for slope stays a constant m. This means that for all points along the line,

$$m = \frac{\text{Change in } y}{\text{Change in } x} = \frac{y - b}{x - 0} = \frac{y - b}{x}.$$

We can clear the fraction by multiplying both sides by x, the least common denominator.

$$m = \frac{y - b}{x} \qquad \text{This is the slope of the line in Figure 2.24.}$$

$$mx = \frac{y - b}{x} \cdot x \qquad \text{Multiply both sides by } x.$$

$$mx = y - b \qquad \text{Simplify: } \frac{y - b}{\cancel{x}} \cdot \cancel{x} = y - b.$$

$$mx + b = y - b + b \qquad \text{Add } b \text{ to both sides and solve for } y.$$

$$mx + b = y \qquad \text{Simplify.}$$

Now, if we reverse the two sides, we obtain the slope-intercept form of the equation of a line.

3 Find a line's slope and y-intercept from its equation. ▶

Great Question!

Which are the variables and which are the constants in the slope-intercept equation $y = mx + b$?

The variables in $y = mx + b$ vary in different ways. The values for slope, m, and y-intercept, b, vary from one line's equation to another. However, they remain constant in the equation of a single line. By contrast, the variables x and y represent the infinitely many points, (x, y), on a single line. Thus, these variables vary in both the equation of a single line, as well as from one equation to another.

Slope-Intercept Form of the Equation of a Line

The **slope-intercept form of the equation** of a nonvertical line with slope m and y-intercept b is

$$y = mx + b.$$

The slope-intercept form of a line's equation, $y = mx + b$, can be expressed in function notation by replacing y with $f(x)$:

$$f(x) = mx + b.$$

We have seen that functions in this form are called **linear functions**. Thus, in the equation of a linear function, the x-coefficient is the line's slope and the constant term is the y-intercept. Here are two examples:

$$y = 2x - 4 \qquad\qquad f(x) = \frac{1}{2}x + 2.$$

The slope is 2.   The y-intercept is −4.   The slope is $\frac{1}{2}$.   The y-intercept is 2.

If a linear function's equation is in standard form, $Ax + By = C$, do you see how we can identify the line's slope and y-intercept? Solve the equation for y and convert to slope-intercept form.

EXAMPLE 3 Converting from Standard Form to Slope-Intercept Form

Give the slope and the y-intercept for the line whose equation is

$$5x + 3y = -12.$$

Solution We convert $5x + 3y = -12$ to slope-intercept form by solving the equation for y. In this form, the coefficient of x is the line's slope and the constant term is the y-intercept.

$5x + 3y = -12$	This is the given equation in standard form, $Ax + By = C$.
Our goal is to isolate y.	
$5x - 5x + 3y = -5x - 12$	Add $-5x$ to both sides.
$3y = -5x - 12$	Simplify.
$\dfrac{3y}{3} = \dfrac{-5x - 12}{3}$	Divide both sides by 3.
$y = -\dfrac{5}{3}x - 4$	Divide each term in the numerator by 3.

The slope is $-\frac{5}{3}$.   The y-intercept is −4. ∎

☑ **CHECK POINT 3** Give the slope and the y-intercept for the line whose equation is $8x - 4y = 20$.

4 Graph linear functions in slope-intercept form.

If a linear function's equation is in slope-intercept form, we can use the y-intercept and the slope to obtain its graph.

Graphing $y = mx + b$ Using the Slope and y-Intercept

1. Plot the point containing the y-intercept on the y-axis. This is the point $(0, b)$.
2. Obtain a second point using the slope, m. Write m as a fraction, and use rise over run, starting at the point on the y-axis, to plot this point.
3. Use a straightedge to draw a line through the two points. Draw arrowheads at the ends of the line to show that the line continues indefinitely in both directions.

EXAMPLE 4   Graphing by Using the Slope and y-Intercept

Graph the line whose equation is $y = 3x - 4$.

Solution   The equation $y = 3x - 4$ is in the form $y = mx + b$. The slope, m, is the coefficient of x. The y-intercept, b, is the constant term.

$$y = 3x + (-4)$$

> The slope is 3.   The y-intercept is -4.

Now that we have identified the slope and the y-intercept, we use the three-step procedure to graph the equation.

Step 1. Plot the point containing the y-intercept on the y-axis. The y-intercept is -4. We plot the point $(0, -4)$, shown in **Figure 2.25(a)**.

Step 2. Obtain a second point using the slope, m. Write m as a fraction, and use rise over run, starting at the point on the y-axis, to plot this point. We express the slope, 3, as a fraction.

$$m = \frac{3}{1} = \frac{\text{Rise}}{\text{Run}}$$

We plot the second point on the line by starting at $(0, -4)$, the first point. Based on the slope, we move 3 units *up* (the rise) and 1 unit to the *right* (the run). This puts us at a second point on the line, $(1, -1)$, shown in **Figure 2.25(b)**.

Step 3. Use a straightedge to draw a line through the two points. The graph of $y = 3x - 4$ is shown in **Figure 2.25(c)**.

Great Question!

If the slope is an integer, such as 3, why should I express it as $\frac{3}{1}$ for graphing purposes?

Writing the slope, m, as a fraction allows you to identify the rise (the fraction's numerator) and the run (the fraction's denominator).

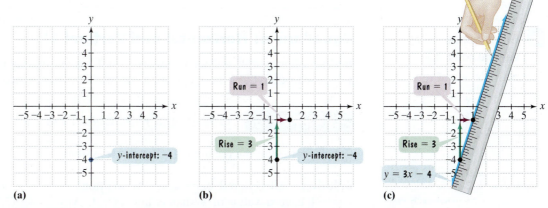

Figure 2.25 Graphing $y = 3x - 4$ using the y-intercept and slope ∎

✓ **CHECK POINT 4**   Graph the line whose equation is $y = 4x - 3$.

EXAMPLE 5 Graphing by Using the Slope and y-Intercept

Graph the linear function: $f(x) = -\dfrac{3}{2}x + 2$.

Solution The equation of the line is in the form $f(x) = mx + b$. We can find the slope, m, by identifying the coefficient of x. We can find the y-intercept, b, by identifying the constant term.

$$f(x) = -\frac{3}{2}x + 2$$

The slope is $-\frac{3}{2}$.   The y-intercept is 2.

Now that we have identified the slope, $-\frac{3}{2}$, and the y-intercept, 2, we use the three-step procedure to graph the equation.

Step 1. Plot the point containing the y-intercept on the y-axis. The y-intercept is 2. We plot $(0, 2)$, shown in **Figure 2.26**.

Step 2. Obtain a second point using the slope, m. Write m as a fraction, and use rise over run, starting at the point on the y-axis, to plot this point. The slope, $-\frac{3}{2}$, is already written as a fraction.

$$m = -\frac{3}{2} = \frac{-3}{2} = \frac{\text{Rise}}{\text{Run}}$$

We plot the second point on the line by starting at $(0, 2)$, the first point. Based on the slope, we move 3 units *down* (the rise) and 2 units to the *right* (the run). This puts us at a second point on the line, $(2, -1)$, shown in **Figure 2.26**.

Step 3. Use a straightedge to draw a line through the two points. The graph of the linear function $f(x) = -\frac{3}{2}x + 2$ is shown as a blue line in **Figure 2.26**. ∎

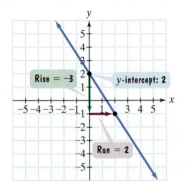

Figure 2.26 The graph of $f(x) = -\dfrac{3}{2}x + 2$

Discover for Yourself

Obtain a second point in Example 5 by writing the slope as follows:

$$m = \frac{3}{-2} = \frac{\text{Rise}}{\text{Run}}.$$

$-\frac{3}{2}$ can be expressed as $\frac{-3}{2}$ or $\frac{3}{-2}$.

Obtain a second point in **Figure 2.26** by moving *up* 3 units and to the *left* 2 units, starting at $(0, 2)$. What do you observe once you draw the line?

✓ **CHECK POINT 5** Graph the linear function: $f(x) = -\dfrac{2}{3}x$.

5 Graph horizontal or vertical lines. ▶

Equations of Horizontal and Vertical Lines

If a line is horizontal, its slope is zero: $m = 0$. Thus, the equation $y = mx + b$ becomes $y = b$, where b is the y-intercept. All horizontal lines have equations of the form $y = b$.

EXAMPLE 6 Graphing a Horizontal Line

Graph $y = -4$ in the rectangular coordinate system.

Figure 2.27 The graph of $y = -4$ or $f(x) = -4$

Solution All ordered pairs that are solutions of $y = -4$ have a value of y that is always -4. Any value can be used for x. In the table on the right, we have selected three of the possible values for x: -2, 0, and 3. The table shows that three ordered pairs that are solutions of $y = -4$ are $(-2, -4)$, $(0, -4)$, and $(3, -4)$. Drawing a line that passes through the three points gives the horizontal line shown in **Figure 2.27**.

x	$y = -4$	(x, y)
-2	-4	$(-2, -4)$
0	-4	$(0, -4)$
3	-4	$(3, -4)$

For all choices of x, y is a constant -4.

∎

✓ **CHECK POINT 6**   Graph $y = 3$ in the rectangular coordinate system.

Equation of a Horizontal Line

A horizontal line is given by an equation of the form

$$y = b,$$

where b is the y-intercept.

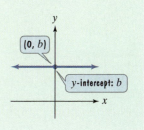

Because any vertical line can intersect the graph of a horizontal line $y = b$ only once, a horizontal line is the graph of a function. Thus, we can express the equation $y = b$ as $f(x) = b$. This linear function is often called a **constant function**.

Next, let's see what we can discover about the graph of an equation of the form $x = a$ by looking at an example.

EXAMPLE 7   Graphing a Vertical Line

Graph the linear equation: $x = 2$.

Solution   All ordered pairs that are solutions of $x = 2$ have a value of x that is always 2. Any value can be used for y. In the table on the right, we have selected three of the possible values for y: -2, 0, and 3. The table shows that three ordered pairs that are solutions of $x = 2$ are $(2, -2)$, $(2, 0)$, and $(2, 3)$. Drawing a line that passes through the three points gives the vertical line shown in **Figure 2.28**. ∎

For all choices of y,

x is always 2.

$x = 2$	y	(x, y)
2	-2	$(2, -2)$
2	0	$(2, 0)$
2	3	$(2, 3)$

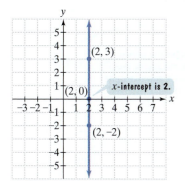

Figure 2.28 The graph of $x = 2$

Equation of a Vertical Line

A vertical line is given by an equation of the form

$$x = a,$$

where a is the x-intercept.

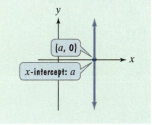

Does a vertical line represent the graph of a linear function? No. Look at the graph of $x = 2$ in **Figure 2.28**. A vertical line drawn through $(2, 0)$ intersects the graph infinitely many times. This shows that infinitely many outputs are associated with the input 2. **No vertical line is a linear function.**

✓ **CHECK POINT 7**   Graph the linear equation: $x = -3$.

Great Question!

How can equations of horizontal and vertical lines fit into the standard form of a line's equation, $Ax + By = C$, which contains two variables?

The linear equations in Examples 6 and 7, $y = -4$ and $x = 2$, each show only one variable.

- $y = -4$ means $0x + 1y = -4$.
- $x = 2$ means $1x + 0y = 2$.

6 Interpret slope as rate of change. ◉

Slope as Rate of Change

Slope is defined as the ratio of a change in y to a corresponding change in x. It describes how fast y is changing with respect to x. For a linear function, slope may be interpreted as the rate of change of the dependent variable per unit change in the independent variable.

Our next example shows how slope can be interpreted as a rate of change in an applied situation. When calculating slope in applied problems, keep track of the units in the numerator and the denominator.

EXAMPLE 8   Slope as a Rate of Change: Garbage Overload

The line graph in **Figure 2.29** shows the average waste, in pounds, produced per person per day in the United States for selected years from 1960 through 2012. Find the slope of the steepest line segment. Describe what this slope represents.

Average Waste Produced per Person per Day in the U.S.

Figure 2.29

Source: Environmental Protection Agency

Solution   We will let x represent a year and y the average waste produced per person per day for that year. The two points shown on the steepest line segment have the following coordinates:

$$(1985, 3.8) \quad \text{and} \quad (1990, 4.6).$$

In 1985, Americans produced 3.8 pounds of waste per day.

In 1990, Americans produced 4.6 pounds of waste per day.

Now we compute the slope.

$$m = \frac{\text{Change in } y}{\text{Change in } x} = \frac{4.6 - 3.8}{1990 - 1985}$$

The unit in the numerator is *pounds per day.*

The unit in the denominator is *year.*

$$= \frac{0.8}{5} = 0.16$$

The slope indicates that from 1985 through 1990, waste production for each American increased by 0.16 pound per day each year. The rate of change per person is an increase of 0.16 pound per day per year. ∎

☑ **CHECK POINT 8**   Use the graph in **Figure 2.29** to find the slope of the line segment that is decreasing the most over a five-year period. Describe what this slope represents.

7 Find a function's average rate of change.

The Average Rate of Change of a Function

If the graph of a function is not a straight line, the **average rate of change** between any two points is the slope of the line containing the two points. For example, **Figure 2.30** shows the graph of a particular man's height, in inches, as a function of his age, in years. Two points on the graph are labeled (13, 57) and (18, 76). At age 13, this man was 57 inches tall, and at age 18, he was 76 inches tall.

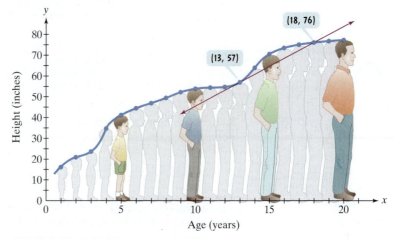

Figure 2.30 Height as a function of age

The man's average growth rate between ages 13 and 18 is the slope of the line containing (13, 57) and (18, 76):

$$m = \frac{\text{Change in } y}{\text{Change in } x} = \frac{76 - 57}{18 - 13} = \frac{19}{5} = 3\frac{4}{5}.$$

This man's average rate of change, or average growth rate, from age 13 to age 18 was $3\frac{4}{5}$, or 3.8, inches per year.

For any function, $y = f(x)$, the slope of the line between any two points on its graph is the **average change in y per unit change in x**.

EXAMPLE 9   Finding the Average Rate of Change

When a person receives a drug injected into a muscle, the concentration of the drug in the body, measured in milligrams per 100 milliliters, is a function of the time elapsed after the injection, measured in hours. **Figure 2.31** shows the graph of such a function, where x represents hours after the injection and $f(x)$ is the drug's concentration at time x. Find the average rate of change in the drug's concentration between 3 and 7 hours.

Solution   At 3 hours, the drug's concentration is 0.05 and at 7 hours, the concentration is 0.02. The average rate of change in its concentration between 3 and 7 hours is the slope of the line connecting the points (3, 0.05) and (7, 0.02).

Figure 2.31 Concentration of a drug as a function of time

$$m = \frac{\text{Change in } y}{\text{Change in } x} = \frac{0.02 - 0.05}{7 - 3} = \frac{-0.03}{4} = -0.0075$$

The average rate of change is -0.0075. This means that the drug's concentration is decreasing at an average rate of 0.0075 milligram per 100 milliliters per hour. ■

Figure 2.31 (repeated)

Great Question!

Can you clarify how you determine the units that you use when describing slope as a rate of change?

Units used to describe *x* and *y* tend to "pile up" when expressing the rate of change of *y* with respect to *x*. The unit used to express the rate of change of *y* with respect to *x* is

In Example 9, the rate of change is described in terms of milligrams per 100 milliliters per hour.

✓ **CHECK POINT 9**   Use **Figure 2.31** to find the average rate of change in the drug's concentration between 1 hour and 3 hours.

Blitzer Bonus ⊙

How Calculus Studies Change

Take a rapid sequence of still photographs of a moving scene and project them onto a screen at thirty shots a second or faster. Our eyes see the results as continuous motion. The small difference between one frame and the next cannot be detected by the human visual system. The idea of calculus likewise regards continuous motion as made up of a sequence of still configurations. Calculus masters the mystery of movement by "freezing the frame" of a continuous changing process, instant by instant. For example, **Figure 2.32** shows a male's changing height over intervals of time. Over the period of time from *P* to *D*, his average rate of growth is his change in height—that is, his height at time *D* minus his height at time *P*— divided by the change in time from *P* to *D*. This is the slope of line *PD*.

The lines *PD, PC, PB,* and *PA* shown in **Figure 2.32** have slopes that show average growth rates for successively shorter periods of time. Calculus makes these time frames so small that they approach a single point—that is, a single instant in time. This point is shown as point *P* in **Figure 2.32**. The slope of the line that touches the graph at *P* gives the male's growth rate at one instant in time, *P*.

Figure 2.32 Analyzing continuous growth over intervals of time and at an instant in time

8 Use slope and *y*-intercept to model data. ⊙

Modeling Data with the Slope-Intercept Form of the Equation of a Line

Our planet has been heating up for more than a century. Most experts have concluded that the increase in the amount of carbon dioxide in the atmosphere has been at least partly responsible for the increase in global surface temperature. The bar graph in **Figure 2.33(a)** shows the average global temperature for selected years from 1900 through 2014. The data are displayed as a set of four points in the scatter plot in **Figure 2.33(b)**.

Also shown on the scatter plot in **Figure 2.33(b)** is a line that passes through or near the four points. Linear functions are useful for modeling data that fall on or near a line. Example 10 illustrates how we can use the equation $y = mx + b$ to obtain a model for the data and make predictions about climate change in the future.

Average Global Temperature

Figure 2.33(a)

Source: National Oceanic and Atmospheric Administration

Figure 2.33(b)

EXAMPLE 10 Modeling with the Slope-Intercept Form of the Equation

a. Use the scatter plot in **Figure 2.33(b)** to find a function in the form $T(x) = mx + b$ that models average global temperature, $T(x)$, in degrees Fahrenheit, x years after 1900.

b. If the trend shown by the data continues, use the model to project the average global temperature in 2050.

Solution

a. We will use the line passing through the points $(0, 56.58)$ and $(114, 58.34)$ to obtain a model. We need values for m, the slope, and b, the *y*-intercept.

$$T(x) = mx + b$$

$$m = \frac{\text{Change in } y}{\text{Change in } x}$$

The point $(0, 56.58)$ lies on the line, so the *y*-intercept is 56.58: $b = 56.58$.

$$m = \frac{58.34 - 56.58}{114 - 0} \approx 0.02$$

Average global temperature, $T(x)$, in degrees Fahrenheit (°F), x years after 1900 can be modeled by the linear function

$$T(x) = 0.02x + 56.58.$$

The slope, approximately 0.02, indicates an increase in average global temperature of 0.02° Fahrenheit per year from 1900 through 2014.

b. Now let's use the function $T(x) = 0.02x + 56.58$ to project average global temperature in 2050. Because 2050 is 150 years after 1900, substitute 150 for x and evaluate the function at 150.

$$T(150) = 0.02(150) + 56.58 = 3 + 56.58 = 59.58$$

Our model projects an average global temperature of 59.58°F in 2050. ∎

☑ **CHECK POINT 10** In our chapter opener, we noted that most scientists believe that there is a strong correlation between carbon dioxide concentration and global temperature. The amount of carbon dioxide in the atmosphere, measured in parts per million, has been increasing as a result of the burning of oil and coal. The buildup of gases and particles is believed to trap heat and raise the planet's temperature. The pre-industrial concentration of atmospheric carbon dioxide was 280 parts per million. The bar graph in **Figure 2.34(a)** shows the average atmospheric concentration of carbon dioxide, in parts per million, for selected years from 1950 through 2010. The data are displayed as a set of four points in the scatter plot in **Figure 2.34(b)**.

Average Atmospheric Concentration of Carbon Dioxide

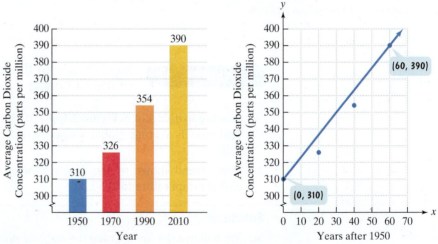

Figure 2.34(a)   Figure 2.34(b)

Source: National Oceanic and Atmospheric Administration

a. Use the line in the scatter plot in **Figure 2.34(b)** to find a function in the form $C(x) = mx + b$ that models average atmospheric concentration of carbon dioxide, $C(x)$, in parts per million, x years after 1950. Round m to two decimal places.

b. If the trend shown by the data continues, use your model from part (a) to project the average atmospheric concentration of carbon dioxide in 2050.

CONCEPT AND VOCABULARY CHECK

Fill in each blank so that the resulting statement is true.

1. Data presented in a visual form as a set of points is called a/an _____. A line that best fits this set of points is called a/an _____ line.

2. The equation $Ax + By = C$, where A and B are not both zero, is called the _____ form of the equation of a line.

3. The x-coordinate of a point where a graph crosses the x-axis is called a/an _____. The y-coordinate of such a point is always _____.

4. The y-coordinate of a point where a graph crosses the y-axis is called a/an _____. The x-coordinate of such a point is always _____.

5. The slope, m, of a line through the distinct points (x_1, y_1) and (x_2, y_2) is given by the formula $m =$ _____.

6. If a line rises from left to right, the line has _____ slope.

7. If a line falls from left to right, the line has _____ slope.

8. The slope of a horizontal line is _____.

9. The slope of a vertical line is _____.

10. The slope-intercept form of the equation of a line is _____, where m represents the _____ and b represents the _____.

11. In order to graph the line whose equation is $y = \frac{2}{5}x + 3$, begin by plotting the point _____. From this point, we move _____ units up (the rise) and _____ units to the right (the run).

12. The graph of the equation $y = 3$ is a/an _____ line.

13. The graph of the equation $x = -2$ is a/an _____ line.

14. The slope of the line through the distinct points (x_1, y_1) and (x_2, y_2) can be interpreted as the rate of change in _____ with respect to _____.

2.4 EXERCISE SET   ▶ MyMathLab®

Practice Exercises

In Exercises 1–14, use intercepts and a checkpoint to graph each linear function.

1. $x + y = 4$
2. $x + y = 2$
3. $x + 3y = 6$
4. $2x + y = 4$
5. $6x - 2y = 12$
6. $6x - 9y = 18$
7. $3x - y = 6$
8. $x - 4y = 8$
9. $x - 3y = 9$
10. $2x - y = 5$
11. $2x = 3y + 6$
12. $3x = 5y - 15$
13. $6x - 3y = 15$
14. $8x - 2y = 12$

In Exercises 15–26, find the slope of the line passing through each pair of points or state that the slope is undefined. Then indicate whether the line through the points rises, falls, is horizontal, or is vertical.

15. $(2, 4)$ and $(3, 8)$
16. $(3, 1)$ and $(5, 4)$
17. $(-1, 4)$ and $(2, 5)$
18. $(-3, -2)$ and $(2, 5)$
19. $(2, 5)$ and $(-1, 5)$
20. $(-6, -3)$ and $(4, -3)$
21. $(-7, 1)$ and $(-4, -3)$
22. $(2, -1)$ and $(-6, 3)$
23. $(-7, -4)$ and $(-3, 6)$
24. $(-3, -4)$ and $(-1, 6)$
25. $\left(\frac{7}{2}, -2\right)$ and $\left(\frac{7}{2}, \frac{1}{4}\right)$
26. $\left(\frac{3}{2}, -6\right)$ and $\left(\frac{3}{2}, \frac{1}{6}\right)$

In Exercises 27–28, find the slope of each line.

27.

28.

In Exercises 29–40, give the slope and y-intercept of each line whose equation is given. Then graph the linear function.

29. $y = 2x + 1$

30. $y = 3x + 2$

31. $y = -2x + 1$

32. $y = -3x + 2$

33. $f(x) = \dfrac{3}{4}x - 2$

34. $f(x) = \dfrac{3}{4}x - 3$

35. $f(x) = -\dfrac{3}{5}x + 7$

36. $f(x) = -\dfrac{2}{5}x + 6$

37. $y = -\dfrac{1}{2}x$

38. $y = -\dfrac{1}{3}x$

39. $y = -\dfrac{1}{2}$

40. $y = -\dfrac{1}{3}$

In Exercises 41–48,

  a. *Rewrite the given equation in slope-intercept form by solving for y.*

  b. *Give the slope and y-intercept.*

  c. *Use the slope and y-intercept to graph the linear function.*

41. $2x + y = 0$

42. $3x + y = 0$

43. $5y = 4x$

44. $4y = 3x$

45. $3x + y = 2$

46. $2x + y = 4$

47. $5x + 3y = 15$

48. $7x + 2y = 14$

In Exercises 49–62, graph each equation in a rectangular coordinate system.

49. $y = 3$

50. $y = 5$

51. $f(x) = -2$

52. $f(x) = -4$

53. $3y = 18$

54. $5y = -30$

55. $f(x) = 2$

56. $f(x) = 1$

57. $x = 5$

58. $x = 4$

59. $3x = -12$

60. $4x = -12$

61. $x = 0$

62. $y = 0$

Practice PLUS

In Exercises 63–66, find the slope of the line passing through each pair of points or state that the slope is undefined. Assume that all variables represent positive real numbers. Then indicate whether the line through the points rises, falls, is horizontal, or is vertical.

63. $(0, a)$ and $(b, 0)$

64. $(-a, 0)$ and $(0, -b)$

65. (a, b) and $(a, b + c)$

66. $(a - b, c)$ and $(a, a + c)$

In Exercises 67–68, give the slope and y-intercept of each line whose equation is given. Assume that $B \neq 0$.

67. $Ax + By = C$

68. $Ax = By - C$

In Exercises 69–70, find the value of y if the line through the two given points is to have the indicated slope.

69. $(3, y)$ and $(1, 4), m = -3$

70. $(-2, y)$ and $(4, -4), m = \dfrac{1}{3}$

In Exercises 71–72, graph each linear function.

71. $3x - 4f(x) = 6$

72. $6x - 5f(x) = 20$

73. If one point on a line is $(3, -1)$ and the line's slope is -2, find the y-intercept.

74. If one point on a line is $(2, -6)$ and the line's slope is $-\dfrac{3}{2}$, find the y-intercept.

Use the figure to make the lists in Exercises 75–76.

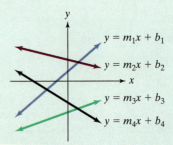

75. List the slopes $m_1, m_2, m_3,$ and m_4 in order of decreasing size.

76. List the y-intercepts $b_1, b_2, b_3,$ and b_4 in order of decreasing size.

Application Exercises

In Exercises 77–80, a linear function that models data is described. Find the slope of each model. Then describe what this means in terms of the rate of change of the dependent variable per unit change in the independent variable.

77. The linear function $f(x) = 55.7x + 60.1$ models the number of smartphones sold in the United States, $f(x)$, in millions, x years after 2006. (*Source: Newsweek, April 11, 2011*)

78. The linear function $f(x) = 2x + 10$ models the amount, $f(x)$, in billions of dollars, that the drug industry spent on marketing information about drugs to doctors x years after 2000. (*Source: IMS Health*)

79. The linear function $f(x) = -0.52x + 24.7$ models the percentage of U.S. adults who smoked cigarettes, $f(x)$, x years after 1997. (*Source: National Center for Health Statistics*)

80. The linear function $f(x) = -0.28x + 1.7$ models the percentage of U.S. taxpayers who were audited by the IRS, $f(x)$, x years after 1996. (*Source: IRS*)

Divorce rates are typically higher for couples who marry in their teens. The graph shows the percentage of marriages ending in divorce by wife's age at marriage. Use the information shown to solve Exercises 81–82.

Percentage of Marriages Ending in Divorce by Wife's Age at Marriage

Source: National Center for Health Statistics

81. **a.** What percentage of marriages in which the wife is under 18 when she marries end in divorce within the first five years?

   b. What percentage of marriages in which the wife is under 18 when she marries end in divorce within the first ten years?

   c. Find the average rate of change in the percentage of marriages ending in divorce between five and ten years of marriage in which the wife is under 18 when she marries.

82. **a.** What percentage of marriages in which the wife is over age 25 when she marries end in divorce within the first five years?

   b. What percentage of marriages in which the wife is over age 25 when she marries end in divorce within the first ten years?

   c. Find the average rate of change in the percentage of marriages ending in divorce between five and ten years of marriage in which the wife is over age 25 when she marries.

83. Shown, again, is the scatter plot that indicates a relationship between the percentage of adult females in a country who are literate and the mortality of children under five. Also shown is a line that passes through or near the points.

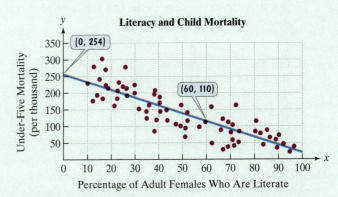

Literacy and Child Mortality

Source: United Nations Development Programme

   a. According to the graph, what is the y-intercept of the line? Describe what this represents in this situation.

   b. Use the coordinates of the two points shown to compute the slope of the line. Describe what this means in terms of the rate of change.

   c. Use the y-intercept from part (a) and the slope from part (b) to write a linear function that models child mortality, $f(x)$, per thousand, for children under five in a country where $x\%$ of adult women are literate.

   d. Use the function from part (c) to predict the mortality rate of children under five in a country where 50% of adult females are literate.

84. A wave of immigration from the Caribbean, Africa, and Latin America is reshaping America's black population. In 2013, approximately one in 11 blacks in America was foreign-born. The figure is projected to rise to one in six by 2060. The bar graph on the left shows the percentage of the U.S. black population that was foreign-born in four selected years from 1980 through 2013. The data are displayed as a set of four points in the scatter plot on the right.

Percentage of U.S. Black Population That Is Foreign-Born

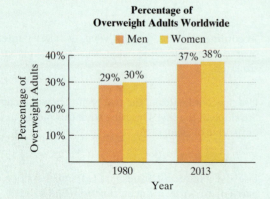

Source: Pew Research Center

a. Use the coordinates of the two points shown in the scatter plot to compute the slope of the line. Describe what this means in terms of the rate of change.

b. Use the *y*-intercept shown in the scatter plot and the slope from part (a) to write a linear function that models the percentage of America's black population, $P(x)$, that was foreign-born *x* years after 1980.

c. Use the function from part (b) to project the percentage of America's black population that will be foreign-born in 2060.

The bar graph shows the percentage of the world's adults who were overweight (a body mass index of 25 or greater) by gender, for two selected years.

Percentage of Overweight Adults Worldwide

■ Men   ■ Women

1980: Men 29%, Women 30%
2013: Men 37%, Women 38%

Year

Source: TheLancet.com

In Exercises 85–86, find a linear function in slope-intercept form that models the given description. Each function should model the percentage of overweight men or women worldwide, P(x), x years after 1980.

85. In 1980, 29% of adult men were overweight. This has increased at an average rate of approximately 0.24% per year since then.

86. In 1980, 30% of adult women were overweight. This has increased at an average rate of approximately 0.24% per year since then.

Explaining the Concepts

87. What is a scatter plot?

88. What is a regression line?

89. What is the standard form of the equation of a line?

90. What is an *x*-intercept of a graph?

91. What is a *y*-intercept of a graph?

92. If you are given the standard form of the equation of a line, explain how to find the *x*-intercept.

93. If you are given the standard form of the equation of a line, explain how to find the *y*-intercept.

94. What is the slope of a line?

95. Describe how to calculate the slope of a line passing through two points.

96. What does it mean if the slope of a line is zero?

97. What does it mean if the slope of a line is undefined?

98. Describe how to find the slope of a line whose equation is given.

99. Describe how to graph a line using the slope and *y*-intercept. Provide an original example with your description.

100. Describe the graph of $y = b$.

101. Describe the graph of $x = a$.

102. If the graph of a function is not a straight line, explain how to find the average rate of change between two points.

103. Take another look at the scatter plot in Exercise 83. Although there is a relationship between literacy and child mortality, we cannot conclude that increased literacy causes child mortality to decrease. Offer two or more possible explanations for the data in the scatter plot.

Technology Exercises

104. Use a graphing utility to verify any three of your hand-drawn graphs in Exercises 1–14. Solve the equation for *y* before entering it.

In Exercises 105–108, use a graphing utility to graph each linear function. Then use the $\boxed{\text{TRACE}}$ *feature to trace along the line and find the coordinates of two points. Use these points to compute the line's slope. Check your result by using the coefficient of x in the line's equation.*

105. $y = 2x + 4$ **106.** $y = -3x + 6$

107. $f(x) = -\dfrac{1}{2}x - 5$

108. $f(x) = \dfrac{3}{4}x - 2$

Critical Thinking Exercises

Make Sense? *In Exercises 109–112, determine whether each statement makes sense or does not make sense, and explain your reasoning.*

109. The graph of my linear function at first rose from left to right, reached a maximum point, and then fell from left to right.

110. A linear function that models tuition and fees at public four-year colleges from 2000 through 2010 has negative slope.

111. The function $S(x) = 49,100x + 1700$ models the average salary for a college professor, $S(x)$, x years after 2000.

112. The federal minimum wage was \$5.15 per hour from 1997 through 2006, so $f(x) = 5.15$ models the minimum wage, $f(x)$, in dollars, for the domain {1997, 1998, 1999, ... , 2006}.

In Exercises 113–116, determine whether each statement is true or false. If the statement is false, make the necessary change(s) to produce a true statement.

113. A linear function with nonnegative slope has a graph that rises from left to right.

114. Every line in the rectangular coordinate system has an equation that can be expressed in slope-intercept form.

115. The graph of the linear function $5x + 6y = 30$ is a line passing through the point $(6, 0)$ with slope $-\dfrac{5}{6}$.

116. The graph of $x = 7$ in the rectangular coordinate system is the single point $(7, 0)$.

In Exercises 117–118, find the coefficients that must be placed in each shaded area so that the function's graph will be a line satisfying the specified conditions.

117. ▮ $x +$ ▮ $y = 12$; x-intercept $= -2$; y-intercept $= 4$

118. ▮ $x +$ ▮ $y = 12$; y-intercept $= -6$; slope $= \dfrac{1}{2}$

119. For the linear function

$$f(x) = mx + b,$$

a. Find $f(x_1 + x_2)$.

b. Find $f(x_1) + f(x_2)$.

c. Is $f(x_1 + x_2) = f(x_1) + f(x_2)$?

Review Exercises

120. Simplify: $\left(\dfrac{4x^2}{y^{-3}}\right)^2$. (Section 1.6, Example 9)

121. Multiply and write the answer in scientific notation:

$$(8 \times 10^{-7})(4 \times 10^3).$$

(Section 1.7, Example 4)

122. Simplify: $5 - [3(x - 4) - 6x]$. (Section 1.2, Example 14)

Preview Exercises

Exercises 123–125 will help you prepare for the material covered in the next section.

123. Write the equation $y - 5 = 7(x + 4)$ in slope-intercept form.

124. Write the equation $y + 3 = -\dfrac{7}{3}(x - 1)$ in slope-intercept form.

125. The equation of a line is $x + 4y - 8 = 0$.

a. Write the equation in slope-intercept form and determine the slope.

b. The product of the line's slope in part (a) and the slope of a second line is -1. What is the slope of the second line?

SECTION

2.5

The Point-Slope Form of the Equation of a Line ▶

What am I supposed to learn?

After studying this section, you should be able to:

1 Use the point-slope form to write equations of a line. ▶

2 Model data with linear functions and make predictions. ▶

3 Find slopes and equations of parallel and perpendicular lines. ▶

If present trends continue, is it possible that our descendants could live to be 200 years of age? To answer this question, we need to develop a function that models life expectancy by birth year. In this section, you will learn to use another form of a line's equation to obtain functions that model data.

Figure 2.35 A line passing through (x_1, y_1) with slope m

Point-Slope Form

We can use the slope of a line to obtain another useful form of the line's equation. Consider a nonvertical line that has slope m and contains the point (x_1, y_1). Now, let (x, y) represent any other point on the line, shown in **Figure 2.35**. Keep in mind that the point (x, y) is arbitrary and is not in one fixed position. By contrast, the point (x_1, y_1) is fixed.

Regardless of where the point (x, y) is located, the steepness of the line in **Figure 2.35** remains the same. Thus, the ratio for slope stays a constant m. This means that for all points (x, y) along the line

$$m = \frac{\text{Change in } y}{\text{Change in } x} = \frac{y - y_1}{x - x_1}.$$

We can clear the fraction by multiplying both sides by $x - x_1$, the least common denominator, where $x - x_1 \neq 0$.

$$m = \frac{y - y_1}{x - x_1} \qquad \text{This is the slope of the line in Figure 2.35.}$$

$$m(x - x_1) = \frac{y - y_1}{x - x_1} \cdot (x - x_1) \qquad \text{Multiply both sides by } x - x_1.$$

$$m(x - x_1) = y - y_1 \qquad \text{Simplify: } \frac{y - y_1}{x - x_1} \cdot (x - x_1) = y - y_1.$$

Now, if we reverse the two sides, we obtain the **point-slope form** of the equation of a line.

Great Question!

When using $y - y_1 = m(x - x_1)$, for which variables do I substitute numbers?

When writing the point-slope form of a line's equation, you will never substitute numbers for x and y. You will substitute values for x_1, y_1, and m.

> ### Point-Slope Form of the Equation of a Line
>
> The **point-slope form of the equation** of a nonvertical line with slope m that passes through the point (x_1, y_1) is
>
> $$y - y_1 = m(x - x_1).$$

For example, the point-slope form of the equation of the line passing through $(1, 5)$ with slope 2 $(m = 2)$ is

$$y - 5 = 2(x - 1).$$

1 Use the point-slope form to write equations of a line. ▶

Using the Point-Slope Form to Write a Line's Equation

If we know the slope of a line and a point not containing the y-intercept through which the line passes, the point-slope form is the equation that we should use. Once we have obtained this equation, it is customary to solve for y and write the equation in slope-intercept form. Examples 1 and 2 illustrate these ideas.

EXAMPLE 1 Writing the Point-Slope Form and the Slope-Intercept Form

Write the point-slope form and the slope-intercept form of the equation of the line with slope 7 that passes through the point $(-4, 5)$.

Solution We begin with the point-slope form of the equation of a line with $m = 7, x_1 = -4$, and $y_1 = 5$.

$$y - y_1 = m(x - x_1) \qquad \text{This is the point-slope form of the equation.}$$

$$y - 5 = 7[x - (-4)] \qquad \text{Substitute the given values.}$$

$$y - 5 = 7(x + 4) \qquad \text{We now have the point-slope form of the equation of the given line.}$$

Now we solve $y - 5 = 7(x + 4)$, the point-slope form, for y and write an equivalent equation in slope-intercept form ($y = mx + b$).

We need to isolate y.	$y - 5 = 7(x + 4)$	This is the point-slope form of the equation.
	$y - 5 = 7x + 28$	Use the distributive property.
	$y = 7x + 33$	Add 5 to both sides.

The slope-intercept form of the line's equation is $y = 7x + 33$. Using function notation, the equation is $f(x) = 7x + 33$. ∎

✓ **CHECK POINT 1**   Write the point-slope form and the slope-intercept form of the equation of the line with slope -2 that passes through the point $(4, -3)$.

EXAMPLE 2   Writing the Point-Slope Form and the Slope-Intercept Form

A line passes through the points $(1, -3)$ and $(-2, 4)$. (See **Figure 2.36**.) Find an equation of the line

a. in point-slope form.   **b.** in slope-intercept form.

Solution

a. To use the point-slope form, we need to find the slope. The slope is the change in the y-coordinates divided by the corresponding change in the x-coordinates.

$$m = \frac{4 - (-3)}{-2 - 1} = \frac{7}{-3} = -\frac{7}{3}$$

This is the definition of slope using $(1, -3)$ and $(-2, 4)$.

We can take either point on the line to be (x_1, y_1). Let's use $(x_1, y_1) = (1, -3)$. Now, we are ready to write the point-slope form of the equation.

$$y - y_1 = m(x - x_1)$$ This is the point-slope form of the equation.

$$y - (-3) = -\frac{7}{3}(x - 1)$$ Substitute: $(x_1, y_1) = (1, -3)$ and $m = -\frac{7}{3}$.

$$y + 3 = -\frac{7}{3}(x - 1)$$ Simplify.

This equation is the point-slope form of the equation of the line shown in **Figure 2.36**.

b. Now, we solve this equation for y and write an equivalent equation in slope-intercept form ($y = mx + b$).

We need to isolate y.	$y + 3 = -\frac{7}{3}(x - 1)$	This is the point-slope form of the equation.
	$y + 3 = -\frac{7}{3}x + \frac{7}{3}$	Use the distributive property.
	$y = -\frac{7}{3}x - \frac{2}{3}$	Subtract 3 from both sides: $\frac{7}{3} - 3 = \frac{7}{3} - \frac{9}{3} = -\frac{2}{3}$.

This equation is the slope-intercept form of the equation of the line shown in **Figure 2.36**. Using function notation, the equation is $f(x) = -\frac{7}{3}x - \frac{2}{3}$. ∎

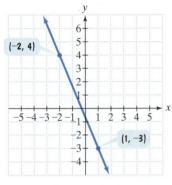

Figure 2.36 Line passing through $(1, -3)$ and $(-2, 4)$

Discover for Yourself

If you are given two points on a line, you can use either point for (x_1, y_1) when you write the point-slope form of its equation. Rework Example 2 using $(-2, 4)$ for (x_1, y_1). Once you solve for y, you should obtain the same slope-intercept form of the equation as the one shown in the last line of the solution to Example 2.

✓ | **CHECK POINT 2**   A line passes through the points $(6, -3)$ and $(2, 5)$. Find an equation of the line

   a. in point-slope form.

   b. in slope-intercept form.

Here is a summary of the various forms for equations of lines:

> ### Equations of Lines
>
> **1.** Standard form: $Ax + By = C$
>
> **2.** Slope-intercept form: $y = mx + b$ or $f(x) = mx + b$
>
> **3.** Horizontal line: $y = b$
>
> **4.** Vertical line: $x = a$
>
> **5.** Point-slope form: $y - y_1 = m(x - x_1)$

In Examples 1 and 2, we eventually wrote a line's equation in slope-intercept form, or in function notation. But where do we start our work?

Starting with $y = mx + b$	Starting with $y - y_1 = m(x - x_1)$
Begin with the slope-intercept form if you know	Begin with the point-slope form if you know
• The slope of the line and the y-intercept	• The slope of the line and a point on the line not containing the y-intercept
or	or
• Two points on the line, one of which contains the y-intercept.	• Two points on the line, neither of which contains the y-intercept.

2 Model data with linear functions and make predictions. ▶

Applications

We have seen that linear functions are useful for modeling data that fall on or near a line. For example, the bar graph in **Figure 2.37(a)** gives the life expectancy for American men and women born in the indicated year. The data for the men are displayed as a set of six points in the scatter plot in **Figure 2.37(b)**.

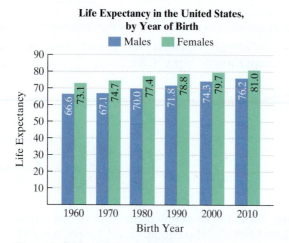

Figure 2.37(a)   Data for men and women

Source: National Center for Health Statistics

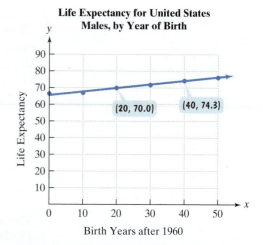

Figure 2.37(b)   A scatter plot for the men's data

Also shown on the scatter plot in **Figure 2.37(b)** is a line that passes through or near the six points. By writing the equation of this line, we can obtain a model for life expectancy and make predictions about how long American men will live in the future.

EXAMPLE 3   Modeling Life Expectancy

Write the slope-intercept form of the equation of the line shown in **Figure 2.37(b)**. Use the equation to predict the life expectancy of an American man born in 2020.

Solution   The line in **Figure 2.37(b)** passes through $(20, 70.0)$ and $(40, 74.3)$. We start by finding its slope.

$$m = \frac{\text{Change in } y}{\text{Change in } x} = \frac{74.3 - 70.0}{40 - 20} = \frac{4.3}{20} = 0.215$$

The slope indicates that for each subsequent birth year, a man's life expectancy is increasing by 0.215 year.

Now we write the line's equation in slope-intercept form.

$y - y_1 = m(x - x_1)$	Begin with the point-slope form.
$y - 70.0 = 0.215(x - 20)$	Either ordered pair can be (x_1, y_1). Let $(x_1, y_1) = (20, 70.0)$. From above, $m = 0.215$.
$y - 70.0 = 0.215x - 4.3$	Apply the distributive property: $0.215(20) = 4.3$.
$y = 0.215x + 65.7$	Add 70 to both sides and solve for y.

A linear function that models life expectancy, $f(x)$, for American men born x years after 1960 is

$$f(x) = 0.215x + 65.7.$$

Now let's use this function to predict the life expectancy of an American man born in 2020. Because 2020 is 60 years after 1960, substitute 60 for x and evaluate the function at 60.

$$f(60) = 0.215(60) + 65.7 = 78.6$$

Our model predicts that American men born in 2020 will have a life expectancy of 78.6 years.   ■

Using Technology

You can use a graphing utility to obtain a model for a scatter plot in which the data points fall on or near a straight line. After entering the data for men in **Figure 2.37(a)**, a graphing utility displays a scatter plot of the data and the regression line, that is, the line that best fits the data.

[0, 50, 10] by [0, 90, 10]

Also displayed is the regression line's equation.

```
        LinReg
y=ax+b
a=.204
b=65.9
```

✓ **CHECK POINT 3**   The data for the life expectancy for American women are displayed as a set of six points in the scatter plot in **Figure 2.38**. Also shown is a line that passes through or near the six points. Use the data points labeled by the voice balloons to write the slope-intercept form of the equation of this line. Round the slope to two decimal places. Then use the linear function to predict the life expectancy of an American woman born in 2020.

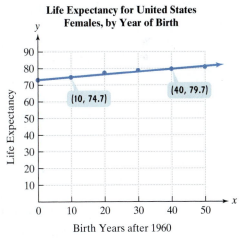

Figure 2.38

Blitzer Bonus ⊙

Predicting Your Own Life Expectancy

The models in Example 3 and Check Point 3 do not take into account your current health, lifestyle, and family history, all of which could increase or decrease your life expectancy. Thomas Perls at Boston University Medical School, who studies centenarians, developed a much more detailed model for life expectancy at livingto100.com. The model takes into account everything from your stress level to your sleep habits and gives you the exact age it predicts you will live to.

3 Find slopes and equations of parallel and perpendicular lines.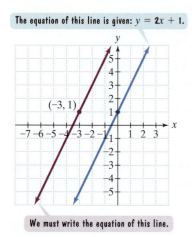

Parallel and Perpendicular Lines

Two nonintersecting lines that lie in the same plane are **parallel**. If two lines do not intersect, the ratio of the vertical change to the horizontal change is the same for each line. Because two parallel lines have the same "steepness," they must have the same slope.

> ### Slope and Parallel Lines
>
> 1. If two nonvertical lines are parallel, then they have the same slope.
> 2. If two distinct nonvertical lines have the same slope, then they are parallel.
> 3. Two distinct vertical lines, both with undefined slopes, are parallel.

The equation of this line is given: $y = 2x + 1$.

(−3, 1)

We must write the equation of this line.

Figure 2.39

EXAMPLE 4   Writing Equations of a Line Parallel to a Given Line

Write an equation of the line passing through $(-3, 1)$ and parallel to the line whose equation is $y = 2x + 1$. Express the equation in point-slope form and slope-intercept form.

Solution   The situation is illustrated in **Figure 2.39**. We are looking for the equation of the red line shown on the left. How do we obtain this equation? Notice that the line passes through the point $(-3, 1)$. Using the point-slope form of the line's equation, we have $x_1 = -3$ and $y_1 = 1$.

$$y - y_1 = m(x - x_1)$$

$$y_1 = 1 \qquad x_1 = -3$$

Now the only thing missing from the equation of the red line is m, the slope. Do we know anything about the slope of either line in **Figure 2.39**? The answer is yes; we know the slope of the blue line on the right, whose equation is given.

$$y = 2x + 1$$

The slope of the blue line on the right in **Figure 2.39** is 2.

Parallel lines have the same slope. Because the slope of the blue line is 2, the slope of the red line, the line whose equation we must write, is also 2: $m = 2$. We now have values for x_1, y_1, and m for the red line.

$$y - y_1 = m(x - x_1)$$

$$y_1 = 1 \qquad m = 2 \qquad x_1 = -3$$

The point-slope form of the red line's equation is

$$y - 1 = 2[x - (-3)] \text{ or}$$
$$y - 1 = 2(x + 3).$$

Solving for y, we obtain the slope-intercept form of the equation.

$$y - 1 = 2x + 6 \qquad \text{Apply the distributive property.}$$
$$y = 2x + 7 \qquad \text{Add 1 to both sides. This is the}$$

slope-intercept form, $y = mx + b$, of the equation. Using function notation, the equation is $f(x) = 2x + 7$. ∎

✓ **CHECK POINT 4**   Write an equation of the line passing through $(-2, 5)$ and parallel to the line whose equation is $y = 3x + 1$. Express the equation in point-slope form and slope-intercept form.

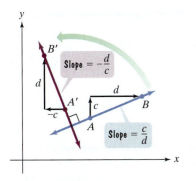

Figure 2.40 Slopes of perpendicular lines

Two lines that intersect at a right angle (90°) are said to be **perpendicular**, shown in **Figure 2.40**. The relationship between the slopes of perpendicular lines is not as obvious as the relationship between parallel lines. **Figure 2.40** shows line AB, with slope $\frac{c}{d}$. Rotate line AB through 90° counterclockwise to obtain line $A'B'$, perpendicular to line AB. The figure indicates that the rise and the run of the new line are reversed from the original line, but the run is now negative. This means that the slope of the new line is $-\frac{d}{c}$. Notice that the product of the slopes of the two perpendicular lines is -1:

$$\left(\frac{c}{d}\right)\left(-\frac{d}{c}\right) = -1.$$

This relationship holds for all nonvertical perpendicular lines and is summarized in the following box:

Slope and Perpendicular Lines

1. If two nonvertical lines are perpendicular, then the product of their slopes is -1.

2. If the product of the slopes of two lines is -1, then the lines are perpendicular.

3. A horizontal line having zero slope is perpendicular to a vertical line having undefined slope.

An equivalent way of stating this relationship between two nonvertical lines is to say that **one line is perpendicular to another line if its slope is the *negative reciprocal* of the slope of the other line.** For example, if a line has slope 5, any line having slope $-\frac{1}{5}$ is perpendicular to it. Similarly, if a line has slope $-\frac{3}{4}$, any line having slope $\frac{4}{3}$ is perpendicular to it.

EXAMPLE 5   Writing Equations of a Line Perpendicular to a Given Line

a. Find the slope of any line that is perpendicular to the line whose equation is $x + 4y = 8$.

b. Write the equation of the line passing through $(3, -5)$ and perpendicular to the line whose equation is $x + 4y = 8$. Express the equation in point-slope form and slope-intercept form.

Solution

a. We begin by writing the equation of the given line, $x + 4y = 8$, in slope-intercept form. Solve for y.

$x + 4y = 8$	This is the given equation.
$4y = -x + 8$	To isolate the y-term, subtract x from both sides.
$y = -\dfrac{1}{4}x + 2$	Divide both sides by 4.

Slope is $-\frac{1}{4}$.

The given line has slope $-\frac{1}{4}$. Any line perpendicular to this line has a slope that is the negative reciprocal of $-\frac{1}{4}$. Thus, the slope of any perpendicular line is 4.

b. Let's begin by writing the point-slope form of the perpendicular line's equation. Because the line passes through the point $(3, -5)$, we have $x_1 = 3$ and $y_1 = -5$. In part (a), we determined that the slope of any line perpendicular to $x + 4y = 8$ is 4, so the slope of this particular perpendicular line must be 4: $m = 4$.

$$y - y_1 = m(x - x_1)$$

$y_1 = -5$ $m = 4$ $x_1 = 3$

The point-slope form of the perpendicular line's equation is

$$y - (-5) = 4(x - 3) \quad \text{or} \quad y + 5 = 4(x - 3).$$

How can we express this equation in slope-intercept form, $y = mx + b$? We need to solve for y.

$$y + 5 = 4(x - 3) \qquad \text{This is the point-slope form of the line's equation.}$$
$$y + 5 = 4x - 12 \qquad \text{Apply the distributive property.}$$
$$y = 4x - 17 \qquad \text{Subtract 5 from both sides of the equation and solve for } y.$$

The slope-intercept form of the perpendicular line's equation is

$$y = 4x - 17 \quad \text{or} \quad f(x) = 4x - 17. \quad \blacksquare$$

✓ **CHECK POINT 5**

    a. Find the slope of any line that is perpendicular to the line whose equation is $x + 3y = 12$.

    b. Write the equation of the line passing through $(-2, -6)$ and perpendicular to the line whose equation is $x + 3y = 12$. Express the equation in point-slope form and slope-intercept form.

Achieving Success

Organizing and creating your own compact chapter summaries can reinforce what you know and help with the retention of this information. Imagine that your professor will permit two index cards of notes (3 by 5; front and back) on all exams. Organize and create such a two-card summary for the test on this chapter. Begin by determining what information you would find most helpful to include on the cards. Take as long as you need to create the summary. Based on how effective you find this strategy, you may decide to use the technique to help prepare for future exams.

CONCEPT AND VOCABULARY CHECK

Fill in each blank so that the resulting statement is true.

1. The point-slope form of the equation of a nonvertical line with slope m that passes through the point (x_1, y_1) is _____.

2. Two parallel lines have _____ slopes.

3. The product of the slopes of two nonvertical perpendicular lines is _____.

4. The negative reciprocal of 5 is _____.

5. The negative reciprocal of $-\dfrac{3}{5}$ is _____.

6. The slope of the line whose equation is $y = -4x + 3$ is _____. The slope of any line parallel to $y = -4x + 3$ is _____.

7. The slope of the line whose equation is $y = \dfrac{1}{2}x - 5$ is _____. The slope of any line perpendicular to $y = \dfrac{1}{2}x - 5$ is _____.

2.5 EXERCISE SET ▷ MyMathLab®

Practice Exercises

Write the point-slope form of the equation of the line satisfying each of the conditions in Exercises 1–28. Then use the point-slope form to write the slope-intercept form of the equation in function notation.

1. Slope = 3, passing through $(2, 5)$

2. Slope = 4, passing through $(3, 1)$

3. Slope = 5, passing through $(-2, 6)$

4. Slope = 8, passing through $(-4, 1)$

5. Slope = -4, passing through $(-3, -2)$

6. Slope = -6, passing through $(-2, -4)$

7. Slope = -5, passing through $(-2, 0)$

8. Slope = -4, passing through $(0, -3)$

9. Slope = -1, passing through $\left(-2, -\frac{1}{2}\right)$

10. Slope = -1, passing through $\left(-\frac{1}{4}, -4\right)$

11. Slope = $\frac{1}{4}$, passing through the origin

12. Slope = $\frac{1}{5}$, passing through the origin

13. Slope = $-\frac{2}{3}$, passing through $(6, -4)$

14. Slope = $-\frac{2}{5}$, passing through $(15, -4)$

15. Passing through $(6, 3)$ and $(5, 2)$

16. Passing through $(1, 3)$ and $(2, 4)$

17. Passing through $(-2, 0)$ and $(0, 4)$

18. Passing through $(2, 0)$ and $(0, -1)$

19. Passing through $(-6, 13)$ and $(-2, 5)$

20. Passing through $(-3, 2)$ and $(2, -8)$

21. Passing through $(1, 9)$ and $(4, -2)$

22. Passing through $(4, -8)$ and $(8, -3)$

23. Passing through $(-2, -5)$ and $(3, -5)$

24. Passing through $(-1, -4)$ and $(3, -4)$

25. Passing through $(7, 8)$ with x-intercept $= 3$

26. Passing through $(-4, 5)$ with y-intercept $= -3$

27. x-intercept $= 2$ and y-intercept $= -1$

28. x-intercept $= -2$ and y-intercept $= 4$

*In Exercises 29–44, the equation of a line is given. Find the slope of a line that is **a.** parallel to the line with the given equation; and **b.** perpendicular to the line with the given equation.*

29. $y = 5x$

30. $y = 3x$

31. $y = -7x$

32. $y = -9x$

33. $y = \frac{1}{2}x + 3$

34. $y = \frac{1}{4}x - 5$

35. $y = -\frac{2}{5}x - 1$

36. $y = -\frac{3}{7}x - 2$

37. $4x + y = 7$

38. $8x + y = 11$

39. $2x + 4y = 8$

40. $3x + 2y = 6$

41. $2x - 3y = 5$

42. $3x - 4y = -7$

43. $x = 6$

44. $y = 9$

In Exercises 45–48, write an equation for line L in point-slope form and slope-intercept form.

45.

L is parallel to y = 2x.

46.

L is parallel to y = -2x.

47.

L is perpendicular to $y = 2x$.

48.

L is perpendicular to $y = -2x$.

In Exercises 49–56, use the given conditions to write an equation for each line in point-slope form and slope-intercept form.

49. Passing through $(-8, -10)$ and parallel to the line whose equation is $y = -4x + 3$

50. Passing through $(-2, -7)$ and parallel to the line whose equation is $y = -5x + 4$

51. Passing through $(2, -3)$ and perpendicular to the line whose equation is $y = \frac{1}{5}x + 6$

52. Passing through $(-4, 2)$ and perpendicular to the line whose equation is $y = \frac{1}{3}x + 7$

53. Passing through $(-2, 2)$ and parallel to the line whose equation is $2x - 3y = 7$

54. Passing through $(-1, 3)$ and parallel to the line whose equation is $3x - 2y = 5$

55. Passing through $(4, -7)$ and perpendicular to the line whose equation is $x - 2y = 3$

56. Passing through $(5, -9)$ and perpendicular to the line whose equation is $x + 7y = 12$

Practice PLUS

In Exercises 57–64, write the slope-intercept form of the equation of a function f whose graph satisfies the given conditions.

57. The graph of *f* passes through $(-1, 5)$ and is perpendicular to the line whose equation is $x = 6$.

58. The graph of *f* passes through $(-2, 6)$ and is perpendicular to the line whose equation is $x = -4$.

59. The graph of *f* passes through $(-6, 4)$ and is perpendicular to the line that has an *x*-intercept of 2 and a *y*-intercept of -4.

60. The graph of *f* passes through $(-5, 6)$ and is perpendicular to the line that has an *x*-intercept of 3 and a *y*-intercept of -9.

61. The graph of *f* is perpendicular to the line whose equation is $3x - 2y = 4$ and has the same *y*-intercept as this line.

62. The graph of *f* is perpendicular to the line whose equation is $4x - y = 6$ and has the same *y*-intercept as this line.

63. The graph of *f* is the graph of $g(x) = 4x - 3$ shifted down 2 units.

64. The graph of *f* is the graph of $g(x) = 2x - 5$ shifted up 3 units.

65. What is the slope of a line that is parallel to the line whose equation is $Ax + By = C, B \neq 0$?

66. What is the slope of a line that is perpendicular to the line whose equation is $Ax + By = C, A \neq 0$ and $B \neq 0$?

Application Exercises

Americans are getting married later in life, or not getting married at all. In 2010, more than half of Americans ages 25 through 29 were unmarried. The bar graph shows the percentage of never-married men and women in this age group for four selected years. The data are displayed as two sets of four points each, one scatter plot for the percentage of never-married American men and one for the percentage of never-married American women. Also shown for each scatter plot is a line that passes through or near the four points. Use these lines to solve Exercises 67–68 on the next page.

Percentage of United States Population Never Married, Ages 25–29

Source: U.S. Census Bureau

67. In this exercise, you will use the blue line for the women shown on the scatter plot on the previous page to develop a model for the percentage of never-married American females ages 25–29.

a. Use the two points whose coordinates are shown by the voice balloons to find the point-slope form of the equation of the line that models the percentage of never-married American females ages 25–29, *y*, *x* years after 1980.

b. Write the equation from part (a) in slope-intercept form. Use function notation.

c. Use the linear function to predict the percentage of never-married American females, ages 25–29, in 2020.

68. In this exercise, you will use the red line for the men shown on the scatter plot on the previous page to develop a model for the percentage of never-married American males ages 25–29.

a. Use the two points whose coordinates are shown by the voice balloons to find the point-slope form of the equation of the line that models the percentage of never-married American males ages 25–29, *y*, *x* years after 1980.

b. Write the equation from part (a) in slope-intercept form. Use function notation.

c. Use the linear function to predict the percentage of never-married American males, ages 25–29, in 2015.

Phone Fight! *Some of the smartphone giants, such as Motorola, Panasonic, Microsoft, and Apple, are suing each other, charging competitors with stealing their patents. Do more phones mean more lawyers? Think about this as you create linear models for the number of smartphones sold (Exercise 69) and the number of infringement filings by smartphone companies (Exercise 70).*

69. The bar graph shows the number of smartphones sold in the United States from 2004 through 2010.

Number of Smartphones Sold in the United States

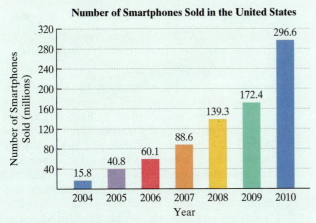

Source: *Newsweek*, April 11, 2011

a. Let *x* represent the number of years after 2004 and let *y* represent the number of smartphones sold, in millions, in the United States. Create a scatter plot that displays the data as a set of seven points in a rectangular coordinate system.

b. Draw a line through the two points that show the number of smartphones sold in 2005 and 2010. Use the coordinates of these points to write the line's

equation in point-slope form and slope-intercept form.

c. Use the slope-intercept form of the equation from part (b) to project the number of smartphones sold in the United States in 2015.

70. The bar graph shows the number of U.S. lawsuits by smartphone companies for patent infringement from 2004 through 2010.

Number of U.S. Lawsuits by Smartphone Companies Charging Competitors with Stealing Their Patents

Source: *Newsweek*, April 11, 2011

a. Let *x* represent the number of years after 2004 and let *y* represent the number of lawsuits. Create a scatter plot that displays the data as a set of seven points in a rectangular coordinate system.

b. Draw a line through the two points that show the number of lawsuits in 2007 and 2010. Use the coordinates of these points to write the line's equation in point-slope form and slope-intercept form.

c. Use the slope-intercept form of the equation from part (b) to project the number of lawsuits by smartphone companies for patent infringement in 2016.

In 2007, the U.S. government faced the prospect of paying out more and more in Social Security, Medicare, and Medicaid benefits. The line graphs show the costs of these entitlement programs, in billions of dollars, from 2007 through 2016 (projected). Use this information to solve Exercises 71–72 on the next page.

Cost, in Billions of Dollars, of the Largest Federal Entitlement Programs

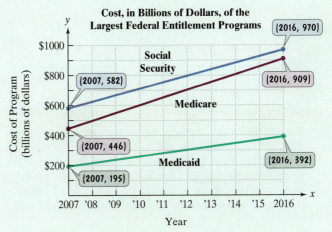

Source: U.S. Congressional Budget Office

(In Exercises 71–72, refer to the line graphs at the bottom of the previous page.)

71. a. Find the slope of the line segment representing Social Security. Round to one decimal place. Describe what this means in terms of rate of change.

b. Find the slope of the line segment representing Medicare. Round to one decimal place. Describe what this means in terms of rate of change.

c. Do the line segments for Social Security and Medicare lie on parallel lines? What does this mean in terms of the rate of change for these entitlement programs?

72. a. Find the slope of the line segment representing Social Security. Round to one decimal place. Describe what this means in terms of rate of change.

b. Find the slope of the line segment representing Medicaid. Round to one decimal place. Describe what this means in terms of rate of change.

c. Do the line segments for Social Security and Medicaid lie on parallel lines? What does this mean in terms of the rate of change for these entitlement programs?

73. Just as money doesn't buy happiness for individuals, the two don't necessarily go together for countries either. However, the scatter plot does show a relationship between a country's annual per capita income and the percentage of people in that country who call themselves "happy."

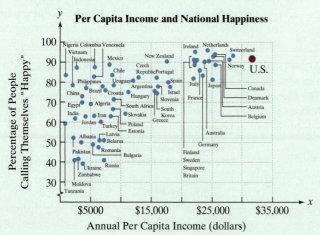

Source: Richard Layard, Happiness: Lessons from a New Science, Penguin, 2005

Draw a line that fits the data so that the spread of the data points around the line is as small as possible. Use the coordinates of two points along your line to write the slope-intercept form of its equation. Express the equation in function notation and use the linear function to make a prediction about national happiness based on per capita income.

Explaining the Concepts

74. Describe how to write the equation of a line if its slope and a point along the line are known.

75. Describe how to write the equation of a line if two points along the line are known.

76. If two lines are parallel, describe the relationship between their slopes.

77. If two lines are perpendicular, describe the relationship between their slopes.

78. If you know a point on a line and you know the equation of a line parallel to this line, explain how to write the line's equation.

79. In Example 3 on page 159, we developed a model that predicted American men born in 2020 will have a life expectancy of 78.6 years. Describe something that might occur that would make this prediction inaccurate.

Technology Exercises

80. The lines whose equations are $y = \frac{1}{3}x + 1$ and $y = -3x - 2$ are perpendicular because the product of their slopes, $\frac{1}{3}$ and -3, respectively, is -1.

a. Use a graphing utility to graph the equations in a $[-10, 10, 1]$ by $[-10, 10, 1]$ viewing rectangle. Do the lines appear to be perpendicular?

b. Now use the zoom square feature of your utility. Describe what happens to the graphs. Explain why this is so.

81. a. Use the statistical menu of your graphing utility to enter the seven data points shown in the scatter plot that you drew in Exercise 69(a).

b. Use the $\boxed{\text{STAT PLOT}}$ menu and the scatter plot capability to draw a scatter plot of the data points.

c. Select the linear regression option. Use your utility to obtain values for a and b for the equation of the regression line, $y = ax + b$. Compare this equation to the one that you obtained by hand in Exercise 69. You may also be given a **correlation coefficient**, r. Values of r close to 1 indicate that the points can be modeled by a linear function and the regression line has a positive slope. Values of r close to -1 indicate that the points can be modeled by a linear function and the regression line has a negative slope. Values of r close to 0 indicate no linear relationship between the variables. In this case, a linear model does not accurately describe the data.

d. Use the appropriate sequence (consult your manual) to graph the regression equation on top of the points in the scatter plot.

82. Repeat Exercise 81 using the seven data points shown in the scatter plot that you drew in Exercise 70(a).

Critical Thinking Exercises

Make Sense?   *In Exercises 83–86, determine whether each statement makes sense or does not make sense, and explain your reasoning.*

83. I can use any two points in a scatter plot to write the point-slope form of the equation of the line through those points. However, the other data points in the scatter plot might not fall on, or even near, this line.

84. I have linear functions that model changes for men and women over the same time period. The functions have the same slope, so their graphs are parallel lines, indicating that the rate of change for men is the same as the rate of change for women.

85. Some of the steel girders in this photo of the Eiffel Tower appear to be perpendicular. I can verify my observation by determining that their slopes are negative reciprocals.

86. When writing equations of lines, it's always easiest to begin by writing the point-slope form of the equation.

In Exercises 87–90, determine whether each statement is true or false. If the statement is false, make the necessary change(s) to produce a true statement.

87. The standard form of the equation of a line passing through $(-3, -1)$ and perpendicular to the line whose equation is $y = -\dfrac{2}{5}x - 4$ is $5x - 2y = -13$.

88. If I change the subtraction signs to addition signs in $y - 12 = 8(x - 2)$, the y-intercept of the corresponding graph will change from -4 to 4.

89. $y - 5 = 2(x - 1)$ is an equation of a line passing through $(4, 11)$.

90. The function $\{(-1, 4), (3, 6), (5, 7), (11, 10)\}$ can be described using $y - 7 = \frac{1}{2}(x - 5)$ with a domain of $\{-1, 3, 5, 11\}$.

91. Determine the value of B so that the line whose equation is $By = 8x - 1$ has slope -2.

92. Determine the value of A so that the line whose equation is $Ax + y = 2$ is perpendicular to the line containing the points $(1, -3)$ and $(-2, 4)$.

93. Consider a line whose x-intercept is -3 and whose y-intercept is -6. Provide the missing coordinate for the following two points that lie on this line: $(-40, \ )$ and $(\ , -200)$.

94. Prove that the equation of a line passing through $(a, 0)$ and $(0, b)$ $(a \neq 0, b \neq 0)$ can be written in the form $\dfrac{x}{a} + \dfrac{y}{b} = 1$. Why is this called the *intercept form* of a line?

Review Exercises

95. If $f(x) = 3x^2 - 8x + 5$, find $f(-2)$. (Section 2.1, Example 3)

96. If $f(x) = x^2 - 3x + 4$ and $g(x) = 2x - 5$, find $(fg)(-1)$. (Section 2.3, Example 4)

97. The sum of the angles of a triangle is $180°$. Find the three angles of a triangle if one angle is $20°$ greater than the smallest angle and the third angle is twice the smallest angle. (Section 1.5, Example 1)

Preview Exercises

Exercises 98–100 will help you prepare for the material covered in the first section of the next chapter.

98. **a.** Does $(-5, -6)$ satisfy $2x - y = -4$?
    b. Does $(-5, -6)$ satisfy $3x - 5y = 15$?

99. Graph $y = -x - 1$ and $4x - 3y = 24$ in the same rectangular coordinate system. At what point do the graphs intersect?

100. Solve: $7x - 2(-2x + 4) = 3$.

Chapter 2 Summary

Definitions and Concepts	Examples

Section 2.1 Introduction to Functions

A relation is any set of ordered pairs. The set of first components of the ordered pairs is the domain and the set of second components is the range. A function is a relation in which each member of the domain corresponds to exactly one member of the range. No two ordered pairs of a function can have the same first component and different second components.

The domain of the relation $\{(1, 2), (3, 4), (3, 7)\}$ is $\{1, 3\}$. The range is $\{2, 4, 7\}$. The relation is not a function: 3, in the domain, corresponds to both 4 and 7 in the range.

If a function is defined by an equation, the notation $f(x)$, read "f of x" or "f at x," describes the value of the function at the number, or input, x.

If $f(x) = 7x - 5$, then
$$f(a + 2) = 7(a + 2) - 5$$
$$= 7a + 14 - 5$$
$$= 7a + 9.$$

Section 2.2 Graphs of Functions

The graph of a function is the graph of its ordered pairs.

The Vertical Line Test for Functions
If any vertical line intersects a graph in more than one point, the graph does not define y as a function of x.

At the left or right of a function's graph, you will often find closed dots, open dots, or arrows. A closed dot shows that the graph ends and the point belongs to the graph. An open dot shows that the graph ends and the point does not belong to the graph. An arrow indicates that the graph extends indefinitely.

The graph of a function can be used to determine the function's domain and its range. To find the domain, look for all the inputs on the x-axis that correspond to points on the graph. To find the range, look for all the outputs on the y-axis that correspond to points on the graph.

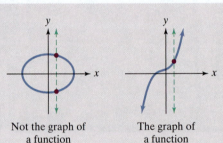

Not the graph of a function　　The graph of a function

(0, 2) belongs to the graph of f; (0, 0) does not.　$y = f(x)$

To find $f(2)$, locate 2 on the x-axis. The graph shows $f(2) = 4$.
Domain of $f = (-\infty, \infty)$
Range of $f = (0, \infty)$

Section 2.3 The Algebra of Functions

A Function's Domain
If a function f does not model data or verbal conditions, its domain is the largest set of real numbers for which the value of $f(x)$ is a real number. Exclude from a function's domain real numbers that cause division by zero and real numbers that result in a square root of a negative number.

$$f(x) = 7x + 13$$
Domain of $f = (-\infty, \infty)$

$$g(x) = \frac{7x}{12 - x}$$
Domain of $g = (-\infty, 12)$ or $(12, \infty)$

Definitions and Concepts	**Examples**

Section 2.3 The Algebra of Functions (continued)

The Algebra of Functions

Let f and g be two functions. The sum $f + g$, the difference $f - g$, the product fg, and the quotient $\dfrac{f}{g}$ are functions whose domains are the set of all real numbers common to the domains of f and g, defined as follows:

1. Sum: $(f + g)(x) = f(x) + g(x)$
2. Difference: $(f - g)(x) = f(x) - g(x)$
3. Product: $(fg)(x) = f(x) \cdot g(x)$
4. Quotient: $\left(\dfrac{f}{g}\right)(x) = \dfrac{f(x)}{g(x)}, g(x) \neq 0$.

Let $f(x) = x^2 + 2x$ and $g(x) = 4 - x$.

- $(f + g)(x) = (x^2 + 2x) + (4 - x) = x^2 + x + 4$
  $(f + g)(-2) = (-2)^2 + (-2) + 4 = 4 - 2 + 4 = 6$
- $(f - g)(x) = (x^2 + 2x) - (4 - x) = x^2 + 2x - 4 + x$
  $\qquad\qquad\qquad = x^2 + 3x - 4$
  $(f - g)(5) = 5^2 + 3 \cdot 5 - 4 = 25 + 15 - 4 = 36$
- $(fg)(1) = f(1) \cdot g(1) = (1^2 + 2 \cdot 1)(4 - 1)$
  $\qquad\qquad = 3(3) = 9$
- $\left(\dfrac{f}{g}\right)(x) = \dfrac{x^2 + 2x}{4 - x}, x \neq 4$

  $\left(\dfrac{f}{g}\right)(3) = \dfrac{3^2 + 2 \cdot 3}{4 - 3} = \dfrac{9 + 6}{1} = 15$

Section 2.4 Linear Functions and Slope

Data presented in a visual form as a set of points is called a scatter plot. A line that best fits the data points is called a regression line.

A function whose graph is a straight line is called a linear function. All linear functions can be written in the form $f(x) = mx + b$.

$f(x) = 3x + 10$ is a linear function.

$g(x) = 3x^2 + 10$ is not a linear function.

If a graph intersects the x-axis at $(a, 0)$, then a is an x-intercept. If a graph intersects the y-axis at $(0, b)$, then b is a y-intercept. The standard form of the equation of a line,

$$Ax + By = C,$$

can be graphed using intercepts and a checkpoint.

Graph using intercepts: $4x + 3y = 12$.

x-intercept:
(Set $y = 0$.)
$\qquad 4x = 12$ — Line passes through **(3, 0)**.
$\qquad\ \ x = 3$

y-intercept:
(Set $x = 0$.)
$\qquad 3y = 12$ — Line passes through **(0, 4)**.
$\qquad\ \ y = 4$

Checkpoint: Let $x = 2$.
$\qquad 4 \cdot 2 + 3y = 12$
$\qquad\ \ \ 8 + 3y = 12$
$\qquad\qquad\ \ 3y = 4$
$\qquad\qquad\ \ \ y = \dfrac{4}{3}$

y-intercept is 4.
Checkpoint: $(2, \frac{4}{3})$
x-intercept is 3.

Definitions and Concepts	**Examples**

Section 2.4 Linear Functions and Slope (continued)

The slope, m, of the line through the points (x_1, y_1) and (x_2, y_2) is

$$m = \frac{y_2 - y_1}{x_2 - x_1}, \quad x_2 - x_1 \neq 0.$$

If the slope is positive, the line rises from left to right. If the slope is negative, the line falls from left to right. The slope of a horizontal line is 0. The slope of a vertical line is undefined.

For points $(-7, 2)$ and $(3, -4)$, the slope of the line through the points is

$$m = \frac{\text{Change in } y}{\text{Change in } x} = \frac{-4 - 2}{3 - (-7)} = \frac{-6}{10} = -\frac{3}{5}.$$

The slope is negative, so the line falls.

For points $(2, -5)$ and $(2, 16)$, the slope of the line through the points is

$$m = \frac{\text{Change in } y}{\text{Change in } x} = \frac{16 - (-5)}{2 - 2} = \frac{21}{0}.$$

The slope is undefined, so the line is vertical.

The slope-intercept form of the equation of a nonvertical line with slope m and y-intercept b is

$$y = mx + b.$$

Using function notation, the equation is

$$f(x) = mx + b.$$

Graph: $f(x) = -\dfrac{3}{4}x + 1$.

Slope is $-\frac{3}{4}$. y-intercept is 1.

Horizontal and Vertical Lines

The graph of $y = b$, or $f(x) = b$, is a horizontal line. The y-intercept is b. The linear function $f(x) = b$ is called a constant function.

The graph of $x = a$ is a vertical line. The x-intercept is a. A vertical line is not a linear function.

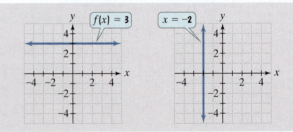

If the graph of a function is not a straight line, the average rate of change between any two points is the slope of the line containing the two points.

For a linear function, slope is the rate of change of the dependent variable per unit change of the independent variable.

The function

$$p(t) = -0.59t + 80.75$$

slope

models the percentage, $p(t)$, of Americans smoking cigarettes t years after 1900. The slope, -0.59, shows that the percentage of smokers is decreasing by 0.59% per year.

Definitions and Concepts	**Examples**

Section 2.5 The Point-Slope Form of the Equation of a Line

The point-slope form of the equation of a nonvertical line with slope m that passes through the point (x_1, y_1) is $$y - y_1 = m(x - x_1).$$	Slope = -4, passing through $(-1, 5)$ $$m = -4 \qquad x_1 = -1 \quad y_1 = 5$$ The point-slope form of the line's equation is $$y - 5 = -4[x - (-1)].$$ Simplify: $$y - 5 = -4(x + 1).$$
To write the point-slope form of the line passing through two points, begin by using the points to compute the slope, m. Use either given point as (x_1, y_1) and write the point-slope equation: $$y - y_1 = m(x - x_1).$$ Solving this equation for y gives the slope-intercept form of the line's equation.	Write equations in point-slope form and in slope-intercept form of the line passing through $(4, 1)$ and $(3, -2)$. $$m = \frac{-2 - 1}{3 - 4} = \frac{-3}{-1} = 3$$ Using $(4, 1)$ as (x_1, y_1), the point-slope form of the equation is $$y - 1 = 3(x - 4).$$ Solve for y to obtain the slope-intercept form. $$y - 1 = 3x - 12$$ $$y = 3x - 11$$ In function notation, $$f(x) = 3x - 11.$$
Nonvertical parallel lines have the same slope. If the product of the slopes of two lines is -1, then the lines are perpendicular. One line is perpendicular to another line if its slope is the negative reciprocal of the slope of the other. A horizontal line having zero slope is perpendicular to a vertical line having undefined slope.	Write equations in point-slope form and in slope-intercept form of the line passing through $(2, -1)$ $$x_1 \quad y_1$$ and perpendicular to $y = -\dfrac{1}{5}x + 6$. $$\text{slope}$$ The slope, m, of the perpendicular line is 5, the negative reciprocal of $-\frac{1}{5}$. $$y - (-1) = 5(x - 2) \qquad \text{Point-slope form of the equation}$$ $$y + 1 = 5(x - 2)$$ $$y + 1 = 5x - 10$$ $$y = 5x - 11 \quad \text{or} \quad f(x) = 5x - 11$$ Slope-intercept form of the equation

CHAPTER 2 REVIEW EXERCISES

2.1 *In Exercises 1–3, determine whether each relation is a function. Give the domain and range for each relation.*

1. $\{(3, 10), (4, 10), (5, 10)\}$

2. $\{(1, 12), (2, 100), (3, \pi), (4, -6)\}$

3. $\{(13, 14), (15, 16), (13, 17)\}$

In Exercises 4–5, find the indicated function values.

4. $f(x) = 7x - 5$
   a. $f(0)$ b. $f(3)$ c. $f(-10)$
   d. $f(2a)$ e. $f(a + 2)$

5. $g(x) = 3x^2 - 5x + 2$
   a. $g(0)$ b. $g(5)$ c. $g(-4)$
   d. $g(b)$ e. $g(4a)$

2.2 *In Exercises 6–7, graph the given functions, f and g, in the same rectangular coordinate system. Select integers for x, starting with −2 and ending with 2. Once you have obtained your graphs, describe how the graph of g is related to the graph of f.*

6. $f(x) = x^2$, $g(x) = x^2 - 1$

7. $f(x) = |x|$, $g(x) = |x| + 2$

In Exercises 8–13, use the vertical line test to identify graphs in which y is a function of x.

8.

9.

10.

11.

12.

13.

Use the graph of f to solve Exercises 14–18.

14. Find $f(-2)$.

15. Find $f(0)$.

16. For what value of x is $f(x) = -5$?

17. Find the domain of f.

18. Find the range of f.

19. The graph shows the height, in meters, of an eagle in terms of its time, in seconds, in flight.

a. Use the graph to explain why the eagle's height is a function of its time in flight.

b. Find $f(15)$. Describe what this means in practical terms.

c. What is a reasonable estimate of the eagle's maximum height?

d. For what values of x is $f(x) = 20$? Describe what this means in practical terms.

e. Use the graph of the function to write a description of the eagle's flight.

2.3 *In Exercises 20–22, find the domain of each function.*

20. $f(x) = 7x - 3$

21. $g(x) = \dfrac{1}{x + 8}$

22. $f(x) = x + \dfrac{3x}{x - 5}$

In Exercises 23–24, find **a.** $(f + g)(x)$ *and* **b.** $(f + g)(3)$.

23. $f(x) = 4x - 5, g(x) = 2x + 1$

24. $f(x) = 5x^2 - x + 4, g(x) = x - 3$

In Exercises 25–26, for each pair of functions, f and g, determine the domain of f + g.

25. $f(x) = 3x + 4, g(x) = \dfrac{5}{4 - x}$

26. $f(x) = \dfrac{7x}{x + 6}, g(x) = \dfrac{4}{x + 1}$

In Exercises 27–34, let

$$f(x) = x^2 - 2x \quad \text{and} \quad g(x) = x - 5.$$

Find each of the following.

27. $(f + g)(x)$ and $(f + g)(-2)$

28. $f(3) + g(3)$

29. $(f - g)(x)$ and $(f - g)(1)$

30. $f(4) - g(4)$

31. $(fg)(-3)$

32. $\left(\dfrac{f}{g}\right)(x)$ and $\left(\dfrac{f}{g}\right)(4)$

33. The domain of $f - g$

34. The domain of $\dfrac{f}{g}$

2.4 *In Exercises 35–37, use intercepts and a checkpoint to graph each linear function.*

35. $x + 2y = 4$

36. $2x - 3y = 12$

37. $4x = 8 - 2y$

In Exercises 38–41, find the slope of the line passing through each pair of points or state that the slope is undefined. Then indicate whether the line through the points rises, falls, is horizontal, or is vertical.

38. $(5, 2)$ and $(2, -4)$

39. $(-2, 3)$ and $(7, -3)$

40. $(3, 2)$ and $(3, -1)$

41. $(-3, 4)$ and $(-1, 4)$

In Exercises 42–44, give the slope and y-intercept of each line whose equation is given. Then graph the linear function.

42. $y = 2x - 1$

43. $f(x) = -\dfrac{1}{2}x + 4$

44. $y = \dfrac{2}{3}x$

In Exercises 45–47, rewrite the equation in slope-intercept form. Give the slope and y-intercept.

45. $2x + y = 4$

46. $-3y = 5x$

47. $5x + 3y = 6$

In Exercises 48–52, graph each equation in a rectangular coordinate system.

48. $y = 2$

49. $7y = -21$

50. $f(x) = -4$

51. $x = 3$

52. $2x = -10$

53. The function $f(t) = -0.27t + 70.45$ models record time, $f(t)$, in seconds, for the women's 400-meter run t years after 1900. What is the slope of this model? Describe what this means in terms of rate of change.

54. The stated intent of the 1994 "don't ask, don't tell" policy was to reduce the number of discharges of gay men and lesbians from the military. Nearly 14,000 active-duty gay servicemembers were dismissed under the policy, which officially ended in 2011, after 18 years. The line graph shows the number of discharges under "don't ask, don't tell" from 1994 through 2010.

Number of Active-Duty Gay Servicemembers Discharged from the Military for Homosexuality

Source: General Accountability Office

a. Find the average rate of change, rounded to the nearest whole number, from 1994 through 1998. Describe what this means.

b. Find the average rate of change, rounded to the nearest whole number, from 2001 through 2006. Describe what this means.

55. The graph shows that a linear function describes the relationship between Fahrenheit temperature, F, and Celsius temperature, C.

a. Use the points labeled by the voice balloons to find a function in the form $F = mC + b$ that expresses Fahrenheit temperature, F, in terms of Celsius temperature, C.

b. Use the function from part (a) to find the Fahrenheit temperature when the Celsius temperature is 30°.

2.5 *In Exercises 56–59, use the given conditions to write an equation for each line in point-slope form and in slope-intercept form.*

56. Passing through $(-3, 2)$ with slope -6

57. Passing through $(1, 6)$ and $(-1, 2)$

58. Passing through $(4, -7)$ and parallel to the line whose equation is $3x + y = 9$

59. Passing through $(-2, 6)$ and perpendicular to the line whose equation is $y = \frac{1}{3}x + 4$

60. The bar graph shows the average age at which men in the United States married for the first time from 2009 through 2013. The data are displayed as five points in a scatter plot. Also shown is a line that passes through or near the points.

Men's Average Age of First Marriage

Source: U.S. Census Bureau

a. Use the two points whose coordinates are shown by the voice balloons to find the point-slope form of the equation of the line that models men's average age of first marriage, y, x years after 2008.

b. Write the equation from part (a) in slope-intercept form. Use function notation.

c. If trends shown from 2009 through 2013 continue, use the linear function to project men's average age of first marriage in 2020.

CHAPTER 2 TEST   Step-by-step test solutions are found on the Chapter Test Prep Videos available in MyMathLab® or on YouTube (search "BlitzerInterAlg7e" and click on "Channels").

In Exercises 1–2, determine whether each relation is a function. Give the domain and range for each relation.

1. $\{(1, 2), (3, 4), (5, 6), (6, 6)\}$

2. $\{(2, 1), (4, 3), (6, 5), (6, 6)\}$

3. If $f(x) = 3x - 2$, find $f(a + 4)$.

4. If $f(x) = 4x^2 - 3x + 6$, find $f(-2)$.

5. Graph $f(x) = x^2 - 1$ and $g(x) = x^2 + 1$ in the same rectangular coordinate system. Select integers for x, starting with -2 and ending with 2. Once you have obtained your graphs, describe how the graph of g is related to the graph of f.

In Exercises 6–7, identify the graph or graphs in which y is a function of x.

6.

7.

Use the graph of f to solve Exercises 8–11.

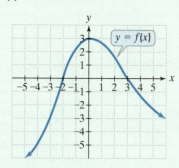

8. Find $f(6)$.

9. List two values of x for which $f(x) = 0$.

10. Find the domain of f.

11. Find the range of f.

12. Find the domain of $f(x) = \dfrac{6}{10 - x}$.

In Exercises 13–17, let
$$f(x) = x^2 + 4x \quad and \quad g(x) = x + 2.$$

Find each of the following.

13. $(f + g)(x)$ and $(f + g)(3)$

14. $(f - g)(x)$ and $(f - g)(-1)$

15. $(fg)(-5)$

16. $\left(\dfrac{f}{g}\right)(x)$ and $\left(\dfrac{f}{g}\right)(2)$

17. The domain of $\dfrac{f}{g}$

In Exercises 18–20, graph each linear function.

18. $4x - 3y = 12$

19. $f(x) = -\dfrac{1}{3}x + 2$

20. $f(x) = 4$

In Exercises 21–22, find the slope of the line passing through each pair of points or state that the slope is undefined. Then indicate whether the line through the points rises, falls, is horizontal, or is vertical.

21. $(5, 2)$ and $(1, 4)$

22. $(4, 5)$ and $(4, -5)$

The function $V(t) = 3.6t + 140$ models the number of Super Bowl viewers, $V(t)$, in millions, t years after 1995. Use the model to solve Exercises 23–24.

23. Find $V(10)$. Describe what this means in terms of the variables in the model.

24. What is the slope of this model? Describe what this means in terms of rate of change.

In Exercises 25–27, use the given conditions to write an equation for each line in point-slope form and slope-intercept form.

25. Passing through $(-1, -3)$ and $(4, 2)$

26. Passing through $(-2, 3)$ and perpendicular to the line whose equation is $y = -\frac{1}{2}x - 4$

27. Passing through $(6, -4)$ and parallel to the line whose equation is $x + 2y = 5$

28. In the United States, it is illegal to drive with a blood alcohol concentration of 0.08 or higher. (A blood alcohol concentration of 0.08 indicates 0.08% alcohol in the blood.) The scatter plot shows the relationship between the number of twelve-ounce beers consumed per hour by an adult man and that person's corresponding blood alcohol concentration. Also shown is a line that passes through or near the data points.

Source: National Highway Traffic Safety Administration

a. Use the two points whose coordinates are shown by the voice balloons to find the point-slope form of the equation of the line that models the blood alcohol concentration of an adult man, y, who consumes x twelve-ounce beers per hour.

b. Write the equation from part (a) in slope-intercept form. Use function notation.

c. Use the linear function to predict the blood alcohol concentration of an adult man who consumes 8 twelve-ounce beers in an hour. (Effects at this level include major impairment of all physical and mental functions.)

CUMULATIVE REVIEW EXERCISES   (CHAPTERS 1–2)

1. Use the roster method to list the elements in the set:

   $\{x \,|\, x \text{ is a whole number less than 4}\}$.

2. Determine whether the following statement is true or false:

   $\pi \notin \{x \,|\, x \text{ is an irrational number}\}$.

In Exercises 3–4, use the order of operations to simplify each expression.

3. $\dfrac{8 - 3^2 \div 9}{|-5| - [5 - (18 \div 6)]^2}$

4. $4 - (2 - 9)^0 + 3^2 \div 1 + 3$

5. Simplify: $3 - [2(x - 2) - 5x]$.

In Exercises 6–8, solve each equation. If the solution set is ∅ or ℝ, classify the equation as an inconsistent equation or an identity.

6. $2 + 3x - 4 = 2(x - 3)$

7. $4x + 12 - 8x = -6(x - 2) + 2x$

8. $\dfrac{x - 2}{4} = \dfrac{2x + 6}{3}$

9. After a 20% reduction, a computer sold for $1800. What was the computer's price before the reduction?

10. Solve for t: $A = p + prt$.

In Exercises 11–12, simplify each exponential expression.

11. $(3x^4 y^{-5})^{-2}$

12. $\left(\dfrac{3x^2 y^{-4}}{x^{-3} y^2}\right)^2$

13. Multiply and write the answer in scientific notation:

   $(7 \times 10^{-8})(3 \times 10^2)$.

14. Is $\{(1, 5), (2, 5), (3, 5), (4, 5), (6, 5)\}$ a function? Give the relation's domain and range.

15. Graph $f(x) = |x| - 1$ and $g(x) = |x| + 2$ in the same rectangular coordinate system. Select integers for x, starting with -2 and ending with 2. Once you have obtained your graphs, describe how the graph of g is related to the graph of f.

16. Find the domain of $f(x) = \dfrac{1}{15 - x}$.

17. If $f(x) = 3x^2 - 4x + 2$ and $g(x) = x^2 - 5x - 3$, find $(f - g)(x)$ and $(f - g)(-1)$.

In Exercises 18–19, graph each linear function.

18. $f(x) = -2x + 4$

19. $x - 2y = 6$

20. Write equations in point-slope form and slope-intercept form for the line passing through $(3, -5)$ and parallel to the line whose equation is $y = 4x + 7$.

Systems of Linear Equations

Held in a different country every five years, with contributions from more than 120 countries, the World Expo is a showcase of cool gizmos, a multinational take on what the future might be. Who needs a human friend when you can connect with an adorable robot that recognizes you, engages in (meaningful?) conversation in any language, takes orders, plays a mean trumpet, and even does algebra? And what do the entrepreneurs who create these robots and other global visions want to do? Generate profit, of course. In this chapter, you'll learn how algebra models every business venture, from a kid selling lemonade to companies producing innovative "personal partner" robots.

Here's where you'll find this application:

- Functions of business, including modeling business ventures with profit functions, are discussed in Section 3.2, pages 201–203.

SECTION

3.1

Systems of Linear Equations in Two Variables

What am I supposed to learn?

After studying this section, you should be able to:

1. Determine whether an ordered pair is a solution of a system of linear equations.

2. Solve systems of linear equations by graphing.

3. Solve systems of linear equations by substitution.

4. Solve systems of linear equations by addition.

5. Select the most efficient method for solving a system of linear equations.

6. Identify systems that do not have exactly one ordered-pair solution.

"My wife and I were happy for twenty years. Then we met."

—Henny Youngman

"Getting married for sex is like buying a 747 for the free peanuts."

—Jeff Foxworthy

Despite an endless array of jokes about marriage, **Figure 3.1** indicates that fewer U.S. adults are getting married. Our changing marital status can be analyzed using a pair of linear models in two variables.

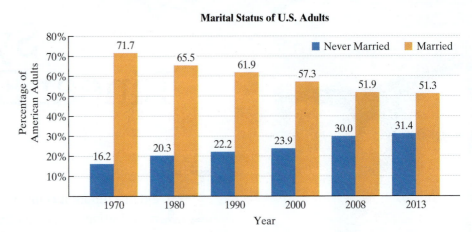

Marital Status of U.S. Adults

Figure 3.1

Source: U.S. Census Bureau

In the first three sections of this chapter, you will learn to model your world with two equations in two variables and three equations in three variables. The methods you learn for solving these systems provide the foundation for solving complex problems involving thousands of equations containing thousands of variables. In this section's Exercise Set (Exercise 91), you will apply these methods to analyze America's changing marital status.

1. Determine whether an ordered pair is a solution of a system of linear equations.

Systems of Linear Equations and Their Solutions

We have seen that all equations in the form $Ax + By = C$ are straight lines when graphed. Two such equations are called a **system of linear equations**, or a **linear system**. A **solution of a system of linear equations** is an ordered pair that satisfies both equations in the system. For example, (3, 4) satisfies the system

$$\begin{cases} x + y = 7 & \text{(3 + 4 is, indeed, 7.)} \\ x - y = -1. & \text{(3 − 4 is, indeed, −1.)} \end{cases}$$

Thus, (3, 4) satisfies both equations and is a solution of the system. The solution can be described by saying that $x = 3$ and $y = 4$. The solution can also be described using set notation. The solution set of the system is $\{(3, 4)\}$—that is, the set consisting of the ordered pair $(3, 4)$.

A system of linear equations can have exactly one solution, no solution, or infinitely many solutions. We begin with systems with exactly one solution.

EXAMPLE 1   Determining Whether Ordered Pairs Are Solutions of a System

Consider the system:

$$\begin{cases} x + 2y = -7 \\ 2x - 3y = 0. \end{cases}$$

Determine if each ordered pair is a solution of the system:

a. $(-3, -2)$ **b.** $(1, -4)$.

Solution

a. We begin by determining whether $(-3, -2)$ is a solution. Because -3 is the x-coordinate and -2 is the y-coordinate of $(-3, -2)$, we replace x with -3 and y with -2.

$$
\begin{array}{ll}
x + 2y = -7 & 2x - 3y = 0 \\
-3 + 2(-2) \stackrel{?}{=} -7 & 2(-3) - 3(-2) \stackrel{?}{=} 0 \\
-3 + (-4) \stackrel{?}{=} -7 & -6 - (-6) \stackrel{?}{=} 0 \\
-7 = -7, \;\; \text{true} & -6 + 6 \stackrel{?}{=} 0 \\
& 0 = 0, \;\; \text{true}
\end{array}
$$

The pair $(-3, -2)$ satisfies both equations: It makes each equation true. Thus, the ordered pair is a solution of the system.

b. To determine whether $(1, -4)$ is a solution, we replace x with 1 and y with -4.

$$
\begin{array}{ll}
x + 2y = -7 & 2x - 3y = 0 \\
1 + 2(-4) \stackrel{?}{=} -7 & 2 \cdot 1 - 3(-4) \stackrel{?}{=} 0 \\
1 + (-8) \stackrel{?}{=} -7 & 2 - (-12) \stackrel{?}{=} 0 \\
-7 = -7, \;\; \text{true} & 2 + 12 \stackrel{?}{=} 0 \\
& 14 = 0, \;\; \text{false}
\end{array}
$$

The pair $(1, -4)$ fails to satisfy *both* equations: It does not make both equations true. Thus, the ordered pair is not a solution of the system. ∎

✓ CHECK POINT 1   Consider the system:

$$\begin{cases} 2x + 5y = -24 \\ 3x - 5y = 14. \end{cases}$$

Determine if each ordered pair is a solution of the system:

a. $(-7, -2)$ **b.** $(-2, -4)$.

2 Solve systems of linear equations by graphing. ▶

Solving Linear Systems by Graphing

The solution of a system of two linear equations in two variables can be found by graphing both of the equations in the same rectangular coordinate system. For a system with one solution, **the coordinates of the point of intersection give the system's solution.**

Great Question!

Can I use a rough sketch on scratch paper to solve a linear system by graphing?

No. When solving linear systems by graphing, neatly drawn graphs are essential for determining points of intersection.

- Use rectangular coordinate graph paper.
- Use a ruler or straightedge.
- Use a pencil with a sharp point.

Solving Systems of Two Linear Equations in Two Variables, *x* and *y*, by Graphing

1. Graph the first equation.
2. Graph the second equation on the same set of axes.
3. If the lines intersect at a point, determine the coordinates of this point of intersection. The ordered pair is the solution to the system.
4. Check the solution in both equations.

EXAMPLE 2 Solving a Linear System by Graphing

Solve by graphing:

$$\begin{cases} y = -x - 1 \\ 4x - 3y = 24. \end{cases}$$

Solution

Step 1. Graph the first equation. We use the *y*-intercept and the slope to graph $y = -x - 1$.

$$y = -x - 1$$

The slope is −1. The *y*-intercept is −1.

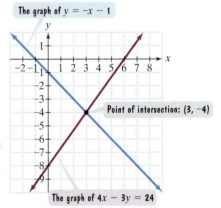

The graph of $y = -x - 1$

Point of intersection: (3, −4)

The graph of $4x - 3y = 24$

Figure 3.2

The graph of the linear function is shown as a blue line in **Figure 3.2**.

Step 2. Graph the second equation on the same axes. We use intercepts to graph $4x - 3y = 24$.

x-intercept (Set $y = 0$.)	*y*-intercept (Set $x = 0$.)
$4x - 3 \cdot 0 = 24$	$4 \cdot 0 - 3y = 24$
$4x = 24$	$-3y = 24$
$x = 6$	$y = -8$

The *x*-intercept is 6, so the line passes through $(6, 0)$. The *y*-intercept is −8, so the line passes through $(0, -8)$. The graph of $4x - 3y = 24$ is shown as a red line in **Figure 3.2**.

Step 3. Determine the coordinates of the intersection point. This ordered pair is the system's solution. Using **Figure 3.2**, it appears that the lines intersect at $(3, -4)$. The "apparent" solution of the system is $(3, -4)$.

Step 4. Check the solution in both equations.

Check $(3, -4)$ in $y = -x - 1$:	Check $(3, -4)$ in $4x - 3y = 24$:
$-4 \overset{?}{=} -3 - 1$	$4(3) - 3(-4) \overset{?}{=} 24$
$-4 = -4,$ true	$12 + 12 \overset{?}{=} 24$
	$24 = 24,$ true

Because both equations are satisfied, $(3, -4)$ is the solution and $\{(3, -4)\}$ is the solution set. ■

Using Technology

A graphing utility can be used to solve the system in Example 2. Solve each equation for *y*, if necessary, graph the equations, and use the intersection feature. The utility displays the solution $(3, -4)$ as $x = 3, y = -4$.

Intersection
X=3   Y=-4

$[-10, 10, 1]$ by $[-10, 10, 1]$

✓ CHECK POINT 2 Solve by graphing:

$$\begin{cases} y = -2x + 6 \\ 2x - y = -2. \end{cases}$$

3 Solve systems of linear equations by substitution. ▶

Eliminating a Variable Using the Substitution Method

Finding the solution to a linear system by graphing equations may not be easy to do. For example, a solution of $\left(-\frac{2}{3}, \frac{157}{29} \right)$ would be difficult to "see" as an intersection point on a graph.

Let's consider a method that does not depend on finding a system's solution visually: the substitution method. This method involves converting the system to one equation in one variable by an appropriate substitution.

Great Question!

In the first step of the substitution method, how do I know which variable to isolate and in which equation?

You can choose both the variable and the equation. If possible, solve for a variable whose coefficient is 1 or −1 to avoid working with fractions.

Solving Linear Systems by Substitution

1. Solve either of the equations for one variable in terms of the other. (If one of the equations is already in this form, you can skip this step.)
2. Substitute the expression found in step 1 into the other equation. This will result in an equation in one variable.
3. Solve the equation containing one variable.
4. Back-substitute the value found in step 3 into one of the original equations. Simplify and find the value of the remaining variable.
5. Check the proposed solution in both of the system's given equations.

EXAMPLE 3   Solving a System by Substitution

Solve by the substitution method:

$$\begin{cases} y = -2x + 4 \\ 7x - 2y = 3. \end{cases}$$

Solution

Step 1. Solve either of the equations for one variable in terms of the other. This step has already been done for us. The first equation, $y = -2x + 4$, is solved for y in terms of x.

Step 2. Substitute the expression from step 1 into the other equation. We substitute the expression $-2x + 4$ for y in the second equation:

$$y = \boxed{-2x + 4} \qquad 7x - 2\boxed{y} = 3. \quad \text{Substitute } -2x + 4 \text{ for } y.$$

This gives us an equation in one variable, namely

$$7x - 2(-2x + 4) = 3.$$

The variable y has been eliminated.

Step 3. Solve the resulting equation containing one variable.

$$7x - 2(-2x + 4) = 3 \qquad \text{This is the equation containing one variable.}$$
$$7x + 4x - 8 = 3 \qquad \text{Apply the distributive property.}$$
$$11x - 8 = 3 \qquad \text{Combine like terms.}$$
$$11x = 11 \qquad \text{Add 8 to both sides.}$$
$$x = 1 \qquad \text{Divide both sides by 11.}$$

Step 4. Back-substitute the obtained value into one of the original equations. We now know that the x-coordinate of the solution is 1. To find the y-coordinate, we back-substitute the x-value into either original equation. We will use

$$y = -2x + 4.$$

Substitute 1 for x.

$$y = -2 \cdot 1 + 4 = -2 + 4 = 2$$

With $x = 1$ and $y = 2$, the proposed solution is $(1, 2)$.

Step 5. Check the proposed solution in both of the system's given equations. Replace x with 1 and y with 2.

$$y = -2x + 4 \qquad\qquad 7x - 2y = 3$$
$$2 \overset{?}{=} -2 \cdot 1 + 4 \qquad\qquad 7(1) - 2(2) \overset{?}{=} 3$$
$$2 \overset{?}{=} -2 + 4 \qquad\qquad 7 - 4 \overset{?}{=} 3$$
$$2 = 2, \text{ true} \qquad\qquad 3 = 3, \text{ true}$$

The pair $(1, 2)$ satisfies both equations. The solution is $(1, 2)$ and the system's solution set is $\{(1, 2)\}$. ∎

✓ **CHECK POINT 3** Solve by the substitution method:

$$\begin{cases} y = 3x - 7 \\ 5x - 2y = 8. \end{cases}$$

EXAMPLE 4 Solving a System by Substitution

Solve by the substitution method:

$$\begin{cases} 5x + 2y = 1 \\ x - 3y = 7. \end{cases}$$

Solution

Step 1. Solve either of the equations for one variable in terms of the other. We begin by isolating one of the variables in either of the equations. By solving for x in the second equation, which has a coefficient of 1, we can avoid fractions.

$$x - 3y = 7 \qquad \text{This is the second equation in the given system.}$$
$$x = 3y + 7 \qquad \text{Solve for } x \text{ by adding } 3y \text{ to both sides.}$$

Step 2. Substitute the expression from step 1 into the other equation. We substitute $3y + 7$ for x in the first equation.

$$x = \boxed{3y + 7} \qquad 5\boxed{x} + 2y = 1$$

This gives us an equation in one variable, namely

$$5(3y + 7) + 2y = 1.$$

The variable x has been eliminated.

Step 3. Solve the resulting equation containing one variable.

$$5(3y + 7) + 2y = 1 \qquad \text{This is the equation containing one variable.}$$
$$15y + 35 + 2y = 1 \qquad \text{Apply the distributive property.}$$
$$17y + 35 = 1 \qquad \text{Combine like terms.}$$
$$17y = -34 \qquad \text{Subtract 35 from both sides.}$$
$$y = -2 \qquad \text{Divide both sides by 17.}$$

Step 4. Back-substitute the obtained value into one of the original equations. We back-substitute -2 for y into one of the original equations to find x. Let's use both equations to show that we obtain the same value for x in either case.

Using the first equation:	Using the second equation:
$5x + 2y = 1$	$x - 3y = 7$
$5x + 2(-2) = 1$	$x - 3(-2) = 7$
$5x - 4 = 1$	$x + 6 = 7$
$5x = 5$	$x = 1$
$x = 1$	

With $x = 1$ and $y = -2$, the proposed solution is $(1, -2)$.

Great Question!

When I back-substitute the value for one of the variables, which equation should I use?

The equation from step 1, in which one variable is expressed in terms of the other, is equivalent to one of the original equations. It is often easiest to back-substitute the obtained value into this equation to find the value of the other variable. After obtaining both values, get into the habit of checking the ordered-pair solution in *both* equations of the system.

Step 5. Check. Take a moment to show that $(1, -2)$ satisfies both given equations. The solution is $(1, -2)$ and the solution set is $\{(1, -2)\}$. ∎

✓ **CHECK POINT 4** Solve by the substitution method:

$$\begin{cases} 3x + 2y = 4 \\ 2x + \ y = 1. \end{cases}$$

4 Solve systems of linear equations by addition. ▶

Eliminating a Variable Using the Addition Method

The substitution method is most useful if one of the given equations has an isolated variable. A third method for solving a linear system is the addition method. Like the substitution method, the addition method involves eliminating a variable and ultimately solving an equation containing only one variable. However, this time we eliminate a variable by adding the equations.

For example, consider the following system of linear equations:

$$\begin{cases} \quad 3x - 4y = 11 \\ -3x + 2y = -7. \end{cases}$$

When we add these two equations, the x-terms are eliminated. This occurs because the coefficients of the x-terms, 3 and -3, are opposites (additive inverses) of each other:

$$\begin{cases} \quad 3x - 4y = 11 \\ \underline{-3x + 2y = -7} \end{cases}$$

Add: $\quad 0x - 2y = \ 4$ The sum is an equation in one variable.

$$-2y = \ 4$$

$$y = -2. \quad \text{Divide both sides by } -2 \text{ and solve for } y.$$

Now we can back-substitute -2 for y into one of the original equations to find x. It does not matter which equation you use; you will obtain the same value for x in either case. If we use either equation, we can show that $x = 1$ and the solution $(1, -2)$ satisfies both equations in the system.

When we use the addition method, we want to obtain two equations whose sum is an equation containing only one variable. The key step is to **obtain, for one of the variables, coefficients that differ only in sign**. To do this, we may need to multiply one or both equations by some nonzero number so that the coefficients of one of the variables, x or y, become opposites. Then when the two equations are added, this variable will be eliminated.

EXAMPLE 5 Solving a System by the Addition Method

Great Question!

Isn't the addition method also called the elimination method?

Although the addition method is also known as the elimination method, variables are eliminated when using both the substitution and addition methods. The name *addition method* specifically tells us that the elimination of a variable is accomplished by adding two equations.

Solve by the addition method:

$$\begin{cases} 3x + 4y = -10 \\ 5x - 2y = \ \ 18. \end{cases}$$

Solution We must rewrite one or both equations in equivalent forms so that the coefficients of the same variable (either x or y) are opposites of each other. Consider the terms in y in each equation, that is, $4y$ and $-2y$. To eliminate y, we can multiply each term of the second equation by 2 and then add equations.

$$\begin{cases} 3x + 4y = -10 \ \xrightarrow{\text{No change}} \\ 5x - 2y = \ \ 18 \ \xrightarrow{\text{Multiply by 2.}} \end{cases} \begin{cases} \ 3x + 4y = -10 \\ \underline{10x - 4y = \ \ \ 36} \end{cases}$$

Add: $13x + 0y = \ \ 26$

$$13x = \ \ 26$$

$$x = \ \ 2 \quad \text{Divide both sides by 13 and solve for } x.$$

Thus, $x = 2$. We back-substitute this value into either one of the given equations. We'll use the first one.

$$3x + 4y = -10 \quad \text{This is the first equation in the given system.}$$
$$3(2) + 4y = -10 \quad \text{Substitute 2 for } x.$$
$$6 + 4y = -10 \quad \text{Multiply.}$$
$$4y = -16 \quad \text{Subtract 6 from both sides.}$$
$$y = -4 \quad \text{Divide both sides by 4.}$$

We see that $x = 2$ and $y = -4$. The ordered pair $(2, -4)$ can be shown to satisfy both equations in the system. Consequently, the solution is $(2, -4)$ and the solution set is $\{(2, -4)\}$. ∎

Solving Linear Systems by Addition

1. If necessary, rewrite both equations in the form $Ax + By = C$.
2. If necessary, multiply either equation or both equations by appropriate nonzero numbers so that the sum of the x-coefficients or the sum of the y-coefficients is 0.
3. Add the equations in step 2. The sum will be an equation in one variable.
4. Solve the equation in one variable.
5. Back-substitute the value obtained in step 4 into either of the given equations and solve for the other variable.
6. Check the solution in both of the original equations.

✓ **CHECK POINT 5**   Solve by the addition method:

$$\begin{cases} 4x - 7y = -16 \\ 2x + 5y = \quad 9. \end{cases}$$

EXAMPLE 6   Solving a System by the Addition Method

Solve by the addition method:

$$\begin{cases} 7x = 5 - 2y \\ 3y = 16 - 2x. \end{cases}$$

Solution

Step 1. Rewrite both equations in the form $Ax + By = C$. We first arrange the system so that variable terms appear on the left and constants appear on the right. We obtain

$$\begin{cases} 7x + 2y = 5 \quad \text{Add } 2y \text{ to both sides of the first equation.} \\ 2x + 3y = 16. \quad \text{Add } 2x \text{ to both sides of the second equation.} \end{cases}$$

Step 2. If necessary, multiply either equation or both equations by appropriate numbers so that the sum of the x-coefficients or the sum of the y-coefficients is 0. We can eliminate x or y. Let's eliminate y by multiplying the first equation by 3 and the second equation by -2.

$$\begin{cases} 7x + 2y = \quad 5 \\ 2x + 3y = 16 \end{cases} \xrightarrow[\text{Multiply by } -2.]{\text{Multiply by 3.}} \begin{cases} 21x + 6y = \quad 15 \\ -4x - 6y = -32 \end{cases}$$

Step 3. Add the equations.   Add: $17x + 0y = -17$
$$17x = -17$$

Step 4. Solve the equation in one variable. We solve $17x = -17$ by dividing both sides by 17.

$$\frac{17x}{17} = \frac{-17}{17} \qquad \text{Divide both sides by 17.}$$

$$x = -1 \qquad \text{Simplify.}$$

Step 5. Back-substitute and find the value of the other variable. We can back-substitute -1 for x into either one of the given equations. We'll use the second one.

$$3y = 16 - 2x \qquad \text{This is the second equation in the given system.}$$

$$3y = 16 - 2(-1) \qquad \text{Substitute } -1 \text{ for } x.$$

$$3y = 16 + 2 \qquad \text{Multiply.}$$

$$3y = 18 \qquad \text{Add.}$$

$$y = 6 \qquad \text{Divide both sides by 3.}$$

We found that $x = -1$ and $y = 6$. The proposed solution is $(-1, 6)$.

Step 6. Check. Take a moment to show that $(-1, 6)$ satisfies both given equations. The solution is $(-1, 6)$ and the solution set is $\{(-1, 6)\}$. ∎

✓ **CHECK POINT 6**   Solve by the addition method:

$$\begin{cases} 3x = 2 - 4y \\ 5y = -1 - 2x. \end{cases}$$

Some linear systems have solutions that are not integers. If the value of one variable turns out to be a "messy" fraction, back-substitution might lead to cumbersome arithmetic. If this happens, you can return to the original system and use the addition method a second time to find the value of the other variable.

EXAMPLE 7   Solving a System by the Addition Method

Solve by the addition method:

$$\begin{cases} \dfrac{x}{2} - 5y = 32 \\[2mm] \dfrac{3x}{2} - 7y = 45. \end{cases}$$

Solution

Step 1. Rewrite both equations in the form $Ax + By = C$. Although each equation is already in this form, the coefficients of x are not integers. There is less chance for error if the coefficients for x and y in $Ax + By = C$ are integers. Consequently, we begin by clearing fractions. Multiply both sides of each equation by 2.

$$\begin{cases} \dfrac{x}{2} - 5y = 32 \\[2mm] \dfrac{3x}{2} - 7y = 45 \end{cases} \xrightarrow[\text{Multiply by 2.}]{\text{Multiply by 2.}} \begin{cases} x - 10y = 64 \\ 3x - 14y = 90 \end{cases}$$

Step 2. If necessary, multiply either equation or both equations by appropriate numbers so that the sum of the x-coefficients or the sum of the y-coefficients is 0. We will eliminate x. Multiply the first equation with integral coefficients by -3 and leave the second equation unchanged.

$$\begin{cases} x - 10y = 64 \\ 3x - 14y = 90 \end{cases} \xrightarrow[\text{No change}]{\text{Multiply by } -3.} \begin{cases} -3x + 30y = -192 \\ \underline{3x - 14y = 90} \end{cases}$$

Step 3. Add the equations.   Add: $0x + 16y = -102$

$$16y = -102$$

Step 4. Solve the equation in one variable. We solve $16y = -102$ by dividing both sides by 16.

$$\frac{16y}{16} = \frac{-102}{16} \qquad \text{Divide both sides by 16.}$$

$$y = -\frac{102}{16} = -\frac{51}{8} \qquad \text{Simplify.}$$

Step 5. Back-substitute and find the value of the other variable. Back-substitution of $-\frac{51}{8}$ for y into either of the given equations results in cumbersome arithmetic. Instead, let's use the addition method on the system with integral coefficients from step 1 to find the value of x. Thus, we eliminate y by multiplying the first equation by -7 and the second equation by 5.

$$\begin{cases} x - 10y = 64 \\ 3x - 14y = 90 \end{cases} \begin{array}{l} \xrightarrow{\text{Multiply by } -7.} \\ \xrightarrow{\text{Multiply by } 5.} \end{array} \begin{cases} -7x + 70y = -448 \\ \underline{15x - 70y = 450} \end{cases}$$

$$\text{Add:} \qquad 8x = 2 = 2$$

$$x = \frac{2}{8} = \frac{1}{4}$$

We found that $x = \frac{1}{4}$ and $y = -\frac{51}{8}$. The proposed solution is $\left(\frac{1}{4}, -\frac{51}{8}\right)$.

Step 6. Check. For this system, a calculator is helpful in showing that $\left(\frac{1}{4}, -\frac{51}{8}\right)$ satisfies both of the original equations of the system. The solution is $\left(\frac{1}{4}, -\frac{51}{8}\right)$ and the solution set is $\left\{\left(\frac{1}{4}, -\frac{51}{8}\right)\right\}$. ∎

☑ **CHECK POINT 7**   Solve by the addition method:

$$\begin{cases} \dfrac{3x}{2} - 2y = \dfrac{5}{2} \\ x - \dfrac{5y}{2} = -\dfrac{3}{2}. \end{cases}$$

5 Select the most efficient method for solving a system of linear equations. ▶

Comparing the Three Solution Methods

The following chart compares the graphing, substitution, and addition methods for solving systems of linear equations in two variables. With increased practice, you will find it easier to select the best method for solving a particular linear system.

Comparing Solution Methods

Method	Advantages	Disadvantages
Graphing	You can see the solutions.	If the solutions do not involve integers or are too large to be seen on the graph, it's impossible to tell exactly what the solutions are.
Substitution	Gives exact solutions. Easy to use if a variable is on one side by itself.	Solutions cannot be seen. Introduces extensive work with fractions when no variable has a coefficient of 1 or -1.
Addition	Gives exact solutions. Easy to use if no variable has a coefficient of 1 or -1.	Solutions cannot be seen.

6 Identify systems that do not have exactly one ordered-pair solution.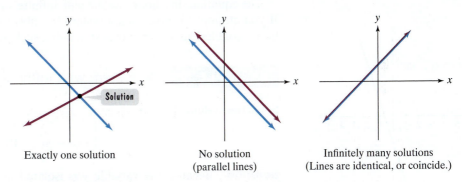

Linear Systems Having No Solution or Infinitely Many Solutions

We have seen that a system of linear equations in two variables represents a pair of lines. The lines either intersect at one point, are parallel, or are identical. Thus, there are three possibilities for the number of solutions to a system of two linear equations.

> ### The Number of Solutions to a System of Two Linear Equations
>
> The number of solutions to a system of two linear equations in two variables is given by one of the following. (See **Figure 3.3**.)
>
> | Number of Solutions | What This Means Graphically |
> |---|---|
> | Exactly one ordered-pair solution | The two lines intersect at one point. |
> | No solution | The two lines are parallel. |
> | Infinitely many solutions | The two lines are identical. |

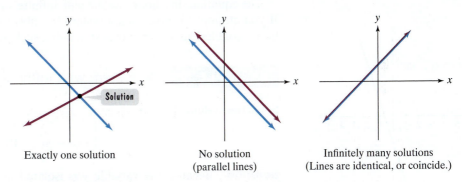

Figure 3.3 Possible graphs for a system of two linear equations in two variables

A linear system with no solution is called an **inconsistent system**. If you attempt to solve such a system by substitution or addition, you will eliminate both variables. A false statement, such as $0 = 6$, will be the result.

EXAMPLE 8 A System with No Solution

Solve the system:

$$\begin{cases} 3x - 2y = 6 \\ 6x - 4y = 18. \end{cases}$$

Solution Because no variable is isolated, we will use the addition method. To obtain coefficients of x that differ only in sign, we multiply the first equation by -2.

$$\begin{cases} 3x - 2y = 6 \xrightarrow{\text{Multiply by } -2.} \\ 6x - 4y = 18 \xrightarrow{\text{No change}} \end{cases} \begin{cases} -6x + 4y = -12 \\ \underline{6x - 4y = 18} \\ \text{Add:} \quad 0 = 6 \end{cases}$$

There are no values of x and y for which $0 = 6$. No values of x and y satisfy $0x + 0y = 6$.

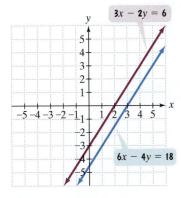

Figure 3.4 The graph of an inconsistent system

The false statement $0 = 6$ indicates that the system is inconsistent and has no solution. The solution set is the empty set, $\varnothing$. ∎

The lines corresponding to the two equations in Example 8 are shown in **Figure 3.4**. The lines are parallel and have no point of intersection.

Discover for Yourself

Show that the graphs of $3x - 2y = 6$ and $6x - 4y = 18$ must be parallel lines by solving each equation for y. What is the slope and the y-intercept for each line? What does this mean? If a linear system is inconsistent, what must be true about the slopes and the y-intercepts for the system's graphs?

✓ **CHECK POINT 8**   Solve the system:

$$\begin{cases} 5x - 2y = 4 \\ -10x + 4y = 7. \end{cases}$$

A linear system that has at least one solution is called a **consistent system**. Lines that intersect and lines that coincide both represent consistent systems. If the lines coincide, then the consistent system has infinitely many solutions, represented by every point on the coinciding lines.

The equations in a linear system with infinitely many solutions are called **dependent**. If you attempt to solve such a system by substitution or addition, you will eliminate both variables. However, a true statement, such as $10 = 10$, will be the result.

EXAMPLE 9   A System with Infinitely Many Solutions

Solve the system:

$$\begin{cases} y = 3x - 2 \\ 15x - 5y = 10. \end{cases}$$

Solution   Because the variable y is isolated in $y = 3x - 2$, the first equation, we can use the substitution method. We substitute the expression for y into the second equation.

$y = \boxed{3x - 2}$ $15x - 5\boxed{y} = 10$ Substitute $3x - 2$ for y.

$15x - 5(3x - 2) = 10$ The substitution results in an equation in one variable.

$15x - 15x + 10 = 10$ Apply the distributive property.

$10 = 10$ Simplify.

> This statement is true for all values of x and y.

In our final step, both variables have been eliminated and the resulting statement, $10 = 10$, is true. This true statement indicates that the system has infinitely many solutions. The solution set consists of all points (x, y) lying on either of the coinciding lines, $y = 3x - 2$ or $15x - 5y = 10$, as shown in **Figure 3.5**.

We express the solution set for the system in one of two equivalent ways:

$$\{(x, y) \mid y = 3x - 2\} \quad \text{or} \quad \{(x, y) \mid 15x - 5y = 10\}.$$

> The set of all ordered pairs (x, y) such that $y = 3x - 2$

> The set of all ordered pairs (x, y) such that $15x - 5y = 10$ ∎

✓ **CHECK POINT 9**   Solve the system:

$$\begin{cases} x = 4y - 8 \\ 5x - 20y = -40. \end{cases}$$

Figure 3.5 The graph of a system with infinitely many solutions

$15x - 5y = 10$

$y = 3x - 2$

Great Question!

The system in Example 9 has infinitely many solutions. Does that mean that any ordered pair of numbers is a solution?

No. Although the system in Example 9 has infinitely many solutions, this does not mean that any ordered pair of numbers you can form will be a solution. The ordered pair (x, y) must satisfy one of the system's equations, $y = 3x - 2$ or $15x - 5y = 10$, and there are infinitely many such ordered pairs. Because the graphs are coinciding lines, the ordered pairs that are solutions of one of the equations are also solutions of the other equation.

Achieving Success

Take some time to consider how you are doing in this course. Check your performance by answering the following questions:

- Are you attending all lectures?
- For each hour of class time, are you spending at least two hours outside of class completing all homework assignments, checking answers, correcting errors, and using all resources to get the help that you need?
- Are you reviewing for quizzes and tests?
- Are you reading the textbook? In all college courses, you are responsible for the information in the text, whether or not it is covered in class.
- Are you keeping an organized notebook? Does each page have the appropriate section number from the text on top? Do the pages contain examples your instructor works during lecture and other relevant class notes? Have you included your worked-out homework exercises? Do you keep a special section for graded exams?
- Are you analyzing your mistakes and learning from your errors?
- Are there ways you can improve how you are doing in the course?

CONCEPT AND VOCABULARY CHECK

Fill in each blank so that the resulting statement is true.

1. A solution to a system of linear equations in two variables is an ordered pair that _____.

2. When solving a system of linear equations by graphing, the system's solution is determined by locating _____.

3. When solving

$$\begin{cases} 3x - 2y = 5 \\ y = 3x - 3 \end{cases}$$

   by the substitution method, we obtain $x = \frac{1}{3}$, so the solution set is _____.

4. When solving

$$\begin{cases} 2x + 10y = 9 \\ 8x + 5y = 7 \end{cases}$$

   by the addition method, we can eliminate y by multiplying the second equation by _____ and then adding the equations.

5. When solving

$$\begin{cases} 4x - 3y = 15 \\ 3x - 2y = 10 \end{cases}$$

   by the addition method, we can eliminate y by multiplying the first equation by 2 and the second equation by _____ and then adding the equations.

6. When solving

$$\begin{cases} 12x - 21y = 24 \\ 4x - 7y = 7 \end{cases}$$

   by the addition method, we obtain $0 = 3$, so the solution set is _____. The linear system is a/an _____ system. If you attempt to solve such a system by graphing, you will obtain two lines that are _____.

7. When solving

$$\begin{cases} x = 3y + 2 \\ 5x - 15y = 10 \end{cases}$$

   by the substitution method, we obtain $10 = 10$, so the solution set is _____. The equations in this system are called _____. If you attempt to solve such a system by graphing, you will obtain two lines that _____.

3.1 EXERCISE SET ▶ MyMathLab®

Practice Exercises

In Exercises 1–6, determine whether the given ordered pair is a solution of the system.

1. $(7, -5)$
$$\begin{cases} x - y = 12 \\ x + y = 2 \end{cases}$$

2. $(-3, 1)$
$$\begin{cases} x - y = -4 \\ 2x + 10y = 4 \end{cases}$$

3. $(2, -1)$
$$\begin{cases} 3x + 4y = 2 \\ 2x + 5y = 1 \end{cases}$$

4. $(4, 2)$
$$\begin{cases} 2x - 5y = -2 \\ 3x + 4y = 18 \end{cases}$$

5. $(5, -3)$
$$\begin{cases} y = 2x - 13 \\ 4x + 9y = -7 \end{cases}$$

6. $(-3, -4)$
$$\begin{cases} y = 3x + 5 \\ 5x - 2y = -7 \end{cases}$$

In Exercises 7–24, solve each system by graphing. Identify systems with no solution and systems with infinitely many solutions, using set notation to express their solution sets.

7. $\begin{cases} x + y = 4 \\ x - y = 2 \end{cases}$

8. $\begin{cases} x + y = 6 \\ x - y = -4 \end{cases}$

9. $\begin{cases} 2x + y = 4 \\ y = 4x + 1 \end{cases}$

10. $\begin{cases} x + 2y = 4 \\ y = -2x - 1 \end{cases}$

11. $\begin{cases} 3x - 2y = 6 \\ x - 4y = -8 \end{cases}$

12. $\begin{cases} 4x + y = 4 \\ 3x - y = 3 \end{cases}$

13. $\begin{cases} 2x + 3y = 6 \\ 4x = -6y + 12 \end{cases}$

14. $\begin{cases} 3x - 3y = 6 \\ 2x = 2y + 4 \end{cases}$

15. $\begin{cases} y = 2x - 2 \\ y = -5x + 5 \end{cases}$

16. $\begin{cases} y = -x + 1 \\ y = 3x + 5 \end{cases}$

17. $\begin{cases} 3x - y = 4 \\ 6x - 2y = 4 \end{cases}$

18. $\begin{cases} 2x - y = -4 \\ 4x - 2y = 6 \end{cases}$

19. $\begin{cases} 2x + y = 4 \\ 4x + 3y = 10 \end{cases}$

20. $\begin{cases} 4x - y = 9 \\ x - 3y = 16 \end{cases}$

21. $\begin{cases} x - y = 2 \\ y = 1 \end{cases}$

22. $\begin{cases} x + 2y = 1 \\ x = 3 \end{cases}$

23. $\begin{cases} 3x + y = 3 \\ 6x + 2y = 12 \end{cases}$

24. $\begin{cases} 2x - 3y = 6 \\ 4x - 6y = 24 \end{cases}$

In Exercises 25–42, solve each system by the substitution method. Identify inconsistent systems and systems with dependent equations, using set notation to express their solution sets.

25. $\begin{cases} x + y = 6 \\ y = 2x \end{cases}$

26. $\begin{cases} x + y = 10 \\ y = 4x \end{cases}$

27. $\begin{cases} 2x + 3y = 9 \\ x = y + 2 \end{cases}$

28. $\begin{cases} 3x - 4y = 18 \\ y = 1 - 2x \end{cases}$

29. $\begin{cases} y = -3x + 7 \\ 5x - 2y = 8 \end{cases}$

30. $\begin{cases} x = 3y + 8 \\ 2x - y = 6 \end{cases}$

31. $\begin{cases} 4x + y = 5 \\ 2x - 3y = 13 \end{cases}$

32. $\begin{cases} x - 3y = 3 \\ 3x + 5y = -19 \end{cases}$

33. $\begin{cases} x - 2y = 4 \\ 2x - 4y = 5 \end{cases}$

34. $\begin{cases} x - 3y = 6 \\ 2x - 6y = 5 \end{cases}$

35. $\begin{cases} 2x + 5y = -4 \\ 3x - y = 11 \end{cases}$

36. $\begin{cases} 2x + 5y = 1 \\ -x + 6y = 8 \end{cases}$

37. $\begin{cases} 2(x - 1) - y = -3 \\ y = 2x + 3 \end{cases}$

38. $\begin{cases} x + y - 1 = 2(y - x) \\ y = 3x - 1 \end{cases}$

39. $\begin{cases} \dfrac{x}{4} - \dfrac{y}{4} = -1 \\ x + 4y = -9 \end{cases}$

40. $\begin{cases} \dfrac{x}{6} - \dfrac{y}{2} = \dfrac{1}{3} \\ x + 2y = -3 \end{cases}$

41. $\begin{cases} y = \dfrac{2}{5}x - 2 \\ 2x - 5y = 10 \end{cases}$

42. $\begin{cases} y = \dfrac{1}{3}x + 4 \\ 3y = x + 12 \end{cases}$

In Exercises 43–58, solve each system by the addition method. Identify inconsistent systems and systems with dependent equations, using set notation to express their solution sets.

43. $\begin{cases} x + y = 7 \\ x - y = 3 \end{cases}$

44. $\begin{cases} 2x + y = 3 \\ x - y = 3 \end{cases}$

45. $\begin{cases} 12x + 3y = 15 \\ 2x - 3y = 13 \end{cases}$

46. $\begin{cases} 4x + 2y = 12 \\ 3x - 2y = 16 \end{cases}$

47. $\begin{cases} x + 3y = 2 \\ 4x + 5y = 1 \end{cases}$

48. $\begin{cases} x + 2y = -1 \\ 2x - y = 3 \end{cases}$

49. $\begin{cases} 6x - y = -5 \\ 4x - 2y = 6 \end{cases}$

50. $\begin{cases} x - 2y = 5 \\ 5x - y = -2 \end{cases}$

51. $\begin{cases} 3x - 5y = 11 \\ 2x - 6y = 2 \end{cases}$

52. $\begin{cases} 4x - 3y = 12 \\ 3x - 4y = 2 \end{cases}$

53. $\begin{cases} 2x - 5y = 13 \\ 5x + 3y = 17 \end{cases}$

54. $\begin{cases} 4x + 5y = -9 \\ 6x - 3y = -3 \end{cases}$

55. $\begin{cases} 2x + 6y = 8 \\ 3x + 9y = 12 \end{cases}$

56. $\begin{cases} x - 3y = -6 \\ 3x - 9y = 9 \end{cases}$

57. $\begin{cases} 2x - 3y = 4 \\ 4x + 5y = 3 \end{cases}$

58. $\begin{cases} 4x - 3y = 8 \\ 2x - 5y = -14 \end{cases}$

In Exercises 59–82, solve each system by the method of your choice. Identify inconsistent systems and systems with dependent equations, using set notation to express solution sets.

59. $\begin{cases} 3x - 7y = 1 \\ 2x - 3y = -1 \end{cases}$

60. $\begin{cases} 2x - 3y = 2 \\ 5x + 4y = 51 \end{cases}$

61. $\begin{cases} x = y + 4 \\ 3x + 7y = -18 \end{cases}$

62. $\begin{cases} y = 3x + 5 \\ 5x - 2y = -7 \end{cases}$

63. $\begin{cases} 9x + \dfrac{4y}{3} = 5 \\ 4x - \dfrac{y}{3} = 5 \end{cases}$

64. $\begin{cases} \dfrac{x}{6} - \dfrac{y}{5} = -4 \\ \dfrac{x}{4} - \dfrac{y}{6} = -2 \end{cases}$

65. $\begin{cases} \dfrac{1}{4}x - \dfrac{1}{9}y = \dfrac{2}{3} \\ \dfrac{1}{2}x - \dfrac{1}{3}y = 1 \end{cases}$

66. $\begin{cases} \dfrac{1}{16}x - \dfrac{3}{4}y = -1 \\ \dfrac{3}{4}x + \dfrac{5}{2}y = 11 \end{cases}$

67. $\begin{cases} x = 3y - 1 \\ 2x - 6y = -2 \end{cases}$

68. $\begin{cases} x = 4y - 1 \\ 2x - 8y = -2 \end{cases}$

69. $\begin{cases} y = 2x + 1 \\ y = 2x - 3 \end{cases}$ **70.** $\begin{cases} y = 2x + 4 \\ y = 2x - 1 \end{cases}$

71. $\begin{cases} 0.4x + 0.3y = 2.3 \\ 0.2x - 0.5y = 0.5 \end{cases}$ **72.** $\begin{cases} 0.2x - y = -1.4 \\ 0.7x - 0.2y = -1.6 \end{cases}$

73. $\begin{cases} 5x - 40 = 6y \\ 2y = 8 - 3x \end{cases}$ **74.** $\begin{cases} 4x - 24 = 3y \\ 9y = 3x - 1 \end{cases}$

75. $\begin{cases} 3(x + y) = 6 \\ 3(x - y) = -36 \end{cases}$ **76.** $\begin{cases} 4(x - y) = -12 \\ 4(x + y) = -20 \end{cases}$

77. $\begin{cases} 3(x - 3) - 2y = 0 \\ 2(x - y) = -x - 3 \end{cases}$ **78.** $\begin{cases} 5x + 2y = -5 \\ 4(x + y) = 6(2 - x) \end{cases}$

79. $\begin{cases} x + 2y - 3 = 0 \\ 12 = 8y + 4x \end{cases}$ **80.** $\begin{cases} 2x - y - 5 = 0 \\ 10 = 4x - 2y \end{cases}$

81. $\begin{cases} 3x + 4y = 0 \\ 7x = 3y \end{cases}$ **82.** $\begin{cases} 5x + 8y = 20 \\ 4y = -5x \end{cases}$

Practice PLUS

In Exercises 83–84, solve each system by the method of your choice.

83. $\begin{cases} \dfrac{x + 2}{2} - \dfrac{y + 4}{3} = 3 \\ \dfrac{x + y}{5} = \dfrac{x - y}{2} - \dfrac{5}{2} \end{cases}$ **84.** $\begin{cases} \dfrac{x - y}{3} = \dfrac{x + y}{2} - \dfrac{1}{2} \\ \dfrac{x + 2}{2} - 4 = \dfrac{y + 4}{3} \end{cases}$

In Exercises 85–86, solve each system for x and y, expressing either value in terms of a or b, if necessary. Assume that $a \neq 0$ and $b \neq 0$.

85. $\begin{cases} 5ax + 4y = 17 \\ ax + 7y = 22 \end{cases}$ **86.** $\begin{cases} 4ax + by = 3 \\ 6ax + 5by = 8 \end{cases}$

87. For the linear function $f(x) = mx + b, f(-2) = 11$ and $f(3) = -9$. Find m and b.

88. For the linear function $f(x) = mx + b, f(-3) = 23$ and $f(2) = -7$. Find m and b.

Use the graphs of the linear functions to solve Exercises 89–90.

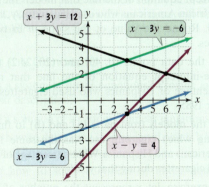

89. Write the linear system whose solution set is $\{(6, 2)\}$. Express each equation in the system in slope-intercept form.

90. Write the linear system whose solution set is $\varnothing$. Express each equation in the system in slope-intercept form.

Application Exercises

91. We opened the section with a bar graph that showed fewer U.S. adults are getting married. (See **Figure 3.1** on page 178.) The data can be modeled by the following system of linear equations:

$\begin{cases} -3x + 10y = 160 \\ x + 2y = 142. \end{cases}$

Percentage of never-married American adults, y, x years after 1970

Percentage of married American adults, y, x years after 1970

a. Use these models to determine the year, rounded to the nearest year, when the percentage of never-married adults will be the same as the percentage of married adults. For that year, approximately what percentage of Americans, rounded to the nearest percent, will belong to each group?

b. How is your approximate solution from part (a) shown by the following graphs?

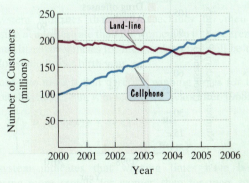

92. The graph shows that from 2000 through 2006, Americans unplugged land lines and switched to cellphones.

Number of Cellphone and Land-Line Customers in the United States

Source: Federal Communications Commission

a. Use the graphs to estimate the point of intersection. In what year was the number of cellphone and land-line customers the same? How many millions of customers were there for each?

b. The function $4.3x + y = 198$ models the number of land-line customers, in millions, x years after 2000.

114. The solution set of the system

$$\begin{cases} 5x - y = 1 \\ 10x - 2y = 2 \end{cases}$$

is $\{(2, 9)\}$.

115. A system of linear equations can have a solution set consisting of precisely two ordered pairs.

116. The solution set of the system

$$\begin{cases} y = 4x - 3 \\ y = 4x + 5 \end{cases}$$

is the empty set.

117. Determine a and b so that $(2, 1)$ is a solution of this system:

$$\begin{cases} ax - by = 4 \\ bx + ay = 7. \end{cases}$$

118. Write a system of equations having $\{(-2, 7)\}$ as a solution set. (More than one system is possible.)

119. Solve the system for x and y in terms of $a_1, b_1, c_1, a_2, b_2,$ and c_2:

$$\begin{cases} a_1x + b_1y = c_1 \\ a_2x + b_2y = c_2. \end{cases}$$

Review Exercises

120. Solve: $6x = 10 + 5(3x - 4)$. (Section 1.4, Example 3)

121. Simplify: $(4x^2y^4)^2(-2x^5y^0)^3$. (Section 1.6, Example 9)

122. If $f(x) = x^2 - 3x + 7$, find $f(-1)$. (Section 2.1, Example 3)

Preview Exercises

Exercises 123–125 will help you prepare for the material covered in the next section.

123. The formula $I = Pr$ is used to find the simple interest, I, earned for one year when the principal, P, is invested at an annual interest rate, r. Write an expression for the total interest earned on a principal of x dollars at a rate of 15% ($r = 0.15$) and a principal of y dollars at a rate of 7% ($r = 0.07$).

124. A chemist working on a flu vaccine needs to obtain 50 milliliters of a 30% sodium-iodine solution. How many milliliters of sodium-iodine are needed in the solution?

125. A company that manufactures running shoes sells them at $80 per pair. Write an expression for the revenue that is generated by selling x pairs of shoes.

SECTION

3.2

Problem Solving and Business Applications Using Systems of Equations ▶

Driving through your neighborhood, you see kids selling lemonade. Would it surprise you to know that this activity can be analyzed using functions and systems of equations? By doing so, you will view profit and loss in the business world in a new way. In this section, we use systems of equations to solve problems and model business ventures.

What am I supposed to learn?

After studying this section, you should be able to:

1 Solve problems using systems of equations. ▶

2 Use functions to model revenue, cost, and profit, and perform a break-even analysis. ▶

1 Solve problems using systems of equations. ▶

A Strategy for Solving Word Problems Using Systems of Equations

When we solved problems in Chapter 1, we let x represent a quantity that was unknown. Problems in this section involve two unknown quantities. We will let x and y represent these quantities. We then translate from the verbal conditions of the problem into a *system* of linear equations.

EXAMPLE 1 Solving a Problem Involving Energy Efficiency of Building Materials

A heat-loss survey by an electric company indicated that a wall of a house containing 40 square feet of glass and 60 square feet of plaster lost 1920 Btu (British thermal units) of heat. A second wall containing 10 square feet of glass and 100 square feet of plaster lost 1160 Btu of heat. Determine the heat lost per square foot for the glass and for the plaster.

Solution

Step 1. Use variables to represent unknown quantities.

Let $x =$ the heat lost per square foot for the glass.

Let $y =$ the heat lost per square foot for the plaster.

Step 2. Write a system of equations that models the problem's conditions. The heat loss for each wall is the heat lost by the glass plus the heat lost by the plaster. One wall containing 40 square feet of glass and 60 square feet of plaster lost 1920 Btu of heat.

Heat lost by the glass	$+$	heat lost by the plaster	$=$	total heat lost.
$\begin{pmatrix} \text{Number} \\ \text{of ft}^2 \end{pmatrix} \cdot \begin{pmatrix} \text{heat lost} \\ \text{per ft}^2 \end{pmatrix}$	$+$	$\begin{pmatrix} \text{number} \\ \text{of ft}^2 \end{pmatrix} \cdot \begin{pmatrix} \text{heat lost} \\ \text{per ft}^2 \end{pmatrix}$	$=$	total heat lost.
$40 \quad \cdot \quad x$	$+$	$60 \quad \cdot \quad y$	$=$	1920

A second wall containing 10 square feet of glass and 100 square feet of plaster lost 1160 Btu of heat.

Heat lost by the glass	$+$	heat lost by the plaster	$=$	total heat lost.
$\begin{pmatrix} \text{Number} \\ \text{of ft}^2 \end{pmatrix} \cdot \begin{pmatrix} \text{heat lost} \\ \text{per ft}^2 \end{pmatrix}$	$+$	$\begin{pmatrix} \text{number} \\ \text{of ft}^2 \end{pmatrix} \cdot \begin{pmatrix} \text{heat lost} \\ \text{per ft}^2 \end{pmatrix}$	$=$	total heat lost.
$10 \quad \cdot \quad x$	$+$	$100 \quad \cdot \quad y$	$=$	1160

Step 3. Solve the system and answer the problem's question. The system

$$\begin{cases} 40x + 60y = 1920 \\ 10x + 100y = 1160 \end{cases}$$

can be solved by addition. We'll multiply the second equation by -4 and then add equations to eliminate x.

$$\begin{cases} 40x + 60y = 1920 \xrightarrow{\text{No change}} \\ 10x + 100y = 1160 \xrightarrow{\text{Multiply by } -4.} \end{cases} \begin{cases} 40x + 60y = 1920 \\ -40x - 400y = -4640 \end{cases}$$

$$\text{Add:} \qquad -340y = -2720$$

$$y = \frac{-2720}{-340} = 8$$

Now we can find the value of x by back-substituting 8 for y in either of the system's equations.

$$10x + 100y = 1160 \qquad \text{We'll use the second equation.}$$
$$10x + 100(8) = 1160 \qquad \text{Back-substitute 8 for } y.$$
$$10x + 800 = 1160 \qquad \text{Multiply.}$$
$$10x = 360 \qquad \text{Subtract 800 from both sides.}$$
$$x = 36 \qquad \text{Divide both sides by 10.}$$

We see that $x = 36$ and $y = 8$. Because x represents heat lost per square foot for the glass and y for the plaster, the glass lost 36 Btu of heat per square foot and the plaster lost 8 Btu per square foot.

Step 4. Check the proposed solution in the original wording of the problem. The problem states that the wall with 40 square feet of glass and 60 square feet of plaster lost 1920 Btu.

$$40(36) + 60(8) = 1440 + 480 = 1920 \text{ Btu of heat}$$

> Proposed solution is 36 Btu per ft^2 for glass and 8 Btu per ft^2 for plaster.

Our proposed solution checks with the first statement. The problem also states that the wall with 10 square feet of glass and 100 square feet of plaster lost 1160 Btu.

$$10(36) + 100(8) = 360 + 800 = 1160 \text{ Btu of heat}$$

Our proposed solution also checks with the second statement. ∎

✓ CHECK POINT 1 University of Arkansas researchers discovered that we underestimate the number of calories in restaurant meals. The next time you eat out, take the number of calories you think you ate and double it. The researchers concluded that this number should be a more accurate estimate. The actual number of calories in one portion of hamburger and fries and two portions of fettuccine Alfredo is 4240. The actual number of calories in two portions of hamburger and fries and one portion of fettuccine Alfredo is 3980. Find the actual number of calories in each of these dishes. (*Source: Consumer Reports*, January/February, 2007)

Test Your Calorie I.Q.

Hamburger and Fries
Average guess:
777 calories

Fettuccine Alfredo
Average guess:
704 calories

Next, we will solve problems involving investments, mixtures, and motion with systems of equations. We will continue using our four-step problem-solving strategy. We will also use tables to help organize the information in the problems.

Dual Investments with Simple Interest

Simple interest involves interest calculated only on the amount of money that we invest, called the **principal**. The formula $I = Pr$ is used to find the simple interest, I, earned for one year when the principal, P, is invested at an annual interest rate, r. Dual investment problems involve different amounts of money in two or more investments, each paying a different rate.

EXAMPLE 2 Solving a Dual Investment Problem

Your grandmother needs your help. She has $50,000 to invest. Part of this money is to be invested in noninsured bonds paying 15% annual interest. The rest of this money is to be invested in a government-insured certificate of deposit paying 7% annual interest. She told you that she requires $6000 per year in extra income from both of these investments. How much money should be placed in each investment?

Solution
Step 1. Use variables to represent unknown quantities.

Let x = the amount invested in the 15% noninsured bonds.

Let y = the amount invested in the 7% certificate of deposit.

Step 2. Write a system of equations that models the problem's conditions. Because Grandma has $50,000 to invest,

The amount invested at 15%	plus	the amount invested at 7%	equals	$50,000.
x	+	y	=	50,000

Furthermore, Grandma requires $6000 in total interest. We can use a table to organize the information in the problem and obtain a second equation.

	Principal (amount invested)	×	Interest rate	=	Interest earned
15% Investment	x		0.15		0.15x
7% Investment	y		0.07		0.07y

The interest for the two investments combined must be $6000.

$$0.15x \quad + \quad 0.07y \quad = \quad 6000$$

Step 3. Solve the system and answer the problem's question. The system

$$\begin{cases} x + y = 50{,}000 \\ 0.15x + 0.07y = 6000 \end{cases}$$

can be solved by substitution or addition. Substitution works well because both variables in the first equation have coefficients of 1. Addition also works well; if we multiply the first equation by -0.15 or -0.07, adding equations will eliminate a variable. We will use addition.

$$\begin{cases} x + y = 50{,}000 \\ 0.15x + 0.07y = 6000 \end{cases} \xrightarrow[\text{No change}]{\text{Multiply by } -0.07.} \begin{cases} -0.07x - 0.07y = -3500 \\ 0.15x + 0.07y = 6000 \end{cases}$$

$$\text{Add:} \quad 0.08x = 2500$$

$$x = \frac{2500}{0.08}$$

$$x = 31{,}250$$

Because x represents the amount that should be invested at 15%, Grandma should place $31,250 in 15% noninsured bonds. Now we can find y, the amount that she should place in the 7% certificate of deposit. We do so by back-substituting 31,250 for x in either of the system's equations.

$x + y = 50{,}000$	We'll use the first equation.
$31{,}250 + y = 50{,}000$	Back-substitute 31,250 for x.
$y = 18{,}750$	Subtract 31,250 from both sides.

Because $x = 31{,}250$ and $y = 18{,}750$, Grandma should invest $31,250 at 15% and $18,750 at 7%.

Step 4. Check the proposed answers in the original wording of the problem. Has Grandma invested $50,000?

$$\$31{,}250 + \$18{,}750 = \$50{,}000$$

Yes, all her money was placed in the dual investments. Can she count on $6000 interest? The interest earned on $31,250 at 15% is ($31,250)(0.15), or $4687.50. The interest earned on $18,750 at 7% is ($18,750)(0.07), or $1312.50. The total interest is $4687.50 + $1312.50, or $6000, exactly as it should be. You've made your grandmother happy. (Now if you would just visit her more often ...) ∎

✓ **CHECK POINT 2**   You inherited $5000 with the stipulation that for the first year the money had to be invested in two funds paying 9% and 11% annual interest. How much did you invest at each rate if the total interest earned for the year was $487?

Problems Involving Mixtures

Chemists and pharmacists often have to change the concentration of solutions and other mixtures. In these situations, the amount of a particular ingredient in the solution or mixture is expressed as a percentage of the total.

EXAMPLE 3 Solving a Mixture Problem

A chemist working on a flu vaccine needs to mix a 10% sodium-iodine solution with a 60% sodium-iodine solution to obtain 50 milliliters of a 30% sodium-iodine solution. How many milliliters of the 10% solution and of the 60% solution should be mixed?

Solution

Step 1. Use variables to represent unknown quantities.

Let x = the number of milliliters of the 10% solution to be used in the mixture.

Let y = the number of milliliters of the 60% solution to be used in the mixture.

Step 2. Write a system of equations that models the problem's conditions. The situation is illustrated in **Figure 3.6**. The chemist needs 50 milliliters of a 30% sodium-iodine solution. We form a table that shows the amount of sodium-iodine in each of the three solutions.

	Volume	x	y	50
	Strength	10% solution	60% solution	30% solution

Figure 3.6

Solution	Number of Milliliters	×	Percent of Sodium-Iodine	=	Amount of Sodium-Iodine
10% Solution	x		10% = 0.1		$0.1x$
60% Solution	y		60% = 0.6		$0.6y$
30% Mixture	50		30% = 0.3		$0.3(50) = 15$

The chemist needs to obtain a 50-milliliter mixture.

$$x + y = 50$$

The 50-milliliter mixture must be 30% sodium-iodine. The amount of sodium-iodine must be 30% of 50, or $(0.3)(50) = 15$ milliliters.

$$0.1x + 0.6y = 15$$

Step 3. Solve the system and answer the problem's question. The system

$$\begin{cases} x + y = 50 \\ 0.1x + 0.6y = 15 \end{cases}$$

can be solved by substitution or addition. Let's use substitution. The first equation can easily be solved for x or y. Solving for y, we obtain $y = 50 - x$.

$$y = \boxed{50 - x} \qquad 0.1x + 0.6\boxed{y} = 15$$

We substitute $50 - x$ for y in the second equation. This gives us an equation in one variable.

$0.1x + 0.6(50 - x) =$	15	This equation contains one variable, x.
$0.1x + 30 - 0.6x =$	15	Apply the distributive property.
$-0.5x + 30 =$	15	Combine like terms.
$-0.5x =$	-15	Subtract 30 from both sides.
$x = \dfrac{-15}{-0.5} =$	30	Divide both sides by -0.5.

Back-substituting 30 for x in either of the system's equations ($x + y = 50$ is easier to use) gives $y = 20$. Because x represents the number of milliliters of the 10% solution and y the number of milliliters of the 60% solution, the chemist should mix 30 milliliters of the 10% solution with 20 milliliters of the 60% solution.

Step 4. Check the proposed solution in the original wording of the problem. The problem states that the chemist needs 50 milliliters of a 30% sodium-iodine solution. The amount of sodium-iodine in this mixture is 0.3(50), or 15 milliliters. The amount of sodium-iodine in 30 milliliters of the 10% solution is 0.1(30), or 3 milliliters. The amount of sodium-iodine in 20 milliliters of the 60% solution is 0.6(20) = 12 milliliters. The amount of sodium-iodine in the two solutions used in the mixture is 3 milliliters + 12 milliliters, or 15 milliliters, exactly as it should be. ∎

✓ **CHECK POINT 3**   A chemist needs to mix a 12% acid solution with a 20% acid solution to obtain 160 ounces of a 15% acid solution. How many ounces of each of the acid solutions must be used?

Great Question!

Are there similarities between dual investment problems and mixture problems?

Problems involving dual investments and problems involving mixtures are both based on the same idea: The total amount times the rate gives the amount.

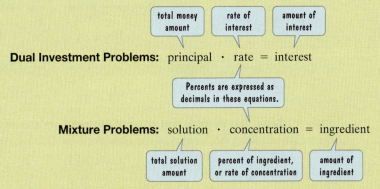

Dual Investment Problems: principal · rate = interest

Percents are expressed as decimals in these equations.

Mixture Problems: solution · concentration = ingredient

Our dual investment problem involved mixing two investments. Our mixture problem involved mixing two liquids. The equations in these problems are obtained from similar conditions:

Dual Investment Problems		**Mixture Problems**	
Interest from investment 1 + interest from investment 2 = amount of interest from mixed investments.		Ingredient amount in solution 1 + ingredient amount in solution 2 = amount of ingredient in mixture.	

Being aware of the similarities between dual investment and mixture problems should make you a better problem solver in a variety of situations that involve mixtures.

Problems Involving Motion

We have seen that the distance, d, covered by any moving body is the product of its average rate, r, and its time in motion, t:

$$d = rt. \quad \text{Distance equals rate times time.}$$

Wind and water current have the effect of increasing or decreasing a traveler's rate.

EXAMPLE 4   Solving a Motion Problem

When a small airplane flies with the wind, it can travel 450 miles in 3 hours. When the same airplane flies in the opposite direction against the wind, it takes 5 hours to fly the same distance. Find the average rate of the plane in still air and the average rate of the wind.

Solution

Step 1. Use variables to represent unknown quantities.

Let $x = $ the average rate of the plane in still air.
Let $y = $ the average rate of the wind.

Step 2. Write a system of equations that models the problem's conditions. As it travels with the wind, the plane's rate is increased. The net rate is its rate in still air, x, plus the rate of the wind, y, given by the expression $x + y$. As it travels against the wind, the plane's rate is decreased. The net rate is its rate in still air, x, minus the rate of the wind, y, given by the expression $x - y$. Here is a chart that summarizes the problem's information and includes the increased and decreased rates.

	Rate	$\times$	Time	$=$	Distance
Trip with the Wind	$x + y$		3		$3(x + y)$
Trip against the Wind	$x - y$		5		$5(x - y)$

The problem states that the distance in each direction is 450 miles. We use this information to write our system of equations.

The distance of the trip with the wind   is   450 miles.

$$3(x + y) = 450$$

The distance of the trip against the wind   is   450 miles.

$$5(x - y) = 450$$

Step 3. Solve the system and answer the problem's question. We can simplify the system by dividing both sides of the equations by 3 and 5, respectively.

$$\begin{cases} 3(x + y) = 450 \\ 5(x - y) = 450 \end{cases} \xrightarrow{\text{Divide by 3.}} \xrightarrow{\text{Divide by 5.}} \begin{cases} x + y = 150 \\ x - y = 90 \end{cases}$$

Solve the system on the right by the addition method.

$$\begin{cases} x + y = 150 \\ \underline{x - y = 90} \end{cases}$$

Add: $2x = 240$

$x = 120$ Divide both sides by 2.

Back-substituting 120 for x in either of the system's equations gives $y = 30$. Because $x = 120$ and $y = 30$, the average rate of the plane in still air is 120 miles per hour and the average rate of the wind is 30 miles per hour.

Step 4. Check the proposed solution in the original wording of the problem. The problem states that the distance in each direction is 450 miles. The average rate of the plane with the wind is $120 + 30 = 150$ miles per hour. In 3 hours, it travels $150 \cdot 3$, or 450 miles, which checks with the stated condition. Furthermore, the average rate of the plane against the wind is $120 - 30 = 90$ miles per hour. In 5 hours, it travels $90 \cdot 5 = 450$ miles, which is the stated distance. ∎

✓ **CHECK POINT 4** With the current, a motorboat can travel 84 miles in 2 hours. Against the current, the same trip takes 3 hours. Find the average rate of the boat in still water and the average rate of the current.

② Use functions to model revenue, cost, and profit, and perform a break-even analysis. ▶

Functions of Business: Break-Even Analysis

Suppose that a company produces and sells x units of a product. Its *revenue* is the money generated by selling x units of the product. Its *cost* is the cost of producing x units of the product.

Revenue and Cost Functions

A company produces and sells x units of a product.

Revenue Function

$$R(x) = (\text{price per unit sold})x$$

Cost Function

$$C(x) = \text{fixed cost} + (\text{cost per unit produced})x$$

The point of intersection of the graphs of the revenue and cost functions is called the **break-even point**. The x-coordinate of the point reveals the number of units that a company must produce and sell so that money coming in, the revenue, is equal to money going out, the cost. The y-coordinate of the break-even point gives the amount of money coming in and going out. Example 5 illustrates the use of the substitution method in determining a company's break-even point.

EXAMPLE 5 Finding a Break-Even Point

Technology is now promising to bring light, fast, and beautiful wheelchairs to millions of disabled people. A company is planning to manufacture these radically different wheelchairs. Fixed cost will be $500,000 and it will cost $400 to produce each wheelchair. Each wheelchair will be sold for $600.

a. Write the cost function, C, of producing x wheelchairs.

b. Write the revenue function, R, from the sale of x wheelchairs.

c. Determine the break-even point. Describe what this means.

Solution

a. The cost function is the sum of the fixed cost and variable cost.

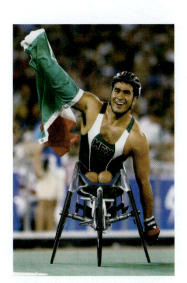

Fixed cost of $500,000   plus   Variable cost: $400 for each chair produced

$$C(x) = 500,000 + 400x$$

b. The revenue function is the money generated from the sale of x wheelchairs.

Revenue per chair, $600, times   the number of chairs sold

$$R(x) = 600x$$

c. The break-even point occurs where the graphs of C and R intersect. Thus, we find this point by solving the system

$$\begin{cases} C(x) = 500{,}000 + 400x \\ R(x) = 600x \end{cases} \quad \text{or} \quad \begin{cases} y = 500{,}000 + 400x \\ y = 600x. \end{cases}$$

Using the substitution method, we can substitute $600x$ for y in the first equation.

$$600x = 500{,}000 + 400x \qquad \text{Substitute } 600x \text{ for } y \text{ in } y = 500{,}000 + 400x$$
$$200x = 500{,}000 \qquad\qquad \text{Subtract } 400x \text{ from both sides.}$$
$$x = 2500 \qquad\qquad\qquad \text{Divide both sides by 200.}$$

Back-substituting 2500 for x in either of the system's equations (or functions), we obtain

$$R(2500) = 600(2500) = 1{,}500{,}000.$$

We used $R(x) = 600x$.

The break-even point is $(2500, 1{,}500{,}000)$. This means that the company will break even if it produces and sells 2500 wheelchairs. At this level, the money coming in is equal to the money going out: \$1,500,000. ∎

    Figure 3.7 shows the graphs of the revenue and cost functions for the wheelchair business. Similar graphs and models apply no matter how small or large a business venture may be.

Figure 3.7

    The intersection point confirms that the company breaks even by producing and selling 2500 wheelchairs. Can you see what happens for $x < 2500$? The red cost graph lies above the blue revenue graph. The cost is greater than the revenue and the business is losing money. Thus, if they sell fewer than 2500 wheelchairs, the result is a *loss*. By contrast, look at what happens for $x > 2500$. The blue revenue graph lies above the red cost graph. The revenue is greater than the cost and the business is making money. Thus, if they sell more than 2500 wheelchairs, the result is a *gain*.

☑ **CHECK POINT 5**   A company that manufactures running shoes has a fixed cost of $300,000. Additionally, it costs $30 to produce each pair of shoes. The shoes are sold at $80 per pair.

    a. Write the cost function, C, of producing x pairs of running shoes.

    b. Write the revenue function, R, from the sale of x pairs of running shoes.

    c. Determine the break-even point. Describe what this means.

What does every entrepreneur, from a kid selling lemonade to Mark Zuckerberg, want to do? Generate profit, of course. The *profit* made is the money taken in, or the revenue, minus the money spent, or the cost. This relationship between revenue and cost allows us to define the *profit function, P(x)*.

> ### The Profit Function
>
> The profit, $P(x)$, generated after producing and selling x units of a product is given by the **profit function**
>
> $$P(x) = R(x) - C(x),$$
>
> where R and C are the revenue and cost functions, respectively.

EXAMPLE 6   Writing a Profit Function

Use the revenue and cost functions for the wheelchair business in Example 5,

$$R(x) = 600x \quad \text{and} \quad C(x) = 500{,}000 + 400x,$$

to write the profit function for producing and selling x wheelchairs.

Solution   The profit function is the difference between the revenue function and the cost function.

$$\begin{aligned} P(x) &= R(x) - C(x) &&\text{This is the definition of the profit function.}\\ &= 600x - (500{,}000 + 400x) &&\text{Substitute the given functions.}\\ &= 600x - 500{,}000 - 400x &&\text{Distribute } -1 \text{ to each term in parentheses.}\\ &= 200x - 500{,}000 &&\text{Simplify: } 600x - 400x = 200x. \end{aligned}$$

The profit function is $P(x) = 200x - 500{,}000.$ ∎

The graph of the profit function for the wheelchair business, $P(x) = 200x - 500{,}000$, is shown in **Figure 3.8**. The red portion lies below the x-axis and shows a loss when fewer than 2500 wheelchairs are sold. The business is "in the red." The black portion lies above the x-axis and shows a gain when more than 2500 wheelchairs are sold. The wheelchair business is "in the black."

☑ **CHECK POINT 6**   Use the revenue and cost functions that you obtained in Check Point 5 to write the profit function for producing and selling x pairs of running shoes.

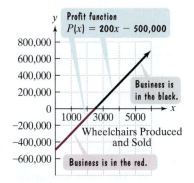

Figure 3.8

Achieving Success

Avoid asking for help on a problem that you have not thought about. This is basically asking your instructor to do the work for you. First try solving the problem on your own!

CONCEPT AND VOCABULARY CHECK

Fill in each blank so that the resulting statement is true.

1. If computers sell for \$1180 per unit and hard drives sell for \$125 per unit, the revenue from *x* computers and *y* hard drives can be represented by _____.

2. The combined yearly interest for *x* dollars invested at 12% and *y* dollars invested at 9% is represented by _____.

3. The total amount of acid in *x* milliliters of a 9% acid solution and *y* milliliters of a 60% acid solution is represented by _____.

4. If *x* represents the average rate of a plane in still air and *y* represents the average rate of the wind, the plane's rate with the wind is represented by _____ and the plane's rate against the wind is represented by _____.

5. If *x* + *y* represents a motorboat's rate with the current, in miles per hour, its distance after 4 hours is represented by _____.

6. A company's _____ function is the money generated by selling *x* units of its product. The difference between this function and the company's cost function is called its _____ function.

7. A company has a graph that shows the money it generates by selling *x* units of its product. It also has a graph that shows its cost of producing *x* units of its product. The point of intersection of these graphs is called the company's _____.

3.2 EXERCISE SET

Practice Exercises

In Exercises 1–4, let x represent one number and let y represent the other number. Use the given conditions to write a system of equations. Solve the system and find the numbers.

1. The sum of two numbers is 7. If one number is subtracted from the other, the result is −1. Find the numbers.

2. The sum of two numbers is 2. If one number is subtracted from the other, the result is 8. Find the numbers.

3. Three times a first number decreased by a second number is 1. The first number increased by twice the second number is 12. Find the numbers.

4. The sum of three times a first number and twice a second number is 8. If the second number is subtracted from twice the first number, the result is 3. Find the numbers.

In Exercises 5–8, cost and revenue functions for producing and selling x units of a product are given. Cost and revenue are expressed in dollars.

  a. *Find the number of units that must be produced and sold to break even. At this level, what is the dollar amount coming in and going out?*

  b. *Write the profit function from producing and selling x units of the product.*

5. $C(x) = 25{,}500 + 15x$
   $R(x) = 32x$

6. $C(x) = 15{,}000 + 12x$
   $R(x) = 32x$

7. $C(x) = 105x + 70{,}000$
   $R(x) = 245x$

8. $C(x) = 1.2x + 1500$
   $R(x) = 1.7x$

Application Exercises

In Exercises 9–40, use the four-step strategy to solve each problem.

9. The current generation of college students grew up playing interactive online games, and many continue to play in college. The bar graph shows the percentage of U.S. college students playing online games, by gender.

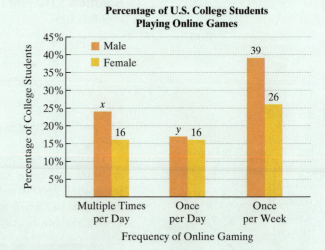

Percentage of U.S. College Students Playing Online Games

Source: statista.com

A total of 41% of college men play online games multiple times per day or once per day. The difference in the percentage who play multiple times per day and once per day is 7%. Find the percentage of college men who play online games multiple times per day and the percentage of college men who play online games once per day.

10. A number of studies have emphasized the importance of sleep for students' success in their academic performance. The bar graph shows the actual sleep hours and the sleep hours to function best for U.S. college students.

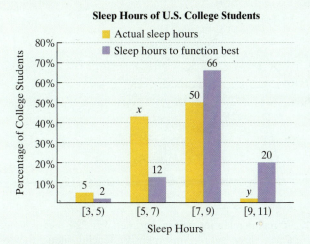

Sleep Hours of U.S. College Students

Source: Chiang et al., "The Effects of Sleep on Academic Performance and Job Performance," *College Student Journal*, Spring 2014

A total of 45% of college students sleep between 5 and 7 hours or between 9 and 11 hours. The difference in the percentage who sleep between 5 and 7 hours and between 9 and 11 hours is 41%. Find the percentage of college students who sleep between 5 and 7 hours and the percentage of college students who sleep between 9 and 11 hours.

11. One week a computer store sold a total of 36 computers and external hard drives. The revenue from these sales was $27,710. If computers sold for $1180 per unit and hard drives for $125 per unit, how many of each did the store sell?

12. There were 180 people at a civic club fundraiser. Members paid $4.50 per ticket and nonmembers paid $8.25 per ticket. If total receipts amounted to $1222.50, how many members and how many nonmembers attended the fundraiser?

13. You invested $7000 in two accounts paying 6% and 8% annual interest. If the total interest earned for the year was $520, how much was invested at each rate?

14. You invested $11,000 in stocks and bonds, paying 5% and 8% annual interest. If the total interest earned for the year was $730, how much was invested in stocks and how much was invested in bonds?

15. You invested money in two funds. Last year, the first fund paid a dividend of 9% and the second a dividend of 3%, and you received a total of $900. This year, the first fund paid a 10% dividend and the second only 1%, and you received a total of $860. How much money did you invest in each fund?

16. You invested money in two funds. Last year, the first fund paid a dividend of 8% and the second a dividend of 5%, and you received a total of $1330. This year, the first fund paid a 12% dividend and the second only 2%, and you received a total of $1500. How much money did you invest in each fund?

17. Things did not go quite as planned. You invested $20,000, part of it in a stock with a 12% annual return. However, the rest of the money suffered a 5% loss. If the total annual income from both investments was $1890, how much was invested at each rate?

18. Things did not go quite as planned. You invested $30,000, part of it in a stock with a 14% annual return. However, the rest of the money suffered a 6% loss. If the total annual income from both investments was $200, how much was invested at each rate?

19. A wine company needs to blend a California wine with a 5% alcohol content and a French wine with a 9% alcohol content to obtain 200 gallons of wine with a 7% alcohol content. How many gallons of each kind of wine must be used?

20. A jeweler needs to mix an alloy with a 16% gold content and an alloy with a 28% gold content to obtain 32 ounces of a new alloy with a 25% gold content. How many ounces of each of the original alloys must be used?

21. For thousands of years, gold has been considered one of Earth's most precious metals. One hundred percent pure gold is 24-karat gold, which is too soft to be made into jewelry. In the United States, most gold jewelry is 14-karat gold, approximately 58% gold. If 18-karat gold is 75% gold and 12-karat gold is 50% gold, how much of each should be used to make a 14-karat gold bracelet weighing 300 grams?

22. In the "Peanuts" cartoon shown, solve the problem that is sending Peppermint Patty into an agitated state. How much cream and how much milk, to the nearest hundredth of a gallon, must be mixed together to obtain 50 gallons of cream that contains 12.5% butterfat?

Peanuts copyright © 1978 Peanuts Worldwide LLC. Distributed by Universal Uclick. Reprinted with permission. All rights reserved.

23. The manager of a candystand at a large multiplex cinema has a popular candy that sells for $1.60 per pound. The manager notices a different candy worth $2.10 per pound that is not selling well. The manager decides to form a mixture of both types of candy to help clear the inventory of the more expensive type. How many pounds of each kind of candy should be used to create a 75-pound mixture selling for $1.90 per pound?

24. A grocer needs to mix raisins at $2.00 per pound with granola at $3.25 per pound to obtain 10 pounds of a mixture that costs $2.50 per pound. How many pounds of raisins and how many pounds of granola must be used?

25. A coin purse contains a mixture of 15 coins in nickels and dimes. The coins have a total value of $1.10. Determine the number of nickels and the number of dimes in the purse.

26. A coin purse contains a mixture of 15 coins in dimes and quarters. The coins have a total value of $3.30. Determine the number of dimes and the number of quarters in the purse.

27. When a small plane flies with the wind, it can travel 800 miles in 5 hours. When the plane flies in the opposite direction, against the wind, it takes 8 hours to fly the same distance. Find the rate of the plane in still air and the rate of the wind.

28. When a plane flies with the wind, it can travel 4200 miles in 6 hours. When the plane flies in the opposite direction, against the wind, it takes 7 hours to fly the same distance. Find the rate of the plane in still air and the rate of the wind.

29. A boat's crew rowed 16 kilometers downstream, with the current, in 2 hours. The return trip upstream, against the current, covered the same distance, but took 4 hours. Find the crew's rowing rate in still water and the rate of the current.

30. A motorboat traveled 36 miles downstream, with the current, in 1.5 hours. The return trip upstream, against the current, covered the same distance, but took 2 hours. Find the boat's rate in still water and the rate of the current.

31. With the current, you can canoe 24 miles in 4 hours. Against the same current, you can canoe only $\frac{3}{4}$ of this distance in 6 hours. Find your rate in still water and the rate of the current.

32. With the current, you can row 24 miles in 3 hours. Against the same current, you can row only $\frac{2}{3}$ of this distance in 4 hours. Find your rowing rate in still water and the rate of the current.

33. A student has two test scores. The difference between the scores is 12 and the mean, or average, of the scores is 80. What are the two test scores?

34. A student has two test scores. The difference between the scores is 8 and the mean, or average, of the scores is 88. What are the two test scores?

In Exercises 35–36, an isosceles triangle containing two angles with equal measure is shown. The degree measure of each triangle's three interior angles and an exterior angle is represented with variables. Find the measure of the three interior angles.

35.

36.

37. A rectangular lot whose perimeter is 220 feet is fenced along three sides. An expensive fencing along the lot's length costs $20 per foot, and an inexpensive fencing along the two side widths costs only $8 per foot. The total cost of the fencing along the three sides comes to $2040. What are the lot's dimensions?

38. A rectangular lot whose perimeter is 260 feet is fenced along three sides. An expensive fencing along the lot's length costs $16 per foot, and an inexpensive fencing along the two side widths costs only $5 per foot. The total cost of the fencing along the three sides comes to $1780. What are the lot's dimensions?

39. A new restaurant is to contain two-seat tables and four-seat tables. Fire codes limit the restaurant's maximum occupancy to 56 customers. If the owners have hired enough servers to handle 17 tables of customers, how many of each kind of table should they purchase?

40. A hotel has 200 rooms. Those with kitchen facilities rent for $100 per night and those without kitchen facilities rent for $80 per night. On a night when the hotel was completely occupied, revenues were $17,000. How many of each type of room does the hotel have?

The figure shows the graphs of the cost and revenue functions for a company that manufactures and sells small radios. Use the information in the figure to solve Exercises 41–46.

41. How many radios must be produced and sold for the company to break even?

42. More than how many radios must be produced and sold for the company to have a profit?

43. Use the formulas shown in the voice balloons to find $R(200) - C(200)$. Describe what this means for the company.

44. Use the formulas shown in the voice balloons to find $R(300) - C(300)$. Describe what this means for the company.

45. a. Use the formulas shown in the voice balloons to write the company's profit function, P, from producing and selling x radios.

    b. Find the company's profit if 10,000 radios are produced and sold.

46. a. Use the formulas shown in the voice balloons to write the company's profit function, P, from producing and selling x radios.

    b. Find the company's profit if 20,000 radios are produced and sold.

Exercises 47–50 describe a number of business ventures. For each exercise,

a. *Write the cost function, C.*

b. *Write the revenue function, R.*

c. *Determine the break-even point. Describe what this means.*

47. A company that manufactures small canoes has a fixed cost of $18,000. It costs $20 to produce each canoe. The selling price is $80 per canoe. (In solving this exercise, let x represent the number of canoes produced and sold.)

48. A company that manufactures bicycles has a fixed cost of $100,000. It costs $100 to produce each bicycle. The selling price is $300 per bike. (In solving this exercise, let x represent the number of bicycles produced and sold.)

49. You invest in a new play. The cost includes an overhead of $30,000, plus production costs of $2500 per performance. A sold-out performance brings in $3125. (In solving this exercise, let x represent the number of sold-out performances.)

50. You invested $30,000 and started a business writing greeting cards. Supplies cost 2¢ per card and you are selling each card for 50¢. (In solving this exercise, let x represent the number of cards produced and sold.)

Explaining the Concepts

51. Describe the conditions in a problem that enable it to be solved using a system of linear equations.

52. Write a word problem that can be solved by translating to a system of linear equations. Then solve the problem.

53. Describe a revenue function for a business venture.

54. Describe a cost function for a business venture. What are the two kinds of costs that are modeled by this function?

55. What is the profit function for a business venture and how is it determined?

56. Describe the break-even point for a business.

57. The law of supply and demand states that, in a free market economy, a commodity tends to be sold at its equilibrium price. At this price, the amount that the seller will supply is the same amount that the consumer will buy. Explain how graphs can be used to determine the equilibrium price.

58. Many students hate mixture problems and decide to ignore them, stating, "I'll just skip that one on the test." If you share this opinion, describe what you find particularly unappealing about this kind of problem.

Technology Exercises

In Exercises 59–60, graph the revenue and cost functions in the same viewing rectangle. Then use the intersection feature to determine the break-even point.

59. $R(x) = 50x, \quad C(x) = 20x + 180$

60. $R(x) = 92.5x, \quad C(x) = 52x + 1782$

61. Use the procedure in Exercises 59–60 to verify your work for any one of the break-even points that you found in Exercises 47–50.

Critical Thinking Exercises

Make Sense? *In Exercises 62–65, determine whether each statement makes sense or does not make sense, and explain your reasoning.*

62. A system of linear equations can be used to model and compare the fees charged by two different taxicab companies.

63. I should mix 6 liters of a 50% acid solution with 4 liters of a 25% acid solution to obtain 10 liters of a 75% acid solution.

64. If I know the perimeter of this rectangle and triangle, each in the same unit of measure, I can use a system of linear equations to determine values for x and y.

 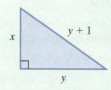

65. You told me that you flew against the wind from Miami to Seattle in 7 hours and, at the same time, your friend flew with the wind from Seattle to Miami in only 5.6 hours. You have not given me enough information to determine the average rate of the wind.

66. The radiator in your car contains 4 gallons of antifreeze and water. The mixture is 45% antifreeze. How much of this mixture should be drained and replaced with pure antifreeze in order to have a 60% antifreeze solution? Round to the nearest tenth of a gallon.

67. A marching band has 52 members, and there are 24 in the pom-pom squad. They wish to form several hexagons and squares like those diagrammed below. Can it be done with no people left over?

B = Band Member
P = Pom-pom Person

68. A boy has as many brothers as he has sisters. Each of his sisters has twice as many brothers as she has sisters. How many boys and girls are in this family?

69. When entering your test score into a computer, your professor accidently reversed the two digits. This error reduced your score by 36 points. Your professor told you that the sum of the digits of your actual score was 14, corrected the error, and agreed to give you extra credit if you could determine the actual score without looking back at the test. What was your actual test score? (*Hint:* Let t = the tens-place digit of your actual score and let u = the units-place digit of your actual score. Thus, $10t + u$ represents your actual test score.)

70. A dealer paid a total of $67 for mangos and avocados. The mangos were sold at a profit of 20% on the dealer's cost, but the avocados started to spoil, resulting in a selling price of a 2% loss on the dealer's cost. The dealer made a profit of $8.56 on the total transaction. How much did the dealer pay for the mangos and for the avocados?

Review Exercises

In Exercises 71–72, use the given conditions to write an equation for each line in point-slope form and slope-intercept form.

71. Passing through $(-2, 5)$ and $(-6, 13)$
(Section 2.5, Example 2)

72. Passing through $(-3, 0)$ and parallel to the line whose equation is $-x + y = 7$
(Section 2.5, Example 4)

73. Find the domain of $g(x) = \dfrac{x - 2}{3 - x}$.
(Section 2.3, Example 1)

Preview Exercises

Exercises 74–76 will help you prepare for the material covered in the next section.

74. If $x = 3$, $y = 2$, and $z = -3$, does the ordered triple (x, y, z) satisfy the equation $2x - y + 4z = -8$?

75. Consider the following equations:

$$\begin{cases} 5x - 2y - 4z = 3 & \text{Equation 1} \\ 3x + 3y + 2z = -3. & \text{Equation 2} \end{cases}$$

Eliminate z by copying Equation 1, multiplying Equation 2 by 2, and then adding the equations.

76. Write an equation involving a, b, and c based on the following description:

When the value of x in $y = ax^2 + bx + c$ is 4, the value of y is 1682.

SECTION

3.3

Systems of Linear Equations in Three Variables ▶

What am I supposed to learn?

After studying this section, you should be able to:

1 Verify the solution of a system of linear equations in three variables. ▶

2 Solve systems of linear equations in three variables. ▶

3 Identify inconsistent and dependent systems. ▶

4 Solve problems using systems in three variables. ▶

All animals sleep, but the length of time they sleep varies widely: Cattle sleep for only a few minutes at a time. We humans seem to need more sleep than other animals, up to eight hours a day. Without enough sleep, we have difficulty concentrating, make mistakes in routine tasks, lose energy, and feel bad-tempered. There is a relationship between hours of sleep and death rate per year per 100,000 people. How many hours of sleep will put you in the group with the minimum death rate? In this section, we will answer this question by solving a system of linear equations with more than two variables.

Point of intersection

P

Figure 3.9

1 Verify the solution of a system of linear equations in three variables. ▶

Systems of Linear Equations in Three Variables and Their Solutions

An equation such as $x + 2y - 3z = 9$ is called a *linear equation in three variables*. In general, any equation of the form

$$Ax + By + Cz = D,$$

where A, B, C, and D are real numbers such that A, B, and C are not all 0, is a **linear equation in three variables: *x*, *y*, and *z*.** The graph of this linear equation in three variables is a plane in three-dimensional space.

The process of solving a system of three linear equations in three variables is geometrically equivalent to finding the point of intersection (assuming that there is one) of three planes in space (see **Figure 3.9**). A **solution** of a system of linear equations in three variables is an ordered triple of real numbers that satisfies all equations in the system. The **solution set** of the system is the set of all its solutions.

EXAMPLE 1 Determining Whether an Ordered Triple Satisfies a System

Show that the ordered triple $(-1, 2, -2)$ is a solution of the system:

$$\begin{cases} x + 2y - 3z = 9. \\ 2x - y + 2z = -8. \\ -x + 3y - 4z = 15. \end{cases}$$

Solution Because -1 is the x-coordinate, 2 is the y-coordinate, and -2 is the z-coordinate of $(-1, 2, -2)$, we replace x with -1, y with 2, and z with -2 in each of the three equations.

$$x + 2y - 3z = 9$$
$$-1 + 2(2) - 3(-2) \stackrel{?}{=} 9$$
$$-1 + 4 + 6 \stackrel{?}{=} 9$$
$$9 = 9, \quad \text{true}$$

$$2x - y + 2z = -8$$
$$2(-1) - 2 + 2(-2) \stackrel{?}{=} -8$$
$$-2 - 2 - 4 \stackrel{?}{=} -8$$
$$-8 = -8, \quad \text{true}$$

$$-x + 3y - 4z = 15$$
$$-(-1) + 3(2) - 4(-2) \stackrel{?}{=} 15$$
$$1 + 6 + 8 \stackrel{?}{=} 15$$
$$15 = 15, \quad \text{true}$$

The ordered triple $(-1, 2, -2)$ satisfies the three equations: It makes each equation true. Thus, the ordered triple is a solution of the system. ∎

✓ CHECK POINT 1 Show that the ordered triple $(-1, -4, 5)$ is a solution of the system:

$$\begin{cases} x - 2y + 3z = 22 \\ 2x - 3y - z = 5 \\ 3x + y - 5z = -32. \end{cases}$$

2 Solve systems of linear equations in three variables. ▶

Solving Systems of Linear Equations in Three Variables by Eliminating Variables

The method for solving a system of linear equations in three variables is similar to that used on systems of linear equations in two variables. We use addition to eliminate any variable, reducing the system to two equations in two variables. Once we obtain a system of two equations in two variables, we use addition or substitution to eliminate a variable. The result is a single equation in one variable. We solve this equation to get the value of the remaining variable. Other variable values are found by back-substitution.

Great Question!

When solving a linear system in three variables, which variable should I eliminate first?

It does not matter which variable you eliminate first, as long as you eliminate the same variable in two different pairs of equations.

Solving Linear Systems in Three Variables by Eliminating Variables

1. Reduce the system to two equations in two variables. This is usually accomplished by taking two different pairs of equations and using the addition method to eliminate the same variable from both pairs.

2. Solve the resulting system of two equations in two variables using addition or substitution. The result is an equation in one variable that gives the value of that variable.

3. Back-substitute the value of the variable found in step 2 into either of the equations in two variables to find the value of the second variable.

4. Use the values of the two variables from steps 2 and 3 to find the value of the third variable by back-substituting into one of the original equations.

5. Check the proposed solution in each of the original equations.

EXAMPLE 2   Solving a System in Three Variables

Solve the system:

$$\begin{cases} 5x - 2y - 4z = 3 & \text{Equation 1} \\ 3x + 3y + 2z = -3 & \text{Equation 2} \\ -2x + 5y + 3z = 3. & \text{Equation 3} \end{cases}$$

Solution   There are many ways to proceed. Because our initial goal is to reduce the system to two equations in two variables, **the central idea is to take two different pairs of equations and eliminate the same variable from both pairs**.

Step 1. Reduce the system to two equations in two variables. We choose any two equations and use the addition method to eliminate a variable. Let's eliminate z using Equations 1 and 2. We do so by multiplying Equation 2 by 2. Then we add equations.

(Equation 1) $\begin{cases} 5x - 2y - 4z = 3 \end{cases}$ $\xrightarrow{\text{No change}}$ $\begin{cases} 5x - 2y - 4z = 3 \end{cases}$
(Equation 2) $\begin{cases} 3x + 3y + 2z = -3 \end{cases}$ $\xrightarrow{\text{Multiply by 2.}}$ $\begin{cases} 6x + 6y + 4z = -6 \end{cases}$

Add: $\quad 11x + 4y \qquad = -3$ Equation 4

Now we must eliminate the *same* variable from another pair of equations. We can eliminate z using Equations 2 and 3. First, we multiply Equation 2 by -3. Next, we multiply Equation 3 by 2. Finally, we add equations.

(Equation 2) $\begin{cases} 3x + 3y + 2z = -3 \end{cases}$ $\xrightarrow{\text{Multiply by }-3.}$ $\begin{cases} -9x - 9y - 6z = 9 \end{cases}$
(Equation 3) $\begin{cases} -2x + 5y + 3z = 3 \end{cases}$ $\xrightarrow{\text{Multiply by 2.}}$ $\begin{cases} -4x + 10y + 6z = 6 \end{cases}$

Add: $\quad -13x + y \qquad = 15$ Equation 5

Equations 4 and 5 give us a system of two equations in two variables.

Step 2. Solve the resulting system of two equations in two variables. We will use the addition method to solve Equations 4 and 5 for x and y. To do so, we multiply Equation 5 by -4 and add this to Equation 4.

(Equation 4) $\begin{cases} 11x + 4y = -3 \end{cases}$ $\xrightarrow{\text{No change}}$ $\begin{cases} 11x + 4y = -3 \end{cases}$
(Equation 5) $\begin{cases} -13x + y = 15 \end{cases}$ $\xrightarrow{\text{Multiply by }-4.}$ $\begin{cases} 52x - 4y = -60 \end{cases}$

Add: $\quad 63x \qquad = -63$

$\qquad\qquad x \qquad = -1$ Divide both sides by 63.

Step 3. Use back-substitution in one of the equations in two variables to find the value of the second variable. We back-substitute -1 for x in either Equation 4 or 5 to find the value of y. We will use Equation 5.

$$-13x + y = 15 \qquad \text{Equation 5}$$
$$-13(-1) + y = 15 \qquad \text{Substitute } -1 \text{ for } x.$$
$$13 + y = 15 \qquad \text{Multiply.}$$
$$y = 2 \qquad \text{Subtract 13 from both sides.}$$

Step 4. Back-substitute the values found for two variables into one of the original equations to find the value of the third variable. We can now use any one of the original equations and back-substitute the values of x and y to find the value for z. We will use Equation 2.

$$3x + 3y + 2z = -3 \qquad \text{Equation 2}$$
$$3(-1) + 3(2) + 2z = -3 \qquad \text{Substitute } -1 \text{ for } x \text{ and } 2 \text{ for } y.$$
$$3 + 2z = -3 \qquad \begin{array}{l}\text{Multiply and then add:} \\ 3(-1) + 3(2) = -3 + 6 = 3.\end{array}$$
$$2z = -6 \qquad \text{Subtract 3 from both sides.}$$
$$z = -3 \qquad \text{Divide both sides by 2.}$$

With $x = -1$, $y = 2$, and $z = -3$, the proposed solution is the ordered triple $(-1, 2, -3)$.

Step 5. Check. Check the proposed solution, $(-1, 2, -3)$, by substituting the values for x, y, and z into each of the three original equations. These substitutions yield three true statements. Thus, the solution is $(-1, 2, -3)$ and the solution set is $\{(-1, 2, -3)\}$. ∎

✓ **CHECK POINT 2**   Solve the system:

$$\begin{cases} x + 4y - z = 20 \\ 3x + 2y + z = 8 \\ 2x - 3y + 2z = -16. \end{cases}$$

In some examples, one of the variables is missing from a given equation. In this case, the missing variable should be eliminated from the other two equations, thereby making it possible to omit one of the elimination steps. We illustrate this idea in Example 3.

EXAMPLE 3   Solving a System of Equations with a Missing Term

Solve the system:

$$\begin{cases} x + + z = 8 & \text{Equation 1} \\ x + y + 2z = 17 & \text{Equation 2} \\ x + 2y + z = 16. & \text{Equation 3} \end{cases}$$

Solution

Step 1. Reduce the system to two equations in two variables. Because Equation 1 contains only x and z, we can omit one of the elimination steps by eliminating y using Equations 2 and 3. This will give us two equations in x and z. To eliminate y using Equations 2 and 3, we multiply Equation 2 by -2 and add Equation 3.

(Equation 2) $\begin{cases} x + y + 2z = 17 \\ (Equation 3) \quad x + 2y + z = 16 \end{cases}$ $\xrightarrow[\text{No change}]{\text{Multiply by } -2.}$ $\begin{cases} -2x - 2y - 4z = -34 \\ \underline{x + 2y + z = 16} \end{cases}$

Add: $\quad -x - 3z = -18$ Equation 4

Equation 4 and the given Equation 1 provide us with a system of two equations in two variables:

$$\begin{cases} x + z = 8 & \text{Equation 1} \\ -x - 3z = -18. & \text{Equation 4} \end{cases}$$

Step 2. Solve the resulting system of two equations in two variables. We will solve Equations 1 and 4 for x and z.

$$\begin{cases} x + z = & 8 & \text{Equation 1} \\ -x - 3z = & -18 & \text{Equation 4} \end{cases}$$

Add: $\begin{aligned} -2z &= -10 \\ z &= 5 \end{aligned}$ Divide both sides by -2.

Step 3. Use back-substitution in one of the equations in two variables to find the value of the second variable. To find x, we back-substitute 5 for z in either Equation 1 or 4. We will use Equation 1.

$$\begin{aligned} x + z &= 8 & \text{Equation 1} \\ x + 5 &= 8 & \text{Substitute 5 for } z. \\ x &= 3 & \text{Subtract 5 from both sides.} \end{aligned}$$

Step 4. Back-substitute the values found for two variables into one of the original equations to find the value of the third variable. To find y, we back-substitute 3 for x and 5 for z into Equation 2, $x + y + 2z = 17$, or Equation 3, $x + 2y + z = 16$. We can't use Equation 1, $x + z = 8$, because y is missing in this equation. We will use Equation 2.

$$\begin{aligned} x + y + 2z &= 17 & \text{Equation 2} \\ 3 + y + 2(5) &= 17 & \text{Substitute 3 for } x \text{ and 5 for } z. \\ y + 13 &= 17 & \text{Multiply and add.} \\ y &= 4 & \text{Subtract 13 from both sides.} \end{aligned}$$

We found that $z = 5$, $x = 3$, and $y = 4$. Thus, the proposed solution is the ordered triple $(3, 4, 5)$.

Step 5. Check. Substituting 3 for x, 4 for y, and 5 for z into each of the three original equations yields three true statements. Consequently, the solution is $(3, 4, 5)$ and the solution set is $\{(3, 4, 5)\}$. ∎

✓ CHECK POINT 3   Solve the system:

$$\begin{cases} 2y - z = & 7 \\ x + 2y + z = & 17 \\ 2x - 3y + 2z = & -1. \end{cases}$$

3 Identify inconsistent and dependent systems. ▶

Inconsistent and Dependent Systems

A system of three linear equations in three variables represents three planes. The three planes need not intersect at one point. The planes may have no common point of intersection and represent an **inconsistent system** with no solution. **Figure 3.10** illustrates some of the geometric possibilities for inconsistent systems.

Three planes are parallel with no common intersection point.

Two planes are parallel with no common intersection point.

Planes intersect two at a time. There is no intersection point common to all three planes.

Figure 3.10 Three planes may have no common point of intersection.

If you attempt to solve an inconsistent system algebraically, at some point in the solution process you will eliminate all three variables. A false statement, such as $0 = -10$, will be the result. For example, consider the system

$$\begin{cases} 2x + 5y + z = 12 & \text{Equation 1} \\ x - 2y + 4z = -10 & \text{Equation 2} \\ -3x + 6y - 12z = 20. & \text{Equation 3} \end{cases}$$

Suppose we reduce the system to two equations in two variables by eliminating x. To eliminate x using Equations 2 and 3, we multiply Equation 2 by 3 and add Equation 3:

$$\begin{cases} x - 2y + 4z = -10 \\ -3x + 6y - 12z = 20 \end{cases} \xrightarrow[\text{No change}]{\text{Multiply by 3.}} \begin{cases} 3x - 6y + 12z = -30 \\ -3x + 6y - 12z = 20 \end{cases}$$

$$\text{Add:} \qquad\qquad 0 = -10$$

There are no values of x, y, and z for which $0 = -10$. The false statement $0 = -10$ indicates that the system is inconsistent and has no solution. The solution set is the empty set, $\varnothing$.

We have seen that a linear system that has at least one solution is called a **consistent system**. Planes that intersect at one point and planes that intersect at infinitely many points both represent consistent systems. **Figure 3.11** illustrates two different cases of three planes that intersect at infinitely many points. The equations in these linear systems with infinitely many solutions are called **dependent**.

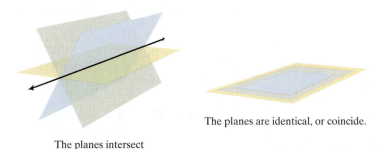

The planes are identical, or coincide.

The planes intersect along a common line.

Figure 3.11 Three planes may intersect at infinitely many points.

If you attempt to solve a system with dependent equations algebraically, at some point in the solution process you will eliminate all three variables. A true statement, such as $0 = 0$, will be the result. If this occurs as you are solving a linear system, simply state that the equations are dependent.

4 Solve problems using systems in three variables. ▶

Applications

Systems of equations may allow us to find models for data without using a graphing utility. Three data points that do not lie on or near a line determine the graph of a function of the form

$$y = ax^2 + bx + c, a \neq 0.$$

Such a function is called a **quadratic function**. If $a > 0$, its graph is shaped like a bowl, making it ideal for modeling situations in which values of y are decreasing and then increasing. In Chapter 8, we'll have lots of interesting things to tell you about quadratic functions and their graphs.

The process of determining a function whose graph contains given points is called **curve fitting**. In our next example, we fit the curve whose equation is $y = ax^2 + bx + c$ to three data points. Using a system of equations, we find values for a, b, and c.

EXAMPLE 4 Modeling Data Relating Sleep and Death Rate

In a study relating sleep and death rate, the following data were obtained. Use the function $y = ax^2 + bx + c$ to model the data.

x (Average Number of Hours of Sleep)	y (Death Rate per Year per 100,000 Males)
4	1682
7	626
9	967

Using Technology

The graph of

$$y = 104.5x^2 - 1501.5x + 6016$$

is displayed in a [3, 12, 1] by [500, 2000, 100] viewing rectangle. The minimum function feature shows that the lowest point on the graph is approximately (7.2, 622.5). Men who average 7.2 hours of sleep are in the group with the lowest death rate, approximately 622.5 deaths per 100,000 males.

Minimum
X=7.1842088   Y=622.45395

Solution We need to find values for a, b, and c in $y = ax^2 + bx + c$. We can do so by solving a system of three linear equations in a, b, and c. We obtain the three equations by using the values of x and y from the data as follows:

$y = ax^2 + bx + c$ *Use the quadratic function to model the data.*

When $x = 4$, $y = 1682$:
When $x = 7$, $y = 626$:
When $x = 9$, $y = 967$:

$$\begin{cases} 1682 = a \cdot 4^2 + b \cdot 4 + c \\ 626 = a \cdot 7^2 + b \cdot 7 + c \\ 967 = a \cdot 9^2 + b \cdot 9 + c \end{cases} \text{ or } \begin{cases} 16a + 4b + c = 1682 \\ 49a + 7b + c = 626 \\ 81a + 9b + c = 967. \end{cases}$$

The easiest way to solve this system is to eliminate c from two pairs of equations, obtaining two equations in a and b. Solving this system gives $a = 104.5$, $b = -1501.5$, and $c = 6016$. We now substitute the values for a, b, and c into $y = ax^2 + bx + c$. The function that models the given data is

$$y = 104.5x^2 - 1501.5x + 6016. \quad \blacksquare$$

We can use the model that we obtained in Example 4 to find the death rate of males who average, say, 6 hours of sleep. First, write the model in function notation:

$$f(x) = 104.5x^2 - 1501.5x + 6016.$$

Substitute 6 for x:

$$f(6) = 104.5(6)^2 - 1501.5(6) + 6016 = 769.$$

According to the model, the death rate for males who average 6 hours of sleep is 769 deaths per 100,000 males.

✓ **CHECK POINT 4** Find the quadratic function $y = ax^2 + bx + c$ whose graph passes through the points $(1, 4)$, $(2, 1)$, and $(3, 4)$.

Problems involving three unknowns can be solved using the same strategy for solving problems with two unknown quantities. You can let x, y, and z represent the unknown quantities. We then translate from the verbal conditions of the problem to a system of three equations in three variables. Problems of this type are included in the Exercise Set that follows.

Achieving Success

Think about finding a tutor.

If you're attending all lectures, taking good class notes, reading the textbook, and doing all the assigned homework, but still having difficulty in a math course, you might want to find a tutor. Many on-campus learning centers and math labs have trained people available to help you. Sometimes a TA who has previously taught the course is available. **Make sure the tutor is both good at math and familiar with the particular course you're taking.** Bring your textbook, class notes, the problems you've done, and information about course policy and tests to each meeting with your tutor. That way he or she can be sure the tutoring sessions address your exact needs.

CONCEPT AND VOCABULARY CHECK

Fill in each blank so that the resulting statement is true.

1. A solution of a system of linear equations in three variables is an ordered _____ of real numbers that satisfies <u>all/some</u> of the equations in the system.

   Circle the correct choice.

2. Consider the following system:

$$\begin{cases} x + \cdot y - z = -1 & \text{Equation 1} \\ 2x - 2y - 5z = 7 & \text{Equation 2} \\ 4x + y - 2z = 7. & \text{Equation 3} \end{cases}$$

   We can eliminate x from Equations 1 and 2 by multiplying Equation 1 by _____ and adding equations. We can eliminate x from Equations 1 and 3 by multiplying Equation 1 by _____ and adding equations.

3. Consider the following system:

$$\begin{cases} x + y + z = 2 & \text{Equation 1} \\ 2x - 3y = 3 & \text{Equation 2} \\ 10y - z = 12. & \text{Equation 3} \end{cases}$$

   Equation 2 does not contain the variable _____. To obtain a second equation that does not contain this variable, we can _____.

4. A function of the form $y = ax^2 + bx + c, a \neq 0$, is called a/an _____ function.

5. The process of determining a function whose graph contains given points is called _____.

3.3 EXERCISE SET ▶ MyMathLab®

Practice Exercises

In Exercises 1–4, determine if the given ordered triple is a solution of the system.

1. $(2, -1, 3)$
$$\begin{cases} x + y + z = 4 \\ x - 2y - z = 1 \\ 2x - y - z = -1 \end{cases}$$

2. $(5, -3, -2)$
$$\begin{cases} x + y + z = 0 \\ x + 2y - 3z = 5 \\ 3x + 4y + 2z = -1 \end{cases}$$

3. $(4, 1, 2)$
$$\begin{cases} x - 2y = 2 \\ 2x + 3y = 11 \\ y - 4z = -7 \end{cases}$$

4. $(-1, 3, 2)$
$$\begin{cases} x - 2z = -5 \\ y - 3z = -3 \\ 2x - z = -4 \end{cases}$$

Solve each system in Exercises 5–22. If there is no solution or if there are infinitely many solutions and a system's equations are dependent, so state.

5. $$\begin{cases} x + y + 2z = 11 \\ x + y + 3z = 14 \\ x + 2y - z = 5 \end{cases}$$

6. $$\begin{cases} 2x + y - 2z = -1 \\ 3x - 3y - z = 5 \\ x - 2y + 3z = 6 \end{cases}$$

7. $$\begin{cases} 4x - y + 2z = 11 \\ x + 2y - z = -1 \\ 2x + 2y - 3z = -1 \end{cases}$$

8. $$\begin{cases} x - y + 3z = 8 \\ 3x + y - 2z = -2 \\ 2x + 4y + z = 0 \end{cases}$$

9. $$\begin{cases} 3x + 2y - 3z = -2 \\ 2x - 5y + 2z = -2 \\ 4x - 3y + 4z = 10 \end{cases}$$

10. $$\begin{cases} 2x + 3y + 7z = 13 \\ 3x + 2y - 5z = -22 \\ 5x + 7y - 3z = -28 \end{cases}$$

11. $$\begin{cases} 2x - 4y + 3z = 17 \\ x + 2y - z = 0 \\ 4x - y - z = 6 \end{cases}$$

12. $$\begin{cases} x + z = 3 \\ x + 2y - z = 1 \\ 2x - y + z = 3 \end{cases}$$

13. $$\begin{cases} 2x + y = 2 \\ x + y - z = 4 \\ 3x + 2y + z = 0 \end{cases}$$

14. $$\begin{cases} x + 3y + 5z = 20 \\ y - 4z = -16 \\ 3x - 2y + 9z = 36 \end{cases}$$

15. $$\begin{cases} x + y = -4 \\ y - z = 1 \\ 2x + y + 3z = -21 \end{cases}$$

40. The bar graph shows the average number of hours U.S. college students study per week for seven selected majors. Study times are rounded to the nearest hour.

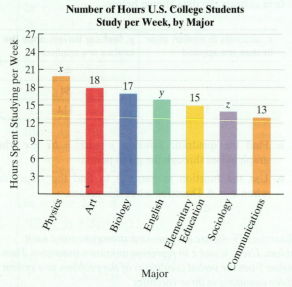

Number of Hours U.S. College Students Study per Week, by Major

Source: statista.com

The combined weekly study time for students majoring in physics, English, and sociology is 50 hours. The difference between weekly study time for physics majors and English majors is 4 hours. The difference between weekly study time for physics majors and sociology majors is 6 hours. Find the average number of hours per week that physics majors, English majors, and sociology majors spend studying.

41. A person invested $6700 for one year, part at 8%, part at 10%, and the remainder at 12%. The total annual income from these investments was $716. The amount of money invested at 12% was $300 more than the amounts invested at 8% and 10% combined. Find the amount invested at each rate.

42. A person invested $17,000 for one year, part at 10%, part at 12%, and the remainder at 15%. The total annual income from these investments was $2110. The amount of money invested at 12% was $1000 less than the amounts invested at 10% and 15% combined. Find the amount invested at each rate.

43. At a college production of *Streetcar Named Desire*, 400 tickets were sold. The ticket prices were $8, $10, and $12, and the total income from ticket sales was $3700. How many tickets of each type were sold if the combined number of $8 and $10 tickets sold was 7 times the number of $12 tickets sold?

44. A certain brand of razor blades comes in packages of 6, 12, and 24 blades, costing $2, $3, and $4 per package, respectively. A store sold 12 packages containing a total of 162 razor blades and took in $35. How many packages of each type were sold?

45. Three foods have the following nutritional content per ounce.

	Calories	Protein (in grams)	Vitamin C (in milligrams)
Food A	40	5	30
Food B	200	2	10
Food C	400	4	300

If a meal consisting of the three foods allows exactly 660 calories, 25 grams of protein, and 425 milligrams of vitamin C, how many ounces of each kind of food should be used?

46. A furniture company produces three types of desks: a children's model, an office model, and a deluxe model. Each desk is manufactured in three stages: cutting, construction, and finishing. The time requirements for each model and manufacturing stage are given in the following table.

	Children's Model	Office Model	Deluxe Model
Cutting	2 hr	3 hr	2 hr
Construction	2 hr	1 hr	3 hr
Finishing	1 hr	1 hr	2 hr

Each week the company has available a maximum of 100 hours for cutting, 100 hours for construction, and 65 hours for finishing. If all available time must be used, how many of each type of desk should be produced each week?

Explaining the Concepts

47. What is a system of linear equations in three variables?

48. How do you determine whether a given ordered triple is a solution of a system of linear equations in three variables?

49. Describe in general terms how to solve a system in three variables.

50. Describe what happens when using algebraic methods to solve an inconsistent system.

51. Describe what happens when using algebraic methods to solve a system with dependent equations.

Technology Exercises

52. Does your graphing utility have a feature that allows you to solve linear systems by entering coefficients and constant terms? If so, use this feature to verify the solutions to any five exercises that you worked by hand from Exercises 5–16.

53. Verify your results in Exercises 23–26 by using a graphing utility to graph the quadratic function. Trace along the curve and convince yourself that the three points given in the exercise lie on the function's graph.

Critical Thinking Exercises

Make Sense? *In Exercises 54–57, determine whether each statement makes sense or does not make sense, and explain your reasoning.*

54. Solving a system in three variables, I found that $x = 3$ and $y = -1$. Because z represents a third variable, z cannot equal 3 or -1.

55. A system of linear equations in three variables, x, y, and z, cannot contain an equation in the form $y = mx + b$.

56. I'm solving a three-variable system in which one of the given equations has a missing term, so it will not be necessary to use any of the original equations twice when I reduce the system to two equations in two variables.

57. Because the percentage of the U.S. population that was foreign-born decreased from 1910 through 1970 and then increased after that, a quadratic function of the form $f(x) = ax^2 + bx + c$, rather than a linear function of the form $f(x) = mx + b$, should be used to model the data.

In Exercises 58–61, determine whether each statement is true or false. If the statement is false, make the necessary change(s) to produce a true statement.

58. The ordered triple (2, 15, 14) is the only solution of the equation $x + y - z = 3$.

59. The equation $x - y - z = -6$ is satisfied by $(2, -3, 5)$.

60. If two equations in a system are $x + y - z = 5$ and $x + y - z = 6$, then the system must be inconsistent.

61. An equation with four variables, such as $x + 2y - 3z + 5w = 2$, cannot be satisfied by real numbers.

62. In the following triangle, the degree measures of the three interior angles and two of the exterior angles are represented with variables. Find the measure of each interior angle.

63. A modernistic painting consists of triangles, rectangles, and pentagons, all drawn so as to not overlap or share sides. Within each rectangle are drawn 2 red roses, and each pentagon contains 5 carnations. How many triangles, rectangles, and pentagons appear in the painting if the painting contains a total of 40 geometric figures, 153 sides of geometric figures, and 72 flowers?

64. Two blocks of wood having the same length and width are placed on the top and bottom of a table, as shown in (a). Length A measures 32 centimeters. The blocks are rearranged as shown in (b). Length B measures 28 centimeters. Determine the height of the table.

(a)   (b)

Review Exercises

In Exercises 65–67, graph each linear function.

65. $f(x) = -\dfrac{3}{4}x + 3$ (Section 2.4, Example 5)

66. $-2x + y = 6$ (Section 2.4, Example 1)

67. $f(x) = -5$ (Section 2.4, Example 6)

Preview Exercises

Exercises 68–70 will help you prepare for the material covered in the next section.

68. Solve the system:

$$\begin{cases} x + 2y = -1 \\ \quad\;\; y = \;\; 1. \end{cases}$$

What makes it fairly easy to find the solution?

69. Solve the system:

$$\begin{cases} x + y + 2z = 19 \\ \quad\;\; y + 2z = 13 \\ \qquad\qquad z = \;\; 5. \end{cases}$$

What makes it fairly easy to find the solution?

70. Consider the following array of numbers:

$$\begin{bmatrix} 1 & 2 & -1 \\ 4 & -3 & -15 \end{bmatrix}.$$

Rewrite the array as follows: Multiply each number in the top row by -4 and add this product to the corresponding number in the bottom row. Do not change the numbers in the top row.

Our goal in solving a linear system using matrices is to produce a matrix with 1s down the diagonal from upper left to lower right on the left side of the vertical bar, called the **main diagonal**, and 0s below the 1s. In general, the matrix will be one of the following forms.

This is the desired form for systems with two equations.
$$\left[\begin{array}{cc|c} 1 & a & b \\ 0 & 1 & c \end{array}\right]$$

$$\left[\begin{array}{ccc|c} 1 & a & b & c \\ 0 & 1 & d & e \\ 0 & 0 & 1 & f \end{array}\right]$$
This is the desired form for systems with three equations.

The last row of these matrices gives us the value of one variable. The values of the other variables can then be found by back-substitution.

2 Perform matrix row operations. ▶

Matrix Row Operations

A matrix with 1s down the main diagonal and 0s below the 1s is said to be in **row-echelon form**. How do we produce a matrix in this form? We use **row operations** on the augmented matrix. These row operations are just like what you did when solving a linear system by the addition method. The difference is that we no longer write the variables, usually represented by x, y, and z.

Matrix Row Operations

The following row operations produce matrices that represent systems with the same solution set:

1. Two rows of a matrix may be interchanged. This is the same as interchanging two equations in a linear system.

2. The elements in any row may be multiplied by a nonzero number. This is the same as multiplying both sides of an equation by a nonzero number.

3. The elements in any row may be multiplied by a nonzero number, and these products may be added to the corresponding elements in any other row. This is the same as multiplying an equation by a nonzero number and then adding equations to eliminate a variable.

Two matrices are **row equivalent** if one can be obtained from the other by a sequence of row operations.

Each matrix row operation in the preceding box can be expressed symbolically as follows:

1. Interchange the elements in the ith and jth rows: $R_i \leftrightarrow R_j$.
2. Multiply each element in the ith row by k: kR_i.
3. Add k times the elements in row i to the corresponding elements in row j: $kR_i + R_j$.

Great Question!

Can you clarify what I'm supposed to do to find $kR_i + R_j$?. Which row do I work with and which row do I replace?

When performing the row operation

$$kR_i + R_j$$

you use row i to find the products. However, **elements in row i do not change. It is the elements in row j that change:** Add k times the elements in row i to the corresponding elements in row j. Replace elements in row j by these sums.

EXAMPLE 1 Performing Matrix Row Operations

Use the matrix

$$\left[\begin{array}{ccc|c} 3 & 18 & -12 & 21 \\ 1 & 2 & -3 & 5 \\ -2 & -3 & 4 & -6 \end{array}\right]$$

and perform each indicated row operation:

a. $R_1 \leftrightarrow R_2$ **b.** $\dfrac{1}{3}R_1$ **c.** $2R_2 + R_3$.

Solution

a. The notation $R_1 \leftrightarrow R_2$ means to interchange the elements in row 1 and row 2. This results in the row-equivalent matrix

$$\begin{bmatrix} 1 & 2 & -3 & | & 5 \\ 3 & 18 & -12 & | & 21 \\ -2 & -3 & 4 & | & -6 \end{bmatrix}.$$

This was row 2; now it's row 1.

This was row 1; now it's row 2.

b. The notation $\frac{1}{3}R_1$ means to multiply each element in row 1 by $\frac{1}{3}$. This results in the row-equivalent matrix

$$\begin{bmatrix} \frac{1}{3}(3) & \frac{1}{3}(18) & \frac{1}{3}(-12) & | & \frac{1}{3}(21) \\ 1 & 2 & -3 & | & 5 \\ -2 & -3 & 4 & | & -6 \end{bmatrix} = \begin{bmatrix} 1 & 6 & -4 & | & 7 \\ 1 & 2 & -3 & | & 5 \\ -2 & -3 & 4 & | & -6 \end{bmatrix}.$$

c. The notation $2R_2 + R_3$ means to add 2 times the elements in row 2 to the corresponding elements in row 3. Replace the elements in row 3 by these sums. First, we find 2 times the elements in row 2, namely, 1, 2, −3, and 5:

$$2(1) \text{ or } 2, \qquad 2(2) \text{ or } 4, \qquad 2(-3) \text{ or } -6, \qquad 2(5) \text{ or } 10.$$

Now we add these products to the corresponding elements in row 3. Although we use row 2 to find the products, row 2 does not change. It is the elements in row 3 that change, resulting in the row-equivalent matrix

Replace row 3 by the sum of itself and 2 times row 2.

$$\begin{bmatrix} 3 & 18 & -12 & | & 21 \\ 1 & 2 & -3 & | & 5 \\ -2+2=0 & -3+4=1 & 4+(-6)=-2 & | & -6+10=4 \end{bmatrix} = \begin{bmatrix} 3 & 18 & -12 & | & 21 \\ 1 & 2 & -3 & | & 5 \\ 0 & 1 & -2 & | & 4 \end{bmatrix}. \quad \blacksquare$$

✓ **CHECK POINT 1** Use the matrix

$$\begin{bmatrix} 4 & 12 & -20 & | & 8 \\ 1 & 6 & -3 & | & 7 \\ -3 & -2 & 1 & | & -9 \end{bmatrix}$$

and perform each indicated row operation:

a. $R_1 \leftrightarrow R_2$

b. $\frac{1}{4}R_1$

c. $3R_2 + R_3$.

③ Use matrices to solve linear systems in two variables. ▶

Solving Linear Systems in Two Variables Using Matrices

The process that we use to solve linear systems using matrix row operations is often called **Gaussian elimination**, after the German mathematician Carl Friedrich Gauss (1777–1855). At the top of the next page are the steps used in solving linear systems in two variables with matrices.

Solution

Step 1. Write the augmented matrix for the system.

| Linear System | Augmented Matrix |

$$\begin{cases} 3x + y + 2z = 31 \\ x + y + 2z = 19 \\ x + 3y + 2z = 25 \end{cases} \qquad \left[\begin{array}{ccc|c} 3 & 1 & 2 & 31 \\ 1 & 1 & 2 & 19 \\ 1 & 3 & 2 & 25 \end{array}\right]$$

Step 2. Use matrix row operations to simplify the matrix to row-echelon form, with 1s down the main diagonal from upper left to lower right, and 0s below the 1s in the first and second columns. Our first step in achieving this goal is to get 1 in the top position of the first column.

We want 1 in this position.
$$\left[\begin{array}{ccc|c} 3 & 1 & 2 & 31 \\ 1 & 1 & 2 & 19 \\ 1 & 3 & 2 & 25 \end{array}\right]$$

To get 1 in this position, we interchange row 1 and row 2: $R_1 \leftrightarrow R_2$.
We could also interchange row 1 and row 3 to get 1 in the upper left-hand corner.

$$\left[\begin{array}{ccc|c} 1 & 1 & 2 & 19 \\ 3 & 1 & 2 & 31 \\ 1 & 3 & 2 & 25 \end{array}\right] \quad \begin{array}{l}\text{This was row 2; now it's row 1.} \\ {} \\ \text{This was row 1; now it's row 2.}\end{array}$$

Now we want to get 0s below the 1 in the first column.

We want 0 in these positions.
$$\left[\begin{array}{ccc|c} 1 & 1 & 2 & 19 \\ 3 & 1 & 2 & 31 \\ 1 & 3 & 2 & 25 \end{array}\right]$$

To get a 0 where there is now a 3, multiply the top row of numbers by -3 and add these products to the second row of numbers: $-3R_1 + R_2$. To get a 0 in the bottom of the first column where there is now a 1, multiply the top row of numbers by -1 and add these products to the third row of numbers: $-1R_1 + R_3$. Although we are using row 1 to find the products, the numbers in row 1 do not change.

Replace row 2 by $-3R_1 + R_2$.
Replace row 3 by $-1R_1 + R_3$.

$$\left[\begin{array}{ccc|c} 1 & 1 & 2 & 19 \\ -3(1)+3 & -3(1)+1 & -3(2)+2 & -3(19)+31 \\ -1(1)+1 & -1(1)+3 & -1(2)+2 & -1(19)+25 \end{array}\right] = \left[\begin{array}{ccc|c} 1 & 1 & 2 & 19 \\ 0 & -2 & -4 & -26 \\ 0 & 2 & 0 & 6 \end{array}\right]$$

We want 1 in this position.

We move on to the second column. To get 1 in the desired position, we multiply -2 by its reciprocal, $-\frac{1}{2}$. Therefore, we multiply all the numbers in the second row by $-\frac{1}{2}$: $-\frac{1}{2}R_2$.

$$-\frac{1}{2}R_2 \quad \left[\begin{array}{ccc|c} 1 & 1 & 2 & 19 \\ -\frac{1}{2}(0) & -\frac{1}{2}(-2) & -\frac{1}{2}(-4) & -\frac{1}{2}(-26) \\ 0 & 2 & 0 & 6 \end{array}\right] = \left[\begin{array}{ccc|c} 1 & 1 & 2 & 19 \\ 0 & 1 & 2 & 13 \\ 0 & 2 & 0 & 6 \end{array}\right]$$

We want 0 in this position.

We are not yet done with the second column. The voice balloon shows that we want to get a 0 where there is now a 2. If we multiply the second row of numbers by -2 and add these products to the third row of numbers, we will get 0 in this position: $-2R_2 + R_3$. Although we are using the numbers in row 2 to find the products, the numbers in row 2 do not change.

Replace row 3 by $-2R_2 + R_3$.

$$\left[\begin{array}{ccc|c} 1 & 1 & 2 & 19 \\ 0 & 1 & 2 & 13 \\ -2(0)+0 & -2(1)+2 & -2(2)+0 & -2(13)+6 \end{array}\right] = \left[\begin{array}{ccc|c} 1 & 1 & 2 & 19 \\ 0 & 1 & 2 & 13 \\ 0 & 0 & -4 & -20 \end{array}\right]$$

We want 1 in this position.

We move on to the third column. To get 1 in the desired position, we multiply -4 by its reciprocal, $-\frac{1}{4}$. Therefore, we multiply all the numbers in the third row by $-\frac{1}{4}$: $-\frac{1}{4}R_3$.

$$-\tfrac{1}{4}R_3 \begin{bmatrix} 1 & 1 & 2 & | & 19 \\ 0 & 1 & 2 & | & 13 \\ -\frac{1}{4}(0) & -\frac{1}{4}(0) & -\frac{1}{4}(-4) & | & -\frac{1}{4}(-20) \end{bmatrix} = \begin{bmatrix} 1 & 1 & 2 & | & 19 \\ 0 & 1 & 2 & | & 13 \\ 0 & 0 & 1 & | & 5 \end{bmatrix}$$

We now have the desired matrix in row-echelon form, with 1s down the main diagonal and 0s below the 1s in the first and second columns.

Step 3. Write the system of linear equations corresponding to the matrix from step 2 and use back-substitution to find the system's solution. The system represented by the matrix from step 2 is

$$\begin{bmatrix} 1 & 1 & 2 & | & 19 \\ 0 & 1 & 2 & | & 13 \\ 0 & 0 & 1 & | & 5 \end{bmatrix} \rightarrow \begin{cases} 1x + 1y + 2z = 19 \\ 0x + 1y + 2z = 13 \\ 0x + 0y + 1z = 5 \end{cases} \text{ or } \begin{cases} x + y + 2z = 19 \quad \text{1} \\ \quad\quad y + 2z = 13. \quad \text{2} \\ \quad\quad\quad\quad z = 5 \quad \text{3} \end{cases}$$

We immediately see from Equation 3 that the value for z is 5. To find y, we back-substitute 5 for z in the second equation.

$$\begin{aligned} y + 2z &= 13 && \text{Equation 2} \\ y + 2(5) &= 13 && \text{Substitute 5 for } z. \\ y &= 3 && \text{Solve for } y. \end{aligned}$$

Finally, back-substitute 3 for y and 5 for z in the first equation.

$$\begin{aligned} x + y + 2z &= 19 && \text{Equation 1} \\ x + 3 + 2(5) &= 19 && \text{Substitute 3 for } y \text{ and 5 for } z. \\ x + 13 &= 19 && \text{Multiply and add.} \\ x &= 6 && \text{Subtract 13 from both sides.} \end{aligned}$$

The solution of the original system is $(6, 3, 5)$ and the solution set is $\{(6, 3, 5)\}$. Check to see that the solution satisfies all three equations in the given system. ■

Using Technology

Most graphing utilities can convert an augmented matrix to row-echelon form, with 1s down the main diagonal and 0s below the 1s. However, row-echelon form is not unique. Your graphing utility might give a row-echelon form different from the one you obtained by hand. However, all row-echelon forms for a given system's augmented matrix produce the same solution to the system. Enter the augmented matrix and name it A. Then use the REF (row-echelon form) command on matrix A.

✓ **CHECK POINT 3**
Use matrices to solve the system:

$$\begin{cases} 2x + y + 2z = 18 \\ x - y + 2z = 9 \\ x + 2y - z = 6. \end{cases}$$

Modern supercomputers are capable of solving systems with more than 600,000 variables. The augmented matrices for such systems are huge, but the solution using matrices is exactly like what we did in Example 3. Work with the augmented matrix, one column at a time. Get 1s down the main diagonal from upper left to lower right, and 0s below the 1s.

5 Use matrices to identify inconsistent and dependent systems.

Inconsistent Systems and Systems with Dependent Equations

When solving a system using matrices, you might obtain a matrix with a row in which the numbers to the left of the vertical bar are all zeros, but a nonzero number appears on the right. In such a case, the system is inconsistent and has no solution. For example, a system of equations that yields the following matrix is an inconsistent system:

$$\begin{bmatrix} 1 & -2 & | & 3 \\ 0 & 0 & | & -4 \end{bmatrix}.$$

The second row of the matrix represents the equation $0x + 0y = -4$, which is false for all values of x and y.

If you obtain a matrix in which a 0 appears across an entire row, the system contains dependent equations and has infinitely many solutions. This row of zeros represents $0x + 0y = 0$ or $0x + 0y + 0z = 0$. These equations are satisfied by infinitely many ordered pairs or triples.

In Exercises 41–42, perform each matrix row operation and write the new matrix.

41.
$$\left[\begin{array}{cccc|c} 1 & -1 & 1 & 1 & 3 \\ 0 & 1 & -2 & -1 & 0 \\ 2 & 0 & 3 & 4 & 11 \\ 5 & 1 & 2 & 4 & 6 \end{array}\right] \begin{array}{l} \\ \\ -2R_1 + R_3 \\ -5R_1 + R_4 \end{array}$$

42.
$$\left[\begin{array}{cccc|c} 1 & -5 & 2 & -2 & 4 \\ 0 & 1 & -3 & -1 & 0 \\ 3 & 0 & 2 & -1 & 6 \\ -4 & 1 & 4 & 2 & -3 \end{array}\right] \begin{array}{l} \\ \\ -3R_1 + R_3 \\ 4R_1 + R_4 \end{array}$$

In Exercises 43–44, solve each system using matrices. You will need to use matrix row operations to obtain matrices like those in Exercises 39 and 40, with 1s down the main diagonal and 0s below the 1s. Express the solution set as $\{(w, x, y, z)\}$.

43.
$$\begin{cases} w + x + y + z = 4 \\ 2w + x - 2y - z = 0 \\ w - 2x - y - 2z = -2 \\ 3w + 2x + y + 3z = 4 \end{cases}$$

44.
$$\begin{cases} w + x + y + z = 5 \\ w + 2x - y - 2z = -1 \\ w - 3x - 3y - z = -1 \\ 2w - x + 2y - z = -2 \end{cases}$$

Application Exercises

45. A ball is thrown straight upward. The graph shows the ball's height, $s(t)$, in feet, after t seconds.

a. Find the quadratic function

$$s(t) = at^2 + bt + c$$

whose graph passes through the three points labeled on the graph. Solve the system of linear equations involving $a, b,$ and c using matrices.

b. Find and interpret $s(3.5)$. Identify your solution as a point on the graph shown.

46. A football is kicked straight upward. The graph shows the football's height, $s(t)$, in feet, after t seconds.

a. Find the quadratic function

$$s(t) = at^2 + bt + c$$

whose graph passes through the three points labeled on the graph. Solve the system of linear equations involving $a, b,$ and c using matrices.

b. Find and interpret $s(7)$. Identify your solution as a point on the graph shown.

Write a system of linear equations in three variables to solve Exercises 47–48. Then use matrices to solve the system.

Exercises 47–48 are based on a Time/CNN telephone poll that included never-married single women between the ages of 18 and 49 and never-married single men between the ages of 18 and 49. The circle graphs show the results for one of the questions in the poll.

If You Couldn't Find the Perfect Mate, Would You Marry Someone Else?

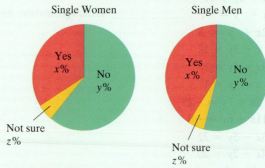

Source: Time/CNN

47. For single women in the poll, the percentage who said no exceeded the combined percentages for those who said yes and those who said not sure by 22%. If the percentage who said yes is doubled, it is 7% more than the percentage who said no. Find the percentage of single women who responded yes, no, and not sure.

48. For single men in the poll, the percentage who said no exceeded the combined percentages for those who said yes and those who said not sure by 8%. If the percentage who said yes is doubled, it is 28% more than the percentage who said no. Find the percentage of single men who responded yes, no, and not sure.

Explaining the Concepts

49. What is a matrix?

50. Describe what is meant by the augmented matrix of a system of linear equations.

51. In your own words, describe each of the three matrix row operations. Give an example of each of the operations.

52. Describe how to use matrices and row operations to solve a system of linear equations.

53. When solving a system using matrices, how do you know if the system has no solution?

54. When solving a system using matrices, how do you know if the system has infinitely many solutions?

Technology Exercises

55. Most graphing utilities can perform row operations on matrices. Consult the owner's manual for your graphing utility to learn proper keystroke sequences for performing these operations. Then duplicate the row operations of any three exercises that you solved from Exercises 3–12.

56. If your graphing utility has a REF (row-echelon form) command, use this feature to verify your work with any five systems from Exercises 15–38.

57. A matrix with 1s down the main diagonal and 0s in every position *above and below* each 1 is said to be in **reduced row-echelon form**.

If your graphing utility has a RREF (reduced row-echelon form) command, use this feature to verify your work with any five systems from Exercises 15–38.

Critical Thinking Exercises

Make Sense? *In Exercises 58–61, determine whether each statement makes sense or does not make sense, and explain your reasoning.*

58. Matrix row operations remind me of what I did when solving a linear system by the addition method, although I no longer write the variables.

59. When I use matrices to solve linear systems, the only arithmetic involves multiplication or a combination of multiplication and addition.

60. When I use matrices to solve linear systems, I spend most of my time using row operations to express the system's augmented matrix in row-echelon form.

61. Using row operations on an augmented matrix, I obtain a row in which 0s appear to the left of the vertical bar, but 6 appears on the right, so the system I'm working with has infinitely many solutions.

In Exercises 62–65, determine whether each statement is true or false. If the statement is false, make the necessary change(s) to produce a true statement.

62. A matrix row operation such as $-\frac{4}{5}R_1 + R_2$ is not permitted because of the negative fraction.

63. The augmented matrix for the system

$$\begin{cases} x - 3y = 5 \\ y - 2z = 7 \\ 2x + z = 4 \end{cases} \text{ is } \begin{bmatrix} 1 & -3 & | & 5 \\ 1 & -2 & | & 7 \\ 2 & 1 & | & 4 \end{bmatrix}.$$

64. In solving a linear system of three equations in three variables, we begin with the augmented matrix and use row operations to obtain a row-equivalent matrix with 0s down the diagonal from left to right and 1s below each 0.

65. The row operation $kR_i + R_j$ indicates that it is the elements in row i that change.

66. The vitamin content per ounce for three foods is given in the following table.

	Milligrams per Ounce		
	Thiamin	Riboflavin	Niacin
Food A	3	7	1
Food B	1	5	3
Food C	3	8	2

a. Use matrices to show that no combination of these foods can provide exactly 14 milligrams of thiamin, 32 milligrams of riboflavin, and 9 milligrams of niacin.

b. Use matrices to describe in practical terms what happens if the riboflavin requirement is increased by 5 milligrams and the other requirements stay the same.

Review Exercises

67. If $f(x) = -3x + 10$, find $f(2a - 1)$. (Section 2.1, Example 3)

68. If $f(x) = 3x$ and $g(x) = 2x - 3$, find $(fg)(-1)$. (Section 2.3, Example 4)

69. Simplify: $\dfrac{-4x^8y^{-12}}{12x^{-3}y^{24}}$. (Section 1.6, Example 9)

Preview Exercises

Exercises 70–72 will help you prepare for the material covered in the next section. Simplify the expression in each exercise.

70. $2(-5) - (-3)(4)$

71. $\dfrac{2(-5) - 1(-4)}{5(-5) - 6(-4)}$

72. $2(-30 - (-3)) - 3(6 - 9) + (-1)(1 - 15)$

3.5

Determinants and Cramer's Rule

After studying this section, you should be able to:

1. Evaluate a second-order determinant.

2. Solve a system of linear equations in two variables using Cramer's rule.

3. Evaluate a third-order determinant.

4. Solve a system of linear equations in three variables using Cramer's rule.

As cyberspace absorbs more and more of our work, play, shopping, and socializing, where will it all end? Which activities will still be offline in 2025?

Our technologically transformed lives can be traced back to the English inventor Charles Babbage (1791–1871). Babbage knew of a method for solving linear systems called *Cramer's rule*, in honor of the Swiss geometer Gabriel Cramer (1704–1752). Cramer's rule was simple, but involved numerous multiplications for large systems. Babbage designed a machine, called the "difference engine," that

A portion of Charles Babbage's unrealized Difference Engine

consisted of toothed wheels on shafts for performing these multiplications. Despite the fact that only one-seventh of the functions ever worked, Babbage's invention demonstrated how complex calculations could be handled mechanically. In 1944, scientists at IBM used the lessons of the difference engine to create the world's first computer.

Those who invented computers hoped to relegate the drudgery of repeated computation to a machine. In this section, we look at a method for solving linear systems that played a critical role in this process. The method uses real numbers, called *determinants*, that are associated with arrays of numbers. As with matrix methods, solutions are obtained by writing down the coefficients and constants of a linear system and performing operations with them.

1. Evaluate a second-order determinant.

The Determinant of a 2 × 2 Matrix

A matrix of **order $m \times n$** has m rows and n columns. If $m = n$, a matrix has the same number of rows as columns and is called a **square matrix**. Associated with every square matrix is a real number, called its **determinant**. The determinant for a 2 × 2 square matrix is defined as follows:

Great Question!

What does the definition of a determinant mean? What am I supposed to do?

To evaluate a second-order determinant, find the difference of the product of the two diagonals.

$$\begin{vmatrix} a_1 & b_1 \\ a_2 & b_2 \end{vmatrix} = a_1 b_2 - a_2 b_1$$

Definition of the Determinant of a 2 × 2 Matrix

The determinant of the matrix $\begin{bmatrix} a_1 & b_1 \\ a_2 & b_2 \end{bmatrix}$ is denoted by $\begin{vmatrix} a_1 & b_1 \\ a_2 & b_2 \end{vmatrix}$ and is defined by

$$\begin{vmatrix} a_1 & b_1 \\ a_2 & b_2 \end{vmatrix} = a_1 b_2 - a_2 b_1.$$

We also say that the **value** of the **second-order determinant** $\begin{vmatrix} a_1 & b_1 \\ a_2 & b_2 \end{vmatrix}$ is $a_1 b_2 - a_2 b_1$.

Example 1 illustrates that the determinant of a matrix may be positive or negative. The determinant can also have 0 as its value.

EXAMPLE 1 Evaluating the Determinant of a 2 × 2 Matrix

Evaluate the determinant of each of the following matrices:

a. $\begin{bmatrix} 5 & 6 \\ 7 & 3 \end{bmatrix}$

b. $\begin{bmatrix} 2 & 4 \\ -3 & -5 \end{bmatrix}$.

Discover for Yourself

Write and then evaluate three determinants, one whose value is positive, one whose value is negative, and one whose value is 0.

Solution We multiply and subtract as indicated.

a. $\begin{vmatrix} 5 & 6 \\ 7 & 3 \end{vmatrix} = 5 \cdot 3 - 7 \cdot 6 = 15 - 42 = -27$ The value of the second-order determinant is -27.

b. $\begin{vmatrix} 2 & 4 \\ -3 & -5 \end{vmatrix} = 2(-5) - (-3)(4) = -10 + 12 = 2$ The value of the second-order determinant is 2. ∎

✓ **CHECK POINT 1** Evaluate the determinant of each of the following matrices:

a. $\begin{bmatrix} 10 & 9 \\ 6 & 5 \end{bmatrix}$ **b.** $\begin{bmatrix} 4 & 3 \\ -5 & -8 \end{bmatrix}$.

---

2 Solve a system of linear equations in two variables using Cramer's rule. ▶

Solving Systems of Linear Equations in Two Variables Using Determinants

Determinants can be used to solve a linear system in two variables. In general, such a system appears as

$$\begin{cases} a_1x + b_1y = c_1 \\ a_2x + b_2y = c_2. \end{cases}$$

Let's first solve this system for x using the addition method. We can solve for x by eliminating y from the equations. Multiply the first equation by b_2 and the second equation by $-b_1$. Then add the two equations:

$$\begin{cases} a_1x + b_1y = c_1 \\ a_2x + b_2y = c_2 \end{cases} \xrightarrow[\text{Multiply by } -b_1.]{\text{Multiply by } b_2.} \begin{cases} a_1b_2x + b_1b_2y = c_1b_2 \\ -a_2b_1x - b_1b_2y = -c_2b_1 \end{cases}$$

$$\text{Add: } (a_1b_2 - a_2b_1)x = c_1b_2 - c_2b_1$$

$$x = \frac{c_1b_2 - c_2b_1}{a_1b_2 - a_2b_1}$$

Because

$$\begin{vmatrix} c_1 & b_1 \\ c_2 & b_2 \end{vmatrix} = c_1b_2 - c_2b_1 \quad \text{and} \quad \begin{vmatrix} a_1 & b_1 \\ a_2 & b_2 \end{vmatrix} = a_1b_2 - a_2b_1,$$

we can express our answer for x as the quotient of two determinants:

$$x = \frac{\begin{vmatrix} c_1 & b_1 \\ c_2 & b_2 \end{vmatrix}}{\begin{vmatrix} a_1 & b_1 \\ a_2 & b_2 \end{vmatrix}}.$$

Similarly, we could use the addition method to solve our system for y, again expressing y as the quotient of two determinants. This method of using determinants to solve the linear system, called **Cramer's rule**, is summarized in the box on the next page.

Solving a Linear System in Two Variables Using Determinants

Cramer's Rule

If
$$\begin{cases} a_1x + b_1y = c_1 \\ a_2x + b_2y = c_2 \end{cases}$$

then

$$x = \frac{\begin{vmatrix} c_1 & b_1 \\ c_2 & b_2 \end{vmatrix}}{\begin{vmatrix} a_1 & b_1 \\ a_2 & b_2 \end{vmatrix}} \quad \text{and} \quad y = \frac{\begin{vmatrix} a_1 & c_1 \\ a_2 & c_2 \end{vmatrix}}{\begin{vmatrix} a_1 & b_1 \\ a_2 & b_2 \end{vmatrix}},$$

where

$$\begin{vmatrix} a_1 & b_1 \\ a_2 & b_2 \end{vmatrix} \neq 0.$$

Here are some helpful tips when solving

$$\begin{cases} a_1x + b_1y = c_1 \\ a_2x + b_2y = c_2 \end{cases}$$

using determinants:

1. Three different determinants are used to find x and y. The determinants in the denominators for x and y are identical. The determinants in the numerators for x and y differ. In abbreviated notation, we write

$$x = \frac{D_x}{D} \quad \text{and} \quad y = \frac{D_y}{D}, \text{where } D \neq 0.$$

2. The elements of D, the determinant in the denominator, are the coefficients of the variables in the system.

$$D = \begin{vmatrix} a_1 & b_1 \\ a_2 & b_2 \end{vmatrix}$$

3. D_x, the determinant in the numerator of x, is obtained by replacing the x-coefficients, in D, a_1 and a_2, with the constants on the right sides of the equations, c_1 and c_2.

$$D = \begin{vmatrix} a_1 & b_1 \\ a_2 & b_2 \end{vmatrix} \quad \text{and} \quad D_x = \begin{vmatrix} c_1 & b_1 \\ c_2 & b_2 \end{vmatrix} \qquad \text{Replace the column with } a_1 \text{ and } a_2 \text{ with the constants } c_1 \text{ and } c_2 \text{ to get } D_x$$

4. D_y, the determinant in the numerator for y, is obtained by replacing the y-coefficients, in D, b_1 and b_2, with the constants on the right sides of the equations, c_1 and c_2.

$$D = \begin{vmatrix} a_1 & b_1 \\ a_2 & b_2 \end{vmatrix} \quad \text{and} \quad D_y = \begin{vmatrix} a_1 & c_1 \\ a_2 & c_2 \end{vmatrix} \qquad \text{Replace the column with } b_1 \text{ and } b_2 \text{ with the constants } c_1 \text{ and } c_2 \text{ to get } D_y$$

EXAMPLE 2 Using Cramer's Rule to Solve a Linear System

Use Cramer's rule to solve the system:

$$\begin{cases} 5x - 4y = 2 \\ 6x - 5y = 1. \end{cases}$$

Solution Because

$$x = \frac{D_x}{D} \quad \text{and} \quad y = \frac{D_y}{D},$$

we will set up and evaluate the three determinants D, D_x, and D_y.

1. D, the determinant in both denominators, consists of the x- and y-coefficients.

$$D = \begin{vmatrix} 5 & -4 \\ 6 & -5 \end{vmatrix} = (5)(-5) - (6)(-4) = -25 + 24 = -1$$

Because this determinant is not zero, we continue to use Cramer's rule to solve the system.

$$\begin{cases} 5x - 4y = 2 \\ 6x - 5y = 1 \end{cases}$$

The system we are interested in solving (repeated so that you don't have to look back)

2. D_x, the determinant in the numerator for x, is obtained by replacing the x-coefficients in D, 5 and 6, by the constants on the right sides of the equations, 2 and 1.

$$D_x = \begin{vmatrix} 2 & -4 \\ 1 & -5 \end{vmatrix} = (2)(-5) - (1)(-4) = -10 + 4 = -6$$

3. D_y, the determinant in the numerator for y, is obtained by replacing the y-coefficients in D, -4 and -5, by the constants on the right sides of the equations, 2 and 1.

$$D_y = \begin{vmatrix} 5 & 2 \\ 6 & 1 \end{vmatrix} = (5)(1) - (6)(2) = 5 - 12 = -7$$

4. Thus,

$$x = \frac{D_x}{D} = \frac{-6}{-1} = 6 \quad \text{and} \quad y = \frac{D_y}{D} = \frac{-7}{-1} = 7.$$

As always, the ordered pair $(6, 7)$ should be checked by substituting these values into the original equations. The solution is $(6, 7)$ and the solution set is $\{(6, 7)\}$. ∎

✓ **CHECK POINT 2** Use Cramer's rule to solve the system:

$$\begin{cases} 5x + 4y = 12 \\ 3x - 6y = 24. \end{cases}$$

3 Evaluate a third-order determinant. ▶

The Determinant of a 3 × 3 Matrix

The determinant for a 3 × 3 matrix is defined in terms of second-order determinants:

> **Definition of the Determinant of a 3 × 3 Matrix**
>
> A third-order determinant is defined by
>
> $$\begin{vmatrix} a_1 & b_1 & c_1 \\ a_2 & b_2 & c_2 \\ a_3 & b_3 & c_3 \end{vmatrix} = a_1 \overset{\text{Subtract.}}{\begin{vmatrix} b_2 & c_2 \\ b_3 & c_3 \end{vmatrix}} - a_2 \begin{vmatrix} b_1 & c_1 \\ b_3 & c_3 \end{vmatrix} + a_3 \overset{\text{Add.}}{\begin{vmatrix} b_1 & c_1 \\ b_2 & c_2 \end{vmatrix}}.$$
>
> Each a on the right comes from the first column.

$$\begin{vmatrix} a_1 & b_1 & c_1 \\ a_2 & b_2 & c_2 \\ a_3 & b_3 & c_3 \end{vmatrix} =$$

$$a_1 \begin{vmatrix} b_2 & c_2 \\ b_3 & c_3 \end{vmatrix} - a_2 \begin{vmatrix} b_1 & c_1 \\ b_3 & c_3 \end{vmatrix} + a_3 \begin{vmatrix} b_1 & c_1 \\ b_2 & c_2 \end{vmatrix}$$

The definition of a third-order determinant (repeated)

Here are some tips that should be helpful when evaluating the determinant of a 3×3 matrix:

Evaluating the Determinant of a 3 × 3 Matrix

1. Each of the three terms in the definition of a third-order determinant contains two factors—a numerical factor and a second-order determinant.

2. The numerical factor in each term is an element from the first column of the third-order determinant.

3. The minus sign precedes the second term.

4. The second-order determinant that appears in each term is obtained by crossing out the row and the column containing the numerical factor.

$$a_1 \begin{vmatrix} b_2 & c_2 \\ b_3 & c_3 \end{vmatrix} - a_2 \begin{vmatrix} b_1 & c_1 \\ b_3 & c_3 \end{vmatrix} + a_3 \begin{vmatrix} b_1 & c_1 \\ b_2 & c_2 \end{vmatrix}$$

$$\begin{vmatrix} a_1 & b_1 & c_1 \\ a_2 & b_2 & c_2 \\ a_3 & b_3 & c_3 \end{vmatrix} \quad \begin{vmatrix} a_1 & b_1 & c_1 \\ a_2 & b_2 & c_2 \\ a_3 & b_3 & c_3 \end{vmatrix} \quad \begin{vmatrix} a_1 & b_1 & c_1 \\ a_2 & b_2 & c_2 \\ a_3 & b_3 & c_3 \end{vmatrix}$$

The **minor** of an element is the determinant that remains after deleting the row and column of that element. For this reason, we call this method **expansion by minors**.

EXAMPLE 3 Evaluating the Determinant of a 3 × 3 Matrix

Evaluate the determinant of the following matrix:

$$\begin{bmatrix} 4 & 1 & 0 \\ -9 & 3 & 4 \\ -3 & 8 & 1 \end{bmatrix}.$$

Solution We know that each of the three terms in the determinant contains a numerical factor and a second-order determinant. The numerical factors are from the first column of the given matrix. They are shown in red in the following matrix:

$$\begin{bmatrix} 4 & 1 & 0 \\ -9 & 3 & 4 \\ -3 & 8 & 1 \end{bmatrix}.$$

We find the minor for each numerical factor by deleting the row and column of that element.

$$\begin{bmatrix} 4 & 1 & 0 \\ -9 & 3 & 4 \\ -3 & 8 & 1 \end{bmatrix} \qquad \begin{bmatrix} 4 & 1 & 0 \\ -9 & 3 & 4 \\ -3 & 8 & 1 \end{bmatrix} \qquad \begin{bmatrix} 4 & 1 & 0 \\ -9 & 3 & 4 \\ -3 & 8 & 1 \end{bmatrix}$$

The minor for 4 is $\begin{vmatrix} 3 & 4 \\ 8 & 1 \end{vmatrix}$.

The minor for -9 is $\begin{vmatrix} 1 & 0 \\ 8 & 1 \end{vmatrix}$.

The minor for -3 is $\begin{vmatrix} 1 & 0 \\ 3 & 4 \end{vmatrix}$.

Now we have three numerical factors, 4, -9, and -3, and three second-order determinants. We multiply each numerical factor by its second-order determinant to find the three terms of the third-order determinant:

$$4 \begin{vmatrix} 3 & 4 \\ 8 & 1 \end{vmatrix}, \quad -9 \begin{vmatrix} 1 & 0 \\ 8 & 1 \end{vmatrix}, \quad -3 \begin{vmatrix} 1 & 0 \\ 3 & 4 \end{vmatrix}.$$

Based on the preceding definition, we subtract the second term from the first term and add the third term.

Don't forget to supply the minus sign.

$$\begin{vmatrix} 4 & 1 & 0 \\ -9 & 3 & 4 \\ -3 & 8 & 1 \end{vmatrix} = 4\begin{vmatrix} 3 & 4 \\ 8 & 1 \end{vmatrix} - (-9)\begin{vmatrix} 1 & 0 \\ 8 & 1 \end{vmatrix} - 3\begin{vmatrix} 1 & 0 \\ 3 & 4 \end{vmatrix}$$

Begin by evaluating the three second-order determinants.

$$= 4(3\cdot1 - 8\cdot4) + 9(1\cdot1 - 8\cdot0) - 3(1\cdot4 - 3\cdot0)$$

$$= 4(3 - 32) + 9(1 - 0) - 3(4 - 0) \quad \text{Multiply within parentheses.}$$

$$= 4(-29) + 9(1) - 3(4) \quad \text{Subtract within parentheses.}$$

$$= -116 + 9 - 12 \quad \text{Multiply.}$$

$$= -119 \quad \text{Add and subtract as indicated.} \blacksquare$$

✓ **CHECK POINT 3** Evaluate the determinant of the following matrix:

$$\begin{bmatrix} 2 & 1 & 7 \\ -5 & 6 & 0 \\ -4 & 3 & 1 \end{bmatrix}.$$

4 Solve a system of linear equations in three variables using Cramer's rule. ▶

Solving Systems of Linear Equations in Three Variables Using Determinants

Cramer's rule can be applied to solving systems of linear equations in three variables. The determinants in the numerator and denominator of the quotients determining each variable are third-order determinants.

Solving Three Equations in Three Variables Using Determinants

Cramer's Rule

If

$$\begin{cases} a_1x + b_1y + c_1z = d_1 \\ a_2x + b_2y + c_2z = d_2 \\ a_3x + b_3y + c_3z = d_3, \end{cases}$$

then

$$x = \frac{D_x}{D}, \quad y = \frac{D_y}{D}, \quad \text{and} \quad z = \frac{D_z}{D}.$$

These four third-order determinants are given by

$$D = \begin{vmatrix} a_1 & b_1 & c_1 \\ a_2 & b_2 & c_2 \\ a_3 & b_3 & c_3 \end{vmatrix}$$

These are the coefficients of the variables x, y, and z.
$D \neq 0$

$$D_x = \begin{vmatrix} d_1 & b_1 & c_1 \\ d_2 & b_2 & c_2 \\ d_3 & b_3 & c_3 \end{vmatrix}$$

Replace x-coefficients in D with the constants on the right of the three equations.

$$D_y = \begin{vmatrix} a_1 & d_1 & c_1 \\ a_2 & d_2 & c_2 \\ a_3 & d_3 & c_3 \end{vmatrix}$$

Replace y-coefficients in D with the constants on the right of the three equations.

$$D_z = \begin{vmatrix} a_1 & b_1 & d_1 \\ a_2 & b_2 & d_2 \\ a_3 & b_3 & d_3 \end{vmatrix}.$$

Replace z-coefficients in D with the constants on the right of the three equations.

EXAMPLE 4 Using Cramer's Rule to Solve a Linear System in Three Variables

Use Cramer's rule to solve:

$$\begin{cases} x + 2y - z = -4 \\ x + 4y - 2z = -6 \\ 2x + 3y + z = 3. \end{cases}$$

Solution Because

$$x = \frac{D_x}{D}, \quad y = \frac{D_y}{D}, \quad \text{and} \quad z = \frac{D_z}{D},$$

we need to set up and evaluate four determinants.

Step 1. Set up the determinants.

1. D, the determinant in all three denominators, consists of the x-, y-, and z-coefficients.

$$D = \begin{vmatrix} 1 & 2 & -1 \\ 1 & 4 & -2 \\ 2 & 3 & 1 \end{vmatrix}$$

2. D_x, the determinant in the numerator for x, is obtained by replacing the x-coefficients in D, 1, 1, and 2, with the constants on the right sides of the equations, -4, -6, and 3.

$$D_x = \begin{vmatrix} -4 & 2 & -1 \\ -6 & 4 & -2 \\ 3 & 3 & 1 \end{vmatrix}$$

3. D_y, the determinant in the numerator for y, is obtained by replacing the y-coefficients in D, 2, 4, and 3, with the constants on the right sides of the equations, -4, -6, and 3.

$$D_y = \begin{vmatrix} 1 & -4 & -1 \\ 1 & -6 & -2 \\ 2 & 3 & 1 \end{vmatrix}$$

4. D_z, the determinant in the numerator for z, is obtained by replacing the z-coefficients in D, -1, -2, and 1, with the constants on the right sides of the equations, -4, -6, and 3.

$$D_z = \begin{vmatrix} 1 & 2 & -4 \\ 1 & 4 & -6 \\ 2 & 3 & 3 \end{vmatrix}$$

Step 2. Evaluate the four determinants.

$$D = \begin{vmatrix} 1 & 2 & -1 \\ 1 & 4 & -2 \\ 2 & 3 & 1 \end{vmatrix} = 1\begin{vmatrix} 4 & -2 \\ 3 & 1 \end{vmatrix} - 1\begin{vmatrix} 2 & -1 \\ 3 & 1 \end{vmatrix} + 2\begin{vmatrix} 2 & -1 \\ 4 & -2 \end{vmatrix}$$

$$= 1(4 + 6) - 1(2 + 3) + 2(-4 + 4)$$

$$= 1(10) - 1(5) + 2(0) = 5$$

Using the same technique to evaluate each determinant, we obtain

$$D_x = -10, \quad D_y = 5, \quad \text{and} \quad D_z = 20.$$

Great Question!

Can I use D, the determinant in each denominator, to find the determinants in the numerators?

To find D_x, D_y, and D_z, you'll need to apply the evaluation process for a 3×3 determinant three times. The values of D_x, D_y, and D_z cannot be obtained from the numbers that occur in the computation of D.

Step 3. Substitute these four values and solve the system.

$$x = \frac{D_x}{D} = \frac{-10}{5} = -2$$

$$y = \frac{D_y}{D} = \frac{5}{5} = 1$$

$$z = \frac{D_z}{D} = \frac{20}{5} = 4$$

The ordered triple $(-2, 1, 4)$ can be checked by substitution into the original three equations. The solution is $(-2, 1, 4)$ and the solution set is $\{(-2, 1, 4)\}$. ∎

✓ **CHECK POINT 4**   Use Cramer's rule to solve the system:

$$\begin{cases} 3x - 2y + z = 16 \\ 2x + 3y - z = -9 \\ x + 4y + 3z = 2. \end{cases}$$

Great Question!

What should I do if D, the determinant in the denominator of Cramer's rule, is zero?

If $D = 0$, the system is inconsistent or contains dependent equations. Use a method other than Cramer's rule to determine the solution set.

Although we have focused on applying determinants to solve linear systems, they have other applications, some of which we consider in the Exercise Set that follows.

CONCEPT AND VOCABULARY CHECK

Fill in each blank so that the resulting statement is true.

1. $\begin{vmatrix} 5 & 4 \\ 2 & 3 \end{vmatrix} = \_\_ \cdot \_\_ - \_\_ \cdot \_\_ = \_\_ - \_\_ = \_\_$

   The value of this second-order _____ is _____.

2. Using Cramer's rule to solve

   $$\begin{cases} x + y = 8 \\ x - y = -2, \end{cases}$$

   we obtain

   $$x = \frac{\begin{vmatrix} \_\_ & \_\_ \\ \_\_ & \_\_ \end{vmatrix}}{\begin{vmatrix} \_\_ & \_\_ \\ \_\_ & \_\_ \end{vmatrix}} \quad \text{and} \quad y = \frac{\begin{vmatrix} \_\_ & \_\_ \\ \_\_ & \_\_ \end{vmatrix}}{\begin{vmatrix} \_\_ & \_\_ \\ \_\_ & \_\_ \end{vmatrix}}.$$

3. $\begin{vmatrix} 3 & 2 & 1 \\ 4 & 3 & 1 \\ 5 & 1 & 1 \end{vmatrix} = 3\begin{vmatrix} \_\_ & \_\_ \\ \_\_ & \_\_ \end{vmatrix} - 4\begin{vmatrix} \_\_ & \_\_ \\ \_\_ & \_\_ \end{vmatrix} + 5\begin{vmatrix} \_\_ & \_\_ \\ \_\_ & \_\_ \end{vmatrix}$

4. Using Cramer's rule to solve

$$\begin{cases} 3x + y + 4z = -8 \\ 2x + 3y - 2z = 11 \\ x - 3y - 2z = 4 \end{cases}$$

for y, we obtain

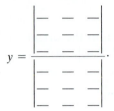

$$y = \frac{\begin{vmatrix} \underline{\quad} & \underline{\quad} & \underline{\quad} \\ \underline{\quad} & \underline{\quad} & \underline{\quad} \\ \underline{\quad} & \underline{\quad} & \underline{\quad} \end{vmatrix}}{\begin{vmatrix} \underline{\quad} & \underline{\quad} & \underline{\quad} \\ \underline{\quad} & \underline{\quad} & \underline{\quad} \\ \underline{\quad} & \underline{\quad} & \underline{\quad} \end{vmatrix}}.$$

3.5 EXERCISE SET   ▶ MyMathLab®

Practice Exercises

Evaluate each determinant in Exercises 1–10.

1. $\begin{vmatrix} 5 & 7 \\ 2 & 3 \end{vmatrix}$

2. $\begin{vmatrix} 4 & 8 \\ 5 & 6 \end{vmatrix}$

3. $\begin{vmatrix} -4 & 1 \\ 5 & 6 \end{vmatrix}$

4. $\begin{vmatrix} 7 & 9 \\ -2 & -5 \end{vmatrix}$

5. $\begin{vmatrix} -7 & 14 \\ 2 & -4 \end{vmatrix}$

6. $\begin{vmatrix} 1 & 3 \\ -8 & 2 \end{vmatrix}$

7. $\begin{vmatrix} -5 & -1 \\ -2 & -7 \end{vmatrix}$

8. $\begin{vmatrix} \frac{1}{5} & \frac{1}{6} \\ -6 & 5 \end{vmatrix}$

9. $\begin{vmatrix} \frac{1}{2} & \frac{1}{2} \\ \frac{1}{8} & -\frac{3}{4} \end{vmatrix}$

10. $\begin{vmatrix} \frac{2}{3} & \frac{1}{3} \\ -\frac{1}{2} & \frac{3}{4} \end{vmatrix}$

For Exercises 11–26, use Cramer's rule to solve each system or use another method to determine that the system is inconsistent or contains dependent equations.

11. $\begin{cases} x + y = 7 \\ x - y = 3 \end{cases}$

12. $\begin{cases} 2x + y = 3 \\ x - y = 3 \end{cases}$

13. $\begin{cases} 12x + 3y = 15 \\ 2x - 3y = 13 \end{cases}$

14. $\begin{cases} x - 2y = 5 \\ 5x - y = -2 \end{cases}$

15. $\begin{cases} 4x - 5y = 17 \\ 2x + 3y = 3 \end{cases}$

16. $\begin{cases} 3x + 2y = 2 \\ 2x + 2y = 3 \end{cases}$

17. $\begin{cases} x - 3y = 4 \\ 3x - 4y = 12 \end{cases}$

18. $\begin{cases} 2x - 9y = 5 \\ 3x - 3y = 11 \end{cases}$

19. $\begin{cases} 3x - 4y = 4 \\ 2x + 2y = 12 \end{cases}$

20. $\begin{cases} 3x = 7y + 1 \\ 2x = 3y - 1 \end{cases}$

21. $\begin{cases} 2x = 3y + 2 \\ 5x = 51 - 4y \end{cases}$

22. $\begin{cases} y = -4x + 2 \\ 2x = 3y + 8 \end{cases}$

23. $\begin{cases} 3x = 2 - 3y \\ 2y = 3 - 2x \end{cases}$

24. $\begin{cases} x + 2y - 3 = 0 \\ 12 = 8y + 4x \end{cases}$

25. $\begin{cases} 4y = 16 - 3x \\ 6x = 32 - 8y \end{cases}$

26. $\begin{cases} 2x = 7 + 3y \\ 4x - 6y = 3 \end{cases}$

Evaluate each determinant in Exercises 27–32.

27. $\begin{vmatrix} 3 & 0 & 0 \\ 2 & 1 & -5 \\ 2 & 5 & -1 \end{vmatrix}$

28. $\begin{vmatrix} 4 & 0 & 0 \\ 3 & -1 & 4 \\ 2 & -3 & 5 \end{vmatrix}$

29. $\begin{vmatrix} 3 & 1 & 0 \\ -3 & 4 & 0 \\ -1 & 3 & -5 \end{vmatrix}$

30. $\begin{vmatrix} 2 & -4 & 2 \\ -1 & 0 & 5 \\ 3 & 0 & 4 \end{vmatrix}$

31. $\begin{vmatrix} 1 & 1 & 1 \\ 2 & 2 & 2 \\ -3 & 4 & -5 \end{vmatrix}$

32. $\begin{vmatrix} 1 & 2 & 3 \\ 2 & 2 & -3 \\ 3 & 2 & 1 \end{vmatrix}$

In Exercises 33–40, use Cramer's rule to solve each system.

33. $\begin{cases} x + y + z = 0 \\ 2x - y + z = -1 \\ -x + 3y - z = -8 \end{cases}$

34. $\begin{cases} x - y + 2z = 3 \\ 2x + 3y + z = 9 \\ -x - y + 3z = 11 \end{cases}$

35. $\begin{cases} 4x - 5y - 6z = -1 \\ x - 2y - 5z = -12 \\ 2x - y = 7 \end{cases}$

36. $\begin{cases} x - 3y + z = -2 \\ x + 2y = 8 \\ 2x - y = 1 \end{cases}$

37. $\begin{cases} x + y + z = 4 \\ x - 2y + z = 7 \\ x + 3y + 2z = 4 \end{cases}$

38. $\begin{cases} 2x + 2y + 3z = 10 \\ 4x - y + z = -5 \\ 5x - 2y + 6z = 1 \end{cases}$

39. $\begin{cases} x + 2z = 4 \\ 2y - z = 5 \\ 2x + 3y = 13 \end{cases}$

40. $\begin{cases} 3x + 2z = 4 \\ 5x - y = -4 \\ 4y + 3z = 22 \end{cases}$

Practice PLUS

In Exercises 41–42, evaluate each determinant.

41. $\begin{vmatrix} \begin{vmatrix} 3 & 1 \\ -2 & 3 \end{vmatrix} & \begin{vmatrix} 7 & 0 \\ 1 & 5 \end{vmatrix} \\ \begin{vmatrix} 3 & 0 \\ 0 & 7 \end{vmatrix} & \begin{vmatrix} 9 & -6 \\ 3 & 5 \end{vmatrix} \end{vmatrix}$

42. $\begin{vmatrix} \begin{vmatrix} 5 & 0 \\ 4 & -3 \end{vmatrix} & \begin{vmatrix} -1 & 0 \\ 0 & -1 \end{vmatrix} \\ \begin{vmatrix} 7 & -5 \\ 4 & 6 \end{vmatrix} & \begin{vmatrix} 4 & 1 \\ -3 & 5 \end{vmatrix} \end{vmatrix}$

In Exercises 43–44, write the system of linear equations for which Cramer's rule yields the given determinants.

43. $D = \begin{vmatrix} 2 & -4 \\ 3 & 5 \end{vmatrix}$, $D_x = \begin{vmatrix} 8 & -4 \\ -10 & 5 \end{vmatrix}$

44. $D = \begin{vmatrix} 2 & -3 \\ 5 & 6 \end{vmatrix}$, $D_x = \begin{vmatrix} 8 & -3 \\ 11 & 6 \end{vmatrix}$

In Exercises 45–48, solve each equation for x.

45. $\begin{vmatrix} -2 & x \\ 4 & 6 \end{vmatrix} = 32$

46. $\begin{vmatrix} x+3 & -6 \\ x-2 & -4 \end{vmatrix} = 28$

47. $\begin{vmatrix} 1 & x & -2 \\ 3 & 1 & 1 \\ 0 & -2 & 2 \end{vmatrix} = -8$

48. $\begin{vmatrix} 2 & x & 1 \\ -3 & 1 & 0 \\ 2 & 1 & 4 \end{vmatrix} = 39$

Application Exercises

Determinants are used to find the area of a triangle whose vertices are given by three points in a rectangular coordinate system. The area of a triangle with vertices (x_1, y_1), (x_2, y_2), and (x_3, y_3) is

$$\text{Area} = \pm\frac{1}{2}\begin{vmatrix} x_1 & y_1 & 1 \\ x_2 & y_2 & 1 \\ x_3 & y_3 & 1 \end{vmatrix},$$

where the $\pm$ symbol indicates that the appropriate sign should be chosen to yield a positive area. Use this information to work Exercises 49–50.

49. Use determinants to find the area of the triangle whose vertices are $(3, -5)$, $(2, 6)$, and $(-3, 5)$.

50. Use determinants to find the area of the triangle whose vertices are $(1, 1)$, $(-2, -3)$, and $(11, -3)$.

Determinants are used to show that three points lie on the same line (are collinear). If $\begin{vmatrix} x_1 & y_1 & 1 \\ x_2 & y_2 & 1 \\ x_3 & y_3 & 1 \end{vmatrix} = 0$,

then the points (x_1, y_1), (x_2, y_2), and (x_3, y_3) are collinear. If the determinant does not equal 0, then the points are not collinear. Use this information to work Exercises 51–52.

51. Are the points $(3, -1)$, $(0, -3)$, and $(12, 5)$ collinear?

52. Are the points $(-4, -6)$, $(1, 0)$, and $(11, 12)$ collinear?

Determinants are used to write an equation of a line passing through two points. An equation of the line passing through the distinct points (x_1, y_1) and (x_2, y_2) is given by

$$\begin{vmatrix} x & y & 1 \\ x_1 & y_1 & 1 \\ x_2 & y_2 & 1 \end{vmatrix} = 0.$$

Use this information to work Exercises 53–54.

53. Use the determinant to write an equation for the line passing through $(3, -5)$ and $(-2, 6)$. Then expand the determinant, expressing the line's equation in slope-intercept form.

54. Use the determinant to write an equation for the line passing through $(-1, 3)$ and $(2, 4)$. Then expand the determinant, expressing the line's equation in slope-intercept form.

Explaining the Concepts

55. Explain how to evaluate a second-order determinant.

56. Describe the determinants D, D_x, and D_y in terms of the coefficients and constants in a system of two equations in two variables.

57. Explain how to evaluate a third-order determinant.

58. When expanding a determinant by minors, when is it necessary to supply a minus sign?

59. Without going into too much detail, describe how to solve a linear system in three variables using Cramer's rule.

60. In applying Cramer's rule, what does it mean if $D = 0$?

61. The process of solving a linear system in three variables using Cramer's rule can involve tedious computation. Is there a way of speeding up this process, perhaps using Cramer's rule to find the value for only one of the variables? Describe how this process might work, presenting a specific example with your description. Remember that your goal is still to find the value for each variable in the system.

62. If you could use only one method to solve linear systems in three variables, which method would you select? Explain why this is so.

Technology Exercises

63. Use the feature of your graphing utility that evaluates the determinant of a square matrix to verify any five of the determinants that you evaluated by hand in Exercises 1–10 or Exercises 27–32.

64. What is the fastest method for solving a linear system with your graphing utility?

Critical Thinking Exercises

Make Sense? *In Exercises 65–68, determine whether each statement makes sense or does not make sense, and explain your reasoning.*

65. I'm solving a linear system using a determinant that contains two rows and three columns.

66. I can speed up the tedious computations required by Cramer's rule by using the value of D to determine the value of D_x.

67. When using Cramer's rule to solve a linear system, the number of determinants that I set up and evaluate is the same as the number of variables in the system.

68. Using Cramer's rule to solve a linear system, I found the value of D_x to be zero, so the system is inconsistent.

In Exercises 69–72, determine whether each statement is true or false. If the statement is false, make the necessary change(s) to produce a true statement.

69. Only one 2×2 determinant is needed to evaluate

$$\begin{vmatrix} 2 & 3 & -2 \\ 0 & 1 & 3 \\ 0 & 4 & -1 \end{vmatrix}.$$

70. If $D = 0$, then every variable has a value of 0.

71. Because there are different determinants in the numerators of x and y, if a system is solved using Cramer's rule, x and y cannot have the same value.

72. Using Cramer's rule, we use $\dfrac{D}{D_y}$ to get the value of y.

73. What happens to the value of a second-order determinant if the two columns are interchanged?

74. Consider the system

$$\begin{cases} a_1 x + b_1 y = c_1 \\ a_2 x + b_2 y = c_2. \end{cases}$$

Use Cramer's rule to prove that if the first equation of the system is replaced by the sum of the two equations, the resulting system has the same solution as the original system.

75. Show that the equation of a line through (x_1, y_1) and (x_2, y_2) is given by the following determinant.

$$\begin{vmatrix} x & y & 1 \\ x_1 & y_1 & 1 \\ x_2 & y_2 & 1 \end{vmatrix} = 0$$

Review Exercises

76. Solve: $6x - 4 = 2 + 6(x - 1)$. (Section 1.4, Example 6)

77. Solve for y: $-2x + 3y = 7$. (Section 1.5, Example 6)

78. Solve: $\dfrac{4x + 1}{3} = \dfrac{x - 3}{6} + \dfrac{x + 5}{6}$. (Section 1.4, Example 4)

Preview Exercises

Exercises 79–81 will help you prepare for the material covered in the first section of the next chapter.

79. Solve: $\dfrac{x + 3}{4} = \dfrac{x - 2}{3} + \dfrac{1}{4}$.

80. Solve: $-2x - 4 = x + 5$.

81. Use interval notation to describe values of x for which $2(x + 4)$ is greater than $2x + 3$.

Chapter 3 Summary

Definitions and Concepts	**Examples**

Section 3.1 Systems of Linear Equations in Two Variables

A system of linear equations in two variables, x and y, consists of two equations that can be written in the form $Ax + By = C$. The solution set is the set of all ordered pairs that satisfy both equations. Using the graphing method, a solution of a linear system is a point common to the graphs of both equations in the system.

Solve by graphing: $\begin{cases} 2x - y = 6 \\ x + y = 6. \end{cases}$

The intersection point gives the solution: $(4, 2)$. The solution set is $\{(4, 2)\}$.

Definitions and Concepts	**Examples**

Section 3.1 Systems of Linear Equations in Two Variables (continued)

To solve a linear system by the substitution method,

1. Solve either equation for one variable in terms of the other.
2. Substitute the expression for that variable into the other equation.
3. Solve the equation in one variable.
4. Back-substitute the value of the variable into one of the original equations and find the value of the other variable.
5. Check the proposed solution in both equations.

Solve by the substitution method:

$$\begin{cases} y = 2x - 3 \\ 4x - 3y = 5. \end{cases}$$

Substitute $2x - 3$ for y in the second equation.

$$4x - 3(2x - 3) = 5$$
$$4x - 6x + 9 = 5$$
$$-2x + 9 = 5$$
$$-2x = -4$$
$$x = 2$$

Find y. Substitute 2 for x in $y = 2x - 3$.

$$y = 2(2) - 3 = 4 - 3 = 1$$

The ordered pair $(2, 1)$ checks. The solution is $(2, 1)$ and $\{(2, 1)\}$ is the solution set.

To solve a linear system by the addition method,

1. Write the equations in $Ax + By = C$ form.
2. Multiply one or both equations by nonzero numbers so that coefficients of one variable are opposites.
3. Add equations.
4. Solve the resulting equation for the variable.
5. Back-substitute the value of the variable into either original equation and find the value of the remaining variable.
6. Check the proposed solution in both equations.

Solve by the addition method:

$$\begin{cases} 2x + y = 10 \\ 3x + 4y = 25. \end{cases}$$

Eliminate y. Multiply the first equation by -4.

$$\begin{cases} -8x - 4y = -40 \\ \underline{3x + 4y = 25} \end{cases}$$
$$\text{Add:} \quad -5x = -15$$
$$x = 3$$

Find y. Back-substitute 3 for x. Use the first equation, $2x + y = 10$.

$$2(3) + y = 10$$
$$6 + y = 10$$
$$y = 4$$

The ordered pair $(3, 4)$ checks. The solution is $(3, 4)$ and $\{(3, 4)\}$ is the solution set.

A linear system with at least one solution is a consistent system. A system that has no solution, with $\varnothing$ as its solution set, is an inconsistent system. A linear system with infinitely many solutions has dependent equations. Solving inconsistent systems by substitution or addition leads to a false statement, such as $0 = 3$. Solving systems with dependent equations leads to a true statement, such as $7 = 7$.

One solution: consistent

No solution: inconsistent

Infinitely many solutions: dependent and consistent

Definitions and Concepts	**Examples**

Section 3.2 Problem Solving and Business Applications Using Systems of Equations

A Problem-Solving Strategy

1. Use variables, usually x and y, to represent unknown quantities.
2. Write a system of equations describing the problem's conditions.
3. Solve the system and answer the problem's question.
4. Check proposed answers in the problem's wording.

You invested $14,000 in two funds paying 7% and 9% interest. Total year-end interest was $1180. How much was invested at each rate?

$$\text{Let } x = \text{amount invested at 7\% and}$$
$$y = \text{amount invested at 9\%.}$$

Solving by substitution or addition, $x = 4000$ and $y = 10,000$. Thus, $4000 was invested at 7% and $10,000 at 9%.

Functions of Business

A company produces and sells x units of a product.

Revenue Function

$$R(x) = (\text{price per unit sold})x$$

Cost Function

$$C(x) = \text{fixed cost} + (\text{cost per unit produced})x$$

Profit Function

$$P(x) = R(x) - C(x)$$

The point of intersection of the graphs of R and C is the break-even point. The x-coordinate of the point reveals the number of units that a company must produce and sell so that the money coming in, the revenue, is equal to the money going out, the cost. The y-coordinate gives the amount of money coming in and going out.

A company that manufactures lamps has a fixed cost of $80,000 and it costs $20 to produce each lamp. Lamps are sold for $70.

a. Write the cost function.

$$C(x) = 80,000 + 20x$$

Fixed cost | Variable cost: $20 per lamp

b. Write the revenue function.

$$R(x) = 70x$$

Revenue per lamp, $70, times number of lamps sold

c. Find the break-even point.

Solve

$$\begin{cases} y = 80,000 + 20x \\ y = 70x \end{cases}$$

by substitution. Solving

$$70x = 80,000 + 20x$$

yields $x = 1600$. Back-substituting, $y = 112,000$. The break-even point is $(1600, 112,000)$: The company breaks even if it sells 1600 lamps. At this level, money coming in equals money going out: $112,000.

Definitions and Concepts	**Examples**

Section 3.3 Systems of Linear Equations in Three Variables

A system of linear equations in three variables, x, y, and z, consists of three equations of the form $Ax + By + Cz = D$. The solution set is the set consisting of the ordered triple that satisfies all three equations. The solution represents the point of intersection of three planes in space.

Is $(2, -1, 3)$ a solution of

$$\begin{cases} 3x + 5y - 2z = -5 \\ 2x + 3y - z = -2 \\ 2x + 4y + 6z = 18? \end{cases}$$

Replace x with 2, y with -1, and z with 3. Using the first equation, we obtain:

$$3 \cdot 2 + 5(-1) - 2(3) \overset{?}{=} -5$$
$$6 - 5 - 6 \overset{?}{=} -5$$
$$-5 = -5, \quad \text{true}$$

The ordered triple $(2, -1, 3)$ satisfies the first equation. In a similar manner, you can verify that it satisfies the other two equations and is a solution.

To solve a linear system in three variables by eliminating variables,

1. Reduce the system to two equations in two variables.
2. Solve the resulting system of two equations in two variables.
3. Use back-substitution in one of the equations in two variables to find the value of the second variable.
4. Back-substitute the values for two variables into one of the original equations to find the value of the third variable.
5. Check.

If all variables are eliminated and a false statement results, the system is inconsistent and has no solution. If a true statement results, the system contains dependent equations and has infinitely many solutions.

Solve

$$\begin{cases} 2x + 3y - 2z = 0 & \text{Equation 1} \\ x + 2y - z = 1 & \text{Equation 2} \\ 3x - y + z = -15. & \text{Equation 3} \end{cases}$$

Add Equations 2 and 3 to eliminate z.

$$4x + y = -14 \quad \text{Equation 4}$$

Eliminate z again. Multiply Equation 3 by 2 and add to Equation 1.

$$8x + y = -30 \quad \text{Equation 5}$$

Multiply Equation 4 by -1 and add to Equation 5.

$$\begin{cases} -4x - y = 14 \\ 8x + y = -30 \end{cases}$$
$$\text{Add:} \quad 4x \quad\;\; = -16$$
$$x = -4$$

Substitute -4 for x in Equation 4.

$$4(-4) + y = -14$$
$$y = 2$$

Substitute -4 for x and 2 for y in Equation 3.

$$3(-4) - 2 + z = -15$$
$$-14 + z = -15$$
$$z = -1$$

Checking verifies that $(-4, 2, -1)$ is the solution and $\{(-4, 2, -1)\}$ is the solution set.

Definitions and Concepts	**Examples**

Section 3.3 Systems of Linear Equations in Three Variables (continued)

Curve Fitting

Curve fitting is determining a function whose graph contains given points. Three points that do not lie on a line determine the graph of a quadratic function

$$y = ax^2 + bx + c.$$

Use the three given points to create a system of three equations. Solve the system to find $a, b,$ and c.

Find the quadratic function $y = ax^2 + bx + c$ whose graph passes through the points $(-1, 2), (1, 8),$ and $(2, 14)$.
Use $y = ax^2 + bx + c.$

When $x = -1, y = 2$: $\quad 2 = a(-1)^2 + b(-1) + c$
When $x = 1, y = 8$: $\quad 8 = a \cdot 1^2 + b \cdot 1 + c$
When $x = 2, y = 14$: $\quad 14 = a \cdot 2^2 + b \cdot 2 + c$

Solving
$$\begin{cases} a - b + c = 2 \\ a + b + c = 8 \\ 4a + 2b + c = 14, \end{cases}$$

$a = 1, b = 3,$ and $c = 4.$ The quadratic function, $y = ax^2 + bx + c,$ is $y = x^2 + 3x + 4.$

Section 3.4 Matrix Solutions to Linear Systems

A matrix is a rectangular array of numbers. The augmented matrix of a linear system is obtained by writing the coefficients of each variable, a vertical bar, and the constants of the system.

$$\begin{cases} x + 4y = 9 \\ 3x + y = 5 \end{cases}$$

The augmented matrix is $\begin{bmatrix} 1 & 4 & | & 9 \\ 3 & 1 & | & 5 \end{bmatrix}.$

The following row operations produce matrices that represent systems with the same solution. Two matrices are row equivalent if one can be obtained from the other by a sequence of these row operations.

1. Interchange the elements in the ith and jth rows:
   $R_i \leftrightarrow R_j.$

2. Multiply each element in the ith row by k: $kR_i.$

3. Add k times the elements in row i to the corresponding elements in row j: $kR_i + R_j.$

Find the result of the row operation $-4R_1 + R_2$:

$$\begin{bmatrix} 1 & 0 & -2 & | & 5 \\ 4 & -1 & 2 & | & 6 \\ 3 & -7 & 9 & | & 10 \end{bmatrix}.$$

Add -4 times the elements in row 1 to the corresponding elements in row 2.

$$\begin{bmatrix} 1 & 0 & -2 & | & 5 \\ -4(1) + 4 & -4(0) + (-1) & -4(-2) + 2 & | & -4(5) + 6 \\ 3 & -7 & 9 & | & 10 \end{bmatrix}$$

$$= \begin{bmatrix} 1 & 0 & -2 & | & 5 \\ 0 & -1 & 10 & | & -14 \\ 3 & -7 & 9 & | & 10 \end{bmatrix}$$

Solving Linear Systems Using Matrices

1. Write the augmented matrix for the system.

2. Use matrix row operations to simplify the matrix to row-echelon form, with 1s down the main diagonal from upper left to lower right, and 0s below the 1s.

3. Write the system of linear equations corresponding to the matrix from step 2 and use back-substitution to find the system's solution.

If you obtain a matrix with a row containing 0s to the left of the vertical bar and a nonzero number on the right, the system is inconsistent. If 0s appear across an entire row, the system contains dependent equations.

Solve using matrices:

$$\begin{cases} 3x + y = 5 \\ x + 4y = 9. \end{cases}$$

$$\begin{bmatrix} 3 & 1 & | & 5 \\ 1 & 4 & | & 9 \end{bmatrix} \xrightarrow{R_1 \leftrightarrow R_2} \begin{bmatrix} 1 & 4 & | & 9 \\ 3 & 1 & | & 5 \end{bmatrix}$$

$$\xrightarrow{-3R_1 + R_2} \begin{bmatrix} 1 & 4 & | & 9 \\ 0 & -11 & | & -22 \end{bmatrix} \xrightarrow{-\frac{1}{11}R_2} \begin{bmatrix} 1 & 4 & | & 9 \\ 0 & 1 & | & 2 \end{bmatrix}$$

$$\longrightarrow \begin{cases} x + 4y = 9 \\ y = 2. \end{cases}$$

When $y = 2, x + 4 \cdot 2 = 9,$ so $x = 1.$ The solution is $(1, 2)$ and the solution set is $\{(1, 2)\}.$

Definitions and Concepts	**Examples**

Section 3.5 Determinants and Cramer's Rule

A square matrix has the same number of rows as columns. A determinant is a real number associated with a square matrix. The determinant is denoted by placing vertical bars about the array of numbers. The value of a second-order determinant is

$$\begin{vmatrix} a_1 & b_1 \\ a_2 & b_2 \end{vmatrix} = a_1 b_2 - a_2 b_1.$$

Evaluate:

$$\begin{vmatrix} 2 & -1 \\ 3 & 4 \end{vmatrix} = 2(4) - 3(-1) = 8 + 3 = 11.$$

Cramer's Rule for Two Linear Equations in Two Variables

If

$$\begin{cases} a_1 x + b_1 y = c_1 \\ a_2 x + b_2 y = c_2, \end{cases}$$

then

$$x = \frac{\begin{vmatrix} c_1 & b_1 \\ c_2 & b_2 \end{vmatrix}}{\begin{vmatrix} a_1 & b_1 \\ a_2 & b_2 \end{vmatrix}} = \frac{D_x}{D} \text{ and } y = \frac{\begin{vmatrix} a_1 & c_1 \\ a_2 & c_2 \end{vmatrix}}{\begin{vmatrix} a_1 & b_1 \\ a_2 & b_2 \end{vmatrix}} = \frac{D_y}{D}, D \neq 0.$$

If $D = 0$, the system is inconsistent or contains dependent equations. Use a method other than Cramer's rule to determine the solution set.

Solve by Cramer's rule:

$$\begin{cases} 5x + 3y = 7 \\ -x + 2y = 9. \end{cases}$$

$$D = \begin{vmatrix} 5 & 3 \\ -1 & 2 \end{vmatrix} = 5(2) - (-1)(3) = 10 + 3 = 13$$

$$D_x = \begin{vmatrix} 7 & 3 \\ 9 & 2 \end{vmatrix} = 7 \cdot 2 - 9 \cdot 3 = 14 - 27 = -13$$

$$D_y = \begin{vmatrix} 5 & 7 \\ -1 & 9 \end{vmatrix} = 5(9) - (-1)(7) = 45 + 7 = 52$$

$$x = \frac{D_x}{D} = \frac{-13}{13} = -1, \quad y = \frac{D_y}{D} = \frac{52}{13} = 4$$

The solution is $(-1, 4)$ and the solution set is $\{(-1, 4)\}$.

The value of a third-order determinant is

$$\begin{vmatrix} a_1 & b_1 & c_1 \\ a_2 & b_2 & c_2 \\ a_3 & b_3 & c_3 \end{vmatrix}$$

$$= a_1 \begin{vmatrix} b_2 & c_2 \\ b_3 & c_3 \end{vmatrix} - a_2 \begin{vmatrix} b_1 & c_1 \\ b_3 & c_3 \end{vmatrix} + a_3 \begin{vmatrix} b_1 & c_1 \\ b_2 & c_2 \end{vmatrix}.$$

Each second-order determinant is called a minor.

Evaluate:

$$\begin{vmatrix} 1 & -2 & 1 \\ 3 & 1 & -2 \\ 5 & 5 & 3 \end{vmatrix}$$

$$= 1\begin{vmatrix} 1 & -2 \\ 5 & 3 \end{vmatrix} - 3\begin{vmatrix} -2 & 1 \\ 5 & 3 \end{vmatrix} + 5\begin{vmatrix} -2 & 1 \\ 1 & -2 \end{vmatrix}$$

$$= 1(3 - (-10)) - 3(-6 - 5) + 5(4 - 1)$$

$$= 1(13) - 3(-11) + 5(3)$$

$$= 13 + 33 + 15 = 61.$$

Definitions and Concepts	**Examples**

Section 3.5 Determinants and Cramer's Rule (continued)

Cramer's Rule for Three Linear Equations in Three Variables

If

$$\begin{cases} a_1x + b_1y + c_1z = d_1 \\ a_2x + b_2y + c_2z = d_2 \\ a_3x + b_3y + c_3z = d_3, \end{cases}$$

then

$$x = \frac{D_x}{D}, \quad y = \frac{D_y}{D}, \quad z = \frac{D_z}{D}.$$

$$D = \begin{vmatrix} a_1 & b_1 & c_1 \\ a_2 & b_2 & c_2 \\ a_3 & b_3 & c_3 \end{vmatrix} \neq 0, \quad D_x = \begin{vmatrix} d_1 & b_1 & c_1 \\ d_2 & b_2 & c_2 \\ d_3 & b_3 & c_3 \end{vmatrix}$$

$$D_y = \begin{vmatrix} a_1 & d_1 & c_1 \\ a_2 & d_2 & c_2 \\ a_3 & d_3 & c_3 \end{vmatrix}, \quad D_z = \begin{vmatrix} a_1 & b_1 & d_1 \\ a_2 & b_2 & d_2 \\ a_3 & b_3 & d_3 \end{vmatrix}$$

If $D = 0$, the system is inconsistent or contains dependent equations. Use a methed other than Cramer's rule to determine the solution set.

Solve by Cramer's rule:

$$\begin{cases} x - 2y + z = 4 \\ 3x + y - 2z = 3 \\ 5x + 5y + 3z = -8. \end{cases}$$

$$D = \begin{vmatrix} 1 & -2 & 1 \\ 3 & 1 & -2 \\ 5 & 5 & 3 \end{vmatrix} = 61 \qquad \text{This evaluation is shown on the bottom of page 247.}$$

$$D_x = \begin{vmatrix} 4 & -2 & 1 \\ 3 & 1 & -2 \\ -8 & 5 & 3 \end{vmatrix} = 61$$

$$D_y = \begin{vmatrix} 1 & 4 & 1 \\ 3 & 3 & -2 \\ 5 & -8 & 3 \end{vmatrix} = -122$$

$$D_z = \begin{vmatrix} 1 & -2 & 4 \\ 3 & 1 & 3 \\ 5 & 5 & -8 \end{vmatrix} = -61$$

$$x = \frac{D_x}{D} = \frac{61}{61} = 1, \quad y = \frac{D_y}{D} = \frac{-122}{61} = -2,$$

$$z = \frac{D_z}{D} = \frac{-61}{61} = -1.$$

The solution is $(1, -2, -1)$ and the solution set is $\{(1, -2, -1)\}$.

CHAPTER 3 REVIEW EXERCISES

3.1 *In Exercises 1–2, determine whether the given ordered pair is a solution of the system.*

1. $(4, 2)$
$$\begin{cases} 2x - 5y = -2 \\ 3x + 4y = 4 \end{cases}$$

2. $(-5, 3)$
$$\begin{cases} \dot{-}x + 2y = 11 \\ y = -\dfrac{x}{3} + \dfrac{4}{3} \end{cases}$$

In Exercises 3–6, solve each system by graphing. Identify systems with no solution and systems with infinitely many solutions, using set notation to express their solution sets.

3. $\begin{cases} x + y = 5 \\ 3x - y = 3 \end{cases}$

4. $\begin{cases} 3x - 2y = 6 \\ 6x - 4y = 12 \end{cases}$

5. $\begin{cases} y = \dfrac{3}{5}x - 3 \\ 2x - y = -4 \end{cases}$

6. $\begin{cases} y = -x + 4 \\ 3x + 3y = -6 \end{cases}$

In Exercises 7–13, solve each system by the substitution method or the addition method. Identify systems with no solution and systems with infinitely many solutions, using set notation to express their solution sets.

7. $\begin{cases} 2x - y = 2 \\ x + 2y = 11 \end{cases}$

8. $\begin{cases} y = -2x + 3 \\ 3x + 2y = -17 \end{cases}$

9. $\begin{cases} 3x + 2y = -8 \\ 2x + 5y = 2 \end{cases}$

10. $\begin{cases} 5x - 2y = 14 \\ 3x + 4y = 11 \end{cases}$

11. $\begin{cases} y = 4 - x \\ 3x + 3y = 12 \end{cases}$

12. $\begin{cases} \dfrac{x}{8} + \dfrac{3y}{4} = \dfrac{19}{8} \\ -\dfrac{x}{2} + \dfrac{3y}{4} = \dfrac{1}{2} \end{cases}$

13. $\begin{cases} x - 2y + 3 = 0 \\ 2x - 4y + 7 = 0 \end{cases}$

3.2 *In Exercises 14–18, use the four-step strategy to solve each problem.*

14. An appliance store is having a sale on small TVs and stereos. One day a salesperson sells 3 of the TVs and 4 stereos for $2530. The next day the salesperson sells 4 of the same TVs and 3 of the same stereos for $2510. What are the prices of a TV and a stereo?

15. You invested $9000 in two funds paying 4% and 7% annual interest. At the end of the year, the total interest from these investments was $555. How much was invested at each rate?

16. A chemist needs to mix a solution that is 34% silver nitrate with one that is 4% silver nitrate to obtain 100 milliliters of a mixture that is 7% silver nitrate. How many milliliters of each of the solutions must be used?

17. When a plane flies with the wind, it can travel 2160 miles in 3 hours. When the plane flies in the opposite direction, against the wind, it takes 4 hours to fly the same distance. Find the rate of the plane in still air and the rate of the wind.

18. The perimeter of a rectangular tabletop is 34 feet. The difference between 4 times the length and 3 times the width is 33 feet. Find the dimensions.

The cost and revenue functions for producing and selling x units of a new graphing calculator are

$$C(x) = 22,500 + 40x \quad \text{and} \quad R(x) = 85x.$$

Use these functions to solve Exercises 19–21.

19. Find the loss or the gain from selling 400 graphing calculators.

20. Determine the break-even point. Describe what this means.

21. Write the profit function, P, from producing and selling x graphing calculators.

22. A company is planning to manufacture computer desks. The fixed cost will be $60,000 and it will cost $200 to produce each desk. Each desk will be sold for $450.

   a. Write the cost function, C, of producing x desks.

   b. Write the revenue function, R, from the sale of x desks.

   c. Determine the break-even point. Describe what this means.

3.3

23. Is $(-3, -2, 5)$ a solution of the system

$$\begin{cases} x + y + z = 0 \\ 2x - 3y + z = 5 \\ 4x + 2y + 4z = 3? \end{cases}$$

Solve each system in Exercises 24–26 by eliminating variables using the addition method. If there is no solution or if there are infinitely many solutions and a system's equations are dependent, so state.

24. $\begin{cases} 2x - y + z = 1 \\ 3x - 3y + 4z = 5 \\ 4x - 2y + 3z = 4 \end{cases}$

25. $\begin{cases} x + 2y - z = 5 \\ 2x - y + 3z = 0 \\ 2y + z = 1 \end{cases}$

26. $\begin{cases} 3x - 4y + 4z = 7 \\ x - y - 2z = 2 \\ 2x - 3y + 6z = 5 \end{cases}$

27. Find the quadratic function $y = ax^2 + bx + c$ whose graph passes through the points $(1, 4)$, $(3, 20)$, and $(-2, 25)$.

28. **20th Century Death** The greatest cause of death in the 20th century was disease, killing 1390 million people. The bar graph shows the five leading causes of death in that century, excluding disease.

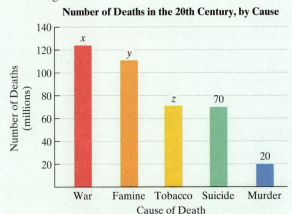

Number of Deaths in the 20th Century, by Cause

Source: Centers for Disease Control and Prevention

War, famine, and tobacco combined resulted in 306 million deaths. The difference between the number of deaths from war and famine was 13 million. The difference between the number of deaths from war and tobacco was 53 million. Find the number of 20th century deaths from war, famine, and tobacco.

3.4   *In Exercises 29–32, perform each matrix row operation and write the new matrix.*

29. $\begin{bmatrix} 1 & -8 & 3 \\ 0 & 7 & -14 \end{bmatrix} \frac{1}{7} R_2$

30. $\begin{bmatrix} 1 & -3 & 1 \\ 2 & 1 & -5 \end{bmatrix} -2R_1 + R_2$

31. $\begin{bmatrix} 2 & -2 & 1 & -1 \\ 1 & 2 & -1 & 2 \\ 6 & 4 & 3 & 5 \end{bmatrix} \frac{1}{2} R_1$

32. $\begin{bmatrix} 1 & 2 & 2 & 2 \\ 0 & 1 & -1 & 2 \\ 0 & 5 & 4 & 1 \end{bmatrix} -5R_2 + R_3$

In Exercises 33–36, solve each system using matrices. If there is no solution or if a system's equations are dependent, so state.

33. $\begin{cases} x + 4y = 7 \\ 3x + 5y = 0 \end{cases}$

34. $\begin{cases} 2x - 3y = 8 \\ -6x + 9y = 4 \end{cases}$

35. $\begin{cases} x + 2y + 3z = -5 \\ 2x + y + z = 1 \\ x + y - z = 8 \end{cases}$

36. $\begin{cases} x - 2y + z = 0 \\ y - 3z = -1 \\ 2y + 5z = -2 \end{cases}$

3.5   *In Exercises 37–40, evaluate each determinant.*

37. $\begin{vmatrix} 3 & 2 \\ -1 & 5 \end{vmatrix}$

38. $\begin{vmatrix} -2 & -3 \\ -4 & -8 \end{vmatrix}$

39. $\begin{vmatrix} 2 & 4 & -3 \\ 1 & -1 & 5 \\ -2 & 4 & 0 \end{vmatrix}$ **40.** $\begin{vmatrix} 4 & 7 & 0 \\ -5 & 6 & 0 \\ 3 & 2 & -4 \end{vmatrix}$

In Exercises 41–44, use Cramer's rule to solve each system. If there is no solution or if a system's equations are dependent, so state.

41. $\begin{cases} x - 2y = 8 \\ 3x + 2y = -1 \end{cases}$

42. $\begin{cases} 7x + 2y = 0 \\ 2x + y = -3 \end{cases}$

43. $\begin{cases} x + 2y + 2z = 5 \\ 2x + 4y + 7z = 19 \\ -2x - 5y - 2z = 8 \end{cases}$

44. $\begin{cases} 2x + y = -4 \\ y - 2z = 0 \\ 3x - 2z = -11 \end{cases}$

45. Use the quadratic function $y = ax^2 + bx + c$ to model the following data:

x (Age of a Driver)	y (Average Number of Automobile Accidents per Day in the United States)
20	400
40	150
60	400

Use Cramer's rule to determine values for a, b, and c. Then use the model to write a statement about the average number of automobile accidents in which 30-year-old drivers and 50-year-old drivers are involved daily.

CHAPTER 3 TEST

Step-by-step test solutions are found on the Chapter Test Prep Videos available in MyMathLab® or on You Tube (search "BlitzerInterAlg7e" and click on "Channels").

1. Solve by graphing

$$\begin{cases} x + y = 6 \\ 4x - y = 4. \end{cases}$$

In Exercises 2–4, solve each system by the substitution method or the addition method. Identify systems with no solution and systems with infinitely many solutions, using set notation to express their solution sets.

2. $\begin{cases} 5x + 4y = 10 \\ 3x + 5y = -7 \end{cases}$

3. $\begin{cases} x = y + 4 \\ 3x + 7y = -18 \end{cases}$

4. $\begin{cases} 4x = 2y + 6 \\ y = 2x - 3 \end{cases}$

In Exercises 5–8, solve each problem.

5. In a new development, 50 one- and two-bedroom condominiums were sold. Each one-bedroom condominium sold for $120 thousand and each two-bedroom condominium sold for $150 thousand. If sales totaled $7050 thousand, how many of each type of unit was sold?

6. You invested $9000 in two funds paying 6% and 7% annual interest. At the end of the year, the total interest from these investments was $610. How much was invested at each rate?

7. You need to mix a 6% peroxide solution with a 9% peroxide solution to obtain 36 ounces of an 8% peroxide solution. How many ounces of each of the solutions must be used?

8. A paddleboat on the Mississippi River travels 48 miles downstream, with the current, in 3 hours. The return trip, against the current, takes the paddleboat 4 hours. Find the boat's rate in still water and the rate of the current.

Use this information to solve Exercises 9–11: A company is planning to produce and sell a new line of computers. The fixed cost will be $360,000 and it will cost $850 to produce each computer. Each computer will be sold for $1150.

9. Write the cost function, C, of producing x computers.

10. Write the revenue function, R, from the sale of x computers.

11. Determine the break-even point. Describe what this means.

12. The cost and revenue functions for producing and selling x units of a toaster oven are

$$C(x) = 40x + 350{,}000 \quad \text{and} \quad R(x) = 125x.$$

Write the profit function, P, from producing and selling x toaster ovens.

13. Solve by eliminating variables using the addition method:

$$\begin{cases} x + y + z = 6 \\ 3x + 4y - 7z = 1 \\ 2x - y + 3z = 5. \end{cases}$$

14. Perform the indicated matrix row operation and write the new matrix.

$$\begin{bmatrix} 1 & 0 & -4 & | & 5 \\ 6 & -1 & 2 & | & 10 \\ 2 & -1 & 4 & | & -3 \end{bmatrix} \quad -6R_1 + R_2$$

In Exercises 15–16, solve each system using matrices.

15. $\begin{cases} 2x + y = 6 \\ 3x - 2y = 16 \end{cases}$

16. $\begin{cases} x - 4y + 4z = -1 \\ 2x - y + 5z = 6 \\ -x + 3y - z = 5 \end{cases}$

In Exercises 17–18, evaluate each determinant.

17. $\begin{vmatrix} -1 & -3 \\ 7 & 4 \end{vmatrix}$

18. $\begin{vmatrix} 3 & 4 & 0 \\ -1 & 0 & -3 \\ 4 & 2 & 5 \end{vmatrix}$

In Exercises 19–20, use Cramer's rule to solve each system.

19. $\begin{cases} 4x - 3y = 14 \\ 3x - y = 3 \end{cases}$

20. $\begin{cases} 2x + 3y + z = 2 \\ 3x + 3y - z = 0 \\ x - 2y - 3z = 1 \end{cases}$

CUMULATIVE REVIEW EXERCISES   (CHAPTERS 1–3)

1. Simplify: $\dfrac{6(8 - 10)^3 + (-2)}{(-5)^2(-2)}$.

2. Simplify: $7x - [5 - 2(4x - 1)]$.

In Exercises 3–5, solve each equation.

3. $5 - 2(3 - x) = 2(2x + 5) + 1$

4. $\dfrac{3x}{5} + 4 = \dfrac{x}{3}$

5. $3x - 4 = 2(3x + 2) - 3x$

6. For a summer sales job, you are choosing between two pay arrangements: a weekly salary of $200 plus 5% commission on sales, or a straight 15% commission. For how many dollars of sales will the earnings be the same regardless of the pay arrangement?

7. Simplify: $\dfrac{-5x^6 y^{-10}}{20x^{-2} y^{20}}$.

8. If $f(x) = -4x + 5$, find $f(a + 2)$.

9. Find the domain of $f(x) = \dfrac{4}{x + 3}$.

10. If $f(x) = 2x^2 - 5x + 2$ and $g(x) = x^2 - 2x + 3$, find $(f - g)(x)$ and $(f - g)(3)$.

In Exercises 11–12, graph each linear function.

11. $f(x) = -\dfrac{2}{3}x + 2$

12. $2x - y = 6$

In Exercises 13–14, use the given conditions to write an equation for each line in point-slope form and slope-intercept form.

13. Passing through $(2, 4)$ and $(4, -2)$

14. Passing through $(-1, 0)$ and parallel to the line whose equation is $3x + y = 6$

In Exercises 15–16, solve each system by eliminating variables using the addition method.

15. $\begin{cases} 3x + 12y = 25 \\ 2x - 6y = 12 \end{cases}$

16. $\begin{cases} x + 3y - z = 5 \\ -x + 2y + 3z = 13 \\ 2x - 5y - z = -8 \end{cases}$

17. If two pads of paper and 19 pens are sold for $5.40 and 7 of the same pads and 4 of the same pens sell for $6.40, find the cost of one pad and one pen.

18. Evaluate:

$$\begin{vmatrix} 0 & 1 & -2 \\ -7 & 0 & -4 \\ 3 & 0 & 5 \end{vmatrix}.$$

19. Solve using matrices:

$$\begin{cases} 2x + 3y - z = -1 \\ x + 2y + 3z = 2 \\ 3x + 5y - 2x = -3. \end{cases}$$

20. Solve using Cramer's rule (determinants):

$$\begin{cases} 3x + 4y = -1 \\ -2x + y = 8. \end{cases}$$

SECTION

4.1

Solving Linear Inequalities

What am I supposed to learn?

After studying this section, you should be able to:

1 Solve linear inequalities.

2 Recognize inequalities with no solution or all real numbers as solutions.

3 Solve applied problems using linear inequalities.

You can go online and obtain a list of cellphone companies and their monthly price plans. You've chosen a plan that has a monthly fee of $80 that includes 1500 voice minutes, 4 GB of data, with a charge of $0.08 per text (regular, photo, or video). Suppose you are limited by how much money you can spend for the month: You can spend at most $100. If we let x represent the number of text messages in a month, we can write an inequality that describes the given conditions:

The monthly fee of $80	plus	the charge of $0.08 per text for x text messages	must be less than or equal to	$100.
80	+	0.08x	≤	100.

Using the commutative property of addition, we can express this inequality as

$$0.08x + 80 \le 100.$$

Placing an inequality symbol between a linear expression ($mx + b$) and a constant results in a *linear inequality in one variable*. In this section, we will study how to solve linear inequalities such as the one shown above. **Solving an inequality** is the process of finding the set of numbers that make the inequality a true statement. These numbers are called the **solutions** of the inequality and we say that they **satisfy** the inequality. The set of all solutions is called the **solution set** of the inequality. We will use interval notation to represent these solution sets.

1 Solve linear inequalities.

Solving Linear Inequalities in One Variable

We know that a linear equation in x can be expressed as $ax + b = 0$. A **linear inequality in x** can be written in one of the following forms: $ax + b < 0, ax + b \le 0, ax + b > 0, ax + b \ge 0$. In each form, $a \ne 0$.

Back to our question from above: How many text messages can you send in a month if you can spend at most $100? We answer the question by solving the linear inequality

$$0.08x + 80 \le 100$$

for x. The solution procedure is nearly identical to that for solving the equation

$$0.08x + 80 = 100.$$

Our goal is to get x by itself on the left side. We do this by first subtracting 80 from both sides to isolate 0.08x:

$0.08x + 80 \le 100$	This is the given inequality.
$0.08x + 80 - 80 \le 100 - 80$	Subtract 80 from both sides.
$0.08x \le 20.$	Simplify.

Finally, we isolate x from $0.08x$ by dividing both sides of the inequality by 0.08:

$$\frac{0.08x}{0.08} \leq \frac{20}{0.08} \qquad \text{Divide both sides by 0.08.}$$

$$x \leq 250. \qquad \text{Simplify.}$$

With at most \$100 per month to spend, you can send no more than 250 text messages each month.

We started with the inequality $0.08x + 80 \leq 100$ and obtained the inequality $x \leq 250$ in the final step. Both of these inequalities have the same solution set. Inequalities such as these, with the same solution set, are said to be **equivalent**.

We isolated x from $0.08x$ by dividing both sides of $0.08x \leq 20$ by 0.08, a positive number. Let's see what happens if we divide both sides of an inequality by a negative number. Consider the inequality $10 < 14$. Divide 10 and 14 by -2:

$$\frac{10}{-2} = -5 \quad \text{and} \quad \frac{14}{-2} = -7.$$

Because -5 lies to the right of -7 on the number line, -5 is greater than -7:

$$-5 > -7.$$

Notice that the direction of the inequality symbol is reversed:

$$10 < 14$$
$$\downarrow$$
$$-5 > -7.$$

Dividing by -2 changes the direction of the inequality symbol.

In general, **when we multiply or divide both sides of an inequality by a negative number, the direction of the inequality symbol is reversed**. When we reverse the direction of the inequality symbol, we say that we change the *sense* of the inequality.

We can isolate a variable in a linear inequality the same way we can isolate a variable in a linear equation. The following properties are used to create equivalent inequalities:

Properties of Inequalities

Property	The Property in Words	Example
The Addition Property of Inequality If $a < b$, then $a + c < b + c$. If $a < b$, then $a - c < b - c$.	If the same quantity is added to or subtracted from both sides of an inequality, the resulting inequality is equivalent to the original one.	$2x + 3 < 7$ Subtract 3: $\quad 2x + 3 - 3 < 7 - 3.$ Simplify: $\quad\quad 2x < 4.$
The Positive Multiplication Property of Inequality If $a < b$ and c is positive, then $ac < bc$. If $a < b$ and c is positive, then $\dfrac{a}{c} < \dfrac{b}{c}$.	If we multiply or divide both sides of an inequality by the same positive quantity, the resulting inequality is equivalent to the original one.	$2x < 4$ Divide by 2: $\quad \dfrac{2x}{2} < \dfrac{4}{2}.$ Simplify: $\quad\quad x < 2.$
The Negative Multiplication Property of Inequality If $a < b$ and c is negative, then $ac > bc$. If $a < b$ and c is negative, then $\dfrac{a}{c} > \dfrac{b}{c}$.	If we multiply or divide both sides of an inequality by the same negative quantity and reverse the direction of the inequality symbol, the resulting inequality is equivalent to the original one.	$-4x < 20$ Divide by -4 and change the sense of the inequality: $\quad \dfrac{-4x}{-4} > \dfrac{20}{-4}.$ Simplify: $\quad\quad x > -5.$

If an inequality does not contain fractions, it can be solved using the following procedure. (In Example 3, we will see how to clear fractions.) Notice, again, how similar this procedure is to the procedure for solving a linear equation.

> ### Solving a Linear Inequality
>
> 1. Simplify the algebraic expression on each side.
> 2. Use the addition property of inequality to collect all the variable terms on one side and all the constant terms on the other side.
> 3. Use the multiplication property of inequality to isolate the variable and solve. Change the sense of the inequality when multiplying or dividing both sides by a negative number.
> 4. Express the solution set in interval notation and graph the solution set on a number line.

EXAMPLE 1 Solving a Linear Inequality

Solve and graph the solution set on a number line:

$$3x - 5 > -17.$$

Solution

Step 1. Simplify each side. Because each side is already simplified, we can skip this step.

Step 2. Collect variable terms on one side and constant terms on the other side. The variable term, $3x$, is already on the left side of $3x - 5 > -17$. We will collect constant terms on the right side by adding 5 to both sides.

$$3x - 5 > -17 \qquad \text{This is the given inequality.}$$
$$3x - 5 + 5 > -17 + 5 \qquad \text{Add 5 to both sides.}$$
$$3x > -12 \qquad \text{Simplify.}$$

Step 3. Isolate the variable and solve. We isolate the variable, x, by dividing both sides by 3. Because we are dividing by a positive number, we do not reverse the direction of the inequality symbol.

$$\frac{3x}{3} > \frac{-12}{3} \qquad \text{Divide both sides by 3.}$$
$$x > -4 \qquad \text{Simplify.}$$

Step 4. Express the solution set in interval notation and graph the set on a number line. The solution set consists of all real numbers that are greater than -4. The interval notation for this solution set is $(-4, \infty)$. The graph of the solution set is shown as follows:

✓ **CHECK POINT 1** Solve and graph the solution set on a number line:

$$4x - 3 > -23.$$

EXAMPLE 2 Solving a Linear Inequality

Solve and graph the solution set on a number line:

$$-2x - 4 > x + 5.$$

Solution

Step 1. Simplify each side. Because each side is already simplified, we can skip this step.

Great Question!

Can I solve the inequality in Example 2 by isolating the variable on the right?

Yes, you can solve

$$-2x - 4 > x + 5$$

by isolating x on the right side. Add $2x$ to both sides.

$$-2x - 4 + 2x > x + 5 + 2x$$
$$-4 > 3x + 5$$

Now subtract 5 from both sides.

$$-4 - 5 > 3x + 5 - 5$$
$$-9 > 3x$$

Finally, divide both sides by 3.

$$\frac{-9}{3} > \frac{3x}{3}$$
$$-3 > x$$

This last inequality means the same thing as $x < -3$.

Step 2. Collect variable terms on one side and constant terms on the other side. We will collect variable terms on the left and constant terms on the right.

$-2x - 4 > x + 5$	This is the given inequality.
$-2x - 4 - x > x + 5 - x$	Subtract x from both sides.
$-3x - 4 > 5$	Simplify.
$-3x - 4 + 4 > 5 + 4$	Add 4 to both sides.
$-3x > 9$	Simplify.

Step 3. Isolate the variable and solve. We isolate the variable, x, by dividing both sides by -3. Because we are dividing by a negative number, we must reverse the direction of the inequality symbol.

$\dfrac{-3x}{-3} < \dfrac{9}{-3}$	Divide both sides by -3 and change the sense of the inequality.
$x < -3$	Simplify.

Step 4. Express the solution set in interval notation and graph the set on a number line. The solution set consists of all real numbers that are less than -3. The interval notation for this solution set is $(-\infty, -3)$. The graph of the solution set is shown as follows:

■

Using Technology

Numeric and Graphic Connections

You can use a graphing utility to check the solution set of a linear inequality. Enter each side of the inequality separately under y_1 and y_2. Then use the table or the graphs. To use the table, first locate the x-value for which the y-values are the same. Then scroll up or down to locate x values for which y_1 is greater than y_2 or for which y_1 is less than y_2. To use the graphs, locate the intersection point and then find the x-values for which the graph of y_1 lies above the graph of y_2 ($y_1 > y_2$) or for which the graph of y_1 lies below the graph of y_2 ($y_1 < y_2$).

Let's verify our work in Example 2 and show that $(-\infty, -3)$ is the solution set of

$$-2x - 4 > x + 5.$$

Enter $y_1 = -2x - 4$ in the $\boxed{y =}$ screen. Enter $y_2 = x + 5$ in the $\boxed{y =}$ screen.

We are looking for values of x for which y_1 is greater than y_2.

Numeric Check

Scrolling through the table shows that $y_1 > y_2$ for values of x that are less than -3 (when $x = -3$, $y_1 = y_2$). This verifies $(-\infty, -3)$ is the solution set of $-2x - 4 > x + 5$.

Graphic Check

Display the graphs for y_1 and y_2. Use the intersection feature. The solution set is the set of x-values for which the graph of y_1 lies above the graph of y_2.

Graphs intersect at $(-3, 2)$. When x is less than -3, the graph of y_1 lies above the graph of y_2. This graphically verifies $(-\infty, -3)$ is the solution set of $-2x - 4 > x + 5$.

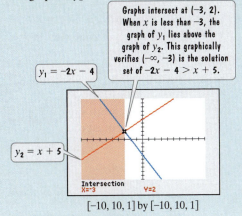

$[-10, 10, 1]$ by $[-10, 10, 1]$

☑ **CHECK POINT 2**   Solve and graph the solution set: $3x + 1 > 7x - 15$.

If an inequality contains fractions, begin by multiplying both sides by the least common denominator. This will clear the inequality of fractions.

EXAMPLE 3   Solving a Linear Inequality Containing Fractions

Solve and graph the solution set on a number line:

$$\frac{x + 3}{4} \geq \frac{x - 2}{3} + \frac{1}{4}.$$

Solution   The denominators are 4, 3, and 4. The least common denominator is 12. We begin by multiplying both sides of the inequality by 12.

$\dfrac{x + 3}{4} \geq \dfrac{x - 2}{3} + \dfrac{1}{4}$	This is the given inequality.
$12\left(\dfrac{x + 3}{4}\right) \geq 12\left(\dfrac{x - 2}{3} + \dfrac{1}{4}\right)$	Multiply both sides by 12. Multiplying by a positive number preserves the sense of the inequality.
$\dfrac{12}{1} \cdot \dfrac{x + 3}{4} \geq \dfrac{12}{1} \cdot \dfrac{x - 2}{3} + \dfrac{12}{1} \cdot \dfrac{1}{4}$	Multiply each term by 12. Use the distributive property on the right side.
$\dfrac{\overset{3}{\cancel{12}}}{1} \cdot \dfrac{x + 3}{\underset{1}{\cancel{4}}} \geq \dfrac{\overset{4}{\cancel{12}}}{1} \cdot \dfrac{x - 2}{\underset{1}{\cancel{3}}} + \dfrac{\overset{3}{\cancel{12}}}{1} \cdot \dfrac{1}{\underset{1}{\cancel{4}}}$	Divide out common factors in each multiplication.
$3(x + 3) \geq 4(x - 2) + 3$	The fractions are now cleared.

Now that the fractions have been cleared, we follow the four steps that we used in the previous examples.

Step 1. Simplify each side.

$3(x + 3) \geq 4(x - 2) + 3$	This is the inequality with the fractions cleared.
$3x + 9 \geq 4x - 8 + 3$	Use the distributive property.
$3x + 9 \geq 4x - 5$	Simplify.

Step 2. Collect variable terms on one side and constant terms on the other side. We will collect variable terms on the left and constant terms on the right.

$3x + 9 - 4x \geq 4x - 5 - 4x$	Subtract 4x from both sides.
$-x + 9 \geq -5$	Simplify.
$-x + 9 - 9 \geq -5 - 9$	Subtract 9 from both sides.
$-x \geq -14$	Simplify.

Step 3. Isolate the variable and solve. To isolate x, we must eliminate the negative sign in front of the x. Because $-x$ means $-1x$, we can do this by multiplying (or dividing) both sides of the inequality by -1. We are multiplying by a negative number. Thus, we must reverse the direction of the inequality symbol.

$(-1)(-x) \leq (-1)(-14)$	Multiply both sides by −1 and change the sense of the inequality.
$x \leq 14$	Simplify.

Step 4. Express the solution set in interval notation and graph the set on a number line. The solution set consists of all real numbers that are less than or equal to 14. The

interval notation for this solution set is $(-\infty, 14]$. The graph of the solution set is shown as follows:

☑ **CHECK POINT 3** Solve and graph the solution set on a number line: $\dfrac{x-4}{2} \geq \dfrac{x-2}{3} + \dfrac{5}{6}$.

---

2 Recognize inequalities with no solution or all real numbers as solutions. ▶

Inequalities with Unusual Solution Sets

We have seen that some equations have no solution. This is also true for some inequalities. An example of such an inequality is

$$x > x + 1.$$

There is no number that is greater than itself plus 1. This inequality has no solution and its solution set is $\varnothing$, the empty set.

By contrast, some inequalities are true for all real numbers. An example of such an inequality is

$$x < x + 1.$$

Every real number is less than itself plus 1. The solution set is $(-\infty, \infty)$.

If you attempt to solve an inequality that has no solution, you will eliminate the variable and obtain a false statement, such as $0 > 1$. If you attempt to solve an inequality that is true for all real numbers, you will eliminate the variable and obtain a true statement, such as $0 < 1$.

EXAMPLE 4 Solving Linear Inequalities

Solve each inequality:

a. $2(x + 4) > 2x + 3$

b. $x + 7 \leq x - 2$.

Using Technology

Graphic Connections
The graphs of

$y_1 = 2(x + 4)$ and $y_2 = 2x + 3$

are parallel lines. The graph of y_1 is always above the graph of y_2. Every value of x satisfies the inequality $y_1 > y_2$. Thus, the solution set of the inequality

$$2(x + 4) > 2x + 3$$

is $(-\infty, \infty)$.

$[-10, 10, 1]$ by $[-10, 10, 1]$

Solution

a.

$2(x + 4) > 2x + 3$	This is the given inequality.
$2x + 8 > 2x + 3$	Apply the distributive property.
$2x + 8 - 2x > 2x + 3 - 2x$	Subtract $2x$ from both sides.
$8 > 3$	Simplify. The statement $8 > 3$ is true.

The inequality $8 > 3$ is true for all values of x. Because this inequality is equivalent to the original inequality, the original inequality is true for all real numbers. The solution set is $(-\infty, \infty)$.

b.

$x + 7 \leq x - 2$	This is the given inequality.
$x + 7 - x \leq x - 2 - x$	Subtract x from both sides.
$7 \leq -2$	Simplify. The statement $7 \leq -2$ is false.

The inequality $7 \leq -2$ is false for all values of x. Because this inequality is equivalent to the original inequality, the original inequality has no solution. The solution set is $\varnothing$. ∎

☑ **CHECK POINT 4** Solve each inequality:

a. $3(x + 1) > 3x + 2$

b. $x + 1 \leq x - 1$.

3 Solve applied problems using linear inequalities. ▶

Applications

Commonly used English phrases such as "at least" and "at most" indicate inequalities. **Table 4.1** lists sentences containing these phrases and their algebraic translations into inequalities.

Table 4.1	English Sentences and Inequalities
English Sentence	**Inequality**
x is at least 5.	$x \geq 5$
x is at most 5.	$x \leq 5$
x is between 5 and 7.	$5 < x < 7$
x is no more than 5.	$x \leq 5$
x is no less than 5.	$x \geq 5$

Our next example shows how to use an inequality to select the better deal when considering two pricing options. We use our strategy for solving word problems, translating from the verbal conditions of the problem to a linear inequality.

EXAMPLE 5   Selecting the Better Deal

Acme Car rental agency charges $4 a day plus $0.15 per mile, whereas Interstate rental agency charges $20 a day and $0.05 per mile. How many miles must be driven to make the daily cost of an Acme rental a better deal than an Interstate rental?

Solution
Step 1. Let x represent one of the unknown quantities. We are looking for the number of miles that must be driven in a day to make Acme the better deal. Thus,

let x = the number of miles driven in a day.

Step 2. Represent other unknown quantities in terms of x. We are not asked to find another quantity, so we can skip this step.

Step 3. Write an inequality in x that models the conditions. Acme is a better deal than Interstate if the daily cost of Acme is less than the daily cost of Interstate.

$$4 \;+\; 0.15 \;\cdot\; x \;<\; 20 \;+\; 0.05 \;\cdot\; x$$

Step 4. Solve the inequality and answer the question.

$4 + 0.15x < 20 + 0.05x$	This is the inequality that models the verbal conditions.
$4 + 0.15x - 0.05x < 20 + 0.05x - 0.05x$	Subtract 0.05x from both sides.
$4 + 0.1x < 20$	Simplify.
$4 + 0.1x - 4 < 20 - 4$	Subtract 4 from both sides.
$0.1x < 16$	Simplify.
$\dfrac{0.1x}{0.1} < \dfrac{16}{0.1}$	Divide both sides by 0.1.
$x < 160$	Simplify.

Thus, driving fewer than 160 miles per day makes Acme the better deal.

Using Technology

Graphic Connections
The graphs of the daily cost models for the car rental agencies

$$y_1 = 4 + 0.15x$$
$$\text{and } y_2 = 20 + 0.05x$$

are shown in a $[0, 300, 10]$ by $[0, 40, 4]$ viewing rectangle. The graphs intersect at $(160, 28)$. To the left of $x = 160$, the graph of Acme's daily cost lies below that of Interstate's daily cost. This shows that for fewer than 160 miles per day, Acme offers the better deal.

Step 5. Check the proposed solution in the original wording of the problem. One way to do this is to take a mileage less than 160 miles per day to see if Acme is the better deal. Suppose that 150 miles are driven in a day.

$$\text{Cost for Acme} = 4 + 0.15(150) = 26.50$$
$$\text{Cost for Interstate} = 20 + 0.05(150) = 27.50$$

Acme has a lower daily cost, making Acme the better deal. ■

✓ **CHECK POINT 5**   A car can be rented from Basic Rental for $260 per week with no extra charge for mileage. Continental charges $80 per week plus 25 cents for each mile driven to rent the same car. How many miles must be driven in a week to make the rental cost for Basic Rental a better deal than Continental's?

Achieving Success

Learn from your mistakes. Being human means making mistakes. By finding and understanding your errors, you will become a better math student.

Source of Error	Remedy
Not Understanding a Concept	Review the concept by finding a similar example in your textbook or class notes. Ask your professor questions to help clarify the concept.
Skipping Steps	Show clear step-by-step solutions. Detailed solution procedures help organize your thoughts and enhance understanding. Doing too many steps mentally often results in preventable mistakes.
Carelessness	Write neatly. Not being able to read your own math writing leads to errors. Avoid writing in pen so you won't have to put huge marks through incorrect work.

"You can achieve your goal if you persistently pursue it."
   —Cha Sa-Soon, a 68-year-old South Korean woman who passed her country's written driver's-license exam on her 950th try (*Source: Newsweek*)

CONCEPT AND VOCABULARY CHECK

Fill in each blank so that the resulting statement is true.

1. The addition property of inequality states that if $a < b$, then $a + c$ _____.

2. The positive multiplication property of inequality states that if $a < b$ and c is positive, then ac _____.

3. The negative multiplication property of inequality states that if $a < b$ and c is negative, then ac _____.

4. The linear inequality $-3x - 4 > 5$ can be solved by first _____ to both sides and then _____ both sides by _____, which changes the _____ of the inequality symbol from _____ to _____.

5. In solving an inequality, if you eliminate the variable and obtain a false statement such as $7 < -2$, the solution set is _____.

6. In solving an inequality, if you eliminate the variable and obtain a true statement such as $8 > 3$, the solution set is _____.

7. The algebraic translation of "x is at least 7" is _____.

8. The algebraic translation of "x is at most 7" is _____.

9. The algebraic translation of "x is no more than 7" is _____.

10. The algebraic translation of "x is no less than 7" is _____.

4.1 EXERCISE SET ▶ MyMathLab®

Practice Exercises

In Exercises 1–32, solve each linear inequality. Other than ∅, graph the solution set on a number line.

1. $5x + 11 < 26$
2. $2x + 5 < 17$
3. $3x - 8 \geq 13$
4. $8x - 2 \geq 14$
5. $-9x \geq 36$
6. $-5x \leq 30$
7. $8x - 11 \leq 3x - 13$
8. $18x + 45 \leq 12x - 8$
9. $4(x + 1) + 2 \geq 3x + 6$
10. $8x + 3 > 3(2x + 1) + x + 5$
11. $2x - 11 < -3(x + 2)$
12. $-4(x + 2) > 3x + 20$
13. $1 - (x + 3) \geq 4 - 2x$
14. $5(3 - x) \leq 3x - 1$
15. $\dfrac{x}{4} - \dfrac{1}{2} \leq \dfrac{x}{2} + 1$
16. $\dfrac{3x}{10} + 1 \geq \dfrac{1}{5} - \dfrac{x}{10}$
17. $1 - \dfrac{x}{2} > 4$
18. $7 - \dfrac{4}{5}x < \dfrac{3}{5}$
19. $\dfrac{x - 4}{6} \geq \dfrac{x - 2}{9} + \dfrac{5}{18}$
20. $\dfrac{4x - 3}{6} + 2 \geq \dfrac{2x - 1}{12}$
21. $4(3x - 2) - 3x < 3(1 + 3x) - 7$
22. $3(x - 8) - 2(10 - x) < 5(x - 1)$
23. $8(x + 1) \leq 7(x + 5) + x$
24. $4(x - 1) \geq 3(x - 2) + x$
25. $3x < 3(x - 2)$
26. $5x < 5(x - 3)$
27. $7(x + 4) - 13 < 12 + 13(3 + x)$
28. $-3[7x - (2x - 3)] > -2(x + 1)$
29. $6 - \dfrac{2}{3}(3x - 12) \leq \dfrac{2}{5}(10x + 50)$
30. $\dfrac{2}{7}(7 - 21x) - 4 > 10 - \dfrac{3}{11}(11x - 11)$
31. $3[3(x + 5) + 8x + 7] + 5[3(x - 6)$
    $-2(3x - 5)] < 2(4x + 3)$
32. $5[3(2 - 3x) - 2(5 - x)] - 6[5(x - 2)$
    $-2(4x - 3)] < 3x + 19$
33. Let $f(x) = 3x + 2$ and $g(x) = 5x - 8$. Find all values of x for which $f(x) > g(x)$.
34. Let $f(x) = 2x - 9$ and $g(x) = 5x + 4$. Find all values of x for which $f(x) > g(x)$.

35. Let $f(x) = \dfrac{2}{5}(10x + 15)$ and $g(x) = \dfrac{1}{4}(8 - 12x)$. Find all values of x for which $g(x) \leq f(x)$.
36. Let $f(x) = \dfrac{3}{5}(10x - 15) + 9$ and $g(x) = \dfrac{3}{8}(16 - 8x) - 7$. Find all values of x for which $g(x) \leq f(x)$.
37. Let $f(x) = 1 - (x + 3) + 2x$. Find all values of x for which $f(x)$ is at least 4.
38. Let $f(x) = 2x - 11 + 3(x + 2)$. Find all values of x for which $f(x)$ is at most 0.

Practice PLUS

In Exercises 39–40, solve each linear inequality and graph the solution set on a number line.

39. $2(x + 3) > 6 - \{4[x - (3x - 4) - x] + 4\}$
40. $3(4x - 6) < 4 - \{5x - [6x - (4x - (3x + 2))]\}$

In Exercises 41–42, write an inequality with x isolated on the left side that is equivalent to the given inequality.

41. $ax + b > c$; Assume $a < 0$.
42. $\dfrac{ax + b}{c} > b$; Assume $a > 0$ and $c < 0$.

In Exercises 43–44, use the graphs of y_1 and y_2 to solve each inequality.

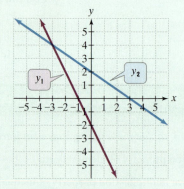

43. $y_1 \geq y_2$
44. $y_1 \leq y_2$

In Exercises 45–46, use the table of values for the linear functions y_1 and y_2 to solve each inequality.

X	Y₁	Y₂
-1.8	9.4	7.4
-1.7	9.1	7.6
-1.6	8.8	7.8
-1.5	8.5	8
-1.4	8.2	8.2
-1.3	7.9	8.4
-1.2	7.6	8.6
-1.1	7.3	8.8
-1	7	9
-.9	6.7	9.2
-.8	6.4	9.4
X=-1.8		

45. $y_1 < y_2$
46. $y_1 > y_2$

Application Exercises

The graphs show that the three components of love, namely passion, intimacy, and commitment, progress differently over time. Passion peaks early in a relationship and then declines. By contrast, intimacy and commitment build gradually. Use the graphs to solve Exercises 47–54.

The Course of Love Over Time

Source: R.J. Sternberg, *A Triangular Theory of Love, Psychological Review, 93,* 119–135.

47. Use interval notation to write an inequality that expresses for which years in a relationship intimacy is greater than commitment.

48. Use interval notation to write an inequality that expresses for which years in a relationship passion is greater than or equal to intimacy.

49. What is the relationship between passion and intimacy on the interval $[5, 7)$?

50. What is the relationship between intimacy and commitment on the interval $[4, 7)$?

51. What is the relationship between passion and commitment on the interval $(6, 8)$?

52. What is the relationship between passion and commitment on the interval $(7, 9)$?

53. What is the maximum level of intensity for passion? After how many years in a relationship does this occur?

54. After approximately how many years do levels of intensity for commitment exceed the maximum level of intensity for passion?

Diversity Index. *What is the chance that the next person I meet will be different from me? The diversity index, from 0 (no diversity) to 100, measures the chance that two randomly selected people are a different race or ethnicity. The diversity index in the United States varies widely from region to region, from as high as 81 in Hawaii to as low as 11 in Vermont. The bar graph at the top of the next column shows the national diversity index for the United States for four years in the period from 1980 through 2010. Exercises 55–56 are based on the data in the graph.*

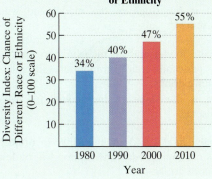

Chance That Two Randomly Selected Americans Are a Different Race or Ethnicity

Source: USA Today

55. The data in the graph can be modeled by the formula
$$D = 0.7x + 34,$$
where D is the national diversity index for the United States x years after 1980.

    a. How well does the formula model the national diversity index for 2010?

    b. Use an inequality to determine in which years the national diversity index will exceed 62.

56. The data in the graph can be modeled by the formula
$$D = 0.7x + 34,$$
where D is the national diversity index for the United States x years after 1980.

    a. Does the formula underestimate or overestimate the national diversity index for 2000? By how many points?

    b. Use an inequality to determine in which years the national diversity index will exceed 69.

Drinking Less Soda. *The bar graph shows annual per-capita consumption of 12-ounce servings of carbonated soda in the United States for three selected years.*

Per-Capita Consumption of Carbonated Soda in the U.S.

Source: Beverage Digest

The data displayed by the graph can be modeled by the formula
$$S = -8x + 583,$$
where S is per-capita consumption of carbonated soda (12-ounce servings) x years after 1998 (the year of U.S. peak soda consumption). Use this model to solve Exercises 57–58.

(*In Exercises 57–58, refer to the model and graph at the bottom of the previous page.*)

57. a. What is the slope of the model and what does it represent?

  b. Use the model and an inequality to project in which years per-capita soda consumption will be less than 415 12-ounce drinks.

58. a. What is the slope of the model and what does it represent?

  b. Use the model and an inequality to project in which years per-capita soda consumption will be less than 383 12-ounce drinks.

In Exercises 59–66, use the strategy for solving word problems, translating from the verbal conditions of the problem to a linear inequality.

59. A truck can be rented from Basic Rental for $50 a day plus $0.20 per mile. Continental charges $20 per day plus $0.50 per mile to rent the same truck. How many miles must be driven in a day to make the rental cost for Basic Rental a better deal than Continental's?

60. You are choosing between two landline telephone plans. Plan A has a monthly fee of $15 with a charge of $0.08 per minute for all calls. Plan B has a monthly fee of $3 with a charge of $0.12 per minute for all calls. How many calling minutes in a month make plan A the better deal?

61. A city commission has proposed two tax bills. The first bill requires that a homeowner pay $1800 plus 3% of the assessed home value in taxes. The second bill requires taxes of $200 plus 8% of the assessed home value. What price range of home assessment would make the first bill a better deal for the homeowner?

62. A local bank charges $8 per month plus 5¢ per check. The credit union charges $2 per month plus 8¢ per check. How many checks should be written each month to make the credit union a better deal?

63. A company manufactures and sells blank DVDs. The weekly fixed cost is $10,000 and it costs $0.40 to produce each DVD. The selling price is $2.00 per DVD. How many DVDs must be produced and sold each week for the company to have a profit?

64. A company manufactures and sells personalized stationery. The weekly fixed cost is $3000 and it costs $3.00 to produce each package of stationery. The selling price is $5.50 per package. How many packages of stationery must be produced and sold each week for the company to have a profit?

65. An elevator at a construction site has a maximum capacity of 3000 pounds. If the elevator operator weighs 200 pounds and each cement bag weighs 70 pounds, how many bags of cement can be safely lifted on the elevator in one trip?

66. An elevator at a construction site has a maximum capacity of 2500 pounds. If the elevator operator weighs 160 pounds and each cement bag weighs 60 pounds, how many bags of cement can be safely lifted on the elevator in one trip?

Explaining the Concepts

67. When graphing the solutions of an inequality, what does a parenthesis signify? What does a bracket signify?

68. When solving an inequality, when is it necessary to change the sense of the inequality? Give an example.

69. Describe ways in which solving a linear inequality is similar to solving a linear equation.

70. Describe ways in which solving a linear inequality is different from solving a linear equation.

71. When solving a linear inequality, describe what happens if the solution set is $(-\infty, \infty)$.

72. When solving a linear inequality, describe what happens if the solution set is $\varnothing$.

73. What is the slope of the model in Exercise 55 or Exercise 56? What does this represent in terms of the 0-to-100 national diversity index?

Technology Exercises

In Exercises 74–75, solve each inequality using a graphing utility. Graph each side separately. Then determine the values of x for which the graph on the left side lies above the graph on the right side.

74. $-3(x - 6) > 2x - 2$

75. $-2(x + 4) > 6x + 16$

76. Use a graphing utility's $\boxed{\text{TABLE}}$ feature to verify your work in Exercises 74–75.

Use the same technique employed in Exercises 74–75 to solve each inequality in Exercises 77–78. In each case, what conclusion can you draw? What happens if you try solving the inequalities algebraically?

77. $12x - 10 > 2(x - 4) + 10x$

78. $2x + 3 > 3(2x - 4) - 4x$

79. A bank offers two checking account plans. Plan A has a base service charge of $4.00 per month plus 10¢ per check. Plan B charges a base service charge of $2.00 per month plus 15¢ per check.

  a. Write models for the total monthly costs for each plan if x checks are written.

  b. Use a graphing utility to graph the models in the same $[0, 50, 10]$ by $[0, 10, 1]$ viewing rectangle.

  c. Use the graphs (and the intersection feature) to determine for what number of checks per month plan A will be better than plan B.

  d. Verify the result of part (c) algebraically by solving an inequality.

Critical Thinking Exercises

Make Sense? *In Exercises 80–83, determine whether each statement makes sense or does not make sense, and explain your reasoning.*

80. I began the solution of $5 - 3(x + 2) > 10x$ by simplifying the left side, obtaining $2x + 4 > 10x$.

81. I have trouble remembering when to reverse the direction of an inequality symbol, so I avoid this difficulty by collecting variable terms on an appropriate side.

82. If you tell me that three times a number is less than two times that number, it's obvious that no number satisfies this condition, and there is no need for me to write and solve an inequality.

83. Whenever I solve a linear inequality in which the coefficients of the variable on each side are the same, the solution set is $\varnothing$ or $(-\infty, \infty)$.

In Exercises 84–87, determine whether each statement is true or false. If the statement is false, make the necessary change(s) to produce a true statement.

84. The inequality $3x > 6$ is equivalent to $2 > x$.

85. The smallest real number in the solution set of $2x > 6$ is 4.

86. If x is at least 7, then $x > 7$.

87. The inequality $-3x > 6$ is equivalent to $-2 > x$.

88. Find a so that the solution set of $ax + 4 \le -12$ is $[8, \infty)$.

89. What's wrong with this argument? Suppose x and y represent two real numbers, where $x > y$.

$2 > 1$	This is a true statement.
$2(y - x) > 1(y - x)$	Multiply both sides by $y - x$.
$2y - 2x > y - x$	Use the distributive property.
$y - 2x > -x$	Subtract y from both sides.
$y > x$	Add $2x$ to both sides.

The final inequality, $y > x$, is impossible because we were initially given $x > y$.

Review Exercises

90. If $f(x) = x^2 - 2x + 5$, find $f(-4)$. (Section 2.1, Example 3)

91. Solve the system:
$$\begin{cases} 2x - y - z = -3 \\ 3x - 2y - 2z = -5 \\ -x + y + 2z = 4. \end{cases}$$
(Section 3.3, Example 2)

92. Simplify: $\left(\dfrac{2x^4 y^{-2}}{4xy^3}\right)^3$. (Section 1.6, Example 9)

Preview Exercises

Exercises 93–95 will help you prepare for the material covered in the next section.

93. Consider the sets $A = \{1, 2, 3, 4\}$ and $B = \{3, 4, 5, 6, 7\}$.

  a. Write the set consisting of elements common to both set A and set B.

  b. Write the set consisting of elements that are members of set A or of set B or of both sets.

94. a. Solve: $x - 3 < 5$.

  b. Solve: $2x + 4 < 14$.

  c. Give an example of a number that satisfies the inequality in part (a) and the inequality in part (b).

  d. Give an example of a number that satisfies the inequality in part (a), but not the inequality in part (b).

95. a. Solve: $2x - 6 \ge -4$.

  b. Solve: $5x + 2 \ge 17$.

  c. Give an example of a number that satisfies the inequality in part (a) and the inequality in part (b).

  d. Give an example of a number that satisfies the inequality in part (a), but not the inequality in part (b).

4.2

Compound Inequalities

What am I supposed to learn?

After studying this section, you should be able to:

1 Find the intersection of two sets.

2 Solve compound inequalities involving *and*.

3 Find the union of two sets.

4 Solve compound inequalities involving *or*.

Among the U.S. presidents, more have had dogs than cats in the White House. Sets of presidents with dogs, cats, dogs and cats, dogs or cats, and neither, with the number of presidents belonging to each of these sets, are shown in **Figure 4.1**.

$17 + 9 + 2 = 28$ presidents had dogs OR cats.

17 presidents had dogs only.

9 presidents had both dogs AND cats.

2 presidents had cats only.

15 presidents had neither dogs nor cats.

Presidents with Dogs   Presidents with Cats

Figure 4.1 Presidential pets in the White House

Source: factmonster.com

Figure 4.1 indicates that 9 presidents had dogs *and* cats in the White House. Furthermore, 28 presidents had dogs *or* cats in the White House. In this section, we'll focus on sets joined with the word *and* and the word *or*. The difference is that we'll be using solution sets of inequalities rather than sets of presidents with pets.

A **compound inequality** is formed by joining two inequalities with the word *and* or the word *or*.

Examples of Compound Inequalities

- $x - 3 < 5$ and $2x + 4 < 14$
- $3x - 5 \leq 13$ or $5x + 2 > -3$

Compound inequalities illustrate the importance of the words *and* and *or* in mathematics, as well as in everyday English.

1 Find the intersection of two sets.

Compound Inequalities Involving *And*

If A and B are sets, we can form a new set consisting of all elements that are in both A and B. This set is called the *intersection* of the two sets.

Definition of the Intersection of Sets

The **intersection** of sets A and B, written $A \cap B$, is the set of elements common to both set A **and** set B. This definition can be expressed in set-builder notation as follows:

$$A \cap B = \{x \,|\, x \in A \text{ and } x \in B\}.$$

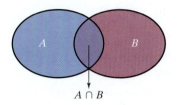

Figure 4.2 Picturing the intersection of two sets

Figure 4.2 shows a useful way of picturing the intersection of sets A and B. The figure indicates that $A \cap B$ contains those elements that belong to both A and B at the same time.

EXAMPLE 1 Finding the Intersection of Two Sets

Find the intersection: $\{7, 8, 9, 10, 11\} \cap \{6, 8, 10, 12\}$.

Solution The elements common to $\{7, 8, 9, 10, 11\}$ and $\{6, 8, 10, 12\}$ are 8 and 10. Thus,

$$\{7, 8, 9, 10, 11\} \cap \{6, 8, 10, 12\} = \{8, 10\}. \quad \blacksquare$$

✓ **CHECK POINT 1** Find the intersection: $\{3, 4, 5, 6, 7\} \cap \{3, 7, 8, 9\}$.

2 Solve compound inequalities involving *and*. ▶

A number is a **solution of a compound inequality formed by the word *and*** if it is a solution of both inequalities. For example, the solution set of the compound inequality

$$x \leq 6 \quad \text{and} \quad x \geq 2$$

is the set of values of x that satisfy both $x \leq 6$ and $x \geq 2$. Thus, the solution set is the intersection of the solution sets of the two inequalities.

What are the numbers that satisfy both $x \leq 6$ and $x \geq 2$? These numbers are easier to see if we graph the solution set to each inequality on a number line. These graphs are shown in **Figure 4.3**. The intersection is shown in the third graph.

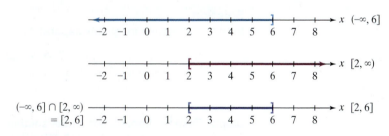

Figure 4.3 Numbers satisfying both $x \leq 6$ and $x \geq 2$

The numbers common to both sets are those that are less than or equal to 6 and greater than or equal to 2. This set in interval notation is $[2, 6]$.

Here is a procedure for finding the solution set of a compound inequality containing the word *and*.

Solving Compound Inequalities Involving *AND*

1. Solve each inequality separately.
2. Graph the solution set to each inequality on a number line and take the intersection of these solution sets. This intersection appears as the portion of the number line that the two graphs have in common.

EXAMPLE 2   Solving a Compound Inequality with *And*

Solve: $x - 3 < 5$ and $2x + 4 < 14$.

Solution
Step 1. Solve each inequality separately.

$$x - 3 < 5 \quad \text{and} \quad 2x + 4 < 14$$
$$x < 8 \qquad\qquad 2x < 10$$
$$x < 5$$

Step 2. Take the intersection of the solution sets of the two inequalities. We graph the solution sets of $x < 8$ and $x < 5$. The intersection is shown in the third graph.

The numbers common to both sets are those that are less than 5. The solution set in interval notation is $(-\infty, 5)$. Take a moment to check that any number in $(-\infty, 5)$ satisfies both of the original inequalities. ■

✓ **CHECK POINT 2**   Solve: $x + 2 < 5$ and $2x - 4 < -2$.

EXAMPLE 3   Solving a Compound Inequality with *And*

Solve: $2x - 7 > 3$ and $5x - 4 < 6$.

Solution
Step 1. Solve each inequality separately.

$$2x - 7 > 3 \quad \text{and} \quad 5x - 4 < 6$$
$$2x > 10 \qquad\qquad 5x < 10$$
$$x > 5 \qquad\qquad x < 2$$

Step 2. Take the intersection of the solution sets of the two inequalities. We graph the solution sets of $x > 5$ and $x < 2$. We use these graphs to find their intersection.

There is no number that is both greater than 5 and at the same time less than 2. Thus, the solution set is the empty set, $\varnothing$. ■

✓ **CHECK POINT 3**   Solve: $4x - 5 > 7$ and $5x - 2 < 3$.

If $a < b$, the compound inequality

$$a < x \text{ and } x < b$$

can be written in the shorter form

$$a < x < b.$$

For example, the compound inequality

$$-3 < 2x + 1 \text{ and } 2x + 1 < 3$$

can be abbreviated

$$-3 < 2x + 1 < 3.$$

The word *and* does not appear when the inequality is written in the shorter form, although it is implied. The shorter form enables us to solve both inequalities at once. By performing the same operations on all three parts of the inequality, our goal is to **isolate x in the middle**.

EXAMPLE 4   Solving a Compound Inequality

Solve and graph the solution set:

$$-3 < 2x + 1 \leq 3.$$

Solution   We would like to isolate x in the middle. We can do this by first subtracting 1 from all three parts of the compound inequality. Then we isolate x from $2x$ by dividing all three parts of the inequality by 2.

$-3 < 2x + 1 \leq 3$	This is the given inequality.
$-3 - 1 < 2x + 1 - 1 \leq 3 - 1$	Subtract 1 from all three parts.
$-4 < 2x \leq 2$	Simplify.
$\dfrac{-4}{2} < \dfrac{2x}{2} \leq \dfrac{2}{2}$	Divide each part by 2.
$-2 < x \leq 1$	Simplify.

The solution set consists of all real numbers greater than -2 and less than or equal to 1, represented by $(-2, 1]$ in interval notation. The graph is shown as follows:

Using Technology

Numeric and Graphic Connections

Let's verify our work in Example 4 and show that $(-2, 1]$ is the solution set of $-3 < 2x + 1 \leq 3$.

Numeric Check

To check numerically, enter $y_1 = 2x + 1$.

The shaded part of the table shows that values of $y_1 = 2x + 1$ are greater than -3 and less than or equal to 3 when x is in the interval $(-2, 1]$.

Graphic Check

To check graphically, graph each part of

$$-3 < 2x + 1 \leq 3.$$

Enter $y_1 = -3.$ Enter $y_2 = 2x + 1.$ Enter $y_3 = 3.$

The figure shows that the graph of $y_2 = 2x + 1$ lies above the graph of $y_1 = -3$ and on or below the graph of $y_3 = 3$ when x is in the interval $(-2, 1]$.

$[-3, 3, 1]$ by $[-5, 5, 1]$

✓ **CHECK POINT 4**   Solve and graph the solution set: $1 \le 2x + 3 < 11$.

3 Find the union of two sets. ▶

Compound Inequalities Involving *Or*

Another set that we can form from sets *A* and *B* consists of elements that are in *A* or *B* or in both sets. This set is called the *union* of the two sets.

> ### Definition of the Union of Sets
>
> The **union** of sets *A* and *B*, written $A \cup B$, is the set of elements that are members of set *A* **or** of set *B* or of both sets. This definition can be expressed in set-builder notation as follows:
>
> $$A \cup B = \{x \,|\, x \in A \text{ or } x \in B\}.$$

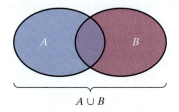

$A \cup B$

Figure 4.4 Picturing the union of two sets

Figure 4.4 shows a useful way of picturing the union of sets *A* and *B*. The figure indicates that $A \cup B$ is formed by joining the sets together.

We can find the union of set *A* and set *B* by listing the elements of set *A*. Then, we include any elements of set *B* that have not already been listed. Enclose all elements that are listed with braces. This shows that the union of two sets is also a set.

EXAMPLE 5   Finding the Union of Two Sets

Find the union: $\{7, 8, 9, 10, 11\} \cup \{6, 8, 10, 12\}$.

Solution   To find $\{7, 8, 9, 10, 11\} \cup \{6, 8, 10, 12\}$, start by listing all the elements from the first set, namely 7, 8, 9, 10, and 11. Now list all the elements from the second set that are not in the first set, namely 6 and 12. The union is the set consisting of all these elements. Thus,

$$\{7, 8, 9, 10, 11\} \cup \{6, 8, 10, 12\} = \{6, 7, 8, 9, 10, 11, 12\}.$$

Although 8 and 10 appear in both sets,   do not list 8 and 10 twice.   ∎

✓ **CHECK POINT 5**   Find the union: $\{3, 4, 5, 6, 7\} \cup \{3, 7, 8, 9\}$.

4 Solve compound inequalities involving *or*. ▶

A number is a **solution of a compound inequality formed by the word *or*** if it is a solution of either inequality. Thus, the solution set of a compound inequality formed by the word *or* is the union of the solution sets of the two inequalities.

> ### Solving Compound Inequalities Involving OR
>
> 1. Solve each inequality separately.
> 2. Graph the solution set to each inequality on a number line and take the union of these solution sets. This union appears as the portion of the number line representing the total collection of numbers in the two graphs.

EXAMPLE 6   Solving a Compound Inequality with *Or*

Solve: $2x - 3 < 7$ or $35 - 4x \leq 3$.

Solution
Step 1. Solve each inequality separately.

$$2x - 3 < 7 \quad \text{or} \quad 35 - 4x \leq 3$$
$$2x < 10 \qquad\qquad -4x \leq -32$$
$$x < 5 \qquad\qquad\quad x \geq 8$$

Step 2. Take the union of the solution sets of the two inequalities. We graph the solution sets of $x < 5$ and $x \geq 8$. We use these graphs to find their union.

The solution set consists of all numbers that are less than 5 or greater than or equal to 8. The solution set in interval notation is $(-\infty, 5) \cup [8, \infty)$. There is no shortcut way to express this union when interval notation is used. ■

✓ **CHECK POINT 6**   Solve: $3x - 5 \leq -2$ or $10 - 2x < 4$.

EXAMPLE 7   Solving a Compound Inequality with *Or*

Solve: $3x - 5 \leq 13$ or $5x + 2 > -3$.

Solution
Step 1. Solve each inequality separately.

$$3x - 5 \leq 13 \quad \text{or} \quad 5x + 2 > -3$$
$$3x \leq 18 \qquad\qquad 5x > -5$$
$$x \leq 6 \qquad\qquad\quad x > -1$$

Step 2. Take the union of the solution sets of the two inequalities. We graph the solution sets of $x \leq 6$ and $x > -1$. We use these graphs to find their union.

Because all real numbers are either less than or equal to 6 or greater than −1 or both, the union of the two sets fills the entire number line. Thus, the solution set in interval notation is $(-\infty, \infty)$. Any real number that you select will satisfy at least one of the original inequalities. ■

✓ **CHECK POINT 7**   Solve: $2x + 5 \geq 3$ or $2x + 3 < 3$.

CONCEPT AND VOCABULARY CHECK

Fill in each blank so that the resulting statement is true.

1. The set of elements common to both set A and set B is called the _____ of sets A and B, and is symbolized by _____.

2. The set of elements that are members of set A or set B or of both sets is called the _____ of sets A and B, and is symbolized by _____.

3. The set of elements common to both $(-\infty, 9)$ and $(-\infty, 12)$ is _____.

4. The set of elements in $(-\infty, 9)$ or $(-\infty, 12)$ or in both sets is _____.

5. The way to solve $-7 < 3x - 4 \le 5$ is to isolate x in the _____.

4.2 EXERCISE SET ▶ MyMathLab®

Practice Exercises

In Exercises 1–6, find the intersection of the sets.

1. $\{1, 2, 3, 4\} \cap \{2, 4, 5\}$
2. $\{1, 3, 7\} \cap \{2, 3, 8\}$
3. $\{1, 3, 5, 7\} \cap \{2, 4, 6, 8, 10\}$
4. $\{0, 1, 3, 5\} \cap \{-5, -3, -1\}$
5. $\{a, b, c, d\} \cap \varnothing$
6. $\{w, y, z\} \cap \varnothing$

In Exercises 7–24, solve each compound inequality. Use graphs to show the solution set to each of the two given inequalities, as well as a third graph that shows the solution set of the compound inequality. Except for the empty set, express the solution set in interval notation.

7. $x > 3$ and $x > 6$
8. $x > 2$ and $x > 4$
9. $x \le 5$ and $x \le 1$
10. $x \le 6$ and $x \le 2$
11. $x < 2$ and $x \ge -1$
12. $x < 3$ and $x \ge -1$
13. $x > 2$ and $x < -1$
14. $x > 3$ and $x < -1$
15. $5x < -20$ and $3x > -18$
16. $3x \le 15$ and $2x > -6$
17. $x - 4 \le 2$ and $3x + 1 > -8$
18. $3x + 2 > -4$ and $2x - 1 < 5$
19. $2x > 5x - 15$ and $7x > 2x + 10$
20. $6 - 5x > 1 - 3x$ and $4x - 3 > x - 9$
21. $4(1 - x) < -6$ and $\dfrac{x - 7}{5} \le -2$
22. $5(x - 2) > 15$ and $\dfrac{x - 6}{4} \le -2$
23. $x - 1 \le 7x - 1$ and $4x - 7 < 3 - x$
24. $2x + 1 > 4x - 3$ and $x - 1 \ge 3x + 5$

In Exercises 25–32, solve each inequality and graph the solution set on a number line. Express the solution set in interval notation.

25. $6 < x + 3 < 8$
26. $7 < x + 5 < 11$
27. $-3 \le x - 2 < 1$
28. $-6 < x - 4 \le 1$

29. $-11 < 2x - 1 \le -5$
30. $3 \le 4x - 3 < 19$
31. $-3 \le \dfrac{2x}{3} - 5 < -1$
32. $-6 \le \dfrac{x}{2} - 4 < -3$

In Exercises 33–38, find the union of the sets.

33. $\{1, 2, 3, 4\} \cup \{2, 4, 5\}$
34. $\{1, 3, 7, 8\} \cup \{2, 3, 8\}$
35. $\{1, 3, 5, 7\} \cup \{2, 4, 6, 8, 10\}$
36. $\{0, 1, 3, 5\} \cup \{2, 4, 6\}$
37. $\{a, e, i, o, u\} \cup \varnothing$
38. $\{e, m, p, t, y\} \cup \varnothing$

In Exercises 39–54, solve each compound inequality. Use graphs to show the solution set to each of the two given inequalities, as well as a third graph that shows the solution set of the compound inequality. Express the solution set in interval notation.

39. $x > 3$ or $x > 6$
40. $x > 2$ or $x > 4$
41. $x \le 5$ or $x \le 1$
42. $x \le 6$ or $x \le 2$
43. $x < 2$ or $x \ge -1$
44. $x < 3$ or $x \ge -1$
45. $x \ge 2$ or $x < -1$
46. $x \ge 3$ or $x < -1$
47. $3x > 12$ or $2x < -6$
48. $3x < 3$ or $2x > 10$
49. $3x + 2 \le 5$ or $5x - 7 \ge 8$
50. $2x - 5 \le -11$ or $5x + 1 \ge 6$
51. $4x + 3 < -1$ or $2x - 3 \ge -11$
52. $2x + 1 < 15$ or $3x - 4 \ge -1$
53. $-2x + 5 > 7$ or $-3x + 10 > 2x$
54. $16 - 3x \ge -8$ or $13 - x > 4x + 3$
55. Let $f(x) = 2x + 3$ and $g(x) = 3x - 1$. Find all values of x for which $f(x) \ge 5$ and $g(x) > 11$.
56. Let $f(x) = 4x + 5$ and $g(x) = 3x - 4$. Find all values of x for which $f(x) \ge 5$ and $g(x) \le 2$.

57. Let $f(x) = 3x - 1$ and $g(x) = 4 - x$. Find all values of x for which $f(x) < -1$ or $g(x) < -2$.

58. Let $f(x) = 2x - 5$ and $g(x) = 3 - x$. Find all values of x for which $f(x) \geq 3$ or $g(x) < 0$.

Practice PLUS

In Exercises 59–60, write an inequality with x isolated in the middle that is equivalent to the given inequality. Assume $a > 0$, $b > 0$, and $c > 0$.

59. $-c < ax - b < c$

60. $-2 < \dfrac{ax - b}{c} < 2$

In Exercises 61–62, use the graphs of y_1, y_2, and y_3 to solve each compound inequality.

61. $-3 \leq 2x - 1 \leq 5$

62. $x - 2 < 2x - 1 < x + 2$

[−5, 5, 1] by [−5, 8, 1]

63. Solve $x - 2 < 2x - 1 < x + 2$, the inequality in Exercise 62, using algebraic methods. (*Hint:* Rewrite the inequality as $2x - 1 > x - 2$ and $2x - 1 < x + 2$.)

64. Use the hint given in Exercise 63 to solve $x \leq 3x - 10 \leq 2x$.

In Exercises 65–66, use the table to solve each inequality.

65. $-2 \leq 5x + 3 < 13$

66. $-3 < 2x - 5 \leq 3$

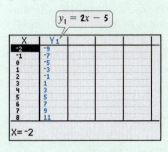

In Exercises 67–68, use the roster method to find the set of negative integers that are solutions of each inequality.

67. $5 - 4x \geq 1$ and $3 - 7x < 31$

68. $-5 < 3x + 4 \leq 16$

Application Exercises

In more U.S. marriages, spouses have different faiths. The bar graph shows the percentage of households with an interfaith marriage in 1988 and 2012. Also shown is the percentage of households in which a person of faith is married to someone with no religion.

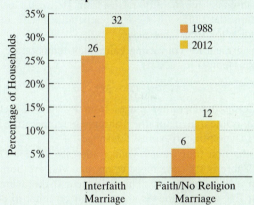

Source: General Social Survey, University of Chicago

The formula

$$I = \frac{1}{4}x + 26$$

models the percentage of U.S. households with an interfaith marriage, I, x years after 1988. The formula

$$N = \frac{1}{4}x + 6$$

models the percentage of U.S. households in which a person of faith is married to someone with no religion, N, x years after 1988. Use these models to solve Exercises 69–70.

69. a. In which years will more than 33% of U.S. households have an interfaith marriage?

b. In which years will more than 14% of U.S. households have a person of faith married to someone with no religion?

c. Based on your answers to parts (a) and (b), in which years will more than 33% of households have an interfaith marriage and more than 14% have a faith/no religion marriage?

d. Based on your answers to parts (a) and (b), in which years will more than 33% of households have an interfaith marriage or more than 14% have a faith/no religion marriage?

(Refer to the models and graph at the bottom of the previous page.)

70. a. In which years will more than 34% of U.S. households have an interfaith marriage?

b. In which years will more than 15% of U.S. households have a person of faith married to someone with no religion?

c. Based on your answers to parts (a) and (b), in which years will more than 34% of households have an interfaith marriage and more than 15% have a faith/no religion marriage?

d. Based on your answers to parts (a) and (b), in which years will more than 34% of households have an interfaith marriage or more than 15% have a faith/no religion marriage?

71. The formula for converting Celsius temperature, C, to Fahrenheit temperature, F, is

$$F = \frac{9}{5}C + 32.$$

If Fahrenheit temperature ranges from 41° to 50°, inclusive, what is the range for Celsius temperature? Use interval notation to express this range.

72. The formula for converting Fahrenheit temperature, F, to Celsius temperature, C, is

$$C = \frac{5}{9}(F - 32).$$

If Celsius temperature ranges from 15° to 35°, inclusive, what is the range for the Fahrenheit temperature? Use interval notation to express this range.

73. On the first four exams, your grades are 70, 75, 87, and 92. There is still one more exam, and you are hoping to earn a B in the course. This will occur if the average of your five exam grades is greater than or equal to 80 and less than 90. What range of grades on the fifth exam will result in earning a B? Use interval notation to express this range.

74. On the first four exams, your grades are 82, 75, 80, and 90. There is still a final exam, and it counts as two grades. You are hoping to earn a B in the course: This will occur if the average of your six exam grades is greater than or equal to 80 and less than 90. What range of grades on the final exam will result in earning a B? Use interval notation to express this range.

75. The toll to a bridge is $3.00. A three-month pass costs $7.50 and reduces the toll to $0.50. A six-month pass costs $30 and permits crossing the bridge for no additional fee. How many crossings per three-month period does it take for the three-month pass to be the best deal?

76. Parts for an automobile repair cost $175. The mechanic charges $34 per hour. If you receive an estimate for at least $226 and at most $294 for fixing the car, what is the time interval that the mechanic will be working on the job?

Explaining the Concepts

77. Describe what is meant by the intersection of two sets. Give an example.

78. Explain how to solve a compound inequality involving *and*.

79. Why is $1 < 2x + 3 < 9$ a compound inequality? What are the two inequalities and what is the word that joins them?

80. Explain how to solve $1 < 2x + 3 < 9$.

81. Describe what is meant by the union of two sets. Give an example.

82. Explain how to solve a compound inequality involving *or*.

Technology Exercises

In Exercises 83–86, solve each inequality using a graphing utility. Graph each of the three parts of the inequality separately in the same viewing rectangle. The solution set consists of all values of x for which the graph of the linear function in the middle lies between the graphs of the constant functions on the left and the right.

83. $1 < x + 3 < 9$

84. $-1 < \dfrac{x + 4}{2} < 3$

85. $1 \le 4x - 7 \le 3$

86. $2 \le 4 - x \le 7$

87. Use a graphing utility's $\boxed{\text{TABLE}}$ feature to verify your work in Exercises 83–86.

Critical Thinking Exercises

Make Sense? *In Exercises 88–91, determine whether each statement makes sense or does not make sense, and explain your reasoning.*

88. I've noticed that when solving some compound inequalities with *or*, there is no way to express the solution set using a single interval, but this does not happen with *and* compound inequalities.

89. Compound inequalities with *and* have solutions that satisfy both inequalities, whereas compound inequalities with *or* have solutions that satisfy at least one of the inequalities.

90. I'm considering the compound inequality $x < 8$ and $x \ge 8$, so the solution set is $\varnothing$.

91. I'm considering the compound inequality $x < 8$ or $x \ge 8$, so the solution set is $(-\infty, \infty)$.

In Exercises 92–95, determine whether each statement is true or false. If the statement is false, make the necessary change(s) to produce a true statement.

92. $(-\infty, -1] \cap [-4, \infty) = [-4, -1]$

93. $(-\infty, 3) \cup (-\infty, -2) = (-\infty, -2)$

94. The union of two sets can never give the same result as the intersection of those same two sets.

95. The solution set of the compound inequality $x < a$ and $x > a$ is the set of all real numbers excluding a.

96. Solve and express the solution set in interval notation: $-7 \le 8 - 3x \le 20$ and $-7 < 6x - 1 < 41$.

The graphs of $f(x) = \sqrt{4 - x}$ and $g(x) = \sqrt{x + 1}$ are shown in a $[-3, 10, 1]$ by $[-2, 5, 1]$ viewing rectangle.

$f(x) = \sqrt{4 - x}$

$g(x) = \sqrt{x + 1}$

In Exercises 97–100, use the graphs and interval notation to express the domain of the given function.

97. The domain of f

98. The domain of g

99. The domain of $f + g$

100. The domain of $\dfrac{f}{g}$

101. At the end of the day, the change machine at a laundrette contained at least \$3.20 and at most \$5.45 in nickels, dimes, and quarters. There were 3 fewer dimes than twice the number of nickels and 2 more quarters than twice the number of nickels. What was the least possible number and the greatest possible number of nickels?

Review Exercises

102. If $f(x) = x^2 - 3x + 4$ and $g(x) = 2x - 5$, find $(g - f)(x)$ and $(g - f)(-1)$. (Section 2.3, Example 4)

103. Use function notation to write the equation of the line passing through $(4, 2)$ and perpendicular to the line whose equation is $4x - 2y = 8$. (Section 2.5, Example 5)

104. Simplify: $4 - [2(x - 4) - 5]$. (Section 1.2, Example 14)

Preview Exercises

Exercises 105–107 will help you prepare for the material covered in the next section.

105. Find all values of x satisfying $1 - 4x = 3$ or $1 - 4x = -3$.

106. Find all values of x satisfying $3x - 1 = x + 5$ or $3x - 1 = -(x + 5)$.

107. **a.** Substitute -5 for x and determine whether -5 satisfies $|2x + 3| \geq 5$.

   b. Does 0 satisfy $|2x + 3| \geq 5$?

SECTION

4.3

Equations and Inequalities Involving Absolute Value ▶

What am I supposed to learn?

After studying this section, you should be able to:

1. Solve absolute value equations. ▶

2. Solve absolute value inequalities of the form $|u| < c$. ▶

3. Solve absolute value inequalities of the form $|u| > c$. ▶

4. Recognize absolute value inequalities with no solution or all real numbers as solutions. ▶

5. Solve problems using absolute value inequalities. ▶

What activities do you dread? Reading math textbooks with bottle-cap designs painted on brick walls on the cover? (Be kind!) No, that's not America's most-dreaded activity. In a random sample of 1000 U.S. adults, 46% of those questioned responded, "Public speaking."

Numerical information, such as the percentage of adults who dread public speaking, is often given with a margin of error. Inequalities involving absolute value are used to describe errors in polling, as well as errors of measurement in manufacturing, engineering, science, and other fields. In this section, you will learn to solve equations and inequalities containing absolute value. With these skills, you will be able to analyze data on the activities U.S. adults say they dread.

1 Solve absolute value equations.

Figure 4.5 If $|x| = 2$, then $x = 2$ or $x = -2$.

Equations Involving Absolute Value

We have seen that the absolute value of a, denoted $|a|$, is the distance from 0 to a on a number line. Now consider **absolute value equations**, such as

$$|x| = 2.$$

This means that we must determine real numbers whose distance from the origin on a number line is 2. **Figure 4.5** shows that there are two numbers such that $|x| = 2$, namely, 2 and -2. We write $x = 2$ or $x = -2$. This observation can be generalized as follows:

Rewriting an Absolute Value Equation Without Absolute Value Bars

If c is a positive real number and u represents any algebraic expression, then $|u| = c$ is equivalent to $u = c$ or $u = -c$.

EXAMPLE 1 Solving an Equation Involving Absolute Value

Solve: $|2x - 3| = 11$.

Solution

$\|2x - 3\| = 11$	This is the given equation.
$2x - 3 = 11 \quad$ or $\quad 2x - 3 = -11$	Rewrite the equation without absolute value bars: $\|u\| = c$ is equivalent to $u = c$ or $u = -c$. In this case $u = 2x - 3$.
$2x = 14 \qquad\qquad 2x = -8$	Add 3 to both sides of each equation.
$x = 7 \qquad\qquad x = -4$	Divide both sides of each equation by 2.

Check 7:	Check -4:	
$\|2x - 3\| = 11$	$\|2x - 3\| = 11$	This is the original equation.
$\|2(7) - 3\| \overset{?}{=} 11$	$\|2(-4) - 3\| \overset{?}{=} 11$	Substitute the proposed solutions.
$\|14 - 3\| \overset{?}{=} 11$	$\|-8 - 3\| \overset{?}{=} 11$	Perform operations inside the absolute value bars.
$\|11\| \overset{?}{=} 11$	$\|-11\| \overset{?}{=} 11$	
$11 = 11, \quad$ true	$11 = 11, \quad$ true	These true statements indicate that 7 and -4 are solutions.

The solutions are -4 and 7. We can also say that the solution set is $\{-4, 7\}$. ■

☑ CHECK POINT 1 Solve: $|2x - 1| = 5$.

EXAMPLE 2 Solving an Equation Involving Absolute Value

Solve: $5|1 - 4x| - 15 = 0$.

Solution

$5\|1 - 4x\| - 15 = 0$	This is the given equation.

We need to isolate $|1 - 4x|$, the absolute value expression.

$5\|1 - 4x\| = 15$	Add 15 to both sides.
$\|1 - 4x\| = 3$	Divide both sides by 5.
$1 - 4x = 3 \quad$ or $\quad 1 - 4x = -3$	Rewrite $\|u\| = c$ as $u = c$ or $u = -c$.
$-4x = 2 \qquad\qquad -4x = -4$	Subtract 1 from both sides of each equation.
$x = -\frac{1}{2} \qquad\qquad x = 1$	Divide both sides of each equation by -4.

Using Technology

Graphic Connections
You can use a graphing utility to verify the solution set of an absolute value equation. Consider, for example,

$$|2x - 3| = 11.$$

Graph $y_1 = |2x - 3|$ and $y_2 = 11$. The graphs are shown in a $[-10, 10, 1]$ by $[-1, 15, 1]$ viewing rectangle. The x-coordinates of the intersection points are -4 and 7, verifying that $\{-4, 7\}$ is the solution set.

$[-10, 10, 1]$ by $[-1, 15, 1]$

Take a moment to check $-\frac{1}{2}$ and 1, the proposed solutions, in the original equation, $5|1 - 4x| - 15 = 0$. In each case, you should obtain the true statement $0 = 0$. The solutions are $-\frac{1}{2}$ and 1, and the solution set is $\left\{-\frac{1}{2}, 1\right\}$. ∎

✓ **CHECK POINT 2**   Solve: $2|1 - 3x| - 28 = 0$.

The absolute value of a number is never negative. Thus, if u is an algebraic expression and c is a negative number, then $|u| = c$ has no solution. For example, the equation $|3x - 6| = -2$ has no solution because $|3x - 6|$ cannot be negative. The solution set is $\varnothing$, the empty set.

The absolute value of 0 is 0. Thus, if u is an algebraic expression and $|u| = 0$, the solution is found by solving $u = 0$. For example, the solution of $|x - 2| = 0$ is obtained by solving $x - 2 = 0$. The solution is 2 and the solution set is {2}.

Some equations have two absolute value expressions, such as

$$|3x - 1| = |x + 5|.$$

These absolute value expressions are equal when the expressions inside the absolute value bars are equal to or opposites of each other.

> **Rewriting an Absolute Value Equation with Two Absolute Values without Absolute Value Bars**
>
> If $|u| = |v|$, then $u = v$ or $u = -v$.

EXAMPLE 3   Solving an Absolute Value Equation with Two Absolute Values

Solve: $|3x - 1| = |x + 5|$.

Solution   We rewrite the equation without absolute value bars.

$$|u| = |v| \quad \text{means} \quad u = v \quad \text{or} \quad u = -v$$

$$|3x - 1| = |x + 5| \quad \text{means} \quad 3x - 1 = x + 5 \quad \text{or} \quad 3x - 1 = -(x + 5).$$

We now solve the two equations that do not contain absolute value bars.

$$
\begin{array}{ll}
3x - 1 = x + 5 \quad \text{or} & 3x - 1 = -(x + 5) \\
2x - 1 = 5 & 3x - 1 = -x - 5 \\
2x = 6 & 4x - 1 = -5 \\
x = 3 & 4x = -4 \\
& x = -1
\end{array}
$$

Take a moment to complete the solution process by checking the two proposed solutions in the original equation. The solutions are -1 and 3, and the solution set is $\{-1, 3\}$. ∎

✓ **CHECK POINT 3**   Solve: $|2x - 7| = |x + 3|$.

2 Solve absolute value inequalities of the form $|u| < c$.

Inequalities Involving Absolute Value

Absolute value can also arise in inequalities. Consider, for example,

$$|x| < 2.$$

This means that the distance from 0 to x on a number line is less than 2, as shown in **Figure 4.6**.

If the distance from 0 to x is less than 2 ($|x| < 2$),
then x is between -2 and 2 ($-2 < x < 2$).

Figure 4.6

Generalizing from the observations in the voice balloons gives us a method for solving inequalities of the form $|u| < c$, where c is a positive number. This method involves rewriting the given inequality without absolute value bars.

> **Solving Absolute Value Inequalities of the Form $|u| < c$**
>
> If c is a positive real number and u represents any algebraic expression, then
>
> $$|u| < c \text{ is equivalent to } -c < u < c.$$
>
> This rule is valid if $<$ is replaced by $\leq$.

EXAMPLE 4 Solving an Absolute Value Inequality of the Form $|u| < c$

Solve and graph the solution set on a number line:

$$|x - 4| < 3.$$

Solution We rewrite the inequality without absolute value bars.

$|u| < c$ means $-c < u < c$.

$$|x - 4| < 3 \text{ means } -3 < x - 4 < 3.$$

We solve the compound inequality by adding 4 to all three parts.

$$-3 < x - 4 < 3$$
$$-3 + 4 < x - 4 + 4 < 3 + 4$$
$$1 < x < 7$$

The solution set is all real numbers greater than 1 and less than 7, denoted in interval notation by (1, 7). The graph of the solution set is shown as follows:

We can use the rectangular coordinate system to visualize the solution set of

$$|x - 4| < 3.$$

Figure 4.7 shows the graphs of $f(x) = |x - 4|$ and $g(x) = 3$. The solution set of $|x - 4| < 3$ consists of all values of x for which the blue graph of f lies below the red graph of g. These x-values make up the interval (1, 7), which is the solution set.

Figure 4.7 The solution set of $|x - 4| < 3$ is (1, 7).

✓ **CHECK POINT 4** Solve and graph the solution set on a number line: $|x - 2| < 5$.

Before rewriting an absolute value inequality without absolute value bars, isolate the absolute value expression on one side of the inequality.

EXAMPLE 5   Solving an Absolute Value Inequality

Solve and graph the solution set on a number line: $-2|3x + 5| + 7 \geq -13$.

Solution

$$-2|3x + 5| + 7 \geq -13 \qquad \text{This is the given inequality.}$$

We need to isolate $|3x + 5|$, the absolute value expression.

$$-2|3x + 5| + 7 - 7 \geq -13 - 7 \qquad \text{Subtract 7 from both sides.}$$

$$-2|3x + 5| \geq -20 \qquad \text{Simplify.}$$

$$\frac{-2|3x + 5|}{-2} \leq \frac{-20}{-2} \qquad \text{Divide both sides by } -2 \text{ and change the sense of the inequality.}$$

$$|3x + 5| \leq 10 \qquad \text{Simplify.}$$

$$-10 \leq 3x + 5 \leq 10 \qquad \text{Rewrite without absolute value bars: } |u| \leq c \text{ means } -c \leq u \leq c.$$

Now we need to isolate x in the middle.

$$-10 - 5 \leq 3x + 5 - 5 \leq 10 - 5 \qquad \text{Subtract 5 from all three parts.}$$

$$-15 \leq 3x \leq 5 \qquad \text{Simplify.}$$

$$\frac{-15}{3} \leq \frac{3x}{3} \leq \frac{5}{3} \qquad \text{Divide each part by 3.}$$

$$-5 \leq x \leq \frac{5}{3} \qquad \text{Simplify.}$$

The solution set is $\left[-5, \frac{5}{3}\right]$ in interval notation. The graph is shown as follows:

✓ CHECK POINT 5   Solve and graph the solution set on a number line: $-3|5x - 2| + 20 \geq -19$.

3 Solve absolute value inequalities of the form $|u| > c$. ▶

Now let's consider absolute value inequalities with greater than symbols, such as

$$|x| > 2.$$

This means that the distance from 0 to x on a number line is greater than 2, as shown in **Figure 4.8**.

If the distance from 0 to x greater than 2 $(|x| > 2)$, then x is less than -2 or greater than 2 $(x < -2 \text{ or } x > 2)$.

Figure 4.8

Generalizing from the observations in the voice balloons gives us a method for solving inequalities of the form $|u| > c$, where c is a positive number. This method once again involves rewriting the given inequality without absolute value bars.

Solving Absolute Value Inequalities of the Form $|u| > c$

If c is a positive real number and u represents any algebraic expression, then

$$|u| > c \text{ is equivalent to } u < -c \text{ or } u > c.$$

This rule is valid if $>$ is replaced by $\geq$.

EXAMPLE 6   Solving an Absolute Value Inequality of the Form $|u| \geq c$

Solve and graph the solution set on a number line:

$$|2x + 3| \geq 5.$$

Solution   We rewrite the inequality without absolute value bars.

$|u| \geq c$ means $u \leq -c$ or $u \geq c.$

$|2x + 3| \geq 5$ means $2x + 3 \leq -5$ or $2x + 3 \geq 5.$

We solve this compound inequality by solving each of these inequalities separately. Then we take the union of their solution sets.

$2x + 3 \leq -5$ or $2x + 3 \geq 5$		These are the inequalities without absolute value bars.
$2x \leq -8$	$2x \geq 2$	Subtract 3 from both sides.
$x \leq -4$	$x \geq 1$	Divide both sides by 2.

The solution set consists of all numbers that are less than or equal to -4 or greater than or equal to 1. The solution set in interval notation is $(-\infty, -4] \cup [1, \infty)$. The graph of the solution set is shown as follows:

$$\xleftarrow{\hspace{1cm}} \; {}^{-7} \;\; {}^{-6} \;\; {}^{-5} \;\; {}^{-4} \;\; {}^{-3} \;\; {}^{-2} \;\; {}^{-1} \;\; 0 \;\; 1 \;\; 2 \;\; 3 \; \xrightarrow{\hspace{1cm}} x$$

■

Great Question!

The graph of the solution set in Example 6 consists of two intervals. When is the graph of the solution set of an absolute value inequality a single interval and when is it divided into two intervals?

If u is a linear expression and $c > 0$, the graph of the solution set for $|u| > c$ will be divided into two intervals whose union cannot be represented as a single interval. The graph of the solution set for $|u| < c$ will be a single interval. Avoid the common error of rewriting $|u| > c$ as $-c < u > c$.

✓ **CHECK POINT 6**   Solve and graph the solution set on a number line: $|2x - 5| \geq 3.$

Great Question!

Please cut to the chase. What do I need to know when rewriting absolute value equations and inequalities without absolute value bars?

Here's a brief summary. If $c > 0$,

- $|u| = c$ is equivalent to $u = c$ or $u = -c.$
- $|u| < c$ is equivalent to $-c < u < c.$
- $|u| > c$ is equivalent to $u < -c$ or $u > c.$

4 Recognize absolute value inequalities with no solution or all real numbers as solutions.

Absolute Value Inequalities with Unusual Solution Sets

We have been working with $|u| < c$ and $|u| > c$, where c is a positive number. Now let's see what happens to these inequalities if c is a negative number. Consider, for example, $|x| < -2$. Because $|x|$ always has a value that is greater than or equal to 0, there is no number whose absolute value is less than -2. The inequality $|x| < -2$ has no solution. The solution set is $\varnothing$.

Now consider the inequality $|x| > -2$. Because $|x|$ is never negative, all numbers have an absolute value greater than -2. All real numbers satisfy the inequality $|x| > -2$. The solution set is $(-\infty, \infty)$.

Absolute Value Inequalities with Unusual Solution Sets

If u is an algebraic expression and c is a negative number,

1. The inequality $|u| < c$ has no solution.
2. The inequality $|u| > c$ is true for all real numbers for which u is defined.

---

5 Solve problems using absolute value inequalities.

Applications

We opened this section with this question:

What activities do you dread?

In a random sample of 1000 U.S. adults, 46% of those questioned responded, "Public speaking." The problem is that this is a single random sample. Do 46% of adults in the entire U.S. population dread public speaking?

If you look at the results of a poll like the one in **Figure 4.9**, you will observe that a margin of error is reported.

Activities U.S. Adults Say They Dread

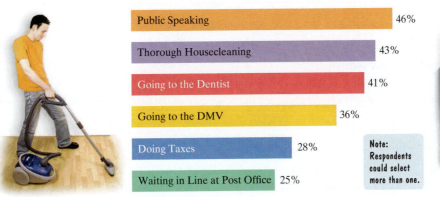

Figure 4.9

Source: TNS survey of 1000 adults, March 2010, Margin of error: ±3.2%

The margin of error in the dreaded-activities poll is ±3.2%. This means that the actual percentage of U.S. adults who dread public speaking is at most 3.2% greater than or less than 46%. If x represents the percentage of U.S. adults in the population who dread public speaking, then the poll's margin of error can be expressed as an absolute value inequality:

$$|x - 46| \le 3.2.$$

EXAMPLE 7 Analyzing a Poll's Margin of Error

The inequality

$$|x - 28| \leq 3.2$$

describes the percentage of U.S. adults in the population who dread doing taxes. Solve the inequality and interpret the solution.

Solution We rewrite the inequality without absolute value bars.

$|u| \leq c$ **means** $-c \leq u \leq c.$

$|x - 28| \leq 3.2$ means $-3.2 \leq x - 28 \leq 3.2.$

We solve the compound inequality by adding 28 to all three parts.

$$-3.2 \leq x - 28 \leq 3.2$$ This is $|x - 28| \leq 3.2$ without absolute value bars.

$$-3.2 + 28 \leq x - 28 + 28 \leq 3.2 + 28$$ Add 28 to all three parts.

$$24.8 \leq x \leq 31.2$$ Simplify. The solution is [24.8, 31.2].

The percentage of U.S. adults in the population who dread doing taxes is somewhere between a low of 24.8% and a high of 31.2%. Notice that these percents are 3.2% above and below the given 28% in **Figure 4.9** on the previous page, and that 3.2% is the poll's margin of error. ■

✓ **CHECK POINT 7**   Solve the inequality:

$$|x - 41| \leq 3.2.$$

Interpret the solution in terms of the information in **Figure 4.9** on the previous page.

Achieving Success

Use index cards to help learn new terms.

Many of the terms, notations, and formulas used in this book will be new to you. Buy a pack of 3 × 5 index cards. On each card, list a new vocabulary word, symbol, or title of a formula. On the other side of the card, put the definition or formula. Here are four examples:

Effective Index Cards

Rewriting |u| = c without absolute value, c > 0
Front

|u| = c is equivalent to u = c or u = -c
Back

Rewriting |u| = |v| without absolute value
Front

|u| = |v| is equivalent to u = v or u = -v
Back

Rewriting |u| < c without absolute value, c > 0
Front

|u| < c is equivalent to -c < u < c
Back

Rewriting |u| > c without absolute value, c > 0
Front

|u| > c is equivalent to u < -c or u > c
Back

Review these cards frequently. Use the cards to quiz yourself and prepare for exams.

CONCEPT AND VOCABULARY CHECK

Fill in each blank so that the resulting statement is true.

1. If $c > 0$, $|u| = c$ is equivalent to $u =$ _____ or $u =$ _____.

2. $|u| = |v|$ is equivalent to $u =$ _____ or $u =$ _____.

3. If $c > 0$, $|u| < c$ is equivalent to _____ $< u <$ _____.

4. If $c > 0$, $|u| > c$ is equivalent to $u <$ _____ or $u >$ _____.

5. $|u| < c$ has no solution if c _____ 0.

6. $|u| > c$ is true for all real numbers if c _____ 0.

Match each absolute value equation or inequality in the left column with an equivalent statement from the right column.

7. $|3x - 1| = 5$

8. $|3x - 1| \geq 5$

9. $|3x - 1| \leq 5$

10. $|3x - 1| = |x|$

11. $|3x - 1| < -5$

12. $|3x - 1| > -5$

A. $-5 \leq 3x - 1 \leq 5$

B. $3x - 1 = x$ or $3x - 1 = -x$

C. $3x - 1 = 5$ or $3x - 1 = -5$

D. $\varnothing$

E. $3x - 1 \leq -5$ or $3x - 1 \geq 5$

F. $(-\infty, \infty)$

4.3 EXERCISE SET   ▶ MyMathLab®

Practice Exercises

In Exercises 1–38, find the solution set for each equation.

1. $|x| = 8$
2. $|x| = 6$
3. $|x - 2| = 7$
4. $|x + 1| = 5$
5. $|2x - 1| = 7$
6. $|2x - 3| = 11$
7. $\left|\dfrac{4x - 2}{3}\right| = 2$
8. $\left|\dfrac{3x - 1}{5}\right| = 1$
9. $|x| = -8$
10. $|x| = -6$
11. $|x + 3| = 0$
12. $|x + 2| = 0$
13. $2|y + 6| = 10$
14. $3|y + 5| = 12$
15. $3|2x - 1| = 21$
16. $2|3x - 2| = 14$
17. $|6y - 2| + 4 = 32$
18. $|3y - 1| + 10 = 25$
19. $7|5x| + 2 = 16$
20. $7|3x| + 2 = 16$
21. $|x + 1| + 5 = 3$
22. $|x + 1| + 6 = 2$
23. $|4y + 1| + 10 = 4$
24. $|3y - 2| + 8 = 1$
25. $|2x - 1| + 3 = 3$
26. $|3x - 2| + 4 = 4$
27. $|5x - 8| = |3x + 2|$
28. $|4x - 9| = |2x + 1|$
29. $|2x - 4| = |x - 1|$
30. $|6x| = |3x - 9|$

31. $|2x - 5| = |2x + 5|$
32. $|3x - 5| = |3x + 5|$
33. $|x - 3| = |5 - x|$
34. $|x - 3| = |6 - x|$
35. $|2y - 6| = |10 - 2y|$
36. $|4y + 3| = |4y + 5|$
37. $\left|\dfrac{2x}{3} - 2\right| = \left|\dfrac{x}{3} + 3\right|$
38. $\left|\dfrac{x}{2} - 2\right| = \left|x - \dfrac{1}{2}\right|$

In Exercises 39–74, solve and graph the solution set on a number line.

39. $|x| < 3$
40. $|x| < 5$
41. $|x - 2| < 1$
42. $|x - 1| < 5$
43. $|x + 2| \leq 1$
44. $|x + 1| \leq 5$
45. $|2x - 6| < 8$
46. $|3x + 5| < 17$
47. $|x| > 3$
48. $|x| > 5$
49. $|x + 3| > 1$
50. $|x - 2| > 5$
51. $|x - 4| \geq 2$
52. $|x - 3| \geq 4$
53. $|3x - 8| > 7$
54. $|5x - 2| > 13$
55. $|2(x - 1) + 4| \leq 8$
56. $|3(x - 1) + 2| \leq 20$

57. $\left| \dfrac{2x + 6}{3} \right| < 2$

58. $\left| \dfrac{3x - 3}{4} \right| < 6$

59. $\left| \dfrac{2x + 2}{4} \right| \geq 2$

60. $\left| \dfrac{3x - 3}{9} \right| \geq 1$

61. $\left| 3 - \dfrac{2x}{3} \right| > 5$

62. $\left| 3 - \dfrac{3x}{4} \right| > 9$

63. $|x - 2| < -1$

64. $|x - 3| < -2$

65. $|x + 6| > -10$

66. $|x + 4| > -12$

67. $|x + 2| + 9 \leq 16$

68. $|x - 2| + 4 \leq 5$

69. $2|2x - 3| + 10 > 12$

70. $3|2x - 1| + 2 > 8$

71. $-4|1 - x| < -16$

72. $-2|5 - x| < -6$

73. $3 \leq |2x - 1|$

74. $9 \leq |4x + 7|$

75. Let $f(x) = |5 - 4x|$. Find all values of x for which $f(x) = 11$.

76. Let $f(x) = |2 - 3x|$. Find all values of x for which $f(x) = 13$.

77. Let $f(x) = |3 - x|$ and $g(x) = |3x + 11|$. Find all values of x for which $f(x) = g(x)$.

78. Let $f(x) = |3x + 1|$ and $g(x) = |6x - 2|$. Find all values of x for which $f(x) = g(x)$.

79. Let $g(x) = |-1 + 3(x + 1)|$. Find all values of x for which $g(x) \leq 5$.

80. Let $g(x) = |-3 + 4(x + 1)|$. Find all values of x for which $g(x) \leq 3$.

81. Let $h(x) = |2x - 3| + 1$. Find all values of x for which $h(x) > 6$.

82. Let $h(x) = |2x - 4| - 6$. Find all values of x for which $h(x) > 18$.

Practice PLUS

83. When 3 times a number is subtracted from 4, the absolute value of the difference is at least 5. Use interval notation to express the set of all real numbers that satisfy this condition.

84. When 4 times a number is subtracted from 5, the absolute value of the difference is at most 13. Use interval notation to express the set of all real numbers that satisfy this condition.

In Exercises 85–86, solve each inequality. Assume that $a > 0$ and $c > 0$.

85. $|ax + b| < c$

86. $|ax + b| \geq c$

In Exercises 87–88, use the graph of $f(x) = |4 - x|$ to solve each equation or inequality.

87. $|4 - x| = 1$ 88. $|4 - x| < 5$

In Exercises 89–90, use the table to solve each inequality.

89. $|2x + 1| \leq 3$

90. $|2x + 1| \geq 3$

Application Exercises

How to Blow Your Job Interview. *The data in the bar graph are from a random survey of 1910 job interviewers. The graph shows the top interviewer turnoffs and the percentage of surveyed interviewers who were offended by each of these behaviors. In Exercises 91–92, let x represent the actual percentage of interviewers in the entire population of job interviewers.*

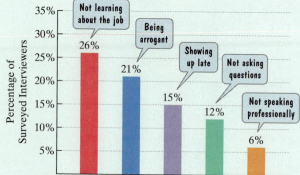

Top Interviewer Turnoffs

Source: Scott Erker, PhD., and Kelli Buczynski, "Are You Failing the Interview? 2009 Survey of Global Interviewing Practices and Perceptions." Development Dimensions International.

91. Solve the inequality: $|x - 21| \leq 3$. Interpret the solution in terms of the information in the graph. What is the margin of error?

92. Solve the inequality: $|x - 15| \leq 3$. Interpret the solution in terms of the information in the graph at the bottom of the previous page. What is the margin of error?

93. The inequality $|T - 57| \leq 7$ describes the range of monthly average temperature, T, in degrees Fahrenheit, for San Francisco, California. Solve the inequality and interpret the solution.

94. The inequality $|T - 50| \leq 22$ describes the range of monthly average temperature, T, in degrees Fahrenheit, for Albany, New York. Solve the inequality and interpret the solution.

The specifications for machine parts are given with tolerance limits that describe a range of measurements for which the part is acceptable. In Exercises 95–96, x represents the length of a machine part, in centimeters. The tolerance limit is 0.01 centimeter.

95. Solve: $|x - 8.6| \leq 0.01$. If the length of the machine part is supposed to be 8.6 centimeters, interpret the solution.

96. Solve: $|x - 9.4| \leq 0.01$. If the length of the machine part is supposed to be 9.4 centimeters, interpret the solution.

97. If a coin is tossed 100 times, we would expect approximately 50 of the outcomes to be heads. It can be demonstrated that a coin is unfair if h, the number of outcomes that result in heads, satisfies $\left|\dfrac{h - 50}{5}\right| \geq 1.645$. Describe the number of outcomes that result in heads that determine an unfair coin that is tossed 100 times.

Explaining the Concepts

98. Explain how to solve an equation containing one absolute value expression.

99. Explain why the procedure that you described in Exercise 98 does not apply to the equation $|x - 5| = -3$. What is the solution set of this equation?

100. Describe how to solve an absolute value equation with two absolute values.

101. Describe how to solve an absolute value inequality of the form $|u| < c$ for $c > 0$.

102. Explain why the procedure that you described in Exercise 101 does not apply to the inequality $|x - 5| < -3$. What is the solution set of this inequality?

103. Describe how to solve an absolute value inequality of the form $|u| > c$ for $c > 0$.

104. Explain why the procedure that you described in Exercise 103 does not apply to the inequality $|x - 5| > -3$. What is the solution set of this inequality?

Technology Exercises

In Exercises 105–107, solve each equation using a graphing utility. Graph each side separately in the same viewing rectangle. The solutions are the x-coordinates of the intersection points.

105. $|x + 1| = 5$

106. $|3(x + 4)| = 12$

107. $|2x - 3| = |9 - 4x|$

In Exercises 108–110, solve each inequality using a graphing utility. Graph each side separately in the same viewing rectangle. The solution set consists of all values of x for which the graph of the left side lies below the graph of the right side.

108. $|2x + 3| < 5$

109. $\left|\dfrac{2x - 1}{3}\right| < \dfrac{5}{3}$

110. $|x + 4| < -1$

In Exercises 111–113, solve each inequality using a graphing utility. Graph each side separately in the same viewing rectangle. The solution set consists of all values of x for which the graph of the left side lies above the graph of the right side.

111. $|2x - 1| > 7$

112. $|0.1x - 0.4| + 0.4 > 0.6$

113. $|x + 4| > -1$

114. Use a graphing utility to verify the solution sets for any five equations or inequalities that you solved by hand in Exercises 1–74.

Critical Thinking Exercises

Make Sense? *In Exercises 115–118, determine whether each statement makes sense or does not make sense, and explain your reasoning.*

115. I solved $|x - 2| = 5$ by rewriting the equation as $x - 2 = 5$ or $x + 2 = 5$.

116. I solved $|x - 2| > 5$ by rewriting the inequality as $-5 < x - 2 > 5$.

117. Because the absolute value of any expression is never less than a negative number, I can immediately conclude that the inequality $|2x - 5| - 9 < -4$ has no solution.

118. I'll win the contest if I can complete the crossword puzzle in 20 minutes plus or minus 5 minutes, so my winning time, x, is modeled by $|x - 20| \leq 5$.

In Exercises 119–122, determine whether each statement is true or false. If the statement is false, make the necessary change(s) to produce a true statement.

119. All absolute value equations have two solutions.

120. The equation $|x| = -6$ is equivalent to $x = 6$ or $x = -6$.

121. Values of -5 and 5 satisfy $|x| = 5$, $|x| \leq 5$, and $|x| \geq -5$.

122. The absolute value of any linear expression is greater than 0 for all real numbers except the number for which the expression is equal to 0.

123. Write an absolute value inequality for which the interval shown is the solution.

124. The percentage, p, of defective products manufactured by a company is given by $|p - 0.3\%| \le 0.2\%$. If 100,000 products are manufactured and the company offers a \$5 refund for each defective product, describe the company's cost for refunds.

125. Solve: $|2x + 5| = 3x + 4$.

Review and Preview Exercises

Exercises 126–128 will enable you to review graphing linear functions. In addition, they will help you prepare for the material covered in the next section. In each exercise, graph the linear function.

126. $3x - 5y = 15$ (Section 2.4, Example 1)

127. $f(x) = -\dfrac{2}{3}x$ (Section 2.4, Example 5)

128. $f(x) = -2$ (Section 2.4, Example 6)

MID-CHAPTER CHECK POINT   Section 4.1–Section 4.3

What You Know: We learned to solve linear inequalities, expressing solution sets in interval notation. We know that it is necessary to change the sense of an inequality when multiplying or dividing both sides by a negative number. We solved compound inequalities with *and* by finding the intersection of solution sets and with *or* by finding the union of solution sets. Finally, we solved equations and inequalities involving absolute value by carefully rewriting the given equation or inequality without absolute value bars. For positive values of c, we wrote $|u| = c$ as $u = c$ or $u = -c$. We wrote $|u| < c$ as $-c < u < c$, and we wrote $|u| > c$ as $u < -c$ or $u > c$.

In Exercises 1–18, solve each inequality or equation.

1. $4 - 3x \ge 12 - x$

2. $5 \le 2x - 1 < 9$

3. $|4x - 7| = 5$

4. $-10 - 3(2x + 1) > 8x + 1$

5. $2x + 7 < -11$ or $-3x - 2 < 13$

6. $|3x - 2| \le 4$

7. $|x + 5| = |5x - 8|$

8. $5 - 2x \ge 9$ and $5x + 3 > -17$

9. $3x - 2 > -8$ or $2x + 1 < 9$

10. $\dfrac{x}{2} + 3 \le \dfrac{x}{3} + \dfrac{5}{2}$

11. $\dfrac{2}{3}(6x - 9) + 4 > 5x + 1$

12. $|5x + 3| > 2$

13. $7 - \left|\dfrac{x}{2} + 2\right| \le 4$

14. $5(x - 2) - 3(x + 4) \ge 2x - 20$

15. $\dfrac{x + 3}{4} < \dfrac{1}{3}$

16. $5x + 1 \ge 4x - 2$ and $2x - 3 > 5$

17. $3 - |2x - 5| = -6$

18. $3 + |2x - 5| = -6$

In Exercises 19–22, solve each problem.

19. A car rental agency rents a certain car for \$40 per day with unlimited mileage or \$24 per day plus \$0.20 per mile. How far can a customer drive this car per day for the \$24 option to cost no more than the unlimited mileage option?

20. To receive a B in a course, you must have an average of at least 80% but less than 90% on five exams. Your grades on the first four exams were 95%, 79%, 91%, and 86%. What range of grades on the fifth exam will result in a B for the course?

21. A retiree requires an annual income of at least \$9000 from an investment paying 7.5% annual interest. How much should the retiree invest to achieve the desired return?

22. A company that manufactures compact discs has fixed monthly overhead costs of \$60,000. Each disc costs \$0.18 to produce and sells for \$0.30. How many discs should be produced and sold each month for the company to have a profit of at least \$30,000?

4.4

Linear Inequalities in Two Variables

What am I supposed to learn?

After studying this section, you should be able to:

1 Graph a linear inequality in two variables.

2 Use mathematical models involving linear inequalities.

3 Graph a system of linear inequalities.

This book was written in Point Reyes National Seashore, 40 miles north of San Francisco. The park consists of 75,000 acres with miles of pristine surf-washed beaches, forested ridges, and bays bordered by white cliffs.

Like your author, many people are kept inspired and energized surrounded by nature's unspoiled beauty. In this section, you will see how systems of inequalities model whether a region's natural beauty manifests itself in forests, grasslands, or deserts.

Linear Inequalities in Two Variables and Their Solutions

We have seen that equations in the form $Ax + By = C$ are straight lines when graphed. If we change the symbol $=$ to $>$, $<$, $\geq$, or $\leq$, we obtain a **linear inequality in two variables**. Some examples of linear inequalities in two variables are $x + y > 2$, $3x - 5y \leq 15$, and $2x - y < 4$.

A **solution of an inequality in two variables**, x and y, is an ordered pair of real numbers with the following property: When the x-coordinate is substituted for x and the y-coordinate is substituted for y in the inequality, we obtain a true statement. For example, $(3, 2)$ is a solution of the inequality $x + y > 1$. When 3 is substituted for x and 2 is substituted for y, we obtain the true statement $3 + 2 > 1$, or $5 > 1$. Because there are infinitely many pairs of numbers that have a sum greater than 1, the inequality $x + y > 1$ has infinitely many solutions. Each ordered-pair solution is said to **satisfy** the inequality. Thus, $(3, 2)$ satisfies the inequality $x + y > 1$.

1 Graph a linear inequality in two variables.

The Graph of a Linear Inequality in Two Variables

We know that the graph of an equation in two variables is the set of all points whose coordinates satisfy the equation. Similarly, the **graph of an inequality in two variables** is the set of all points whose coordinates satisfy the inequality.

Let's use **Figure 4.10** to get an idea of what the graph of a linear inequality in two variables looks like. Part of the figure shows the graph of the linear equation $x + y = 2$. The line divides the points in the rectangular coordinate system into three sets. First, there is the set of points along the line, satisfying $x + y = 2$. Next, there is the set of points in the green region above the line. Points in the green region satisfy the linear inequality $x + y > 2$. Finally, there is the set of points in the purple region below the line. Points in the purple region satisfy the linear inequality $x + y < 2$.

A **half-plane** is the set of all the points on one side of a line. In **Figure 4.10**, the green region is a half-plane. The purple region is also a half-plane. A half-plane is the graph of a linear inequality that involves $>$ or $<$. The graph of a linear inequality that involves $\geq$ or $\leq$ is a half-plane and a line. A solid line is used to show that a line is part of a graph. A dashed line is used to show that a line is not part of a graph.

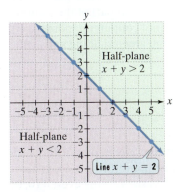

Figure 4.10

Graphing a Linear Inequality in Two Variables

1. Replace the inequality symbol with an equal sign and graph the corresponding linear equation. Draw a solid line if the original inequality contains a ≤ or ≥ symbol. Draw a dashed line if the original inequality contains a < or > symbol.

2. Choose a test point from one of the half-planes. (Do not choose a point on the line.) Substitute the coordinates of the test point into the inequality.

3. If a true statement results, shade the half-plane containing this test point. If a false statement results, shade the half-plane not containing this test point.

EXAMPLE 1 Graphing a Linear Inequality in Two Variables

Graph: $2x - 3y \geq 6$.

Solution

Step 1. Replace the inequality symbol by = and graph the linear equation. We need to graph $2x - 3y = 6$. We can use intercepts to graph this line.

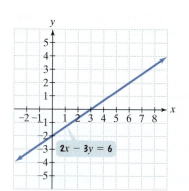

Figure 4.11 Preparing to graph $2x - 3y \geq 6$

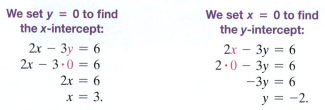

We set $y = 0$ to find the x-intercept:	We set $x = 0$ to find the y-intercept:
$2x - 3y = 6$	$2x - 3y = 6$
$2x - 3 \cdot 0 = 6$	$2 \cdot 0 - 3y = 6$
$2x = 6$	$-3y = 6$
$x = 3.$	$y = -2.$

The x-intercept is 3, so the line passes through $(3, 0)$. The y-intercept is -2, so the line passes through $(0, -2)$. Using the intercepts, the line is shown in **Figure 4.11** as a solid line. This is because the inequality $2x - 3y \geq 6$ contains a ≥ symbol, in which equality is included.

Step 2. Choose a test point from one of the half-planes and not from the line. Substitute its coordinates into the inequality. Figure 4.11 shows that the line $2x - 3y = 6$ divides the plane into three parts—the line itself and two half-planes. The points in one half-plane satisfy $2x - 3y > 6$. The points in the other half-plane satisfy $2x - 3y < 6$. We need to find which half-plane belongs to the solution of $2x - 3y \geq 6$. To do so, we test a point from either half-plane. The origin, $(0, 0)$, is the easiest point to test.

$2x - 3y \geq 6$	This is the given inequality.
$2 \cdot 0 - 3 \cdot 0 \overset{?}{\geq} 6$	Test (0, 0) by substituting 0 for x and 0 for y.
$0 - 0 \overset{?}{\geq} 6$	Multiply.
$0 \geq 6$	This statement is false.

Step 3. If a false statement results, shade the half-plane not containing the test point. Because 0 is not greater than or equal to 6, the test point, $(0, 0)$, is not part of the solution set. Thus, the half-plane below the solid line $2x - 3y = 6$ is part of the solution set. The solution set is the line and the half-plane that does not contain the point $(0, 0)$, indicated by shading this half-plane. The graph is shown using green shading and a blue line in **Figure 4.12**. ∎

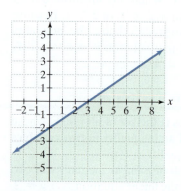

Figure 4.12 The graph of $2x - 3y \geq 6$

✓ CHECK POINT 1 Graph: $4x - 2y \geq 8$.

When graphing a linear inequality, choose a test point that lies in one of the half-planes and *not on the line dividing the half-planes*. The test point $(0, 0)$ is convenient because it is easy to calculate when 0 is substituted for each variable. However, if $(0, 0)$ lies on the dividing line and not in a half-plane, a different test point must be selected.

EXAMPLE 2 Graphing a Linear Inequality in Two Variables

Graph: $y > -\dfrac{2}{3}x$.

Solution

Step 1. Replace the inequality symbol by = and graph the linear equation. Because we are interested in graphing $y > -\frac{2}{3}x$, we begin by graphing $y = -\frac{2}{3}x$. We can use the slope and the y-intercept to graph this linear function.

$$y = -\frac{2}{3}x + 0$$

Slope $= \dfrac{-2}{3} = \dfrac{\text{rise}}{\text{run}}$ y-intercept $= 0$

The y-intercept is 0, so the line passes through $(0, 0)$. Using the y-intercept and the slope, the line is shown in **Figure 4.13** as a dashed line. This is because the inequality $y > -\frac{2}{3}x$ contains a $>$ symbol, in which equality is not included.

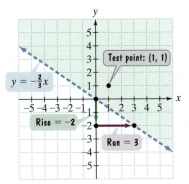

Test point: (1, 1)

$y = -\frac{2}{3}x$

Rise = -2

Run = 3

Figure 4.13 The graph of $y > -\frac{2}{3}x$

Step 2. Choose a test point from one of the half-planes and not from the line. Substitute its coordinates into the inequality. We cannot use $(0, 0)$ as a test point because it lies on the line and not in a half-plane. Let's use $(1, 1)$, which lies in the half-plane above the line.

$$y > -\frac{2}{3}x \qquad \textcolor{blue}{\text{This is the given inequality.}}$$

$$1 \overset{?}{>} -\frac{2}{3}\cdot 1 \qquad \textcolor{blue}{\text{Test (1, 1) by substituting 1 for } x \text{ and 1 for } y.}$$

$$1 > -\frac{2}{3} \qquad \textcolor{blue}{\text{This statement is true.}}$$

Step 3. If a true statement results, shade the half-plane containing the test point. Because 1 is greater than $-\frac{2}{3}$, the test point $(1, 1)$ is part of the solution set. All the points on the same side of the line $y = -\frac{2}{3}x$ as the point $(1, 1)$ are members of the solution set. The solution set is the half-plane that contains the point $(1, 1)$, indicated by shading this half-plane. The graph is shown using green shading and a dashed blue line in **Figure 4.13**. ■

Using Technology

Most graphing utilities can graph inequalities in two variables with the $\boxed{\text{SHADE}}$ feature. The procedure varies by model, so consult your manual. For most graphing utilities, you must first solve for y if it is not already isolated. The figure shows the graph of $y > -\frac{2}{3}x$. Most displays do not distinguish between dashed and solid boundary lines.

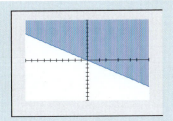

☑ **CHECK POINT 2** Graph: $y > -\dfrac{3}{4}x$.

Graphing Linear Inequalities without Using Test Points

You can graph inequalities in the form $y > mx + b$ or $y < mx + b$ without using test points. The inequality symbol indicates which half-plane to shade.

- If $y > mx + b$, shade the half-plane above the line $y = mx + b$.
- If $y < mx + b$, shade the half-plane below the line $y = mx + b$.

Observe how this is illustrated in **Figure 4.13**. The graph of $y > -\frac{2}{3}x$ is the half-plane above the line $y = -\frac{2}{3}x$.

Great Question!

When is it important to use test points to graph linear inequalities?

Continue using test points to graph inequalities in the form $Ax + By > C$ or $Ax + By < C$. The graph of $Ax + By > C$ can lie above or below the line given by $Ax + By = C$, depending on the value of B. The same comment applies to the graph of $Ax + By < C$.

It is also not necessary to use test points when graphing inequalities involving half-planes on one side of a vertical or a horizontal line.

For the Vertical Line $x = a$:	For the Horizontal Line $y = b$:
• If $x > a$, shade the half-plane to the right of $x = a$.	• If $y > b$, shade the half-plane above $y = b$.
• If $x < a$, shade the half-plane to the left of $x = a$.	• If $y < b$, shade the half-plane below $y = b$.

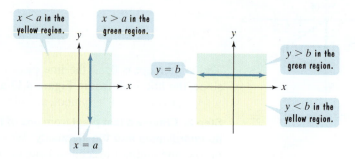

EXAMPLE 3 Graphing Inequalities without Using Test Points

Graph each inequality in a rectangular coordinate system:

a. $y \leq -3$ **b.** $x > 2$.

Solution

a. $y \leq -3$

Graph $y = -3$, a horizontal line with y-intercept −3. The line is solid because equality is included in $y \leq -3$. Because of the less than part of $\leq$, shade the half-plane below the horizontal line.

b. $x > 2$

Graph $x = 2$, a vertical line with x-intercept 2. The line is dashed because equality is not included in $x > 2$. Because of $>$, the greater than symbol, shade the half-plane to the right of the vertical line.

✓ **CHECK POINT 3** Graph each inequality in a rectangular coordinate system:

a. $y > 1$ **b.** $x \leq -2$.

---

2 Use mathematical models involving linear inequalities. ▶

Modeling with Systems of Linear Inequalities

Just as two or more linear equations make up a system of linear equations, two or more linear inequalities make up a **system of linear inequalities**. A **solution of a system of linear inequalities** in two variables is an ordered pair that satisfies each inequality in the system.

EXAMPLE 4   Forests, Grasslands, Deserts, and Systems of Inequalities

Temperature and precipitation affect whether or not trees and forests can grow. At certain levels of precipitation and temperature, only grasslands and deserts will exist. **Figure 4.14** shows three kinds of regions—deserts, grasslands, and forests—that result from various ranges of temperature, T, and precipitation, P.

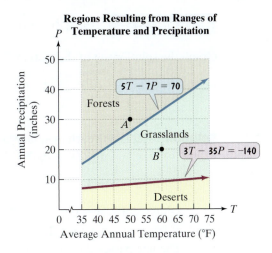

Figure 4.14

Source: Albert Miller et al., *Elements of Meteorology.* © 1983 Merrill Publishing Company

Systems of inequalities can be used to model where forests, grasslands, and deserts occur. Because these regions occur when the average annual temperature, T, is 35°F or greater, each system contains the inequality $T \geq 35$.

Forests occur if	Grasslands occur if	Deserts occur if
$\begin{cases} T \geq 35 \\ 5T - 7P < 70. \end{cases}$	$\begin{cases} T \geq 35 \\ 5T - 7P \geq 70 \\ 3T - 35P \leq -140. \end{cases}$	$\begin{cases} T \geq 35 \\ 3T - 35P > -140. \end{cases}$

Show that point A in **Figure 4.14** is a solution of the system of inequalities that models where forests occur.

Solution   Point A has coordinates $(50, 30)$. This means that if a region has an average annual temperature of 50°F and an average annual precipitation of 30 inches, a forest occurs. We can show that $(50, 30)$ satisfies the system of inequalities for forests by substituting 50 for T and 30 for P in each inequality in the system.

$$T \geq 35 \qquad\qquad 5T - 7P < 70$$
$$50 \geq 35, \quad \text{true} \qquad 5 \cdot 50 - 7 \cdot 30 \overset{?}{<} 70$$
$$250 - 210 \overset{?}{<} 70$$
$$40 < 70, \quad \text{true}$$

The coordinates $(50, 30)$ make each inequality true. Thus, $(50, 30)$ satisfies the system for forests. ∎

✓ **CHECK POINT 4**   Show that point B in **Figure 4.14** is a solution of the system of inequalities that models where grasslands occur.

3 Graph a system of linear inequalities. ▶

Graphing Systems of Linear Inequalities

The **solution set of a system of linear inequalities in two variables** is the set of all ordered pairs that satisfy each inequality in the system. Thus, to graph a system of inequalities in two variables, begin by graphing each individual inequality in the same rectangular coordinate system. Then find the region, if there is one, that is common to every graph in the system. This region of intersection gives a picture of the system's solution set.

EXAMPLE 5 Graphing a System of Linear Inequalities

Graph the solution set of the system:

$$\begin{cases} x - y < 1 \\ 2x + 3y \geq 12. \end{cases}$$

Solution Replacing each inequality symbol with an equal sign indicates that we need to graph $x - y = 1$ and $2x + 3y = 12$. We can use intercepts to graph these lines.

$x - y = 1$		$2x + 3y = 12$

x-intercept: $x - 0 = 1$ *Set $y = 0$ in each equation.* x-intercept: $2x + 3 \cdot 0 = 12$
$x = 1$ $2x = 12$
The line passes through $(1, 0)$. $x = 6$
                                                                  The line passes through $(6, 0)$.

y-intercept: $0 - y = 1$ *Set $x = 0$ in each equation.* y-intercept: $2 \cdot 0 + 3y = 12$
$-y = 1$ $3y = 12$
$y = -1$ $y = 4$
The line passes through $(0, -1)$.   The line passes through $(0, 4)$.

Now we are ready to graph the solution set of the system of linear inequalities.

Graph $x - y < 1$. The blue line, $x - y = 1$, is dashed: Equality is not included in $x - y < 1$. Because $(0, 0)$ makes the inequality true $(0 - 0 < 1$, or $0 < 1$, is true), shade the half-plane containing $(0, 0)$ in yellow.

$x - y = 1$: passes through $(1, 0)$ and $(0, -1)$

The graph of $x - y < 1$

Add the graph of $2x + 3y \geq 12$. The red line, $2x + 3y = 12$, is solid: Equality is included in $2x + 3y \geq 12$. Because $(0, 0)$ makes the inequality false $(2 \cdot 0 + 3 \cdot 0 \geq 12$, or $0 \geq 12$, is false), shade the half-plane not containing $(0, 0)$ using green vertical shading.

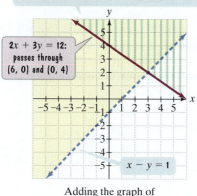

$2x + 3y = 12$: passes through $(6, 0)$ and $(0, 4)$

$x - y = 1$

Adding the graph of $2x + 3y \geq 12$

The solution set of the system is graphed as the intersection (the overlap) of the two half-planes. This is the region in which the yellow shading and the green vertical shading overlap.

This open dot shows $(3, 2)$ is not in the solution set. It does not satisfy $x - y < 1$.

The graph of $x - y < 1$ and $2x + 3y \geq 12$

■

✓ **CHECK POINT 5** Graph the solution set of the system:

$$\begin{cases} x - 3y < 6 \\ 2x + 3y \geq -6. \end{cases}$$

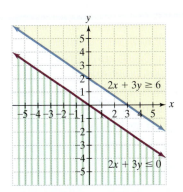

Figure 4.15 A system of inequalities with no solution

A system of inequalities has no solution if there are no points in the rectangular coordinate system that simultaneously satisfy each inequality in the system. For example, the system

$$\begin{cases} 2x + 3y \geq 6 \\ 2x + 3y \leq 0, \end{cases}$$

whose separate graphs are shown in **Figure 4.15**, has no overlapping region. Thus, the system has no solution. The solution set is $\varnothing$, the empty set.

EXAMPLE 6 Graphing a System of Inequalities

Graph the solution set of the system:

$$\begin{cases} x - y < 2 \\ -2 \leq x < 4 \\ y < 3. \end{cases}$$

Solution We begin by graphing $x - y < 2$, the first given inequality. The line $x - y = 2$ has an x-intercept of 2 and a y-intercept of -2. The test point $(0, 0)$ makes the inequality $x - y < 2$ true. The graph of $x - y < 2$ is shown in **Figure 4.16**.

Now, let's consider the second given inequality, $-2 \leq x < 4$. Replacing the inequality symbols by $=$, we obtain $x = -2$ and $x = 4$, graphed as red vertical lines in **Figure 4.17**. The line $x = 4$ is not included. Because x is between -2 and 4, we shade the region between the vertical lines. We must intersect this region with the yellow region in **Figure 4.16**. The resulting region is shown in yellow and green vertical shading in **Figure 4.17**.

Finally, let's consider the third given inequality, $y < 3$. Replacing the inequality symbol by $=$, we obtain $y = 3$, which graphs as a horizontal line. Because of the less than symbol in $y < 3$, the graph consists of the half-plane below the line $y = 3$. We must intersect this half-plane with the region in **Figure 4.17**. The resulting region is shown in yellow and green vertical shading in **Figure 4.18**. This region represents the graph of the solution set of the given system.

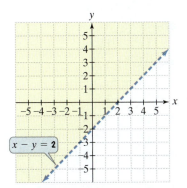

Figure 4.16 The graph of $x - y < 2$

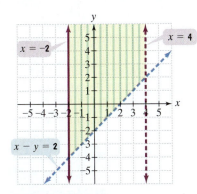

Figure 4.17 The graph of $x - y < 2$ and $-2 \leq x < 4$

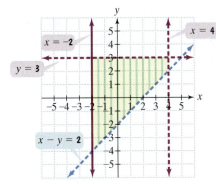

Figure 4.18 The graph of $x - y < 2$ and $-2 \leq x < 4$ and $y < 3$ ∎

☑ **CHECK POINT 6** Graph the solution set of the system:

$$\begin{cases} x + y < 2 \\ -2 \leq x < 1 \\ y > -3. \end{cases}$$

CONCEPT AND VOCABULARY CHECK

Fill in each blank so that the resulting statement is true.

1. The ordered pair (3, 2) is a/an _____ of the inequality $x + y > 1$ because when 3 is substituted for _____ and 2 is substituted for _____, the true statement _____ is obtained.

2. The set of all points that satisfy a linear inequality in two variables is called the _____ of the inequality.

3. The set of all points on one side of a line is called a/an _____.

4. True or false: The graph of $2x - 3y > 6$ includes the line $2x - 3y = 6$. _____

5. True or false: The graph of the linear equation $2x - 3y = 6$ is used to graph the linear inequality $2x - 3y > 6$. _____

6. True or false: When graphing $4x - 2y \geq 8$, to determine which side of the line to shade, choose a test point on $4x - 2y = 8$. _____

7. The solution set of the system

$$\begin{cases} x - y < 1 \\ 2x + 3y \geq 12 \end{cases}$$

   is the set of ordered pairs that satisfy _____ and _____.

8. True or false: The graph of the solution set of the system

$$\begin{cases} x - 3y < 6 \\ 2x + 3y \geq -6 \end{cases}$$

   includes the intersection point of $x - 3y = 6$ and $2x + 3y = -6$. _____

4.4 EXERCISE SET ▶ MyMathLab®

Practice Exercises

In Exercises 1–22, graph each inequality.

1. $x + y \geq 3$
2. $x + y \geq 2$
3. $x - y < 5$
4. $x - y < 6$
5. $x + 2y > 4$
6. $2x + y > 6$
7. $3x - y \leq 6$
8. $x - 3y \leq 6$
9. $\dfrac{x}{2} + \dfrac{y}{3} < 1$
10. $\dfrac{x}{4} + \dfrac{y}{2} < 1$
11. $y > \dfrac{1}{3}x$
12. $y > \dfrac{1}{4}x$
13. $y \leq 3x + 2$
14. $y \leq 2x - 1$
15. $y < -\dfrac{1}{4}x$
16. $y < -\dfrac{1}{3}x$
17. $x \leq 2$
18. $x \leq -4$
19. $y > -4$
20. $y > -2$
21. $y \geq 0$
22. $x \leq 0$

In Exercises 23–46, graph the solution set of each system of inequalities or indicate that the system has no solution.

23. $\begin{cases} 3x + 6y \leq 6 \\ 2x + y \leq 8 \end{cases}$
24. $\begin{cases} x - y \geq 4 \\ x + y \leq 6 \end{cases}$
25. $\begin{cases} 2x - 5y \leq 10 \\ 3x - 2y > 6 \end{cases}$
26. $\begin{cases} 2x - y \leq 4 \\ 3x + 2y > -6 \end{cases}$

27. $\begin{cases} y > 2x - 3 \\ y < -x + 6 \end{cases}$
28. $\begin{cases} y < -2x + 4 \\ y < x - 4 \end{cases}$
29. $\begin{cases} x + 2y \leq 4 \\ y \geq x - 3 \end{cases}$
30. $\begin{cases} x + y \leq 4 \\ y \geq 2x - 4 \end{cases}$
31. $\begin{cases} x \leq 2 \\ y \geq -1 \end{cases}$
32. $\begin{cases} x \leq 3 \\ y \leq -1 \end{cases}$
33. $-2 \leq x < 5$
34. $-2 < y \leq 5$
35. $\begin{cases} x - y \leq 1 \\ x \geq 2 \end{cases}$
36. $\begin{cases} 4x - 5y \geq -20 \\ x \geq -3 \end{cases}$
37. $\begin{cases} x + y > 4 \\ x + y < -1 \end{cases}$
38. $\begin{cases} x + y > 3 \\ x + y < -2 \end{cases}$
39. $\begin{cases} x + y > 4 \\ x + y > -1 \end{cases}$
40. $\begin{cases} x + y > 3 \\ x + y > -2 \end{cases}$
41. $\begin{cases} x - y \leq 2 \\ x \geq -2 \\ y \leq 3 \end{cases}$
42. $\begin{cases} 3x + y \leq 6 \\ x \geq -2 \\ y \leq 4 \end{cases}$
43. $\begin{cases} x \geq 0 \\ y \geq 0 \\ 2x + 5y \leq 10 \\ 3x + 4y \leq 12 \end{cases}$
44. $\begin{cases} x \geq 0 \\ y \geq 0 \\ 2x + y \leq 4 \\ 2x - 3y \leq 6 \end{cases}$
45. $\begin{cases} 3x + y \leq 6 \\ 2x - y \leq -1 \\ x \geq -2 \\ y \leq 4 \end{cases}$
46. $\begin{cases} 2x + y \leq 6 \\ x + y \geq 2 \\ 1 \leq x \leq 2 \\ y \leq 3 \end{cases}$

Practice PLUS

In Exercises 47–48, write each sentence as a linear inequality in two variables. Then graph the inequality.

47. The *y*-variable is at least 4 more than the product of −2 and the *x*-variable.

48. The *y*-variable is at least 2 more than the product of −3 and the *x*-variable.

In Exercises 49–50, write the given sentences as a system of linear inequalities in two variables. Then graph the system.

49. The sum of the *x*-variable and the *y*-variable is at most 4. The *y*-variable added to the product of 3 and the *x*-variable does not exceed 6.

50. The sum of the *x*-variable and the *y*-variable is at most 3. The *y*-variable added to the product of 4 and the *x*-variable does not exceed 6.

In Exercises 51–52, rewrite each inequality in the system without absolute value bars. Then graph the rewritten system in rectangular coordinates.

51. $\begin{cases} |x| \le 2 \\ |y| \le 3 \end{cases}$

52. $\begin{cases} |x| \le 1 \\ |y| \le 2 \end{cases}$

*The graphs of solution sets of systems of inequalities involve finding the intersection of the solution sets of two or more inequalities. By contrast, in Exercises 53–54 you will be graphing the **union** of the solution sets of two inequalities.*

53. Graph the union of $y > \frac{3}{2}x - 2$ and $y < 4$.

54. Graph the union of $x - y \ge -1$ and $5x - 2y \le 10$.

Without graphing, in Exercises 55–58, determine if each system has no solution or infinitely many solutions.

55. $\begin{cases} 3x + y < 9 \\ 3x + y > 9 \end{cases}$

56. $\begin{cases} 6x - y \le 24 \\ 6x - y > 24 \end{cases}$

57. $\begin{cases} 3x + y \le 9 \\ 3x + y \ge 9 \end{cases}$

58. $\begin{cases} 6x - y \le 24 \\ 6x - y \ge 24 \end{cases}$

Application Exercises

Maximum heart rate, H, in beats per minute is a function of age, a, modeled by the formula

$$H = 220 - a,$$

where $10 \le a \le 70$. The bar graph at the top of the next column shows the target heart rate ranges for four types of exercise goals in terms of maximum heart rate.

Target Heart Rate Ranges for Exercise Goals

Fraction of Maximum Heart Rate, $220 - a$

Source: Vitality Magazine

In Exercises 59–62, systems of inequalities will be used to model three of the target heart rate ranges shown in the bar graph. We begin with the target heart rate range for cardiovascular conditioning, modeled by the following system of inequalities:

$10 \le a \le 70$ — Heart rate ranges apply to ages 10 through 70, inclusive.

$H \ge 0.7(220 - a)$ — Target heart rate range is greater than or equal to 70% of maximum heart rate

$H \le 0.8(220 - a)$. — and less than or equal to 80% of maximum heart rate.

The graph of this system is shown in the figure. Use the graph to solve Exercises 59–60.

Target Heart Rate Range for Cardiovascular Conditioning

59. a. What are the coordinates of point *A* and what does this mean in terms of age and heart rate?

b. Show that point *A* is a solution of the system of inequalities.

60. a. What are the coordinates of point *B* and what does this mean in terms of age and heart rate?

b. Show that point *B* is a solution of the system of inequalities.

(*In Exercises 61–62, refer to the models and graphs on the previous page.*)

61. Write a system of inequalities that models the target heart rate range for the goal of losing weight.

62. Write a system of inequalities that models the target heart rate range for improving overall health.

63. On your next vacation, you will divide lodging between large resorts and small inns. Let x represent the number of nights spent in large resorts. Let y represent the number of nights spent in small inns.

    a. Write a system of inequalities that models the following conditions:

        You want to stay at least 5 nights. At least one night should be spent at a large resort. Large resorts average $200 per night and small inns average $100 per night. Your budget permits no more than $700 for lodging.

    b. Graph the solution set of the system of inequalities in part (a).

    c. Based on your graph in part (b), how many nights could you spend at a large resort and still stay within your budget?

64. **a.** An elevator can hold no more than 2000 pounds. If children average 80 pounds and adults average 160 pounds, write a system of inequalities that models when the elevator holding x children and y adults is overloaded.

    b. Graph the solution set of the system of inequalities in part (a).

Explaining the Concepts

65. What is a linear inequality in two variables? Provide an example with your description.

66. How do you determine if an ordered pair is a solution of an inequality in two variables, x and y?

67. What is a half-plane?

68. What does a solid line mean in the graph of an inequality?

69. What does a dashed line mean in the graph of an inequality?

70. Explain how to graph $x - 2y < 4$.

71. What is a system of linear inequalities?

72. What is a solution of a system of linear inequalities?

73. Explain how to graph the solution set of a system of inequalities.

74. What does it mean if a system of linear inequalities has no solution?

Technology Exercises

Graphing utilities can be used to shade regions in the rectangular coordinate system, thereby graphing an inequality in two variables. Read the section of the user's manual for your graphing utility that describes how to shade a region. Then use your graphing utility to graph the inequalities in Exercises 75–78.

75. $y \leq 4x + 4$

76. $y \geq \dfrac{2}{3}x - 2$

77. $2x + y \leq 6$

78. $3x - 2y \geq 6$

79. Does your graphing utility have any limitations in terms of graphing inequalities? If so, what are they?

80. Use a graphing utility with a $\boxed{\text{SHADE}}$ feature to verify any five of the graphs that you drew by hand in Exercises 1–22.

81. Use a graphing utility with a $\boxed{\text{SHADE}}$ feature to verify any five of the graphs that you drew by hand for the systems in Exercises 23–46.

Critical Thinking Exercises

Make Sense? *In Exercises 82–85, determine whether each statement makes sense or does not make sense, and explain your reasoning.*

82. When graphing a linear inequality, I should always use $(0, 0)$ as a test point because it's easy to perform the calculations when 0 is substituted for each variable.

83. If you want me to graph $x < 3$, you need to tell me whether to use a number line or a rectangular coordinate system.

84. When graphing $3x - 4y < 12$, it's not necessary for me to graph the linear equation $3x - 4y = 12$ because the inequality contains a $<$ symbol, in which equality is not included.

85. Linear inequalities can model situations in which I'm interested in purchasing two items at different costs, I can spend no more than a specified amount on both items, and I want to know how many of each item I can purchase.

In Exercises 86–89, determine whether each statement is true or false. If the statement is false, make the necessary change(s) to produce a true statement.

86. The graph of $3x - 5y < 10$ consists of a dashed line and a shaded half-plane below the line.

87. The graph of $y \geq -x + 1$ consists of a solid line that rises from left to right and a shaded half-plane above the line.

88. The ordered pair $(-2, 40)$ satisfies the following system:

$$\begin{cases} y \geq 9x + 11 \\ 13x + y > 14. \end{cases}$$

89. For the graph of $y < x - 3$, the points $(0, -3)$ and $(8, 5)$ lie on the graph of the corresponding linear equation, but neither point is a solution of the inequality.

90. Write a linear inequality in two variables whose graph is shown.

In Exercises 91–92, write a system of inequalities for each graph.

91.

92.

93. Write a linear inequality in two variables satisfying the following conditions: The points $(-3, -8)$ and $(4, 6)$ lie on the graph of the corresponding linear equation and each point is a solution of the inequality. The point $(1, 1)$ is also a solution.

94. Write a system of inequalities whose solution set includes every point in the rectangular coordinate system.

95. Sketch the graph of the solution set for the following system of inequalities:

$$\begin{cases} y \geq nx + b \ (n < 0, b > 0) \\ y \leq mx + b \ (m > 0, b > 0). \end{cases}$$

Review Exercises

96. Solve using matrices:

$$\begin{cases} 3x - y = 8 \\ x - 5y = -2. \end{cases}$$

(Section 3.4, Example 2)

97. Solve by graphing:

$$\begin{cases} y = 3x - 2 \\ y = -2x + 8. \end{cases}$$

(Section 3.1, Example 2)

98. Evaluate:

$$\begin{vmatrix} 8 & 2 & -1 \\ 3 & 0 & 5 \\ 6 & -3 & 4 \end{vmatrix}.$$

(Section 3.5, Example 3)

Preview Exercises

Exercises 99–101 will help you prepare for the material covered in the next section.

99. a. Graph the solution set of the system:

$$\begin{cases} x + y \geq 6 \\ x \leq 8 \\ y \leq 5. \end{cases}$$

b. List the points that form the corners of the graphed region in part (a).

c. Evaluate $3x + 2y$ at each of the points obtained in part (b).

100. a. Graph the solution set of the system:

$$\begin{cases} x \geq 0 \\ y \geq 0 \\ 3x - 2y \leq 6 \\ y \leq -x + 7. \end{cases}$$

b. List the points that form the corners of the graphed region in part (a).

c. Evaluate $2x + 5y$ at each of the points obtained in part (b).

101. Bottled water and medical supplies are to be shipped to survivors of an earthquake by plane. The bottled water weighs 20 pounds per container and medical kits weigh 10 pounds per kit. Each plane can carry no more than 80,000 pounds. If x represents the number of bottles of water to be shipped per plane and y represents the number of medical kits per plane, write an inequality that models each plane's 80,000-pound weight restriction.

4.5

Linear Programming

What am I supposed to learn?

After studying this section, you should be able to:

1 Write an objective function modeling a quantity that must be maximized or minimized.

2 Use inequalities to model limitations in a situation.

3 Use linear programming to solve problems.

West Berlin children at Tempelhof airport watch fleets of U.S. airplanes bringing in supplies to circumvent the Soviet blockade. The airlift began June 28, 1948, and continued for 15 months.

The Berlin Airlift (1948–1949) was an operation by the United States and Great Britain in response to military action by the former Soviet Union: Soviet troops closed all roads and rail lines between West Germany and Berlin, cutting off supply routes to the city. The Allies used a mathematical technique developed during World War II to maximize the quantities of supplies transported. During the 15-month airlift, 278,228 flights provided basic necessities to blockaded Berlin, saving one of the world's great cities.

In this section, we will look at an important application of systems of linear inequalities. Such systems arise in **linear programming**, a method for solving problems in which a particular quantity that must be maximized or minimized is limited by other factors. Linear programming is one of the most widely used tools in management science. It helps businesses allocate resources to manufacture products in a way that will maxmize profit. Linear programming accounts for more than 50% and perhaps as much as 90% of all computing time used for management decisions in business. The Allies used linear programming to save Berlin.

Objective Functions in Linear Programming

1 Write an objective function modeling a quantity that must be maximized or minimized.

Many problems involve quantities that must be maximized or minimized. Businesses are interested in maximizing profit. A relief operation in which bottled water and medical kits are shipped to earthquake survivors needs to maximize the number of survivors helped by this shipment. An **objective function** is an algebraic expression in two or more variables describing a quantity that must be maximized or minimized.

EXAMPLE 1 Writing an Objective Function

Bottled water and medical supplies are to be shipped to survivors of an earthquake by plane. Each container of bottled water will serve 10 people and each medical kit will aid 6 people. If x represents the number of bottles of water to be shipped and y represents the number of medical kits, write the objective function that models the number of people that can be helped.

Solution Because each bottle of water serves 10 people and each medical kit aids 6 people, we have

The number of people helped	is	10 times the number of bottles of water	plus	6 times the number of medical kits.
$=$		$10x$	$+$	$6y.$

Using z to represent the number of people helped, the objective function is

$$z = 10x + 6y.$$

Unlike the functions that we have seen so far, the objective function is an equation in three variables. For a value of x and a value of y, there is one and only one value of z. Thus, z is a function of x and y. ■

☑ **CHECK POINT 1**   A company manufactures bookshelves and desks for computers. Let x represent the number of bookshelves manufactured daily and y the number of desks manufactured daily. The company's profits are $25 per bookshelf and $55 per desk. Write the objective function that models the company's total daily profit, z, from x bookshelves and y desks. (Check Points 2 through 4 are related to this situation, so keep track of your answers.)

---

2 Use inequalities to model limitations in a situation. ▶

Constraints in Linear Programming

Ideally, the number of earthquake survivors helped in Example 1 should increase without restriction so that every survivor receives water and medical supplies. However, the planes that ship these supplies are subject to weight and volume restrictions. In linear programming problems, such restrictions are called **constraints**. Each constraint is expressed as a linear inequality. The list of constraints forms a system of linear inequalities.

EXAMPLE 2   Writing a Constraint

Each plane can carry no more than 80,000 pounds. The bottled water weighs 20 pounds per container and each medical kit weighs 10 pounds. Let x represent the number of bottles of water to be shipped and y the number of medical kits. Write an inequality that models this constraint.

Solution   Because each plane can carry no more than 80,000 pounds, we have

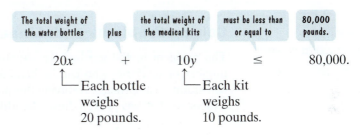

| The total weight of the water bottles | plus | the total weight of the medical kits | must be less than or equal to | 80,000 pounds. |

$$20x \quad + \quad 10y \quad \leq \quad 80,000.$$

└─ Each bottle weighs 20 pounds.   └─ Each kit weighs 10 pounds.

The plane weight constraint is modeled by the inequality

$$20x + 10y \leq 80,000. \quad ■$$

☑ **CHECK POINT 2**   To maintain high quality, the company in Check Point 1 should not manufacture more than a total of 80 bookshelves and desks per day. Write an inequality that models this constraint.

In addition to a weight constraint on its cargo, each plane has a limited amount of space in which to carry supplies. Example 3 demonstrates how to express this constraint.

EXAMPLE 3   Writing a Constraint

Each plane can carry a total volume of supplies that does not exceed 6000 cubic feet. Each water bottle is 1 cubic foot and each medical kit also has a volume of 1 cubic foot. With x still representing the number of water bottles and y the number of medical kits, write an inequality that models this second constraint.

Solution   Because each plane can carry a volume of supplies that does not exceed 6000 cubic feet, we have

$$1x \quad + \quad 1y \quad \le \quad 6000.$$

Each bottle is 1 cubic foot.   Each kit is 1 cubic foot.

The plane volume constraint is modeled by the inequality $x + y \le 6000$. ∎

In summary, here's what we have described so far in this aid-to-earthquake-survivors situation:

$$z = 10x + 6y$$ This is the objective function modeling the number of people helped with x bottles of water and y medical kits.

$$\begin{cases} 20x + 10y \le 80{,}000 \\ \quad\; x + y \le 6000. \end{cases}$$ These are the constraints based on each plane's weight and volume limitations.

☑ **CHECK POINT 3**   To meet customer demand, the company in Check Point 1 must manufacture between 30 and 80 bookshelves per day, inclusive. Furthermore, the company must manufacture at least 10 and no more than 30 desks per day. Write an inequality that models each of these sentences. Then summarize what you have described about this company by writing the objective function for its profits and the three constraints.

---

③ Use linear programming to solve problems. ⏵

Solving Problems with Linear Programming

The problem in the earthquake situation described previously is to maximize the number of survivors who can be helped, subject to each plane's weight and volume constraints. The process of solving this problem is called *linear programming*, based on a theorem that was proven during World War II.

Solving a Linear Programming Problem

Let $z = ax + by$ be an objective function that depends on x and y. Furthermore, z is subject to a number of linear constraints on x and y. If a maximum or minimum value of z exists, it can be determined as follows:

1. Graph the system of inequalities representing the constraints.
2. Find the value of the objective function at each corner, or **vertex**, of the graphed region. The maximum and minimum of the objective function occur at one or more of the corner points.

EXAMPLE 4   Solving a Linear Programming Problem

Determine how many bottles of water and how many medical kits should be sent on each plane to maximize the number of earthquake survivors who can be helped.

Solution   We must maximize $z = 10x + 6y$ subject to the following constraints:

$$\begin{cases} 20x + 10y \le 80{,}000 \\ \quad\; x + y \le 6000. \end{cases}$$

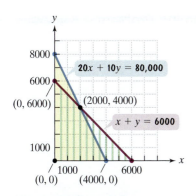

Figure 4.19 The region in quadrant I representing the constraints $20x + 10y \leq 80,000$ and $x + y \leq 6000$

Step 1. Graph the system of inequalities representing the constraints. Because x (the number of bottles of water per plane) and y (the number of medical kits per plane) must be nonnegative, we need to graph the system of inequalities in quadrant I and its boundary only.

To graph the inequality $20x + 10y \leq 80,000$, we graph the equation $20x + 10y = 80,000$ as a solid blue line (**Figure 4.19**). Setting $y = 0$, the x-intercept is 4000 and setting $x = 0$, the y-intercept is 8000. Using $(0,0)$ as a test point, the inequality is satisfied, so we shade below the blue line, as shown in yellow in **Figure 4.19**.

Now we graph $x + y \leq 6000$ by first graphing $x + y = 6000$ as a solid red line. Setting $y = 0$, the x-intercept is 6000. Setting $x = 0$, the y-intercept is 6000. Using $(0,0)$ as a test point, the inequality is satisfied, so we shade below the red line, as shown using green vertical shading in **Figure 4.19**.

We use the addition method to find where the lines $20x + 10y = 80,000$ and $x + y = 6000$ intersect.

$$\begin{cases} 20x + 10y = 80,000 \\ x + y = 6000 \end{cases} \xrightarrow[\text{Multiply by }-10.]{\text{No change}} \begin{cases} 20x + 10y = 80,000 \\ -10x - 10y = -60,000 \end{cases}$$

$$\text{Add:} \quad 10x \qquad = 20,000$$
$$x \qquad = 2000$$

Back-substituting 2000 for x in $x + y = 6000$, we find $y = 4000$, so the intersection point is $(2000, 4000)$.

The system of inequalities representing the constraints is shown by the region in which the yellow shading and the green vertical shading overlap in **Figure 4.19**. The graph of the system of inequalities is shown again in **Figure 4.20**. The red and blue line segments are included in the graph.

Step 2. Find the value of the objective function at each corner of the graphed region. The maximum and minimum of the objective function occur at one or more of the corner points. We must evaluate the objective function, $z = 10x + 6y$, at the four corners, or vertices, of the region in **Figure 4.20**.

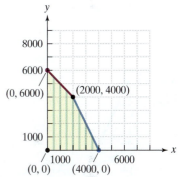

Figure 4.20

Corner (x, y)	Objective Function $z = 10x + 6y$
$(0, 0)$	$z = 10(0) + 6(0) = 0$
$(4000, 0)$	$z = 10(4000) + 6(0) = 40,000$
$(2000, 4000)$	$z = 10(2000) + 6(4000) = 44,000$ ← maximum
$(0, 6000)$	$z = 10(0) + 6(6000) = 36,000$

Thus, the maximum value of z is 44,000 and this occurs when $x = 2000$ and $y = 4000$. In practical terms, this means that the maximum number of earthquake survivors who can be helped with each plane shipment is 44,000. This can be accomplished by sending 2000 water bottles and 4000 medical kits per plane. ∎

☑ **CHECK POINT 4**   For the company in Check Points 1–3, how many bookshelves and how many desks should be manufactured per day to obtain maximum profit? What is the maximum daily profit?

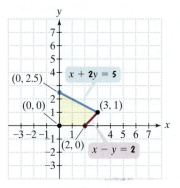

Figure 4.21 The graph of $x + 2y \leq 5$ and $x - y \leq 2$ in quadrant I

EXAMPLE 5   Solving a Linear Programming Problem

Find the maximum value of the objective function

$$z = 2x + y$$

subject to the following constraints:

$$\begin{cases} x \geq 0, y \geq 0 \\ x + 2y \leq 5 \\ x - y \leq 2. \end{cases}$$

Solution   We begin by graphing the region in quadrant I ($x \geq 0, y \geq 0$) formed by the constraints. The graph is shown in **Figure 4.21**.

Now we evaluate the objective function at the four vertices of this region.

Objective function: $z = 2x + y$

At $(0, 0)$: $z = 2 \cdot 0 + 0 = 0$

At $(2, 0)$: $z = 2 \cdot 2 + 0 = 4$

At $(3, 1)$: $z = 2 \cdot 3 + 1 = 7$ ← Maximum value of z

At $(0, 2.5)$: $z = 2 \cdot 0 + 2.5 = 2.5$

Thus, the maximum value of z is 7, and this occurs when $x = 3$ and $y = 1$. ∎

We can see why the objective function in Example 5 has a maximum value that occurs at a vertex by solving the equation for y.

$z = 2x + y$ This is the objective function of Example 5.

$y = -2x + z$ Solve for y. Recall that the slope-intercept

     form of the equation of a line is $y = mx + b$.

Slope $= -2$ y-intercept $= z$

In this form, z represents the y-intercept of the objective function. The equation describes infinitely many parallel lines (one for each value of z), each with slope -2. The process in linear programming involves finding the maximum z-value for all lines that intersect the region determined by the constraints. Of all the lines whose slope is -2, we're looking for the one with the greatest y-intercept that intersects the given region. As we see in **Figure 4.22**, such a line will pass through one (or possibly more) of the vertices of the region.

Figure 4.22 The line with slope -2 with the greatest y-intercept that intersects the shaded region passes through one of the vertices of the region.

✓ **CHECK POINT 5**   Find the maximum value of the objective function $z = 3x + 5y$ subject to the constraints $x \geq 0, y \geq 0, x + y \geq 1, x + y \leq 6$.

CONCEPT AND VOCABULARY CHECK

Fill in each blank so that the resulting statement is true.

1. A method for finding the maximum or minimum value of a quantity that is subject to various limitations is called _____.

2. An algebraic expression in two or more variables describing a quantity that must be maximized or minimized is called a/an _____ function.

3. A system of linear inequalities is used to represent restrictions, or _____, on a function that must be maximized or minimized. Using the graph of such a system of inequalities, the maximum and minimum values of the function occur at one or more of the _____ points.

4.5 EXERCISE SET   ▶ **MyMathLab**®

Practice Exercises

In Exercises 1–4, find the value of the objective function at each corner of the graphed region. What is the maximum value of the objective function? What is the minimum value of the objective function?

1. Objective Function
$z = 5x + 6y$

2. Objective Function
$z = 3x + 2y$

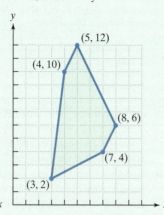

3. Objective Function
$z = 40x + 50y$

4. Objective Function
$z = 30x + 45y$

In Exercises 5–14, an objective function and a system of linear inequalities representing constraints are given.

a. *Graph the system of inequalities representing the constraints.*

b. *Find the value of the objective function at each corner of the graphed region.*

c. *Use the values in part (b) to determine the maximum value of the objective function and the values of x and y for which the maximum occurs.*

5. Objective Function $z = 3x + 2y$
Constraints
$$\begin{cases} x \geq 0, y \geq 0 \\ 2x + y \leq 8 \\ x + y \geq 4 \end{cases}$$

6. Objective Function $z = 2x + 3y$
Constraints
$$\begin{cases} x \geq 0, y \geq 0 \\ 2x + y \leq 8 \\ 2x + 3y \leq 12 \end{cases}$$

7. Objective Function $z = 4x + y$
Constraints
$$\begin{cases} x \geq 0, y \geq 0 \\ 2x + 3y \leq 12 \\ x + y \geq 3 \end{cases}$$

8. Objective Function $z = x + 6y$
Constraints
$$\begin{cases} x \geq 0, y \geq 0 \\ 2x + y \leq 10 \\ x - 2y \geq -10 \end{cases}$$

9. Objective Function $z = 3x - 2y$
Constraints
$$\begin{cases} 1 \leq x \leq 5 \\ y \geq 2 \\ x - y \geq -3 \end{cases}$$

10. Objective Function $z = 5x - 2y$
Constraints
$$\begin{cases} 0 \leq x \leq 5 \\ 0 \leq y \leq 3 \\ x + y \geq 2 \end{cases}$$

11. Objective Function $z = 4x + 2y$
Constraints
$$\begin{cases} x \geq 0, y \geq 0 \\ 2x + 3y \leq 12 \\ 3x + 2y \leq 12 \\ x + y \geq 2 \end{cases}$$

12. Objective Function $z = 2x + 4y$
Constraints
$$\begin{cases} x \geq 0, y \geq 0 \\ x + 3y \geq 6 \\ x + y \geq 3 \\ x + y \leq 9 \end{cases}$$

13. Objective Function $z = 10x + 12y$
Constraints
$$\begin{cases} x \geq 0, y \geq 0 \\ 2x + y \leq 10 \\ 2x + 3y \leq 18 \end{cases}$$

14. Objective Function $z = 5x + 6y$
Constraints
$$\begin{cases} x \geq 0, y \geq 0 \\ 2x + y \geq 10 \\ x + 2y \geq 10 \\ x + y \leq 10 \end{cases}$$

Application Exercises

15. A television manufacturer makes rear-projection and plasma televisions. The profit per unit is $125 for the rear-projection televisions and $200 for the plasma televisions.

a. Let $x =$ the number of rear-projection televisions manufactured in a month and $y =$ the number of plasma televisions manufactured in a month. Write the objective function that models the total monthly profit.

b. The manufacturer is bound by the following constraints:

- Equipment in the factory allows for making at most 450 rear-projection televisions in one month.

- Equipment in the factory allows for making at most 200 plasma televisions in one month.

- The cost to the manufacturer per unit is $600 for the rear-projection televisions and $900 for the plasma televisions. Total monthly costs cannot exceed $360,000.

Write a system of three inequalities that models these constraints.

c. Graph the system of inequalities in part (b). Use only the first quadrant and its boundary, because x and y must both be nonnegative.

d. Evaluate the objective function for total monthly profit at each of the five vertices of the graphed region. [The vertices should occur at $(0, 0)$, $(0, 200)$, $(300, 200)$, $(450, 100)$, and $(450, 0)$.]

e. Complete the missing portions of this statement: The television manufacturer will make the greatest profit by manufacturing ___ rear-projection televisions each month and ___ plasma televisions each month. The maximum monthly profit is $_____.

16. a. A student earns $15 per hour for tutoring and $10 per hour as a teacher's aide. Let x = the number of hours each week spent tutoring and y = the number of hours each week spent as a teacher's aide. Write the objective function that models total weekly earnings.

b. The student is bound by the following constraints:

- To have enough time for studies, the student can work no more than 20 hours a week.

- The tutoring center requires that each tutor spend at least three hours a week tutoring.

- The tutoring center requires that each tutor spend no more than eight hours a week tutoring.

Write a system of three inequalities that models these constraints.

c. Graph the system of inequalities in part (b). Use only the first quadrant and its boundary, because x and y are nonnegative.

d. Evaluate the objective function for total weekly earnings at each of the four vertices of the graphed region. [The vertices should occur at $(3, 0)$, $(8, 0)$, $(3, 17)$, and $(8, 12)$.]

e. Complete the missing portions of this statement: The student can earn the maximum amount per week by tutoring for ___ hours per week and working as a teacher's aide for ___ hours per week. The maximum amount that the student can earn each week is $___.

Use the two steps for solving a linear programming problem, given in the box on page 300, to solve the problems in Exercises 17–23.

17. A manufacturer produces two models of mountain bicycles. The times (in hours) required for assembling and painting each model are given in the following table:

	Model A	Model B
Assembling	5	4
Painting	2	3

The maximum total weekly hours available in the assembly department and the paint department are 200 hours and 108 hours, respectively. The profits per unit are $25 for model A and $15 for model B. How many of each type should be produced to maximize profit?

18. A large institution is preparing lunch menus containing foods A and B. The specifications for the two foods are given in the following table:

Food	Units of Fat per Ounce	Units of Carbohydrates per Ounce	Units of Protein per Ounce
A	1	2	1
B	1	1	1

Each lunch must provide at least 6 units of fat per serving, no more than 7 units of protein, and at least 10 units of carbohydrates. The institution can purchase food A for $0.12 per ounce and food B for $0.08 per ounce. How many ounces of each food should a serving contain to meet the dietary requirement at the least cost?

19. Food and clothing are shipped to survivors of a hurricane. Each carton of food will feed 12 people, while each carton of clothing will help 5 people. Each 20-cubic-foot box of food weighs 50 pounds and each 10-cubic-foot box of clothing weighs 20 pounds. The commercial carriers transporting food and clothing are bound by the following constraints:

- The total weight per carrier cannot exceed 19,000 pounds.
- The total volume must be less than 8000 cubic feet.

How many cartons of food and how many cartons of clothing should be sent with each plane shipment to maximize the number of people who can be helped?

20. On June 24, 1948, the former Soviet Union blocked all land and water routes through East Germany to Berlin. A gigantic airlift was organized using American and British planes to bring food, clothing, and other supplies to the more than 2 million people in West Berlin. The cargo capacity was 30,000 cubic feet for an American plane and 20,000 cubic feet for a British plane. To break the Soviet blockade, the Western Allies had to maximize cargo capacity, but were subject to the following restrictions:

- No more than 44 planes could be used.
- The larger American planes required 16 personnel per flight, double that of the requirement for the British planes. The total number of personnel available could not exceed 512.
- The cost of an American flight was $9000 and the cost of a British flight was $5000. Total weekly costs could not exceed $300,000.

Find the number of American planes and the number of British planes that were used to maximize cargo capacity.

21. A theater is presenting a program on drinking and driving for students and their parents. The proceeds will be donated to a local alcohol information center. Admission is $2.00 for parents and $1.00 for students. However, the situation has two constraints: The theater can hold no more than 150 people and every two parents must bring at least one student. How many parents and students should attend to raise the maximum amount of money?

22. You are about to take a test that contains computation problems worth 6 points each and word problems worth 10 points each. You can do a computation problem in 2 minutes and a word problem in 4 minutes. You have 40 minutes to take the test and may answer no more than 12 problems. Assuming you answer all the problems attempted correctly, how many of each type of problem must you do to maximize your score? What is the maximum score?

23. In 1978, a ruling by the Civil Aeronautics Board allowed Federal Express to purchase larger aircraft. Federal Express's options included 20 Boeing 727s that United Airlines was retiring and/or the French-built Dassault Fanjet Falcon 20. To aid in their decision, executives at Federal Express analyzed the following data:

	Boeing 727	Falcon 20
Direct Operating Cost	$1400 per hour	$500 per hour
Payload	42,000 pounds	6000 pounds

Federal Express was faced with the following constraints:

- Hourly operating cost was limited to $35,000.
- Total payload had to be at least 672,000 pounds.
- Only twenty 727s were available.

Given the constraints, how many of each kind of aircraft should Federal Express have purchased to maximize the number of aircraft?

Explaining the Concepts

24. What kinds of problems are solved using the linear programming method?

25. What is an objective function in a linear programming problem?

26. What is a constraint in a linear programming problem? How is a constraint represented?

27. In your own words, describe how to solve a linear programming problem.

28. Describe a situation in your life in which you would like to maximize something, but are limited by at least two constraints. Can linear programming be used in this situation? Explain your answer.

Critical Thinking Exercises

Make Sense? *In Exercises 29–32, determine whether each statement makes sense or does not make sense, and explain your reasoning.*

29. In order to solve a linear programming problem, I use the graph representing the constraints and the graph of the objective function.

30. I use the coordinates of each vertex from my graph representing the constraints to find the values that maximize or minimize an objective function.

31. I need to be able to graph systems of linear inequalities in order to solve linear programming problems.

32. An important application of linear programming for businesses involves maximizing profit.

33. Suppose that you inherit $10,000. The will states how you must invest the money. Some (or all) of the money must be invested in stocks and bonds. The requirements are that at least $3000 be invested in bonds, with expected returns of $0.08 per dollar, and at least $2000 be invested in stocks, with expected returns of $0.12 per dollar. Because the stocks are medium risk, the final stipulation requires that the investment in bonds should never be less than the investment in stocks. How should the money be invested so as to maximize your expected returns?

34. Consider the objective function $z = Ax + By$ ($A > 0$ and $B > 0$) subject to the following constraints: $2x + 3y \le 9$, $x - y \le 2, x \ge 0$, and $y \ge 0$. Prove that the objective function will have the same maximum value at the vertices $(3, 1)$ and $(0, 3)$ if $A = \frac{2}{3}B$.

Review Exercises

35. Simplify: $(2x^4y^3)(3xy^4)^3$. (Section 1.6, Example 9)

36. Solve for L: $3P = \dfrac{2L - W}{4}$. (Section 1.5, Example 8)

37. If $f(x) = x^3 + 2x^2 - 5x + 4$, find $f(-1)$. (Section 2.1, Example 3)

Preview Exercises

Exercises 38–40 will help you prepare for the material covered in the first section of the next chapter.

In Exercises 38–39, simplify each algebraic expression.

38. $(-9x^3 + 7x^2 - 5x + 3) + (13x^3 + 2x^2 - 8x - 6)$

39. $(7x^3 - 8x^2 + 9x - 6) - (2x^3 - 6x^2 - 3x + 9)$

40. The figures show the graphs of two functions.

a. Which function, f or g, has a graph that rises to the left and falls to the right?

b. Which function, f or g, has a graph that falls to the left and rises to the right?

Chapter 4 Summary

Definitions and Concepts	**Examples**

Section 4.1 Solving Linear Inequalities

A linear inequality in one variable can be written in the form $ax + b < 0$, $ax + b \le 0$, $ax + b > 0$, or $ax + b \ge 0$. The set of all numbers that make the inequality a true statement is its solution set, represented using interval notation.

Solving a Linear Inequality

1. Simplify each side.

2. Collect variable terms on one side and constant terms on the other side.

3. Isolate the variable and solve.

If an inequality is multiplied or divided by a negative number, the direction of the inequality symbol must be reversed.

Solve: $2(x + 3) - 5x \le 15$.

$$2x + 6 - 5x \le 15$$
$$-3x + 6 \le 15$$
$$-3x \le 9$$
$$\frac{-3x}{-3} \ge \frac{9}{-3}$$
$$x \ge -3$$

Solution set: $[-3, \infty)$

Section 4.2 Compound Inequalities

Intersection ($\cap$) and Union ($\cup$)

$A \cap B$ is the set of elements common to both set A and set B.
$A \cup B$ is the set of elements that are members of set A or set B or of both sets.

$\{1, 3, 5, 7\} \cap \{5, 7, 9, 11\} = \{5, 7\}$
$\{1, 3, 5, 7\} \cup \{5, 7, 9, 11\} = \{1, 3, 5, 7, 9, 11\}$

A compound inequality is formed by joining two inequalities with the word *and* or *or*.

When the connecting word is *and*, graph each inequality separately and take the intersection of their solution sets.

Solve: $x + 1 > 3$ and $x + 4 \le 8$.

$$x > 2 \qquad \text{and} \qquad x \le 4$$

Solution set: $(2, 4]$

The compound inequality $a < x < b$ means $a < x$ and $x < b$. Solve by isolating the variable in the middle.

Solve: $-1 < \dfrac{2x + 1}{3} \le 2$.

$$-3 < 2x + 1 \le 6 \quad \text{Multiply by 3.}$$
$$-4 < 2x \le 5 \quad \text{Subtract 1.}$$
$$-2 < x \le \frac{5}{2} \quad \text{Divide by 2.}$$

Solution set: $\left(-2, \dfrac{5}{2}\right]$

When the connecting word in a compound inequality is *or*, graph each inequality separately and take the union of their solution sets.

Solve: $x - 2 > -3$ or $2x \le -6$.

$$x > -1 \quad \text{or} \quad x \le -3$$

Solution set: $(-\infty, -3] \cup (-1, \infty)$

Section 4.3 Equations and Inequalities Involving Absolute Value

Absolute Value Equations

1. If $c > 0$, then $|u| = c$ means $u = c$ or $u = -c$.

2. If $c < 0$, then $|u| = c$ has no solution.

3. If $c = 0$, then $|u| = 0$ means $u = 0$.

Solve: $|2x - 7| = 3$.

$$2x - 7 = 3 \quad \text{or} \quad 2x - 7 = -3$$
$$2x = 10 \qquad\qquad 2x = 4$$
$$x = 5 \qquad\qquad\quad x = 2$$

The solution set is $\{2, 5\}$.

Definitions and Concepts	Examples

Section 4.3 Equations and Inequalities Involving Absolute Value (continued)

Absolute Value Equations with Two Absolute Value Bars

If $|u| = |v|$, then $u = v$ or $u = -v$.

Solve: $|x - 6| = |2x + 1|$.

$$x - 6 = 2x + 1 \quad \text{or} \quad x - 6 = -(2x + 1)$$
$$-x - 6 = 1 \qquad\qquad\quad x - 6 = -2x - 1$$
$$-x = 7 \qquad\qquad\qquad 3x - 6 = -1$$
$$x = -7 \qquad\qquad\qquad\quad 3x = 5$$
$$x = \frac{5}{3}$$

The solutions are -7 and $\frac{5}{3}$, and the solution set is $\left\{-7, \frac{5}{3}\right\}$.

Solving Absolute Value Inequalities

If c is a positive number,

1. $|u| < c$ is equivalent to $-c < u < c$.
2. $|u| > c$ is equivalent to $u < -c$ or $u > c$.

In each case, the absolute value inequality is rewritten as an equivalent compound inequality without absolute value bars.

Solve: $|x - 4| < 3$.
$$-3 < x - 4 < 3$$
$$1 < x < 7 \quad \text{Add 4.}$$
The solution set is $(1, 7)$.

$$\underset{0\ \ 1\ \ 2\ \ 3\ \ 4\ \ 5\ \ 6\ \ 7\ \ 8}{\xrightarrow{\hspace{0.5cm}(\ \ \ \ \ \ \ \ \ )\hspace{0.5cm}}} x$$

Solve: $\left|\dfrac{x}{3} - 1\right| \geq 2$.

$$\frac{x}{3} - 1 \leq -2 \quad \text{or} \quad \frac{x}{3} - 1 \geq 2.$$
$$x - 3 \leq -6 \quad \text{or} \quad x - 3 \geq 6 \quad \text{Multiply by 3.}$$
$$x \leq -3 \quad \text{or} \qquad\quad x \geq 9 \quad \text{Add 3.}$$

The solution set is $(-\infty, -3] \cup [9, \infty)$.

$$\underset{-6\ -4\ -2\ \ 0\ \ 2\ \ 4\ \ 6\ \ 8\ 10}{\xleftarrow{\hspace{0.3cm}]\hspace{2cm}[\hspace{0.3cm}}} x$$

Absolute Value Inequalities with Unusual Solution Sets

If c is a negative number,

1. $|u| < c$ has no solution.
2. $|u| > c$ is true for all real numbers for which u is defined.

- $|x - 4| < -3$ has no solution. The solution set is $\varnothing$.
- $|3x + 6| > -12$ is true for all real numbers. The solution set is $(-\infty, \infty)$.

Section 4.4 Linear Inequalities in Two Variables

If the equal sign in $Ax + By = C$ is replaced with an inequality symbol, the result is a linear inequality in two variables. Its graph is the set of all points whose coordinates satisfy the inequality. To obtain the graph,

1. Replace the inequality symbol with an equal sign and graph the boundary line. Use a solid line for $\leq$ or $\geq$ and a dashed line for $<$ or $>$.
2. Choose a test point not on the line and substitute its coordinates into the inequality.
3. If a true statement results, shade the half-plane containing the test point. If a false statement results, shade the half-plane not containing the test point.

Graph: $x - 2y \leq 4$.

1. Graph $x - 2y = 4$. Use a solid line because the inequality symbol is $\leq$.
2. Test $(0, 0)$.
$$x - 2y \leq 4$$
$$0 - 2 \cdot 0 \overset{?}{\leq} 4$$
$$0 \leq 4, \quad \text{true}$$
3. The inequality is true. Shade the half-plane containing $(0, 0)$.

Definitions and Concepts	**Examples**

Section 4.4 Linear Inequalities in Two Variables (continued)

Two or more linear inequalities make up a system of linear inequalities. A solution is an ordered pair satisfying all inequalities in the system. To graph a system of inequalities, graph each inequality in the system. The overlapping region, if there is one, represents the solutions of the system. If there is no overlapping region, the system has no solution.	Graph the solutions of the system: $$\begin{cases} y \le -2x \\ x - y \ge 3. \end{cases}$$

Section 4.5 Linear Programming

Linear programming is a method for solving problems in which a particular quantity that must be maximized or minimized is limited. An objective function is an algebraic expression in three variables modeling a quantity that must be maximized or minimized. Constraints are restrictions, expressed as linear inequalities. **Solving a Linear Programming Problem** 1. Graph the system of inequalities representing the constraints. 2. Find the value of the objective function at each corner, or vertex, of the graphed region. The maximum and minimum of the objective function occur at one or more vertices.	Find the maximum value of the objective function $z = 3x + 2y$ subject to the following constraints: $x \ge 0, y \ge 0, 2x + 3y \le 18, 2x + y \le 10$. 1. Graph the system of inequalities representing the constraints. 2. Evaluate the objective function at each vertex.

2. Evaluate the objective function at each vertex.

Vertex	$z = 3x + 2y$
$A(0, 0)$	$z = 3(0) + 2(0) = 0$
$B(0, 6)$	$z = 3(0) + 2(6) = 12$
$C(3, 4)$	$z = 3(3) + 2(4) = 17$
$D(5, 0)$	$z = 3(5) + 2(0) = 15$

The maximum value of the objective function is 17.

CHAPTER 4 REVIEW EXERCISES

4.1 *In Exercises 1–6, solve each linear inequality. Other than ∅, graph the solution set on a number line.*

1. $-6x + 3 \le 15$

2. $6x - 9 \ge -4x - 3$

3. $\dfrac{x}{3} - \dfrac{3}{4} - 1 > \dfrac{x}{2}$

4. $6x + 5 > -2(x - 3) - 25$

5. $3(2x - 1) - 2(x - 4) \ge 7 + 2(3 + 4x)$

6. $2x + 7 \le 5x - 6 - 3x$

7. A person can choose between two charges on a checking account. The first method involves a fixed cost of $11 per month plus 6¢ for each check written. The second method involves a fixed cost of $4 per month plus 20¢ for each check written. How many checks should be written to make the first method a better deal?

8. A salesperson earns $500 per month plus a commission of 20% of sales. Describe the sales needed to receive a total income that exceeds $3200 per month.

4.2   *In Exercises 9–12, let $A = \{a, b, c\}$, $B = \{a, c, d, e\}$, and $C = \{a, d, f, g\}$. Find the indicated set.*

9. $A \cap B$

10. $A \cap C$

11. $A \cup B$

12. $A \cup C$

In Exercises 13–23, solve each compound inequality. Other than $\varnothing$, graph the solution set on a number line.

13. $x \le 3$ and $x < 6$

14. $x \le 3$ or $x < 6$

15. $-2x < -12$ and $x - 3 < 5$

16. $5x + 3 \le 18$ and $2x - 7 \le -5$

17. $2x - 5 > -1$ and $3x < 3$

18. $2x - 5 > -1$ or $3x < 3$

19. $x + 1 \le -3$ or $-4x + 3 < -5$

20. $5x - 2 \le -22$ or $-3x - 2 > 4$

21. $5x + 4 \ge -11$ or $1 - 4x \ge 9$

22. $-3 < x + 2 \le 4$

23. $-1 \le 4x + 2 \le 6$

24. To receive a B in a course, you must have an average of at least 80% but less than 90% on five exams. Your grades on the first four exams were 95%, 79%, 91%, and 86%. What range of grades on the fifth exam will result in a B for the course? Use interval notation to express this range.

4.3   *In Exercises 25–28, find the solution set for each equation.*

25. $|2x + 1| = 7$

26. $|3x + 2| = -5$

27. $2|x - 3| - 7 = 10$

28. $|4x - 3| = |7x + 9|$

In Exercises 29–33, solve each absolute value inequality. Other than $\varnothing$, graph the solution set on a number line.

29. $|2x + 3| \le 15$

30. $\left| \dfrac{2x + 6}{3} \right| > 2$

31. $|2x + 5| - 7 < -6$

32. $-4|x + 2| + 5 \le -7$

33. $|2x - 3| + 4 \le -10$

34. Approximately 90% of the population sleeps h hours daily, where h is modeled by the inequality $|h - 6.5| \le 1$. Write a sentence describing the range for the number of hours that most people sleep. Do *not* use the phrase "absolute value" in your description.

4.4   *In Exercises 35–40, graph each inequality in a rectangular coordinate system.*

35. $3x - 4y > 12$

36. $x - 3y \le 6$

37. $y \le -\dfrac{1}{2}x + 2$

38. $y > \dfrac{3}{5}x$

39. $x \le 2$

40. $y > -3$

In Exercises 41–49, graph the solution set of each system of inequalities or indicate that the system has no solution.

41. $\begin{cases} 2x - y \le 4 \\ x + y \ge 5 \end{cases}$

42. $\begin{cases} y < -x + 4 \\ y > x - 4 \end{cases}$

43. $-3 \le x < 5$

44. $-2 < y \le 6$

45. $\begin{cases} x \ge 3 \\ y \le 0 \end{cases}$

46. $\begin{cases} 2x - y > -4 \\ x \ge 0 \end{cases}$

47. $\begin{cases} x + y \le 6 \\ y \ge 2x - 3 \end{cases}$

48. $\begin{cases} 3x + 2y \ge 4 \\ x - y \le 3 \\ x \ge 0, y \ge 0 \end{cases}$

49. $\begin{cases} 2x - y > 2 \\ 2x - y < -2 \end{cases}$

4.5

50. Find the value of the objective function $z = 2x + 3y$ at each corner of the graphed region shown. What is the maximum value of the objective function? What is the minimum value of the objective function?

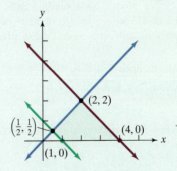

In Exercises 51–53, graph the region determined by the constraints. Then find the maximum value of the given objective function, subject to the constraints.

51. Objective Function $z = 2x + 3y$

Constraints $\begin{cases} x \ge 0, y \ge 0 \\ x + y \le 8 \\ 3x + 2y \ge 6 \end{cases}$

52. Objective Function $z = x + 4y$

Constraints $\begin{cases} 0 \le x \le 5, 0 \le y \le 7 \\ x + y \ge 3 \end{cases}$

53. Objective Function $\quad z = 5x + 6y$
Constraints $\qquad \begin{cases} x \geq 0, y \geq 0 \\ y \leq x \\ 2x + y \leq 12 \\ 2x + 3y \geq 6 \end{cases}$

54. A paper manufacturing company converts wood pulp to writing paper and newsprint. The profit on a unit of writing paper is $500 and the profit on a unit of newsprint is $350.

   a. Let x represent the number of units of writing paper produced daily. Let y represent the number of units of newsprint produced daily. Write the objective function that models total daily profit.

   b. The manufacturer is bound by the following constraints:
- Equipment in the factory allows for making at most 200 units of paper (writing paper and newsprint) in a day.
- Regular customers require at least 10 units of writing paper and at least 80 units of newsprint daily.

   Write a system of inequalities that models these constraints.

   c. Graph the inequalities in part (b). Use only the first quadrant, because x and y must both be positive. (*Suggestion*: Let each unit along the x- and y-axes represent 20.)

   d. Evaluate the objective function at each of the three vertices of the graphed region.

   e. Complete the missing portions of this statement: The company will make the greatest profit by producing ___ units of writing paper and ___ units of newsprint each day. The maximum daily profit is $_____.

55. A manufacturer of lightweight tents makes two models whose specifications are given in the following table.

	Cutting Time per Tent	Assembly Time per Tent
Model A	0.9 hour	0.8 hour
Model B	1.8 hours	1.2 hours

Each month, the manufacturer has no more than 864 hours of labor available in the cutting department and at most 672 hours in the assembly division. The profits come to $25 per tent for model A and $40 per tent for model B. How many of each should be manufactured monthly to maximize the profit?

CHAPTER 4 TEST

Step-by-step test solutions are found on the Chapter Test Prep Videos available in MyMathLab® or on YouTube (search "BlitzerInterAlg7e" and click on "Channels").

In Exercises 1–2, solve and graph the solution set on a number line.

1. $3(x + 4) \geq 5x - 12$

2. $\dfrac{x}{6} + \dfrac{1}{8} \leq \dfrac{x}{2} - \dfrac{3}{4}$

3. You are choosing between two landline phone plans. Plan A charges $25 per month for unlimited calls. Plan B has a monthly fee of $13 with a charge of $0.06 per minute. How many minutes of calls in a month make plan A the better deal?

4. Find the intersection: $\{2, 4, 6, 8, 10\} \cap \{4, 6, 12, 14\}$.

5. Find the union: $\{2, 4, 6, 8, 10\} \cup \{4, 6, 12, 14\}$.

In Exercises 6–10, solve each compound inequality. Other than $\varnothing$, graph the solution set on a number line.

6. $2x + 4 < 2$ and $x - 3 > -5$

7. $x + 6 \geq 4$ and $2x + 3 \geq -2$

8. $2x - 3 < 5$ or $3x - 6 \leq 4$

9. $x + 3 \leq -1$ or $-4x + 3 < -5$

10. $-3 \leq \dfrac{2x + 5}{3} < 6$

In Exercises 11–12, find the solution set for each equation.

11. $|5x + 3| = 7$

12. $|6x + 1| = |4x + 15|$

In Exercises 13–14, solve and graph the solution set on a number line.

13. $|2x - 1| < 7$

14. $|2x - 3| \geq 5$

15. The inequality $|b - 98.6| > 8$ describes a person's body temperature, b, in degrees Fahrenheit, when hyperthermia (extremely high body temperature) or hypothermia (extremely low body temperature) occurs. Solve the inequality and interpret the solution.

In Exercises 16–18, graph each inequality in a rectangular coordinate system.

16. $3x - 2y < 6$

17. $y \geq \dfrac{1}{2}x - 1$

18. $y \leq -1$

In Exercises 19–21, graph the solution set of each system of inequalities.

19. $\begin{cases} x + y \geq 2 \\ x - y \geq 4 \end{cases}$ **20.** $\begin{cases} 3x + y \leq 9 \\ 2x + 3y \geq 6 \\ x \geq 0, y \geq 0 \end{cases}$ **21.** $-2 < x \leq 4$

22. Find the maximum value of the objective function $z = 3x + 5y$ subject to the following constraints: $x \geq 0, y \geq 0, x + y \leq 6, x \geq 2$.

23. A manufacturer makes two types of jet skis, regular and deluxe. The profit on a regular jet ski is $200 and the profit on the deluxe model is $250. To meet customer demand, the company must manufacture at least 50 regular jet skis per week and at least 75 deluxe models. To maintain high quality, the total number of both models of jet skis manufactured by the company should not exceed 150 per week. How many jet skis of each type should be manufactured per week to obtain maximum profit? What is the maximum weekly profit?

CUMULATIVE REVIEW EXERCISES   (CHAPTERS 1–4)

In Exercises 1–2, solve each equation.

1. $5(x + 1) + 2 = x - 3(2x + 1)$

2. $\dfrac{2(x + 6)}{3} = 1 + \dfrac{4x - 7}{3}$

3. Simplify: $\dfrac{-10x^2 y^4}{15x^7 y^{-3}}$.

4. If $f(x) = x^2 - 3x + 4$, find $f(-3)$ and $f(2a)$.

5. If $f(x) = 3x^2 - 4x + 1$ and $g(x) = x^2 - 5x - 1$, find $(f - g)(x)$ and $(f - g)(2)$.

6. Use function notation to write the equation of the line passing through $(2, 3)$ and perpendicular to the line whose equation is $y = 2x - 3$.

In Exercises 7–10, graph each equation or inequality in a rectangular coordinate system.

7. $f(x) = 2x + 1$

8. $y > 2x$

9. $2x - y \geq 6$

10. $f(x) = -1$

11. Solve the system:

$$\begin{cases} 3x - y + z = -15 \\ x + 2y - z = 1. \\ 2x + 3y - 2z = 0 \end{cases}$$

12. Solve using matrices:

$$\begin{cases} 2x - y = -4 \\ x + 3y = 5. \end{cases}$$

13. Evaluate: $\begin{vmatrix} 4 & 3 \\ -1 & -5 \end{vmatrix}$.

14. A motel with 60 rooms charges $90 per night for rooms with kitchen facilities and $80 per night for rooms without kitchen facilities. When all rooms are occupied, the nightly revenue is $5260. How many rooms of each kind are there?

15. Which of the following are functions?

In Exercises 16–20, solve and graph the solution set on a number line.

16. $\dfrac{x}{4} - \dfrac{3}{4} - 1 \leq \dfrac{x}{2}$

17. $2x + 5 \leq 11$ and $-3x > 18$

18. $x - 4 \geq 1$ or $-3x + 1 \geq -5 - x$

19. $|2x + 3| \leq 17$

20. $|3x - 8| > 7$

Solution Consider each term of the polynomial $7x^2y^3 - 17x^4y^2 + xy - 6y^2 + 9$.

Term	Coefficient	Degree (Sum of Exponents on the Variables)
$7x^2y^3$	7	$2 + 3 = 5$
$-17x^4y^2$	-17	$4 + 2 = 6$
xy	1	$1 + 1 = 2$
$-6y^2$	-6	$0 + 2 = 2$
9	9	$0 + 0 = 0$

Think of xy as $1x^1y^1$.
Think of $-6y^2$ as $-6x^0y^2$.
Think of 9 as $9x^0y^0$.

The degree of the polynomial is the greatest degree of any term of the polynomial, which is 6. The leading term is the term of the greatest degree, which is $-17x^4y^2$. Its coefficient, -17, is the leading coefficient. ■

✓ **CHECK POINT 1** Determine the coefficient of each term, the degree of each term, the degree of the polynomial, the leading term, and the leading coefficient of the polynomial

$$8x^4y^5 - 7x^3y^2 - x^2y - 5x + 11.$$

If a polynomial contains three or more variables, the degree of a term is the sum of the exponents of all the variables. Here is an example of a polynomial in three variables, x, y, and z:

The coefficients are $\frac{1}{4}$, -2, 6, and 5.

$$\frac{1}{4}xy^2z^4 - 2xyz + 6x^2 + 5.$$

Degree: $1 + 2 + 4 = 7$ Degree: $1 + 1 + 1 = 3$ Degree: 2   Degree: 0

The degree of this polynomial is the greatest degree of any term of the polynomial, which is 7.

Polynomial Functions

2 Evaluate polynomial functions.

The expression $4x^3 - 5x^2 + 3$ is a polynomial. If we write

$$f(x) = 4x^3 - 5x^2 + 3,$$

then we have a **polynomial function**. In a polynomial function, the expression that defines the function is a polynomial. How do we evaluate a polynomial function? Use substitution, just as we did to evaluate other functions in Chapter 2.

EXAMPLE 2 Evaluating a Polynomial Function

The polynomial function

$$f(x) = -49x^3 + 806x^2 + 3776x + 2503$$

models the number of AIDS cases diagnosed in the United States, $f(x)$, x years after 1983, where $0 \le x \le 8$. Find $f(6)$ and describe what this means in practical terms.

Wha
to le

After s
you sh

1 U
p

2 E
fu

3 D
b

4 A

5 S
p

Number of Cases Diagnosed

80,000
70,000
60,000
50,000
40,000
30,000
20,000
10,000

Figure

Source: U

$f(x) =$

Cases Diagnosed

[(

Figure
modelin
from 19

Solution   To find $f(6)$, or f of 6, we replace each occurrence of x in the function's formula with 6.

$$f(x) = -49x^3 + 806x^2 + 3776x + 2503 \qquad \text{This is the given function.}$$

$$f(6) = -49(6)^3 + 806(6)^2 + 3776(6) + 2503 \qquad \text{Replace each occurrence of } x \text{ with 6.}$$

$$= -49(216) + 806(36) + 3776(6) + 2503 \qquad \text{Evaluate exponential expressions.}$$

$$= -10{,}584 + 29{,}016 + 22{,}656 + 2503 \qquad \text{Multiply.}$$

$$= 43{,}591 \qquad \text{Add.}$$

Thus, $f(6) = 43{,}591$. According to the model, this means that 6 years after 1983, in 1989, there were 43,591 AIDS cases diagnosed in the United States. (The actual number, shown in **Figure 5.1**, is 43,499.) ∎

Using Technology

Once each occurrence of x in $f(x) = -49x^3 + 806x^2 + 3776x + 2503$ is replaced with 6, the resulting computation can be performed using a scientific calculator or a graphing calculator.

$$-49(6)^3 + 806(6)^2 + 3776(6) + 2503$$

Many Scientific Calculators

49 $\boxed{+/-}$ $\boxed{\times}$ 6 $\boxed{y^x}$ 3 $\boxed{+}$ 806 $\boxed{\times}$ 6 $\boxed{y^x}$ 2 $\boxed{+}$ 3776 $\boxed{\times}$ 6 $\boxed{+}$ 2503 $\boxed{=}$

Many Graphing Calculators

$\boxed{(-)}$ 49 $\boxed{\times}$ 6 $\boxed{\wedge}$ 3 $\boxed{+}$ 806 $\boxed{\times}$ 6 $\boxed{\wedge}$ 2 $\boxed{+}$ 3776 $\boxed{\times}$ 6 $\boxed{+}$ 2503 $\boxed{\text{ENTER}}$

The display should be 43591. This number can also be obtained by using a graphing utility's feature that evaluates a function or by using its table feature.

✓ CHECK POINT 2   For the polynomial function

$$f(x) = 4x^3 - 3x^2 - 5x + 6,$$

find $f(2)$.

Smooth, Continuous Graphs

Polynomial functions of degree 2 or higher have graphs that are *smooth* and *continuous*. By **smooth**, we mean that the graph contains only rounded curves with no sharp corners. By **continuous**, we mean that the graph has no breaks and can be drawn without lifting your pencil from the rectangular coordinate system. These ideas are illustrated in **Figure 5.3**.

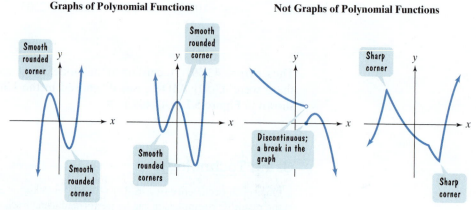

Figure 5.3 Recognizing graphs of polynomial functions

3 Determine end behavior. ▶

End Behavior of Polynomial Functions

Figure 5.4 shows the graph of the function

$$f(x) = -49x^3 + 806x^2 + 3776x + 2503,$$

which models U.S. AIDS cases from 1983 through 1991. Look at what happens to the graph when we extend the year up through 2005. By year 21 (2004), the values of *y* are negative and the function no longer models AIDS cases. We've added an arrow to the graph at the far right to emphasize that it continues to decrease without bound. This far-right *end behavior* of the graph is one reason that this function is inappropriate for modeling AIDS cases into the future.

Years after 1983
[0, 22, 1] by [−10,000, 85,000, 5000]

Figure 5.4 By extending the viewing rectangle, we see that *y* is eventually negative and the function no longer models the number of AIDS cases.

The behavior of the graph of a function to the far left or the far right is called its **end behavior**. Although the graph of a polynomial function may have intervals where it increases and intervals where it decreases, the graph will eventually rise or fall without bound as it moves far to the left or far to the right.

How can you determine whether the graph of a polynomial function goes up or down at each end? **The end behavior depends upon the leading term.** In particular, the sign of the leading coefficient and the degree of the polynomial reveal the graph's end behavior. With regard to end behavior, only the leading term—that is, the term of the greatest degree—counts, as summarized by the **Leading Coefficient Test**.

The Leading Coefficient Test

As *x* increases or decreases without bound, the graph of a polynomial function eventually rises or falls. In particular,

1. For odd-degree polynomials:

If the leading coefficient is positive, the graph falls to the left and rises to the right. ($\swarrow$, $\nearrow$)

If the leading coefficient is negative, the graph rises to the left and falls to the right. ($\nwarrow$, $\searrow$)

2. For even-degree polynomials:

If the leading coefficient is positive, the graph rises to the left and rises to the right. ($\nwarrow$, $\nearrow$)

If the leading coefficient is negative, the graph falls to the left and falls to the right. ($\swarrow$, $\searrow$)

Great Question!

What's the bottom line on the Leading Coefficient Test?

Odd-degree polynomial functions have graphs with opposite behavior at each end. Even-degree polynomial functions have graphs with the same behavior at each end.

Figure 5.5 The graph of $f(x) = x^3 + 3x^2 - x - 3$

EXAMPLE 3   Using the Leading Coefficient Test

Use the Leading Coefficient Test to determine the end behavior of the graph of

$$f(x) = x^3 + 3x^2 - x - 3.$$

Solution   We begin by identifying the sign of the leading coefficient and the degree of the polynomial.

$$f(x) = x^3 + 3x^2 - x - 3$$

The leading coefficient, 1, is positive.

The degree of the polynomial, 3, is odd.

The degree of the function *f* is 3, which is odd. Odd-degree polynomial functions have graphs with opposite behavior at each end. The leading coefficient, 1, is positive. Thus, the graph falls to the left and rises to the right ($\swarrow$, $\nearrow$). The graph of *f* is shown in **Figure 5.5**. ∎

☑ **CHECK POINT 3**   Use the Leading Coefficient Test to determine the end behavior of the graph of $f(x) = x^4 - 4x^2$.

EXAMPLE 4   Using the Leading Coefficient Test

Use end behavior to explain why

$$f(x) = -49x^3 + 806x^2 + 3776x + 2503$$

is only an appropriate model for AIDS cases for a limited time period.

Solution   We begin by identifying the sign of the leading coefficient and the degree of the polynomial.

$$f(x) = -49x^3 + 806x^2 + 3776x + 2503$$

The leading coefficient, −49, is negative.

The degree of the polynomial, 3, is odd.

The degree of f in $f(x) = -49x^3 + 806x^2 + 3776x + 2053$ is 3, which is odd. Odd-degree polynomial functions have graphs with opposite behavior at each end. The leading coefficient, -49, is negative. Thus, the graph rises to the left and falls to the right ($\nwarrow$, $\searrow$). The fact that the graph falls to the right indicates that at some point the number of AIDS cases will be negative, an impossibility. If a function has a graph that decreases without bound over time, it will not be capable of modeling nonnegative phenomena over long time periods. Model breakdown will eventually occur. ■

✓ **CHECK POINT 4**   The polynomial function

$$f(x) = -0.27x^3 + 9.2x^2 - 102.9x + 400$$

models the ratio of students to computers in U.S. public schools x years after 1980. Use end behavior to determine whether this function could be an appropriate model for computers in the classroom well into the twenty-first century. Explain your answer.

If you use a graphing utility to graph a polynomial function, it is important to select a viewing rectangle that accurately reveals the graph's end behavior. If the viewing rectangle, or window, is too small, it may not accurately show a complete graph with the appropriate end behavior.

EXAMPLE 5   Using the Leading Coefficient Test

The graph of $f(x) = -x^4 + 8x^3 + 4x^2 + 2$ was obtained with a graphing utility using a $[-8, 8, 1]$ by $[-10, 10, 1]$ viewing rectangle. The graph is shown in **Figure 5.6**. Is this a complete graph that shows the end behavior of the function?

Solution   We begin by identifying the sign of the leading coefficient and the degree of the polynomial.

$$f(x) = -x^4 + 8x^3 + 4x^2 + 2$$

| The leading coefficient, -1, is negative. | The degree of the polynomial, 4, is even. |

[−8, 8, 1] by [−10, 10, 1]

Figure 5.6

[−10, 10, 1] by [−1000, 750, 250]

Figure 5.7

The degree of f is 4, which is even. Even-degree polynomial functions have graphs with the same behavior at each end. The leading coefficient, -1, is negative. Thus, the graph should fall to the left and fall to the right ($\swarrow$, $\searrow$). The graph in **Figure 5.6** is falling to the left, but it is not falling to the right. Therefore, the graph is not complete enough to show end behavior. A more complete graph of the function is shown in a larger viewing rectangle in **Figure 5.7**. ■

✓ **CHECK POINT 5**   The graph of $f(x) = x^3 + 13x^2 + 10x - 4$ is shown in a standard viewing rectangle in **Figure 5.8**. Use the Leading Coefficient Test to determine whether this is a complete graph that shows the end behavior of the function. Explain your answer.

Figure 5.8

4 Add polynomials. ▶

Adding Polynomials

Polynomials are added by combining like terms. Here are two examples that illustrate the use of the distributive property in adding monomials and combining like terms:

$$-9x^3 + 13x^3 = (-9 + 13)x^3 = 4x^3$$

Add coefficients and keep the same variable factor(s).

$$-7x^3y^2 + 4x^3y^2 = (-7 + 4)x^3y^2 = -3x^3y^2.$$

EXAMPLE 6   Adding Polynomials

Add: $(-6x^3 + 5x^2 + 4) + (2x^3 + 7x^2 - 10)$.

Solution

$$(-6x^3 + 5x^2 + 4) + (2x^3 + 7x^2 - 10)$$

$= -6x^3 + 5x^2 + 4 + 2x^3 + 7x^2 - 10$ Remove the parentheses. Like terms are shown in the same color.

$= \underbrace{-6x^3 + 2x^3} + \underbrace{5x^2 + 7x^2} + \underbrace{4 - 10}$ Rearrange the terms so that like terms are adjacent.

$= \quad -4x^3 \qquad + 12x^2 \qquad -6$ Combine like terms.

$= -4x^3 + 12x^2 - 6$ This is the same sum as above, written more concisely. ∎

✓ **CHECK POINT 6**   Add: $(-7x^3 + 4x^2 + 3) + (4x^3 + 6x^2 - 13)$.

EXAMPLE 7   Adding Polynomials

Add: $(5x^3y - 4x^2y - 7y) + (2x^3y + 6x^2y - 4y - 5)$.

Solution

$(5x^3y - 4x^2y - 7y) + (2x^3y + 6x^2y - 4y - 5)$ The given problem involves adding polynomials in two variables.

$= 5x^3y - 4x^2y - 7y + 2x^3y + 6x^2y - 4y - 5$ Remove the parentheses. Like terms are shown in the same color.

$= \underbrace{5x^3y + 2x^3y} - \underbrace{4x^2y + 6x^2y} - \underbrace{7y - 4y} - \underbrace{5}$ Rearrange the terms so that like terms are adjacent.

$= \quad 7x^3y \qquad + 2x^2y \qquad - 11y \quad - 5$ Combine like terms. ∎

Polynomials can be added by arranging like terms in columns. Then combine like terms, column by column. Here's the solution to Example 7 using columns and a vertical format:

$$\begin{array}{r} 5x^3y - 4x^2y - \ 7y \\ 2x^3y + 6x^2y - \ 4y - 5 \\ \hline 7x^3y + 2x^2y - 11y - 5 \end{array}$$

✓ **CHECK POINT 7**   Add: $(7xy^3 - 5xy^2 - 3y) + (2xy^3 + 8xy^2 - 12y - 9)$.

5 Subtract polynomials. ▶

Subtracting Polynomials

We subtract real numbers by adding the opposite, or additive inverse, of the number being subtracted. For example,

$$8 - 3 = 8 + (-3) = 5.$$

We can apply this idea to polynomial subtraction. For example,

> Change the subtraction to addition.

$$(6x^2 + 5x - 3) - (4x^2 - 2x - 7) = (6x^2 + 5x - 3) + (-4x^2 + 2x + 7)$$

> Replace $4x^2 - 2x - 7$ with its opposite, or additive inverse.

$$= 6x^2 + 5x - 3 - 4x^2 + 2x + 7 \qquad \text{Remove the parentheses.}$$
$$= 6x^2 - 4x^2 + 5x + 2x - 3 + 7 \qquad \text{Rearrange the terms.}$$
$$= 2x^2 + 7x + 4 \qquad \text{Combine like terms.}$$

Great Question!

How do I find the opposite, or additive inverse, of the polynomial being subtracted?

To write the additive inverse, change the sign of each term of the polynomial.

> **Subtracting Polynomials**
>
> To subtract one polynomial from another, add the opposite, or additive inverse, of the polynomial being subtracted.

Great Question!

Can I use a vertical format to subtract polynomials, like we did with addition?

Yes. Here's the solution to Example 8 with a vertical format. Notice that you still distribute the negative sign, thereby adding the opposite.

$$
\begin{array}{r}
7x^3 - 8x^2 + 9x - 6 \\
-(2x^3 - 6x^2 - 3x + 9) \\
\hline
\end{array}
$$

$$
\begin{array}{r}
7x^3 - 8x^2 + 9x - 6 \\
+ \; -2x^3 + 6x^2 + 3x - 9 \\
\hline
5x^3 - 2x^2 + 12x - 15
\end{array}
$$

EXAMPLE 8 Subtracting Polynomials

Subtract: $(7x^3 - 8x^2 + 9x - 6) - (2x^3 - 6x^2 - 3x + 9)$.

Solution

$(7x^3 - 8x^2 + 9x - 6) - (2x^3 - 6x^2 - 3x + 9)$

$= (7x^3 - 8x^2 + 9x - 6) + (-2x^3 + 6x^2 + 3x - 9)$ Add the opposite of the polynomial being subtracted.

$= 7x^3 - 8x^2 + 9x - 6 - 2x^3 + 6x^2 + 3x - 9$ Remove the parentheses. Like terms are shown in the same color.

$= \underbrace{7x^3 - 2x^3}_{5x^3} \underbrace{- 8x^2 + 6x^2}_{-2x^2} \underbrace{+ 9x + 3x}_{+12x} \underbrace{- 6 - 9}_{-15}$ Rearrange terms.

$=$ Combine like terms. ∎

✓ **CHECK POINT 8**   Subtract: $(14x^3 - 5x^2 + x - 9) - (4x^3 - 3x^2 - 7x + 1)$.

EXAMPLE 9 Subtracting Polynomials

Subtract $-2x^5y^2 - 3x^3y + 7$ from $3x^5y^2 - 4x^3y - 3$.

Solution

$(3x^5y^2 - 4x^3y - 3) - (-2x^5y^2 - 3x^3y + 7)$

$= (3x^5y^2 - 4x^3y - 3) + (2x^5y^2 + 3x^3y - 7)$ Add the opposite of the polynomial being subtracted.

$= 3x^5y^2 - 4x^3y - 3 + 2x^5y^2 + 3x^3y - 7$ Remove the parentheses. Like terms are shown in the same color.

$= \underbrace{3x^5y^2 + 2x^5y^2}_{5x^5y^2} \underbrace{- 4x^3y + 3x^3y}_{-x^3y} \underbrace{- 3 - 7}_{-10}$ Rearrange terms.

$=$ Combine like terms. ∎

Great Question!

When I subtract one polynomial from another, as in Example 9, how can I tell which polynomial is being subtracted?

Be careful of the order in Example 9. For example, subtracting 2 from 5 is equivalent to $5 - 2$. In general, subtracting B from A means $A - B$. The order of the resulting algebraic expression is not the same as the order in English.

✓ **CHECK POINT 9**   Subtract $-7x^2y^5 - 4xy^3 + 2$ from $6x^2y^5 - 2xy^3 - 8$.

Achieving Success

Form a study group with other students in your class. Working in small groups often serves as an excellent way to learn and reinforce new material. Set up helpful procedures and guidelines for the group. "Talk" math by discussing and explaining the concepts and exercises to one another.

CONCEPT AND VOCABULARY CHECK

Fill in each blank so that the resulting statement is true.

1. A polynomial is a single term or the sum of two or more terms containing variables with exponents that are _____ numbers.

2. It is customary to write the terms of a polynomial in the order of descending powers of the variable. This is called the _____ form of a polynomial.

3. A simplified polynomial that has exactly one term is called a/an _____.

4. A simplified polynomial that has two terms is called a/an _____.

5. A simplified polynomial that has three terms is called a/an _____.

6. If $a \neq 0$, the degree of ax^n is _____.

7. If $a \neq 0$, the degree of $ax^n y^m$ is _____.

8. The degree of a polynomial is the _____ degree of all the terms of the polynomial. If there is precisely one term of the greatest degree, it is called the _____ term. Its coefficient is called the _____ coefficient.

9. True or false: Polynomial functions of degree 2 or higher have graphs that contain only rounded curves with no sharp corners. _____

10. True or false: Polynomial functions of degree 2 or higher have breaks in their graphs. _____

11. The behavior of the graph of a polynomial function to the far left or the far right is called its _____ behavior, which depends upon the _____ term.

12. The graph of $f(x) = x^3$ _____ to the left and _____ to the right.

13. The graph of $f(x) = -x^3$ _____ to the left and _____ to the right.

14. The graph of $f(x) = x^2$ _____ to the left and _____ to the right.

15. The graph of $f(x) = -x^2$ _____ to the left and _____ to the right.

16. True or false: Odd-degree polynomial functions have graphs with opposite behavior at each end. _____

17. True or false: Even-degree polynomial functions have graphs with the same behavior at each end. _____

18. Terms of a polynomial that contain the same variables raised to the same powers are called _____ terms.

Add or subtract, if possible.

19. $-7x^3 + 4x^3$

20. $-4x^3 y + x^3 y$

21. $x^5 + x^5$

22. $7x^5 y^2 - (-3x^5 y^2)$

23. $12xy^2 - 12y^2$

5.1 EXERCISE SET   MyMathLab®

Practice Exercises

In Exercises 1–10, determine the coefficient of each term, the degree of each term, the degree of the polynomial, the leading term, and the leading coefficient of the polynomial.

1. $-x^4 + x^2$

2. $x^3 - 4x^2$

3. $5x^3 + 7x^2 - x + 9$

4. $11x^3 - 6x^2 + x + 3$

5. $3x^2 - 7x^4 - x + 6$

6. $2x^2 - 9x^4 - x + 5$

7. $x^3y^2 - 5x^2y^7 + 6y^2 - 3$

8. $12x^4y - 5x^3y^7 - x^2 + 4$

9. $x^5 + 3x^2y^4 + 7xy + 9x - 2$

10. $3x^6 + 4x^4y^4 - x^3y + 4x^2 - 5$

In Exercises 11–20, let

$$f(x) = x^2 - 5x + 6 \quad \text{and} \quad g(x) = 2x^3 - x^2 + 4x - 1.$$

Find the indicated function values.

11. $f(3)$ **12.** $f(4)$ **13.** $f(-1)$

14. $f(-2)$ **15.** $g(3)$ **16.** $g(2)$

17. $g(-2)$ **18.** $g(-3)$

19. $g(0)$ **20.** $f(0)$

In Exercises 21–24, identify which graphs are not those of polynomial functions.

21.

22.

23.

24.

In Exercises 25–28, use the Leading Coefficient Test to determine the end behavior of the graph of the given polynomial function. Then use this end behavior to match the polynomial function with its graph. [The graphs are labeled (a) through (d).]

25. $f(x) = -x^4 + x^2$

26. $f(x) = x^3 - 4x^2$

27. $f(x) = x^2 - 6x + 9$

28. $f(x) = -x^3 - x^2 + 5x - 3$

a.

b.

c.

d.

In Exercises 29–40, add the polynomials. Assume that all variable exponents represent whole numbers.

29. $(-6x^3 + 5x^2 - 8x + 9) + (17x^3 + 2x^2 - 4x - 13)$

30. $(-7x^3 + 6x^2 - 11x + 13) + (19x^3 - 11x^2 + 7x - 17)$

31. $\left(\frac{2}{5}x^4 + \frac{2}{3}x^3 + \frac{5}{8}x^2 + 7\right) + \left(-\frac{4}{5}x^4 + \frac{1}{3}x^3 - \frac{1}{4}x^2 - 7\right)$

32. $\left(\frac{1}{5}x^4 + \frac{1}{3}x^3 + \frac{3}{8}x^2 + 6\right) + \left(-\frac{3}{5}x^4 + \frac{2}{3}x^3 - \frac{1}{2}x^2 - 6\right)$

33. $(7x^2y - 5xy) + (2x^2y - xy)$

34. $(-4x^2y + xy) + (7x^2y + 8xy)$

35. $(5x^2y + 9xy + 12) + (-3x^2y + 6xy + 3)$

36. $(8x^2y + 12xy + 14) + (-2x^2y + 7xy + 4)$

37. $(9x^4y^2 - 6x^2y^2 + 3xy) + (-18x^4y^2 - 5x^2y - xy)$

38. $(10x^4y^2 - 3x^2y^2 + 2xy) + (-16x^4y^2 - 4x^2y - xy)$

39. $(x^{2n} + 5x^n - 8) + (4x^{2n} - 7x^n + 2)$

40. $(6y^{2n} + y^n + 5) + (3y^{2n} - 4y^n - 15)$

In Exercises 41–50, subtract the polynomials. Assume that all variable exponents represent whole numbers.

41. $(17x^3 - 5x^2 + 4x - 3) - (5x^3 - 9x^2 - 8x + 11)$

42. $(18x^3 - 2x^2 - 7x + 8) - (9x^3 - 6x^2 - 5x + 7)$

43. $(13y^5 + 9y^4 - 5y^2 + 3y + 6) - (-9y^5 - 7y^3 + 8y^2 + 11)$

44. $(12y^5 + 7y^4 - 3y^2 + 6y + 7) - (-10y^5 - 8y^3 + 3y^2 + 14)$

45. $(x^3 + 7xy - 5y^2) - (6x^3 - xy + 4y^2)$

46. $(x^4 - 7xy - 5y^3) - (6x^4 - 3xy + 4y^3)$

47. $(3x^4y^2 + 5x^3y - 3y) - (2x^4y^2 - 3x^3y - 4y + 6x)$

48. $(5x^4y^2 + 6x^3y - 7y) - (3x^4y^2 - 5x^3y - 6y + 8x)$

49. $(7y^{2n} + y^n - 4) - (6y^{2n} - y^n - 1)$

50. $(8x^{2n} + x^n - 4) - (9x^{2n} - x^n - 2)$

51. Subtract $-5a^2b^4 - 8ab^2 - ab$ from $3a^2b^4 - 5ab^2 + 7ab$.

52. Subtract $-7a^2b^4 - 8ab^2 - ab$ from $13a^2b^4 - 17ab^2 + ab$.

53. Subtract $-4x^3 - x^2y + xy^2 + 3y^3$ from $x^3 + 2x^2y - y^3$.

54. Subtract $-6x^3 + x^2y - xy^2 + 2y^3$ from $x^3 + 2xy^2 - y^3$.

Practice PLUS

55. Add $6x^4 - 5x^3 + 2x$ to the difference between $4x^3 + 3x^2 - 1$ and $x^4 - 2x^2 + 7x - 3$.

56. Add $5x^4 - 2x^3 + 7x$ to the difference between $2x^3 + 5x^2 - 3$ and $-x^4 - x^2 - x - 1$.

57. Subtract $9x^2y^2 - 3x^2 - 5$ from the sum of $-6x^2y^2 - x^2 - 1$ and $5x^2y^2 + 2x^2 - 1$.

58. Subtract $6x^2y^3 - 2x^2 - 7$ from the sum of $-5x^2y^3 + 3x^2 - 4$ and $4x^2y^3 - 2x^2 - 6$.

In Exercises 59–64, let

$$f(x) = -3x^3 - 2x^2 - x + 4$$
$$g(x) = x^3 - x^2 - 5x - 4$$
$$h(x) = -2x^3 + 5x^2 - 4x + 1.$$

Find the indicated function, function value, or polynomial.

59. $(f - g)(x)$ and $(f - g)(-1)$

60. $(g - h)(x)$ and $(g - h)(-1)$

61. $(f + g - h)(x)$ and $(f + g - h)(-2)$

62. $(g + h - f)(x)$ and $(g + h - f)(-2)$

63. $2f(x) - 3g(x)$

64. $-2g(x) - 3h(x)$

Application Exercises

As we noted in the chapter opener, experts fear that without conservation efforts, tigers could disappear from the wild by 2022. Just one hundred years ago, there were at least 100,000 wild tigers. By 2010, the estimated world tiger population was 3200. The bar graph shows the estimated world tiger population for selected years from 1970 through 2010. Also shown is a polynomial function, with its graph, that models the data. Use this information to solve Exercises 65–68.

Vanishing Tigers

Estimated World Tiger Population

Graph of a Polynomial Model for the Data

$f(x) = 0.76x^3 - 30x^2 - 882x + 37{,}807$

Source: World Wildlife Fund

(In Exercises 65–68, be sure to refer to the information on vanishing tigers at the bottom of the previous page.)

65. a. Find and interpret $f(40)$. Identify this information as a point on the graph of f.

   b. Does $f(40)$ overestimate or underestimate the actual data shown by the bar graph? By how much?

66. a. Find and interpret $f(10)$. Identify this information as a point on the graph of f.

   b. Does $f(10)$ overestimate or underestimate the actual data shown by the bar graph? By how much?

67. Use the Leading Coefficient Test to determine the end behavior to the right for the graph of f. Will this function be useful in modeling the world tiger population if conservation efforts to save wild tigers fail? Explain your answer.

68. Use the Leading Coefficient Test to determine the end behavior to the right for the graph of f. Might this function be useful in modeling the world tiger population if conservation efforts to save wild tigers are successful? Explain your answer.

69. The common cold is caused by a rhinovirus. After x days of invasion by the viral particles, the number of particles in our bodies, $f(x)$, in billions, can be modeled by the polynomial function

$$f(x) = -0.75x^4 + 3x^3 + 5.$$

Use the Leading Coefficient Test to determine the graph's end behavior to the right. What does this mean about the number of viral particles in our bodies over time?

70. The polynomial function

$$f(x) = -0.87x^3 + 0.35x^2 + 81.62x + 7684.94$$

models the number of thefts, $f(x)$, in thousands, in the United States x years after 1987. Will this function be useful in modeling the number of thefts over an extended period of time? Explain your answer.

71. A herd of 100 elk is introduced to a small island. The number of elk, $f(x)$, after x years is modeled by the polynomial function

$$f(x) = -x^4 + 21x^2 + 100.$$

Use the Leading Coefficient Test to determine the graph's end behavior to the right. What does this mean about what will eventually happen to the elk population?

Explaining the Concepts

72. What is a polynomial?

73. Explain how to determine the degree of each term of a polynomial.

74. Explain how to determine the degree of a polynomial.

75. Explain how to determine the leading coefficient of a polynomial.

76. What is a polynomial function?

77. What do we mean when we describe the graph of a polynomial function as smooth and continuous?

78. What is meant by the end behavior of a polynomial function?

79. Explain how to use the Leading Coefficient Test to determine the end behavior of a polynomial function.

80. Why is a polynomial function of degree 3 with a negative leading coefficient not appropriate for modeling nonnegative real-world phenomena over a long period of time?

81. Explain how to add polynomials.

82. Explain how to subtract polynomials.

83. In a favorable habitat and without natural predators, a population of reindeer is introduced to an island preserve. The reindeer population t years after their introduction is modeled by the polynomial function

$$f(t) = -0.125t^5 + 3.125t^4 + 4000.$$

Discuss the growth and decline of the reindeer population. Describe the factors that might contribute to this population model.

Technology Exercises

Write a polynomial function that imitates the end behavior of each graph in Exercises 84–87. The dashed portions of the graphs indicate that you should focus only on imitating the left and right end behavior of the graph. You can be flexible about what occurs between the left and right ends. Then use your graphing utility to graph the polynomial function and verify that you imitated the end behavior shown in the given graph.

84.

85.

86. **87.**

In Exercises 88–91, use a graphing utility with a viewing rectangle large enough to show end behavior to graph each polynomial function.

88. $f(x) = x^3 + 13x^2 + 10x - 4$

89. $f(x) = -2x^3 + 6x^2 + 3x - 1$

90. $f(x) = -x^4 + 8x^3 + 4x^2 + 2$

91. $f(x) = -x^5 + 5x^4 - 6x^3 + 2x + 20$

In Exercises 92–93, use a graphing utility to graph f and g in the same viewing rectangle. Then use the ZOOM OUT *feature to show that f and g have identical end behavior.*

92. $f(x) = x^3 - 6x + 1, \quad g(x) = x^3$

93. $f(x) = -x^4 + 2x^3 - 6x, \quad g(x) = -x^4$

Critical Thinking Exercises

Make Sense? *In Exercises 94–97, determine whether each statement makes sense or does not make sense, and explain your reasoning.*

94. Many English words have prefixes with meanings similar to those used to describe polynomials, such as *monologue*, *binocular*, and *tricuspid*.

95. I can determine a polynomial's leading coefficient by inspecting the coefficient of the first term.

96. When I'm trying to determine end behavior, it's the coefficient of the first term of a polynomial function written in standard form that I should inspect.

97. When I rearrange the terms of a polynomial, it's important that I move the sign in front of a term with that term.

In Exercises 98–101, determine whether each statement is true or false. If the statement is false, make the necessary change(s) to produce a true statement.

98. $4x^3 + 7x^2 - 5x + \dfrac{2}{x}$ is a polynomial containing four terms.

99. If two polynomials of degree 2 are added, the sum must be a polynomial of degree 2.

100. $(x^2 - 7x) - (x^2 - 4x) = -11x$ for all values of x.

101. All terms of a polynomial are monomials.

In Exercises 102–103, perform the indicated operations. Assume that exponents represent whole numbers.

102. $(x^{2n} - 3x^n + 5) + (4x^{2n} - 3x^n - 4) - (2x^{2n} - 5x^n - 3)$

103. $(y^{3n} - 7y^{2n} + 3) - (-3y^{3n} - 2y^{2n} - 1) + (6y^{3n} - y^{2n} + 1)$

104. From what polynomial must $4x^2 + 2x - 3$ be subtracted to obtain $5x^2 - 5x + 8$?

Review Exercises

105. Solve: $9(x - 1) = 1 + 3(x - 2)$.
(Section 1.4, Example 3)

106. Graph: $2x - 3y < -6$. (Section 4.4, Example 1)

107. Write the point-slope form and slope-intercept form of equations of a line passing through the point $(-2, 5)$ and parallel to the line whose equation is $3x - y = 9$. (Section 2.5, Example 4)

Preview Exercises

Exercises 108–110 will help you prepare for the material covered in the next section.

108. Multiply: $(2x^3 y^2)(5x^4 y^7)$.

109. Use the distributive property to multiply: $2x^4(8x^4 + 3x)$.

110. Simplify and express the polynomial in standard form:

$$3x(x^2 + 4x + 5) + 7(x^2 + 4x + 5).$$

SECTION

5.2

Multiplication of Polynomials

Can that be Axl, your author's yellow lab, sharing a special moment with a baby chick? And if it is (it is), what possible relevance can this have to multiplying polynomials? An answer is promised before you reach the Exercise Set. For now, let's begin by reviewing how to multiply monomials, a skill that you will apply in every polynomial multiplication problem.

Old Dog ... New Chicks

What am I supposed to learn?

After studying this section, you should be able to:

1 Multiply monomials. ▶

2 Multiply a monomial and a polynomial. ▶

3 Multiply polynomials when neither is a monomial. ▶

4 Use FOIL in polynomial multiplication. ▶

5 Square binomials. ▶

6 Multiply the sum and difference of two terms. ▶

7 Find the product of functions. ▶

8 Use polynomial multiplication to evaluate functions. ▶

Multiplying Monomials

To multiply monomials, begin by multiplying the coefficients. Then multiply the variables. Use the product rule for exponents to multiply the variables: Retain the variable and add the exponents.

1 Multiply monomials. ▶

| **EXAMPLE 1** | Multiplying Monomials |

Multiply:

a. $(5x^3y^4)(-6x^7y^8)$ **b.** $(4x^3y^2z^5)(2x^5y^2z^4)$.

Solution

a.
$$(5x^3y^4)(-6x^7y^8) = 5(-6)x^3 \cdot x^7 \cdot y^4 \cdot y^8 \qquad \text{Rearrange factors. This step is usually done mentally.}$$
$$= -30x^{3+7}y^{4+8} \qquad \text{Multiply coefficients and add exponents.}$$
$$= -30x^{10}y^{12} \qquad \text{Simplify.}$$

b.
$$(4x^3y^2z^5)(2x^5y^2z^4) = 4 \cdot 2 \cdot x^3 \cdot x^5 \cdot y^2 \cdot y^2 \cdot z^5 \cdot z^4 \qquad \text{Rearrange factors.}$$
$$= 8x^{3+5}y^{2+2}z^{5+4} \qquad \text{Multiply coefficients and add exponents.}$$
$$= 8x^8y^4z^9 \qquad \text{Simplify.} ∎$$

✓ **CHECK POINT 1**   Multiply:

a. $(6x^5y^7)(-3x^2y^4)$ **b.** $(10x^4y^3z^6)(3x^6y^3z^2)$.

2 Multiply a monomial and a polynomial. ▶

Multiplying a Monomial and a Polynomial That Is Not a Monomial

We use the distributive property to multiply a monomial and a polynomial that is not a monomial. For example,

$$3x^2(2x^3 + 5x) = 3x^2 \cdot 2x^3 + 3x^2 \cdot 5x = 3 \cdot 2x^{2+3} + 3 \cdot 5x^{2+1} = 6x^5 + 15x^3.$$

    Monomial   Binomial   Multiply coefficients and add exponents.

To multiply a monomial and a polynomial, multiply each term of the polynomial by the monomial. Once the monomial factor is distributed, we multiply the resulting monomials using the procedure shown in Example 1.

EXAMPLE 2   Multiplying a Monomial and a Trinomial

Multiply:

a. $4x^3(6x^5 - 2x^2 + 3)$ **b.** $5x^3y^4(2x^7y - 6x^4y^3 - 3)$.

Solution

a. $4x^3(6x^5 - 2x^2 + 3) = 4x^3 \cdot 6x^5 - 4x^3 \cdot 2x^2 + 4x^3 \cdot 3$ Use the distributive property.

$= 24x^8 - 8x^5 + 12x^3$ Multiply coefficients and add exponents.

b. $5x^3y^4(2x^7y - 6x^4y^3 - 3)$

$= 5x^3y^4 \cdot 2x^7y - 5x^3y^4 \cdot 6x^4y^3 - 5x^3y^4 \cdot 3$ Use the distributive property.

$= 10x^{10}y^5 - 30x^7y^7 - 15x^3y^4$ Multiply coefficients and add exponents. ∎

✓ CHECK POINT 2   Multiply:

a. $6x^4(2x^5 - 3x^2 + 4)$

b. $2x^4y^3(5xy^6 - 4x^3y^4 - 5)$.

---

3 Multiply polynomials when neither is a monomial. ⏵

Multiplying Polynomials When Neither Is a Monomial

How do we multiply two polynomials if neither is a monomial? For example, consider

$$(3x + 7)(x^2 + 4x + 5).$$

Binomial   Trinomial

One way to perform this multiplication is to distribute $3x$ throughout the trinomial

$$3x(x^2 + 4x + 5)$$

and 7 throughout the trinomial

$$7(x^2 + 4x + 5).$$

Then combine the like terms that result.

> **Multiplying Polynomials When Neither Is a Monomial**
>
> Multiply each term of one polynomial by each term of the other polynomial. Then combine like terms.

EXAMPLE 3   Multiplying a Binomial and a Trinomial

Multiply: $(3x + 7)(x^2 + 4x + 5)$.

Solution

$(3x + 7)(x^2 + 4x + 5)$

$= 3x(x^2 + 4x + 5) + 7(x^2 + 4x + 5)$ Multiply the trinomial by each term of the binomial.

$= 3x \cdot x^2 + 3x \cdot 4x + 3x \cdot 5 + 7x^2 + 7 \cdot 4x + 7 \cdot 5$ Use the distributive property.

$= 3x^3 + 12x^2 + 15x + 7x^2 + 28x + 35$ Multiply monomials: Multiply coefficients and add exponents.

$= 3x^3 + 19x^2 + 43x + 35$ Combine like terms:

$12x^2 + 7x^2 = 19x^2$ and

$15x + 28x = 43x$. ∎

☑ **CHECK POINT 3** Multiply: $(3x + 2)(2x^2 - 2x + 1)$.

Another method for solving Example 3 is to use a vertical format similar to that used for multiplying whole numbers.

$$
\begin{array}{r}
x^2 + 4x + 5 \\
3x + 7 \\
\hline
7x^2 + 28x + 35 \\
3x^3 + 12x^2 + 15x \\
\hline
3x^3 + 19x^2 + 43x + 35
\end{array}
$$

Write like terms in the same column.

$7(x^2 + 4x + 5)$

$3x(x^2 + 4x + 5)$

Combine like terms.

EXAMPLE 4 Multiplying a Binomial and a Trinomial

Multiply: $(2x^2y + 3y)(5x^4y - 4x^2y + y)$.

Solution

$(2x^2y + 3y)(5x^4y - 4x^2y + y)$

$= 2x^2y(5x^4y - 4x^2y + y) + 3y(5x^4y - 4x^2y + y)$ Multiply the trinomial by each term of the binomial.

$= 2x^2y \cdot 5x^4y - 2x^2y \cdot 4x^2y + 2x^2y \cdot y + 3y \cdot 5x^4y - 3y \cdot 4x^2y + 3y \cdot y$

Use the distributive property.

$= 10x^6y^2 - 8x^4y^2 + 2x^2y^2 + 15x^4y^2 - 12x^2y^2 + 3y^2$ Multiply coefficients and add exponents.

$= 10x^6y^2 + 7x^4y^2 - 10x^2y^2 + 3y^2$ Combine like terms:

$-8x^4y^2 + 15x^4y^2 = 7x^4y^2$
$2x^2y^2 - 12x^2y^2 = -10x^2y^2$. ∎

☑ **CHECK POINT 4** Multiply: $(4xy^2 + 2y)(3xy^4 - 2xy^2 + y)$.

④ Use FOIL in polynomial multiplication. ▶

The Product of Two Binomials: FOIL

Frequently we need to find the product of two binomials. One way to perform this multiplication is to distribute each term in the first binomial throughout the second binomial. For example, we can find the product of the binomials $7x + 2$ and $4x + 5$ as follows:

$$(7x + 2)(4x + 5) = 7x(4x + 5) + 2(4x + 5)$$
$$= 7x(4x) + 7x(5) + 2(4x) + 2(5)$$
$$= 28x^2 + 35x + 8x + 10.$$

Distribute $7x$ over $4x + 5$.

Distribute 2 over $4x + 5$.

We'll combine these like terms later. For now, our interest is in how to obtain *each* of these four terms.

We can also find the product of $7x + 2$ and $4x + 5$ using a method called FOIL, which is based on our work shown above. Any two binomials can be quickly multiplied using the FOIL method, in which **F** represents the product of the **first** terms in each binomial, **O** represents the product of the **outside** terms, **I** represents the product of the two **inside** terms, and **L** represents the product of the **last**, or second, terms in each binomial. For example, we can use the FOIL method to find the product of the binomials $7x + 2$ and $4x + 5$ as follows:

$$(7x + 2)(4x + 5) = 28x^2 + 35x + 8x + 10$$

$$= 28x^2 + 43x + 10 \qquad \text{Combine like terms.}$$

In general, here's how to use the FOIL method to find the product of $ax + b$ and $cx + d$:

Using the FOIL Method to Multiply Binomials

$$(ax + b)(cx + d) = ax \cdot cx + ax \cdot d + b \cdot cx + b \cdot d$$

EXAMPLE 5   Using the FOIL Method

Multiply:

a. $(x + 3)(x + 2)$ **b.** $(3x + 5y)(x - 2y)$ **c.** $(5x^3 - 6)(4x^3 - x)$.

Solution

a. $(x + 3)(x + 2) = x \cdot x + x \cdot 2 + 3 \cdot x + 3 \cdot 2$

$\qquad\qquad\qquad = x^2 + 2x + 3x + 6$

$\qquad\qquad\qquad = x^2 + 5x + 6 \qquad$ Combine like terms.

b. $(3x + 5y)(x - 2y) = 3x \cdot x + 3x(-2y) + 5y \cdot x + 5y(-2y)$

$\qquad\qquad\qquad\qquad = 3x^2 - 6xy + 5xy - 10y^2$

$\qquad\qquad\qquad\qquad = 3x^2 - xy - 10y^2 \qquad$ Combine like terms.

c. $(5x^3 - 6)(4x^3 - x) = 5x^3 \cdot 4x^3 + 5x^3(-x) + (-6)(4x^3) + (-6)(-x)$

$\qquad\qquad\qquad\qquad = 20x^6 - 5x^4 - 24x^3 + 6x \qquad$ There are no like terms to combine. ∎

☑ **CHECK POINT 5**   Multiply:

a. $(x + 5)(x + 3)$ b. $(7x + 4y)(2x - y)$ c. $(4x^3 - 5)(x^3 - 3x)$.

5 Square binomials.

The Square of a Binomial

Let us find $(A + B)^2$, the square of a binomial sum. To do so, we begin with the FOIL method and look for a general rule.

$$(A + B)^2 = (A + B)(A + B) = \overset{F}{A \cdot A} + \overset{O}{A \cdot B} + \overset{I}{A \cdot B} + \overset{L}{B \cdot B}$$
$$= A^2 + 2AB + B^2$$

This result implies the following rule, which is often called a **special-product formula**:

The Square of a Binomial Sum

$$(A + B)^2 = A^2 + 2AB + B^2$$

| The square of a binomial sum | is | first term squared | plus | 2 times the product of the terms | plus | last term squared. |

Great Question!

When finding $(x + 5)^2$, why can't I just write $x^2 + 5^2$, or $x^2 + 25$?

Caution! The square of a sum is *not* the sum of the squares.

$$(A + B)^2 \neq A^2 + B^2$$

The middle term $2AB$ is missing.

$$(x + 3)^2 \neq x^2 + 9$$

Incorrect!

Show that $(x + 5)^2$ and $x^2 + 25$ are not equal by substituting 3 for x in each expression and simplifying.

EXAMPLE 6   Finding the Square of a Binomial Sum

Multiply:

a. $(x + 5)^2$ b. $(3x + 2y)^2$.

Solution   Use the special-product formula shown.

$$(A + B)^2 = A^2 + 2AB + B^2$$

	(First Term)²	+	2 · Product of the Terms	+	(Last Term)²	= Product
a. $(x + 5)^2 =$	x^2	+	$2 \cdot x \cdot 5$	+	5^2	$= x^2 + 10x + 25$
b. $(3x + 2y)^2 =$	$(3x)^2$	+	$2 \cdot 3x \cdot 2y$	+	$(2y)^2$	$= 9x^2 + 12xy + 4y^2$

■

☑ **CHECK POINT 6**   Multiply:

a. $(x + 8)^2$ b. $(4x + 5y)^2$.

The formula for the square of a binomial sum can be interpreted geometrically by analyzing the areas in **Figure 5.9**.

Area of the large square $(A + B)^2$

Sum of the areas of the four smaller rectangles inside the large square $A^2 + AB + AB + B^2$
$= A^2 + 2AB + B^2$

Conclusion:
$$(A + B)^2 = A^2 + 2AB + B^2$$

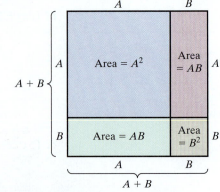

Figure 5.9

A similar pattern occurs for $(A - B)^2$, the square of a binomial difference. Using the FOIL method on $(A - B)^2$, we obtain the following rule:

The Square of a Binomial Difference

$$(A - B)^2 = A^2 - 2AB + B^2$$

| The square of a binomial difference | is | first term squared | minus | 2 times the product of the terms | plus | last term squared. |

EXAMPLE 7 Finding the Square of a Binomial Difference

Multiply:

a. $(x - 8)^2$ **b.** $\left(\frac{1}{2}x - 4y^3\right)^2$.

Solution Use the special-product formula shown.

$$(A - B)^2 = A^2 - 2AB + B^2$$

	(First Term)^2	$-$	2 · Product of the Terms	$+$	(Last Term)^2	= Product
a. $(x - 8)^2 =$	x^2	$-$	$2 \cdot x \cdot 8$	$+$	8^2	$= x^2 - 16x + 64$
b. $\left(\frac{1}{2}x - 4y^3\right)^2 =$	$\left(\frac{1}{2}x\right)^2$	$-$	$2 \cdot \frac{1}{2}x \cdot 4y^3$	$+$	$(4y^3)^2$	$= \frac{1}{4}x^2 - 4xy^3 + 16y^6$

∎

☑ **CHECK POINT 7** Multiply:
a. $(x - 5)^2$ **b.** $(2x - 6y^4)^2$.

---

6 Multiply the sum and difference of two terms. ▶

Multiplying the Sum and Difference of Two Terms

We can use the FOIL method to multiply $A + B$ and $A - B$ as follows:

F O I L

$$(A + B)(A - B) = A^2 - AB + AB - B^2 = A^2 - B^2.$$

Notice that the outside and inside products have a sum of 0 and the terms cancel. The FOIL multiplication provides us with a quick rule for multiplying the sum and difference of two terms, which is another example of a special-product formula.

The Product of the Sum and Difference of Two Terms

$$(A + B)(A - B) = A^2 - B^2$$

| The product of the sum and the difference of the same two terms | is | the square of the first term minus the square of the second term. |

EXAMPLE 8 Finding the Product of the Sum and Difference of Two Terms

Multiply:

a. $(x + 8)(x - 8)$ **b.** $(9x + 5y)(9x - 5y)$ **c.** $(6a^2b - 3b)(6a^2b + 3b)$.

Solution Use the special-product formula shown.

$$(A + B)(A - B) \qquad = \qquad A^2 \quad - \quad B^2$$

First term squared $-$ Second term squared $=$ Product

a. $(x + 8)(x - 8) \qquad = \qquad x^2 \quad - \quad 8^2 \quad = \quad x^2 - 64$

b. $(9x + 5y)(9x - 5y) \qquad = \quad (9x)^2 \quad - \quad (5y)^2 = 81x^2 - 25y^2$

c. $(6a^2b - 3b)(6a^2b + 3b) = (6a^2b)^2 \quad - \quad (3b)^2 = 36a^4b^2 - 9b^2$ ∎

✓ **CHECK POINT 8** Multiply:

a. $(x + 3)(x - 3)$

b. $(5x + 7y)(5x - 7y)$

c. $(5ab^2 - 4a)(5ab^2 + 4a)$.

Special products can sometimes be used to find the products of certain trinomials, as illustrated in Example 9.

EXAMPLE 9 Using the Special Products

Multiply:

a. $(7x + 5 + 4y)(7x + 5 - 4y)$ **b.** $(3x + y + 1)^2$.

Solution

a. By grouping the first two terms within each set of parentheses, we can find the product using the form for the sum and difference of two terms.

$$(A \quad + \quad B) \quad \cdot \quad (A \quad - \quad B) \quad = \quad A^2 \quad - \quad B^2$$

$$[(7x + 5) + 4y] \cdot [(7x + 5) - 4y] = (7x + 5)^2 - (4y)^2$$
$$= (7x)^2 + 2 \cdot 7x \cdot 5 + 5^2 - (4y)^2$$
$$= 49x^2 + 70x + 25 - 16y^2$$

b. We can group the terms so that the formula for the square of a binomial can be applied.

$$(A \quad + \quad B)^2 \quad = \quad A^2 \quad + \quad 2 \quad \cdot \quad A \quad \cdot \quad B \quad + \quad B^2$$

$$[(3x + y) + 1]^2 = (3x + y)^2 + 2 \cdot (3x + y) \cdot 1 + 1^2$$
$$= 9x^2 + 6xy + y^2 + 6x + 2y + 1$$ ∎

Discover for Yourself

Group $(3x + y + 1)^2$ as $[3x + (y + 1)]^2$. Verify that you get the same product as we obtained in Example 9(b).

✓ **CHECK POINT 9** Multiply:

a. $(3x + 2 + 5y)(3x + 2 - 5y)$

b. $(2x + y + 3)^2$.

---

7 Find the product of functions. ▶

Multiplication of Polynomial Functions

In Chapter 2, we developed an algebra of functions, defining the product of functions f and g as follows:

$$(fg)(x) = f(x) \cdot g(x).$$

Now that we know how to multiply polynomials, we can find the product of functions.

EXAMPLE 10 Using the Algebra of Functions

Let $f(x) = x - 5$ and $g(x) = x - 2$. Find:

a. $(fg)(x)$ **b.** $(fg)(1)$.

Solution

a. $(fg)(x) = f(x) \cdot g(x)$ This is the definition of the product function, fg.

$\qquad\qquad = (x - 5)(x - 2)$ Substitute the given functions.

$\qquad$ F   O   I   L

$\qquad\qquad = x^2 - 2x - 5x + 10$ Multiply by the FOIL method.

$\qquad\qquad = x^2 - 7x + 10$ Combine like terms.

Thus,

$$(fg)(x) = x^2 - 7x + 10.$$

b. We use the product function to find $(fg)(1)$—that is, the value of the function fg at 1. Replace x with 1.

$$(fg)(1) = 1^2 - 7 \cdot 1 + 10 = 4 \quad \blacksquare$$

Example 10 involved linear and quadratic functions.

$$f(x) = x - 5 \qquad g(x) = x - 2 \qquad (fg)(x) = x^2 - 7x + 10$$

These are linear functions of the form $f(x) = mx + b$.

This is a quadratic function of the form $f(x) = ax^2 + bx + c$.

All three of these functions are polynomial functions. A linear function is a first-degree polynomial function. A quadratic function is a second-degree polynomial function.

✓ **CHECK POINT 10** Let $f(x) = x - 3$ and $g(x) = x - 7$. Find:

a. $(fg)(x)$

b. $(fg)(2)$.

---

8 Use polynomial multiplication to evaluate functions. ▶

If you are given a function, f, calculus can reveal how it is changing at any instant in time. The algebraic expression

$$\frac{f(a + h) - f(a)}{h}$$

plays an important role in this process. Our work with polynomial multiplication can be used to evaluate the numerator of this expression.

EXAMPLE 11 Using Polynomial Multiplication to Evaluate Functions

Given $f(x) = x^2 - 7x + 3$, find and simplify each of the following:

a. $f(a + 4)$ **b.** $f(a + h) - f(a)$.

Solution

a. We find $f(a + 4)$, read "f at a plus 4," by replacing x with $a + 4$ each time that x appears in the polynomial.

$$f(x) \quad = \quad x^2 \qquad - 7x \qquad + 3$$

| Replace x with $a + 4$. | Replace x with $a + 4$. | Replace x with $a + 4$. | Copy the 3. There is no x in this term. |

$$f(a + 4) \quad = \quad (a + 4)^2 \quad - 7(a + 4) \qquad + 3$$

$$= a^2 + 8a + 16 - 7a - 28 + 3 \quad \text{Multiply as indicated.}$$

$$= a^2 + a - 9 \qquad \text{Combine like terms: } 8a - 7a = a$$
$$\text{and } 16 - 28 + 3 = -9.$$

b. To find $f(a + h) - f(a)$, we first replace each occurrence of x in $f(x) = x^2 - 7x + 3$ with $a + h$ and then replace each occurrence of x with a. Then we perform the resulting operations and simplify.

| This is $f(a + h)$. Use $f(x) = x^2 - 7x + 3$ and replace x with $a + h$. | This is $f(a)$. Use $f(x) = x^2 - 7x + 3$ and replace x with a. |

$$f(a + h) - f(a) = \boxed{(a + h)^2 - 7(a + h) + 3} - (a^2 - 7a + 3)$$

$$= (a^2 + 2ah + h^2 - 7a - 7h + 3) - (a^2 - 7a + 3) \quad \text{Perform the multiplications required by } f(a + h).$$

$$= (a^2 + 2ah + h^2 - 7a - 7h + 3) + (-a^2 + 7a - 3) \quad \text{Add the opposite of the polynomial being subtracted. Like terms are shown in the same color.}$$

$$= a^2 - a^2 - 7a + 7a + 3 - 3 + 2ah + h^2 - 7h \quad \text{Group like terms.}$$

$$= 2ah + h^2 - 7h \qquad \text{Simplify. Observe that}$$
$$a^2 - a^2 = 0, -7a + 7a = 0,$$
$$\text{and } 3 - 3 = 0. \ \blacksquare$$

✓ **CHECK POINT 11**   Given $f(x) = x^2 - 5x + 4$, find and simplify each of the following:

a. $f(a + 3)$ **b.** $f(a + h) - f(a)$.

Blitzer Bonus ⊙

Labrador Retrievers and Polynomial Multiplication

The color of a Labrador retriever is determined by its pair of genes. A single gene is inherited at random from each parent. The black-fur gene, B, is dominant. The yellow-fur gene, Y, is recessive. This means that labs with at least one black-fur gene (BB or BY) have black coats. Only labs with two yellow-fur genes (YY) have yellow coats.

Axl, your author's yellow lab, inherited his genetic makeup from two black BY parents.

Second BY parent, a black lab with a recessive yellow-fur gene

	B	Y
B	BB	BY
Y	BY	YY

First BY parent, a black lab with a recessive yellow-fur gene

The table shows the four possible combinations of color genes that BY parents can pass to their offspring.

Because YY is one of four possible outcomes, the probability that a yellow lab like Axl will be the offspring of these black parents is $\frac{1}{4}$. The probabilities suggested by the table can be modeled by the expression $\left(\frac{1}{2}B + \frac{1}{2}Y\right)^2$.

$$\left(\frac{1}{2}B + \frac{1}{2}Y\right)^2 = \left(\frac{1}{2}B\right)^2 + 2\left(\frac{1}{2}B\right)\left(\frac{1}{2}Y\right) + \left(\frac{1}{2}Y\right)^2$$

$$= \frac{1}{4}BB \quad + \quad \frac{1}{2}BY \quad + \quad \frac{1}{4}YY$$

| The probability of a black lab with two dominant black genes is $\frac{1}{4}$. | The probability of a black lab with a recessive yellow gene is $\frac{1}{2}$. | The probability of a yellow lab with two recessive yellow genes is $\frac{1}{4}$. |

CONCEPT AND VOCABULARY CHECK

Fill in each blank so that the resulting statement is true.

1. To multiply the variables in monomials, retain each variable and _____ the exponents.

2. To multiply $7x^3(4x^5 - 8x^2 + 6)$, use the _____ property to multiply each term of the trinomial _____ by the monomial _____.

3. To multiply $(5x + 3)(x^2 + 8x + 7)$, begin by multiplying each term of $x^2 + 8x + 7$ by _____. Then multiply each term of $x^2 + 8x + 7$ by _____. Then combine _____ terms.

4. When using the FOIL method to find $(x + 7)(3x + 5)$, the product of the first terms is _____, the product of the outside terms is _____, the product of the inside terms is _____, and the product of the last terms is _____.

5. $(A + B)^2 = $ _____. The square of a binomial sum is the first term _____ plus 2 times the _____ plus the last term _____.

6. $(A - B)^2 = $ _____. The square of a binomial difference is the first term squared _____ 2 times the _____ _____ the last term squared.

> plus or minus?

7. $(A + B)(A - B) = $ _____. The product of the sum and difference of the same two terms is the square of the first term _____ the square of the second term.

8. If $f(x) = x^2 - 4x + 7$, we find $f(a + h)$ by replacing each occurrence of _____ with _____.

5.2 EXERCISE SET   MyMathLab®

Practice Exercises

Throughout the practice exercises, assume that any variable exponents represent whole numbers.

In Exercises 1–8, multiply the monomials.

1. $(3x^2)(5x^4)$
2. $(4x^2)(6x^4)$
3. $(3x^2y^4)(5xy^7)$
4. $(6x^4y^2)(3x^7y)$
5. $(-3xy^2z^5)(2xy^7z^4)$
6. $(11x^2yz^4)(-3xy^5z^6)$
7. $(-8x^{2n}y^{n-5})\left(-\dfrac{1}{4}x^ny^3\right)$
8. $(-9x^{3n}y^{n-3})\left(-\dfrac{1}{3}x^ny^2\right)$

In Exercises 9–22, multiply the monomial and the polynomial.

9. $4x^2(3x + 2)$
10. $5x^2(6x + 7)$
11. $2y(y^2 - 5y)$
12. $3y(y^2 - 4y)$
13. $5x^3(2x^5 - 4x^2 + 9)$
14. $6x^3(3x^5 - 5x^2 + 7)$
15. $4xy(7x + 3y)$
16. $5xy(8x + 3y)$
17. $3ab^2(6a^2b^3 + 5ab)$
18. $5ab^2(10a^2b^3 + 7ab)$

19. $-4x^2y(3x^4y^2 - 7xy^3 + 6)$
20. $-3x^2y(10x^2y^4 - 2xy^3 + 7)$
21. $-4x^n\left(3x^{2n} - 5x^n + \dfrac{1}{2}x\right)$
22. $-10x^n\left(4x^{2n} - 3x^n + \dfrac{1}{5}x\right)$

In Exercises 23–34, find each product using either a horizontal or a vertical format.

23. $(x - 3)(x^2 + 2x + 5)$
24. $(x + 4)(x^2 - 5x + 8)$
25. $(x - 1)(x^2 + x + 1)$
26. $(x - 2)(x^2 + 2x + 4)$
27. $(a - b)(a^2 + ab + b^2)$
28. $(a + b)(a^2 - ab + b^2)$
29. $(x^2 + 2x - 1)(x^2 + 3x - 4)$
30. $(x^2 - 2x + 3)(x^2 + x + 1)$
31. $(x - y)(x^2 - 3xy + y^2)$
32. $(x - y)(x^2 - 4xy + y^2)$
33. $(xy + 2)(x^2y^2 - 2xy + 4)$
34. $(xy + 3)(x^2y^2 - 2xy + 5)$

In Exercises 35–54, use the FOIL method to multiply the binomials.

35. $(x + 4)(x + 7)$
36. $(x + 5)(x + 8)$
37. $(y + 5)(y - 6)$
38. $(y + 5)(y - 8)$

39. $(5x + 3)(2x + 1)$

40. $(4x + 3)(5x + 1)$

41. $(3y - 4)(2y - 1)$

42. $(5y - 2)(3y - 1)$

43. $(3x - 2)(5x - 4)$

44. $(2x - 3)(4x - 5)$

45. $(x - 3y)(2x + 7y)$

46. $(3x - y)(2x + 5y)$

47. $(7xy + 1)(2xy - 3)$

48. $(3xy - 1)(5xy + 2)$

49. $(x - 4)(x^2 - 5)$

50. $(x - 5)(x^2 - 3)$

51. $(8x^3 + 3)(x^2 - 5)$

52. $(7x^3 + 5)(x^2 - 2)$

53. $(3x^n - y^n)(x^n + 2y^n)$

54. $(5x^n - y^n)(x^n + 4y^n)$

In Exercises 55–68, multiply using one of the rules for the square of a binomial.

55. $(x + 3)^2$

56. $(x + 4)^2$

57. $(y - 5)^2$

58. $(y - 6)^2$

59. $(2x + y)^2$

60. $(4x + y)^2$

61. $(5x - 3y)^2$

62. $(3x - 4y)^2$

63. $(2x^2 + 3y)^2$

64. $(4x^2 + 5y)^2$

65. $(4xy^2 - xy)^2$

66. $(5xy^2 - xy)^2$

67. $(a^n + 4b^n)^2$

68. $(3a^n - b^n)^2$

In Exercises 69–82, multiply using the rule for the product of the sum and difference of two terms.

69. $(x + 4)(x - 4)$

70. $(x + 5)(x - 5)$

71. $(5x + 3)(5x - 3)$

72. $(3x + 2)(3x - 2)$

73. $(4x + 7y)(4x - 7y)$

74. $(8x + 7y)(8x - 7y)$

75. $(y^3 + 2)(y^3 - 2)$

76. $(y^3 + 3)(y^3 - 3)$

77. $(1 - y^5)(1 + y^5)$

78. $(2 - y^5)(2 + y^5)$

79. $(7xy^2 - 10y)(7xy^2 + 10y)$

80. $(3xy^2 - 4y)(3xy^2 + 4y)$

81. $(5a^n - 7)(5a^n + 7)$

82. $(10b^n - 3)(10b^n + 3)$

In Exercises 83–94, find each product.

83. $[(2x + 3) + 4y][(2x + 3) - 4y]$

84. $[(3x + 2) + 5y][(3x + 2) - 5y]$

85. $(x + y + 3)(x + y - 3)$

86. $(x + y + 4)(x + y - 4)$

87. $(5x + 7y - 2)(5x + 7y + 2)$

88. $(7x + 5y - 2)(7x + 5y + 2)$

89. $[5y + (2x + 3)][5y - (2x + 3)]$

90. $[8y + (3x + 2)][8y - (3x + 2)]$

91. $(x + y + 1)^2$

92. $(x + y + 2)^2$

93. $(x + 1)(x - 1)(x^2 + 1)$

94. $(x + 2)(x - 2)(x^2 + 4)$

95. Let $f(x) = x - 2$ and $g(x) = x + 6$. Find each of the following.

   a. $(fg)(x)$

   b. $(fg)(-1)$

   c. $(fg)(0)$

96. Let $f(x) = x - 4$ and $g(x) = x + 10$. Find each of the following.

   a. $(fg)(x)$

   b. $(fg)(-1)$

   c. $(fg)(0)$

97. Let $f(x) = x - 3$ and $g(x) = x^2 + 3x + 9$. Find each of the following.

   a. $(fg)(x)$

   b. $(fg)(-2)$

   c. $(fg)(0)$

98. Let $f(x) = x + 3$ and $g(x) = x^2 - 3x + 9$. Find each of the following.

   a. $(fg)(x)$

   b. $(fg)(-2)$

   c. $(fg)(0)$

In Exercises 99–102, find each of the following and simplify:

   a. $f(a + 2)$ **b.** $f(a + h) - f(a)$.

99. $f(x) = x^2 - 3x + 7$

100. $f(x) = x^2 - 4x + 9$

101. $f(x) = 3x^2 + 2x - 1$

102. $f(x) = 4x^2 + 5x - 1$

Practice PLUS

In Exercises 103–112, perform the indicated operation or operations.

103. $(3x + 4y)^2 - (3x - 4y)^2$

104. $(5x + 2y)^2 - (5x - 2y)^2$

105. $(5x - 7)(3x - 2) - (4x - 5)(6x - 1)$

106. $(3x + 5)(2x - 9) - (7x - 2)(x - 1)$

107. $(2x + 5)(2x - 5)(4x^2 + 25)$

108. $(3x + 4)(3x - 4)(9x^2 + 16)$

109. $(x - 1)^3$

110. $(x - 2)^3$

111. $\dfrac{(2x - 7)^5}{(2x - 7)^3}$

112. $\dfrac{(5x - 3)^6}{(5x - 3)^4}$

Application Exercises

In Exercises 113–114, find the area of the large rectangle in two ways:

   a. *Find the sum of the areas of the four smaller rectangles.*

   b. *Multiply the length and the width of the large rectangle using the FOIL method. Compare this product with your answer to part (a).*

113.

114.

In Exercises 115–116, express each polynomial in standard form—that is, in descending powers of x.

   a. *Write a polynomial that represents the area of the large rectangle.*

   b. *Write a polynomial that represents the area of the small, unshaded rectangle.*

   c. *Write a polynomial that represents the area of the shaded blue region.*

115.

116.

In Exercises 117–118, express each polynomial in standard form.

   a. *Write a polynomial that represents the area of the rectangular base of the open box.*

   b. *Write a polynomial that represents the volume of the open box.*

117.

118.

119. A popular model of carry-on luggage has a length that is 10 inches greater than its depth. Airline regulations require that the sum of the length, width, and depth cannot exceed 40 inches. These conditions, with the assumption that this sum *is* 40 inches, can be modeled by a function that gives the volume of the luggage, V, in cubic inches, in terms of its depth, x, in inches.

$$\boxed{\text{Volume}} = \boxed{\text{depth}} \cdot \boxed{\text{length}} \cdot \boxed{\text{width: } 40 - (\text{depth} + \text{length})}$$

$$V(x) = x \cdot (x + 10) \cdot [40 - (x + x + 10)]$$

$$V(x) = x(x + 10)(30 - 2x)$$

   a. Perform the multiplications in the formula for $V(x)$ and express the formula in standard form.

   b. Use the function's formula from part (a) and the Leading Coefficient Test to determine the end behavior of its graph.

   c. Does the end behavior to the right make this function useful in modeling the volume of carry-on luggage as its depth continues to increase?

   d. Use the formula from part (a) to find $V(10)$. Describe what this means in practical terms.

   e. The graph of the function modeling the volume of carry-on luggage is shown below. Identify your answer from part (d) as a point on the graph.

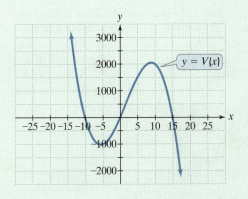

   f. Use the graph to describe a realistic domain, x, for the volume function, where x represents the depth of the carry-on luggage. Use interval notation to express this realistic domain.

120. Before working this exercise, be sure that you have read the Blitzer Bonus on page 336. The table shows the four combinations of color genes that a YY yellow lab and a BY black lab can pass to their offspring.

	B	Y
Y	BY	YY
Y	BY	YY

a. How many combinations result in a yellow lab with two recessive yellow genes? What is the probability of a yellow lab?

b. How many combinations result in a black lab with a recessive yellow gene? What is the probability of a black lab?

c. Find the product of Y and $\frac{1}{2}$B + $\frac{1}{2}$Y. How does this product model the probabilities that you determined in parts (a) and (b)?

Explaining the Concepts

121. Explain how to multiply monomials. Give an example.

122. Explain how to multiply a monomial and a polynomial that is not a monomial. Give an example.

123. Explain how to multiply a binomial and a trinomial.

124. What is the FOIL method and when is it used? Give an example of the method.

125. Explain how to square a binomial sum. Give an example.

126. Explain how to square a binomial difference. Give an example.

127. Explain how to find the product of the sum and difference of two terms. Give an example with your explanation.

128. How can the graph of function fg be obtained from the graphs of functions f and g?

129. Explain how to find $f(a + h) - f(a)$ for a given function f.

Technology Exercises

In Exercises 130–133, use a graphing utility to graph the functions y_1 and y_2. Select a viewing rectangle that is large enough to show the end behavior of y_2. What can you conclude? Verify your conclusions using polynomial multiplication.

130. $y_1 = (x - 2)^2$
$y_2 = x^2 - 4x + 4$

131. $y_1 = (x - 4)(x^2 - 3x + 2)$
$y_2 = x^3 - 7x^2 + 14x - 8$

132. $y_1 = (x - 1)(x^2 + x + 1)$
$y_2 = x^3 - 1$

133. $y_1 = (x + 1.5)(x - 1.5)$
$y_2 = x^2 - 2.25$

134. Graph $f(x) = x + 4, g(x) = x - 2$, and the product function, fg, in a $[-6, 6, 1]$ by $[-10, 10, 1]$ viewing rectangle. Trace along the curves and show that $(fg)(1) = f(1) \cdot g(1)$.

Critical Thinking Exercises

Make Sense? *In Exercises 135–138, determine whether each statement makes sense or does not make sense, and explain your reasoning.*

135. Knowing the difference between factors and terms is important: In $(3x^2y)^2$, I can distribute the exponent 2 on each factor, but in $(3x^2 + y)^2$, I cannot do the same thing on each term.

136. I used the FOIL method to find the product of $x + 5$ and $x^2 + 2x + 1$.

137. Instead of using the formula for the square of a binomial sum, I prefer to write the binomial sum twice and then apply the FOIL method.

138. Special-product formulas have patterns that make their multiplications quicker than using the FOIL method.

In Exercises 139–142, determine whether each statement is true or false. If the statement is false, make the necessary change(s) to produce a true statement.

139. If f is a polynomial function, then
$f(a + h) - f(a) = f(a) + f(h) - f(a) = f(h)$.

140. $(x - 5)^2 = x^2 - 5x + 25$

141. $(x + 1)^2 = x^2 + 1$

142. Suppose a square garden has an area represented by $9x^2$ square feet. If one side is made 7 feet longer and the other side is made 2 feet shorter, then the trinomial that represents the area of the larger garden is $9x^2 + 15x - 14$ square feet.

143. Express the area of the plane figure shown as a polynomial in standard form.

In Exercises 144–145, represent the volume of each figure as a polynomial in standard form.

144.

145.

146. Simplify: $(y^n + 2)(y^n - 2) - (y^n - 3)^2$.

147. The product of two consecutive odd integers is 22 less than the square of the greater integer. Find the integers.

Review Exercises

148. Solve: $|3x + 4| \geq 10$. (Section 4.3, Example 6)

149. Solve: $2 - 6x \leq 20$. (Section 4.1, Example 2)

150. Write in scientific notation: $8,034,000,000$. (Section 1.7, Example 2)

Preview Exercises

Exercises 151–153 will help you prepare for the material covered in the next section.

151. Replace each boxed question mark with a polynomial that results in the given product.
   a. $3x^3 \cdot \boxed{?} = 9x^5$
   b. $2x^3y^2 \cdot \boxed{?} = 12x^5y^4$

In Exercises 152–153, a polynomial is given in factored form. Use multiplication to find the product of the factors.

152. $(x - 5)(x^2 + 3)$

153. $(x + 4)(3x - 2y)$

SECTION

5.3

Greatest Common Factors and Factoring by Grouping

What am I supposed to learn?

After studying this section, you should be able to:

1. Factor out the greatest common factor of a polynomial.

2. Factor out a common factor with a negative coefficient.

3. Factor by grouping.

A two-year-old girl is asked, "Do you have a sister?" She answers, "Yes." "What is your sister's name?" "Hannah." Asked if Hannah has a sister, the two-year-old replies, "No." The child can go in the direction from self to sister but she cannot reverse this direction and move from sister back to self.

As our intellects develop, we learn to reverse the direction of our thinking. Reversibility of thought is found throughout algebra. For example, we can multiply polynomials and show that

$$7x(3x + 4) = 21x^2 + 28x.$$

We can also reverse $7x(3x + 4) = 21x^2 + 28x$ and express the resulting polynomial as

$$21x^2 + 28x = 7x(3x + 4).$$

Factoring a polynomial consisting of the sum of monomials means finding an equivalent expression that is a product.

Factoring $21x^2 + 28x$

Sum of monomials   Equivalent expression that is a product

$$21x^2 + 28x = 7x(3x + 4)$$

The factors of $21x^2 + 28x$ are $7x$ and $3x + 4$.

In this chapter, we will be factoring over the set of integers, meaning that the coefficients in the factors are integers. Polynomials that cannot be factored using integer coefficients are called **prime polynomials over the set of integers**.

1 Factor out the greatest common factor of a polynomial. ▶

Factoring Out the Greatest Common Factor

In any factoring problem, the first step is to look for the *greatest common factor*. The **greatest common factor**, abbreviated GCF, is an expression with the greatest coefficient and of the highest degree that divides each term of the polynomial. Can you see that $7x$ is the greatest common factor of $21x^2 + 28x$? 7 is the greatest integer that divides both 21 and 28. Furthermore, x is the greatest power of x that divides x^2 and x.

The variable part of the greatest common factor always contains the *smallest* power of a variable that appears in all terms of the polynomial. For example, consider the polynomial

$$21x^2 + 28x.$$

x^1, or x, is the variable raised to the smallest exponent.

We see that x is the variable part of the greatest common factor, $7x$.

When factoring a monomial from a polynomial, determine the greatest common factor of all terms in the polynomial. Sometimes there may not be a GCF other than 1. When a GCF other than 1 exists, we use the following procedure:

> ### Factoring a Monomial from a Polynomial
>
> 1. Determine the greatest common factor of all terms in the polynomial.
> 2. Express each term as the product of the GCF and its other factor.
> 3. Use the distributive property to factor out the GCF.

EXAMPLE 1   Factoring Out the Greatest Common Factor

Factor: $21x^2 + 28x$.

Solution   The GCF of the two terms of the polynomial is $7x$.

$$21x^2 + 28x$$

$$= 7x(3x) + 7x(4) \qquad \text{Express each term as the product of the GCF and its other factor.}$$

$$= 7x(3x + 4) \qquad \text{Factor out the GCF.}$$

We can check this factorization by multiplying $7x$ and $3x + 4$, obtaining the original polynomial as the answer. ∎

☑ **CHECK POINT 1**   Factor: $20x^2 + 30x$.

EXAMPLE 2 Factoring Out the Greatest Common Factor

Factor:

a. $9x^5 + 15x^3$ **b.** $16x^2y^3 - 24x^3y^4$ **c.** $12x^5y^4 - 4x^4y^3 + 2x^3y^2$.

Solution

a. First, determine the greatest common factor.

3 is the greatest integer that divides 9 and 15.

$$9x^5 + 15x^3$$

x^3 is the variable raised to the smallest exponent.

The GCF of the two terms of the polynomial is $3x^3$.

$$9x^5 + 15x^3$$
$$= 3x^3 \cdot 3x^2 + 3x^3 \cdot 5 \qquad \text{Express each term as the product of the GCF and its other factor.}$$
$$= 3x^3(3x^2 + 5) \qquad \text{Factor out the GCF.}$$

b. Begin by determining the greatest common factor.

8 is the greatest integer that divides 16 and 24.

$$16x^2y^3 - 24x^3y^4$$

The variables raised to the smallest exponents are x^2 and y^3.

The GCF of the two terms of the polynomial is $8x^2y^3$.

$$16x^2y^3 - 24x^3y^4$$
$$= 8x^2y^3 \cdot 2 - 8x^2y^3 \cdot 3xy \qquad \text{Express each term as the product of the GCF and its other factor.}$$
$$= 8x^2y^3(2 - 3xy) \qquad \text{Factor out the GCF.}$$

c. First, determine the greatest common factor of the three terms.

2 is the greatest integer that divides 12, 4, and 2.

$$12x^5y^4 - 4x^4y^3 + 2x^3y^2$$

The variables raised to the smallest exponents are x^3 and y^2.

The GCF is $2x^3y^2$.

$$12x^5y^4 - 4x^4y^3 + 2x^3y^2$$
$$= 2x^3y^2 \cdot 6x^2y^2 - 2x^3y^2 \cdot 2xy + 2x^3y^2 \cdot 1 \qquad \text{Express each term as the product of the GCF and its other factor.}$$

You can obtain the factors shown in black by dividing each term of the given polynomial by $2x^3y^2$, the GCF.

$$\frac{12x^5y^4}{2x^3y^2} = 6x^2y^2 \qquad \frac{4x^4y^3}{2x^3y^2} = 2xy \qquad \frac{2x^3y^2}{2x^3y^2} = 1$$

$$= 2x^3y^2(6x^2y^2 - 2xy + 1) \qquad \text{Factor out the GCF.} \blacksquare$$

Because factoring reverses the process of multiplication, all factorizations can be checked by multiplying. Take a few minutes to check each of the three factorizations in Example 2. Use the distributive property to multiply the factors. This should give the original polynomial.

☑ **CHECK POINT 2** Factor:

a. $9x^4 + 21x^2$

b. $15x^3y^2 - 25x^4y^3$

c. $16x^4y^5 - 8x^3y^4 + 4x^2y^3$.

---

2 Factor out a common factor with a negative coefficient. ⓒ

When the leading coefficient of a polynomial is negative, it is often desirable to factor out a common factor with a negative coefficient. The common factor is the GCF preceded by a negative sign.

EXAMPLE 3 Using a Common Factor with a Negative Coefficient

Factor: $-3x^3 + 12x^2 - 15x$.

Solution The GCF is $3x$. Because the leading coefficient, -3, is negative, we factor out a common factor with a negative coefficient. We will factor out the opposite of the GCF, or $-3x$.

$$-3x^3 + 12x^2 - 15x$$

$$= -3x(x^2) - 3x(-4x) - 3x(5) \quad \text{Express each term as the product of the common factor and its other factor.}$$

$$= -3x(x^2 - 4x + 5) \quad \text{Factor out the opposite of the GCF.} \quad ■$$

☑ **CHECK POINT 3** Factor out a common factor with a negative coefficient: $-2x^3 + 10x^2 - 6x$.

Factoring by Grouping

Up to now, we have factored a monomial from a polynomial. By contrast, in our next example, the greatest common factor of the polynomial is a binomial.

EXAMPLE 4 Factoring Out the Greatest Common Binomial Factor

Factor:

a. $2(x - 7) + 9a(x - 7)$ **b.** $5y(a - b) - (a - b)$.

Solution Let's identify the common binomial factor in each part of the problem.

$$2(x - 7) + 9a(x - 7) \qquad 5y(a - b) - (a - b)$$

The GCF, a binomial, is $x - 7$.   The GCF, a binomial, is $a - b$.

We factor out each common binomial factor as follows.

a. $2(x - 7) + 9a(x - 7)$

$\quad = (x - 7)2 + (x - 7)9a$ This step, usually omitted, shows each term as the product of the GCF and its other factor, in that order.

$\quad = (x - 7)(2 + 9a)$ Factor out the GCF.

b. $5y(a - b) - (a - b)$

$\quad = 5y(a - b) - 1(a - b)$ Write $-(a - b)$ as $-1(a - b)$ to aid in the factoring.

$\quad = (a - b)(5y - 1)$ Factor out the GCF. ∎

☑ **CHECK POINT 4**   Factor:

a. $3(x - 4) + 7a(x - 4)$

b. $7x(a + b) - (a + b)$.

---

3 Factor by grouping.

Some polynomials have only a greatest common factor of 1. However, by a suitable grouping of the terms, it still may be possible to factor. This process, called **factoring by grouping**, is illustrated in Example 5.

EXAMPLE 5   Factoring by Grouping

Factor: $x^3 - 5x^2 + 3x - 15$.

Solution   There is no factor other than 1 common to all four terms. However, we can group terms that have a common factor:

$$\boxed{x^3 - 5x^2} \; + \; \boxed{3x - 15}$$

Common factor is x^2.   Common factor is 3.

We now factor the given polynomial as follows:

$x^3 - 5x^2 + 3x - 15$

$= (x^3 - 5x^2) + (3x - 15)$ Group terms with common factors.

$= x^2(x - 5) + 3(x - 5)$ Factor out the greatest common factor from the grouped terms. The remaining two terms have $x - 5$ as a common binomial factor.

$= (x - 5)(x^2 + 3)$.   Factor out the GCF.

Thus, $x^3 - 5x^2 + 3x - 15 = (x - 5)(x^2 + 3)$. Check the factorization by multiplying the right side of the equation using the FOIL method. Because the factorization is correct, you should obtain the original polynomial. ∎

Discover for Yourself

In Example 5, group the terms as follows:

$x^2 + 3x - 5x^2 - 15.$

Factor out the common factor from each group and complete the factoring process. Describe what happens. What can you conclude?

☑ **CHECK POINT 5**   Factor: $x^3 - 4x^2 + 5x - 20$.

Factoring by Grouping

1. Group terms that have a common monomial factor. There will usually be two groups. Sometimes the terms must be rearranged.

2. Factor out the common monomial factor from each group.

3. Factor out the remaining common binomial factor (if one exists).

EXAMPLE 6 Factoring by Grouping

Factor: $3x^2 + 12x - 2xy - 8y$.

Solution There is no factor other than 1 common to all four terms. However, we can group terms that have a common factor:

$$\boxed{3x^2 + 12x} \; + \; \boxed{-2xy - 8y} \,.$$

Common factor is $3x$:
$3x^2 + 12x = 3x(x + 4)$.

Use $-2y$, rather than $2y$, as the common factor:
$-2xy - 8y = -2y(x + 4)$. In this way, the common binomial factor, $x + 4$, appears.

The voice balloons illustrate that it is sometimes necessary to use a factor with a negative coefficient to obtain a common binomial factor for the two groupings. We now factor the given polynomial as follows:

$$
\begin{aligned}
3x^2 &+ 12x - 2xy - 8y \\
&= (3x^2 + 12x) + (-2xy - 8y) &&\text{Group terms with common factors.} \\
&= 3x(x + 4) - 2y(x + 4) &&\text{Factor out the common factors} \\
& &&\text{from the grouped terms.} \\
&= (x + 4)(3x - 2y). &&\text{Factor out the GCF.}
\end{aligned}
$$

Thus, $3x^2 + 12x - 2xy - 8y = (x + 4)(3x - 2y)$. Using the commutative property of multiplication, the factorization can also be expressed as $(3x - 2y)(x + 4)$. Verify the factorization by showing that, regardless of the order, FOIL multiplication gives the original polynomial. ∎

✓ **CHECK POINT 6** Factor: $4x^2 + 20x - 3xy - 15y$.

Factoring by grouping sometimes requires that the terms be rearranged before the groupings are made. For example, consider the polynomial

$$3x^2 - 8y + 12x - 2xy.$$

The first two terms have no common factor other than 1. We must rearrange the terms and try a different grouping. Example 6 showed one such rearrangement of two groupings.

Achieving Success

When using your professor's office hours, show up prepared. If you are having difficulty with a concept or problem, bring your work so that your instructor can determine where you are having trouble. If you miss a lecture, read the appropriate section in the textbook, borrow class notes, and attempt the assigned homework before your office visit. Because this text has video lectures for every section, you might find it helpful to view them in MyMathLab to review the material you missed. It is not realistic to expect your professor to rehash all or part of a class lecture during office hours.

CONCEPT AND VOCABULARY CHECK

Fill in each blank so that the resulting statement is true.

1. The process of writing a polynomial containing the sum of monomials as a product is called _____.

2. An expression with the greatest coefficient and of the highest degree that divides each term of a polynomial is called the _____. The variable part of this expression contains the _____ power of a variable that appears in all terms of the polynomial.

3. True or false: The factorization of $12x^4 + 21x^2$ is $3x^2 \cdot 4x^2 + 3x^2 \cdot 7$. _____

4. We factor $-2x^3 + 10x^2 - 6x$ by factoring out _____.

5. True or false: The factorization of $x^3 - 4x^2 + 5x - 20$ is $x^2(x - 4) + 5(x - 4)$. _____

5.3 EXERCISE SET   ▶ MyMathLab®

Practice Exercises

Throughout the practice exercises, assume that any variable exponents represent whole numbers.

In Exercises 1–22, factor the greatest common factor from each polynomial.

1. $10x^2 + 4x$
2. $12x^2 + 9x$
3. $y^2 - 4y$
4. $y^2 - 7y$
5. $x^3 + 5x^2$
6. $x^3 + 7x^2$
7. $12x^4 - 8x^2$
8. $20x^4 - 8x^2$
9. $32x^4 + 2x^3 + 8x^2$
10. $9x^4 + 18x^3 + 6x^2$
11. $4x^2y^3 + 6xy$
12. $6x^3y^2 + 9xy$
13. $30x^2y^3 - 10xy^2$
14. $27x^2y^3 - 18xy^2$
15. $12xy - 6xz + 4xw$
16. $14xy - 10xz + 8xw$
17. $15x^3y^6 - 9x^4y^4 + 12x^2y^5$
18. $15x^4y^6 - 3x^3y^5 + 12x^4y^4$
19. $25x^3y^6z^2 - 15x^4y^4z^4 + 25x^2y^5z^3$
20. $49x^4y^3z^5 - 70x^3y^5z^4 + 35x^4y^4z^3$
21. $15x^{2n} - 25x^n$
22. $12x^{3n} - 9x^{2n}$

In Exercises 23–34, factor out the negative of the greatest common factor.

23. $-4x + 12$
24. $-5x + 20$
25. $-8x - 48$
26. $-7x - 63$
27. $-2x^2 + 6x - 14$
28. $-2x^2 + 8x - 12$
29. $-5y^2 + 40x$
30. $-9y^2 + 45x$
31. $-4x^3 + 32x^2 - 20x$
32. $-5x^3 + 50x^2 - 10x$

33. $-x^2 - 7x + 5$
34. $-x^2 - 8x + 8$

In Exercises 35–44, factor the greatest common binomial factor from each polynomial.

35. $4(x + 3) + a(x + 3)$
36. $5(x + 4) + a(x + 4)$
37. $x(y - 6) - 7(y - 6)$
38. $x(y - 9) - 5(y - 9)$
39. $3x(x + y) - (x + y)$
40. $7x(x + y) - (x + y)$
41. $4x^2(3x - 1) + 3x - 1$
42. $6x^2(5x - 1) + 5x - 1$
43. $(x + 2)(x + 3) + (x - 1)(x + 3)$
44. $(x + 4)(x + 5) + (x - 1)(x + 5)$

In Exercises 45–68, factor by grouping.

45. $x^2 + 3x + 5x + 15$
46. $x^2 + 2x + 4x + 8$
47. $x^2 + 7x - 4x - 28$
48. $x^2 + 3x - 5x - 15$
49. $x^3 - 3x^2 + 4x - 12$
50. $x^3 - 2x^2 + 5x - 10$
51. $xy - 6x + 2y - 12$
52. $xy - 5x + 9y - 45$
53. $xy + x - 7y - 7$
54. $xy + x - 5y - 5$
55. $10x^2 - 12xy + 35xy - 42y^2$
56. $3x^2 - 6xy + 5xy - 10y^2$
57. $4x^3 - x^2 - 12x + 3$
58. $3x^3 - 2x^2 - 6x + 4$
59. $x^2 - ax - bx + ab$
60. $x^2 + ax - bx - ab$
61. $x^3 - 12 - 3x^2 + 4x$
62. $2x^3 - 10 + 4x^2 - 5x$
63. $ay - by + bx - ax$
64. $cx - dx + dy - cy$
65. $ay^2 + 2by^2 - 3ax - 6bx$
66. $3a^2x + 6a^2y - 2bx - 4by$
67. $x^ny^n + 3x^n + y^n + 3$
68. $x^ny^n - x^n + 2y^n - 2$

Practice PLUS

In Exercises 69–78, factor each polynomial.

69. $ab - c - ac + b$

70. $ab - 3c - ac + 3b$

71. $x^3 - 5 + 4x^3y - 20y$

72. $x^3 - 2 + 3x^3y - 6y$

73. $2y^7(3x - 1)^5 - 7y^6(3x - 1)^4$

74. $3y^9(3x - 2)^7 - 5y^8(3x - 2)^6$

75. $ax^2 + 5ax - 2a + bx^2 + 5bx - 2b$

76. $ax^2 + 3ax - 11a + bx^2 + 3bx - 11b$

77. $ax + ay + az - bx - by - bz + cx + cy + cz$

78. $ax^2 + ay^2 - az^2 + bx^2 + by^2 - bz^2 + cx^2 + cy^2 - cz^2$

Application Exercises

79. A ball is thrown straight upward. The function

$$f(t) = -16t^2 + 40t$$

describes the ball's height above the ground, $f(t)$, in feet, t seconds after it is thrown.

  a. Find and interpret $f(2)$.

  b. Find and interpret $f(2.5)$.

  c. Factor the polynomial $-16t^2 + 40t$ and write the function in factored form.

  d. Use the factored form of the function to find $f(2)$ and $f(2.5)$. Do you get the same answers as you did in parts (a) and (b)? If so, does this prove that your factorization is correct? Explain.

80. An explosion causes debris to rise vertically. The function

$$f(t) = -16t^2 + 72t$$

describes the height of the debris above the ground, $f(t)$, in feet, t seconds after the explosion.

  a. Find and interpret $f(2)$.

  b. Find and interpret $f(4.5)$.

  c. Factor the polynomial $-16t^2 + 72t$ and write the function in factored form.

  d. Use the factored form of the function to find $f(2)$ and $f(4.5)$. Do you get the same answers as you did in parts (a) and (b)? If so, does this prove that your factorization is correct? Explain.

81. Your computer store is having an incredible sale. The price on one model is reduced by 40%. Then the sale price is reduced by another 40%. If x is the computer's original price, the sale price can be represented by

$$(x - 0.4x) - 0.4(x - 0.4x).$$

  a. Factor out $(x - 0.4x)$ from each term. Then simplify the resulting expression.

  b. Use the simplified expression from part (a) to answer these questions. With a 40% reduction followed by a 40% reduction, is the computer selling at 20% of its original price? If not, at what percentage of the original price is it selling?

82. Your local electronics store is having an end-of-the-year sale. The price on a plasma television had been reduced by 30%. Now the sale price is reduced by another 30%. If x is the television's original price, the sale price can be represented by

$$(x - 0.3x) - 0.3(x - 0.3x).$$

  a. Factor out $(x - 0.3x)$ from each term. Then simplify the resulting expression.

  b. Use the simplified expression from part (a) to answer these questions. With a 30% reduction followed by a 30% reduction, is the television selling at 40% of its original price? If not, at what percentage of the original price is it selling?

Exercises 83–84 involve compound interest. **Compound interest** *is interest computed on your original savings as well as on any accumulated interest.*

83. After 2 years, the balance, A, in an account with principal P and interest rate r compounded annually is given by the formula

$$A = P + Pr + (P + Pr)r.$$

Use factoring by grouping to express the formula as $A = P(1 + r)^2$.

84. After 3 years, the balance, A, in an account with principal P and interest rate r compounded annually is given by the formula

$$A = P(1 + r)^2 + P(1 + r)^2r.$$

Use factoring by grouping to express the formula as $A = P(1 + r)^3$.

85. The area of the skating rink with semicircular ends shown is $A = \pi r^2 + 2rl$. Express the area, A, in factored form.

86. The amount of sheet metal needed to manufacture a cylindrical tin can, that is, its surface area, S, is $S = 2\pi r^2 + 2\pi rh$. Express the surface area, S, in factored form.

Explaining the Concepts

87. What is factoring?

88. If a polynomial has a greatest common factor other than 1, explain how to find its GCF.

89. Using an example, explain how to factor out the greatest common factor of a polynomial.

90. Suppose that a polynomial contains four terms and can be factored by grouping. Explain how to obtain the factorization.

91. Use two different groupings to factor

$$ac - ad + bd - bc$$

in two ways. Then explain why the two factorizations are the same.

92. Write a sentence that uses the word *factor* as a noun. Then write a sentence that uses the word *factor* as a verb.

Technology Exercises

In Exercises 93–96, use a graphing utility to graph the function on each side of the equation in the same viewing rectangle. Use end behavior to show a complete picture of the polynomial function on the left side. Do the graphs coincide? If so, this means that the polynomial on the left side has been factored correctly. If not, factor the polynomial correctly and then use your graphing utility to verify the factorization.

93. $x^2 - 4x = x(x - 4)$

94. $x^2 - 2x + 5x - 10 = (x - 2)(x - 5)$

95. $x^2 + 2x + x + 2 = x(x + 2) + 1$

96. $x^3 - 3x^2 + 4x - 12 = (x^2 + 4)(x - 3)$

Critical Thinking Exercises

Make Sense? *In Exercises 97–100, determine whether each statement makes sense or does not make sense, and explain your reasoning.*

97. After I've factored a polynomial, my answer cannot always be checked by multiplication.

98. The word *greatest* in greatest common factor is helpful because it tells me to look for the greatest power of a variable appearing in all terms.

99. Although $20x^3$ appears as a term in both $20x^3 + 8x^2$ and $20x^3 + 10x$, I'll need to factor $20x^3$ in different ways to obtain each polynomial's factorization.

100. You grouped the polynomial's terms using different groupings than I did, yet we both obtained the same factorization.

In Exercises 101–104, determine whether each statement is true or false. If the statement is false, make the necessary change(s) to produce a true statement.

101. Because the GCF of $9x^3 + 6x^2 + 3x$ is $3x$, it is not necessary to write the 1 when $3x$ is factored from the last term.

102. Some polynomials with four terms, such as $x^3 + x^2 + 4x - 4$, cannot be factored by grouping.

103. The polynomial $28x^3 - 7x^2 + 36x - 9$ can be factored by grouping terms as follows:

$$(28x^3 + 36x) + (-7x^2 - 9).$$

104. $x^2 - 2$ is a factor of $2 - 50x - x^2 + 25x^3$.

In Exercises 105–107, factor each polynomial. Assume that all variable exponents represent whole numbers.

105. $x^{4n} + x^{2n} + x^{3n}$

106. $3x^{3m}y^m - 6x^{2m}y^{2m}$

107. $8y^{2n+4} + 16y^{2n+3} - 12y^{2n}$

In Exercises 108–109, write a polynomial that fits the given description. Do not use a polynomial that appeared in this section or in the Exercise Set.

108. The polynomial has three terms and can be factored using a greatest common factor that has both a negative coefficient and a variable.

109. The polynomial has four terms and can be factored by grouping.

Review Exercises

110. Solve by Cramer's rule:

$$\begin{cases} 3x - 2y = 8 \\ 2x - 5y = 10. \end{cases}$$

(Section 3.5, Example 2)

111. Determine whether each relation is a function.

   a. $\{(0, 5), (3, -5), (5, 5), (7, -5)\}$

   b. $\{(1, 2), (3, 4), (5, 5), (5, 6)\}$ (Section 2.1, Example 2)

112. The length of a rectangle is 2 feet greater than twice its width. If the rectangle's perimeter is 22 feet, find the length and width. (Section 1.5, Example 5)

Preview Exercises

Exercises 113–115 will help you prepare for the material covered in the next section. In each exercise, replace the boxed question mark with an integer that results in the given product. Some trial and error may be necessary.

113. $(x + 3)(x + \boxed{?}) = x^2 + 7x + 12$

114. $(x - \boxed{?})(x - 12) = x^2 - 14x + 24$

115. $(x + 3y)(x - \boxed{?}y) = x^2 - 4xy - 21y^2$

5.4 Factoring Trinomials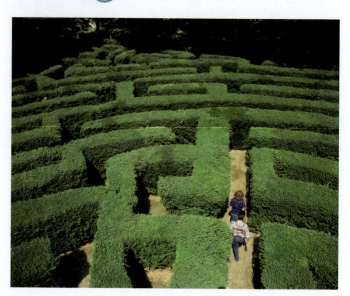

What am I supposed to learn?

After studying this section, you should be able to:

1 Factor a trinomial whose leading coefficient is 1.

2 Factor using a substitution.

3 Factor a trinomial whose leading coefficient is not 1.

4 Factor trinomials by grouping.

---

1 Factor a trinomial whose leading coefficient is 1.

A great deal of trial and error is involved in finding your way out of this maze. Trial and error play an important role in problem solving and can be helpful in leading to correct solutions. In this section, you will use trial and error to factor trinomials following a problem-solving process that is not very different from learning to traverse the maze.

Factoring a Trinomial Whose Leading Coefficient Is 1

In Section 5.2, we used the FOIL method to multiply two binomials. The product was often a trinomial. The following are some examples:

Factored Form	F	O	I	L	Trinomial Form
$(x + 3)(x + 4) = x^2 + 4x + 3x + 12 = x^2 + 7x + 12$
$(x - 3)(x - 4) = x^2 - 4x - 3x + 12 = x^2 - 7x + 12$
$(x + 3)(x - 5) = x^2 - 5x + 3x - 15 = x^2 - 2x - 15$

Observe that each trinomial is of the form $x^2 + bx + c$, where the coefficient of the squared term is 1. Our goal in the first part of this section is to start with the trinomial form and, assuming that it is factorable, return to the factored form.

The first FOIL multiplication shown in our list indicates that

$$(x + 3)(x + 4) = x^2 + 7x + 12.$$

Let's reverse the sides of this equation:

$$x^2 + 7x + 12 = (x + 3)(x + 4).$$

We can use $x^2 + 7x + 12 = (x + 3)(x + 4)$ to make several important observations about the factors on the right side.

$$x^2 + 7x + 12 = (x + 3)(x + 4)$$

The first term of each factor is x. The product of the First terms is $x \cdot x = x^2$.

$$x^2 + 7x + 12 = (x + 3)(x + 4)$$

3 and 4 are factors of 12. The product of the Last terms is $3 \cdot 4 = 12$.

$$x^2 + 7x + 12 = (x + 3)(x + 4)$$

I: $3x$

O: $4x$

The sum of the Outside and Inside products is $4x + 3x = 7x$.

These observations provide us with a procedure for factoring $x^2 + bx + c$.

A Strategy for Factoring $x^2 + bx + c$

1. Enter x as the first term of each factor.

$$(x \quad)(x \quad) = x^2 + bx + c$$

2. List pairs of factors of the constant c.

3. Try various combinations of these factors. Select the combination in which the sum of the Outside and Inside products is equal to bx.

4. Check your work by multiplying the factors using the FOIL method. You should obtain the original trinomial.

If none of the possible combinations yield an Outside product and an Inside product whose sum is equal to bx, the trinomial cannot be factored using integers and is called **prime** over the set of integers.

EXAMPLE 1 Factoring a Trinomial Whose Leading Coefficient Is 1

Factor: $x^2 + 5x + 6$.

Solution

Step 1. Enter x as the first term of each factor.

$$x^2 + 5x + 6 = (x \quad)(x \quad)$$

Step 2. List pairs of factors of the constant, 6.

| Factors of 6 | 6, 1 | 3, 2 | −6, −1 | −3, −2 |

Step 3. Try various combinations of these factors. The correct factorization of $x^2 + 5x + 6$ is the one in which the sum of the Outside and Inside products is equal to $5x$. Here is a list of the possible factorizations.

Possible Factorizations of $x^2 + 5x + 6$	Sum of Outside and Inside Products (Should Equal 5x)
$(x + 6)(x + 1)$	$x + 6x = 7x$
$(x + 3)(x + 2)$	$2x + 3x = 5x$
$(x - 6)(x - 1)$	$-x - 6x = -7x$
$(x - 3)(x - 2)$	$-2x - 3x = -5x$

This is the required middle term.

Thus,

$$x^2 + 5x + 6 = (x + 3)(x + 2).$$

Step 4. Check this result by multiplying the right side using the FOIL method. You should obtain the original trinomial. Because of the commutative property, the factorization can also be expressed as

$$x^2 + 5x + 6 = (x + 2)(x + 3). \quad \blacksquare$$

In factoring a trinomial of the form $x^2 + bx + c$, you can speed things up by listing the factors of c and then finding their sums. We are interested in a sum of b. For example, in factoring $x^2 + 5x + 6$, we are interested in the factors of 6 whose sum is 5.

Factors of 6	6, 1	3, 2	−6, −1	−3, −2
Sum of Factors	7	5	−7	−5

This is the desired sum.

Thus, $x^2 + 5x + 6 = (x + 3)(x + 2)$.

Using Technology

Numeric and Graphic Connections

If a polynomial contains one variable, a graphing utility can be used to check its factorization. For example, the factorization in Example 1 can be checked graphically or numerically.

Numeric Check

Use the TABLE feature.

Graphic Check

Use the GRAPH feature to display graphs for y_1 and y_2.

$[-7, 5, 1]$ by $[-1, 10, 1]$

Notice that the graph of the quadratic function is shaped like a bowl. The graph of the even-degree quadratic function exhibits the same behavior at each end, rising to the left and rising to the right (↖, ↗).

✓ **CHECK POINT 1**   Factor: $x^2 + 6x + 8$.

EXAMPLE 2   Factoring a Trinomial Whose Leading Coefficient Is 1

Factor: $x^2 - 14x + 24$.

Solution

Step 1. Enter *x* as the first term of each factor.

$$x^2 - 14x + 24 = (x \quad)(x \quad)$$

To find the second term of each factor, we must find two integers whose product is 24 and whose sum is −14.

Step 2. List pairs of factors of the constant, 24. Because the desired sum, −14, is negative, we will list only the negative pairs of factors of 24.

Negative Factors of 24	−24, −1	−12, −2	−8, −3	−6, −4

Step 3. Try various combinations of these factors. We are interested in the factors whose sum is -14.

Negative Factors of 24	$-24, -1$	$-12, -2$	$-8, -3$	$-6, -4$
Sum of Factors	-25	-14	-11	-10

This is the desired sum.

Thus, $x^2 - 14x + 24 = (x - 12)(x - 2)$. ∎

Great Question!

Is there a way to eliminate some of the combinations of factors for $x^2 + bx + c$ when c is positive?

Yes. To factor $x^2 + bx + c$ when c is positive, find two numbers with the same sign as the middle term.

$$x^2 + 5x + 6 = (x + 3)(x + 2)$$

Same signs

$$x^2 - 14x + 24 = (x - 12)(x - 2)$$

Same signs

✓ CHECK POINT 2   Factor: $x^2 - 9x + 20$.

EXAMPLE 3   Factoring a Trinomial Whose Leading Coefficient Is 1

Factor: $y^2 + 7y - 60$.

Solution

Step 1. Enter y as the first term of each factor.

$$y^2 + 7y - 60 = (y \quad)(y \quad)$$

To find the second term of each factor, we must find two integers whose product is -60 and whose sum is 7.

Steps 2 and 3. List pairs of factors of the constant, -60, and try various combinations of these factors. Because the desired sum, 7, is positive, the positive factor of -60 must be farther from 0 than the negative factor is. Thus, we will only list pairs of factors of -60 in which the positive factor has the larger absolute value.

Some Factors of -60	$60, -1$	$30, -2$	$20, -3$	$15, -4$	$12, -5$	$10, -6$
Sum of Factors	59	28	17	11	7	4

This is the desired sum.

Thus, $y^2 + 7y - 60 = (y + 12)(y - 5)$. ∎

Great Question!

Is there a way to eliminate some of the combinations of factors for $x^2 + bx + c$ when c is negative?

Yes. To factor $x^2 + bx + c$ when c is negative, find two numbers with opposite signs whose sum is the coefficient of the middle term.

$$y^2 + 7y - 60 = (y + 12)(y - 5)$$

Negative — Opposite signs

✓ **CHECK POINT 3**   Factor: $y^2 + 19y - 66$.

EXAMPLE 4   Factoring a Trinomial in Two Variables

Factor: $x^2 - 4xy - 21y^2$.

Solution

Step 1. Enter x as the first term of each factor. Because the last term of the trinomial contains y^2, the second term of each factor must contain y.

$$x^2 - 4xy - 21y^2 = (x \quad ?y)(x \quad ?y)$$

The question marks indicate that we are looking for the coefficients of y in each factor. To find these coefficients, we must find two integers whose product is -21 and whose sum is -4.

Steps 2 and 3. List pairs of factors of the coefficient of the last term, -21, and try various combinations of these factors. We are interested in the factors whose sum is -4.

Factors of -21	1, -21	3, -7	-1, 21	-3, 7
Sum of Factors	-20	-4	20	4

This is the desired sum.

Thus, $x^2 - 4xy - 21y^2 = (x + 3y)(x - 7y)$ or $(x - 7y)(x + 3y)$.

Step 4. Verify the factorization using the FOIL method.

$$(x + 3y)(x - 7y) = x^2 - 7xy + 3xy - 21y^2 = x^2 - 4xy - 21y^2$$

Because the product of the factors is the original polynomial, the factorization is correct. ■

✓ **CHECK POINT 4**   Factor: $x^2 - 5xy + 6y^2$.

Can every trinomial be factored? The answer is no. For example, consider

$$x^2 + x - 5 = (x \qquad)(x \qquad).$$

To find the second term of each factor, we must find two integers whose product is -5 and whose sum is 1. Because no such integers exist, $x^2 + x - 5$ cannot be factored. This trinomial is prime.

To factor some polynomials, more than one technique must be used. **Always begin by trying to factor out the greatest common factor.** A polynomial is **factored completely** when it is written as the product of prime polynomials.

Great Question!

Can my graphing utility verify that a polynomial is factored completely?

The technology box on page 352 shows how to use a graphing utility to check a polynomial's factorization graphically or numerically. The graphing utility verifies that the factorization is equivalent to the original polynomial. However, a graphing utility cannot verify that the factorization is complete.

EXAMPLE 5 Factoring Completely

Factor: $8x^3 - 40x^2 - 48x$.

Solution The GCF of the three terms of the polynomial is $8x$. We begin by factoring out $8x$. Then we factor the remaining trinomial.

$$8x^3 - 40x^2 - 48x$$
$$= 8x(x^2 - 5x - 6) \quad \text{Factor out the GCF.}$$
$$= 8x(x \quad)(x \quad) \quad \text{Begin factoring } x^2 - 5x - 6. \text{ Find two integers}$$
$$\qquad\qquad\qquad\qquad \text{whose product is } -6 \text{ and whose sum is } -5.$$
$$= 8x(x - 6)(x + 1) \quad \text{The integers are } -6 \text{ and } 1.$$

Thus,

$$8x^3 - 40x^2 - 48x = 8x(x - 6)(x + 1).$$

Be sure to include the GCF in the factorization.

You can check this factorization by multiplying the binomials using the FOIL method. Then use the distributive property and multiply each term in this product by $8x$. Try doing this now. Because the factorization is correct, you should obtain the original polynomial. ■

✓ **CHECK POINT 5** Factor: $3x^3 - 15x^2 - 42x$.

Some trinomials, such as $-x^2 + 5x + 6$, have a leading coefficient of -1. Because it is easier to factor a trinomial with a positive leading coefficient, begin by factoring out -1. For example,

$$-x^2 + 5x + 6 = -1(x^2 - 5x - 6) = -(x - 6)(x + 1).$$

2 Factor using a substitution.

In some trinomials, the highest power is greater than 2, and the exponent in one of the terms is half that of the other term. By letting u equal the variable to the smaller power, the trinomial can be written in a form that makes its possible factorization more obvious. Here are some examples:

Given Trinomial	Substitution	New Trinomial
$x^6 - 8x^3 + 15$ or $(x^3)^2 - 8x^3 + 15$	$u = x^3$	$u^2 - 8u + 15$
$x^4 - 8x^2 - 9$ or $(x^2)^2 - 8x^2 - 9$	$u = x^2$	$u^2 - 8u - 9$

In each case, we factor the given trinomial by working with the new trinomial on the right. If a factorization is found, we replace all occurrences of u in the factorization with the substitution shown in the middle column.

EXAMPLE 6   Factoring by Substitution

Factor: $x^6 - 8x^3 + 15$.

Solution   Notice that the exponent on x^3 is half that of the exponent on x^6. We will let u equal the variable to the power that is half of 6. Thus, let $u = x^3$.

$$(x^3)^2 - 8x^3 + 15 \qquad \textcolor{blue}{\text{This is the given polynomial, with } x^6 \text{ written as } (x^3)^2.}$$

$$= u^2 - 8u + 15 \qquad \textcolor{blue}{\text{Let } u = x^3. \text{ Rewrite the trinomial in terms of } u.}$$

$$= (u - 5)(u - 3) \qquad \textcolor{blue}{\text{Factor.}}$$

$$= (x^3 - 5)(x^3 - 3) \qquad \textcolor{blue}{\text{Now substitute } x^3 \text{ for } u.}$$

Thus, the given trinomial can be factored as

$$x^6 - 8x^3 + 15 = (x^3 - 5)(x^3 - 3).$$

Check this result using FOIL multiplication on the right. ■

✓ **CHECK POINT 6**   Factor: $x^6 - 7x^3 + 10$.

---

3 Factor a trinomial whose leading coefficient is not 1. ▶

Factoring a Trinomial Whose Leading Coefficient Is Not 1

How do we factor a trinomial such as $5x^2 - 14x + 8$? Notice that the leading coefficient is 5. We must find two binomials whose product is $5x^2 - 14x + 8$. The product of the First terms must be $5x^2$:

$$(5x \qquad)(x \qquad).$$

From this point on, the factoring strategy is exactly the same as the one we use to factor a trinomial whose leading coefficient is 1.

A Strategy for Factoring $ax^2 + bx + c$

Assume, for the moment, that there is no greatest common factor.

1. Find two First terms whose product is ax^2:

$$(\Box x + \quad)(\Box x + \quad) = ax^2 + bx + c.$$

2. Find two Last terms whose product is c:

$$(\Box x + \Box)(\Box x + \Box) = ax^2 + bx + c.$$

3. By trial and error, perform steps 1 and 2 until the sum of the Outside product and Inside product is bx:

$$(\Box x + \Box)(\Box x + \Box) = ax^2 + bx + c.$$
$$\text{Sum of O + I}$$

If no such combinations exist, the polynomial is prime.

Great Question!

Should I feel discouraged if it takes me a while to get the correct factorization?

The *error* part of the factoring strategy plays an important role in the process. If you do not get the correct factorization the first time, this is not a bad thing. This error is often helpful in leading you to the correct factorization.

EXAMPLE 7 Factoring a Trinomial Whose Leading Coefficient Is Not 1

Factor: $5x^2 - 14x + 8$.

Solution

Step 1. Find two First terms whose product is $5x^2$.

$$5x^2 - 14x + 8 = (5x \quad)(x \quad)$$

Great Question!

When factoring trinomials, must I list every possible factorization before getting the correct one?

With practice, you will find that you don't have to list every possible factorization of the trinomial. As you practice factoring, you will be able to narrow down the list of possible factorizations to just a few. When it comes to factoring, practice makes perfect.

Step 2. Find two Last terms whose product is 8. The number 8 has pairs of factors that are either both positive or both negative. Because the middle term, $-14x$, is negative, both factors must be negative. The negative factorizations of 8 are $-1(-8)$ and $-2(-4)$.

Step 3. Try various combinations of these factors. The correct factorization of $5x^2 - 14x + 8$ is the one in which the sum of the Outside and Inside products is equal to $-14x$. Here is a list of the possible factorizations:

Possible Factorizations of $5x^2 - 14x + 8$	Sum of Outside and Inside Products (Should Equal $-14x$)	
$(5x - 1)(x - 8)$	$-40x - x = -41x$	
$(5x - 8)(x - 1)$	$-5x - 8x = -13x$	
$(5x - 2)(x - 4)$	$-20x - 2x = -22x$	
$(5x - 4)(x - 2)$	$-10x - 4x = -14x$	This is the required middle term.

Thus,

$$5x^2 - 14x + 8 = (5x - 4)(x - 2).$$

Show that this factorization is correct by multiplying the factors using the FOIL method. You should obtain the original trinomial. ∎

✓ **CHECK POINT 7** Factor: $3x^2 - 20x + 28$.

EXAMPLE 8 Factoring a Trinomial Whose Leading Coefficient Is Not 1

Factor: $8x^6 - 10x^5 - 3x^4$.

Solution The GCF of the three terms of the polynomial is x^4. We begin by factoring out x^4.

$$8x^6 - 10x^5 - 3x^4 = x^4(8x^2 - 10x - 3)$$

Now we factor the remaining trinomial, $8x^2 - 10x - 3$.

Step 1. Find two First terms whose product is $8x^2$.

$$8x^2 - 10x - 3 \overset{?}{=} (8x \quad)(x \quad)$$
$$8x^2 - 10x - 3 \overset{?}{=} (4x \quad)(2x \quad)$$

Step 2. Find two Last terms whose product is -3. The possible factorizations are $1(-3)$ and $-1(3)$.

Step 3. Try various combinations of these factors. The correct factorization of $8x^2 - 10x - 3$ is the one in which the sum of the Outside and Inside products is equal to $-10x$. At the top of the next page is a list of the possible factorizations.

Great Question!

I space out reading your long lists of possible factorizations. Are there any rules for shortening these lists?

Here are some suggestions for reducing the list of possible factorizations for $ax^2 + bx + c$.

1. If b is relatively small, avoid the larger factors of a.

2. If c is positive, the signs in both binomial factors must match the sign of b.

3. If the trinomial has no common factor, no binomial factor can have a common factor.

4. Reversing the signs in the binomial factors reverses the sign of bx, the middle term.

Possible Factorizations of $8x^2 - 10x - 3$	Sum of Outside and Inside Products (Should Equal $-10x$)
$(8x + 1)(x - 3)$	$-24x + x = -23x$
$(8x - 3)(x + 1)$	$8x - 3x = 5x$
$(8x - 1)(x + 3)$	$24x - x = 23x$
$(8x + 3)(x - 1)$	$-8x + 3x = -5x$
$(4x + 1)(2x - 3)$	$-12x + 2x = -10x$ ← This is the required middle term.
$(4x - 3)(2x + 1)$	$4x - 6x = -2x$
$(4x - 1)(2x + 3)$	$12x - 2x = 10x$
$(4x + 3)(2x - 1)$	$-4x + 6x = 2x$

The factorization of $8x^2 - 10x - 3$ is $(4x + 1)(2x - 3)$. Now we include the GCF in the complete factorization of the given polynomial. Thus,

$$8x^6 - 10x^5 - 3x^4 = x^4(8x^2 - 10x - 3) = x^4(4x + 1)(2x - 3).$$

This is the complete factorization with the GCF, x^4, included. ∎

✓ **CHECK POINT 8** Factor: $6x^6 + 19x^5 - 7x^4$.

We have seen that not every trinomial can be factored. For example, consider

$$6x^2 + 14x + 7 = (6x + \square)(x + \square)$$

$$6x^2 + 14x + 7 = (3x + \square)(2x + \square).$$

The possible factors for the last term are 1 and 7. However, regardless of how these factors are placed in the boxes shown, the sum of the Outside and Inside products is not equal to $14x$. Thus, the trinomial $6x^2 + 14x + 7$ cannot be factored and is prime.

EXAMPLE 9 Factoring a Trinomial in Two Variables

Factor: $3x^2 - 13xy + 4y^2$.

Solution

Step 1. Find two First terms whose product is $3x^2$.

$$3x^2 - 13xy + 4y^2 = (3x \quad ?y)(x \quad ?y)$$

The question marks indicate that we are looking for the coefficients of y in each factor. **Steps 2 and 3. List pairs of factors of the coefficient of the last term, 4, and try various combinations of these factors.** The correct factorization is the one in which the sum of the Outside and Inside products is equal to $-13xy$. Because of the negative coefficient, -13, we will consider only the negative pairs of factors of 4. The possible factorizations are $-1(-4)$ and $-2(-2)$.

Possible Factorizations of $3x^2 - 13xy + 4y^2$	Sum of Outside and Inside Products (Should Equal $-13xy$)
$(3x - y)(x - 4y)$	$-12xy - xy = -13xy$ ← This is the required middle term.
$(3x - 4y)(x - y)$	$-3xy - 4xy = -7xy$
$(3x - 2y)(x - 2y)$	$-6xy - 2xy = -8xy$

Thus,

$$3x^2 - 13xy + 4y^2 = (3x - y)(x - 4y). \quad ∎$$

✓ **CHECK POINT 9**   Factor: $2x^2 - 7xy + 3y^2$.

EXAMPLE 10   Factoring by Substitution

Factor: $6y^4 + 13y^2 + 6$.

Solution   Notice that the exponent on y^2 is half that of the exponent on y^4. We will let u equal the variable to the smaller power. Thus, let $u = y^2$.

$$6(y^2)^2 + 13y^2 + 6 \qquad \text{\color{teal}This is the given polynomial, with } y^4 \text{ written as } (y^2)^2.$$

$$= 6u^2 + 13u + 6 \qquad \text{\color{teal}Let } u = y^2. \text{ Rewrite the trinomial in terms of } u.$$

$$= (3u + 2)(2u + 3) \qquad \text{\color{teal}Factor the trinomial.}$$

$$= (3y^2 + 2)(2y^2 + 3) \qquad \text{\color{teal}Now substitute } y^2 \text{ for } u.$$

Therefore, $6y^4 + 13y^2 + 6 = (3y^2 + 2)(2y^2 + 3)$. Check using FOIL multiplication.   ∎

✓ **CHECK POINT 10**   Factor: $3y^4 + 10y^2 - 8$.

---

4 Factor trinomials by grouping. ▶

Factoring Trinomials by Grouping

A second method for factoring $ax^2 + bx + c, a \neq 1$, is called the **grouping method**. This method involves both trial and error, as well as grouping. The trial and error in factoring $ax^2 + bx + c$ depends upon finding two numbers, p and q, for which $p + q = b$. Then we factor $ax^2 + px + qx + c$ using grouping.

Let's see how this works by looking at a particular factorization:

$$15x^2 - 7x - 2 = (3x - 2)(5x + 1).$$

If we multiply using FOIL on the right, we obtain

$$(3x - 2)(5x + 1) = 15x^2 + 3x - 10x - 2.$$

In this case, the desired numbers, p and q, are $p = 3$ and $q = -10$. Compare these numbers to ac and b in the given polynomial.

$$\boxed{a = 15} \quad \boxed{b = -7} \quad \boxed{c = -2}$$

$$15x^2 - 7x - 2$$

$$\boxed{ac = 15(-2) = -30}$$

Can you see that p and q, 3 and -10, are factors of ac, or -30? Furthermore, p and q have a sum of b, namely -7. By expressing the middle term, $-7x$, in terms of p and q, we can factor by grouping as follows:

$$15x^2 - 7x - 2$$

$$= 15x^2 + (3x - 10x) - 2 \qquad \text{\color{teal}Rewrite } -7x \text{ as } 3x - 10x.$$

$$= (15x^2 + 3x) + (-10x - 2) \qquad \text{\color{teal}Group terms.}$$

$$= 3x(5x + 1) - 2(5x + 1) \qquad \text{\color{teal}Factor from each group.}$$

$$= (5x + 1)(3x - 2). \qquad \text{\color{teal}Factor out } 5x + 1, \text{ the common binomial factor.}$$

> **Factoring $ax^2 + bx + c$ Using Grouping ($a \neq 1$)**
>
> 1. Multiply the leading coefficient, a, and the constant, c.
> 2. Find the factors of ac whose sum is b.
> 3. Rewrite the middle term, bx, as a sum or difference using the factors from step 2.
> 4. Factor by grouping.

EXAMPLE 11   Factoring a Trinomial by Grouping

Factor by grouping: $12x^2 - 5x - 2$.

Solution   The trinomial is of the form $ax^2 + bx + c$.

$$12x^2 - 5x - 2$$

$a = 12$ $b = -5$ $c = -2$

Discover for Yourself

In step 2, we found that the desired numbers were -8 and 3. We wrote $-5x$ as $-8x + 3x$. What happens if we write $-5x$ as $3x - 8x$? Use factoring by grouping on

$$12x^2 - 5x - 2$$
$$= 12x^2 + 3x - 8x - 2.$$

Describe what happens.

Step 1.   Multiply the leading coefficient, a, and the constant, c. Using $a = 12$ and $c = -2$,

$$ac = 12(-2) = -24.$$

Step 2.   Find the factors of ac whose sum is b. We want the factors of -24 whose sum is b, or -5. The factors of -24 whose sum is -5 are -8 and 3.

Step 3.   Rewrite the middle term, $-5x$, as a sum or difference using the factors from step 2, -8 and 3.

$$12x^2 - 5x - 2 = 12x^2 - 8x + 3x - 2$$

Step 4.   Factor by grouping.

$$
\begin{aligned}
&= (12x^2 - 8x) + (3x - 2) && \text{Group terms.}\\
&= 4x(3x - 2) + 1(3x - 2) && \text{Factor from each group.}\\
&= (3x - 2)(4x + 1) && \text{Factor out } 3x - 2, \text{ the}\\
& && \text{common binomial factor.}
\end{aligned}
$$

Thus,

$$12x^2 - 5x - 2 = (3x - 2)(4x + 1). \;\blacksquare$$

✓ **CHECK POINT 11**   Factor by grouping: $8x^2 - 22x + 5$.

CONCEPT AND VOCABULARY CHECK

Fill in each blank so that the resulting statement is true.

1. A polynomial is factored _____ when it is written as a product of prime polynomials.

2. We begin the process of factoring a polynomial by first factoring out the _____, assuming that there is one other than 1.

3. $x^2 + 8x + 15 = (x + 3)(x \underline{\qquad})$

4. $x^2 - 9x + 20 = (x - 5)(x \underline{\qquad})$

5. $x^2 + 13x - 48 = (x - 3)(x \underline{\qquad})$

6. $x^2 + 17xy - 60y^2 = (x + 20y)(x \underline{\qquad})$

7. $2x^3 - 30x^2 - 108x = \underline{\qquad}(x^2 - 15x - 54) = \underline{\qquad}(x + 3)(x \underline{\qquad})$

8. $3x^2 - 14x + 11 = (x - 1)(3x \underline{\qquad})$

9. $6x^2 + 13x - 63 = (3x - 7)(\underline{\qquad})$

10. $63x^2 - 89xy + 30y^2 = (9x - 5y)(7x \underline{\qquad})$

5.4 EXERCISE SET ⊙ MyMathLab®

Practice Exercises

In Exercises 1–30, factor each trinomial, or state that the trinomial is prime. Check each factorization using FOIL multiplication.

1. $x^2 + 5x + 6$

2. $x^2 + 10x + 9$

3. $x^2 + 8x + 12$

4. $x^2 + 8x + 15$

5. $x^2 + 9x + 20$

6. $x^2 + 11x + 24$

7. $y^2 + 10y + 16$

8. $y^2 + 9y + 18$

9. $x^2 - 8x + 15$

10. $x^2 - 5x + 6$

11. $y^2 - 12y + 20$

12. $y^2 - 25y + 24$

13. $a^2 + 5a - 14$

14. $a^2 + a - 12$

15. $x^2 + x - 30$

16. $x^2 + 14x - 32$

17. $x^2 - 3x - 28$

18. $x^2 - 4x - 21$

19. $y^2 - 5y - 36$

20. $y^2 - 3y - 40$

21. $x^2 - x + 7$

22. $x^2 + 3x + 8$

23. $x^2 - 9xy + 14y^2$

24. $x^2 - 8xy + 15y^2$

25. $x^2 - xy - 30y^2$

26. $x^2 - 3xy - 18y^2$

27. $x^2 + xy + y^2$

28. $x^2 - xy + y^2$

29. $a^2 - 18ab + 80b^2$

30. $a^2 - 18ab + 45b^2$

In Exercises 31–38, factor completely.

31. $3x^2 + 3x - 18$

32. $4x^2 - 4x - 8$

33. $2x^3 - 14x^2 + 24x$

34. $2x^3 + 6x^2 + 4x$

35. $3y^3 - 15y^2 + 18y$

36. $4y^3 + 12y^2 - 72y$

37. $2x^4 - 26x^3 - 96x^2$

38. $3x^4 + 54x^3 + 135x^2$

In Exercises 39–44, factor by introducing an appropriate substitution.

39. $x^6 - x^3 - 6$

40. $x^6 + x^3 - 6$

41. $x^4 - 5x^2 - 6$

42. $x^4 - 4x^2 - 5$

43. $(x + 1)^2 + 6(x + 1) + 5$ (Let $u = x + 1$.)

44. $(x + 1)^2 + 8(x + 1) + 7$ (Let $u = x + 1$.)

In Exercises 45–68, use the method of your choice to factor each trinomial, or state that the trinomial is prime. Check each factorization using FOIL multiplication.

45. $3x^2 + 8x + 5$

46. $2x^2 + 9x + 7$

47. $5x^2 + 56x + 11$

48. $5x^2 - 16x + 3$

49. $3y^2 + 22y - 16$

50. $5y^2 + 33y - 14$

51. $4y^2 + 9y + 2$

52. $8y^2 + 10y + 3$

53. $10x^2 + 19x + 6$

54. $6x^2 + 19x + 15$

55. $8x^2 - 18x + 9$

56. $4x^2 - 27x + 18$

57. $6y^2 - 23y + 15$

58. $16y^2 - 6y - 27$

59. $6y^2 + 14y + 3$

60. $4y^2 + 22y - 5$

61. $3x^2 + 4xy + y^2$

62. $2x^2 + 3xy + y^2$

63. $6x^2 - 7xy - 5y^2$

64. $6x^2 - 5xy - 6y^2$

65. $15x^2 - 31xy + 10y^2$

66. $15x^2 + 11xy - 14y^2$

67. $3a^2 - ab - 14b^2$

68. $15a^2 - ab - 6b^2$

In Exercises 69–82, factor completely.

69. $15x^3 - 25x^2 + 10x$

70. $10x^3 + 24x^2 + 14x$

71. $24x^4 + 10x^3 - 4x^2$

72. $15x^4 - 39x^3 + 18x^2$

73. $15y^5 - 2y^4 - y^3$

74. $10y^5 - 17y^4 + 3y^3$

75. $24x^2 + 3xy - 27y^2$

76. $12x^2 + 10xy - 8y^2$

77. $6a^2b - 2ab - 60b$

78. $8a^2b + 34ab - 84b$

79. $12x^2y - 34xy^2 + 14y^3$

80. $12x^2y - 46xy^2 + 14y^3$

81. $13x^3y^3 + 39x^3y^2 - 52x^3y$

82. $4x^3y^5 + 24x^2y^5 - 64xy^5$

In Exercises 83–92, factor by introducing an appropriate substitution.

83. $2x^4 - x^2 - 3$

84. $5x^4 + 2x^2 - 3$

85. $2x^6 + 11x^3 + 15$

86. $2x^6 + 13x^3 + 15$

87. $2y^{10} + 7y^5 + 3$

88. $5y^{10} + 29y^5 - 42$

89. $5(x + 1)^2 + 12(x + 1) + 7$ (Let $u = x + 1$.)

90. $3(x + 1)^2 - 5(x + 1) + 2$ (Let $u = x + 1$.)

91. $2(x - 3)^2 - 5(x - 3) - 7$

92. $3(x - 2)^2 - 5(x - 2) - 2$

Practice PLUS

In Exercises 93–100, factor completely.

93. $x^2 - 0.5x + 0.06$

94. $x^2 + 0.3x - 0.04$

95. $x^2 - \dfrac{3}{49} + \dfrac{2}{7}x$

96. $x^2 - \dfrac{6}{25} + \dfrac{1}{5}x$

97. $acx^2 - bcx + adx - bd$

98. $acx^2 - bcx - adx + bd$

99. $-4x^5y^2 + 7x^4y^3 - 3x^3y^4$

100. $-5x^4y^3 + 7x^3y^4 - 2x^2y^5$

101. If $(fg)(x) = 3x^2 - 22x + 39$, find f and g.

102. If $(fg)(x) = 4x^2 - x - 5$, find f and g.

In Exercises 103–104, a large rectangle formed by a number of smaller rectangles is shown. Factor the sum of the areas of the smaller rectangles to determine the dimensions of the large rectangle.

103.   **104.**

Application Exercises

105. A diver jumps directly upward from a board that is 32 feet high. The function

$$f(t) = -16t^2 + 16t + 32$$

describes the diver's height above the water, $f(t)$, in feet, after t seconds.

a. Find and interpret $f(1)$.

b. Find and interpret $f(2)$.

c. Factor the expression for $f(t)$ and write the function in completely factored form.

d. Use the factored form of the function to find $f(1)$ and $f(2)$.

106. The function $V(x) = 3x^3 - 2x^2 - 8x$ describes the volume, $V(x)$, in cubic inches, of the box shown whose height is x inches.

a. Find and interpret $V(4)$.

b. Factor the expression for $V(x)$ and write the function in completely factored form.

c. Use the factored form of the function to find $V(4)$ and $V(5)$.

107. Find the area of the large rectangle shown below in two ways.

a. Find the sum of the areas of the six smaller rectangles and squares.

b. Express the area of the large rectangle as the product of its length and width.

c. Explain how the figure serves as a geometric model for the factorization of the sum that you wrote in part (a).

108. If x represents a positive integer, factor $x^3 + 3x^2 + 2x$ to show that the trinomial represents the product of three consecutive integers.

Explaining the Concepts

109. Explain how to factor $x^2 + 8x + 15$.

110. Give two helpful suggestions for factoring $x^2 - 5x + 6$.

111. In factoring $x^2 + bx + c$, describe how the last term in each binomial factor is related to b and to c.

112. Describe the first thing that you should try doing when factoring a polynomial.

113. What does it mean to factor completely?

114. Explain how to factor $x^6 - 7x^3 + 10$ by substitution.

115. Is it possible to factor $x^6 - 7x^3 + 10$ without using substitution? How might this be done?

116. Explain how to factor $2x^2 - x - 1$.

Technology Exercises

In Exercises 117–120, use a graphing utility to graph the function on each side of the equation in the same viewing rectangle. Use end behavior to show a complete picture of the polynomial function on the left side. Do the graphs coincide? If so, this means that the polynomial on the left side has been factored correctly. If not, factor the polynomial correctly and then use your graphing utility to verify the factorization.

117. $x^2 + 7x + 12 = (x + 4)(x + 3)$

118. $x^2 - 7x + 6 = (x - 2)(x - 3)$

119. $6x^3 + 5x^2 - 4x = x(3x + 4)(2x - 1)$

120. $x^4 - x^2 - 20 = (x^2 + 5)(x^2 - 4)$

121. Use the $\boxed{\text{TABLE}}$ feature of a graphing utility to verify any two of your factorizations in Exercises 39–44.

Critical Thinking Exercises

Make Sense?   *In Exercises 122–125, determine whether each statement makes sense or does not make sense, and explain your reasoning.*

122. Although $(x + 2)(x - 5)$ is the same as $(x - 5)(x + 2)$, the factorization $(2 - x)(2 + x)$ is not the same as $-(x - 2)(x + 2)$.

123. I'm often able to use an incorrect factorization to lead me to the correct factorization.

124. My graphing calculator showed the same graph for $y_1 = 20x^3 - 70x^2 + 60x$ and $y_2 = 10x(2x^2 - 7x + 6)$, so I can conclude that the complete factorization of $20x^3 - 70x^2 + 60x$ is $10x(2x^2 - 7x + 6)$.

125. First factoring out the greatest common factor makes it easier for me to determine how to factor the remaining factor, assuming that it is not prime.

In Exercises 126–129, determine whether each statement is true or false. If the statement is false, make the necessary change(s) to produce a true statement.

126. Once a GCF is factored from $6y^6 - 19y^5 + 10y^4$, the remaining trinomial factor is prime.

127. One factor of $8y^2 - 51y + 18$ is $8y - 3$.

128. We can immediately tell that $6x^2 - 11xy - 10y^2$ is prime because 11 is a prime number and the polynomial contains two variables.

129. A factor of $12x^2 - 19xy + 5y^2$ is $4x - y$.

In Exercises 130–131, find all integers b so that the trinomial can be factored.

130. $4x^2 + bx - 1$

131. $3x^2 + bx + 5$

In Exercises 132–137, factor each polynomial. Assume that all variable exponents represent whole numbers.

132. $9x^{2n} + x^n - 8$

133. $4x^{2n} - 9x^n + 5$

134. $a^{2n+2} - a^{n+2} - 6a^2$

135. $b^{2n+2} + 3b^{n+2} - 10b^2$

136. $3c^{n+2} - 10c^{n+1} + 3c^n$

137. $2d^{n+2} - 5d^{n+1} + 3d^n$

Review Exercises

138. Solve: $-2x \leq 6$ and $-2x + 3 < -7$. (Section 4.2, Example 2)

139. Solve the system:
$$\begin{cases} 2x - y - 2z = -1 \\ x - 2y - z = 1 \\ x + y + z = 4. \end{cases}$$
(Section 3.3, Example 2)

140. Factor: $4x^3 + 8x^2 - 5x - 10$. (Section 5.3, Example 5)

Preview Exercises

Exercises 141–143 will help you prepare for the material covered in the next section. In each exercise, factor the polynomial. (You'll soon be learning techniques that will shorten the factoring process.)

141. $x^2 + 14x + 49$

142. $x^2 - 8x + 16$

143. $x^2 - 25$ (or $x^2 + 0x - 25$)

MID-CHAPTER CHECK POINT   Section 5.1–Section 5.4

What You Know: We learned the vocabulary of polynomials and observed the smooth, continuous graphs of polynomial functions. We used the Leading Coefficient Test to describe the end behavior of these graphs. We learned to add, subtract, and multiply polynomials. We used a number of fast methods for finding products of polynomials, including the FOIL method for multiplying binomials, special-product formulas for squaring binomials $[(A + B)^2 = A^2 + 2AB + B^2; (A - B)^2 = A^2 - 2AB + B^2]$, and a special-product formula for the product of the sum and difference of two terms $[(A + B)(A - B) = A^2 - B^2]$. We learned to factor out a polynomial's greatest common factor and to use grouping to factor polynomials with more than three terms. We factored polynomials with three terms, beginning with trinomials with leading coefficient 1 and moving on to $ax^2 + bx + c$, with $a \neq 1$. We saw that the factoring process should begin by looking for a GCF and, if there is one, factoring it out first.

In Exercises 1–18, perform the indicated operations.

1. $(-8x^3 + 6x^2 - x + 5) - (-7x^3 + 2x^2 - 7x - 12)$

2. $(6x^2yz^4)\left(-\dfrac{1}{3}x^5y^2z\right)$

3. $5x^2y\left(6x^3y^2 - 7xy - \dfrac{2}{5}\right)$

4. $(3x - 5)(x^2 + 3x - 8)$

5. $(x^2 - 2x + 1)(2x^2 + 3x - 4)$

6. $(x^2 - 2x + 1) - (2x^2 + 3x - 4)$

7. $(6x^3y - 11x^2y - 4y) + (-11x^3y + 5x^2y - y - 6)$
   $-(-x^3y + 2y - 1)$

8. $(2x + 5)(4x - 1)$

9. $(2xy - 3)(5xy + 2)$

10. $(3x - 2y)(3x + 2y)$

11. $(3xy + 1)(2x^2 - 3y)$

12. $(7x^3y + 5x)(7x^3y - 5x)$

13. $3(x + h)^2 - 2(x + h) + 5 - (3x^2 - 2x + 5)$

14. $(x^2 - 3)^2$

15. $(x^2 - 3)(x^3 + 5x + 2)$

16. $(2x + 5y)^2$

17. $(x + 6 + 3y)(x + 6 - 3y)$

18. $(x + y + 5)^2$

In Exercises 19–30, factor completely, or state that the polynomial is prime.

19. $x^2 - 5x - 24$

20. $15xy + 5x + 6y + 2$

21. $5x^2 + 8x - 4$

22. $35x^2 + 10x - 50$

23. $9x^2 - 9x - 18$

24. $10x^3y^2 - 20x^2y^2 + 35x^2y$

25. $18x^2 + 21x + 5$

26. $12x^2 - 9xy - 16x + 12y$

27. $9x^2 - 15x + 4$

28. $3x^6 + 11x^3 + 10$

29. $25x^3 + 25x^2 - 14x$

30. $2x^4 - 6x - x^3y + 3y$

SECTION

5.5

Factoring Special Forms ▶

What am I supposed to learn?

After studying this section, you should be able to:

1. Factor the difference of two squares. ▶

2. Factor perfect square trinomials. ▶

3. Use grouping to obtain the difference of two squares. ▶

4. Factor the sum or difference of two cubes. ▶

Bees use honeycombs to store honey and house larvae. They construct honey storage cells from wax. Each cell has the shape of a six-sided figure whose sides are all the same length and whose angles all have the same measure, called a regular hexagon. The cells fit together perfectly, preventing dirt or predators from entering. Squares or equilateral triangles would fit equally well, but regular hexagons provide the largest storage space for the amount of wax used.

In this section, we develop factoring techniques by reversing the formulas for special products discussed in Section 5.2. Like the construction of honeycombs, these factorizations can be visualized by perfectly fitting together "cells" of squares and rectangles to form larger rectangles.

Factoring the Difference of Two Squares

A method for factoring the difference of two squares is obtained by reversing the special product for the sum and difference of two terms.

1 Factor the difference of two squares.

The Difference of Two Squares

If A and B are real numbers, variables, or algebraic expressions, then

$$A^2 - B^2 = (A + B)(A - B).$$

In words: The difference of the squares of two terms factors as the product of a sum and a difference of those terms.

EXAMPLE 1 Factoring the Difference of Two Squares

Factor:

a. $9x^2 - 100$ **b.** $36y^6 - 49x^4$.

Solution We must express each term as the square of some monomial. Then we use the formula for factoring $A^2 - B^2$.

a. $9x^2 - 100 = (3x)^2 - 10^2 = (3x + 10)(3x - 10)$

$$A^2 \quad - \quad B^2 \quad = \quad (A \quad + \quad B) \quad (A \quad - \quad B)$$

b. $36y^6 - 49x^4 = (6y^3)^2 - (7x^2)^2 = (6y^3 + 7x^2)(6y^3 - 7x^2)$ ∎

In order to apply the factoring formula for $A^2 - B^2$, each term must be the square of an integer or a polynomial.

- A number that is the square of an integer is called a **perfect square**. For example, 100 is a perfect square because $100 = 10^2$.

- Any exponential expression involving a perfect-square coefficient and variables to even powers is a perfect square. For example, $100y^6$ is a perfect square because $100y^6 = (10y^3)^2$.

Great Question!

You mentioned that because $100 = 10^2$, 100 is a perfect square. What are some other perfect squares that I should recognize?

It's helpful to identify perfect squares. Here are 16 perfect squares, each printed in boldface.

1 $= 1^2$	**25** $= 5^2$	**81** $= 9^2$	**169** $= 13^2$
4 $= 2^2$	**36** $= 6^2$	**100** $= 10^2$	**196** $= 14^2$
9 $= 3^2$	**49** $= 7^2$	**121** $= 11^2$	**225** $= 15^2$
16 $= 4^2$	**64** $= 8^2$	**144** $= 12^2$	**256** $= 16^2$

✓ **CHECK POINT 1** Factor:

a. $16x^2 - 25$ **b.** $100y^6 - 9x^4$.

Be careful when determining whether or not to apply the factoring formula for the difference of two squares.

Prime Over the Integers

- $x^2 - 5$ - $x^7 - 25$

5 is not a perfect square.

7 is an odd power. x^7 is not the square of any integer power of x.

Factorable

Even powers

- $1 - x^6y^4$

Perfect square: $1 = 1^2$

Perfect square: $x^6y^4 = (x^3y^2)^2$

When factoring, always check first for common factors. If there are common factors, factor out the GCF and then factor the resulting polynomial.

EXAMPLE 2   Factoring Out the GCF and Then Factoring the Difference of Two Squares

Factor: $3y - 3x^6y^5$.

Solution   The GCF of the two terms of the polynomial is $3y$. We begin by factoring out $3y$.

$$3y - 3x^6y^5 = 3y(1 - x^6y^4) = 3y[1^2 - (x^3y^2)^2] = 3y(1 + x^3y^2)(1 - x^3y^2)$$

Factor out the GCF. $A^2 \ - \ B^2 \ = \ (A \ + \ B) \ (A \ - \ B)$ ∎

✓ **CHECK POINT 2**   Factor: $6y - 6x^2y^7$.

We have seen that a polynomial is factored completely when it is written as the product of prime polynomials. To be sure that you have factored completely, check to see whether any factors with more than one term in the factored polynomial can be factored further. If so, continue factoring.

EXAMPLE 3   A Repeated Factorization

Factor completely: $81x^4 - 16$.

Solution

$$81x^4 - 16 = (9x^2)^2 - 4^2$$ Express as the difference of two squares.

$$= (9x^2 + 4)(9x^2 - 4)$$ The factors are the sum and difference of the expressions being squared.

$$= (9x^2 + 4)[(3x)^2 - 2^2]$$ The factor $9x^2 - 4$ is the difference of two squares and can be factored.

$$= (9x^2 + 4)(3x + 2)(3x - 2)$$ The factors of $9x^2 - 4$ are the sum and difference of the expressions being squared. ∎

Are you tempted to further factor $9x^2 + 4$, the sum of two squares, in Example 3? Resist the temptation! **The sum of two squares, $A^2 + B^2$, with no common factor other than 1 is a prime polynomial.**

✓ **CHECK POINT 3**   Factor completely: $16x^4 - 81$.

In our next example, we begin with factoring by grouping. We can then factor further using the difference of two squares.

EXAMPLE 4   Factoring Completely

Factor completely: $x^3 + 5x^2 - 9x - 45$.

Solution

$$x^3 + 5x^2 - 9x - 45$$

$$= (x^3 + 5x^2) + (-9x - 45)$$ Group terms with common factors.

$$= x^2(x + 5) - 9(x + 5)$$ Factor out the common factor from each group.

$$= (x + 5)(x^2 - 9)$$ Factor out $x + 5$, the common binomial factor, from both terms.

$$= (x + 5)(x + 3)(x - 3)$$ Factor $x^2 - 3^2$, the difference of two squares. ∎

Great Question!

Why isn't factoring $81x^4 - 16$ as

$$(9x^2 + 4)(9x^2 - 4)$$

a complete factorization?

The second factor, $9x^2 - 4$, is itself a difference of two squares and can be factored.

✓ **CHECK POINT 4**   Factor completely: $x^3 + 7x^2 - 4x - 28$.

In Examples 1–4, we used the formula for factoring the difference of two squares. Although we obtained the formula by reversing the special product for the sum and difference of two terms, it can also be obtained geometrically.

The combined area of the three purple rectangles is $A^2 - B^2$.

We've rearranged the three purple rectangles. Their combined area is $(A + B)(A - B)$.

Because the three purple rectangles make up the same combined area in both figures,

$$A^2 - B^2 = (A + B)(A - B).$$

2 Factor perfect square trinomials. ⓘ

Factoring Perfect Square Trinomials

Our next factoring technique is obtained by reversing the special products for squaring binomials. The trinomials that are factored using this technique are called **perfect square trinomials.**

> ### Factoring Perfect Square Trinomials
>
> Let A and B be real numbers, variables, or algebraic expressions.
>
> **1.** $A^2 + 2AB + B^2 = (A + B)^2$ **2.** $A^2 - 2AB + B^2 = (A - B)^2$
>
>   Same sign   Same sign

The two items in the box show that perfect square trinomials, $A^2 + 2AB + B^2$ and $A^2 - 2AB + B^2$, come in two forms: one in which the coefficient of the middle term is positive and one in which the coefficient of the middle term is negative. Here's how to recognize a perfect square trinomial:

1. The first and last terms are squares of monomials or integers.

2. The middle term is twice the product of the expressions being squared in the first and last terms.

EXAMPLE 5   Factoring Perfect Square Trinomials

Factor:

a. $x^2 + 14x + 49$ **b.** $4x^2 + 12xy + 9y^2$ **c.** $9y^4 - 12y^2 + 4$.

Solution

a. $x^2 + 14x + 49 = x^2 + 2 \cdot x \cdot 7 + 7^2 = (x + 7)^2$ The middle term has a positive sign.

$A^2 + 2AB + B^2 = (A + B)^2$

b. We suspect that $4x^2 + 12xy + 9y^2$ is a perfect square trinomial because $4x^2 = (2x)^2$ and $9y^2 = (3y)^2$. The middle term can be expressed as twice the product of $2x$ and $3y$.

$$4x^2 + 12xy + 9y^2 = (2x)^2 + 2 \cdot 2x \cdot 3y + (3y)^2 = (2x + 3y)^2$$

$$A^2 \quad + \quad 2AB \quad + \quad B^2 \quad = \quad (A + B)^2$$

c. $9y^4 - 12y^2 + 4 = (3y^2)^2 - 2 \cdot 3y^2 \cdot 2 + 2^2 = (3y^2 - 2)^2$ The middle term has a negative sign. ∎

$$A^2 \quad - \quad 2AB \quad + \quad B^2 \quad = \quad (A - B)^2$$

✓ **CHECK POINT 5** Factor:

a. $x^2 + 6x + 9$

b. $16x^2 + 40xy + 25y^2$

c. $4y^4 - 20y^2 + 25.$

---

3 Use grouping to obtain the difference of two squares. ▶

Using Special Forms When Factoring by Grouping

If a polynomial contains four terms, try factoring by grouping. In the next example, we group the terms to obtain the difference of two squares. One of the squares is a perfect square trinomial.

EXAMPLE 6   Using Grouping to Obtain the Difference of Two Squares

Factor: $x^2 - 8x + 16 - y^2.$

Solution

$$x^2 - 8x + 16 - y^2$$

$$= (x^2 - 8x + 16) - y^2$$ Group as a perfect square trinomial minus y^2 to obtain a difference of two squares.

$$= (x - 4)^2 - y^2$$ Factor the perfect square trinomial.

$$= (x - 4 + y)(x - 4 - y)$$ Factor the difference of two squares. The factors are the sum and difference of the expressions being squared. ∎

✓ **CHECK POINT 6**   Factor: $x^2 + 10x + 25 - y^2.$

EXAMPLE 7   Using Grouping to Obtain the Difference of Two Squares

Factor: $a^2 - b^2 + 10b - 25.$

Solution   Grouping into two groups of two terms does not result in a common binomial factor. Let's look for a perfect square trinomial. Can you see that the perfect square trinomial is the expression being subtracted from a^2?

$$a^2 - b^2 + 10b - 25$$

$$= a^2 - (b^2 - 10b + 25)$$ Factor out -1 and group as $a^2 -$ (perfect square trinomial) to obtain a difference of two squares.

$$= a^2 - (b - 5)^2$$ Factor the perfect square trinomial.

$$= [a + (b - 5)][a - (b - 5)]$$ Factor the difference of squares. The factors are the sum and difference of the expressions being squared.

$$= (a + b - 5)(a - b + 5)$$ Simplify. ∎

> ☑ **CHECK POINT 7** Factor: $a^2 - b^2 + 4b - 4$.

4 Factor the sum or difference of two cubes. ▶

Factoring the Sum or Difference of Two Cubes

Here are two multiplications that lead to factoring formulas for the sum of two cubes and the difference of two cubes:

$$(A + B)(A^2 - AB + B^2) = A(A^2 - AB + B^2) + B(A^2 - AB + B^2)$$
$$= A^3 - A^2B + AB^2 + A^2B - AB^2 + B^3$$
$$= A^3 + B^3$$

The product results in the sum of two cubes.

Combine like terms:
$-A^2B + A^2B = 0$ and
$AB^2 - AB^2 = 0$.

and

$$(A - B)(A^2 + AB + B^2) = A(A^2 + AB + B^2) - B(A^2 + AB + B^2)$$
$$= A^3 + A^2B + AB^2 - A^2B - AB^2 - B^3$$
$$= A^3 - B^3.$$

The product results in the difference of two cubes.

Combine like terms:
$A^2B - A^2B = 0$ and
$AB^2 - AB^2 = 0$.

By reversing the two sides of these equations, we obtain formulas that allow us to factor a sum or difference of two cubes. These formulas should be memorized.

Factoring the Sum or Difference of Two Cubes

1. Factoring the Sum of Two Cubes

$$A^3 + B^3 = (A + B)(A^2 - AB + B^2)$$

Same signs   Opposite signs

2. Factoring the Difference of Two Cubes

$$A^3 - B^3 = (A - B)(A^2 + AB + B^2)$$

Same signs   Opposite signs

Great Question!

What are some cubes that I should be able to identify?

When factoring the sum or difference of cubes, it is helpful to recognize the following cubes:

$$1 = 1^3$$
$$8 = 2^3$$
$$27 = 3^3$$
$$64 = 4^3$$
$$125 = 5^3$$
$$216 = 6^3$$
$$1000 = 10^3.$$

EXAMPLE 8 Factoring the Sum of Two Cubes

Factor:

a. $x^3 + 125$ **b.** $x^6 + 64y^3$.

Solution We must express each term as the cube of some monomial. Then we use the formula for factoring $A^3 + B^3$.

a. $x^3 + 125 = x^3 + 5^3 = (x + 5)(x^2 - x \cdot 5 + 5^2) = (x + 5)(x^2 - 5x + 25)$

$A^3 + B^3 = (A + B)(A^2 - AB + B^2)$

b. $x^6 + 64y^3 = (x^2)^3 + (4y)^3 = (x^2 + 4y)[(x^2)^2 - x^2 \cdot 4y + (4y)^2]$

$A^3 + B^3 = (A + B)(A^2 - AB + B^2)$

$$= (x^2 + 4y)(x^4 - 4x^2y + 16y^2)$$ ∎

☑ **CHECK POINT 8**   Factor:

a. $x^3 + 27$

b. $x^6 + 1000y^3$.

EXAMPLE 9   Factoring the Difference of Two Cubes

Factor:

a. $x^3 - 216$ **b.** $8 - 125x^3y^3$.

Solution   We must express each term as the cube of some monomial. Then we use the formula for factoring $A^3 - B^3$.

a. $x^3 - 216 = x^3 - 6^3 = (x - 6)(x^2 + x \cdot 6 + 6^2) = (x - 6)(x^2 + 6x + 36)$

$$A^3 - B^3 = (A - B)(A^2 + AB + B^2)$$

b. $8 - 125x^3y^3 = 2^3 - (5xy)^3 = (2 - 5xy)[2^2 + 2 \cdot 5xy + (5xy)^2]$

$$A^3 - B^3 = (A - B)(A^2 + AB + B^2)$$

$$= (2 - 5xy)(4 + 10xy + 25x^2y^2) \quad ■$$

☑ **CHECK POINT 9**   Factor:

a. $x^3 - 8$

b. $1 - 27x^3y^3$.

Great Question!

The formulas for factoring $A^3 + B^3$ and $A^3 - B^3$ are difficult to remember and easy to confuse. Can you help me out?

A Cube of SOAP

When factoring sums or differences of cubes, observe the sign patterns.

Same signs

$$A^3 + B^3 = (A + B)(A^2 - AB + B^2)$$

Opposite signs   Always positive

Same signs

$$A^3 - B^3 = (A - B)(A^2 + AB + B^2)$$

Opposite signs   Always positive

The word *SOAP* is a way to remember these patterns:

S   O   A   P.

Same signs   Opposite signs   Always Positive

CONCEPT AND VOCABULARY CHECK

Fill in each blank so that the resulting statement is true.

1. The formula for factoring the difference of two squares is $A^2 - B^2 =$ _____.

2. A formula for factoring a perfect square trinomial is $A^2 + 2AB + B^2 =$ _____.

3. A formula for factoring a perfect square trinomial is $A^2 - 2AB + B^2 =$ _____.

4. The formula for factoring the sum of two cubes is $A^3 + B^3 =$ _____.

5. The formula for factoring the difference of two cubes is $A^3 - B^3 =$ _____.

6. $16x^2 - 49 = ($ _____ $+ 7)($ _____ $- 7)$

7. $a^2 - (b + 3)^2 = [a +$ _____ $][a -$ _____ $]$

8. $x^2 - 14x + 49 = (x$ _____ $)^2$

9. $16x^2 + 40xy + 25y^2 = ($ _____ $+ 5y)^2$

10. $x^3 + 27 = (x$ _____ $)(x^2$ _____ $+ 9)$

11. $x^3 - 1000 = (x$ _____ $)(x^2 + 10x$ _____ $)$

12. True or false: $x^2 - 10$ is the difference of two perfect squares. _____

13. True or false: $x^2 + 8x + 16$ is a perfect square trinomial. _____

14. True or false: $x^2 - 5x + 25$ is a perfect square trinomial. _____

15. True or false: $x^6 + 1000y^3$ is the sum of two cubes. _____

16. True or false: $x^3 - 100$ is the difference of two cubes. _____

5.5 EXERCISE SET ▶ MyMathLab®

Practice Exercises

In Exercises 1–22, factor each difference of two squares. Assume that any variable exponents represent whole numbers.

1. $x^2 - 4$

2. $x^2 - 16$

3. $9x^2 - 25$

4. $4x^2 - 9$

5. $9 - 25y^2$

6. $16 - 49y^2$

7. $36x^2 - 49y^2$

8. $64x^2 - 25y^2$

9. $x^2y^2 - 1$

10. $x^2y^2 - 100$

11. $9x^4 - 25y^6$

12. $25x^4 - 9y^6$

13. $x^{14} - y^4$

14. $x^4 - y^{10}$

15. $(x - 3)^2 - y^2$

16. $(x - 6)^2 - y^2$

17. $a^2 - (b - 2)^2$

18. $a^2 - (b - 3)^2$

19. $x^{2n} - 25$

20. $x^{2n} - 36$

21. $1 - a^{2n}$

22. $4 - b^{2n}$

In Exercises 23–48, factor completely, or state that the polynomial is prime.

23. $2x^3 - 8x$

24. $2x^3 - 72x$

25. $50 - 2y^2$

26. $72 - 2y^2$

27. $8x^2 - 8y^2$

28. $6x^2 - 6y^2$

29. $2x^3y - 18xy$

30. $2x^3y - 32xy$

31. $a^3b^2 - 49ac^2$

32. $4a^3c^2 - 16ax^2y^2$

33. $5y - 5x^2y^7$

34. $2y - 2x^6y^3$

35. $8x^2 + 8y^2$

36. $6x^2 + 6y^2$

37. $x^2 + 25y^2$

38. $x^2 + 36y^2$

39. $x^4 - 16$

40. $x^4 - 1$

41. $81x^4 - 1$

42. $1 - 81x^4$

43. $2x^5 - 2xy^4$

44. $3x^5 - 3xy^4$

45. $x^3 + 3x^2 - 4x - 12$

46. $x^3 + 3x^2 - 9x - 27$

47. $x^3 - 7x^2 - x + 7$

48. $x^3 - 6x^2 - x + 6$

In Exercises 49–64, factor any perfect square trinomials, or state that the polynomial is prime.

49. $x^2 + 4x + 4$

50. $x^2 + 2x + 1$

51. $x^2 - 10x + 25$

52. $x^2 - 14x + 49$

53. $x^4 - 4x^2 + 4$

54. $x^4 - 6x^2 + 9$

55. $9y^2 + 6y + 1$

56. $4y^2 + 4y + 1$

57. $64y^2 - 16y + 1$

58. $25y^2 - 10y + 1$

59. $x^2 - 12xy + 36y^2$

60. $x^2 + 16xy + 64y^2$

61. $x^2 - 8xy + 64y^2$

62. $x^2 - 9xy + 81y^2$

63. $9x^2 + 48xy + 64y^2$

64. $16x^2 - 40xy + 25y^2$

In Exercises 65–74, factor by grouping to obtain the difference of two squares.

65. $x^2 - 6x + 9 - y^2$

66. $x^2 - 12x + 36 - y^2$

67. $x^2 + 20x + 100 - x^4$

68. $x^2 + 16x + 64 - x^4$

69. $9x^2 - 30x + 25 - 36y^2$

70. $25x^2 - 20x + 4 - 81y^2$

71. $x^4 - x^2 - 2x - 1$

72. $x^4 - x^2 - 6x - 9$

73. $z^2 - x^2 + 4xy - 4y^2$

74. $z^2 - x^2 + 10xy - 25y^2$

In Exercises 75–94, factor using the formula for the sum or difference of two cubes.

75. $x^3 + 64$

76. $x^3 + 1$

77. $x^3 - 27$

78. $x^3 - 1000$

79. $8y^3 + 1$

80. $27y^3 + 1$

81. $125x^3 - 8$

82. $27x^3 - 8$

83. $x^3y^3 + 27$

84. $x^3y^3 + 64$

85. $64x - x^4$

86. $216x - x^4$

87. $x^6 + 27y^3$

88. $x^6 + 8y^3$

89. $125x^6 - 64y^6$

90. $125x^6 - y^6$

91. $x^9 + 1$

92. $x^9 - 1$

93. $(x - y)^3 - y^3$

94. $x^3 + (x + y)^3$

Practice PLUS

In Exercises 95–104, factor completely.

95. $0.04x^2 + 0.12x + 0.09$

96. $0.09x^2 - 0.12x + 0.04$

97. $8x^4 - \dfrac{x}{8}$

98. $27x^4 + \dfrac{x}{27}$

99. $x^6 - 9x^3 + 8$

100. $x^6 + 9x^3 + 8$

101. $x^8 - 15x^4 - 16$

102. $x^8 + 15x^4 - 16$

103. $x^5 - x^3 - 8x^2 + 8$

104. $x^5 - x^3 + 27x^2 - 27$

105. The figure shows four purple rectangles that fit together to form a large square.

a. Express the area of the large square in terms of one of its sides, $A + B$.

b. Write an expression for the area of each of the four rectangles that form the large square.

c. Use the sum of the areas from part (b) to write a second expression for the area of the large square.

d. Set the expression from part (c) equal to the expression from part (a). What factoring technique have you established?

Application Exercises

In Exercises 106–109, find the formula for the area of the shaded blue region and express it in factored form.

106.

107.

108.

109.

In Exercises 110–111, find the formula for the volume of the region outside the smaller rectangular solid and inside the larger rectangular solid. Then express the volume in factored form.

110.

111.

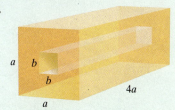

Explaining the Concepts

112. Explain how to factor the difference of two squares. Provide an example with your explanation.

113. What is a perfect square trinomial and how is it factored?

114. Explain how to factor $x^2 - y^2 + 8x - 16$. Should the expression be grouped into two groups of two terms? If not, why not, and what sort of grouping should be used?

115. Explain how to factor $x^3 + 1$.

Technology Exercises

In Exercises 116–123, use a graphing utility to graph the function on each side of the equation in the same viewing rectangle. Use end behavior to show a complete picture of the polynomial function on the left side. Do the graphs coincide? If so, this means that the polynomial on the left side has been factored correctly. If not, factor the polynomial correctly and then use your graphing utility to verify the factorization.

116. $9x^2 - 4 = (3x + 2)(3x - 2)$

117. $x^2 + 4x + 4 = (x + 4)^2$

118. $9x^2 + 12x + 4 = (3x + 2)^2$

119. $25 - (x^2 + 4x + 4) = (x + 7)(x - 3)$

120. $(2x + 3)^2 - 9 = 4x(x + 3)$

121. $(x - 3)^2 + 8(x - 3) + 16 = (x - 1)^2$

122. $x^3 - 1 = (x - 1)(x^2 - x + 1)$

123. $(x + 1)^3 + 1 = (x + 1)(x^2 + x + 1)$

124. Use the $\boxed{\text{TABLE}}$ feature of a graphing utility to verify any two of your factorizations in Exercises 67–68 or 77–78.

Critical Thinking Exercises

Make Sense? *In Exercises 125–128, determine whether each statement makes sense or does not make sense, and explain your reasoning.*

125. Although I can factor the difference of squares and perfect square trinomials using trial and error associated with FOIL, recognizing these special forms shortens the process.

126. Although $x^3 + 2x^2 - 5x - 6$ can be factored as $(x + 1)(x + 3)(x - 2)$, I have not yet learned techniques to obtain this factorization.

127. I factored $4x^2 - 100$ completely and obtained $(2x + 10)(2x - 10)$.

128. You told me that the area of a square is represented by $9x^2 + 12x + 4$ square inches, so I factored and concluded that the length of one side must be $3x + 2$ inches.

In Exercises 129–132, determine whether each statement is true or false. If the statement is false, make the necessary change(s) to produce a true statement.

129. $9x^2 + 15x + 25 = (3x + 5)^2$

130. $x^3 - 27 = (x - 3)(x^2 + 6x + 9)$

131. $x^3 - 64 = (x - 4)^3$

132. $4x^2 - 121 = (2x - 11)^2$

In Exercises 133–136, factor each polynomial completely. Assume that any variable exponents represent whole numbers.

133. $y^3 + x + x^3 + y$

134. $36x^{2n} - y^{2n}$

135. $x^{3n} + y^{12n}$

136. $4x^{2n} + 20x^n y^m + 25y^{2m}$

137. Factor $x^6 - y^6$ first as the difference of squares and then as the difference of cubes. From these two factorizations, determine a factorization for $x^4 + x^2y^2 + y^4$.

In Exercises 138–139, find all integers k so that the trinomial is a perfect square trinomial.

138. $kx^2 + 8xy + y^2$

139. $64x^2 - 16x + k$

Review Exercises

140. Solve: $2x + 2 \geq 12$ and $\dfrac{2x - 1}{3} \leq 7$. (Section 4.2, Example 2)

141. Solve using matrices:
$$\begin{cases} 3x - 2y = -8 \\ x + 6y = 4. \end{cases}$$
(Section 3.4, Example 2)

142. Factor: $3x^2 + 21x - xy - 7y$. (Section 5.3, Example 6)

Preview Exercises

Exercises 143–145 will help you prepare for the material covered in the next section. In each exercise, factor completely.

143. $2x^3 + 8x^2 + 8x$

144. $5x^3 - 40x^2y + 35xy^2$

145. $9b^2x + 9b^2y - 16x - 16y$

5.6

A General Factoring Strategy

What am I supposed to learn?

After studying this section, you should be able to:

1. Use a general strategy for factoring polynomials.

Successful problem solving involves understanding the problem, devising a plan for solving it, and then carrying out the plan. In this section, you will learn a step-by-step strategy that provides a plan and direction for solving factoring problems.

A Strategy for Factoring Polynomials

It is important to practice factoring a wide variety of polynomials so that you can quickly select the appropriate technique. The polynomial is factored completely when all its polynomial factors, except possibly for monomial factors, are prime. Because of the commutative property, the order of the factors does not matter.

Here is a general strategy for factoring polynomials:

1. Use a general strategy for factoring polynomials.

A Strategy for Factoring a Polynomial

1. If there is a common factor, factor out the GCF or factor out a common factor with a negative coefficient.

2. Determine the number of terms in the polynomial and try factoring as follows:

   a. If there are two terms, can the binomial be factored by using one of the following special forms?

   Difference of two squares: $A^2 - B^2 = (A + B)(A - B)$

   Sum of two cubes: $A^3 + B^3 = (A + B)(A^2 - AB + B^2)$

   Difference of two cubes: $A^3 - B^3 = (A - B)(A^2 + AB + B^2)$

b. If there are three terms, is the trinomial a perfect square trinomial? If so, factor by using one of the following special forms:

$$A^2 + 2AB + B^2 = (A + B)^2$$
$$A^2 - 2AB + B^2 = (A - B)^2.$$

If the trinomial is not a perfect square trinomial, try factoring by trial and error or grouping.

c. If there are four or more terms, try factoring by grouping.

3. Check to see if any factors with more than one term in the factored polynomial can be factored further. If so, factor completely.

Remember to check the factored form by multiplying or by using the ⌷TABLE⌷ or ⌷GRAPH⌷ feature of a graphing utility.

The following examples and those in the Exercise Set are similar to the previous factoring problems. However, these factorizations are not all of the same type. They are intentionally mixed to promote the development of a general factoring strategy.

Using Technology

Graphic Connections
The polynomial functions
$y_1 = 2x^3 + 8x^2 + 8x$
and $y_2 = 2x(x + 2)^2$ have identical graphs. This verifies that

$$2x^3 + 8x^2 + 8x = 2x(x + 2)^2.$$

[−4, 2, 1] by [−10, 10, 1]

The degree of y_1 is 3, which is odd. Odd-degree polynomial functions have graphs with opposite behavior at each end. The leading coefficient, 2, is positive. The graph should fall to the left and rise to the right (↙, ↗). The viewing rectangle used is complete enough to show this end behavior.

EXAMPLE 1 Factoring a Polynomial

Factor: $2x^3 + 8x^2 + 8x$.

Solution
Step 1. If there is a common factor, factor out the GCF. Because $2x$ is common to all terms, we factor it out.

$$2x^3 + 8x^2 + 8x = 2x(x^2 + 4x + 4) \quad \text{Factor out the GCF.}$$

Step 2. Determine the number of terms and factor accordingly. The factor $x^2 + 4x + 4$ has three terms and is a perfect square trinomial. We factor using $A^2 + 2AB + B^2 = (A + B)^2$.

$$2x^3 + 8x^2 + 8x = 2x(x^2 + 4x + 4)$$
$$= 2x(x^2 + 2 \cdot x \cdot 2 + 2^2)$$

$$\underbrace{\qquad\quad A^2 \quad + \quad 2AB \quad + \quad B^2 \qquad}$$

$$= 2x(x + 2)^2 \qquad A^2 + 2AB + B^2 = (A + B)^2$$

Step 3. Check to see if factors can be factored further. In this problem, they cannot. Thus,

$$2x^3 + 8x^2 + 8x = 2x(x + 2)^2. \quad \blacksquare$$

☑ **CHECK POINT 1** Factor: $3x^3 - 30x^2 + 75x$.

EXAMPLE 2 Factoring a Polynomial

Factor: $4x^2y - 16xy - 20y$.

Solution
Step 1. If there is a common factor, factor out the GCF. Because $4y$ is common to all terms, we factor it out.

$$4x^2y - 16xy - 20y = 4y(x^2 - 4x - 5) \quad \text{Factor out the GCF.}$$

Step 2. Determine the number of terms and factor accordingly. The factor $x^2 - 4x - 5$ has three terms, but it is not a perfect square trinomial. We factor it using trial and error.

$$4x^2y - 16xy - 20y = 4y(x^2 - 4x - 5) = 4y(x + 1)(x - 5)$$

Step 3. Check to see if factors can be factored further. In this case, they cannot, so we have factored completely. ■

✓ **CHECK POINT 2**   Factor: $3x^2y - 12xy - 36y$.

EXAMPLE 3   Factoring a Polynomial

Factor: $9b^2x - 16y - 16x + 9b^2y$.

Solution

Step 1. If there is a common factor, factor out the GCF. Other than 1 or -1, there is no common factor.

Step 2. Determine the number of terms and factor accordingly. There are four terms. We try factoring by grouping. Notice that the first and last terms have a common factor of $9b^2$ and the two middle terms have a common factor of -16. Thus, we begin by rearranging the terms.

$$
\begin{aligned}
&9b^2x - 16y - 16x + 9b^2y \\
&= (9b^2x + 9b^2y) + (-16x - 16y) \qquad \text{Rearrange terms and group terms with common factors.} \\
&= 9b^2(x + y) - 16(x + y) \qquad \text{Factor from each group.} \\
&= (x + y)(9b^2 - 16) \qquad \text{Factor out the common binomial factor, } x + y.
\end{aligned}
$$

Step 3. Check to see if factors can be factored further. We note that $9b^2 - 16$ is the difference of two squares, $(3b)^2 - 4^2$, so we continue factoring.

$$
\begin{aligned}
&9b^2x - 16y - 16x + 9b^2y \\
&= (x + y)[(3b)^2 - 4^2] \qquad \text{Express } 9b^2 - 16 \text{ as the difference of squares.} \\
&= (x + y)(3b + 4)(3b - 4) \qquad \text{The factors of } 9b^2 - 16 \text{ are the sum and difference of the expressions being squared.} \ ■
\end{aligned}
$$

✓ **CHECK POINT 3**   Factor: $16a^2x - 25y - 25x + 16a^2y$.

EXAMPLE 4   Factoring a Polynomial

Factor: $x^2 - 25a^2 + 8x + 16$.

Solution

Step 1. If there is a common factor, factor out the GCF. Other than 1 or -1, there is no common factor.

Step 2. Determine the number of terms and factor accordingly. There are four terms. We try factoring by grouping. Grouping into two groups of two terms does not result in a common binomial factor. Let's try grouping as a difference of squares.

$$
\begin{aligned}
&x^2 - 25a^2 + 8x + 16 \\
&= (x^2 + 8x + 16) - 25a^2 \qquad \text{Rearrange terms and group as a perfect square trinomial minus } 25a^2 \text{ to obtain a difference of squares.} \\
&= (x + 4)^2 - (5a)^2 \qquad \text{Factor the perfect square trinomial.} \\
&= (x + 4 + 5a)(x + 4 - 5a) \qquad \text{Factor the difference of squares. The factors are the sum and difference of the expressions being squared.}
\end{aligned}
$$

Step 3. Check to see if factors can be factored further. In this case, they cannot, so we have factored completely. ■

✓ **CHECK POINT 4**   Factor: $x^2 - 36a^2 + 20x + 100$.

EXAMPLE 5   Factoring a Polynomial

Factor: $3x^{10} + 3x$.

Solution

Step 1.   If there is a common factor, factor out the GCF.   Because $3x$ is common to both terms, we factor it out.

$$3x^{10} + 3x = 3x(x^9 + 1) \qquad \text{Factor out the GCF.}$$

Step 2.   Determine the number of terms and factor accordingly.   The factor $x^9 + 1$ has two terms. This binomial can be expressed as $(x^3)^3 + 1^3$, so it can be factored as the sum of two cubes.

$$3x^{10} + 3x = 3x(x^9 + 1)$$
$$= 3x[(x^3)^3 + 1^3] = 3x(x^3 + 1)[(x^3)^2 - x^3 \cdot 1 + 1^2]$$

$$A^3 + B^3 \quad = \quad (A + B)\ (A^2 - AB + B^2)$$

$$= 3x(x^3 + 1)(x^6 - x^3 + 1) \quad \text{Simplify.}$$

Step 3.   Check to see if factors can be factored further.   We note that $x^3 + 1$ is the sum of two cubes, $x^3 + 1^3$, so we continue factoring.

$$3x^{10} + 3x$$
$$= 3x(x^3 + 1)(x^6 - x^3 + 1) \qquad \text{This is our factorization in the previous step.}$$

$$A^3 + B^3 = (A + B)\ (A^2 - AB + B^2) \qquad \text{Factor completely by factoring } x^3 + 1^3, \text{ the sum of cubes.}$$

$$= 3x(x + 1)(x^2 - x + 1)(x^6 - x^3 + 1) \quad ■$$

✓ **CHECK POINT 5**   Factor: $x^{10} + 512x$. *Hint*: $512 = 8^3$.

Achieving Success

In the next chapter, you will see how mathematical models involving quotients of polynomials describe environmental issues. Factoring is an essential skill for working with such models. **Success in mathematics cannot be achieved without a complete understanding of factoring.** Be sure to work all the assigned exercises in Exercise Set 5.6 so that you can apply each of the factoring techniques discussed in this chapter. The more deeply you force your brain to think about factoring by working many exercises, the better will be your chances of achieving success in the next chapter and in future mathematics courses.

CONCEPT AND VOCABULARY CHECK

Here is a list of the factoring techniques that we have discussed.

a. Factoring out the GCF

b. Factoring out the negative of the GCF

c. Factoring by grouping

d. Factoring trinomials by trial and error or grouping

e. Factoring the difference of two squares

$A^2 - B^2 = (A + B)(A - B)$

f. Factoring perfect square trinomials

$A^2 + 2AB + B^2 = (A + B)^2$

$A^2 - 2AB + B^2 = (A - B)^2$

g. Factoring the sum of two cubes

$A^3 + B^3 = (A + B)(A^2 - AB + B^2)$

h. Factoring the difference of two cubes

$A^3 - B^3 = (A - B)(A^2 + AB + B^2)$

Fill in each blank by writing the letter of the technique (a through h) for factoring the polynomial.

1. $-4x^2 + 20x$ _____

2. $9x^2 - 16$ _____

3. $125x^3 - 1$ _____

4. $x^2 + 7x + xy + 7y$ _____

5. $x^2 - 4x - 5$ _____

6. $x^2 + 4x + 4$ _____

7. $3x^6 + 6x$ _____

8. $x^9 + 1$ _____

5.6 EXERCISE SET   ▶ MyMathLab®

Practice Exercises

In Exercises 1–68, factor completely, or state that the polynomial is prime.

1. $x^3 - 16x$

2. $x^3 - x$

3. $3x^2 + 18x + 27$

4. $8x^2 + 40x + 50$

5. $81x^3 - 3$

6. $24x^3 - 3$

7. $x^2y - 16y + 32 - 2x^2$

8. $12x^2y - 27y - 4x^2 + 9$

9. $4a^2b - 2ab - 30b$

10. $32y^2 - 48y + 18$

11. $ay^2 - 4a - 4y^2 + 16$

12. $ax^2 - 16a - 2x^2 + 32$

13. $11x^5 - 11xy^2$

14. $4x^9 - 400x$

15. $4x^5 - 64x$

16. $7x^5 - 7x$

17. $x^3 - 4x^2 - 9x + 36$

18. $x^3 - 5x^2 - 4x + 20$

19. $2x^5 + 54x^2$

20. $3x^5 + 24x^2$

21. $3x^4y - 48y^5$

22. $32x^4y - 2y^5$

23. $12x^3 + 36x^2y + 27xy^2$

24. $18x^3 + 48x^2y + 32xy^2$

25. $x^2 - 12x + 36 - 49y^2$

26. $x^2 - 10x + 25 - 36y^2$

27. $4x^2 + 25y^2$

28. $16x^2 + 49y^2$

29. $12x^3y - 12xy^3$

30. $9x^2y^2 - 36y^2$

31. $6bx^2 + 6by^2$

32. $6x^2 - 66$

33. $x^4 - xy^3 + x^3y - y^4$

34. $x^3 - xy^2 + x^2y - y^3$

35. $x^2 - 4a^2 + 12x + 36$

36. $x^2 - 49a^2 + 14x + 49$

37. $5x^3 + x^6 - 14$

38. $6x^3 + x^6 - 16$

39. $4x - 14 + 2x^3 - 7x^2$

40. $3x^3 + 8x + 9x^2 + 24$

41. $54x^3 - 16y^3$

42. $54x^3 - 250y^3$

43. $x^2 + 10x - y^2 + 25$

44. $x^2 + 6x - y^2 + 9$

45. $x^8 - y^8$

46. $x^8 - 1$

47. $x^3y - 16xy^3$

48. $x^3y - 100xy^3$

49. $x + 8x^4$

50. $x + 27x^4$

51. $16y^2 - 4y - 2$

52. $32y^2 + 4y - 6$

53. $14y^3 + 7y^2 - 10y$

54. $5y^3 - 45y^2 + 70y$

55. $27x^2 + 36xy + 12y^2$

56. $125x^2 + 50xy + 5y^2$

57. $12x^3 + 3xy^2$

58. $3x^4 + 27x^2$

59. $x^6y^6 - x^3y^3$

60. $x^3 - 2x^2 - x + 2$

61. $(x + 5)(x - 3) + (x + 5)(x - 7)$

62. $(x + 4)(x - 9) + (x + 4)(2x - 3)$

63. $a^2(x - y) + 4(y - x)$

64. $b^2(x - 3) + c^2(3 - x)$

65. $(c + d)^4 - 1$

66. $(c + d)^4 - 16$

67. $p^3 - pq^2 + p^2q - q^3$

68. $p^3 - pq^2 - p^2q + q^3$

Practice PLUS

In Exercises 69–80, factor completely.

69. $x^4 - 5x^2y^2 + 4y^4$

70. $x^4 - 10x^2y^2 + 9y^4$

71. $(x + y)^2 + 6(x + y) + 9$

72. $(x - y)^2 - 8(x - y) + 16$

73. $(x - y)^4 - 4(x - y)^2$

74. $(x + y)^4 - 100(x + y)^2$

75. $2x^2 - 7xy^2 + 3y^4$

76. $3x^2 + 5xy^2 + 2y^4$

77. $x^3 - y^3 - x + y$

78. $x^3 + y^3 + x^2 - y^2$

79. $x^6y^3 + x^3 - 8x^3y^3 - 8$

80. $x^6y^3 - x^3 + x^3y^3 - 1$

Application Exercises

In Exercises 81–86,

a. *Write an expression for the area of the shaded blue region.*

b. *Write the expression in factored form.*

81.

82.

83.

84.

85.

86.

Explaining the Concepts

87. Describe a strategy that can be used to factor polynomials.

88. Describe some of the difficulties in factoring polynomials. What suggestions can you offer to overcome these difficulties?

Technology Exercises

In Exercises 89–92, use a graphing utility to graph the function on each side of the equation in the same viewing rectangle. Use end behavior to show a complete picture of the polynomial on the left side. Do the graphs coincide? If so, the factorization is correct. If not, factor correctly and then use your graphing utility to verify the factorization.

89. $4x^2 - 12x + 9 = (4x - 3)^2$

90. $2x^3 + 10x^2 - 2x - 10 = 2(x + 5)(x^2 + 1)$

91. $x^4 - 16 = (x^2 + 4)(x + 2)(x - 2)$

92. $x^3 + 1 = (x + 1)^3$

93. Use the $\boxed{\text{TABLE}}$ feature of a graphing utility to verify any two of your complete factorizations in Exercises 15–20.

Critical Thinking Exercises

Make Sense? *In Exercises 94–97, determine whether each statement makes sense or does not make sense, and explain your reasoning.*

94. It takes a great deal of practice to get good at factoring a wide variety of polynomials.

95. Multiplying polynomials is relatively mechanical, but factoring often requires a great deal of thought.

96. The factorable trinomial $4x^2 + 8x + 3$ and the prime trinomial $4x^2 + 8x + 1$ are in the form $ax^2 + bx + c$, but $b^2 - 4ac$ is a perfect square only in the case of the factorable trinomial.

97. You told me that the volume of a rectangular solid is represented by $5x^3 + 30x^2 + 40x$ cubic inches, so I factored completely and concluded that the dimensions are $5x$ inches, $x + 2$ inches, and $x + 5$ inches.

In Exercises 98–101, determine whether each statement is true or false. If the statement is false, make the necessary change(s) to produce a true statement.

98. $x^4 - 16$ is factored completely as $(x^2 + 4)(x^2 - 4)$.

99. The trinomial $x^2 - 4x - 4$ is a prime polynomial.

100. $x^2 + 36 = (x + 6)^2$

101. $x^3 - 64 = (x + 4)(x^2 + 4x - 16)$

In Exercises 102–104, factor completely. Assume that variable exponents represent whole numbers.

102. $x^{2n+3} - 10x^{n+3} + 25x^3$

103. $3x^{n+2} - 13x^{n+1} + 4x^n$

104. $x^{4n+1} - xy^{4n}$

105. In certain circumstances, the sum of two perfect squares can be factored by adding and subtracting the same perfect square. For example,

$x^4 + 4 = x^4 + 4x^2 + 4 - 4x^2.$ Add and subtract $4x^2$.

Use this first step to factor $x^4 + 4$.

106. Express $x^3 + x + 2x^4 + 4x^2 + 2$ as the product of two polynomials of degree 2.

Review Exercises

107. Solve: $\dfrac{3x - 1}{5} + \dfrac{x + 2}{2} = -\dfrac{3}{10}$.

(Section 1.4, Example 4)

108. Simplify: $(4x^3y^{-1})^2(2x^{-3}y)^{-1}$.

(Section 1.6, Example 9)

109. Evaluate: $\begin{vmatrix} 0 & -3 & 2 \\ 1 & 5 & 3 \\ -2 & 1 & 4 \end{vmatrix}$.

(Section 3.5, Example 3)

Preview Exercises

Exercises 110–112 will help you prepare for the material covered in the next section.

110. Evaluate $(2x + 3)(x - 4)$ in your head for $x = 4$.

111. Evaluate $-16(t - 6)(t + 4)$ in your head for $t = 6$.

112. Express as an equivalent equation with a factored trinomial on the left side and zero on the right side:

$x^2 + (x + 7)^2 = (x + 8)^2.$

Polynomial Equations and Their Applications

What am I supposed to learn?

After studying this section, you should be able to:

1 Solve quadratic equations by factoring.

2 Solve higher-degree polynomial equations by factoring.

3 Solve problems using polynomial equations.

Motion and change are the very essence of life. Moving air brushes against our faces; rain falls on our heads; birds fly past us; plants spring from the earth, grow, and then die; and rocks thrown upward reach a maximum height before falling to the ground. In this section, you will use quadratic functions and factoring strategies to model and visualize motion. Analyzing the where and when of moving objects involves equations in which the highest exponent on the variable is 2, called *quadratic equations*.

The Standard Form of a Quadratic Equation

We begin by defining a quadratic equation.

> ### Definition of a Quadratic Equation
>
> A **quadratic equation** in x is an equation that can be written in the **standard form**
>
> $$ax^2 + bx + c = 0,$$
>
> where a, b, and c are real numbers, with $a \neq 0$. A quadratic equation in x is also called a **second-degree polynomial equation** in x.

Here is an example of a quadratic equation in standard form:

$$x^2 - 12x + 27 = 0.$$

$a = 1$ $b = -12$ $c = 27$

1 Solve quadratic equations by factoring.

Solving Quadratic Equations by Factoring

We can factor the left side of the quadratic equation $x^2 - 12x + 27 = 0$. We obtain $(x - 3)(x - 9) = 0$. If a quadratic equation has zero on one side and a factored expression on the other side, it can be solved using the **zero-product principle**.

The Zero-Product Principle

If the product of two algebraic expressions is zero, then at least one of the factors is equal to zero.

$$\text{If } AB = 0, \text{ then } A = 0 \text{ or } B = 0.$$

For example, consider the equation $(x - 3)(x - 9) = 0$. According to the zero-product principle, this product can be zero only if at least one of the factors is zero. We set each individual factor equal to zero and solve the resulting equations for x.

$$(x - 3)(x - 9) = 0$$
$$x - 3 = 0 \quad \text{or} \quad x - 9 = 0$$
$$x = 3 \qquad\qquad x = 9$$

The solutions of the original quadratic equation, $x^2 - 12x + 27 = 0$, are 3 and 9. The solution set is $\{3, 9\}$.

Solving a Quadratic Equation by Factoring

1. If necessary, rewrite the equation in the standard form $ax^2 + bx + c = 0$, moving all terms to one side, thereby obtaining zero on the other side.
2. Factor completely.
3. Apply the zero-product principle, setting each factor containing a variable equal to zero.
4. Solve the equations in step 3.
5. Check the solutions in the original equation.

EXAMPLE 1 Solving a Quadratic Equation by Factoring

Solve: $2x^2 - 5x = 12$.

Solution

Step 1. Move all terms to one side and obtain zero on the other side. Subtract 12 from both sides and write the equation in standard form.

$$2x^2 - 5x - 12 = 12 - 12$$
$$2x^2 - 5x - 12 = 0$$

Step 2. Factor.

$$(2x + 3)(x - 4) = 0$$

Steps 3 and 4. Set each factor equal to zero and solve the resulting equations.

$$2x + 3 = 0 \quad \text{or} \quad x - 4 = 0$$
$$2x = -3 \qquad\qquad x = 4$$
$$x = -\frac{3}{2}$$

Step 5. Check the solutions in the original equation.

Check $-\dfrac{3}{2}$:	Check 4:

$$2x^2 - 5x = 12$$

$$2\left(-\frac{3}{2}\right)^2 - 5\left(-\frac{3}{2}\right) \overset{?}{=} 12$$

$$2\left(\frac{9}{4}\right) - 5\left(-\frac{3}{2}\right) \overset{?}{=} 12$$

$$\frac{9}{2} + \frac{15}{2} \overset{?}{=} 12$$

$$\frac{24}{2} \overset{?}{=} 12$$

$$12 = 12, \;\; \text{true}$$

$$2x^2 - 5x = 12$$

$$2(4)^2 - 5(4) \overset{?}{=} 12$$

$$2(16) - 5(4) \overset{?}{=} 12$$

$$32 - 20 \overset{?}{=} 12$$

$$12 = 12, \;\; \text{true}$$

The solutions are $-\frac{3}{2}$ and 4, and the solution set is $\left\{-\frac{3}{2}, 4\right\}$. ■

✓ **CHECK POINT 1**   Solve: $2x^2 - 9x = 5$.

Great Question!

After factoring a polynomial, should I set each factor equal to zero?

No. Do not confuse factoring a polynomial with solving a quadratic equation by factoring.

Factoring a Polynomial	**Solving a Quadratic Equation**

Factor: $2x^2 - 5x - 12$.

> This is not an equation. There is no equal sign.

Solution: $(2x + 3)(x - 4)$

> Stop! Avoid the common error of setting each factor equal to zero.

Solve: $2x^2 - 5x - 12 = 0$.

> This is an equation. There is an equal sign.

Solution: $(2x + 3)(x - 4) = 0$

$2x + 3 = 0$ or $x - 4 = 0$

$x = -\dfrac{3}{2}$ $\qquad\qquad$ $x = 4$

The solution set is $\left\{-\dfrac{3}{2}, 4\right\}$.

There is an important relationship between a quadratic equation in standard form, such as

$$2x^2 - 5x - 12 = 0$$

and a quadratic function, such as

$$y = 2x^2 - 5x - 12.$$

The solutions of $ax^2 + bx + c = 0$ correspond to the x-intercepts of the graph of the quadratic function $y = ax^2 + bx + c$. For example, you can visualize the solutions of $2x^2 - 5x - 12 = 0$ by looking at the x-intercepts of the graph of the quadratic function $y = 2x^2 - 5x - 12$. The graph, shaped like a bowl, is shown in **Figure 5.10**. The solutions of the equation $2x^2 - 5x - 12 = 0$, $-\frac{3}{2}$ and 4, appear as the graph's x-intercepts.

x-intercept is $-\frac{3}{2}$. x-intercept is 4.

$[-4, 6, 1]$ by $[-16, 4, 1]$

Figure 5.10

EXAMPLE 2 Solving Quadratic Equations by Factoring

Solve:

a. $5x^2 = 20x$　　**b.** $x^2 + 4 = 8x - 12$　　**c.** $(x - 7)(x + 5) = -20$.

Solution

a.

$$5x^2 = 20x$$ This is the given equation.

$$5x^2 - 20x = 0$$ Subtract 20x from both sides and write the equation in standard form.

$$5x(x - 4) = 0$$ Factor.

$$5x = 0 \quad \text{or} \quad x - 4 = 0$$ Set each factor equal to O.

$$x = 0 \qquad\qquad x = 4$$ Solve the resulting equations.

Check by substituting 0 and 4 into the given equation. The graph of $y = 5x^2 - 20x$, obtained with a graphing utility, is shown in **Figure 5.11**. The x-intercepts are 0 and 4. This verifies that the solutions are 0 and 4, and the solution set is {0, 4}.

b.

$$x^2 + 4 = 8x - 12$$ This is the given equation.

$$x^2 - 8x + 16 = 0$$ Write the equation in standard form by subtracting 8x and adding 12 on both sides.

$$(x - 4)(x - 4) = 0$$ Factor.

$$x - 4 = 0 \quad \text{or} \quad x - 4 = 0$$ Set each factor equal to O.

$$x = 4 \qquad\qquad x = 4$$ Solve the resulting equations.

Notice that there is only one solution (or, if you prefer, a repeated solution). The trinomial $x^2 - 8x + 16$ is a perfect square trinomial that could have been factored as $(x - 4)^2$. The graph of $y = x^2 - 8x + 16$, obtained with a graphing utility, is shown in **Figure 5.12**. The graph has only one x-intercept at 4. This verifies that the equation's solution is 4 and the solution set is {4}.

c. Be careful! Although the left side of $(x - 7)(x + 5) = -20$ is factored, we cannot use the zero-product principle. Why not? The right side of the equation is not 0. So we begin by multiplying the factors on the left side of the equation. Then we add 20 to both sides to obtain 0 on the right side.

$$(x - 7)(x + 5) = -20$$ This is the given equation.

$$x^2 - 2x - 35 = -20$$ Use the FOIL method to multiply on the left side.

$$x^2 - 2x - 15 = 0$$ Add 20 to both sides.

$$(x + 3)(x - 5) = 0$$ Factor.

$$x + 3 = 0 \quad \text{or} \quad x - 5 = 0$$ Set each factor equal to O.

$$x = -3 \qquad\qquad x = 5$$ Solve the resulting equations.

Check by substituting -3 and 5 into the given equation. The graph of $y = x^2 - 2x - 15$, obtained with a graphing utility, is shown in **Figure 5.13**. The x-intercepts are -3 and 5. This verifies that the solutions are -3 and 5, and the solution set is {−3, 5}.

Figure 5.11 The solution set of $5x^2 = 20x$, or $5x^2 - 20x = 0$, is {0, 4}.

[−2, 6, 1] by [−20, 10, 1]

x-intercept: 0　*x*-intercept: 4　$y = 5x^2 - 20x$

Figure 5.12 The solution set of $x^2 + 4 = 8x - 12$, or $x^2 - 8x + 16 = 0$, is {4}.

[−1, 10, 1] by [0, 20, 1]

$y = x^2 - 8x + 16$　*x*-intercept: 4

[−5, 7, 1] by [−18, 5, 1]

x-intercept: −3　*x*-intercept: 5　$y = x^2 - 2x - 15$

Figure 5.13 The solution set of $(x - 7)(x + 5) = -20$, or $x^2 - 2x - 15 = 0$, is {−3, 5}. ■

Great Questions!

In Example 2(a), can I simplify $5x^2 = 20x$ by dividing both sides by x? In Example 2(c), can I solve $(x - 7)(x + 5) = -20$ by using the zero-product principle in the first step?

No and no. Avoid the following errors:

$5x^2 = 20x$

$\dfrac{5x^2}{x} = \dfrac{20x}{x}$ Never divide both sides of an equation by x. Division by zero is undefined

$5x = 20$ and x may be zero. Indeed, the solutions for this equation (Example 2a) are O

$x = 4$ and 4. Dividing both sides by x does not permit us to find both solutions.

$(x - 7)(x + 5) = -20$

$x - 7 = -20$ or $x + 5 = -20$ The zero-product principle cannot be used because the

$x = -13$ or $x = -25$ right side of the equation is not equal to O.

✓ **CHECK POINT 2**   Solve:

   a. $3x^2 = 2x$

   b. $x^2 + 7 = 10x - 18$

   c. $(x - 2)(x + 3) = 6.$

---

2 Solve higher-degree polynomial equations by factoring. ⏵

Polynomial Equations

A **polynomial equation** is the result of setting two polynomials equal to each other. The equation is in **standard form** if one side is 0 and the polynomial on the other side is in standard form, that is, in descending powers of the variable. The **degree of a polynomial equation** is the same as the highest degree of any term in the equation. Here are examples of three polynomial equations:

$$3x + 5 = 14 \qquad 2x^2 + 7x = 4 \qquad x^3 + x^2 = 4x + 4.$$

| This equation is of degree 1 because 1 is the highest degree. | This equation is of degree 2 because 2 is the highest degree. | This equation is of degree 3 because 3 is the highest degree. |

Notice that a polynomial equation of degree 1 is a linear equation. A polynomial equation of degree 2 is a quadratic equation.

   Some polynomial equations of degree 3 or higher can be solved by moving all terms to one side, thereby obtaining 0 on the other side. Once the equation is in standard form, factor and then set each factor equal to 0.

EXAMPLE 3   Solving a Polynomial Equation by Factoring

Solve by factoring: $x^3 + x^2 = 4x + 4.$

Solution

Step 1. Move all terms to one side and obtain zero on the other side. Subtract $4x$ and subtract 4 from both sides.

$$x^3 + x^2 - 4x - 4 = 4x + 4 - 4x - 4$$

$$x^3 + x^2 - 4x - 4 = 0$$

Step 2. Factor. Use factoring by grouping. Group terms that have a common factor.

$$\boxed{x^3 + x^2} + \boxed{-4x - 4} = 0$$

Common factor is x^2.   Common factor is -4.

Using Technology

Numeric Connections

A graphing utility's ⎡TABLE⎤ feature can be used to numerically verify that $\{-2, -1, 2\}$ is the solution set of

$$x^3 + x^2 = 4x + 4.$$

Discover for Yourself

Suggest a method involving intersecting graphs that can be used with a graphing utility to verify that $\{-2, -1, 2\}$ is the solution set of

$$x^3 + x^2 = 4x + 4.$$

Apply this method to verify the solution set.

$$x^3 + x^2 - 4x - 4 = 0 \qquad \text{We have repeated the equation from the bottom of the previous page.}$$

$$x^2(x + 1) - 4(x + 1) = 0 \qquad \text{Factor } x^2 \text{ from the first two terms and } -4 \text{ from the last two terms.}$$

$$(x + 1)(x^2 - 4) = 0 \qquad \text{Factor out the common binomial, } x + 1, \text{ from each term.}$$

$$(x + 1)(x + 2)(x - 2) = 0 \qquad \text{Factor completely by factoring } x^2 - 4 \text{ as the difference of two squares.}$$

Steps 3 and 4. Set each factor equal to zero and solve the resulting equations.

$$x + 1 = 0 \quad \text{or} \quad x + 2 = 0 \quad \text{or} \quad x - 2 = 0$$
$$x = -1 \qquad\qquad x = -2 \qquad\qquad x = 2$$

Step 5. Check the solutions in the original equation. Check the three solutions, -1, -2, and 2, by substituting them into the original equation. Can you verify that the solutions are -1, -2, and 2, and the solution set is $\{-2, -1, 2\}$? ∎

Using Technology

Graphic Connections

You can use a graphing utility to check the solutions to $x^3 + x^2 - 4x - 4 = 0$. Graph $y = x^3 + x^2 - 4x - 4$, as shown on the right. Is the graph complete? Because the degree, 3, is odd, and the leading coefficient, 1, is positive, it should fall to the left and rise to the right ($\swarrow$, $\nearrow$). The graph shows this end behavior and is therefore complete. The x-intercepts are -2, -1, and 2, corresponding to the equation's solutions.

$[-5, 5, 1]$ by $[-8, 2, 1]$

☑ **CHECK POINT 3** Solve by factoring: $2x^3 + 3x^2 = 8x + 12$.

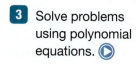 Solve problems using polynomial equations.

Applications of Polynomial Equations

Solving polynomial equations by factoring can be used to answer questions about variables contained in mathematical models.

EXAMPLE 4 Modeling Motion

You throw a ball straight up from a rooftop 384 feet high with an initial speed of 32 feet per second. The function

$$s(t) = -16t^2 + 32t + 384$$

describes the ball's height above the ground, $s(t)$, in feet, t seconds after you throw it. The ball misses the rooftop on its way down and eventually strikes the ground. How long will it take for the ball to hit the ground?

Solution The ball hits the ground when $s(t)$, its height above the ground, is 0 feet. Thus, we substitute 0 for $s(t)$ in the given function and solve for t.

$$s(t) = -16t^2 + 32t + 384 \qquad \text{This is the function that models the ball's height.}$$

$$0 = -16t^2 + 32t + 384 \qquad \text{Substitute O for } s(t).$$

$$0 = -16(t^2 - 2t - 24) \qquad \text{Factor out } -16, \text{ the negative of the GCF.}$$

$$0 = -16(t - 6)(t + 4)$$ Factor $t^2 - 2t - 24$, the trinomial.

> Do not set the constant, -16, equal to zero: $-16 \neq 0$.

$$t - 6 = 0 \quad \text{or} \quad t + 4 = 0$$ Set each variable factor equal to 0.

$$t = 6 \qquad\qquad t = -4$$ Solve for t.

Because we begin describing the ball's height at $t = 0$, we discard the solution $t = -4$. The ball hits the ground after 6 seconds. ■

Figure 5.14 shows the graph of the function $s(t) = -16t^2 + 32t + 384$. The horizontal axis is labeled t, for the ball's time in motion. The vertical axis is labeled $s(t)$, for the ball's height above the ground at time t. Because time and height are both positive, the function is graphed in quadrant I only.

The graph visually shows what we discovered algebraically: The ball hits the ground after 6 seconds. The graph also reveals that the ball reaches its maximum height, 400 feet, after 1 second. Then the ball begins to fall.

Figure 5.14

✓ **CHECK POINT 4** Use the function $s(t) = -16t^2 + 32t + 384$ to determine when the ball's height is 336 feet. Identify your meaningful solution as a point on the graph in **Figure 5.14**.

In our next example, we use our five-step strategy for solving word problems.

EXAMPLE 5 Solving a Problem Involving Landscape Design

A rectangular garden measures 80 feet by 60 feet. A large path of uniform width is to be added along both shorter sides and one longer side of the garden. The landscape designer doing the work wants to double the garden's area with the addition of this path. How wide should the path be?

Solution

Step 1. Let x represent one of the unknown quantities. We will let

$$x = \text{the width of the path.}$$

The situation is illustrated in **Figure 5.15**. The figure shows the original 80-by-60 foot rectangular garden and the path of width x added along both shorter sides and one longer side.

Figure 5.15 The garden's area is to be doubled by adding the path.

Step 2. Represent other unknown quantities in terms of x. Because the path is added along both shorter sides and one longer side, **Figure 5.15** shows that

$$80 + 2x = \text{the length of the new, expanded rectangle}$$

$$60 + x = \text{the width of the new, expanded rectangle.}$$

Step 3. Write an equation that models the conditions. The area of the rectangle must be doubled by the addition of the path.

The area, or length times width, of the new, expanded rectangle	must be	twice that of	the area of the garden.

$$(80 + 2x)(60 + x) \ = \ 2 \ \cdot \ 80 \ \cdot \ 60$$

Figure 5.15 (repeated) The garden's area is to be doubled by adding the path.

Step 4. Solve the equation and answer the question.

$$(80 + 2x)(60 + x) = 2 \cdot 80 \cdot 60$$

This is the equation that models the problem's conditions.

$$4800 + 200x + 2x^2 = 9600$$

Multiply. Use FOIL on the left side.

$$2x^2 + 200x - 4800 = 0$$

Subtract 9600 from both sides and write the equation in standard form.

$$2(x^2 + 100x - 2400) = 0$$

Factor out 2, the GCF.

$$2(x - 20)(x + 120) = 0$$

Factor the trinomial.

$$x - 20 = 0 \quad \text{or} \quad x + 120 = 0$$

Set each variable factor equal to 0.

$$x = 20 \quad \text{or} \quad x = -120$$

Solve for x.

The path cannot have a negative width. Because -120 is geometrically impossible, we use $x = 20$. The width of the path should be 20 feet.

Step 5. Check the proposed solution in the original wording of the problem. Has the landscape architect doubled the garden's area with the 20-foot-wide path? The area of the garden is 80 feet times 60 feet, or 4800 square feet. Because $80 + 2x$ and $60 + x$ represent the length and width of the expanded rectangle,

$$80 + 2x = 80 + 2 \cdot 20 = 120 \text{ feet is the expanded rectangle's length.}$$

$$60 + x = 60 + 20 = 80 \text{ feet is the expanded rectangle's width.}$$

The area of the expanded rectangle is 120 feet times 80 feet, or 9600 square feet. This is double the area of the garden, 4800 square feet, as specified by the problem's conditions. ∎

✓ **CHECK POINT 5** A rectangular garden measures 16 feet by 12 feet. A path of uniform width is to be added so as to surround the entire garden. The landscape artist doing the work wants the garden and path to cover an area of 320 square feet. How wide should the path be?

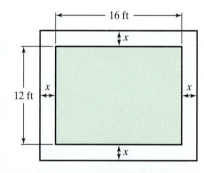

The solution to our next problem relies on knowing the **Pythagorean Theorem**. The theorem relates the lengths of the three sides of a **right triangle**, a triangle with one angle measuring 90°. The side opposite the 90° angle is called the **hypotenuse**. The other sides are called **legs**. The legs form the two sides of the right angle.

The Pythagorean Theorem

The sum of the squares of the lengths of the legs of a right triangle equals the square of the length of the hypotenuse.

If the legs have lengths a and b, and the hypotenuse has length c, then

$$a^2 + b^2 = c^2.$$

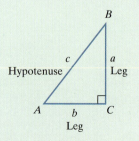

EXAMPLE 6 Using the Pythagorean Theorem to Obtain a Polynomial Equation

Figure 5.16 shows a tent with wires attached to help stabilize it. The length of each wire is 8 feet greater than the distance from the ground to where it is attached to the tent. The distance from the base of the tent to where the wire is anchored exceeds this height by 7 feet. Find the length of each wire used to stabilize the tent.

$x + 8$ x

$x + 7$

Figure 5.16

Solution **Figure 5.16** shows a right triangle. The lengths of the legs are x and $x + 7$. The length of the hypotenuse, $x + 8$, represents the length of the wire. We use the Pythagorean Theorem to find this length.

$$\text{leg}^2 + \text{leg}^2 = \text{hypotenuse}^2$$

$$x^2 + (x + 7)^2 = (x + 8)^2$$

This is the equation arising from the Pythagorean Theorem.

$$x^2 + x^2 + 14x + 49 = x^2 + 16x + 64$$

Square $x + 7$ and $x + 8$.

$$2x^2 + 14x + 49 = x^2 + 16x + 64$$

Combine like terms: $x^2 + x^2 = 2x^2$.

$$x^2 - 2x - 15 = 0$$

Subtract $x^2 + 16x + 64$ from both sides and write the quadratic equation in standard form.

$$(x - 5)(x + 3) = 0$$

Factor the trinomial.

$$x - 5 = 0 \quad \text{or} \quad x + 3 = 0$$

Set each factor equal to 0.

$$x = 5 \qquad\qquad x = -3$$

Solve for x.

Because x represents the distance from the ground to where the wire is attached, x cannot be negative. Thus, we only use $x = 5$. **Figure 5.16** shows that the length of the wire is $x + 8$ feet. The length of the wire is $5 + 8$ feet, or 13 feet.

We can check to see that the lengths of the three sides of the right triangle, x, $x + 7$, and $x + 8$, satisfy the Pythagorean Theorem when $x = 5$. The lengths are 5 feet, 12 feet, and 13 feet.

$$\text{leg}^2 + \text{leg}^2 = \text{hypotenuse}^2$$

$$5^2 + 12^2 \stackrel{?}{=} 13^2$$

$$25 + 144 \stackrel{?}{=} 169$$

$$169 \stackrel{?}{=} 169, \quad \text{true} \quad \blacksquare$$

✓ **CHECK POINT 6** A guy wire is attached to a tree to help it grow straight. The situation is illustrated in **Figure 5.17**. The length of the wire is 2 feet greater than the distance from the base of the tree to the stake. Find the length of the wire.

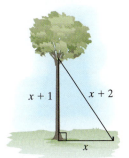

$x + 1$ $x + 2$

x

Figure 5.17

Achieving Success

Be sure to use the chapter test prep video on YouTube to prepare for each chapter test. These videos contain worked-out solutions to every exercise in the chapter test and let you review any exercises you miss.

Are you using any of the other textbook supplements for help and additional study? These include:

- The Student Solutions Manual. This contains fully worked solutions to the odd-numbered section exercises plus all Check Points, Concept and Vocabulary Checks, Review/Preview Exercises, Mid-Chapter Check Points, Chapter Reviews, Chapter Tests, and Cumulative Reviews.

- Lecture Videos on DVD. These are keyed to each section of the text and contain short video clips of an instructor working key text examples and exercises.

- MyMathLab is a text-specific online course.

CONCEPT AND VOCABULARY CHECK

Fill in each blank so that the resulting statement is true.

1. An equation that can be written in the standard form $ax^2 + bx + c = 0, a \neq 0$, is called a/an _____ equation.

2. The zero-product principle states that if $AB = 0$, then _____.

3. The solutions of $ax^2 + bx + c = 0$ correspond to the _____ for the graph of $y = ax^2 + bx + c$.

4. The equation $5x^2 = 20x$ can be written in standard form by _____ on both sides.

5. The equation $x^2 + 4 = 8x - 12$ can be written in standard form by _____ and _____ on both sides.

6. The result of setting two polynomials equal to each other is called a/an _____ equation. The equation is in standard form if one side is _____ and the polynomial on the other side is in standard form, or in _____ powers of the variable. The degree of the equation is the _____ degree of any term in the equation.

7. A triangle with one angle measuring 90° is called a/an _____ triangle. The side opposite the 90° angle is called the _____. The other sides are called _____.

8. The Pythagorean Theorem states that in any _____ triangle, the sum of the squares of the lengths of the _____ equals _____.

5.7 EXERCISE SET

Practice Exercises

In Exercises 1–36, use factoring to solve each quadratic equation. Check by substitution or by using a graphing utility and identifying x-intercepts.

1. $x^2 + x - 12 = 0$
2. $x^2 - 2x - 15 = 0$
3. $x^2 + 6x = 7$
4. $x^2 - 4x = 45$
5. $3x^2 + 10x - 8 = 0$
6. $2x^2 - 5x - 3 = 0$
7. $5x^2 = 8x - 3$
8. $7x^2 = 30x - 8$
9. $3x^2 = 2 - 5x$
10. $5x^2 = 2 + 3x$
11. $x^2 = 8x$
12. $x^2 = 4x$
13. $3x^2 = 5x$
14. $2x^2 = 5x$
15. $x^2 + 4x + 4 = 0$
16. $x^2 + 6x + 9 = 0$
17. $x^2 = 14x - 49$
18. $x^2 = 12x - 36$
19. $9x^2 = 30x - 25$
20. $4x^2 = 12x - 9$
21. $x^2 - 25 = 0$
22. $x^2 - 49 = 0$
23. $9x^2 = 100$
24. $4x^2 = 25$

25. $x(x - 3) = 18$
26. $x(x - 4) = 21$
27. $(x - 3)(x + 8) = -30$
28. $(x - 1)(x + 4) = 14$
29. $x(x + 8) = 16(x - 1)$
30. $x(x + 9) = 4(2x + 5)$
31. $(x + 1)^2 - 5(x + 2) = 3x + 7$
32. $(x + 1)^2 = 2(x + 5)$
33. $x(8x + 1) = 3x^2 - 2x + 2$
34. $2x(x + 3) = -5x - 15$
35. $\dfrac{x^2}{18} + \dfrac{x}{2} + 1 = 0$
36. $\dfrac{x^2}{4} - \dfrac{5x}{2} + 6 = 0$

In Exercises 37–46, use factoring to solve each polynomial equation. Check by substitution or by using a graphing utility and identifying x-intercepts.

37. $x^3 + 4x^2 - 25x - 100 = 0$
38. $x^3 - 2x^2 - x + 2 = 0$
39. $x^3 - x^2 = 25x - 25$
40. $x^3 + 2x^2 = 16x + 32$
41. $3x^4 - 48x^2 = 0$
42. $5x^4 - 20x^2 = 0$
43. $x^4 - 4x^3 + 4x^2 = 0$
44. $x^4 - 6x^3 + 9x^2 = 0$
45. $2x^3 + 16x^2 + 30x = 0$
46. $3x^3 - 9x^2 - 30x = 0$

In Exercises 47–50, determine the x-intercepts of the graph of each quadratic function. Then match the function with its graph, labeled (a)–(d). Each graph is shown in a $[-10, 10, 1]$ *by* $[-10, 10, 1]$ *viewing rectangle.*

47. $y = x^2 - 6x + 8$

48. $y = x^2 - 2x - 8$

49. $y = x^2 + 6x + 8$

50. $y = x^2 + 2x - 8$

a.

b.

c.

d.

Practice PLUS

In Exercises 51–54, solve each polynomial equation.

51. $x(x + 1)^3 - 42(x + 1)^2 = 0$

52. $x(x - 2)^3 - 35(x - 2)^2 = 0$

53. $-4x[x(3x - 2) - 8](25x^2 - 40x + 16) = 0$

54. $-7x[x(2x - 5) - 12](9x^2 + 30x + 25) = 0$

In Exercises 55–58, find all values of c satisfying the given conditions.

55. $f(x) = x^2 - 4x - 27$ and $f(c) = 5$.

56. $f(x) = 5x^2 - 11x + 6$ and $f(c) = 4$.

57. $f(x) = 2x^3 + x^2 - 8x + 2$ and $f(c) = 6$.

58. $f(x) = x^3 + 4x^2 - x + 6$ and $f(c) = 10$.

In Exercises 59–62, find all numbers satisfying the given conditions.

59. The product of the number decreased by 1 and increased by 4 is 24.

60. The product of the number decreased by 6 and increased by 2 is 20.

61. If 5 is subtracted from 3 times the number, the result is the square of 1 less than the number.

62. If the square of the number is subtracted from 61, the result is the square of 1 more than the number.

In Exercises 63–64, list all numbers that must be excluded from the domain of the given function.

63. $f(x) = \dfrac{3}{x^2 + 4x - 45}$

64. $f(x) = \dfrac{7}{x^2 - 3x - 28}$

Application Exercises

A gymnast dismounts the uneven parallel bars at a height of 8 feet with an initial upward velocity of 8 feet per second. The function

$$s(t) = -16t^2 + 8t + 8$$

describes the height of the gymnast's feet above the ground, s(t), in feet, t seconds after dismounting. The graph of the function is shown, with unlabeled tick marks along the horizontal axis. Use the function to solve Exercises 65–66.

Time (seconds)

65. How long will it take the gymnast to reach the ground? Use this information to provide a number on each tick mark along the horizontal axis in the figure shown.

66. When will the gymnast be 8 feet above the ground? Identify the solution(s) as one or more points on the graph.

In a round-robin chess tournament, each player is paired with every other player once. The function

$$f(x) = \frac{x^2 - x}{2}$$

models the number of chess games, $f(x)$, that must be played in a round-robin tournament with x chess players. Use this function to solve Exercises 67–68.

67. In a round-robin chess tournament, 21 games were played. How many players were entered in the tournament?

68. In a round-robin chess tournament, 36 games were played. How many players were entered in the tournament?

The graph of the quadratic function in Exercises 67–68 is shown. Use the graph to solve Exercises 69–70.

69. Identify your solution to Exercise 67 as a point on the graph.

70. Identify your solution to Exercise 68 as a point on the graph.

71. The length of a rectangular sign is 3 feet longer than the width. If the sign's area is 54 square feet, find its length and width.

72. A rectangular parking lot has a length that is 3 yards greater than the width. The area of the parking lot is 180 square yards. Find the length and the width.

73. Each side of a square is lengthened by 3 inches. The area of this new, larger square is 64 square inches. Find the length of a side of the original square.

74. Each side of a square is lengthened by 2 inches. The area of this new, larger square is 36 square inches. Find the length of a side of the original square.

75. A pool measuring 10 meters by 20 meters is surrounded by a path of uniform width, as shown in the figure. If the area of the pool and the path combined is 600 square meters, what is the width of the path?

76. A vacant rectangular lot is being turned into a community vegetable garden measuring 15 meters by 12 meters. A path of uniform width is to surround the garden. If the area of the lot is 378 square meters, find the width of the path surrounding the garden.

77. As part of a landscaping project, you put in a flower bed measuring 10 feet by 12 feet. You plan to surround the bed with a uniform border of low-growing plants.

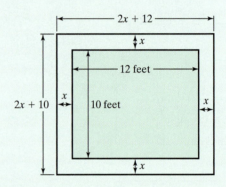

a. Write a polynomial that describes the area of the uniform border that surrounds your flower bed. *Hint:* The area of the border is the area of the large rectangle shown in the figure minus the area of the flower bed.

b. The low growing plants surrounding the flower bed require 1 square foot each when mature. If you have 168 of these plants, how wide a strip around the flower bed should you prepare for the border?

78. As part of a landscaping project, you put in a flower bed measuring 20 feet by 30 feet. To finish off the project, you are putting in a uniform border of pine bark around the outside of the rectangular garden. You have enough pine bark to cover 336 square feet. How wide should the border be?

79. A machine produces open boxes using square sheets of metal. The figure illustrates that the machine cuts equal-sized squares measuring 2 inches on a side from the corners and then shapes the metal into an open box by turning up the sides. If each box must have a volume of 200 cubic inches, find the length and width of the open box.

80. A machine produces open boxes using square sheets of metal. The machine cuts equal-sized squares measuring 3 inches on a side from the corners and then shapes the metal into an open box by turning up the sides. If each box must have a volume of 75 cubic inches, find the length and width of the open box.

81. The rectangular floor of a closet is divided into two right triangles by drawing a diagonal, as shown in the figure. One leg of the right triangle is 2 feet more than twice the other leg. The hypotenuse is 13 feet. Determine the closet's length and width.

82. A piece of wire measuring 20 feet is attached to a telephone pole as a guy wire. The distance along the ground from the bottom of the pole to the end of the wire is 4 feet greater than the height where the wire is attached to the pole. How far up the pole does the guy wire reach?

83. A tree is supported by a wire anchored in the ground 15 feet from its base. The wire is 4 feet longer than the height that it reaches on the tree. Find the length of the wire.

84. A tree is supported by a wire anchored in the ground 5 feet from its base. The wire is 1 foot longer than the height that it reaches on the tree. Find the length of the wire.

Explaining the Concepts

85. What is a quadratic equation?

86. What is the zero-product principle?

87. Explain how to solve $x^2 - x = 6$.

88. Describe the relationship between the solutions of a quadratic equation and the graph of the corresponding quadratic function.

89. What is a polynomial equation? When is it in standard form?

90. What is the degree of a polynomial equation? What are polynomial equations of degree 1 and degree 2, respectively, called?

91. Explain how to solve $x^3 + x^2 = x + 1$.

92. If something is thrown straight up, or possibly dropped, describe a situation in which it is important to know how long it will take the object to hit the ground or possibly the water.

93. A toy rocket is launched vertically upward. Using a quadratic equation, we find that the rocket will reach a height of 220 feet at 2.5 seconds and again at 5.5 seconds. How can this be?

94. Describe a situation in which a landscape designer might use polynomials and polynomial equations.

95. In your own words, state the Pythagorean Theorem.

Technology Exercises

In Exercises 96–99, use a graphing utility with a viewing rectangle large enough to show end behavior to graph each polynomial function. Then use the x-intercepts for the graph to solve the polynomial equation. Check by substitution.

96. Use the graph of $y = x^2 + 3x - 4$ to solve
$x^2 + 3x - 4 = 0$.

97. Use the graph of $y = x^3 + 3x^2 - x - 3$ to solve
$x^3 + 3x^2 - x - 3 = 0$.

98. Use the graph of $y = 2x^3 - 3x^2 - 11x + 6$ to solve
$2x^3 - 3x^2 - 11x + 6 = 0$.

99. Use the graph of $y = -x^4 + 4x^3 - 4x^2$ to solve
$-x^4 + 4x^3 - 4x^2 = 0$.

100. Use the $\boxed{\text{TABLE}}$ feature of a graphing utility to verify the solution sets for any two equations in Exercises 31–32 or 39–40.

Critical Thinking Exercises

Make Sense? *In Exercises 101–104, determine whether each statement makes sense or does not make sense, and explain your reasoning.*

101. I'm working with a quadratic function that describes the length of time a ball has been thrown into the air and its height above the ground, and I find the function's graph more meaningful than its equation.

102. I set the quadratic equation $2x^2 - 5x = 12$ equal to zero and obtained $2x^2 - 5x = 0$.

103. Because some trinomials are prime, some quadratic equations cannot be solved by factoring.

104. I'm looking at a graph with one x-intercept, so it must be the graph of a linear function or a vertical line.

In Exercises 105–108, determine whether each statement is true or false. If the statement is false, make the necessary change(s) to produce a true statement.

105. Quadratic equations solved by factoring always have two different solutions.

106. If $4x(x^2 + 49) = 0$, then

$4x = 0 \quad$ or $\quad x^2 + 49 = 0$

$x = 0 \quad$ or $\qquad x = 7 \quad$ or $\quad x = -7$.

107. If -4 is a solution of $7y^2 + (2k - 5)y - 20 = 0$, then k must equal 14.

108. Some quadratic equations have more than two solutions.

109. Write a quadratic equation in standard form whose solutions are -3 and 7.

110. Solve: $|x^2 + 2x - 36| = 12$.

Review Exercises

111. Solve: $|3x - 2| = 8$.
(Section 4.3, Example 1)

112. Simplify: $3(5 - 7)^2 + \sqrt{16} + 12 \div (-3)$.
(Section 1.2, Example 7)

113. You invested \$3000 in two accounts paying 5% and 8% annual interest. If the total interest earned for the year is \$189, how much was invested at each rate? (Section 3.2, Example 2)

Preview Exercises

Exercises 114–116 will help you prepare for the material covered in the first section of the next chapter.

114. If $f(x) = \dfrac{120x}{100 - x}$, find $f(20)$.

115. Find the domain of $f(x) = \dfrac{4}{x - 2}$.

116. Factor the numerator and the denominator. Then simplify by dividing out the common factor in the numerator and the denominator.

$$\frac{x^2 - 7x - 18}{2x^2 + 3x - 2}$$

Chapter 5 Summary

Definitions and Concepts	Examples

Section 5.1 Introduction to Polynomials and Polynomial Functions

A polynomial is a single term or the sum of two or more terms containing variables with whole-number exponents. A monomial is a polynomial with exactly one term; a binomial has exactly two terms; a trinomial has exactly three terms. If $a \neq 0$, the degree of ax^n is n and the degree of $ax^n y^m$ is $n + m$. The degree of a nonzero constant is 0. The constant 0 has no defined degree. The degree of a polynomial is the greatest degree of any term. The leading term is the term of greatest degree. Its coefficient is called the leading coefficient.

$$7x^3 y \;-\; 4x^5 y^4 \;-\; 2x^4 y$$

Degree is $3 + 1 = 4.$ Degree is $5 + 4 = 9.$ Degree is $4 + 1 = 5.$

The degree of the polynomial is 9. The leading term is $-4x^5 y^4$. The leading coefficient is -4. This polynomial is a trinomial.

In a polynomial function, the expression that defines the function is a polynomial. Polynomial functions have graphs that are smooth and continuous. The behavior of the graph of a polynomial function to the far left or the far right is called its end behavior.

The Leading Coefficient Test

1. Odd-degree polynomial functions have graphs with opposite behavior at each end. If the leading coefficient is positive, the graph falls to the left and rises to the right. If the leading coefficient is negative, the graph rises to the left and falls to the right.

2. Even-degree polynomial functions have graphs with the same behavior at each end. If the leading coefficient is positive, the graph rises to the left and rises to the right. If the leading coefficient is negative, the graph falls to the left and falls to the right.

Describe the end behavior of the graph of each polynomial function:

- $f(x) = -2x^3 + 3x^2 + 11x - 6$

  The degree, 3, is odd. The leading coefficient, -2, is negative. The graph rises to the left and falls to the right.

- $f(x) = x^2 - 4$

  The degree, 2, is even. The leading coefficient, 1, is positive. The graph rises to the left and rises to the right.

To add polynomials, add like terms.

$(6x^3 y + 5x^2 y - 7y) + (-9x^3 y + x^2 y + 6y)$
$= (6x^3 y - 9x^3 y) + (5x^2 y + x^2 y) + (-7y + 6y)$
$= -3x^3 y + 6x^2 y - y$

To subtract two polynomials, add the opposite, or additive inverse, of the polynomial being subtracted.

$(5y^3 - 9y^2 - 4) - (3y^3 - 12y^2 - 5)$
$= (5y^3 - 9y^2 - 4) + (-3y^3 + 12y^2 + 5)$
$= (5y^3 - 3y^3) + (-9y^2 + 12y^2) + (-4 + 5)$
$= 2y^3 + 3y^2 + 1$

Definitions and Concepts	**Examples**

Section 5.2 Multiplication of Polynomials

To multiply monomials, multiply coefficients and add exponents.	$(-2x^2y^4)(-3x^3y)$ $= (-2)(-3)x^{2+3}y^{4+1} = 6x^5y^5$
To multiply a monomial and a polynomial, multiply each term of the polynomial by the monomial.	$7x^2y(4x^3y^5 - 2xy - 1)$ $= 7x^2y \cdot 4x^3y^5 - 7x^2y \cdot 2xy - 7x^2y \cdot 1$ $= 28x^5y^6 - 14x^3y^2 - 7x^2y$
To multiply polynomials if neither is a monomial, multiply each term of one by each term of the other.	$(x^3 + 2x)(5x^2 - 3x + 4)$ $= x^3(5x^2 - 3x + 4) + 2x(5x^2 - 3x + 4)$ $= 5x^5 - 3x^4 + 4x^3 + 10x^3 - 6x^2 + 8x$ $= 5x^5 - 3x^4 + 14x^3 - 6x^2 + 8x$
The FOIL method may be used when multiplying two binomials: First terms multiplied. Outside terms multiplied. Inside terms multiplied. Last terms multiplied.	$\boxed{F}$ $\boxed{O}$ $\boxed{I}$ $\boxed{L}$ $(5x - 3y)(2x + y) = 5x \cdot 2x + 5x \cdot y + (-3y) \cdot 2x + (-3y) \cdot y$ $= 10x^2 + 5xy - 6xy - 3y^2$ $= 10x^2 - xy - 3y^2$
The Square of a Binomial Sum $(A + B)^2 = A^2 + 2AB + B^2$	$(x^2 + 6)^2 = (x^2)^2 + 2 \cdot x^2 \cdot 6 + 6^2$ $= x^4 + 12x^2 + 36$
The Square of a Binomial Difference $(A - B)^2 = A^2 - 2AB + B^2$	$(4x - 5)^2 = (4x)^2 - 2 \cdot 4x \cdot 5 + 5^2$ $= 16x^2 - 40x + 25$
The Product of the Sum and Difference of Two Terms $(A + B)(A - B) = A^2 - B^2$	• $(3x + 7y)(3x - 7y) = (3x)^2 - (7y)^2$ $= 9x^2 - 49y^2$ • $[(x + 2) - 4y][(x + 2) + 4y]$ $= (x + 2)^2 - (4y)^2 = x^2 + 4x + 4 - 16y^2$

Section 5.3 Greatest Common Factors and Factoring by Grouping

Factoring a polynomial consisting of the sum of monomials means finding an equivalent expression that is a product. Polynomials that cannot be factored using integer coefficients are called prime polynomials over the integers. The greatest common factor, GCF, is an expression that divides every term of the polynomial. The GCF is the product of the largest common numerical factor and the variable of lowest degree common to every term of the polynomial. To factor a monomial from a polynomial, express each term as the product of the GCF and its other factor. Then use the distributive property to factor out the GCF. When the leading coefficient of a polynomial is negative, it is often desirable to factor out a common factor with a negative coefficient.	Factor: $4x^4y^2 - 12x^2y^3 + 20xy^2$. (GCF is $4xy^2$.) $= 4xy^2 \cdot x^3 - 4xy^2 \cdot 3xy + 4xy^2 \cdot 5$ $= 4xy^2(x^3 - 3xy + 5)$ Factor: $-25x^3 + 10x^2 - 15x$. (Use $-5x$ as a common factor.) $= -5x(5x^2) - 5x(-2x) - 5x(3)$ $= -5x(5x^2 - 2x + 3)$

Definitions and Concepts	**Examples**

Section 5.3 Greatest Common Factors and Factoring by Grouping (continued)

To factor by grouping, factor out the GCF from each group. Then factor out the remaining factor.	$xy + 7x - 2y - 14$ $= x(y + 7) - 2(y + 7)$ $= (y + 7)(x - 2)$

Section 5.4 Factoring Trinomials

To factor a trinomial of the form $x^2 + bx + c$, find two numbers whose product is c and whose sum is b. The factorization is $(x + \text{one number})(x + \text{other number}).$	Factor: $x^2 + 9x + 20$. Find two numbers whose product is 20 and whose sum is 9. The numbers are 4 and 5. $x^2 + 9x + 20 = (x + 4)(x + 5)$
In some trinomials, the highest power is greater than 2, and the exponent in one of the terms is half that of the other term. Factor by introducing a substitution. Let u equal the variable to the smaller power.	Factor: $x^6 - 7x^3 + 12$. $\qquad = (x^3)^2 - 7x^3 + 12$ $\qquad = u^2 - 7u + 12 \quad$ Let $u = x^3$. $\qquad = (u - 4)(u - 3)$ $\qquad = (x^3 - 4)(x^3 - 3)$
To factor $ax^2 + bx + c$ by trial and error, try various combinations of factors of ax^2 and c until a middle term of bx is obtained for the sum of the outside and inside products.	Factor: $2x^2 + 7x - 15$. Factors of $2x^2$: $2x, x$ Factors of -15: 1 and -15, -1 and 15, 3 and -5, -3 and 5 $(2x - 3)(x + 5)$ Sum of outside and inside products should equal $7x$. $10x - 3x = 7x$ Thus, $2x^2 + 7x - 15 = (2x - 3)(x + 5)$.
To factor $ax^2 + bx + c$ by grouping, find the factors of ac whose sum is b. Write bx as a sum or difference using these factors. Then factor by grouping.	Factor: $2x^2 + 7x - 15$. Find the factors of $2(-15)$, or -30, whose sum is 7. They are 10 and -3. $\quad 2x^2 + 7x - 15$ $\quad = 2x^2 + 10x - 3x - 15$ $\quad = 2x(x + 5) - 3(x + 5) = (x + 5)(2x - 3)$

Section 5.5 Factoring Special Forms

The Difference of Two Squares $A^2 - B^2 = (A + B)(A - B)$	$16x^2 - 9y^2$ $= (4x)^2 - (3y)^2 = (4x + 3y)(4x - 3y)$
Perfect Square Trinomials $A^2 + 2AB + B^2 = (A + B)^2$ $A^2 - 2AB + B^2 = (A - B)^2$	• $x^2 + 20x + 100 = x^2 + 2 \cdot x \cdot 10 + 10^2 = (x + 10)^2$ • $9x^2 - 30x + 25 = (3x)^2 - 2 \cdot 3x \cdot 5 + 5^2 = (3x - 5)^2$
Sum or Difference of Cubes $A^3 + B^3 = (A + B)(A^2 - AB + B^2)$ $A^3 - B^3 = (A - B)(A^2 + AB + B^2)$	$125x^3 - 8 = (5x)^3 - 2^3$ $\qquad = (5x - 2)[(5x)^2 + 5x \cdot 2 + 2^2]$ $\qquad = (5x - 2)(25x^2 + 10x + 4)$

Definitions and Concepts	**Examples**

Section 5.5 Factoring Special Forms (continued)

| When using factoring by grouping, terms can sometimes be grouped to obtain the difference of two squares. One of the squares is a perfect square trinomial. | $x^2 + 18x + 81 - 25y^2$
 $= (x + 9)^2 - (5y)^2$
 $= (x + 9 + 5y)(x + 9 - 5y)$ |

Section 5.6 A General Factoring Strategy

| **A Factoring Strategy**
 1. Factor out the GCF or a common factor with a negative coefficient.
 2. **a.** If two terms, try
 $A^2 - B^2 = (A + B)(A - B)$
 $A^3 + B^3 = (A + B)(A^2 - AB + B^2)$
 $A^3 - B^3 = (A - B)(A^2 + AB + B^2)$.
 b. If three terms, try
 $A^2 + 2AB + B^2 = (A + B)^2$
 $A^2 - 2AB + B^2 = (A - B)^2$.
 If not a perfect square trinomial, try trial and error or grouping.
 c. If four terms, try factoring by grouping.
 3. See if any factors can be factored further. | Factor: $3x^4 + 12x^3 - 3x^2 - 12x$.
 The GCF is $3x$.
 $3x^4 + 12x^3 - 3x^2 - 12x$
 $= 3x(x^3 + 4x^2 - x - 4)$
 Four terms: Try grouping.
 $= 3x[x^2(x + 4) - 1(x + 4)]$
 $= 3x(x + 4)(x^2 - 1)$
 This can be factored further.
 $= 3x(x + 4)(x + 1)(x - 1)$ |

Section 5.7 Polynomial Equations and Their Applications

| A quadratic equation in x can be written in the standard form
 $ax^2 + bx + c = 0, a \neq 0$.
 A polynomial equation is the result of setting two polynomials equal to each other. The equation is in standard form if one side is 0 and the polynomial on the other side is in standard form, that is, in descending powers of the variable. In standard form, its degree is the highest degree of any term in the equation. A polynomial equation of degree 1 is a linear equation and of degree 2 a quadratic equation. Some polynomial equations can be solved by writing the equation in standard form, factoring, and then using the zero-product principle: If a product is 0, then at least one of the factors is equal to 0. | Solve: $5x^2 + 7x = 6$.
 $5x^2 + 7x - 6 = 0$
 $(5x - 3)(x + 2) = 0$
 $5x - 3 = 0$ or $x + 2 = 0$
 $5x = 3$ $x = -2$
 $x = \dfrac{3}{5}$
 The solutions are -2 and $\frac{3}{5}$, and the solution set is $\left\{-2, \frac{3}{5}\right\}$. (The solutions are the x-intercepts of the graph of $y = 5x^2 + 7x - 6$.) |

CHAPTER 5 REVIEW EXERCISES

5.1 *In Exercises 1–2, determine the coefficient of each term, the degree of each term, the degree of the polynomial, the leading term, and the leading coefficient of the polynomial.*

1. $-5x^3 + 7x^2 - x + 2$

2. $8x^4y^2 - 7xy^6 - x^3y$

3. If $f(x) = x^3 - 4x^2 + 3x - 1$, find $f(-2)$.

4. The bar graph shows the number of record daily high temperatures set across the United States, by decade.

Number of Record Daily High Temperatures in the United States, by Decade

Source: University Corporation for Atmospheric Research, UCAR MAGAZINE

The data can be modeled by the function

$$f(x) = 10x^3 - 200x^2 + 1230x - 2016,$$

where $f(x)$ is the number of record daily high temperatures in the United States, in hundreds of thousands, in decade x, with $x = 6 = 1960s, 7 = 1970s, 8 = 1980s, 9 = 1990s,$ and $10 = 2000s.$

a. Find and interpret $f(10)$.

b. Does your answer in part (a) underestimate or overestimate the number of record highs shown by the graph? By how much?

In Exercises 5–8, use the Leading Coefficient Test to determine the end behavior of the graph of the given polynomial function. Then use this end behavior to match the polynomial function with its graph. [The graphs are labeled (a) through (d).]

5. $f(x) = -x^3 + x^2 + 2x$

6. $f(x) = x^6 - 6x^4 + 9x^2$

7. $f(x) = x^5 - 5x^3 + 4x$

8. $f(x) = -x^4 + 1$

a.

b.

c.

d.

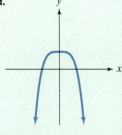

9. The Brazilian Amazon rain forest is the world's largest tropical rain forest, with some of the greatest biodiversity of any region. In 2009, the number of trees cut down in the Amazon dropped to its lowest level in 20 years. The line graph shows the number of square kilometers cleared from 2001 through 2009.

Amazon Deforestation

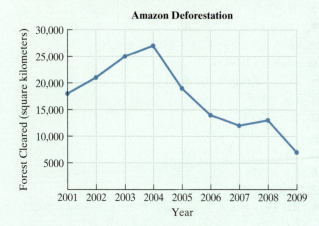

Source: Brazil's National Institute for Space Research

The data in the line graph can be modeled by the following third- and fourth-degree polynomial functions:

$$f(x) = 158x^3 - 2845x^2 + 12{,}926x + 7175$$
$$g(x) = -17x^4 + 508x^3 - 5180x^2 + 18{,}795x + 2889.$$

Amazon deforestation, in square kilometers, x years after **2000**

a. Use the Leading Coefficient Test to determine the end behavior to the right for the graph of f.

b. Assume that the rate at which the Amazon rain forest is being cut down continues to decline. Based on your answer to part (a), will f be useful in modeling Amazon deforestation over an extended period of time? Explain your answer.

c. Use the Leading Coefficient Test to determine the end behavior to the right for the graph of g.

d. Assume that the rate at which the Amazon rain forest is being cut down continues to decline. Based on your answer to part (c), will g be useful in modeling Amazon deforestation over an extended period of time? Explain your answer.

In Exercises 10–11, add the polynomials.

10. $(-8x^3 + 5x^2 - 7x + 4) + (9x^3 - 11x^2 + 6x - 13)$

11. $(7x^3y - 13x^2y - 6y) + (5x^3y + 11x^2y - 8y - 17)$

In Exercises 12–13, subtract the polynomials.

12. $(7x^3 - 6x^2 + 5x - 11) - (-8x^3 + 4x^2 - 6x - 3)$

13. $(4x^3y^2 - 7x^3y - 4) - (6x^3y^2 - 3x^3y + 4)$

14. Subtract $-2x^3 - x^2y + xy^2 + 7y^3$ from $x^3 + 4x^2y - y^3$.

5.2 *In Exercises 15–27, multiply the polynomials.*

15. $(4x^2yz^5)(-3x^4yz^2)$

16. $6x^3\left(\dfrac{1}{3}x^5 - 4x^2 - 2\right)$

17. $7xy^2(3x^4y^2 - 5xy - 1)$

18. $(2x + 5)(3x^2 + 7x - 4)$

19. $(x^2 + x - 1)(x^2 + 3x + 2)$

20. $(4x - 1)(3x - 5)$

21. $(3xy - 2)(5xy + 4)$

22. $(3x + 7y)^2$

23. $(x^2 - 5y)^2$

24. $(2x + 7y)(2x - 7y)$

25. $(3xy^2 - 4x)(3xy^2 + 4x)$

26. $[(x + 3) + 5y][(x + 3) - 5y]$

27. $(x + y + 4)^2$

28. Let $f(x) = x - 3$ and $g(x) = 2x + 5$. Find $(fg)(x)$ and $(fg)(-4)$.

29. Let $f(x) = x^2 - 7x + 2$. Find each of the following and simplify:

    a. $f(a - 1)$

    b. $f(a + h) - f(a)$.

5.3 *In Exercises 30–33, factor the greatest common factor from each polynomial.*

30. $16x^3 + 24x^2$

31. $2x - 36x^2$

32. $21x^2y^2 - 14xy^2 + 7xy$

33. $18x^3y^2 - 27x^2y$

In Exercises 34–35, factor out a common factor with a negative coefficient.

34. $-12x^2 + 8x - 48$

35. $-x^2 - 11x + 14$

In Exercises 36–38, factor by grouping.

36. $x^3 - x^2 - 2x + 2$

37. $xy - 3x - 5y + 15$

38. $5ax - 15ay + 2bx - 6by$

5.4 *In Exercises 39–47, factor each trinomial completely, or state that the trinomial is prime.*

39. $x^2 + 8x + 15$

40. $x^2 + 16x - 80$

41. $x^2 + 16xy - 17y^2$

42. $3x^3 - 36x^2 + 33x$

43. $3x^2 + 22x + 7$

44. $6x^2 - 13x + 6$

45. $5x^2 - 6xy - 8y^2$

46. $6x^3 + 5x^2 - 4x$

47. $2x^2 + 11x + 15$

In Exercises 48–51, factor by introducing an appropriate substitution.

48. $x^6 + x^3 - 30$

49. $x^4 - 10x^2 - 39$

50. $(x + 5)^2 + 10(x + 5) + 24$

51. $5x^6 + 17x^3 + 6$

5.5 *In Exercises 52–55, factor each difference of two squares.*

52. $4x^2 - 25$

53. $1 - 81x^2y^2$

54. $x^8 - y^6$

55. $(x - 1)^2 - y^2$

In Exercises 56–60, factor any perfect square trinomials, or state that the polynomial is prime.

56. $x^2 + 16x + 64$

57. $9x^2 - 6x + 1$

58. $25x^2 + 20xy + 4y^2$

59. $49x^2 + 7x + 1$

60. $25x^2 - 40xy + 16y^2$

In Exercises 61–62, factor by grouping to obtain the difference of two squares.

61. $x^2 + 18x + 81 - y^2$

62. $z^2 - 25x^2 + 10x - 1$

In Exercises 63–65, factor using the formula for the sum or difference of two cubes.

63. $64x^3 + 27$

64. $125x^3 - 8$

65. $x^3y^3 + 1$

5.6 *In Exercises 66–90, factor completely, or state that the polynomial is prime.*

66. $15x^2 + 3x$

67. $12x^4 - 3x^2$

68. $20x^4 - 24x^3 + 28x^2 - 12x$

69. $x^3 - 15x^2 + 26x$

70. $-2y^4 + 24y^3 - 54y^2$

71. $9x^2 - 30x + 25$

72. $5x^2 - 45$

73. $2x^3 - x^2 - 18x + 9$

74. $6x^2 - 23xy + 7y^2$

75. $2y^3 + 12y^2 + 18y$

76. $x^2 + 6x + 9 - 4a^2$

77. $8x^3 - 27$

78. $x^5 - x$

79. $x^4 - 6x^2 + 9$

80. $x^2 + xy + y^2$

81. $4a^3 + 32$

82. $x^4 - 81$

83. $ax + 3bx - ay - 3by$

84. $27x^3 - 125y^3$

85. $10x^3y + 22x^2y - 24xy$

86. $6x^6 + 13x^3 - 5$

87. $2x + 10 + x^2y + 5xy$

88. $y^3 + 2y^2 - 25y - 50$

89. $a^8 - 1$

90. $9(x - 4) + y^2(4 - x)$

In Exercises 91–92,

    a. *Write an expression for the area of the shaded blue region.*

    b. *Write the expression in factored form.*

91.

92.

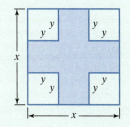

5.7 *In Exercises 93–97, use factoring to solve each polynomial equation.*

93. $x^2 + 6x + 5 = 0$

94. $3x^2 = 22x - 7$

95. $(x + 3)(x - 2) = 50$

96. $3x^2 = 12x$

97. $x^3 + 5x^2 = 9x + 45$

98. A model rocket is launched from the top of a cliff 144 feet above sea level. The function

$$s(t) = -16t^2 + 128t + 144$$

describes the rocket's height above the water, $s(t)$, in feet, t seconds after it is launched. The rocket misses the edge of the cliff on its way down and eventually lands in the ocean. How long will it take for the rocket to hit the water?

99. How much distance do you need to bring your car to a complete stop? A function used by those who study automobile safety is

$$d(x) = \frac{x^2}{20} + x,$$

where $d(x)$ is the stopping distance, in feet, for a car traveling at x miles per hour.

    a. If it takes you 40 feet to come to a complete stop, how fast was your car traveling?

    b. The graph of the quadratic function that models stopping distance is shown. Identify your solution from part (a) as a point on the graph.

Stopping Distance of a Car

$d(x) = \dfrac{x^2}{20} + x$

    c. Describe the trend shown by the graph.

100. The length of a rectangular sign is 3 feet longer than the width. If the sign has space for 54 square feet of advertising, find its length and its width.

101. A painting measuring 10 inches by 16 inches is surrounded by a frame of uniform width. If the combined area of the painting and frame is 280 square inches, determine the width of the frame.

102. A lot is in the shape of a right triangle. The longer leg of the triangle is 20 yards longer than twice the length of the shorter leg. The hypotenuse is 30 yards longer than twice the length of the shorter leg. What are the lengths of the three sides?

CHAPTER 5 TEST

Step-by-step test solutions are found on the Chapter Test Prep Videos available in MyMathLab® or on YouTube (search "BlitzerInterAlg7e" and click on "Channels").

In Exercises 1–2, give the degree and the leading coefficient of the polynomial.

1. $7x - 5 + x^2 - 6x^3$

2. $4xy^3 + 7x^4y^5 - 3xy^4$

3. If $f(x) = 3x^3 + 5x^2 - x + 6$, find $f(0)$ and $f(-2)$.

In Exercises 4–5, use the Leading Coefficient Test to describe the end behavior of the graph of the polynomial function.

4. $f(x) = -16x^2 + 160x$

5. $f(x) = 4x^3 + 12x^2 - x - 3$

In Exercises 6–13, perform the indicated operations.

6. $(4x^3y - 19x^2y - 7y) + (3x^3y + x^2y + 6y - 9)$

7. $(6x^2 - 7x - 9) - (-5x^2 + 6x - 3)$

8. $(-7x^3y)(-5x^4y^2)$

9. $(x - y)(x^2 - 3xy - y^2)$

10. $(7x - 9y)(3x + y)$

11. $(2x - 5y)(2x + 5y)$

12. $(4y - 7)^2$

13. $[(x + 2) + 3y][(x + 2) - 3y]$

14. Let $f(x) = x + 2$ and $g(x) = 3x - 5$. Find $(fg)(x)$ and $(fg)(-5)$.

15. Let $f(x) = x^2 - 5x + 3$. Find $f(a + h) - f(a)$ and simplify.

In Exercises 16–33, factor completely, or state that the polynomial is prime.

16. $14x^3 - 15x^2$

17. $81y^2 - 25$

18. $x^3 + 3x^2 - 25x - 75$

19. $25x^2 - 30x + 9$

20. $x^2 + 10x + 25 - 9y^2$

21. $x^4 + 1$

22. $y^2 - 16y - 36$

23. $14x^2 + 41x + 15$

24. $5x^3 - 5$

25. $12x^2 - 3y^2$

26. $12x^2 - 34x + 10$

27. $3x^4 - 3$

28. $x^8 - y^8$

29. $12x^2y^4 + 8x^3y^2 - 36x^2y$

30. $x^6 - 12x^3 - 28$

31. $x^4 - 2x^2 - 24$

32. $12x^2y - 27xy + 6y$

33. $y^4 - 3y^3 + 2y^2 - 6y$

In Exercises 34–37, solve each polynomial equation.

34. $3x^2 = 5x + 2$

35. $(5x + 4)(x - 1) = 2$

36. $15x^2 - 5x = 0$

37. $x^3 - 4x^2 - x + 4 = 0$

38. A baseball is thrown straight up from a rooftop 448 feet high. The function

$$s(t) = -16t^2 + 48t + 448$$

describes the ball's height above the ground, $s(t)$, in feet, t seconds after it is thrown. How long will it take for the ball to hit the ground?

39. An architect is allowed 15 square yards of floor space to add a small bedroom to a house. Because of the room's design in relationship to the existing structure, the width of the rectangular floor must be 7 yards less than two times the length. Find the length and width of the rectangular floor that the architect is permitted.

40. Find the lengths of the three sides of the right triangle in the figure shown.

CUMULATIVE REVIEW EXERCISES (CHAPTERS 1–5)

In Exercises 1–7, solve each equation, inequality, or system of equations.

1. $8(x + 2) - 3(2 - x) = 4(2x + 6) - 2$

2. $\begin{cases} 2x + 4y = -6 \\ x = 2y - 5 \end{cases}$

3. $\begin{cases} 2x - y + 3z = 0 \\ 2y + z = 1 \\ x + 2y - z = 5 \end{cases}$

4. $2x + 4 < 10$ and $3x - 1 > 5$

5. $|2x - 5| \geq 9$

6. $2x^2 = 7x - 5$

7. $2x^3 + 6x^2 = 20x$

8. Solve for x: $x = \dfrac{ax + b}{c}$.

9. Use function notation to write the equation of the line passing through $(-2, -3)$ and $(2, 5)$.

10. In a campuswide election for student government president, 2800 votes were cast for the two candidates. If the winner had 160 more votes than the loser, how many votes were cast for each candidate?

In Exercises 11–13, graph each equation or inequality in a rectangular coordinate system.

11. $f(x) = -\dfrac{1}{3}x + 1$

12. $4x - 5y < 20$

13. $y \leq -1$

14. Simplify: $-\dfrac{8x^3y^6}{16x^9y^{-4}}$.

15. Write in scientific notation: 0.0000706.

In Exercises 16–17, perform the indicated operations.

16. $(3x^2 - y)^2$

17. $(3x^2 - y)(3x^2 + y)$

In Exercises 18–20, factor completely.

18. $x^3 - 3x^2 - 9x + 27$

19. $x^6 - x^2$

20. $14x^3y^2 - 28x^4y^2$

Rational Expressions, Functions, and Equations

I s the newest route to college through video games? A few collegiate gaming factoids:

- Hundreds of colleges have formed video-gaming teams and some have made gaming a varsity sport. Top Nexus destroyers can receive thousands of dollars in e-sports scholarships.

- College e-sports are predicted to galvanize the video-game industry. The numbers are impressive: In 2014, an estimated 10,000 students competed in the video game *Collegiate StarLeague*. More than 130,000 viewers watched the *League of Legends* college finals.

- Video-game companies, recognizing a great market opportunity when they see one, are helping to underwrite the college gaming explosion. In 2015, Riot Games, creator of *League of Legends*, offered $360,000 in scholarship money to gifted players. Blizzard Entertainment ran a 64-team collegiate tournament when they introduced *Heroes of the Storm* (dubbed Heroes of the Dorm). ESPN televised the finals and each member of the winning team received up to three years of tuition. Teams from over 650 colleges tried to qualify.

But wait, there's much more! We are on the cusp of a revolution in the way people play video games. Virtual reality headsets will enable players to experience video games as immersive three-dimensional environments, basically putting the gamer inside the game.

In this chapter, you will see how functions involving fractional expressions provide insights into phenomena as diverse as virtual reality video games, the aftereffects of candy on your mouth, the cost of environmental cleanup, the relationship between heart rate and life span, and even our ongoing processes of learning and forgetting.

Here's where you'll find these applications:

- Virtual reality video games: Section 6.7 opener and Section 6.7, Example 3
- Sugar and your mouth: Exercise Set 6.1, Exercises 109–112
- Environmental cleanup: Section 6.1, Example 1, and Section 6.6, Example 6
- Heart rate and life span: Exercise Set 6.8, Exercises 29–32
- Learning and forgetting: Exercise Set 6.6, Exercises 53–60.

Rational Expressions and Functions: Multiplying and Dividing ▶

Environmental scientists and municipal planners often make decisions using **cost-benefit models**. These mathematical models estimate the cost of removing a pollutant from the atmosphere as a function of the percentage of pollutant removed. What kinds of functions describe the cost of reducing environmental pollution? In this section, we introduce this new category of functions, called *rational functions*.

Rational Expressions

A **rational expression** consists of a polynomial divided by a nonzero polynomial. Here are some examples of rational expressions:

$$\frac{120x}{100 - x}, \quad \frac{2x + 1}{2x^2 - x - 1}, \quad \frac{3x^2 + 12xy - 15y^2}{6x^3 - 6xy^2}.$$

Rational Functions

A **rational function** is a function defined by a formula that is a rational expression.

1. Evaluate rational functions. ▶

> **EXAMPLE 1** Using a Cost-Benefit Model

The rational function

$$f(x) = \frac{120x}{100 - x}$$

models the cost, $f(x)$, in thousands of dollars, to remove $x\%$ of the pollutants that a city has discharged into a lake. Find and interpret each of the following:

a. $f(20)$ **b.** $f(80)$.

Solution We use substitution to evaluate a rational function, just as we did to evaluate other functions in Chapter 2.

a. $f(x) = \dfrac{120x}{100 - x}$ This is the given rational function.

$f(20) = \dfrac{120(20)}{100 - 20}$ To find f of 20, replace x with 20.

$= \dfrac{2400}{80} = 30$ Perform the indicated operations.

Thus, $f(20) = 30$. This means that the cost to remove 20% of the lake's pollutants is $30 thousand. **Figure 6.1** illustrates the solution by the point $(20, 30)$ on the graph of the rational function.

b. $f(x) = \dfrac{120x}{100 - x}$ This is the given rational function.

$$f(80) = \dfrac{120(80)}{100 - 80} = \dfrac{9600}{20} = 480$$ To find f of 80, replace x with 80.

Thus, $f(80) = 480$. This means that the cost to remove 80% of the lake's pollutants is $480 thousand. **Figure 6.1** illustrates the solution by the point $(80, 480)$ on the graph of the cost-benefit model. The graph illustrates that costs rise steeply as the percentage of pollutants removed increases. ∎

✓ **CHECK POINT 1**   Use the rational function in Example 1 to find and interpret: **a.** $f(40)$ **b.** $f(60)$. Identify each solution as a point on the graph of the rational function in **Figure 6.1**.

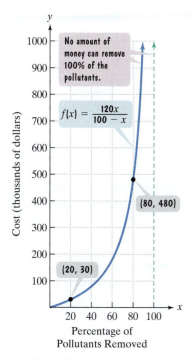

No amount of money can remove 100% of the pollutants.

$f(x) = \dfrac{120x}{100 - x}$

$(80, 480)$

$(20, 30)$

Percentage of Pollutants Removed

Figure 6.1

2 Find the domain of a rational function. ▶

The Domain of a Rational Function

Does the cost-benefit model

$$f(x) = \dfrac{120x}{100 - x}$$

indicate that the city can clean up its lake completely? To do this, the city must remove 100% of the pollutants. The problem is that the rational function is undefined for $x = 100$.

$$f(x) = \dfrac{120x}{100 - x}$$ If $x = 100$, the value of the denominator is 0.

Notice how the graph of the rational function in **Figure 6.1** approaches, but never touches, the dashed green vertical line whose equation is $x = 100$. The graph continues to rise more and more steeply, visually showing the escalating costs. By never touching the dashed vertical line, the graph illustrates that no amount of money will be enough to remove all pollutants from the lake.

In Chapter 2, we learned to exclude from a function's domain real numbers that cause division by zero. Thus, the **domain of a rational function** is the set of all real numbers except those for which the denominator is zero. We can find the domain by determining when the denominator is zero. For the cost-benefit model, the denominator is zero when $x = 100$. Furthermore, for this model, negative values of x and values of x greater than 100 are not meaningful. The domain of the function is $[0, 100)$ and excludes 100.

Inspection can sometimes be used to find a rational function's domain. Here are two examples.

This numerator can be zero, so there is no need to exclude 3 from the domain.

$$f(x) = \dfrac{4}{x - 2} \qquad g(x) = \dfrac{x - 3}{(x + 1)(x - 1)}$$

This denominator would equal zero if $x = 2$.   This factor would equal zero if $x = -1$.   This factor would equal zero if $x = 1$.

The domain of f, which excludes 2, can be expressed in interval notation:

$$\text{Domain of } f = (-\infty, 2) \cup (2, \infty).$$

$$g(x) = \frac{x - 3}{(x + 1)(x - 1)}$$

This factor would equal zero if $x = -1$.	This factor would equal zero if $x = 1$.

Function g (repeated)

Likewise, the domain of g, which excludes both -1 and 1, can be expressed in interval notation:

$$\text{Domain of } g = (-\infty, -1) \cup (-1, 1) \cup (1, \infty).$$

EXAMPLE 2 Finding the Domain of a Rational Function

Find the domain of f if

$$f(x) = \frac{2x + 1}{2x^2 - x - 1}.$$

Solution The domain of f is the set of all real numbers except those for which the denominator is zero. We can identify such numbers by setting the denominator equal to zero and solving for x.

$$2x^2 - x - 1 = 0 \qquad \text{Set the denominator equal to O.}$$
$$(2x + 1)(x - 1) = 0 \qquad \text{Factor.}$$
$$2x + 1 = 0 \quad \text{or} \quad x - 1 = 0 \qquad \text{Set each factor equal to O.}$$
$$2x = -1 \qquad\qquad x = 1 \qquad \text{Solve the resulting equations.}$$
$$x = -\frac{1}{2}$$

Because $-\frac{1}{2}$ and 1 make the denominator zero, these are the values to exclude. Thus,

$$\text{Domain of } f = \left(-\infty, -\frac{1}{2}\right) \cup \left(-\frac{1}{2}, 1\right) \cup (1, \infty). \quad \blacksquare$$

Using Technology

We can use the $\boxed{\text{TABLE}}$ feature of a graphing utility (with $\Delta\text{Tbl} = .5$) to verify our work with

$$f(x) = \frac{2x + 1}{2x^2 - x - 1}.$$

Enter

$$y_1 = \boxed{(}\,\boxed{2}\,\boxed{x}\,\boxed{+}\,\boxed{1}\,\boxed{)}\,\boxed{\div}$$

$$\boxed{(}\,\boxed{2}\,\boxed{x}\,\boxed{\wedge}\,\boxed{2}\,\boxed{-}\,\boxed{x}\,\boxed{-}\,\boxed{1}\,\boxed{)}\cdot$$

and press $\boxed{\text{TABLE}}$.
On some calculators, it may be necessary to press the right arrow key to exit the exponent after entering the exponent, 2.

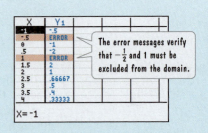

X	Y₁
-1	-.5
-.5	ERROR
0	-1
.5	-2
1	ERROR
1.5	2
2	1
2.5	.66667
3	.5
3.5	.4
4	.33333

X= -1

The error messages verify that $-\frac{1}{2}$ and 1 must be excluded from the domain.

$\checkmark$ **CHECK POINT 2** Find the domain of f if

$$f(x) = \frac{x - 5}{2x^2 + 5x - 3}.$$

Does every rational function have values to exclude? The answer is no. For example, consider

$$f(x) = \frac{2}{x^2 + 1}.$$

No real-number values of x cause this denominator to equal zero.

The domain of f consists of the set of all real numbers. Recall that this set can be expressed in three different ways:

$$\text{Domain of } f = \{x \mid x \text{ is a real number}\} \text{ or } \mathbb{R} \text{ or } (-\infty, \infty).$$

3 Interpret information given by the graph of a rational function.

What Graphs of Rational Functions Look Like

In everyday speech, a continuous process is one that goes on without interruption and without abrupt changes. In mathematics, a continuous function has much the same meaning. The graph of a continuous function does not have interrupting breaks, such as holes, gaps, or jumps. Thus, the graph of a continuous function can be drawn without lifting a pencil off the paper.

Most rational functions are not continuous functions. For example, the graph of the rational function

$$f(x) = \frac{2x + 1}{2x^2 - x - 1}$$

Using Technology

When using a graphing utility to graph a rational function, remember to enclose both the numerator and denominator in parentheses. The graph of the rational function

$$y = \frac{2x + 1}{2x^2 - x - 1}$$

is shown in a $[-6.6, 6.6, 1]$ by $[-6.6, 6.6, 1]$ viewing rectangle.

Some graphing utilities incorrectly connect the two pieces of the graph shown above, resulting in a nearly vertical line appearing on the screen. If your graphing utility does this, we recommend that you use DOT mode. Consult your owner's manual.

is shown in **Figure 6.2**. In Example 2, we excluded $-\frac{1}{2}$ and 1 from this function's domain. Unlike the graph of a polynomial function, this graph has two breaks in it—one at each of the excluded values. At $-\frac{1}{2}$, a hole in the graph appears. The graph is composed of two distinct branches. Each branch approaches, but never touches, the dashed vertical line drawn at 1, the other excluded value.

A vertical line that the graph of a function approaches, but does not touch, is said to be a **vertical asymptote** of the graph. In **Figure 6.2**, the line $x = 1$ is a vertical asymptote of the graph of f. A rational function may have no vertical asymptotes, one vertical asymptote, or several vertical asymptotes. The graph of a rational function never intersects a vertical asymptote.

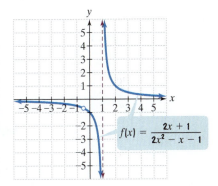

Figure 6.2

Unlike the graph of a polynomial function, the graph of the rational function in **Figure 6.2** does not go up or down at each end. At the far left and the far right, the graph is getting closer to, but not actually reaching, the x-axis. This shows that as x increases or decreases without bound, the function values are approaching 0. The line $y = 0$ (that is, the x-axis) is a **horizontal asymptote** of the graph. Many, but not all, rational functions have horizontal asymptotes.

For the remainder of this section, we will focus on the rational expressions that define rational functions. Operations with these expressions should remind you of those performed in arithmetic with fractions.

4 Simplify rational expressions.

Simplifying Rational Expressions

A rational expression is **simplified** if its numerator and denominator have no common factors other than 1 or -1. The following procedure can be used to simplify rational expressions:

Simplifying Rational Expressions

1. Factor the numerator and the denominator completely.
2. Divide both the numerator and the denominator by any common factors.

EXAMPLE 3 Simplifying a Rational Expression

Simplify: $\dfrac{x^2 + 4x + 3}{x + 1}$.

Solution

$$\dfrac{x^2 + 4x + 3}{x + 1} = \dfrac{(x + 1)(x + 3)}{1(x + 1)} \qquad \text{Factor the numerator and denominator.}$$

$$= \dfrac{(x + 1)(x + 3)}{1(x + 1)} \qquad \text{Divide out the common factor, } x + 1.$$

$$= x + 3 \quad \blacksquare$$

Simplifying a rational expression can change the numbers that make it undefined. For example, we just showed that

$$\dfrac{x^2 + 4x + 3}{x + 1} = x + 3.$$

| This is undefined for $x = -1$. | The simplified form is defined for all real numbers. |

Thus, to equate the two expressions, we must restrict the values of x in the simplified expression to exclude -1. We can write

$$\dfrac{x^2 + 4x + 3}{x + 1} = x + 3, x \neq -1.$$

Without this restriction, the expressions are not equal. The slight difference between them is illustrated using the graphs of

$$f(x) = \dfrac{x^2 + 4x + 3}{x + 1} \quad \text{and} \quad g(x) = x + 3$$

in **Figure 6.3**.

Figure 6.3 Visualizing the difference between $\dfrac{x^2 + 4x + 3}{x + 1}$ and $x + 3$

Hereafter, we will assume that a simplified rational expression is equal to the original rational expression for all real numbers except those for which either denominator is 0.

✓ **CHECK POINT 3** Simplify: $\dfrac{x^2 + 7x + 10}{x + 2}$.

EXAMPLE 4 Simplifying Rational Expressions

Simplify:

a. $\dfrac{x^2 - 7x - 18}{2x^2 + 3x - 2}$ **b.** $\dfrac{3x^2 + 12xy - 15y^2}{6x^3 - 6xy^2}$.

Solution

a. $\dfrac{x^2 - 7x - 18}{2x^2 + 3x - 2} = \dfrac{(x - 9)(x + 2)}{(2x - 1)(x + 2)}$ Factor the numerator and denominator.

$= \dfrac{(x - 9)\cancel{(x + 2)}}{(2x - 1)\cancel{(x + 2)}}$ Divide out the common factor, $x + 2$.

$= \dfrac{x - 9}{2x - 1}$

b. $\dfrac{3x^2 + 12xy - 15y^2}{6x^3 - 6xy^2} = \dfrac{3(x^2 + 4xy - 5y^2)}{6x(x^2 - y^2)}$ Factor out the GCF in the numerator and denominator.

$= \dfrac{3(x + 5y)(x - y)}{6x(x + y)(x - y)}$ Factor the numerator and denominator completely.

$= \dfrac{\overset{1}{\cancel{3}}(x + 5y)\cancel{(x - y)}}{\underset{2}{\cancel{6}}x(x + y)\cancel{(x - y)}}$ Divide out the common factor, $3(x - y)$.

$= \dfrac{x + 5y}{2x(x + y)}$ It is not necessary to carry out the multiplication in the denominator. ∎

☑ **CHECK POINT 4**   Simplify:

a. $\dfrac{x^2 - 2x - 15}{3x^2 + 8x - 3}$ **b.** $\dfrac{3x^2 + 9xy - 12y^2}{9x^3 - 9xy^2}$.

Great Question!

Can I simplify $\dfrac{x + 4}{x}$ **by dividing the numerator and the denominator by** x**?**

No. When simplifying rational expressions, you can only divide out, or cancel, *factors* common to the numerator and denominator. **It is incorrect to divide out common terms from the numerator and denominator.**

Incorrect!

$$\cancel{\dfrac{x + 4}{x} = 4} \qquad \cancel{\dfrac{x^2 - 4}{4} = x^2 - 1} \qquad \cancel{\dfrac{x^{\overset{x}{2}} - 9^{\overset{3}{}}}{x_1 - 3_1} = x - 3}$$

The first two expressions have no common factors in their numerators and denominators. Only when expressions are multiplied can they be factors. **If you can't factor, then don't try to cancel.** The third rational expression can be simplified as follows:

Correct

$$\dfrac{x^2 - 9}{x - 3} = \dfrac{(x + 3)\cancel{(x - 3)}}{1\cancel{(x - 3)}} = x + 3.$$

Divide out the common factor, $x - 3$.

5 Multiply rational expressions. 🅾️

Multiplying Rational Expressions

The product of two rational expressions is the product of their numerators divided by the product of their denominators. For example,

$$\frac{x^2}{y+3} \cdot \frac{x+5}{y-7} = \frac{x^2(x+5)}{(y+3)(y-7)}.$$

> Multiply numerators.

> Multiply denominators.

Here is a step-by-step procedure for multiplying rational expressions. Before multiplying, divide out any factors common to both a numerator and a denominator.

> **Multiplying Rational Expressions**
>
> 1. Factor all numerators and denominators completely.
> 2. Divide numerators and denominators by common factors.
> 3. Multiply the remaining factors in the numerators and multiply the remaining factors in the denominators.

> **EXAMPLE 5** Multiplying Rational Expressions

Multiply: $\dfrac{x+3}{x-4} \cdot \dfrac{x^2-2x-8}{x^2-9}$.

Solution

$$\frac{x+3}{x-4} \cdot \frac{x^2-2x-8}{x^2-9}$$

$$= \frac{1(x+3)}{1(x-4)} \cdot \frac{(x-4)(x+2)}{(x+3)(x-3)} \qquad \text{Factor all numerators and denominators completely.}$$

$$= \frac{1(x+3)}{1(x-4)} \cdot \frac{(x-4)(x+2)}{(x+3)(x-3)} \qquad \text{Divide numerators and denominators by common factors.}$$

$$= \frac{x+2}{x-3} \qquad \text{Multiply the remaining factors in the numerators and in the denominators.} \blacksquare$$

> ✓ **CHECK POINT 5** Multiply: $\dfrac{x+4}{x-7} \cdot \dfrac{x^2-4x-21}{x^2-16}$.

Some rational expressions contain factors in the numerator and denominator that are opposites, or additive inverses. Here is an example of such an expression:

$$\frac{(2x+5)(2x-5)}{3(5-2x)}.$$

> The factors $2x - 5$ and $5 - 2x$ are opposites. They differ only in sign.

Although you can factor out -1 from the numerator or the denominator and then divide out the common factor, there is an even faster way to simplify this rational expression.

> **Simplifying Rational Expressions with Opposite Factors in the Numerator and Denominator**
>
> The quotient of two polynomials that have opposite signs and are additive inverses is -1.

For example,

$$\frac{(2x + 5)(2x - 5)}{3(5 - 2x)} = \frac{(2x + 5)(2x - 5)^{(-1)}}{3(5 - 2x)} = \frac{-(2x + 5)}{3} \quad \text{or} \quad -\frac{2x + 5}{3} \quad \text{or} \quad \frac{-2x - 5}{3}.$$

Factoring out −1 is done mentally:
$$\frac{(2x + 5)(-1)(5 - 2x)}{3(5 - 2x)}.$$

EXAMPLE 6 Multiplying Rational Expressions

Multiply: $\dfrac{5x + 5}{7x - 7x^2} \cdot \dfrac{2x^2 + x - 3}{4x^2 - 9}.$

Solution

$$\frac{5x + 5}{7x - 7x^2} \cdot \frac{2x^2 + x - 3}{4x^2 - 9}$$

$$= \frac{5(x + 1)}{7x(1 - x)} \cdot \frac{(2x + 3)(x - 1)}{(2x + 3)(2x - 3)}$$ Factor all numerators and denominators completely.

$$= \frac{5(x + 1)}{7x(1 - x)} \cdot \frac{(2x + 3)(x - 1)^{(-1)}}{(2x + 3)(2x - 3)}$$ Divide numerators and denominators by common factors. Because $1 - x$ and $x - 1$ are opposites, their quotient is -1.

$$= \frac{-5(x + 1)}{7x(2x - 3)} \quad \text{or} \quad -\frac{5(x + 1)}{7x(2x - 3)}$$ Multiply the remaining factors in the numerators and in the denominators. ∎

✓ **CHECK POINT 6** Multiply: $\dfrac{4x + 8}{6x - 3x^2} \cdot \dfrac{3x^2 - 4x - 4}{9x^2 - 4}.$

---

6 Divide rational expressions. ▶

Dividing Rational Expressions

The quotient of two rational expressions is the product of the first expression and the multiplicative inverse, or reciprocal, of the second expression. The reciprocal is found by interchanging the numerator and the denominator of the expression.

Dividing Rational Expressions

If $P, Q, R,$ and S are polynomials, where $Q \neq 0, R \neq 0,$ and $S \neq 0,$ then

$$\frac{P}{Q} \div \frac{R}{S} = \frac{P}{Q} \cdot \frac{S}{R} = \frac{PS}{QR}.$$

Change division to multiplication.

Replace $\dfrac{R}{S}$ with its reciprocal by interchanging numerator and denominator.

Thus, **we find the quotient of two rational expressions by inverting the divisor and multiplying**. For example,

$$\frac{x}{7} \div \frac{6}{y} = \frac{x}{7} \cdot \frac{y}{6} = \frac{xy}{42}.$$

Change the division to multiplication.

Replace $\dfrac{6}{y}$ with its reciprocal by interchanging numerator and denominator.

Great Question!

If I'm multiplying or dividing rational expressions and an expression appears without a denominator, what should I do?

When performing operations with rational expressions, if a rational expression is written without a denominator, it is helpful to write the expression with a denominator of 1. In Example 7(a), we wrote

$$4x^2 - 25 \quad \text{as} \quad \frac{4x^2 - 25}{1}.$$

EXAMPLE 7 Dividing Rational Expressions

Divide:

a. $(4x^2 - 25) \div \dfrac{2x + 5}{14}$

b. $\dfrac{x^2 + 3x - 10}{2x} \div \dfrac{x^2 - 5x + 6}{x^2 - 3x}.$

Solution

a. $(4x^2 - 25) \div \dfrac{2x + 5}{14}$

$= \dfrac{4x^2 - 25}{1} \div \dfrac{2x + 5}{14}$ Write $4x^2 - 25$ with a denominator of 1.

$= \dfrac{4x^2 - 25}{1} \cdot \dfrac{14}{2x + 5}$ Invert the divisor and multiply.

$= \dfrac{(2x + 5)(2x - 5)}{1} \cdot \dfrac{14}{1(2x + 5)}$ Factor.

$= \dfrac{(\cancel{2x + 5})(2x - 5)}{1} \cdot \dfrac{14}{1(\cancel{2x + 5})}$ Divide the numerator and denominator by the common factor, $2x + 5$.

$= 14(2x - 5)$ Multiply the remaining factors in the numerators and in the denominators.

b. $\dfrac{x^2 + 3x - 10}{2x} \div \dfrac{x^2 - 5x + 6}{x^2 - 3x}$

$= \dfrac{x^2 + 3x - 10}{2x} \cdot \dfrac{x^2 - 3x}{x^2 - 5x + 6}$ Invert the divisor and multiply.

$= \dfrac{(x + 5)(x - 2)}{2x} \cdot \dfrac{x(x - 3)}{(x - 3)(x - 2)}$ Factor.

$= \dfrac{(x + 5)(\cancel{x - 2})}{2\cancel{x}} \cdot \dfrac{\cancel{x}(\cancel{x - 3})}{(\cancel{x - 3})(\cancel{x - 2})}$ Divide numerators and denominators by common factors.

$= \dfrac{x + 5}{2}$ Multiply the remaining factors in the numerators and in the denominators. ∎

✓ **CHECK POINT 7** Divide:

a. $(9x^2 - 49) \div \dfrac{3x - 7}{9}$

b. $\dfrac{x^2 - x - 12}{5x} \div \dfrac{x^2 - 10x + 24}{x^2 - 6x}.$

Achieving Success

Analyze the errors you make on quizzes and tests. For each error, write out the correct solution along with a description of the concept needed to solve the problem correctly. Do your mistakes indicate gaps in understanding concepts or do you at times believe that you are just not a good test taker? Are you repeatedly making the same kinds of mistakes on tests? Keeping track of errors should increase your understanding of the material, resulting in improved test scores.

CONCEPT AND VOCABULARY CHECK

Fill in each blank so that the resulting statement is true.

1. A rational expression consists of a/an _____ divided by a nonzero _____.

2. The domain of a rational function is the set of all real numbers except those for which the denominator is _____.

3. A vertical line that the graph of a function approaches, but does not touch, is said to be a vertical _____ of the graph.

4. A horizontal line that the graph of a rational function approaches at the far left or the far right is called a horizontal _____ of the graph.

5. We simplify a rational expression by _____ the numerator and the denominator completely. Then we divide the numerator and the denominator by any _____.

6. The rational expression $\dfrac{(x + 3)(x + 5)}{x + 3}$ simplifies to _____ if $x \neq -3$.

7. True or false: $\dfrac{x^2 - 9}{9}$ simplifies to $x^2 - 1$. _____

8. The rational expression $\dfrac{x - 5}{5 - x}$ simplifies to _____.

9. The product of two rational expressions is the product of their _____ divided by the product of their _____.

10. The quotient of two rational expressions is the product of the first expression and the _____ of the second:

$$\frac{P}{Q} \div \frac{R}{S} = \frac{P}{Q} \cdot \frac{}{} = \underline{}, Q \neq 0, R \neq 0, S \neq 0.$$

11. $\dfrac{x}{7} \cdot \dfrac{x}{10} = \underline{}$

12. $\dfrac{x}{7} \div \dfrac{x}{10} = \underline{}$

6.1 EXERCISE SET ▶ MyMathLab®

Practice Exercises

In Exercises 1–6, use the given rational function to find the indicated function values. If a function value does not exist, so state.

1. $f(x) = \dfrac{x^2 - 9}{x + 3}; f(-2), f(0), f(5)$

2. $f(x) = \dfrac{x^2 - 16}{x + 4}; f(-2), f(0), f(5)$

3. $f(x) = \dfrac{x^2 - 2x - 3}{4 - x}; f(-1), f(4), f(6)$

4. $f(x) = \dfrac{x^2 - 3x - 4}{3 - x}; f(-1), f(3), f(5)$

5. $g(t) = \dfrac{2t^3 - 5}{t^2 + 1}; g(-1), g(0), g(2)$

6. $g(t) = \dfrac{2t^3 - 1}{t^2 + 4}; g(-1), g(0), g(2)$

In Exercises 7–16, find the domain of the given rational function.

7. $f(x) = \dfrac{x - 2}{x - 5}$

8. $f(x) = \dfrac{x - 3}{x - 6}$

9. $f(x) = \dfrac{x - 4}{(x - 1)(x + 3)}$

10. $f(x) = \dfrac{x - 5}{(x - 2)(x + 4)}$

11. $f(x) = \dfrac{2x}{(x + 5)^2}$

12. $f(x) = \dfrac{2x}{(x + 7)^2}$

13. $f(x) = \dfrac{3x}{x^2 - 8x + 15}$

14. $f(x) = \dfrac{3x}{x^2 - 13x + 36}$

15. $f(x) = \dfrac{(x - 1)^2}{3x^2 - 2x - 8}$

16. $f(x) = \dfrac{(x - 1)^2}{4x^2 - 13x + 3}$

The graph of a rational function, f, is shown in the figure. Use the graph to solve Exercises 17–26.

(In Exercises 17–26, use the graph on the previous page.)

17. Find $f(4)$. **18.** Find $f(1)$.

19. What is the domain of f? What is the range of f?

20. What are the equations of the vertical asymptotes of the graph of f?

21. Describe the end behavior of the graph at the far left. What is the equation of the horizontal asymptote?

22. Describe the end behavior of the graph at the far right. What is the equation of the horizontal asymptote?

23. Explain how the graph shows that $f(-2)$ does not exist.

24. Explain how the graph shows that $f(2)$ does not exist.

25. How can you tell that this is not the graph of a polynomial function?

26. List two real numbers that are not function values of f.

In Exercises 27–50, simplify each rational expression. If the rational expression cannot be simplified, so state.

27. $\dfrac{x^2 - 4}{x - 2}$

28. $\dfrac{x^2 - 25}{x - 5}$

29. $\dfrac{x + 2}{x^2 - x - 6}$

30. $\dfrac{x + 1}{x^2 - 2x - 3}$

31. $\dfrac{4x + 20}{x^2 + 5x}$

32. $\dfrac{5x + 30}{x^2 + 6x}$

33. $\dfrac{4y - 20}{y^2 - 25}$

34. $\dfrac{6y - 42}{y^2 - 49}$

35. $\dfrac{3x - 5}{25 - 9x^2}$

36. $\dfrac{5x - 2}{4 - 25x^2}$

37. $\dfrac{y^2 - 49}{y^2 - 14y + 49}$

38. $\dfrac{y^2 - 9}{y^2 - 6y + 9}$

39. $\dfrac{x^2 + 7x - 18}{x^2 - 3x + 2}$

40. $\dfrac{x^2 - 4x - 5}{x^2 + 5x + 4}$

41. $\dfrac{3x + 7}{3x + 10}$

42. $\dfrac{2x + 3}{2x + 5}$

43. $\dfrac{x^2 - x - 12}{16 - x^2}$

44. $\dfrac{x^2 - 7x + 12}{9 - x^2}$

45. $\dfrac{x^2 + 3xy - 10y^2}{3x^2 - 7xy + 2y^2}$

46. $\dfrac{x^2 + 2xy - 3y^2}{2x^2 + 5xy - 3y^2}$

47. $\dfrac{x^3 - 8}{x^2 - 4}$

48. $\dfrac{x^3 - 1}{x^2 - 1}$

49. $\dfrac{x^3 + 4x^2 - 3x - 12}{x + 4}$

50. $\dfrac{x^3 - 2x^2 + x - 2}{x - 2}$

In Exercises 51–72, multiply as indicated.

51. $\dfrac{x - 3}{x + 7} \cdot \dfrac{3x + 21}{2x - 6}$

52. $\dfrac{x - 2}{x + 3} \cdot \dfrac{2x + 6}{5x - 10}$

53. $\dfrac{x^2 - 49}{x^2 - 4x - 21} \cdot \dfrac{x + 3}{x}$

54. $\dfrac{x^2 - 25}{x^2 - 3x - 10} \cdot \dfrac{x + 2}{x}$

55. $\dfrac{x^2 - 9}{x^2 - x - 6} \cdot \dfrac{x^2 + 5x + 6}{x^2 + x - 6}$

56. $\dfrac{x^2 - 1}{x^2 - 4} \cdot \dfrac{x^2 - 5x + 6}{x^2 - 2x - 3}$

57. $\dfrac{x^2 + 4x + 4}{x^2 + 8x + 16} \cdot \dfrac{(x + 4)^3}{(x + 2)^3}$

58. $\dfrac{x^2 - 2x + 1}{x^2 - 4x + 4} \cdot \dfrac{(x - 2)^3}{(x - 1)^3}$

59. $\dfrac{8y + 2}{y^2 - 9} \cdot \dfrac{3 - y}{4y^2 + y}$

60. $\dfrac{6y + 2}{y^2 - 1} \cdot \dfrac{1 - y}{3y^2 + y}$

61. $\dfrac{y^3 - 8}{y^2 - 4} \cdot \dfrac{y + 2}{2y}$

62. $\dfrac{y^2 + 6y + 9}{y^3 + 27} \cdot \dfrac{1}{y + 3}$

63. $(x - 3) \cdot \dfrac{x^2 + x + 1}{x^2 - 5x + 6}$

64. $(x + 1) \cdot \dfrac{x + 2}{x^2 + 7x + 6}$

65. $\dfrac{x^2 + xy}{x^2 - y^2} \cdot \dfrac{4x - 4y}{x}$

66. $\dfrac{x^2 - y^2}{x} \cdot \dfrac{x^2 + xy}{x + y}$

67. $\dfrac{x^2 + 2xy + y^2}{x^2 - 2xy + y^2} \cdot \dfrac{4x - 4y}{3x + 3y}$

68. $\dfrac{2x^2 - 3xy - 2y^2}{3x^2 - 4xy + y^2} \cdot \dfrac{3x^2 - 2xy - y^2}{x^2 + xy - 6y^2}$

69. $\dfrac{4a^2 + 2ab + b^2}{2a + b} \cdot \dfrac{4a^2 - b^2}{8a^3 - b^3}$

70. $\dfrac{27a^3 - 8b^3}{b^2 - b - 6} \cdot \dfrac{bc - b - 3c + 3}{3ac - 2bc - 3a + 2b}$

71. $\dfrac{10z^2 + 13z - 3}{3z^2 - 8z + 5} \cdot \dfrac{2z^2 - 3z - 2}{25z^2 - 10z + 1} \cdot \dfrac{15z^2 - 28z + 5}{4z^2 - 9}$

72. $\dfrac{2z^2 - 2z - 12}{z^2 - 49} \cdot \dfrac{4z^2 - 1}{2z^2 + 5z + 2} \cdot \dfrac{2z^2 - 13z - 7}{2z^2 - 7z + 3}$

In Exercises 73–90, divide as indicated.

73. $\dfrac{x + 5}{7} \div \dfrac{4x + 20}{9}$

74. $\dfrac{x + 1}{3} \div \dfrac{3x + 3}{7}$

75. $\dfrac{4}{y - 6} \div \dfrac{40}{7y - 42}$

76. $\dfrac{7}{y - 5} \div \dfrac{28}{3y - 15}$

77. $\dfrac{x^2 - 2x}{15} \div \dfrac{x - 2}{5}$

78. $\dfrac{x^2 - x}{15} \div \dfrac{x - 1}{5}$

79. $\dfrac{y^2 - 25}{2y - 2} \div \dfrac{y^2 + 10y + 25}{y^2 + 4y - 5}$

80. $\dfrac{y^2 + y}{y^2 - 4} \div \dfrac{y^2 - 1}{y^2 + 5y + 6}$

81. $(x^2 - 16) \div \dfrac{x^2 + 3x - 4}{x^2 + 4}$

82. $(x^2 + 4x - 5) \div \dfrac{x^2 - 25}{x + 7}$

83. $\dfrac{y^2 - 4y - 21}{y^2 - 10y + 25} \div \dfrac{y^2 + 2y - 3}{y^2 - 6y + 5}$

84. $\dfrac{y^2 + 4y - 21}{y^2 + 3y - 28} \div \dfrac{y^2 + 14y + 48}{y^2 + 4y - 32}$

85. $\dfrac{8x^3 - 1}{4x^2 + 2x + 1} \div \dfrac{x - 1}{(x - 1)^2}$

86. $\dfrac{x^2 - 9}{x^3 - 27} \div \dfrac{x^2 + 6x + 9}{x^2 + 3x + 9}$

87. $\dfrac{x^2 - 4y^2}{x^2 + 3xy + 2y^2} \div \dfrac{x^2 - 4xy + 4y^2}{x + y}$

88. $\dfrac{xy - y^2}{x^2 + 2x + 1} \div \dfrac{2x^2 + xy - 3y^2}{2x^2 + 5xy + 3y^2}$

89. $\dfrac{x^4 - y^8}{x^2 + y^4} \div \dfrac{x^2 - y^4}{3x^2}$

90. $\dfrac{(x - y)^3}{x^3 - y^3} \div \dfrac{x^2 - 2xy + y^2}{x^2 - y^2}$

Practice PLUS

In Exercises 91–98, perform the indicated operation or operations.

91. $\dfrac{x^3 - 4x^2 + x - 4}{2x^3 - 8x^2 + x - 4} \cdot \dfrac{2x^3 + 2x^2 + x + 1}{x^4 - x^3 + x^2 - x}$

92. $\dfrac{y^3 + y^2 + yz^2 + z^2}{y^3 + y + y^2 + 1} \cdot \dfrac{y^3 + y + y^2z + z}{2y^2 + 2yz - yz^2 - z^3}$

93. $\dfrac{ax - ay + 3x - 3y}{x^3 + y^3} \div \dfrac{ab + 3b + ac + 3c}{xy - x^2 - y^2}$

94. $\dfrac{a^3 + b^3}{ac - ad - bc + bd} \div \dfrac{ab - a^2 - b^2}{ac - ad + bc - bd}$

95. $\dfrac{a^2b + b}{3a^2 - 4a - 20} \cdot \dfrac{a^2 + 5a}{2a^2 + 11a + 5} \div \dfrac{ab^2}{6a^2 - 17a - 10}$

96. $\dfrac{a^2 - 8a + 15}{2a^3 - 10a^2} \cdot \dfrac{2a^2 + 3a}{3a^3 - 27a} \div \dfrac{14a + 21}{a^2 - 6a - 27}$

97. $\dfrac{a - b}{4c} \div \left(\dfrac{b - a}{c} \div \dfrac{a - b}{c^2} \right)$

98. $\left(\dfrac{a - b}{4c} \div \dfrac{b - a}{c} \right) \div \dfrac{a - b}{c^2}$

In Exercises 99–102, find $\dfrac{f(a + h) - f(a)}{h}$ *and simplify.*

99. $f(x) = 7x - 4$

100. $f(x) = -3x + 5$

101. $f(x) = x^2 - 5x + 3$

102. $f(x) = 3x^2 - 4x + 7$

In Exercises 103–104, let

$$f(x) = \dfrac{(x + 2)^2}{1 - 2x} \quad \text{and} \quad g(x) = \dfrac{x + 2}{2x - 1}.$$

103. Find $\left(\dfrac{f}{g} \right)(x)$ and the domain of $\dfrac{f}{g}$.

104. Find $\left(\dfrac{g}{f} \right)(x)$ and the domain of $\dfrac{g}{f}$.

Application Exercises

The rational function

$$f(x) = \dfrac{130x}{100 - x}$$

models the cost, $f(x)$, in millions of dollars, to inoculate x% of the population against a particular strain of flu. The graph of the rational function is shown. Use the function's equation to solve Exercises 105–108.

105. Find and interpret $f(60)$. Identify your solution as a point on the graph.

106. Find and interpret $f(80)$. Identify your solution as a point on the graph.

107. What value of x must be excluded from the rational function's domain? Does the cost model indicate that we can inoculate all of the population against the flu? Explain.

108. What happens to the cost as x approaches 100%? How is this shown by the graph? Explain what this means.

The function

$$f(x) = \frac{6.5x^2 - 20.4x + 234}{x^2 + 36}$$

models the pH level, f(x), of the human mouth x minutes after a person eats food containing sugar. The graph of this function is shown in the figure. Use the function's equation or graph, as specified, to solve Exercises 109–112.

Number of Minutes after
Eating Food Containing Sugar

109. After eating sugar, when is the pH level the lowest? Use the function's equation to determine the pH level, to the nearest tenth, at this time.

110. Use the graph to obtain a reasonable estimate, to the nearest tenth, of the pH level of the human mouth 42 minutes after a person eats food containing sugar.

111. According to the graph, what is the normal pH level of the human mouth? What does the end behavior at the far right of the graph indicate in terms of the mouth's pH level over time?

112. Use the graph to describe what happens to the pH level during the first hour.

*Among all deaths from a particular disease, the percentage that is smoking related (21–39 cigarettes per day) is a function of the disease's **incidence ratio**. The incidence ratio describes the number of times more likely smokers are than nonsmokers to die from the disease. The following table shows the incidence ratios for heart disease and lung cancer for two age groups.*

Incidence Ratios

	Heart Disease	Lung Cancer
Ages 55–64	1.9	10
Ages 65–74	1.7	9

Source: Alexander M. Walker, *Observation and Inference*, Epidemiology Resources Inc., 1991.

For example, the incidence ratio of 9 in the table means that smokers between the ages of 65 and 74 are 9 times more likely than nonsmokers in the same age group to die from lung cancer.

The rational function

$$P(x) = \frac{100(x - 1)}{x}$$

models the percentage of smoking-related deaths among all deaths from a disease, P(x), in terms of the disease's incidence ratio, x. The graph of the rational function is shown. Use this function to solve Exercises 113–116.

The Disease's Incidence Ratio:
The number of times more likely smokers are
than nonsmokers to die from the disease

113. Find $P(10)$. Describe what this means in terms of the incidence ratio, 10, given in the table. Identify your solution as a point on the graph.

114. Find $P(9)$. Round to the nearest percent. Describe what this means in terms of the incidence ratio, 9, given in the table. Identify your solution as a point on the graph.

115. What is the horizontal asymptote of the graph? Describe what this means about the percentage of deaths caused by smoking with increasing incidence ratios.

116. According to the model and its graph, is there a disease for which all deaths are caused by smoking? Explain your answer.

Explaining the Concepts

117. What is a rational expression? Give an example with your explanation.

118. What is a rational function? Provide an example.

119. What is the domain of a rational function?

120. If you are given the equation of a rational function, explain how to determine its domain.

121. Describe two ways the graph of a rational function differs from the graph of a polynomial function.

122. What is a vertical asymptote?

123. What is a horizontal asymptote?

124. Explain how to simplify a rational expression.

125. Explain how to simplify a rational expression with opposite factors in the numerator and the denominator.

126. Explain how to multiply rational expressions.

127. Explain how to divide rational expressions.

128. Although your friend has a family history of heart disease, he smokes, on average, 25 cigarettes per day. He sees the table showing incidence ratios for heart disease (see Exercises 113–116) and feels comfortable that they are less than 2, compared to 9 and 10 for lung cancer. He claims that all family deaths have been from heart disease and decides not to give up smoking. Use the given function and its graph to describe some additional information not given in the table that might influence his decision.

Technology Exercises

In Exercises 129–132, determine if the multiplication or division has been performed correctly by graphing the function on each side of the equation in the same viewing rectangle. If the graphs do not coincide, correct the expression on the right side and then verify your correction using the graphing utility.

129. $\dfrac{x^2 + x}{3x} \cdot \dfrac{6x}{x + 1} = 2x$

130. $\dfrac{x^3 - 25x}{x^2 - 3x - 10} \cdot \dfrac{x + 2}{x} = x + 5$

131. $\dfrac{x^2 - 9}{x + 4} \div \dfrac{x - 3}{x + 4} = x - 3$

132. $(x - 5) \div \dfrac{2x^2 - 11x + 5}{4x^2 - 1} = 2x - 1$

133. Use the $\boxed{\text{TABLE}}$ feature of a graphing utility to verify the domains that you determined for any two functions in Exercises 7–16.

134. a. Graph $f(x) = \dfrac{x^2 - x - 2}{x - 2}$ and $g(x) = x + 1$ in the same viewing rectangle. What do you observe?

b. Simplify the formula in the definition of function f. Do f and g represent exactly the same function? Explain.

c. Display the graphs of f and g separately in the utility's viewing rectangle. $\boxed{\text{TRACE}}$ along each of the curves until you get to $x = 2$. What difference do you observe? What does this mean?

Critical Thinking Exercises

Make Sense? *In Exercises 135–138, determine whether each statement makes sense or does not make sense, and explain your reasoning.*

135. I cannot simplify, multiply, or divide rational expressions without knowing how to factor polynomials.

136. I simplified $\dfrac{2(x + 2) - 5(x + 1)}{(x + 2)(x + 1)}$ by dividing the numerator and the denominator by $(x + 2)(x + 1)$.

137. The values to exclude from the domain of $f(x) = \dfrac{x - 3}{x - 7}$ are 3 and 7.

138. When performing the division
$$\dfrac{3x}{x + 2} \div \dfrac{(x + 2)^2}{x - 4},$$
I began by dividing the numerator and the denominator by the common factor, $x + 2$.

In Exercises 139–142, determine whether each statement is true or false. If the statement is false, make the necessary change(s) to produce a true statement.

139. $\dfrac{x^2 - 25}{x - 5} = x - 5$

140. $\dfrac{x^2 + 7}{7} = x^2 + 1$

141. The domain of $f(x) = \dfrac{7}{x(x - 3) + 5(x - 3)}$ is $(-\infty, 3) \cup (3, \infty)$.

142. The restrictions on the values of x when performing the division
$$\dfrac{f(x)}{g(x)} \div \dfrac{h(x)}{k(x)}$$
are $g(x) \neq 0$, $k(x) \neq 0$, and $h(x) \neq 0$.

143. Graph: $f(x) = \dfrac{x^2 - x - 2}{x - 2}$.

144. Simplify: $\dfrac{6x^{3n} + 6x^{2n}y^n}{x^{2n} - y^{2n}}$.

145. Divide: $\dfrac{y^{2n} - 1}{y^{2n} + 3y^n + 2} \div \dfrac{y^{2n} + y^n - 12}{y^{2n} - y^n - 6}$.

146. Solve for x in terms of a and write the resulting rational expression in simplified form. Include any necessary restrictions on a.
$$a^2(x - 1) = 4(x - 1) + 5a + 10$$

Review Exercises

147. Graph: $4x - 5y \geq 20$. (Section 4.4, Example 1)

148. Multiply: $(2x - 5)(x^2 - 3x - 6)$. (Section 5.2, Example 3)

149. Simplify: $\left(\dfrac{ab^{-3}c^{-4}}{4a^5b^{10}c^{-3}}\right)^{-2}$. (Section 1.6, Example 9)

Preview Exercises

Exercises 150–152 will help you prepare for the material covered in the next section. In each exercise, perform the indicated operation. Where possible, reduce the answer to its lowest terms.

150. $\dfrac{7}{10} - \dfrac{3}{10}$ **151.** $\dfrac{1}{2} + \dfrac{2}{3}$ **152.** $\dfrac{7}{15} - \dfrac{3}{10}$

SECTION

6.2 Adding and Subtracting Rational Expressions

What am I supposed to learn?

After studying this section, you should be able to:

1. Add rational expressions with the same denominator.

2. Subtract rational expressions with the same denominator.

3. Find the least common denominator.

4. Add and subtract rational expressions with different denominators.

5. Add and subtract rational expressions with opposite denominators.

Did you know that people in California are at far greater risk of injury or death from drunk drivers than from earthquakes? According to the U.S. Bureau of Justice Statistics, half the arrests for driving under the influence of alcohol involve drivers ages 25 through 34. The rational function

$$f(x) = \frac{27{,}725(x - 14)}{x^2 + 9} - 5x$$

models the number of arrests for driving under the influence, $f(x)$, per 100,000 drivers, as a function of a driver's age, x.

The formula for function f involves the subtraction of two expressions. It is possible to perform this subtraction and express the function's formula as a single rational expression. In this section, we will draw on your experience from arithmetic to add and subtract rational expressions. We return to the driving-under-the-influence model and its graph in Exercise 83 in the section's Exercise Set.

1. Add rational expressions with the same denominator.

Addition and Subtraction When Denominators Are the Same

To add rational numbers having the same denominators, such as $\frac{2}{9}$ and $\frac{5}{9}$, we add the numerators and place the sum over the common denominator:

$$\frac{2}{9} + \frac{5}{9} = \frac{2 + 5}{9} = \frac{7}{9}.$$

We add rational expressions with the same denominator in an identical manner.

Adding Rational Expressions with Common Denominators

If $\dfrac{P}{R}$ and $\dfrac{Q}{R}$ are rational expressions, then

$$\frac{P}{R} + \frac{Q}{R} = \frac{P + Q}{R}.$$

To add rational expressions with the same denominator, add numerators and place the sum over the common denominator. If possible, factor and simplify the result.

EXAMPLE 1 Adding Rational Expressions When Denominators Are the Same

Add: $\dfrac{x^2 + 2x - 2}{x^2 + 3x - 10} + \dfrac{5x + 12}{x^2 + 3x - 10}$.

Solution

$$\dfrac{x^2 + 2x - 2}{x^2 + 3x - 10} + \dfrac{5x + 12}{x^2 + 3x - 10}$$

$$= \dfrac{x^2 + 2x - 2 + 5x + 12}{x^2 + 3x - 10}$$ Add numerators. Place this sum over the common denominator.

$$= \dfrac{x^2 + 7x + 10}{x^2 + 3x - 10}$$ Combine like terms: $2x + 5x = 7x$ and $-2 + 12 = 10$.

$$= \dfrac{(x + 2)(x + 5)}{(x - 2)(x + 5)}$$ Factor and simplify by dividing out the common factor, $x + 5$.

$$= \dfrac{x + 2}{x - 2}$$ ∎

☑ **CHECK POINT 1**   Add: $\dfrac{x^2 - 5x - 15}{x^2 + 5x + 6} + \dfrac{2x + 5}{x^2 + 5x + 6}$.

The following box shows how to subtract rational expressions with the same denominator:

2 Subtract rational expressions with the same denominator. ▶

> **Subtracting Rational Expressions with Common Denominators**
>
> If $\dfrac{P}{R}$ and $\dfrac{Q}{R}$ are rational expressions, then
>
> $$\dfrac{P}{R} - \dfrac{Q}{R} = \dfrac{P - Q}{R}.$$
>
> To subtract rational expressions with the same denominator, subtract numerators and place the difference over the common denominator. If possible, factor and simplify the result.

EXAMPLE 2 Subtracting Rational Expressions When Denominators Are the Same

Subtract: $\dfrac{3y^3 - 5x^3}{x^2 - y^2} - \dfrac{4y^3 - 6x^3}{x^2 - y^2}$.

Solution

$$\dfrac{3y^3 - 5x^3}{x^2 - y^2} - \dfrac{4y^3 - 6x^3}{x^2 - y^2} = \dfrac{3y^3 - 5x^3 - (4y^3 - 6x^3)}{x^2 - y^2}$$ Subtract numerators and include parentheses to indicate that both terms are subtracted. Place this difference over the common denominator.

$$= \dfrac{3y^3 - 5x^3 - 4y^3 + 6x^3}{x^2 - y^2}$$ Remove parentheses and then distribute the minus to change the sign of each term.

$$= \dfrac{x^3 - y^3}{x^2 - y^2}$$ Combine like terms: $-5x^3 + 6x^3 = x^3$ and $3y^3 - 4y^3 = -y^3$.

$$= \dfrac{(x - y)(x^2 + xy + y^2)}{(x + y)(x - y)}$$ Factor and simplify by dividing out the common factor, $x - y$.

$$= \dfrac{x^2 + xy + y^2}{x + y}$$ ∎

Great Question!

When subtracting a numerator containing more than one term, do I just subtract the first term?

No. When a numerator is being subtracted, be sure to **subtract every term in that expression**.

The − sign applies to the entire numerator, $4y^3 - 6x^3$.

Insert parentheses to indicate this.

The sign of every term of $4y^3 - 6x^3$ changes.

$$\frac{3y^3 - 5x^3}{x^2 - y^2} - \frac{4y^3 - 6x^3}{x^2 - y^2} = \frac{3y^3 - 5x^3 - (4y^3 - 6x^3)}{x^2 - y^2} = \frac{3y^3 - 5x^3 - 4y^3 + 6x^3}{x^2 - y^2}$$

The entire numerator of the second rational expression must be subtracted. Avoid the common error of subtracting only the first term.

$-6x^3$ must also be subtracted.

Incorrect!

$$\frac{3y^3 - 5x^3}{x^2 - y^2} - \frac{4y^3 - 6x^3}{x^2 - y^2} \quad \frac{3y^3 - 5x^3 - 4y^3 - 6x^3}{x^2 - y^2}$$

✓ **CHECK POINT 2**   Subtract: $\dfrac{5x - y}{x^2 - y^2} - \dfrac{4x - 2y}{x^2 - y^2}$.

---

3 Find the least common denominator. ▶

Finding the Least Common Denominator

We can gain insight into adding rational expressions with different denominators by looking closely at what we do when adding fractions with different denominators. For example, suppose that we want to add $\frac{1}{2}$ and $\frac{2}{3}$. We must first write the fractions with the same denominator. We look for the smallest number that contains both 2 and 3 as factors. This number, 6, is then used as the *least common denominator*, or LCD.

The **least common denominator** of several rational expressions is a polynomial consisting of the product of all prime factors in the denominators, with each factor raised to the greatest power of its occurrence in any denominator.

Finding the Least Common Denominator

1. Factor each denominator completely.
2. List the factors of the first denominator.
3. Add to the list in step 2 any factors of the second denominator that do not appear in the list.
4. Form the product of each different factor from the list in step 3. This product is the least common denominator.

EXAMPLE 3   Finding the Least Common Denominator

Find the LCD of

$$\frac{3}{10x^2} \quad \text{and} \quad \frac{7}{15x}.$$

Solution

Step 1. Factor each denominator completely.

$$10x^2 = 5 \cdot 2x^2 \quad (\text{or } 5 \cdot 2 \cdot x \cdot x)$$
$$15x = 5 \cdot 3x$$

Great Question!

What's the relationship between the powers in the factored denominators and the LCD?

After factoring denominators completely, the LCD can be determined by taking each factor to the highest power it appears in any factorization. The *Mathematics Teacher* magazine accused the LCD of keeping up with the Joneses. The LCD wants everything (all of the factors) the other denominators have.

Step 2. List the factors of the first denominator.

$$5, 2, x^2 \quad (\text{or } 5, 2, x, x)$$

Step 3. Add any unlisted factors from the second denominator. Two factors from $5 \cdot 3x$ are already in our list. These factors include 5 and x. We add the unlisted factor, 3, to our list. We have

$$3, 5, 2, x^2.$$

Step 4. The least common denominator is the product of all factors in the final list. Thus,

$$3 \cdot 5 \cdot 2 \cdot x^2,$$

or $30x^2$, is the least common denominator. ■

✓ **CHECK POINT 3** Find the LCD of

$$\frac{7}{6x^2} \quad \text{and} \quad \frac{2}{9x}.$$

EXAMPLE 4 Finding the Least Common Denominator

Find the LCD of

$$\frac{9}{7x^2 + 28x} \quad \text{and} \quad \frac{11}{x^2 + 8x + 16}.$$

Solution

Step 1. Factor each denominator completely.

$$7x^2 + 28x = 7x(x + 4)$$
$$x^2 + 8x + 16 = (x + 4)^2$$

Step 2. List the factors of the first denominator.

$$7, x, (x + 4)$$

Step 3. Add any unlisted factors from the second denominator. The second denominator is $(x + 4)^2$, or $(x + 4)(x + 4)$. One factor of $x + 4$ is already in our list, but the other factor is not. We add a second factor of $x + 4$ to the list. We have

$$7, x, (x + 4), (x + 4).$$

Step 4. The least common denominator is the product of all factors in the final list. Thus,

$$7x(x + 4)(x + 4), \text{ or } 7x(x + 4)^2,$$

is the least common denominator. ■

✓ **CHECK POINT 4** Find the LCD of

$$\frac{7}{5x^2 + 15x} \quad \text{and} \quad \frac{9}{x^2 + 6x + 9}.$$

④ Add and subtract rational expressions with different denominators. ▶

Addition and Subtraction When Denominators Are Different

Finding the least common denominator for two (or more) rational expressions is the first step needed to add or subtract the expressions. For example, to add $\frac{1}{2}$ and $\frac{2}{3}$, we first determine that the LCD is 6. Then we write each fraction in terms of the LCD.

$$\frac{1}{2} + \frac{2}{3} = \frac{1}{2} \cdot \frac{3}{3} + \frac{2}{3} \cdot \frac{2}{2}$$

Multiply the numerator and denominator of each fraction by whatever extra factors are required to form 6, the LCD.

$\frac{3}{3} = 1$ and $\frac{2}{2} = 1$. Multiplying by 1 does not change a fraction's value.

$$= \frac{3}{6} + \frac{4}{6}$$

Perform the required multiplications.

$$= \frac{3+4}{6} = \frac{7}{6}$$

Add numerators. Place this sum over the LCD. Simplify.

We follow the same steps in adding or subtracting rational expressions with different denominators.

> ### Adding and Subtracting Rational Expressions That Have Different Denominators
>
> 1. Find the LCD of the rational expressions.
> 2. Rewrite each rational expression as an equivalent expression whose denominator is the LCD. To do so, multiply the numerator and the denominator of each rational expression by any factor(s) needed to convert the denominator into the LCD.
> 3. Add or subtract numerators, placing the resulting expression over the LCD.
> 4. If possible, simplify the resulting rational expression.

EXAMPLE 5   Adding Rational Expressions with Different Denominators

Add: $\dfrac{3}{10x^2} + \dfrac{7}{15x}$.

Solution

Step 1. Find the least common denominator. In Example 3, we found that the LCD for these rational expressions is $30x^2$.

Step 2. Write equivalent expressions with the LCD as denominators. We must rewrite each rational expression with a denominator of $30x^2$.

$$\frac{3}{10x^2} \cdot \frac{3}{3} = \frac{9}{30x^2} \qquad \frac{7}{15x} \cdot \frac{2x}{2x} = \frac{14x}{30x^2}$$

Multiply the numerator and denominator by 3 to get $30x^2$, the LCD.

Multiply the numerator and denominator by $2x$ to get $30x^2$, the LCD.

Because $\frac{3}{3} = 1$ and $\frac{2x}{2x} = 1$, we are not changing the value of either rational expression, only its appearance. In summary, we have

$$\frac{3}{10x^2} + \frac{7}{15x}$$

The LCD is $30x^2$.

$$= \frac{3}{10x^2} \cdot \frac{3}{3} + \frac{7}{15x} \cdot \frac{2x}{2x}$$

Write equivalent expressions with the LCD.

$$= \frac{9}{30x^2} + \frac{14x}{30x^2}.$$

Perform the required multiplications.

Steps 3 and 4. Add numerators, putting this sum over the LCD. Simplify, if possible.

$$= \frac{9+14x}{30x^2} \quad \text{or} \quad \frac{14x+9}{30x^2}$$

The numerator is prime and further simplification is not possible. ∎

Great Question!

Can't I just simplify things and add rational expressions by adding numerators and adding denominators?

No. It is incorrect to add rational expressions by adding numerators and adding denominators. Avoid this common error.

Incorrect!

$$\frac{3}{10x^2} + \frac{7}{15x}$$

$$= \frac{3+7}{10x^2 + 15x}$$

$$= \frac{10}{10x^2 + 15x}$$

✓ **CHECK POINT 5** Add: $\dfrac{7}{6x^2} + \dfrac{2}{9x}$.

EXAMPLE 6 Adding Rational Expressions with Different Denominators

Add: $\dfrac{x}{x-3} + \dfrac{x-1}{x+3}$.

Solution

Step 1. Find the least common denominator. Begin by factoring the denominators.

$$x - 3 = 1(x - 3)$$
$$x + 3 = 1(x + 3)$$

The factors of the first denominator are 1 and $x - 3$. The only factor from the second denominator that is unlisted is $x + 3$. Thus, the least common denominator is $1(x - 3)(x + 3)$, or $(x - 3)(x + 3)$.

Step 2. Write equivalent expressions with the LCD as denominators.

$$\frac{x}{x-3} + \frac{x-1}{x+3}$$

$$= \frac{x(x+3)}{(x-3)(x+3)} + \frac{(x-1)(x-3)}{(x-3)(x+3)}$$ Multiply each numerator and denominator by the extra factor required to form $(x-3)(x+3)$, the LCD.

Steps 3 and 4. Add numerators, putting this sum over the LCD. Simplify, if possible.

$$= \frac{x(x+3) + (x-1)(x-3)}{(x-3)(x+3)}$$

$$= \frac{x^2 + 3x + x^2 - 4x + 3}{(x-3)(x+3)}$$ Perform the multiplications using the distributive property and FOIL.

$$= \frac{2x^2 - x + 3}{(x-3)(x+3)}$$ Combine like terms: $x^2 + x^2 = 2x^2$ and $3x - 4x = -x$.

The numerator is prime and further simplification is not possible. ■

✓ **CHECK POINT 6** Add: $\dfrac{x}{x-4} + \dfrac{x-2}{x+4}$.

EXAMPLE 7 Subtracting Rational Expressions with Different Denominators

Subtract: $\dfrac{x-1}{x^2+x-6} - \dfrac{x-2}{x^2+4x+3}$.

Solution

Step 1. Find the least common denominator. Begin by factoring the denominators.

$$x^2 + x - 6 = (x + 3)(x - 2)$$
$$x^2 + 4x + 3 = (x + 3)(x + 1)$$

The factors of the first denominator are $x + 3$ and $x - 2$. The only factor from the second denominator that is unlisted is $x + 1$. Thus, the least common denominator is $(x + 3)(x - 2)(x + 1)$.

Step 2. Write equivalent expressions with the LCD as denominators.

$$\frac{x-1}{x^2+x-6} - \frac{x-2}{x^2+4x+3}$$

$$= \frac{x-1}{(x+3)(x-2)} - \frac{x-2}{(x+3)(x+1)}$$

Factor denominators.
The LCD is
$(x+3)(x-2)(x+1)$.

$$= \frac{(x-1)(x+1)}{(x+3)(x-2)(x+1)} - \frac{(x-2)(x-2)}{(x+3)(x-2)(x+1)}$$

Multiply each numerator
and denominator by the
extra factor required to form
$(x+3)(x-2)(x+1)$, the LCD.

Steps 3 and 4. Subtract numerators, putting this difference over the LCD. Simplify, if possible.

$$= \frac{(x-1)(x+1) - (x-2)(x-2)}{(x+3)(x-2)(x+1)}$$

$$= \frac{x^2-1-(x^2-4x+4)}{(x+3)(x-2)(x+1)}$$

Perform the multiplications in the numerator.
Don't forget the parentheses.

$$= \frac{x^2-1-x^2+4x-4}{(x+3)(x-2)(x+1)}$$

Remove parentheses and change the sign of
each term in parentheses.

$$= \frac{4x-5}{(x+3)(x-2)(x+1)}$$

Combine like terms.

The numerator is prime and further simplification is not possible. ∎

☑ **CHECK POINT 7** Subtract: $\dfrac{2x-3}{x^2-5x+6} - \dfrac{x+4}{x^2-2x-3}$.

EXAMPLE 8 Adding and Subtracting Rational Expressions with Different Denominators

Perform the indicated operations:

$$\frac{3y+2}{y-5} + \frac{4}{3y+4} - \frac{7y^2+24y+28}{3y^2-11y-20}.$$

Solution

Step 1. Find the least common denominator. Begin by factoring the denominators.

$$y-5 = 1(y-5)$$
$$3y+4 = 1(3y+4)$$
$$3y^2-11y-20 = (3y+4)(y-5)$$

The factors of the first denominator are 1 and $y-5$. The only factor from the second denominator that is unlisted is $3y+4$. Adding this factor to our list, we have $1, y-5,$ and $3y+4$. We have listed all factors from the third denominator. Thus, the least common denominator is $1(y-5)(3y+4)$, or $(y-5)(3y+4)$.

Step 2. Write equivalent expressions with the LCD as denominators.

$$\frac{3y+2}{y-5} + \frac{4}{3y+4} - \frac{7y^2+24y+28}{3y^2-11y-20}$$

$$= \frac{3y+2}{y-5} + \frac{4}{3y+4} - \frac{7y^2+24y+28}{(3y+4)(y-5)}$$

Factor denominators. The LCD is
$(y-5)(3y+4)$.

$$= \frac{(3y+2)(3y+4)}{(y-5)(3y+4)} + \frac{4(y-5)}{(y-5)(3y+4)} - \frac{7y^2+24y+28}{(3y+4)(y-5)}$$

Multiply the first
two numerators and
denominators by the
extra factor required
to form the LCD.

Steps 3 and 4. Add and subtract numerators, putting this result over the LCD. Simplify, if possible.

$$= \frac{(3y + 2)(3y + 4) + 4(y - 5) - (7y^2 + 24y + 28)}{(y - 5)(3y + 4)}$$

$$= \frac{9y^2 + 18y + 8 + 4y - 20 - 7y^2 - 24y - 28}{(y - 5)(3y + 4)}$$

Perform multiplications. Remove parentheses and change the sign of each term.

$$= \frac{2y^2 - 2y - 40}{(y - 5)(3y + 4)}$$

Combine like terms:
$9y^2 - 7y^2 = 2y^2$, $18y + 4y - 24y = -2y$,
and $8 - 20 - 28 = -40$.

$$= \frac{2(y^2 - y - 20)}{(y - 5)(3y + 4)}$$

Factor out the GCF in the numerator.

$$= \frac{2(y + 4)(y - 5)}{(y - 5)(3y + 4)}$$

Factor completely and simplify.

$$= \frac{2(y + 4)}{3y + 4} \quad \blacksquare$$

✓ **CHECK POINT 8**   Perform the indicated operations:

$$\frac{y - 1}{y - 2} + \frac{y - 6}{y^2 - 4} - \frac{y + 1}{y + 2}.$$

5 Add and subtract rational expressions with opposite denominators. ▶

In some situations, we need to add or subtract rational expressions with denominators that are opposites, or additive inverses. Multiply the numerator and the denominator of either of the rational expressions by -1. Then they will have the same denominators.

EXAMPLE 9   Adding Rational Expressions When Denominators Are Opposites

Add: $\dfrac{4x - 16y}{x - 5y} + \dfrac{x - 6y}{5y - x}$.

Solution

$$\frac{4x - 16y}{x - 5y} + \frac{x - 6y}{5y - x}$$

The denominators, $x - 5y$ and $5y - x$, are opposites, or additive inverses.

$$= \frac{4x - 16y}{x - 5y} + \frac{(-1)}{(-1)} \cdot \frac{x - 6y}{5y - x}$$

Multiply the numerator and denominator of the second rational expression by -1.

$$= \frac{4x - 16y}{x - 5y} + \frac{-x + 6y}{-5y + x}$$

Perform the multiplications by -1.

$$= \frac{4x - 16y}{x - 5y} + \frac{-x + 6y}{x - 5y}$$

Rewrite $-5y + x$ as $x - 5y$. Both rational expressions have the same denominator.

$$= \frac{4x - 16y - x + 6y}{x - 5y}$$

Add numerators. Place this sum over the common denominator.

$$= \frac{3x - 10y}{x - 5y}$$

Combine like terms. ■

✓ **CHECK POINT 9**   Add: $\dfrac{4x - 7y}{x - 3y} + \dfrac{x - 2y}{3y - x}$.

CONCEPT AND VOCABULARY CHECK

Fill in each blank so that the resulting statement is true.

1. $\dfrac{P}{R} + \dfrac{Q}{R} = $ _____ : To add rational expressions with the same denominator, add _____ and place the sum over the _____ .

2. $\dfrac{P}{R} - \dfrac{Q}{R} = $ _____ : To subtract rational expressions with the same denominator, subtract _____ and place the difference over the _____ .

3. $\dfrac{x}{3} - \dfrac{5 - y}{3} = $ _____

4. When adding or subtracting rational expressions with denominators that are opposites, or additive inverses, multiply either expression by _____ to obtain a common denominator.

5. The first step in finding the least common denominator of $\dfrac{7}{5x^2 + 15x}$ and $\dfrac{9}{x^2 + 6x + 9}$ is to _____ .

6. Consider the following subtraction problem:
   $$\frac{x - 1}{x^2 + x - 6} - \frac{x - 2}{x^2 + 4x + 3}.$$
   The factors of the first denominator are _____ . The factors of the second denominator are _____ . The LCD is _____ .

7. An equivalent expression for $\dfrac{7}{15x}$ with a denominator of $30x^2$ can be obtained by multiplying the numerator and denominator by _____ .

8. An equivalent expression for $\dfrac{3y + 2}{y - 5}$ with a denominator of $(3y + 4)(y - 5)$ can be obtained by multiplying the numerator and denominator by _____ .

9. An equivalent expression for $\dfrac{x - 2y}{3y - x}$ with a denominator of $x - 3y$ can be obtained by multiplying the numerator and denominator by _____ .

6.2 EXERCISE SET ▶ MyMathLab®

Practice Exercises

In Exercises 1–16, perform the indicated operations. These exercises involve addition and subtraction when denominators are the same. Simplify the result, if possible.

1. $\dfrac{2}{9x} + \dfrac{4}{9x}$

2. $\dfrac{11}{6x} + \dfrac{4}{6x}$

3. $\dfrac{x}{x - 5} + \dfrac{9x + 3}{x - 5}$

4. $\dfrac{x}{x - 3} + \dfrac{11x + 5}{x - 3}$

5. $\dfrac{x^2 - 2x}{x^2 + 3x} + \dfrac{x^2 + x}{x^2 + 3x}$

6. $\dfrac{x^2 + 7x}{x^2 - 5x} + \dfrac{x^2 - 4x}{x^2 - 5x}$

7. $\dfrac{y^2}{y^2 - 9} + \dfrac{9 - 6y}{y^2 - 9}$

8. $\dfrac{y^2}{y^2 - 25} + \dfrac{25 - 10y}{y^2 - 25}$

9. $\dfrac{3x}{4x - 3} - \dfrac{2x - 1}{4x - 3}$

10. $\dfrac{3x}{7x - 4} - \dfrac{2x - 1}{7x - 4}$

11. $\dfrac{x^2 - 2}{x^2 + 6x - 7} - \dfrac{19 - 4x}{x^2 + 6x - 7}$

12. $\dfrac{x^2 + 6x + 2}{x^2 + x - 6} - \dfrac{2x - 1}{x^2 + x - 6}$

13. $\dfrac{20y^2 + 5y + 1}{6y^2 + y - 2} - \dfrac{8y^2 - 12y - 5}{6y^2 + y - 2}$

14. $\dfrac{y^2 + 3y - 6}{y^2 - 5y + 4} - \dfrac{4y - 4 - 2y^2}{y^2 - 5y + 4}$

15. $\dfrac{2x^3 - 3y^3}{x^2 - y^2} - \dfrac{x^3 - 2y^3}{x^2 - y^2}$

16. $\dfrac{4y^3 - 3x^3}{y^2 - x^2} - \dfrac{3y^3 - 2x^3}{y^2 - x^2}$

In Exercises 17–28, find the least common denominator of the rational expressions.

17. $\dfrac{11}{25x^2}$ and $\dfrac{14}{35x}$

18. $\dfrac{7}{15x^2}$ and $\dfrac{9}{24x}$

19. $\dfrac{2}{x-5}$ and $\dfrac{3}{x^2-25}$

20. $\dfrac{2}{x+3}$ and $\dfrac{5}{x^2-9}$

21. $\dfrac{7}{y^2-100}$ and $\dfrac{13}{y(y-10)}$

22. $\dfrac{7}{y^2-4}$ and $\dfrac{15}{y(y+2)}$

23. $\dfrac{8}{x^2-16}$ and $\dfrac{x}{x^2-8x+16}$

24. $\dfrac{3}{x^2-25}$ and $\dfrac{x}{x^2-10x+25}$

25. $\dfrac{7}{y^2-5y-6}$ and $\dfrac{y}{y^2-4y-5}$

26. $\dfrac{3}{y^2-y-20}$ and $\dfrac{y}{2y^2+7y-4}$

27. $\dfrac{7y}{2y^2+7y+6}$, $\dfrac{3}{y^2-4}$, and $\dfrac{-7y}{2y^2-3y-2}$

28. $\dfrac{5y}{y^2-9}$, $\dfrac{8}{y^2+6y+9}$, and $\dfrac{-5y}{2y^2+5y-3}$

In Exercises 29–66, perform the indicated operations. These exercises involve addition and subtraction when denominators are different. Simplify the result, if possible.

29. $\dfrac{3}{5x^2}+\dfrac{10}{x}$

30. $\dfrac{7}{2x^2}+\dfrac{4}{x}$

31. $\dfrac{4}{x-2}+\dfrac{3}{x+1}$

32. $\dfrac{2}{x-3}+\dfrac{7}{x+2}$

33. $\dfrac{3x}{x^2+x-2}+\dfrac{2}{x^2-4x+3}$

34. $\dfrac{7x}{x^2+2x-8}+\dfrac{3}{x^2-3x+2}$

35. $\dfrac{x-6}{x+5}+\dfrac{x+5}{x-6}$

36. $\dfrac{x-2}{x+7}+\dfrac{x+7}{x-2}$

37. $\dfrac{3x}{x^2-25}-\dfrac{4}{x+5}$

38. $\dfrac{8x}{x^2-16}-\dfrac{5}{x+4}$

39. $\dfrac{3y+7}{y^2-5y+6}-\dfrac{3}{y-3}$

40. $\dfrac{2y+9}{y^2-7y+12}-\dfrac{2}{y-3}$

41. $\dfrac{x^2-6}{x^2+9x+18}-\dfrac{x-4}{x+6}$

42. $\dfrac{x^2-39}{x^2+3x-10}-\dfrac{x-7}{x-2}$

43. $\dfrac{4x+1}{x^2+7x+12}+\dfrac{2x+3}{x^2+5x+4}$

44. $\dfrac{3x-2}{x^2-x-6}+\dfrac{4x-3}{x^2-9}$

45. $\dfrac{x+4}{x^2-x-2}-\dfrac{2x+3}{x^2+2x-8}$

46. $\dfrac{2x+1}{x^2-7x+6}-\dfrac{x+3}{x^2-5x-6}$

47. $4+\dfrac{1}{x-3}$

48. $7+\dfrac{1}{x-5}$

49. $\dfrac{y-7}{y^2-16}+\dfrac{7-y}{16-y^2}$

50. $\dfrac{y-3}{y^2-25}+\dfrac{y-3}{25-y^2}$

51. $\dfrac{x+7}{3x+6}+\dfrac{x}{4-x^2}$

52. $\dfrac{x+5}{4x+12}+\dfrac{x}{9-x^2}$

53. $\dfrac{2x}{x-4}+\dfrac{64}{x^2-16}-\dfrac{2x}{x+4}$

54. $\dfrac{x}{x-3}+\dfrac{x+2}{x^2-2x-3}-\dfrac{4}{x+1}$

55. $\dfrac{5x}{x^2-y^2}-\dfrac{7}{y-x}$

56. $\dfrac{9x}{x^2-y^2}-\dfrac{10}{y-x}$

57. $\dfrac{3}{5x+6}-\dfrac{4}{x-2}+\dfrac{x^2-x}{5x^2-4x-12}$

58. $\dfrac{x-1}{x^2+2x+1}-\dfrac{3}{2x-2}+\dfrac{x}{x^2-1}$

59. $\dfrac{3x-y}{x^2-9xy+20y^2}+\dfrac{2y}{x^2-25y^2}$

60. $\dfrac{x+2y}{x^2+4xy+4y^2}-\dfrac{2x}{x^2-4y^2}$

61. $\dfrac{3x}{x^2-4}+\dfrac{5x}{x^2+x-2}-\dfrac{3}{x^2-4x+4}$

62. $\dfrac{1}{x} + \dfrac{4}{x^2 - 4} - \dfrac{2}{x^2 - 2x}$

63. $\dfrac{6a + 5b}{6a^2 + 5ab - 4b^2} - \dfrac{a + 2b}{9a^2 - 16b^2}$

64. $\dfrac{5a - b}{a^2 + ab - 2b^2} - \dfrac{3a + 2b}{a^2 + 5ab - 6b^2}$

65. $\dfrac{1}{m^2 + m - 2} - \dfrac{3}{2m^2 + 3m - 2} + \dfrac{2}{2m^2 - 3m + 1}$

66. $\dfrac{5}{2m^2 - 5m - 3} + \dfrac{3}{2m^2 + 5m + 2} - \dfrac{1}{m^2 - m - 6}$

Practice PLUS

In Exercises 67–74, perform the indicated operations. Simplify the result, if possible.

67. $\left(\dfrac{2x + 3}{x + 1} \cdot \dfrac{x^2 + 4x - 5}{2x^2 + x - 3}\right) - \dfrac{2}{x + 2}$

68. $\dfrac{1}{x^2 - 2x - 8} \div \left(\dfrac{1}{x - 4} - \dfrac{1}{x + 2}\right)$

69. $\left(2 - \dfrac{6}{x + 1}\right)\left(1 + \dfrac{3}{x - 2}\right)$

70. $\left(4 - \dfrac{3}{x + 2}\right)\left(1 + \dfrac{5}{x - 1}\right)$

71. $\left(\dfrac{1}{x + h} - \dfrac{1}{x}\right) \div h$

72. $\left(\dfrac{5}{x - 5} - \dfrac{2}{x + 3}\right) \div (3x + 25)$

73. $\left(\dfrac{1}{a^3 - b^3} \cdot \dfrac{ac + ad - bc - bd}{1}\right) - \dfrac{c - d}{a^2 + ab + b^2}$

74. $\dfrac{ab}{a^2 + ab + b^2} + \left(\dfrac{ac - ad - bc + bd}{ac - ad + bc - bd} \div \dfrac{a^3 - b^3}{a^3 + b^3}\right)$

75. If $f(x) = \dfrac{2x - 3}{x + 5}$ and $g(x) = \dfrac{x^2 - 4x - 19}{x^2 + 8x + 15}$, find $(f - g)(x)$ and the domain of $f - g$.

76. If $f(x) = \dfrac{2x - 1}{x^2 + x - 6}$ and $g(x) = \dfrac{x + 2}{x^2 + 5x + 6}$, find $(f - g)(x)$ and the domain of $f - g$.

Application Exercises

You plan to drive from Miami, Florida, to Atlanta, Georgia. Your trip involves approximately 470 miles of travel in Florida and 250 miles in Georgia. The speed limit is 70 miles per hour in Florida and 65 miles per hour in Georgia. If you average x miles per hour over these speed limits, the total driving time, T(x), in hours, is given by the function

$$T(x) = \dfrac{470}{x + 70} + \dfrac{250}{x + 65}.$$

The graph of T is shown in the figure. Use the function's equation to solve Exercises 77–82.

77. Find and interpret $T(0)$. Round to the nearest hour. Identify your solution as a point on the graph.

78. Find and interpret $T(5)$. Round to the nearest hour. Identify your solution as a point on the graph.

79. Find a simplified form of $T(x)$ by adding the rational expressions in the function's formula. Then use this form of the function to find $T(0)$.

80. Find a simplified form of $T(x)$ by adding the rational expressions in the function's formula. Then use this form of the function to find $T(5)$.

81. Use the graph to answer this question. If you want the driving time to be 9 hours, how much over the speed limits do you need to drive? Round to the nearest mile per hour. Does this seem like a realistic driving time for the trip? Explain your answer.

82. Use the graph to answer this question. If you want the driving time to be 8 hours, how much over the speed limits do you need to drive? Round to the nearest mile per hour. Does this seem like a realistic driving time for the trip? Explain your answer.

83. In the section opener, we saw that the rational function

$$f(x) = \frac{27,725(x - 14)}{x^2 + 9} - 5x$$

models the number of arrests, $f(x)$, per 100,000 drivers, for driving under the influence of alcohol as a function of a driver's age, x. The graph of f for an appropriate domain is shown in the figure.

Driving Under the Influence of Alcohol and a Driver's Age

$$f(x) = \frac{27,725(x - 14)}{x^2 + 9} - 5x$$

a. Use the function's equation to find and interpret $f(20)$. Round to the nearest whole number. Identify your solution as a point on the graph.

b. Find a simplified form of $f(x)$ by subtracting the rational expressions in the function's formula and writing the equation as a single rational expression.

c. Use the graph to determine the age, to the nearest five years, that corresponds to the greatest number of arrests. Then use the form of the function that you obtained in part (b) to determine the number of arrests, per 100,000 drivers, for this age group. Round to the nearest whole number.

In Exercises 84–85, express the perimeter of each rectangle as a single rational expression.

84.

$\frac{x}{x + 5}$

$\frac{x}{x + 4}$

85.

$\frac{x}{x + 8}$

$\frac{x}{x + 7}$

Explaining the Concepts

86. Explain how to add rational expressions when denominators are the same. Give an example with your explanation.

87. Explain how to subtract rational expressions when denominators are the same. Give an example with your explanation.

88. Explain how to find the least common denominator for denominators of $x^2 - 100$ and $x^2 - 20x + 100$.

89. Explain how to add rational expressions that have different denominators. Use $\frac{5}{x + 1} + \frac{3}{x + 4}$ in your explanation.

90. Explain how to add rational expressions when denominators are opposites. Use an example to support your explanation.

Explain the error in Exercises 91–92. Then rewrite the right side of the equation to correct the error that now exists.

91. $\dfrac{1}{a} + \dfrac{1}{b} = \dfrac{1}{a + b}$

92. $\dfrac{1}{x} + \dfrac{3}{7} = \dfrac{4}{x + 7}$

Critical Thinking Exercises

Make Sense? *In Exercises 93–96, determine whether each statement makes sense or does not make sense, and explain your reasoning.*

93. When a numerator is being subtracted, I find that inserting parentheses helps me to distribute the negative sign to every term.

94. The reason I can rewrite rational expressions with a common denominator is that 1 is the multiplicative identity.

95. The fastest way for me to add $\dfrac{5}{x - 7} + \dfrac{3}{7 - x}$ is by using $(x - 7)(7 - x)$ as the LCD.

96. Although $\dfrac{2x^3 + 11x^2}{x + 3} + \dfrac{5x^3 + 4x^2}{x + 3}$ looks more complicated than $\dfrac{2}{x + 3} + \dfrac{5}{x - 3}$, it takes me more steps to perform the less complicated-looking addition.

In Exercises 97–100, determine whether each statement is true or false. If the statement is false, make the necessary change(s) to produce a true statement.

97. $\dfrac{2}{x + 3} + \dfrac{3}{x + 4} = \dfrac{5}{2x + 7}$

98. $\dfrac{a}{b} + \dfrac{a}{c} = \dfrac{a}{b + c}$

99. $6 + \dfrac{1}{x} = \dfrac{7}{x}$

100. $\dfrac{1}{x + 3} + \dfrac{x + 3}{2} = \dfrac{1}{(x + 3)} + \dfrac{(x + 3)}{2} = 1 + \dfrac{1}{2} = \dfrac{3}{2}$

In Exercises 101–103, perform the indicated operations.

101. $\dfrac{1}{x^n - 1} - \dfrac{1}{x^n + 1} - \dfrac{1}{x^{2n} - 1}$

102. $\left(1 - \dfrac{1}{x}\right)\left(1 - \dfrac{1}{x + 1}\right)\left(1 - \dfrac{1}{x + 2}\right)\left(1 - \dfrac{1}{x + 3}\right)$

103. $(x - y)^{-1} + (x - y)^{-2}$

Review Exercises

104. Simplify: $\left(\dfrac{3x^2 y^{-2}}{y^3}\right)^{-2}$. (Section 1.6, Example 9)

105. Solve: $|3x - 1| \le 14$. (Section 4.3, Example 4)

106. Factor completely: $50x^3 - 18x$. (Section 5.5, Example 2)

Preview Exercises

Exercises 107–109 will help you prepare for the material covered in the next section.

107. Multiply and simplify: $x^2 y^2\left(\dfrac{1}{x} + \dfrac{y}{x^2}\right)$.

108. Multiply and simplify: $x(x + h)\left(\dfrac{1}{x + h} - \dfrac{1}{x}\right)$.

109. Divide: $\dfrac{x^2 - 1}{x^2} \div \dfrac{x^2 - 4x + 3}{x^2}$.

SECTION

6.3

Complex Rational Expressions

What am I supposed to learn?

After studying this section, you should be able to:

1. Simplify complex rational expressions by multiplying by 1.

2. Simplify complex rational expressions by dividing.

One area in finance of great interest to us ordinary folks when we buy a new car is the amount of each monthly payment. If P is the principal, or the amount borrowed, i is the monthly interest rate, and n is the number of monthly payments, then the amount, A, of each monthly payment is

$$A = \frac{Pi}{1 - \dfrac{1}{(1 + i)^n}}.$$

Do you notice anything unusual about the rational expression for the amount of each payment? It has a separate rational expression in its denominator.

Numerator ——— Pi ——— Main fraction bar

Denominator ——— $1 - \dfrac{1}{(1 + i)^n}$

A separate rational expression occurs in the denominator.

Complex rational expressions, also called **complex fractions**, have numerators or denominators containing one or more rational expressions. Here is another example of such an expression:

Numerator

Main fraction bar

$$\frac{\dfrac{1}{x} + \dfrac{y}{x^2}}{\dfrac{1}{y} + \dfrac{x}{y^2}}.$$

Denominator

In this section, we study two methods for simplifying complex rational expressions.

1 Simplify complex rational expressions by multiplying by 1. ▶

Simplifying Complex Rational Expressions by Multiplying by 1

One method for simplifying a complex rational expression is to find the least common denominator of all the rational expressions in its numerator and denominator. Then multiply each term in its numerator and denominator by this least common denominator. Because we are multiplying by a form of 1, we will obtain an equivalent expression that does not contain fractions in the numerator or denominator.

> **Simplifying a Complex Rational Expression by Multiplying by 1 in the Form $\dfrac{\text{LCD}}{\text{LCD}}$**
>
> 1. Find the LCD of all rational expressions within the complex rational expression.
> 2. Multiply both the numerator and the denominator of the complex rational expression by this LCD.
> 3. Use the distributive property and multiply each term in the numerator and denominator by this LCD. Simplify each term. No fractional expressions should remain within the numerator or denominator of the main fraction.
> 4. If possible, factor and simplify.

EXAMPLE 1   Simplifying a Complex Rational Expression

Simplify:

$$\frac{\dfrac{1}{x} + \dfrac{y}{x^2}}{\dfrac{1}{y} + \dfrac{x}{y^2}}.$$

Solution   The denominators in the complex rational expression are $x, x^2, y,$ and y^2. The LCD is x^2y^2. Multiply both the numerator and the denominator of the complex rational expression by x^2y^2.

$$\frac{\dfrac{1}{x} + \dfrac{y}{x^2}}{\dfrac{1}{y} + \dfrac{x}{y^2}} = \frac{x^2y^2}{x^2y^2} \cdot \frac{\left(\dfrac{1}{x} + \dfrac{y}{x^2}\right)}{\left(\dfrac{1}{y} + \dfrac{x}{y^2}\right)}$$

Multiply the numerator and the denominator by x^2y^2.

$$= \frac{\boxed{x^2 y^2} \cdot \frac{1}{x} + \boxed{x^2 y^2} \cdot \frac{y}{x^2}}{\boxed{x^2 y^2} \cdot \frac{1}{y} + \boxed{x^2 y^2} \cdot \frac{x}{y^2}}$$

> In all four rational expressions, we have divided numerators and denominators by common boxed factors.

Use the distributive property on

$$\frac{x^2 y^2 \cdot \left(\frac{1}{x} + \frac{y}{x^2} \right)}{x^2 y^2 \cdot \left(\frac{1}{y} + \frac{x}{y^2} \right)}.$$

$$= \frac{xy^2 + y^3}{x^2 y + x^3}$$

Simplify.

$$= \frac{y^2(x + y)}{x^2(y + x)}$$

Factor and simplify.

$$= \frac{y^2}{x^2} \blacksquare$$

☑ **CHECK POINT 1**   Simplify:

$$\frac{\frac{x}{y} - 1}{\frac{x^2}{y^2} - 1}.$$

Great Question!

It's been a while since I've seen negative exponents. Can you refresh my memory and tell me what they mean?

In Section 1.6, we introduced the negative-exponent rule:

$$b^{-n} = \frac{1}{b^n}, b \neq 0.$$

See pages 72–74 if you need to review negative integers as exponents.

Complex rational expressions are often written with negative exponents. For example,

$$\frac{x^{-1} + x^{-2} y}{y^{-1} + xy^{-2}} \quad \text{means} \quad \frac{\frac{1}{x} + \frac{y}{x^2}}{\frac{1}{y} + \frac{x}{y^2}}.$$

This is the expression that we simplified in Example 1. If an expression contains negative exponents, first rewrite it as an equivalent expression with positive exponents. Then simplify by multiplying the numerator and the denominator by the LCD.

EXAMPLE 2   Simplifying a Complex Rational Expression

Simplify:

$$\frac{\frac{1}{x + h} - \frac{1}{x}}{h}.$$

Solution   The denominators in the complex rational expression are $x + h$ and x. The LCD is $x(x + h)$. Multiply both the numerator and the denominator of the complex rational expression by $x(x + h)$.

$$\frac{\frac{1}{x + h} - \frac{1}{x}}{h} = \frac{x(x + h)}{x(x + h)} \cdot \frac{\left(\frac{1}{x + h} - \frac{1}{x} \right)}{h}$$

Multiply the numerator and the denominator by $x(x + h)$.

$$= \frac{x(x + h) \cdot \frac{1}{x + h} - x(x + h) \cdot \frac{1}{x}}{x(x + h)h}$$

Use the distributive property in the numerator to multiply every term by the LCD.

$$= \frac{x - (x + h)}{x(x + h)h}$$

Simplify: $x(x + h) \cdot \frac{1}{(x + h)} = x$

and $x(x + h) \cdot \frac{1}{x} = x + h.$

$$= \frac{x - x - h}{x(x + h)h}$$

Remove parentheses and change the sign of each term.

$$= \frac{-h}{x(x + h)h}$$ Simplify: $x - x - h = -h$.

$$= \frac{-\cancel{h}}{x(x + h)\cancel{h}}$$ Divide the numerator and the denominator by the common factor, h.

$$= -\frac{1}{x(x + h)} \quad \blacksquare$$

☑ **CHECK POINT 2** Simplify:

$$\frac{\dfrac{1}{x + 7} - \dfrac{1}{x}}{7}.$$

---

2 Simplify complex rational expressions by dividing. ▶

Simplifying Complex Rational Expressions by Dividing

A second method for simplifying a complex rational expression is to combine its numerator into a single rational expression and combine its denominator into a single rational expression. Then perform the division by inverting the denominator and multiplying.

> ### Simplifying a Complex Rational Expression by Dividing
>
> 1. If necessary, add or subtract to get a single rational expression in the numerator.
> 2. If necessary, add or subtract to get a single rational expression in the denominator.
> 3. Perform the division indicated by the main fraction bar: Invert the denominator of the complex rational expression and multiply.
> 4. If possible, simplify.

EXAMPLE 3 Simplifying a Complex Rational Expression

Simplify:

$$\frac{\dfrac{x + 1}{x} + \dfrac{x + 1}{x - 1}}{\dfrac{x + 2}{x} - \dfrac{2}{x - 1}}.$$

Solution

Step 1. Add to get a single rational expression in the numerator.

$$\frac{x + 1}{x} + \frac{x + 1}{x - 1}$$

The LCD is $x(x - 1)$.

$$= \frac{(x + 1)(x - 1)}{x(x - 1)} + \frac{x(x + 1)}{x(x - 1)} = \frac{(x + 1)(x - 1) + x(x + 1)}{x(x - 1)} = \frac{x^2 - 1 + x^2 + x}{x(x - 1)} = \frac{2x^2 + x - 1}{x(x - 1)}$$

Step 2. Subtract to get a single rational expression in the denominator.

$$\frac{x + 2}{x} - \frac{2}{x - 1}$$

The LCD is $x(x - 1)$.

$$= \frac{(x + 2)(x - 1)}{x(x - 1)} - \frac{2x}{x(x - 1)} = \frac{(x + 2)(x - 1) - 2x}{x(x - 1)} = \frac{x^2 + x - 2 - 2x}{x(x - 1)} = \frac{x^2 - x - 2}{x(x - 1)}$$

Steps 3 and 4. Perform the division indicated by the main fraction bar: Invert and multiply. If possible, simplify.

$$\frac{\dfrac{x+1}{x}+\dfrac{x+1}{x-1}}{\dfrac{x+2}{x}-\dfrac{2}{x-1}}=\frac{\dfrac{2x^2+x-1}{x(x-1)}}{\dfrac{x^2-x-2}{x(x-1)}}$$

> These are the single rational expressions from steps 1 and 2.

$$=\frac{2x^2+x-1}{x(x-1)}\cdot\frac{x(x-1)}{x^2-x-2}=\frac{(2x-1)(x+1)}{x(x-1)}\cdot\frac{x(x-1)}{(x-2)(x+1)}=\frac{2x-1}{x-2}$$

> Invert and multiply.

✓ **CHECK POINT 3** Simplify:

$$\frac{\dfrac{x+1}{x-1}-\dfrac{x-1}{x+1}}{\dfrac{x-1}{x+1}+\dfrac{x+1}{x-1}}.$$

Which of the two methods do you prefer? Let's try them both in Example 4.

EXAMPLE 4 Simplifying a Complex Rational Expression: Comparing Methods

Simplify:

$$\frac{1-x^{-2}}{1-4x^{-1}+3x^{-2}}.$$

Solution First rewrite the expression without negative exponents.

$$\frac{1-x^{-2}}{1-4x^{-1}+3x^{-2}}=\frac{1-\dfrac{1}{x^2}}{1-\dfrac{4}{x}+\dfrac{3}{x^2}}$$

> The negative exponents affect only the variables and not the constants.

Method 1 Multiplying by 1		Method 2 Dividing	
$\dfrac{1-\dfrac{1}{x^2}}{1-\dfrac{4}{x}+\dfrac{3}{x^2}}=\dfrac{x^2}{x^2}\cdot\dfrac{\left(1-\dfrac{1}{x^2}\right)}{\left(1-\dfrac{4}{x}+\dfrac{3}{x^2}\right)}$	Multiply the numerator and denominator by x^2, the LCD of all fractions.	$\dfrac{1-\dfrac{1}{x^2}}{1-\dfrac{4}{x}+\dfrac{3}{x^2}}=\dfrac{\dfrac{x^2}{x^2}-\dfrac{1}{x^2}}{\dfrac{x^2}{x^2}-\dfrac{4}{x}\cdot\dfrac{x}{x}+\dfrac{3}{x^2}}$	Get a single rational expression in the numerator and in the denominator.
$=\dfrac{x^2\cdot1-x^2\cdot\dfrac{1}{x^2}}{x^2\cdot1-x^2\cdot\dfrac{4}{x}+x^2\cdot\dfrac{3}{x^2}}$	Apply the distributive property.	$=\dfrac{\dfrac{x^2-1}{x^2}}{\dfrac{x^2-4x+3}{x^2}}$	
$=\dfrac{x^2-1}{x^2-4x+3}$	Simplify.	$=\dfrac{x^2-1}{x^2}\cdot\dfrac{x^2}{x^2-4x+3}$	Invert and multiply.
$=\dfrac{(x+1)(x-1)}{(x-3)(x-1)}$	Factor and simplify.	$=\dfrac{(x+1)(x-1)}{x^2}\cdot\dfrac{x^2}{(x-1)(x-3)}$	Factor and simplify.
$=\dfrac{x+1}{x-3}$		$=\dfrac{x+1}{x-3}$	

☑ **CHECK POINT 4**   Simplify by the method of your choice:

$$\frac{1 - 4x^{-2}}{1 - 7x^{-1} + 10x^{-2}}.$$

Achieving Success

According to the Ebbinghaus retention model, you forget 50% of processed information within one hour of leaving the classroom. You lose 60% to 70% within 24 hours. After 30 days, 70% is gone. Reviewing and rewriting class notes is an effective way to counteract this phenomenon. At the very least, read your lecture notes at the end of each day. The more you engage with the material, the more you retain.

CONCEPT AND VOCABULARY CHECK

Fill in each blank so that the resulting statement is true.

1. A rational expression whose numerator or denominator or both contains rational expressions is called a/an _____ rational expression or a/an _____ fraction.

2. $\dfrac{\dfrac{7}{x} + \dfrac{5}{x^2}}{\dfrac{5}{x} + 1} = \dfrac{x^2}{x^2} \cdot \dfrac{\left(\dfrac{7}{x} + \dfrac{5}{x^2}\right)}{\left(\dfrac{5}{x} + 1\right)} = \dfrac{x^2 \cdot \dfrac{7}{x} + x^2 \cdot \dfrac{5}{x^2}}{x^2 \cdot \dfrac{5}{x} + x^2 \cdot 1} = \dfrac{\underline{} + \underline{}}{\underline{} + \underline{}}$

3. $\dfrac{\dfrac{1}{x + 3} - \dfrac{1}{x}}{3} = \dfrac{x(x + 3)}{x(x + 3)} \cdot \dfrac{\left(\dfrac{1}{x + 3} - \dfrac{1}{x}\right)}{3} = \dfrac{x(x + 3) \cdot \underline{} - x(x + 3) \cdot \underline{}}{3x(x + 3)} = \dfrac{\underline{} - (\underline{})}{3x(x + 3)} = \dfrac{\underline{}}{3x(x + 3)} = \underline{}$

6.3 EXERCISE SET   ▶ MyMathLab®

Practice Exercises

In Exercises 1–40, simplify each complex rational expression by the method of your choice.

1. $\dfrac{4 + \dfrac{2}{x}}{1 - \dfrac{3}{x}}$

2. $\dfrac{5 - \dfrac{2}{x}}{3 + \dfrac{1}{x}}$

3. $\dfrac{\dfrac{3}{x} + \dfrac{x}{3}}{\dfrac{x}{3} - \dfrac{3}{x}}$

4. $\dfrac{\dfrac{x}{5} - \dfrac{5}{x}}{\dfrac{1}{5} + \dfrac{1}{x}}$

5. $\dfrac{\dfrac{1}{x} + \dfrac{1}{y}}{\dfrac{1}{x} - \dfrac{1}{y}}$

6. $\dfrac{\dfrac{x}{y} + \dfrac{1}{x}}{\dfrac{y}{x} + \dfrac{1}{x}}$

7. $\dfrac{8x^{-2} - 2x^{-1}}{10x^{-1} - 6x^{-2}}$

8. $\dfrac{12x^{-2} - 3x^{-1}}{15x^{-1} - 9x^{-2}}$

9. $\dfrac{\dfrac{1}{x - 2}}{1 - \dfrac{1}{x - 2}}$

10. $\dfrac{\dfrac{1}{x + 2}}{1 + \dfrac{1}{x + 2}}$

11. $\dfrac{\dfrac{1}{x + 5} - \dfrac{1}{x}}{5}$

12. $\dfrac{\dfrac{1}{x + 6} - \dfrac{1}{x}}{6}$

13. $\dfrac{\dfrac{4}{x + 4}}{\dfrac{1}{x + 4} - \dfrac{1}{x}}$

14. $\dfrac{\dfrac{7}{x + 7}}{\dfrac{1}{x + 7} - \dfrac{1}{x}}$

15. $\dfrac{\dfrac{1}{x - 1} + 1}{\dfrac{1}{x + 1} - 1}$

16. $\dfrac{\dfrac{1}{x + 1} - 1}{\dfrac{1}{x - 1} + 1}$

17. $\dfrac{x^{-1} + y^{-1}}{(x + y)^{-1}}$

18. $(x^{-1} + y^{-1})^{-1}$

19. $\dfrac{\dfrac{x + 2}{x - 2} - \dfrac{x - 2}{x + 2}}{\dfrac{x - 2}{x + 2} + \dfrac{x + 2}{x - 2}}$

20. $\dfrac{\dfrac{x + 1}{x - 1} - \dfrac{x - 1}{x + 1}}{\dfrac{x - 1}{x + 1} + \dfrac{x + 2}{x - 1}}$

21. $\dfrac{\dfrac{2}{x^3 y} + \dfrac{5}{x y^4}}{\dfrac{5}{x^3 y} - \dfrac{3}{x y}}$

22. $\dfrac{\dfrac{3}{x y^2} + \dfrac{2}{x^2 y}}{\dfrac{1}{x^2 y} + \dfrac{2}{x y^3}}$

23. $\dfrac{\dfrac{3}{x+2} - \dfrac{3}{x-2}}{\dfrac{5}{x^2-4}}$

24. $\dfrac{\dfrac{3}{x+1} - \dfrac{3}{x-1}}{\dfrac{5}{x^2-1}}$

25. $\dfrac{3a^{-1} + 3b^{-1}}{4a^{-2} - 9b^{-2}}$

26. $\dfrac{5a^{-1} - 2b^{-1}}{25a^{-2} - 4b^{-2}}$

27. $\dfrac{\dfrac{4x}{x^2-4} - \dfrac{5}{x-2}}{\dfrac{2}{x-2} + \dfrac{3}{x+2}}$

28. $\dfrac{\dfrac{2}{x+3} + \dfrac{5x}{x^2-9}}{\dfrac{4}{x+3} + \dfrac{2}{x-3}}$

29. $\dfrac{\dfrac{2y}{y^2+4y+3}}{\dfrac{1}{y+3} + \dfrac{2}{y+1}}$

30. $\dfrac{\dfrac{5y}{y^2-5y+6}}{\dfrac{3}{y-3} + \dfrac{2}{y-2}}$

31. $\dfrac{\dfrac{2}{a^2} - \dfrac{1}{ab} - \dfrac{1}{b^2}}{\dfrac{1}{a^2} - \dfrac{3}{ab} + \dfrac{2}{b^2}}$

32. $\dfrac{\dfrac{2}{b^2} - \dfrac{5}{ab} - \dfrac{3}{a^2}}{\dfrac{2}{b^2} + \dfrac{7}{ab} + \dfrac{3}{a^2}}$

33. $\dfrac{\dfrac{2x}{x^2-25} + \dfrac{1}{3x-15}}{\dfrac{5}{x-5} + \dfrac{3}{4x-20}}$

34. $\dfrac{\dfrac{7x}{2x-2} + \dfrac{x}{x^2-1}}{\dfrac{4}{x+1} - \dfrac{1}{3x+3}}$

35. $\dfrac{\dfrac{3}{x+2y} - \dfrac{2y}{x^2+2xy}}{\dfrac{3y}{x^2+2xy} + \dfrac{5}{x}}$

36. $\dfrac{\dfrac{1}{x^3-y^3}}{\dfrac{1}{x-y} - \dfrac{1}{x^2+xy+y^2}}$

37. $\dfrac{\dfrac{2}{m^2-3m+2} + \dfrac{2}{m^2-m-2}}{\dfrac{2}{m^2-1} + \dfrac{2}{m^2+4m+3}}$

38. $\dfrac{\dfrac{m}{m^2-9} - \dfrac{2}{m^2-4m+4}}{\dfrac{3}{m^2-5m+6} + \dfrac{m}{m^2+m-6}}$

39. $\dfrac{\dfrac{2}{a^2+2a-8} + \dfrac{1}{a^2+5a+4}}{\dfrac{1}{a^2-5a+6} + \dfrac{2}{a^2-a-2}}$

40. $\dfrac{\dfrac{3}{a^2+10a+25} - \dfrac{1}{a^2-a-2}}{\dfrac{4}{a^2+6a+5} - \dfrac{2}{a^2+3a-10}}$

Practice PLUS

In Exercises 41–46, perform the indicated operations. Simplify the result, if possible.

41. $\dfrac{\dfrac{x-1}{x^2-4}}{1 + \dfrac{1}{x-2}} - \dfrac{1}{x-2}$

42. $\dfrac{\dfrac{x-3}{x^2-16}}{1 + \dfrac{1}{x-4}} - \dfrac{1}{x-4}$

43. $\dfrac{3}{1 - \dfrac{3}{3+x}} - \dfrac{3}{\dfrac{3}{3-x} - 1}$

44. $\dfrac{5}{1 - \dfrac{5}{5+x}} - \dfrac{5}{\dfrac{5}{5-x} - 1}$

45. $\dfrac{x}{1 - \dfrac{1}{1 + \dfrac{1}{x}}}$

46. $\dfrac{\dfrac{1}{x+1}}{x - \dfrac{1}{x + \dfrac{1}{x}}}$

In Exercises 47–48, let $f(x) = \dfrac{1+x}{1-x}$.

47. Find $f\left(\dfrac{1}{x+3}\right)$ and simplify.

48. Find $f\left(\dfrac{1}{x-6}\right)$ and simplify.

In Exercises 49–50, use the given rational function to find and simplify

$$\dfrac{f(a+h) - f(a)}{h}.$$

49. $f(x) = \dfrac{3}{x}$

50. $f(x) = \dfrac{1}{x^2}$

Application Exercises

51. How much are your monthly payments on a loan? If P is the principal, or amount borrowed, i is the monthly interest rate (as a decimal), and n is the number of monthly payments, then the amount, A, of each monthly payment is

$$A = \dfrac{Pi}{1 - \dfrac{1}{(1+i)^n}}.$$

a. Simplify the complex rational expression for the amount of each payment.

b. You purchase a $20,000 automobile at 1% monthly interest to be paid over 48 months. How much do you pay each month? Use the simplified rational expression from part (a) and a calculator. Round to the nearest dollar.

52. The average rate on a round-trip commute having a one-way distance d is given by the complex rational expression

$$\frac{2d}{\dfrac{d}{r_1} + \dfrac{d}{r_2}},$$

in which r_1 and r_2 are the rates on the outgoing and return trips, respectively.

a. Simplify the complex rational expression.

b. Find your average rate if you drive to campus averaging 30 miles per hour and return home on the same route averaging 40 miles per hour. Use the simplified rational expression from part (a).

53. If three resistors with resistances $R_1, R_2,$ and R_3 are connected in parallel, their combined resistance, R, is given by the formula

$$R = \frac{1}{\dfrac{1}{R_1} + \dfrac{1}{R_2} + \dfrac{1}{R_3}}.$$

Simplify the complex rational expression on the right side of the formula. Then find R, to the nearest hundredth of an ohm, when R_1 is 4 ohms, R_2 is 8 ohms, and R_3 is 12 ohms.

54. A camera lens has a measurement called its focal length, f. When an object is in focus, its distance from the lens, p, and its image distance from the lens, q, satisfy the formula

$$f = \frac{1}{\dfrac{1}{p} + \dfrac{1}{q}}.$$

Simplify the complex rational expression on the right side of the formula.

Explaining the Concepts

55. What is a complex rational expression? Give an example with your explanation.

56. Describe two ways to simplify $\dfrac{\dfrac{2}{x} + \dfrac{2}{y}}{\dfrac{2}{x} - \dfrac{2}{y}}$.

57. Which method do you prefer for simplifying complex rational expressions? Why?

58. Of the four complex rational expressions in Exercises 51–54, which one do you find most useful? Explain how you might use this expression in a practical situation.

Technology Exercises

In Exercises 59–62, use a graphing utility to determine if the simplification is correct by graphing the function on each side of the equation in the same viewing rectangle. If the graphs do not coincide, correct the expression on the right side and then verify your correction using the graphing utility.

59. $\dfrac{x - \dfrac{1}{2x + 1}}{1 - \dfrac{x}{2x + 1}} = 2x - 1$

60. $\dfrac{\dfrac{1}{x} + 1}{\dfrac{1}{x}} = 2$

61. $\dfrac{\dfrac{1}{x} + \dfrac{1}{3}}{\dfrac{1}{3x}} = x + \dfrac{1}{3}$

62. $\dfrac{\dfrac{x}{3}}{\dfrac{2}{x + 1}} = \dfrac{3(x + 1)}{2}$

Critical Thinking Exercises

Make Sense? *In Exercises 63–66, determine whether each statement makes sense or does not make sense, and explain your reasoning.*

63. I simplified

$$\frac{\dfrac{1 + 3x}{xy}}{5 + 4y}$$

by multiplying the numerator by xy.

64. By noticing that

$$\frac{\dfrac{1}{x + 7} - \dfrac{1}{x}}{7}$$

repeats x and 7 twice, it's fairly easy to simplify the complex fraction in my head without showing all the steps.

65. I simplified

$$\frac{3 - \dfrac{6}{x + 5}}{1 + \dfrac{7}{x - 4}}$$

by multiplying by 1 and obtained $\dfrac{3 - 6(x - 4)}{1 + 7(x + 5)}$.

66. Before simplifying $\dfrac{1 - x^{-2}}{1 - 5x^{-3}}$, I wrote the complex fraction without negative exponents as

$$\frac{1 - \dfrac{1}{x^2}}{1 - \dfrac{1}{5x^3}}.$$

67. Simplify:

$$\frac{\dfrac{x + h}{x + h + 1} - \dfrac{x}{x + 1}}{h}.$$

68. Simplify:

$$x + \cfrac{1}{x + \cfrac{1}{x + \cfrac{1}{x}}}.$$

69. If $f(x) = \dfrac{1}{x + 1}$, find $f(f(a))$ and simplify.

70. Let x represent the first of two consecutive integers. Find a simplified expression that represents the reciprocal of the sum of the reciprocals of the two integers.

Review Exercises

71. Solve: $x^2 + 27 = 12x$. (Section 5.7, Example 2)

72. Multiply: $(4x^2 - y)^2$. (Section 5.2, Example 7)

73. Solve: $-4 < 3x - 7 < 8$. (Section 4.2, Example 4)

Preview Exercises

Exercises 74–76 will help you prepare for the material covered in the next section.

74. Simplify: $\dfrac{8x^4y^5}{4x^3y^2}$.

75. Divide 737 by 21 without using a calculator. Write the answer as

$$\text{quotient} + \frac{\text{remainder}}{\text{divisor}}.$$

76. Simplify: $6x^2 + 3x - (6x^2 - 4x)$.

SECTION

6.4   Division of Polynomials

What am I supposed to learn?

After studying this section, you should be able to:

1. Divide a polynomial by a monomial. ▶

2. Use long division to divide by a polynomial containing more than one term. ▶

During the 1980s, the controversial economist Arthur Laffer promoted the idea that tax *increases* lead to a *reduction* in government revenue. Called supply-side economics, the theory uses rational functions as models. One such function

$$f(x) = \frac{80x - 8000}{x - 110}, \quad 30 \leq x \leq 100$$

models the government tax revenue, $f(x)$, in tens of billions of dollars, as a function of the tax rate, x. The graph of the rational function is shown in **Figure 6.4**. The graph shows tax revenue decreasing quite dramatically as the tax rate increases. At a tax rate of (gasp) 100%, the government takes all our money and no one has an incentive to work. With no income earned, zero dollars in tax revenue is generated.

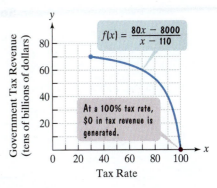

Figure 6.4

Like all rational functions, the Laffer model consists of the quotient of polynomials. Although the rational expression in the model cannot be simplified, it is possible to perform the division, thereby expressing the function in another form. In this section, you will learn to divide polynomials.

1 Divide a polynomial by a monomial. ▶

Dividing a Polynomial by a Monomial

We have seen that to divide monomials, we divide the coefficients and subtract the exponents when bases are the same. For example,

$$\frac{25x^{12}}{5x^4} = \frac{25}{5}x^{12-4} = 5x^8 \quad \text{and} \quad \frac{60x^4y^3}{-30x^2y} = \frac{60}{-30}x^{4-2}y^{3-1} = -2x^2y^2.$$

How do we divide a polynomial that is not a monomial by a monomial? We divide each term of the polynomial by the monomial. For example,

Polynomial dividend

Monomial divisor

$$\frac{10x^8 + 15x^6}{5x^3} = \frac{10x^8}{5x^3} + \frac{15x^6}{5x^3} = \frac{10}{5}x^{8-3} + \frac{15}{5}x^{6-3} = 2x^5 + 3x^3.$$

Divide the first term by $5x^3$.

Divide the second term by $5x^3$.

> ### Dividing a Polynomial That Is Not a Monomial by a Monomial
>
> To divide a polynomial by a monomial, divide each term of the polynomial by the monomial.

EXAMPLE 1   Dividing a Polynomial by a Monomial

Divide: $(15x^3 - 5x^2 + x + 5) \div (5x)$.

Solution

$$\frac{15x^3 - 5x^2 + x + 5}{5x}$$

Rewrite the division in a vertical format.

$$= \frac{15x^3}{5x} - \frac{5x^2}{5x} + \frac{x}{5x} + \frac{5}{5x}$$

Divide each term of the polynomial by the monomial.

$$= 3x^2 - x + \frac{1}{5} + \frac{1}{x}$$

Simplify each quotient. ■

✓ **CHECK POINT 1**   Divide: $(16x^3 - 32x^2 + 2x + 4) \div 4x$.

EXAMPLE 2   Dividing a Polynomial by a Monomial

Divide $8x^4y^5 - 10x^4y^3 + 12x^2y^3$ by $4x^3y^2$.

Solution

$$\frac{8x^4y^5 - 10x^4y^3 + 12x^2y^3}{4x^3y^2}$$

Express the division of $8x^4y^5 - 10x^4y^3 + 12x^2y^3$ by $4x^3y^2$ in a vertical format.

$$= \frac{8x^4y^5}{4x^3y^2} - \frac{10x^4y^3}{4x^3y^2} + \frac{12x^2y^3}{4x^3y^2}$$

Divide each term of the polynomial by the monomial.

$$= 2xy^3 - \frac{5}{2}xy + \frac{3y}{x}$$

Simplify each quotient. ■

✓ **CHECK POINT 2**   Divide $15x^4y^5 - 5x^3y^4 + 10x^2y^2$ by $5x^2y^3$.

2 Use long division to divide by a polynomial containing more than one term. ▶

Dividing by a Polynomial Containing More Than One Term

We now look at division by a polynomial containing more than one term, such as

$$x - 2\overline{)x^2 - 14x + 24}.$$

Divisor has two terms and is a binomial.

The polynomial dividend has three terms and is a trinomial.

When a divisor has more than one term, the four steps used to divide whole numbers—**divide, multiply, subtract, bring down the next term**—form the repetitive procedure for polynomial long division.

EXAMPLE 3   Dividing a Polynomial by a Binomial

Divide $x^2 - 14x + 24$ by $x - 2$.

Solution   The following steps illustrate how polynomial division is very similar to numerical division.

$$x - 2\overline{)x^2 - 14x + 24}$$

Arrange the terms of the dividend $(x^2 - 14x + 24)$ and the divisor $(x - 2)$ in descending powers of x.

$$\begin{array}{r} x \\ x - 2\overline{)x^2 - 14x + 24} \end{array}$$

DIVIDE x^2 (the first term in the dividend) by x (the first term in the divisor): $\frac{x^2}{x} = x$. Align like terms.

$x(x - 2) = x^2 - 2x$
$$\begin{array}{r} x \\ x - 2\overline{)x^2 - 14x + 24} \\ x^2 - 2x \end{array}$$

MULTIPLY each term in the divisor $(x - 2)$ by x, aligning terms of the product under like terms in the dividend.

Change signs of the polynomial being subtracted.
$$\begin{array}{r} x \\ x - 2\overline{)x^2 - 14x + 24} \\ \ominus\,x^2 \oplus 2x \\ \hline -12x \end{array}$$

SUBTRACT $x^2 - 2x$ from $x^2 - 14x$ by changing the sign of each term in the lower expression and adding.

$$\begin{array}{r} x \\ x - 2\overline{)x^2 - 14x + 24} \\ x^2 - 2x \downarrow \\ \hline -12x + 24 \end{array}$$

BRING DOWN 24 from the original dividend and add algebraically to form a new dividend.

$$\begin{array}{r} x - 12 \\ x - 2\overline{)x^2 - 14x + 24} \\ x^2 - 2x \\ \hline -12x + 24 \end{array}$$

Find the second term of the quotient. DIVIDE the first term of $-12x + 24$ by x, the first term of the divisor: $\frac{-12x}{x} = -12$.

$$-12(x - 2) = -12x + 24 \qquad \begin{array}{r} x - 12 \\ x - 2 \overline{)x^2 - 14x + 24} \\ \underline{x^2 - 2x} \\ -12x + 24 \\ \underline{-12x + 24} \\ 0 \end{array}$$

MULTIPLY the divisor $(x - 2)$ by -12, aligning under like terms in the new dividend. Then subtract to obtain the remainder of 0.

Remainder

The quotient is $x - 12$. Because the remainder is 0, we can conclude that $x - 2$ is a factor of $x^2 - 14x + 24$ and

$$\frac{x^2 - 14x + 24}{x - 2} = x - 12. \quad \blacksquare$$

After performing polynomial long division, the answer can be checked. Find the product of the divisor and the quotient, and add the remainder. If the result is the dividend, the answer to the division problem is correct. For example, let's check our work in Example 3.

Dividend   **Divisor**

$$\frac{x^2 - 14x + 24}{x - 2} = x - 12$$

Quotient to be checked

Multiply the divisor and the quotient, and add the remainder, 0:

$$(x - 2)(x - 12) + 0 = x^2 - 12x - 2x + 24 + 0 = x^2 - 14x + 24.$$

Divisor   **Quotient**   **Remainder**   **This is the dividend.**

Because we obtained the dividend, the quotient is correct.

☑ **CHECK POINT 3**   Divide $3x^2 - 14x + 16$ by $x - 2$.

Before considering additional examples, let's summarize the general procedure for dividing by a polynomial that contains more than one term.

> ### Long Division of Polynomials
>
> 1. Arrange the terms of both the dividend and the divisor in descending powers of any variable.
> 2. **Divide** the first term in the dividend by the first term in the divisor. The result is the first term of the quotient.
> 3. **Multiply** every term in the divisor by the first term in the quotient. Write the resulting product beneath the dividend with like terms lined up.
> 4. **Subtract** the product from the dividend.
> 5. **Bring down** the next term in the original dividend and write it next to the remainder to form a new dividend.
> 6. Use this new expression as the dividend and repeat this process until the remainder can no longer be divided. This will occur when the degree of the remainder (the highest exponent on a variable in the remainder) is less than the degree of the divisor.

In our next long division, we will obtain a nonzero remainder.

EXAMPLE 4   Long Division of Polynomials

Divide $4 - 5x - x^2 + 6x^3$ by $3x - 2$.

Solution In order to divide $4 - 5x - x^2 + 6x^3$ by $3x - 2$, we begin by writing the dividend in descending powers of x.

$$4 - 5x - x^2 + 6x^3 = 6x^3 - x^2 - 5x + 4$$

Now we divide $3x^2$ by $3x$ to obtain x, multiply x and the divisor, and subtract.

Now we divide $-3x$ by $3x$ to obtain -1, multiply -1 and the divisor, and subtract.

The quotient is $2x^2 + x - 1$ and the remainder is 2. When there is a nonzero remainder, as in this example, list the quotient, plus the remainder above the divisor. Thus,

$$\frac{6x^3 - x^2 - 5x + 4}{3x - 2} = \underbrace{2x^2 + x - 1}_{\text{Quotient}} + \underbrace{\frac{2}{3x - 2}}_{\substack{\text{Remainder} \\ \text{above divisor}}}.$$

Check this result by showing that the product of the divisor and the quotient,

$$(3x - 2)(2x^2 + x - 1),$$

plus the remainder, 2, is the dividend, $6x^3 - x^2 - 5x + 4$. ∎

✓ **CHECK POINT 4** Divide $-9 + 7x - 4x^2 + 4x^3$ by $2x - 1$.

If a power of x is missing in either a dividend or a divisor, add that power of x with a coefficient of 0 and then divide. In this way, like terms will be aligned as you carry out the long division.

EXAMPLE 5   Long Division of Polynomials

Divide $6x^4 + 5x^3 + 3x - 5$ by $3x^2 - 2x$.

Solution   We write the dividend, $6x^4 + 5x^3 + 3x - 5$, as $6x^4 + 5x^3 + 0x^2 + 3x - 5$ to keep all like terms aligned.

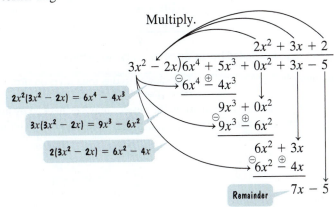

The division process is finished because the degree of $7x - 5$, which is 1, is less than the degree of the divisor $3x^2 - 2x$, which is 2.

We see that the answer is

$$\frac{6x^4 + 5x^3 + 3x - 5}{3x^2 - 2x} = 2x^2 + 3x + 2 + \frac{7x - 5}{3x^2 - 2x}. \quad \blacksquare$$

✓ **CHECK POINT 5**   Divide $2x^4 + 3x^3 - 7x - 10$ by $x^2 - 2x$.

CONCEPT AND VOCABULARY CHECK

Fill in each blank so that the resulting statement is true.

1. To divide $16x^3 - 32x^2 + 2x + 4$ by $4x$, divide each term of _____ by _____.

2. Consider the following long division problem:

$$x + 4\overline{)6x - 4 + 2x^3}.$$

We begin the division process by rewriting the dividend as _____.

3. Consider the following long division problem:

$$3x - 1\overline{)6x^3 + 7x^2 + 12x - 5}.$$

We begin the division process by dividing _____ by _____. We obtain _____. We write this result above _____ in the dividend.

4. In the following long division problem, the first step has been completed:

$$5x - 2\overline{)10x^3 + 6x^2 - 9x + 10}. \quad \overset{2x^2}{}$$

The next step is to multiply _____ and _____. We obtain _____. We write this result below _____.

5. In the following long division problem, the first two steps have been completed:

$$3x - 5 \overline{) 6x^2 + 8x - 4}$$
$$\underline{6x^2 - 10x}$$

with $2x$ above.

The next step is to subtract _____ from _____. We obtain _____. Then we bring down _____ and form the new dividend _____.

6. In the following long division problem, most of the steps have been completed:

$$2x + 1 \overline{) 6x^2 - 7x + 4}$$
$$\underline{6x^2 + 3x}$$
$$-10x + 4$$
$$\underline{-10x - 5}$$
$$?$$

with $3x - 5$ above.

Completing the step designated by the question mark, we obtain _____. Thus, the quotient is _____ and the remainder is _____. The answer to this long division problem is _____.

7. After performing polynomial long division, the answer may be checked by multiplying the _____ by the _____, and then adding the _____. You should obtain the _____.

6.4 EXERCISE SET ▶ MyMathLab®

Practice Exercises

In Exercises 1–12, divide the polynomial by the monomial.

1. $\dfrac{25x^7 - 15x^5 + 10x^3}{5x^3}$

2. $\dfrac{49x^7 - 28x^5 + 14x^3}{7x^3}$

3. $\dfrac{18x^3 + 6x^2 - 9x - 6}{3x}$

4. $\dfrac{25x^3 + 50x^2 - 40x - 10}{5x}$

5. $(28x^3 - 7x^2 - 16x) \div (4x^2)$

6. $(70x^3 - 10x^2 - 14x) \div (7x^2)$

7. $(25x^8 - 50x^7 + 3x^6 - 40x^5) \div (-5x^5)$

8. $(18x^7 - 9x^6 + 20x^5 - 10x^4) \div (-2x^4)$

9. $(18a^3b^2 - 9a^2b - 27ab^2) \div (9ab)$

10. $(12a^2b^2 + 6a^2b - 15ab^2) \div (3ab)$

11. $(36x^4y^3 - 18x^3y^2 - 12x^2y) \div (6x^3y^3)$

12. $(40x^4y^3 - 20x^3y^2 - 50x^2y) \div (10x^3y^3)$

In Exercises 13–36, divide as indicated. Check at least five of your answers by showing that the product of the divisor and the quotient, plus the remainder, is the dividend.

13. $(x^2 + 8x + 15) \div (x + 5)$

14. $(x^2 + 3x - 10) \div (x - 2)$

15. $(x^3 - 2x^2 - 5x + 6) \div (x - 3)$

16. $(x^3 + 5x^2 + 7x + 2) \div (x + 2)$

17. $(x^2 - 7x + 12) \div (x - 5)$

18. $(2x^2 + x - 9) \div (x - 2)$

19. $(2x^2 + 13x + 5) \div (2x + 3)$

20. $(8x^2 + 6x - 25) \div (4x + 9)$

21. $(x^3 + 3x^2 + 5x + 4) \div (x + 1)$

22. $(x^3 + 6x^2 - 2x + 3) \div (x - 1)$

23. $(4y^3 + 12y^2 + 7y - 3) \div (2y + 3)$

24. $(6y^3 + 7y^2 + 12y - 5) \div (3y - 1)$

25. $(9x^3 - 3x^2 - 3x + 4) \div (3x + 2)$

26. $(2x^3 + 13x^2 + 9x - 6) \div (2x + 3)$

27. $(4x^3 - 6x - 11) \div (2x - 4)$

28. $(2x^3 + 6x - 4) \div (x + 4)$

29. $(4y^3 - 5y) \div (2y - 1)$

30. $(6y^3 - 5y) \div (2y - 1)$

31. $(4y^4 - 17y^2 + 14y - 3) \div (2y - 3)$

32. $(2y^4 - y^3 + 16y^2 - 4) \div (2y - 1)$

33. $(4x^4 + 3x^3 + 4x^2 + 9x - 6) \div (x^2 + 3)$

34. $(3x^5 - x^3 + 4x^2 - 12x - 8) \div (x^2 - 2)$

35. $(15x^4 + 3x^3 + 4x^2 + 4) \div (3x^2 - 1)$

36. $(18x^4 + 9x^3 + 3x^2) \div (3x^2 + 1)$

In Exercises 37–40, find a simplified expression for $\left(\dfrac{f}{g}\right)(x)$.

37. $f(x) = 8x^3 - 38x^2 + 49x - 10,$

$g(x) = 4x - 1$

38. $f(x) = 2x^3 - 9x^2 - 17x + 39,$

$g(x) = 2x - 3$

39. $f(x) = 2x^4 - 7x^3 + 7x^2 - 9x + 10,$

$g(x) = 2x - 5$

40. $f(x) = 4x^4 + 6x^3 + 3x - 1,$

$g(x) = 2x^2 + 1$

Practice PLUS

In Exercises 41–50, divide as indicated.

41. $\dfrac{x^4 + y^4}{x + y}$

42. $\dfrac{x^5 + y^5}{x + y}$

43. $\dfrac{3x^4 + 5x^3 + 7x^2 + 3x - 2}{x^2 + x + 2}$

44. $\dfrac{x^4 - x^3 - 7x^2 - 7x - 2}{x^2 - 3x - 2}$

45. $\dfrac{4x^3 - 3x^2 + x + 1}{x^2 + x + 1}$

46. $\dfrac{x^4 - x^2 + 1}{x^2 + x + 1}$

47. $\dfrac{x^5 - 1}{x^2 - x + 2}$

48. $\dfrac{5x^5 - 7x^4 + 3x^3 - 20x^2 + 28x - 12}{x^3 - 4}$

49. $\dfrac{4x^3 - 7x^2y - 16xy^2 + 3y^3}{x - 3y}$

50. $\dfrac{12x^3 - 19x^2y + 13xy^2 - 10y^3}{4x - 5y}$

In Exercises 51–52, find $\left(\dfrac{f - g}{h}\right)(x)$ and the domain of $\dfrac{f - g}{h}$.

51. $f(x) = 3x^3 + 4x^2 - x - 4, g(x) = -5x^3 + 22x^2 - 28x - 12,$

$h(x) = 4x + 1$

52. $f(x) = x^3 + 9x^2 - 6x + 25, g(x) = -3x^3 + 2x^2 - 14x + 5,$

$h(x) = 2x + 4$

In Exercises 53–54, solve each equation for x in terms of a and simplify.

53. $ax + 2x + 4 = 3a^3 + 10a^2 + 6a$

54. $ax - 3x + 6 = a^3 - 6a^2 + 11a$

Application Exercises

In the section opener, we saw that

$$f(x) = \dfrac{80x - 8000}{x - 110}, \quad 30 \le x \le 100$$

models the government tax revenue, $f(x)$, in tens of billions of dollars, as a function of the tax rate percentage, x. Use this function to solve Exercises 55–58. Round to the nearest ten billion dollars.

55. Find and interpret $f(30)$. Identify the solution as a point on the graph of the function in **Figure 6.4** on page 439.

56. Find and interpret $f(70)$. Identify the solution as a point on the graph of the function in **Figure 6.4** on page 439.

57. Rewrite the function by using long division to perform

$$(80x - 8000) \div (x - 110).$$

Then use this new form of the function to find $f(30)$. Do you obtain the same answer as you did in Exercise 55? Which form of the function do you find easier to use?

58. Rewrite the function by using long division to perform

$$(80x - 8000) \div (x - 110).$$

Then use this new form of the function to find $f(70)$. Do you obtain the same answer as you did in Exercise 56? Which form of the function do you find easier to use?

Explaining the Concepts

59. Explain how to divide a polynomial that is not a monomial by a monomial. Give an example.

60. In your own words, explain how to divide by a polynomial containing more than one term. Use $\dfrac{x^2 + 4}{x + 2}$ in your explanation.

61. When performing polynomial long division, explain when to stop dividing.

62. After performing polynomial long division, explain how to check the answer.

63. When performing polynomial long division with missing terms, explain the advantage of writing the missing terms with zero coefficients.

64. The idea of supply-side economics is that an increase in the tax rate may actually reduce government revenue. What explanation can you offer for this theory?

Technology Exercises

In Exercises 65–67, use a graphing utility to determine if the division has been performed correctly. Graph the function on each side of the equation in the same viewing rectangle. If the graphs do not coincide, correct the expression on the right side by using polynomial long division. Then verify your correction using the graphing utility.

65. $(6x^2 + 16x + 8) \div (3x + 2) = 2x + 4$

66. $(4x^3 + 7x^2 + 8x + 20) \div (2x + 4) = 2x^2 - \dfrac{1}{2}x + 3$

67. $(3x^4 + 4x^3 - 32x^2 - 5x - 20) \div (x + 4) = 3x^3 - 8x^2 + 5$

68. Use the ⟨TABLE⟩ feature of a graphing utility to verify any two division results that you obtained in Exercises 13–36.

Critical Thinking Exercises

Make Sense? *In Exercises 69–72, determine whether each statement makes sense or does not make sense, and explain your reasoning.*

69. When performing the division

$$(2x^3 + 13x^2 + 9x - 6) \div (2x + 3),$$

I mentally cover up the +3, the second term of the binomial divisor, before dividing into the dividend.

70. When performing the division $(x^5 + 1) \div (x + 1)$, there's no need for me to follow all the steps involved in polynomial long division because I can work the problem in my head and see that the quotient must be $x^4 + 1$.

71. Because of exponential properties, the degree of the quotient must be the difference between the degree of the dividend and the degree of the divisor.

72. When performing the division $(x^3 + 1) \div (x + 2)$, the purpose of rewriting $x^3 + 1$ as $x^3 + 0x^2 + 0x + 1$ is to keep all like terms aligned.

In Exercises 73–76, determine whether each statement is true or false. If the statement is false, make the necessary change(s) to produce a true statement.

73. All long-division problems can be done by the alternative method of factoring the dividend and canceling identical factors in the dividend and the divisor.

74. Polynomial long division always shows that the answer is a polynomial.

75. The long division process should be continued until the degree of the remainder is the same as the degree of the divisor.

76. If a polynomial long-division problem results in a remainder that is zero, then the divisor is a factor of the dividend.

In Exercises 77–78, divide as indicated.

77. $(x^{3n} - 4x^{2n} - 2x^n - 12) \div (x^n - 5)$

78. $(x^{3n} + 1) \div (x^n + 1)$

79. When $2x^2 - 7x + 9$ is divided by a polynomial, the quotient is $2x - 3$ and the remainder is 3. Find the polynomial.

80. Find k so that the remainder is 0:
$(20x^3 + 23x^2 - 10x + k) \div (4x + 3)$.

Review Exercises

81. Solve: $|2x - 3| > 4$. (Section 4.3, Example 6)

82. Write 40,610,000 in scientific notation. (Section 1.7, Example 2)

83. Simplify: $2x - 4[x - 3(2x + 1)]$. (Section 1.2, Example 14)

Preview Exercises

Exercises 84–86 will help you prepare for the material covered in the next section.

84. a. Divide: $\dfrac{5x^3 + 6x + 8}{x + 2}$.

b. Find the sum of the numbers in each column, designated by ☐, in the following array of numbers. Then describe the relationship between the four numbers in the bottom row of the array and your answer to the division problem in part (a).

$$
\begin{array}{r|rrrr}
-2 & 5 & 0 & 6 & 8 \\
   &   & -10 & 20 & -52 \\
\hline
   & 5 & \square & \square & \square
\end{array}
$$

85. a. Divide: $\dfrac{3x^3 - 4x^2 + 2x - 1}{x + 1}$.

   b. Find the sum of the numbers in each column, designated by $\square$, in the following array of numbers. Then describe the relationship between the four numbers in the bottom row of the array and your answer to the division problem in part (a).

$$
\begin{array}{r|rrrr}
-1 & 3 & -4 & 2 & -1 \\
   &   & -3 & 7 & -9 \\
\hline
   & 3 & \square & \square & \square
\end{array}
$$

86. Divide $2x^3 - 3x^2 - 11x + 6$ by $x - 3$. Use your answer to factor $2x^3 - 3x^2 - 11x + 6$ completely.

MID-CHAPTER CHECK POINT   Section 6.1–Section 6.4

 What You Know: We learned that the domain of a rational function is the set of all real numbers except those for which the denominator is zero. We saw that graphs of rational functions have vertical asymptotes or breaks at these excluded values. We learned to simplify rational expressions by dividing the numerator and the denominator by common factors. We performed a variety of operations with rational expressions, including multiplication, division, addition, and subtraction. We used two methods (multiplying by 1 and dividing) to simplify complex rational expressions. Finally, we used long division when dividing by a polynomial with more than one term.

1. Simplify: $\dfrac{x^2 - x - 6}{x^2 + 3x - 18}$.

In Exercises 2–19, perform the indicated operation(s) and, if possible, simplify.

2. $\dfrac{2x^2 - 8x - 11}{x^2 + 3x - 4} + \dfrac{x^2 + 14x - 13}{x^2 + 3x - 4}$

3. $\dfrac{x^3 - 27}{4x^2 - 4x} \cdot \dfrac{4x}{x - 3}$

4. $5 + \dfrac{7}{x - 2}$

5. $\dfrac{x - \dfrac{4}{x + 6}}{\dfrac{1}{x + 6} + x}$

6. $(2x^4 - 13x^3 + 17x^2 + 18x - 24) \div (x - 4)$

7. $\dfrac{x^3 y - y^3 x}{x^2 y - xy^2}$

8. $(28x^8 y^3 - 14x^6 y^2 + 3x^2 y^2) \div (7x^2 y)$

9. $\dfrac{2x - 1}{x + 6} - \dfrac{x + 3}{x - 2}$

10. $\dfrac{3}{x - 2} - \dfrac{2}{x + 2} - \dfrac{x}{x^2 - 4}$

11. $\dfrac{3x^2 - 7x - 6}{3x^2 - 13x - 10} \div \dfrac{2x^2 - x - 1}{4x^2 - 18x - 10}$

12. $\dfrac{3}{7 - x} + \dfrac{x - 2}{x - 7}$

13. $(6x^4 - 3x^3 - 11x^2 + 2x + 4) \div (3x^2 - 1)$

14. $\dfrac{5 + \dfrac{2}{x}}{3 - \dfrac{1}{x}}$

15. $\dfrac{x}{x^2 - 7x + 6} - \dfrac{x}{x^2 - 2x - 24}$

16. $\dfrac{\dfrac{3}{x + 1} + \dfrac{4}{x}}{\dfrac{4}{x}}$

17. $\dfrac{x^2 - x - 6}{x + 1} \div \left(\dfrac{x^2 - 9}{x^2 - 1} \cdot \dfrac{x - 1}{x + 3} \right)$

18. $(64x^3 + 4) \div (4x + 2)$

19. $\dfrac{x + 1}{x^2 + x - 2} - \dfrac{1}{x^2 - 3x + 2} + \dfrac{2x}{x^2 - 4}$

20. Find the domain of $f(x) = \dfrac{5x - 10}{x^2 + 5x - 14}$. Then simplify the function's equation.

SECTION

6.5 Synthetic Division and the Remainder Theorem

What am I supposed to learn?

After studying this section, you should be able to:

1. Divide polynomials using synthetic division.

2. Evaluate a polynomial function using the Remainder Theorem.

3. Show that a number is a solution of a polynomial equation using the Remainder Theorem.

A moth has moved into your closet. She appeared in your bedroom at night, but somehow her relatively stout body escaped your clutches. Within a few weeks, swarms of moths in your tattered wardrobe suggest that Mama Moth was in the family way. There must be at least 200 critters nesting in every crevice of your clothing.

Two hundred plus moth-tykes from one female moth— is this possible? Indeed it is. The number of eggs, $f(x)$, in a female moth is a function of her abdominal width, x, in millimeters, modeled by

$$f(x) = 14x^3 - 17x^2 - 16x + 34, \quad 1.5 \le x \le 3.5.$$

Because there are 200 moths feasting on your favorite sweaters, Mama's abdominal width can be estimated by finding the solutions of the polynomial equation

$$14x^3 - 17x^2 - 16x + 34 = 200.$$

With mathematics present even in your quickly disappearing attire, we move from rags to a shortcut for long division, called *synthetic division*. In the Exercise Set (Exercise 41), you will use this shortcut to find Mama Moth's abdominal width.

1. Divide polynomials using synthetic division.

Dividing Polynomials Using Synthetic Division

We can use **synthetic division** to divide polynomials if the divisor is of the form $x - c$. This method provides a quotient more quickly than long division. Let's compare the two methods showing $x^3 + 4x^2 - 5x + 5$ divided by $x - 3$.

Notice the relationship between the polynomials in the long division process and the numbers that appear in synthetic division.

> These are the coefficients of the dividend $x^3 + 4x^2 - 5x + 5$.

> The divisor is $x - 3$.
> This is 3, or c, in $x - c$.

$$
\begin{array}{r|rrrr}
3 & 1 & 4 & -5 & 5 \\
  &   & 3 & 21 & 48 \\
\hline
  & 1 & 7 & 16 & 53 \\
\end{array}
$$

> These are the coefficients of the quotient $x^2 + 7x + 16$.

> This is the remainder.

Now let's look at the steps involved in synthetic division.

Synthetic Division To divide a polynomial by $x - c$:

Example

1. Arrange polynomials in descending powers, with a 0 coefficient for any missing term.

$$x - 3 \overline{)\, x^3 + 4x^2 - 5x + 5}$$

2. Write c for the divisor, $x - c$. To the right, write the coefficients of the dividend.

$$
\begin{array}{r|rrrr}
3 & 1 & 4 & -5 & 5 \\
\end{array}
$$

3. Write the leading coefficient of the dividend on the bottom row.

$$
\begin{array}{r|rrrr}
3 & 1 & 4 & -5 & 5 \\
\hline
  & 1 \\
\end{array}
$$
Bring down 1.

4. Multiply c (in this case, 3) times the value just written on the bottom row. Write the product in the next column in the second row.

$$
\begin{array}{r|rrrr}
3 & 1 & 4 & -5 & 5 \\
  &   & 3 \\
\hline
  & 1 \\
\end{array}
$$
Multiply by 3: $3 \cdot 1 = 3$.

5. Add the values in this new column, writing the sum in the bottom row.

$$
\begin{array}{r|rrrr}
3 & 1 & 4 & -5 & 5 \\
  &   & 3 & \text{Add.} \\
\hline
  & 1 & 7 \\
\end{array}
$$

6. Repeat this series of multiplications and additions until all columns are filled in.

$$
\begin{array}{r|rrrr}
3 & 1 & 4 & -5 & 5 \\
  &   & 3 & 21 & \text{Add.} \\
\hline
  & 1 & 7 & 16 \\
\end{array}
$$
Multiply by 3: $3 \cdot 7 = 21$.

$$
\begin{array}{r|rrrr}
3 & 1 & 4 & -5 & 5 \\
  &   & 3 & 21 & 48 & \text{Add.} \\
\hline
  & 1 & 7 & 16 & 53 \\
\end{array}
$$
Multiply by 3: $3 \cdot 16 = 48$.

7. Use the numbers in the last row to write the quotient, plus the remainder above the divisor. **The degree of the first term of the quotient is one less than the degree of the first term of the dividend.** The final value in this row is the remainder.

> Written from
> 1   7   16   53
> the last row of the synthetic division

$$x^2 + 7x + 16 + \frac{53}{x - 3}$$
$$x - 3 \overline{)\, x^3 + 4x^2 - 5x + 5}$$

EXAMPLE 1 Using Synthetic Division

Use synthetic division to divide $5x^3 + 6x + 8$ by $x + 2$.

Solution The divisor must be in the form $x - c$. Thus, we write $x + 2$ as $x - (-2)$. This means that $c = -2$. Writing a 0 coefficient for the missing x^2-term in the dividend, we can express the division as follows:

$$x - (-2)\,\overline{)5x^3 + 0x^2 + 6x + 8.}$$

Now we are ready to set up the problem so that we can use synthetic division.

Use the coefficients of the dividend
$5x^3 + 0x^2 + 6x + 8$ in descending powers of x.

This is c in
$x - (-2)$. $\underline{-2|}$ 5 0 6 8

We begin the synthetic division process by bringing down 5. This is followed by a series of multiplications and additions.

1. Bring down 5.

$\underline{-2|}$ 5 0 6 8

5

2. Multiply: $-2(5) = -10$.

$\underline{-2|}$ 5 0 6 8
     -10

5

Multiply 5 by -2.

3. Add: $0 + (-10) = -10$.

$\underline{-2|}$ 5 0 6 8
     -10 Add.

5 -10

4. Multiply: $-2(-10) = 20$.

$\underline{-2|}$ 5 0 6 8
     -10 20

5 -10

Multiply -10 by -2.

5. Add: $6 + 20 = 26$.

$\underline{-2|}$ 5 0 6 8
     -10 20 Add.

5 -10 26

6. Multiply: $-2(26) = -52$.

$\underline{-2|}$ 5 0 6 8
     -10 20 -52

5 -10 26

Multiply 26 by -2.

7. Add: $8 + (-52) = -44$.

$\underline{-2|}$ 5 0 6 8
     -10 20 -52 Add.

5 -10 26 -44

The numbers in the last row represent the coefficients of the quotient and the remainder. The degree of the first term of the quotient is one less than that of the dividend. Because the degree of the dividend, $5x^3 + 6x + 8$, is 3, the degree of the quotient is 2. This means that the 5 in the last row represents $5x^2$.

$\underline{-2|}$ 5   0   6   8
      -10 20 -52

5 -10 26 -44

The quotient is
$5x^2 - 10x + 26$.

The remainder
is -44.

Thus,

$$x + 2\,\overline{)5x^3 + 6x + 8} \qquad 5x^2 - 10x + 26 - \dfrac{44}{x + 2}$$

✓ **CHECK POINT 1**   Use synthetic division to divide

$$x^3 - 7x - 6 \text{ by } x + 2.$$

2 Evaluate a polynomial function using the Remainder Theorem. ▶

The Remainder Theorem

We have seen that the answer to a long division problem can be checked: Find the product of the divisor and the quotient and add the remainder. The result should be the dividend. If the divisor is $x - c$, we can express this idea symbolically:

$$f(x) = (x - c)q(x) + r.$$

Dividend   Divisor   Quotient   The remainder, r, is a constant when dividing by $x - c$.

Now let's evaluate f at c.

$$f(c) = (c - c)q(c) + r \quad \text{Find } f(c) \text{ by letting } x = c \text{ in } f(x) = (x - c)q(x) + r.$$
This will give an expression for r.

$$f(c) = 0 \cdot q(c) + r \qquad c - c = 0$$

$$f(c) = r \qquad 0 \cdot q(c) = 0 \text{ and } 0 + r = r.$$

What does this last equation mean? If a polynomial is divided by $x - c$, the remainder is the value of the polynomial at c. This result is called the **Remainder Theorem**.

The Remainder Theorem

If the polynomial $f(x)$ is divided by $x - c$, then the remainder is $f(c)$.

Example 2 shows how we can use the Remainder Theorem to evaluate a polynomial function at 2. Rather than substituting 2 for x, we divide the function by $x - 2$. The remainder is $f(2)$.

EXAMPLE 2   Using the Remainder Theorem to Evaluate a Polynomial Function

Given $f(x) = x^3 - 4x^2 + 5x + 3$, use the Remainder Theorem to find $f(2)$.

Solution   By the Remainder Theorem, if $f(x)$ is divided by $x - 2$, then the remainder is $f(2)$. We'll use synthetic division to divide.

$$
\begin{array}{r|rrrr}
2 & 1 & -4 & 5 & 3 \\
  &   & 2 & -4 & 2 \\
\hline
  & 1 & -2 & 1 & 5 \\
\end{array}
$$
Remainder

The remainder, 5, is the value of $f(2)$. Thus, $f(2) = 5$. We can verify that this is correct by evaluating $f(2)$ directly. Using $f(x) = x^3 - 4x^2 + 5x + 3$, we obtain

$$f(2) = 2^3 - 4 \cdot 2^2 + 5 \cdot 2 + 3 = 8 - 16 + 10 + 3 = 5. \quad \blacksquare$$

✓ **CHECK POINT 2**   Given $f(x) = 3x^3 + 4x^2 - 5x + 3$, use the Remainder Theorem to find $f(-4)$.

3 Show that a number is a solution of a polynomial equation using the Remainder Theorem. ▶

If the polynomial $f(x)$ is divided by $x - c$ and the remainder is zero, then $f(c) = 0$. This means that c is a solution of the polynomial equation $f(x) = 0$.

EXAMPLE 3 Using the Remainder Theorem

Show that 3 is a solution of the equation

$$2x^3 - 3x^2 - 11x + 6 = 0.$$

Then solve the polynomial equation.

Solution One way to show that 3 is a solution is to substitute 3 for x in the equation and obtain 0. An easier way is to use synthetic division and the Remainder Theorem.

Proposed solution

$$\begin{array}{r|rrrr} 3 & 2 & -3 & -11 & 6 \\ & & 6 & 9 & -6 \\ \hline & 2 & 3 & -2 & 0 \end{array}$$

Remainder

$$x - 3 \overline{)2x^3 - 3x^2 - 11x + 6} \quad \text{with quotient } 2x^2 + 3x - 2$$

Equivalently,

$$2x^3 - 3x^2 - 11x + 6 = (x - 3)(2x^2 + 3x - 2).$$

Because the remainder is 0, the polynomial has a value of 0 when $x = 3$. Thus, 3 is a solution of the given equation.

The synthetic division also shows that $x - 3$ divides the polynomial with a zero remainder. Thus, $x - 3$ is a factor of the polynomial, as shown to the right of the synthetic division. The other factor is the quotient found in the last row of the synthetic division. Now we can solve the polynomial equation.

$$2x^3 - 3x^2 - 11x + 6 = 0 \qquad \text{This is the given equation.}$$
$$(x - 3)(2x^2 + 3x - 2) = 0 \qquad \text{Factor using the result from the synthetic division.}$$
$$(x - 3)(2x - 1)(x + 2) = 0 \qquad \text{Factor the trinomial.}$$
$$x - 3 = 0 \quad \text{or} \quad 2x - 1 = 0 \quad \text{or} \quad x + 2 = 0 \qquad \text{Set each factor equal to 0.}$$
$$x = 3 \qquad\qquad x = \tfrac{1}{2} \qquad\qquad x = -2 \qquad \text{Solve for } x.$$

The solutions are $-2, \tfrac{1}{2}$, and 3, and the solution set is $\left\{ -2, \tfrac{1}{2}, 3 \right\}$. ■

Using Technology

Graphic Connections
Because the solution set of

$$2x^3 - 3x^2 - 11x + 6 = 0$$

is $\left\{ -2, \tfrac{1}{2}, 3 \right\}$, this implies that the polynomial function

$$f(x) = 2x^3 - 3x^2 - 11x + 6$$

has x-intercepts at $-2, \tfrac{1}{2}$, and 3. This is verified by the graph of f.

x-intercept: $\tfrac{1}{2}$
x-intercept: -2
x-intercept: 3

$[-10, 10, 1]$ by $[-15, 15, 1]$

✓ **CHECK POINT 3** Use synthetic division to show that -1 is a solution of the equation

$$15x^3 + 14x^2 - 3x - 2 = 0.$$

Then solve the polynomial equation.

CONCEPT AND VOCABULARY CHECK

Fill in each blank so that the resulting statement is true.

1. To divide $x^3 + 5x^2 - 7x + 1$ by $x - 4$ using synthetic division, the first step is to write

$$\underline{\quad}\big|\ \underline{\quad}\ \underline{\quad}\ \underline{\quad}\ \underline{\quad}.$$

2. To divide $4x^3 - 8x - 2$ by $x + 5$ using synthetic division, the first step is to write

$$\underline{\quad}\big|\ \underline{\quad}\ \underline{\quad}\ \underline{\quad}\ \underline{\quad}.$$

3. True or false:

$$\begin{array}{r|rrrr} -1 & 3 & -4 & 2 & -1 \\ & & -3 & 7 & -9 \\ \hline & 3 & -7 & 9 & -10 \end{array} \quad \text{means} \quad \frac{3x^3 - 4x^2 + 2x - 1}{x + 1} = 3x^2 - 7x + 9 - \frac{10}{x + 1}. \ \underline{\quad}$$

4. The Remainder Theorem states that if the polynomial $f(x)$ is divided by $x - c$, then the remainder is $\underline{\quad}$.

6.5 EXERCISE SET ▶ MyMathLab®

Practice Exercises

In Exercises 1–18, divide using synthetic division. In the first two exercises, begin the process as shown.

1. $(2x^2 + x - 10) \div (x - 2)$ $2\rfloor$ 2   1   −10
2. $(x^2 + x - 2) \div (x - 1)$ $1\rfloor$ 1   1   −2
3. $(3x^2 + 7x - 20) \div (x + 5)$
4. $(5x^2 - 12x - 8) \div (x + 3)$
5. $(4x^3 - 3x^2 + 3x - 1) \div (x - 1)$
6. $(5x^3 - 6x^2 + 3x + 11) \div (x - 2)$
7. $(6x^5 - 2x^3 + 4x^2 - 3x + 1) \div (x - 2)$
8. $(x^5 + 4x^4 - 3x^2 + 2x + 3) \div (x - 3)$
9. $(x^2 - 5x - 5x^3 + x^4) \div (5 + x)$
10. $(x^2 - 6x - 6x^3 + x^4) \div (6 + x)$
11. $(3x^3 + 2x^2 - 4x + 1) \div \left(x - \dfrac{1}{3}\right)$
12. $(2x^4 - x^3 + 2x^2 - 3x + 1) \div \left(x - \dfrac{1}{2}\right)$
13. $\dfrac{x^5 + x^3 - 2}{x - 1}$
14. $\dfrac{x^7 + x^5 - 10x^3 + 12}{x + 2}$
15. $\dfrac{x^4 - 256}{x - 4}$
16. $\dfrac{x^7 - 128}{x - 2}$
17. $\dfrac{2x^5 - 3x^4 + x^3 - x^2 + 2x - 1}{x + 2}$
18. $\dfrac{x^5 - 2x^4 - x^3 + 3x^2 - x + 1}{x - 2}$

In Exercises 19–26, use synthetic division and the Remainder Theorem to find the indicated function value.

19. $f(x) = 2x^3 - 11x^2 + 7x - 5;$ $f(4)$
20. $f(x) = x^3 - 7x^2 + 5x - 6;$ $f(3)$
21. $f(x) = 3x^3 - 7x^2 - 2x + 5;$ $f(-3)$
22. $f(x) = 4x^3 + 5x^2 - 6x - 4;$ $f(-2)$
23. $f(x) = x^4 + 5x^3 + 5x^2 - 5x - 6;$ $f(3)$
24. $f(x) = x^4 - 5x^3 + 5x^2 + 5x - 6;$ $f(2)$
25. $f(x) = 2x^4 - 5x^3 - x^2 + 3x + 2;$ $f\left(-\dfrac{1}{2}\right)$
26. $f(x) = 6x^4 + 10x^3 + 5x^2 + x + 1;$ $f\left(-\dfrac{2}{3}\right)$

In Exercises 27–32, use synthetic division to show that the number given to the right of each equation is a solution of the equation. Then solve the polynomial equation.

27. $x^3 - 4x^2 + x + 6 = 0;$ -1
28. $x^3 - 2x^2 - x + 2 = 0;$ -1
29. $2x^3 - 5x^2 + x + 2 = 0;$ 2
30. $2x^3 - 3x^2 - 11x + 6 = 0;$ -2
31. $6x^3 + 25x^2 - 24x + 5 = 0;$ -5
32. $3x^3 + 7x^2 - 22x - 8 = 0;$ -4

Practice PLUS

In Exercises 33–36, use the graph or the table to determine a solution of each equation. Use synthetic division to verify that this number is a solution of the equation. Then solve the polynomial equation.

33. $x^3 + 2x^2 - 5x - 6 = 0$

$y = x^3 + 2x^2 - 5x - 6$

$[0, 4, 1]$ by $[-25, 25, 5]$

34. $2x^3 + x^2 - 13x + 6 = 0$

$y = 2x^3 + x^2 - 13x + 6$

$[-4, 0, 1]$ by $[-25, 25, 5]$

35. $6x^3 - 11x^2 + 6x - 1 = 0$

$y_1 = 6x^3 - 11x^2 + 6x - 1$

X	Y1
-3	-280
-2	-105
-1	-24
0	-1
1	0
2	15
3	80
4	231
5	504
6	935
7	1560

X= -3

36. $2x^3 + 11x^2 - 7x - 6 = 0$

In Exercises 37–38, perform the given operations.

37. $(22x - 24 + 7x^3 - 29x^2 + 4x^4)(x + 4)^{-1}$

38. $(9 - x^2 + 6x + 2x^3)(x + 1)^{-1}$

In Exercises 39–40, write a polynomial that represents the length of each rectangle.

39.

The width is $x + 0.2$ units.

The area is $0.5x^3 - 0.3x^2 + 0.22x + 0.06$ square units.

40.

The width is $x + \frac{3}{4}$ units.

The area is $8x^3 - 6x^2 - 5x + 3$ square units.

Application Exercises

41. a. Use synthetic division to show that 3 is a solution of the polynomial equation

$$14x^3 - 17x^2 - 16x - 177 = 0.$$

b. Use the solution from part (a) to solve this problem. The number of eggs, $f(x)$, in a female moth is a function of her abdominal width, x, in millimeters, modeled by

$$f(x) = 14x^3 - 17x^2 - 16x + 34.$$

What is the abdominal width when there are 211 eggs?

42. a. Use synthetic division to show that 2 is a solution of the polynomial equation

$$2h^3 + 14h^2 - 72 = 0.$$

b. Use the solution from part (a) to solve this problem. The width of a rectangular box is twice the height and the length is 7 inches more than the height. If the volume is 72 cubic inches, find the dimensions of the box.

Explaining the Concepts

43. Explain how to perform synthetic division. Use the division problem

$$(2x^3 - 3x^2 - 11x + 7) \div (x - 3)$$

to support your explanation.

44. State the Remainder Theorem.

45. Explain how the Remainder Theorem can be used to find $f(-6)$ if $f(x) = x^4 + 7x^3 + 8x^2 + 11x + 5$. What advantage is there to using the Remainder Theorem in this situation rather than evaluating $f(-6)$ directly?

46. Explain how the Remainder Theorem and synthetic division can be used to determine whether -4 is a solution of the following equation:

$$5x^3 + 22x^2 + x - 28 = 0.$$

Technology Exercise

47. For each equation that you solved in Exercises 27–32, use a graphing utility to graph the polynomial function on the left side of the equation. Use end behavior to obtain a complete graph. Then use the graph's x-intercepts to verify your solutions.

Critical Thinking Exercises

Make Sense? In Exercises 48–51, determine whether each statement makes sense or does not make sense, and explain your reasoning.

48. There are certain kinds of polynomial divisions that I can perform using long division, but not using synthetic division.

49. Every time I divide polynomials using synthetic division, I am using a highly condensed form of the long division procedure where omitting the variables and exponents does not involve the loss of any essential data.

50. The only nongraphic method that I have for evaluating a function at a given value is to substitute that value into the function's equation.

51. The Remainder Theorem gives me a method for factoring certain polynomials that I could not factor using the factoring strategies from the previous chapter.

52. Synthetic division is a process for dividing a polynomial by $x - c$. The coefficient of x is 1. How might synthetic division be used if you are dividing by $2x - 4$?

53. Use synthetic division to show that 5 is a solution of

$$x^4 - 4x^3 - 9x^2 + 16x + 20 = 0.$$

Then solve the polynomial equation. *Hint:* Use factoring by grouping when working with the quotient factor.

Review Exercises

54. Solve: $4x + 3 - 13x - 7 < 2(3 - 4x)$.
(Section 4.1, Example 3)

55. Solve: $2x(x + 3) + 6(x - 3) = -28$.
(Section 5.7, Example 2)

56. Solve by Cramer's rule:

$$\begin{cases} 7x - 6y = 17 \\ 3x + \ y = 18. \end{cases}$$

(Section 3.5, Example 2)

Preview Exercises

Exercises 57–59 will help you prepare for the material covered in the next section. In each exercise, find the LCD of the rational expressions.

57. $\dfrac{x + 4}{2x}, \dfrac{x + 20}{3x}$

58. $\dfrac{x}{3}, \dfrac{9}{x}, 4$

59. $\dfrac{2x}{x - 3}, \dfrac{6}{x + 3}, -\dfrac{28}{x^2 - 9}$

SECTION

6.6

Rational Equations

What am I supposed to learn?

After studying this section, you should be able to:

1 Solve rational equations.

2 Solve problems involving rational functions that model applied situations.

The lake is in one of the city's favorite parks and the time has come to clean it up. Voters in the city have committed $80 thousand for the cleanup. We know that

$$f(x) = \frac{120x}{100 - x}$$

models the cost, $f(x)$, in thousands of dollars, to remove $x\%$ of the lake's pollutants. What percentage of the pollutants can be removed for $80 thousand?

To determine the percentage, we use the given cost-benefit model. Voters have committed $80 thousand, so substitute 80 for $f(x)$:

$$80 = \frac{120x}{100 - x}.$$

> This equation contains a rational expression.

Now we have to solve the equation and find the value for x. This variable represents the percentage of the pollutants that can be removed for $80 thousand.

A **rational equation**, also called a **fractional equation**, is an equation containing one or more rational expressions. The equation shown above is an example of a rational equation. Do you see that there is a variable in a denominator? This is a characteristic of many rational equations. In this section, you will learn a procedure for solving such equations.

1 Solve rational equations.

Solving Rational Equations

We have seen that the LCD is used to add and subtract rational expressions. By contrast, when solving rational equations, **the LCD is used as a multiplier that clears an equation of fractions**.

EXAMPLE 1   Solving a Rational Equation

Solve: $\dfrac{x + 4}{2x} + \dfrac{x + 20}{3x} = 3$.

Solution   Notice that the variable x appears in both denominators. We must avoid any values of the variable that make a denominator zero.

$$\frac{x + 4}{2x} + \frac{x + 20}{3x} = 3$$

This denominator would equal zero if $x = 0$.   This denominator would equal zero if $x = 0$.

We see that x cannot equal zero.

The denominators are $2x$ and $3x$. The least common denominator is $6x$. We begin by multiplying both sides of the equation by $6x$. We will also write the restriction that x cannot equal zero to the right of the equation.

$$\frac{x + 4}{2x} + \frac{x + 20}{3x} = 3, x \neq 0 \qquad \text{This is the given equation.}$$

$$6x\left(\frac{x + 4}{2x} + \frac{x + 20}{3x}\right) = 6x \cdot 3 \qquad \text{Multiply both sides by } 6x, \text{ the LCD.}$$

$$\frac{6x}{1} \cdot \frac{x + 4}{2x} + \frac{6x}{1} \cdot \frac{x + 20}{3x} = 18x \qquad \text{Use the distributive property.}$$

$$3(x + 4) + 2(x + 20) = 18x \qquad \text{Divide out common factors in the multiplications.}$$

Observe that the equation is now cleared of fractions.

$$3x + 12 + 2x + 40 = 18x \qquad \text{Use the distributive property.}$$

$$5x + 52 = 18x \qquad \text{Combine like terms.}$$

$$52 = 13x \qquad \text{Subtract } 5x \text{ from both sides.}$$

$$4 = x \qquad \text{Divide both sides by 13.}$$

The proposed solution, 4, is not part of the restriction $x \neq 0$. It should check in the original equation.

Check 4:

$$\frac{x + 4}{2x} + \frac{x + 20}{3x} = 3$$

$$\frac{4 + 4}{2 \cdot 4} + \frac{4 + 20}{3 \cdot 4} \stackrel{?}{=} 3$$

$$\frac{8}{8} + \frac{24}{12} \stackrel{?}{=} 3$$

$$1 + 2 \stackrel{?}{=} 3$$

$$3 = 3, \quad \text{true}$$

This true statement verifies that the solution is 4 and the solution set is {4}. ∎

 CHECK POINT 1   Solve: $\dfrac{x + 6}{2x} + \dfrac{x + 24}{5x} = 2$.

The following steps may be used to solve a rational equation:

Solving Rational Equations

1. List restrictions on the variable. Avoid any values of the variable that make a denominator zero.

2. Clear the equation of fractions by multiplying both sides by the LCD of all rational expressions in the equation.

3. Solve the resulting equation.

4. Reject any proposed solution that is in the list of restrictions on the variable. Check other proposed solutions in the original equation.

EXAMPLE 2   Solving a Rational Equation

Solve: $\dfrac{x+1}{x+10} = \dfrac{x-2}{x+4}$.

Solution
Step 1. List restrictions on the variable.

This denominator would equal zero if $x = -10$.

$$\dfrac{x+1}{x+10} = \dfrac{x-2}{x+4}$$

This denominator would equal zero if $x = -4$.

The restrictions are $x \ne -10$ and $x \ne -4$.

Step 2. Multiply both sides by the LCD. The denominators are $x+10$ and $x+4$. Thus, the LCD is $(x+10)(x+4)$.

$$\dfrac{x+1}{x+10} = \dfrac{x-2}{x+4}, \quad x \ne -10, x \ne -4 \qquad \text{This is the given equation.}$$

$$(x+10)(x+4)\cdot\left(\dfrac{x+1}{x+10}\right) = (x+10)(x+4)\cdot\left(\dfrac{x-2}{x+4}\right) \qquad \begin{array}{l}\text{Multiply both sides by}\\ \text{the LCD.}\end{array}$$

$$(x+4)(x+1) = (x+10)(x-2) \qquad \text{Simplify.}$$

Step 3. Solve the resulting equation.

$$\begin{array}{ll}(x+4)(x+1) = (x+10)(x-2) & \text{This is the equation cleared of fractions.}\\ x^2 + 5x + 4 = x^2 + 8x - 20 & \text{Use FOIL multiplication on each side.}\\ 5x + 4 = 8x - 20 & \text{Subtract } x^2 \text{ from both sides.}\\ -3x + 4 = -20 & \text{Subtract } 8x \text{ from both sides.}\\ -3x = -24 & \text{Subtract 4 from both sides.}\\ x = 8 & \text{Divide both sides by } -3.\end{array}$$

Step 4. Check the proposed solution in the original equation. The proposed solution, 8, is not part of the restriction that $x \ne -10$ and $x \ne -4$. Substitute 8 for x in the given equation. You should obtain the true statement $\frac{1}{2} = \frac{1}{2}$. The solution is 8 and the solution set is $\{8\}$. ■

☑ **CHECK POINT 2**   Solve: $\dfrac{x-3}{x+1} = \dfrac{x-2}{x+6}$.

Great Question!

Earlier in the chapter, I learned how to add and subtract rational expressions. In this section, I'm solving equations that contain addition and subtraction of rational expressions. Can you do a side-by-side comparison of the two procedures?

We simplify rational expressions. We solve rational equations. Notice the differences between the procedures.

Simplifying a Rational Expression	Solving a Rational Equation
Simplify: $\dfrac{9}{4x} - \dfrac{5}{2x} - \dfrac{3}{4}$.	Solve: $\dfrac{9}{4x} - \dfrac{5}{2x} = \dfrac{3}{4}$.
This is not an equation. There is no equal sign.	This is an equation. There is an equal sign.
Solution The LCD is $4x$. Rewrite each expression with this LCD and retain the LCD.	**Solution** The LCD is $4x$. Multiply both sides by this LCD and clear the fractions.
$\dfrac{9}{4x} - \dfrac{5}{2x} - \dfrac{3}{4}$	$4x\left(\dfrac{9}{4x} - \dfrac{5}{2x}\right) = 4x\cdot\dfrac{3}{4}$
$= \dfrac{9}{4x} - \dfrac{5\cdot2}{2x\cdot2} - \dfrac{3\cdot x}{4\cdot x}$	$\dfrac{4x}{1}\cdot\dfrac{9}{4x} - \dfrac{4x}{1}\cdot\dfrac{5}{2x} = 4x\cdot\dfrac{3}{4}$
$= \dfrac{9}{4x} - \dfrac{10}{4x} - \dfrac{3x}{4x}$	$9 - 10 = 3x$
$= \dfrac{9-10-3x}{4x} = \dfrac{-1-3x}{4x}$	$-1 = 3x$
	$-\dfrac{1}{3} = x$
	The solution set is $\left\{-\frac{1}{3}\right\}$.

You only eliminate the denominators when solving a rational equation with an equal sign. You should never begin by eliminating the denominators when simplifying a rational expression involving addition or subtraction with no equal sign.

EXAMPLE 3 Solving a Rational Equation

Solve: $\dfrac{x}{x - 3} = \dfrac{3}{x - 3} + 9$.

Solution

Step 1. List restrictions on the variable.

$$\dfrac{x}{x - 3} = \dfrac{3}{x - 3} + 9$$

> These denominators are zero if $x = 3$.

The restriction is $x \ne 3$.

Step 2. Multiply both sides by the LCD. The LCD is $x - 3$.

$$\dfrac{x}{x - 3} = \dfrac{3}{x - 3} + 9, \quad x \ne 3 \qquad \text{This is the given equation.}$$

$$(x - 3) \cdot \dfrac{x}{x - 3} = (x - 3)\left(\dfrac{3}{x - 3} + 9\right) \qquad \text{Multiply both sides by the LCD.}$$

$$(x - 3) \cdot \dfrac{x}{x - 3} = (x - 3) \cdot \dfrac{3}{x - 3} + 9(x - 3) \qquad \begin{array}{l}\text{Use the distributive property on}\\ \text{the right side.}\end{array}$$

$$x = 3 + 9(x - 3) \qquad \text{Simplify.}$$

Step 3. Solve the resulting equation.

$$x = 3 + 9(x - 3) \qquad \text{This is the equation cleared of fractions.}$$
$$x = 3 + 9x - 27 \qquad \text{Use the distributive property on the right side.}$$
$$x = 9x - 24 \qquad \text{Combine numerical terms.}$$
$$-8x = -24 \qquad \text{Subtract 9x from both sides.}$$
$$x = 3 \qquad \text{Divide both sides by } -8.$$

Great Question!

Cut to the chase: When do I get rid of proposed solutions in rational equations?

Reject any proposed solution that causes any denominator in a rational equation to equal 0.

Step 4. Check proposed solutions. The proposed solution, 3, is *not* a solution because of the restriction that $x \ne 3$. Notice that 3 makes both of the denominators zero in the original equation. There is no solution for this equation. The solution set is $\varnothing$, the empty set. ∎

✓ **CHECK POINT 3** Solve: $\dfrac{8x}{x + 1} = 4 - \dfrac{8}{x + 1}$.

Examples 4 and 5 involve rational equations that become quadratic equations after clearing fractions.

EXAMPLE 4 Solving a Rational Equation

Solve: $\dfrac{x}{3} + \dfrac{9}{x} = 4$.

Solution

Step 1. List restrictions on the variable.

$$\frac{x}{3} + \frac{9}{x} = 4$$

This denominator would equal zero if $x = 0$.

The restriction is $x \neq 0$.

Step 2. Multiply both sides by the LCD. The denominators are 3 and x. Thus, the LCD is $3x$.

Using Technology

Graphic Connections

The graphs of

$$y_1 = \frac{x}{3} + \frac{9}{x}$$

and

$$y_2 = 4$$

have two intersection points. The x-coordinates of the points are 3 and 9. This verifies that

$$\frac{x}{3} + \frac{9}{x} = 4$$

has both 3 and 9 as solutions.

$[-2, 12, 1]$ by $[-1, 5, 1]$

$$\frac{x}{3} + \frac{9}{x} = 4, \quad x \neq 0 \qquad \text{This is the given equation.}$$

$$3x\left(\frac{x}{3} + \frac{9}{x}\right) = 3x \cdot 4 \qquad \text{Multiply both sides by the LCD.}$$

$$3x \cdot \frac{x}{3} + 3x \cdot \frac{9}{x} = 12x \qquad \text{Use the distributive property on the left side.}$$

$$x^2 + 27 = 12x \qquad \text{Simplify.}$$

Step 3. Solve the resulting equation. Can you see that we have a quadratic equation? Write the equation in standard form and solve for x.

$$x^2 + 27 = 12x \qquad \text{This is the equation cleared of fractions.}$$

$$x^2 - 12x + 27 = 0 \qquad \text{Subtract 12x from both sides.}$$

$$(x - 9)(x - 3) = 0 \qquad \text{Factor.}$$

$$x - 9 = 0 \quad \text{or} \quad x - 3 = 0 \qquad \text{Set each factor equal to 0.}$$

$$x = 9 \qquad\qquad x = 3 \qquad \text{Solve the resulting equations.}$$

Step 4. Check proposed solutions in the original equation. The proposed solutions, 9 and 3, are not part of the restriction that $x \neq 0$. Substitute 9 for x, and then 3 for x, in the given equation. In each case, you should obtain the true statement $4 = 4$. The solutions are 3 and 9, and the solution set is $\{3, 9\}$. ■

✓ **CHECK POINT 4** Solve: $\dfrac{x}{2} + \dfrac{12}{x} = 5$.

EXAMPLE 5 Solving a Rational Equation

Solve: $\dfrac{2x}{x - 3} + \dfrac{6}{x + 3} = -\dfrac{28}{x^2 - 9}$.

Solution

Step 1. List restrictions on the variable. By factoring denominators, it makes it easier to see values that make the denominators zero.

$$\frac{2x}{x - 3} + \frac{6}{x + 3} = -\frac{28}{(x + 3)(x - 3)}$$

This denominator is zero if $x = 3$.

This denominator is zero if $x = -3$.

This denominator is zero if $x = -3$ or $x = 3$.

The restrictions are $x \neq -3$ and $x \neq 3$.

Step 2. Multiply both sides by the LCD. The LCD is $(x + 3)(x - 3)$.

$$\frac{2x}{x - 3} + \frac{6}{x + 3} = -\frac{28}{(x + 3)(x - 3)}, \quad x \neq -3, x \neq 3$$

This is the given equation with a denominator factored.

$$(x + 3)(x - 3)\left(\frac{2x}{x - 3} + \frac{6}{x + 3}\right) = (x + 3)(x - 3)\left(-\frac{28}{(x + 3)(x - 3)}\right)$$

Multiply both sides by the LCD.

$$(x + 3)(x - 3) \cdot \frac{2x}{x - 3} + (x + 3)(x - 3) \cdot \frac{6}{x + 3} = (x + 3)(x - 3)\left(-\frac{28}{(x + 3)(x - 3)}\right)$$

Use the distributive property on the left side.

$$2x(x + 3) + 6(x - 3) = -28$$

Simplify.

Step 3. Solve the resulting equation.

$$2x(x + 3) + 6(x - 3) = -28$$

This is the equation cleared of fractions.

$$2x^2 + 6x + 6x - 18 = -28$$

Use the distributive property twice on the left side.

$$2x^2 + 12x - 18 = -28$$

Combine like terms.

$$2x^2 + 12x + 10 = 0$$

Add 28 to both sides and write the quadratic equation in standard form.

$$2(x^2 + 6x + 5) = 0$$

Factor out the GCF.

$$2(x + 5)(x + 1) = 0$$

Factor the trinomial.

$$x + 5 = 0 \quad \text{or} \quad x + 1 = 0$$

Set each variable factor equal to 0.

$$x = -5 \qquad x = -1$$

Solve for x.

Step 4. Check the proposed solutions in the original equation. The proposed solutions, -5 and -1, are not part of the restriction that $x \neq -3$ and $x \neq 3$. Substitute -5 for x, and then -1 for x, in the given equation. The resulting true statements verify that -5 and -1 are the solutions, and $\{-5, -1\}$ is the solution set. ∎

✓ **CHECK POINT 5** Solve: $\dfrac{3}{x - 3} + \dfrac{5}{x - 4} = \dfrac{x^2 - 20}{x^2 - 7x + 12}$.

---

2 Solve problems involving rational functions that model applied situations. ⏵

Applications of Rational Equations

Solving rational equations can be used to answer questions about variables contained in rational functions.

EXAMPLE 6 Using a Cost-Benefit Model

The function

$$f(x) = \frac{120x}{100 - x}$$

models the cost, $f(x)$, in thousands of dollars, to remove $x\%$ of a lake's pollutants. If voters commit $80 thousand for this project, what percentage of the pollutants can be removed?

Solution   Substitute 80, the cost in thousands of dollars, for $f(x)$ and solve the resulting rational equation for x.

$$80 = \frac{120x}{100 - x}$$ The LCD is 100 − x.

$$(100 - x)80 = (100 - x) \cdot \frac{120x}{100 - x}$$ Multiply both sides by the LCD.

$$80(100 - x) = 120x$$ Simplify.

$$8000 - 80x = 120x$$ Use the distributive property on the left side.

$$8000 = 200x$$ Add 80x to both sides.

$$40 = x$$ Divide both sides by 200.

If voters commit $80 thousand, 40% of the lake's pollutants can be removed.   ∎

✓ **CHECK POINT 6**   Use the cost-benefit model in Example 6 to answer this question: If voters in the city commit $120 thousand for the project, what percentage of the lake's pollutants can be removed?

Achieving Success

Assuming that you have done very well preparing for an exam, **there are certain things you can do that will make you a better test taker**.

- Get a good sleep the night before the exam.
- Have a good breakfast that balances protein, carbohydrates, and fruit.
- Just before the exam, briefly review the relevant material in the chapter summary.
- Bring everything you need to the exam, including two pencils, an eraser, scratch paper (if permitted), a calculator (if you're allowed to use one), water, and a watch.
- Survey the entire exam quickly to get an idea of its length.
- Read the directions to each problem carefully. Make sure that you have answered the specific question asked.
- Work the easy problems first. Then return to the hard problems you are not sure of. Doing the easy problems first will build your confidence. If you get bogged down on any one problem, you may not be able to complete the exam and receive credit for the questions you can easily answer.
- Attempt every problem. There may be partial credit even if you do not obtain the correct answer.
- Work carefully. Show your step-by-step solutions neatly. Check your work and answers.
- Watch the time. Pace yourself and be aware of when half the time is up. Determine how much of the exam you have completed. This will indicate if you're moving at a good pace or need to speed up. Prepare to spend more time on problems worth more points.
- Never turn in a test early. Use every available minute you are given for the test. If you have extra time, double check your arithmetic and look over your solutions.

CONCEPT AND VOCABULARY CHECK

Fill in each blank so that the resulting statement is true.

1. We clear a rational equation of fractions by multiplying both sides by the _____ of all rational expressions in the equation.

2. We reject any proposed solution of a rational equation that causes a denominator to equal _____.

3. The first step in solving

$$\frac{4}{x} + \frac{1}{2} = \frac{5}{x}$$

is to multiply both sides by _____.

4. The first step in solving

$$\frac{x-6}{x+5} = \frac{x-3}{x+1}$$

is to multiply both sides by _____.

5. The restrictions on the variable in the rational equation

$$\frac{1}{x-2} - \frac{2}{x+4} = \frac{2x-1}{x^2+2x-8}$$

are _____ and _____.

6.

$$\frac{5}{x+4} + \frac{3}{x+3} = \frac{12x+9}{(x+4)(x+3)}$$

$$(x+4)(x+3)\left(\frac{5}{x+4} + \frac{3}{x+3}\right) = (x+4)(x+3)\left(\frac{12x+9}{(x+4)(x+3)}\right)$$

The resulting equation cleared of fractions is _____.

7. True or false: A rational equation can have no solution. _____

6.6 EXERCISE SET

Practice Exercises

In Exercises 1–34, solve each rational equation. If an equation has no solution, so state.

1. $\dfrac{1}{x} + 2 = \dfrac{3}{x}$

2. $\dfrac{1}{x} - 3 = \dfrac{4}{x}$

3. $\dfrac{5}{x} + \dfrac{1}{3} = \dfrac{6}{x}$

4. $\dfrac{4}{x} + \dfrac{1}{2} = \dfrac{5}{x}$

5. $\dfrac{x-2}{2x} + 1 = \dfrac{x+1}{x}$

6. $\dfrac{7x-4}{5x} = \dfrac{9}{5} - \dfrac{4}{x}$

7. $\dfrac{3}{x+1} = \dfrac{5}{x-1}$

8. $\dfrac{6}{x+3} = \dfrac{4}{x-3}$

9. $\dfrac{x-6}{x+5} = \dfrac{x-3}{x+1}$

10. $\dfrac{x+2}{x+10} = \dfrac{x-3}{x+4}$

11. $\dfrac{x+6}{x+3} = \dfrac{3}{x+3} + 2$

12. $\dfrac{3x+1}{x-4} = \dfrac{6x+5}{2x-7}$

13. $1 - \dfrac{4}{x+7} = \dfrac{5}{x+7}$

14. $5 - \dfrac{2}{x-5} = \dfrac{3}{x-5}$

15. $\dfrac{4x}{x+2} + \dfrac{2}{x-1} = 4$

16. $\dfrac{3x}{x+1} + \dfrac{4}{x-2} = 3$

17. $\dfrac{8}{x^2-9} + \dfrac{4}{x+3} = \dfrac{2}{x-3}$

18. $\dfrac{32}{x^2-25} = \dfrac{4}{x+5} + \dfrac{2}{x-5}$

19. $x + \dfrac{7}{x} = -8$

20. $x + \dfrac{6}{x} = -7$

21. $\dfrac{6}{x} - \dfrac{x}{3} = 1$

22. $\dfrac{x}{2} - \dfrac{12}{x} = 1$

23. $\dfrac{x+6}{3x-12} = \dfrac{5}{x-4} + \dfrac{2}{3}$

24. $\dfrac{1}{5x+5} = \dfrac{3}{x+1} - \dfrac{7}{5}$

25. $\dfrac{1}{x-1} + \dfrac{1}{x+1} = \dfrac{2}{x^2-1}$

26. $\dfrac{1}{x-2} + \dfrac{1}{x+2} = \dfrac{4}{x^2-4}$

27. $\dfrac{5}{x+4} + \dfrac{3}{x+3} = \dfrac{12x+19}{x^2+7x+12}$

28. $\dfrac{2x-1}{x^2+2x-8} + \dfrac{2}{x+4} = \dfrac{1}{x-2}$

29. $\dfrac{4x}{x+3} - \dfrac{12}{x-3} = \dfrac{4x^2+36}{x^2-9}$

30. $\dfrac{2}{x+3} - \dfrac{5}{x+1} = \dfrac{3x+5}{x^2+4x+3}$

31. $\dfrac{4}{x^2+3x-10} + \dfrac{1}{x^2+9x+20} = \dfrac{2}{x^2+2x-8}$

32. $\dfrac{4}{x^2+3x-10} - \dfrac{1}{x^2+x-6} = \dfrac{3}{x^2-x-12}$

33. $\dfrac{3y}{y^2+5y+6} + \dfrac{2}{y^2+y-2} = \dfrac{5y}{y^2+2y-3}$

34. $\dfrac{y-1}{y^2-4} + \dfrac{y}{y^2-y-2} = \dfrac{2y-1}{y^2+3y+2}$

In Exercises 35–38, a rational function g is given. Find all values of a for which g(a) is the indicated value.

35. $g(x) = \dfrac{x}{2} + \dfrac{20}{x}; g(a) = 7$

36. $g(x) = \dfrac{x}{4} + \dfrac{5}{x}; g(a) = 3$

37. $g(x) = \dfrac{5}{x+2} + \dfrac{25}{x^2+4x+4}; g(a) = 20$

38. $g(x) = \dfrac{3x-2}{x+1} + \dfrac{x+2}{x-1}; g(a) = 4$

Practice PLUS

In Exercises 39–46, solve or simplify, whichever is appropriate.

39. $\dfrac{x+2}{x^2-x} - \dfrac{6}{x^2-1}$

40. $\dfrac{x+3}{x^2-x} - \dfrac{8}{x^2-1}$

41. $\dfrac{x+2}{x^2-x} - \dfrac{6}{x^2-1} = 0$

42. $\dfrac{x+3}{x^2-x} - \dfrac{8}{x^2-1} = 0$

43. $\dfrac{1}{x^3-8} + \dfrac{3}{(x-2)(x^2+2x+4)} = \dfrac{2}{x^2+2x+4}$

44. $\dfrac{2}{x^3-1} + \dfrac{4}{(x-1)(x^2+x+1)} = -\dfrac{1}{x^2+x+1}$

45. $\dfrac{1}{x^3-8} + \dfrac{3}{(x-2)(x^2+2x+4)} - \dfrac{2}{x^2+2x+4}$

46. $\dfrac{2}{x^3-1} + \dfrac{4}{(x-1)(x^2+x+1)} + \dfrac{1}{x^2+x+1}$

In Exercises 47–48, find all values of a for which
$f(a) = g(a) + 1.$

47. $f(x) = \dfrac{x+2}{x+3}, g(x) = \dfrac{x+1}{x^2+2x-3}$

48. $f(x) = \dfrac{4}{x-3}, g(x) = \dfrac{10}{x^2+x-12}$

In Exercises 49–50, find all values of a for which
$(f+g)(a) = h(a).$

49. $f(x) = \dfrac{5}{x-4}, g(x) = \dfrac{3}{x-3}, h(x) = \dfrac{x^2-20}{x^2-7x+12}$

50. $f(x) = \dfrac{6}{x+3}, g(x) = \dfrac{2x}{x-3}, h(x) = -\dfrac{28}{x^2-9}$

Application Exercises

The function
$$f(x) = \dfrac{250x}{100-x}$$

models the cost, f(x), in millions of dollars, to remove x% of a river's pollutants. Use this function to solve Exercises 51–52.

51. If the government commits $375 million for this project, what percentage of the pollutants can be removed?

52. If the government commits $750 million for this project, what percentage of the pollutants can be removed?

In an experiment about memory, students in a language class are asked to memorize 40 vocabulary words in Latin, a language with which they are not familiar. After studying the words for one day, students are tested each day thereafter to see how many words they remember. The class average is then found. The function
$$f(x) = \dfrac{5x+30}{x}$$

models the average number of Latin words remembered by the students, f(x), after x days. The graph of the rational function is shown. Use the function to solve Exercises 53–56.

Average Number of Words Remembered Over Time

$f(x) = \dfrac{5x+30}{x}$

(In Exercises 53–56, be sure to refer to the rational function

$$f(x) = \frac{5x + 30}{x} \text{ and its graph on the previous page.})$$

53. After how many days do the students remember 8 words? Identify your solution as a point on the graph.

54. After how many days do the students remember 7 words? Identify your solution as a point on the graph.

55. What is the horizontal asymptote of the graph? Describe what this means about the average number of Latin words remembered by the students over an extended period of time.

56. According to the graph, between which two days do students forget the most? Describe the trend shown by the memory function's graph.

Rational functions can be used to model learning. Many of these functions model the proportion of correct responses as a function of the number of trials of a particular task. One such model, called a learning curve, is

$$f(x) = \frac{0.9x - 0.4}{0.9x + 0.1},$$

where $f(x)$ is the proportion of correct responses after x trials. If $f(x) = 0$, there are no correct responses. If $f(x) = 1$, all responses are correct. The graph of the rational function is shown. Use the function to solve Exercises 57–60.

A Learning Curve

Number of Learning Trials

57. How many learning trials are necessary for 0.95 of the responses to be correct? Identify your solution as a point on the graph.

58. How many learning trials are necessary for 0.5 of the responses to be correct? Identify your solution as a point on the graph.

59. Describe the trend shown by the graph in terms of learning new tasks. What happens initially and what happens as time increases?

60. What is the horizontal asymptote of the graph? Once the performance level approaches peak efficiency, what

effect does additional practice have on performance? Describe how this is shown by the graph.

61. A company wants to increase the 10% peroxide content of its product by adding pure peroxide (100% peroxide). If x liters of pure peroxide are added to 500 liters of its 10% solution, the concentration, $C(x)$, of the new mixture is given by

$$C(x) = \frac{x + 0.1(500)}{x + 500}.$$

How many liters of pure peroxide should be added to produce a new product that is 28% peroxide?

62. Suppose that x liters of pure acid are added to 200 liters of a 35% acid solution.

    a. Write a function that gives the concentration, $C(x)$, of the new mixture. (*Hint*: See Exercise 61.)

    b. How many liters of pure acid should be added to produce a new mixture that is 74% acid?

Explaining the Concepts

63. What is a rational equation?

64. Explain how to solve a rational equation.

65. Explain how to find restrictions on the variable in a rational equation.

66. Why should restrictions on the variable in a rational equation be listed before you begin solving the equation?

67. Describe similarities and differences between the procedures needed to solve the following problems:

$$\text{Add: } \frac{2}{x} + \frac{3}{4}. \qquad \text{Solve: } \frac{2}{x} + \frac{3}{4} = 1.$$

68. Rational functions model learning and forgetting. Use the graphs in Exercises 53–56 and in Exercises 57–60 to describe one similarity and one difference between learning and forgetting over time.

69. Does the graph of the learning curve shown in Exercises 57–60 indicate that practice makes perfect? Explain. Does this have anything to do with what psychologists call "the curse of perfection"?

Technology Exercises

In Exercises 70–74, use a graphing utility to solve each rational equation. Graph each side of the equation in the given viewing rectangle. The solution is the first coordinate of the point(s) of intersection. Check by direct substitution.

70. $\dfrac{x}{2} + \dfrac{x}{4} = 6$; $[-5, 10, 1]$ by $[-5, 10, 1]$

71. $\dfrac{50}{x} = 2x$; $[-10, 10, 1]$ by $[-20, 20, 2]$

72. $x + \dfrac{6}{x} = -5; [-10, 10, 1]$ by $[-7, 10, 1]$

73. $\dfrac{2}{x} = x + 1; [-5, 5, 1]$ by $[-5, 5, 1]$

74. $\dfrac{3}{x} - \dfrac{x + 21}{3x} = \dfrac{5}{3}; [-5, 5, 1]$ by $[-5, 5, 1]$

Critical Thinking Exercises

Make Sense? In Exercises 75–78, determine whether each statement makes sense or does not make sense, and explain your reasoning.

75. I must have made an error if a rational equation produces no solution.

76. I added two rational expressions and found the solution set.

77. I can solve the equation $\dfrac{40}{x} = \dfrac{15}{x - 20}$ by multiplying both sides by the LCD or by setting the cross product of 40 and $x - 20$ equal to the cross product of 15 and x.

78. I'm solving a rational equation that became a quadratic equation, so my rational equation will have two solutions.

In Exercises 79–82, determine whether each statement is true or false. If the statement is false, make the necessary change(s) to produce a true statement.

79. Once a rational equation is cleared of fractions, all solutions of the resulting equation are also solutions of the rational equation.

80. We find

$$\frac{4}{x} - \frac{2}{x + 1}$$

by multiplying each term by the LCD, $x(x + 1)$, thereby clearing fractions.

81. All real numbers satisfy the equation $\dfrac{7}{x} - \dfrac{2}{x} = \dfrac{5}{x}$.

82. In order to find a number to add to the numerator and denominator of $\frac{3}{16}$ to result in $\frac{1}{2}$, we could solve the following rational equation:

$$\frac{3 + x}{16 + x} = \frac{1}{2}.$$

83. Solve: $\left(\dfrac{1}{x + 1} + \dfrac{x}{1 - x} \right) \div \left(\dfrac{x}{x + 1} - \dfrac{1}{x - 1} \right) = -1.$

84. Solve: $\left| \dfrac{x + 1}{x + 8} \right| = \dfrac{2}{3}.$

85. Write an original rational equation that has no solution.

86. Solve

$$\left(\frac{4}{x - 1} \right)^2 + 2\left(\frac{4}{x - 1} \right) + 1 = 0$$

by introducing the substitution $u = \dfrac{4}{x - 1}$.

Review Exercises

87. Graph the solution set:

$$\begin{cases} x + 2y \geq 2 \\ x - y \geq -4. \end{cases}$$

(Section 4.4, Example 5)

88. Solve:

$$\frac{x - 4}{2} - \frac{1}{5} = \frac{7x + 1}{20}.$$

(Section 1.4, Example 4)

89. Solve for F:

$$C = \frac{5F - 160}{9}.$$

(Section 1.5, Example 8)

Preview Exercises

Exercises 90–92 will help you prepare for the material covered in the next section.

90. Solve for p: $qf + pf = pq$.

91. Solve: $\dfrac{40}{x} + \dfrac{40}{x + 30} = 2.$

92. A plane flies at an average rate of 450 miles per hour. It can travel 980 miles with the wind in the same amount of time as it travels 820 miles against the wind. Solve the equation

$$\frac{980}{450 + x} = \frac{820}{450 - x}$$

to find the average rate of the wind, x, in miles per hour.

SECTION

6.7

Formulas and Applications of Rational Equations

What am I supposed to learn?

After studying this section, you should be able to:

1. Solve a formula with a rational expression for a variable.

2. Solve business problems involving average cost.

3. Solve problems involving time in motion.

4. Solve problems involving work.

The current generation of college students grew up playing interactive online games and many continue to play in school. Hundreds of colleges have formed organized gaming teams, many as campus clubs. Enter the Oculus Rift, revolutionizing the way people experience gaming. The Oculus Rift is a virtual reality headset that enables users to experience video games as immersive three-dimensional environments. Basically, it puts the gamer inside the game.

The cost of manufacturing Oculus Rift headsets can be modeled by rational functions. In this section, you will see that high production levels of the Oculus Rift can keep the price of this amazing invention low, perhaps making this the device that brings home virtual reality to reality.

1. Solve a formula with a rational expression for a variable.

Solving a Formula for a Variable

Formulas and mathematical models frequently contain rational expressions. We solve for a specified variable using the procedure for solving rational equations. The goal is to get the specified variable alone on one side of the equation. To do so, collect all terms with this variable on one side and all other terms on the other side. It is sometimes necessary to factor out the variable you are solving for.

EXAMPLE 1   Solving for a Variable in a Formula

The formula

$$S = \frac{C}{1 - r}$$

describes a product's selling price, S, in terms of its cost to the seller, C, and its markup rate, r, in decimal form.

a. Solve the formula for r.

b. What is the markup rate on a textbook costing the campus bookstore $140 and selling for $200?

Solution

a. Our goal is to isolate the variable r.

$$S = \frac{C}{1 - r}$$

> We need r by itself on one side of the formula.

We begin by multiplying both sides by the least common denominator, $1 - r$, to clear the equation of fractions.

$$S = \frac{C}{1 - r}$$ This is the given formula.

$$(1 - r)S = (1 - r)\left(\frac{C}{1 - r}\right)$$ Multiply both sides by $1 - r$, the LCD.

$$S - Sr = C$$ Use the distributive property on the left side. On the right side,

$$(1 - r)\left(\frac{C}{1 - r}\right) = C.$$

Observe that the formula is now cleared of fractions. The term with r, the specified variable, is already on the left side of the equation. To isolate this term, subtract S from both sides.

$$S - Sr = C$$ This is the equation cleared of fractions.

$$-Sr = C - S$$ Subtract S from both sides.

$$\frac{-Sr}{-S} = \frac{C - S}{-S}$$ Divide both sides by $-S$ and solve for r.

$$r = \frac{C - S}{-S}$$ Simplify.

Great Question!

Now that we've solved the formula for r, it looks weird with a negative sign in the denominator. Is there another way to write this formula?

Yes. Multiply the numerator and the denominator by -1.

$$r = \frac{C - S}{-S} = \frac{(-1)}{(-1)} \cdot \frac{C - S}{-S} = \frac{-C + S}{S} = \frac{S - C}{S}$$

b. Now we find the markup rate on a textbook costing the campus bookstore $140 and selling for $200.

$$C = 140$$

$$r = \frac{C - S}{-S} \quad S = 200 \qquad \text{This is the formula solved for } r \text{ from part (a).}$$

$$r = \frac{140 - 200}{-200} = \frac{-60}{-200} = \frac{3}{10} = 0.3 = 30\%$$

The bookstore's markup rate is 30%. ∎

✓ **CHECK POINT 1** Solve for x: $a = \dfrac{b}{x + 2}$.

EXAMPLE 2   Solving for a Variable in a Formula

If you wear glasses, did you know that each lens has a measurement called its focal length, f? When an object is in focus, its distance from the lens, p, and the distance from the lens to your retina, q, satisfy the formula

$$\frac{1}{p} + \frac{1}{q} = \frac{1}{f}.$$

Figure 6.5

(See **Figure 6.5**.) Solve this formula for p.

Solution Our goal is to isolate the variable p. We begin by multiplying both sides by the least common denominator, pqf, to clear the equation of fractions.

$$\frac{1}{p} + \frac{1}{q} = \frac{1}{f}$$ This is the given formula.

$$pqf\left(\frac{1}{p} + \frac{1}{q}\right) = pqf\left(\frac{1}{f}\right)$$ Multiply both sides by pqf, the LCD.

$$pqf\left(\frac{1}{p}\right) + pqf\left(\frac{1}{q}\right) = pqf\left(\frac{1}{f}\right)$$ Use the distributive property on the left side.

$$qf + pf = pq$$ Simplify.

Observe that the formula is now cleared of fractions. Collect terms with p, the specified variable, on one side of the equation. To do so, subtract pf from both sides.

$$qf + pf = pq$$ This is the equation cleared of fractions.

$$qf = pq - pf$$ Subtract pf from both sides.

$$qf = p(q - f)$$ Factor out p, the specified variable.

$$\frac{qf}{q - f} = \frac{p(q - f)}{q - f}$$ Divide both sides by $q - f$ and solve for p.

$$\frac{qf}{q - f} = p$$ Simplify. ■

☑ **CHECK POINT 2** Solve $\dfrac{1}{x} + \dfrac{1}{y} = \dfrac{1}{z}$ for x.

---

2 Solve business problems involving average cost. ▶

Business Problems Involving Average Cost

We have seen that the cost function for a business is the sum of its fixed and variable costs:

$$C(x) = (\text{fixed cost}) + cx$$

> Cost per unit times the number of units produced, x

The **average cost** per unit for a company to produce x units is the sum of its fixed and variable costs divided by the number of units produced. The **average cost function** is a rational function that is denoted by $\overline{C}$. Thus,

> Cost of producing x units: fixed plus variable costs

$$\overline{C}(x) = \frac{(\text{fixed cost}) + cx}{x}.$$

> Number of units produced

EXAMPLE 3 Putting the Video-Game Player Inside the Game

We return to the Oculus Rift, a virtual reality headset that enables users to experience video games as immersive three-dimensional environments described in the section opener. Suppose the company that manufactures this invention has a fixed monthly cost of $600,000 and that it costs $50 to produce each headset.

a. Write the cost function, C, of producing x headsets.

b. Write the average cost function, $\overline{C}$, of producing x headsets.

c. How many headsets must be produced each month for the company to reach an average cost of $350 per headset?

Solution

a. The cost function, C, is the sum of the fixed cost and the variable costs.

$$C(x) = 600{,}000 + 50x$$

Fixed cost is $600,000.

Variable cost: $50 for each headset

b. The average cost function, $\overline{C}$, is the sum of fixed and variable costs divided by the number of headsets produced.

$$\overline{C}(x) = \frac{600{,}000 + 50x}{x}$$

c. We are interested in the company's production level that results in an average cost of $350 per headset. Substitute 350, the average cost, for $\overline{C}(x)$ and solve the resulting rational equation for x.

$$350 = \frac{600{,}000 + 50x}{x} \qquad \text{Substitute 350 for } \overline{C}(x).$$

$$350x = 600{,}000 + 50x \qquad \text{Multiply both sides by the LCD, } x.$$

$$300x = 600{,}000 \qquad \text{Subtract 50x from both sides.}$$

$$x = 2000 \qquad \text{Divide both sides by 300.}$$

The company must produce 2000 headsets each month for an average cost of $350 per headset. ∎

Figure 6.6 shows the graph of the average cost function in Example 3. As the production level increases, the average cost of producing each headset decreases. The horizontal asymptote, $y = 50$, is also shown in the figure. This means that the more headsets produced each month, the closer the average cost per system for the company comes to $50. The least possible cost per headset is approaching $50. Competitively low prices take place with high production levels, posing a major problem for small businesses.

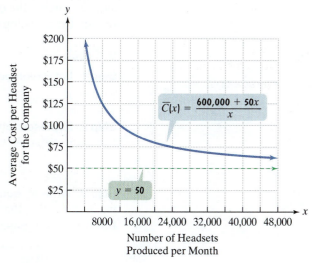

$$\overline{C}(x) = \frac{600{,}000 + 50x}{x}$$

$y = 50$

Average Cost per Headset for the Company

Number of Headsets Produced per Month

Figure 6.6

✓ **CHECK POINT 3** A company is planning to manufacture wheelchairs that are light, fast, and beautiful. Fixed monthly cost will be $500,000 and it will cost $400 to produce each radically innovative chair.

a. Write the cost function, C, of producing x wheelchairs.

b. Write the average cost function, $\overline{C}$, of producing x wheelchairs.

c. How many wheelchairs must be produced each month for the company to reach an average cost of $450 per chair?

3 Solve problems involving time in motion. ▶

Problems Involving Motion

We have seen that the distance, d, covered by any moving body is the product of its average rate, r, and its time in motion, t: $d = rt$. Rational expressions appear in motion problems when the conditions of the problem involve the time traveled. We can obtain an expression for t, the time traveled, by dividing both sides of $d = rt$ by r.

$$d = rt \qquad \text{Distance equals rate times time.}$$

$$\frac{d}{r} = \frac{rt}{r} \qquad \text{Divide both sides by } r.$$

$$\frac{d}{r} = t \qquad \text{Simplify.}$$

Time in Motion

$$t = \frac{d}{r}$$

$$\text{Time traveled} = \frac{\text{Distance traveled}}{\text{Rate of travel}}$$

EXAMPLE 4   A Motion Problem Involving Time

You commute to work a distance of 40 miles and return on the same route at the end of the day. Your average rate on the return trip is 30 miles per hour faster than your average rate on the outgoing trip. If the round trip takes 2 hours, what is your average rate on the outgoing trip to work?

Solution

Step 1. Let x represent one of the unknown quantities. Let

$$x = \text{the rate on the outgoing trip.}$$

Step 2. Represent other unknown quantities in terms of x. Because the average rate on the return trip is 30 miles per hour faster than the average rate on the outgoing trip, let

$$x + 30 = \text{the rate on the return trip.}$$

Step 3. Write an equation that models the conditions. By reading the problem again, we discover that the crucial idea is that the time for the round trip is 2 hours. Thus, the time on the outgoing trip plus the time on the return trip is 2 hours.

	Distance	Rate	Time = $\dfrac{\text{Distance}}{\text{Rate}}$	
Outgoing Trip	40	x	$\dfrac{40}{x}$	The sum of these times is 2 hours.
Return Trip	40	$x + 30$	$\dfrac{40}{x + 30}$	

We are now ready to write an equation that models the problem's conditions.

Time on the outgoing trip *plus* time on the return trip *equals* 2 hours.

$$\frac{40}{x} + \frac{40}{x + 30} = 2$$

Step 4. Solve the equation and answer the question.

$$\frac{40}{x} + \frac{40}{x + 30} = 2$$ This is the equation that models the problem's conditions.

$$x(x + 30)\left(\frac{40}{x} + \frac{40}{x + 30}\right) = 2x(x + 30)$$ Multiply both sides by the LCD, $x(x + 30)$.

$$\cancel{x}(x + 30) \cdot \frac{40}{\cancel{x}} + x\cancel{(x + 30)} \cdot \frac{40}{\cancel{x + 30}} = 2x^2 + 60x$$ Use the distributive property on each side.

$$40(x + 30) + 40x = 2x^2 + 60x$$ Simplify.

$$40x + 1200 + 40x = 2x^2 + 60x$$ Use the distributive property.

$$80x + 1200 = 2x^2 + 60x$$ Combine like terms: $40x + 40x = 80x$.

$$0 = 2x^2 - 20x - 1200$$ Subtract $80x + 1200$ from both sides.

$$0 = 2(x^2 - 10x - 600)$$ Factor out the GCF.

$$0 = 2(x - 30)(x + 20)$$ Factor completely.

$$x - 30 = 0 \quad \text{or} \quad x + 20 = 0$$ Set each variable factor equal to 0.

$$x = 30 \qquad\qquad x = -20$$ Solve for x.

Because x represents the rate on the outgoing trip, we reject the negative value, -20. The rate on the outgoing trip is 30 miles per hour. At an outgoing rate of 30 miles per hour, the round trip should take 2 hours.

Step 5. Check the proposed solution in the original wording of the problem. Does the round trip take 2 hours? Because the rate on the return trip is 30 miles per hour faster than the rate on the outgoing trip, the rate on the return trip is $30 + 30$, or 60 miles per hour.

$$\text{Time on the outgoing trip} = \frac{\text{Distance}}{\text{Rate}} = \frac{40}{30} = \frac{4}{3} \text{ hours}$$

$$\text{Time on the return trip} = \frac{\text{Distance}}{\text{Rate}} = \frac{40}{60} = \frac{2}{3} \text{ hour}$$

The total time for the round trip is $\frac{4}{3} + \frac{2}{3} = \frac{6}{3}$, or 2 hours. This checks with the original conditions of the problem. ∎

Using Technology

Graphic Connections
The graph of the rational function for the time for the round trip,

$$f(x) = \frac{40}{x} + \frac{40}{x + 30},$$

is shown in a $[0, 60, 30]$ by $[0, 10, 1]$ viewing rectangle.

At an outgoing rate of 30 miles per hour, the trip takes 2 hours.

Rate on Outgoing Trip (miles per hour)
$[0, 60, 30]$ by $[0, 10, 1]$

The graph is falling from left to right. This shows that the time for the round trip decreases as the rate increases.

✓ **CHECK POINT 4**   After riding at a steady speed for 40 miles, a bicyclist had a flat tire and walked 5 miles to a repair shop. The cycling rate was 4 times faster than the walking rate. If the time spent cycling and walking was 5 hours, at what rate was the cyclist riding?

4 Solve problems involving work. ▶

Problems Involving Work

You are thinking of designing your own Web site. You estimate that it will take 15 hours to do the job. In 1 hour, $\frac{1}{15}$ of the job is completed. In 2 hours, $\frac{2}{15}$ of the job is completed. In 3 hours, the fractional part of the job done is $\frac{3}{15}$, or $\frac{1}{5}$. In x hours, the fractional part of the job that you can complete is $\frac{x}{15}$.

Your friend, who has experience developing Web sites, took 10 hours working on her own to design an impressive site. You wonder about the possibility of working together. How long would it take both of you to design your Web site?

Problems involving work usually have two (or more) people or machines working together to complete a job. The amount of time it takes each person to do the job working alone is frequently known. The question deals with how long it will take both people working together to complete the job.

In work problems, **the number 1 represents one whole job completed**. For example, the completion of your Web site is represented by 1. Equations in work problems are often based on the following condition:

$$\begin{array}{ccc} \boxed{\begin{matrix}\text{Fractional part of}\\\text{the job done by}\\\text{the first person}\end{matrix}} & + & \boxed{\begin{matrix}\text{fractional part of}\\\text{the job done by}\\\text{the second person}\end{matrix}} & = & \boxed{\begin{matrix}1 \text{ (one whole job}\\\text{completed).}\end{matrix}} \end{array}$$

EXAMPLE 5 Solving a Problem Involving Work

You can design a Web site in 15 hours. Your friend can design the same site in 10 hours. How long will it take to design the Web site if you both work together?

Solution

Step 1. Let *x* represent one of the unknown quantities. Let

$x = $ the time, in hours, for you and your friend working together to design the Web site.

Step 2. Represent other unknown quantities in terms of *x*. Because there are no other unknown quantities, we can skip this step.

Step 3. Write an equation that models the conditions. We construct a table to help find the fractional parts of the task completed by you and your friend in x hours.

		Fractional part of job completed in 1 hour	Time working together	Fractional part of job completed in x hours
You can design the site in 15 hours.	You	$\dfrac{1}{15}$	x	$\dfrac{x}{15}$
Your friend can design the site in 10 hours.	Your friend	$\dfrac{1}{10}$	x	$\dfrac{x}{10}$

$$\begin{array}{ccccc} \boxed{\begin{matrix}\text{Fractional part of}\\\text{the job done by you}\end{matrix}} & + & \boxed{\begin{matrix}\text{fractional part of the}\\\text{job done by your friend}\end{matrix}} & = & \boxed{\begin{matrix}\text{one whole}\\\text{job.}\end{matrix}} \\ \dfrac{x}{15} & + & \dfrac{x}{10} & = & 1 \end{array}$$

Step 4. Solve the equation and answer the question.

$$\dfrac{x}{15} + \dfrac{x}{10} = 1 \qquad \text{This is the equation that models the problem's conditions.}$$

$$30\left(\dfrac{x}{15} + \dfrac{x}{10}\right) = 30 \cdot 1 \qquad \text{Multiply both sides by 30, the LCD.}$$

$$30 \cdot \dfrac{x}{15} + 30 \cdot \dfrac{x}{10} = 30 \qquad \text{Use the distributive property on the left side.}$$

$$2x + 3x = 30 \qquad \text{Simplify: } \overset{2}{\underset{1}{\cancel{30}}} \cdot \dfrac{x}{\underset{1}{\cancel{15}}} = 2x \text{ and } \overset{3}{\underset{1}{\cancel{30}}} \cdot \dfrac{x}{\underset{1}{\cancel{10}}} = 3x.$$

$$5x = 30 \qquad \text{Combine like terms.}$$

$$x = 6 \qquad \text{Divide both sides by 5.}$$

If you both work together, you can design your Web site in 6 hours.

Step 5. Check the proposed solution in the original wording of the problem. Will you both complete the job in 6 hours? Because you can design the site in 15 hours, in 6 hours, you can complete $\frac{6}{15}$, or $\frac{2}{5}$, of the job. Because your friend can design the site in 10 hours, in 6 hours, she can complete $\frac{6}{10}$, or $\frac{3}{5}$, of the job. Notice that $\frac{2}{5} + \frac{3}{5} = 1$, which represents the completion of the entire job, or one whole job. ∎

Great Question!

Is there an equation that models all the work problems in this section?
Yes. Let

a = the time it takes person A to do a job working alone, and

b = the time it takes person B to do the same job working alone.

If x represents the time it takes for A and B to complete the entire job working together, then the situation can be modeled by the rational equation

$$\frac{x}{a} + \frac{x}{b} = 1.$$

☑ **CHECK POINT 5**   A new underwater tunnel is being built using tunnel-boring machines that begin at opposite ends of the tunnel. One tunnel-boring machine can complete the tunnel in 18 months. A faster machine can tunnel to the other side in 9 months. If both machines start at opposite ends and work at the same time, in how many months will the tunnel be finished?

EXAMPLE 6   Solving a Problem Involving Work

After designing your Web site, you and your friend decide to go into business setting up sites for others. With lots of practice, you can now work together and design a modest site in 4 hours. Your friend is still a faster worker. Working alone, you require 6 more hours than she does to design a site for a client. How many hours does it take your friend to design a Web site if she works alone?

Solution

Step 1. Let x represent one of the unknown quantities. Let

x = the time, in hours, for your friend to design a Web site working alone.

Step 2. Represent other unknown quantities in terms of x. Because you require 6 more hours than your friend to do the job, let

$x + 6$ = the time, in hours, for you to design a Web site working alone.

Step 3. Write an equation that models the conditions. Working together, you and your friend can complete the job in 4 hours. We construct a table to find the fractional part of the task completed by you and your friend in 4 hours.

		Fractional part of job completed in 1 hour	Time working together	Fractional part of job completed in 4 hours
Your friend can design the site in x hours. →	Your friend	$\dfrac{1}{x}$	4	$\dfrac{4}{x}$
You can design the site in $x + 6$ hours. →	You	$\dfrac{1}{x + 6}$	4	$\dfrac{4}{x + 6}$

Because you can both complete the job in 4 hours,

Fractional part of the job done by your friend	+	fractional part of the job done by you	=	one whole job.

$$\frac{4}{x} + \frac{4}{x + 6} = 1.$$

Step 4. Solve the equation and answer the question.

$$\frac{4}{x} + \frac{4}{x + 6} = 1$$

This is the equation that models the problem's conditions.

$$x(x + 6)\left(\frac{4}{x} + \frac{4}{x + 6}\right) = x(x + 6) \cdot 1$$

Multiply both sides by $x(x + 6)$, the LCD.

$$x(x + 6) \cdot \frac{4}{x} + x(x + 6) \cdot \frac{4}{(x + 6)} = x^2 + 6x$$

Use the distributive property on each side.

$$4(x + 6) + 4x = x^2 + 6x$$

Simplify.

$$4x + 24 + 4x = x^2 + 6x$$

Use the distributive property.

$$8x + 24 = x^2 + 6x$$

Combine like terms: $4x + 4x = 8x$.

$$0 = x^2 - 2x - 24$$

Subtract $8x + 24$ from both sides and write the quadratic equation in standard form.

$$0 = (x - 6)(x + 4)$$

Factor.

$$x - 6 = 0 \quad \text{or} \quad x + 4 = 0$$

Set each factor equal to O.

$$x = 6 \qquad x = -4$$

Solve for x.

Because x represents the time for your friend to design a Web site working alone, we reject the negative value, -4. Your friend can design a Web site in 6 hours.

Step 5. Check the proposed solution in the original wording of the problem. Will you both complete the job in 4 hours? Working alone, your friend takes 6 hours. Because you require 6 more hours than your friend, you require 12 hours to complete the job on your own.

In 4 hours, your friend completes $\frac{4}{6} = \frac{2}{3}$ of the job.

In 4 hours, you complete $\frac{4}{12} = \frac{1}{3}$ of the job.

Notice that $\frac{2}{3} + \frac{1}{3} = 1$, which represents the completion of the entire job, or the design of one Web site. ∎

✓ **CHECK POINT 6** An experienced carpenter can panel a room 3 times faster than an apprentice can. Working together, they can panel the room in 6 hours. How long would it take each person working alone to do the job?

CONCEPT AND VOCABULARY CHECK

Fill in each blank so that the resulting statement is true.

1. The first step in solving $\frac{1}{x} + \frac{1}{y} = \frac{1}{z}$ for z is to multiply both sides of the equation by the least common denominator, which is _____.

2. The cost function for a business, $C(x)$, is the sum of the _____ cost and the _____ costs.

3. The average cost function for a company, $\overline{C}(x)$, is its cost function divided by the _____.

4. The formula $t = \frac{d}{r}$ states that time traveled is _____ divided by _____.

5. In work problems, the number _____ represents the whole job completed.

6. If you can complete a job in 19 hours, the fractional part of the job that you can complete in x hours is represented by _____.

6.7 EXERCISE SET ▶ MyMathLab®

Practice Exercises

In Exercises 1–14, solve each formula for the specified variable.

1. $\dfrac{V_1}{V_2} = \dfrac{P_2}{P_1}$ for P_1 (chemistry)

2. $\dfrac{V_1}{V_2} = \dfrac{P_2}{P_1}$ for V_2 (chemistry)

3. $\dfrac{1}{p} + \dfrac{1}{q} = \dfrac{1}{f}$ for f (optics)

4. $\dfrac{1}{p} + \dfrac{1}{q} = \dfrac{1}{f}$ for q (optics)

5. $P = \dfrac{A}{1 + r}$ for r (investment)

6. $S = \dfrac{a}{1 - r}$ for r (mathematics)

7. $F = \dfrac{Gm_1m_2}{d^2}$ for m_1 (physics)

8. $F = \dfrac{Gm_1m_2}{d^2}$ for m_2 (physics)

9. $z = \dfrac{x - \bar{x}}{s}$ for x (statistics)

10. $z = \dfrac{x - \bar{x}}{s}$ for s (statistics)

11. $I = \dfrac{E}{R + r}$ for R (electronics)

12. $I = \dfrac{E}{R + r}$ for r (electronics)

13. $f = \dfrac{f_1 f_2}{f_1 + f_2}$ for f_1 (optics)

14. $f = \dfrac{f_1 f_2}{f_1 + f_2}$ for f_2 (optics)

Application Exercises

The figure shows the graph of the average cost function for the company described in Check Point 3 that manufactures wheelchairs. Use the graph to solve Exercises 15–18.

Number of Wheelchairs Produced per Month

15. How many wheelchairs must be produced each month for the company to have an average cost of $410 per chair?

16. How many wheelchairs must be produced each month for the company to have an average cost of $425 per chair?

17. What is the equation of the horizontal asymptote shown by the dashed green line? What is the meaning of the horizontal asymptote as production level increases?

18. Describe the end behavior of the graph at the far right. Is there a production level that results in an average cost of $400 per chair? Explain your answer.

19. A company is planning to manufacture mountain bikes. Fixed monthly cost will be $100,000 and it will cost $100 to produce each bicycle.
   a. Write the cost function, C, of producing x mountain bikes.
   b. Write the average cost function, $\overline{C}$, of producing x mountain bikes.
   c. How many mountain bikes must be produced each month for the company to have an average cost of $300 per bike?

20. A company is planning to manufacture small canoes. Fixed monthly cost will be $20,000 and it will cost $20 to produce each canoe.
   a. Write the cost function, C, of producing x canoes.
   b. Write the average cost function, $\overline{C}$, of producing x canoes.
   c. How many canoes must be produced each month for the company to have an average cost of $40 per canoe?

It's vacation time. You drive 90 miles along a scenic highway and then take a 5-mile run along a hiking trail. Your driving rate is nine times that of your running rate. The graph shows the total time you spend driving and running, $f(x)$, as a function of your running rate, x. Use the graph of this rational function to solve Exercises 21–24.

$$f(x) = \frac{90}{9x} + \frac{5}{x}$$

Running Rate (miles per hour)

(Be sure to use the graph of the rational function at the bottom of the previous page to solve Exercises 21–24.)

21. If the total time for driving and running is 3 hours, what is your running rate?

22. If the total time for driving and running is 5 hours, what is your running rate?

23. Describe the behavior of the graph as x approaches 0. What does this show about the time driving and running as your running rate is close to zero miles per hour (a stupefied crawl)?

24. The graph is falling from left to right. What does this show?

Use a rational equation to solve Exercises 25–36. Each exercise is a problem involving motion.

25. A car can travel 300 miles in the same amount of time it takes a bus to travel 180 miles. If the rate of the bus is 20 miles per hour slower than the rate of the car, find the average rate for each.

26. A passenger train can travel 240 miles in the same amount of time it takes a freight train to travel 160 miles. If the rate of the freight train is 20 miles per hour slower than the rate of the passenger train, find the average rate of each.

27. You ride your bike to campus a distance of 5 miles and return home on the same route. Going to campus, you ride mostly downhill and average 9 miles per hour faster than on your return trip home. If the round trip takes one hour and ten minutes—that is, $\frac{7}{6}$ hours—what is your average rate on the return trip?

28. An engine pulls a train 140 miles. Then a second engine, whose average rate is 5 miles per hour faster than the first engine, takes over and pulls the train 200 miles. The total time required for both engines is 9 hours. Find the average rate of each engine.

29. In still water, a boat averages 7 miles per hour. It takes the same amount of time to travel 20 miles downstream, with the current, as 8 miles upstream, against the current. What is the rate of the water's current?

	Distance	Rate	Time = $\dfrac{\text{Distance}}{\text{Rate}}$
With the current	20	$7 + x$	
Against the current	8	$7 - x$	

These times are equal.

30. In still water, a boat averages 8 miles per hour. It takes the same amount of time to travel 30 miles downstream, with the current, as 18 miles upstream, against the current. What is the rate of the water's current?

	Distance	Rate	Time = $\dfrac{\text{Distance}}{\text{Rate}}$
With the current	30	$8 + x$	
Against the current	18	$8 - x$	

These times are equal.

31. The rate of the jet stream is 100 miles per hour. Traveling with the jet stream, an airplane can fly 2400 miles in the same amount of time as it takes to fly 1600 miles against the jet stream. What is the airplane's average rate in calm air?

32. The wind is blowing at an average rate of 10 miles per hour. Riding with the wind, a bicyclist can cycle 75 miles in the same amount of time it takes to cycle 15 miles against the wind. What is the bicyclist's average rate in calm air?

33. A moving sidewalk at an airport glides at a rate of 1.8 feet per second. Walking on the moving sidewalk, you travel 100 feet forward in the same time it takes to travel 40 feet in the opposite direction. Find your walking speed on a nonmoving sidewalk.

34. A moving sidewalk at an airport glides at a rate of 1.8 feet per second. Walking on the moving sidewalk, you travel 105 feet forward in the same time it takes to travel 50 feet in the opposite direction. Find your walking speed on a nonmoving sidewalk. Round to the nearest tenth.

35. Two runners, one averaging 8 miles per hour and the other 6 miles per hour, start at the same place and run along the same trail. The slower runner arrives at the end of the trail a half hour after the faster runner. How far did each person run?

36. Two sailboats, one averaging 20 miles per hour and the other 18 miles per hour, start at the same place and follow the same course. The slower boat arrives at the end of the course $\frac{1}{6}$ of an hour after the faster sailboat. How far did each boat travel?

Use a rational equation to solve Exercises 37–48. Each exercise is a problem involving work.

37. You promised your parents that you would wash the family car. You have not started the job and they are due home in 20 minutes. You can wash the car in 45 minutes and your sister claims she can do it in 30 minutes. If you work together, how long will it take to do the job? Will this give you enough time before your parents return?

38. You must leave for campus in half an hour, or you will be late for class. Unfortunately, you are snowed in. You can shovel the driveway in 45 minutes and your brother claims he can do it in 36 minutes. If you shovel together, how long will it take to clear the driveway? Will this give you enough time before you have to leave?

39. A pool can be filled by one pipe in 6 hours and by a second pipe in 12 hours. How long will it take using both pipes to fill the pool?

40. A pond can be filled by one pipe in 8 hours and by a second pipe in 24 hours. How long will it take using both pipes to fill the pond?

41. Working with your cousin, you can refinish a table in 3 hours. Working alone, your cousin can complete the job in 4 hours. How long would it take you to refinish the table working alone?

42. Working with your cousin, you can split a cord of firewood in 5 hours. Working alone, your cousin can complete the job in 7 hours. How long would it take you to split the firewood working alone?

43. An earthquake strikes and an isolated area is without food or water. Three crews arrive. One can dispense needed supplies in 20 hours, a second in 30 hours, and a third in 60 hours. How long will it take all three crews working together to dispense food and water?

44. A hurricane strikes and a rural area is without food or water. Three crews arrive. One can dispense needed supplies in 10 hours, a second in 15 hours, and a third in 20 hours. How long will it take all three crews working together to dispense food and water?

45. An office has an old copying machine and a new one. Working together, it takes both machines 6 hours to make all the copies of the annual financial report. Working alone, it takes the old copying machine 5 hours longer than the new one to make all the copies of the report. How long would it take the new copying machine to make all the copies working alone?

46. A demolition company wants to build a brick wall to hide from public view the area where they store wrecked cars. Working together, an experienced bricklayer and an apprentice can build the wall in 12 hours. Working alone, it takes the apprentice 10 hours longer than the experienced bricklayer to do the job. How long would it take the experienced bricklayer to build the wall working alone?

47. A faucet can fill a sink in 5 minutes. It takes twice as long for the drain to empty the sink. How long will it take to fill the sink if the drain is open and the faucet is on?

48. A pool can be filled by a pipe in 3 hours. It takes 3 times as long for another pipe to empty the pool. How long will it take to fill the pool if both pipes are open?

Exercises 49–56 contain a variety of problems. Use a rational equation to solve each exercise.

49. What number multiplied by the numerator and added to the denominator of $\frac{4}{5}$ makes the resulting fraction equivalent to $\frac{3}{2}$?

50. What number multiplied by the numerator and subtracted from the denominator of $\frac{9}{11}$ makes the resulting fraction equivalent to $-\frac{12}{5}$?

51. The sum of 2 times a number and twice its reciprocal is $\frac{20}{3}$. Find the number(s).

52. If 2 times the reciprocal of a number is subtracted from 3 times the number, the difference is 1. Find the number(s).

53. You have 35 hits in 140 times at bat. Your batting average is $\frac{35}{140}$, or 0.25. How many consecutive hits must you get to increase your batting average to 0.30?

54. You have 30 hits in 120 times at bat. Your batting average is $\frac{30}{120}$, or 0.25. How many consecutive hits must you get to increase your batting average to 0.28?

55. If one pipe can fill a pool in a hours and a second pipe can fill the pool in b hours, write a formula for the time, x, in terms of a and b, for the number of hours it takes both pipes, working together, to fill the pool.

56. If one pipe can fill a pool in a hours and a second pipe can empty the pool in b hours, write a formula for the time, x, in terms of a and b, for the number of hours it takes to fill the pool with both of these pipes open.

Explaining the Concepts

57. Without showing the details, explain how to solve the formula

$$\frac{1}{R} = \frac{1}{R_1} + \frac{1}{R_2}$$

for R_1. (The formula is used in electronics.)

58. Explain how to find the average cost function for a business.

59. How does the average cost function illustrate a problem for small businesses?

60. What is the relationship among time traveled, distance traveled, and rate of travel?

61. If you know how many hours it takes for you to do a job, explain how to find the fractional part of the job you can complete in x hours.

62. If you can do a job in 6 hours and your friend can do the same job in 3 hours, explain how to find how long it takes to complete the job working together. It is not necessary to solve the problem.

63. When two people work together to complete a job, describe one factor that can result in more or less time than the time given by the rational equations we have been using.

Technology Exercises

64. For Exercises 19–20, use a graphing utility to graph the average cost function described by the problem's conditions. Then TRACE along the curve and find the point that visually shows the solution in part (c).

65. For Exercises 45–46, use a graphing utility to graph the function representing the sum of the fractional parts of the job done by the two machines or the two people. Then TRACE along the curve and find the point that visually shows the problem's solution.

66. A boat can travel 10 miles per hour in still water. The boat travels 24 miles upstream, against the current, and then 24 miles downstream, with the current.

    a. Let x = the rate of the current. Write a function in terms of x that models the total time for the boat to travel upstream and downstream.

    b. Use a graphing utility to graph the rational function in part (a).

    c. TRACE along the curve and determine the current's rate if the trip's time is 5 hours. Then verify this result algebraically.

Critical Thinking Exercises

Make Sense? *Read this excerpt from an advertisement for bikes with aerodynamic coverings.*

> Our high-performance bicycles have the aerodynamic design of custom racing bikes, but are practical for everyday riding. The aerodynamic covering will increase your average speed by 10 miles per hour. Cyclists using our bicycles, versus bikes without aerodynamic coverings, reduced time on a 75-mile test run by 2 hours.

Now you are interested in finding the average rate of the bikes with the aerodynamic coverings on the 75-mile test run. With this goal in mind, determine whether each statement in Exercises 67–70 makes sense or does not make sense, and explain your reasoning.

67. I decided to organize the critical information from the advertisement in a table with the following entries:

	Distance	Rate	Time
With covering	75	$x + 10$	$\dfrac{75}{x + 10}$
Without covering	75	x	$\dfrac{75}{x}$

68. The ad stated that bikes with coverings reduced time on the 75-mile test run by 2 hours, so I used my table from Exercise 67 and modeled this condition with the rational equation
$$\frac{75}{x} = \frac{75}{x + 10} - 2.$$

69. The equation in x that modeled the conditions had a positive and a negative value for x, so I rejected the negative solution.

70. My professor verified that 15 is the correct value for x in the equation modeling the conditions, so I used my table from Exercise 67 and concluded that the average rate of the covered bikes on the 75-mile test run was 15 miles per hour.

In Exercises 71–74, determine whether each statement is true or false. If the statement is false, make the necessary change(s) to produce a true statement.

71. As production level increases, the average cost for a company to produce each unit of its product also increases.

72. To solve $qf + pf = pq$ for p, subtract qf from both sides and then divide by f.

73. If you plan a theater trip that costs \$300 to rent a limousine and \$25 per ticket, the cost per person, $f(x)$, for a group of x people is modeled by the rational function
$$f(x) = \frac{300 + 25x}{x}.$$

74. If you can clean the house in 3 hours and your sloppy friend can completely mess it up in 6 hours, then $\dfrac{x}{3} - \dfrac{x}{6} = 1$ can be used to find how long it takes to clean the house if you both "work" together.

75. Solve $\dfrac{1}{s} = f + \dfrac{1 - f}{p}$ for f.

76. A new schedule for a train requires it to travel 351 miles in $\frac{1}{4}$ hour less time than before. To accomplish this, the rate of the train must be increased by 2 miles per hour. What should the average rate of the train be so that it can keep on the new schedule?

77. It takes Mr. Todd 4 hours longer to prepare an order of pies than it takes Mrs. Lovett. They bake together for 2 hours when Mrs. Lovett leaves. Mr. Todd takes 7 additional hours to complete the work. Working alone, how long does it take Mrs. Lovett to prepare the pies?

Review Exercises

78. Factor: $x^2 + 4x + 4 - 9y^2$. (Section 5.6, Example 4)

79. Solve using matrices:
$$\begin{cases} 2x + 5y = -5 \\ x + 2y = -1. \end{cases}$$
(Section 3.4, Example 2)

80. Solve the system:
$$\begin{cases} x + y + z = 4 \\ 2x + 5y = 1 \\ x - y - 2x = 0. \end{cases}$$
(Section 3.3, Example 3)

Preview Exercises

Exercises 81–83 will help you prepare for the material covered in the next section.

81. a. If $y = kx^2$, find the value of k using $x = 2$ and $y = 64$.

   b. Substitute the value for k into $y = kx^2$ and write the resulting equation.

   c. Use the equation from part (b) to find y when $x = 5$.

82. a. If $y = \dfrac{k}{x}$, find the value of k using $x = 8$ and $y = 12$.

   b. Substitute the value for k into $y = \dfrac{k}{x}$ and write the resulting equation.

   c. Use the equation from part (b) to find y when $x = 3$.

83. If $S = \dfrac{kA}{P}$, find the value of k using $A = 60{,}000$, $P = 40$, and $S = 12{,}000$.

SECTION

6.8

Modeling Using Variation

What am I supposed to learn?

After studying this section, you should be able to:

1 Solve direct variation problems.

2 Solve inverse variation problems.

3 Solve combined variation problems.

4 Solve problems involving joint variation.

Have you ever wondered how telecommunication companies estimate the number of phone calls expected per day between two cities? The formula

$$C = \frac{0.02P_1P_2}{d^2}$$

shows that the daily number of phone calls, C, increases as the populations of the cities, P_1 and P_2, in thousands, increase, and decreases as the distance, d, between the cities increases.

Certain formulas occur so frequently in applied situations that they are given special names. Variation formulas show how one quantity changes in relation to other quantities. Quantities can vary *directly, inversely*, or *jointly*. In this section, we look at situations that can be modeled by each of these kinds of variation.

1 Solve direct variation problems.

Direct Variation

When you swim underwater, the pressure in your ears depends on the depth at which you are swimming. The formula

$$p = 0.43d$$

describes the water pressure, p, in pounds per square inch, at a depth of d feet. We can use this linear function to determine the pressure in your ears at various depths.

In each case, use $p = 0.43d$:

If $d = 20$, $p = 0.43(20) = 8.6$.   At a depth of 20 feet, water pressure is 8.6 pounds per square inch.

Doubling the depth doubles the pressure.

If $d = 40$, $p = 0.43(40) = 17.2$.   At a depth of 40 feet, water pressure is 17.2 pounds per square inch.

Doubling the depth doubles the pressure.

If $d = 80$, $p = 0.43(80) = 34.4$.   At a depth of 80 feet, water pressure is 34.4 pounds per square inch.

The formula $p = 0.43d$ illustrates that water pressure is a constant multiple of your underwater depth. If your depth is doubled, the pressure is doubled; if your depth is tripled, the pressure is tripled; and so on. Because of this, the pressure in your ears is said to **vary directly** as your underwater depth. The **equation of variation** is

$$p = 0.43d.$$

Generalizing, we obtain the following statement:

Direct Variation

If a situation is described by an equation in the form

$$y = kx,$$

where k is a nonzero constant, we say that **y varies directly as x** or **y is directly proportional to x**. The number k is called the **constant of variation** or the **constant of proportionality**.

Can you see that **the direct variation equation, $y = kx$, is a special case of the linear function $y = mx + b$?** When $m = k$ and $b = 0$, $y = mx + b$ becomes $y = kx$. Thus, the slope of a direct variation equation is k, the constant of variation. Because b, the y-intercept, is 0, the graph of a direct variation equation is a line passing through the origin. This is illustrated in **Figure 6.7**, which shows the graph of $p = 0.43d$: Water pressure varies directly as depth.

Figure 6.7 Water pressure at various depths

Problems involving direct variation can be solved using the following procedure. This procedure applies to direct variation problems, as well as to the other kinds of variation problems that we will discuss.

Solving Variation Problems

1. Write an equation that models the given English statement.
2. Substitute the given pair of values into the equation in step 1 and find the value of k, the constant of variation.
3. Substitute the value of k into the equation in step 1.
4. Use the equation from step 3 to answer the problem's question.

EXAMPLE 1   Solving a Direct Variation Problem

Many areas of Northern California depend on the snowpack of the Sierra Nevada mountain range for their water supply. The volume of water produced from melting snow varies directly as the volume of snow. Meteorologists have determined that 250 cubic centimeters of snow will melt to 28 cubic centimeters of water. How much water does 1200 cubic centimeters of melting snow produce?

Solution

Step 1. Write an equation. We know that *y varies directly as x* is expressed as

$$y = kx.$$

By changing letters, we can write an equation that models the following English statement: Volume of water, W, varies directly as volume of snow, S.

$$W = kS$$

Step 2. Use the given values to find k. We are told that 250 cubic centimeters of snow will melt to 28 cubic centimeters of water. Substitute 28 for W and 250 for S in the direct variation equation. Then solve for k.

$$W = kS \qquad \text{Volume of water varies directly as volume of melting snow.}$$

$$28 = k(250) \qquad \text{250 cubic centimeters of snow melt to 28 cubic centimeters of water.}$$

$$\frac{28}{250} = \frac{k(250)}{250} \qquad \text{Divide both sides by 250.}$$

$$0.112 = k \qquad \text{Simplify.}$$

Step 3. Substitute the value of k into the equation.

$$W = kS \qquad \text{This is the equation from step 1.}$$

$$W = 0.112S \qquad \text{Replace } k, \text{ the constant of variation, with 0.112.}$$

Step 4. Answer the problem's question. How much water does 1200 cubic centimeters of melting snow produce? Substitute 1200 for S in $W = 0.112S$ and solve for W.

$$W = 0.112S \qquad \text{Use the equation from step 3.}$$

$$W = 0.112(1200) \qquad \text{Substitute 1200 for } S.$$

$$W = 134.4 \qquad \text{Multiply.}$$

A snowpack measuring 1200 cubic centimeters will produce 134.4 cubic centimeters of water. ∎

✓ **CHECK POINT 1**   The number of gallons of water, W, used when taking a shower varies directly as the time, t, in minutes, in the shower. A shower lasting 5 minutes uses 30 gallons of water. How much water is used in a shower lasting 11 minutes?

The direct variation equation $y = kx$ is a linear function. If $k > 0$, then the slope of the line is positive. Consequently, as x increases, y also increases.

A direct variation situation can involve variables to higher powers. For example, y can vary directly as $x^2 (y = kx^2)$ or as $x^3 (y = kx^3)$.

Direct Variation with Powers

y varies directly as the nth power of x if there exists some nonzero constant k such that

$$y = kx^n.$$

We also say that **y is directly proportional to the nth power of x.**

Direct variation with exponents that are whole numbers is modeled by polynomial functions. In our next example, the graph of the variation equation is the graph of a quadratic function.

EXAMPLE 2   Solving a Direct Variation Problem

The distance, s, that a body falls from rest varies directly as the square of the time, t, of the fall. If skydivers fall 64 feet in 2 seconds, how far will they fall in 4.5 seconds?

Solution

Step 1. Write an equation. We know that *y varies directly as the square of x* is expressed as

$$y = kx^2.$$

By changing letters, we can write an equation that models the following English statement: Distance, s, varies directly as the square of time, t, of the fall.

$$s = kt^2$$

Step 2. Use the given values to find k. Skydivers fall 64 feet in 2 seconds. Substitute 64 for s and 2 for t in the direct variation equation. Then solve for k.

$s = kt^2$	Distance varies directly as the square of time.
$64 = k \cdot 2^2$	Skydivers fall 64 feet in 2 seconds.
$64 = 4k$	Simplify: $2^2 = 4$.
$\dfrac{64}{4} = \dfrac{4k}{4}$	Divide both sides by 4.
$16 = k$	Simplify.

Step 3. Substitute the value of k into the equation.

$s = kt^2$	Use the equation from step 1.
$s = 16t^2$	Replace k, the constant of variation, with 16.

Step 4. Answer the problem's question. How far will the skydivers fall in 4.5 seconds? Substitute 4.5 for t in $s = 16t^2$ and solve for s.

$$s = 16(4.5)^2 = 16(20.25) = 324$$

Thus, in 4.5 seconds, the skydivers will fall 324 feet. ■

We can express the variation equation from Example 2 in function notation, writing

$$s(t) = 16t^2.$$

The distance that a body falls from rest is a function of the time, t, of the fall. The graph of this quadratic function is shown in **Figure 6.8**. The graph increases rapidly from left to right, showing the effects of the acceleration of gravity.

Distance Skydivers Fall over Time

Distance the Skydivers Fall (feet)

Time the Skydivers Fall (seconds)

Figure 6.8 The graph of $s(t) = 16t^2$

✓ CHECK POINT 2   The distance required to stop a car varies directly as the square of its speed. If it requires 200 feet to stop a car traveling 60 miles per hour, how many feet are required to stop a car traveling 100 miles per hour?

2 Solve inverse variation problems. ▶

Inverse Variation

The distance from San Francisco to Los Angeles is 420 miles. The time that it takes to drive from San Francisco to Los Angeles depends on the average rate at which one drives and is given by

$$\text{Time} = \frac{420}{\text{Rate}}.$$

For example, if you average 30 miles per hour, the time for the drive is

$$\text{Time} = \frac{420}{30} = 14,$$

or 14 hours. If you average 50 miles per hour, the time for the drive is

$$\text{Time} = \frac{420}{50} = 8.4,$$

or 8.4 hours. As your rate (or speed) increases, the time for the trip decreases and vice versa. This is illustrated by the graph in **Figure 6.9**.

Figure 6.9

We can express the time for the San Francisco–Los Angeles trip using t for time and r for rate:

$$t = \frac{420}{r}.$$

This equation is an example of an **inverse variation** equation. Time, t, **varies inversely** as rate, r. When two quantities vary inversely, one quantity increases as the other decreases, and vice versa.

Generalizing, we obtain the following statement:

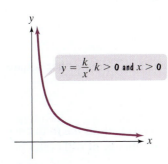

Figure 6.10 The graph of the inverse variation equation

Inverse Variation

If a situation is described by an equation in the form

$$y = \frac{k}{x},$$

where k is a nonzero constant, we say that **y varies inversely as x** or **y is inversely proportional to x.** The number k is called the **constant of variation**.

Notice that **the inverse variation equation**

$$y = \frac{k}{x}, \quad \text{or} \quad f(x) = \frac{k}{x},$$

is a **rational function.** For $k > 0$ and $x > 0$, the graph of the function takes on the shape shown in **Figure 6.10**.

We use the same procedure to solve inverse variation problems as we did to solve direct variation problems. Example 3 illustrates this procedure.

Doubling the pressure halves the volume.

EXAMPLE 3   Solving an Inverse Variation Problem

When you use a spray can and press the valve at the top, you decrease the pressure of the gas in the can. This decrease of pressure causes the volume of the gas in the can to increase. Because the gas needs more room than is provided in the can, it expands in spray form through the small hole near the valve. In general, if the temperature is constant, the pressure, P, of a gas in a container varies inversely as the volume, V, of the container. The pressure of a gas sample in a container whose volume is 8 cubic inches is 12 pounds per square inch. If the sample expands to a volume of 22 cubic inches, what is the new pressure of the gas?

Solution

Step 1. Write an equation. We know that y *varies inversely as x* is expressed as

$$y = \frac{k}{x}.$$

By changing letters, we can write an equation that models the following English statement: The pressure, P, of a gas in a container varies inversely as the volume, V.

$$P = \frac{k}{V}$$

Step 2. Use the given values to find *k*. The pressure of a gas sample in a container whose volume is 8 cubic inches is 12 pounds per square inch. Substitute 12 for P and 8 for V in the inverse variation equation. Then solve for k.

$$P = \frac{k}{V} \qquad \text{Pressure varies inversely as volume.}$$

$$12 = \frac{k}{8} \qquad \text{The pressure in an 8-cubic-inch container is 12 pounds per square inch.}$$

$$12 \cdot 8 = \frac{k}{8} \cdot 8 \qquad \text{Multiply both sides by 8.}$$

$$96 = k \qquad \text{Simplify.}$$

Step 3. Substitute the value of *k* into the equation.

$$P = \frac{k}{V} \qquad \text{Use the equation from step 1.}$$

$$P = \frac{96}{V} \qquad \text{Replace } k, \text{ the constant of variation, with 96.}$$

Step 4. Answer the problem's question. We need to find the pressure when the volume expands to 22 cubic inches. Substitute 22 for V and solve for P.

$$P = \frac{96}{V} = \frac{96}{22} = 4\frac{4}{11}$$

When the volume is 22 cubic inches, the pressure of the gas is $4\frac{4}{11}$ pounds per square inch. ∎

✓ **CHECK POINT 3**   The length of a violin string varies inversely as the frequency of its vibrations. A violin string 8 inches long vibrates at a frequency of 640 cycles per second. What is the frequency of a 10-inch string?

3 Solve combined variation problems. ▶

Combined Variation

In **combined variation**, direct variation and inverse variation occur at the same time. For example, as the advertising budget, A, of a company increases, its monthly sales, S, also increase. Monthly sales vary directly as the advertising budget:

$$S = kA.$$

By contrast, as the price of the company's product, P, increases, its monthly sales, S, decrease. Monthly sales vary inversely as the price of the product:

$$S = \frac{k}{P}.$$

We can combine these two variation equations into one equation:

$$S = \frac{kA}{P}.$$

Monthly sales, S, vary directly as the advertising budget, A, and inversely as the price of the product, P.

The following example illustrates an application of combined variation.

EXAMPLE 4   Solving a Combined Variation Problem

The owners of Rollerblades Now determined that the monthly sales, S, of its skates vary directly as its advertising budget, A, and inversely as the price of the skates, P. When \$60,000 is spent on advertising and the price of the skates is \$40, the monthly sales are 12,000 pairs of rollerblades.

a. Write an equation of variation that models this situation.

b. Determine monthly sales if the amount of the advertising budget is increased to \$70,000.

Solution

a. Write an equation:

$$S = \frac{kA}{P}.$$

Translate "sales vary directly as the advertising budget and inversely as the skates' price."

Use the given values to find k.

$$12,000 = \frac{k(60,000)}{40}$$

When \$60,000 is spent on advertising ($A = 60,000$) and the price is \$40 ($P = 40$), monthly sales are 12,000 units ($S = 12,000$).

$$12,000 = k \cdot 1500$$

Divide 60,000 by 40.

$$\frac{12,000}{1500} = \frac{k \cdot 1500}{1500}$$

Divide both sides of the equation by 1500.

$$8 = k$$

Simplify.

Therefore, the equation of variation that models monthly sales is

$$S = \frac{8A}{P}.$$

Substitute 8 for k in $S = \frac{kA}{P}$.

b. The advertising budget is increased to \$70,000, so $A = 70,000$. The skates' price is still \$40, so $P = 40$.

$$S = \frac{8A}{P}$$

This is the equation from part (a).

$$S = \frac{8(70,000)}{40}$$

Substitute 70,000 for A and 40 for P.

$$S = 14,000$$

Simplify.

With a \$70,000 advertising budget and \$40 price, the company can expect to sell 14,000 pairs of rollerblades in a month (up from 12,000). ∎

✓ **CHECK POINT 4**   The number of minutes needed to solve an Exercise Set of variation problems varies directly as the number of problems and inversely as the number of people working to solve the problems. It takes 4 people 32 minutes to solve 16 problems. How many minutes will it take 8 people to solve 24 problems?

4 Solve problems involving joint variation.

Joint Variation

Joint variation is a variation in which a variable varies directly as the product of two or more other variables. Thus, the equation $y = kxz$ is read "y varies jointly as x and z."

Joint variation plays a critical role in Isaac Newton's formula for gravitation:

$$F = G\frac{m_1 m_2}{d^2}.$$

The formula states that the force of gravitation, F, between two bodies varies jointly as the product of their masses, m_1 and m_2, and inversely as the square of the distance between them, d. (G is the gravitational constant.) The formula indicates that gravitational force exists between any two objects in the universe, increasing as the distance between the bodies decreases. One practical result is that the pull of the moon on the oceans is greater on the side of Earth closer to the moon. This gravitational imbalance is what produces tides.

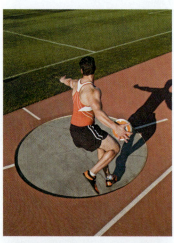

EXAMPLE 5 Modeling Centrifugal Force

The centrifugal force, C, of a body moving in a circle varies jointly with the radius of the circular path, r, and the body's mass, m, and inversely with the square of the time, t, it takes to move about one full circle. A 6-gram body moving in a circle with radius 100 centimeters at a rate of 1 revolution in 2 seconds has a centrifugal force of 6000 dynes. Find the centrifugal force of an 18-gram body moving in a circle with radius 100 centimeters at a rate of 1 revolution in 3 seconds.

Solution

$$C = \frac{krm}{t^2}$$

Translate "Centrifugal force, C, varies jointly with radius, r, and mass, m, and inversely with the square of time, t."

$$6000 = \frac{k(100)(6)}{2^2}$$

A 6-gram body ($m = 6$) moving in a circle with radius 100 centimeters ($r = 100$) at 1 revolution in 2 seconds ($t = 2$) has a centrifugal force of 6000 dynes ($C = 6000$).

$$6000 = 150k$$

Simplify: $\frac{100(6)}{2^2} = \frac{600}{4} = 150.$

$$40 = k$$

Divide both sides by 150 and solve for k.

$$C = \frac{40rm}{t^2}$$

Substitute 40 for k in the model for centrifugal force.

$$C = \frac{40(100)(18)}{3^2}$$

Find centrifugal force, C, of an 18-gram body ($m = 18$) moving in a circle with radius 100 centimeters ($r = 100$) at 1 revolution in 3 seconds ($t = 3$).

$$= 8000$$

Simplify.

The centrifugal force is 8000 dynes. ∎

✓ **CHECK POINT 5** The volume of a cone, V, varies jointly as its height, h, and the square of its radius, r. A cone with a radius measuring 6 feet and a height measuring 10 feet has a volume of 120π cubic feet. Find the volume of a cone having a radius of 12 feet and a height of 2 feet.

CONCEPT AND VOCABULARY CHECK

Fill in each blank so that the resulting statement is true.

1. y varies directly as x can be modeled by the equation _____, where k is called the _____.

2. y varies directly as the nth power of x can be modeled by the equation _____.

3. y varies inversely as x can be modeled by the equation _____.

4. y varies directly as x and inversely as z can be modeled by the equation _____.

5. y varies jointly as x and z can be modeled by the equation _____.

6. In the equation $S = \dfrac{8A}{P}$, S varies _____ as A and _____ as P.

7. In the equation $C = \dfrac{0.02P_1P_2}{d^2}$, C varies _____ as P_1 and P_2 and _____ as the square of d.

6.8 EXERCISE SET   MyMathLab®

Practice Exercises

Use the four-step procedure for solving variation problems given on page 480 to solve Exercises 1–10.

1. y varies directly as x. $y = 65$ when $x = 5$. Find y when $x = 12$.

2. y varies directly as x. $y = 45$ when $x = 5$. Find y when $x = 13$.

3. y varies inversely as x. $y = 12$ when $x = 5$. Find y when $x = 2$.

4. y varies inversely as x. $y = 6$ when $x = 3$. Find y when $x = 9$.

5. y varies directly as x and inversely as the square of z. $y = 20$ when $x = 50$ and $z = 5$. Find y when $x = 3$ and $z = 6$.

6. a varies directly as b and inversely as the square of c. $a = 7$ when $b = 9$ and $c = 6$. Find a when $b = 4$ and $c = 8$.

7. y varies jointly as x and z. $y = 25$ when $x = 2$ and $z = 5$. Find y when $x = 8$ and $z = 12$.

8. C varies jointly as A and T. $C = 175$ when $A = 2100$ and $T = 4$. Find C when $A = 2400$ and $T = 6$.

9. y varies jointly as a and b, and inversely as the square root of c. $y = 12$ when $a = 3$, $b = 2$, and $c = 25$. Find y when $a = 5$, $b = 3$, and $c = 9$.

10. y varies jointly as m and the square of n, and inversely as p. $y = 15$ when $m = 2, n = 1$, and $p = 6$. Find y when $m = 3, n = 4$, and $p = 10$.

Practice PLUS

In Exercises 11–20, write an equation that expresses each relationship. Then solve the equation for y.

11. x varies jointly as y and z.

12. x varies jointly as y and the square of z.

13. x varies directly as the cube of z and inversely as y.

14. x varies directly as the cube root of z and inversely as y.

15. x varies jointly as y and z and inversely as the square root of w.

16. x varies jointly as y and z and inversely as the square of w.

17. x varies jointly as z and the sum of y and w.

18. x varies jointly as z and the difference between y and w.

19. x varies directly as z and inversely as the difference between y and w.

20. x varies directly as z and inversely as the sum of y and w.

Application Exercises

Use the four-step procedure for solving variation problems given on page 480 to solve Exercises 21–28.

21. An alligator's tail length, T, varies directly as its body length, B. An alligator with a body length of 4 feet has a tail length of 3.6 feet. What is the tail length of an alligator whose body length is 6 feet?

|← —— Body length, B —— →|← —— Tail length, T —— →|

22. An object's weight on the moon, M, varies directly as its weight on Earth, E. Neil Armstrong, the first person to step on the moon on July 20, 1969, weighed 360 pounds on Earth (with all of his equipment on) and 60 pounds on the moon. What is the moon weight of a person who weighs 186 pounds on Earth?

23. The height that a ball bounces varies directly as the height from which it was dropped. A tennis ball dropped from 12 inches bounces 8.4 inches. From what height was the tennis ball dropped if it bounces 56 inches?

24. The distance that a spring will stretch varies directly as the force applied to the spring. A force of 12 pounds is needed to stretch a spring 9 inches. What force is required to stretch the spring 15 inches?

25. If all men had identical body types, their weight would vary directly as the cube of their height. Shown below is Robert Wadlow, who reached a record height of 8 feet 11 inches (107 inches) before his death at age 22. If a man who is 5 feet 10 inches tall (70 inches) with the same body type as Mr. Wadlow weighs 170 pounds, what was Robert Wadlow's weight shortly before his death?

26. On a dry asphalt road, a car's stopping distance varies directly as the square of its speed. A car traveling at 45 miles per hour can stop in 67.5 feet. What is the stopping distance for a car traveling at 60 miles per hour?

27. The figure shows that a bicyclist tips the cycle when making a turn. The angle B, formed by the vertical direction and the bicycle, is called the banking angle. The banking angle varies inversely as the cycle's turning radius. When the turning radius is 4 feet, the banking angle is 28°. What is the banking angle when the turning radius is 3.5 feet?

28. The water temperature of the Pacific Ocean varies inversely as the water's depth. At a depth of 1000 meters, the water temperature is 4.4° Celsius. What is the water temperature at a depth of 5000 meters?

Heart rates and life spans of most mammals can be modeled using inverse variation. The bar graph shows the average heart rate and the average life span of five mammals. You will use the data to solve Exercises 29–30.

Heart Rate and Life Span

29. a. A mammal's average life span, L, in years, varies inversely as its average heart rate, R, in beats per minute. Use the data shown for horses to write the equation that models this relationship.

b. Is the inverse variation equation in part (a) an exact model or an approximate model for the data shown for lions?

c. Elephants have an average heart rate of 27 beats per minute. Determine their average life span.

30. **a.** A mammal's average life span, L, in years, varies inversely as its average heart rate, R, in beats per minute. Use the data shown for cats at the bottom of the previous page to write the equation that models this relationship.

b. Is the inverse variation equation in part (a) an exact model or an approximate model for the data shown for squirrels?

c. Mice have an average heart rate of 634 beats per minute. Determine their average life span, rounded to the nearest year.

The figure shows the graph of the inverse variation model that you wrote in Exercise 29(a) or Exercise 30(a). Use the graph of this rational function to solve Exercises 31–32.

The Graph of an Inverse Variation Model for Heart Rate and Life Span

31. **a.** If a mammal has a life span of 20 years, use the graph to estimate its heart rate, rounded to the nearest 10 beats per minute.

b. Use the inverse variation equation that you wrote in Exercise 29(a) or Exercise 30(a) to determine the heart rate, rounded to the nearest beat per minute, for a mammal with a life span of 20 years.

c. The bar graph at the bottom of the previous page uses two bars to display the data for horses. How is this information shown on the graph of the inverse variation model?

32. **a.** If a mammal has a life span of 50 years, use the graph to estimate its heart rate, rounded to the nearest 10 beats per minute.

b. Use the inverse variation equation that you wrote in Exercise 29(a) or Exercise 30(a) to determine the heart rate, rounded to the nearest beat per minute, for a mammal with a life span of 50 years.

c. The bar graph at the bottom of the previous page uses two bars to display the data for lions. How is this information shown on the graph of the inverse variation model?

Continue to use the four-step procedure for solving variation problems given on page 480 to solve Exercises 33–40.

33. Radiation machines, used to treat tumors, produce an intensity of radiation that varies inversely as the square of the distance from the machine. At 3 meters, the radiation intensity is 62.5 milliroentgens per hour. What is the intensity at a distance of 2.5 meters?

34. The illumination provided by a car's headlight varies inversely as the square of the distance from the headlight. A car's headlight produces an illumination of 3.75 footcandles at a distance of 40 feet. What is the illumination when the distance is 50 feet?

35. Body-mass index, or BMI, takes both weight and height into account when assessing whether an individual is underweight or overweight. BMI varies directly as one's weight, in pounds, and inversely as the square of one's height, in inches. In adults, normal values for the BMI are between 20 and 25, inclusive. Values below 20 indicate that an individual is underweight and values above 30 indicate that an individual is obese. A person who weighs 180 pounds and is 5 feet, or 60 inches, tall has a BMI of 35.15. What is the BMI, to the nearest tenth, for a 170 pound person who is 5 feet 10 inches tall. Is this person overweight?

36. One's intelligence quotient, or IQ, varies directly as a person's mental age and inversely as that person's chronological age. A person with a mental age of 25 and a chronological age of 20 has an IQ of 125. What is the chronological age of a person with a mental age of 40 and an IQ of 80?

37. The heat loss of a glass window varies jointly as the window's area and the difference between the outside and inside temperatures. A window 3 feet wide by 6 feet long loses 1200 Btu per hour when the temperature outside is 20° colder than the temperature inside. Find the heat loss through a glass window that is 6 feet wide by 9 feet long when the temperature outside is 10° colder than the temperature inside.

38. Kinetic energy varies jointly as the mass and the square of the velocity. A mass of 8 grams and velocity of 3 centimeters per second has a kinetic energy of 36 ergs. Find the kinetic energy for a mass of 4 grams and velocity of 6 centimeters per second.

39. Sound intensity varies inversely as the square of the distance from the sound source. If you are in a movie theater and you change your seat to one that is twice as far from the speakers, how does the new sound intensity compare to that of your original seat?

40. Many people claim that as they get older, time seems to pass more quickly. Suppose that the perceived length of a period of time is inversely proportional to your age. How long will a year seem to be when you are three times as old as you are now?

41. The average number of daily phone calls, C, between two cities varies jointly as the product of their populations, P_1 and P_2, and inversely as the square of the distance, d, between them.

a. Write an equation that expresses this relationship.

b. The distance between San Francisco (population: 777,000) and Los Angeles (population: 3,695,000) is 420 miles. If the average number of daily phone calls between the cities is 326,000, find the value of k to two decimal places and write the equation of variation.

c. Memphis (population: 650,000) is 400 miles from New Orleans (population: 490,000). Find the average number of daily phone calls, to the nearest whole number, between these cities.

42. The force of wind blowing on a window positioned at a right angle to the direction of the wind varies jointly as the area of the window and the square of the wind's speed. It is known that a wind of 30 miles per hour blowing on a window measuring 4 feet by 5 feet exerts a force of 150 pounds. During a storm with winds of 60 miles per hour, should hurricane shutters be placed on a window that measures 3 feet by 4 feet and is capable of withstanding 300 pounds of force?

43. The table shows the values for the current, I, in an electric circuit and the resistance, R, of the circuit.

I (amperes)	0.5	1.0	1.5	2.0	2.5	3.0	4.0	5.0
R (ohms)	12	6.0	4.0	3.0	2.4	2.0	1.5	1.2

a. Graph the ordered pairs in the table of values, with values of I along the x-axis and values of R along the y-axis. Connect the eight points with a smooth curve.

b. Does current vary directly or inversely as resistance? Use your graph and explain how you arrived at your answer.

c. Write an equation of variation for I and R, using one of the ordered pairs in the table to find the constant of variation. Then use your variation equation to verify the other seven ordered pairs in the table.

Explaining the Concepts

44. What does it mean if two quantities vary directly?

45. In your own words, explain how to solve a variation problem.

46. What does it mean if two quantities vary inversely?

47. Explain what is meant by combined variation. Give an example with your explanation.

48. Explain what is meant by joint variation. Give an example with your explanation.

In Exercises 49–50, describe in words the variation shown by the given equation.

49. $z = \dfrac{k\sqrt{x}}{y^2}$

50. $z = kx^2\sqrt{y}$

51. We have seen that the daily number of phone calls between two cities varies jointly as their populations and inversely as the square of the distance between them. This model, used by telecommunication companies to estimate the line capacities needed among various cities, is called the *gravity model*. Compare the model to Newton's formula for gravitation on page 486 and describe why the name *gravity model* is appropriate.

Technology Exercises

52. Use a graphing utility to graph any three of the variation equations in Exercises 21–28. Then ┃TRACE┃ along each curve and identify the point that corresponds to the problem's solution.

Critical Thinking Exercises

Make Sense? *In Exercises 53–56, determine whether each statement makes sense or does not make sense, and explain your reasoning.*

53. I'm using an inverse variation equation and I need to determine the value of the dependent variable when the independent variable is zero.

54. The graph of this direct variation equation has a positive constant of variation and shows one variable increasing as the other variable decreases.

55. When all is said and done, it seems to me that direct variation equations are special kinds of linear functions and inverse variation equations are special kinds of rational functions.

56. Using the language of variation, I can now state the formula for the area of a trapezoid, $A = \frac{1}{2}h(b_1 + b_2)$, as, "A trapezoid's area varies jointly with its height and the sum of its bases."

57. In a hurricane, the wind pressure varies directly as the square of the wind velocity. If wind pressure is a measure of a hurricane's destructive capacity, what happens to this destructive power when the wind speed doubles?

58. The heat generated by a stove element varies directly as the square of the voltage and inversely as the resistance. If the voltage remains constant, what needs to be done to triple the amount of heat generated?

59. Galileo's telescope brought about revolutionary changes in astronomy. A comparable leap in our ability to observe the universe took place as a result of the Hubble Space Telescope. The space telescope can see stars and galaxies whose brightness is $\frac{1}{50}$ of the faintest objects now observable using ground-based telescopes. Use the fact that the brightness of a point source, such as a star, varies inversely as the square of its distance from an observer to show that the space telescope can see about seven times farther than a ground-based telescope.

Review Exercises

60. Evaluate:
$$\begin{vmatrix} -1 & 2 \\ 3 & -4 \end{vmatrix}.$$

(Section 3.5, Example 1)

61. Factor completely:
$$x^2y - 9y - 3x^2 + 27.$$

(Section 5.6, Example 3)

62. Find the degree of
$$7xy + x^2y^2 - 5x^3 - 7.$$

(Section 5.1, Example 1)

Preview Exercises

Exercises 63–65 will help you prepare for the material covered in the first section of the next chapter.

63. If $f(x) = \sqrt{3x + 12}$, find $f(-1)$.

64. If $f(x) = \sqrt{3x + 12}$, find $f(8)$.

65. Use the graph of $f(x) = \sqrt{3x + 12}$ to identify the function's domain and its range.

$f(x) = \sqrt{3x + 12}$

Chapter 6 Summary

Definitions and Concepts	Examples

Section 6.1 Rational Expressions and Functions: Multiplying and Dividing

A rational expression consists of a polynomial divided by a nonzero polynomial. A rational function is defined by a formula that is a rational expression. The domain of a rational function is the set of all real numbers except those for which the denominator is zero. Graphs of rational functions often contain disconnected branches. The graphs often approach, but do not touch, vertical or horizontal lines, called vertical asymptotes and horizontal asymptotes, respectively.

Let $f(x) = \dfrac{x + 2}{x^2 + x - 6}$. Find the domain of f.

$$x^2 + x - 6 = 0$$
$$(x + 3)(x - 2) = 0$$
$$x + 3 = 0 \quad \text{or} \quad x - 2 = 0$$
$$x = -3 \qquad\qquad x = 2$$

These values of x make the denominator zero. Exclude them from the domain.

Domain of $f = (-\infty, -3) \cup (-3, 2) \cup (2, \infty)$

The branches of the graph of f break at -3 and 2.

Vertical asymptote

Vertical asymptote

Simplifying Rational Expressions

1. Factor the numerator and the denominator completely.

2. Divide both the numerator and the denominator by any common factors.

Simplify: $\dfrac{4x + 28}{x^2 - 49}$.

$$\dfrac{4x + 28}{x^2 - 49} = \dfrac{4(x + 7)}{(x + 7)(x - 7)} = \dfrac{4}{x - 7}$$

Definitions and Concepts	**Examples**

Section 6.6 Rational Equations

A rational equation is an equation containing one or more rational expressions.

Solving Rational Equations

1. List restrictions on the variable.
2. Clear fractions by multiplying both sides by the LCD.
3. Solve the resulting equation.
4. Reject any proposed solution in the list of restrictions. Check other proposed solutions in the original equation.

Solve: $\dfrac{7x}{x^2 - 4} + \dfrac{5}{x - 2} = \dfrac{2x}{x^2 - 4}$.

$$\frac{7x}{(x + 2)(x - 2)} + \frac{5}{x - 2} = \frac{2x}{(x + 2)(x - 2)}$$

Denominators would equal 0 if $x = -2$ or $x = 2$. Restrictions: $x \neq -2$ and $x \neq 2$.

LCD is $(x + 2)(x - 2)$.

$$(x + 2)(x - 2)\left[\frac{7x}{(x + 2)(x - 2)} + \frac{5}{x - 2}\right] =$$

$$(x + 2)(x - 2) \cdot \frac{2x}{(x + 2)(x - 2)}$$

$$7x + 5(x + 2) = 2x$$

$$7x + 5x + 10 = 2x$$

$$12x + 10 = 2x$$

$$10 = -10x$$

$$-1 = x$$

The proposed solution, -1, is not part of the restriction $x \neq -2$ and $x \neq 2$. It checks. The solution is -1 and the solution set is $\{-1\}$.

Section 6.7 Formulas and Applications of Rational Equations

To solve a formula for a variable, get the specified variable alone on one side of the formula. When working with formulas containing rational expressions, it is sometimes necessary to factor out the variable you are solving for.

Solve: $\dfrac{e}{E} = \dfrac{r}{r + R}$ for r.

$$E(r + R) \cdot \frac{e}{E} = E(r + R) \cdot \frac{r}{r + R} \qquad \text{LCD is } E(r + R).$$

$$e(r + R) = Er$$

$$er + eR = Er$$

$$eR = Er - er$$

$$eR = (E - e)r$$

$$\frac{eR}{E - e} = r$$

The average cost function, $\overline{C}$, for a business is a rational function representing the average cost for the company to produce each unit of its product. The function consists of the sum of fixed and variable costs divided by the number of units produced. As production level increases, the average cost to produce each unit of a product decreases.

A company has a fixed cost of \$80,000 monthly and it costs \$20 to produce each unit of its product. Its cost function, C, of producing x units is

$$C(x) = 80,000 + 20x.$$

Its average cost function, $\overline{C}$, is

$$\overline{C}(x) = \frac{80,000 + 20x}{x}.$$

Definitions and Concepts	**Examples**

Section 6.7 Formulas and Applications of Rational Equations (continued)

Problems involving time in motion and problems involving work translate into rational equations. Motion problems involving time are solved using

$$\text{Time traveled} = \frac{\text{Distance traveled}}{\text{Rate of travel}}.$$

Work problems are solved using the following condition:

$$\boxed{\text{Fraction of job done by first}} + \boxed{\text{fraction of job done by second}} = \boxed{1}.$$

It takes a cyclist who averages 16 miles per hour in still air the same time to travel 48 miles with the wind as 16 miles against the wind. What is the wind's rate?

$$x = \text{wind's rate}$$
$$16 + x = \text{cyclist's rate with wind}$$
$$16 - x = \text{cyclist's rate against wind}$$

	Distance	Rate	Time = $\dfrac{\text{Distance}}{\text{Rate}}$
With wind	48	$16 + x$	$\dfrac{48}{16 + x}$
Against wind	16	$16 - x$	$\dfrac{16}{16 - x}$

These times are equal.

$$(16 + x)(16 - x) \cdot \frac{48}{16 + x} = \frac{16}{16 - x} \cdot (16 + x)(16 - x)$$
$$48(16 - x) = 16(16 + x)$$

Solving this equation, $x = 8$. The wind's rate is 8 miles per hour.

Section 6.8 Modeling Using Variation

English Statement	**Equation**
y varies directly as x. y is directly proportional to x.	$y = kx$
y varies directly as x^n. y is directly proportional to x^n.	$y = kx^n$
y varies inversely as x. y is inversely proportional to x.	$y = \dfrac{k}{x}$
y varies inversely as x^n. y is inversely proportional to x^n.	$y = \dfrac{k}{x^n}$
y varies directly as x and inversely as z.	$y = \dfrac{kx}{z}$
y varies jointly as x and z.	$y = kxz$

Solving Variation Problems

1. Write an equation that models the given English statement.
2. Substitute the pair of values into the equation in step 1 and find k.
3. Substitute k into the equation in step 1.
4. Use the equation in step 3 to answer the problem's question.

The time that it takes you to drive a certain distance varies inversely as your driving rate. Averaging 40 miles per hour, it takes you 10 hours to drive the distance. How long would the trip take averaging 50 miles per hour?

1.
$$t = \frac{k}{r}$$
Time, t, varies inversely as rate, r.

2. It takes 10 hours at 40 miles per hour.

$$10 = \frac{k}{40}$$
$$k = 10(40) = 400$$

3. $t = \dfrac{400}{r}$

4. How long at 50 miles per hour? Substitute 50 for r.

$$t = \frac{400}{50} = 8$$

It takes 8 hours at 50 miles per hour.

6.7 *In Exercises 57–61, solve each formula for the specified variable.*

57. $P = \dfrac{R - C}{n}$ for C

58. $\dfrac{P_1 V_1}{T_1} = \dfrac{P_2 V_2}{T_2}$ for T_1

59. $T = \dfrac{A - P}{Pr}$ for P

60. $\dfrac{1}{R} = \dfrac{1}{R_1} + \dfrac{1}{R_2}$ for R

61. $I = \dfrac{nE}{R + nr}$ for n

62. A company is planning to manufacture affordable graphing calculators. Fixed monthly cost will be $50,000, and it will cost $25 to produce each calculator.

   a. Write the cost function, C, of producing x graphing calculators.

   b. Write the average cost function, $\overline{C}$, of producing x graphing calculators.

   c. How many graphing calculators must be produced each month for the company to have an average cost of $35 per graphing calculator?

63. After riding at a steady rate for 60 miles, a bicyclist had a flat tire and walked 8 miles to a repair shop. The cycling rate was 3 times faster than the walking rate. If the time spent cycling and walking was 7 hours, at what rate was the cyclist riding?

64. The current of a river moves at 3 miles per hour. It takes a boat a total of 3 hours to travel 12 miles upstream, against the current, and return the same distance traveling downstream, with the current. What is the boat's rate in still water?

65. Working alone, two people can clean their house in 3 hours and 6 hours, respectively. They have agreed to clean together so that they can finish in time to watch a TV program that begins in $1\frac{1}{2}$ hours. How long will it take them to clean the house working together? Can they finish before the program starts?

66. Working together, two crews can clear snow from the city's streets in 20 hours. Working alone, the faster crew requires 9 hours less time than the slower crew. How many hours would it take each crew to clear the streets working alone?

67. An inlet faucet can fill a small pond in 60 minutes. The pond can be emptied by an outlet pipe in 80 minutes. You begin filling the empty pond. By accident, the outlet pipe that empties the pond is left open. Under these conditions, how long will it take for the pond to fill?

6.8 *Solve the variation problems in Exercises 68–73.*

68. A company's profit varies directly as the number of products it sells. The company makes a profit of $1175 on the sale of 25 products. What is the company's profit when it sells 105 products?

69. The distance that a body falls from rest varies directly as the square of the time of the fall. If skydivers fall 144 feet in 3 seconds, how far will they fall in 10 seconds?

70. The pitch of a musical tone varies inversely as its wavelength. A tone has a pitch of 660 vibrations per second and a wavelength of 1.6 feet. What is the pitch of a tone that has a wavelength of 2.4 feet?

71. The loudness of a stereo speaker, measured in decibels, varies inversely as the square of your distance from the speaker. When you are 8 feet from the speaker, the loudness is 28 decibels. What is the loudness when you are 4 feet from the speaker?

72. The time required to assemble computers varies directly as the number of computers assembled and inversely as the number of workers. If 30 computers can be assembled by 6 workers in 10 hours, how long would it take 5 workers to assemble 40 computers?

73. The volume of a pyramid varies jointly as its height and the area of its base. A pyramid with a height of 15 feet and a base with an area of 35 square feet has a volume of 175 cubic feet. Find the volume of a pyramid with a height of 20 feet and a base with an area of 120 square feet.

CHAPTER 6 TEST

Step-by-step test solutions are found on the Chapter Test Prep Videos available in MyMathLab® or on YouTube (search "BlitzerInterAlg7e" and click on "Channels").

1. Find the domain of $f(x) = \dfrac{x^2 - 2x}{x^2 - 7x + 10}$. Then simplify the right side of the function's equation.

In Exercises 2–11, perform the indicated operations. Simplify where possible.

2. $\dfrac{x^2}{x^2 - 16} \cdot \dfrac{x^2 + 7x + 12}{x^2 + 3x}$

3. $\dfrac{x^3 + 27}{x^2 - 1} \div \dfrac{x^2 - 3x + 9}{x^2 - 2x + 1}$

4. $\dfrac{x^2 + 3x - 10}{x^2 + 4x + 3} \cdot \dfrac{x^2 + x - 6}{x^2 + 10x + 25} \cdot \dfrac{x + 1}{x - 2}$

5. $\dfrac{x^2 - 6x - 16}{x^3 + 3x^2 + 2x} \cdot (x^2 - 3x - 4) \div \dfrac{x^2 - 7x + 12}{3x}$

6. $\dfrac{x^2 - 5x - 2}{6x^2 - 11x - 35} - \dfrac{x^2 - 7x + 5}{6x^2 - 11x - 35}$

7. $\dfrac{x}{x + 3} + \dfrac{5}{x - 3}$

8. $\dfrac{2}{x^2 - 4x + 3} + \dfrac{3x}{x^2 + x - 2}$

9. $\dfrac{5x}{x^2 - 4} - \dfrac{2}{x^2 + x - 2}$

10. $\dfrac{x - 4}{x - 5} - \dfrac{3}{x + 5} - \dfrac{10}{x^2 - 25}$

11. $\dfrac{1}{10 - x} + \dfrac{x - 1}{x - 10}$

In Exercises 12–13, simplify each rational expression.

12. $\dfrac{\dfrac{x}{4} - \dfrac{1}{x}}{1 + \dfrac{x + 4}{x}}$

13. $\dfrac{\dfrac{1}{x} - \dfrac{3}{x + 2}}{\dfrac{2}{x^2 + 2x}}$

In Exercises 14–16, divide as indicated.

14. $(12x^4y^3 + 16x^2y^3 - 10x^2y^2) \div (4x^2y)$

15. $(9x^3 - 3x^2 - 3x + 4) \div (3x + 2)$

16. $(3x^4 + 2x^3 - 8x + 6) \div (x^2 - 1)$

17. Divide using synthetic division:
$$(3x^4 + 11x^3 - 20x^2 + 7x + 35) \div (x + 5).$$

18. Given that
$$f(x) = x^4 - 2x^3 - 11x^2 + 5x + 34,$$
use synthetic division and the Remainder Theorem to find $f(-2)$.

19. Use synthetic division to decide whether -2 is a solution of $2x^3 - 3x^2 - 11x + 6 = 0$.

In Exercises 20–21, solve each rational equation.

20. $\dfrac{x}{x + 4} = \dfrac{11}{x^2 - 16} + 2$

21. $\dfrac{x + 1}{x^2 + 2x - 3} - \dfrac{1}{x + 3} = \dfrac{1}{x - 1}$

22. Park rangers introduce 50 elk into a wildlife preserve. The function
$$f(t) = \dfrac{250(3t + 5)}{t + 25}$$
models the elk population, $f(t)$, after t years. How many years will it take for the population to increase to 125 elk?

23. Solve for a: $R = \dfrac{as}{a + s}$.

24. A company is planning to manufacture portable satellite radio players. Fixed monthly cost will be \$300,000 and it will cost \$10 to produce each player.
    a. Write the cost function, C, of producing x players.
    b. Write the average cost function, $\overline{C}$, of producing x players.
    c. How many portable satellite radio players must be produced each month for the company to have an average cost of \$25 per player?

25. It takes one pipe 3 hours to fill a pool and a second pipe 4 hours to drain the pool. The pool is empty and the first pipe begins to fill it. The second pipe is accidently left open, so the water is also draining out of the pool. Under these conditions, how long will it take to fill the pool?

26. A motorboat averages 20 miles per hour in still water. It takes the boat the same amount of time to travel 3 miles with the current as it does to travel 2 miles against the current. What is the current's rate?

27. The intensity of light received at a source varies inversely as the square of the distance from the source. A particular light has an intensity of 20 footcandles at 15 feet. What is the light's intensity at 10 feet?

CUMULATIVE REVIEW EXERCISES (CHAPTERS 1–6)

In Exercises 1–5, solve each equation, inequality, or system.

1. $2x + 5 \leq 11$ and $-3x > 18$

2. $2x^2 = 7x + 4$

3. $\begin{cases} 4x + 3y + 3z = 4 \\ 3x \quad\;\; + 2z = 2 \\ 2x - 5y \quad\;\; = -4 \end{cases}$

4. $|3x - 4| \leq 10$

5. $\dfrac{x}{x - 8} + \dfrac{6}{x - 2} = \dfrac{x^2}{x^2 - 10x + 16}$

6. Solve for s: $I = \dfrac{2R}{w + 2s}$.

7. Solve by graphing: $\begin{cases} 2x - y = 4 \\ x + y = 5. \end{cases}$

8. Use function notation to write the equation of the line with slope -3 and passing through the point $(1, -5)$.

In Exercises 9–11, graph each equation, inequality, or system in a rectangular coordinate system.

9. $y = |x| + 2$

10. $\begin{cases} y \geq 2x - 1 \\ x \geq 1 \end{cases}$

11. $2x - y < 4$

In Exercises 12–15, perform the indicated operations.

12. $[(x + 2) + 3y][(x + 2) - 3y]$

13. $\dfrac{2x^2 + x - 1}{2x^2 - 9x + 4} \div \dfrac{6x + 15}{3x^2 - 12x}$

14. $\dfrac{3x}{x^2 - 9x + 20} - \dfrac{5}{2x - 8}$

15. $(3x^2 + 10x + 10) \div (x + 2)$

In Exercises 16–17, factor completely.

16. $xy - 6x + 2y - 12$

17. $24x^3y + 16x^2y - 30xy$

18. A baseball is thrown straight up from a height of 64 feet. The function
$$s(t) = -16t^2 + 48t + 64$$
describes the ball's height above the ground, $s(t)$, in feet, t seconds after it is thrown. How long will it take for the ball to hit the ground?

19. The local cable television company offers two deals. Basic cable service with one movie channel costs $35 per month. Basic service with two movie channels costs $45 per month. Find the charge for the basic cable service and the charge for each movie channel.

20. A rectangular garden 10 feet wide and 12 feet long is surrounded by a rock border of uniform width. The area of the garden and rock border combined is 168 square feet. What is the width of the rock border?

Radicals, Radical Functions, and Rational Exponents

C an mathematical models be created for events that appear to involve random behavior, such as stock market fluctuations or air turbulence? Chaos theory, a new frontier of mathematics, offers models and computer-generated images that reveal order and underlying patterns where only the erratic and the unpredictable had been observed. Because most behavior is chaotic, the computer has become a canvas that looks more like the real world than anything previously seen. Magnified portions of these computer images yield repetitions of the original structure, as well as new and unexpected patterns. The computer generates these visualizations of chaos by plotting large numbers of points for functions whose domains are nonreal numbers involving the square root of negative one.

Here's where you'll find this application:

- We define $\sqrt{-1}$ in Section 7.7 and hint at chaotic possibilities in the Blitzer Bonus on page 569. If you are intrigued by how the operations of nonreal numbers in Section 7.7 reveal that the world is not random [rather, the underlying patterns are far more intricate than we had previously assumed], we suggest reading *Chaos* by James Gleick, published by Penguin Books.

7.1   Radical Expressions and Functions

What am I supposed to learn?

After studying this section, you should be able to:

1. Evaluate square roots.
2. Evaluate square root functions.
3. Find the domain of square root functions.
4. Use models that are square root functions.
5. Simplify expressions of the form $\sqrt{a^2}$.
6. Evaluate cube root functions.
7. Simplify expressions of the form $\sqrt[3]{a^3}$.
8. Find even and odd roots.
9. Simplify expressions of the form $\sqrt[n]{a^n}$.

S = Sail area
L = Length
D = Displacement

The America's Cup is the supreme event in ocean sailing. Competition is fierce and the costs are huge. Competitors look to mathematics to provide the critical innovation that can make the difference between winning and losing. The basic dimensions of competitors' yachts must satisfy an inequality containing square roots and cube roots:

$$L + 1.25\sqrt{S} - 9.8\sqrt[3]{D} \le 16.296.$$

In the inequality, L is the yacht's length, in meters, S is its sail area, in square meters, and D is its displacement, in cubic meters.

In this section, we introduce a new category of expressions and functions that contain roots. You will see why square root functions are used to describe phenomena that are continuing to grow but whose growth is leveling off.

Square Roots

From our earlier work with exponents, we are aware that the square of 5 and the square of −5 are both 25:

$$5^2 = 25 \quad \text{and} \quad (-5)^2 = 25.$$

The reverse operation of squaring a number is finding the *square root* of the number. For example,

- One square root of 25 is 5 because $5^2 = 25$.
- Another square root of 25 is −5 because $(-5)^2 = 25$.

In general, **if $b^2 = a$, then b is a square root of a.**

The symbol $\sqrt{}$ is used to denote the *positive* or *principal square root* of a number. For example,

- $\sqrt{25} = 5$ because $5^2 = 25$ and 5 is positive.
- $\sqrt{100} = 10$ because $10^2 = 100$ and 10 is positive.

1. Evaluate square roots.

Radical sign $\sqrt{a}$ Radicand

Radical expression

The symbol $\sqrt{}$ that we use to denote the principal square root is called a **radical sign**. The number under the radical sign is called the **radicand**. Together we refer to the radical sign and its radicand as a **radical expression**.

> ### Definition of the Principal Square Root
>
> If a is a nonnegative real number, the nonnegative number b such that $b^2 = a$, denoted by $b = \sqrt{a}$, is the **principal square root** of a.

The symbol $-\sqrt{}$ is used to denote the negative square root of a number. For example,

- $-\sqrt{25} = -5$ because $(-5)^2 = 25$ and -5 is negative.
- $-\sqrt{100} = -10$ because $(-10)^2 = 100$ and -10 is negative.

EXAMPLE 1   Evaluating Square Roots

Evaluate:

a. $\sqrt{81}$ b. $-\sqrt{9}$ c. $\sqrt{\dfrac{4}{49}}$

d. $\sqrt{0.0064}$ e. $\sqrt{36 + 64}$ f. $\sqrt{36} + \sqrt{64}$.

Great Question!

Is $\sqrt{a + b}$ equal to $\sqrt{a} + \sqrt{b}$?

No. In Example 1, parts (e) and (f), observe that $\sqrt{36 + 64}$ is not equal to $\sqrt{36} + \sqrt{64}$. In general,

$$\sqrt{a + b} \neq \sqrt{a} + \sqrt{b}$$

and

$$\sqrt{a - b} \neq \sqrt{a} - \sqrt{b}.$$

Solution

a. $\sqrt{81} = 9$ The principal square root of 81 is 9 because $9^2 = 81$.

b. $-\sqrt{9} = -3$ The negative square root of 9 is -3 because $(-3)^2 = 9$.

c. $\sqrt{\dfrac{4}{49}} = \dfrac{2}{7}$ The principal square root of $\dfrac{4}{49}$ is $\dfrac{2}{7}$ because $\left(\dfrac{2}{7}\right)^2 = \dfrac{4}{49}$.

d. $\sqrt{0.0064} = 0.08$ The principal square root of 0.0064 is 0.08 because $(0.08)^2 = (0.08)(0.08) = 0.0064$.

e. $\sqrt{36 + 64} = \sqrt{100}$ Simplify the radicand.
$= 10$ Take the principal square root of 100, which is 10.

f. $\sqrt{36} + \sqrt{64} = 6 + 8$ $\sqrt{36} = 6$ because $6^2 = 36$. $\sqrt{64} = 8$ because $8^2 = 64$.
$= 14$ ∎

☑ **CHECK POINT 1**   Evaluate:

a. $\sqrt{64}$ b. $-\sqrt{49}$ c. $\sqrt{\dfrac{16}{25}}$

d. $\sqrt{0.0081}$ e. $\sqrt{9 + 16}$ f. $\sqrt{9} + \sqrt{16}$.

Let's see what happens to the radical expression $\sqrt{x}$ if x is a negative number. Is the square root of a negative number a real number? For example, consider $\sqrt{-25}$. Is there a real number whose square is -25? No. Thus, $\sqrt{-25}$ is not a real number. In general, **a square root of a negative number is not a real number**.

2 Evaluate square root functions. ▶

Square Root Functions

Because each nonnegative real number, x, has precisely one principal square root, $\sqrt{x}$, there is a **square root function** defined by

$$f(x) = \sqrt{x}.$$

The domain of this function is $[0, \infty)$. We can graph $f(x) = \sqrt{x}$ by selecting nonnegative real numbers for x. It is easiest to choose perfect squares, numbers that have rational square roots. **Table 7.1** shows five such choices for x and the calculations for the corresponding outputs. We plot these ordered pairs as points in the rectangular coordinate system and connect the points with a smooth curve. The graph of $f(x) = \sqrt{x}$ is shown in **Figure 7.1**.

Table 7.1		
x	$f(x) = \sqrt{x}$	(x, y) or $(x, f(x))$
0	$f(0) = \sqrt{0} = 0$	$(0, 0)$
1	$f(1) = \sqrt{1} = 1$	$(1, 1)$
4	$f(4) = \sqrt{4} = 2$	$(4, 2)$
9	$f(9) = \sqrt{9} = 3$	$(9, 3)$
16	$f(16) = \sqrt{16} = 4$	$(16, 4)$

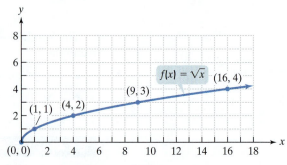

Figure 7.1 The graph of the square root function $f(x) = \sqrt{x}$

Is it possible to choose values of x for **Table 7.1** that are not squares of integers, or perfect squares? Yes. For example, we can let $x = 3$. Thus, $f(3) = \sqrt{3}$. Because 3 is not a perfect square, $\sqrt{3}$ is an irrational number, one that cannot be expressed as a quotient of integers. We can use a calculator to find a decimal approximation of $\sqrt{3}$.

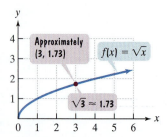

Figure 7.2 Visualizing $\sqrt{3}$ as a point on the graph of $f(x) = \sqrt{x}$

Many Scientific Calculators

$3\ \boxed{\sqrt{}}$

Many Graphing Calculators

$\boxed{\sqrt{}}\ 3\ \boxed{\text{ENTER}}$

Rounding the displayed number to two decimal places, $\sqrt{3} \approx 1.73$. This information is shown visually as a point, approximately $(3, 1.73)$, on the graph of $f(x) = \sqrt{x}$ in **Figure 7.2**.

To evaluate a square root function, we use substitution, just as we have done to evaluate other functions.

EXAMPLE 2 Evaluating Square Root Functions

For each function, find the indicated function value:

a. $f(x) = \sqrt{5x - 6}; f(2)$ **b.** $g(x) = -\sqrt{64 - 8x}; g(-3)$.

Solution

a. $f(2) = \sqrt{5 \cdot 2 - 6}$ Substitute 2 for x in $f(x) = \sqrt{5x - 6}$.

         $= \sqrt{4} = 2$ Simplify the radicand and take the square root.

b. $g(-3) = -\sqrt{64 - 8(-3)}$ Substitute -3 for x in $g(x) = -\sqrt{64 - 8x}$.

         $= -\sqrt{88} \approx -9.38$ Simplify the radicand:
                                     $64 - 8(-3) = 64 - (-24) = 64 + 24 = 88$.
                                       Then use a calculator to approximate $\sqrt{88}$. ∎

✓ **CHECK POINT 2** For each function, find the indicated function value:
     a. $f(x) = \sqrt{12x - 20}; f(3)$
     b. $g(x) = -\sqrt{9 - 3x}; g(-5)$.

3 Find the domain of square root functions. ▶

We have seen that the domain of a function f is the largest set of real numbers for which the value of $f(x)$ is a real number. Because only nonnegative numbers have real square roots, the domain of a square root function is the set of real numbers for which the radicand is nonnegative.

EXAMPLE 3   Finding the Domain of a Square Root Function

Find the domain of

$$f(x) = \sqrt{3x + 12}.$$

Solution   The domain is the set of real numbers, x, for which the radicand, $3x + 12$, is nonnegative. We set the radicand greater than or equal to 0 and solve the resulting inequality.

$$3x + 12 \geq 0$$
$$3x \geq -12$$
$$x \geq -4$$

The domain of f is $[-4, \infty)$. ∎

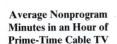

Figure 7.3

Figure 7.3 shows the graph of $f(x) = \sqrt{3x + 12}$ in a $[-10, 10, 1]$ by $[-10, 10, 1]$ viewing rectangle. The graph appears only for $x \geq -4$, verifying $[-4, \infty)$ as the domain. Can you see how the graph also illustrates this square root function's range? The graph only appears for nonnegative values of y. Thus, the range is $[0, \infty)$.

☑ **CHECK POINT 3**   Find the domain of

$$f(x) = \sqrt{9x - 27}.$$

4 Use models that are square root functions. ▶

The graph of the square root function $f(x) = \sqrt{x}$ is increasing from left to right. However, the rate of increase is slowing down as the graph moves to the right. This is why square root functions are often used to model growing phenomena with growth that is leveling off.

EXAMPLE 4   Modeling with a Square Root Function

By 2011, the amount of "clutter," including commercials and plugs for other shows, had increased to the point where an "hour-long" drama on cable TV was 44.8 minutes. The graph in **Figure 7.4** shows the average number of nonprogram minutes in an hour of prime-time cable television. Although the minutes of clutter grew from 1996 through 2011, the growth was leveling off. The data can be modeled by the function

$$M(x) = 0.7\sqrt{x} + 12.5,$$

where $M(x)$ is the average number of nonprogram minutes in an hour of prime-time cable x years after 1996. According to the model, in 2002, how many cluttered minutes disrupted cable TV action in an hour? Round to the nearest tenth of a minute. What is the difference between the actual data and the number of minutes that you obtained?

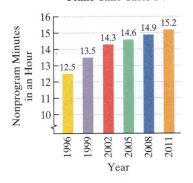

Average Nonprogram Minutes in an Hour of Prime-Time Cable TV

Figure 7.4

Source: The Nielsen Media Company, Monitor-Plus

Solution   Because 2002 is 6 years after 1996, we substitute 6 for x and evaluate the function at 6.

$$M(x) = 0.7\sqrt{x} + 12.5 \qquad \text{Use the given function.}$$
$$M(6) = 0.7\sqrt{6} + 12.5 \qquad \text{Substitute 6 for } x.$$
$$\approx 14.2 \qquad \text{Use a calculator.}$$

The model indicates that there were approximately 14.2 nonprogram minutes in an hour of prime-time cable in 2002. **Figure 7.4** shows 14.3 minutes, so the difference is $14.3 - 14.2$, or 0.1 minute. ∎

☑ **CHECK POINT 4**   If the trend from 1996 through 2011 continues, use the square root function in Example 4 to project how many cluttered minutes, rounded to the nearest tenth, there will be in an hour in 2017.

Simplify expressions of the form $\sqrt{a^2}$.

Simplifying Expressions of the Form $\sqrt{a^2}$

You may think that $\sqrt{a^2} = a$. However, this is not necessarily true. Consider the following examples:

$$\sqrt{4^2} = \sqrt{16} = 4$$
$$\sqrt{(-4)^2} = \sqrt{16} = 4.$$

> The result is not −4, but rather the absolute value of −4, or 4.

Here is a rule for simplifying expressions of the form $\sqrt{a^2}$:

Simplifying $\sqrt{a^2}$

For any real number a,

$$\sqrt{a^2} = |a|.$$

In words, the principal square root of a^2 is the absolute value of a.

Using Technology

Graphic Connections

The graphs of $f(x) = \sqrt{x^2}$ and $g(x) = |x|$ are shown in a $[-10, 10, 1]$ by $[-2, 10, 1]$ viewing rectangle. The graphs are the same. Thus,

$$\sqrt{x^2} = |x|.$$

$f(x) = \sqrt{x^2}$

$g(x) = |x|$

EXAMPLE 5 Simplifying Radical Expressions

Simplify each expression:

a. $\sqrt{(-6)^2}$ **b.** $\sqrt{(x+5)^2}$ **c.** $\sqrt{25x^6}$ **d.** $\sqrt{x^2 - 4x + 4}.$

Solution The principal square root of an expression squared is the absolute value of that expression. In parts (a) and (b), we are given squared radicands. In parts (c) and (d), it will first be necessary to express the radicand as an expression that is squared.

a. $\sqrt{(-6)^2} = |-6| = 6$

b. $\sqrt{(x+5)^2} = |x+5|$

c. To simplify $\sqrt{25x^6}$, first write $25x^6$ as an expression that is squared: $25x^6 = (5x^3)^2$. Then simplify.

$$\sqrt{25x^6} = \sqrt{(5x^3)^2} = |5x^3| \quad \text{or} \quad 5|x^3|$$

d. To simplify $\sqrt{x^2 - 4x + 4}$, first write $x^2 - 4x + 4$ as an expression that is squared by factoring the perfect square trinomial: $x^2 - 4x + 4 = (x - 2)^2$. Then simplify.

$$\sqrt{x^2 - 4x + 4} = \sqrt{(x - 2)^2} = |x - 2| \quad \blacksquare$$

☑ **CHECK POINT 5** Simplify each expression:

a. $\sqrt{(-7)^2}$

b. $\sqrt{(x+8)^2}$

c. $\sqrt{49x^{10}}$

d. $\sqrt{x^2 - 6x + 9}.$

In some situations, we are told that no radicands involve negative quantities raised to even powers. When the expression being squared is nonnegative, it is not necessary to use absolute value when simplifying $\sqrt{a^2}$. For example, assuming that no radicands contain negative quantities that are squared,

$$\sqrt{x^6} = \sqrt{(x^3)^2} = x^3$$
$$\sqrt{25x^2 + 10x + 1} = \sqrt{(5x + 1)^2} = 5x + 1.$$

Evaluate cube root functions.

Cube Roots and Cube Root Functions

Finding the square root of a number reverses the process of squaring a number. Similarly, finding the cube root of a number reverses the process of cubing a number. For example, $2^3 = 8$, and so the cube root of 8 is 2. The notation that we use is $\sqrt[3]{8} = 2$.

> ### Definition of the Cube Root of a Number
> The **cube root** of a real number a is written $\sqrt[3]{a}$.
> $$\sqrt[3]{a} = b \quad \text{means that} \quad b^3 = a.$$

Great Question!

Should I know the cube roots of certain numbers by heart?

Some cube roots occur so frequently that you might want to memorize them.

$$\sqrt[3]{1} = 1$$
$$\sqrt[3]{8} = 2$$
$$\sqrt[3]{27} = 3$$
$$\sqrt[3]{64} = 4$$
$$\sqrt[3]{125} = 5$$
$$\sqrt[3]{216} = 6$$
$$\sqrt[3]{1000} = 10$$

For example,

$$\sqrt[3]{64} = 4 \quad \text{because} \quad 4^3 = 64.$$
$$\sqrt[3]{-27} = -3 \quad \text{because} \quad (-3)^3 = -27.$$

In contrast to square roots, the cube root of a negative number is a real number. All real numbers have cube roots. The cube root of a positive number is positive. The cube root of a negative number is negative.

Because every real number, x, has precisely one cube root, $\sqrt[3]{x}$, there is a **cube root function** defined by

$$f(x) = \sqrt[3]{x}.$$

The domain of this function is the set of all real numbers. We can graph $f(x) = \sqrt[3]{x}$ by selecting perfect cubes, numbers that have rational cube roots, for x. **Table 7.2** shows five such choices for x and the calculations for the corresponding outputs. We plot these ordered pairs as points in the rectangular coordinate system and connect the points with a smooth curve. The graph of $f(x) = \sqrt[3]{x}$ is shown in **Figure 7.5**.

Table 7.2

x	$f(x) = \sqrt[3]{x}$	(x, y) or $(x, f(x))$
-8	$f(-8) = \sqrt[3]{-8} = -2$	$(-8, -2)$
-1	$f(-1) = \sqrt[3]{-1} = -1$	$(-1, -1)$
0	$f(0) = \sqrt[3]{0} = 0$	$(0, 0)$
1	$f(1) = \sqrt[3]{1} = 1$	$(1, 1)$
8	$f(8) = \sqrt[3]{8} = 2$	$(8, 2)$

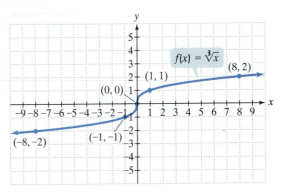

Figure 7.5 The graph of the cube root function $f(x) = \sqrt[3]{x}$

Notice that both the domain and the range of $f(x) = \sqrt[3]{x}$ are the set of all real numbers, $(-\infty, \infty)$.

EXAMPLE 6   Evaluating Cube Root Functions

For each function, find the indicated function value:

a. $f(x) = \sqrt[3]{x - 2}; \quad f(127)$

b. $g(x) = \sqrt[3]{8x - 8}; \quad g(-7).$

Solution

a.
$$f(x) = \sqrt[3]{x} - 2 \qquad \text{This is the given function.}$$
$$f(127) = \sqrt[3]{127} - 2 \qquad \text{Substitute 127 for } x.$$
$$= \sqrt[3]{125} \qquad \text{Simplify the radicand.}$$
$$= 5 \qquad \sqrt[3]{125} = 5 \text{ because } 5^3 = 125.$$

b.
$$g(x) = \sqrt[3]{8x} - 8 \qquad \text{This is the given function.}$$
$$g(-7) = \sqrt[3]{8(-7)} - 8 \qquad \text{Substitute } -7 \text{ for } x.$$
$$= \sqrt[3]{-64} \qquad \text{Simplify the radicand: } 8(-7) - 8 = -56 - 8 = -64.$$
$$= -4 \qquad \sqrt[3]{-64} = -4 \text{ because } (-4)^3 = -64. \quad \blacksquare$$

✓ **CHECK POINT 6** For each function, find the indicated function value:

a. $f(x) = \sqrt[3]{x} - 6; \quad f(33)$

b. $g(x) = \sqrt[3]{2x} + 2; \quad g(-5).$

---

7 Simplify expressions of the form $\sqrt[3]{a^3}$. ▶

Because the cube root of a positive number is positive and the cube root of a negative number is negative, absolute value is not needed to simplify expressions of the form $\sqrt[3]{a^3}$.

> **Simplifying $\sqrt[3]{a^3}$**
>
> For any real number a,
> $$\sqrt[3]{a^3} = a.$$
>
> In words, the cube root of any expression cubed is that expression.

EXAMPLE 7 Simplifying a Cube Root

Simplify: $\sqrt[3]{-64x^3}$.

Solution Begin by expressing the radicand as an expression that is cubed: $-64x^3 = (-4x)^3$. Then simplify.

$$\sqrt[3]{-64x^3} = \sqrt[3]{(-4x)^3} = -4x$$

We can check our answer by cubing $-4x$:

$$(-4x)^3 = (-4)^3 x^3 = -64x^3.$$

By obtaining the original radicand, we know that our simplification is correct. ∎

✓ **CHECK POINT 7** Simplify: $\sqrt[3]{-27x^3}$.

---

8 Find even and odd roots. ▶

Even and Odd *n*th Roots

Up to this point, we have focused on square roots and cube roots. Other radical expressions have different roots. For example, the fifth root of a, written $\sqrt[5]{a}$, is the number b for which $b^5 = a$. Thus,

$$\sqrt[5]{32} = 2 \qquad \text{because} \qquad 2^5 = 2 \cdot 2 \cdot 2 \cdot 2 \cdot 2 = 32.$$

The radical expression $\sqrt[n]{a}$ represents the **nth root** of a. The number n is called the **index**. An index of 2 represents a square root and is not written. An index of 3 represents a cube root.

If the index n in $\sqrt[n]{a}$ is an odd number, a root is said to be an **odd root**. A cube root is an odd root. Other odd roots have the same characteristics as cube roots.

- Every real number has exactly one real root when n is odd. An odd root of a positive number is positive and an odd root of a negative number is negative.

$$3^5 = 3 \cdot 3 \cdot 3 \cdot 3 \cdot 3 = 243, \text{ so the fifth root of 243 is 3.}$$
$$(-3)^5 = (-3)(-3)(-3)(-3)(-3) = -243, \text{ so the fifth root of } -243 \text{ is } -3.$$

- The (odd) nth root of a, $\sqrt[n]{a}$, is the number b for which $b^n = a$.

$$\sqrt[5]{243} = 3 \qquad\qquad \sqrt[5]{-243} = -3$$

$3^5 = 243$ | $(-3)^5 = -243$

Great Question!

Should I know the higher roots of certain numbers by heart?

Some higher even and odd roots occur so frequently that you might want to memorize them.

Fourth Roots	Fifth Roots
$\sqrt[4]{1} = 1$	$\sqrt[5]{1} = 1$
$\sqrt[4]{16} = 2$	$\sqrt[5]{32} = 2$
$\sqrt[4]{81} = 3$	$\sqrt[5]{243} = 3$
$\sqrt[4]{256} = 4$	
$\sqrt[4]{625} = 5$	

If the index n in $\sqrt[n]{a}$ is an even number, a root is said to be an **even root**. A square root is an even root. Other even roots have the same characteristics as square roots.

- Every positive real number has two real roots when n is even. One root is positive and one is negative.

$$2^4 = 2 \cdot 2 \cdot 2 \cdot 2 = 16 \qquad \text{and} \qquad (-2)^4 = (-2)(-2)(-2)(-2) = 16,$$

so both 2 and -2 are fourth roots of 16.

- The positive root, called the **principal nth root** and represented by $\sqrt[n]{a}$, is the nonnegative number b for which $b^n = a$. The symbol $-\sqrt[n]{a}$ is used to denote the negative nth root.

$$\sqrt[4]{16} = 2 \qquad\qquad -\sqrt[4]{16} = -2$$

$2^4 = 16$ | $(-2)^4 = 16$

- **An even root of a negative number is not a real number.**

$$\sqrt[4]{-16} \text{ is not a real number.}$$

EXAMPLE 8   Finding Even and Odd Roots

Find the indicated root, or state that the expression is not a real number:

a. $\sqrt[4]{81}$ **b.** $-\sqrt[4]{81}$ **c.** $\sqrt[4]{-81}$ **d.** $\sqrt[5]{-32}$.

Solution

a. $\sqrt[4]{81} = 3$

The principal fourth root of 81 is 3 because $3^4 = 3 \cdot 3 \cdot 3 \cdot 3 = 81$.

b. $-\sqrt[4]{81} = -3$

The negative fourth root of 81 is -3 because $(-3)^4 = (-3)(-3)(-3)(-3) = 81$.

c. $\sqrt[4]{-81}$ is not a real number because the index, 4, is even and the radicand, -81, is negative. No real number can be raised to the fourth power to give a negative result such as -81. Real numbers to even powers can only result in nonnegative numbers.

d. $\sqrt[5]{-32} = -2$ because $(-2)^5 = (-2)(-2)(-2)(-2)(-2) = -32$. An odd root of a negative real number is always negative. ∎

☑ **CHECK POINT 8**   Find the indicated root, or state that the expression is not a real number:

a. $\sqrt[4]{16}$ **b.** $-\sqrt[4]{16}$ **c.** $\sqrt[4]{-16}$ **d.** $\sqrt[5]{-1}$.

---

9 Simplify expressions of the form $\sqrt[n]{a^n}$. ▶

Simplifying Expressions of the Form $\sqrt[n]{a^n}$

We have seen that

$$\sqrt{a^2} = |a| \qquad \text{and} \qquad \sqrt[3]{a^3} = a.$$

Expressions of the form $\sqrt[n]{a^n}$ can be simplified in the same manner. Unless a is known to be nonnegative, absolute value notation is needed when n is even. When the index is odd, absolute value bars are not used.

Simplifying $\sqrt[n]{a^n}$

For any real number a,

1. If n is even, $\sqrt[n]{a^n} = |a|$.
2. If n is odd, $\sqrt[n]{a^n} = a$.

EXAMPLE 9   Simplifying Radical Expressions

Simplify:

a. $\sqrt[4]{(x-3)^4}$ **b.** $\sqrt[5]{(2x+7)^5}$ **c.** $\sqrt[6]{(-5)^6}$.

Solution   Each expression involves the nth root of a radicand raised to the nth power. Thus, each radical expression can be simplified. Absolute value bars are necessary in parts (a) and (c) because the index, n, is even.

a. $\sqrt[4]{(x-3)^4} = |x-3|$ $\sqrt[n]{a^n} = |a|$ if n is even.
b. $\sqrt[5]{(2x+7)^5} = 2x+7$ $\sqrt[n]{a^n} = a$ if n is odd.
c. $\sqrt[6]{(-5)^6} = |-5| = 5$ $\sqrt[n]{a^n} = |a|$ if n is even. ∎

☑ **CHECK POINT 9**   Simplify:

a. $\sqrt[4]{(x+6)^4}$ **b.** $\sqrt[5]{(3x-2)^5}$ **c.** $\sqrt[6]{(-8)^6}$.

Achieving Success

The Secret of Math Success

What's the secret of math success? The bar graph in **Figure 7.6** shows that Japanese teachers and students are more likely than their American counterparts to believe that the key to doing well in math is working hard. Americans tend to think that either you have mathematical intelligence or you don't. Alan Bass, author of *Math Study Skills* (Pearson Education, 2008), strongly disagrees with this American perspective:

Figure 7.6

Source: Wade and Tavris, *Psychology*, Ninth Edition, Pearson, 2008.

"Human beings are easily intelligent enough to understand the basic principles of math. I cannot repeat this enough, but I'll try... **Poor performance in math is not due to a lack of intelligence!** The fact is that the key to success in math is in taking an intelligent approach. Students come up to me and say, 'I'm just not good at math.' Then I ask to see their class notebooks and they show me a chaotic mess of papers jammed into a folder. In math, that's a lot like taking apart your car's engine, dumping the heap of disconnected parts back under the hood, and then going to a mechanic to ask why it won't run. Students come to me and say, 'I'm just not good at math.' Then I ask them about their study habits and they say, 'I have to do all my studying on the weekend.' In math, that's a lot like trying to do all your eating or sleeping on the weekends and wondering why you're so tired and hungry all the time. **How you approach math is much more important than how smart you are.**"

—Alan Bass

CONCEPT AND VOCABULARY CHECK

Fill in each blank so that the resulting statement is true.

1. The symbol $\sqrt{}$ is used to denote the nonnegative, or _____, square root of a number.

2. $\sqrt{64} = 8$ because _____ $= 64$.

3. The domain of $f(x) = \sqrt{x}$ is _____.

4. The domain of $f(x) = \sqrt{5x - 20}$ can be found by solving the inequality _____.

5. For any real number a, $\sqrt{a^2} =$ _____.

6. $\sqrt[3]{1000} = 10$ because _____ $= 1000$.

7. $\sqrt[3]{-125} = -5$ because _____ $= -125$.

8. For any real number a, $\sqrt[3]{a^3} =$ _____.

9. The domain of $f(x) = \sqrt[3]{x}$ is _____.

10. The radical expression $\sqrt[n]{a}$ represents the _____ root of a. The number n is called the _____.

11. If n is even, $\sqrt[n]{a^n} =$ _____. If n is odd, $\sqrt[n]{a^n} =$ _____.

12. True or false: $-\sqrt{25}$ is a real number. _____

13. True or false: $\sqrt{-25}$ is a real number. _____

14. True or false: $\sqrt[3]{-1}$ is a real number. _____

15. True or false: $\sqrt[4]{-1}$ is a real number. _____

7.1 EXERCISE SET ▶ MyMathLab®

Practice Exercises

In Exercises 1–20, evaluate each expression, or state that the expression is not a real number.

1. $\sqrt{36}$
2. $\sqrt{16}$
3. $-\sqrt{36}$
4. $-\sqrt{16}$
5. $\sqrt{-36}$
6. $\sqrt{-16}$
7. $\sqrt{\dfrac{1}{25}}$
8. $\sqrt{\dfrac{1}{49}}$
9. $-\sqrt{\dfrac{9}{16}}$
10. $-\sqrt{\dfrac{4}{25}}$
11. $\sqrt{0.81}$
12. $\sqrt{0.49}$
13. $-\sqrt{0.04}$
14. $-\sqrt{0.64}$
15. $\sqrt{25 - 16}$
16. $\sqrt{144 + 25}$
17. $\sqrt{25} - \sqrt{16}$
18. $\sqrt{144} + \sqrt{25}$
19. $\sqrt{16 - 25}$
20. $\sqrt{25 - 144}$

In Exercises 21–26, find the indicated function values for each function. If necessary, round to two decimal places. If the function value is not a real number and does not exist, so state.

21. $f(x) = \sqrt{x - 2};\ \ f(18), f(3), f(2), f(-2)$

22. $f(x) = \sqrt{x - 3};\ \ f(28), f(4), f(3), f(-1)$

23. $g(x) = -\sqrt{2x + 3};\ \ g(11), g(1), g(-1), g(-2)$

24. $g(x) = -\sqrt{2x + 1};\ \ g(4), g(1), g\left(-\dfrac{1}{2}\right), g(-1)$

25. $h(x) = \sqrt{(x - 1)^2};\ \ h(5), h(3), h(0), h(-5)$

26. $h(x) = \sqrt{(x - 2)^2};\ \ h(5), h(3), h(0), h(-5)$

In Exercises 27–32, find the domain of each square root function. Then use the domain to match the radical function with its graph. [The graphs are labeled (a) through (f) and are shown in $[-10, 10, 1]$ by $[-10, 10, 1]$ viewing rectangles below and on the next page.]

27. $f(x) = \sqrt{x - 3}$
28. $f(x) = \sqrt{x + 2}$
29. $f(x) = \sqrt{3x + 15}$
30. $f(x) = \sqrt{3x - 15}$
31. $f(x) = \sqrt{6 - 2x}$
32. $f(x) = \sqrt{8 - 2x}$

a.

104. A motorist is involved in an accident. A police officer measures the car's skid marks to be 45 feet long. Use the function described at the bottom of the previous page to estimate the speed at which the motorist was traveling before braking. If the posted speed limit is 35 miles per hour and the motorist tells the officer she was not speeding, should the officer believe her? Explain.

Explaining the Concepts

105. What are the square roots of 36? Explain why each of these numbers is a square root.

106. What does the symbol $\sqrt{\ }$ denote? Which of your answers in Exercise 105 is given by this symbol? Write the symbol needed to obtain the other answer.

107. Explain why $\sqrt{-1}$ is not a real number.

108. Explain how to find the domain of a square root function.

109. Explain how to simplify $\sqrt{a^2}$. Give an example with your explanation.

110. Explain why $\sqrt[3]{8}$ is 2. Then describe what is meant by the cube root of a real number.

111. Describe two differences between odd and even roots.

112. Explain how to simplify $\sqrt[n]{a^n}$ if n is even and if n is odd. Give examples with your explanations.

113. Explain the meaning of the words *radical*, *radicand*, and *index*. Give an example with your explanation.

114. Describe the trend in a boy's growth from birth through five years, shown in the graph for Exercise 101. Why is a square root function a useful model for the data?

Technology Exercises

115. Use a graphing utility to graph $y_1 = \sqrt{x}$, $y_2 = \sqrt{x + 4}$, and $y_3 = \sqrt{x - 3}$ in the same $[-5, 10, 1]$ by $[0, 6, 1]$ viewing rectangle. Describe one similarity and one difference that you observe among the graphs. Use the word *shift* in your response.

116. Use a graphing utility to graph $y = \sqrt{x}$, $y = \sqrt{x} + 4$, and $y = \sqrt{x} - 3$ in the same $[-1, 10, 1]$ by $[-10, 10, 1]$ viewing rectangle. Describe one similarity and one difference that you observe among the graphs.

117. Use a graphing utility to graph $f(x) = \sqrt{x}$, $g(x) = -\sqrt{x}$, $h(x) = \sqrt{-x}$, and $k(x) = -\sqrt{-x}$ in the same $[-10, 10, 1]$ by $[-4, 4, 1]$ viewing rectangle. Use the graphs to describe the domains and the ranges of functions f, g, h, and k.

118. Use a graphing utility to graph $y_1 = \sqrt{x^2}$ and $y_2 = -x$ in the same viewing rectangle.
  a. For what values of x is $\sqrt{x^2} = -x$?
  b. For what values of x is $\sqrt{x^2} \neq -x$?

Critical Thinking Exercises

Make Sense? *In Exercises 119–122, determine whether each statement makes sense or does not make sense, and explain your reasoning.*

119. $\sqrt[4]{(-8)^4}$ cannot be positive 8 because the power and the index cancel each other.

120. If I am given any real number, that number has exactly one odd root and two even roots.

121. I need to restrict the domains of radical functions with even indices, but these restrictions are not necessary when indices are odd.

122. Using my calculator, I determined that $5^5 = 3125$, so 5 must be the fifth root of 3125.

In Exercises 123–126, determine whether each statement is true or false. If the statement is false, make the necessary change(s) to produce a true statement.

123. The domain of $f(x) = \sqrt[3]{x - 4}$ is $[4, \infty)$.

124. If n is odd and b is negative, then $\sqrt[n]{b}$ is not a real number.

125. If $x = -2$, then $\sqrt{x^6} = x^3$.

126. The expression $\sqrt[n]{4}$ represents increasingly larger numbers for $n = 2, 3, 4, 5, 6$, and so on.

127. Write a function whose domain is $(-\infty, 5]$.

128. Let $f(x) = \sqrt{x - 3}$ and $g(x) = \sqrt{x + 1}$. Find the domain of $f + g$ and $\dfrac{f}{g}$.

129. Simplify: $\sqrt{(2x + 3)^{10}}$.

In Exercises 130–131, graph each function by hand. Then describe the relationship between the function that you graphed and the graph of $f(x) = \sqrt{x}$.

130. $g(x) = \sqrt{x} + 2$

131. $h(x) = \sqrt{x + 3}$

Review Exercises

132. Simplify: $3x - 2[x - 3(x + 5)]$. (Section 1.2, Example 14)

133. Simplify: $(-3x^{-4}y^3)^{-2}$. (Section 1.6, Example 7c)

134. Solve: $|3x - 4| > 11$. (Section 4.3, Example 6)

Preview Exercises

Exercises 135–137 will help you prepare for the material covered in the next section. In each exercise, use properties of exponents to simplify the expression. Be sure that no negative exponents appear in your simplified expression. (If you have forgotten how to simplify an exponential expression, see the box on page 77.)

135. $(2^3 x^5)(2^4 x^{-6})$

136. $\dfrac{32x^2}{16x^5}$

137. $(x^{-2}y^3)^4$

SECTION

7.2

What am I supposed to learn?

After studying this section, you should be able to:

1. Use the definition of $a^{\frac{1}{n}}$.

2. Use the definition of $a^{\frac{m}{n}}$.

3. Use the definition of $a^{-\frac{m}{n}}$.

4. Simplify expressions with rational exponents.

5. Simplify radical expressions using rational exponents.

Rational Exponents

The Galápagos Islands are a chain of volcanic islands lying 600 miles west of Ecuador. They are famed for over 5000 species of plants and animals, including a rare flightless cormorant, marine iguanas, and giant tortoises weighing more than 600 pounds. Early in 2001, the plants and wildlife that live in the Galápagos were at risk from a massive oil spill that flooded 150,000 gallons of toxic fuel into one of the world's most fragile ecosystems. The long-term danger of the accident is that

Marine iguana of the Galápagos Islands

fuel sinking to the ocean floor will destroy algae that is vital to the food chain. Any imbalance in the food chain could threaten the rare Galápagos plant and animal species that have evolved for thousands of years in isolation with little human intervention.

At risk on these ecologically vulnerable islands are unique flora and fauna that helped to inspire Charles Darwin's theory of evolution. Darwin made an enormous collection of the island's plant species. The function

$$f(x) = 29x^{\frac{1}{3}}$$

models the number of plant species, $f(x)$, on the various islands of the Galápagos in terms of the area, x, in square miles, of a particular island. But x to the *what* power? How can we interpret the information given by this function? In this section, we turn our attention to rational exponents such as $\frac{1}{3}$ and their relationship to roots of real numbers.

Defining Rational Exponents

We define rational exponents so that their properties are the same as the properties for integer exponents. For example, suppose that $x = 7^{\frac{1}{3}}$. We know that exponents are multiplied when an exponential expression is raised to a power. For this to be true,

$$x^3 = \left(7^{\frac{1}{3}}\right)^3 = 7^{\frac{1}{3} \cdot 3} = 7^1 = 7.$$

We see that $x^3 = 7$. This means that x is the number whose cube is 7. Thus, $x = \sqrt[3]{7}$.

Remember that we began with $x = 7^{\frac{1}{3}}$. This means that

$$7^{\frac{1}{3}} = \sqrt[3]{7}.$$

We can generalize $7^{\frac{1}{3}} = \sqrt[3]{7}$ with the following definition:

1 Use the definition of $a^{\frac{1}{n}}$. ▶

The Definition of $a^{\frac{1}{n}}$

If $\sqrt[n]{a}$ represents a real number and $n \geq 2$ is an integer, then

$$a^{\frac{1}{n}} = \sqrt[n]{a}.$$

> The denominator of the rational exponent is the radical's index.

If n is even, a must be nonnegative. If n is odd, a can be any real number.

Using Technology

This graphing utility screen shows that
$\sqrt{64} = 8$ and $64^{\frac{1}{2}} = 8$.

```
√64
                    8
64^1/2
                    8
```

EXAMPLE 1 Using the Definition of $a^{\frac{1}{n}}$

Use radical notation to rewrite each expression. Simplify, if possible:

a. $64^{\frac{1}{2}}$ **b.** $(-125)^{\frac{1}{3}}$ **c.** $(6x^2y)^{\frac{1}{5}}$.

Solution

a. $64^{\frac{1}{2}} = \sqrt{64} = 8$

> The denominator is the index.

b. $(-125)^{\frac{1}{3}} = \sqrt[3]{-125} = -5$

c. $(6x^2y)^{\frac{1}{5}} = \sqrt[5]{6x^2y}$ ■

☑ **CHECK POINT 1** Use radical notation to rewrite each expression. Simplify, if possible:

a. $25^{\frac{1}{2}}$ **b.** $(-8)^{\frac{1}{3}}$ **c.** $(5xy^2)^{\frac{1}{4}}$.

In our next example, we begin with radical notation and rewrite the expression with rational exponents.

> The radical's index becomes the exponent's denominator.

$$\sqrt[n]{a} = a^{\frac{1}{n}}$$

> The radicand becomes the base.

EXAMPLE 2 Using the Definition of $a^{\frac{1}{n}}$

Rewrite with rational exponents:

a. $\sqrt[5]{13ab}$ **b.** $\sqrt[7]{\dfrac{xy^2}{17}}$.

Solution Parentheses are needed to show that the entire radicand becomes the base.

a. $\sqrt[5]{13ab} = (13ab)^{\frac{1}{5}}$

> The index is the exponent's denominator.

b. $\sqrt[7]{\dfrac{xy^2}{17}} = \left(\dfrac{xy^2}{17}\right)^{\frac{1}{7}}$ ■

✓ **CHECK POINT 2**   Rewrite with rational exponents:

a. $\sqrt[4]{5xy}$

b. $\sqrt[5]{\dfrac{a^3 b}{2}}$.

---

2 Use the definition of $a^{\frac{m}{n}}$. ▶

Can rational exponents have numerators other than 1? The answer is yes. If the numerator is some other integer, we still want to multiply exponents when raising a power to a power. For this reason,

$$a^{\frac{2}{3}} = \left(a^{\frac{1}{3}}\right)^2 \quad \text{and} \quad a^{\frac{2}{3}} = \left(a^2\right)^{\frac{1}{3}}.$$

> This means $(\sqrt[3]{a})^2$.

> This means $\sqrt[3]{a^2}$.

Thus,

$$a^{\frac{2}{3}} = \left(\sqrt[3]{a}\right)^2 = \sqrt[3]{a^2}.$$

Do you see that the denominator, 3, of the rational exponent is the same as the index of the radical? The numerator, 2, of the rational exponent serves as an exponent in each of the two radical forms. We generalize these ideas with the following definition:

> **The Definition of $a^{\frac{m}{n}}$**
>
> If $\sqrt[n]{a}$ represents a real number, $\dfrac{m}{n}$ is a positive rational number reduced to lowest terms, and $n \geq 2$ is an integer, then
>
> $$a^{\frac{m}{n}} = \left(\sqrt[n]{a}\right)^m.$$
>
> > First take the nth root of a.
>
> and
>
> $$a^{\frac{m}{n}} = \sqrt[n]{a^m}.$$
>
> > First raise a to the m power.

The first form of the definition shown in the box involves taking the root first. This form is often preferable because smaller numbers are involved. Notice that the rational exponent consists of two parts, indicated by the following voice balloons:

> The numerator is the exponent.

$$a^{\frac{m}{n}} = \left(\sqrt[n]{a}\right)^m.$$

> The denominator is the radical's index.

EXAMPLE 3   Using the Definition of $a^{\frac{m}{n}}$

Use radical notation to rewrite each expression and simplify:

a. $1000^{\frac{2}{3}}$ **b.** $16^{\frac{3}{2}}$ **c.** $-32^{\frac{3}{5}}$.

Solution

a. $(1000)^{\frac{2}{3}} = \left(\sqrt[3]{1000}\right)^2 = 10^2 = 100$

> The denominator of $\frac{2}{3}$ is the root and the numerator is the exponent.

b. $16^{\frac{3}{2}} = \left(\sqrt{16}\right)^3 = 4^3 = 64$

c. $-32^{\frac{3}{5}} = -\left(\sqrt[5]{32}\right)^3 = -2^3 = -8$

> The base is 32 and the negative sign is not affected by the exponent.

■

Using Technology

Here are the calculator keystroke sequences for $1000^{\frac{2}{3}}$:

Many Scientific Calculators

1000 y^x $($ 2 $\div$ 3 $)$ $=$

Many Graphing Calculators

1000 $\wedge$ $($ 2 $\div$ 3 $)$ ENTER

✓ **CHECK POINT 3**   Use radical notation to rewrite each expression and simplify:

a. $8^{\frac{4}{3}}$ **b.** $25^{\frac{3}{2}}$ **c.** $-81^{\frac{3}{4}}$.

In our next example, we begin with radical notation and rewrite the expression with rational exponents. When changing from radical form to exponential form, the index becomes the denominator of the rational exponent.

EXAMPLE 4   Using the Definition of $a^{\frac{m}{n}}$

Rewrite with rational exponents:

a. $\sqrt[3]{7^5}$ **b.** $\left(\sqrt[4]{13xy}\right)^9$.

Solution

a. $\sqrt[3]{7^5} = 7^{\frac{5}{3}}$

> The index is the exponent's denominator.

b. $\left(\sqrt[4]{13xy}\right)^9 = (13xy)^{\frac{9}{4}}$ ∎

✓ **CHECK POINT 4**   Rewrite with rational exponents:

a. $\sqrt[3]{6^4}$ **b.** $\left(\sqrt[5]{2xy}\right)^7$.

3 Use the definition of $a^{-\frac{m}{n}}$. ▶

Can a rational exponent be negative? Yes. The way that negative rational exponents are defined is similar to the way that negative integer exponents are defined.

> **The Definition of $a^{-\frac{m}{n}}$**
>
> If $a^{\frac{m}{n}}$ is a nonzero real number, then
>
> $$a^{-\frac{m}{n}} = \frac{1}{a^{\frac{m}{n}}}.$$

EXAMPLE 5   Using the Definition of $a^{-\frac{m}{n}}$

Rewrite each expression with a positive exponent. Simplify, if possible:

a. $36^{-\frac{1}{2}}$ **b.** $125^{-\frac{1}{3}}$ **c.** $16^{-\frac{3}{4}}$ **d.** $(7xy)^{-\frac{4}{7}}$.

Solution

a. $36^{-\frac{1}{2}} = \dfrac{1}{36^{\frac{1}{2}}} = \dfrac{1}{\sqrt{36}} = \dfrac{1}{6}$

b. $125^{-\frac{1}{3}} = \dfrac{1}{125^{\frac{1}{3}}} = \dfrac{1}{\sqrt[3]{125}} = \dfrac{1}{5}$

c. $16^{-\frac{3}{4}} = \dfrac{1}{16^{\frac{3}{4}}} = \dfrac{1}{\left(\sqrt[4]{16}\right)^3} = \dfrac{1}{2^3} = \dfrac{1}{8}$

d. $(7xy)^{-\frac{4}{7}} = \dfrac{1}{(7xy)^{\frac{4}{7}}}$ ∎

Using Technology

Here are the calculator keystroke sequences for $16^{-\frac{3}{4}}$:

Many Scientific Calculators

16 y^x (3 +/− ÷ 4) =

Many Graphing Calculators

16 ^ ((−) 3 ÷ 4) ENTER

> ☑ **CHECK POINT 5** Rewrite each expression with a positive exponent. Simplify, if possible:
>
> **a.** $100^{-\frac{1}{2}}$ **b.** $8^{-\frac{1}{3}}$
>
> **c.** $32^{-\frac{3}{5}}$ **d.** $(3xy)^{-\frac{5}{9}}$.

Properties of Rational Exponents

The same properties apply to rational exponents as to integer exponents. The following is a summary of these properties:

Properties of Rational Exponents

If m and n are rational exponents, and a and b are real numbers for which the following expressions are defined, then

1. $b^m \cdot b^n = b^{m+n}$ When multiplying exponential expressions with the same base, add the exponents. Use this sum as the exponent of the common base.

2. $\dfrac{b^m}{b^n} = b^{m-n}$ When dividing exponential expressions with the same base, subtract the exponents. Use this difference as the exponent of the common base.

3. $(b^m)^n = b^{mn}$ When an exponential expression is raised to a power, multiply the exponents. Place the product of the exponents on the base and remove the parentheses.

4. $(ab)^n = a^n b^n$ When a product is raised to a power, raise each factor to that power and multiply.

5. $\left(\dfrac{a}{b}\right)^n = \dfrac{a^n}{b^n}$ When a quotient is raised to a power, raise the numerator to that power and divide by the denominator to that power.

④ Simplify expressions with rational exponents. ▶

We can use these properties to simplify exponential expressions with rational exponents. As with integer exponents, an expression with rational exponents is **simplified** when:

- No parentheses appear.
- No powers are raised to powers.
- Each base occurs only once.
- No negative or zero exponents appear.

Great Question!

Because $6 \cdot 6 = 36$, should I multiply 6 and 6 when simplifying $6^{\frac{1}{7}} \cdot 6^{\frac{4}{7}}$?

No. To simplify $6^{\frac{1}{7}} \cdot 6^{\frac{4}{7}}$, **do not multiply the numerical bases.**

Incorrect!

$6^{\frac{1}{7}} \cdot 6^{\frac{4}{7}} = 36^{\frac{1}{7}+\frac{4}{7}}$

EXAMPLE 6   Simplifying Expressions with Rational Exponents

Simplify:

a. $6^{\frac{1}{7}} \cdot 6^{\frac{4}{7}}$ **b.** $\dfrac{32x^{\frac{1}{2}}}{16x^{\frac{3}{4}}}$ **c.** $\left(8.3^{\frac{3}{4}}\right)^{\frac{2}{3}}$ **d.** $\left(x^{-\frac{2}{5}} y^{\frac{1}{3}}\right)^{\frac{1}{2}}$.

Solution

a. $6^{\frac{1}{7}} \cdot 6^{\frac{4}{7}} = 6^{\frac{1}{7}+\frac{4}{7}}$ To multiply with the same base, add exponents.

$= 6^{\frac{5}{7}}$ Simplify: $\dfrac{1}{7} + \dfrac{4}{7} = \dfrac{5}{7}$.

b. $\dfrac{32x^{\frac{1}{2}}}{16x^{\frac{3}{4}}} = \dfrac{32}{16}x^{\frac{1}{2}-\frac{3}{4}}$ Divide coefficients. To divide with the same base, subtract exponents.

$= 2x^{\frac{2}{4}-\frac{3}{4}}$ Write exponents in terms of the LCD, 4.

$= 2x^{-\frac{1}{4}}$ Subtract: $\dfrac{2}{4} - \dfrac{3}{4} = -\dfrac{1}{4}$.

$= \dfrac{2}{x^{\frac{1}{4}}}$ Rewrite with a positive exponent: $a^{-\frac{m}{n}} = \dfrac{1}{a^{\frac{m}{n}}}$.

c. $\left(8.3^{\frac{3}{4}}\right)^{\frac{2}{3}} = 8.3^{\left(\frac{3}{4}\right)\left(\frac{2}{3}\right)}$ To raise a power to a power, multiply exponents.

$= 8.3^{\frac{1}{2}}$ Multiply: $\dfrac{3}{4} \cdot \dfrac{2}{3} = \dfrac{6}{12} = \dfrac{1}{2}$.

d. $\left(x^{-\frac{2}{5}}y^{\frac{1}{3}}\right)^{\frac{1}{2}} = \left(x^{-\frac{2}{5}}\right)^{\frac{1}{2}}\left(y^{\frac{1}{3}}\right)^{\frac{1}{2}}$ To raise a product to a power, raise each factor to the power.

$= x^{-\frac{1}{5}}y^{\frac{1}{6}}$ Multiply: $-\dfrac{2}{5} \cdot \dfrac{1}{2} = -\dfrac{1}{5}$ and $\dfrac{1}{3} \cdot \dfrac{1}{2} = \dfrac{1}{6}$.

$= \dfrac{y^{\frac{1}{6}}}{x^{\frac{1}{5}}}$ Rewrite with positive exponents. ∎

✓ **CHECK POINT 6**   Simplify:

a. $7^{\frac{1}{2}} \cdot 7^{\frac{1}{3}}$ **b.** $\dfrac{50x^{\frac{1}{3}}}{10x^{\frac{4}{3}}}$ **c.** $\left(9.1^{\frac{2}{5}}\right)^{\frac{3}{4}}$ **d.** $\left(x^{-\frac{3}{5}}y^{\frac{1}{4}}\right)^{\frac{1}{3}}$.

5 Simplify radical expressions using rational exponents. ▶

Using Rational Exponents to Simplify Radical Expressions

Some radical expressions can be simplified using rational exponents. We will use the following procedure:

> ### Simplifying Radical Expressions Using Rational Exponents
> 1. Rewrite each radical expression as an exponential expression with a rational exponent.
> 2. Simplify using properties of rational exponents.
> 3. Rewrite in radical notation if rational exponents still appear.

EXAMPLE 7   Simplifying Radical Expressions Using Rational Exponents

Use rational exponents to simplify. Assume that all variables represent nonnegative numbers.

a. $\sqrt[10]{x^5}$ **b.** $\sqrt[3]{27a^{15}}$ **c.** $\sqrt[4]{x^6y^2}$ **d.** $\sqrt{x} \cdot \sqrt[3]{x}$ **e.** $\sqrt[3]{\sqrt{x}}$

Solution

a. $\sqrt[10]{x^5} = x^{\frac{5}{10}}$ Rewrite as an exponential expression.

$= x^{\frac{1}{2}}$ Simplify the exponent.

$= \sqrt{x}$ Rewrite in radical notation.

b. $\sqrt[3]{27a^{15}} = \left(27a^{15}\right)^{\frac{1}{3}}$ Rewrite as an exponential expression.

$= 27^{\frac{1}{3}}\left(a^{15}\right)^{\frac{1}{3}}$ Raise each factor in parentheses to the $\frac{1}{3}$ power.

$= \sqrt[3]{27} \cdot a^{15\left(\frac{1}{3}\right)}$ To raise a power to a power, multiply exponents.

$= 3a^5$ $\sqrt[3]{27} = 3$. Multiply exponents: $15 \cdot \frac{1}{3} = 5$.

c. $\sqrt[4]{x^6 y^2} = (x^6 y^2)^{\frac{1}{4}}$ Rewrite as an exponential expression.

$= (x^6)^{\frac{1}{4}}(y^2)^{\frac{1}{4}}$ Raise each factor in parentheses to the $\frac{1}{4}$ power.

$= x^{\frac{6}{4}} y^{\frac{2}{4}}$ To raise powers to powers, multiply.

$= x^{\frac{3}{2}} y^{\frac{1}{2}}$ Simplify.

$= (x^3 y)^{\frac{1}{2}}$ $a^n b^n = (ab)^n$

$= \sqrt{x^3 y}$ Rewrite in radical notation.

d. $\sqrt{x} \cdot \sqrt[3]{x} = x^{\frac{1}{2}} \cdot x^{\frac{1}{3}}$ Rewrite as exponential expressions.

$= x^{\frac{1}{2}+\frac{1}{3}}$ To multiply with the same base, add exponents.

$= x^{\frac{3}{6}+\frac{2}{6}}$ Write exponents in terms of the LCD, 6.

$= x^{\frac{5}{6}}$ Add: $\frac{3}{6} + \frac{2}{6} = \frac{5}{6}$.

$= \sqrt[6]{x^5}$ Rewrite in radical notation.

e. $\sqrt[3]{\sqrt{x}} = \sqrt[3]{x^{\frac{1}{2}}}$ Write the radicand as an exponential expression.

$= \left(x^{\frac{1}{2}}\right)^{\frac{1}{3}}$ Write the entire expression in exponential form.

$= x^{\frac{1}{6}}$ To raise powers to powers, multiply: $\frac{1}{2} \cdot \frac{1}{3} = \frac{1}{6}$.

$= \sqrt[6]{x}$ Rewrite in radical notation. ■

✓ **CHECK POINT 7**   Use rational exponents to simplify:

a. $\sqrt[6]{x^3}$ **b.** $\sqrt[3]{8a^{12}}$ **c.** $\sqrt[8]{x^4 y^2}$

d. $\dfrac{\sqrt{x}}{\sqrt[3]{x}}$ **e.** $\sqrt{\sqrt[3]{x}}$.

CONCEPT AND VOCABULARY CHECK

Fill in each blank so that the resulting statement is true.

1. $36^{\frac{1}{2}} = \sqrt{\underline{}} = \underline{}$ **2.** $8^{\frac{1}{3}} = \sqrt[3]{\underline{}} = \underline{}$ **3.** $a^{\frac{1}{n}} = \underline{}$ **4.** $16^{\frac{3}{4}} = (\sqrt[4]{\underline{}})^3 = (\underline{})^3 = \underline{}$

5. $a^{\frac{m}{n}} = \underline{}$ **6.** $\sqrt[3]{2^5} = 2^{-}$ **7.** $16^{-\frac{3}{2}} = \dfrac{1}{\underline{}} = \dfrac{1}{(\sqrt{\underline{}})^-} = \dfrac{1}{(\underline{})^-} = \dfrac{1}{\underline{}}$

7.2 EXERCISE SET   ▶ MyMathLab®

Practice Exercises

In Exercises 1–20, use radical notation to rewrite each expression. Simplify, if possible.

1. $49^{\frac{1}{2}}$ **2.** $100^{\frac{1}{2}}$

3. $(-27)^{\frac{1}{3}}$ **4.** $(-64)^{\frac{1}{3}}$

5. $-16^{\frac{1}{4}}$ **6.** $-81^{\frac{1}{4}}$

7. $(xy)^{\frac{1}{3}}$ **8.** $(xy)^{\frac{1}{4}}$

9. $(2xy^3)^{\frac{1}{5}}$ **10.** $(3xy^4)^{\frac{1}{5}}$

11. $81^{\frac{3}{2}}$ **12.** $25^{\frac{3}{2}}$

13. $125^{\frac{2}{3}}$ **14.** $1000^{\frac{2}{3}}$

15. $(-32)^{\frac{3}{5}}$ **16.** $(-27)^{\frac{2}{3}}$

17. $27^{\frac{2}{3}} + 16^{\frac{3}{4}}$

18. $4^{\frac{5}{2}} - 8^{\frac{2}{3}}$

19. $(xy)^{\frac{4}{7}}$ **20.** $(xy)^{\frac{4}{9}}$

In Exercises 21–38, rewrite each expression with rational exponents.

21. $\sqrt{7}$ **22.** $\sqrt{13}$

23. $\sqrt[3]{5}$ **24.** $\sqrt[3]{6}$

25. $\sqrt[5]{11x}$ **26.** $\sqrt[5]{13x}$

27. $\sqrt{x^3}$ **28.** $\sqrt{x^5}$

29. $\sqrt[5]{x^3}$ **30.** $\sqrt[7]{x^4}$

31. $\sqrt[5]{x^2 y}$ **32.** $\sqrt[7]{xy^3}$

33. $(\sqrt{19xy})^3$ **34.** $(\sqrt{11xy})^3$

35. $(\sqrt[6]{7xy^2})^5$ **36.** $(\sqrt[6]{9x^2 y})^5$

37. $2x\sqrt[3]{y^2}$ **38.** $4x\sqrt[5]{y^2}$

In Exercises 39–54, rewrite each expression with a positive rational exponent. Simplify, if possible.

39. $49^{-\frac{1}{2}}$ **40.** $9^{-\frac{1}{2}}$

41. $27^{-\frac{1}{3}}$

42. $125^{-\frac{1}{3}}$

43. $16^{-\frac{3}{4}}$

44. $81^{-\frac{5}{4}}$

45. $8^{-\frac{2}{3}}$

46. $32^{-\frac{4}{5}}$

47. $\left(\dfrac{8}{27}\right)^{-\frac{1}{3}}$

48. $\left(\dfrac{8}{125}\right)^{-\frac{1}{3}}$

49. $(-64)^{-\frac{2}{3}}$

50. $(-8)^{-\frac{2}{3}}$

51. $(2xy)^{-\frac{7}{10}}$

52. $(4xy)^{-\frac{4}{7}}$

53. $5xz^{-\frac{1}{3}}$

54. $7xz^{-\frac{1}{4}}$

In Exercises 55–78, use properties of rational exponents to simplify each expression. Assume that all variables represent positive numbers.

55. $3^{\frac{3}{4}} \cdot 3^{\frac{1}{4}}$

56. $5^{\frac{2}{3}} \cdot 5^{\frac{1}{3}}$

57. $\dfrac{16^{\frac{3}{4}}}{16^{\frac{1}{4}}}$

58. $\dfrac{100^{\frac{3}{4}}}{100^{\frac{31}{4}}}$

59. $x^{\frac{1}{2}} \cdot x^{\frac{1}{3}}$

60. $x^{\frac{1}{2}} \cdot x^{\frac{2}{3}}$

61. $\dfrac{x^{\frac{4}{5}}}{x^{\frac{1}{5}}}$

62. $\dfrac{x^{\frac{3}{7}}}{x^{\frac{1}{7}}}$

63. $\dfrac{x^{\frac{1}{3}}}{x^{\frac{3}{4}}}$

64. $\dfrac{x^{\frac{1}{4}}}{x^{\frac{3}{5}}}$

65. $\left(5^{\frac{2}{3}}\right)^3$

66. $\left(3^{\frac{4}{5}}\right)^5$

67. $\left(y^{-\frac{2}{3}}\right)^{\frac{1}{4}}$

68. $\left(y^{-\frac{3}{4}}\right)^{\frac{1}{6}}$

69. $\left(2x^{\frac{1}{5}}\right)^5$

70. $\left(2x^{\frac{1}{4}}\right)^4$

71. $\left(25x^4y^6\right)^{\frac{1}{2}}$

72. $\left(125x^9y^6\right)^{\frac{1}{3}}$

73. $\left(x^{\frac{1}{2}}y^{-\frac{3}{5}}\right)^{\frac{1}{2}}$

74. $\left(x^{\frac{1}{4}}y^{-\frac{2}{5}}\right)^{\frac{1}{3}}$

75. $\dfrac{3^{\frac{1}{2}} \cdot 3^{\frac{3}{4}}}{3^{\frac{1}{4}}}$

76. $\dfrac{5^{\frac{1}{4}} \cdot 5^{\frac{1}{2}}}{5^{\frac{1}{4}}}$

77. $\dfrac{\left(3y^{\frac{1}{4}}\right)^3}{y^{\frac{1}{12}}}$

78. $\dfrac{\left(2y^{\frac{1}{5}}\right)^4}{y^{\frac{3}{10}}}$

In Exercises 79–112, use rational exponents to simplify each expression. If rational exponents appear after simplifying, write the answer in radical notation. Assume that all variables represent positive numbers.

79. $\sqrt[8]{x^2}$

80. $\sqrt[10]{x^2}$

81. $\sqrt[3]{8a^6}$

82. $\sqrt[3]{27a^{12}}$

83. $\sqrt[5]{x^{10}y^{15}}$

84. $\sqrt[5]{x^{15}y^{20}}$

85. $\left(\sqrt[3]{xy}\right)^{18}$

86. $\left(\sqrt[3]{xy}\right)^{21}$

87. $\sqrt[10]{(3y)^2}$

88. $\sqrt[12]{(3y)^2}$

89. $\left(\sqrt[6]{2a}\right)^4$

90. $\left(\sqrt[8]{2a}\right)^6$

91. $\sqrt[9]{x^6y^3}$

92. $\sqrt[4]{x^2y^6}$

93. $\sqrt{2} \cdot \sqrt[3]{2}$

94. $\sqrt{3} \cdot \sqrt[3]{3}$

95. $\sqrt[5]{x^2} \cdot \sqrt{x}$

96. $\sqrt[7]{x^2} \cdot \sqrt{x}$

97. $\sqrt[4]{a^2b} \cdot \sqrt[3]{ab}$

98. $\sqrt[6]{ab^2} \cdot \sqrt[3]{a^2b}$

99. $\dfrac{\sqrt[4]{x}}{\sqrt[5]{x}}$

100. $\dfrac{\sqrt[3]{x}}{\sqrt[4]{x}}$

101. $\dfrac{\sqrt[3]{y^2}}{\sqrt[6]{y}}$

102. $\dfrac{\sqrt[5]{y^2}}{\sqrt[10]{y^3}}$

103. $\sqrt[4]{\sqrt{x}}$

104. $\sqrt[5]{\sqrt{x}}$

105. $\sqrt{\sqrt{x^2y}}$

106. $\sqrt{\sqrt{xy^2}}$

107. $\sqrt[4]{\sqrt[3]{2x}}$

108. $\sqrt[5]{\sqrt[3]{2x}}$

109. $\left(\sqrt[4]{x^3y^5}\right)^{12}$

110. $\left(\sqrt[5]{x^4y^2}\right)^{20}$

111. $\dfrac{\sqrt[4]{a^5b^5}}{\sqrt{ab}}$

112. $\dfrac{\sqrt[4]{a^3b^3}}{\sqrt{ab}}$

Practice PLUS

In Exercises 113–116, use the distributive property or the FOIL method to perform each multiplication.

113. $x^{\frac{1}{3}}\left(x^{\frac{1}{3}} - x^{\frac{2}{3}}\right)$

114. $x^{-\frac{1}{4}}\left(x^{\frac{9}{4}} - x^{\frac{1}{4}}\right)$

115. $\left(x^{\frac{1}{2}} - 3\right)\left(x^{\frac{1}{2}} + 5\right)$

116. $\left(x^{\frac{1}{3}} - 2\right)\left(x^{\frac{1}{3}} + 6\right)$

In Exercises 117–120, factor out the greatest common factor from each expression.

117. $6x^{\frac{1}{2}} + 2x^{\frac{3}{2}}$

118. $8x^{\frac{1}{4}} + 4x^{\frac{5}{4}}$

119. $15x^{\frac{1}{3}} - 60x$

120. $7x^{\frac{1}{3}} - 70x$

In Exercises 121–124, simplify each expression. Assume that all variables represent positive numbers.

121. $\left(49x^{-2}y^4\right)^{-\frac{1}{2}}\left(xy^{\frac{1}{2}}\right)$

122. $\left(8x^{-6}y^3\right)^{\frac{1}{3}}\left(x^{\frac{5}{6}}y^{-\frac{1}{3}}\right)^6$

123. $\left(\dfrac{x^{-\frac{5}{4}}y^{\frac{1}{3}}}{x^{-\frac{3}{4}}}\right)^{-6}$

124. $\left(\dfrac{x^{\frac{1}{2}}y^{-\frac{7}{4}}}{y^{-\frac{5}{4}}}\right)^{-4}$

Application Exercises

The Galápagos Islands, lying 600 miles west of Ecuador, are famed for their extraordinary wildlife. The function

$$f(x) = 29x^{\frac{1}{3}}$$

models the number of plant species, $f(x)$, on the various islands of the Galápagos chain in terms of the area, x, in square miles, of a particular island. Use the function to solve Exercises 125–126.

125. How many species of plants are on a Galápagos island that has an area of 8 square miles?

126. How many species of plants are on a Galápagos island that has an area of 27 square miles?

The function

$$f(x) = 70x^{\frac{3}{4}}$$

models the number of calories per day, f(x), a person needs to maintain life in terms of that person's weight, x, in kilograms. (1 kilogram is approximately 2.2 pounds.) Use this model and a calculator to solve Exercises 127–128. Round answers to the nearest calorie.

127. How many calories per day does a person who weighs 80 kilograms (approximately 176 pounds) need to maintain life?

128. How many calories per day does a person who weighs 70 kilograms (approximately 154 pounds) need to maintain life?

The way that we perceive the temperature on a cold day depends on both air temperature and wind speed. The windchill is what the air temperature would have to be with no wind to achieve the same chilling effect on the skin. In 2002, the National Weather Service issued new windchill temperatures, shown in the table below. (One reason for this new windchill index is that the wind speed is now calculated at 5 feet, the average height of the human body's face, rather than 33 feet, the height of the standard anemometer, an instrument that calculates wind speed.)

New Windchill Temperature Index

Air Temperature (°F)

	30	25	20	15	10	5	0	−5	−10	−15	−20	−25
5	25	19	13	7	1	−5	−11	−16	−22	−28	−34	−40
10	21	15	9	3	−4	−10	−16	−22	−28	−35	−41	−47
15	19	13	6	0	−7	−13	−19	−26	−32	−39	−45	−51
20	17	11	4	−2	−9	−15	−22	−29	−35	−42	−48	−55
25	16	9	3	−4	−11	−17	−24	−31	−37	−44	−51	−58
30	15	8	1	−5	−12	−19	−26	−33	−39	−46	−53	−60
35	14	7	0	−7	−14	−21	−27	−34	−41	−48	−55	−62
40	13	6	−1	−8	−15	−22	−29	−36	−43	−50	−57	−64
45	12	5	−2	−9	−16	−23	−30	−37	−44	−51	−58	−65
50	12	4	−3	−10	−17	−24	−31	−38	−45	−52	−60	−67
55	11	4	−3	−11	−18	−25	−32	−39	−46	−54	−61	−68
60	10	3	−4	−11	−19	−26	−33	−40	−48	−55	−62	−69

Wind Speed (miles per hour)

Frostbite occurs in 15 minutes or less.

Source: National Weather Service

The windchill temperatures shown in the table can be calculated using

$$C = 35.74 + 0.6215t - 35.74\sqrt[25]{v^4} + 0.4275t\sqrt[25]{v^4},$$

in which C is the windchill, in degrees Fahrenheit, t is the air temperature, in degrees Fahrenheit, and v is the wind speed, in miles per hour. Use the formula to solve Exercises 129–132.

129. a. Rewrite the equation for calculating windchill temperatures using rational exponents.

b. Use the form of the equation in part (a) and a calculator to find the windchill temperature, to the nearest degree, when the air temperature is 25°F and the wind speed is 30 miles per hour.

130. a. Rewrite the equation for calculating windchill temperatures using rational exponents.

b. Use the form of the equation in part (a) and a calculator to find the windchill temperature, to the nearest degree, when the air temperature is 35°F and the wind speed is 15 miles per hour.

131. a. Substitute 0 for t in the equation with rational exponents from Exercise 129(a) and write a function C(v) that gives the windchill temperature as a function of wind speed for an air temperature of 0°F.

b. Find and interpret C(25). Use a calculator and round to the nearest degree.

c. Identify your solution to part (b) on the graph shown.

[0, 60, 5] by [−35, 0, 5]

132. a. Substitute 30 for t in the equation with rational exponents from Exercise 130(a) and write a function C(v) that gives the windchill temperature as a function of wind speed for an air temperature of 30°F. Simplify the function's formula so that it contains exactly two terms.

b. Find and interpret C(40). Use a calculator and round to the nearest degree.

c. Identify your solution to part (b) on the graph shown.

Air Speed (miles per hour)
[0, 60, 5] by [0, 25, 5]

Your job is to determine whether or not yachts are eligible for the America's Cup, the supreme event in ocean sailing. The basic dimensions of competitors' yachts must satisfy

$$L + 1.25\sqrt{S} - 9.8\sqrt[3]{D} \le 16.296,$$

where L is the yacht's length, in meters, S is its sail area, in square meters, and D is its displacement, in cubic meters. Use this information to solve Exercises 133–134.

133. a. Rewrite the inequality using rational exponents.

b. Use your calculator to determine if a yacht with length 20.85 meters, sail area 276.4 square meters, and displacement 18.55 cubic meters is eligible for the America's Cup.

134. a. Rewrite the inequality using rational exponents.

b. Use your calculator to determine if a yacht with length 22.85 meters, sail area 312.5 square meters, and displacement 22.34 cubic meters is eligible for the America's Cup.

Explaining the Concepts

135. What is the meaning of $a^{\frac{1}{n}}$? Give an example to support your explanation.

136. What is the meaning of $a^{\frac{m}{n}}$? Give an example.

137. What is the meaning of $a^{-\frac{m}{n}}$? Give an example.

138. Explain why $a^{\frac{1}{n}}$ is negative when n is odd and a is negative. What happens if n is even and a is negative? Why?

139. In simplifying $36^{\frac{3}{2}}$, is it better to use $a^{\frac{m}{n}} = \sqrt[n]{a^m}$ or $a^{\frac{m}{n}} = (\sqrt[n]{a})^m$? Explain.

140. How can you tell if an expression with rational exponents is simplified?

141. Explain how to simplify $\sqrt[3]{x} \cdot \sqrt{x}$.

142. Explain how to simplify $\sqrt[3]{\sqrt{x}}$.

Technology Exercises

143. Use a scientific or graphing calculator to verify your results in Exercises 15–18.

144. Use a scientific or graphing calculator to verify your results in Exercises 45–50.

Exercises 145–147 show a number of simplifications, not all of which are correct. Enter the left side of each equation as y_1 and the right side as y_2. Then use your graphing utility's TABLE *feature to determine if the simplification is correct. If it is not, correct the right side and use the* TABLE *feature to verify your simplification.*

145. $x^{\frac{3}{5}} \cdot x^{-\frac{1}{10}} = x^{\frac{1}{2}}$

146. $\left(x^{-\frac{1}{2}} \cdot x^{\frac{3}{4}}\right)^{-2} = x^{\frac{1}{2}}$

147. $\dfrac{x^{\frac{1}{4}}}{x^{\frac{1}{2}} \cdot x^{-\frac{3}{4}}} = \dfrac{1}{x^{\frac{1}{2}}}$

Critical Thinking Exercises

Make Sense? *In Exercises 148–151, determine whether each statement makes sense or does not make sense, and explain your reasoning.*

148. By adding the exponents, I simplified $7^{\frac{1}{2}} \cdot 7^{\frac{1}{2}}$ and obtained 49.

149. When I use the definition for $a^{\frac{m}{n}}$, I usually prefer to first raise a to the m power because smaller numbers are involved.

150. There's no question that $(-64)^{\frac{1}{3}} = -64^{\frac{1}{3}}$, so I can conclude that $(-64)^{\frac{1}{2}} = -64^{\frac{1}{2}}$.

151. I checked the following simplification and every step is correct:

$$5\left(4 - 5^{\frac{1}{2}}\right) = 5 \cdot 4 - 5 \cdot 5^{\frac{1}{2}}$$
$$= 20 - 25^{\frac{1}{2}}$$
$$= 20 - \sqrt{25}$$
$$= 20 - 5 = 15.$$

In Exercises 152–155, determine whether each statement is true or false. If the statement is false, make the necessary change(s) to produce a true statement.

152. If n is odd, then $(-b)^{\frac{1}{n}} = -b^{\frac{1}{n}}$.

153. $(a + b)^{\frac{1}{n}} = a^{\frac{1}{n}} + b^{\frac{1}{n}}$

154. $8^{-\frac{2}{3}} = -4$

155. $4^{-3.5} = \dfrac{1}{128}$

156. A mathematics professor recently purchased a birthday cake for her son with the inscription

Happy $\left(2^{\frac{5}{2}} \cdot 2^{\frac{3}{4}} \div 2^{\frac{1}{4}}\right)$ th Birthday.

How old is the son?

157. The birthday boy in Exercise 156, excited by the inscription on the cake, tried to wolf down the whole thing. Professor Mom, concerned about the possible metamorphosis of her son into a blimp, exclaimed, "Hold on! It is your birthday, so why not take $\dfrac{8^{-\frac{4}{3}} + 2^{-2}}{16^{-\frac{3}{4}} + 2^{-1}}$ of the cake? I'll eat half of what's left over." How much of the cake did the professor eat?

158. Simplify: $\left[3 + \left(27^{\frac{2}{3}} + 32^{\frac{2}{5}}\right)\right]^{\frac{3}{2}} - 9^{\frac{1}{2}}$.

159. Find the domain of $f(x) = (x - 3)^{\frac{1}{2}}(x + 4)^{-\frac{1}{2}}$.

Review Exercises

160. Write the equation of the linear function whose graph passes through (5, 1) and (4, 3). (Section 2.5, Example 2)

161. Graph $y \le -\dfrac{3}{2}x + 3$. (Section 4.4, Example 2)

162. Solve by Cramer's rule:

$$\begin{cases} 5x - 3y = 3 \\ 7x + y = 25. \end{cases}$$

(Section 3.5, Example 2)

Preview Exercises

Exercises 163–165 will help you prepare for the material covered in the next section.

163. a. Find $\sqrt{16} \cdot \sqrt{4}$.

    b. Find $\sqrt{16 \cdot 4}$.

    c. Based on your answers to parts (a) and (b), what can you conclude?

164. a. Use a calculator to approximate $\sqrt{300}$ to two decimal places.

    b. Use a calculator to approximate $10\sqrt{3}$ to two decimal places.

    c. Based on your answers to parts (a) and (b), what can you conclude?

165. Simplify: **a.** $\sqrt[3]{x^{21}}$ **b.** $\sqrt[6]{y^{24}}$.

SECTION
7.3

Multiplying and Simplifying Radical Expressions ▶

What am I supposed to learn?

After studying this section, you should be able to:

1 Use the product rule to multiply radicals. ▶

2 Use factoring and the product rule to simplify radicals. ▶

3 Multiply radicals and then simplify. ▶

Mirror II (1963), George Tooker. Addison Gallery, Phillips Academy, MA. © George Tooker.

We opened this book with a model that described our improving emotional health as we age. Unfortunately, not everything gets better. The aging process is also accompanied by a number of physical transformations, including changes in vision that require glasses for reading, the onset of wrinkles and sagging skin, and a decrease in heart response. A change in heart response occurs fairly early; after 20, our hearts become less adept at accelerating in response to exercise. In this section's Exercise Set (Exercises 95–96), you will see how a radical function models changes in heart function throughout the aging process, as we turn to multiplying and simplifying radical expressions.

1 Use the product rule to multiply radicals. ▶

The Product Rule for Radicals

A rule for multiplying radicals can be generalized by comparing $\sqrt{25} \cdot \sqrt{4}$ and $\sqrt{25 \cdot 4}$. Notice that

$$\sqrt{25} \cdot \sqrt{4} = 5 \cdot 2 = 10 \quad \text{and} \quad \sqrt{25 \cdot 4} = \sqrt{100} = 10.$$

Because we obtain 10 in both situations, the original radical expressions must be equal. That is,

$$\sqrt{25} \cdot \sqrt{4} = \sqrt{25 \cdot 4}.$$

This result is a special case of the **product rule for radicals** that can be generalized as follows:

The Product Rule for Radicals

If $\sqrt[n]{a}$ and $\sqrt[n]{b}$ are real numbers, then

$$\sqrt[n]{a} \cdot \sqrt[n]{b} = \sqrt[n]{ab}.$$

The product of two nth roots is the nth root of the product of the radicands.

EXAMPLE 1   Using the Product Rule for Radicals

Multiply:

a. $\sqrt{3} \cdot \sqrt{7}$ **b.** $\sqrt{x+7} \cdot \sqrt{x-7}$ **c.** $\sqrt[3]{7} \cdot \sqrt[3]{9}$ **d.** $\sqrt[8]{10x} \cdot \sqrt[8]{8x^4}$.

Solution   In each problem, the indices are the same. Thus, we multiply by multiplying the radicands.

a. $\sqrt{3} \cdot \sqrt{7} = \sqrt{3 \cdot 7} = \sqrt{21}$

b. $\sqrt{x+7} \cdot \sqrt{x-7} = \sqrt{(x+7)(x-7)} = \sqrt{x^2 - 49}$

> This is not equal to $\sqrt{x^2} - \sqrt{49}$.

c. $\sqrt[3]{7} \cdot \sqrt[3]{9} = \sqrt[3]{7 \cdot 9} = \sqrt[3]{63}$

d. $\sqrt[8]{10x} \cdot \sqrt[8]{8x^4} = \sqrt[8]{10x \cdot 8x^4} = \sqrt[8]{80x^5}$ ■

✓ **CHECK POINT 1**   Multiply:

a. $\sqrt{5} \cdot \sqrt{11}$ **b.** $\sqrt{x+4} \cdot \sqrt{x-4}$

c. $\sqrt[3]{6} \cdot \sqrt[3]{10}$ **d.** $\sqrt[7]{2x} \cdot \sqrt[7]{6x^3}$.

Great Question!

Can I use the product rule to simplify radicals with different indices, such as $\sqrt{x} \cdot \sqrt[3]{x}$?

No. The product rule can be used only when the radicals have the same index. If indices differ, rational exponents can be used, as in $\sqrt{x} \cdot \sqrt[3]{x}$, which was Example 7(d) in the previous section.

2 Use factoring and the product rule to simplify radicals. ▶

Using Factoring and the Product Rule to Simplify Radicals

In Chapter 5, we saw that a number that is the square of an integer is a **perfect square**. For example, 100 is a perfect square because $100 = 10^2$. A number is a **perfect cube** if it is the cube of an integer. Thus, 125 is a perfect cube because $125 = 5^3$. In general, a number is a **perfect nth power** if it is the nth power of an integer. Thus, p is a perfect nth power if there is an integer q such that $p = q^n$.

A radical of index n is **simplified** when its radicand has no factors other than 1 that are perfect nth powers. For example, $\sqrt{300}$ is not simplified because it can be expressed as $\sqrt{100 \cdot 3}$ and 100 is a perfect square. We can use the product rule in the form

$$\sqrt[n]{ab} = \sqrt[n]{a} \cdot \sqrt[n]{b}$$

to simplify $\sqrt[n]{ab}$ when a or b is a perfect nth power. Consider $\sqrt{300}$. To simplify, we factor 300 so that one of its factors is the greatest perfect square possible.

Using Technology

You can use a calculator to provide numerical support that $\sqrt{300} = 10\sqrt{3}$. First find an approximation for $\sqrt{300}$:

$$300 \boxed{\sqrt{}} \approx 17.32$$

or

$$\boxed{\sqrt{}} \, 300 \boxed{\text{ENTER}} \approx 17.32.$$

Now find an aproximation for $10\sqrt{3}$:

$$10 \boxed{\times} 3 \boxed{\sqrt{}} \approx 17.32$$

or

$$10 \boxed{\sqrt{}} \, 3 \boxed{\text{ENTER}} \approx 17.32.$$

Correct to two decimal places,

$$\sqrt{300} \approx 17.32 \quad \text{and} \quad 10\sqrt{3} \approx 17.32.$$

This verifies that

$$\sqrt{300} = 10\sqrt{3}.$$

Use this technique to support the numerical results for the answers in this section. *Caution:* A simplified radical does not mean a decimal approximation.

$$\sqrt{300} = \sqrt{100 \cdot 3} \qquad \text{Factor 300: 100 is the greatest perfect square factor.}$$
$$= \sqrt{100} \cdot \sqrt{3} \qquad \text{Use the product rule: } \sqrt[n]{ab} = \sqrt[n]{a} \cdot \sqrt[n]{b}.$$
$$= 10\sqrt{3} \qquad \text{Write } \sqrt{100} \text{ as 10. We read } 10\sqrt{3} \text{ as "ten times the square root of three."}$$

Simplifying Radical Expressions by Factoring

A radical expression whose index is n is **simplified** when its radicand has no factors that are perfect nth powers. To simplify, use the following procedure:

1. Write the radicand as the product of two factors, one of which is the greatest perfect nth power.
2. Use the product rule to take the nth root of each factor.
3. Find the nth root of the perfect nth power.

EXAMPLE 2 Simplifying Radicals by Factoring

Simplify by factoring:

a. $\sqrt{75}$ **b.** $\sqrt[3]{54}$ **c.** $\sqrt[5]{64}$ **d.** $\sqrt{500xy^2}$.

Solution

a. $\sqrt{75} = \sqrt{25 \cdot 3}$ 25 is the greatest perfect square that is a factor of 75.
$\quad\quad\;\; = \sqrt{25} \cdot \sqrt{3}$ Take the square root of each factor: $\sqrt[n]{ab} = \sqrt[n]{a} \cdot \sqrt[n]{b}$
$\quad\quad\;\; = 5\sqrt{3}$ Write $\sqrt{25}$ as 5.

b. $\sqrt[3]{54} = \sqrt[3]{27 \cdot 2}$ 27 is the greatest perfect cube that is a factor of 54: $27 = 3^3$.
$\quad\quad\;\; = \sqrt[3]{27} \cdot \sqrt[3]{2}$ Take the cube root of each factor: $\sqrt[n]{ab} = \sqrt[n]{a} \cdot \sqrt[n]{b}$.
$\quad\quad\;\; = 3\sqrt[3]{2}$ Write $\sqrt[3]{27}$ as 3.

c. $\sqrt[5]{64} = \sqrt[5]{32 \cdot 2}$ 32 is the greatest perfect fifth power that is a factor of 64: $32 = 2^5$.
$\quad\quad\;\; = \sqrt[5]{32} \cdot \sqrt[5]{2}$ Take the fifth root of each factor: $\sqrt[n]{ab} = \sqrt[n]{a} \cdot \sqrt[n]{b}$
$\quad\quad\;\; = 2\sqrt[5]{2}$ Write $\sqrt[5]{32}$ as 2.

d. $\sqrt{500xy^2} = \sqrt{100y^2 \cdot 5x}$ $100y^2$ is the greatest perfect square that is a factor of $500xy^2$: $100y^2 = (10y)^2$.
$\quad\quad\;\; = \sqrt{100y^2} \cdot \sqrt{5x}$ Factor into two radicals.
$\quad\quad\;\; = 10|y|\sqrt{5x}$ Take the square root of $100y^2$. ∎

✓ CHECK POINT 2 Simplify by factoring:

a. $\sqrt{80}$ **b.** $\sqrt[3]{40}$
c. $\sqrt[4]{32}$ **d.** $\sqrt{200x^2 y}$.

EXAMPLE 3 Simplifying a Radical Function

If

$$f(x) = \sqrt{2x^2 + 4x + 2},$$

express the function, f, in simplified form.

Solution Begin by factoring the radicand. The GCF is 2. Simplification is possible if we obtain a factor that is a perfect square.

$$f(x) = \sqrt{2x^2 + 4x + 2}$$ This is the given function.

$$= \sqrt{2(x^2 + 2x + 1)}$$ Factor out the GCF.

$$= \sqrt{2(x + 1)^2}$$ Factor the perfect square trinomial: $A^2 + 2AB + B^2 = (A + B)^2$.

$$= \sqrt{2} \cdot \sqrt{(x + 1)^2}$$ Take the square root of each factor. The factor $(x + 1)^2$. is a perfect square.

$$= \sqrt{2}|x + 1|$$ Take the square root of $(x + 1)^2$.

In simplified form,

$$f(x) = \sqrt{2}|x + 1|. \quad \blacksquare$$

Using Technology

Graphic Connections

The graphs of

$$f(x) = \sqrt{2x^2 + 4x + 2}, \quad g(x) = \sqrt{2}|x + 1|, \quad \text{and} \quad h(x) = \sqrt{2}(x + 1)$$

are shown in **Figure 7.7** in three separate $[-5, 5, 1]$ by $[-5, 5, 1]$ viewing rectangles. The graphs in **Figure 7.7 (a)** and **(b)** are identical. This verifies that our simplification in Example 3 is correct: $\sqrt{2x^2 + 4x + 2} = \sqrt{2}|x + 1|$. Now compare the graphs in **Figure 7.7 (a)** and **(c)**. Can you see that they are not the same? This illustrates the importance of not leaving out absolute value bars:

$$\sqrt{2x^2 + 4x + 2} \neq \sqrt{2}(x + 1).$$

Figure 7.7 (a)   **(b)**   **(c)**

✅ **CHECK POINT 3**   If $f(x) = \sqrt{3x^2 - 12x + 12}$, express the function, f, in simplified form.

For the remainder of this chapter, in situations that do not involve functions, we will **assume that no radicands involve negative quantities raised to even powers. Based upon this assumption, absolute value bars are not necessary when taking even roots.**

> ### Simplifying When Variables to Even Powers in a Radicand Are Nonnegative Quantities
>
> For any nonnegative real number a,
>
> $$\sqrt[n]{a^n} = a.$$

In simplifying an nth root, how do we find variable factors in the radicand that are perfect nth powers? **The perfect nth powers have exponents that are divisible by n.** Simplification is possible by observation or by using rational exponents. Here are some examples:

- $\sqrt{x^6} = \sqrt{(x^3)^2} = x^3$ or $\sqrt{x^6} = (x^6)^{\frac{1}{2}} = x^3$

  6 is divisible by the index, 2. Thus, x^6 is a perfect square.

- $\sqrt[3]{y^{21}} = \sqrt[3]{(y^7)^3} = y^7$ or $\sqrt[3]{y^{21}} = (y^{21})^{\frac{1}{3}} = y^7$

  21 is divisible by the index, 3. Thus, y^{21} is a perfect cube.

- $\sqrt[6]{z^{24}} = \sqrt[6]{(z^4)^6} = z^4$ or $\sqrt[6]{z^{24}} = (z^{24})^{\frac{1}{6}} = z^4$.

  24 is divisible by the index, 6. Thus, z^{24} is a perfect 6th power.

EXAMPLE 4 Simplifying a Radical by Factoring

Simplify: $\sqrt{x^5 y^{13} z^7}$.

Solution We write the radicand as the product of the greatest perfect square factor and another factor. Because the index is 2, variables that have exponents that are divisible by 2 are part of the perfect square factor. We use the greatest exponents that are divisible by 2.

Discover for Yourself

Square the answer in Example 4 and show that it is correct. If it is a square root, you should obtain the given radicand, $x^5 y^{13} z^7$.

$$\sqrt{x^5 y^{13} z^7} = \sqrt{x^4 \cdot x \cdot y^{12} \cdot y \cdot z^6 \cdot z} \quad \text{Use the greatest even power of each variable.}$$

$$= \sqrt{(x^4 y^{12} z^6)(xyz)} \quad \text{Group the perfect square factors.}$$

$$= \sqrt{x^4 y^{12} z^6} \cdot \sqrt{xyz} \quad \text{Factor into two radicals.}$$

$$= x^2 y^6 z^3 \sqrt{xyz} \quad \sqrt{x^4 y^{12} z^6} = (x^4 y^{12} z^6)^{\frac{1}{2}} = x^2 y^6 z^3 \quad \blacksquare$$

✓ **CHECK POINT 4** Simplify: $\sqrt{x^9 y^{11} z^3}$.

EXAMPLE 5 Simplifying a Radical by Factoring

Simplify: $\sqrt[3]{32 x^8 y^{16}}$.

Solution We write the radicand as the product of the greatest perfect cube factor and another factor. Because the index is 3, variables that have exponents that are divisible by 3 are part of the perfect cube factor. We use the greatest exponents that are divisible by 3.

$$\sqrt[3]{32 x^8 y^{16}} = \sqrt[3]{8 \cdot 4 \cdot x^6 \cdot x^2 \cdot y^{15} \cdot y} \quad \text{Identify perfect cube factors.}$$

$$= \sqrt[3]{(8 x^6 y^{15})(4 x^2 y)} \quad \text{Group the perfect cube factors.}$$

$$= \sqrt[3]{8 x^6 y^{15}} \cdot \sqrt[3]{4 x^2 y} \quad \text{Factor into two radicals.}$$

$$= 2 x^2 y^5 \sqrt[3]{4 x^2 y} \quad \sqrt[3]{8} = 2 \text{ and} \\ \sqrt[3]{x^6 y^{15}} = (x^6 y^{15})^{\frac{1}{3}} = x^2 y^5. \quad \blacksquare$$

✓ **CHECK POINT 5** Simplify: $\sqrt[3]{40 x^{10} y^{14}}$.

EXAMPLE 6 Simplifying a Radical by Factoring

Simplify: $\sqrt[5]{64 x^3 y^7 z^{29}}$.

Solution We write the radicand as the product of the greatest perfect 5th power and another factor. Because the index is 5, variables that have exponents that are divisible by 5 are part of the perfect fifth factor. We use the greatest exponents that are divisible by 5.

$$\sqrt[5]{64 x^3 y^7 z^{29}} = \sqrt[5]{32 \cdot 2 \cdot x^3 \cdot y^5 \cdot y^2 \cdot z^{25} \cdot z^4} \quad \text{Identify perfect fifth factors.}$$

$$= \sqrt[5]{(32 y^5 z^{25})(2 x^3 y^2 z^4)} \quad \text{Group the perfect fifth factors.}$$

$$= \sqrt[5]{32 y^5 z^{25}} \cdot \sqrt[5]{2 x^3 y^2 z^4} \quad \text{Factor into two radicals.}$$

$$= 2 y z^5 \sqrt[5]{2 x^3 y^2 z^4} \quad \sqrt[5]{32} = 2 \text{ and } \sqrt[5]{y^5 z^{25}} = (y^5 z^{25})^{\frac{1}{5}} = y z^5. \quad \blacksquare$$

✓ **CHECK POINT 6** Simplify: $\sqrt[5]{32 x^{12} y^2 z^8}$.

---

3 Multiply radicals and then simplify. ▶

Multiplying and Simplifying Radicals

We have seen how to use the product rule when multiplying radicals with the same index. Sometimes after multiplying, we can simplify the resulting radical.

EXAMPLE 7 Multiplying Radicals and Then Simplifying

Multiply and simplify:

a. $\sqrt{15} \cdot \sqrt{3}$ **b.** $7\sqrt[3]{4} \cdot 5\sqrt[3]{6}$ **c.** $\sqrt[4]{8x^3y^2} \cdot \sqrt[4]{8x^5y^3}$.

Great Question!

When should I write an expression under one radical and when should I separate the radicals?

- Use $\sqrt[n]{a} \cdot \sqrt[n]{b} = \sqrt[n]{ab}$, writing under one radical, when *multiplying*.
- Use $\sqrt[n]{ab} = \sqrt[n]{a} \cdot \sqrt[n]{b}$, factoring into two radicals, when *simplifying*.

Solution

a. $\sqrt{15} \cdot \sqrt{3} = \sqrt{15 \cdot 3}$ Use the product rule.

$= \sqrt{45} = \sqrt{9 \cdot 5}$ 9 is the greatest perfect square factor of 45.

$= \sqrt{9} \cdot \sqrt{5} = 3\sqrt{5}$

b. $7\sqrt[3]{4} \cdot 5\sqrt[3]{6} = 35\sqrt[3]{4 \cdot 6}$ Use the product rule.

$= 35\sqrt[3]{24} = 35\sqrt[3]{8 \cdot 3}$ 8 is the greatest perfect cube factor of 24.

$= 35\sqrt[3]{8} \cdot \sqrt[3]{3} = 35 \cdot 2 \cdot \sqrt[3]{3}$

$= 70\sqrt[3]{3}$

c. $\sqrt[4]{8x^3y^2} \cdot \sqrt[4]{8x^5y^3} = \sqrt[4]{8x^3y^2 \cdot 8x^5y^3}$ Use the product rule.

$= \sqrt[4]{64x^8y^5}$ Multiply.

$= \sqrt[4]{16 \cdot 4 \cdot x^8 \cdot y^4 \cdot y}$ Identify perfect fourth factors.

$= \sqrt[4]{(16x^8y^4)(4y)}$ Group the perfect fourth factors.

$= \sqrt[4]{16x^8y^4} \cdot \sqrt[4]{4y}$ Factor into two radicals.

$= 2x^2y\sqrt[4]{4y}$ $\sqrt[4]{16} = 2$ and $\sqrt[4]{x^8y^4} = (x^8y^4)^{\frac{1}{4}} = x^2y$. ∎

☑ **CHECK POINT 7** Multiply and simplify:

a. $\sqrt{6} \cdot \sqrt{2}$

b. $10\sqrt[3]{16} \cdot 5\sqrt[3]{2}$

c. $\sqrt[4]{4x^2y} \cdot \sqrt[4]{8x^6y^3}$.

CONCEPT AND VOCABULARY CHECK

Fill in each blank so that the resulting statement is true.

1. If $\sqrt[n]{a}$ and $\sqrt[n]{b}$ are real numbers, then $\sqrt[n]{a} \cdot \sqrt[n]{b} = $ _____.

2. $\sqrt[3]{7} \cdot \sqrt[3]{11} = \sqrt[3]{\_\_}$

3. $\sqrt[3]{8 \cdot 5} = \sqrt[3]{__} \cdot \sqrt[3]{__} = $ _____

4. $\sqrt{5(x + 1)^2} = $ _____

5. Variable factors in a radicand that are perfect *n*th powers can be simplified by observation. For example, $\sqrt{x^{10}} = $ _____, $\sqrt[3]{x^{12}} = $ _____, and $\sqrt[6]{x^{30}} = $ _____.

7.3 EXERCISE SET   ▶ MyMathLab®

Practice Exercises

In Exercises 1–20, use the product rule to multiply.

1. $\sqrt{3} \cdot \sqrt{5}$ 2. $\sqrt{7} \cdot \sqrt{5}$

3. $\sqrt[3]{2} \cdot \sqrt[3]{9}$ 4. $\sqrt[3]{5} \cdot \sqrt[3]{4}$

5. $\sqrt[4]{11} \cdot \sqrt[4]{3}$ 6. $\sqrt[5]{9} \cdot \sqrt[5]{3}$

7. $\sqrt{3x} \cdot \sqrt{11y}$ 8. $\sqrt{5x} \cdot \sqrt{11y}$

9. $\sqrt[5]{6x^3} \cdot \sqrt[5]{4x}$ 10. $\sqrt[4]{6x^2} \cdot \sqrt[4]{3x}$

11. $\sqrt{x + 3} \cdot \sqrt{x - 3}$

12. $\sqrt{x + 6} \cdot \sqrt{x - 6}$

13. $\sqrt[6]{x - 4} \cdot \sqrt[6]{(x - 4)^4}$

14. $\sqrt[6]{x - 5} \cdot \sqrt[6]{(x - 5)^4}$

15. $\sqrt{\dfrac{2x}{3}} \cdot \sqrt{\dfrac{3}{2}}$ 16. $\sqrt{\dfrac{2x}{5}} \cdot \sqrt{\dfrac{5}{2}}$

17. $\sqrt[4]{\dfrac{x}{7}} \cdot \sqrt[4]{\dfrac{3}{y}}$

18. $\sqrt[4]{\dfrac{x}{3}} \cdot \sqrt[4]{\dfrac{7}{y}}$

19. $\sqrt[5]{7x^2y} \cdot \sqrt[5]{11x^3y^2}$

20. $\sqrt[9]{12x^2y^3} \cdot \sqrt[9]{3x^3y^4}$

In Exercises 21–32, simplify by factoring.

21. $\sqrt{50}$ **22.** $\sqrt{27}$

23. $\sqrt{45}$ **24.** $\sqrt{28}$

25. $\sqrt{75x}$ **26.** $\sqrt{40x}$

27. $\sqrt[3]{16}$ **28.** $\sqrt[3]{54}$

29. $\sqrt[3]{27x^3}$ **30.** $\sqrt[3]{250x^3}$

31. $\sqrt[3]{-16x^2y^3}$ **32.** $\sqrt[3]{-32x^2y^3}$

In Exercises 33–38, express the function, f, in simplified form. Assume that x can be any real number.

33. $f(x) = \sqrt{36(x+2)^2}$

34. $f(x) = \sqrt{81(x-2)^2}$

35. $f(x) = \sqrt[3]{32(x+2)^3}$

36. $f(x) = \sqrt[3]{48(x-2)^3}$

37. $f(x) = \sqrt{3x^2 - 6x + 3}$

38. $f(x) = \sqrt{5x^2 - 10x + 5}$

In Exercises 39–60, simplify by factoring. Assume that all variables in a radicand represent positive real numbers and no radicands involve negative quantities raised to even powers.

39. $\sqrt{x^7}$ **40.** $\sqrt{x^5}$

41. $\sqrt{x^8y^9}$ **42.** $\sqrt{x^6y^7}$

43. $\sqrt{48x^3}$ **44.** $\sqrt{40x^3}$

45. $\sqrt[3]{y^8}$ **46.** $\sqrt[3]{y^{11}}$

47. $\sqrt[3]{x^{14}y^3z}$ **48.** $\sqrt[3]{x^3y^{17}z^2}$

49. $\sqrt[3]{81x^8y^6}$ **50.** $\sqrt[3]{32x^9y^{17}}$

51. $\sqrt[3]{(x+y)^5}$

52. $\sqrt[3]{(x+y)^4}$

53. $\sqrt[5]{y^{17}}$ **54.** $\sqrt[5]{y^{18}}$

55. $\sqrt[5]{64x^6y^{17}}$ **56.** $\sqrt[5]{64x^7y^{16}}$

57. $\sqrt[4]{80x^{10}}$ **58.** $\sqrt[4]{96x^{11}}$

59. $\sqrt[4]{(x-3)^{10}}$

60. $\sqrt[4]{(x-2)^{14}}$

In Exercises 61–82, multiply and simplify. Assume that all variables in a radicand represent positive real numbers and no radicands involve negative quantities raised to even powers.

61. $\sqrt{12} \cdot \sqrt{2}$

62. $\sqrt{3} \cdot \sqrt{6}$

63. $\sqrt{5x} \cdot \sqrt{10y}$

64. $\sqrt{8x} \cdot \sqrt{10y}$

65. $\sqrt{12x} \cdot \sqrt{3x}$

66. $\sqrt{20x} \cdot \sqrt{5x}$

67. $\sqrt{50xy} \cdot \sqrt{4xy^2}$

68. $\sqrt{5xy} \cdot \sqrt{10xy^2}$

69. $2\sqrt{5} \cdot 3\sqrt{40}$

70. $3\sqrt{15} \cdot 5\sqrt{6}$

71. $\sqrt[3]{12} \cdot \sqrt[3]{4}$

72. $\sqrt[4]{4} \cdot \sqrt[4]{8}$

73. $\sqrt{5x^3} \cdot \sqrt{8x^2}$

74. $\sqrt{2x^7} \cdot \sqrt{12x^4}$

75. $\sqrt[3]{25x^4y^2} \cdot \sqrt[3]{5xy^{12}}$

76. $\sqrt[3]{6x^7y} \cdot \sqrt[3]{9x^4y^{12}}$

77. $\sqrt[4]{8x^2y^3z^6} \cdot \sqrt[4]{2x^4yz}$

78. $\sqrt[4]{4x^2y^3z^3} \cdot \sqrt[4]{8x^4yz^6}$

79. $\sqrt[5]{8x^4y^6z^2} \cdot \sqrt[5]{8xy^7z^4}$

80. $\sqrt[5]{8x^4y^3z^3} \cdot \sqrt[5]{8xy^9z^8}$

81. $\sqrt[3]{x-y} \cdot \sqrt[3]{(x-y)^7}$

82. $\sqrt[3]{x-6} \cdot \sqrt[3]{(x-6)^7}$

Practice PLUS

In Exercises 83–90, simplify each expression. Assume that all variables in a radicand represent positive real numbers and no radicands involve negative quantities raised to even powers.

83. $-2x^2y\left(\sqrt[3]{54x^3y^7z^2}\right)$

84. $\dfrac{-x^2y^7}{2}\left(\sqrt[3]{-32x^4y^9z^7}\right)$

85. $-3y\left(\sqrt[5]{64x^3y^6}\right)$

86. $-4x^2y^7\left(\sqrt[5]{-32x^{11}y^{17}}\right)$

87. $\left(-2xy^2\sqrt{3x}\right)\left(xy\sqrt{6x}\right)$

88. $\left(-5x^2y^3z\sqrt{2xyz}\right)\left(-x^4z\sqrt{10xz}\right)$

89. $\left(2x^2y\sqrt[4]{8xy}\right)\left(-3xy^2\sqrt[4]{2x^2y^3}\right)$

90. $\left(5a^2b\sqrt[4]{8a^2b}\right)\left(4ab\sqrt[4]{4a^3b^2}\right)$

Application Exercises

The function

$$d(x) = \sqrt{\dfrac{3x}{2}}$$

models the distance, $d(x)$, in miles, that a person x feet high can see to the horizon. Use this function to solve Exercises 91–92.

91. The pool deck on a cruise ship is 72 feet above the water. How far can passengers on the pool deck see? Write the answer in simplified radical form. Then use the simplified radical form and a calculator to express the answer to the nearest tenth of a mile.

92. The captain of a cruise ship is on the star deck, which is 120 feet above the water. How far can the captain see? Write the answer in simplified radical form. Then use the simplified radical form and a calculator to express the answer to the nearest tenth of a mile.

Paleontologists use the function

$$W(x) = 4\sqrt{2x}$$

to estimate the walking speed of a dinosaur, W(x), in feet per second, where x is the length, in feet, of the dinosaur's leg. The graph of W is shown in the figure. Use this information to solve Exercises 93–94.

Dinosaur Walking Speeds

Leg Length (feet)
[0, 12, 1] by [0, 20, 2]

93. What is the walking speed of a dinosaur whose leg length is 6 feet? Use the function's equation and express the answer in simplified radical form. Then use the function's graph to estimate the answer to the nearest foot per second.

94. What is the walking speed of a dinosaur whose leg length is 10 feet? Use the function's equation and express the answer in simplified radical form. Then use the function's graph to estimate the answer to the nearest foot per second.

Your **cardiac index** *is your heart's output, in liters of blood per minute, divided by your body's surface area, in square meters. The cardiac index, C(x), can be modeled by*

$$C(x) = \frac{7.644}{\sqrt[4]{x}}, \quad 10 \le x \le 80,$$

where x is an individual's age, in years. The graph of the function is shown. Use the function to solve Exercises 95–96.

95. a. Find the cardiac index of a 32-year-old. Express the denominator in simplified radical form and reduce the fraction.

  b. Use the form of the answer in part (a) and a calculator to express the cardiac index to the nearest hundredth. Identify your solution as a point on the graph.

96. a. Find the cardiac index of an 80-year-old. Express the denominator in simplified radical form and reduce the fraction.

  b. Use the form of the answer in part (a) and a calculator to express the cardiac index to the nearest hundredth. Identify your solution as a point on the graph.

Explaining the Concepts

97. What is the product rule for radicals? Give an example to show how it is used.

98. Explain why $\sqrt{50}$ is not simplified. What do we mean when we say a radical expression is simplified?

99. In simplifying an *n*th root, explain how to find variable factors in the radicand that are perfect *n*th powers.

100. Without showing all the details, explain how to simplify $\sqrt[3]{16x^{14}}$.

101. As you get older, what would you expect to happen to your heart's output? Explain how this is shown in the graph for Exercises 95–96. Is this trend taking place progressively more rapidly or more slowly over the entire interval? What does this mean about this aspect of aging?

Technology Exercises

102. Use a calculator to provide numerical support for your simplifications in Exercises 21–24 and 27–28. In each case, find a decimal approximation for the given expression. Then find a decimal approximation for your simplified expression. The approximations should be the same.

In Exercises 103–106, determine if each simplification is correct by graphing the function on each side of the equation with your graphing utility. Use the given viewing rectangle. The graphs should be the same. If they are not, correct the right side of the equation and then use your graphing utility to verify the simplification.

103. $\sqrt{x^4} = x^2$; [0, 5, 1] by [0, 20, 1]

104. $\sqrt{8x^2} = 4x\sqrt{2}$; [−5, 5, 1] by [−5, 20, 1]

105. $\sqrt{3x^2 - 6x + 3} = (x - 1)\sqrt{3}$; [−5, 5, 1] by [−5, 5, 1]

106. $\sqrt[3]{2x} \cdot \sqrt[3]{4x^2} = 4x$; [−10, 10, 1] by [−10, 10, 1]

Critical Thinking Exercises

Make Sense? *In Exercises 107–110, determine whether each statement makes sense or does not make sense, and explain your reasoning.*

107. Because the product rule for radicals applies when $\sqrt[n]{a}$ and $\sqrt[n]{b}$ are real numbers, I can use it to find $\sqrt[3]{16} \cdot \sqrt[3]{-4}$, but not to find $\sqrt{8} \cdot \sqrt{-2}$.

108. I multiply *n*th roots by taking the *n*th root of the product of the radicands.

109. I need to know how to factor a trinomial to simplify $\sqrt{x^2 - 10x + 25}$.

110. I know that I've simplified a radical expression when it contains a single radical.

In Exercises 111–114, determine whether each statement is true or false. If the statement is false, make the necessary change(s) to produce a true statement.

111. $2\sqrt{5} \cdot 6\sqrt{5} = 12\sqrt{5}$

112. $\sqrt[3]{4} \cdot \sqrt[3]{4} = 4$

113. $\sqrt{12} = 2\sqrt{6}$

114. $\sqrt[3]{3^{15}} = 243$

115. If a number is tripled, what happens to its square root?

116. What must be done to a number so that its cube root is tripled?

117. If $f(x) = \sqrt[3]{2x}$ and $(fg)(x) = 2x$, find $g(x)$.

118. Graph $f(x) = \sqrt{(x-1)^2}$ by hand.

Review Exercises

119. Solve: $2x - 1 \le 21$ and $2x + 2 \ge 12$.
(Section 4.2, Example 2)

120. Solve:
$$\begin{cases} 5x + 2y = 2 \\ 4x + 3y = -4. \end{cases}$$
(Section 3.1, Example 6)

121. Factor: $64x^3 - 27$. (Section 5.5, Example 9)

Preview Exercises

Exercises 122–124 will help you prepare for the material covered in the next section.

122. a. Simplify: $21x + 10x$.
 b. Simplify: $21\sqrt{2} + 10\sqrt{2}$.

123. a. Simplify: $4x - 12x$.
 b. Simplify: $4\sqrt[3]{2} - 12\sqrt[3]{2}$.

124. Simplify: $\dfrac{\sqrt[4]{7y^5}}{\sqrt[4]{x^{12}}}$.

SECTION

7.4

Adding, Subtracting, and Dividing Radical Expressions ▶

What am I supposed to learn?

After studying this section, you should be able to:

1 Add and subtract radical expressions. ▶

2 Use the quotient rule to simplify radical expressions. ▶

3 Use the quotient rule to divide radical expressions. ▶

The future is now: You have the opportunity to explore the cosmos in a starship traveling near the speed of light. The experience will enable you to understand the mysteries of the universe firsthand, transporting you to unimagined levels of knowing and being. The downside: According to Einstein's theory of relativity, close to the speed of light, your aging rate relative to friends on Earth is nearly zero. You will return from your two-year journey to an unknown futuristic world. In this section's Exercise Set (Exercise 86), we provide an expression involving radical division that models your return to this unrecognizable world. To make sense of the model, we turn to various operations with radicals, including addition, subtraction, and division.

1 Add and subtract radical expressions. ▶

Adding and Subtracting Radical Expressions

We know that like terms have exactly the same variable factors and can be combined. For example,

$$7x + 6x = (7 + 6)x = 13x.$$

Two or more radical expressions that have the same indices *and* the same radicands are called **like radicals**. Like radicals are combined in exactly the same way that we combine like terms. For example,

$$7\sqrt{11} + 6\sqrt{11} = (7 + 6)\sqrt{11} = 13\sqrt{11}.$$

7 square roots of 11 plus 6 square roots of 11 result in 13 square roots of 11.

EXAMPLE 1   Adding and Subtracting Like Radicals

Simplify (add or subtract) by combining like radical terms:

a. $7\sqrt{2} + 8\sqrt{2}$

b. $\sqrt[3]{5} - 4x\sqrt[3]{5} + 8\sqrt[3]{5}$

c. $8\sqrt[6]{5x} - 5\sqrt[6]{5x} + 4\sqrt[3]{5x}.$

Solution

a. $7\sqrt{2} + 8\sqrt{2}$

$= (7 + 8)\sqrt{2}$ Apply the distributive property.

$= 15\sqrt{2}$ Simplify.

b. $\sqrt[3]{5} - 4x\sqrt[3]{5} + 8\sqrt[3]{5}$

$= (1 - 4x + 8)\sqrt[3]{5}$ Apply the distributive property.

$= (9 - 4x)\sqrt[3]{5}$ Simplify.

c. $8\sqrt[6]{5x} - 5\sqrt[6]{5x} + 4\sqrt[3]{5x}$

$= (8 - 5)\sqrt[6]{5x} + 4\sqrt[3]{5x}$ Apply the distributive property to the two terms with like radicals.

$= 3\sqrt[6]{5x} + 4\sqrt[3]{5x}$ The indices, 6 and 3, differ. These are not like radicals and cannot be combined. ■

✓ **CHECK POINT 1**   Simplify by combining like radical terms:

a. $8\sqrt{13} + 2\sqrt{13}$

b. $9\sqrt[3]{7} - 6x\sqrt[3]{7} + 12\sqrt[3]{7}$

c. $7\sqrt[4]{3x} - 2\sqrt[4]{3x} + 2\sqrt[3]{3x}.$

In some cases, radical expressions can be combined once they have been simplified. For example, to add $\sqrt{2}$ and $\sqrt{8}$, we can write $\sqrt{8}$ as $\sqrt{4 \cdot 2}$ because 4 is a perfect square factor of 8.

$$\sqrt{2} + \sqrt{8} = \sqrt{2} + \sqrt{4 \cdot 2} = 1\sqrt{2} + 2\sqrt{2} = (1 + 2)\sqrt{2} = 3\sqrt{2}$$

Always begin by simplifying radical terms. This makes it possible to identify and combine any like radicals.

EXAMPLE 2 Combining Radicals That First Require Simplification

Simplify by combining like radical terms, if possible:

a. $7\sqrt{18} + 5\sqrt{8}$ **b.** $4\sqrt{27x} - 8\sqrt{12x}$ **c.** $7\sqrt{3} - 2\sqrt{5}.$

Solution

a. $7\sqrt{18} + 5\sqrt{8}$

$= 7\sqrt{9 \cdot 2} + 5\sqrt{4 \cdot 2}$ Factor the radicands using the greatest perfect square factors.

$= 7\sqrt{9} \cdot \sqrt{2} + 5\sqrt{4} \cdot \sqrt{2}$ Take the square root of each factor.

$= 7 \cdot 3 \cdot \sqrt{2} + 5 \cdot 2 \cdot \sqrt{2}$ $\sqrt{9} = 3$ and $\sqrt{4} = 2.$

$= 21\sqrt{2} + 10\sqrt{2}$ Multiply.

$= (21 + 10)\sqrt{2}$ Apply the distributive property.

$= 31\sqrt{2}$ Simplify.

b. $4\sqrt{27x} - 8\sqrt{12x}$

$= 4\sqrt{9 \cdot 3x} - 8\sqrt{4 \cdot 3x}$ Factor the radicands using the greatest perfect square factors.

$= 4\sqrt{9} \cdot \sqrt{3x} - 8\sqrt{4} \cdot \sqrt{3x}$ Take the square root of each factor.

$= 4 \cdot 3 \cdot \sqrt{3x} - 8 \cdot 2 \cdot \sqrt{3x}$ $\sqrt{9} = 3$ and $\sqrt{4} = 2.$

$= 12\sqrt{3x} - 16\sqrt{3x}$ Multiply.

$= (12 - 16)\sqrt{3x}$ Apply the distributive property.

$= -4\sqrt{3x}$ Simplify.

c. $7\sqrt{3} - 2\sqrt{5}$ cannot be simplified. The radical expressions have different radicands, namely 3 and 5, and are not like terms. ■

☑ **CHECK POINT 2**   Simplify by combining like radical terms, if possible:

a. $3\sqrt{20} + 5\sqrt{45}$

b. $3\sqrt{12x} - 6\sqrt{27x}$

c. $8\sqrt{5} - 6\sqrt{2}.$

EXAMPLE 3 Adding and Subtracting with Higher Indices

Simplify by combining like radical terms, if possible:

a. $2\sqrt[3]{16} - 4\sqrt[3]{54}$ **b.** $5\sqrt[3]{xy^2} + \sqrt[3]{8x^4y^5}.$

Solution

a. $2\sqrt[3]{16} - 4\sqrt[3]{54}$

$= 2\sqrt[3]{8 \cdot 2} - 4\sqrt[3]{27 \cdot 2}$ Factor the radicands using the greatest perfect cube factors.

$= 2\sqrt[3]{8} \cdot \sqrt[3]{2} - 4\sqrt[3]{27} \cdot \sqrt[3]{2}$ Take the cube root of each factor.

$= 2 \cdot 2 \cdot \sqrt[3]{2} - 4 \cdot 3 \cdot \sqrt[3]{2}$ $\sqrt[3]{8} = 2$ and $\sqrt[3]{27} = 3.$

$= 4\sqrt[3]{2} - 12\sqrt[3]{2}$ Multiply.

$= (4 - 12)\sqrt[3]{2}$ Apply the distributive property.

$= -8\sqrt[3]{2}$ Simplify.

b. $5\sqrt[3]{xy^2} + \sqrt[3]{8x^4y^5}$

$= 5\sqrt[3]{xy^2} + \sqrt[3]{(8x^3y^3)xy^2}$ Factor the second radicand using the greatest perfect cube factor.

$= 5\sqrt[3]{xy^2} + \sqrt[3]{8x^3y^3} \cdot \sqrt[3]{xy^2}$ Take the cube root of each factor.

$= 5\sqrt[3]{xy^2} + 2xy\sqrt[3]{xy^2}$ $\sqrt[3]{8} = 2$ and $\sqrt[3]{x^3y^3} = (x^3y^3)^{\frac{1}{3}} = xy.$

$= (5 + 2xy)\sqrt[3]{xy^2}$ Apply the distributive property. ■

✓ **CHECK POINT 3**   Simplify by combining like radical terms, if possible:

a. $3\sqrt[3]{24} - 5\sqrt[3]{81}$

b. $5\sqrt[3]{x^2y} + \sqrt[3]{27x^5y^4}.$

Dividing Radical Expressions

We have seen that the root of a product is the product of the roots. The root of a quotient can also be expressed as the quotient of roots. Here is an example:

$$\sqrt{\frac{64}{4}} = \sqrt{16} = 4 \quad \text{and} \quad \frac{\sqrt{64}}{\sqrt{4}} = \frac{8}{2} = 4.$$

This expression is the square root of a quotient.

This expression is the quotient of two square roots.

The two procedures produce the same result, 4. This is a special case of the **quotient rule for radicals**.

2 Use the quotient rule to simplify radical expressions. ⊙

> ### The Quotient Rule for Radicals
> If $\sqrt[n]{a}$ and $\sqrt[n]{b}$ are real numbers and $b \neq 0$, then
>
> $$\sqrt[n]{\frac{a}{b}} = \frac{\sqrt[n]{a}}{\sqrt[n]{b}}.$$
>
> The *nth* root of a quotient is the quotient of the *nth* roots of the numerator and denominator.

We know that a radical is simplified when its radicand has no factors other than 1 that are perfect *nth* powers. The quotient rule can be used to simplify some radicals. Keep in mind that all variables in radicands represent positive real numbers.

EXAMPLE 4   Using the Quotient Rule to Simplify Radicals

Simplify using the quotient rule:

a. $\sqrt[3]{\frac{16}{27}}$ **b.** $\sqrt{\frac{x^2}{25y^6}}$ **c.** $\sqrt[4]{\frac{7y^5}{x^{12}}}.$

Solution   We simplify each expression by taking the roots of the numerator and the denominator. Then we use factoring to simplify the resulting radicals, if possible.

a. $\sqrt[3]{\dfrac{16}{27}} = \dfrac{\sqrt[3]{16}}{\sqrt[3]{27}} = \dfrac{\sqrt[3]{8 \cdot 2}}{3} = \dfrac{\sqrt[3]{8} \cdot \sqrt[3]{2}}{3} = \dfrac{2\sqrt[3]{2}}{3}$

b. $\sqrt{\dfrac{x^2}{25y^6}} = \dfrac{\sqrt{x^2}}{\sqrt{25y^6}} = \dfrac{x}{5(y^6)^{\frac{1}{2}}} = \dfrac{x}{5y^3}$

> Try to do this step mentally.

c. $\sqrt[4]{\dfrac{7y^5}{x^{12}}} = \dfrac{\sqrt[4]{7y^5}}{\sqrt[4]{x^{12}}} = \dfrac{\sqrt[4]{y^4 \cdot 7y}}{\sqrt[4]{x^{12}}} = \dfrac{y\sqrt[4]{7y}}{x^3}$ ∎

✓ **CHECK POINT 4**   Simplify using the quotient rule:

a. $\sqrt[3]{\dfrac{24}{125}}$ **b.** $\sqrt{\dfrac{9x^3}{y^{10}}}$ **c.** $\sqrt[3]{\dfrac{8y^7}{x^{12}}}.$

By reversing the two sides of the quotient rule, we obtain a procedure for dividing radical expressions.

3 Use the quotient rule to divide radical expressions. ▶

> **Dividing Radical Expressions**
>
> If $\sqrt[n]{a}$ and $\sqrt[n]{b}$ are real numbers and $b \neq 0$, then
>
> $$\frac{\sqrt[n]{a}}{\sqrt[n]{b}} = \sqrt[n]{\frac{a}{b}}.$$
>
> To divide two radical expressions with the same index, divide the radicands and retain the common index.

Great Question!

When should I write a quotient under a single radical and when should I use separate radicals for a quotient's numerator and denominator?

- Use
  $$\frac{\sqrt[n]{a}}{\sqrt[n]{b}} = \sqrt[n]{\frac{a}{b}},$$
  writing under one radical (as in Example 5), when *dividing*.

- Use
  $$\sqrt[n]{\frac{a}{b}} = \frac{\sqrt[n]{a}}{\sqrt[n]{b}},$$
  with separate radicals for the numerator and the denominator (as in Example 4), when *simplifying* a quotient.

EXAMPLE 5   Dividing Radical Expressions

Divide and, if possible, simplify:

a. $\dfrac{\sqrt{48x^3}}{\sqrt{6x}}$ **b.** $\dfrac{\sqrt{45xy}}{2\sqrt{5}}$ **c.** $\dfrac{\sqrt[3]{16x^5y^2}}{\sqrt[3]{2xy^{-1}}}.$

Solution   In each part of this problem, the indices in the numerator and the denominator are the same. Perform each division by dividing the radicands and retaining the common index.

a. $\dfrac{\sqrt{48x^3}}{\sqrt{6x}} = \sqrt{\dfrac{48x^3}{6x}} = \sqrt{8x^2} = \sqrt{4x^2 \cdot 2} = \sqrt{4x^2} \cdot \sqrt{2} = 2x\sqrt{2}$

b. $\dfrac{\sqrt{45xy}}{2\sqrt{5}} = \dfrac{1}{2} \cdot \sqrt{\dfrac{45xy}{5}} = \dfrac{1}{2} \cdot \sqrt{9xy} = \dfrac{1}{2} \cdot 3\sqrt{xy}$ or $\dfrac{3\sqrt{xy}}{2}$

c. $\dfrac{\sqrt[3]{16x^5y^2}}{\sqrt[3]{2xy^{-1}}} = \sqrt[3]{\dfrac{16x^5y^2}{2xy^{-1}}}$ Divide the radicands and retain the common index.

$= \sqrt[3]{8x^{5-1}y^{2-(-1)}}$ Divide factors in the radicand. Subtract exponents on common bases.

$= \sqrt[3]{8x^4y^3}$ Simplify.

$= \sqrt[3]{(8x^3y^3)x}$ Factor using the greatest perfect cube factor.

$= \sqrt[3]{8x^3y^3} \cdot \sqrt[3]{x}$ Factor into two radicals.

$= 2xy\sqrt[3]{x}$ Simplify. ∎

✓ **CHECK POINT 5** Divide and, if possible, simplify:

a. $\dfrac{\sqrt{40x^5}}{\sqrt{2x}}$ b. $\dfrac{\sqrt{50xy}}{2\sqrt{2}}$ c. $\dfrac{\sqrt[3]{48x^7y}}{\sqrt[3]{6xy^{-2}}}$.

CONCEPT AND VOCABULARY CHECK

Fill in each blank so that the resulting statement is true.

1. $5\sqrt{3} + 8\sqrt{3} = (\_ + \_)\sqrt{3} = \underline{\hspace{1cm}}$

2. $\sqrt{27} - \sqrt{12} = \sqrt{\_\cdot 3} - \sqrt{\_\cdot 3} = \_\sqrt{3} - \_\sqrt{3} = \_$

3. $\sqrt[3]{54} + \sqrt[3]{16} = \sqrt[3]{\_\cdot 2} + \sqrt[3]{\_\cdot 2} = \_\sqrt[3]{2} + \_\sqrt[3]{2} = \underline{\hspace{1cm}}$

4. If $\sqrt[n]{a}$ and $\sqrt[n]{b}$ are real numbers and $b \neq 0$, the quotient rule for radicals states that

$$\sqrt[n]{\dfrac{a}{b}} = \underline{\hspace{1cm}}.$$

5. $\sqrt[3]{\dfrac{8}{27}} = \dfrac{\sqrt[3]{\_\_}}{\sqrt[3]{\_\_}} = \underline{\hspace{0.6cm}}$

6. If $x > 0$,

$$\dfrac{\sqrt{72x^3}}{\sqrt{2x}} = \sqrt{\dfrac{\_\_}{\_\_}} = \sqrt{\_\_} = \_.$$

7.4 EXERCISE SET   ▶ MyMathLab®

Practice Exercises

In this Exercise Set, assume that all variables represent positive real numbers.

In Exercises 1–10, add or subtract as indicated.

1. $8\sqrt{5} + 3\sqrt{5}$
2. $7\sqrt{3} + 2\sqrt{3}$
3. $9\sqrt[3]{6} - 2\sqrt[3]{6}$
4. $9\sqrt[3]{7} - 4\sqrt[3]{7}$
5. $4\sqrt[5]{2} + 3\sqrt[5]{2} - 5\sqrt[5]{2}$
6. $6\sqrt[5]{3} + 2\sqrt[5]{3} - 3\sqrt[5]{3}$
7. $3\sqrt{13} - 2\sqrt{5} - 2\sqrt{13} + 4\sqrt{5}$
8. $8\sqrt{17} - 5\sqrt{19} - 6\sqrt{17} + 4\sqrt{19}$
9. $3\sqrt{5} - \sqrt[3]{x} + 4\sqrt{5} + 3\sqrt[3]{x}$
10. $6\sqrt{7} - \sqrt[3]{x} + 2\sqrt{7} + 5\sqrt[3]{x}$

In Exercises 11–28, add or subtract as indicated. You will need to simplify terms to identify the like radicals.

11. $\sqrt{3} + \sqrt{27}$
12. $\sqrt{5} + \sqrt{20}$
13. $7\sqrt{12} + \sqrt{75}$
14. $5\sqrt{12} + \sqrt{75}$
15. $3\sqrt{32x} - 2\sqrt{18x}$
16. $5\sqrt{45x} - 2\sqrt{20x}$
17. $5\sqrt[3]{16} + \sqrt[3]{54}$
18. $3\sqrt[3]{24} + \sqrt[3]{81}$
19. $3\sqrt{45x^3} + \sqrt{5x}$
20. $8\sqrt{45x^3} + \sqrt{5x}$
21. $\sqrt[3]{54xy^3} + y\sqrt[3]{128x}$
22. $\sqrt[3]{24xy^3} + y\sqrt[3]{81x}$
23. $\sqrt[3]{54x^4} - \sqrt[3]{16x}$
24. $\sqrt[3]{81x^4} - \sqrt[3]{24x}$
25. $\sqrt{9x - 18} + \sqrt{x - 2}$
26. $\sqrt{4x - 12} + \sqrt{x - 3}$
27. $2\sqrt[3]{x^4y^2} + 3x\sqrt[3]{xy^2}$
28. $4\sqrt[3]{x^4y^2} + 5x\sqrt[3]{xy^2}$

In Exercises 29–44, simplify using the quotient rule.

29. $\sqrt{\dfrac{11}{4}}$
30. $\sqrt{\dfrac{19}{25}}$
31. $\sqrt[3]{\dfrac{19}{27}}$
32. $\sqrt[3]{\dfrac{11}{64}}$
33. $\sqrt{\dfrac{x^2}{36y^8}}$
34. $\sqrt{\dfrac{x^2}{144y^{12}}}$
35. $\sqrt{\dfrac{8x^3}{25y^6}}$
36. $\sqrt{\dfrac{50x^3}{81y^8}}$
37. $\sqrt[3]{\dfrac{x^4}{8y^3}}$
38. $\sqrt[3]{\dfrac{x^5}{125y^3}}$
39. $\sqrt[3]{\dfrac{50x^8}{27y^{12}}}$
40. $\sqrt[3]{\dfrac{81x^8}{8y^{15}}}$
41. $\sqrt[4]{\dfrac{9y^6}{x^8}}$
42. $\sqrt[4]{\dfrac{13y^7}{x^{12}}}$
43. $\sqrt[5]{\dfrac{64x^{13}}{y^{20}}}$
44. $\sqrt[5]{\dfrac{64x^{14}}{y^{15}}}$

In Exercises 45–66, divide and, if possible, simplify.

45. $\dfrac{\sqrt{40}}{\sqrt{5}}$

46. $\dfrac{\sqrt{200}}{\sqrt{10}}$

47. $\dfrac{\sqrt[3]{48}}{\sqrt[3]{6}}$

48. $\dfrac{\sqrt[3]{54}}{\sqrt[3]{2}}$

49. $\dfrac{\sqrt{54x^3}}{\sqrt{6x}}$

50. $\dfrac{\sqrt{72x^3}}{\sqrt{2x}}$

51. $\dfrac{\sqrt{x^5y^3}}{\sqrt{xy}}$

52. $\dfrac{\sqrt{x^7y^6}}{\sqrt{x^3y^2}}$

53. $\dfrac{\sqrt{200x^3}}{\sqrt{10x^{-1}}}$

54. $\dfrac{\sqrt{500x^3}}{\sqrt{10x^{-1}}}$

55. $\dfrac{\sqrt{48a^8b^7}}{\sqrt{3a^{-2}b^{-3}}}$

56. $\dfrac{\sqrt{54a^7b^{11}}}{\sqrt{3a^{-4}b^{-2}}}$

57. $\dfrac{\sqrt{72xy}}{2\sqrt{2}}$

58. $\dfrac{\sqrt{50xy}}{2\sqrt{2}}$

59. $\dfrac{\sqrt[3]{24x^3y^5}}{\sqrt[3]{3y^2}}$

60. $\dfrac{\sqrt[3]{250x^5y^3}}{\sqrt[3]{2x^3}}$

61. $\dfrac{\sqrt[4]{32x^{10}y^8}}{\sqrt[4]{2x^2y^{-2}}}$

62. $\dfrac{\sqrt[5]{96x^{12}y^{11}}}{\sqrt[5]{3x^2y^{-2}}}$

63. $\dfrac{\sqrt[3]{x^2+5x+6}}{\sqrt[3]{x+2}}$

64. $\dfrac{\sqrt[3]{x^2+7x+12}}{\sqrt[3]{x+3}}$

65. $\dfrac{\sqrt[3]{a^3+b^3}}{\sqrt[3]{a+b}}$

66. $\dfrac{\sqrt[3]{a^3-b^3}}{\sqrt[3]{a-b}}$

Practice PLUS

In Exercises 67–76, perform the indicated operations.

67. $\dfrac{\sqrt{32}}{5}+\dfrac{\sqrt{18}}{7}$

68. $\dfrac{\sqrt{27}}{2}+\dfrac{\sqrt{75}}{7}$

69. $3x\sqrt{8xy^2}-5y\sqrt{32x^3}+\sqrt{18x^3y^2}$

70. $6x\sqrt{3xy^2}-4x^2\sqrt{27xy}-5\sqrt{75x^5y}$

71. $5\sqrt{2x^3}+\dfrac{30x^3\sqrt{24x^2}}{3x^2\sqrt{3x}}$

72. $7\sqrt{2x^3}+\dfrac{40x^3\sqrt{150x^2}}{5x^2\sqrt{3x}}$

73. $2x\sqrt{75xy}-\dfrac{\sqrt{81xy^2}}{\sqrt{3x^{-2}y}}$

74. $5\sqrt{8x^2y^3}-\dfrac{9x^2\sqrt{64y}}{3x\sqrt{2y^{-2}}}$

75. $\dfrac{15x^4\sqrt[3]{80x^3y^2}}{5x^3\sqrt[3]{2x^2y}}-\dfrac{75\sqrt[3]{5x^3y}}{25\sqrt[3]{x^{-1}}}$

76. $\dfrac{16x^4\sqrt[3]{48x^3y^2}}{8x^3\sqrt[3]{3x^2y}}-\dfrac{20\sqrt[3]{2x^3y}}{4\sqrt[3]{x^{-1}}}$

In Exercises 77–80, find $\left(\frac{f}{g}\right)(x)$ and the domain of $\frac{f}{g}$. Express each quotient function in simplified form.

77. $f(x)=\sqrt{48x^5}, g(x)=\sqrt{3x^2}$

78. $f(x)=\sqrt{x^2-25}, g(x)=\sqrt{x+5}$

79. $f(x)=\sqrt[3]{32x^6}, g(x)=\sqrt[3]{2x^2}$

80. $f(x)=\sqrt[3]{2x^6}, g(x)=\sqrt[3]{16x}$

Application Exercises

*Exercises 81–84 involve the perimeter and area of various geometric figures. Refer to **Table 1.6** on page 60 if you've forgotten any of the formulas.*

In Exercises 81–82, find the perimeter and area of each rectangle. Express answers in simplified radical form.

81.

$\sqrt{125}$ feet

$2\sqrt{20}$ feet

82.

$4\sqrt{20}$ feet

$\sqrt{80}$ feet

83. Find the perimeter of the triangle in simplified radical form.

$\sqrt{45}$ m

$\sqrt{80}$ m

$\sqrt{125}$ m

84. Find the area of the trapezoid in simplified radical form.

$\sqrt{2}$ feet

$\sqrt{6}$ feet

$\sqrt{8}$ feet

85. America is getting older. The graph shows the elderly U.S. population for ages 65–84 and for ages 85 and older in 2010, with projections for 2020 and beyond.

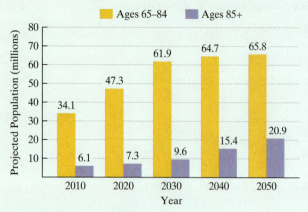

Projected Elderly United States Population

Source: U.S. Census Bureau

The function $f(x) = 5\sqrt{x} + 34.1$ models the projected number of Americans ages 65–84, $f(x)$, in millions, x years after 2010.

a. Use the function to find $f(40) - f(10)$. Express this difference in simplified radical form. What does this simplified radical represent?

b. Use a calculator and write your answer in part (a) to the nearest tenth. Does this rounded decimal overestimate or underestimate the difference in the projected data shown by the bar graph? By how much?

86. What does travel in space have to do with radicals? Imagine that in the future we will be able to travel in starships at velocities approaching the speed of light (approximately 186,000 miles per second). According to Einstein's theory of relativity, time would pass more quickly on Earth than it would in the moving starship. The radical expression

$$R_f \frac{\sqrt{c^2 - v^2}}{\sqrt{c^2}}$$

gives the aging rate of an astronaut relative to the aging rate of a friend, R_f, on Earth. In the expression, v is the astronaut's velocity and c is the speed of light.

a. Use the quotient rule and simplify the expression that shows your aging rate relative to a friend on Earth. Working in a step-by-step manner, express your aging rate as

$$R_f\sqrt{1 - \left(\frac{v}{c}\right)^2}.$$

b. You are moving at velocities approaching the speed of light. Substitute c, the speed of light, for v in the simplified expression from part (a). Simplify

completely. Close to the speed of light, what is your aging rate relative to a friend on Earth? What does this mean?

Explaining the Concepts

87. What are like radicals? Give an example with your explanation.

88. Explain how to add like radicals. Give an example with your explanation.

89. If only like radicals can be combined, why is it possible to add $\sqrt{2}$ and $\sqrt{8}$?

90. Explain how to simplify a radical expression using the quotient rule. Provide an example.

91. Explain how to divide radical expressions with the same index.

92. In Exercise 85, use the data displayed by the bar graph to explain why we used a square root function to model the projected population for the 65–84 age group, but not for the 85+ group.

Technology Exercises

93. Use a calculator to provide numerical support for any four exercises that you worked from Exercises 1–66 that do not contain variables. Begin by finding a decimal approximation for the given expression. Then find a decimal approximation for your answer. The two decimal approximations should be the same.

In Exercises 94–96, determine if each operation is performed correctly by graphing the function on each side of the equation with your graphing utility. Use the given viewing rectangle. The graphs should be the same. If they are not, correct the right side of the equation and then use your graphing utility to verify the correction.

94. $\sqrt{4x} + \sqrt{9x} = 5\sqrt{x}$

[0, 5, 1] by [0, 10, 1]

95. $\sqrt{16x} - \sqrt{9x} = \sqrt{7x}$

[0, 5, 1] by [0, 5, 1]

96. $x\sqrt{8} + x\sqrt{2} = x\sqrt{10}$

[−5, 5, 1] by [−15, 15, 1]

Critical Thinking Exercises

Make Sense? *In Exercises 97–100, determine whether each statement makes sense or does not make sense, and explain your reasoning.*

97. I divide nth roots by taking the nth root of the quotient of the radicands.

98. The unlike radicals $3\sqrt{2}$ and $5\sqrt{3}$ remind me of the unlike terms $3x$ and $5y$ that cannot be combined by addition or subtraction.

99. I simplified the terms of $3\sqrt[3]{81} + 2\sqrt[3]{54}$, and then I was able to identify and add the like radicals.

100. Without using a calculator, <u>it's easier for me to estimate</u> the decimal value of $\sqrt{72} + \sqrt{32} + \sqrt{18}$ by first simplifying.

In Exercises 101–104, determine whether each statement is true or false. If the statement is false, make the necessary change(s) to produce a true statement.

101. $\sqrt{5} + \sqrt{5} = \sqrt{10}$

102. $4\sqrt{3} + 5\sqrt{3} = 9\sqrt{6}$

103. If any two radical expressions are completely simplified, they can then be combined through addition or subtraction.

104. $\dfrac{\sqrt{-8}}{\sqrt{2}} = \sqrt{\dfrac{-8}{2}} = \sqrt{-4} = -2$

105. If an irrational number is decreased by $2\sqrt{18} - \sqrt{50}$, the result is $\sqrt{2}$. What is the number?

106. Simplify: $\dfrac{\sqrt{20}}{3} + \dfrac{\sqrt{45}}{4} - \sqrt{80}$.

107. Simplify: $\dfrac{6\sqrt{49xy}\,\sqrt{ab^2}}{7\sqrt{36x^{-3}y^{-5}}\sqrt{a^{-9}b^{-1}}}$.

Review Exercises

108. Solve: $2(3x - 1) - 4 = 2x - (6 - x)$.
(Section 1.4, Example 3)

109. Factor: $x^2 - 8xy + 12y^2$.
(Section 5.4, Example 4)

110. Add: $\dfrac{2}{x^2 + 5x + 6} + \dfrac{3x}{x^2 + 6x + 9}$.
(Section 6.2, Example 6)

Preview Exercises

Exercises 111–113 will help you prepare for the material covered in the next section.

111. **a.** Multiply: $7(x + 5)$.
  b. Multiply: $\sqrt{7}\,(x + \sqrt{5})$.

112. **a.** Multiply: $(x + 5)(6x + 3)$.
  b. Multiply: $(\sqrt{2} + 5)(6\sqrt{2} + 3)$.

113. Multiply and simplify:
$$\frac{10y}{\sqrt[5]{4x^3y}} \cdot \frac{\sqrt[5]{8x^2y^4}}{\sqrt[5]{8x^2y^4}}.$$

MID-CHAPTER CHECK POINT   Section 7.1–Section 7.4

What You Know: We learned to find roots of numbers. We saw that the domain of a square root function is the set of real numbers for which the radicand is nonnegative. We learned to simplify radical expressions, using $\sqrt[n]{a^n} = |a|$ if n is even and $\sqrt[n]{a^n} = a$ if n is odd. The definition $a^{\frac{m}{n}} = (\sqrt[n]{a})^m = \sqrt[n]{a^m}$ connected rational exponents and radicals. Finally, we performed various operations with radicals, including multiplication, addition, subtraction, and division.

In Exercises 1–23, simplify the given expression or perform the indicated operation(s) and, if possible, simplify. Assume that all variables represent positive real numbers.

1. $\sqrt{100} - \sqrt[3]{-27}$

2. $\sqrt{8x^5y^7}$

3. $3\sqrt[3]{4x^2} + 2\sqrt[3]{4x^2}$

4. $\left(3\sqrt[3]{4x^2}\right)\left(2\sqrt[3]{4x^2}\right)$

5. $27^{\frac{2}{3}} + (-32)^{\frac{3}{5}}$

6. $\left(64x^3y^{\frac{1}{4}}\right)^{\frac{1}{3}}$

7. $5\sqrt{27} - 4\sqrt{48}$

8. $\sqrt{\dfrac{500x^3}{4y^4}}$

9. $\dfrac{x}{\sqrt[4]{x}}$

10. $\sqrt[3]{54x^5}$

11. $\dfrac{\sqrt[3]{160}}{\sqrt[3]{2}}$

12. $\sqrt[5]{\dfrac{x^{10}}{y^{20}}}$

13. $\dfrac{\left(x^{\frac{2}{3}}\right)^2}{\left(x^{\frac{1}{4}}\right)^3}$

14. $\sqrt[6]{x^6y^4}$

15. $\sqrt[7]{(x - 2)^3} \cdot \sqrt[7]{(x - 2)^6}$

16. $\sqrt[4]{32x^{11}y^{17}}$

17. $4\sqrt[3]{16} + 2\sqrt[3]{54}$

18. $\dfrac{\sqrt[7]{x^4y^9}}{\sqrt[7]{x^{-5}y^7}}$

19. $(-125)^{-\frac{2}{3}}$

20. $\sqrt{2} \cdot \sqrt[3]{2}$

21. $\sqrt[3]{\dfrac{32x}{y^4}} \cdot \sqrt[3]{\dfrac{2x^2}{y^2}}$

22. $\sqrt{32xy^2} \cdot \sqrt{2x^3y^5}$

23. $4x\sqrt{6x^4y^3} - 7y\sqrt{24x^6y}$

In Exercises 24–25, find the domain of each function.

24. $f(x) = \sqrt{30 - 5x}$

25. $g(x) = \sqrt[3]{3x - 15}$

SECTION

7.5 Multiplying with More Than One Term and Rationalizing Denominators

What am I supposed to learn?

After studying this section, you should be able to:

1 Multiply radical expressions with more than one term. ▶

2 Use polynomial special products to multiply radicals. ▶

3 Rationalize denominators containing one term. ▶

4 Rationalize denominators containing two terms. ▶

5 Rationalize numerators. ▶

1 Multiply radical expressions with more than one term. ▶

$$\frac{7\sqrt{2}}{\sqrt{6}} \cdot \frac{\sqrt{6}}{\sqrt{6}} = \frac{7\sqrt{2 \cdot 3}}{6} = \frac{7}{3}\sqrt{3}$$

Peanuts copyright © 1979 Peanuts Worldwide LLC. Distributed by Universal Uclick. Reprinted with permission. All rights reserved

The late Charles Schulz, creator of the "Peanuts" comic strip, transfixed 350 million readers worldwide with the joys and angst of his hapless Charlie Brown and Snoopy, a romantic self-deluded beagle. In 18,250 comic strips that spanned nearly 50 years, mathematics was often featured. Is the discussion of radicals shown on the left the real thing, or is it just an algebraic scam? By the time you complete this section on multiplying and dividing radicals, you will be able to decide.

Multiplying Radical Expressions with More Than One Term

Radical expressions with more than one term are multiplied in much the same way that polynomials with more than one term are multiplied. Example 1 uses the distributive property and the FOIL method to perform multiplications.

EXAMPLE 1 Multiplying Radicals

Multiply:

a. $\sqrt{7}(x + \sqrt{2})$ **b.** $\sqrt[3]{x}(\sqrt[3]{6} - \sqrt[3]{x^2})$ **c.** $(5\sqrt{2} + 2\sqrt{3})(4\sqrt{2} - 3\sqrt{3})$.

Solution

a. $\sqrt{7}(x + \sqrt{2})$

$= \sqrt{7} \cdot x + \sqrt{7} \cdot \sqrt{2}$ Use the distributive property.

$= x\sqrt{7} + \sqrt{14}$ Multiply the radicals.

b. $\sqrt[3]{x}(\sqrt[3]{6} - \sqrt[3]{x^2})$

$= \sqrt[3]{x} \cdot \sqrt[3]{6} - \sqrt[3]{x} \cdot \sqrt[3]{x^2}$ Use the distributive property.

$= \sqrt[3]{6x} - \sqrt[3]{x^3}$ Multiply the radicals: $\sqrt[n]{a} \cdot \sqrt[n]{b} = \sqrt[n]{ab}$.

$= \sqrt[3]{6x} - x$ Simplify: $\sqrt[3]{x^3} = x$.

c. $\left(5\sqrt{2} + 2\sqrt{3}\right)\left(4\sqrt{2} - 3\sqrt{3}\right)$ Use the FOIL method.

$$= \left(5\sqrt{2}\right)\left(4\sqrt{2}\right) + \left(5\sqrt{2}\right)\left(-3\sqrt{3}\right) + \left(2\sqrt{3}\right)\left(4\sqrt{2}\right) + \left(2\sqrt{3}\right)\left(-3\sqrt{3}\right)$$

$$= 20 \cdot 2 - 15\sqrt{6} + 8\sqrt{6} - 6 \cdot 3 \qquad \text{Multiply. Note that } \sqrt{2} \cdot \sqrt{2} = \sqrt{4} = 2 \text{ and } \sqrt{3} \cdot \sqrt{3} = \sqrt{9} = 3.$$

$$= 40 - 15\sqrt{6} + 8\sqrt{6} - 18 \qquad \text{Complete the multiplications.}$$

$$= (40 - 18) + \left(-15\sqrt{6} + 8\sqrt{6}\right) \qquad \text{Group like terms. Try to do this step mentally.}$$

$$= 22 - 7\sqrt{6} \qquad \text{Combine numerical terms and like radicals.} \blacksquare$$

✓ **CHECK POINT 1**   Multiply:

   a. $\sqrt{6}\left(x + \sqrt{10}\right)$
   b. $\sqrt[3]{y}\left(\sqrt[3]{y^2} - \sqrt[3]{7}\right)$
   c. $\left(6\sqrt{5} + 3\sqrt{2}\right)\left(2\sqrt{5} - 4\sqrt{2}\right).$

---

2 Use polynomial special products to multiply radicals. ▶

Some radicals can be multiplied using the special products for multiplying polynomials.

EXAMPLE 2   Using Special Products to Multiply Radicals

Multiply:

   a. $\left(\sqrt{3} + \sqrt{7}\right)^2$ **b.** $\left(\sqrt{7} + \sqrt{3}\right)\left(\sqrt{7} - \sqrt{3}\right)$ **c.** $\left(\sqrt{a} - \sqrt{b}\right)\left(\sqrt{a} + \sqrt{b}\right).$

Solution   Use the special-product formulas shown.

$$(A + B)^2 = A^2 + 2 \cdot A \cdot B + B^2$$

a. $\left(\sqrt{3} + \sqrt{7}\right)^2 = \left(\sqrt{3}\right)^2 + 2 \cdot \sqrt{3} \cdot \sqrt{7} + \left(\sqrt{7}\right)^2$ Use the special product for $(A + B)^2$.

$$= 3 + 2\sqrt{21} + 7 \qquad \text{Multiply the radicals.}$$

$$= 10 + 2\sqrt{21} \qquad \text{Simplify.}$$

$$(A + B) \cdot (A - B) = A^2 - B^2$$

b. $\left(\sqrt{7} + \sqrt{3}\right)\left(\sqrt{7} - \sqrt{3}\right) = \left(\sqrt{7}\right)^2 - \left(\sqrt{3}\right)^2$ Use the special product for $(A + B)(A - B)$.

$$= 7 - 3 \qquad \text{Simplify: } \left(\sqrt{a}\right)^2 = a.$$

$$= 4$$

$$(A - B) \cdot (A + B) = A^2 - B^2$$

c. $\left(\sqrt{a} - \sqrt{b}\right)\left(\sqrt{a} + \sqrt{b}\right) = \left(\sqrt{a}\right)^2 - \left(\sqrt{b}\right)^2 = a - b$ ∎

Radical expressions that involve the sum and difference of the same two terms are called **conjugates**. For example,

$$\sqrt{7} + \sqrt{3} \quad \text{and} \quad \sqrt{7} - \sqrt{3}$$

are conjugates of each other. Parts (b) and (c) of Example 2 illustrate that the product of two radical expressions need not be a radical expression:

$$(\sqrt{7} + \sqrt{3})(\sqrt{7} - \sqrt{3}) = 4$$
$$(\sqrt{a} - \sqrt{b})(\sqrt{a} + \sqrt{b}) = a - b.$$

> The product of conjugates does not contain a radical.

Later in this section, we will use conjugates to simplify quotients.

☑ **CHECK POINT 2**   Multiply:

a. $\left(\sqrt{5} + \sqrt{6}\right)^2$ **b.** $\left(\sqrt{6} + \sqrt{5}\right)\left(\sqrt{6} - \sqrt{5}\right)$

c. $\left(\sqrt{a} - \sqrt{7}\right)\left(\sqrt{a} + \sqrt{7}\right).$

3 Rationalize denominators containing one term. ▶

Rationalizing Denominators Containing One Term

You can use a calculator to compare the approximate values for $\dfrac{1}{\sqrt{3}}$ and $\dfrac{\sqrt{3}}{3}$. The two approximations are the same. This is not a coincidence:

$$\frac{1}{\sqrt{3}} = \frac{1}{\sqrt{3}} \cdot \boxed{\frac{\sqrt{3}}{\sqrt{3}}} = \frac{\sqrt{3}}{\sqrt{9}} = \frac{\sqrt{3}}{3}.$$

> Any nonzero number divided by itself is 1. Multiplication by 1 does not change the value of $\dfrac{1}{\sqrt{3}}$.

This process involves rewriting a radical expression as an equivalent expression in which the denominator no longer contains any radicals. The process is called **rationalizing the denominator**. When the denominator contains a single radical with an nth root, **multiply the numerator and the denominator by a radical of index n that produces a perfect nth power in the denominator's radicand**.

EXAMPLE 3   Rationalizing Denominators

Rationalize each denominator:

a. $\dfrac{\sqrt{5}}{\sqrt{6}}$ **b.** $\sqrt[3]{\dfrac{7}{25}}.$

Solution

a. If we multiply the numerator and the denominator of $\dfrac{\sqrt{5}}{\sqrt{6}}$ by $\sqrt{6}$, the denominator becomes $\sqrt{6} \cdot \sqrt{6} = \sqrt{36} = 6$. The denominator's radicand, 36, is a perfect square. The denominator no longer contains a radical. Therefore, we multiply by 1, choosing $\dfrac{\sqrt{6}}{\sqrt{6}}$ for 1.

$$\frac{\sqrt{5}}{\sqrt{6}} = \frac{\sqrt{5}}{\sqrt{6}} \cdot \frac{\sqrt{6}}{\sqrt{6}}$$

> Multiply the numerator and denominator by $\sqrt{6}$ to remove the radical in the denominator.

$$= \frac{\sqrt{30}}{\sqrt{36}}$$

> Multiply numerators and multiply denominators. The denominator's radicand, 36, is a perfect square.

$$= \frac{\sqrt{30}}{6}$$

> Simplify: $\sqrt{36} = 6.$

b. Using the quotient rule, we can express $\sqrt[3]{\dfrac{7}{25}}$ as $\dfrac{\sqrt[3]{7}}{\sqrt[3]{25}}$. We have cube roots, so we want the denominator's radicand to be a perfect cube. Right now, the denominator's radicand is 25 or 5^2. We know that $\sqrt[3]{5^3} = 5$. If we multiply the numerator and the denominator of $\dfrac{\sqrt[3]{7}}{\sqrt[3]{25}}$ by $\sqrt[3]{5}$, the denominator becomes

$$\sqrt[3]{25} \cdot \sqrt[3]{5} = \sqrt[3]{5^2} \cdot \sqrt[3]{5} = \sqrt[3]{5^3} = 5.$$

The denominator's radicand, 5^3, is a perfect cube. The denominator no longer contains a radical. Therefore, we multiply by 1, choosing $\dfrac{\sqrt[3]{5}}{\sqrt[3]{5}}$ for 1.

$$\sqrt[3]{\dfrac{7}{25}} = \dfrac{\sqrt[3]{7}}{\sqrt[3]{25}}$$

Use the quotient rule and rewrite as the quotient of radicals.

$$= \dfrac{\sqrt[3]{7}}{\sqrt[3]{5^2}}$$

Write the denominator's radicand as an exponential expression.

$$= \dfrac{\sqrt[3]{7}}{\sqrt[3]{5^2}} \cdot \dfrac{\sqrt[3]{5}}{\sqrt[3]{5}}$$

Multiply numerator and denominator by $\sqrt[3]{5}$ to remove the radical in the denominator.

$$= \dfrac{\sqrt[3]{35}}{\sqrt[3]{5^3}}$$

Multiply numerators and denominators. The denominator's radicand, 5^3, is a perfect cube.

$$= \dfrac{\sqrt[3]{35}}{5}$$

Simplify: $\sqrt[3]{5^3} = 5$. ■

Great Question!

What exactly does rationalizing a denominator do to an irrational number in the denominator?

Rationalizing a numerical denominator makes that denominator a rational number.

✓ **CHECK POINT 3**   Rationalize each denominator:

a. $\dfrac{\sqrt{3}}{\sqrt{7}}$

b. $\sqrt[3]{\dfrac{2}{9}}$.

Example 3 showed that it is helpful to express the denominator's radicand using exponents. In this way, we can easily find the extra factor or factors needed to produce a perfect nth power. For example, suppose that $\sqrt[5]{8}$ appears in the denominator. We want a perfect fifth power. By expressing $\sqrt[5]{8}$ as $\sqrt[5]{2^3}$, we would multiply the numerator and denominator by $\sqrt[5]{2^2}$ because

$$\sqrt[5]{2^3} \cdot \sqrt[5]{2^2} = \sqrt[5]{2^5} = 2.$$

EXAMPLE 4   Rationalizing Denominators

Rationalize each denominator:

a. $\sqrt{\dfrac{3x}{5y}}$

b. $\dfrac{\sqrt[3]{x}}{\sqrt[3]{36y}}$

c. $\dfrac{10y}{\sqrt[5]{4x^3y}}$.

Solution   By examining each denominator, you can determine how to multiply by 1. For the square root, we must produce exponents of 2 in the radicand. For the cube root, we need exponents of 3, and for the fifth root, we want exponents of 5.

Examine the denominators to determine how to remove each radical.

- $\sqrt{5y}$ - $\sqrt[3]{36y}$ or $\sqrt[3]{6^2y}$ - $\sqrt[5]{4x^3y}$ or $\sqrt[5]{2^2x^3y}$

| Multiply by $\sqrt{5y}$: $\sqrt{5y} \cdot \sqrt{5y} = \sqrt{25y^2} = 5y.$ | Multiply by $\sqrt[3]{6y^2}$: $\sqrt[3]{6^2y} \cdot \sqrt[3]{6y^2} = \sqrt[3]{6^3y^3} = 6y.$ | Multiply by $\sqrt[5]{2^3x^2y^4}$: $\sqrt[5]{2^2x^3y} \cdot \sqrt[5]{2^3x^2y^4} = \sqrt[5]{2^5x^5y^5} = 2xy.$ |

a. $\sqrt{\dfrac{3x}{5y}} = \dfrac{\sqrt{3x}}{\sqrt{5y}} = \dfrac{\sqrt{3x}}{\sqrt{5y}} \cdot \dfrac{\sqrt{5y}}{\sqrt{5y}} = \dfrac{\sqrt{15xy}}{\sqrt{25y^2}} = \dfrac{\sqrt{15xy}}{5y}$

Multiply by 1. $25y^2$ is a perfect square.

b. $\dfrac{\sqrt[3]{x}}{\sqrt[3]{36y}} = \dfrac{\sqrt[3]{x}}{\sqrt[3]{6^2y}} = \dfrac{\sqrt[3]{x}}{\sqrt[3]{6^2y}} \cdot \dfrac{\sqrt[3]{6y^2}}{\sqrt[3]{6y^2}} = \dfrac{\sqrt[3]{6xy^2}}{\sqrt[3]{6^3y^3}} = \dfrac{\sqrt[3]{6xy^2}}{6y}$

Multiply by 1. 6^3y^3 is a perfect cube.

c. $\dfrac{10y}{\sqrt[5]{4x^3y}} = \dfrac{10y}{\sqrt[5]{2^2x^3y}} = \dfrac{10y}{\sqrt[5]{2^2x^3y}} \cdot \dfrac{\sqrt[5]{2^3x^2y^4}}{\sqrt[5]{2^3x^2y^4}}$

Multiply by 1.

$= \dfrac{10y\sqrt[5]{2^3x^2y^4}}{\sqrt[5]{2^5x^5y^5}} = \dfrac{10y\sqrt[5]{8x^2y^4}}{2xy} = \dfrac{5\sqrt[5]{8x^2y^4}}{x}$

$2^5x^5y^5$ is a perfect 5th power.   Simplify: Divide numerator and denominator by $2y$. ∎

✓ **CHECK POINT 4**   Rationalize each denominator:

a. $\sqrt{\dfrac{2x}{7y}}$ **b.** $\dfrac{\sqrt[3]{x}}{\sqrt[3]{9y}}$ **c.** $\dfrac{6x}{\sqrt[5]{8x^2y^4}}$.

---

4 Rationalize denominators containing two terms. ⏵

Rationalizing Denominators Containing Two Terms

How can we rationalize a denominator if the denominator contains two terms with one or more square roots? **Multiply the numerator and the denominator by the conjugate of the denominator.** Here are three examples of such expressions:

- $\dfrac{8}{3\sqrt{2} + 4}$ - $\dfrac{2 + \sqrt{5}}{\sqrt{6} - \sqrt{3}}$ - $\dfrac{h}{\sqrt{x + h} - \sqrt{x}}$

| The conjugate of the denominator is $3\sqrt{2} - 4$. | The conjugate of the denominator is $\sqrt{6} + \sqrt{3}$. | The conjugate of the denominator is $\sqrt{x + h} + \sqrt{x}$. |

The product of the denominator and its conjugate is found using the formula

$$(A + B)(A - B) = A^2 - B^2.$$

The simplified product will not contain a radical.

EXAMPLE 5   Rationalizing a Denominator Containing Two Terms

Rationalize the denominator: $\dfrac{8}{3\sqrt{2} + 4}$.

Solution The conjugate of the denominator is $3\sqrt{2} - 4$. If we multiply the numerator and the denominator by $3\sqrt{2} - 4$, the simplified denominator will not contain a radical. Therefore, we multiply by 1, choosing $\dfrac{3\sqrt{2} - 4}{3\sqrt{2} - 4}$ for 1.

$$\frac{8}{3\sqrt{2} + 4} = \frac{8}{3\sqrt{2} + 4} \cdot \frac{3\sqrt{2} - 4}{3\sqrt{2} - 4} \qquad \text{Multiply by 1.}$$

$$= \frac{8(3\sqrt{2} - 4)}{(3\sqrt{2})^2 - 4^2} \qquad (A + B)(A - B) = A^2 - B^2$$

$$= \frac{8(3\sqrt{2} - 4)}{18 - 16} \qquad (3\sqrt{2})^2 = 9 \cdot 2 = 18$$

Leave the numerator in factored form. This helps simplify, if possible.

$$= \frac{8(3\sqrt{2} - 4)}{2} \qquad \text{This expression can still be simplified.}$$

$$= \frac{\overset{4}{8}(3\sqrt{2} - 4)}{\underset{1}{2}} \qquad \text{Divide the numerator and denominator by 2.}$$

$$= 4(3\sqrt{2} - 4) \quad \text{or} \quad 12\sqrt{2} - 16 \quad \blacksquare$$

✓ **CHECK POINT 5** Rationalize the denominator: $\dfrac{18}{2\sqrt{3} + 3}$.

EXAMPLE 6 Rationalizing a Denominator Containing Two Terms

Rationalize the denominator: $\dfrac{2 + \sqrt{5}}{\sqrt{6} - \sqrt{3}}$.

Solution The conjugate of the denominator is $\sqrt{6} + \sqrt{3}$. Multiplication of both the numerator and denominator by $\sqrt{6} + \sqrt{3}$ will rationalize the denominator. This will produce a rational number in the denominator.

$$\frac{2 + \sqrt{5}}{\sqrt{6} - \sqrt{3}} = \frac{2 + \sqrt{5}}{\sqrt{6} - \sqrt{3}} \cdot \frac{\sqrt{6} + \sqrt{3}}{\sqrt{6} + \sqrt{3}} \qquad \text{Multiply by 1.}$$

$$= \frac{\overset{F}{2\sqrt{6}} + \overset{O}{2\sqrt{3}} + \overset{I}{\sqrt{5} \cdot \sqrt{6}} + \overset{L}{\sqrt{5} \cdot \sqrt{3}}}{(\sqrt{6})^2 - (\sqrt{3})^2} \qquad \begin{array}{l}\text{Use FOIL in the numerator and} \\ (A - B)(A + B) = A^2 - B^2 \\ \text{in the denominator.}\end{array}$$

$$= \frac{2\sqrt{6} + 2\sqrt{3} + \sqrt{30} + \sqrt{15}}{6 - 3}$$

$$= \frac{2\sqrt{6} + 2\sqrt{3} + \sqrt{30} + \sqrt{15}}{3} \qquad \begin{array}{l}\text{Further simplification is} \\ \text{not possible.} \quad \blacksquare\end{array}$$

✓ **CHECK POINT 6** Rationalize the denominator: $\dfrac{3 + \sqrt{7}}{\sqrt{5} - \sqrt{2}}$.

Rationalizing Numerators

We have seen that square root functions are often used to model growing phenomena with growth that is leveling off. **Figure 7.8** shows a male's height as a function of his age. The pattern of his growth suggests modeling with a square root function.

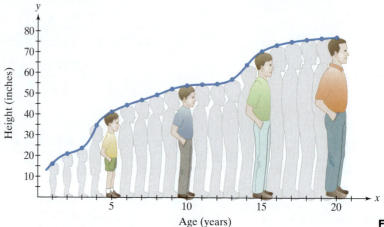

Figure 7.8

If we use $f(x) = \sqrt{x}$ to model height, $f(x)$, at age x, the expression

$$\frac{f(a + h) - f(a)}{h} = \frac{\sqrt{a + h} - \sqrt{a}}{h}$$

describes the man's average growth rate from age a to age $a + h$. Can you see that this expression is not defined if $h = 0$? However, to explore the man's average growth rates for successively shorter periods of time, we need to know what happens to the expression as h takes on values that get closer and closer to 0.

What happens to growth near the instant in time that the man is age a? The question is answered in calculus by **rationalizing the numerator**. The procedure is similar to rationalizing the denominator. **To rationalize a numerator, multiply by 1 to eliminate the radical in the *numerator*.**

EXAMPLE 7   Rationalizing a Numerator

Rationalize the numerator:

$$\frac{\sqrt{a + h} - \sqrt{a}}{h}.$$

Solution   The conjugate of the numerator is $\sqrt{a + h} + \sqrt{a}$. If we multiply the numerator and the denominator by $\sqrt{a + h} + \sqrt{a}$, the simplified numerator will not contain radicals. Therefore, we multiply by 1, choosing $\dfrac{\sqrt{a + h} + \sqrt{a}}{\sqrt{a + h} + \sqrt{a}}$ for 1.

$$\frac{\sqrt{a + h} - \sqrt{a}}{h} = \frac{\sqrt{a + h} - \sqrt{a}}{h} \cdot \frac{\sqrt{a + h} + \sqrt{a}}{\sqrt{a + h} + \sqrt{a}} \qquad \text{Multiply by 1.}$$

$$= \frac{(\sqrt{a + h})^2 - (\sqrt{a})^2}{h(\sqrt{a + h} + \sqrt{a})} \qquad \begin{array}{l} (A - B)(A + B) = A^2 - B^2 \\ \text{Leave the denominator in} \\ \text{factored form.} \end{array}$$

$$= \frac{a + h - a}{h(\sqrt{a + h} + \sqrt{a})} \qquad \begin{array}{l} (\sqrt{a + h})^2 = a + h \\ \text{and } (\sqrt{a})^2 = a. \end{array}$$

$$= \frac{h}{h(\sqrt{a + h} + \sqrt{a})} \qquad \text{Simplify the numerator.}$$

$$= \frac{1}{\sqrt{a + h} + \sqrt{a}} \qquad \begin{array}{l} \text{Simplify by dividing the numerator} \\ \text{and the denominator by } h. \blacksquare \end{array}$$

✓ **CHECK POINT 7** Rationalize the numerator: $\dfrac{\sqrt{x+3} - \sqrt{x}}{3}$.

Achieving Success

Do you get to choose your seat during an exam? If so, select a desk that minimizes distractions and puts you in the right frame of mind. Many students can focus better if they do not sit near friends. If possible, sit near a window, next to a wall, or in the front row. In these locations, you can glance up from time to time without looking at another student's work.

CONCEPT AND VOCABULARY CHECK

Fill in each blank so that the resulting statement is true.

1. Consider the following multiplication problem:

$$\left(7\sqrt{5} + 3\sqrt{2}\right)\left(10\sqrt{5} - 6\sqrt{2}\right).$$

Using the FOIL method, the product of the first terms is _____, the product of the outside terms is _____, the product of the inside terms is _____, and the product of the last terms is _____.

2. $(\sqrt{10} + \sqrt{5})(\sqrt{10} - \sqrt{5}) = (\_\_)^2 - (\_\_)^2 = \_\_ - \_\_ = \_\_$

3. The process of rewriting a radical expression as an equivalent expression in which the denominator no longer contains any radicals is called _____.

4. The number $\dfrac{\sqrt{7}}{\sqrt{5}}$ can be rewritten without a radical in the denominator by multiplying the numerator and denominator by _____.

5. The number $\sqrt[3]{\dfrac{2}{3}}$ can be rewritten without a radical in the denominator by multiplying the numerator and the denominator by _____.

6. The conjugate of $7\sqrt{2} + 5$ is _____.

7. The number $\dfrac{2\sqrt{6} + \sqrt{5}}{3\sqrt{6} - \sqrt{5}}$ can be rewritten without a radical in the denominator by multiplying the numerator and denominator by _____.

7.5 EXERCISE SET ▶ MyMathLab®

Practice Exercises

In this Exercise Set, assume that all variables represent positive real numbers.

In Exercises 1–38, multiply as indicated. If possible, simplify any radical expressions that appear in the product.

1. $\sqrt{2}(x + \sqrt{7})$
2. $\sqrt{5}(x + \sqrt{3})$
3. $\sqrt{6}(7 - \sqrt{6})$
4. $\sqrt{3}(5 - \sqrt{3})$
5. $\sqrt{3}(4\sqrt{6} - 2\sqrt{3})$
6. $\sqrt{6}(4\sqrt{6} - 3\sqrt{2})$
7. $\sqrt[3]{2}(\sqrt[3]{6} + 4\sqrt[3]{5})$
8. $\sqrt[3]{3}(\sqrt[3]{6} + 7\sqrt[3]{4})$
9. $\sqrt[3]{x}(\sqrt[3]{16x^2} - \sqrt[3]{x})$
10. $\sqrt[3]{x}(\sqrt[3]{24x^2} - \sqrt[3]{x})$
11. $(5 + \sqrt{2})(6 + \sqrt{2})$
12. $(7 + \sqrt{2})(8 + \sqrt{2})$
13. $(6 + \sqrt{5})(9 - 4\sqrt{5})$
14. $(4 + \sqrt{5})(10 - 3\sqrt{5})$
15. $(6 - 3\sqrt{7})(2 - 5\sqrt{7})$
16. $(7 - 2\sqrt{7})(5 - 3\sqrt{7})$
17. $(\sqrt{2} + \sqrt{7})(\sqrt{3} + \sqrt{5})$

18. $\left(\sqrt{3} + \sqrt{2}\right)\left(\sqrt{10} + \sqrt{11}\right)$
19. $\left(\sqrt{2} - \sqrt{7}\right)\left(\sqrt{3} - \sqrt{5}\right)$
20. $\left(\sqrt{3} - \sqrt{2}\right)\left(\sqrt{10} - \sqrt{11}\right)$
21. $\left(3\sqrt{2} - 4\sqrt{3}\right)\left(2\sqrt{2} + 5\sqrt{3}\right)$
22. $\left(3\sqrt{5} - 2\sqrt{3}\right)\left(4\sqrt{5} + 5\sqrt{3}\right)$
23. $\left(\sqrt{3} + \sqrt{5}\right)^2$
24. $\left(\sqrt{2} + \sqrt{7}\right)^2$
25. $\left(\sqrt{3x} - \sqrt{y}\right)^2$
26. $\left(\sqrt{2x} - \sqrt{y}\right)^2$
27. $\left(\sqrt{5} + 7\right)\left(\sqrt{5} - 7\right)$
28. $\left(\sqrt{6} + 2\right)\left(\sqrt{6} - 2\right)$
29. $\left(2 - 5\sqrt{3}\right)\left(2 + 5\sqrt{3}\right)$
30. $\left(3 - 5\sqrt{2}\right)\left(3 + 5\sqrt{2}\right)$
31. $\left(3\sqrt{2} + 2\sqrt{3}\right)\left(3\sqrt{2} - 2\sqrt{3}\right)$
32. $\left(4\sqrt{3} + 3\sqrt{2}\right)\left(4\sqrt{3} - 3\sqrt{2}\right)$
33. $\left(3 - \sqrt{x}\right)\left(2 - \sqrt{x}\right)$
34. $\left(4 - \sqrt{x}\right)\left(3 - \sqrt{x}\right)$
35. $\left(\sqrt[3]{x} - 4\right)\left(\sqrt[3]{x} + 5\right)$
36. $\left(\sqrt[3]{x} - 3\right)\left(\sqrt[3]{x} + 7\right)$
37. $\left(x + \sqrt[3]{y^2}\right)\left(2x - \sqrt[3]{y^2}\right)$
38. $\left(x - \sqrt[5]{y^3}\right)\left(2x + \sqrt[5]{y^3}\right)$

In Exercises 39–64, rationalize each denominator.

39. $\dfrac{\sqrt{2}}{\sqrt{5}}$
40. $\dfrac{\sqrt{7}}{\sqrt{3}}$
41. $\sqrt{\dfrac{11}{x}}$
42. $\sqrt{\dfrac{6}{x}}$
43. $\dfrac{9}{\sqrt{3y}}$
44. $\dfrac{12}{\sqrt{3y}}$
45. $\dfrac{1}{\sqrt[3]{2}}$
46. $\dfrac{1}{\sqrt[3]{3}}$
47. $\dfrac{6}{\sqrt[3]{4}}$
48. $\dfrac{10}{\sqrt[3]{5}}$
49. $\sqrt[3]{\dfrac{2}{3}}$
50. $\sqrt[3]{\dfrac{3}{4}}$
51. $\dfrac{4}{\sqrt[3]{x}}$
52. $\dfrac{7}{\sqrt[3]{x}}$
53. $\sqrt[3]{\dfrac{2}{y^2}}$
54. $\sqrt[3]{\dfrac{5}{y^2}}$
55. $\dfrac{7}{\sqrt[3]{2x^2}}$
56. $\dfrac{10}{\sqrt[3]{4x^2}}$
57. $\sqrt[3]{\dfrac{2}{xy^2}}$
58. $\sqrt[3]{\dfrac{3}{xy^2}}$
59. $\dfrac{3}{\sqrt[4]{x}}$
60. $\dfrac{5}{\sqrt[4]{x}}$
61. $\dfrac{6}{\sqrt[5]{8x^3}}$
62. $\dfrac{10}{\sqrt[5]{16x^2}}$
63. $\dfrac{2x^2y}{\sqrt[5]{4x^2y^4}}$
64. $\dfrac{3xy^2}{\sqrt[5]{8xy^3}}$

In Exercises 65–74, simplify each radical expression and then rationalize the denominator.

65. $\dfrac{9}{\sqrt{3x^2y}}$
66. $\dfrac{25}{\sqrt{5x^2y}}$
67. $-\sqrt{\dfrac{75a^5}{b^3}}$
68. $-\sqrt{\dfrac{150a^3}{b^5}}$
69. $\sqrt{\dfrac{7m^2n^3}{14m^3n^2}}$
70. $\sqrt{\dfrac{5m^4n^6}{15m^3n^4}}$
71. $\dfrac{3}{\sqrt[4]{x^5y^3}}$
72. $\dfrac{5}{\sqrt[4]{x^2y^7}}$
73. $\dfrac{12}{\sqrt[3]{-8x^5y^8}}$
74. $\dfrac{15}{\sqrt[3]{-27x^4y^{11}}}$

In Exercises 75–92, rationalize each denominator. Simplify, if possible.

75. $\dfrac{8}{\sqrt{5} + 2}$
76. $\dfrac{15}{\sqrt{6} + 1}$
77. $\dfrac{13}{\sqrt{11} - 3}$
78. $\dfrac{17}{\sqrt{10} - 2}$
79. $\dfrac{6}{\sqrt{5} + \sqrt{3}}$
80. $\dfrac{12}{\sqrt{7} + \sqrt{3}}$
81. $\dfrac{\sqrt{a}}{\sqrt{a} - \sqrt{b}}$
82. $\dfrac{\sqrt{b}}{\sqrt{a} - \sqrt{b}}$
83. $\dfrac{25}{5\sqrt{2} - 3\sqrt{5}}$
84. $\dfrac{35}{5\sqrt{2} - 3\sqrt{5}}$
85. $\dfrac{\sqrt{5} + \sqrt{3}}{\sqrt{5} - \sqrt{3}}$
86. $\dfrac{\sqrt{11} - \sqrt{5}}{\sqrt{11} + \sqrt{5}}$
87. $\dfrac{\sqrt{x} + 1}{\sqrt{x} + 3}$
88. $\dfrac{\sqrt{x} - 2}{\sqrt{x} - 5}$
89. $\dfrac{5\sqrt{3} - 3\sqrt{2}}{3\sqrt{2} - 2\sqrt{3}}$
90. $\dfrac{2\sqrt{6} + \sqrt{5}}{3\sqrt{6} - \sqrt{5}}$
91. $\dfrac{2\sqrt{x} + \sqrt{y}}{\sqrt{y} - 2\sqrt{x}}$
92. $\dfrac{3\sqrt{x} + \sqrt{y}}{\sqrt{y} - 3\sqrt{x}}$

In Exercises 93–104, rationalize each numerator. Simplify, if possible.

93. $\sqrt{\dfrac{3}{2}}$
94. $\sqrt{\dfrac{5}{3}}$
95. $\dfrac{\sqrt[3]{4x}}{\sqrt[3]{y}}$
96. $\dfrac{\sqrt[3]{2x}}{\sqrt[3]{y}}$
97. $\dfrac{\sqrt{x} + 3}{\sqrt{x}}$
98. $\dfrac{\sqrt{x} + 4}{\sqrt{x}}$
99. $\dfrac{\sqrt{a} + \sqrt{b}}{\sqrt{a} - \sqrt{b}}$
100. $\dfrac{\sqrt{a} - \sqrt{b}}{\sqrt{a} + \sqrt{b}}$
101. $\dfrac{\sqrt{x + 5} - \sqrt{x}}{5}$

102. $\dfrac{\sqrt{x+7} - \sqrt{x}}{7}$

103. $\dfrac{\sqrt{x} + \sqrt{y}}{x^2 - y^2}$

104. $\dfrac{\sqrt{x} - \sqrt{y}}{x^2 - y^2}$

Practice PLUS

In Exercises 105–112, add or subtract as indicated. Begin by rationalizing denominators for all terms in which denominators contain radicals.

105. $\sqrt{2} + \dfrac{1}{\sqrt{2}}$

106. $\sqrt{5} + \dfrac{1}{\sqrt{5}}$

107. $\sqrt[3]{25} - \dfrac{15}{\sqrt[3]{5}}$

108. $\sqrt[4]{8} - \dfrac{20}{\sqrt[3]{2}}$

109. $\sqrt{6} - \sqrt{\dfrac{1}{6}} + \sqrt{\dfrac{2}{3}}$

110. $\sqrt{15} - \sqrt{\dfrac{5}{3}} + \sqrt{\dfrac{3}{5}}$

111. $\dfrac{2}{\sqrt{2} + \sqrt{3}} + \sqrt{75} - \sqrt{50}$

112. $\dfrac{5}{\sqrt{2} + \sqrt{7}} - 2\sqrt{32} + \sqrt{28}$

113. Let $f(x) = x^2 - 6x - 4$. Find $f\left(3 - \sqrt{13}\right)$.

114. Let $f(x) = x^2 + 4x - 2$. Find $f\left(-2 + \sqrt{6}\right)$.

115. Let $f(x) = \sqrt{9 + x}$. Find $f\left(3\sqrt{5}\right) \cdot f\left(-3\sqrt{5}\right)$.

116. Let $f(x) = x^2$. Find $f\left(\sqrt{a+1} - \sqrt{a-1}\right)$.

Application Exercises

117. The early Greeks believed that the most pleasing of all rectangles were **golden rectangles**, whose ratio of width to height is

$$\frac{w}{h} = \frac{2}{\sqrt{5} - 1}.$$

The Parthenon at Athens fits into a golden rectangle once the triangular pediment is reconstructed.

Rationalize the denominator of the golden ratio. Then use a calculator and find the ratio of width to height, correct to the nearest hundredth, in golden rectangles.

118. In the "Peanuts" cartoon shown in the section opener on page 542, Woodstock appears to be working steps mentally.

Fill in the missing steps that show how to go from $\dfrac{7\sqrt{2 \cdot 2 \cdot 3}}{6}$ to $\dfrac{7}{3}\sqrt{3}$.

In Exercises 119–120, write expressions for the perimeter and area of each figure. Then simplify these expressions. Assume that all measures are given in inches.

119.

120.

The Pythagorean Theorem for right triangles tells us that the length of the hypotenuse is the square root of the sum of the squares of the lengths of the legs. In Exercises 121–122, use the Pythagorean Theorem to find the length of each hypotenuse in simplified radical form. Assume that all measures are given in inches.

121.

122.

Explaining the Concepts

123. Explain how to perform this multiplication: $\sqrt{2}\left(\sqrt{7} + \sqrt{10}\right)$.

124. Explain how to perform this multiplication: $\left(2 + \sqrt{3}\right)\left(4 + \sqrt{3}\right)$.

125. Explain how to perform this multiplication: $\left(2 + \sqrt{3}\right)^2$.

126. What are conjugates? Give an example with your explanation.

127. Describe how to multiply conjugates.

128. Describe what it means to rationalize a denominator. Use both $\dfrac{1}{\sqrt{5}}$ and $\dfrac{1}{5 + \sqrt{5}}$ in your explanation.

129. When a radical expression has its denominator rationalized, we change the denominator so that it no longer contains any radicals. Doesn't this change the value of the radical expression? Explain.

130. Square the real number $\dfrac{2}{\sqrt{3}}$. Observe that the radical is eliminated from the denominator. Explain whether this process is equivalent to rationalizing the denominator.

Technology Exercises

In Exercises 131–134, determine if each operation is performed correctly by graphing the function on each side of the equation with your graphing utility. Use the given viewing rectangle. The graphs should be the same. If they are not, correct the right side of the equation and then use your graphing utility to verify the correction.

131. $(\sqrt{x} - 1)(\sqrt{x} - 1) = x + 1$
$[0, 5, 1]$ by $[-1, 2, 1]$

132. $(\sqrt{x} + 2)(\sqrt{x} - 2) = x^2 - 4$ for $x \geq 0$
$[0, 10, 1]$ by $[-10, 10, 1]$

133. $(\sqrt{x} + 1)^2 = x + 1$
$[0, 8, 1]$ by $[0, 15, 1]$

134. $\dfrac{3}{\sqrt{x + 3} - \sqrt{x}} = \sqrt{x + 3} + \sqrt{x}$
$[0, 8, 1]$ by $[0, 6, 1]$

Critical Thinking Exercises

Make Sense? *In Exercises 135–138, determine whether each statement makes sense or does not make sense, and explain your reasoning.*

135. I use the same ideas to multiply $(\sqrt{2} + 5)(\sqrt{2} + 4)$ that I did to find the binomial product $(x + 5)(x + 4)$.

136. I used a special-product formula and simplified as follows:
$(\sqrt{2} + \sqrt{5})^2 = 2 + 5 = 7$.

137. In some cases when I multiply a square root expression and its conjugate, the simplified product contains a radical.

138. I use the fact that 1 is the multiplicative identity to both rationalize denominators and rewrite rational expressions with a common denominator.

In Exercises 139–142, determine whether each statement is true or false. If the statement is false, make the necessary change(s) to produce a true statement.

139. $\dfrac{\sqrt{3} + 7}{\sqrt{3} - 2} = -\dfrac{7}{2}$

140. $\dfrac{4}{\sqrt{x + y}} = \dfrac{4\sqrt{x - y}}{x - y}$

141. $\dfrac{4\sqrt{x}}{\sqrt{x} - y} = \dfrac{4x + 4y\sqrt{x}}{x - y^2}$

142. $(\sqrt{x} - 7)^2 = x - 49$

143. Solve:
$$7[(2x - 5) - (x + 1)] = (\sqrt{7} + 2)(\sqrt{7} - 2).$$

144. Simplify: $(\sqrt{2 + \sqrt{3}} + \sqrt{2 - \sqrt{3}})^2$.

145. Rationalize the denominator: $\dfrac{1}{\sqrt{2} + \sqrt{3} + \sqrt{4}}$.

Review Exercises

146. Add: $\dfrac{2}{x - 2} + \dfrac{3}{x^2 - 4}$.
(Section 6.2, Example 6)

147. Solve: $\dfrac{2}{x - 2} + \dfrac{3}{x^2 - 4} = 0$.
(Section 6.6, Example 5)

148. If $f(x) = x^4 - 3x^2 - 2x + 5$, use synthetic division and the Remainder Theorem to find $f(-2)$. (Section 6.5, Example 2)

Preview Exercises

Exercises 149–151 will help you prepare for the material covered in the next section.

149. Multiply: $(\sqrt{x + 4} + 1)^2$.

150. Solve: $4x^2 - 16x + 16 = 4(x + 4)$.

151. Solve: $26 - 11x = 16 - 8x + x^2$.

SECTION

7.6

Radical Equations

What am I supposed to learn?

After studying this section, you should be able to:

1 Solve radical equations.

2 Use models that are radical functions to solve problems.

One of the most dramatic developments in the U.S. work force has been the increase in the number of women. In 1960, approximately 38% of women belonged to the labor force. By 2000, that number had increased to over 60%. With higher levels of education than ever before, most women now choose careers in the labor force rather than homemaking.

The function

$$f(x) = 3.5\sqrt{x} + 38$$

models the percentage of U.S. women in the labor force, $f(x)$, x years after 1960. How can we predict the year when, say, 70% of women will participate in the U.S. work force? Substitute 70 for $f(x)$ in $f(x) = 3.5\sqrt{x} + 38$ and solve for x:

$$70 = 3.5\sqrt{x} + 38.$$

The resulting equation contains a variable in the radicand and is called a *radical equation*. A **radical equation** is an equation in which the variable occurs in a square root, cube root, or any higher root. Some examples of radical equations are

$$\sqrt{2x+3} = 5, \quad \sqrt{3x+1} - \sqrt{x+4} = 1, \quad \text{and} \quad \sqrt[3]{3x-1} + 4 = 0.$$

Variables occur in radicands.

In this section, you will learn how to solve radical equations. Solving such equations will enable you to solve new kinds of problems using radical functions.

1 Solve radical equations. ⓓ

Solving Radical Equations

Consider the following radical equation:

$$\sqrt{x} = 9.$$

We solve the equation by squaring both sides:

Squaring both sides eliminates the square root.

$$(\sqrt{x})^2 = 9^2$$
$$x = 81.$$

The proposed solution, 81, can be checked in the original equation, $\sqrt{x} = 9$. Because $\sqrt{81} = 9$, the solution is 81 and the solution set is {81}.

In general, we solve radical equations with square roots by squaring both sides of the equation. We solve radical equations with nth roots by raising both sides of the equation to the nth power. Unfortunately, if n is even, all the solutions of the equation raised to the even power may not be solutions of the original equation. Consider, for example, the equation

$$x = 4.$$

If we square both sides, we obtain

$$x^2 = 16.$$

$$x^2 - 16 = 0 \qquad \text{Subtract 16 from both sides and write the quadratic equation in standard form.}$$

$$(x+4)(x-4) = 0 \qquad \text{Factor.}$$

$$x + 4 = 0 \quad \text{or} \quad x - 4 = 0 \qquad \text{Set each factor equal to O.}$$

$$x = -4 \qquad\qquad x = 4 \qquad \text{Solve the resulting equations.}$$

The equation $x^2 = 16$ has two solutions, -4 and 4. By contrast, only 4 is a solution of the original equation, $x = 4$. For this reason, **when raising both sides of an equation to an even power, always check proposed solutions in the original equation**.

Here is a general method for solving radical equations with nth roots:

Solving Radical Equations Containing nth Roots

1. If necessary, arrange terms so that one radical is isolated on one side of the equation.
2. Raise both sides of the equation to the nth power to eliminate the nth root.
3. Solve the resulting equation. If this equation still contains radicals, repeat steps 1 and 2.
4. Check all proposed solutions in the original equation.

> **EXAMPLE 1** Solving a Radical Equation

Solve: $\sqrt{2x + 3} = 5$.

Solution

Step 1. Isolate a radical on one side. The radical, $\sqrt{2x + 3}$, is already isolated on the left side of the equation, so we can skip this step.

Step 2. Raise both sides to the nth power. Because n, the index, is 2, we square both sides.

$$\sqrt{2x + 3} = 5 \qquad \text{This is the given equation.}$$
$$\left(\sqrt{2x + 3}\right)^2 = 5^2 \qquad \text{Square both sides to eliminate the radical.}$$
$$2x + 3 = 25 \qquad \text{Simplify.}$$

Step 3. Solve the resulting equation.

$$2x + 3 = 25 \qquad \text{The resulting equation is a linear equation.}$$
$$2x = 22 \qquad \text{Subtract 3 from both sides.}$$
$$x = 11 \qquad \text{Divide both sides by 2.}$$

Step 4. Check the proposed solution in the original equation. Because both sides were raised to an even power, this check is essential.

Check 11:
$$\sqrt{2x + 3} = 5$$
$$\sqrt{2 \cdot 11 + 3} \stackrel{?}{=} 5$$
$$\sqrt{25} \stackrel{?}{=} 5$$
$$5 = 5, \qquad \text{true}$$

The solution is 11 and the solution set is {11}. ∎

> ☑ **CHECK POINT 1** Solve: $\sqrt{3x + 4} = 8$.

> **EXAMPLE 2** Solving a Radical Equation

Solve: $\sqrt{x - 3} + 6 = 5$.

Solution

Step 1. Isolate a radical on one side. The radical, $\sqrt{x - 3}$, can be isolated by subtracting 6 from both sides. We obtain

$$\sqrt{x - 3} = -1.$$

> A principal square root cannot be negative. This equation has no solution. Let's continue the solution procedure to see what happens.

Step 2. Raise both sides to the nth power. Because n, the index, is 2, we square both sides.

$$\left(\sqrt{x - 3}\right)^2 = (-1)^2$$
$$x - 3 = 1 \qquad \text{Simplify.}$$

Step 3. Solve the resulting equation.

$$x - 3 = 1 \qquad \text{The resulting equation is a linear equation.}$$
$$x = 4 \qquad \text{Add 3 to both sides.}$$

Step 4. Check the proposed solution in the original equation.

Check 4:

$$\sqrt{x-3} + 6 = 5$$
$$\sqrt{4-3} + 6 \overset{?}{=} 5$$
$$\sqrt{1} + 6 \overset{?}{=} 5$$
$$1 + 6 \overset{?}{=} 5$$
$$7 = 5, \quad \text{false}$$

This false statement indicates that 4 is not a solution. Thus, the equation has no solution. The solution set is $\varnothing$, the empty set. ∎

Example 2 illustrates that extra solutions may be introduced when you raise both sides of a radical equation to an even power. Such solutions, which are not solutions of the given equation, are called **extraneous solutions**. Thus, 4 is an extraneous solution of $\sqrt{x-3} + 6 = 5$.

☑ **CHECK POINT 2** Solve: $\sqrt{x-1} + 7 = 2$.

EXAMPLE 3 Solving a Radical Equation

Solve:

$$x + \sqrt{26 - 11x} = 4.$$

Solution

Step 1. Isolate a radical on one side. We isolate the radical, $\sqrt{26-11x}$, by subtracting x from both sides.

$$x + \sqrt{26 - 11x} = 4 \qquad \text{This is the given equation.}$$
$$\sqrt{26 - 11x} = 4 - x \qquad \text{Subtract } x \text{ from both sides.}$$

Step 2. Square both sides.

$$\left(\sqrt{26 - 11x}\right)^2 = (4 - x)^2$$
$$26 - 11x = 16 - 8x + x^2 \qquad \text{Simplify. Use the special-product formula } (A-B)^2 = A^2 - 2AB + B^2 \text{ to square } 4 - x.$$

Step 3. Solve the resulting equation. Because of the x^2-term, the resulting equation is a quadratic equation. We need to write this quadratic equation in standard form. We can obtain zero on the left side by subtracting 26 and adding $11x$ on both sides.

$$26 - 26 - 11x + 11x = 16 - 26 - 8x + 11x + x^2$$
$$0 = x^2 + 3x - 10 \qquad \text{Simplify.}$$
$$0 = (x + 5)(x - 2) \qquad \text{Factor.}$$
$$x + 5 = 0 \quad \text{or} \quad x - 2 = 0 \qquad \text{Set each factor equal to zero.}$$
$$x = -5 \qquad\qquad x = 2 \qquad \text{Solve for } x.$$

Step 4. Check the proposed solutions in the original equation.

Check −5:
$$x + \sqrt{26 - 11x} = 4$$
$$-5 + \sqrt{26 - 11(-5)} \overset{?}{=} 4$$
$$-5 + \sqrt{81} \overset{?}{=} 4$$
$$-5 + 9 \overset{?}{=} 4$$
$$4 = 4, \quad \text{true}$$

Check 2:
$$x + \sqrt{26 - 11x} = 4$$
$$2 + \sqrt{26 - 11 \cdot 2} \overset{?}{=} 4$$
$$2 + \sqrt{4} \overset{?}{=} 4$$
$$2 + 2 \overset{?}{=} 4$$
$$4 = 4, \quad \text{true}$$

The solutions are −5 and 2, and the solution set is $\{-5, 2\}$. ∎

Great Question!

Can I square the right side of $\sqrt{26 - 11x} = 4 - x$ by first squaring 4 and then squaring x?

No. Be sure to square *both sides* of an equation. Do *not* square each term.

Correct:
$$\left(\sqrt{26 - 11x}\right)^2 = (4 - x)^2$$

Incorrect!
$$\left(\sqrt{26 - 11x}\right)^2 = 4^2 - x^2$$

Using Technology

Graphic Connections

You can use a graphing utility to provide a graphic check that $\{-5, 2\}$ is the solution set of $x + \sqrt{26 - 11x} = 4$.

Use the given equation

$$x + \sqrt{26 - 11x} = 4.$$

Enter $y_1 = x + \sqrt{26 - 11x}$ in the $\boxed{y=}$ screen.

Enter $y_2 = 4$ in the $\boxed{y=}$ screen.

Use the equivalent equation

$$x + \sqrt{26 - 11x} - 4 = 0.$$

Enter $y_1 = x + \sqrt{26 - 11x} - 4$ in the $\boxed{y=}$ screen.

Display graphs for y_1 and y_2. The solutions are the x-coordinates of the intersection points. These x-coordinates are -5 and 2. This verifies $\{-5, 2\}$ as the solution set of $x + \sqrt{26 - 11x} = 4$.

Display the graph for y_1. The solutions are the x-intercepts. The x-intercepts are -5 and 2. This verifies $\{-5, 2\}$ as the solution set of $x + \sqrt{26 - 11x} - 4 = 0$.

$y_1 = x + \sqrt{26 - 11x}$

x-coordinate of intersection point is -5.

x-coordinate of intersection point is 2.

$y_2 = 4$

-5 2

$[-8, 5, 1]$ by $[0, 7, 1]$

$y_1 = x + \sqrt{26 - 11x} - 4$

x-intercept: 2

x-intercept: -5

$[-10, 3, 1]$ by $[-4, 3, 1]$

✓ **CHECK POINT 3**   Solve: $\sqrt{6x + 7} - x = 2$.

The solution of a radical equation with two or more square root expressions involves isolating a radical, squaring both sides, and then repeating this process. Let's consider an equation containing two square root expressions.

EXAMPLE 4   Solving an Equation That Has Two Radicals

Solve: $\sqrt{3x + 1} - \sqrt{x + 4} = 1$.

Solution

Step 1. Isolate a radical on one side. We can isolate the radical $\sqrt{3x + 1}$ by adding $\sqrt{x + 4}$ to both sides. We obtain

$$\sqrt{3x + 1} = \sqrt{x + 4} + 1.$$

Step 2. Square both sides.

$$(\sqrt{3x + 1})^2 = (\sqrt{x + 4} + 1)^2$$

Squaring the expression on the right side of the equation can be a bit tricky. We have to use the formula

$$(A + B)^2 = A^2 + 2AB + B^2.$$

Focusing on just the right side, here is how the squaring is done:

$$(A + B)^2 \quad = \quad A^2 \ + \ 2 \ \cdot \ A \ \cdot \ B \ + \ B^2$$

$$(\sqrt{x + 4} + 1)^2 = (\sqrt{x + 4})^2 + 2 \cdot \sqrt{x + 4} \cdot 1 + 1^2 = x + 4 + 2\sqrt{x + 4} + 1.$$

Now let's return to squaring both sides.

$$\left(\sqrt{3x+1}\right)^2 = \left(\sqrt{x+4}+1\right)^2$$ Square both sides of the equation with an isolated radical.

$$3x + 1 = x + 4 + 2\sqrt{x+4} + 1$$ $\left(\sqrt{3x+1}\right)^2 = 3x + 1$; square the right side using the formula for $(A+B)^2$.

$$3x + 1 = x + 5 + 2\sqrt{x+4}$$ Combine numerical terms on the right side: $4 + 1 = 5$.

Can you see that the resulting equation still contains a radical, namely $\sqrt{x+4}$? Thus, we need to repeat the first two steps.

Repeat Step 1. Isolate a radical on one side. We isolate $2\sqrt{x+4}$, the radical term, by subtracting $x + 5$ from both sides. We obtain

$$3x + 1 = x + 5 + 2\sqrt{x+4}$$ This is the equation from our last step.

$$2x - 4 = 2\sqrt{x+4}.$$ Subtract x and subtract 5 from both sides.

Although we can simplify the equation by dividing both sides by 2, this sort of simplification is not always helpful. Thus, we will work with the equation in this form.

Repeat Step 2. Square both sides.

> Be careful in squaring both sides. Use $(A - B)^2 = A^2 - 2AB + B^2$ to square the left side. Use $(AB)^2 = A^2B^2$ to square the right side.

$$(2x - 4)^2 = \left(2\sqrt{x+4}\right)^2$$ Square both sides.

$$4x^2 - 16x + 16 = 4(x + 4)$$ Square both 2 and $\sqrt{x+4}$ on the right side.

Step 3. Solve the resulting equation. We solve this quadratic equation by writing it in standard form.

$$4x^2 - 16x + 16 = 4x + 16$$ Use the distributive property.

$$4x^2 - 20x = 0$$ Subtract $4x + 16$ from both sides.

$$4x(x - 5) = 0$$ Factor.

$$4x = 0 \quad \text{or} \quad x - 5 = 0$$ Set each factor equal to zero.

$$x = 0 \qquad\qquad x = 5$$ Solve for x.

Step 4. Check the proposed solutions in the original equation.

Check 0:	Check 5:
$\sqrt{3x+1} - \sqrt{x+4} = 1$	$\sqrt{3x+1} - \sqrt{x+4} = 1$
$\sqrt{3\cdot 0 + 1} - \sqrt{0+4} \overset{?}{=} 1$	$\sqrt{3\cdot 5 + 1} - \sqrt{5+4} \overset{?}{=} 1$
$\sqrt{1} - \sqrt{4} \overset{?}{=} 1$	$\sqrt{16} - \sqrt{9} \overset{?}{=} 1$
$1 - 2 \overset{?}{=} 1$	$4 - 3 \overset{?}{=} 1$
$-1 = 1,$ false	$1 = 1,$ true

The check indicates that 0 is not a solution. It is an extraneous solution brought about by squaring each side of the equation. The only solution is 5 and the solution set is {5}. ■

☑ **CHECK POINT 4**   Solve: $\sqrt{x+5} - \sqrt{x-3} = 2.$

EXAMPLE 5   Solving a Radical Equation

Solve: $(3x - 1)^{\frac{1}{3}} + 4 = 0$.

Solution   Although we can rewrite the equation in radical form

$$\sqrt[3]{3x - 1} + 4 = 0,$$

it is not necessary to do so. Because the equation involves a cube root, we isolate the radical term—that is, the term with the rational exponent—and cube both sides.

$(3x - 1)^{\frac{1}{3}} + 4 = 0$ This is the given equation.

$(3x - 1)^{\frac{1}{3}} = -4$ Subtract 4 from both sides and isolate the term with the rational exponent.

$\left[(3x - 1)^{\frac{1}{3}} \right]^3 = (-4)^3$ Cube both sides.

$3x - 1 = -64$ Multiply exponents on the left side and simplify.

$3x = -63$ Add 1 to both sides.

$x = -21$ Divide both sides by 3.

Because both sides were raised to an odd power, it is not essential to check the proposed solution, -21. However, checking is always a good idea. Do so now and verify that -21 is the solution and the solution set is $\{-21\}$.   ■

Example 5 illustrates that a radical equation with rational exponents can be solved by

1. isolating the expression with the rational exponent, and
2. raising both sides of the equation to a power that is the reciprocal of the rational exponent.

Keep in mind that it is essential to check proposed solutions when both sides have been raised to even powers. Thus, equations with rational exponents such as $\frac{1}{2}$ and $\frac{1}{4}$ must be checked.

✓ **CHECK POINT 5**   Solve: $(2x - 3)^{\frac{1}{3}} + 3 = 0$.

---

2 Use models that are radical functions to solve problems.

Applications of Radical Equations

Radical equations can be solved to answer questions about variables contained in radical functions.

EXAMPLE 6   Women in the Labor Force

The bar graph in **Figure 7.9** shows the percentage of U.S. women in the labor force from 1960 through 2010. The function $f(x) = 3.5\sqrt{x} + 38$ models the percentage of U.S. women in the labor force, $f(x)$, x years after 1960. According to the model, when will 70% of U.S. women participate in the work force?

Percentage of United States Women in the Labor Force

Figure 7.9

Source: U.S. Department of Labor

Solution   To find when 70% of U.S. women will be in the work force, substitute 70 for $f(x)$ in the given function. Then solve for x, the number of years after 1960.

$$f(x) = 3.5\sqrt{x} + 38 \qquad \text{This is the given function.}$$

$$70 = 3.5\sqrt{x} + 38 \qquad \text{Substitute 70 for } f(x).$$

$$32 = 3.5\sqrt{x} \qquad \text{Subtract 38 from both sides.}$$

$$\frac{32}{3.5} = \sqrt{x} \qquad \text{Divide both sides by 3.5.}$$

$$\left(\frac{32}{3.5}\right)^2 = \left(\sqrt{x}\right)^2 \qquad \text{Square both sides.}$$

$$84 \approx x \qquad \text{Use a calculator.}$$

The model indicates that 70% of U.S. women will be in the labor force approximately 84 years after 1960. Because $1960 + 84 = 2044$, this is projected to occur in 2044. ∎

☑ **CHECK POINT 6**   Use the function in Example 6 to project when 73% of U.S. women will participate in the work force.

CONCEPT AND VOCABULARY CHECK

Fill in each blank so that the resulting statement is true.

1. An equation in which the variable occurs in a square root, cube root, or any higher root is called a/an _____ equation.

2. Solutions of a squared equation that are not solutions of the original equation are called _____ solutions.

3. Consider the equation

   $$\sqrt{2x + 1} = x - 7.$$

   Squaring the left side and simplifying results in _____. Squaring the right side and simplifying results in

   _____.

4. Consider the equation

   $$\sqrt{x + 2} = 3 - \sqrt{x - 1}.$$

   Squaring the left side and simplifying results in _____. Squaring the right side and simplifying results in

   _____.

5. Consider the equation

   $$(2x + 3)^{\frac{1}{3}} = 2.$$

   Cubing the left side and simplifying results in _____. Cubing the right side and simplifying results in _____.

6. True or false: 4 is a solution of $\sqrt{5x + 16} = x + 2.$ _____

7. True or false: -3 is a solution of $\sqrt{5x + 16} = x + 2.$ _____

7.6 EXERCISE SET   ▶ MyMathLab®

Practice Exercises

In Exercises 1–38, solve each radical equation.

1. $\sqrt{3x - 2} = 4$
2. $\sqrt{5x - 1} = 8$
3. $\sqrt{5x - 4} - 9 = 0$
4. $\sqrt{3x - 2} - 5 = 0$
5. $\sqrt{3x + 7} + 10 = 4$
6. $\sqrt{2x + 5} + 11 = 6$
7. $x = \sqrt{7x + 8}$
8. $x = \sqrt{6x + 7}$
9. $\sqrt{5x + 1} = x + 1$
10. $\sqrt{2x + 1} = x - 7$
11. $x = \sqrt{2x - 2} + 1$
12. $x = \sqrt{3x + 7} - 3$
13. $x - 2\sqrt{x - 3} = 3$
14. $3x - \sqrt{3x + 7} = -5$
15. $\sqrt{2x - 5} = \sqrt{x + 4}$
16. $\sqrt{6x + 2} = \sqrt{5x + 3}$
17. $\sqrt[3]{2x + 11} = 3$
18. $\sqrt[3]{6x - 3} = 3$
19. $\sqrt[3]{2x - 6} - 4 = 0$

CORRECT:

$$\sqrt{-25} \cdot \sqrt{-4} = \sqrt{25}\sqrt{-1} \cdot \sqrt{4}\sqrt{-1}$$
$$= 5i \cdot 2i$$
$$= 10i^2 = 10(-1) = -10$$

INCORRECT:

$$\sqrt{-25} \cdot \sqrt{-4} = \sqrt{(-25)(-4)}$$
$$= \sqrt{100}$$
$$= 10$$

EXAMPLE 4 Multiplying Square Roots of Negative Numbers

Multiply: $\sqrt{-3} \cdot \sqrt{-5}$.

Solution

$$\sqrt{-3} \cdot \sqrt{-5} = \sqrt{3}\sqrt{-1} \cdot \sqrt{5}\sqrt{-1}$$
$$= i\sqrt{3} \cdot i\sqrt{5} \qquad \text{Express square roots in terms of } i.$$
$$= i^2\sqrt{15} \qquad \sqrt{3} \cdot \sqrt{5} = \sqrt{15} \text{ and } i \cdot i = i^2.$$
$$= (-1)\sqrt{15} \qquad i^2 = -1$$
$$= -\sqrt{15} \quad \blacksquare$$

✓ **CHECK POINT 4** Multiply: $\sqrt{-5} \cdot \sqrt{-7}$.

---

4 Divide complex numbers. ▶

Conjugates and Division

It is possible to multiply imaginary numbers and obtain a real number. Here is an example:

$$\overset{\text{F} \quad \text{O} \quad \text{I} \quad \text{L}}{(4 + 7i)(4 - 7i)} = 16 - 28i + 28i - 49i^2$$
$$= 16 - 49i^2 = 16 - 49(-1) = 65.$$

Replace i^2 with -1.

You can also perform $(4 + 7i)(4 - 7i)$ using the formula

$$(A + B)(A - B) = A^2 - B^2.$$

A real number is obtained even faster:

$$(4 + 7i)(4 - 7i) = 4^2 - (7i)^2 = 16 - 49i^2 = 16 - 49(-1) = 65.$$

The **conjugate** of the complex number $a + bi$ is $a - bi$. The **conjugate** of the complex number $a - bi$ is $a + bi$. The multiplication problem that we just performed involved conjugates. The multiplication of conjugates always results in a real number:

$$(a + bi)(a - bi) = a^2 - (bi)^2 = a^2 - b^2i^2 = a^2 - b^2(-1) = a^2 + b^2.$$

The product eliminates i.

Conjugates are used to divide complex numbers. The goal of the division procedure is to obtain a real number in the denominator. This real number becomes the denominator of a and b in the quotient $a + bi$. By multiplying the numerator and the denominator of the division by the conjugate of the denominator, you will obtain this real number in the denominator. Here are two examples of such divisions:

- $\dfrac{7 + 4i}{2 - 5i}$

  The conjugate of the denominator is 2 + 5i.

- $\dfrac{5i - 4}{3i}$ or $\dfrac{5i - 4}{0 + 3i}$.

  The conjugate of the denominator is 0 − 3i, or −3i.

The procedure for dividing complex numbers, illustrated in Examples 5 and 6, should remind you of rationalizing denominators.

EXAMPLE 5   Using Conjugates to Divide Complex Numbers

Divide and simplify to the form $a + bi$:

$$\frac{7 + 4i}{2 - 5i}.$$

Solution The conjugate of the denominator is $2 + 5i$. Multiplication of both the numerator and the denominator by $2 + 5i$ will eliminate i from the denominator while maintaining the value of the expression.

$$\frac{7 + 4i}{2 - 5i} = \frac{7 + 4i}{2 - 5i} \cdot \frac{2 + 5i}{2 + 5i} \qquad \text{Multiply by 1.}$$

$$\begin{array}{ll} = \dfrac{14 + 35i + 8i + 20i^2}{2^2 - (5i)^2} & \text{Use FOIL in the numerator and } (A - B)(A + B) = \\ & A^2 - B^2 \text{ in the denominator.} \end{array}$$

$$= \frac{14 + 43i + 20i^2}{4 - 25i^2} \qquad \text{Simplify.}$$

$$= \frac{14 + 43i + 20(-1)}{4 - 25(-1)} \qquad i^2 = -1$$

$$= \frac{14 + 43i - 20}{4 + 25} \qquad \text{Perform the multiplications involving } -1.$$

$$\begin{array}{ll} = \dfrac{-6 + 43i}{29} & \text{Combine real terms in the numerator and} \\ & \text{denominator.} \end{array}$$

$$= -\frac{6}{29} + \frac{43}{29}i \qquad \text{Express the answer in the form } a + bi. \ \blacksquare$$

☑ **CHECK POINT 5**   Divide and simplify to the form $a + bi$:

$$\frac{6 + 2i}{4 - 3i}.$$

EXAMPLE 6   Using Conjugates to Divide Complex Numbers

Divide and simplify to the form $a + bi$:

$$\frac{5i - 4}{3i}.$$

7.7 EXERCISE SET ▶ MyMathLab®

Practice Exercises

In Exercises 1–16, express each number in terms of i and simplify, if possible.

1. $\sqrt{-100}$

2. $\sqrt{-49}$

3. $\sqrt{-23}$

4. $\sqrt{-21}$

5. $\sqrt{-18}$

6. $\sqrt{-125}$

7. $\sqrt{-63}$

8. $\sqrt{-28}$

9. $-\sqrt{-108}$

10. $-\sqrt{-300}$

11. $5 + \sqrt{-36}$

12. $7 + \sqrt{-4}$

13. $15 + \sqrt{-3}$

14. $20 + \sqrt{-5}$

15. $-2 - \sqrt{-18}$

16. $-3 - \sqrt{-27}$

In Exercises 17–32, add or subtract as indicated. Write the result in the form a + bi.

17. $(3 + 2i) + (5 + i)$

18. $(6 + 5i) + (4 + 3i)$

19. $(7 + 2i) + (1 - 4i)$

20. $(-2 + 6i) + (4 - i)$

21. $(10 + 7i) - (5 + 4i)$

22. $(11 + 8i) - (2 + 5i)$

23. $(9 - 4i) - (10 + 3i)$

24. $(8 - 5i) - (6 + 2i)$

25. $(3 + 2i) - (5 - 7i)$

26. $(-7 + 5i) - (9 - 11i)$

27. $(-5 + 4i) - (-13 - 11i)$

28. $(-9 + 2i) - (-17 - 6i)$

29. $8i - (14 - 9i)$

30. $15i - (12 - 11i)$

31. $\left(2 + i\sqrt{3}\right) + \left(7 + 4i\sqrt{3}\right)$

32. $\left(4 + i\sqrt{5}\right) + \left(8 + 6i\sqrt{5}\right)$

In Exercises 33–62, find each product. Write imaginary results in the form a + bi.

33. $2i(5 + 3i)$

34. $5i(4 + 7i)$

35. $3i(7i - 5)$

36. $8i(4i - 3)$

37. $-7i(2 - 5i)$

38. $-6i(3 - 5i)$

39. $(3 + i)(4 + 5i)$

40. $(4 + i)(5 + 6i)$

41. $(7 - 5i)(2 - 3i)$

42. $(8 - 4i)(3 - 2i)$

43. $(6 - 3i)(-2 + 5i)$

44. $(7 - 2i)(-3 + 6i)$

45. $(3 + 5i)(3 - 5i)$

46. $(2 + 7i)(2 - 7i)$

47. $(-5 + 3i)(-5 - 3i)$

48. $(-4 + 2i)(-4 - 2i)$

49. $\left(3 - i\sqrt{2}\right)\left(3 + i\sqrt{2}\right)$

50. $\left(5 - i\sqrt{3}\right)\left(5 + i\sqrt{3}\right)$

51. $(2 + 3i)^2$

52. $(3 + 2i)^2$

53. $(5 - 2i)^2$

54. $(5 - 3i)^2$

55. $\sqrt{-7} \cdot \sqrt{-2}$

56. $\sqrt{-7} \cdot \sqrt{-3}$

57. $\sqrt{-9} \cdot \sqrt{-4}$

58. $\sqrt{-16} \cdot \sqrt{-4}$

59. $\sqrt{-7} \cdot \sqrt{-25}$

60. $\sqrt{-3} \cdot \sqrt{-36}$

61. $\sqrt{-8} \cdot \sqrt{-3}$

62. $\sqrt{-9} \cdot \sqrt{-5}$

In Exercises 63–84, divide and simplify to the form a + bi.

63. $\dfrac{2}{3 + i}$

64. $\dfrac{3}{4 + i}$

65. $\dfrac{2i}{1 + i}$

66. $\dfrac{5i}{2 + i}$

67. $\dfrac{7}{4 - 3i}$

68. $\dfrac{9}{1 - 2i}$

69. $\dfrac{6i}{3 - 2i}$

70. $\dfrac{5i}{2 - 3i}$

71. $\dfrac{1 + i}{1 - i}$

72. $\dfrac{1 - i}{1 + i}$

73. $\dfrac{2 - 3i}{3 + i}$

74. $\dfrac{2 + 3i}{3 - i}$

75. $\dfrac{5 - 2i}{3 + 2i}$

76. $\dfrac{6 - 3i}{4 + 2i}$

77. $\dfrac{4 + 5i}{3 - 7i}$

78. $\dfrac{5 - i}{3 - 2i}$

79. $\dfrac{7}{3i}$

80. $\dfrac{5}{7i}$

81. $\dfrac{8 - 5i}{2i}$

82. $\dfrac{3 + 4i}{5i}$

83. $\dfrac{4 + 7i}{-3i}$

84. $\dfrac{5 + i}{-4i}$

In Exercises 85–100, simplify each expression.

85. i^{10}

86. i^{14}

87. i^{11}

88. i^{15}

89. i^{22}

90. i^{46}

91. i^{200}

92. i^{400}

93. i^{17}

94. i^{21}

95. $(-i)^4$

96. $(-i)^6$

97. $(-i)^9$

98. $(-i)^{13}$

99. $i^{24} + i^2$

100. $i^{28} + i^{30}$

Practice PLUS

In Exercises 101–108, perform the indicated operation(s) and write the result in the form a + bi.

101. $(2 - 3i)(1 - i) - (3 - i)(3 + i)$

102. $(8 + 9i)(2 - i) - (1 - i)(1 + i)$

103. $(2 + i)^2 - (3 - i)^2$

104. $(4 - i)^2 - (1 + 2i)^2$

105. $5\sqrt{-16} + 3\sqrt{-81}$

106. $5\sqrt{-8} + 3\sqrt{-18}$

107. $\dfrac{i^4 + i^{12}}{i^8 - i^7}$

108. $\dfrac{i^8 + i^{40}}{i^4 + i^3}$

109. Let $f(x) = x^2 - 2x + 2$. Find $f(1 + i)$.

110. Let $f(x) = x^2 - 2x + 5$. Find $f(1 - 2i)$.

In Exercises 111–114, simplify each evaluation to the form $a + bi$.

111. Let $f(x) = x - 3i$ and $g(x) = 4x + 2i$. Find $(fg)(-1)$.

112. Let $f(x) = 12x - i$ and $g(x) = 6x + 3i$. Find $(fg)\left(-\dfrac{1}{3}\right)$.

113. Let $f(x) = \dfrac{x^2 + 19}{2 - x}$. Find $f(3i)$.

114. Let $f(x) = \dfrac{x^2 + 11}{3 - x}$. Find $f(4i)$.

Application Exercises

Complex numbers are used in electronics to describe the current in an electric circuit. Ohm's law relates the current in a circuit, I, in amperes, the voltage of the circuit, E, in volts, and the resistance of the circuit, R, in ohms, by the formula $E = IR$. Use this formula to solve Exercises 115–116.

115. Find E, the voltage of a circuit, if $I = (4 - 5i)$ amperes and $R = (3 + 7i)$ ohms.

116. Find E, the voltage of a circuit, if $I = (2 - 3i)$ amperes and $R = (3 + 5i)$ ohms.

117. The mathematician Girolamo Cardano is credited with the first use (in 1545) of negative square roots in solving the now-famous problem, "Find two numbers whose sum is 10 and whose product is 40." Show that the complex numbers $5 + i\sqrt{15}$ and $5 - i\sqrt{15}$ satisfy the conditions of the problem. (Cardano did not use the symbolism $i\sqrt{15}$ or even $\sqrt{-15}$. He wrote R.m 15 for $\sqrt{-15}$, meaning "radix minus 15." He regarded the numbers $5 + $ R.m 15 and $5 - $ R.m 15 as "fictitious" or "ghost numbers," and considered the problem "manifestly impossible." But in a mathematically adventurous spirit, he exclaimed, "Nevertheless, we will operate.")

Explaining the Concepts

118. What is the imaginary unit i?

119. Explain how to write $\sqrt{-64}$ as a multiple of i.

120. What is a complex number? Explain when a complex number is a real number and when it is an imaginary number. Provide examples with your explanation.

121. Explain how to add complex numbers. Give an example.

122. Explain how to subtract complex numbers. Give an example.

123. Explain how to find the product of $2i$ and $5 + 3i$.

124. Explain how to find the product of $3 + 2i$ and $5 + 3i$.

125. Explain how to find the product of $3 + 2i$ and $3 - 2i$.

126. Explain how to find the product of $\sqrt{-1}$ and $\sqrt{-4}$. Describe a common error in the multiplication that needs to be avoided.

127. What is the conjugate of $2 + 3i$? What happens when you multiply this complex number by its conjugate?

128. Explain how to divide complex numbers. Provide an example with your explanation.

129. Explain each of the three jokes in the cartoon on page 562.

130. A stand-up comedian uses algebra in some jokes, including one about a telephone recording that announces "You have just reached an imaginary number. Please multiply by i and dial again." Explain the joke.

Explain the error in Exercises 131–132.

131. $\sqrt{-9} + \sqrt{-16} = \sqrt{-25} = i\sqrt{25} = 5i$

132. $\left(\sqrt{-9}\right)^2 = \sqrt{-9} \cdot \sqrt{-9} = \sqrt{81} = 9$

Critical Thinking Exercises

Make Sense? *In Exercises 133–136, determine whether each statement makes sense or does not make sense, and explain your reasoning.*

133. The humor in this cartoon is based on the fact that "rational" and "real" have different meanings in mathematics and in everyday speech.

© 2007 GJ Caulkins

134. The word *imaginary* in imaginary numbers tells me that these numbers are undefined.

135. By writing the imaginary number $5i$, I can immediately see that 5 is the constant and i is the variable.

136. When I add or subtract complex numbers, I am basically combining like terms.

In Exercises 137–140, determine whether each statement is true or false. If the statement is false, make the necessary change(s) to produce a true statement.

137. Some irrational numbers are not complex numbers.

138. $(3 + 7i)(3 - 7i)$ is an imaginary number.

139. $\dfrac{7 + 3i}{5 + 3i} = \dfrac{7}{5}$

140. In the complex number system, $x^2 + y^2$ (the sum of two squares) can be factored as $(x + yi)(x - yi)$.

In Exercises 141–143, perform the indicated operations and write the result in the form $a + bi$.

141. $\dfrac{4}{(2 + i)(3 - i)}$

142. $\dfrac{1 + i}{1 + 2i} + \dfrac{1 - i}{1 - 2i}$

143. $\dfrac{8}{1 + \dfrac{2}{i}}$

Review Exercises

144. Simplify:

$$\dfrac{\dfrac{x}{y^2} + \dfrac{1}{y}}{\dfrac{y}{x^2} + \dfrac{1}{x}}.$$

(Section 6.3, Example 1)

145. Solve for x: $\dfrac{1}{x} + \dfrac{1}{y} = \dfrac{1}{z}$.

(Section 6.7, Example 1)

146. Solve:

$$2x - \dfrac{x - 3}{8} = \dfrac{1}{2} + \dfrac{x + 5}{2}.$$

(Section 1.4, Example 4)

Preview Exercises

Exercises 147–149 will help you prepare for the material covered in the first section of the next chapter.

147. Solve by factoring: $2x^2 + 7x - 4 = 0$.

148. Solve by factoring: $x^2 = 9$.

149. Use substitution to determine if $-\sqrt{6}$ is a solution of the quadratic equation $3x^2 = 18$.

Chapter 7 Summary

Definitions and Concepts	**Examples**

Section 7.1 Radical Expressions and Functions

If $b^2 = a$, then b is a square root of a. The principal square root of a, designated $\sqrt{a}$, is the nonnegative number satisfying $b^2 = a$. The negative square root of a is written $-\sqrt{a}$. A square root of a negative number is not a real number.

A radical function in x is a function defined by an expression containing a root of x. The domain of a square root function is the set of real numbers for which the radicand is nonnegative.

Let $f(x) = \sqrt{6 - 2x}$.
$f(-15) = \sqrt{6 - 2(-15)} = \sqrt{6 + 30} = \sqrt{36} = 6$
$f(5) = \sqrt{6 - 2 \cdot 5} = \sqrt{6 - 10} = \sqrt{-4}$, not a real number
Domain of f: Set the radicand greater than or equal to zero.
$$6 - 2x \geq 0$$
$$-2x \geq -6$$
$$x \leq 3$$
Domain of $f = (-\infty, 3]$

The cube root of a real number a is written $\sqrt[3]{a}$.
$$\sqrt[3]{a} = b \quad \text{means that} \quad b^3 = a.$$

The nth root of a real number a is written $\sqrt[n]{a}$. The number n is the index. Every real number has one root when n is odd. The odd nth root of a, $\sqrt[n]{a}$, is the number b for which $b^n = a$. Every positive real number has two real roots when n is even. An even root of a negative number is not a real number.

- If n is even, then $\sqrt[n]{a^n} = |a|$.
- If n is odd, then $\sqrt[n]{a^n} = a$.

- $\sqrt[3]{-8} = -2$ because $(-2)^3 = -8$.
- $\sqrt[4]{-16}$ is not a real number.
- $\sqrt{x^2 - 14x + 49} = \sqrt{(x - 7)^2} = |x - 7|$
- $\sqrt[3]{125(x + 6)^3} = 5(x + 6)$

Definitions and Concepts	**Examples**

Section 7.2 Rational Exponents

- $a^{\frac{1}{n}} = \sqrt[n]{a}$
- $a^{\frac{m}{n}} = \left(\sqrt[n]{a}\right)^m$ or $\sqrt[n]{a^m}$
- $a^{-\frac{m}{n}} = \dfrac{1}{a^{\frac{m}{n}}}$

- $121^{\frac{1}{2}} = \sqrt{121} = 11$
- $64^{\frac{1}{3}} = \sqrt[3]{64} = 4$
- $27^{\frac{5}{3}} = \left(\sqrt[3]{27}\right)^5 = 3^5 = 3 \cdot 3 \cdot 3 \cdot 3 \cdot 3 = 243$
- $16^{-\frac{3}{4}} = \dfrac{1}{16^{\frac{3}{4}}} = \dfrac{1}{\left(\sqrt[4]{16}\right)^3} = \dfrac{1}{2^3} = \dfrac{1}{8}$
- $\left(\sqrt[3]{7xy}\right)^4 = (7xy)^{\frac{4}{3}}$

Properties of integer exponents are true for rational exponents. An expression with rational exponents is simplified when no parentheses appear, no powers are raised to powers, each base occurs once, and no negative or zero exponents appear.

Simplify: $\left(8x^{\frac{1}{3}}y^{-\frac{1}{2}}\right)^{\frac{1}{3}}$.

$= 8^{\frac{1}{3}}\left(x^{\frac{1}{3}}\right)^{\frac{1}{3}}\left(y^{-\frac{1}{2}}\right)^{\frac{1}{3}}$

$= 2x^{\frac{1}{9}}y^{-\frac{1}{6}} = \dfrac{2x^{\frac{1}{9}}}{y^{\frac{1}{6}}}$

Some radical expressions can be simplifed using rational exponents. Rewrite the expression using rational exponents, simplify, and rewrite in radical notation if rational exponents still appear.

- $\sqrt[9]{x^3} = x^{\frac{3}{9}} = x^{\frac{1}{3}} = \sqrt[3]{x}$
- $\sqrt[5]{x^2} \cdot \sqrt[4]{x} = x^{\frac{2}{5}} \cdot x^{\frac{1}{4}} = x^{\frac{2}{5}+\frac{1}{4}}$

$= x^{\frac{8}{20}+\frac{5}{20}} = x^{\frac{13}{20}} = \sqrt[20]{x^{13}}$

Section 7.3 Multiplying and Simplifying Radical Expressions

The product rule for radicals can be used to multiply radicals: $\sqrt[n]{a} \cdot \sqrt[n]{b} = \sqrt[n]{ab}$.

$\sqrt[3]{7x} \cdot \sqrt[3]{10y^2} = \sqrt[3]{7x \cdot 10y^2} = \sqrt[3]{70xy^2}$

The product rule for radicals can be used to simplify radicals: $\sqrt[n]{ab} = \sqrt[n]{a} \cdot \sqrt[n]{b}$.

A radical expression with index n is simplified when its radicand has no factors that are perfect nth powers. To simplify, write the radicand as the product of two factors, one of which is the greatest perfect nth power. Then use the product rule to take the nth root of each factor. If all variables in a radicand are positive, then

$$\sqrt[n]{a^n} = a.$$

Some radicals can be simplified after multiplication is performed.

- Simplify: $\sqrt[3]{54x^7y^{11}}$.

$= \sqrt[3]{27 \cdot 2 \cdot x^6 \cdot x \cdot y^9 \cdot y^2}$

$= \sqrt[3]{(27x^6y^9)(2xy^2)}$

$= \sqrt[3]{27x^6y^9} \cdot \sqrt[3]{2xy^2} = 3x^2y^3\sqrt[3]{2xy^2}$

- Assuming positive variables, multiply and simplify:

$\sqrt[4]{4x^2y} \cdot \sqrt[4]{4xy^3}$.

$= \sqrt[4]{4x^2y \cdot 4xy^3} = \sqrt[4]{16x^3y^4}$

$= \sqrt[4]{16y^4} \cdot \sqrt[4]{x^3} = 2y\sqrt[4]{x^3}$

Section 7.4 Adding, Subtracting, and Dividing Radical Expressions

Like radicals have the same indices and radicands. Like radicals can be added or subtracted using the distributive property. In some cases, radicals can be combined once they have been simplified.

$4\sqrt{18} - 6\sqrt{50}$

$= 4\sqrt{9 \cdot 2} - 6\sqrt{25 \cdot 2} = 4 \cdot 3\sqrt{2} - 6 \cdot 5\sqrt{2}$

$= 12\sqrt{2} - 30\sqrt{2} = -18\sqrt{2}$

Definitions and Concepts	**Examples**

Section 7.4 Adding, Subtracting, and Dividing Radical Expressions (continued)

The quotient rule for radicals can be used to simplify radicals:

$$\sqrt[n]{\frac{a}{b}} = \frac{\sqrt[n]{a}}{\sqrt[n]{b}}.$$

$$\sqrt[3]{-\frac{8}{x^{12}}} = \frac{\sqrt[3]{-8}}{\sqrt[3]{x^{12}}} = -\frac{2}{x^4}$$

$$\sqrt[3]{x^{12}} = (x^{12})^{\frac{1}{3}} = x^4$$

The quotient rule for radicals can be used to divide radicals with the same indices:

$$\frac{\sqrt[n]{a}}{\sqrt[n]{b}} = \sqrt[n]{\frac{a}{b}}.$$

Some radicals can be simplified after the division is performed.

Assuming a positive variable, divide and simplify:

$$\frac{\sqrt[4]{64x^5}}{\sqrt[4]{2x^{-2}}} = \sqrt[4]{32x^{5-(-2)}} = \sqrt[4]{32x^7}$$

$$= \sqrt[4]{16 \cdot 2 \cdot x^4 \cdot x^3} = \sqrt[4]{16x^4} \cdot \sqrt[4]{2x^3}$$

$$= 2x\sqrt[4]{2x^3}.$$

Section 7.5 Multiplying with More Than One Term and Rationalizing Denominators

Radical expressions with more than one term are multiplied in much the same way that polynomials with more than one term are multiplied.

- $\sqrt{5}(2\sqrt{6} - \sqrt{3}) = 2\sqrt{30} - \sqrt{15}$
- $(4\sqrt{3} - 2\sqrt{2})(\sqrt{3} + \sqrt{2})$

  F   O   I   L

$$= 4\sqrt{3} \cdot \sqrt{3} + 4\sqrt{3} \cdot \sqrt{2} - 2\sqrt{2} \cdot \sqrt{3} - 2\sqrt{2} \cdot \sqrt{2}$$
$$= 4 \cdot 3 + 4\sqrt{6} - 2\sqrt{6} - 2 \cdot 2$$
$$= 12 + 4\sqrt{6} - 2\sqrt{6} - 4 = 8 + 2\sqrt{6}$$

Radical expressions that involve the sum and difference of the same two terms are called conjugates. Use

$$(A + B)(A - B) = A^2 - B^2$$

to multiply conjugates.

$$(8 + 2\sqrt{5})(8 - 2\sqrt{5})$$
$$= 8^2 - (2\sqrt{5})^2$$
$$= 64 - 4 \cdot 5$$
$$= 64 - 20 = 44$$

The process of rewriting a radical expression as an equivalent expression without any radicals in the denominator is called rationalizing the denominator. When the denominator contains a single radical with an nth root, multiply the numerator and the denominator by a radical of index n that produces a perfect nth power in the denominator's radicand.

Rationalize the denominator: $\dfrac{7}{\sqrt[3]{2x}}$.

$$= \frac{7}{\sqrt[3]{2x}} \cdot \frac{\sqrt[3]{4x^2}}{\sqrt[3]{4x^2}} = \frac{7\sqrt[3]{4x^2}}{\sqrt[3]{8x^3}} = \frac{7\sqrt[3]{4x^2}}{2x}$$

If the denominator contains two terms, rationalize the denominator by multiplying the numerator and the denominator by the conjugate of the denominator.

$$\frac{13}{5 - \sqrt{3}} = \frac{13}{5 - \sqrt{3}} \cdot \frac{5 + \sqrt{3}}{5 + \sqrt{3}}$$
$$= \frac{13(5 + \sqrt{3})}{5^2 - (\sqrt{3})^2}$$
$$= \frac{13(5 + \sqrt{3})}{25 - 3} = \frac{13(5 + \sqrt{3})}{22}$$

| **Definitions and Concepts** | **Examples** |

Section 7.6 Radical Equations

A radical equation is an equation in which the variable occurs in a radicand.

Solving Radical Equations Containing *n*th Roots

1. Isolate one radical on one side of the equation.
2. Raise both sides to the *n*th power.
3. Solve the resulting equation.
4. Check proposed solutions in the original equation. Solutions of an equation to an even power that is radical-free, but not the original equation, are called extraneous solutions.

Solve: $\sqrt{6x + 13} - 2x = 1$.

$$\sqrt{6x + 13} = 2x + 1 \qquad \text{Isolate the radical.}$$
$$(\sqrt{6x + 13})^2 = (2x + 1)^2 \qquad \text{Square both sides.}$$
$$6x + 13 = 4x^2 + 4x + 1 \qquad \begin{array}{l}(A + B)^2 = \\ A^2 + 2AB + B^2\end{array}$$
$$0 = 4x^2 - 2x - 12 \qquad \begin{array}{l}\text{Subtract } 6x + 13 \\ \text{from both sides.}\end{array}$$
$$0 = 2(2x^2 - x - 6) \qquad \text{Factor out the GCF.}$$
$$0 = 2(2x + 3)(x - 2) \qquad \text{Factor completely.}$$
$$2x + 3 = 0 \quad \text{or} \quad x - 2 = 0 \qquad \begin{array}{l}\text{Set variable factors} \\ \text{equal to zero.}\end{array}$$
$$2x = -3 \qquad\qquad x = 2 \quad \text{Solve for } x.$$
$$x = -\frac{3}{2}$$

Check both proposed solutions. 2 checks, but $-\frac{3}{2}$ is extraneous.

The solution is 2 and the solution set is {2}.

Section 7.7 Complex Numbers

The imaginary unit *i* is defined as
$$i = \sqrt{-1}, \quad \text{where} \quad i^2 = -1.$$
The set of numbers in the form $a + bi$ is called the set of complex numbers; *a* is the real part and *b* is the imaginary part. If $b = 0$, the complex number is a real number. If $b \neq 0$, the complex number is an imaginary number.

- $\sqrt{-81} = \sqrt{81(-1)} = \sqrt{81}\sqrt{-1} = 9i$
- $\sqrt{-75} = \sqrt{75(-1)} = \sqrt{25 \cdot 3}\sqrt{-1} = 5i\sqrt{3}$

To add or subtract complex numbers, add or subtract their real parts and add or subtract their imaginary parts.

$$(2 - 4i) - (7 - 10i)$$
$$= 2 - 4i - 7 + 10i$$
$$= (2 - 7) + (-4 + 10)i = -5 + 6i$$

To multiply complex numbers, multiply as if they were polynomials. After completing the multiplication, replace i^2 with -1. When performing operations with square roots of negative numbers, begin by expressing all square roots in terms of *i*. Then multiply.

F O I L

- $(2 - 3i)(4 + 5i) = 8 + 10i - 12i - 15i^2$
$$= 8 + 10i - 12i - 15(-1)$$
$$= 23 - 2i$$
- $\sqrt{-36} \cdot \sqrt{-100} = \sqrt{36(-1)} \cdot \sqrt{100(-1)}$
$$= 6i \cdot 10i = 60i^2 = 60(-1) = -60$$

The complex numbers $a + bi$ and $a - bi$ are conjugates. Conjugates can be multiplied using the formula
$$(A + B)(A - B) = A^2 - B^2.$$
The multiplication of conjugates results in a real number.

$$(3 + 5i)(3 - 5i) = 3^2 - (5i)^2$$
$$= 9 - 25i^2$$
$$= 9 - 25(-1) = 34$$

If you solve a quadratic equation by completing the square and the solutions are rational numbers, the equation can also be solved by factoring. By contrast, quadratic equations with irrational solutions cannot be solved by factoring. However, all quadratic equations can be solved by completing the square.

✓ **CHECK POINT 6**   Solve by completing the square: $x^2 + 4x - 1 = 0$.

We have seen that the leading coefficient must be 1 in order to complete the square. If the coefficient of the x^2-term in a quadratic equation is not 1, you must divide each side of the equation by this coefficient before completing the square. For example, to solve $9x^2 - 6x - 4 = 0$ by completing the square, first divide every term by 9:

$$\frac{9x^2}{9} - \frac{6x}{9} - \frac{4}{9} = \frac{0}{9}$$

$$x^2 - \frac{6}{9}x - \frac{4}{9} = 0$$

$$x^2 - \frac{2}{3}x - \frac{4}{9} = 0.$$

Now that the coefficient of the x^2-term is 1, we can solve by completing the square.

EXAMPLE 7   Solving a Quadratic Equation by Completing the Square

Solve by completing the square: $9x^2 - 6x - 4 = 0$.

Solution

$9x^2 - 6x - 4 = 0$	This is the original equation.
$x^2 - \dfrac{2}{3}x - \dfrac{4}{9} = 0$	Divide both sides by 9.
$x^2 - \dfrac{2}{3}x = \dfrac{4}{9}$	Add $\frac{4}{9}$ to both sides to isolate the binomial.
$x^2 - \dfrac{2}{3}x + \dfrac{1}{9} = \dfrac{4}{9} + \dfrac{1}{9}$	Complete the square: Half of $-\frac{2}{3}$ is $-\frac{2}{6}$, or $-\frac{1}{3}$, and $\left(-\frac{1}{3}\right)^2 = \frac{1}{9}$.
$\left(x - \dfrac{1}{3}\right)^2 = \dfrac{5}{9}$	Factor and simplify.
$x - \dfrac{1}{3} = \sqrt{\dfrac{5}{9}}$ or $x - \dfrac{1}{3} = -\sqrt{\dfrac{5}{9}}$	Apply the square root property.
$x - \dfrac{1}{3} = \dfrac{\sqrt{5}}{3}$ $x - \dfrac{1}{3} = -\dfrac{\sqrt{5}}{3}$	$\sqrt{\dfrac{5}{9}} = \dfrac{\sqrt{5}}{\sqrt{9}} = \dfrac{\sqrt{5}}{3}$
$x = \dfrac{1}{3} + \dfrac{\sqrt{5}}{3}$ $x = \dfrac{1}{3} - \dfrac{\sqrt{5}}{3}$	Add $\frac{1}{3}$ to both sides and solve for x.
$x = \dfrac{1 + \sqrt{5}}{3}$ $x = \dfrac{1 - \sqrt{5}}{3}$	Express solutions with a common denominator.

The solutions are $\dfrac{1 \pm \sqrt{5}}{3}$ and the solution set is $\left\{\dfrac{1 \pm \sqrt{5}}{3}\right\}$. ∎

✓ **CHECK POINT 7**   Solve by completing the square: $2x^2 + 3x - 4 = 0$.

Using Technology

Graphic Connections

Obtain a decimal approximation for each solution of

$$9x^2 - 6x - 4 = 0,$$

the equation in Example 7.

$$\frac{1 + \sqrt{5}}{3} \approx 1.1$$

$$\frac{1 - \sqrt{5}}{3} \approx -0.4$$

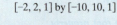

$f(x) = 9x^2 - 6x - 4$

x-intercept ≈ −0.4 x-intercept ≈ 1.1

$[-2, 2, 1]$ by $[-10, 10, 1]$

The x-intercepts of $f(x) = 9x^2 - 6x - 4$ verify the solutions.

EXAMPLE 8 Solving a Quadratic Equation by Completing the Square

Solve by completing the square: $2x^2 - x + 6 = 0$.

Solution

$$2x^2 - x + 6 = 0 \qquad \text{This is the original equation.}$$

$$x^2 - \frac{1}{2}x + 3 = 0 \qquad \text{Divide both sides by 2.}$$

$$x^2 - \frac{1}{2}x = -3 \qquad \text{Subtract 3 from both sides to isolate the binomial.}$$

$$x^2 - \frac{1}{2}x + \frac{1}{16} = -3 + \frac{1}{16} \qquad \text{Complete the square: Half of } -\frac{1}{2} \text{ is } -\frac{1}{4} \text{ and } \left(-\frac{1}{4}\right)^2 = \frac{1}{16}.$$

$$\left(x - \frac{1}{4}\right)^2 = \frac{-47}{16} \qquad \text{Factor and simplify: } -3 + \frac{1}{16} = \frac{-48}{16} + \frac{1}{16} = \frac{-47}{16}.$$

$$x - \frac{1}{4} = \pm\sqrt{\frac{-47}{16}} \qquad \text{Apply the square root property.}$$

$$x - \frac{1}{4} = \pm i\frac{\sqrt{47}}{4} \qquad \pm\sqrt{\frac{-47}{16}} = \pm\sqrt{\frac{47(-1)}{16}} = \pm i\sqrt{\frac{47}{16}}$$

$$= \pm i\frac{\sqrt{47}}{\sqrt{16}} = \pm i\frac{\sqrt{47}}{4}$$

$$x = \frac{1}{4} \pm i\frac{\sqrt{47}}{4} \qquad \text{Add } \frac{1}{4} \text{ to both sides and solve for } x.$$

The solutions are $\frac{1}{4} \pm i\frac{\sqrt{47}}{4}$ and the solution set is $\left\{\frac{1}{4} \pm i\frac{\sqrt{47}}{4}\right\}$. ∎

✓ **CHECK POINT 8** Solve by completing the square: $3x^2 - 9x + 8 = 0$.

---

4 Solve problems using the square root property. ▶

Applications

We all want a wonderful life with fulfilling work, good health, and loving relationships. And let's be honest: Financial security wouldn't hurt! Achieving this goal depends on understanding how money in a savings account grows in remarkable ways as a result of *compound interest*. **Compound interest** is interest computed on your original investment as well as on any accumulated interest. For example, suppose you deposit $1000, the principal, in a savings account at a rate of 5%. **Table 8.1** shows how the investment grows if the interest earned is automatically added on to the principal.

Table 8.1	Compound Interest on $1000		
Year	**Starting Balance**	**Interest Earned: $I = Pr$**	**New Balance**
1	$1000	$1000 × 0.05 = $50	$1050
2	$1050	$1050 × 0.05 = $52.50	$1102.50
3	$1102.50	$1102.50 × 0.05 ≈ $55.13	$1157.63

A faster way to determine the amount, A, in an account subject to compound interest is to use the following formula:

> ### A Formula for Compound Interest
> Suppose that an amount of money, P, is invested at interest rate r, compounded annually. In t years, the amount, A, or balance, in the account is given by the formula
> $$A = P(1 + r)^t.$$

Some compound interest problems can be solved using quadratic equations.

EXAMPLE 9   Solving a Compound Interest Problem

You invested $1000 in an account whose interest is compounded annually. After 2 years, the amount, or balance, in the account is $1210. Find the annual interest rate.

Solution   We are given that

P (the amount invested) $= \$1000$

t (the time of the investment) $= 2$ years

A (the amount, or balance, in the account) $= \$1210$.

We are asked to find the annual interest rate, r. We substitute the three given values into the compound interest formula and solve for r.

$A = P(1 + r)^t$	Use the compound interest formula.
$1210 = 1000(1 + r)^2$	Substitute the given values.
$\dfrac{1210}{1000} = (1 + r)^2$	Divide both sides by 1000.
$\dfrac{121}{100} = (1 + r)^2$	Simplify the fraction.
$1 + r = \sqrt{\dfrac{121}{100}} \quad \text{or} \quad 1 + r = -\sqrt{\dfrac{121}{100}}$	Apply the square root property.
$1 + r = \dfrac{11}{10} \qquad\qquad 1 + r = -\dfrac{11}{10}$	$\sqrt{\dfrac{121}{100}} = \dfrac{\sqrt{121}}{\sqrt{100}} = \dfrac{11}{10}$
$r = \dfrac{11}{10} - 1 \qquad\quad r = -\dfrac{11}{10} - 1$	Subtract 1 from both sides and solve for r.
$r = \dfrac{1}{10} \qquad\qquad\quad r = -\dfrac{21}{10}$	$\dfrac{11}{10} - 1 = \dfrac{11}{10} - \dfrac{10}{10} = \dfrac{1}{10}$ and $-\dfrac{11}{10} - 1 = -\dfrac{11}{10} - \dfrac{10}{10} = -\dfrac{21}{10}$.

Because the interest rate cannot be negative, we reject $-\dfrac{21}{10}$. Thus, the annual interest rate is $\dfrac{1}{10} = 0.10 = 10\%$.

We can check this answer using the formula $A = P(1 + r)^t$. If $1000 is invested for 2 years at 10% interest, compounded annually, the balance in the account is

$$A = \$1000(1 + 0.10)^2 = \$1000(1.10)^2 = \$1210.$$

Because this is precisely the amount given by the problem's conditions, the annual interest rate is, indeed, 10% compounded annually. ■

CHECK POINT 9   You invested $3000 in an account whose interest is compounded annually. After 2 years, the amount, or balance, in the account is $4320. Find the annual interest rate.

In Chapter 5, we solved problems using the Pythagorean Theorem. Recall that in a right triangle, the side opposite the 90° angle is the hypotenuse and the other sides are legs. The Pythagorean Theorem states that the sum of the squares of the lengths of the legs equals the square of the length of the hypotenuse. Some problems that involve the Pythagorean Theorem can be solved using the square root property.

EXAMPLE 10   Using the Pythagorean Theorem and the Square Root Property

a. A wheelchair ramp with a length of 122 inches has a horizontal distance of 120 inches. What is the ramp's vertical distance?

b. Construction laws are very specific when it comes to access ramps for the disabled. Every vertical rise of 1 inch requires a horizontal run of 12 inches. Does this ramp satisfy the requirement?

Solution

Figure 8.2

a. **Figure 8.2** shows the right triangle that is formed by the ramp, the wall, and the ground. We can find x, the ramp's vertical distance, using the Pythagorean Theorem.

$$\underbrace{x^2}_{(\text{leg})^2} \underbrace{+}_{\text{plus}} \underbrace{120^2}_{(\text{leg})^2} \underbrace{=}_{\text{equals}} \underbrace{122^2}_{(\text{hypotenuse})^2}$$

We solve this equation using the square root property.

$x^2 + 120^2 = 122^2$ This is the equation resulting from the Pythagorean Theorem.

$x^2 + 14{,}400 = 14{,}884$ Square 120 and 122.

$x^2 = 484$ Isolate x^2 by subtracting 14,400 from both sides.

$x = \sqrt{484}$ or $x = -\sqrt{484}$ Apply the square root property.

$x = 22$ $x = -22$

Because x represents the ramp's vertical distance, we reject the negative value. Thus, the ramp's vertical distance is 22 inches.

b. Every vertical rise of 1 inch requires a horizontal run of 12 inches. Because the ramp has a vertical distance of 22 inches, it requires a horizontal distance of 22(12) inches, or 264 inches. The horizontal distance is only 120 inches, so this ramp does not satisfy construction laws for access ramps for the disabled. ∎

CHECK POINT 10   A 50-foot supporting wire is to be attached to an antenna. The wire is anchored 20 feet from the base of the antenna. How high up the antenna is the wire attached? Express the answer in simplified radical form. Then find a decimal approximation to the nearest tenth of a foot.

Achieving Success

Many intermediate algebra courses cover only selected sections from this chapter and the book's remaining chapters. Regardless of the content requirements for your course, **it's never too early to start thinking about a final exam**. Here are some strategies to help you prepare for your final:

- Review your back exams. Be sure you understand any error that you made. Seek help with any concepts that are still unclear.
- Ask your professor if there are additional materials to help students review for the final. This includes review sheets and final exams from previous semesters.
- Attend any review sessions conducted by your professor or by the math department.
- Use the strategy first introduced on page 162: Imagine that your professor will permit two 3 by 5 index cards of notes on the final. Organize and create such a two-card summary for the most vital information in the course, including all important formulas. Refer to the chapter summaries in the textbook to prepare your personalized summary.
- For further review, work the relevant exercises in the Cumulative Review at the end of this chapter. The 42 exercises in the Cumulative Review for Chapters 1 through 8 cover most of the objectives in the book's first eight chapters.
- Write your own final exam with detailed solutions for each item. You can use test questions from back exams in mixed order, worked examples from the textbook's chapter summaries, exercises in the Cumulative Reviews, and problems from course handouts. Use your test as a practice final exam.

CONCEPT AND VOCABULARY CHECK

Fill in each blank so that the resulting statement is true.

1. The square root property states that if $u^2 = d$, then $u =$ _____.

2. If $x^2 = 7$, then $x =$ _____.

3. If $x^2 = \dfrac{11}{2}$, then $x =$ _____. Rationalizing denominators, we obtain $x =$ _____.

4. If $x^2 = -9$, then $x =$ _____.

5. To complete the square on $x^2 + 10x$, add _____.

6. To complete the square on $x^2 - 3x$, add _____.

7. To complete the square on $x^2 - \dfrac{4}{5}x$, add _____.

8. To solve $x^2 + 6x = 7$ by completing the square, add _____ to both sides of the equation.

9. To solve $x^2 - \dfrac{2}{3}x = \dfrac{4}{9}$ by completing the square, add _____ to both sides of the equation.

8.1 EXERCISE SET   ▶ MyMathLab®

Practice Exercises

In Exercises 1–22, solve each equation by the square root property. If possible, simplify radicals or rationalize denominators. Express imaginary solutions in the form a + bi.

1. $3x^2 = 75$

2. $5x^2 = 20$

3. $7x^2 = 42$

4. $8x^2 = 40$

5. $16x^2 = 25$

6. $4x^2 = 49$

7. $3x^2 - 2 = 0$

8. $3x^2 - 5 = 0$

9. $25x^2 + 16 = 0$

10. $4x^2 + 49 = 0$

11. $(x + 7)^2 = 9$

12. $(x + 3)^2 = 64$

13. $(x - 3)^2 = 5$

14. $(x - 4)^2 = 3$

15. $2(x + 2)^2 = 16$

16. $3(x + 2)^2 = 36$

17. $(x - 5)^2 = -9$

18. $(x - 5)^2 = -4$

19. $\left(x + \dfrac{3}{4}\right)^2 = \dfrac{11}{16}$

20. $\left(x + \dfrac{2}{5}\right)^2 = \dfrac{7}{25}$

21. $x^2 - 6x + 9 = 36$

22. $x^2 - 6x + 9 = 49$

In Exercises 23–34, determine the constant that should be added to the binomial so that it becomes a perfect square trinomial. Then write and factor the trinomial.

23. $x^2 + 2x$

24. $x^2 + 4x$

25. $x^2 - 14x$

26. $x^2 - 10x$

27. $x^2 + 7x$

28. $x^2 + 9x$

29. $x^2 - \dfrac{1}{2}x$

30. $x^2 - \dfrac{1}{3}x$

31. $x^2 + \dfrac{4}{3}x$

32. $x^2 + \dfrac{4}{5}x$

33. $x^2 - \dfrac{9}{4}x$

34. $x^2 - \dfrac{9}{5}x$

In Exercises 35–58, solve each quadratic equation by completing the square.

35. $x^2 + 4x = 32$

36. $x^2 + 6x = 7$

37. $x^2 + 6x = -2$

38. $x^2 + 2x = 5$

39. $x^2 - 8x + 1 = 0$

40. $x^2 + 8x - 5 = 0$

41. $x^2 + 2x + 2 = 0$

42. $x^2 - 4x + 8 = 0$

43. $x^2 + 3x - 1 = 0$

44. $x^2 - 3x - 5 = 0$

45. $x^2 + \dfrac{4}{7}x + \dfrac{3}{49} = 0$

46. $x^2 + \dfrac{6}{5}x + \dfrac{8}{25} = 0$

47. $x^2 + x - 1 = 0$

48. $x^2 - 7x + 3 = 0$

49. $2x^2 + 3x - 5 = 0$

50. $2x^2 + 5x - 3 = 0$

51. $3x^2 + 6x + 1 = 0$

52. $3x^2 - 6x + 2 = 0$

53. $3x^2 - 8x + 1 = 0$

54. $2x^2 + 3x - 4 = 0$

55. $8x^2 - 4x + 1 = 0$

56. $9x^2 - 6x + 5 = 0$

57. $2x^2 - 5x + 7 = 0$

58. $4x^2 - 2x + 5 = 0$

59. If $g(x) = \left(x - \dfrac{2}{5}\right)^2$, find all values of x for which $g(x) = \dfrac{9}{25}$.

60. If $g(x) = \left(x + \dfrac{1}{3}\right)^2$, find all values of x for which $g(x) = \dfrac{4}{9}$.

61. If $h(x) = 5(x + 2)^2$, find all values of x for which $h(x) = -125$.

62. If $h(x) = 3(x - 4)^2$, find all values of x for which $h(x) = -12$.

Practice PLUS

63. Three times the square of the difference between a number and 2 is -12. Find the number(s).

64. Three times the square of the difference between a number and 9 is -27. Find the number(s).

In Exercises 65–68, solve the formula for the specified variable. Because each variable is nonnegative, list only the principal square root. If possible, simplify radicals or rationalize denominators.

65. $h = \dfrac{v^2}{2g}$ for v

66. $s = \dfrac{kwd^2}{l}$ for d

67. $A = P(1 + r)^2$ for r

68. $C = \dfrac{kP_1P_2}{d^2}$ for d

In Exercises 69–72, solve each quadratic equation by completing the square.

69. $\dfrac{x^2}{3} + \dfrac{x}{9} - \dfrac{1}{6} = 0$

70. $\dfrac{x^2}{2} - \dfrac{x}{6} - \dfrac{3}{4} = 0$

71. $x^2 - bx = 2b^2$

72. $x^2 - bx = 6b^2$

73. The ancient Greeks used a geometric method for completing the square in which they literally transformed a figure into a square.

a. Write a binomial in x that represents the combined area of the small square and the eight rectangular stripes that make up the incomplete square on the left.

b. What is the area of the region in the bottom-right corner that literally completes the square?

c. Write a trinomial in x that represents the combined area of the small square, the eight rectangular stripes, and the bottom-right corner that make up the complete square on the right.

d. Use the length of each side of the complete square on the right to express its area as the square of a binomial.

74. An **isosceles right triangle** has legs that are the same length and acute angles each measuring $45°$.

a. Write an expression in terms of a that represents the length of the hypotenuse.

b. Use your result from part (a) to write a sentence that describes the length of the hypotenuse of an isosceles right triangle in terms of the length of a leg.

Application Exercises

In Exercises 75–78, use the compound interest formula

$$A = P(1 + r)^t$$

to find the annual interest rate, r.

75. In 2 years, an investment of $2000 grows to $2880.

76. In 2 years, an investment of $2000 grows to $2420.

77. In 2 years, an investment of $1280 grows to $1445.

78. In 2 years, an investment of $80,000 grows to $101,250.

The bar graph shows the number of billionaires worldwide for six selected years from 1987 through 2012.

Number of Billionaires Worldwide

Source: Forbes

The data can be modeled by the function

$$f(x) = 2x^2 + 140,$$

where $f(x)$ represents the number of billionaires worldwide x years after 1987. Use this information to solve Exercises 79–80.

79. a. According to the model, how many billionaires were there in 2007? Does this underestimate or overestimate the number displayed by the graph? By how much?

b. According to the model, in which year will there be 1940 billionaires worldwide?

80. a. According to the model, how many billionaires were there in 2012? Does this underestimate or overestimate the number displayed by the graph? By how much?

b. According to the model, in which year will there be 2062 billionaires worldwide?

The function $s(t) = 16t^2$ models the distance, $s(t)$, in feet, that an object falls in t seconds. Use this function and the square root property to solve Exercises 81–82. Express answers in simplified radical form. Then use your calculator to find a decimal approximation to the nearest tenth of a second.

81. A sky diver jumps from an airplane and falls for 4800 feet before opening a parachute. For how many seconds was the diver in a free fall?

82. A sky diver jumps from an airplane and falls for 3200 feet before opening a parachute. For how many seconds was the diver in a free fall?

Use the Pythagorean Theorem and the square root property to solve Exercises 83–88. Express answers in simplified radical form. Then find a decimal approximation to the nearest tenth.

83. A rectangular park is 6 miles long and 3 miles wide. How long is a pedestrian route that runs diagonally across the park?

84. A rectangular park is 4 miles long and 2 miles wide. How long is a pedestrian route that runs diagonally across the park?

85. The base of a 30-foot ladder is 10 feet from the building. If the ladder reaches the flat roof, how tall is the building?

30 ft

x

10 ft

86. The doorway into a room is 4 feet wide and 8 feet high. What is the diameter of the largest circular tabletop that can be taken through this doorway diagonally?

←4 ft→

8 ft

87. A supporting wire is to be attached to the top of a 50-foot antenna. If the wire must be anchored 50 feet from the base of the antenna, what length of wire is required?

88. A supporting wire is to be attached to the top of a 70-foot antenna. If the wire must be anchored 70 feet from the base of the antenna, what length of wire is required?

89. A square flower bed is to be enlarged by adding 2 meters on each side. If the larger square has an area of 196 square meters, what is the length of a side of the original square?

2 meters

2 meters   2 meters

2 meters

x meters

90. A square flower bed is to be enlarged by adding 4 feet on each side. If the larger square has an area of 225 square feet, what is the length of a side of the original square?

Explaining the Concepts

91. What is the square root property?

92. Explain how to solve $(x - 1)^2 = 16$ using the square root property.

93. Explain how to complete the square for a binomial. Use $x^2 + 6x$ to illustrate your explanation.

94. Explain how to solve $x^2 + 6x + 8 = 0$ by completing the square.

95. What is compound interest?

96. In your own words, describe the compound interest formula
$$A = P(1 + r)^t.$$

Technology Exercises

97. Use a graphing utility to solve $4 - (x + 1)^2 = 0$. Graph $y = 4 - (x + 1)^2$ in a $[-5, 5, 1]$ by $[-5, 5, 1]$ viewing rectangle. The equation's solutions are the graph's x-intercepts. Check by substitution in the given equation.

98. Use a graphing utility to solve $(x - 1)^2 - 9 = 0$. Graph $y = (x - 1)^2 - 9$ in a $[-5, 5, 1]$ by $[-9, 3, 1]$ viewing rectangle. The equation's solutions are the graph's x-intercepts. Check by substitution in the given equation.

99. Use a graphing utility and x-intercepts to verify any of the real solutions that you obtained for five of the quadratic equations in Exercises 35–54.

Critical Thinking Exercises

Make Sense?   *In Exercises 100–103, determine whether each statement makes sense or does not make sense, and explain your reasoning.*

100. When the coefficient of the x-term in a quadratic equation is negative and I'm solving by completing the square, I add a negative constant to each side of the equation.

101. When I complete the square for the binomial $x^2 + bx$, I obtain a different polynomial, but when I solve a quadratic equation by completing the square, I obtain an equation with the same solution set.

102. When I use the square root property to determine the length of a right triangle's side, I don't even bother to list the negative square root.

103. When I solved $4x^2 + 10x = 0$ by completing the square, I added 25 to both sides of the equation.

In Exercises 104–107, determine whether each statement is true or false. If the statement is false, make the necessary change(s) to produce a true statement.

104. The graph of $y = (x - 2)^2 + 3$ cannot have x-intercepts.

105. The equation $(x - 5)^2 = 12$ is equivalent to $x - 5 = 2\sqrt{3}$.

106. In completing the square for $2x^2 - 6x = 5$, we should add 9 to both sides.

107. Although not every quadratic equation can be solved by completing the square, they can all be solved by factoring.

108. Solve for y: $\dfrac{x^2}{a^2} + \dfrac{y^2}{b^2} = 1$.

109. Solve by completing the square:

$$x^2 + x + c = 0.$$

110. Solve by completing the square:

$$x^2 + bx + c = 0.$$

111. Solve: $x^4 - 8x^2 + 15 = 0$.

Review Exercises

112. Simplify: $4x - 2 - 3[4 - 2(3 - x)]$.
(Section 1.2, Example 14)

113. Factor: $1 - 8x^3$. (Section 5.5, Example 9)

114. Divide: $(x^4 - 5x^3 + 2x^2 - 6) \div (x - 3)$.
(Section 6.5, Example 1)

Preview Exercises

Exercises 115–117 will help you prepare for the material covered in the next section.

115. a. Solve by factoring: $8x^2 + 2x - 1 = 0$.

 b. The quadratic equation in part (a) is in the standard form $ax^2 + bx + c = 0$. Compute $b^2 - 4ac$. Is $b^2 - 4ac$ a perfect square?

116. a. Solve by factoring: $9x^2 - 6x + 1 = 0$.

 b. The quadratic equation in part (a) is in the standard form $ax^2 + bx + c = 0$. Compute $b^2 - 4ac$.

117. a. Clear fractions in the following equation and write in the form $ax^2 + bx + c = 0$:

$$3 + \frac{4}{x} = -\frac{2}{x^2}.$$

 b. For the equation you wrote in part (a), compute $b^2 - 4ac$.

SECTION

8.2

The Quadratic Formula

What am I supposed to learn?

After studying this section, you should be able to:

1 Solve quadratic equations using the quadratic formula.

2 Use the discriminant to determine the number and type of solutions.

3 Determine the most efficient method to use when solving a quadratic equation.

4 Write quadratic equations from solutions.

5 Use the quadratic formula to solve problems.

Until fairly recently, many doctors believed that your blood pressure was theirs to know and yours to worry about. Today, however, people are encouraged to find out their blood pressure. That pumped-up cuff that squeezes against your upper arm measures blood pressure in millimeters (mm) of mercury (Hg). Blood pressure is given in two numbers: systolic pressure over diastolic pressure, such as 120 over 80. Systolic pressure is the pressure of blood against the artery walls when the heart contracts. Diastolic pressure is the pressure of blood against the artery walls when the heart is at rest.

In this section, we will derive a formula that will enable you to solve quadratic equations more quickly than the method of completing the square. Using this formula, we will work with functions that model changing systolic pressure for men and women with age.

1 Solve quadratic equations using the quadratic formula. ▶

Solving Quadratic Equations Using the Quadratic Formula

We can use the method of completing the square to derive a formula that can be used to solve all quadratic equations. The derivation given below also shows a particular quadratic equation, $3x^2 - 2x - 4 = 0$, to specifically illustrate each of the steps.

Deriving the Quadratic Formula

Standard Form of a Quadratic Equation	Comment	A Specific Example
$ax^2 + bx + c = 0, a > 0$	This is the given equation.	$3x^2 - 2x - 4 = 0$
$x^2 + \dfrac{b}{a}x + \dfrac{c}{a} = 0$	Divide both sides by the coefficient of x^2.	$x^2 - \dfrac{2}{3}x - \dfrac{4}{3} = 0$
$x^2 + \dfrac{b}{a}x = -\dfrac{c}{a}$	Isolate the binomial by adding $-\dfrac{c}{a}$ on both sides.	$x^2 - \dfrac{2}{3}x = \dfrac{4}{3}$
$x^2 + \dfrac{b}{a}x + \left(\dfrac{b}{2a}\right)^2 = -\dfrac{c}{a} + \left(\dfrac{b}{2a}\right)^2$ (half)^2	Complete the square. Add the square of half the coefficient of x to both sides.	$x^2 - \dfrac{2}{3}x + \left(-\dfrac{1}{3}\right)^2 = \dfrac{4}{3} + \left(-\dfrac{1}{3}\right)^2$ (half)^2
$x^2 + \dfrac{b}{a}x + \dfrac{b^2}{4a^2} = -\dfrac{c}{a} + \dfrac{b^2}{4a^2}$		$x^2 - \dfrac{2}{3}x + \dfrac{1}{9} = \dfrac{4}{3} + \dfrac{1}{9}$
$\left(x + \dfrac{b}{2a}\right)^2 = -\dfrac{c}{a}\cdot\dfrac{4a}{4a} + \dfrac{b^2}{4a^2}$	Factor on the left side and obtain a common denominator on the right side.	$\left(x - \dfrac{1}{3}\right)^2 = \dfrac{4}{3}\cdot\dfrac{3}{3} + \dfrac{1}{9}$
$\left(x + \dfrac{b}{2a}\right)^2 = \dfrac{-4ac + b^2}{4a^2}$	Add fractions on the right side.	$\left(x - \dfrac{1}{3}\right)^2 = \dfrac{12 + 1}{9}$
$\left(x + \dfrac{b}{2a}\right)^2 = \dfrac{b^2 - 4ac}{4a^2}$		$\left(x - \dfrac{1}{3}\right)^2 = \dfrac{13}{9}$
$x + \dfrac{b}{2a} = \pm\sqrt{\dfrac{b^2 - 4ac}{4a^2}}$	Apply the square root property.	$x - \dfrac{1}{3} = \pm\sqrt{\dfrac{13}{9}}$
$x + \dfrac{b}{2a} = \pm\dfrac{\sqrt{b^2 - 4ac}}{2a}$	Take the square root of the quotient, simplifying the denominator.	$x - \dfrac{1}{3} = \pm\dfrac{\sqrt{13}}{3}$
$x = \dfrac{-b}{2a} \pm \dfrac{\sqrt{b^2 - 4ac}}{2a}$	Solve for x by subtracting $\dfrac{b}{2a}$ from both sides.	$x = \dfrac{1}{3} \pm \dfrac{\sqrt{13}}{3}$
$x = \dfrac{-b \pm \sqrt{b^2 - 4ac}}{2a}$	Combine fractions on the right side.	$x = \dfrac{1 \pm \sqrt{13}}{3}$

The formula shown at the bottom of the left column is called the *quadratic formula*. A similar proof shows that the same formula can be used to solve quadratic equations if a, the coefficient of the x^2-term, is negative.

The Quadratic Formula

The solutions of a quadratic equation in standard form $ax^2 + bx + c = 0$, with $a \neq 0$, are given by the **quadratic formula**:

$$x = \frac{-b \pm \sqrt{b^2 - 4ac}}{2a}.$$

x equals negative b plus or minus the square root of $b^2 - 4ac$, all divided by 2a.

To use the quadratic formula, write the quadratic equation in standard form if necessary. Then determine the numerical values for a (the coefficient of the x^2-term), b (the coefficient of the x-term), and c (the constant term). Substitute the values of a, b, and c into the quadratic formula and evaluate the expression. The $\pm$ sign indicates that there are two (not necessarily distinct) solutions of the equation.

EXAMPLE 1 Solving a Quadratic Equation Using the Quadratic Formula

Solve using the quadratic formula: $8x^2 + 2x - 1 = 0$.

Solution The given equation is in standard form. Begin by identifying the values for a, b, and c.

$$8x^2 + 2x - 1 = 0$$

$a = 8 \qquad b = 2 \qquad c = -1$

Substituting these values into the quadratic formula and simplifying gives the equation's solutions.

$$x = \frac{-b \pm \sqrt{b^2 - 4ac}}{2a} \qquad \text{Use the quadratic formula.}$$

$$x = \frac{-2 \pm \sqrt{2^2 - 4(8)(-1)}}{2(8)} \qquad \text{Substitute the values for } a, b, \text{ and } c: \\ a = 8, b = 2, \text{ and } c = -1.$$

$$= \frac{-2 \pm \sqrt{4 - (-32)}}{16} \qquad 2^2 - 4(8)(-1) = 4 - (-32)$$

$$= \frac{-2 \pm \sqrt{36}}{16} \qquad 4 - (-32) = 4 + 32 = 36$$

$$= \frac{-2 \pm 6}{16} \qquad \sqrt{36} = 6$$

Now we will evaluate this expression in two different ways to obtain the two solutions. On the left, we will *add* 6 to -2. On the right, we will *subtract* 6 from -2.

$$x = \frac{-2 + 6}{16} \quad \text{or} \quad x = \frac{-2 - 6}{16}$$

$$= \frac{4}{16} = \frac{1}{4} \qquad\qquad = \frac{-8}{16} = -\frac{1}{2}$$

The solutions are $-\frac{1}{2}$ and $\frac{1}{4}$, and the solution set is $\left\{-\frac{1}{2}, \frac{1}{4}\right\}$. ∎

Using Technology

Graphic Connections

The graph of the quadratic function

$$y = 8x^2 + 2x - 1$$

has x-intercepts at $-\frac{1}{2}$ and $\frac{1}{4}$. This verifies that $\left\{-\frac{1}{2}, \frac{1}{4}\right\}$ is the solution set of the quadratic equation

$$8x^2 + 2x - 1 = 0.$$

x-intercept is $-\frac{1}{2}$. x-intercept is $\frac{1}{4}$.

$[-2, 2, 1]$ by $[-3, 10, 1]$

In Example 1, the solutions of $8x^2 + 2x - 1 = 0$ are rational numbers. This means that the equation can also be solved by factoring. The reason that the solutions are rational numbers is that $b^2 - 4ac$, the radicand in the quadratic formula, is 36, which is a perfect square. If a, b, and c are rational numbers, all quadratic equations for which $b^2 - 4ac$ is a perfect square have rational solutions.

☑ **CHECK POINT 1** Solve using the quadratic formula: $2x^2 + 9x - 5 = 0$.

EXAMPLE 2 Solving a Quadratic Equation Using the Quadratic Formula

Solve using the quadratic formula:

$$2x^2 = 4x + 1.$$

Solution The quadratic equation must be in standard form to identify the values for a, b, and c. To move all terms to one side and obtain zero on the right, we subtract $4x + 1$ from both sides. Then we can identify the values for a, b, and c.

$$2x^2 = 4x + 1 \qquad \text{This is the given equation.}$$
$$2x^2 - 4x - 1 = 0 \qquad \text{Subtract } 4x + 1 \text{ from both sides.}$$

$$a = 2 \quad b = -4 \quad c = -1$$

Substituting these values into the quadratic formula and simplifying gives the equation's solutions.

$$x = \frac{-b \pm \sqrt{b^2 - 4ac}}{2a} \qquad \text{Use the quadratic formula.}$$

$$x = \frac{-(-4) \pm \sqrt{(-4)^2 - 4(2)(-1)}}{2(2)} \qquad \begin{array}{l}\text{Substitute the values for } a, b, \text{ and } c: \\ a = 2, b = -4, \text{ and } c = -1.\end{array}$$

$$= \frac{4 \pm \sqrt{16 - (-8)}}{4} \qquad (-4)^2 - 4(2)(-1) = 16 - (-8)$$

$$= \frac{4 \pm \sqrt{24}}{4} \qquad 16 - (-8) = 16 + 8 = 24$$

$$= \frac{4 \pm 2\sqrt{6}}{4} \qquad \sqrt{24} = \sqrt{4 \cdot 6} = \sqrt{4} \cdot \sqrt{6} = 2\sqrt{6}$$

$$= \frac{2(2 \pm \sqrt{6})}{4} \qquad \text{Factor out 2 from the numerator.}$$

$$= \frac{2 \pm \sqrt{6}}{2} \qquad \text{Divide the numerator and denominator by 2.}$$

The solutions are $\dfrac{2 \pm \sqrt{6}}{2}$, and the solution set is $\left\{ \dfrac{2 + \sqrt{6}}{2}, \dfrac{2 - \sqrt{6}}{2} \right\}$ or $\left\{ \dfrac{2 \pm \sqrt{6}}{2} \right\}$. ■

Using Technology

You can use a graphing utility to verify that the solutions of $2x^2 - 4x - 1 = 0$ are $\dfrac{2 \pm \sqrt{6}}{2}$. Begin by entering $y_1 = 2x^2 - 4x - 1$ in the $\boxed{Y=}$ screen. Then evaluate this function at each of the proposed solutions.

In each case, the function value is 0, verifying that the solutions satisfy $2x^2 - 4x - 1 = 0$.

In Example 2, the solutions of $2x^2 = 4x + 1$ are irrational numbers. This means that the equation cannot be solved by factoring. The reason that the solutions are irrational numbers is that $b^2 - 4ac$, the radicand in the quadratic formula, is 24, which is not a perfect square. Notice, too, that the solutions, $\dfrac{2 + \sqrt{6}}{2}$ and $\dfrac{2 - \sqrt{6}}{2}$, are conjugates.

Great Question!

The simplification of the irrational solutions in Example 2 was kind of tricky. Any suggestions to guide the process?

Many students use the quadratic formula correctly until the last step, where they make an error in simplifying the solutions. Be sure to factor the numerator before dividing the numerator and the denominator by the greatest common factor.

$$\frac{4 \pm 2\sqrt{6}}{4} = \frac{2(2 \pm \sqrt{6})}{4} = \frac{\overset{1}{\cancel{2}}(2 \pm \sqrt{6})}{\underset{2}{\cancel{4}}} = \frac{2 \pm \sqrt{6}}{2}$$

You cannot divide just one term in the numerator and the denominator by their greatest common factor.

Incorrect!

$$\frac{\overset{1}{\cancel{4}} \pm 2\sqrt{6}}{\underset{1}{\cancel{4}}} = 1 \pm 2\sqrt{6} \qquad \frac{4 \pm \overset{1}{\cancel{2}}\sqrt{6}}{\underset{2}{\cancel{4}}} = \frac{4 \pm \sqrt{6}}{2}$$

Can all irrational solutions of quadratic equations be simplified? No. The following solutions cannot be simplified:

$$\frac{5 \pm 2\sqrt{7}}{2} \qquad \frac{-4 \pm 3\sqrt{7}}{2}.$$

> Other than 1, terms in each numerator have no common factor.

✓ **CHECK POINT 2** Solve using the quadratic formula: $2x^2 = 6x - 1$.

EXAMPLE 3 Solving a Quadratic Equation Using the Quadratic Formula

Solve using the quadratic formula:

$$3x^2 + 2 = -4x.$$

Solution Begin by writing the quadratic equation in standard form.

$$3x^2 + 2 = -4x \qquad \text{This is the given equation.}$$

$$3x^2 + 4x + 2 = 0 \qquad \text{Add } 4x \text{ to both sides.}$$

$$a = 3 \quad b = 4 \quad c = 2$$

Substituting these values into the quadratic formula and simplifying gives the equation's solutions.

$$x = \frac{-b \pm \sqrt{b^2 - 4ac}}{2a} \qquad \text{Use the quadratic formula.}$$

$$x = \frac{-4 \pm \sqrt{4^2 - 4\cdot3\cdot2}}{2\cdot3} \qquad \begin{array}{l}\text{Substitute the values for } a, b, \text{ and } c:\\ a = 3, b = 4, \text{ and } c = 2.\end{array}$$

$$= \frac{-4 \pm \sqrt{16 - 24}}{6} \qquad \text{Multiply under the radical.}$$

$$= \frac{-4 \pm \sqrt{-8}}{6} \qquad \text{Subtract under the radical.}$$

$$= \frac{-4 \pm 2i\sqrt{2}}{6} \qquad \begin{array}{l}\sqrt{-8} = \sqrt{8(-1)} = \sqrt{8}\sqrt{-1} = i\sqrt{8}\\ = i\sqrt{4\cdot2} = 2i\sqrt{2}\end{array}$$

$$= \frac{2(-2 \pm i\sqrt{2})}{6} \qquad \text{Factor out 2 from the numerator.}$$

$$= \frac{-2 \pm i\sqrt{2}}{3} \qquad \text{Divide the numerator and denominator by 2.}$$

$$= -\frac{2}{3} \pm i\frac{\sqrt{2}}{3} \qquad \begin{array}{l}\text{Express in the form } a + bi,\\ \text{writing } i \text{ before the square root.}\end{array}$$

The solutions are $-\dfrac{2}{3} \pm i\dfrac{\sqrt{2}}{3}$, and the solution set is $\left\{ -\dfrac{2}{3} + i\dfrac{\sqrt{2}}{3}, -\dfrac{2}{3} - i\dfrac{\sqrt{2}}{3} \right\}$ or $\left\{ -\dfrac{2}{3} \pm i\dfrac{\sqrt{2}}{3} \right\}$. ∎

In Example 3, the solutions of $3x^2 + 2 = -4x$ are imaginary numbers. This means that the equation cannot be solved using factoring. The reason that the solutions are imaginary numbers is that $b^2 - 4ac$, the radicand in the quadratic formula, is -8, which is negative. Notice, too, that the solutions are complex conjugates.

Using Technology

Graphic Connections

The graph of the quadratic function

$$y = 3x^2 + 4x + 2$$

has no x-intercepts. This verifies that the equation in Example 3

$$3x^2 + 2 = -4x, \quad \text{or}$$

$$3x^2 + 4x + 2 = 0$$

has imaginary solutions.

$y = 3x^2 + 4x + 2$

$[-3, 3, 1]$ by $[-2, 10, 1]$

Great Question!

Should I check irrational and imaginary solutions by substitution in the given quadratic equation?

No. Checking irrational and imaginary solutions can be time-consuming. The solutions given by the quadratic formula are always correct, unless you have made a careless error. Checking for computational errors or errors in simplification is sufficient.

☑ **CHECK POINT 3** Solve using the quadratic formula: $3x^2 + 5 = -6x$.

Some rational equations can be solved using the quadratic formula. For example, consider the equation

$$3 + \frac{4}{x} = -\frac{2}{x^2}.$$

The denominators are x and x^2. The least common denominator is x^2. We clear fractions by multiplying both sides of the equation by x^2. Notice that x cannot equal zero.

$$x^2\left(3 + \frac{4}{x}\right) = x^2\left(-\frac{2}{x^2}\right), \quad x \neq 0$$

$$3x^2 + \frac{4}{x} \cdot x^2 = x^2\left(-\frac{2}{x^2}\right) \quad \text{Use the distributive property.}$$

$$3x^2 + 4x = -2 \quad \text{Simplify.}$$

By adding 2 to both sides of $3x^2 + 4x = -2$, we obtain the standard form of the quadratic equation:

$$3x^2 + 4x + 2 = 0.$$

This is the equation that we solved in Example 3. The two imaginary solutions are not part of the restriction that $x \neq 0$.

The Discriminant

2 Use the discriminant to determine the number and type of solutions. ▶

The quantity $b^2 - 4ac$, which appears under the radical sign in the quadratic formula, is called the **discriminant**. **Table 8.2** shows how the discriminant of the quadratic equation $ax^2 + bx + c = 0$ determines the number and type of solutions.

Table 8.2 The Discriminant and the Kinds of Solutions to $ax^2 + bx + c = 0$

Discriminant $b^2 - 4ac$	Kinds of Solutions to $ax^2 + bx + c = 0$	Graph of $y = ax^2 + bx + c$
$b^2 - 4ac > 0$	**Two unequal real solutions:** If a, b, and c are rational numbers and the discriminant is a perfect square, the solutions are rational. If the discriminant is not a perfect square, the solutions are irrational conjugates.	Two x-intercepts
$b^2 - 4ac = 0$	**One solution (a repeated solution) that is a real number:** If a, b, and c are rational numbers, the repeated solution is also a rational number.	One x-intercept
$b^2 - 4ac < 0$	**No real solution; two imaginary solutions:** The solutions are complex conjugates.	No x-intercepts

EXAMPLE 4 Using the Discriminant

For each equation, compute the discriminant. Then determine the number and type of solutions:

a. $3x^2 + 4x - 5 = 0$ **b.** $9x^2 - 6x + 1 = 0$ **c.** $3x^2 - 8x + 7 = 0.$

Solution Begin by identifying the values for a, b, and c in each equation. Then compute $b^2 - 4ac$, the discriminant.

a. $3x^2 + 4x - 5 = 0$

$a = 3$ $b = 4$ $c = -5$

Substitute and compute the discriminant:

$$b^2 - 4ac = 4^2 - 4 \cdot 3(-5) = 16 - (-60) = 16 + 60 = 76.$$

The discriminant, 76, is a positive number that is not a perfect square. Thus, there are two real irrational solutions. (These solutions are conjugates of each other.)

b. $9x^2 - 6x + 1 = 0$

$a = 9$ $b = -6$ $c = 1$

Substitute and compute the discriminant:

$$b^2 - 4ac = (-6)^2 - 4 \cdot 9 \cdot 1 = 36 - 36 = 0.$$

The discriminant, 0, shows that there is only one real solution. This real solution is a rational number.

c. $3x^2 - 8x + 7 = 0$

$a = 3$ $b = -8$ $c = 7$

$$b^2 - 4ac = (-8)^2 - 4 \cdot 3 \cdot 7 = 64 - 84 = -20$$

The negative discriminant, -20, shows that there are two imaginary solutions. (These solutions are complex conjugates of each other.) ∎

✓ **CHECK POINT 4** For each equation, compute the discriminant. Then determine the number and type of solutions:

a. $x^2 + 6x + 9 = 0$

b. $2x^2 - 7x - 4 = 0$

c. $3x^2 - 2x + 4 = 0.$

> **Great Question!**
>
> **Is the square root sign part of the discriminant?**
>
> No. The discriminant is $b^2 - 4ac$. It is not $\sqrt{b^2 - 4ac}$, so do not give the discriminant as a radical.

③ Determine the most efficient method to use when solving a quadratic equation.

Determining Which Method to Use

All quadratic equations can be solved by the quadratic formula. However, if an equation is in the form $u^2 = d$, such as $x^2 = 5$ or $(2x + 3)^2 = 8$, it is faster to use the square root property, taking the square root of both sides. If the equation is not in the form $u^2 = d$, write the quadratic equation in standard form ($ax^2 + bx + c = 0$). Try to solve the equation by factoring. If $ax^2 + bx + c$ cannot be factored, then solve the quadratic equation by the quadratic formula.

Because we used the method of completing the square to derive the quadratic formula, we no longer need it for solving quadratic equations. However, we will use completing the square in Chapter 10 to help graph certain kinds of equations.

Table 8.3 summarizes our observations about which technique to use when solving a quadratic equation.

Table 8.3	Determining the Most Efficient Technique to Use When Solving a Quadratic Equation	
Description and Form of the Quadratic Equation	**Most Efficient Solution Method**	**Example**
$ax^2 + bx + c = 0$ and $ax^2 + bx + c$ can be factored easily.	Factor and use the zero-product principle.	$3x^2 + 5x - 2 = 0$ $(3x - 1)(x + 2) = 0$ $3x - 1 = 0$ or $x + 2 = 0$ $x = \dfrac{1}{3}$ $\qquad x = -2$
$ax^2 + c = 0$ The quadratic equation has no x-term. $(b = 0)$	Solve for x^2 and apply the square root property.	$4x^2 - 7 = 0$ $4x^2 = 7$ $x^2 = \dfrac{7}{4}$ $x = \pm \dfrac{\sqrt{7}}{2}$
$u^2 = d$; u is a first-degree polynomial.	Use the square root property.	$(x + 4)^2 = 5$ $x + 4 = \pm\sqrt{5}$ $x = -4 \pm\sqrt{5}$
$ax^2 + bx + c = 0$ and $ax^2 + bx + c$ cannot be factored or the factoring is too difficult.	Use the quadratic formula: $x = \dfrac{-b \pm \sqrt{b^2 - 4ac}}{2a}.$	$x^2 - 2x - 6 = 0$ $a = 1 \quad b = -2 \quad c = -6$ $x = \dfrac{-(-2) \pm \sqrt{(-2)^2 - 4(1)(-6)}}{2(1)}$ $= \dfrac{2 \pm \sqrt{4 + 24}}{2}$ $= \dfrac{2 \pm \sqrt{28}}{2} = \dfrac{2 \pm \sqrt{4}\sqrt{7}}{2}$ $= \dfrac{2 \pm 2\sqrt{7}}{2} = \dfrac{2(1 \pm \sqrt{7})}{2}$ $= 1 \pm \sqrt{7}$

4 Write quadratic equations from solutions. ▶

Writing Quadratic Equations from Solutions

Using the zero-product principle, the equation $(x - 3)(x + 5) = 0$ has two solutions, 3 and -5. By applying the zero-product principle in reverse, we can find a quadratic equation that has two given numbers as its solutions.

> **The Zero-Product Principle in Reverse**
> If $A = 0$ or $B = 0$, then $AB = 0$.

EXAMPLE 5 Writing Equations from Solutions

Write a quadratic equation with the given solution set:

a. $\left\{-\dfrac{5}{3}, \dfrac{1}{2}\right\}$ **b.** $\left\{-2\sqrt{3}, 2\sqrt{3}\right\}$ **c.** $\{-5i, 5i\}$.

Solution

a. Because the solution set is $\left\{-\dfrac{5}{3}, \dfrac{1}{2}\right\}$, then

$$x = -\frac{5}{3} \quad \text{or} \qquad x = \frac{1}{2}.$$

$x + \dfrac{5}{3} = 0$	or	$x - \dfrac{1}{2} = 0$

Obtain zero on one side of each equation.

$3x + 5 = 0$	or	$2x - 1 = 0$

Clear fractions, multiplying by 3 and 2, respectively.

$$(3x + 5)(2x - 1) = 0$$

Use the zero-product principle in reverse: If $A = O$ or $B = O$, then $AB = O$.

$$6x^2 - 3x + 10x - 5 = 0$$

Use the FOIL method to multiply.

$$6x^2 + 7x - 5 = 0$$

Combine like terms.

Thus, one equation is $6x^2 + 7x - 5 = 0$. Many other quadratic equations have $\left\{-\frac{5}{3}, \frac{1}{2}\right\}$ for their solution sets. These equations can be obtained by multiplying both sides of $6x^2 + 7x - 5 = 0$ by any nonzero real number.

b. Because the solution set is $\left\{-2\sqrt{3}, 2\sqrt{3}\right\}$, then

$$x = -2\sqrt{3} \quad \text{or} \qquad x = 2\sqrt{3}.$$

$x + 2\sqrt{3} = 0$	or	$x - 2\sqrt{3} = 0$

Obtain zero on one side of each equation.

$$\left(x + 2\sqrt{3}\right)\left(x - 2\sqrt{3}\right) = 0$$

Use the zero-product principle in reverse: If $A = O$ or $B = O$, then $AB = O$.

$$x^2 - \left(2\sqrt{3}\right)^2 = 0$$

Multiply conjugates using $(A + B)(A - B) = A^2 - B^2$.

$$x^2 - 12 = 0$$

$(2\sqrt{3})^2 = 2^2(\sqrt{3})^2 = 4 \cdot 3 = 12$

Thus, one equation is $x^2 - 12 = 0$.

c. Because the solution set is $\{-5i, 5i\}$, then

$$x = -5i \quad \text{or} \qquad x = 5i.$$

$x + 5i = 0$	or	$x - 5i = 0$

Obtain zero on one side of each equation.

$$(x + 5i)(x - 5i) = 0$$

Use the zero-product principle in reverse: If $A = O$ or $B = O$, then $AB = O$.

$$x^2 - (5i)^2 = 0$$

Multiply conjugates using $(A + B)(A - B) = A^2 - B^2$.

$$x^2 - 25i^2 = 0$$

$(5i)^2 = 5^2 i^2 = 25i^2$

$$x^2 - 25(-1) = 0$$

$i^2 = -1$

$$x^2 + 25 = 0$$

This is the required equation. ∎

✓ **CHECK POINT 5**   Write a quadratic equation with the given solution set:
a. $\left\{-\frac{3}{5}, \frac{1}{4}\right\}$ **b.** $\left\{-5\sqrt{2}, 5\sqrt{2}\right\}$
c. $\{-7i, 7i\}$.

---

5 Use the quadratic formula to solve problems.

Applications

Quadratic equations can be solved to answer questions about variables contained in quadratic functions.

Blood Pressure and Age

The graphs in **Figure 8.3** illustrate that a person's normal systolic blood pressure, measured in millimeters of mercury (mm Hg), depends on his or her age. The function

$$P(A) = 0.006A^2 - 0.02A + 120$$

models a man's normal systolic pressure, $P(A)$, at age A.

a. Find the age, to the nearest year, of a man whose normal systolic blood pressure is 125 mm Hg.

b. Use the graphs in **Figure 8.3** to describe the differences between the normal systolic blood pressures of men and women as they age.

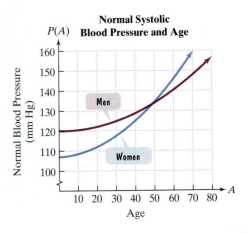

Figure 8.3

Solution

a. We are interested in the age of a man with a normal systolic blood pressure of 125 millimeters of mercury. Thus, we substitute 125 for $P(A)$ in the given function for men. Then we solve for A, the man's age.

$$P(A) = 0.006A^2 - 0.02A + 120$$ This is the given function for men.

$$125 = 0.006A^2 - 0.02A + 120$$ Substitute 125 for $P(A)$.

$$0 = 0.006A^2 - 0.02A - 5$$ Subtract 125 from both sides and write the quadratic equation in standard form.

$$a = 0.006 \quad b = -0.02 \quad c = -5$$

Because the trinomial on the right side of the equation is prime, we solve using the quadratic formula.

Using Technology

On most calculators, here is how to approximate

$$\frac{0.02 + \sqrt{0.1204}}{0.012}.$$

Many Scientific Calculators

(.02 + .1204 √)

÷ .012 =

Many Graphing Calculators

(.02 + √ .1204)

÷ .012 ENTER

If your calculator displays an open parenthesis after √, you'll need to enter another closed parenthesis here.

Notice that the variable is A, rather than the usual x.

$$A = \frac{-b \pm \sqrt{b^2 - 4ac}}{2a}$$ Use the quadratic formula.

$$= \frac{-(-0.02) \pm \sqrt{(-0.02)^2 - 4(0.006)(-5)}}{2(0.006)}$$ Substitute the values for a, b, and c: $a = 0.006$, $b = -0.02$, and $c = -5$.

$$= \frac{0.02 \pm \sqrt{0.1204}}{0.012}$$ Use a calculator to simplify the expression under the square root.

$$\approx \frac{0.02 \pm 0.347}{0.012}$$ Use a calculator: $\sqrt{0.1204} \approx 0.347$.

$$A \approx \frac{0.02 + 0.347}{0.012} \quad \text{or} \quad A \approx \frac{0.02 - 0.347}{0.012}$$

$$A \approx 31 \qquad\qquad\qquad A \approx -27$$ Use a calculator and round to the nearest integer.

Reject this solution. Age cannot be negative.

The positive solution, $A \approx 31$, indicates that 31 is the approximate age of a man whose normal systolic blood pressure is 125 mm Hg. This is illustrated by the black lines with the arrows on the red graph representing men in **Figure 8.4**.

b. Take a second look at the graphs in **Figure 8.4**. Before approximately age 50, the blue graph representing women's normal systolic blood pressure lies below the red graph representing men's normal systolic blood pressure. Thus, up to age 50, women's normal systolic blood pressure is lower than

Figure 8.4

men's, although it is increasing at a faster rate. After age 50, women's normal systolic blood pressure is higher than men's. ∎

✓ **CHECK POINT 6** The function $P(A) = 0.01A^2 + 0.05A + 107$ models a woman's normal systolic blood pressure, $P(A)$, at age A. Use this function to find the age, to the nearest year, of a woman whose normal systolic blood pressure is 115 mm Hg. Use the blue graph in **Figure 8.4** to verify your solution.

Achieving Success

A test-taking tip: Go for partial credit. Always show your work. If worse comes to worst, write something down, anything, even if it's a formula that you think might solve a problem or a possible idea or procedure for solving the problem. Here's an example:

Test Question: Solve by the quadratic formula:

$$2x^2 = -4x + 5.$$

Student Solution:

> $2x^2 = -4x + 5$
> $2x^2 + 4x - 5 = 0$
> $a = 2 \quad b = 4 \quad c = -5$
>
> Egad! I forgot the formula.
>
> $$x = \frac{-b \pm \sqrt{?}}{2a} = \frac{-4 \pm \sqrt{?}}{2 \cdot 2}$$
>
> I know I need to find ?.
> That will give me the solutions.
> Perhaps I can simplify after that.
> ???

CONCEPT AND VOCABULARY CHECK

Fill in each blank so that the resulting statement is true.

1. The solutions of a quadratic equation in standard form $ax^2 + bx + c = 0, a \neq 0$, are given by the quadratic formula

    $x =$ _____.

2. In order to solve $2x^2 + 9x - 5 = 0$ by the quadratic formula, we use $a =$ _____, $b =$ _____, and $c =$ _____.

3. In order to solve $x^2 = 4x + 1$ by the quadratic formula, we use $a =$ _____, $b =$ _____, and $c =$ _____.

4. $x = \dfrac{-(-4) \pm \sqrt{(-4)^2 - 4(1)(2)}}{2(1)}$ simplifies to $x =$ _____.

5. $x = \dfrac{-4 \pm \sqrt{4^2 - 4 \cdot 2 \cdot 5}}{2 \cdot 2}$ simplifies to $x =$ _____.

6. The discriminant of $ax^2 + bx + c = 0$ is defined by _____.

7. If the discriminant of $ax^2 + bx + c = 0$ is negative, the quadratic equation has _____ real solutions.

8. If the discriminant of $ax^2 + bx + c = 0$ is positive, the quadratic equation has _____ real solutions.

9. The most efficient technique for solving $(2x + 7)^2 = 25$ is by using _____.

10. The most efficient technique for solving $x^2 + 5x - 10 = 0$ is by using _____.

11. The most efficient technique for solving $x^2 + 8x + 15 = 0$ is by using _____.

12. True or false: An equation with the solution set $\{2, 5\}$ is $(x + 2)(x + 5) = 0$. _____

8.2 EXERCISE SET   ▶ MyMathLab®

Practice Exercises

In Exercises 1–18, solve each equation using the quadratic formula. Simplify solutions, if possible.

1. $x^2 + 8x + 12 = 0$

2. $x^2 + 8x + 15 = 0$

3. $2x^2 - 7x = -5$

4. $5x^2 + 8x = -3$

5. $x^2 + 3x - 20 = 0$

6. $x^2 + 5x - 10 = 0$

7. $3x^2 - 7x = 3$

8. $4x^2 + 3x = 2$

9. $6x^2 = 2x + 1$

10. $2x^2 = -4x + 5$

11. $4x^2 - 3x = -6$

12. $9x^2 + x = -2$

13. $x^2 - 4x + 8 = 0$

14. $x^2 + 6x + 13 = 0$

15. $3x^2 = 8x - 7$

16. $3x^2 = 4x - 6$

17. $2x(x - 2) = x + 12$

18. $2x(x + 4) = 3x - 3$

In Exercises 19–30, compute the discriminant. Then determine the number and type of solutions for the given equation.

19. $x^2 + 8x + 3 = 0$

20. $x^2 + 7x + 4 = 0$

21. $x^2 + 6x + 8 = 0$

22. $x^2 + 2x - 3 = 0$

23. $2x^2 + x + 3 = 0$

24. $2x^2 - 4x + 3 = 0$

25. $2x^2 + 6x = 0$

26. $3x^2 - 5x = 0$

27. $5x^2 + 3 = 0$

28. $5x^2 + 4 = 0$

29. $9x^2 = 12x - 4$

30. $4x^2 = 20x - 25$

In Exercises 31–50, solve each equation by the method of your choice. Simplify solutions, if possible.

31. $3x^2 - 4x = 4$

32. $2x^2 - x = 1$

33. $x^2 - 2x = 1$

34. $2x^2 + 3x = 1$

35. $3x^2 = x - 9$

36. $2x^2 = -6x - 7$

37. $(2x - 5)(x + 1) = 2$

38. $(2x + 3)(x + 4) = 1$

39. $(3x - 4)^2 = 16$

40. $(2x + 7)^2 = 25$

41. $\dfrac{x^2}{2} + 2x + \dfrac{2}{3} = 0$

42. $\dfrac{x^2}{3} - x - \dfrac{1}{6} = 0$

43. $(3x - 2)^2 = 10$

44. $(4x - 1)^2 = 15$

45. $\dfrac{1}{x} + \dfrac{1}{x + 2} = \dfrac{1}{3}$

46. $\dfrac{1}{x} + \dfrac{1}{x + 3} = \dfrac{1}{4}$

47. $(2x - 6)(x + 2) = 5(x - 1) - 12$

48. $7x(x - 2) = 3 - 2(x + 4)$

49. $x^2 + 10 = 2(2x - 1)$

50. $x(x + 6) = -12$

In Exercises 51–64, write a quadratic equation in standard form with the given solution set.

51. $\{-3, 5\}$

52. $\{-2, 6\}$

53. $\left\{-\dfrac{2}{3}, \dfrac{1}{4}\right\}$

54. $\left\{-\dfrac{5}{6}, \dfrac{1}{3}\right\}$

55. $\{-\sqrt{2}, \sqrt{2}\}$

56. $\{-\sqrt{3}, \sqrt{3}\}$

57. $\{-2\sqrt{5}, 2\sqrt{5}\}$

58. $\{-3\sqrt{5}, 3\sqrt{5}\}$

59. $\{-6i, 6i\}$

60. $\{-8i, 8i\}$

61. $\{1 + i, 1 - i\}$

62. $\{2 + i, 2 - i\}$

63. $\{1 + \sqrt{2}, 1 - \sqrt{2}\}$

64. $\{1 + \sqrt{3}, 1 - \sqrt{3}\}$

Practice PLUS

Exercises 65–68 describe quadratic equations. Match each description with the graph of the corresponding quadratic function. Each graph is shown in a $[-10, 10, 1]$ by $[-10, 10, 1]$ viewing rectangle.

65. A quadratic equation whose solution set contains imaginary numbers

66. A quadratic equation whose discriminant is 0

67. A quadratic equation whose solution set is $\{3 \pm \sqrt{2}\}$

68. A quadratic equation whose solution set contains integers

a.

b.

c.

d.

69. When the sum of 6 and twice a positive number is subtracted from the square of the number, 0 results. Find the number.

70. When the sum of 1 and twice a negative number is subtracted from twice the square of the number, 0 results. Find the number.

In Exercises 71–76, solve each equation by the method of your choice.

71. $\dfrac{1}{x^2 - 3x + 2} = \dfrac{1}{x + 2} + \dfrac{5}{x^2 - 4}$

72. $\dfrac{x - 1}{x - 2} + \dfrac{x}{x - 3} = \dfrac{1}{x^2 - 5x + 6}$

73. $\sqrt{2}x^2 + 3x - 2\sqrt{2} = 0$

74. $\sqrt{3}x^2 + 6x + 7\sqrt{3} = 0$

75. $|x^2 + 2x| = 3$

76. $|x^2 + 3x| = 2$

Application Exercises

A driver's age has something to do with his or her chance of getting into a fatal car crash. The bar graph shows the number of fatal vehicle crashes per 100 million miles driven for drivers of various age groups. For example, 25-year-old drivers are involved in 4.1 fatal crashes per 100 million miles driven. Thus, when a group of 25-year-old Americans have driven a total of 100 million miles, approximately 4 have been in accidents in which someone died.

Age of United States Drivers and Fatal Crashes

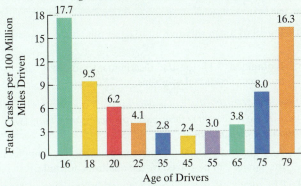

Source: Insurance Institute for Highway Safety

The number of fatal vehicle crashes per 100 million miles, $f(x)$, for drivers of age x can be modeled by the quadratic function

$$f(x) = 0.013x^2 - 1.19x + 28.24.$$

Use the function to solve Exercises 77–78.

77. What age groups are expected to be involved in 3 fatal crashes per 100 million miles driven? How well does the function model the trend in the actual data shown in the bar graph?

78. What age groups are expected to be involved in 10 fatal crashes per 100 million miles driven? How well does the function model the trend in the actual data shown in the bar graph?

Throwing events in track and field include the shot put, the discus throw, the hammer throw, and the javelin throw. The distance that an athlete can achieve depends on the initial velocity of the object thrown and the angle above the horizontal at which the object leaves the hand.

In Exercises 79–80, an athlete whose event is the shot put releases the shot with the same initial velocity, but at different angles.

79. When the shot is released at an angle of 35°, its path can be modeled by the function

$$f(x) = -0.01x^2 + 0.7x + 6.1,$$

in which x is the shot's horizontal distance, in feet, and $f(x)$ is its height, in feet. This function is shown by one of the graphs, (a) or (b), in the figure. Use the function to determine the shot's maximum distance. Use a calculator and round to the nearest tenth of a foot. Which graph, (a) or (b), shows the shot's path?

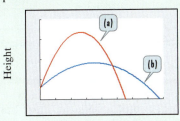

Horizontal Distance
[0, 80, 10] by [0, 40, 10]

80. When the shot is released at an angle of 65°, its path can be modeled by the function

$$f(x) = -0.04x^2 + 2.1x + 6.1,$$

in which x is the shot's horizontal distance, in feet, and $f(x)$ is its height, in feet. This function is shown by one of the graphs, (a) or (b), in the figure above. Use the function to determine the shot's maximum distance. Use a calculator and round to the nearest tenth of a foot. Which graph, (a) or (b), shows the shot's path?

81. The length of a rectangle is 4 meters longer than the width. If the area is 8 square meters, find the rectangle's dimensions. Round to the nearest tenth of a meter.

82. The length of a rectangle exceeds twice its width by 3 inches. If the area is 10 square inches, find the rectangle's dimensions. Round to the nearest tenth of an inch.

83. The longer leg of a right triangle exceeds the shorter leg by 1 inch, and the hypotenuse exceeds the longer leg by 7 inches. Find the lengths of the legs. Round to the nearest tenth of a inch.

84. The hypotenuse of a right triangle is 6 feet long. One leg is 2 feet shorter than the other. Find the lengths of the legs. Round to the nearest tenth of a foot.

85. A rain gutter is made from sheets of aluminum that are 20 inches wide. As shown in the figure, the edges are turned up to form right angles. Determine the depth of the gutter that will allow a cross-sectional area of 13 square inches. Show that there are two different solutions to the problem. Round to the nearest tenth of an inch.

86. A piece of wire is 8 inches long. The wire is cut into two pieces and then each piece is bent into a square. Find the length of each piece if the sum of the areas of these squares is to be 2 square inches.

87. Working together, two people can mow a large lawn in 4 hours. One person can do the job alone 1 hour faster than the other person. How long does it take each person working alone to mow the lawn? Round to the nearest tenth of an hour.

88. A pool has an inlet pipe to fill it and an outlet pipe to empty it. It takes 2 hours longer to empty the pool than it does to fill it. The inlet pipe is turned on to fill the pool, but the outlet pipe is accidentally left open. Despite this, the pool fills in 8 hours. How long does it take the outlet pipe to empty the pool? Round to the nearest tenth of an hour.

Explaining the Concepts

89. What is the quadratic formula and why is it useful?

90. Without going into specific details for every step, describe how the quadratic formula is derived.

91. Explain how to solve $x^2 + 6x + 8 = 0$ using the quadratic formula.

92. If a quadratic equation has imaginary solutions, how is this shown on the graph of the corresponding quadratic function?

93. What is the discriminant and what information does it provide about a quadratic equation?

94. If you are given a quadratic equation, how do you determine which method to use to solve it?

95. Explain how to write a quadratic equation from its solution set. Give an example with your explanation.

Technology Exercises

96. Use a graphing utility to graph the quadratic function related to any five of the quadratic equations in Exercises 19–30. How does each graph illustrate what you determined algebraically using the discriminant?

97. Reread Exercise 85. The cross-sectional area of the gutter is given by the quadratic function

$$f(x) = x(20 - 2x).$$

Graph the function in a $[0, 10, 1]$ by $[0, 60, 5]$ viewing rectangle. Then TRACE along the curve or use the maximum function feature to determine the depth of

the gutter that will maximize its cross-sectional area and allow the greatest amount of water to flow. What is the maximum area? Does the situation described in Exercise 85 take full advantage of the sheets of aluminum?

Critical Thinking Exercises

Make Sense? *In Exercises 98–101, determine whether each statement makes sense or does not make sense, and explain your reasoning.*

98. Because I want to solve $25x^2 - 169 = 0$ fairly quickly, I'll use the quadratic formula.

99. I simplified $\dfrac{3 + 2\sqrt{3}}{2}$ to $3 + \sqrt{3}$ because 2 is a factor of $2\sqrt{3}$.

100. I need to find a square root to determine the discriminant.

101. I obtained -17 for the discriminant, so there are two imaginary irrational solutions.

In Exercises 102–105, determine whether each statement is true or false. If the statement is false, make the necessary change(s) to produce a true statement.

102. Any quadratic equation that can be solved by completing the square can be solved by the quadratic formula.

103. The quadratic formula is developed by applying factoring and the zero-product principle to the quadratic equation $ax^2 + bx + c = 0$.

104. In using the quadratic formula to solve the quadratic equation $5x^2 = 2x - 7$, we have $a = 5, b = 2$, and $c = -7$.

105. The quadratic formula can be used to solve the equation $x^2 - 9 = 0$.

106. Solve for t: $\quad s = -16t^2 + v_0 t$.

107. A rectangular swimming pool is 12 meters long and 8 meters wide. A tile border of uniform width is to be built around the pool using 120 square meters of tile. The tile is from a discontinued stock (so no additional materials are available) and all 120 square meters are to be used. How wide should the border be? Round to the nearest tenth of a meter. If zoning laws require at least a 2-meter-wide border around the pool, can this be done with the available tile?

108. The area of the shaded green region outside the rectangle and inside the triangle is 10 square yards. Find the triangle's height, represented by $2x$. Round to the nearest tenth of a yard.

Review Exercises

109. Solve: $|5x + 2| = |4 - 3x|$. (Section 4.3, Example 3)

110. Solve: $\sqrt{2x - 5} - \sqrt{x - 3} = 1$.
(Section 7.6, Example 4)

111. Rationalize the denominator: $\dfrac{5}{\sqrt{3} + x}$. (Section 7.5, Example 5)

Preview Exercises

Exercises 112–114 will help you prepare for the material covered in the next section.

112. Use point plotting to graph $f(x) = x^2$ and $g(x) = x^2 + 2$ in the same rectangular coordinate system.

113. Use point plotting to graph $f(x) = x^2$ and $g(x) = (x + 2)^2$ in the same rectangular coordinate system.

114. Find the x-intercepts for the graph of $f(x) = -2(x - 3)^2 + 8$.

SECTION

8.3

Quadratic Functions and Their Graphs

What am I supposed to learn?

After studying this section, you should be able to:

1 Recognize characteristics of parabolas. ▶

2 Graph parabolas in the form $f(x) = a(x - h)^2 + k$. ▶

3 Graph parabolas in the form $f(x) = ax^2 + bx + c$. ▶

4 Determine a quadratic function's minimum or maximum value. ▶

5 Solve problems involving a quadratic function's minimum or maximum value. ▶

We have a long history of throwing things. Before 400 B.C., the Greeks competed in games that included discus throwing. In the seventeenth century, English soldiers organized cannonball-throwing competitions. In 1827, a Yale University student, disappointed over failing an exam, took out his frustrations at the passing of a collection plate in chapel. Seizing the monetary tray, he flung it in the direction of a large open space on campus. Yale students see this act of frustration as the origin of the Frisbee.

In this section, we study quadratic functions and their graphs. By graphing functions that model the paths of the things we throw, you will be able to determine both the maximum height these objects attain and the distance these objects travel.

Graphs of Quadratic Functions

The graph of any quadratic function

$$f(x) = ax^2 + bx + c, \quad a \neq 0,$$

1 Recognize characteristics of parabolas. ▶

is called a **parabola**. Parabolas are shaped like bowls or inverted bowls, as shown in **Figure 8.5** at the top of the next page. If the coefficient of x^2 (the value of a in $ax^2 + bx + c$) is positive, the parabola opens upward. If the coefficient of x^2 is negative, the parabola opens downward. The **vertex** (or turning point) of the parabola is the lowest point on the graph when it opens upward and the highest point on the graph when it opens downward.

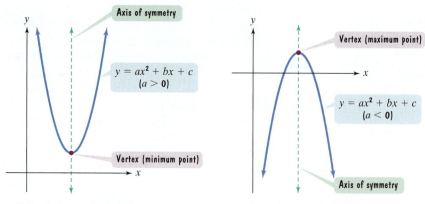

$a > 0$: Parabola opens upward. $a < 0$: Parabola opens downward.

Figure 8.5 Characteristics of graphs of quadratic functions

The two halves of a parabola are mirror images of each other. A "mirror line" through the vertex, called the **axis of symmetry**, divides the figure in half. If a parabola is folded along its axis of symmetry, the two halves match exactly.

2 Graph parabolas in the form $f(x) = a(x - h)^2 + k$. ⊙

Graphing Quadratic Functions in the Form $f(x) = a(x - h)^2 + k$

One way to obtain the graph of a quadratic function is to use point plotting. Let's begin by graphing the functions $f(x) = x^2$, $g(x) = 2x^2$, and $h(x) = \frac{1}{2}x^2$ in the same rectangular coordinate system. Select integers for x, starting with -3 and ending with 3. A partial table of coordinates for each function is shown below. The three parabolas are shown in **Figure 8.6**.

x	$f(x) = x^2$	(x, y) or $(x, f(x))$	x	$g(x) = 2x^2$	(x, y) or $(x, g(x))$
-3	$f(-3) = (-3)^2 = 9$	$(-3, 9)$	-3	$g(-3) = 2(-3)^2 = 18$	$(-3, 18)$
-2	$f(-2) = (-2)^2 = 4$	$(-2, 4)$	-2	$g(-2) = 2(-2)^2 = 8$	$(-2, 8)$
-1	$f(-1) = (-1)^2 = 1$	$(-1, 1)$	-1	$g(-1) = 2(-1)^2 = 2$	$(-1, 2)$
0	$f(0) = 0^2 = 0$	$(0, 0)$	0	$g(0) = 2 \cdot 0^2 = 0$	$(0, 0)$
1	$f(1) = 1^2 = 1$	$(1, 1)$	1	$g(1) = 2 \cdot 1^2 = 2$	$(1, 2)$
2	$f(2) = 2^2 = 4$	$(2, 4)$	2	$g(2) = 2 \cdot 2^2 = 8$	$(2, 8)$
3	$f(3) = 3^2 = 9$	$(3, 9)$	3	$g(3) = 2 \cdot 3^2 = 18$	$(3, 18)$

x	$h(x) = \frac{1}{2}x^2$	(x, y) or $(x, h(x))$
-3	$h(-3) = \frac{1}{2}(-3)^2 = \frac{9}{2}$	$\left(-3, \frac{9}{2}\right)$
-2	$h(-2) = \frac{1}{2}(-2)^2 = 2$	$(-2, 2)$
-1	$h(-1) = \frac{1}{2}(-1)^2 = \frac{1}{2}$	$\left(-1, \frac{1}{2}\right)$
0	$h(0) = \frac{1}{2} \cdot 0^2 = 0$	$(0, 0)$
1	$h(1) = \frac{1}{2} \cdot 1^2 = \frac{1}{2}$	$\left(1, \frac{1}{2}\right)$
2	$h(2) = \frac{1}{2} \cdot 2^2 = 2$	$(2, 2)$
3	$h(3) = \frac{1}{2} \cdot 3^2 = \frac{9}{2}$	$\left(3, \frac{9}{2}\right)$

Figure 8.6

Can you see that the graphs of f, g, and h in **Figure 8.6** all have the same vertex, $(0, 0)$? They also have the same axis of symmetry, the y-axis, or $x = 0$. This is true for all graphs of the form $f(x) = ax^2$. However, the blue graph of $g(x) = 2x^2$ is a narrower parabola than the red graph of $f(x) = x^2$. By contrast, the green graph of $h(x) = \frac{1}{2}x^2$ is a flatter parabola than the red graph of $f(x) = x^2$.

Is there a more efficient method than point plotting to obtain the graph of a quadratic function? The answer is yes. The method is based on comparing graphs of the form $g(x) = a(x - h)^2 + k$ to those of the form $f(x) = ax^2$.

In **Figure 8.7(a)**, the graph of $f(x) = ax^2$ for $a > 0$ is shown in black. The parabola's vertex is $(0, 0)$ and it opens upward. In **Figure 8.7(b)**, the graph of $f(x) = ax^2$ for $a < 0$ is shown in black. The parabola's vertex is $(0, 0)$ and it opens downward.

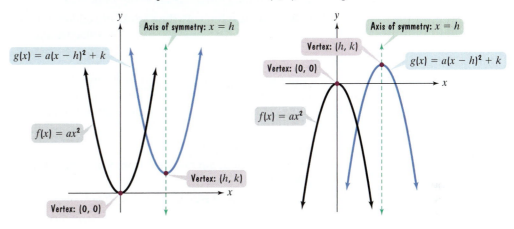

(a) $a > 0$: Parabola opens upward.

(b) $a < 0$: Parabola opens downward.

Figure 8.7 Moving, or shifting, the graph of $f(x) = ax^2$

Figures 8.7(a) and **8.7(b)** also show the graph of $g(x) = a(x - h)^2 + k$ in blue. Compare these graphs to those of $f(x) = ax^2$. Observe that h determines a horizontal move, or shift, and k determines a vertical move, or shift, of the graph of $f(x) = ax^2$:

$$g(x) = a(x - h)^2 + k.$$

| If $h > 0$, the graph of $f(x) = ax^2$ is shifted h units to the right. | If $k > 0$, the graph of $y = a(x - h)^2$ is shifted k units up. |

Consequently, the vertex $(0, 0)$ on the black graph of $f(x) = ax^2$ moves to the point (h, k) on the blue graph of $g(x) = a(x - h)^2 + k$. The axis of symmetry is the vertical line whose equation is $x = h$.

The form of the expression for g is convenient because it immediately identifies the vertex of the parabola as (h, k).

Quadratic Functions in the Form $f(x) = a(x - h)^2 + k$

The graph of

$$f(x) = a(x - h)^2 + k, \quad a \neq 0$$

is a parabola whose vertex is the point (h, k). The parabola is symmetric with respect to the line $x = h$. If $a > 0$, the parabola opens upward; if $a < 0$, the parabola opens downward.

The sign of a in $f(x) = a(x - h)^2 + k$ determines whether the parabola opens upward or downward. Furthermore, if $|a|$ is small, the parabola opens more flatly than if $|a|$ is large. On the next page is a general procedure for graphing parabolas whose equations are in this form.

Graphing Quadratic Functions with Equations in the Form $f(x) = a(x - h)^2 + k$

To graph $f(x) = a(x - h)^2 + k$,

1. Determine whether the parabola opens upward or downward. If $a > 0$, it opens upward. If $a < 0$, it opens downward.

2. Determine the vertex of the parabola. The vertex is (h, k).

3. Find any x-intercepts by solving $f(x) = 0$. The equation's real solutions are the x-intercepts.

4. Find the y-intercept by computing $f(0)$.

5. Plot the intercepts, the vertex, and additional points as necessary. Connect these points with a smooth curve that is shaped like a bowl or an inverted bowl.

In the graphs that follow, we will show each axis of symmetry as a dashed vertical line. Because this vertical line passes through the vertex, (h, k), its equation is $x = h$. The line is dashed because it is not part of the parabola.

Great Question!

I'm confused about finding h from the equation $f(x) = a(x - h)^2 + k$. Can you help me out?

It's easy to make a sign error when finding h, the x-coordinate of the vertex. In

$$f(x) = a(x - h)^2 + k,$$

h is the number that follows the subtraction sign.

- $f(x) = -2(x - 3)^2 + 8$

  The number after the subtraction is 3: $h = 3$.

- $f(x) = (x + 3)^2 + 1$
  $= (x - (-3))^2 + 1$

  The number after the subtraction is −3: $h = -3$.

EXAMPLE 1 Graphing a Quadratic Function in the Form $f(x) = a(x - h)^2 + k$

Graph the quadratic function $f(x) = -2(x - 3)^2 + 8$.

Solution We can graph this function by following the steps in the preceding box. We begin by identifying values for a, h, and k.

$$f(x) = a(x - h)^2 + k$$

$$a = -2 \quad h = 3 \quad k = 8$$

$$f(x) = -2(x - 3)^2 + 8$$

Step 1. Determine how the parabola opens. Note that a, the coefficient of x^2, is -2. Thus, $a < 0$; this negative value tells us that the parabola opens downward.

Step 2. Find the vertex. The vertex of the parabola is at (h, k). Because $h = 3$ and $k = 8$, the parabola has its vertex at $(3, 8)$.

Step 3. Find the x-intercepts by solving $f(x) = 0$. Replace $f(x)$ with 0 in $f(x) = -2(x - 3)^2 + 8$.

$$0 = -2(x - 3)^2 + 8 \qquad \text{Find } x\text{-intercepts, setting } f(x) \text{ equal to 0.}$$

$$2(x - 3)^2 = 8 \qquad \text{Solve for } x. \text{ Add } 2(x - 3)^2 \text{ to both sides of the equation.}$$

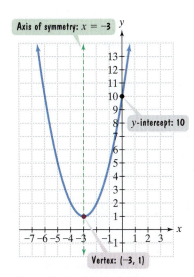

Axis of symmetry: $x = 3$

Vertex: (3, 8)

x-intercept: 1

x-intercept: 5

Figure 8.8 The graph of $f(x) = -2(x - 3)^2 + 8$

$$(x - 3)^2 = 4 \qquad \text{Divide both sides by 2.}$$

$$x - 3 = \sqrt{4} \quad \text{or} \quad x - 3 = -\sqrt{4} \qquad \text{Apply the square root property.}$$

$$x - 3 = 2 \qquad\qquad x - 3 = -2 \qquad \sqrt{4} = 2$$

$$x = 5 \qquad\qquad x = 1 \qquad \text{Add 3 to both sides in each equation.}$$

The x-intercepts are 5 and 1. The parabola passes through $(5, 0)$ and $(1, 0)$.

Step 4. Find the y-intercept by computing $f(0)$. Replace x with 0 in $f(x) = -2(x - 3)^2 + 8$.

$$f(0) = -2(0 - 3)^2 + 8 = -2(-3)^2 + 8 = -2(9) + 8 = -10$$

The y-intercept is -10. The parabola passes through $(0, -10)$.

Step 5. Graph the parabola. With a vertex at $(3, 8)$, x-intercepts at 5 and 1, and a y-intercept at -10, the graph of f is shown in **Figure 8.8**. The axis of symmetry is the vertical line whose equation is $x = 3$. ∎

☑ **CHECK POINT 1**   Graph the quadratic function $f(x) = -(x - 1)^2 + 4$.

EXAMPLE 2   Graphing a Quadratic Function in the Form $f(x) = a(x - h)^2 + k$

Graph the quadratic function $f(x) = (x + 3)^2 + 1$.

Solution   We begin by finding values for a, h, and k.

$$f(x) = a(x - h)^2 + k \qquad \text{Form of quadratic function}$$
$$f(x) = (x + 3)^2 + 1 \qquad \text{Given function}$$
$$f(x) = 1(x - (-3))^2 + 1$$

$a = 1 \qquad h = -3 \qquad k = 1$

Step 1. Determine how the parabola opens. Note that a, the coefficient of x^2, is 1. Thus, $a > 0$; this positive value tells us that the parabola opens upward.

Step 2. Find the vertex. The vertex of the parabola is at (h, k). Because $h = -3$ and $k = 1$, the parabola has its vertex at $(-3, 1)$. See **Figure 8.9**.

Step 3. Find the x-intercepts by solving $f(x) = 0$. Replace $f(x)$ with 0 in $f(x) = (x + 3)^2 + 1$. Because the vertex is at $(-3, 1)$, which lies above the x-axis, and the parabola opens upward, it appears that this parabola has no x-intercepts. We can verify this observation algebraically.

$$0 = (x + 3)^2 + 1 \qquad \text{Find possible } x\text{-intercepts, setting } f(x) \text{ equal to O.}$$

$$-1 = (x + 3)^2 \qquad \text{Solve for } x. \text{ Subtract 1 from both sides.}$$

$$x + 3 = \sqrt{-1} \quad \text{or} \quad x + 3 = -\sqrt{-1} \qquad \text{Apply the square root property.}$$

$$x + 3 = i \qquad\qquad x + 3 = -i \qquad \sqrt{-1} = i$$

$$x = -3 + i \qquad\qquad x = -3 - i \qquad \text{The solutions are } -3 \pm i.$$

Because this equation has no real solutions, the parabola has no x-intercepts.

Step 4. Find the y-intercept by computing $f(0)$. Replace x with 0 in $f(x) = (x + 3)^2 + 1$.

$$f(0) = (0 + 3)^2 + 1 = 3^2 + 1 = 9 + 1 = 10$$

The y-intercept is 10. The parabola passes through $(0, 10)$.

Axis of symmetry: $x = -3$

y-intercept: 10

Vertex: (-3, 1)

Figure 8.9 The graph of $f(x) = (x + 3)^2 + 1$

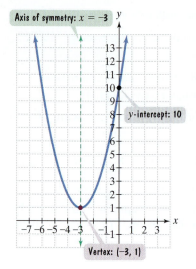

Axis of symmetry: $x = -3$

y-intercept: 10

Vertex: $(-3, 1)$

Figure 8.9 (repeated)

Step 5. Graph the parabola. With a vertex at $(-3, 1)$, no x-intercepts, and a y-intercept at 10, the graph of f is shown in **Figure 8.9**. The axis of symmetry is the vertical line whose equation is $x = -3$. ∎

Great Question!

You graphed the parabola in Figure 8.9 using only two points. Is there a way I can find some additional points to feel more secure about the graph?

Yes. You can find an additional point or two by choosing a value of x between the vertex and the y-intercept. For example, let $x = -1$. Replace x with -1 in $f(x) = (x + 3)^2 + 1$.

$$f(-1) = (-1 + 3)^2 + 1 = 2^2 + 1 = 4 + 1 = 5$$

The parabola passes through $(-1, 5)$.

More points? You can use the axis of symmetry and mirror both $(-1, 5)$ and $(0, 10)$. Mirroring $(-1, 5)$ gives the point $(-5, 5)$. Mirroring $(0, 10)$ gives the point $(-6, 10)$. Take a moment to identify these points in **Figure 8.9**.

✓ **CHECK POINT 2**   Graph the quadratic function $f(x) = (x - 2)^2 + 1$.

---

3 Graph parabolas in the form $f(x) = ax^2 + bx + c$. ▶

Graphing Quadratic Functions in the Form $f(x) = ax^2 + bx + c$

Quadratic functions are frequently expressed in the form $f(x) = ax^2 + bx + c$. How can we identify the vertex of a parabola whose equation is in this form? Completing the square provides the answer to this question.

$f(x) = ax^2 + bx + c$

$= a\left(x^2 + \dfrac{b}{a}x\right) + c$ Factor out a from $ax^2 + bx$.

$= a\left(x^2 + \dfrac{b}{a}x + \dfrac{b^2}{4a^2}\right) + c - a\left(\dfrac{b^2}{4a^2}\right)$

Complete the square by adding the square of half the coefficient of x.

By completing the square, we added $a \cdot \dfrac{b^2}{4a^2}$. To avoid changing the function's equation, we must subtract this term.

$= a\left(x + \dfrac{b}{2a}\right)^2 + c - \dfrac{b^2}{4a}$ Write the trinomial as the square of a binomial and simplify the constant term.

Now let's compare the form of this equation with a quadratic function in the form $f(x) = a(x - h)^2 + k$.

The form we know how to graph

$$f(x) = a(x - h)^2 + k$$

$h = -\dfrac{b}{2a}$ $k = c - \dfrac{b^2}{4a}$

Equation under discussion

$$f(x) = a\left(x - \left(-\dfrac{b}{2a}\right)\right)^2 + c - \dfrac{b^2}{4a}$$

The important part of this observation is that h, the x-coordinate of the vertex, is $-\dfrac{b}{2a}$. The y-coordinate can be found by evaluating the function at $-\dfrac{b}{2a}$.

The Vertex of a Parabola Whose Equation Is $f(x) = ax^2 + bx + c$

Consider the parabola defined by the quadratic function $f(x) = ax^2 + bx + c$. The parabola's vertex is $\left(-\dfrac{b}{2a}, f\left(-\dfrac{b}{2a}\right)\right)$. The x-coordinate is $-\dfrac{b}{2a}$. The y-coordinate is found by substituting the x-coordinate into the parabola's equation and evaluating the function at this value of x.

EXAMPLE 3   Finding a Parabola's Vertex

Find the vertex for the parabola whose equation is $f(x) = 3x^2 + 12x + 8$.

Solution   We know that the x-coordinate of the vertex is $x = -\dfrac{b}{2a}$. Let's identify the numbers a, b, and c in the given equation, which is in the form $f(x) = ax^2 + bx + c$.

$$f(x) = 3x^2 + 12x + 8$$

$$a = 3 \qquad b = 12 \qquad c = 8$$

Substitute the values of a and b into the equation for the x-coordinate:

$$x = -\frac{b}{2a} = -\frac{12}{2 \cdot 3} = -\frac{12}{6} = -2.$$

The x-coordinate of the vertex is -2. We substitute -2 for x into the equation of the function, $f(x) = 3x^2 + 12x + 8$, to find the y-coordinate:

$$f(-2) = 3(-2)^2 + 12(-2) + 8 = 3(4) + 12(-2) + 8 = 12 - 24 + 8 = -4.$$

The vertex is $(-2, -4)$. ■

✓ **CHECK POINT 3**   Find the vertex for the parabola whose equation is $f(x) = 2x^2 + 8x - 1$.

We can apply our five-step procedure and graph parabolas in the form $f(x) = ax^2 + bx + c$.

Graphing Quadratic Functions with Equations in the Form $f(x) = ax^2 + bx + c$

To graph $f(x) = ax^2 + bx + c$,

1. Determine whether the parabola opens upward or downward. If $a > 0$, it opens upward. If $a < 0$, it opens downward.

2. Determine the vertex of the parabola. The vertex is $\left(-\dfrac{b}{2a}, f\left(-\dfrac{b}{2a}\right)\right)$.

3. Find any x-intercepts by solving $f(x) = 0$. The real solutions of $ax^2 + bx + c = 0$ are the x-intercepts.

4. Find the y-intercept by computing $f(0)$. Because $f(0) = c$ (the constant term in the function's equation), the y-intercept is c and the parabola passes through $(0, c)$.

5. Plot the intercepts, the vertex, and additional points as necessary. Connect these points with a smooth curve.

☑ **CHECK POINT 6**   Use function g, whose equation and graph are shown in **Figure 8.12** on the previous page, to find the maximum height, to the nearest tenth of a foot, when the shot is released at an angle of 65°.

Quadratic functions can also be modeled from verbal conditions. Once we have obtained a quadratic function, we can then use the x-coordinate of the vertex to determine its maximum or minimum value. Here is a step-by-step strategy for solving these kinds of problems:

Strategy for Solving Problems Involving Maximizing or Minimizing Quadratic Functions

1. Read the problem carefully and decide which quantity is to be maximized or minimized.
2. Use the conditions of the problem to express the quantity as a function in one variable.
3. Rewrite the function in the form $f(x) = ax^2 + bx + c$.
4. Calculate $-\dfrac{b}{2a}$. If $a > 0$, f has a minimum at $x = -\dfrac{b}{2a}$. This minimum value is $f\left(-\dfrac{b}{2a}\right)$. If $a < 0$, f has a maximum at $x = -\dfrac{b}{2a}$. This maximum value is $f\left(-\dfrac{b}{2a}\right)$.
5. Answer the question posed in the problem.

EXAMPLE 7   Minimizing a Product

Among all pairs of numbers whose difference is 10, find a pair whose product is as small as possible. What is the minimum product?

Solution

Step 1. Decide what must be maximized or minimized. We must minimize the product of two numbers. Calling the numbers x and y, and calling the product P, we must minimize

$$P = xy.$$

Step 2. Express this quantity as a function in one variable. In the formula $P = xy$, P is expressed in terms of two variables, x and y. However, because the difference of the numbers is 10, we can write

$$x - y = 10.$$

We can solve this equation for y in terms of x (or vice versa), substitute the result into $P = xy$, and obtain P as a function of one variable.

$$-y = -x + 10 \qquad \text{Subtract } x \text{ from both sides of } x - y = 10.$$

$$y = x - 10 \qquad \text{Multiply both sides of the equation by } -1 \text{ and solve for } y.$$

Now we substitute $x - 10$ for y in $P = xy$.

$$P = xy = x(x - 10).$$

Because P is now a function of x, we can write

$$P(x) = x(x - 10).$$

Step 3. Write the function in the form $f(x) = ax^2 + bx + c$. We apply the distributive property to obtain

$$P(x) = x(x - 10) = x^2 - 10x.$$

$a = 1$ $b = -10$

Using Technology

Numeric Connections

The $\boxed{\text{TABLE}}$ feature of a graphing utility can be used to verify our work in Example 7.

Enter $y_1 = x^2 - 10x$, the function for the product, when one of the numbers is x.

X	Y₁
0	0
1	-9
2	-16
3	-21
4	-24
5	-25
6	-24
7	-21
8	-16
9	-9
10	0

X=0

The product is a minimum, −25, when one of the numbers is 5.

Step 4. Calculate $-\dfrac{b}{2a}$. If $a > 0$, the function has a minimum at this value. The voice balloons show that $a = 1$ and $b = -10$.

$$x = -\frac{b}{2a} = -\frac{-10}{2(1)} = -(-5) = 5$$

This means that the product, P, of two numbers whose difference is 10 is a minimum when one of the numbers, x, is 5.

Step 5. Answer the question posed by the problem. The problem asks for the two numbers and the minimum product. We found that one of the numbers, x, is 5. Now we must find the second number, y.

$$y = x - 10 = 5 - 10 = -5.$$

The number pair whose difference is 10 and whose product is as small as possible is $5, -5$. The minimum product is $5(-5)$, or -25. ∎

$\boxed{\checkmark}$ **CHECK POINT 7**   Among all pairs of numbers whose difference is 8, find a pair whose product is as small as possible. What is the minimum product?

EXAMPLE 8   Maximizing Area

You have 100 yards of fencing to enclose a rectangular region. Find the dimensions of the rectangle that maximize the enclosed area. What is the maximum area?

Solution

Step 1. Decide what must be maximized or minimized. We must maximize area. What we do not know are the rectangle's dimensions, x and y.

Step 2. Express this quantity as a function in one variable. Because we must maximize area, we have $A = xy$. We need to transform this into a function in which A is represented by one variable. Because you have 100 yards of fencing, the perimeter of the rectangle is 100 yards. This means that

$$2x + 2y = 100.$$

We can solve this equation for y in terms of x, substitute the result into $A = xy$, and obtain A as a function in one variable. We begin by solving for y.

$$2y = 100 - 2x \qquad \text{Subtract } 2x \text{ from both sides.}$$

$$y = \frac{100 - 2x}{2} \qquad \text{Divide both sides by 2.}$$

$$y = 50 - x \qquad \text{Divide each term in the numerator by 2.}$$

7.

8.

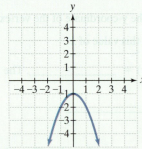

In Exercises 9–16, find the coordinates of the vertex for the parabola defined by the given quadratic function.

9. $f(x) = 2(x - 3)^2 + 1$

10. $f(x) = -3(x - 2)^2 + 12$

11. $f(x) = -2(x + 1)^2 + 5$

12. $f(x) = -2(x + 4)^2 - 8$

13. $f(x) = 2x^2 - 8x + 3$

14. $f(x) = 3x^2 - 12x + 1$

15. $f(x) = -x^2 - 2x + 8$

16. $f(x) = -2x^2 + 8x - 1$

In Exercises 17–38, use the vertex and intercepts to sketch the graph of each quadratic function. Use the graph to identify the function's range.

17. $f(x) = (x - 4)^2 - 1$

18. $f(x) = (x - 1)^2 - 2$

19. $f(x) = (x - 1)^2 + 2$

20. $f(x) = (x - 3)^2 + 2$

21. $y - 1 = (x - 3)^2$

22. $y - 3 = (x - 1)^2$

23. $f(x) = 2(x + 2)^2 - 1$

24. $f(x) = \dfrac{5}{4} - \left(x - \dfrac{1}{2}\right)^2$

25. $f(x) = 4 - (x - 1)^2$

26. $f(x) = 1 - (x - 3)^2$

27. $f(x) = x^2 - 2x - 3$

28. $f(x) = x^2 - 2x - 15$

29. $f(x) = x^2 + 3x - 10$

30. $f(x) = 2x^2 - 7x - 4$

31. $f(x) = 2x - x^2 + 3$

32. $f(x) = 5 - 4x - x^2$

33. $f(x) = x^2 + 6x + 3$

34. $f(x) = x^2 + 4x - 1$

35. $f(x) = 2x^2 + 4x - 3$

36. $f(x) = 3x^2 - 2x - 4$

37. $f(x) = 2x - x^2 - 2$

38. $f(x) = 6 - 4x + x^2$

In Exercises 39–44, an equation of a quadratic function is given.

 a. *Determine, without graphing, whether the function has a minimum value or a maximum value.*

 b. *Find the minimum or maximum value and determine where it occurs.*

 c. *Identify the function's domain and its range.*

39. $f(x) = 3x^2 - 12x - 1$

40. $f(x) = 2x^2 - 8x - 3$

41. $f(x) = -4x^2 + 8x - 3$

42. $f(x) = -2x^2 - 12x + 3$

43. $f(x) = 5x^2 - 5x$

44. $f(x) = 6x^2 - 6x$

Practice PLUS

In Exercises 45–48, give the domain and the range of each quadratic function whose graph is described.

45. The vertex is $(-1, -2)$ and the parabola opens up.

46. The vertex is $(-3, -4)$ and the parabola opens down.

47. Maximum $= -6$ at $x = 10$

48. Minimum $= 18$ at $x = -6$

In Exercises 49–52, write an equation of the parabola that has the same shape as the graph of $f(x) = 2x^2$, but with the given point as the vertex.

49. $(5, 3)$

50. $(7, 4)$

51. $(-10, -5)$

52. $(-8, -6)$

In Exercises 53–56, write an equation of the parabola that has the same shape as the graph of $f(x) = 3x^2$ or $g(x) = -3x^2$, but with the given maximum or minimum.

53. Maximum $= 4$ at $x = -2$

54. Maximum $= -7$ at $x = 5$

55. Minimum $= 0$ at $x = 11$

56. Minimum $= 0$ at $x = 9$

Application Exercises

57. A person standing close to the edge on the top of a 160-foot building throws a baseball vertically upward. The quadratic function

$$s(t) = -16t^2 + 64t + 160$$

models the ball's height above the ground, $s(t)$, in feet, t seconds after it was thrown.

a. After how many seconds does the ball reach its maximum height? What is the maximum height?

b. How many seconds does it take until the ball finally hits the ground? Round to the nearest tenth of a second.

c. Find $s(0)$ and describe what this means.

d. Use your results from parts (a) through (c) to graph the quadratic function. Begin the graph with $t = 0$ and end with the value of t for which the ball hits the ground.

58. A person standing close to the edge on the top of a 200-foot building throws a baseball vertically upward. The quadratic function

$$s(t) = -16t^2 + 64t + 200$$

models the ball's height above the ground, $s(t)$, in feet, t seconds after it was thrown.

a. After how many seconds does the ball reach its maximum height? What is the maximum height?

b. How many seconds does it take until the ball finally hits the ground? Round to the nearest tenth of a second.

c. Find $s(0)$ and describe what this means.

d. Use your results from parts (a) through (c) to graph the quadratic function. Begin the graph with $t = 0$ and end with the value of t for which the ball hits the ground.

59. Among all pairs of numbers whose sum is 16, find a pair whose product is as large as possible. What is the maximum product?

60. Among all pairs of numbers whose sum is 20, find a pair whose product is as large as possible. What is the maximum product?

61. Among all pairs of numbers whose difference is 16, find a pair whose product is as small as possible. What is the minimum product?

62. Among all pairs of numbers whose difference is 24, find a pair whose product is as small as possible. What is the minimum product?

63. You have 600 feet of fencing to enclose a rectangular plot that borders on a river. If you do not fence the side along the river, find the length and width of the plot that will maximize the area. What is the largest area that can be enclosed?

64. You have 200 feet of fencing to enclose a rectangular plot that borders on a river. If you do not fence the side along the river, find the length and width of the plot that will maximize the area. What is the largest area that can be enclosed?

65. You have 50 yards of fencing to enclose a rectangular region. Find the dimensions of the rectangle that maximize the enclosed area. What is the maximum area?

66. You have 80 yards of fencing to enclose a rectangular region. Find the dimensions of the rectangle that maximize the enclosed area. What is the maximum area?

67. A rain gutter is made from sheets of aluminum that are 20 inches wide by turning up the edges to form right angles. Determine the depth of the gutter that will maximize its cross-sectional area and allow the greatest amount of water to flow. What is the maximum cross-sectional area?

68. A rain gutter is made from sheets of aluminum that are 12 inches wide by turning up the edges to form right angles. Determine the depth of the gutter that will maximize its cross-sectional area and allow the greatest amount of water to flow. What is the maximum cross-sectional area?

Using Technology

Graphic Connections

The solution set for

$$4x^2 \leq 1 - 2x$$

or, equivalently,

$$4x^2 + 2x - 1 \leq 0$$

can be verified with a graphing utility. The graph of $f(x) = 4x^2 + 2x - 1$ was obtained using a $[-2, 2, 1]$ by $[-10, 10, 1]$ viewing rectangle. The graph lies on or below the x-axis, representing $\leq$, for all x in

$$\left[\frac{-1 - \sqrt{5}}{4}, \frac{-1 + \sqrt{5}}{4}\right]$$

$$\approx [-0.8, 0.3].$$

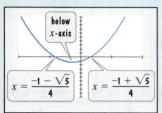

$[-2, 2, 1]$ by $[-10, 10, 1]$

Step 5. Write the solution set, selecting the interval or intervals that satisfy the given inequality. We are interested in solving $f(x) \leq 0$, where $f(x) = 4x^2 + 2x - 1$. Based on our work in step 4, we see that $f(x) < 0$ for all x in $\left(\frac{-1 - \sqrt{5}}{4}, \frac{-1 + \sqrt{5}}{4}\right)$. However, because the inequality involves $\leq$ (less than or *equal to*), we must also include the solutions of $4x^2 + 2x - 1 = 0$, namely $\frac{-1 - \sqrt{5}}{4}$ and $\frac{-1 + \sqrt{5}}{4}$, in the solution set. Thus, the solution set of the given inequality $4x^2 \leq 1 - 2x$, or, equivalently, $4x^2 + 2x - 1 \leq 0$, is

$$\left[\frac{-1 - \sqrt{5}}{4}, \frac{-1 + \sqrt{5}}{4}\right].$$

The graph of the solution set on a number line is shown as follows:

$$\xleftarrow{\quad\quad} \underset{-5 \;-4 \;-3 \;-2 \;-1 \quad 0 \quad 1 \quad 2 \quad 3 \quad 4 \quad 5}{\longrightarrow} x$$ ■

✓ **CHECK POINT 2** Solve and graph the solution set on a real number line: $2x^2 \leq -6x - 1$.

EXAMPLE 3 Solving a Polynomial Inequality

Solve and graph the solution set on a real number line: $x^3 + x^2 \leq 4x + 4$.

Solution

Step 1. Express the inequality in the form $f(x) \leq 0$ or $f(x) \geq 0$. We begin by rewriting the inequality so that 0 is on the right side.

$$\begin{array}{ll}
x^3 + x^2 \leq 4x + 4 & \text{This is the given inequality.} \\
x^3 + x^2 - 4x - 4 \leq 4x + 4 - 4x - 4 & \text{Subtract } 4x + 4 \text{ from both sides.} \\
x^3 + x^2 - 4x - 4 \leq 0 & \text{Simplify.}
\end{array}$$

This inequality is equivalent to the one we wish to solve. It is in the form $f(x) \leq 0$, where $f(x) = x^3 + x^2 - 4x - 4$.

Step 2. Solve the equation $f(x) = 0$. We find the x-intercepts of $f(x) = x^3 + x^2 - 4x - 4$ by solving the equation $x^3 + x^2 - 4x - 4 = 0$.

$$\begin{array}{ll}
x^3 + x^2 - 4x - 4 = 0 & \text{This polynomial equation is of degree 3.} \\
x^2(x + 1) - 4(x + 1) = 0 & \text{Factor } x^2 \text{ from the first two terms and} \\
& -4 \text{ from the last two terms.} \\
(x + 1)(x^2 - 4) = 0 & \text{A common factor of } x + 1 \text{ is factored from} \\
& \text{the expression.} \\
(x + 1)(x + 2)(x - 2) = 0 & \text{Factor completely.} \\
x + 1 = 0 \quad \text{or} \quad x + 2 = 0 \quad \text{or} \quad x - 2 = 0 & \text{Set each factor equal to O.} \\
x = -1 \qquad\qquad x = -2 \qquad\qquad x = 2 & \text{Solve for } x.
\end{array}$$

The x-intercepts of f are $-2, -1$, and 2. We will use these x-intercepts as boundary points on a number line.

Step 3. Locate the boundary points on a number line and separate the line into intervals. The number line with the boundary points is shown as follows:

$$\xleftarrow{\quad\quad} \underset{-5 \;-4 \;-3 \;-2 \;-1 \quad 0 \quad 1 \quad 2 \quad 3 \quad 4 \quad 5}{\longrightarrow} x$$

The boundary points divide the number line into four intervals:

$$(-\infty, -2) \quad (-2, -1) \quad (-1, 2) \quad (2, \infty).$$

Step 4. Choose one test value within each interval and evaluate f at that number.

Interval	Test Value	Substitute into $f(x) = x^3 + x^2 - 4x - 4$	Conclusion
$(-\infty, -2)$	-3	$f(-3) = (-3)^3 + (-3)^2 - 4(-3) - 4$ $= -10$, negative	$f(x) < 0$ for all x in $(-\infty, -2)$.
$(-2, -1)$	-1.5	$f(-1.5) = (-1.5)^3 + (-1.5)^2 - 4(-1.5) - 4$ $= 0.875$, positive	$f(x) > 0$ for all x in $(-2, -1)$.
$(-1, 2)$	0	$f(0) = 0^3 + 0^2 - 4 \cdot 0 - 4$ $= -4$, negative	$f(x) < 0$ for all x in $(-1, 2)$.
$(2, \infty)$	3	$f(3) = 3^3 + 3^2 - 4 \cdot 3 - 4$ $= 20$, positive	$f(x) > 0$ for all x in $(2, \infty)$.

Using Technology

Graphic Connections

The solution set for

$$x^3 + x^2 \le 4x + 4$$

or, equivalently,

$$x^3 + x^2 - 4x - 4 \le 0$$

can be verified with a graphing utility. The graph of $f(x) = x^3 + x^2 - 4x - 4$ lies on or below the x-axis, representing $\le$, for all x in $(-\infty, -2]$ or $[-1, 2]$.

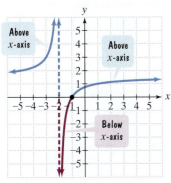

$[-4, 4, 1]$ by $[-7, 3, 1]$

2 Solve rational inequalities.

Figure 8.17 The graph of $f(x) = \dfrac{3x + 3}{2x + 4}$

Step 5. Write the solution set, selecting the interval or intervals that satisfy the given inequality. We are interested in solving $f(x) \le 0$, where $f(x) = x^3 + x^2 - 4x - 4$. Based on our work in step 4, we see that $f(x) < 0$ for all x in $(-\infty, -2)$ or $(-1, 2)$. However, because the inequality involves $\le$ (less than or equal to), we must also include the solutions of $x^3 + x^2 - 4x - 4 = 0$, namely -2, -1, and 2, in the solution set. Thus, the solution set of the given inequality, $x^3 + x^2 \le 4x + 4$, or, equivalently, $x^3 + x^2 - 4x - 4 \le 0$, is

$$(-\infty, -2] \cup [-1, 2].$$

The graph of the solution set on a number line is shown as follows:

☑ **CHECK POINT 3** Solve and graph the solution set on a real number line: $x^3 + 3x^2 \le x + 3$.

Solving Rational Inequalities

A **rational inequality** is any inequality that can be put in one of the forms

$$f(x) < 0, \quad f(x) > 0, \quad f(x) \le 0, \quad \text{or} \quad f(x) \ge 0,$$

where f is a rational function. An example of a rational inequality is

$$\frac{3x + 3}{2x + 4} > 0.$$

This inequality is in the form $f(x) > 0$, where f is the rational function given by

$$f(x) = \frac{3x + 3}{2x + 4}.$$

The graph of f is shown in **Figure 8.17**.

We can find the x-intercept of f by setting the numerator equal to 0:

$$3x + 3 = 0$$
$$3x = -3$$
$$x = -1. \quad \begin{array}{l} f \text{ has an } x\text{-intercept} \\ \text{at } -1 \text{ and passes} \\ \text{through } (-1, 0). \end{array}$$

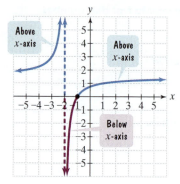

Figure 8.17 The graph of $f(x) = \dfrac{3x + 3}{2x + 4}$ (repeated)

We can determine where f is undefined by setting the denominator of $f(x) = \dfrac{3x + 3}{2x + 4}$ equal to 0:

$$2x + 4 = 0$$
$$2x = -4$$
$$x = -2.$$

> f is undefined at **-2**. **Figure 8.17** shows that the function's vertical asymptote is $x = $ **-2**.

By setting both the numerator and the denominator of f equal to 0, we obtained the solutions -2 and -1. These numbers separate the x-axis into three intervals: $(-\infty, -2), (-2, -1)$, and $(-1, \infty)$. On each interval, the graph of f is either above the x-axis $[f(x) > 0]$ or below the x-axis $[f(x) < 0]$.

Examine the graph in **Figure 8.17** carefully. Can you see that it is above the x-axis for all x in $(-\infty, -2)$ or $(-1, \infty)$, shown in blue? Thus, the solution set of $\dfrac{3x + 3}{2x + 4} > 0$ is $(-\infty, -2) \cup (-1, \infty)$. By contrast, the graph of f lies below the x-axis for all x in $(-2, -1)$, shown in red. Thus, the solution set of $\dfrac{3x + 3}{2x + 4} < 0$ is $(-2, -1)$.

The first step in solving a rational inequality is to bring all terms to one side, obtaining zero on the other side. Then express the rational function on the nonzero side as a single quotient. The second step is to set the numerator and the denominator of f equal to zero. The solutions of these equations serve as boundary points that separate the real number line into intervals. At this point, the procedure is the same as the one we used for solving polynomial inequalities.

EXAMPLE 4 Solving a Rational Inequality

Solve and graph the solution set: $\dfrac{x + 3}{x - 7} < 0$.

Solution

Step 1. Express the inequality so that one side is zero and the other side is a single quotient. The given inequality is already in this form. The form is $f(x) < 0$, where $f(x) = \dfrac{x + 3}{x - 7}$.

Step 2. Set the numerator and the denominator of f equal to zero. The real solutions are the boundary points.

$$x + 3 = 0 \qquad x - 7 = 0$$

> Set the numerator and denominator equal to 0. These are the values that make the quotient zero or undefined.

$$x = -3 \qquad x = 7$$

> Solve for x.

We will use these solutions as boundary points on a number line.

Step 3. Locate the boundary points on a number line and separate the line into intervals. The number line with the boundary points is shown as follows:

The boundary points divide the number line into three intervals:

$$(-\infty, -3) \quad (-3, 7) \quad (7, \infty).$$

Step 4. Choose one test value within each interval and evaluate f **at that number.**

Interval	Test Value	Substitute into $f(x) = \dfrac{x+3}{x-7}$	Conclusion
$(-\infty, -3)$	-4	$f(-4) = \dfrac{-4+3}{-4-7}$ $= \dfrac{-1}{-11} = \dfrac{1}{11}$, positive	$f(x) > 0$ for all x in $(-\infty, -3)$.
$(-3, 7)$	0	$f(0) = \dfrac{0+3}{0-7}$ $= -\dfrac{3}{7}$, negative	$f(x) < 0$ for all x in $(-3, 7)$.
$(7, \infty)$	8	$f(8) = \dfrac{8+3}{8-7}$ $= 11$, positive	$f(x) > 0$ for all x in $(7, \infty)$.

Step 5. Write the solution set, selecting the interval or intervals that satisfy the given inequality. We are interested in solving $f(x) < 0$, where $f(x) = \dfrac{x+3}{x-7}$. Based on our work in step 4, we see that $f(x) < 0$ for all x in $(-3, 7)$.

Because $f(x) < 0$ for all x in $(-3, 7)$, the solution set of the given inequality, $\dfrac{x+3}{x-7} < 0$, is $(-3, 7)$.

The graph of the solution set on a number line is shown as follows:

☑ **CHECK POINT 4** Solve and graph the solution set: $\dfrac{x-5}{x+2} < 0$.

EXAMPLE 5 Solving a Rational Inequality

Solve and graph the solution set: $\dfrac{x+1}{x+3} \geq 2$.

Solution

Step 1. Express the inequality so that one side is zero and the other side is a single quotient. We subtract 2 from both sides to obtain zero on the right.

$$\dfrac{x+1}{x+3} \geq 2 \qquad \text{This is the given inequality.}$$

$$\dfrac{x+1}{x+3} - 2 \geq 0 \qquad \text{Subtract 2 from both sides, obtaining 0 on the right.}$$

$$\dfrac{x+1}{x+3} - \dfrac{2(x+3)}{x+3} \geq 0 \qquad \text{The least common denominator is } x+3. \text{ Express 2 in terms of this denominator.}$$

$$\dfrac{x+1-2(x+3)}{x+3} \geq 0 \qquad \text{Subtract rational expressions.}$$

$$\dfrac{x+1-2x-6}{x+3} \geq 0 \qquad \text{Apply the distributive property.}$$

$$\dfrac{-x-5}{x+3} \geq 0 \qquad \text{Simplify.}$$

This inequality is equivalent to the one we wish to solve. It is in the form $f(x) \geq 0$, where $f(x) = \dfrac{-x-5}{x+3}$.

Great Question!

Can I begin solving $\dfrac{x+1}{x+3} \geq 2$ **by multiplying both sides by** $x + 3$?

No. We do not know if $x + 3$ is positive or negative. Thus, we do not know whether or not to change the sense of the inequality.

Step 2. Set the numerator and the denominator of f equal to zero. We need to solve $f(x) \geq 0$, where $f(x) = \dfrac{-x - 5}{x + 3}$. We use $f(x) = \dfrac{-x - 5}{x + 3}$. The real solutions obtained by setting the numerator and the denominator equal to zero are the boundary points.

$$-x - 5 = 0 \qquad x + 3 = 0$$ Set the numerator and denominator equal to O. These are the values that make the quotient zero or undefined.

$$x = -5 \qquad x = -3$$ Solve for x.

We will use these solutions as boundary points on a number line.

Step 3. Locate the boundary points on a number line and separate the line into intervals. The number line with the boundary points is shown as follows:

$$\begin{array}{c} \qquad -5 \qquad -3 \\ \xleftarrow{\;\;\;\;+\;\;+\;\;+\;\;\bullet\;\;+\;\;+\;\;\bullet\;\;+\;\;+\;\;+\;\;+\;\;+\;\;+\;\;+\;\;} x \\ -7 \;\; -6 \;\; -5 \;\; -4 \;\; -3 \;\; -2 \;\; -1 \;\; 0 \;\; 1 \;\; 2 \;\; 3 \end{array}$$

The boundary points divide the number line into three intervals:

$$(-\infty, -5) \quad (-5, -3) \quad (-3, \infty).$$

Step 4. Choose one test value within each interval and evaluate f at that number.

Interval	Test Value	Substitute into $f(x) = \dfrac{-x - 5}{x + 3}$	Conclusion
$(-\infty, -5)$	-6	$f(-6) = \dfrac{-(-6) - 5}{-6 + 3}$ $= -\tfrac{1}{3}$, negative	$f(x) < 0$ for all x in $(-\infty, -5)$.
$(-5, -3)$	-4	$f(-4) = \dfrac{-(-4) - 5}{-4 + 3}$ $= 1$, positive	$f(x) > 0$ for all x in $(-5, -3)$.
$(-3, \infty)$	0	$f(0) = \dfrac{-0 - 5}{0 + 3}$ $= -\tfrac{5}{3}$, negative	$f(x) < 0$ for all x in $(-3, \infty)$.

Step 5. Write the solution set, selecting the interval or intervals that satisfy the given inequality. We are interested in solving $f(x) \geq 0$, where $f(x) = \dfrac{-x - 5}{x + 3}$. Based on our work in step 4, we see that $f(x) > 0$ for all x in $(-5, -3)$. However, because the inequality involves $\geq$ (greater than or equal to), we must also include the solution of $f(x) = 0$, namely the value that we obtained when we set the numerator of f equal to zero. Thus, we must include -5 in the solution set. The solution set of the given inequality is $[-5, -3)$.

The graph of the solution set on a number line is shown as follows:

$$\begin{array}{c} \xleftarrow{\;\;\;+\;\;+\;\;[\;\;\text{———}\;\;)\;\;+\;\;+\;\;+\;\;+\;\;+\;\;+\;\;+\;\;} x \\ -7 \;\; -6 \;\; -5 \;\; -4 \;\; -3 \;\; -2 \;\; -1 \;\; 0 \;\; 1 \;\; 2 \;\; 3 \end{array} \quad \blacksquare$$

Great Question!

Which boundary points must I always exclude from the solution set of a rational inequality?

Never include values that cause a rational function's denominator to equal zero. Division by zero is undefined.

Using Technology

Graphic Connections

The solution set for

$$\frac{x + 1}{x + 3} \geq 2$$

or, equivalently,

$$\frac{-x - 5}{x + 3} \geq 0$$

can be verified with a graphing utility. The graph of

$$f(x) = \frac{-x - 5}{x + 3}, \text{ shown in blue, lies on or above the}$$

x-axis, representing $\geq$, for all x in $[-5, -3)$. We also graphed the vertical asymptote, $x = -3$, to remind us that f is undefined for $x = -3$.

$[-8, 8, 1]$ by $[-3, 3, 1]$

✓ **CHECK POINT 5** Solve and graph the solution set: $\dfrac{2x}{x+1} \geq 1$.

---

3 Solve problems modeled by polynomial or rational inequalities. ▶

Applications

Polynomial inequalities can be solved to answer questions about variables contained in polynomial functions.

EXAMPLE 6 Modeling the Position of a Free-Falling Object

A ball is thrown vertically upward from the top of the Leaning Tower of Pisa (190 feet high) with an initial velocity of 96 feet per second (**Figure 8.18**). The function

$$s(t) = -16t^2 + 96t + 190$$

models the ball's height above the ground, $s(t)$, in feet, t seconds after it was thrown. During which time period will the ball's height exceed that of the tower?

Solution Using the problem's question and the given model for the ball's height, $s(t) = -16t^2 + 96t + 190$, we obtain a polynomial inequality.

$$-16t^2 + 96t + 190 > 190$$

When will the ball's height   exceed that   of the tower?

Figure 8.18 Throwing a ball from the top of the Leaning Tower of Pisa

190 feet

$-16t^2 + 96t + 190 > 190$	This is the inequality that models the problem's question.
$-16t^2 + 96t > 0$	Subtract 190 from both sides. This inequality is in the form $f(t) > 0$, where $f(t) = -16t^2 + 96t$.
$-16t^2 + 96t = 0$	Solve the equation $f(t) = 0$.
$-16t(t - 6) = 0$	Factor.
$-16t = 0$ or $t - 6 = 0$	Set each factor equal to 0.
$t = 0$ $t = 6$	Solve for t. The boundary points are 0 and 6.

```
 ——+——+——+——●——+——+——+——+——+——●——+——+——▶ t
  -2  -1   0   1   2   3   4   5   6   7   8
```

Locate these values on a number line.

The intervals are $(-\infty, 0)$, $(0, 6)$, and $(6, \infty)$. For our purposes, the mathematical model is useful only from $t = 0$ until the ball hits the ground. (By setting $-16t^2 + 96t + 190$ equal to zero, we find $t \approx 7.57$; the ball hits the ground after approximately 7.57 seconds.) Thus, we use $(0, 6)$ and $(6, 7.57)$ for our intervals.

Interval	Test Value	Substitute into $f(t) = -16t^2 + 96t$	Conclusion
$(0, 6)$	1	$f(1) = -16 \cdot 1^2 + 96 \cdot 1$ $= 80$, positive	$f(t) > 0$ for all t in $(0, 6)$.
$(6, 7.57)$	7	$f(7) = -16 \cdot 7^2 + 96 \cdot 7$ $= -112$, negative	$f(t) < 0$ for all t in $(6, 7.57)$.

We are interested in solving $f(t) > 0$, where $f(t) = -16t^2 + 96t$. We see that $f(t) > 0$ for all t in $(0, 6)$. This means that the ball's height exceeds that of the tower between 0 and 6 seconds. ∎

Using Technology

Graphic Connections

The graphs of

$$y_1 = -16x^2 + 96x + 190$$

and

$$y_2 = 190$$

are shown in a

$$[0, 8, 1] \text{ by } [0, 360, 36]$$

viewing rectangle. The graphs show that the ball's height exceeds that of the tower between 0 and 6 seconds.

✓ **CHECK POINT 6** An object is propelled straight up from ground level with an initial velocity of 80 feet per second. Its height at time t is modeled by

$$s(t) = -16t^2 + 80t,$$

where the height, $s(t)$, is measured in feet and the time, t, is measured in seconds. In which time interval will the object be more than 64 feet above the ground?

CONCEPT AND VOCABULARY CHECK

Fill in each blank so that the resulting statement is true.

1. We solve the polynomial inequality $x^2 + 8x + 15 > 0$ by first solving the equation _____. The real solutions of this equation, -5 and -3, shown on the number line, are called _____ points.

2. The points at -5 and -3 shown above divide the number line into three intervals:

   _____, _____, _____.

3. True or false: A test value for the leftmost interval on the number line shown above could be -10. _____

4. True or false: A test value for the rightmost interval on the number line shown above could be 0. _____

5. Consider the rational inequality

$$\frac{x-1}{x+2} \geq 0.$$

Setting the numerator and the denominator of $\frac{x-1}{x+2}$ equal to zero, we obtain $x = 1$ and $x = -2$. These values are shown as points on the number line. Also shown is information about three test values.

Based on the information shown above, the solution set of $\frac{x-1}{x+2} \geq 0$ is _____.

8.5 EXERCISE SET   ▶ MyMathLab®

Practice Exercises

Solve each polynomial inequality in Exercises 1–40 and graph the solution set on a real number line.

1. $(x - 4)(x + 2) > 0$
2. $(x + 3)(x - 5) > 0$
3. $(x - 7)(x + 3) \leq 0$
4. $(x + 1)(x - 7) \leq 0$
5. $x^2 - 5x + 4 > 0$
6. $x^2 - 4x + 3 < 0$
7. $x^2 + 5x + 4 > 0$
8. $x^2 + x - 6 > 0$
9. $x^2 - 6x + 8 \leq 0$
10. $x^2 - 2x - 3 \geq 0$
11. $3x^2 + 10x - 8 \leq 0$
12. $9x^2 + 3x - 2 \geq 0$
13. $2x^2 + x < 15$
14. $6x^2 + x > 1$
15. $4x^2 + 7x < -3$
16. $3x^2 + 16x < -5$
17. $x^2 - 4x \geq 0$
18. $x^2 + 2x < 0$
19. $2x^2 + 3x > 0$
20. $3x^2 - 5x \leq 0$
21. $-x^2 + x \geq 0$
22. $-x^2 + 2x \geq 0$
23. $x^2 \leq 4x - 2$
24. $x^2 \leq 2x + 2$
25. $3x^2 > 4x + 2$
26. $3x^2 > 10x - 5$
27. $2x^2 - 5x \geq 1$
28. $5x^2 + 8x \geq 11$
29. $x^2 - 6x + 9 < 0$
30. $4x^2 - 4x + 1 \geq 0$
31. $(x - 1)(x - 2)(x - 3) \geq 0$
32. $(x + 1)(x + 2)(x + 3) \geq 0$
33. $x^3 + 2x^2 - x - 2 \geq 0$
34. $x^3 + 2x^2 - 4x - 8 \geq 0$
35. $x^3 - 3x^2 - 9x + 27 < 0$
36. $x^3 + 7x^2 - x - 7 < 0$
37. $x^3 + x^2 + 4x + 4 > 0$
38. $x^3 - x^2 + 9x - 9 > 0$
39. $x^3 \geq 9x^2$
40. $x^3 \leq 4x^2$

Solve each rational inequality in Exercises 41–56 and graph the solution set on a real number line.

41. $\dfrac{x - 4}{x + 3} > 0$
42. $\dfrac{x + 5}{x - 2} > 0$
43. $\dfrac{x + 3}{x + 4} < 0$
44. $\dfrac{x + 5}{x + 2} < 0$
45. $\dfrac{-x + 2}{x - 4} \geq 0$
46. $\dfrac{-x - 3}{x + 2} \leq 0$
47. $\dfrac{4 - 2x}{3x + 4} \leq 0$
48. $\dfrac{3x + 5}{6 - 2x} \geq 0$
49. $\dfrac{x}{x - 3} > 0$
50. $\dfrac{x + 4}{x} > 0$
51. $\dfrac{x + 1}{x + 3} < 2$
52. $\dfrac{x}{x - 1} > 2$
53. $\dfrac{x + 4}{2x - 1} \leq 3$
54. $\dfrac{1}{x - 3} < 1$
55. $\dfrac{x - 2}{x + 2} \leq 2$
56. $\dfrac{x}{x + 2} \geq 2$

In Exercises 57–60, use the given functions to find all values of x that satisfy the required inequality.

57. $f(x) = 2x^2, g(x) = 5x - 2; f(x) \geq g(x)$
58. $f(x) = 4x^2, g(x) = 9x - 2; f(x) < g(x)$
59. $f(x) = \dfrac{2x}{x + 1}, g(x) = 1; f(x) < g(x)$
60. $f(x) = \dfrac{x}{2x - 1}, g(x) = 1; f(x) \geq g(x)$

Practice PLUS

Solve each inequality in Exercises 61–66 and graph the solution set on a real number line.

61. $|x^2 + 2x - 36| > 12$
62. $|x^2 + 6x + 1| > 8$
63. $\dfrac{3}{x + 3} > \dfrac{3}{x - 2}$

64. $\dfrac{1}{x + 1} > \dfrac{2}{x - 1}$

65. $\dfrac{x^2 - x - 2}{x^2 - 4x + 3} > 0$

66. $\dfrac{x^2 - 3x + 2}{x^2 - 2x - 3} > 0$

In Exercises 67–68, use the graph of the polynomial function to solve each inequality.

$f(x) = 2x^3 + 11x^2 - 7x - 6$

$[-7, 3, 1]$ by $[-10, 70, 10]$

67. $2x^3 + 11x^2 \geq 7x + 6$

68. $2x^3 + 11x^2 < 7x + 6$

In Exercises 69–70, use the graph of the rational function to solve each inequality.

$f(x) = \dfrac{x + 1}{x^2 - 4}$

$[-4, 4, 1]$ by $[-4, 4, 1]$

69. $\dfrac{1}{4(x + 2)} \leq -\dfrac{3}{4(x - 2)}$

70. $\dfrac{1}{4(x + 2)} > -\dfrac{3}{4(x - 2)}$

Application Exercises

71. You throw a ball straight up from a rooftop 160 feet high with an initial speed of 48 feet per second. The function

$$s(t) = -16t^2 + 48t + 160$$

models the ball's height above the ground, $s(t)$, in feet, t seconds after it was thrown. During which time period will the ball's height exceed that of the rooftop?

72. Divers in Acapulco, Mexico, dive headfirst from the top of a cliff 87 feet above the Pacific Ocean. The function

$$s(t) = -16t^2 + 8t + 87$$

models a diver's height above the ocean, $s(t)$, in feet, t seconds after leaping. During which time period will the diver's height exceed that of the cliff?

The functions

$$f(x) = 0.0875x^2 - 0.4x + 66.6$$

Dry pavement

and

Wet pavement

$$g(x) = 0.0875x^2 + 1.9x + 11.6$$

model a car's stopping distance, $f(x)$ or $g(x)$, in feet, traveling at x miles per hour. Function f models stopping distance on dry pavement and function g models stopping distance on wet pavement. The graphs of these functions are shown for speeds of 30 miles per hour and greater. Notice that the figure does not specify which graph is the model for dry roads and which is the model for wet roads. Use this information to solve Exercises 73–74.

Graphs of Models for Stopping Distances

73. a. Use the given functions to find the stopping distance on dry pavement and the stopping distance on wet pavement for a car traveling at 35 miles per hour. Round to the nearest foot.

b. Based on your answers to part (a), which rectangular coordinate graph shows stopping distances on dry pavement and which shows stopping distances on wet pavement?

c. How well do your answers to part (a) model the actual stopping distances shown in **Figure 8.15** on page 638?

d. Determine speeds on dry pavement requiring stopping distances that exceed the length of one and one-half football fields, or 540 feet. Round to the nearest mile per hour. How is this shown on the appropriate graph of the models?

74. a. Use the given functions to find the stopping distance on dry pavement and the stopping distance on wet pavement for a car traveling at 55 miles per hour. Round to the nearest foot.

b. Based on your answers to part (a), which rectangular coordinate graph shows stopping distances on dry pavement and which shows stopping distances on wet pavement?

c. How well do your answers to part (a) model the actual stopping distances shown in **Figure 8.15** on page 638?

d. Determine speeds on wet pavement requiring stopping distances that exceed the length of one and one-half football fields, or 540 feet. Round to the nearest mile per hour. How is this shown on the appropriate graph of the models on page 650?

A company manufactures wheelchairs. The average cost function, $\overline{C}$, of producing x wheelchairs per month is given by

$$\overline{C}(x) = \frac{500{,}000 + 400x}{x}.$$

The graph of the rational function is shown. Use the function to solve Exercises 75–76.

Number of Wheelchairs
Produced per Month

75. Describe the company's production level so that the average cost of producing each wheelchair does not exceed $425. Use a rational inequality to solve the problem. Then explain how your solution is shown on the graph.

76. Describe the company's production level so that the average cost of producing each wheelchair does not exceed $410. Use a rational inequality to solve the problem. Then explain how your solution is shown on the graph.

77. The perimeter of a rectangle is 50 feet. Describe the possible length of a side if the area of the rectangle is not to exceed 114 square feet.

78. The perimeter of a rectangle is 180 feet. Describe the possible lengths of a side if the area of the rectangle is not to exceed 800 square feet.

Explaining the Concepts

79. What is a polynomial inequality?

80. What is a rational inequality?

81. Describe similarities and differences between the solutions of

$$(x - 2)(x + 5) \geq 0 \quad \text{and} \quad \frac{x - 2}{x + 5} \geq 0.$$

Technology Exercises

Solve each inequality in Exercises 82–87 using a graphing utility.

82. $x^2 + 3x - 10 > 0$

83. $2x^2 + 5x - 3 \leq 0$

84. $\dfrac{x - 4}{x - 1} \leq 0$

85. $\dfrac{x + 2}{x - 3} \leq 2$

86. $\dfrac{1}{x + 1} \leq \dfrac{2}{x + 4}$

87. $x^3 + 2x^2 - 5x - 6 > 0$

The graph shows stopping distances for trucks at various speeds on dry roads and on wet roads. Use this information to solve Exercises 88–89.

Source: National Highway Traffic Safety Administration

88. a. Use the statistical menu of your graphing utility and the quadratic regression program to obtain the quadratic function that models a truck's stopping distance, $f(x)$, in feet, on dry pavement traveling at x miles per hour. Round the constant term to one decimal place.

   b. Use the function from part (a) to determine speeds on dry pavement requiring stopping distances that exceed 455 feet. Round to the nearest mile per hour.

89. a. Use the statistical menu of your graphing utility and the quadratic regression program to obtain the quadratic function that models a truck's stopping distance, $f(x)$, in feet, on wet pavement traveling at x miles per hour. Round the constant term to one decimal place.

   b. Use the function from part (a) to determine speeds on wet pavement requiring stopping distances that exceed 446 feet.

Critical Thinking Exercises

Make Sense? *In Exercises 90–93, determine whether each statement makes sense or does not make sense, and explain your reasoning.*

90. When solving $f(x) > 0$, where f is a polynomial function, I only pay attention to the sign of f at each test value and not the actual function value.

91. I'm solving a polynomial inequality that has a value for which the polynomial function is undefined.

92. Because it takes me longer to come to a stop on a wet road than on a dry road, graph (a) for Exercises 73–74 is the model for stopping distances on wet pavement and graph (b) is the model for stopping distances on dry pavement.

93. I began the solution of the rational inequality $\dfrac{x+1}{x+3} \geq 2$ by setting both $x+1$ and $x+3$ equal to zero.

In Exercises 94–97, determine whether each statement is true or false. If the statement is false, make the necessary change(s) to produce a true statement.

94. The solution set of $x^2 > 25$ is $(5, \infty)$.

95. The inequality $\dfrac{x-2}{x+3} < 2$ can be solved by multiplying both sides by $x+3$, resulting in the equivalent inequality $x-2 < 2(x+3)$.

96. $(x+3)(x-1) \geq 0$ and $\dfrac{x+3}{x-1} \geq 0$ have the same solution set.

97. The inequality $\dfrac{x-2}{x+3} < 2$ can be solved by multiplying both sides by $(x+3)^2, x \neq -3$, resulting in the equivalent inequality $(x-2)(x+3) < 2(x+3)^2$.

98. Write a quadratic inequality whose solution set is $[-3, 5]$.

99. Write a rational inequality whose solution set is $(-\infty, -4) \cup [3, \infty)$.

In Exercises 100–103, use inspection to describe each inequality's solution set. Do not solve any of the inequalities.

100. $(x-2)^2 > 0$

101. $(x-2)^2 \leq 0$ **102.** $(x-2)^2 < -1$

103. $\dfrac{1}{(x-2)^2} > 0$

104. The graphing calculator screen shows the graph of $y = 4x^2 - 8x + 7$.

$[-2, 6, 1]$ by $[-2, 8, 1]$

a. Use the graph to describe the solution set for $4x^2 - 8x + 7 > 0$.

b. Use the graph to describe the solution set for $4x^2 - 8x + 7 < 0$.

c. Use an algebraic approach to verify each of your descriptions in parts (a) and (b).

105. The graphing calculator screen shows the graph of $y = \sqrt{27 - 3x^2}$. Write and solve a quadratic inequality that explains why the graph only appears for $-3 \leq x \leq 3$.

$[-5, 5, 1]$ by $[0, 6, 1]$

Review Exercises

106. Solve: $\left| \dfrac{x-5}{3} \right| < 8$. (Section 4.3, Example 4)

107. Divide:

$$\frac{2x+6}{x^2+8x+16} \div \frac{x^2-9}{x^2+3x-4}.$$

(Section 6.1, Example 7)

108. Factor completely: $x^4 - 16y^4$. (Section 5.5, Example 3)

Preview Exercises

Exercises 109–111 will help you prepare for the material covered in the first section of the next chapter. In each exercise, use point plotting to graph the function. Begin by setting up a table of coordinates, selecting integers from -3 to 3, inclusive, for x.

109. $f(x) = 2^x$

110. $f(x) = 2^{-x}$

111. $f(x) = 2^x + 1$

Achieving Success

Avoid coursus interruptus.

Now that you've almost completed intermediate algebra, don't interrupt the sequence until you have completed all your required math classes. You'll have better results if you take your math courses without a break. If you start, stop, and start again, it's easy to forget what you've learned and lose your momentum.

Chapter 8 Summary

Definitions and Concepts	**Examples**

Section 8.1 The Square Root Property and Completing the Square

The Square Root Property

If u is an algebraic expression and d is a real number, then

If $u^2 = d$, then $u = \sqrt{d}$ or $u = -\sqrt{d}$.

Equivalently,

If $u^2 = d$, then $u = \pm\sqrt{d}$.

Solve: $\quad (x - 6)^2 = 50.$

$$x - 6 = \pm\sqrt{50}$$
$$x - 6 = \pm\sqrt{25 \cdot 2}$$
$$x - 6 = \pm 5\sqrt{2}$$
$$x = 6 \pm 5\sqrt{2}$$

The solutions are $6 \pm 5\sqrt{2}$ and the solution set is $\{6 \pm 5\sqrt{2}\}$.

Completing the Square

If $x^2 + bx$ is a binomial, then by adding $\left(\dfrac{b}{2}\right)^2$, the square of half the coefficient of x, a perfect square trinomial will result. That is,

$$x^2 + bx + \left(\frac{b}{2}\right)^2 = \left(x + \frac{b}{2}\right)^2.$$

Complete the square:

$$x^2 + \frac{2}{7}x.$$

Half of $\frac{2}{7}$ is $\frac{1}{2} \cdot \frac{2}{7} = \frac{1}{7}$ and $\left(\frac{1}{7}\right)^2 = \frac{1}{49}$.

$$x^2 + \frac{2}{7}x + \frac{1}{49} = \left(x + \frac{1}{7}\right)^2$$

Solving Quadratic Equations by Completing the Square

1. If the coefficient of x^2 is not 1, divide both sides by this coefficient.
2. Isolate variable terms on one side.
3. Complete the square by adding the square of half the coefficient of x to both sides.
4. Factor the perfect square trinomial.
5. Solve by applying the square root property.

Solve by completing the square:

$$2x^2 + 16x - 6 = 0.$$

$$\frac{2x^2}{2} + \frac{16x}{2} - \frac{6}{2} = \frac{0}{2} \quad \text{Divide by 2.}$$
$$x^2 + 8x - 3 = 0 \quad \text{Simplify.}$$
$$x^2 + 8x = 3 \quad \text{Add 3.}$$

The coefficient of x is 8. Half of 8 is 4 and $4^2 = 16$. Add 16 to both sides.

$$x^2 + 8x + 16 = 3 + 16$$
$$(x + 4)^2 = 19$$
$$x + 4 = \pm\sqrt{19}$$
$$x = -4 \pm \sqrt{19}$$

Section 8.2 The Quadratic Formula

The solutions of a quadratic equation in standard form

$$ax^2 + bx + c = 0, \quad a \neq 0,$$

are given by the quadratic formula

$$x = \frac{-b \pm \sqrt{b^2 - 4ac}}{2a}.$$

Solve using the quadratic formula:

$$2x^2 = 6x - 3.$$

First write the equation in standard form by subtracting $6x$ and adding 3 on both sides.

$$2x^2 - 6x + 3 = 0$$

$a = 2$ $\quad$ $b = -6$ $\quad$ $c = 3$

$$x = \frac{-(-6) \pm \sqrt{(-6)^2 - 4 \cdot 2 \cdot 3}}{2 \cdot 2} = \frac{6 \pm \sqrt{36 - 24}}{4}$$
$$= \frac{6 \pm \sqrt{12}}{4} = \frac{6 \pm \sqrt{4 \cdot 3}}{4} = \frac{6 \pm 2\sqrt{3}}{4}$$
$$= \frac{2(3 \pm \sqrt{3})}{2 \cdot 2} = \frac{3 \pm \sqrt{3}}{2}$$

| **Definitions and Concepts** | **Examples** |

Section 8.2 The Quadratic Formula (continued)

The Discriminant

The discriminant, $b^2 - 4ac$, of the quadratic equation $ax^2 + bx + c = 0$ determines the number and type of solutions.

Discriminant	Solutions
Positive perfect square, with a, b, and c rational numbers	2 real rational solutions
Positive and not a perfect square	2 real irrational solutions
Zero, with a, b, and c rational numbers	1 real rational solution
Negative	2 imaginary solutions

$$2x^2 - 7x - 4 = 0$$

$a = 2$ $b = -7$ $c = -4$

$$b^2 - 4ac = (-7)^2 - 4(2)(-4)$$
$$= 49 - (-32) = 49 + 32 = 81$$

Positive perfect square

The equation has two real rational solutions.

Writing Quadratic Equations from Solutions

The zero-product principle in reverse makes it possible to write a quadratic equation from solutions:

If $A = 0$ or $B = 0$, then $AB = 0$.

Write a quadratic equation with the solution set $\left\{-2\sqrt{3}, 2\sqrt{3}\right\}$.

$$x = -2\sqrt{3} \qquad\qquad x = 2\sqrt{3}$$
$$x + 2\sqrt{3} = 0 \quad \text{or} \quad x - 2\sqrt{3} = 0$$
$$\left(x + 2\sqrt{3}\right)\left(x - 2\sqrt{3}\right) = 0$$
$$x^2 - \left(2\sqrt{3}\right)^2 = 0$$
$$x^2 - 12 = 0$$

Section 8.3 Quadratic Functions and Their Graphs

The graph of the quadratic function

$$f(x) = a(x - h)^2 + k, \quad a \neq 0,$$

is called a parabola. The vertex, or turning point, is (h, k). The graph opens upward if a is positive and downward if a is negative. The axis of symmetry is a vertical line passing through the vertex. The graph can be obtained using the vertex, x-intercepts, if any, (set $f(x)$ equal to zero and solve), and the y-intercept (set $x = 0$).

Graph: $f(x) = -(x + 3)^2 + 1$.

$$f(x) = -1(x - (-3))^2 + 1$$

$a = -1$ $h = -3$ $k = 1$

- Vertex $(h, k) = (-3, 1)$
- Opens downward because $a < 0$
- x-intercepts: Set $f(x) = 0$.

$$0 = -(x + 3)^2 + 1$$
$$(x + 3)^2 = 1$$
$$x + 3 = \pm\sqrt{1}$$
$$x + 3 = 1 \quad \text{or} \quad x + 3 = -1$$
$$x = -2 \qquad\qquad x = -4$$

- y-intercept: Set $x = 0$.

$$f(0) = -(0 + 3)^2 + 1 = -9 + 1 = -8$$

Vertex: $(-3, 1)$
$(-4, 0)$ $(-2, 0)$
$(0, -8)$
Axis of symmetry: $x = -3$

Definitions and Concepts	**Examples**

Section 8.3 Quadratic Functions and Their Graphs (continued)

A parabola whose equation is in the form

$$f(x) = ax^2 + bx + c, \quad a \neq 0,$$

has its vertex at

$$\left(-\frac{b}{2a}, f\left(-\frac{b}{2a} \right) \right).$$

The parabola is graphed as described in the left column at the bottom of the previous page. The only difference is how we determine the vertex. If $a > 0$, then f has a minimum that occurs at $x = -\dfrac{b}{2a}$. This minimum value is $f\left(-\dfrac{b}{2a} \right)$. If $a < 0$, then f has a maximum that occurs at $x = -\dfrac{b}{2a}$. This maximum value is $f\left(-\dfrac{b}{2a} \right)$.

Graph:

$$f(x) = x^2 - 6x + 5.$$

$a = 1$ $b = -6$ $c = 5$

Axis of symmetry: $x = 3$

Vertex: $(3, -4)$ $(0, 5)$ $(1, 0)$ $(5, 0)$

- Vertex:
$$x = -\frac{b}{2a} = -\frac{-6}{2 \cdot 1} = 3$$
$$f(3) = 3^2 - 6 \cdot 3 + 5 = -4$$
  Vertex is at $(3, -4)$.
- Opens upward because $a > 0$.
- x-intercepts: Set $f(x) = 0$.
$$x^2 - 6x + 5 = 0$$
$$(x - 1)(x - 5) = 0$$
$$x = 1 \quad \text{or} \quad x = 5$$
- y-intercept: Set $x = 0$.
$$f(0) = 0^2 - 6 \cdot 0 + 5 = 5$$

Section 8.4 Equations Quadratic in Form

An equation that is quadratic in form is one that can be expressed as a quadratic equation using an appropriate substitution. In these equations, the variable factor in one term is the square of the variable factor in the other variable term. Let $u =$ the variable factor that reappears squared. If at any point in the solution process both sides of an equation are raised to an even power, a check is required.

Solve:

$$x^{\frac{2}{3}} - 3x^{\frac{1}{3}} + 2 = 0.$$
$$\left(x^{\frac{1}{3}} \right)^2 - 3x^{\frac{1}{3}} + 2 = 0$$

Let $u = x^{\frac{1}{3}}$.

$$u^2 - 3u + 2 = 0$$
$$(u - 1)(u - 2) = 0$$
$$u - 1 = 0 \quad \text{or} \quad u - 2 = 0$$
$$u = 1 \qquad\qquad u = 2$$
$$x^{\frac{1}{3}} = 1 \qquad\qquad x^{\frac{1}{3}} = 2$$
$$\left(x^{\frac{1}{3}} \right)^3 = 1^3 \qquad \left(x^{\frac{1}{3}} \right)^3 = 2^3$$
$$x = 1 \qquad\qquad x = 8$$

The solutions are 1 and 8, and the solution set is $\{1, 8\}$.

Definitions and Concepts	**Examples**

Section 8.5 Polynomial and Rational Inequalities

Solving Polynomial Inequalities

1. Express the inequality in the form
$$f(x) < 0 \quad \text{or} \quad f(x) > 0,$$
where f is a polynomial function.

2. Solve the equation $f(x) = 0$. The real solutions are the boundary points.

3. Locate these boundary points on a number line, thereby dividing the number line into intervals.

4. Choose one representative number, called a test value, within each interval and evaluate f at that number.

   a. If the value of f is positive, then $f(x) > 0$ for all x in the interval.

   b. If the value of f is negative, then $f(x) < 0$ for all x in the interval.

5. Write the solution set, selecting the interval or intervals that satisfy the given inequality.

This procedure is valid if $<$ is replaced by $\leq$ and $>$ is replaced by $\geq$. In these cases, include the boundary points in the solution set.

Solve: $2x^2 + x - 6 > 0$.

The form of the inequality is $f(x) > 0$ with $f(x) = 2x^2 + x - 6$. Solve $f(x) = 0$.
$$2x^2 + x - 6 = 0$$
$$(2x - 3)(x + 2) = 0$$
$$2x - 3 = 0 \quad \text{or} \quad x + 2 = 0$$
$$x = \frac{3}{2} \qquad\qquad x = -2$$

Use $-3, 0,$ and 2 as test values.
$$f(-3) = 2(-3)^2 + (-3) - 6 = 9, \text{positive}$$
$$f(x) > 0 \text{ for all } x \text{ in } (-\infty, -2).$$
$$f(0) = 2 \cdot 0^2 + 0 - 6 = -6, \text{negative}$$
$$f(x) < 0 \text{ for all } x \text{ in } \left(-2, \frac{3}{2}\right).$$
$$f(2) = 2 \cdot 2^2 + 2 - 6 = 4, \text{positive}$$
$$f(x) > 0 \text{ for all } x \text{ in } \left(\frac{3}{2}, \infty\right).$$

The solution set is $(-\infty, -2) \cup \left(\frac{3}{2}, \infty\right)$.

Solving Rational Inequalities

1. Express the inequality in the form
$$f(x) < 0 \quad \text{or} \quad f(x) > 0,$$
where f is a rational function written as a single quotient.

2. Set the numerator and the denominator of f equal to zero. The real solutions are the boundary points.

3. Locate these boundary points on a number line, thereby dividing the number line into intervals.

4. Choose one representative number, called a test value, within each interval and evaluate f at that number.

   a. If the value of f is positive, then $f(x) > 0$ for all x in the interval.

   b. If the value of f is negative, then $f(x) < 0$ for all x in the interval.

5. Write the solution set, selecting the interval or intervals that satisfy the given inequality.

This procedure is valid if $<$ is replaced by $\leq$ and $>$ is replaced by $\geq$. In these cases, include any values that make the numerator of f zero. Always exclude any values that make the denominator zero.

Solve: $\dfrac{x}{x + 4} \geq 2$.

$$\frac{x}{x + 4} - \frac{2(x + 4)}{x + 4} \geq 0$$
$$\frac{-x - 8}{x + 4} \geq 0$$

The form of the inequality is $f(x) \geq 0$ with $f(x) = \dfrac{-x - 8}{x + 4}$.

Set the numerator and the denominator equal to zero.
$$-x - 8 = 0 \qquad x + 4 = 0$$
$$-8 = x \qquad\qquad x = -4$$

Use $-9, -7,$ and -3 as test values.
$$f(-9) = \frac{-(-9) - 8}{-9 + 4} = \frac{1}{-5}, \text{negative}$$
$$f(x) < 0 \text{ for all } x \text{ in } (-\infty, -8).$$
$$f(-7) = \frac{-(-7) - 8}{-7 + 4} = \frac{-1}{-3} = \frac{1}{3}, \text{positive}$$
$$f(x) > 0 \text{ for all } x \text{ in } (-8, -4).$$
$$f(-3) = \frac{-(-3) - 8}{-3 + 4} = \frac{-5}{1} = -5, \text{negative}$$
$$f(x) < 0 \text{ for all } x \text{ in } (-4, \infty).$$

Because of $\geq$, include -8, the value that makes the numerator zero, in the solution set.
The solution set is $[-8, -4)$.

CHAPTER 8 REVIEW EXERCISES

8.1   *In Exercises 1–5, solve each equation by the square root property. If possible, simplify radicals or rationalize denominators. Express imaginary solutions in the form a + bi.*

1. $2x^2 - 3 = 125$

2. $3x^2 - 150 = 0$

3. $3x^2 - 2 = 0$

4. $(x - 4)^2 = 18$

5. $(x + 7)^2 = -36$

In Exercises 6–7, determine the constant that should be added to the binomial so that it becomes a perfect square trinomial. Then write and factor the trinomial.

6. $x^2 + 20x$

7. $x^2 - 3x$

In Exercises 8–10, solve each quadratic equation by completing the square.

8. $x^2 - 12x + 27 = 0$

9. $x^2 - 7x - 1 = 0$

10. $2x^2 + 3x - 4 = 0$

11. In 2 years, an investment of \$2500 grows to \$2916. Use the compound interest formula
$$A = P(1 + r)^t$$
to find the annual interest rate, r.

12. The function $W(t) = 3t^2$ models the weight of a human fetus, $W(t)$, in grams, after t weeks, where $0 \le t \le 39$. After how many weeks does the fetus weigh 588 grams?

13. A building casts a shadow that is double the length of the building's height. If the distance from the end of the shadow to the top of the building is 300 meters, how high is the building? Express the answer in simplified radical form. Then find a decimal approximation to the nearest tenth of a meter.

8.2   *In Exercises 14–16, solve each equation using the quadratic formula. Simplify solutions, if possible.*

14. $x^2 = 2x + 4$

15. $x^2 - 2x + 19 = 0$

16. $2x^2 = 3 - 4x$

In Exercises 17–19, without solving the given quadratic equation, determine the number and type of solutions.

17. $x^2 - 4x + 13 = 0$

18. $9x^2 = 2 - 3x$

19. $2x^2 + 4x = 3$

In Exercises 20–26, solve each equation by the method of your choice. Simplify solutions, if possible.

20. $3x^2 - 10x - 8 = 0$

21. $(2x - 3)(x + 2) = x^2 - 2x + 4$

22. $5x^2 - x - 1 = 0$

23. $x^2 - 16 = 0$

24. $(x - 3)^2 - 8 = 0$

25. $3x^2 - x + 2 = 0$

26. $\dfrac{5}{x + 1} + \dfrac{x - 1}{4} = 2$

In Exercises 27–29, write a quadratic equation in standard form with the given solution set.

27. $\left\{ -\dfrac{1}{3}, \dfrac{3}{5} \right\}$

28. $\{-9i, 9i\}$

29. $\left\{ -4\sqrt{3}, 4\sqrt{3} \right\}$

30. The graph shows stopping distances for motorcycles at various speeds on dry roads and on wet roads.

Stopping Distances for Motorcycles at Selected Speeds

Source: National Highway Traffic Safety Administration

The functions
$$f(x) = 0.125x^2 - 0.8x + 99$$

Dry pavement

and

Wet pavement

$$g(x) = 0.125x^2 + 2.3x + 27$$

model a motorcycle's stopping distance, $f(x)$ or $g(x)$, in feet traveling at x miles per hour. Function f models stopping distance on dry pavement and function g models stopping distance on wet pavement.

a. Use function g to find the stopping distance on wet pavement for a motorcycle traveling at 35 miles per hour. Round to the nearest foot. Does your rounded answer overestimate or underestimate the stopping distance shown by the graph? By how many feet?

b. Use function f to determine a motorcycle's speed requiring a stopping distance on dry pavement of 267 feet.

31. The graphs of the functions in Exercise 30 are shown for speeds of 30 miles per hour and greater.

Graphs of Models for a Motorcycle's Stopping Distances

$g(x) = 0.125x^2 + 2.3x + 27$

$f(x) = 0.125x^2 - 0.8x + 99$

Stopping Distance (feet)

Speed (miles per hour)

a. How is your answer to Exercise 30(a) shown on the graph of g?

b. How is your answer to Exercise 30(b) shown on the graph of f?

32. A baseball is hit by a batter. The function

$$s(t) = -16t^2 + 140t + 3$$

models the ball's height above the ground, $s(t)$, in feet, t seconds after it is hit. How long will it take for the ball to strike the ground? Round to the nearest tenth of a second.

8.3   *In Exercises 33–36, use the vertex and intercepts to sketch the graph of each quadratic function.*

33. $f(x) = -(x + 1)^2 + 4$ **34.** $f(x) = (x + 4)^2 - 2$

35. $f(x) = -x^2 + 2x + 3$ **36.** $f(x) = 2x^2 - 4x - 6$

37. The function

$$f(x) = -0.02x^2 + x + 1$$

models the yearly growth of a young redwood tree, $f(x)$, in inches, with x inches of rainfall per year. How many inches of rainfall per year result in maximum tree growth? What is the maximum yearly growth?

38. A model rocket is launched upward from a platform 40 feet above the ground. The quadratic function

$$s(t) = -16t^2 + 400t + 40$$

models the rocket's height above the ground, $s(t)$, in feet, t seconds after it was launched. After how many seconds does the rocket reach its maximum height? What is the maximum height?

39. The function

$$f(x) = 104.5x^2 - 1501.5x + 6016$$

models the death rate per year per 100,000 males, $f(x)$, for U.S. men who average x hours of sleep each night. How many hours of sleep, to the nearest tenth of an hour, corresponds to the minimum death rate? What is this minimum death rate, to the nearest whole number?

40. A field bordering a straight stream is to be enclosed. The side bordering the stream is not to be fenced. If 1000 yards of fencing material is to be used, what are the dimensions of the largest rectangular field that can be fenced? What is the maximum area?

41. Among all pairs of numbers whose difference is 14, find a pair whose product is as small as possible. What is the minimum product?

8.4   *In Exercises 42–47, solve each equation by making an appropriate substitution. When necessary, check proposed solutions.*

42. $x^4 - 6x^2 + 8 = 0$

43. $x + 7\sqrt{x} - 8 = 0$

44. $(x^2 + 2x)^2 - 14(x^2 + 2x) = 15$

45. $x^{-2} + x^{-1} - 56 = 0$

46. $x^{\frac{2}{3}} - x^{\frac{1}{3}} - 12 = 0$

47. $x^{\frac{1}{2}} + 3x^{\frac{1}{4}} - 10 = 0$

8.5   *In Exercises 48–52, solve each inequality and graph the solution set on a real number line.*

48. $2x^2 + 5x - 3 < 0$

49. $2x^2 + 9x + 4 \geq 0$

50. $x^3 + 2x^2 > 3x$

51. $\dfrac{x - 6}{x + 2} > 0$ **52.** $\dfrac{x + 3}{x - 4} \leq 5$

53. A model rocket is launched from ground level. The function

$$s(t) = -16t^2 + 48t$$

models the rocket's height above the ground, $s(t)$, in feet, t seconds after it was launched. During which time period will the rocket's height exceed 32 feet?

54. The function

$$H(x) = \frac{15}{8}x^2 - 30x + 200$$

models heart rate, $H(x)$, in beats per minute, x minutes after a strenuous workout.

a. What is the heart rate immediately following the workout?

b. According to the model, during which intervals of time after the strenuous workout does the heart rate exceed 110 beats per minute? For which of these intervals has model breakdown occurred? Which interval provides a more realistic answer? How did you determine this?

Express solutions to all equations in simplified form. Rationalize denominators, if possible.

In Exercises 1–2, solve each equation by the square root property.

1. $2x^2 - 5 = 0$

2. $(x - 3)^2 = 20$

In Exercises 3–4, determine the constant that should be added to the binomial so that it becomes a perfect square trinomial. Then write and factor the trinomial.

3. $x^2 - 16x$

4. $x^2 + \dfrac{2}{5}x$

5. Solve by completing the square: $x^2 - 6x + 7 = 0$.

6. Use the measurements determined by the surveyor to find the width of the pond. Express the answer in simplified radical form.

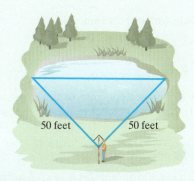

50 feet   50 feet

In Exercises 7–8, without solving the given quadratic equation, determine the number and type of solutions.

7. $3x^2 + 4x - 2 = 0$

8. $x^2 = 4x - 8$

In Exercises 9–12, solve each equation by the method of your choice.

9. $2x^2 + 9x = 5$

10. $x^2 + 8x + 5 = 0$

11. $(x + 2)^2 + 25 = 0$

12. $2x^2 - 6x + 5 = 0$

In Exercises 13–14, write a quadratic equation in standard form with the given solution set.

13. $\{-3, 7\}$

14. $\{-10i, 10i\}$

15. As gas prices surge, more Americans are cycling as a way to save money, stay fit, or both. The bar graph in the next column shows the number of bicycle-friendly U.S. communities, as designated by the League of American Bicyclists, for selected years from 2003 through 2014.

Number of U.S. Communities Designated "Bicycle Friendly" by the League of American Bicyclists

Source: League of American Bicyclists

The function
$$f(x) = 2.4x^2 + 0.7x + 29$$
models the number of bicycle-friendly communities, $f(x)$, x years after 2003.

a. Use the function to find the number of bicycle-friendly communities in 2014. Round to the nearest whole number. Does this rounded value underestimate or overestimate the number shown by the graph? By how much?

b. Use the function to determine the year in which 1206 U.S. communities will be bicycle friendly.

In Exercises 16–17, use the vertex and intercepts to sketch the graph of each quadratic function.

16. $f(x) = (x + 1)^2 + 4$ 17. $f(x) = x^2 - 2x - 3$

A baseball player hits a pop fly into the air. The function
$$s(t) = -16t^2 + 64t + 5$$
models the ball's height above the ground, $s(t)$, in feet, t seconds after it is hit. Use the function to solve Exercises 18–19.

18. When does the baseball reach its maximum height? What is that height?

19. After how many seconds does the baseball hit the ground? Round to the nearest tenth of a second.

20. The function $f(x) = -x^2 + 46x - 360$ models the daily profit, $f(x)$, in hundreds of dollars, for a company that manufactures x computers daily. How many computers should be manufactured each day to maximize profit? What is the maximum daily profit?

In Exercises 21–23, solve each equation by making an appropriate substitution. When necessary, check proposed solutions.

21. $(2x - 5)^2 + 4(2x - 5) + 3 = 0$
22. $x^4 - 13x^2 + 36 = 0$
23. $x^{\frac{2}{3}} - 9x^{\frac{1}{3}} + 8 = 0$

In Exercises 24–25, solve each inequality and graph the solution set on a real number line.

24. $x^2 - x - 12 < 0$ 25. $\dfrac{2x + 1}{x - 3} \le 3$

CUMULATIVE REVIEW EXERCISES (CHAPTERS 1–8)

In Exercises 1–13, solve each equation, inequality, or system.

1. $9(x - 1) = 1 + 3(x - 2)$

2. $\begin{cases} 3x + 4y = -7 \\ x - 2y = -9 \end{cases}$

3. $\begin{cases} x - y + 3z = -9 \\ 2x + 3y - z = 16 \\ 5x + 2y - z = 15 \end{cases}$

4. $7x + 18 \leq 9x - 2$

5. $4x - 3 < 13$ and $-3x - 4 \geq 8$

6. $2x + 4 > 8$ or $x - 7 \geq 3$

7. $|2x - 1| < 5$

8. $\left| \dfrac{2}{3}x - 4 \right| = 2$

9. $\dfrac{4}{x - 3} - \dfrac{6}{x + 3} = \dfrac{24}{x^2 - 9}$

10. $\sqrt{x + 4} - \sqrt{x - 3} = 1$

11. $2x^2 = 5 - 4x$

12. $x^{\frac{2}{3}} - 5x^{\frac{1}{3}} + 6 = 0$

13. $2x^2 + x - 6 \leq 0$

In Exercises 14–17, graph each function, equation, or inequality in a rectangular coordinate system.

14. $x - 3y = 6$

15. $f(x) = \dfrac{1}{2}x - 1$

16. $3x - 2y > -6$

17. $f(x) = -2(x - 3)^2 + 2$

In Exercises 18–28, perform the indicated operations, and simplify, if possible.

18. $4[2x - 6(x - y)]$

19. $(-5x^3y^2)(4x^4y^{-6})$

20. $(8x^2 - 9xy - 11y^2) - (7x^2 - 4xy + 5y^2)$

21. $(3x - 1)(2x + 5)$

22. $(3x^2 - 4y)^2$

23. $\dfrac{3x}{x + 5} - \dfrac{2}{x^2 + 7x + 10}$

24. $\dfrac{1 - \dfrac{9}{x^2}}{1 + \dfrac{3}{x}}$

25. $\dfrac{x^2 - 6x + 8}{3x + 9} \div \dfrac{x^2 - 4}{x + 3}$

26. $\sqrt{5xy} \cdot \sqrt{10x^2y}$

27. $4\sqrt{72} - 3\sqrt{50}$

28. $(5 + 3i)(7 - 3i)$

In Exercises 29–31, factor completely.

29. $81x^4 - 1$

30. $24x^3 - 22x^2 + 4x$

31. $x^3 + 27y^3$

In Exercises 32–34, let $f(x) = x^2 + 3x - 15$ and $g(x) = x - 2$. Find each indicated expression.

32. $(f - g)(x)$ and $(f - g)(5)$

33. $\left(\dfrac{f}{g} \right)(x)$ and the domain of $\dfrac{f}{g}$

34. $\dfrac{f(a + h) - f(a)}{h}$

35. Divide using synthetic division:

$(3x^3 - x^2 + 4x + 8) \div (x + 2)$.

36. Solve for R: $I = \dfrac{R}{R + r}$.

37. Write the slope-intercept form of the equation of the line through $(-2, 5)$ and parallel to the line whose equation is $3x + y = 9$.

38. Evaluate the determinant: $\begin{vmatrix} -2 & -4 \\ 5 & 7 \end{vmatrix}$.

39. The price of a computer is reduced by 30% to $434. What was the original price?

40. The area of a rectangle is 52 square yards. The length of the rectangle is 1 yard longer than 3 times its width. Find the rectangle's dimensions.

41. You invested $4000 in two stocks paying 12% and 14% annual interest. At the end of the year, the total interest from these investments was $508. How much was invested at each rate?

42. The current, I, in amperes, flowing in an electrical circuit varies inversely as the resistance, R, in ohms, in the circuit. When the resistance of an electric percolator is 22 ohms, it draws 5 amperes of current. How much current is needed when the resistance is 10 ohms?

Exponential and Logarithmic Functions

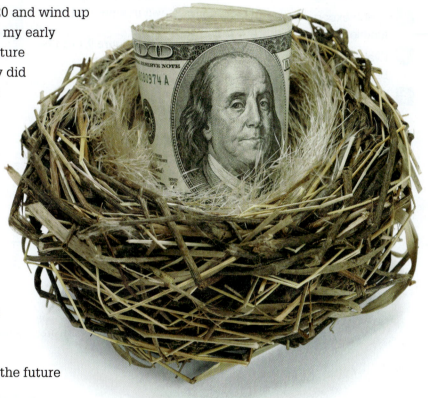

Can I put aside $25,000 when I'm 20 and wind up sitting on half a million dollars by my early fifties? Will population growth lead to a future without comfort or individual choice? Why did I feel I was walking too slowly on my visit to New York City? Are Californians at greater risk from drunk drivers than from earthquakes? What is the difference between earthquakes measuring 6 and 7 on the Richter scale? Can I live forever? And what can possibly be causing merchants at our local shopping mall to grin from ear to ear as they watch the browsers?

The functions that you will be learning about in this chapter will provide you with the mathematics for answering these questions. You will see how these remarkable functions enable us to predict the future and rediscover the past.

Here's where you'll find these applications:

- You'll be sitting on $500,000 in Example 8 of Section 9.5.
- World population growth: Section 9.6, Examples 4 and 5
- Population and walking speed: Section 9.6, Check Point 3
- Alcohol and risk of a car accident: Section 9.5, Example 7
- Earthquake intensity: Section 9.3, Example 9
- Immortality: Blitzer Bonus, page 668

We open the chapter with those grinning merchants and the sound of ka-ching!

9.1

What am I supposed to learn?

After studying this section, you should be able to:

1 Evaluate exponential functions.

2 Graph exponential functions.

3 Evaluate functions with base *e*.

4 Use compound interest formulas.

Exponential Functions

Just browsing? Take your time. Researchers know, to the dollar, the average amount the typical consumer spends per minute at the shopping mall. And the longer you stay, the more you spend. So if you say you're just browsing, that's just fine with the mall merchants. Browsing is time and, as shown in **Figure 9.1**, time is money.

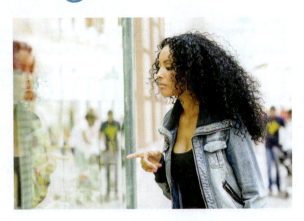

The data in **Figure 9.1** can be modeled by the function

$$f(x) = 42.2(1.56)^x,$$

where $f(x)$ is the average amount spent, in dollars, at a shopping mall after x hours. Can you see how this function is different from polynomial functions? The variable x is in the exponent. Functions whose equations contain a variable in the exponent are called **exponential functions**. Many real-life situations, including population growth, growth of epidemics, radioactive decay, and other changes that involve rapid increase or decrease, can be described using exponential functions.

Mall Browsing Time and Average Amount Spent

Figure 9.1

Source: International Council of Shopping Centers Research, 2006

Definition of an Exponential Function

The **exponential function** f **with base** b is defined by

$$f(x) = b^x \quad \text{or} \quad y = b^x,$$

where b is a positive constant other than 1 ($b > 0$ and $b \neq 1$) and x is any real number.

Here are some examples of exponential functions:

$$f(x) = 2^x \qquad g(x) = 10^x \qquad h(x) = 3^{x+1} \qquad j(x) = \left(\frac{1}{2}\right)^{x-1}.$$

Base is 2.   Base is 10.   Base is 3.   Base is $\frac{1}{2}$.

Each of these functions has a constant base and a variable exponent.

By contrast, the following functions are not exponential functions:

$$F(x) = x^2 \qquad G(x) = 1^x \qquad H(x) = (-1)^x \qquad J(x) = x^x.$$

Variable is the base and not the exponent.	The base of an exponential function must be a positive constant other than 1.	The base of an exponential function must be positive.	Variable is both the base and the exponent.

Why is $G(x) = 1^x$ not classified as an exponential function? The number 1 raised to any power is 1. Thus, the function G can be written as $G(x) = 1$, which is a constant function.

Why is $H(x) = (-1)^x$ not an exponential function? The base of an exponential function must be positive to avoid having to exclude many values of x from the domain that result in nonreal numbers in the range:

$$H(x) = (-1)^x \qquad H\left(\frac{1}{2}\right) = (-1)^{\frac{1}{2}} = \sqrt{-1} = i.$$

Not an exponential function

All values of x resulting in even roots of negative numbers produce nonreal numbers.

1 Evaluate exponential functions.

You will need a calculator to evaluate exponential expressions. Most scientific calculators have a $\boxed{y^x}$ key. Graphing calculators have a $\boxed{\wedge}$ key. To evaluate expressions of the form b^x, enter the base b, press $\boxed{y^x}$ or $\boxed{\wedge}$, enter the exponent x, and finally press $\boxed{=}$ or $\boxed{\text{ENTER}}$.

EXAMPLE 1 Evaluating an Exponential Function

The exponential function $f(x) = 42.2(1.56)^x$ models the average amount spent, $f(x)$, in dollars, at a shopping mall after x hours. What is the average amount spent, to the nearest dollar, after four hours?

Solution Because we are interested in the amount spent after four hours, substitute 4 for x and evaluate the function.

$$f(x) = 42.2(1.56)^x \qquad \text{This is the given function.}$$
$$f(4) = 42.2(1.56)^4 \qquad \text{Substitute 4 for } x.$$

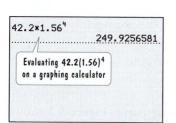

Evaluating $42.2(1.56)^4$ on a graphing calculator

Use a scientific or graphing calculator to evaluate $f(4)$. Press the following keys on your calculator to do this:

Scientific calculator: $42.2 \boxed{\times} 1.56 \boxed{y^x} 4 \boxed{=}$

Graphing calculator: $42.2 \boxed{\times} 1.56 \boxed{\wedge} 4 \boxed{\text{ENTER}}$.

The display should be approximately 249.92566.

$$f(4) = 42.2(1.56)^4 \approx 249.92566 \approx 250$$

Thus, the average amount spent after four hours at a mall is approximately $250. ∎

☑ **CHECK POINT 1** Use the exponential function in Example 1 to find the average amount spent, to the nearest dollar, after three hours at a shopping mall. Does this rounded function value underestimate or overestimate the amount shown in **Figure 9.1**? By how much?

Graphing Exponential Functions

We are familiar with expressions involving b^x where x is a rational number. For example,

$$b^{1.7} = b^{\frac{17}{10}} = \sqrt[10]{b^{17}} \quad \text{and} \quad b^{1.73} = b^{\frac{173}{100}} = \sqrt[100]{b^{173}}.$$

However, note that the definition of $f(x) = b^x$ includes all real numbers for the domain x. You may wonder what b^x means when x is an irrational number, such as $b^{\sqrt{3}}$ or b^{π}. Using closer and closer approximations for $\sqrt{3}$ $(\sqrt{3} \approx 1.73205)$, we can think of $b^{\sqrt{3}}$ as the value that has the successively closer approximations

$$b^{1.7}, b^{1.73}, b^{1.732}, b^{1.73205}, \ldots .$$

In this way, we can graph the exponential function with no holes, or points of discontinuity, at the irrational domain values.

EXAMPLE 2   Graphing an Exponential Function

Graph: $f(x) = 2^x$.

Solution   We begin by setting up a table of coordinates.

x	$f(x) = 2^x$
-3	$f(-3) = 2^{-3} = \dfrac{1}{8}$
-2	$f(-2) = 2^{-2} = \dfrac{1}{4}$
-1	$f(-1) = 2^{-1} = \dfrac{1}{2}$
0	$f(0) = 2^0 = 1$
1	$f(1) = 2^1 = 2$
2	$f(2) = 2^2 = 4$
3	$f(3) = 2^3 = 8$

> We selected integers from -3 to 3, inclusive, to include three negative numbers, 0, and three positive numbers. We also wanted to keep the resulting computations for y relatively simple.

Figure 9.2 The graph of $f(x) = 2^x$

We plot these points, connecting them with a continuous curve. **Figure 9.2** shows the graph of $f(x) = 2^x$. Observe that the graph approaches, but never touches, the negative portion of the x-axis. Thus, the x-axis, or $y = 0$, is a horizontal asymptote. The range is the set of all positive real numbers. Although we used integers for x in our table of coordinates, you can use a calculator to find additional points. For example, $f(0.3) = 2^{0.3} \approx 1.231$ and $f(0.95) = 2^{0.95} \approx 1.932$. The points $(0.3, 1.231)$ and $(0.95, 1.932)$ approximately fit the graph. ■

✓ **CHECK POINT 2**   Graph: $f(x) = 3^x$.

EXAMPLE 3 Graphing an Exponential Function

Graph: $g(x) = \left(\dfrac{1}{2}\right)^x$.

Solution We begin by setting up a table of coordinates. We compute the function values by noting that

$$g(x) = \left(\frac{1}{2}\right)^x = (2^{-1})^x = 2^{-x}.$$

x	$g(x) = \left(\dfrac{1}{2}\right)^x$ or 2^{-x}
-3	$g(-3) = 2^{-(-3)} = 2^3 = 8$
-2	$g(-2) = 2^{-(-2)} = 2^2 = 4$
-1	$g(-1) = 2^{-(-1)} = 2^1 = 2$
0	$g(0) = 2^{-0} = 1$
1	$g(1) = 2^{-1} = \dfrac{1}{2^1} = \dfrac{1}{2}$
2	$g(2) = 2^{-2} = \dfrac{1}{2^2} = \dfrac{1}{4}$
3	$g(3) = 2^{-3} = \dfrac{1}{2^3} = \dfrac{1}{8}$

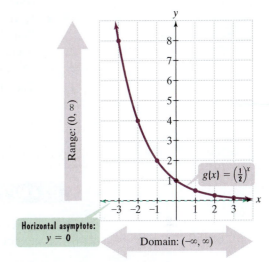

Figure 9.3 The graph of $g(x) = \left(\dfrac{1}{2}\right)^x$

We plot these points, connecting them with a continuous curve. **Figure 9.3** shows the graph of $g(x) = \left(\frac{1}{2}\right)^x$. This time the graph approaches, but never touches, the *positive* portion of the x-axis. Once again, the x-axis, or $y = 0$, is a horizontal asymptote. The range consists of all positive real numbers. ■

Do you notice a relationship between the graphs of $f(x) = 2^x$ and $g(x) = \left(\frac{1}{2}\right)^x$ in **Figures 9.2** and **9.3**? The graph of $g(x) = \left(\frac{1}{2}\right)^x$ is a mirror image, or reflection, of the graph of $f(x) = 2^x$ about the y-axis.

✓ **CHECK POINT 3** Graph: $f(x) = \left(\frac{1}{3}\right)^x$. Note that $f(x) = \left(\frac{1}{3}\right)^x = (3^{-1})^x = 3^{-x}$.

Four exponential functions have been graphed in **Figure 9.4**. Compare the black and green graphs, where $b > 1$, to those in blue and red, where $b < 1$. When $b > 1$, the value of y increases as the value of x increases. When $b < 1$, the value of y decreases as the value of x increases. Notice that all four graphs pass through $(0, 1)$. These graphs illustrate general characteristics of exponential functions, listed in the box on the next page.

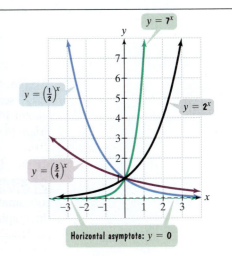

Figure 9.4 Graphs of four exponential functions

Characteristics of Exponential Functions of the Form $f(x) = b^x$

1. The domain of $f(x) = b^x$ consists of all real numbers: $(-\infty, \infty)$. The range of $f(x) = b^x$ consists of all positive real numbers: $(0, \infty)$.

2. The graphs of all exponential functions of the form $f(x) = b^x$ pass through the point $(0, 1)$ because $f(0) = b^0 = 1$ $(b \neq 0)$. The y-intercept is 1.

3. If $b > 1$, $f(x) = b^x$ has a graph that goes up to the right and is an increasing function. The greater the value of b, the steeper the increase.

4. If $0 < b < 1$, $f(x) = b^x$ has a graph that goes down to the right and is a decreasing function. The smaller the value of b, the steeper the decrease.

5. The graph of $f(x) = b^x$ approaches, but does not touch, the x-axis. The x-axis, or $y = 0$, is a horizontal asymptote.

$f(x) = b^x$
$0 < b < 1$

$f(x) = b^x$
$b > 1$

$(0, 1)$

Horizontal asymptote: $y = 0$

EXAMPLE 4 Graphing Exponential Functions

Graph $f(x) = 3^x$ and $g(x) = 3^{x+1}$ in the same rectangular coordinate system. How is the graph of g related to the graph of f?

Solution We begin by setting up a table showing some of the coordinates for f and g, selecting integers from -2 to 2 for x. Notice that $x + 1$ is the exponent for $g(x) = 3^{x+1}$.

x	$f(x) = 3^x$	$g(x) = 3^{x+1}$
-2	$f(-2) = 3^{-2} = \frac{1}{9}$	$g(-2) = 3^{-2+1} = 3^{-1} = \frac{1}{3}$
-1	$f(-1) = 3^{-1} = \frac{1}{3}$	$g(-1) = 3^{-1+1} = 3^0 = 1$
0	$f(0) = 3^0 = 1$	$g(0) = 3^{0+1} = 3^1 = 3$
1	$f(1) = 3^1 = 3$	$g(1) = 3^{1+1} = 3^2 = 9$
2	$f(2) = 3^2 = 9$	$g(2) = 3^{2+1} = 3^3 = 27$

$g(x) = 3^{x+1}$

$f(x) = 3^x$

$(-1, 1)$

$(0, 1)$

Horizontal asymptote: $y = 0$

Figure 9.5

We plot the points for each function and connect them with a smooth curve. Because of the scale on the y-axis, the points on each function corresponding to $x = 2$ are not shown. **Figure 9.5** shows the graphs of $f(x) = 3^x$ and $g(x) = 3^{x+1}$. The graph of g is the graph of f shifted one unit to the left. ∎

✓ **CHECK POINT 4**   Graph $f(x) = 3^x$ and $g(x) = 3^{x-1}$ in the same rectangular coordinate system. Select integers from -2 to 2 for x. How is the graph of g related to the graph of f?

EXAMPLE 5   Graphing Exponential Functions

Graph $f(x) = 2^x$ and $g(x) = 2^x - 3$ in the same rectangular coordinate system. How is the graph of g related to the graph of f?

Solution   We begin by setting up a table showing some of the coordinates for f and g, selecting integers from -2 to 2 for x.

x	$f(x) = 2^x$	$g(x) = 2^x - 3$
-2	$f(-2) = 2^{-2} = \frac{1}{4}$	$g(-2) = 2^{-2} - 3 = \frac{1}{4} - 3 = -2\frac{3}{4}$
-1	$f(-1) = 2^{-1} = \frac{1}{2}$	$g(-1) = 2^{-1} - 3 = \frac{1}{2} - 3 = -2\frac{1}{2}$
0	$f(0) = 2^0 = 1$	$g(0) = 2^0 - 3 = 1 - 3 = -2$
1	$f(1) = 2^1 = 2$	$g(1) = 2^1 - 3 = 2 - 3 = -1$
2	$f(2) = 2^2 = 4$	$g(2) = 2^2 - 3 = 4 - 3 = 1$

Figure 9.6

We plot the points for each function and connect them with a smooth curve. **Figure 9.6** shows the graphs of $f(x) = 2^x$ and $g(x) = 2^x - 3$. The graph of g is the graph of f shifted down three units. As a result, $y = -3$ is the horizontal asymptote for g. ∎

✓ **CHECK POINT 5**   Graph $f(x) = 2^x$ and $g(x) = 2^x + 3$ in the same rectangular coordinate system. Select integers from -2 to 2 for x. How is the graph of g related to the graph of f?

3 Evaluate functions with base e. ▶

Using Technology

Graphic Connections

As n increases, the graph of $y = \left(1 + \frac{1}{n}\right)^n$ approaches the graph of $y = e$.

$[0, 15, 1]$ by $[0, 3, 1]$

The Natural Base e

An irrational number, symbolized by the letter e, appears as the base in many applied exponential functions. The number e is defined as the value that $\left(1 + \frac{1}{n}\right)^n$ approaches as n gets larger and larger. **Table 9.1** shows values of $\left(1 + \frac{1}{n}\right)^n$ for increasingly large values of n. As n increases, the approximate value of e to nine decimal places is

$$e \approx 2.718281827.$$

The irrational number e, approximately 2.72, is called the **natural base**. The function $f(x) = e^x$ is called the **natural exponential function**.

Table 9.1

n	$\left(1 + \frac{1}{n}\right)^n$
1	2
2	2.25
5	2.48832
10	2.59374246
100	2.704813829
1000	2.716923932
10,000	2.718145927
100,000	2.718268237
1,000,000	2.718280469
1,000,000,000	2.718281827

As n takes on increasingly large values, the expression $\left(1 + \frac{1}{n}\right)^n$ approaches e.

Blitzer Bonus ◉

Exponential Growth: The Year Humans Become Immortal

In 2011, *Jeopardy!* aired a three-night match between a personable computer named Watson and the show's two most successful players. The winner: Watson. In the time it took each human contestant to respond to one trivia question, Watson was able to scan the content of one million books. It was also trained to understand the puns and twists of phrases unique to *Jeopardy!* clues.

Watson's remarkable accomplishments can be thought of as a single data point on an exponential curve that models growth in computing power. According to inventor, author, and computer scientist Ray Kurzweil (1948–), computer technology is progressing exponentially, doubling in power each year. What does this mean in terms of the accelerating pace of the graph of $y = 2^x$ that starts slowly and then rockets skyward toward infinity? According to Kurzweil, by 2023, a supercomputer will surpass the brainpower of a human. As progress accelerates exponentially and every hour brings a century's worth of scientific breakthroughs, by 2045, computers will surpass the brainpower equivalent to that of all human brains combined. Here's where it gets exponentially weird: In that year (says Kurzweil), we will be able to scan our consciousness into computers and enter a virtual existence, or swap our bodies for immortal robots. Indefinite life extension will become a reality and people will die only if they choose to.

Use a scientific or graphing calculator with an e^x key to evaluate e to various powers. For example, to find e^2, press the following keys on most calculators:

Scientific calculator: 2 e^x

Graphing calculator: e^x 2 ENTER .

The display should be approximately 7.389.

$$e^2 \approx 7.389$$

The number e lies between 2 and 3. Because $2^2 = 4$ and $3^2 = 9$, it makes sense that e^2, approximately 7.389, lies between 4 and 9.

Because $2 < e < 3$, the graph of $y = e^x$ is between the graphs of $y = 2^x$ and $y = 3^x$, shown in **Figure 9.7**.

Figure 9.7 Graphs of three exponential functions

EXAMPLE 6   Gray Wolf Population

Insatiable killer. That's the reputation the gray wolf acquired in the United States in the nineteenth and early twentieth centuries. Although the label was undeserved, an estimated two million wolves were shot, trapped, or poisoned. By 1960, the population was reduced to 800 wolves. **Figure 9.8** on the next page shows the rebounding population in two recovery areas after the gray wolf was declared an endangered species and received federal protection.

Gray Wolf Population in Two Recovery Areas for Selected Years

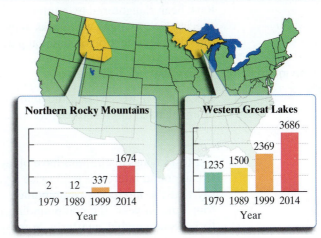

Figure 9.8

Source: U.S. Fish and Wildlife Service

The exponential function

$$f(x) = 1145e^{0.0325x}$$

models the gray wolf population of the Western Great Lakes, $f(x)$, x years after 1978.

a. According to the model, what was the gray wolf population, rounded to the nearest whole number, of the Western Great Lakes in 2014?

b. Does the model underestimate or overestimate the gray wolf population of the Western Great Lakes in 2014? By how much?

Solution

a. Because 2014 is 36 years after 1978 (2014 − 1978 = 36), we substitute 36 for x in the given function.

$$f(x) = 1145e^{0.0325x} \qquad \text{This is the given function.}$$

$$f(36) = 1145e^{0.0325(36)} \qquad \text{Substitute 36 fox x.}$$

Perform this computation on your calculator.

Scientific calculator: 1145 $\times$ $($ .0325 $\times$ 36 $)$ e^x $=$

Graphing calculator: 1145 $\times$ e^x $($ .0325 $\times$ 36 $)$ ENTER

The display should be approximately 3689.181571. Thus,

> This parenthesis is given on some calculators.

$$f(36) = 1145e^{0.0325(36)} \approx 3689.$$

According to the model, the gray wolf population of the Western Great Lakes in 2014 was approximately 3689 wolves.

b. **Figure 9.8** shows that the gray wolf population of the Western Great Lakes in 2014 was 3686 wolves. The model in part (a) indicates a population of 3689 wolves. Thus, the model overestimates the population by 3689 – 3686, or by 3 wolves. ∎

✓ **CHECK POINT 6** Use $f(x) = 1145e^{0.0325x}$, the model described in Example 6, to project the gray wolf population of the Western Great Lakes, rounded to the nearest whole number, in 2017.

Using exponential functions and projections like those in Example 6 and Check Point 6, in 2011, the U.S. Fish and Wildlife Service removed the Northern Rocky

Mountain gray wolf from the endangered species list. Since then, hundreds of wolves have been slaughtered by hunters and trappers. Without protection as an endangered species, free-roaming wolf packs in the United States will be lucky to survive anywhere outside the protection of national parks.

4 Use compound interest formulas. ▶

Compound Interest

In Chapter 8, we saw that the amount of money, A, that a principal, P, will be worth after t years at interest rate r, compounded annually, is given by the formula

$$A = P(1 + r)^t.$$

Most savings institutions have plans in which interest is paid more than once a year. If compound interest is paid twice a year, the compounding period is six months. We say that the interest is **compounded semiannually**. When compound interest is paid four times a year, the compounding period is three months and the interest is said to be **compounded quarterly**. Some plans allow for monthly compounding or daily compounding.

In general, when compound interest is paid n times a year, we say that there are **n compounding periods per year**. The formula $A = P(1 + r)^t$ can be adjusted to take into account the number of compounding periods in a year. If there are n compounding periods per year, the formula becomes

$$A = P\left(1 + \frac{r}{n}\right)^{nt}.$$

Some banks use **continuous compounding**, where the number of compounding periods increases infinitely (compounding interest every trillionth of a second, every quadrillionth of a second, etc.). As n, the number of compounding periods in a year, increases without bound, the expression $\left(1 + \frac{1}{n}\right)^n$ approaches e. As a result, the formula for continuous compounding is $A = Pe^{rt}$. Although continuous compounding sounds terrific, it yields only a fraction of a percent more interest over a year than daily compounding.

Formulas for Compound Interest

After t years, the balance, A, in an account with principal P and annual interest rate r (in decimal form) is given by the following formulas:

1. For n compounding periods per year: $A = P\left(1 + \dfrac{r}{n}\right)^{nt}$

2. For continuous compounding: $A = Pe^{rt}$.

EXAMPLE 7 Choosing Between Investments

You decide to invest $8000 for 6 years and you have a choice between two accounts. The first pays 7% per year, compounded monthly. The second pays 6.85% per year, compounded continuously. Which is the better investment?

Solution The better investment is the one with the greater balance in the account after 6 years. Let's begin with the account with monthly compounding. We use the compound interest model with $P = 8000, r = 7\% = 0.07, n = 12$ (monthly compounding means 12 compounding periods per year), and $t = 6$.

$$A = P\left(1 + \frac{r}{n}\right)^{nt} = 8000\left(1 + \frac{0.07}{12}\right)^{12 \cdot 6} \approx 12{,}160.84$$

The balance in this account after 6 years is $12,160.84.

For the second investment option, we use the model for continuous compounding with $P = 8000$, $r = 6.85\% = 0.0685$, and $t = 6$.

$$A = Pe^{rt} = 8000e^{0.0685(6)} \approx 12{,}066.60$$

The balance in this account after 6 years is $12,066.60, slightly less than the previous amount. Thus, the better investment is the 7% monthly compounding option. ■

✓ **CHECK POINT 7** A sum of $10,000 is invested at an annual rate of 8%. Find the balance in the account after 5 years subject to **a.** quarterly compounding and **b.** continuous compounding.

CONCEPT AND VOCABULARY CHECK

Fill in each blank so that the resulting statement is true.

1. The exponential function f with base b is defined by $f(x) = $ _____, $b > 0$ and $b \neq 1$. Using interval notation, the domain of this function is _____ and the range is _____.

2. The graph of the exponential function f with base b approaches, but does not touch, the _____-axis. This axis, whose equation is _____, is a/an _____ asymptote.

3. The value that $\left(1 + \dfrac{1}{n}\right)^n$ approaches as n gets larger and larger is the irrational number _____, called the _____ base. This irrational number is approximately equal to _____.

4. Consider the compound interest formula
$$A = P\left(1 + \frac{r}{n}\right)^{nt}.$$
This formula gives the balance, _____, in an account with principal _____ and annual interest rate _____, in decimal form, subject to compound interest paid _____ times per year.

5. If compound interest is paid twice a year, we say that the interest is compounded _____. If compound interest is paid four times a year, we say that the interest is compounded _____. If the number of compounding periods increases infinitely, we call this _____ compounding.

9.1 EXERCISE SET ▶ MyMathLab®

Practice Exercises

In Exercises 1–10, approximate each number using a calculator. Round your answer to three decimal places.

1. $2^{3.4}$
2. $3^{2.4}$
3. $3^{\sqrt{5}}$
4. $5^{\sqrt{3}}$
5. $4^{-1.5}$
6. $6^{-1.2}$
7. $e^{2.3}$
8. $e^{3.4}$
9. $e^{-0.95}$
10. $e^{-0.75}$

In Exercises 11–16, set up a table of coordinates for each function. Select integers from −2 to 2, inclusive, for x. Then use the table of coordinates to match the function with its graph. [The graphs are on the next page and labeled (a) through (f).]

11. $f(x) = 3^x$
12. $f(x) = 3^{x-1}$
13. $f(x) = 3^x - 1$
14. $f(x) = -3^x$

The data in the bar graph at the bottom of the previous page can be modeled by

$$f(x) = x + 31 \quad \text{and} \quad g(x) = 32.7e^{0.0217x},$$

in which f(x) and g(x) represent the percentage of high school seniors who applied to more than three colleges x years after 1980. Use these functions to solve Exercises 57–58. Where necessary, round answers to the nearest percent.

57. **a.** According to the linear model, what percentage of high school seniors applied to more than three colleges in 2013?

   b. According to the exponential model, what percentage of high school seniors applied to more than three colleges in 2013?

   c. Which function is a better model for the data shown by the bar graph in 2013?

58. **a.** According to the linear model, what percentage of high school seniors applied to more than three colleges in 2010?

   b. According to the exponential model, what percentage of high school seniors applied to more than three colleges in 2010?

   c. Which function is a better model for the data shown by the bar graph in 2010?

59. In college, we study large volumes of information— information that, unfortunately, we do not often retain for very long. The function

   $$f(x) = 80e^{-0.5x} + 20$$

   describes the percentage of information, $f(x)$, that a particular person remembers x weeks after learning the information.

   a. Substitute 0 for x and, without using a calculator, find the percentage of information remembered at the moment it is first learned.

   b. Substitute 1 for x and find the percentage of information that is remembered after 1 week.

   c. Find the percentage of information that is remembered after 4 weeks.

   d. Find the percentage of information that is remembered after one year (52 weeks).

60. In 1626, Peter Minuit persuaded the Wappinger Indians to sell him Manhattan Island for $24. If the Native Americans had put the $24 into a bank account paying 5% interest, how much would the investment have been worth in the year 2005 if interest were compounded

   a. monthly?

   b. continuously?

The function

$$f(x) = \frac{90}{1 + 270e^{-0.122x}}$$

models the percentage, f(x), of people x years old with some coronary heart disease. Use this function to solve Exercises 61–62. Round answers to the nearest tenth of a percent.

61. Evaluate $f(30)$ and describe what this means in practical terms.

62. Evaluate $f(70)$ and describe what this means in practical terms.

63. The function

   $$N(t) = \frac{30,000}{1 + 20e^{-1.5t}}$$

   describes the number of people, $N(t)$, who become ill with influenza t weeks after its initial outbreak in a town with 30,000 inhabitants. The horizontal asymptote in the graph indicates that there is a limit to the epidemic's growth.

   a. How many people became ill with the flu when the epidemic began? (When the epidemic began, $t = 0$.)

   b. How many people were ill by the end of the third week?

   c. Why can't the spread of an epidemic simply grow indefinitely? What does the horizontal asymptote shown in the graph indicate about the limiting size of the population that becomes ill?

Explaining the Concepts

64. What is an exponential function?

65. What is the natural exponential function?

66. Use a calculator to obtain an approximate value for e to as many decimal places as the display permits. Then use the calculator to evaluate $\left(1 + \dfrac{1}{x}\right)^x$ for $x = 10, 100, 1000, 10,000, 100,000,$ and $1,000,000$. Describe what happens to the expression as x increases.

67. Write an example similar to Example 7 on page 670 in which continuous compounding at a slightly lower yearly interest rate is a better investment than compounding n times per year.

68. Describe how you could use the graph of $f(x) = 2^x$ to obtain a decimal approximation for $\sqrt{2}$.

Technology Exercises

69. You have $10,000 to invest. One bank pays 5% interest compounded quarterly and the other pays 4.5% interest compounded monthly.

a. Use the formula for compound interest to write a function for the balance in each account at any time t in years.

b. Use a graphing utility to graph both functions in an appropriate viewing rectangle. Based on the graphs, which bank offers the better return on your money?

70. a. Graph $y = e^x$ and $y = 1 + x + \dfrac{x^2}{2}$ in the same viewing rectangle.

b. Graph $y = e^x$ and $y = 1 + x + \dfrac{x^2}{2} + \dfrac{x^3}{6}$ in the same viewing rectangle.

c. Graph $y = e^x$ and $y = 1 + x + \dfrac{x^2}{2} + \dfrac{x^3}{6} + \dfrac{x^4}{24}$ in the same viewing rectangle.

d. Describe what you observe in parts (a)–(c). Try generalizing this observation.

Critical Thinking Exercises

Make Sense? *In Exercises 71–74, determine whether each statement makes sense or does not make sense, and explain your reasoning.*

71. My graph of $f(x) = 3 \cdot 2^x$ shows that the horizontal asymptote for f is $x = 3$.

72. I'm using a photocopier to reduce an image over and over by 50%, so the exponential function $f(x) = \left(\frac{1}{2}\right)^x$ models the new image size, where x is the number of reductions.

73. Taxing thoughts: I'm looking at data that show the number of pages in the publication that explains the U.S. tax code for selected years from 1945 through 2013. A linear function appears to be a better choice than an exponential function for modeling the number of pages in the tax code during this period.

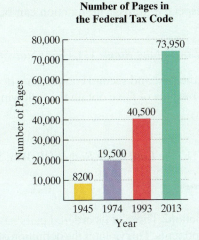

Number of Pages in the Federal Tax Code

Source: CCH Inc.

74. I use the natural base e when determining how much money I'd have in a bank account that earns compound interest subject to continuous compounding.

In Exercises 75–78, determine whether each statement is true or false. If the statement is false, make the necessary change(s) to produce a true statement.

75. As the number of compounding periods increases on a fixed investment, the amount of money in the account over a fixed interval of time will increase without bound.

76. The functions $f(x) = 3^{-x}$ and $g(x) = -3^x$ have the same graph.

77. If $f(x) = 2^x$, then $f(a + b) = f(a) + f(b)$.

78. The functions $f(x) = \left(\frac{1}{3}\right)^x$ and $g(x) = 3^{-x}$ have the same graph.

79. The graphs labeled (a)–(d) in the figure represent $y = 3^x$, $y = 5^x$, $y = \left(\frac{1}{3}\right)^x$, and $y = \left(\frac{1}{5}\right)^x$, but not necessarily in that order. Which is which? Describe the process that enables you to make this decision.

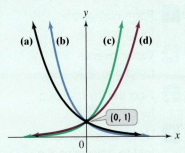

80. The hyperbolic cosine and hyperbolic sine functions are defined by

$$\cosh x = \frac{e^x + e^{-x}}{2} \quad \text{and} \quad \sinh x = \frac{e^x - e^{-x}}{2}.$$

Prove that $(\cosh x)^2 - (\sinh x)^2 = 1$.

Review Exercises

81. Solve for b: $D = \dfrac{ab}{a + b}$.

(Section 6.7, Example 1)

82. Evaluate: $\begin{vmatrix} 3 & -2 \\ 7 & -5 \end{vmatrix}$.

(Section 3.5, Example 1)

83. Solve: $x(x - 3) = 10$.

(Section 5.7, Example 2)

Preview Exercises

Exercises 84–86 will help you prepare for the material covered in the next section.

84. Let $f(x) = 3x - 4$ and $g(x) = x^2 + 6$.

a. Find $f(5)$.

b. Find $g(f(5))$.

85. Simplify: $3\left(\dfrac{x - 2}{3}\right) + 2$.

86. Solve for y: $x = 7y - 5$.

9.2

Composite and Inverse Functions

What am I supposed to learn?

After studying this section, you should be able to:

1 Form composite functions.

2 Verify inverse functions.

3 Find the inverse of a function.

4 Use the horizontal line test to determine if a function has an inverse function.

5 Use the graph of a one-to-one function to graph its inverse function.

Based on Shakespeare's *Romeo and Juliet*, the film *West Side Story* swept the 1961 Academy Awards with ten Oscars. The top four movies to win the most Oscars are shown in **Table 9.2**.

Table 9.2	Films Winning the Most Oscars	
Movie	**Year**	**Number of Academy Awards**
Ben-Hur	1960	11
Titanic	1998	11
The Lord of the Rings: The Return of the King	2003	11
West Side Story	1961	10

Source: Russell Ash, The Top 10 of Everything, 2011

We can use the information in **Table 9.2** to define a function. Let the domain of the function be the set of four movies shown in the table. Let the range be the number of Academy Awards for each of the respective films. The function can be written as follows:

f: {(*Ben-Hur*, 11), (*Titanic*, 11), (*The Lord of the Rings*, 11), (*West Side Story*, 10)}.

Now let's "undo" f by interchanging the first and second components in each of the ordered pairs. Switching the inputs and outputs of f, we obtain the following relation:

Same first component

Undoing f: {(11, *Ben-Hur*), (11, *Titanic*), (11, *The Lord of the Rings*), (10, *West Side Story*)}.

Different second components

Can you see that this relation is not a function? Three of its ordered pairs have the same first component and different second components. This violates the definition of a function.

If a function f is a set of ordered pairs, (x, y), then the changes produced by f can be "undone" by reversing the components of all the ordered pairs. The resulting relation, (y, x), may or may not be a function. In the next section, we will reverse the components of the ordered pairs of exponential functions. To understand what occurs when we interchange components, we turn to the topics of *composite* and *inverse functions*.

① Form composite functions. ▶

Composite Functions

In Chapter 2, we saw that functions could be combined using addition, subtraction, multiplication, and division. Now let's consider another way of combining two functions. To help understand this new combination, suppose that your local computer store is having a sale. The models that are on sale cost either $300 less than the regular price or 85% of the regular price. If x represents the computer's regular price, the discounts can be described with the following functions:

$$f(x) = x - 300 \qquad g(x) = 0.85x.$$

> The computer is on sale for $300 less than its regular price.

> The computer is on sale for 85% of its regular price.

At the store, you bargain with the salesperson. Eventually, she makes an offer you can't refuse. The sale price will be 85% of the regular price followed by a $300 reduction:

$$0.85x - 300.$$

> 85% of the regular price

> followed by a $300 reduction

In terms of functions f and g, this offer can be obtained by taking the output of $g(x) = 0.85x$, namely $0.85x$, and using it as the input of f:

$$f(x) = x - 300$$

> Replace x with $0.85x$, the output of $g(x) = 0.85x$.

$$f(0.85x) = 0.85x - 300.$$

Because $0.85x$ is $g(x)$, we can write this last equation as

$$f(g(x)) = 0.85x - 300.$$

We read this equation as "f of g of x is equal to $0.85x - 300$." We call $f(g(x))$ the **composition of the function f with g**, or a **composite function**. This composite function is written $f \circ g$. Thus,

$$(f \circ g)(x) = f(g(x)) = 0.85x - 300.$$

> This can be read "f of g of x" or "f composed with g of x."

Like all functions, we can evaluate $f \circ g$ for a specified value of x in the function's domain. For example, here's how to find the value of the composite function describing the offer you cannot refuse at 1400:

$$(f \circ g)(x) = 0.85x - 300$$

> Replace x with 1400.

$$(f \circ g)(1400) = 0.85(1400) - 300 = 1190 - 300 = 890.$$

Because $(f \circ g)(1400) = 890$, this means that a computer that regularly sells for $1400 is on sale for $890 subject to both discounts. We can use a partial table of coordinates for each of the discount functions, g and f, to numerically verify this result.

Computer's regular price	85% of the regular price		85% of the regular price	$300 reduction
x	$g(x) = 0.85x$		x	$f(x) = x - 300$
1200	1020		1020	720
1300	1105		1105	805
1400	1190		1190	890

Using these tables, we can find $(f \circ g)(1400)$:

$$(f \circ g)(1400) = f(g(1400)) = f(1190) = 890.$$

The table for g shows that $g(1400) = 1190$.

The table for f shows that $f(1190) = 890$.

This verifies that a computer that regularly sells for $1400 is on sale for $890 subject to both discounts.

Before you run out to buy a computer, let's generalize our discussion of the computer's double discount and define the composition of any two functions.

The Composition of Functions

The **composition of the function f with g** is denoted by $f \circ g$ and is defined by the equation

$$(f \circ g)(x) = f(g(x)).$$

The **domain of the composite function $f \circ g$** is the set of all x such that

1. x is in the domain of g and
2. $g(x)$ is in the domain of f.

EXAMPLE 1   Forming Composite Functions

Given $f(x) = 3x - 4$ and $g(x) = x^2 + 6$, find each of the following composite functions:

a. $(f \circ g)(x)$ **b.** $(g \circ f)(x)$.

Solution

a. We begin with $(f \circ g)(x)$, the composition of f with g. Because $(f \circ g)(x)$ means $f(g(x))$, we must replace each occurrence of x in the equation for f with $g(x)$.

$$f(x) = 3x - 4 \qquad \text{This is the given equation for } f.$$

Replace x with $g(x)$.

$$(f \circ g)(x) = f(g(x)) = 3g(x) - 4$$
$$= 3(x^2 + 6) - 4 \qquad \text{Because } g(x) = x^2 + 6, \text{ replace } g(x) \text{ with } x^2 + 6.$$
$$= 3x^2 + 18 - 4 \qquad \text{Use the distributive property.}$$
$$= 3x^2 + 14 \qquad \text{Simplify.}$$

Thus, $(f \circ g)(x) = 3x^2 + 14$.

b. Next, we find $(g \circ f)(x)$, the composition of g with f. Because $(g \circ f)(x)$, means $g(f(x))$, we must replace each occurrence of x in the equation for g with $f(x)$.

$$g(x) = x^2 + 6 \qquad \text{This is the given equation for } g.$$

Replace x with $f(x)$.

$$(g \circ f)(x) = g(f(x)) = (f(x))^2 + 6$$

$$= (3x - 4)^2 + 6 \qquad \text{Because } f(x) = 3x - 4, \text{ replace } f(x) \text{ with } 3x - 4.$$

$$= 9x^2 - 24x + 16 + 6 \qquad \text{Use } (A - B)^2 = A^2 - 2AB + B^2 \text{ to square } 3x - 4.$$

$$= 9x^2 - 24x + 22 \qquad \text{Simplify.}$$

Thus, $(g \circ f)(x) = 9x^2 - 24x + 22$.

Notice that $f \circ g$ is not the same function as $g \circ f$. ■

✓ **CHECK POINT 1** Given $f(x) = 5x + 6$ and $g(x) = x^2 - 1$, find each of the following composite functions:

a. $(f \circ g)(x)$ **b.** $(g \circ f)(x)$.

Inverse Functions

Here are two functions that describe situations related to the price of a computer, x:

$$f(x) = x - 300 \qquad g(x) = x + 300.$$

Function f subtracts \$300 from the computer's price and function g adds \$300 to the computer's price. Let's see what $f(g(x))$ does. Put $g(x)$ into f:

$$f(x) = x - 300 \qquad \text{This is the given equation for } f.$$

Replace x with $g(x)$.

$$f(g(x)) = g(x) - 300$$

$$= x + 300 - 300 \qquad \text{Because } g(x) = x + 300,$$

$$= x. \qquad \text{replace } g(x) \text{ with } x + 300.$$

This is the computer's original price.

By putting $g(x)$ into f and finding $f(g(x))$, we see that the computer's price, x, went through two changes: the first, an increase; the second, a decrease:

$$x + 300 - 300.$$

The final price of the computer, x, is identical to its starting price, x.

In general, if the changes made to x by function g are undone by the changes made by function f, then

$$f(g(x)) = x.$$

Assume, also, that this "undoing" takes place in the other direction:

$$g(f(x)) = x.$$

Under these conditions, we say that each function is the *inverse function* of the other. The fact that g is the inverse of f is expressed by renaming g as f^{-1}, read "f-inverse." For example, the inverse functions

$$f(x) = x - 300 \qquad g(x) = x + 300$$

are usually named as follows:

$$f(x) = x - 300 \qquad f^{-1}(x) = x + 300.$$

We can use partial tables of coordinates for f and f^{-1} to gain numerical insight into the relationship between a function and its inverse function.

Computer's regular price	$300 reduction		Price with $300 reduction	$300 price increase
x	$f(x) = x - 300$		x	$f^{-1}(x) = x + 300$
1200	900		900	1200
1300	1000		1000	1300
1400	1100		1100	1400

Ordered pairs for f:
(1200, 900), (1300, 1000), (1400, 1100)

Ordered pairs for f^{-1}:
(900, 1200), (1000, 1300), (1100, 1400)

The tables illustrate that if a function f is the set of ordered pairs (x, y), then the inverse, f^{-1}, is the set of ordered pairs (y, x). Using these tables, we can see how one function's changes to x are undone by the other function:

$$(f^{-1} \circ f)(1300) = f^{-1}(f(1300)) = f^{-1}(1000) = 1300.$$

The table for f shows that $f(1300) = 1000$.

The table for f^{-1} shows that $f^{-1}(1000) = 1300$.

The final price of the computer, $1300, is identical to its starting price, $1300.

With these ideas in mind, we present the formal definition of the inverse of a function:

Great Question!

Is the −1 in f^{-1} an exponent?

The notation f^{-1} represents the inverse function of f. The −1 is not an exponent. The notation f^{-1} does *not* mean $\dfrac{1}{f}$:

$$f^{-1} \neq \frac{1}{f}.$$

Definition of the Inverse of a Function

Let f and g be two functions such that

$$f(g(x)) = x \qquad \text{for every } x \text{ in the domain of } g$$

and

$$g(f(x)) = x \qquad \text{for every } x \text{ in the domain of } f.$$

The function g is the **inverse of the function f**, and is denoted by f^{-1} (read "f-inverse"). Thus, $f(f^{-1}(x)) = x$ and $f^{-1}(f(x)) = x$. The domain of f is equal to the range of f^{-1}, and vice versa.

2 Verify inverse functions. ▶

EXAMPLE 2 Verifying Inverse Functions

Show that each function is the inverse of the other:

$$f(x) = 5x \qquad \text{and} \qquad g(x) = \frac{x}{5}.$$

Solution To show that f and g are inverses of each other, we must show that $f(g(x)) = x$ and $g(f(x)) = x$. We begin with $f(g(x))$.

$$f(x) = 5x \qquad \text{This is the given equation for } f.$$

Replace x with $g(x)$.

$$f(g(x)) = 5g(x) = 5\left(\frac{x}{5}\right) = x$$

Because $g(x) = \dfrac{x}{5}$, replace $g(x)$ with $\dfrac{x}{5}$.

Then simplify.

Next, we find $g(f(x))$.

$$g(x) = \frac{x}{5}$$ This is the given equation for g.

Replace x with $f(x)$.

$$g(f(x)) = \frac{f(x)}{5} = \frac{5x}{5} = x$$ Because $f(x) = 5x$, replace $f(x)$ with $5x$. Then simplify.

Because g is the inverse of f (and vice versa), we can use inverse notation and write

$$f(x) = 5x \quad \text{and} \quad f^{-1}(x) = \frac{x}{5}.$$

Notice how f^{-1} undoes the change produced by f: f changes x by multiplying by 5 and f^{-1} undoes this change by dividing by 5. ∎

✓ **CHECK POINT 2** Show that each function is the inverse of the other:

$$f(x) = 7x \quad \text{and} \quad g(x) = \frac{x}{7}.$$

EXAMPLE 3 Verifying Inverse Functions

Show that each function is the inverse of the other:

$$f(x) = 3x + 2 \quad \text{and} \quad g(x) = \frac{x - 2}{3}.$$

Solution To show that f and g are inverses of each other, we must show that $f(g(x)) = x$ and $g(f(x)) = x$. We begin with $f(g(x))$.

$$f(x) = 3x + 2$$

Replace x with $g(x)$.

$$f(g(x)) = 3g(x) + 2 = 3\left(\frac{x - 2}{3}\right) + 2 = (x - 2) + 2 = x$$

$$g(x) = \frac{x - 2}{3}$$

Next, we find $g(f(x))$.

$$g(x) = \frac{x - 2}{3}$$

Replace x with $f(x)$.

$$g(f(x)) = \frac{f(x) - 2}{3} = \frac{(3x + 2) - 2}{3} = \frac{3x}{3} = x$$

$$f(x) = 3x + 2$$

Because g is the inverse of f (and vice versa), we can use inverse notation and write

$$f(x) = 3x + 2 \quad \text{and} \quad f^{-1}(x) = \frac{x - 2}{3}.$$

Notice how f^{-1} undoes the changes produced by f: f changes x by *multiplying* by 3 and *adding* 2, and f^{-1} undoes this by *subtracting* 2 and *dividing* by 3. This "undoing" process is illustrated in **Figure 9.9**. ∎

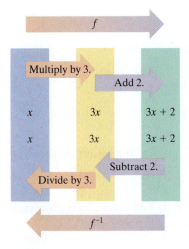

Figure 9.9 f^{-1} undoes the changes produced by f.

☑ **CHECK POINT 3**   Show that each function is the inverse of the other:

$$f(x) = 4x - 7 \quad \text{and} \quad g(x) = \frac{x + 7}{4}.$$

③ Find the inverse of a function. ▶

Finding the Inverse of a Function

The definition of the inverse of a function tells us that the domain of f is equal to the range of f^{-1}, and vice versa. This means that if the function f is the set of ordered pairs (x, y), then the inverse of f is the set of ordered pairs (y, x). If a function is defined by an equation, we can obtain the equation for f^{-1}, the inverse of f, by interchanging the role of x and y in the equation for the function f.

Finding the Inverse of a Function

The equation for the inverse of a function f can be found as follows:

1. Replace $f(x)$ with y in the equation for $f(x)$.
2. Interchange x and y.
3. Solve for y. If this equation does not define y as a function of x, the function f does not have an inverse function and this procedure ends. If this equation does define y as a function of x, the function f has an inverse function.
4. If f has an inverse function, replace y in step 3 with $f^{-1}(x)$. We can verify our result by showing that $f(f^{-1}(x)) = x$ and $f^{-1}(f(x)) = x$.

The procedure for finding a function's inverse uses a *switch-and-solve* strategy. Switch x and y, then solve for y.

EXAMPLE 4   Finding the Inverse of a Function

Find the inverse of $f(x) = 7x - 5$.

Solution

Step 1. Replace $f(x)$ with y:
$$y = 7x - 5.$$

Step 2. Interchange x and y:
$$x = 7y - 5. \quad \text{This is the inverse function.}$$

Step 3. Solve for y:
$$x + 5 = 7y \qquad \text{Add 5 to both sides.}$$
$$\frac{x + 5}{7} = y. \qquad \text{Divide both sides by 7.}$$

Step 4. Replace y with $f^{-1}(x)$:

$$f^{-1}(x) = \frac{x + 5}{7}. \quad \text{The equation is written with } f^{-1} \text{ on the left.}$$

Thus, the inverse of $f(x) = 7x - 5$ is $f^{-1}(x) = \dfrac{x + 5}{7}$. (Verify this result by showing that $f(f^{-1}(x)) = x$ and $f^{-1}(f(x)) = x$.)

The inverse function, f^{-1}, undoes the changes produced by f. f changes x by multiplying by 7 and subtracting 5. f^{-1} undoes this by adding 5 and dividing by 7. ∎

☑ **CHECK POINT 4**   Find the inverse of $f(x) = 2x + 7$.

EXAMPLE 5 Finding the Inverse of a Function

Find the inverse of $f(x) = x^3 + 1$.

Solution

Step 1. Replace $f(x)$ with y: $y = x^3 + 1$.

Step 2. Interchange x and y: $x = y^3 + 1$.

Step 3. Solve for y. We need to solve $x = y^3 + 1$ for y.

Our goal is to isolate y. Because $\sqrt[3]{y^3} = y$, we will take the cube root of both sides of the equation.

$$x - 1 = y^3 \qquad \text{Subtract 1 from both sides of } x = y^3 + 1.$$

$$\sqrt[3]{x - 1} = \sqrt[3]{y^3} \qquad \text{Take the cube root on both sides.}$$

$$\sqrt[3]{x - 1} = y \qquad \text{Simplify.}$$

Step 4. Replace y with $f^{-1}(x)$: $f^{-1}(x) = \sqrt[3]{x - 1}$.

Thus, the inverse of $f(x) = x^3 + 1$ is $f^{-1}(x) = \sqrt[3]{x - 1}$. ∎

✓ **CHECK POINT 5** Find the inverse of $f(x) = 4x^3 - 1$.

---

4 Use the horizontal line test to determine if a function has an inverse function. ▶

The Horizontal Line Test and One-to-One Functions

Let's see what happens if we try to find the inverse of the quadratic function $f(x) = x^2$.

Step 1. Replace $f(x)$ with y: $y = x^2$.

Step 2. Interchange x and y: $x = y^2$.

Step 3. Solve for y: We apply the square root property to solve $y^2 = x$ for y. We obtain

$$y = \pm\sqrt{x}.$$

The $\pm$ in this last equation shows that for certain values of x (all positive real numbers), there are two values of y. Because this equation does not represent y as a function of x, the quadratic function $f(x) = x^2$ does not have an inverse function.

We can use a few of the solutions of $y = x^2$ to illustrate numerically that this function does not have an inverse:

Four solutions of $y = x^2$. $(-2, 4), \quad (-1, 1), \quad (1, 1), \quad (2, 4),$

Interchange x and y in each ordered pair. $(4, -2), \quad (1, -1), \quad (1, 1), \quad (4, 2).$

The input 1 is associated with two outputs, −1 and 1.

The input 4 is associated with two outputs, −2 and 2.

A function provides exactly one output for each input. Thus, the ordered pairs in the bottom row do not define a function.

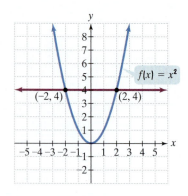

Figure 9.10 The horizontal line intersects the graph twice.

Can we look at the graph of a function and tell if it represents a function with an inverse? Yes. The graph of the quadratic function $f(x) = x^2$ is shown in **Figure 9.10**. Four units above the x-axis, a horizontal line is drawn. This line intersects the graph at two of its points, $(-2, 4)$ and $(2, 4)$. Inverse functions have ordered pairs with the coordinates reversed. We just saw what happened when we interchanged x and y. We obtained $(4, -2)$ and $(4, 2)$, and these ordered pairs do not define a function.

If any horizontal line, such as the one in **Figure 9.10**, intersects a graph at two or more points, the set of these points will not define a function when their coordinates are reversed. This suggests the **horizontal line test** for inverse functions.

The Horizontal Line Test for Inverse Functions

A function f has an inverse that is a function, f^{-1}, if there is no horizontal line that intersects the graph of the function f at more than one point.

EXAMPLE 6   Applying the Horizontal Line Test

Which of the following graphs represent functions that have inverse functions?

Discover for Yourself

How might you restrict the domain of $f(x) = x^2$, graphed in **Figure 9.10**, so that the remaining portion of the graph passes the horizontal line test?

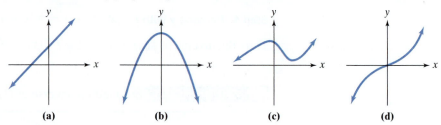

Solution   Notice that horizontal lines can be drawn in graphs **(b)** and **(c)** that intersect the graphs more than once. These graphs do not pass the horizontal line test. These are not the graphs of functions with inverse functions. By contrast, no horizontal line can be drawn in graphs **(a)** and **(d)** that intersects the graphs more than once. These graphs pass the horizontal line test. Thus, the graphs in parts **(a)** and **(d)** represent functions that have inverse functions.

✓ CHECK POINT 6   Which of the following graphs represent functions that have inverse functions?

A function passes the horizontal line test when no two different ordered pairs have the same second component. This means that if $x_1 \neq x_2$, then $f(x_1) \neq f(x_2)$. Such a function is called a **one-to-one function**. Thus, **a one-to-one function is a function in which no two different ordered pairs have the same second component. Only**

one-to-one functions have inverse functions. Any function that passes the horizontal line test is a one-to-one function. Any one-to-one function has a graph that passes the horizontal line test.

5 Use the graph of a one-to-one function to graph its inverse function. ▶

Graphs of *f* and *f*⁻¹

There is a relationship between the graph of a one-to-one function, *f*, and its inverse, *f*⁻¹. Because inverse functions have ordered pairs with the coordinates reversed, if the point (a, b) is on the graph of *f*, then the point (b, a) is on the graph of *f*⁻¹. The points (a, b) and (b, a) are symmetric with respect to the line $y = x$. Thus, **the graph of *f*⁻¹ is a reflection of the graph of *f* about the line $y = x$.** This is illustrated in **Figure 9.11**.

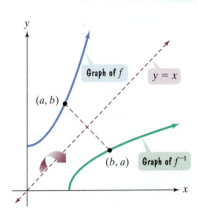

Figure 9.11 The graph of *f*⁻¹ is a reflection of the graph of *f* about $y = x$.

EXAMPLE 7 Graphing the Inverse Function

Use the graph of *f* in **Figure 9.12** to draw the graph of its inverse function.

Solution We begin by noting that no horizontal line intersects the graph of *f*, shown again in blue in **Figure 9.13**, at more than one point, so *f* does have an inverse function. Because the points $(-3, -2), (-1, 0)$, and $(4, 2)$ are on the graph of *f*, the graph of the inverse function, *f*⁻¹, has points with these ordered pairs reversed. Thus, $(-2, -3), (0, -1)$, and $(2, 4)$ are on the graph of *f*⁻¹. We can use these points to graph *f*⁻¹. The graph of *f*⁻¹ is shown in green in **Figure 9.13**. Note that the green graph of *f*⁻¹ is the reflection of the blue graph of *f* about the line $y = x$.

Figure 9.12

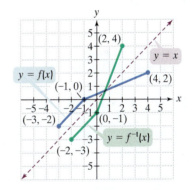

Figure 9.13 The graphs of *f* and *f*⁻¹ ∎

✅ **CHECK POINT 7** The graph of function *f* consists of two line segments, one segment from $(-2, -2)$ to $(-1, 0)$ and a second segment from $(-1, 0)$ to $(1, 2)$. Graph *f* and use the graph to draw the graph of its inverse function.

Achieving Success

Warm up your brain before starting the assigned homework. Researchers say the mind can be strengthened, just like your muscles, with regular training and rigorous practice. Think of the book's Exercise Sets as brain calisthenics. If you're feeling a bit sluggish before any of your mental workouts, try this warmup:

In the list below say the color the word is printed in, not the word itself. Once you can do this in 15 seconds without an error, the warmup is over and it's time to move on to the assigned exercises.

Blue Yellow Red Green Yellow Green Blue Red Yellow Red

CONCEPT AND VOCABULARY CHECK

Fill in each blank so that the resulting statement is true.

1. The notation $f \circ g$, called the _____ of the function f with g, is defined by $(f \circ g)(x) =$ _____.

2. I find $(f \circ g)(x)$ by replacing each occurrence of x in the equation for _____ with _____.

3. The notation $g \circ f$, called the _____ of the function g with f, is defined by $(g \circ f)(x) =$ _____.

4. I find $(g \circ f)(x)$ by replacing each occurrence of x in the equation for _____ with _____.

5. True or false: $f \circ g$ is the same function as $g \circ f$. _____

6. True or false: $f(g(x)) = f(x) \cdot g(x)$ _____

7. The notation f^{-1} means the _____ of the function f.

8. If the function g is the inverse of the function f, then $f(g(x)) =$ _____ and $g(f(x)) =$ _____.

9. A function f has an inverse that is a function if there is no _____ line that intersects the graph of f at more than one point. Such a function is called a/an _____ function.

10. The graph of f^{-1} is a reflection of the graph of f about the line whose equation is _____.

9.2 EXERCISE SET ▶ MyMathLab®

Practice Exercises

In Exercises 1–14, find

a. $(f \circ g)(x)$;   b. $(g \circ f)(x)$;   c. $(f \circ g)(2)$.

1. $f(x) = 2x$, $g(x) = x + 7$

2. $f(x) = 3x$, $g(x) = x - 5$

3. $f(x) = x + 4$, $g(x) = 2x + 1$

4. $f(x) = 5x + 2$, $g(x) = 3x - 4$

5. $f(x) = 4x - 3$, $g(x) = 5x^2 - 2$

6. $f(x) = 7x + 1$, $g(x) = 2x^2 - 9$

7. $f(x) = x^2 + 2$, $g(x) = x^2 - 2$

8. $f(x) = x^2 + 1$, $g(x) = x^2 - 3$

9. $f(x) = \sqrt{x}$, $g(x) = x - 1$

10. $f(x) = \sqrt{x}$, $g(x) = x + 2$

11. $f(x) = 2x - 3$, $g(x) = \dfrac{x + 3}{2}$

12. $f(x) = 6x - 3$, $g(x) = \dfrac{x + 3}{6}$

13. $f(x) = \dfrac{1}{x}$, $g(x) = \dfrac{1}{x}$

14. $f(x) = \dfrac{1}{x}$, $g(x) = \dfrac{2}{x}$

In Exercises 15–24, find $f(g(x))$ and $g(f(x))$ and determine whether each pair of functions f and g are inverses of each other.

15. $f(x) = 4x$ and $g(x) = \dfrac{x}{4}$

16. $f(x) = 6x$ and $g(x) = \dfrac{x}{6}$

17. $f(x) = 3x + 8$ and $g(x) = \dfrac{x - 8}{3}$

18. $f(x) = 4x + 9$ and $g(x) = \dfrac{x - 9}{4}$

19. $f(x) = 5x - 9$ and $g(x) = \dfrac{x + 5}{9}$

20. $f(x) = 3x - 7$ and $g(x) = \dfrac{x + 3}{7}$

21. $f(x) = \dfrac{3}{x - 4}$ and $g(x) = \dfrac{3}{x} + 4$

22. $f(x) = \dfrac{2}{x - 5}$ and $g(x) = \dfrac{2}{x} + 5$

23. $f(x) = -x$ and $g(x) = -x$

24. $f(x) = \sqrt[3]{x - 4}$ and $g(x) = x^3 + 4$

The functions in Exercises 25–44 are all one-to-one. For each function,

a. *Find an equation for $f^{-1}(x)$, the inverse function.*

b. *Verify that your equation is correct by showing that $f(f^{-1}(x)) = x$ and $f^{-1}(f(x)) = x$.*

25. $f(x) = x + 3$

26. $f(x) = x + 5$

27. $f(x) = 2x$

28. $f(x) = 4x$

29. $f(x) = 2x + 3$

30. $f(x) = 3x - 1$

31. $f(x) = x^3 + 2$

32. $f(x) = x^3 - 1$

33. $f(x) = (x + 2)^3$

34. $f(x) = (x - 1)^3$

35. $f(x) = \dfrac{1}{x}$

36. $f(x) = \dfrac{2}{x}$

37. $f(x) = \sqrt{x}$

38. $f(x) = \sqrt[3]{x}$

39. $f(x) = x^2 + 1$, for $x \geq 0$

40. $f(x) = x^2 - 1$, for $x \geq 0$

41. $f(x) = \dfrac{2x + 1}{x - 3}$

42. $f(x) = \dfrac{2x - 3}{x + 1}$

43. $f(x) = \sqrt[3]{x - 4} + 3$

44. $f(x) = x^{\frac{3}{5}}$

Which graphs in Exercises 45–50 represent functions that have inverse functions?

45.

46.

47.

48.

49.

50.

In Exercises 51–54, use the graph of f to draw the graph of its inverse function.

51.

52.

53.

54.

Practice PLUS

In Exercises 55–60, f and g are defined by the following tables. Use the tables to evaluate each composite function.

x	f(x)
−1	1
0	4
1	5
2	−1

x	g(x)
−1	0
1	1
4	2
10	−1

55. $f(g(1))$

56. $f(g(4))$

57. $(g \circ f)(-1)$

58. $(g \circ f)(0)$

59. $f^{-1}(g(10))$

60. $f^{-1}(g(1))$

In Exercises 61–64, use the graphs of f and g to evaluate each composite function.

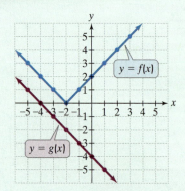

61. $(f \circ g)(-1)$

62. $(f \circ g)(1)$

63. $(g \circ f)(0)$

64. $(g \circ f)(-1)$

In Exercises 65–70, let

$$f(x) = 2x - 5$$
$$g(x) = 4x - 1$$
$$h(x) = x^2 + x + 2.$$

Evaluate the indicated function without finding an equation for the function.

65. $(f \circ g)(0)$

66. $(g \circ f)(0)$

67. $f^{-1}(1)$

68. $g^{-1}(7)$

69. $g(f[h(1)])$

70. $f(g[h(1)])$

Application Exercises

71. The regular price of a computer is x dollars. Let $f(x) = x - 400$ and $g(x) = 0.75x$.

    a. Describe what the functions f and g model in terms of the price of the computer.

    b. Find $(f \circ g)(x)$ and describe what this models in terms of the price of the computer.

    c. Repeat part (b) for $(g \circ f)(x)$.

    d. Which composite function models the greater discount on the computer, $f \circ g$ or $g \circ f$? Explain.

    e. Find f^{-1} and describe what this models in terms of the price of the computer.

72. The regular price of a pair of jeans is x dollars. Let $f(x) = x - 5$ and $g(x) = 0.6x$.

    a. Describe what functions f and g model in terms of the price of the jeans.

    b. Find $(f \circ g)(x)$ and describe what this models in terms of the price of the jeans.

    c. Repeat part (b) for $(g \circ f)(x)$.

    d. Which composite function models the greater discount on the jeans, $f \circ g$ or $g \circ f$? Explain.

    e. Find f^{-1} and describe what this models in terms of the price of the jeans.

Way to Go *Holland was the first country to establish an official bicycle policy. It currently has over 12,000 miles of paths and lanes exclusively for bicycles. The graph at the top of the next page shows the percentage of travel by bike and by car in Holland, as well as in four other selected countries. Use the information in the graph to solve Exercises 73–74.*

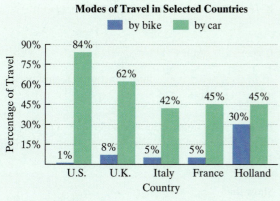

Modes of Travel in Selected Countries

Source: EUROSTAT

73. a. Consider a function, f, whose domain is the set of the five countries shown in the graph. Let the range be the percentage of travel by bike in each of the respective countries. Write function f as a set of ordered pairs.

b. Write the relation that is the inverse of f as a set of ordered pairs. Is this relation a function? Explain your answer.

74. a. Consider a function, f, whose domain is the set of the five countries shown in the graph. Let the range be the percentage of travel by car in each of the respective countries. Write function f as a set of ordered pairs.

b. Write the relation that is the inverse of f as a set of ordered pairs. Is this relation a function? Explain your answer.

75. The graph represents the probability that two people in the same room share a birthday as a function of the number of people in the room. Call the function f.

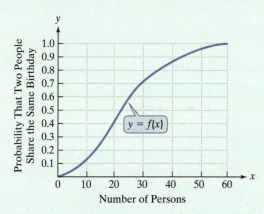

a. Explain why f has an inverse that is a function.

b. Describe in practical terms the meanings of $f^{-1}(0.25)$, $f^{-1}(0.5)$, and $f^{-1}(0.7)$.

76. A study of 900 working women in Texas showed that their feelings changed throughout the day. As the graph indicates, the women felt better as time passed, except for a blip at lunchtime.

Source: D. Kahneman et al., "A Survey Method for Characterizing Daily Life Experience." *Science,* Vol. 306, No. 5702, Dec. 3, 2004, pp. 1776–1780.

a. Does the graph have an inverse that is a function? Explain your answer.

b. Identify two or more times of day when the average happiness level is 3. Express your answers as ordered pairs.

c. Do the ordered pairs in part (b) indicate that the graph represents a one-to-one function? Explain your answer.

77. The formula

$$y = f(x) = \frac{9}{5}x + 32$$

is used to convert from x degrees Celsius to y degrees Fahrenheit. The formula

$$y = g(x) = \frac{5}{9}(x - 32)$$

is used to convert from x degrees Fahrenheit to y degrees Celsius. Show that f and g are inverse functions.

Explaining the Concepts

78. Describe a procedure for finding $(f \circ g)(x)$.

79. Explain how to determine if two functions are inverses of each other.

80. Describe how to find the inverse of a one-to-one function.

81. What is the horizontal line test and what does it indicate?

82. Describe how to use the graph of a one-to-one function to draw the graph of its inverse function.

83. How can a graphing utility be used to visually determine if two functions are inverses of each other?

Technology Exercises

In Exercises 84–91, use a graphing utility to graph each function. Use the graph to determine whether the function has an inverse that is a function (that is, whether the function is one-to-one).

84. $f(x) = x^2 - 1$

85. $f(x) = \sqrt[3]{2 - x}$

86. $f(x) = \dfrac{x^3}{2}$

87. $f(x) = \dfrac{x^4}{4}$

88. $f(x) = |x - 2|$

89. $f(x) = (x - 1)^3$

90. $f(x) = -\sqrt{16 - x^2}$

91. $f(x) = x^3 + x + 1$

In Exercises 92–94, use a graphing utility to graph f and g in the same viewing rectangle. In addition, graph the line y = x and visually determine if f and g are inverses.

92. $f(x) = 4x + 4, \quad g(x) = 0.25x - 1$

93. $f(x) = \dfrac{1}{x} + 2, \quad g(x) = \dfrac{1}{x - 2}$

94. $f(x) = \sqrt[3]{x} - 2, \quad g(x) = (x + 2)^3$

Critical Thinking Exercises

Make Sense?　*In Exercises 95–98, determine whether each statement makes sense or does not make sense, and explain your reasoning.*

95. This diagram illustrates that $f(g(x)) = x^2 + 4$.

96. I must have made a mistake in finding the composite functions $f \circ g$ and $g \circ f$, because I notice that $f \circ g$ is the same function as $g \circ f$.

97. Regardless of what exponential function I graph, its shape indicates that it always has an inverse function.

98. I'm working with the linear function $f(x) = 3x + 5$ and I do not need to find f^{-1} in order to determine the value of $(f \circ f^{-1})(17)$.

In Exercises 99–102, determine whether each statement is true or false. If the statement is false, make the necessary change(s) to produce a true statement.

99. The inverse of $\{(1, 4), (2, 7)\}$ is $\{(2, 7), (1, 4)\}$.

100. The function $f(x) = 5$ is one-to-one.

101. If $f(x) = \sqrt{x}$ and $g(x) = 2x - 1$, then $(f \circ g)(5) = g(2)$.

102. If $f(x) = 3x$, then $f^{-1}(x) = \dfrac{1}{3x}$.

103. If $h(x) = \sqrt{3x^2 + 5}$, find functions f and g so that $h(x) = (f \circ g)(x)$.

104. If $f(x) = 3x$ and $g(x) = x + 5$, find $(f \circ g)^{-1}(x)$ and $(g^{-1} \circ f^{-1})(x)$.

105. Show that

$$f(x) = \frac{3x - 2}{5x - 3}$$

is its own inverse.

106. Consider the two functions defined by $f(x) = m_1 x + b_1$ and $g(x) = m_2 x + b_2$. Prove that the slope of the composite function of f with g is equal to the product of the slopes of the two functions.

Review Exercises

107. Divide and write the quotient in scientific notation:

$$\frac{4.3 \times 10^5}{8.6 \times 10^{-4}}.$$

(Section 1.7, Example 4)

108. Graph: $f(x) = x^2 - 4x + 3$.

(Section 8.3, Example 4)

109. Solve: $\sqrt{x + 4} - \sqrt{x - 1} = 1$.

(Section 7.6, Example 4)

Preview Exercises

Exercises 110–112 will help you prepare for the material covered in the next section.

110. What problem do you encounter when using the switch-and-solve strategy to find the inverse of $f(x) = 2^x$?

111. 25 to what power gives 5? $(25^? = 5)$

112. Solve: $(x - 3)^2 > 0$.

9.3

Logarithmic Functions

The earthquake that ripped through northern California on October 17, 1989, measured 7.1 on the Richter scale, killed more than 60 people, and injured more than 2400. Shown here is San Francisco's Marina district, where shock waves tossed houses off their foundations and into the street.

A higher measure on the Richter scale is more devastating than it seems because for each increase in one unit on the scale, there is a tenfold increase in the intensity of an earthquake. In this section, our focus is on the inverse of the exponential function, called the logarithmic function. The logarithmic function will help you to understand diverse phenomena, including earthquake intensity, human memory, and the pace of life in large cities.

What am I supposed to learn?

After studying this section, you should be able to:

1 Change from logarithmic to exponential form.

2 Change from exponential to logarithmic form.

3 Evaluate logarithms.

4 Use basic logarithmic properties.

5 Graph logarithmic functions.

6 Find the domain of a logarithmic function.

7 Use common logarithms.

8 Use natural logarithms.

The Definition of Logarithmic Functions

No horizontal line can be drawn that intersects the graph of an exponential function at more than one point. This means that the exponential function is one-to-one and has an inverse. Let's use our switch-and-solve strategy to find the inverse.

> All exponential functions have inverse functions.

$$f(x) = b^x$$

Step 1. Replace *f(x)* with *y*: $y = b^x$.

Step 2. Interchange *x* and *y*: $x = b^y$.

Step 3. Solve for *y*:   ?

The question mark indicates that we do not have a method for solving $b^y = x$ for y. To isolate the exponent y, a new notation, called *logarithmic notation*, is needed. This notation gives us a way to name the inverse of $f(x) = b^x$. **The inverse function of the exponential function with base *b* is called the *logarithmic function with base b.***

Definition of the Logarithmic Function

For $x > 0$ and $b > 0, b \neq 1$,

$$y = \log_b x \text{ is equivalent to } b^y = x.$$

The function $f(x) = \log_b x$ is the **logarithmic function with base *b*.**

The equations

$$y = \log_b x \quad \text{and} \quad b^y = x$$

are different ways of expressing the same thing. The first equation is in **logarithmic form** and the second equivalent equation is in **exponential form.**

Notice that a **logarithm, *y*, is an exponent. Logarithmic form allows us to isolate this exponent.** You should learn the location of the base and exponent in each form.

Location of Base and Exponent in Exponential and Logarithmic Forms

Logarithmic Form: $y = \log_b x$ Exponential Form: $b^y = x$

Exponent ⌐ $y = \log_b x$ Base ⌐

Exponent ⌐ $b^y = x$ Base ⌐

Great Question!

Much of what you've discussed so far involves changing from logarithmic form to the more familiar exponential form. Is there a pattern I can use to help me remember how to do this?

Yes. To change from logarithmic form to exponential form, use this pattern:

$$y = \log_b x \qquad \text{means} \qquad b^y = x.$$

1 Change from logarithmic to exponential form. ▶

EXAMPLE 1 Changing from Logarithmic to Exponential Form

Write each equation in its equivalent exponential form:

a. $2 = \log_5 x$ **b.** $3 = \log_b 64$ **c.** $\log_3 7 = y$.

Solution We use the fact that $y = \log_b x$ means $b^y = x$.

a. $2 = \log_5 x$ means $5^2 = x$.   **b.** $3 = \log_b 64$ means $b^3 = 64$.

Logarithms are exponents.   Logarithms are exponents.

c. $\log_3 7 = y$ or $y = \log_3 7$ means $3^y = 7$. ■

✓ **CHECK POINT 1** Write each equation in its equivalent exponential form:
  a. $3 = \log_7 x$ **b.** $2 = \log_b 25$ **c.** $\log_4 26 = y$.

2 Change from exponential to logarithmic form. ▶

EXAMPLE 2 Changing from Exponential to Logarithmic Form

Write each equation in its equivalent logarithmic form:

a. $12^2 = x$ **b.** $b^3 = 8$ **c.** $e^y = 9$.

Solution We use the fact that $b^y = x$ means $y = \log_b x$. In logarithmic form, the exponent is isolated on one side of the equal sign.

a. $12^2 = x$ means $2 = \log_{12} x$.   **b.** $b^3 = 8$ means $3 = \log_b 8$.

Exponents are logarithms.   Exponents are logarithms.

c. $e^y = 9$ means $y = \log_e 9$. ■

✓ **CHECK POINT 2** Write each equation in its equivalent logarithmic form:
  a. $2^5 = x$ **b.** $b^3 = 27$ **c.** $e^y = 33$.

3 Evaluate logarithms. ▶

Remembering that logarithms are exponents makes it possible to evaluate some logarithms by inspection. The logarithm of x with base b, $\log_b x$, is the exponent to which b must be raised to get x. For example, suppose we want to evaluate $\log_2 32$. We ask, 2 to what power gives 32? Because $2^5 = 32$, we can conclude that $\log_2 32 = 5$.

EXAMPLE 3 Evaluating Logarithms

Evaluate:

a. $\log_2 16$ **b.** $\log_3 9$ **c.** $\log_{25} 5$.

Solution

Logarithmic Expression	Question Needed for Evaluation	Logarithmic Expression Evaluated
a. $\log_2 16$	2 to what power gives 16?	$\log_2 16 = 4$ because $2^4 = 16$.
b. $\log_3 9$	3 to what power gives 9?	$\log_3 9 = 2$ because $3^2 = 9$.
c. $\log_{25} 5$	25 to what power gives 5?	$\log_{25} 5 = \frac{1}{2}$ because $25^{\frac{1}{2}} = \sqrt{25} = 5$.

■

✓ CHECK POINT 3 Evaluate:

a. $\log_{10} 100$ **b.** $\log_3 3$ **c.** $\log_{36} 6$.

4 Use basic logarithmic properties. ▶

Basic Logarithmic Properties

Because logarithms are exponents, they have properties that can be verified using the properties of exponents.

Basic Logarithmic Properties Involving 1

1. $\log_b b = 1$ because 1 is the exponent to which b must be raised to obtain b. ($b^1 = b$)

2. $\log_b 1 = 0$ because 0 is the exponent to which b must be raised to obtain 1. ($b^0 = 1$)

EXAMPLE 4 Using Properties of Logarithms

Evaluate:

a. $\log_7 7$ **b.** $\log_5 1$.

Solution

a. Because $\log_b b = 1$, we conclude $\log_7 7 = 1$.

b. Because $\log_b 1 = 0$, we conclude $\log_5 1 = 0$. ■

✓ CHECK POINT 4 Evaluate:

a. $\log_9 9$ **b.** $\log_8 1$.

Now that we are familiar with logarithmic notation, let's resume and finish the switch-and-solve strategy for finding the inverse of $f(x) = b^x$.

Step 1. Replace $f(x)$ with y: $y = b^x$.

Step 2. Interchange x and y: $x = b^y$.

Step 3. Solve for y: $y = \log_b x$.

Step 4. Replace y with $f^{-1}(x)$: $f^{-1}(x) = \log_b x$.

The completed switch-and-solve strategy illustrates that if $f(x) = b^x$, then $f^{-1}(x) = \log_b x$. The inverse of an exponential function is the logarithmic function with the same base.

In Section 9.2, we saw how inverse functions "undo" one another. In particular,

$$f(f^{-1}(x)) = x \quad \text{and} \quad f^{-1}(f(x)) = x.$$

Applying these relationships to exponential and logarithmic functions, we obtain the following **inverse properties of logarithms:**

Inverse Properties of Logarithms

For $b > 0$ and $b \neq 1$,

$\log_b b^x = x$ The logarithm with base b of b raised to a power equals that power.

$b^{\log_b x} = x$ b raised to the logarithm with base b of a number equals that number.

EXAMPLE 5   Using Inverse Properties of Logarithms

Evaluate:

a. $\log_4 4^5$ **b.** $6^{\log_6 9}$.

Solution

a. Because $\log_b b^x = x$, we conclude $\log_4 4^5 = 5$.

b. Because $b^{\log_b x} = x$, we conclude $6^{\log_6 9} = 9$. ∎

✓ **CHECK POINT 5**   Evaluate:
  a. $\log_7 7^8$ **b.** $3^{\log_3 17}$.

---

5 Graph logarithmic functions.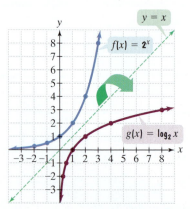

Graphs of Logarithmic Functions

How do we graph logarithmic functions? We use the fact that a logarithmic function is the inverse of an exponential function. This means that the logarithmic function reverses the coordinates of the exponential function. It also means that the graph of the logarithmic function is a reflection of the graph of the exponential function about the line $y = x$.

EXAMPLE 6   Graphs of Exponential and Logarithmic Functions

Graph $f(x) = 2^x$ and $g(x) = \log_2 x$ in the same rectangular coordinate system.

Solution   We first set up a table of coordinates for $f(x) = 2^x$. Reversing these coordinates gives the coordinates for the inverse function $g(x) = \log_2 x$.

x	-2	-1	0	1	2	3
$f(x) = 2^x$	$\frac{1}{4}$	$\frac{1}{2}$	1	2	4	8

Reverse coordinates.

x	$\frac{1}{4}$	$\frac{1}{2}$	1	2	4	8
$g(x) = \log_2 x$	-2	-1	0	1	2	3

We now plot the ordered pairs from each table, connecting them with smooth curves. **Figure 9.14** shows the graphs of $f(x) = 2^x$ and its inverse function $g(x) = \log_2 x$. The graph of the inverse can also be drawn by reflecting the graph of $f(x) = 2^x$ about the line $y = x$. ∎

Figure 9.14 The graphs of $f(x) = 2^x$ and its inverse function

Great Question!

You found the coordinates of $g(x) = \log_2 x$ by reversing the coordinates of $f(x) = 2^x$. Do I have to do it that way?

Not necessarily. You can obtain a partial table of coordinates for $g(x) = \log_2 x$ without having to obtain and reverse coordinates for $f(x) = 2^x$. Because $g(x) = \log_2 x$ means $2^{g(x)} = x$, we begin with values for $g(x)$ and compute corresponding values for x:

Use $x = 2^{g(x)}$ to compute x. For example,
if $g(x) = -2$, $x = 2^{-2} = \frac{1}{2^2} = \frac{1}{4}$.

Start with values for $g(x)$.

x	$\frac{1}{4}$	$\frac{1}{2}$	1	2	4	8
$g(x) = \log_2 x$	-2	-1	0	1	2	3

✓ CHECK POINT 6   Graph $f(x) = 3^x$ and $g(x) = \log_3 x$ in the same rectangular coordinate system.

Figure 9.15 illustrates the relationship between the graph of an exponential function, shown in blue, and its inverse, a logarithmic function, shown in red, for bases greater than 1 and for bases between 0 and 1. Also shown and labeled are the exponential function's horizontal asymptote ($y = 0$) and the logarithmic function's vertical asymptote ($x = 0$).

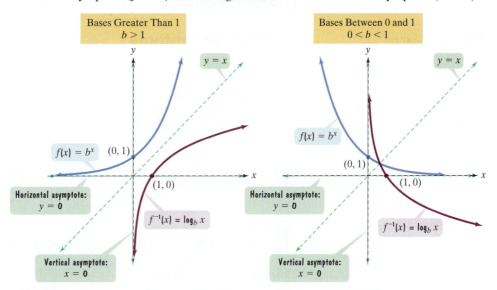

Figure 9.15 Graphs of exponential and logarithmic functions

The red graphs in **Figure 9.15** illustrate the following general characteristics of logarithmic functions:

Characteristics of Logarithmic Functions of the Form $f(x) = \log_b x$

1. The domain of $f(x) = \log_b x$ consists of all positive real numbers: $(0, \infty)$. The range of $f(x) = \log_b x$ consists of all real numbers: $(-\infty, \infty)$.

2. The graphs of all logarithmic functions of the form $f(x) = \log_b x$ pass through the point $(1, 0)$ because $f(1) = \log_b 1 = 0$. The x-intercept is 1. There is no y-intercept.

3. If $b > 1$, $f(x) = \log_b x$ has a graph that goes up to the right and is an increasing function.

4. If $0 < b < 1$, $f(x) = \log_b x$ has a graph that goes down to the right and is a decreasing function.

5. The graph of $f(x) = \log_b x$ approaches, but does not touch, the y-axis. The y-axis, or $x = 0$, is a vertical asymptote.

6 Find the domain of a logarithmic function.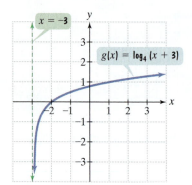

Figure 9.16 The domain of $g(x) = \log_4(x + 3)$ is $(-3, \infty)$.

The Domain of a Logarithmic Function

In Section 9.1, we learned that the domain of an exponential function of the form $f(x) = b^x$ includes all real numbers and its range is the set of positive real numbers. Because the logarithmic function reverses the domain and the range of the exponential function, the **domain of a logarithmic function of the form $f(x) = \log_b x$ is the set of all positive real numbers.** Thus, $\log_2 8$ is defined because the value of x in the logarithmic expression, 8, is greater than zero and therefore is included in the domain of the logarithmic function $f(x) = \log_2 x$. However, $\log_2 0$ and $\log_2(-8)$ are not defined because 0 and -8 are not positive real numbers and therefore are excluded from the domain of the logarithmic function $f(x) = \log_2 x$. In general, **the domain of $f(x) = \log_b g(x)$ consists of all x for which $g(x) > 0$.**

EXAMPLE 7   Finding the Domain of a Logarithmic Function

Find the domain of $g(x) = \log_4(x + 3)$.

Solution   The domain of g consists of all x for which $x + 3 > 0$. Solving this inequality for x, we obtain $x > -3$. Thus, the domain of g is $(-3, \infty)$. This is illustrated in **Figure 9.16**. The vertical asymptote is $x = -3$ and all points on the graph of g have x-coordinates that are greater than -3. ■

☑ **CHECK POINT 7**   Find the domain of $h(x) = \log_4(x - 5)$.

7 Use common logarithms.

Common Logarithms

The logarithmic function with base 10 is called the **common logarithmic function**. The function $f(x) = \log_{10} x$ is usually expressed as $f(x) = \log x$. A calculator with a $\boxed{\text{LOG}}$ key can be used to evaluate common logarithms. Here are some examples.

Logarithm	Most Scientific Calculator Keystrokes	Most Graphing Calculator Keystrokes	Display (or Approximate Display)
$\log 1000$	1000 $\boxed{\text{LOG}}$	$\boxed{\text{LOG}}$ 1000 $\boxed{\text{ENTER}}$	3
$\log \dfrac{5}{2}$	$\boxed{(}$ 5 $\boxed{\div}$ 2 $\boxed{)}$ $\boxed{\text{LOG}}$	$\boxed{\text{LOG}}$ $\boxed{(}$ 5 $\boxed{\div}$ 2 $\boxed{)}$ $\boxed{\text{ENTER}}$	0.39794
$\dfrac{\log 5}{\log 2}$	5 $\boxed{\text{LOG}}$ $\boxed{\div}$ 2 $\boxed{\text{LOG}}$ $\boxed{=}$	$\boxed{\text{LOG}}$ 5 $\boxed{\div}$ $\boxed{\text{LOG}}$ 2 $\boxed{\text{ENTER}}$	2.32193
$\log(-3)$	3 $\boxed{+/-}$ $\boxed{\text{LOG}}$	$\boxed{\text{LOG}}$ $\boxed{(-)}$ 3 $\boxed{\text{ENTER}}$	$\boxed{\text{ERROR}}$

Some graphing calculators display an open parenthesis when the $\boxed{\text{LOG}}$ key is pressed. In this case, remember to close the set of parentheses after entering the function's domain value: $\boxed{\text{LOG}}$ 5 $\boxed{)}$ $\boxed{\div}$ $\boxed{\text{LOG}}$ 2 $\boxed{)}$ $\boxed{\text{ENTER}}$.

The error message or $\boxed{\text{NONREAL ANSWERS}}$ message given by many calculators for $\log(-3)$ is a reminder that the domain of the common logarithmic function, $f(x) = \log x$, is the set of positive real numbers. In general, the domain of $f(x) = \log g(x)$ consists of all x for which $g(x) > 0$.

Many real-life phenomena start with rapid growth and then the growth begins to level off. This type of behavior can be modeled by logarithmic functions.

EXAMPLE 8   Modeling Heights of Children

The percentage of adult height attained by a boy who is x years old can be modeled by

$$f(x) = 29 + 48.8 \log(x + 1),$$

where x represents the boy's age and $f(x)$ represents the percentage of his adult height. Approximately what percentage of his adult height has a boy attained at age eight?

Solution   We substitute the boy's age, 8, for x and evaluate the function at 8.

$f(x) = 29 + 48.8 \log(x + 1)$ This is the given function.

$f(8) = 29 + 48.8 \log(8 + 1)$ Substitute 8 for x.

$= 29 + 48.8 \log 9$ Graphing calculator keystrokes:

29 $+$ 48.8 $\boxed{\text{LOG}}$ 9 $\boxed{\text{ENTER}}$

≈ 76

Thus, an 8-year-old boy has attained approximately 76% of his adult height.   ■

☑ **CHECK POINT 8**   Use the function in Example 8 to answer this question: Approximately what percentage of his adult height has a boy attained at age ten?

The basic properties of logarithms that were listed earlier in this section can be applied to common logarithms.

Properties of Common Logarithms

General Properties		Common Logarithms
1. $\log_b 1 = 0$		**1.** $\log 1 = 0$
2. $\log_b b = 1$		**2.** $\log 10 = 1$
3. $\log_b b^x = x$	Inverse	**3.** $\log 10^x = x$
4. $b^{\log_b x} = x$	properties	**4.** $10^{\log x} = x$

The property $\log 10^x = x$ can be used to evaluate common logarithms involving powers of 10. For example,

$\log 100 = \log 10^2 = 2$, $\log 1000 = \log 10^3 = 3$,   and $\log 10^{7.1} = 7.1$.

EXAMPLE 9   Earthquake Intensity

The magnitude, R, on the Richter scale of an earthquake of intensity I is given by

$$R = \log \frac{I}{I_0},$$

where I_0 is the intensity of a barely felt zero-level earthquake. The earthquake that destroyed San Francisco in 1906 was $10^{8.3}$ times as intense as a zero-level earthquake. What was its magnitude on the Richter scale?

Solution   Because the earthquake was $10^{8.3}$ times as intense as a zero-level earthquake, the intensity, I, is $10^{8.3} I_0$.

$R = \log \dfrac{I}{I_0}$ This is the formula for magnitude on the Richter scale.

$R = \log \dfrac{10^{8.3} I_0}{I_0}$ Substitute $10^{8.3} I_0$ for I.

$= \log 10^{8.3}$ Simplify.

$= 8.3$ Use the property $\log 10^x = x$.

San Francisco's 1906 earthquake registered 8.3 on the Richter scale.   ■

☑ **CHECK POINT 9**   Use the formula in Example 9 to solve this problem. If an earthquake is 10,000 times as intense as a zero-level quake ($I = 10,000 I_0$), what is its magnitude on the Richter scale?

8 Use natural logarithms.

Natural Logarithms

The logarithmic function with base e is called the **natural logarithmic function**. The function $f(x) = \log_e x$ is usually expressed as $f(x) = \ln x$, read "el en of x." A calculator with an $\boxed{\text{LN}}$ key can be used to evaluate natural logarithms. Keystrokes are identical to those shown for common logarithmic evaluations on page 696.

Like the domain of all logarithmic functions, the domain of the natural logarithmic function $f(x) = \ln x$ is the set of all positive real numbers. Thus, the domain of $f(x) = \ln g(x)$ consists of all x for which $g(x) > 0$.

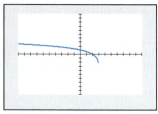

$[-10, 10, 1]$ by $[-10, 10, 1]$

Figure 9.17 The domain of $f(x) = \ln(3 - x)$ is $(-\infty, 3)$.

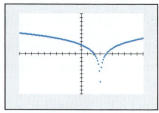

$[-10, 10, 1]$ by $[-10, 10, 1]$

Figure 9.18 The value 3 is excluded from the domain of $g(x) = \ln(x - 3)^2$.

EXAMPLE 10 Finding Domains of Natural Logarithmic Functions

Find the domain of each function:

a. $f(x) = \ln(3 - x)$ **b.** $g(x) = \ln(x - 3)^2$.

Solution

a. The domain of $f(x) = \ln(3 - x)$ consists of all x for which $3 - x > 0$. Solving this inequality for x, we obtain $x < 3$. Thus, the domain of f is $(-\infty, 3)$. This is verified by the graph in **Figure 9.17**.

b. The domain of $g(x) = \ln(x - 3)^2$ consists of all x for which $(x - 3)^2 > 0$. It follows that the domain of g is all real numbers except 3. Thus, the domain of g is $(-\infty, 3) \cup (3, \infty)$. This is shown by the graph in **Figure 9.18**. To make it more obvious that 3 is excluded from the domain, we used a $\boxed{\text{DOT}}$ format. ∎

☑ **CHECK POINT 10** Find the domain of each function:

a. $f(x) = \ln(4 - x)$

b. $g(x) = \ln x^2$.

The basic properties of logarithms that were listed earlier in this section can be applied to natural logarithms.

Properties of Natural Logarithms

General Properties	Natural Logarithms
1. $\log_b 1 = 0$	**1.** $\ln 1 = 0$
2. $\log_b b = 1$	**2.** $\ln e = 1$
3. $\log_b b^x = x$ Inverse	**3.** $\ln e^x = x$
4. $b^{\log_b x} = x$ properties	**4.** $e^{\ln x} = x$

Examine the inverse properties, $\ln e^x = x$ and $e^{\ln x} = x$. Can you see how $\ln$ and e "undo" one another? For example,

$$\ln e^2 = 2, \quad \ln e^{7x^2} = 7x^2, \quad e^{\ln 2} = 2, \quad \text{and} \quad e^{\ln 7x^2} = 7x^2.$$

| **EXAMPLE 11** | Dangerous Heat: Temperature in an Enclosed Vehicle |

When the outside air temperature is anywhere from 72° to 96° Fahrenheit, the temperature in an enclosed vehicle climbs by 43° in the first hour. The bar graph in **Figure 9.19** shows the temperature increase throughout the hour. The function

$$f(x) = 13.4 \ln x - 11.6$$

models the temperature increase, $f(x)$, in degrees Fahrenheit, after x minutes. Use the function to find the temperature increase, to the nearest degree, after 50 minutes. How well does the function model the actual increase shown in **Figure 9.19**?

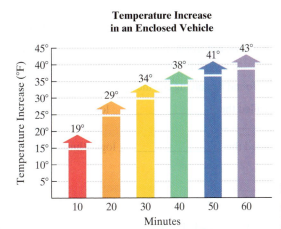

Temperature Increase in an Enclosed Vehicle

Figure 9.19

Source: Lynn I. Gibbs and David W. Lawrence, "Heat Exposure in an Enclosed Automobile." *Journal of the Lousiana State Medical Society*, Volume 147(12), 1995

Solution   We find the temperature increase after 50 minutes by substituting 50 for x and evaluating the function at 50.

$f(x) = 13.4 \ln x - 11.6$	This is the given function.
$f(50) = 13.4 \ln 50 - 11.6$	Substitute 50 for x.
≈ 41	Graphing calculator keystrokes: 13.4 |ln| 50 |−| 11.6 |ENTER|. On some calculators, a parenthesis is needed after 50.

According to the function, the temperature will increase by approximately 41° after 50 minutes. Because the increase shown in **Figure 9.19** is 41°, the function models the actual increase extremely well.   ■

| ✓ **CHECK POINT 11** | Use the function in Example 11 to find the temperature increase, to the nearest degree, after 30 minutes. How well does the function model the actual increase shown in **Figure 9.19**?

Achieving Success

We have seen that **the best way to achieve success in math is through practice**. Keeping up with your homework, preparing for tests, asking questions of your professor, reading your textbook, and attending all classes will help you learn the material and boost your confidence. Use language in a proactive way that reflects a sense of responsibility for your own successes and failures.

Reactive Language	Proactive Language
I'll try.	I'll do it.
That's just the way I am.	I can do better than that.
There's not a thing I can do.	I have options for improvement.
I have to.	I choose to.
I can't.	I can find a way.

SECTION

9.4

Properties of Logarithms

What am I supposed to learn?

After studying this section, you should be able to:

1 Use the product rule. ▶

2 Use the quotient rule. ▶

3 Use the power rule. ▶

4 Expand logarithmic expressions. ▶

5 Condense logarithmic expressions. ▶

6 Use the change-of-base property. ▶

We all learn new things in different ways. In this section, we consider important properties of logarithms. What would be the most effective way for you to learn these properties? Would it be helpful to use your graphing utility and discover one of these properties for yourself? To do so, work Exercise 107 in Exercise Set 9.3 before continuing. Would it be helpful to evaluate certain logarithmic expressions that suggest three of the properties? If this is the case, work Preview Exercises 123–125 in Exercise Set 9.3 before continuing. Would the properties become more meaningful if you could see exactly where they come from? If so, you will find details of the proofs of many of these properties in the appendix. The remainder of our work in this chapter will be based on the properties of logarithms that you learn in this section.

1 Use the product rule. ▶

The Product Rule

Properties of exponents correspond to properties of logarithms. For example, when we multiply exponential expressions with the same base, we add exponents:

$$b^m \cdot b^n = b^{m+n}.$$

This property of exponents, coupled with an awareness that a logarithm is an exponent, suggests the following property, called the **product rule**:

> ### The Product Rule
>
> Let b, M, and N be positive real numbers with $b \neq 1$.
>
> $$\log_b(MN) = \log_b M + \log_b N$$
>
> The logarithm of a product is the sum of the logarithms.

Discover for Yourself

We know that log 100,000 = 5. Show that you get the same result by writing 100,000 as 1000 · 100 and then using the product rule. Then verify the product rule by using other numbers whose logarithms are easy to find.

When we use the product rule to write a single logarithm as the sum of two logarithms, we say that we are **expanding a logarithmic expression**. For example, we can use the product rule to expand $\ln(7x)$:

$$\ln(7x) = \ln 7 + \ln x.$$

The logarithm of a product | is | the sum of the logarithms.

EXAMPLE 1 Using the Product Rule

Use the product rule to expand each logarithmic expression:

a. $\log_4(7 \cdot 5)$ **b.** $\log(10x)$.

Solution

a. $\log_4(7 \cdot 5) = \log_4 7 + \log_4 5$ The logarithm of a product is the sum of the logarithms.

b. $\log(10x) = \log 10 + \log x$ The logarithm of a product is the sum of the logarithms. These are common logarithms with base 10 understood.

$$= 1 + \log x \qquad \text{Because } \log_b b = 1, \text{ then log 10} = 1. \ \blacksquare$$

✓ CHECK POINT 1   Use the product rule to expand each logarithmic expression:

a. $\log_6(7 \cdot 11)$ b. $\log(100x)$.

---

2 Use the quotient rule. ▶

The Quotient Rule

When we divide exponential expressions with the same base, we subtract exponents:

$$\frac{b^m}{b^n} = b^{m-n}.$$

This property suggests the following property of logarithms, called the **quotient rule:**

Discover for Yourself

We know that $\log_2 16 = 4$. Show that you get the same result by writing 16 as $\dfrac{32}{2}$ and then using the quotient rule. Then verify the quotient rule using other numbers whose logarithms are easy to find.

> ### The Quotient Rule
>
> Let b, M, and N be positive real numbers with $b \neq 1$.
>
> $$\log_b\left(\frac{M}{N}\right) = \log_b M - \log_b N$$
>
> The logarithm of a quotient is the difference of the logarithms.

When we use the quotient rule to write a single logarithm as the difference of two logarithms, we say that we are **expanding a logarithmic expression**. For example, we can use the quotient rule to expand $\log \dfrac{x}{2}$:

$$\log\left(\frac{x}{2}\right) = \log x - \log 2.$$

The logarithm of a quotient   is   the difference of the logarithms.

EXAMPLE 2   Using the Quotient Rule

Use the quotient rule to expand each logarithmic expression:

a. $\log_7\left(\dfrac{19}{x}\right)$ b. $\ln\left(\dfrac{e^3}{7}\right)$.

Solution

a. $\log_7\left(\dfrac{19}{x}\right) = \log_7 19 - \log_7 x$ The logarithm of a quotient is the difference of the logarithms.

b. $\ln\left(\dfrac{e^3}{7}\right) = \ln e^3 - \ln 7$ The logarithm of a quotient is the difference of the logarithms. These are natural logarithms with base e understood.

$$= 3 - \ln 7 \qquad \text{Because ln } e^x = x, \text{ then ln } e^3 = 3. \ \blacksquare$$

✓ CHECK POINT 2   Use the quotient rule to expand each logarithmic expression:

a. $\log_8\left(\dfrac{23}{x}\right)$ b. $\ln\left(\dfrac{e^5}{11}\right)$.

3 Use the power rule. ▶

The Power Rule

When an exponential expression is raised to a power, we multiply exponents:

$$(b^m)^n = b^{mn}.$$

This property suggests the following property of logarithms, called the **power rule**:

> ### The Power Rule
>
> Let b and M be positive real numbers with $b \neq 1$, and let p be any real number.
>
> $$\log_b M^p = p \log_b M$$
>
> The logarithm of a number with an exponent is the product of the exponent and the logarithm of that number.

When we use the power rule to "pull the exponent to the front," we say that we are **expanding a logarithmic expression**. For example, we can use the power rule to expand $\ln x^2$:

$$\ln x^2 = 2 \ln x.$$

Figure 9.20 shows the graphs of $y = \ln x^2$ and $y = 2 \ln x$ in $[-5, 5, 1]$ by $[-5, 5, 1]$ viewing rectangles. Are $\ln x^2$ and $2 \ln x$ the same? The graphs illustrate that $y = \ln x^2$ and $y = 2 \ln x$ have different domains. The graphs are only the same if $x > 0$. Thus, we should write

$$\ln x^2 = 2 \ln x \quad \text{for} \quad x > 0.$$

Domain: $(-\infty, 0) \cup (0, \infty)$

Domain: $(0, \infty)$

Figure 9.20
$\ln x^2$ and $2 \ln x$ have different domains.

When expanding a logarithmic expression, you might want to determine whether the rewriting has changed the domain of the expression. For the rest of this section, assume that all variables and variable expressions represent positive numbers.

EXAMPLE 3 Using the Power Rule

Use the power rule to expand each logarithmic expression:

a. $\log_5 7^4$ **b.** $\ln \sqrt{x}$ **c.** $\log(4x)^5$.

Solution

a. $\log_5 7^4 = 4 \log_5 7$ The logarithm of a number with an exponent is the exponent times the logarithm of that number.

b. $\ln \sqrt{x} = \ln x^{\frac{1}{2}}$ Rewrite the radical using a rational exponent.

$= \dfrac{1}{2} \ln x$ Use the power rule to bring the exponent to the front.

c. $\log(4x)^5 = 5 \log(4x)$ We immediately apply the power rule because the entire variable expression, 4x, is raised to the 5th power. ∎

☑ **CHECK POINT 3**   Use the power rule to expand each logarithmic expression:

a. $\log_6 8^9$ **b.** $\ln \sqrt[3]{x}$ **c.** $\log(x + 4)^2$.

---

4 Expand logarithmic expressions. ▶

Expanding Logarithmic Expressions

It is sometimes necessary to use more than one property of logarithms when you expand a logarithmic expression. Properties for expanding logarithmic expressions are as follows:

> **Properties for Expanding Logarithmic Expressions**
>
> For $M > 0$ and $N > 0$:
>
> **1.** $\log_b(MN) = \log_b M + \log_b N$ Product rule
>
> **2.** $\log_b\left(\dfrac{M}{N}\right) = \log_b M - \log_b N$ Quotient rule
>
> **3.** $\log_b M^p = p \log_b M$ Power rule

Great Question!

Are there some common blunders that I can avoid when using properties of logarithms?

The graphs show

$y_1 = \ln(x + 3)$ and $y_2 = \ln x + \ln 3$.

The graphs are not the same:

$\ln(x + 3) \neq \ln x + \ln 3$.

In general,

$\log_b(M + N) \neq \log_b M + \log_b N$.

$y_1 = \ln (x + 3)$

$y_2 = \ln x + \ln 3$

$[-4, 5, 1]$ by $[-3, 3, 1]$

Try to avoid the following errors:

Incorrect!

$\log_b(M + N) = \log_b M + \log_b N$

$\log_b(M - N) = \log_b M - \log_b N$

$\log_b(M \cdot N) = \log_b M \cdot \log_b N$

$\log_b\left(\dfrac{M}{N}\right) = \dfrac{\log_b M}{\log_b N}$

$\dfrac{\log_b M}{\log_b N} = \log_b M - \log_b N$

$\log_b(MN^p) = p \log_b(MN)$

EXAMPLE 4   Expanding Logarithmic Expressions

Use logarithmic properties to expand each expression as much as possible:

a. $\log_b\left(x^2 \sqrt{y}\right)$ **b.** $\log_6\left(\dfrac{\sqrt[3]{x}}{36y^4}\right)$.

Solution   We will have to use two or more of the properties for expanding logarithms in each part of this example.

a. $\log_b\left(x^2\sqrt{y}\right) = \log_b\left(x^2y^{\frac{1}{2}}\right)$ Use exponential notation.

$= \log_b x^2 + \log_b y^{\frac{1}{2}}$ Use the product rule.

$= 2\log_b x + \frac{1}{2}\log_b y$ Use the power rule.

b. $\log_6\left(\dfrac{\sqrt[3]{x}}{36y^4}\right) = \log_6\left(\dfrac{x^{\frac{1}{3}}}{36y^4}\right)$ Use exponential notation.

$= \log_6 x^{\frac{1}{3}} - \log_6(36y^4)$ Use the quotient rule.

$= \log_6 x^{\frac{1}{3}} - \left(\log_6 36 + \log_6 y^4\right)$ Use the product rule on $\log_6(36y^4)$.

$= \frac{1}{3}\log_6 x - \left(\log_6 36 + 4\log_6 y\right)$ Use the power rule.

$= \frac{1}{3}\log_6 x - \log_6 36 - 4\log_6 y$ Apply the distributive property.

$= \frac{1}{3}\log_6 x - 2 - 4\log_6 y$ $\log_6 36 = 2$ because 2 is the power to which we must raise 6 to get 36. $(6^2 = 36)$ ∎

✓ **CHECK POINT 4**   Use logarithmic properties to expand each expression as much as possible:

a. $\log_b\left(x^4\sqrt[3]{y}\right)$ **b.** $\log_5\left(\dfrac{\sqrt{x}}{25y^3}\right)$.

---

5 Condense logarithmic expressions. ▶

Condensing Logarithmic Expressions

To **condense a logarithmic expression**, we write the sum or difference of two or more logarithmic expressions as a single logarithmic expression. We use the properties of logarithms to do so:

Great Question!

Are the properties listed on the right the same as those in the box on page 707?

Yes. The only difference is that we've reversed the sides in each property from the previous box.

Properties for Condensing Logarithmic Expressions

For $M > 0$ and $N > 0$:

1. $\log_b M + \log_b N = \log_b(MN)$ Product rule

2. $\log_b M - \log_b N = \log_b\left(\dfrac{M}{N}\right)$ Quotient rule

3. $p\log_b M = \log_b M^p$ Power rule

EXAMPLE 5   Condensing Logarithmic Expressions

Write as a single logarithm:

a. $\log_4 2 + \log_4 32$

b. $\log(4x - 3) - \log x$.

48. $\frac{1}{3}\ln x + \ln y$

49. $2\log_b x + 3\log_b y$

50. $5\log_b x + 6\log_b y$

51. $5\ln x - 2\ln y$ **52.**

53. $3\ln x - \frac{1}{3}\ln y$ **54.**

55. $4\ln(x + 6) - 3\ln x$

56. $8\ln(x + 9) - 4\ln x$

57. $3\ln x + 5\ln y - 6\ln z$

58. $4\ln x + 7\ln y - 3\ln z$

59. $\frac{1}{2}(\log_5 x + \log_5 y) - 2\log_5(x -$

60. $\frac{1}{3}(\log_4 x - \log_4 y) + 2\log_4(x -$

In Exercises 61–68, use common lo
and a calculator to evaluate to four

61. $\log_5 13$ **62.**

63. $\log_{14} 87.5$ **64.**

65. $\log_{0.1} 17$ **66.**

67. $\log_\pi 63$ **68.**

Practice PLUS

In Exercises 69–74, let $\log_b 2 = A$ a
expression in terms of A and C.

69. $\log_b \frac{3}{2}$ **70.**

71. $\log_b 8$ **72.**

73. $\log_b \sqrt{\frac{2}{27}}$ **74.**

In Exercises 75–88, determine whet
false. Where possible, show work to
the statement is false, make the nece
a true statement.

75. $\ln e = 0$

76. $\ln 0 = e$

77. $\log_4(2x^3) = 3\log_4(2x)$

78. $\ln(8x^3) = 3\ln(2x)$

79. $x\log 10^x = x^2$

80. $\ln(x + 1) = \ln x + \ln 1$

81. $\ln(5x) + \ln 1 = \ln(5x)$

82. $\ln x + \ln(2x) = \ln(3x)$

83. $\log(x + 3) - \log(2x) = \frac{\log(x}{\log(}$

84. $\frac{\log(x + 2)}{\log(x - 1)} = \log(x + 2) - \text{lo}$

Solution

a. $\log_4 2 + \log_4 32 = \log_4(2 \cdot 32)$ Use the product rule.

$\qquad\qquad\qquad\quad = \log_4 64$ We now have a single logarithm. However, we can simplify.

$\qquad\qquad\qquad\quad = 3$ $\log_4 64 = 3$ because $4^3 = 64$.

b. $\log(4x - 3) - \log x = \log\left(\dfrac{4x - 3}{x}\right)$ Use the quotient rule. ■

✓ **CHECK POINT 5**   Write as a single logarithm:

   a. $\log 25 + \log 4$ **b.** $\log(7x + 6) - \log x.$

Coefficients of logarithms must be 1 before you can condense them using the product and quotient rules. For example, to condense

$$2\ln x + \ln(x + 1),$$

the coefficient of the first term must be 1. We use the power rule to rewrite the coefficient as an exponent:

1. Use the power rule to make the number in front an exponent.

$$2\ln x + \ln(x + 1) = \ln x^2 + \ln(x + 1) = \ln[x^2(x + 1)].$$

2. Use the product rule. The sum of logarithms with coefficients of 1 is the logarithm of the product.

EXAMPLE 6   Condensing Logarithmic Expressions

Write as a single logarithm:

a. $\dfrac{1}{2}\log x + 4\log(x - 1)$ **b.** $3\ln(x + 7) - \ln x$

c. $4\log_b x - 2\log_b 6 - \frac{1}{2}\log_b y.$

Solution

a. $\frac{1}{2}\log x + 4\log(x - 1)$

$= \log x^{\frac{1}{2}} + \log(x - 1)^4$ Use the power rule so that all coefficients are 1.

$= \log\left[x^{\frac{1}{2}}(x - 1)^4\right]$ Use the product rule. The condensed form can be expressed as $\log\left[\sqrt{x}\,(x - 1)^4\right]$.

b. $3\ln(x + 7) - \ln x$

$= \ln(x + 7)^3 - \ln x$ Use the power rule so that all coefficients are 1.

$= \ln\left[\dfrac{(x + 7)^3}{x}\right]$ Use the quotient rule.

c. $4\log_b x - 2\log_b 6 - \frac{1}{2}\log_b y$

$= \log_b x^4 - \log_b 6^2 - \log_b y^{\frac{1}{2}}$ Use the power rule so that all coefficients are 1.

$= \log_b x^4 - \left(\log_b 36 + \log_b y^{\frac{1}{2}}\right)$ Rewrite as a single subtraction.

$= \log_b x^4 - \log_b\left(36y^{\frac{1}{2}}\right)$ Use the product rule.

$= \log_b\left(\dfrac{x^4}{36y^{\frac{1}{2}}}\right)$ or $\log_b\left(\dfrac{x^4}{36\sqrt{y}}\right)$ Use the quotient rule. ■

CONCEPT AND V

Fill in each blank so that the re

1. The product rule for loga
   the logarithms.

2. The quotient rule for loga
   the logarithms.

3. The power rule for logari
   _____ of the ex

4. The change-of-base prope
   the property states that

$$\log_b$$

9.4 EXERCISE SE

Practice Exercises

In all exercises, assume that all
represent positive numbers.

In Exercises 1–36, use properti
logarithmic expression as muc
evaluate logarithmic expression

1. $\log_5(7 \cdot 3)$

3. $\log_7(7x)$

5. $\log(1000x)$

7. $\log_7\left(\dfrac{7}{x}\right)$

9. $\log\left(\dfrac{x}{100}\right)$

11. $\log_4\left(\dfrac{64}{y}\right)$

13. $\ln\left(\dfrac{e^2}{5}\right)$

15. $\log_b x^3$

17. $\log N^{-6}$

19. $\ln\sqrt[5]{x}$

21. $\log_b(x^2 y)$

23. $\log_4\left(\dfrac{\sqrt{x}}{64}\right)$

25. $\log_6\left(\dfrac{36}{\sqrt{x+1}}\right)$

26. $\log_8\left(\dfrac{64}{\sqrt{x+1}}\right)$

27. $\log_b\left(\dfrac{x^2 y}{z^2}\right)$

28. $\log_b\left(\dfrac{x^3 y}{z^2}\right)$

---

a. Express the formula so that the expression in brackets is written as a single logarithm.

b. The formula is also used to determine how long it will take chimpanzees and apes to master a task. For example, a typical chimpanzee learning sign language can master a maximum of 65 signs. Use the form of the formula from part (a) to answer this question. How many weeks will it take a chimpanzee to master 30 signs if c for that chimp is 0.03?

Explaining the Concepts

95. Describe the product rule for logarithms and give an example.

96. Describe the quotient rule for logarithms and give an example.

97. Describe the power rule for logarithms and give an example.

98. Without showing the details, explain how to condense $\ln x - 2\ln(x + 1)$.

99. Describe the change-of-base property and give an example.

100. Explain how to use your calculator to find $\log_{14} 283$.

101. You overhear a student talking about a property of logarithms in which division becomes subtraction. Explain what the student means by this.

102. Find $\ln 2$ using a calculator. Then calculate each of the following: $1 - \frac{1}{2}$; $\quad 1 - \frac{1}{2} + \frac{1}{3}$; $\quad 1 - \frac{1}{2} + \frac{1}{3} - \frac{1}{4}$; $1 - \frac{1}{2} + \frac{1}{3} - \frac{1}{4} + \frac{1}{5}$;.... Describe what you observe.

Technology Exercises

103. a. Use a graphing utility (and the change-of-base property) to graph $y = \log_3 x$.

   b. Graph $\quad y = 2 + \log_3 x$, $\quad y = \log_3(x + 2)$, and $y = -\log_3 x$ in the same viewing rectangle as $y = \log_3 x$. Then describe the change or changes that need to be made to the graph of $y = \log_3 x$ to obtain each of these three graphs.

104. Graph $y = \log x$, $y = \log(10x)$, and $y = \log(0.1x)$ in the same viewing rectangle. Describe the relationship among the three graphs. What logarithmic property accounts for this relationship?

105. Use a graphing utility and the change-of-base property to graph $y = \log_3 x$, $y = \log_{25} x$, and $y = \log_{100} x$ in the same viewing rectangle.

   a. Which graph is on the top in the interval $(0, 1)$? Which is on the bottom?

   b. Which graph is on the top in the interval $(1, \infty)$? Which is on the bottom?

---

c. Generalize by writing a statement about which graph is on top, which is on the bottom, and in which intervals, using $y = \log_b x$ where $b > 1$.

Disprove each statement in Exercises 106–110 by

   a. *letting y equal a positive constant of your choice, and*

   b. *using a graphing utility to graph the function on each side of the equal sign. The two functions should have different graphs, showing that the equation is not true in general.*

106. $\log(x + y) = \log x + \log y$

107. $\log\dfrac{x}{y} = \dfrac{\log x}{\log y}$

108. $\ln(x - y) = \ln x - \ln y$

109. $\ln(xy) = (\ln x)(\ln y)$

110. $\dfrac{\ln x}{\ln y} = \ln x - \ln y$

Critical Thinking Exercises

Make Sense? *In Exercises 111–114, determine whether each statement makes sense or does not make sense, and explain your reasoning.*

111. Because I cannot simplify the expression $b^m + b^n$ by adding exponents, there is no property for the logarithm of a sum.

112. Because logarithms are exponents, the product, quotient, and power rules remind me of properties for operations with exponents.

113. I can use any positive number other than 1 in the change-of-base property, but the only practical bases are 10 and e because my calculator gives logarithms for these two bases.

114. I expanded $\log_4 \sqrt{\dfrac{x}{y}}$ by writing the radical using a rational exponent and then applying the quotient rule, obtaining $\dfrac{1}{2}\log_4 x - \log_4 y$.

In Exercises 115–118, determine whether each statement is true or false. If the statement is false, make the necessary change(s) to produce a true statement.

115. $\ln \sqrt{2} = \dfrac{\ln 2}{2}$

116. $\dfrac{\log_7 49}{\log_7 7} = \log_7 49 - \log_7 7$

117. $\log_b(x^3 + y^3) = 3\log_b x + 3\log_b y$

118. $\log_b(xy)^5 = (\log_b x + \log_b y)^5$

119. Use the change-of-base property to prove that
$$\log e = \dfrac{1}{\ln 10}.$$

120. If $\log 3 = A$ and $\log 7 = B$, find $\log_7 9$ in terms of A and B.

121. Write as a single term that does not contain a logarithm:
$$e^{\ln 8x^5 - \ln 2x^2}.$$

Review Exercises

122. Graph: $5x - 2y > 10$. (Section 4.4, Example 1)

123. Solve: $x - 2(3x - 2) > 2x - 3$.
(Section 4.1, Example 2)

124. Divide and simplify: $\dfrac{\sqrt[3]{40x^2y^6}}{\sqrt[3]{5xy}}$.

(Section 7.4, Example 5)

Preview Exercises

Exercises 125–127 will help you prepare for the material covered in the next section.

125. Simplify: $16^{\frac{3}{2}}$.

126. Evaluate $3\ln(2x)$ if $x = \dfrac{e^4}{2}$.

127. Solve: $\dfrac{x + 2}{4x + 3} = \dfrac{1}{x}$.

MID-CHAPTER CHECK POINT   Section 9.1–Section 9.4

 What You Know: We evaluated and graphed exponential functions $[f(x) = b^x, b > 0 \text{ and } b \neq 1]$, including the natural exponential function $[f(x) = e^x, e \approx 2.718]$. We studied composite and inverse functions, noting that a function has an inverse that is a function if there is no horizontal line that intersects the function's graph more than once. The exponential function passes this horizontal line test and we called the inverse of the exponential function with base b the logarithmic function with base b. We learned that $y = \log_b x$ is equivalent to $b^y = x$. We evaluated and graphed logarithmic functions, including the common logarithmic function $[f(x) = \log_{10} x$ or $f(x) = \log x]$ and the natural logarithmic function $[f(x) = \log_e x$ or $f(x) = \ln x]$. Finally, we used properties of logarithms to expand and condense logarithmic expressions.

In Exercises 1–3, find $(f \circ g)(x)$ and $(g \circ f)(x)$. Are f and g inverses of each other?

1. $f(x) = 3x + 2, g(x) = 4x - 5$

2. $f(x) = \sqrt[3]{7x + 5}, g(x) = \dfrac{x^3 - 5}{7}$

3. $f(x) = \log_5 x, g(x) = 5^x$

4. Let $f(x) = \dfrac{x - 1}{x}$ and $g(x) = \sqrt{x + 3}$.

    a. Find $(f \circ g)(6)$.   **b.** Find $(g \circ f)(-1)$.

    c. Find $(f \circ f)(5)$.   **d.** Find $(g \circ g)(-2)$.

In Exercises 5–7, find f^{-1}.

5. $f(x) = \dfrac{2x + 5}{4}$

6. $f(x) = 10x^3 - 7$

7. $f = \{(2, 5), (10, -7), (11, -10)\}$

In Exercises 8–10, which graphs represent functions? Among these graphs, which have inverse functions?

8.

9.

10.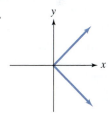

In Exercises 11–14, graph the given function. Give each function's domain and range.

11. $f(x) = 2^x - 3$

12. $f(x) = \left(\frac{1}{3}\right)^x$

13. $f(x) = \log_2 x$

14. $f(x) = \log_2 x + 1$

In Exercises 15–18, find the domain of each function.

15. $f(x) = \log_3(x + 6)$

16. $g(x) = \log_3 x + 6$

17. $h(x) = \log_3(x + 6)^2$

18. $f(x) = 3^{x+6}$

In Exercises 19–29, evaluate each expression without using a calculator. If evaluation is not possible, state the reason.

19. $\log_2 8 + \log_5 25$

20. $\log_3 \frac{1}{9}$

21. $\log_{100} 10$

22. $\log \sqrt[3]{10}$

23. $\log_2(\log_3 81)$

24. $\log_3\left(\log_2 \frac{1}{8}\right)$

25. $6^{\log_6 5}$ **26.** $\ln e^{\sqrt{7}}$

27. $10^{\log 13}$ **28.** $\log_{100} 0.1$

29. $\log_\pi \pi^{\sqrt{\pi}}$

In Exercises 30–31, expand and evaluate numerical terms.

30. $\log\left(\dfrac{\sqrt{xy}}{1000}\right)$

31. $\ln(e^{19}x^{20})$

In Exercises 32–34, write each expression as a single logarithm.

32. $8\log_7 x - \dfrac{1}{3}\log_7 y$

33. $7\log_5 x + 2\log_5 x$

34. $\dfrac{1}{2}\ln x - 3\ln y - \ln(z-2)$

35. Use the formulas

$$A = P\left(1 + \frac{r}{n}\right)^{nt} \quad \text{and} \quad A = Pe^{rt}$$

to solve this exercise. You plan to invest $8000 for 3 years at an annual rate of 8%. How much more is the return if the interest is compounded continuously than monthly? Round to the nearest dollar.

SECTION 9.5

Exponential and Logarithmic Equations

What am I supposed to learn?

After studying this section, you should be able to:

1. Use like bases to solve exponential equations.
2. Use logarithms to solve exponential equations.
3. Use exponential form to solve logarithmic equations.
4. Use the one-to-one property of logarithms to solve logarithmic equations.
5. Solve applied problems involving exponential and logarithmic equations.

---

1. Use like bases to solve exponential equations.

At age 20, you inherit $30,000. You'd like to put aside $25,000 and eventually have over half a million dollars for early retirement. Is this possible? In this section, you will see how techniques for solving equations with variable exponents provide an answer to this question.

Exponential Equations

An **exponential equation** is an equation containing a variable in an exponent. Examples of exponential equations include

$$2^{3x-8} = 16, \quad 4^x = 15, \quad \text{and} \quad 40e^{0.6x} = 240.$$

Some exponential equations can be solved by expressing each side of the equation as a power of the same base. All exponential functions are one-to-one—that is, no two different ordered pairs have the same second component. Thus, if b is a positive number other than 1 and $b^M = b^N$, then $M = N$.

> **Solving Exponential Equations by Expressing Each Side as a Power of the Same Base**
>
> $$\text{If} \quad b^M = b^N, \text{ then } M = N.$$
>
> Express each side as a power of the same base. Set the exponents equal to each other.
>
> 1. Rewrite the equation in the form $b^M = b^N$.
> 2. Set $M = N$.
> 3. Solve for the variable.

EXAMPLE 1 Solving Exponential Equations

Solve: **a.** $2^{3x-8} = 16$ **b.** $16^x = 64$.

Solution In each equation, express both sides as a power of the same base. Then set the exponents equal to each other.

a. Because 16 is 2^4, we express each side of $2^{3x-8} = 16$ in terms of base 2.

$$2^{3x-8} = 16 \quad \text{This is the given equation.}$$
$$2^{3x-8} = 2^4 \quad \text{Write each side as a power of the same base.}$$
$$3x - 8 = 4 \quad \text{If } b^M = b^N, b > 0 \text{ and } b \neq 1, \text{ then } M = N.$$
$$3x = 12 \quad \text{Add 8 to both sides.}$$
$$x = 4 \quad \text{Divide both sides by 3.}$$

Check 4:

$$2^{3x-8} = 16$$
$$2^{3\cdot4-8} \stackrel{?}{=} 16$$
$$2^4 \stackrel{?}{=} 16$$
$$16 = 16, \quad \text{true}$$

The solution is 4 and the solution set is {4}.

b. Because $16 = 4^2$ and $64 = 4^3$, we express each side of $16^x = 64$ in terms of base 4.

$$16^x = 64 \quad \text{This is the given equation.}$$
$$(4^2)^x = 4^3 \quad \text{Write each side as a power of the same base.}$$
$$4^{2x} = 4^3 \quad \text{When an exponential expression is raised to a power, multiply exponents.}$$
$$2x = 3 \quad \text{If two powers of the same base are equal, then the exponents are equal.}$$
$$x = \frac{3}{2} \quad \text{Divide both sides by 2.}$$

Check $\frac{3}{2}$:

$$16^x = 64$$
$$16^{\frac{3}{2}} \stackrel{?}{=} 64$$
$$(\sqrt{16})^3 \stackrel{?}{=} 64 \quad b^{\frac{m}{n}} = (\sqrt[n]{b})^m$$
$$4^3 \stackrel{?}{=} 64$$
$$64 = 64, \quad \text{true}$$

The solution is $\frac{3}{2}$ and the solution set is $\left\{\frac{3}{2}\right\}$. ∎

✓ **CHECK POINT 1** Solve:

a. $5^{3x-6} = 125$ **b.** $4^x = 32$.

Most exponential equations cannot be rewritten so that each side has the same base. Here are two examples:

$$4^x = 15 \qquad\qquad 10^x = 120,000.$$

We cannot rewrite both sides in terms of base 2 or base 4.

We cannot rewrite both sides in terms of base 10.

Using Technology

Graphic Connections

The graphs of

$$y_1 = 2^{3x-8}$$

and $y_2 = 16$

have an intersection point whose x-coordinate is 4. This verifies that {4} is the solution set of $2^{3x-8} = 16$.

$y_2 = 16$

$y_1 = 2^{3x-8}$

$x = 4$

$[-1, 5, 1]$ by $[0, 20, 1]$

Discover for Yourself

The equation $16^x = 64$ can also be solved by writing each side in terms of base 2. Do this. Which solution method do you prefer?

2 Use logarithms to solve exponential equations. ▶

Logarithms are extremely useful in solving equations such as $4^x = 15$ and $10^x = 120{,}000$. The solution begins with isolating the exponential expression. Notice that the exponential expression is already isolated in both $4^x = 15$ and $10^x = 120{,}000$. Then we take the logarithm on both sides. Why can we do this? All logarithmic relations are functions. Thus, if M and N are positive real numbers and $M = N$, then $\log_b M = \log_b N$.

The base that is used when taking the logarithm on both sides of an equation can be any base at all. If the exponential equation involves base 10, as in $10^x = 120{,}000$, we'll take the common logarithm on both sides. If the exponential equation involves any other base, as in $4^x = 15$, we'll take the natural logarithm on both sides.

Using Logarithms to Solve Exponential Equations

1. Isolate the exponential expression.

2. Take the common logarithm on both sides of the equation for base 10. Take the natural logarithm on both sides of the equation for bases other than 10.

3. Simplify using one of the following properties:

$$\ln b^x = x \ln b \quad \text{or} \quad \ln e^x = x \quad \text{or} \quad \log 10^x = x.$$

4. Solve for the variable.

EXAMPLE 2 Solving Exponential Equations

Solve: **a.** $4^x = 15$ **b.** $10^x = 120{,}000$.

Solution We will use the natural logarithmic function to solve $4^x = 15$ and the common logarithmic function to solve $10^x = 120{,}000$.

a. Because the exponential expression, 4^x, is already isolated on the left side of $4^x = 15$, we begin by taking the natural logarithm on both sides of the equation.

$4^x = 15$	This is the given equation.
$\ln 4^x = \ln 15$	Take the natural logarithm on both sides.
$x \ln 4 = \ln 15$	Use the power rule and bring the variable exponent to the front: $\ln b^x = x \ln b$.
$x = \dfrac{\ln 15}{\ln 4}$	Solve for x by dividing both sides by $\ln 4$.

We now have an exact value for x. We use the exact value for x in the equation's solution set. Thus, the equation's solution is $\dfrac{\ln 15}{\ln 4}$ and the solution set is $\left\{ \dfrac{\ln 15}{\ln 4} \right\}$.
We can obtain a decimal approximation by using a calculator: $x \approx 1.95$. Because $4^2 = 16$, it seems reasonable that the solution to $4^x = 15$ is approximately 1.95.

b. Because the exponential expression, 10^x, is already isolated on the left side of $10^x = 120{,}000$, we begin by taking the common logarithm on both sides of the equation.

$10^x = 120{,}000$	This is the given equation.
$\log 10^x = \log 120{,}000$	Take the common logarithm on both sides.
$x = \log 120{,}000$	Use the inverse property $\log 10^x = x$ on the left.

The equation's solution is $\log 120{,}000$ and the solution set is $\{\log 120{,}000\}$. We can obtain a decimal approximation by using a calculator: $x \approx 5.08$. Because $10^5 = 100{,}000$, it seems reasonable that the solution to $10^x = 120{,}000$ is approximately 5.08. ∎

Discover for Yourself

Keep in mind that the base used when taking the logarithm on both sides of an equation can be any base at all. Solve $4^x = 15$ by taking the common logarithm on both sides. Solve again, this time taking the logarithm with base 4 on both sides. Use the change-of-base property to show that the solutions are the same as the one obtained in Example 2(a).

☑ **CHECK POINT 2**   Solve:

a. $5^x = 134$ **b.** $10^x = 8000$.

Find each solution set and then use a calculator to obtain a decimal approximation to two decimal places for the solution.

EXAMPLE 3   Solving an Exponential Equation

Solve: $40e^{0.6x} - 3 = 237$.

Solution   We begin by adding 3 to both sides and dividing both sides by 40 to isolate the exponential expression, $e^{0.6x}$. Then we take the natural logarithm on both sides of the equation.

$40e^{0.6x} - 3 = 237$	This is the given equation.
$40e^{0.6x} = 240$	Add 3 to both sides.
$e^{0.6x} = 6$	Isolate the exponential factor by dividing both sides by 40.
$\ln e^{0.6x} = \ln 6$	Take the natural logarithm on both sides.
$0.6x = \ln 6$	Use the inverse property $\ln e^x = x$ on the left.
$x = \dfrac{\ln 6}{0.6} \approx 2.99$	Divide both sides by 0.6 and solve for x.

Thus, the solution of the equation is $\dfrac{\ln 6}{0.6} \approx 2.99$. Try checking this approximate solution in the original equation to verify that $\left\{ \dfrac{\ln 6}{0.6} \right\}$ is the solution set.   ■

☑ **CHECK POINT 3**   Solve: $7e^{2x} - 5 = 58$. Find the solution set and then use a calculator to obtain a decimal approximation to two decimal places for the solution.

---

3 Use exponential form to solve logarithmic equations. ⊙

Logarithmic Equations

A **logarithmic equation** is an equation containing a variable in a logarithmic expression. Examples of logarithmic equations include

$$\log_4(x + 3) = 2 \quad \text{and} \quad \ln(x + 2) - \ln(4x + 3) = \ln\left(\frac{1}{x}\right).$$

Some logarithmic equations can be expressed in the form $\log_b M = c$. We can solve such equations by rewriting them in exponential form.

Using Exponential Form to Solve Logarithmic Equations

1. Express the equation in the form $\log_b M = c$.
2. Use the definition of a logarithm to rewrite the equation in exponential form:

$$\log_b M = c \quad \text{means} \quad b^c = M.$$

Logarithms are exponents.

3. Solve for the variable.
4. Check proposed solutions in the original equation. Include in the solution set only values for which $M > 0$.

EXAMPLE 4 Solving Logarithmic Equations

Solve: **a.** $\log_4(x + 3) = 2$ **b.** $3 \ln(2x) = 12$.

Solution The form $\log_b M = c$ involves a single logarithm whose coefficient is 1 on one side and a constant on the other side. Equation (a) is already in this form. We will need to divide both sides of equation (b) by 3 to obtain this form.

a.

$\log_4(x + 3) = 2$	This is the given equation.
$4^2 = x + 3$	Rewrite in exponential form: $\log_b M = c$ means $b^c = M$.
$16 = x + 3$	Square 4.
$13 = x$	Subtract 3 from both sides.

Check 13:

$\log_4(x + 3) = 2$	This is the given logarithmic equation.
$\log_4(13 + 3) \overset{?}{=} 2$	Substitute 13 for x.
$\log_4 16 \overset{?}{=} 2$	
$2 = 2,$ true	$\log_4 16 = 2$ because $4^2 = 16$.

This true statement indicates that the solution is 13 and the solution set is {13}.

b.

$3 \ln(2x) = 12$	This is the given equation.
$\ln(2x) = 4$	Divide both sides by 3.
$\log_e(2x) = 4$	Rewrite the natural logarithm showing base e. This step is optional.
$e^4 = 2x$	Rewrite in exponential form: $\log_b M = c$ means $b^c = M$.
$\dfrac{e^4}{2} = x$	Divide both sides by 2.

Check $\dfrac{e^4}{2}$:

$3 \ln(2x) = 12$	This is the given logarithmic equation.
$3 \ln\left[2\left(\dfrac{e^4}{2}\right) \right] \overset{?}{=} 12$	Substitute $\dfrac{e^4}{2}$ for x.
$3 \ln e^4 \overset{?}{=} 12$	Simplify: $\dfrac{\cancel{2}}{\cancel{2}}$
$3 \cdot 4 \overset{?}{=} 12$	Because $\ln e^x = x$, we conclude $\ln e^4 = 4$.
$12 = 12,$ true	

This true statement indicates that the solution is $\dfrac{e^4}{2}$ and the solution set is $\left\{ \dfrac{e^4}{2} \right\}$. ∎

Using Technology

Graphic Connections

The graphs of

$y_1 = \log_4(x + 3)$ and $y_2 = 2$

have an intersection point whose x-coordinate is 13. This verifies that {13} is the solution set for $\log_4(x + 3) = 2$.

$[-3, 17, 1]$ by $[-2, 3, 1]$

✓ **CHECK POINT 4** Solve:

a. $\log_2(x - 4) = 3$ **b.** $4 \ln(3x) = 8$.

Logarithmic expressions are defined only for logarithms of positive real numbers. **Always check proposed solutions of a logarithmic equation in the original equation. Exclude from the solution set any proposed solution that produces the logarithm of a negative number or the logarithm of 0.**

Great Question!

Can a negative number belong to the solution set of a logarithmic equation?

Yes. Here's an example.

$\log_2(x + 20) = 3$	Solve this equation.
$2^3 = x + 20$	Rewrite in exponential form.
$8 = x + 20$	Cube 2.
$-12 = x$	Subtract 20 from both sides.

Check −12:

$\log_2(-12 + 20) \stackrel{?}{=} 3$	Substitute −12 for x.
$\log_2 8 \stackrel{?}{=} 3$	
$3 = 3$, true	$\log_2 8 = 3$ because $2^3 = 8$.

The solution set is $\{-12\}$. Although -12 is negative, it does not produce the logarithm of a negative number in $\log_2(x + 20) = 3$, the given equation. Note that the domain of the expression $\log_2(x + 20)$ is $(-20, \infty)$, which includes negative numbers such as -12.

To rewrite the logarithmic equation $\log_b M = c$ in the equivalent exponential form $b^c = M$, we need a single logarithm whose coefficient is one. It is sometimes necessary to use properties of logarithms to condense logarithms into a single logarithm. In the next example, we use the product rule for logarithms to obtain a single logarithmic expression on the left side.

EXAMPLE 5 Solving a Logarithmic Equation

Solve: $\log_2 x + \log_2(x - 7) = 3$.

Solution

$\log_2 x + \log_2(x - 7) = 3$	This is the given equation.
$\log_2[x(x - 7)] = 3$	Use the product rule to obtain a single logarithm: $\log_b M + \log_b N = \log_b(MN)$.
$2^3 = x(x - 7)$	Rewrite in exponential form.
$8 = x^2 - 7x$	Evaluate 2^3 on the left and apply the distributive property on the right.
$0 = x^2 - 7x - 8$	Set the equation equal to 0.
$0 = (x - 8)(x + 1)$	Factor.
$x - 8 = 0$ or $x + 1 = 0$	Set each factor equal to 0.
$x = 8 \qquad\qquad x = -1$	Solve for x.

Check 8:

$\log_2 x + \log_2(x - 7) = 3$

$\log_2 8 + \log_2(8 - 7) \stackrel{?}{=} 3$

$\log_2 8 + \log_2 1 \stackrel{?}{=} 3$

$3 + 0 \stackrel{?}{=} 3$

$3 = 3$, true

Check −1:

$\log_2 x + \log_2(x - 7) = 3$

$\log_2(-1) + \log_2(-1 - 7) \stackrel{?}{=} 3$

The number −1 does not check. It produces logarithms of negative numbers. Neither −1 nor −8 are in the domain of a logarithmic function.

The solution is 8 and the solution set is $\{8\}$. ■

✓ CHECK POINT 5 Solve: $\log x + \log(x - 3) = 1$.

4 Use the one-to-one property of logarithms to solve logarithmic equations. ▶

Some logarithmic equations can be expressed in the form $\log_b M = \log_b N$. Because all logarithmic functions are one-to-one, we can conclude that $M = N$.

Using the One-to-One Property of Logarithms to Solve Logarithmic Equations

1. Express the equation in the form $\log_b M = \log_b N$. This form involves a single logarithm whose coefficient is 1 on each side of the equation.

2. Use the one-to-one property to rewrite the equation without logarithms: If $\log_b M = \log_b N$, then $M = N$.

3. Solve for the variable.

4. Check proposed solutions in the original equation. Include in the solution set only values for which $M > 0$ and $N > 0$.

EXAMPLE 6 Solving a Logarithmic Equation

Solve: $\ln(x + 2) - \ln(4x + 3) = \ln\left(\dfrac{1}{x}\right)$.

Solution In order to apply the one-to-one property of logarithms, we need a single logarithm whose coefficient is 1 on each side of the equation. The right side is already in this form. We can obtain a single logarithm on the left side by applying the quotient rule.

$$\ln(x + 2) - \ln(4x + 3) = \ln\left(\frac{1}{x}\right)$$
This is the given equation.

$$\ln\left(\frac{x + 2}{4x + 3}\right) = \ln\left(\frac{1}{x}\right)$$
Use the quotient rule to obtain a single logarithm on the left side:
$\log_b M - \log_b N = \log_b\left(\dfrac{M}{N}\right)$.

$$\frac{x + 2}{4x + 3} = \frac{1}{x}$$
Use the one-to-one property:
If $\log_b M = \log_b N$, then $M = N$.

$$x(4x + 3)\left(\frac{x + 2}{4x + 3}\right) = x(4x + 3)\left(\frac{1}{x}\right)$$
Multiply both sides by $x(4x + 3)$, the LCD.

$$x(x + 2) = 4x + 3$$
Simplify.

$$x^2 + 2x = 4x + 3$$
Apply the distributive property.

$$x^2 - 2x - 3 = 0$$
Subtract $4x + 3$ from both sides and set the equation equal to O.

$$(x - 3)(x + 1) = 0$$
Factor.

$$x - 3 = 0 \quad \text{or} \quad x + 1 = 0$$
Set each factor equal to O.

$$x = 3 \qquad\qquad x = -1$$
Solve for x.

Substituting 3 for x into the original equation produces the true statement $\ln\left(\frac{1}{3}\right) = \ln\left(\frac{1}{3}\right)$. However, substituting -1 produces logarithms of negative numbers. Thus, -1 is not a solution. The solution is 3 and the solution set is $\{3\}$. ∎

Using Technology

Numeric Connections
A graphing utility's $\boxed{\text{TABLE}}$ feature can be used to verify that $\{3\}$ is the solution set of

$$\ln(x + 2) - \ln(4x + 3) = \ln\left(\frac{1}{x}\right).$$

$y_1 = \ln(x + 2) - \ln(4x + 3)$

$y_2 = \ln\left(\frac{1}{x}\right)$

y_1 and y_2 are equal when $x = 3$.

> ☑ **CHECK POINT 6**   Solve: $\ln(x - 3) = \ln(7x - 23) - \ln(x + 1)$.

---

5 Solve applied problems involving exponential and logarithmic equations. ▶

Applications

Our first applied example provides a mathematical perspective on the old slogan "Alcohol and driving don't mix." In California, where 38% of fatal traffic crashes involve drunk drivers, it is illegal to drive with a blood alcohol concentration of 0.08 or higher. At these levels, drivers may be arrested and charged with driving under the influence.

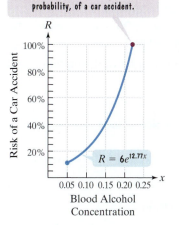

A blood alcohol concentration of 0.22 corresponds to near certainty, or a 100% probability, of a car accident.

Figure 9.22

EXAMPLE 7   Alcohol and Risk of a Car Accident

Medical research indicates that the risk of having a car accident increases exponentially as the concentration of alcohol in the blood increases. The risk is modeled by

$$R = 6e^{12.77x},$$

where x is the blood alcohol concentration and R, given as a percent, is the risk of having a car accident. What blood alcohol concentration corresponds to a 17% risk of a car accident? How is this shown on the graph of R in **Figure 9.22**?

Solution   For a risk of 17%, we let $R = 17$ in the equation and solve for x, the blood alcohol concentration.

$R = 6e^{12.77x}$	This is the given equation.
$6e^{12.77x} = 17$	Substitute 17 for R and (optional) reverse the two sides of the equation.
$e^{12.77x} = \dfrac{17}{6}$	Isolate the exponential factor by dividing both sides by 6.
$\ln e^{12.77x} = \ln\left(\dfrac{17}{6}\right)$	Take the natural logarithm on both sides.
$12.77x = \ln\left(\dfrac{17}{6}\right)$	Use the inverse property $\ln e^x = x$ on the left side.
$x = \dfrac{\ln\left(\dfrac{17}{6}\right)}{12.77} \approx 0.08$	Divide both sides by 12.77.

For a blood alcohol concentration of 0.08, the risk of a car accident is 17%. This is shown on the graph of R in **Figure 9.22** by the point (0.08, 17) that lies on the blue curve. Take a moment to locate this point on the curve. In many states, it is illegal to drive with a blood alcohol concentration of 0.08. ■

> ☑ **CHECK POINT 7**   Use the formula in Example 7 to solve this problem. What blood alcohol concentration corresponds to a 7% risk of a car accident? (In many states, drivers under the age of 21 can lose their licenses for driving at this level.)

Suppose that you inherit $30,000 at age 20. Is it possible to invest $25,000 and have over half a million dollars for early retirement? Our next example illustrates the power of compound interest.

Blitzer Bonus ⏵

Playing Doubles: Interest Rates and Doubling Time

One way to calculate what your savings will be worth at some point in the future is to consider doubling time. The following table shows how long it takes for your money to double at different annual interest rates subject to continuous compounding.

Annual Interest Rate	Years to Double
5%	13.9 years
7%	9.9 years
9%	7.7 years
11%	6.3 years

Of course, the first problem is collecting some money to invest. The second problem is finding a reasonably safe investment with a return of 9% or more.

EXAMPLE 8   Revisiting the Formula for Compound Interest

The formula

$$A = P\left(1 + \frac{r}{n}\right)^{nt}$$

describes the accumulated value, A, of a sum of money, P, the principal, after t years at annual percentage rate r (in decimal form) compounded n times a year. How long will it take $25,000 to grow to $500,000 at 9% annual interest compounded monthly?

Solution

$$A = P\left(1 + \frac{r}{n}\right)^{nt}$$ This is the given formula.

$$500,000 = 25,000\left(1 + \frac{0.09}{12}\right)^{12t}$$ A(the desired accumulated value) = 500,000, P(the principal) = 25,000, r(the interest rate) = 9% = 0.09, and n = 12 (monthly compounding).

Our goal is to solve the equation for t. Let's reverse the two sides of the equation and then simplify within parentheses.

$$25,000\left(1 + \frac{0.09}{12}\right)^{12t} = 500,000$$ Reverse the two sides of the previous equation.

$$25,000(1 + 0.0075)^{12t} = 500,000$$ Divide within parentheses: $\frac{0.09}{12}$ = 0.0075.

$$25,000(1.0075)^{12t} = 500,000$$ Add within parentheses.

$$(1.0075)^{12t} = 20$$ Divide both sides by 25,000.

$$\ln(1.0075)^{12t} = \ln 20$$ Take the natural logarithm on both sides.

$$12t \ln(1.0075) = \ln 20$$ Use the power rule to bring the exponent to the front: $\ln b^x = x \ln b$.

$$t = \frac{\ln 20}{12 \ln 1.0075}$$ Solve for t, dividing both sides by 12 ln 1.0075.

$$\approx 33.4$$ Use a calculator.

After approximately 33.4 years, the $25,000 will grow to an accumulated value of $500,000. If you set aside the money at age 20, you can begin enjoying a life of leisure at about age 53. ∎

✓ **CHECK POINT 8**   How long, to the nearest tenth of a year, will it take $1000 to grow to $3600 at 8% annual interest compounded quarterly?

EXAMPLE 9   Revisiting the Model for Heights of Children

We have seen that the percentage of adult height attained by a boy who is x years old can be modeled by

$$f(x) = 29 + 48.8 \log(x + 1),$$

where x represents the boy's age (from 5 to 15) and $f(x)$ represents the percentage of his adult height. At what age, rounded to the nearest year, has a boy attained 85% of his adult height?

Solution To find at what age a boy has attained 85% of his adult height, we substitute 85 for $f(x)$ and solve for x, the boy's age.

$$f(x) = 29 + 48.8 \log(x + 1)$$ This is the given function.

$$85 = 29 + 48.8 \log(x + 1)$$ Substitute 85 for $f(x)$.

> Our goal is to isolate $\log(x + 1)$ and then rewrite the equation in exponential form.

$$56 = 48.8 \log(x + 1)$$ Subtract 29 from both sides.

$$\frac{56}{48.8} = \log(x + 1)$$ Divide both sides by 48.8.

$$\frac{56}{48.8} = \log_{10}(x + 1)$$ Rewrite the common logarithm showing base 10. This step is optional.

$$10^{\frac{56}{48.8}} = x + 1$$ Rewrite in exponential form.

$$10^{\frac{56}{48.8}} - 1 = x$$ Subtract 1 from both sides.

$$13 \approx x$$ Use a calculator:
10 $\boxed{\wedge}$ $\boxed{(}$ 56 $\boxed{\div}$ 48.8 $\boxed{)}$ $\boxed{-}$ 1 $\boxed{\text{ENTER}}$.
Some calculators require that you use the right arrow key to exit the exponent before you subtract 1.

At approximately age 13, a boy has attained 85% of his adult height. This is shown on the graph of the model in **Figure 9.23** by the point (13, 85).

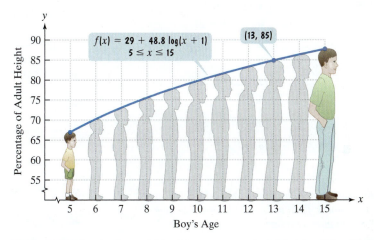

Figure 9.23 Graph of a model for the percentage of adult height attained by a boy ∎

✓ **CHECK POINT 9** The percentage of adult height attained by a girl who is x years old can be modeled by

$$f(x) = 62 + 35 \log(x - 4),$$

where x represents the girl's age (from 5 to 15) and $f(x)$ represents the percentage of her adult height. At what age has a girl attained 97% of her adult height?

CONCEPT AND VOCABULARY CHECK

Fill in each blank so that the resulting statement is true.

1. If $b^M = b^N$, then _____.

2. If $2^{4x-1} = 2^7$, then _____ = 7.

3. If $x \ln 9 = \ln 20$, then $x = $ _____.

4. If $e^{0.6x} = 6$, then $0.6x = $ _____.

5. If $\log_5(x + 1) = 3$, then _____ $= x + 1$.

6. If $\log_3 x + \log_3(x + 1) = 2$, then $\log_3$ _____ $= 2$.

7. If $\ln\left(\dfrac{7x - 23}{x + 1}\right) = \ln(x - 3)$, then _____ $= x - 3$.

8. True or false: $x^4 = 15$ is an exponential equation. _____

9. True or false: $4^x = 15$ is an exponential equation. _____

10. True or false: -3 is a solution of $\log_5 9 = 2 \log_5 x$. _____

9.5 EXERCISE SET ▶ MyMathLab®

Practice Exercises

Solve each exponential equation in Exercises 1–18 by expressing each side as a power of the same base and then equating exponents.

1. $2^x = 64$
2. $3^x = 81$
3. $5^x = 125$
4. $5^x = 625$
5. $2^{2x-1} = 32$
6. $3^{2x+1} = 27$
7. $4^{2x-1} = 64$
8. $5^{3x-1} = 125$
9. $32^x = 8$
10. $4^x = 32$
11. $9^x = 27$
12. $125^x = 625$
13. $3^{1-x} = \frac{1}{27}$
14. $5^{2-x} = \frac{1}{125}$
15. $6^{\frac{x-3}{4}} = \sqrt{6}$
16. $7^{\frac{x-2}{6}} = \sqrt{7}$
17. $4^x = \dfrac{1}{\sqrt{2}}$
18. $9^x = \dfrac{1}{\sqrt[3]{3}}$

Solve each exponential equation in Exercises 19–40 by taking the logarithm on both sides. Express the solution set in terms of logarithms. Then use a calculator to obtain a decimal approximation, correct to two decimal places, for the solution.

19. $e^x = 5.7$
20. $e^x = 0.83$
21. $10^x = 3.91$
22. $10^x = 8.07$
23. $5^x = 17$
24. $19^x = 143$
25. $5e^x = 25$
26. $9e^x = 99$
27. $3e^{5x} = 1977$
28. $4e^{7x} = 10{,}273$
29. $e^{0.7x} = 13$

30. $e^{0.08x} = 4$
31. $1250e^{0.055x} = 3750$
32. $1250e^{0.065x} = 6250$
33. $30 - (1.4)^x = 0$
34. $135 - (4.7)^x = 0$
35. $e^{1-5x} = 793$
36. $e^{1-8x} = 7957$
37. $7^{x+2} = 410$
38. $5^{x-3} = 137$
39. $2^{x+1} = 5^x$
40. $4^{x+1} = 9^x$

Solve each logarithmic equation in Exercises 41–90. Be sure to reject any value of x that is not in the domain of the original logarithmic expressions. Give the exact answer. Then, where necessary, use a calculator to obtain a decimal approximation, correct to two decimal places, for the solution.

41. $\log_3 x = 4$
42. $\log_5 x = 3$
43. $\log_2 x = -4$
44. $\log_2 x = -5$
45. $\log_9 x = \dfrac{1}{2}$
46. $\log_{25} x = \dfrac{1}{2}$
47. $\log x = 2$
48. $\log x = 3$

49. $\log_4(x + 5) = 3$

50. $\log_5(x - 7) = 2$

51. $\log_3(x - 4) = -3$

52. $\log_7(x + 2) = -2$

53. $\log_4(3x + 2) = 3$

54. $\log_2(4x + 1) = 5$

55. $\ln x = 2$

56. $\ln x = 3$

57. $\ln x = -3$

58. $\ln x = -4$

59. $5 \ln(2x) = 20$

60. $6 \ln(2x) = 30$

61. $6 + 2 \ln x = 5$

62. $7 + 3 \ln x = 6$

63. $\ln \sqrt{x + 3} = 1$

64. $\ln \sqrt{x + 4} = 1$

65. $\log_5 x + \log_5(4x - 1) = 1$

66. $\log_6(x + 5) + \log_6 x = 2$

67. $\log_3(x - 5) + \log_3(x + 3) = 2$

68. $\log_2(x - 1) + \log_2(x + 1) = 3$

69. $\log_2(x + 2) - \log_2(x - 5) = 3$

70. $\log_4(x + 2) - \log_4(x - 1) = 1$

71. $\log(3x - 5) - \log(5x) = 2$

72. $\log(2x - 1) - \log x = 2$

73. $\ln(x + 1) - \ln x = 1$

74. $\ln(x + 2) - \ln x = 2$

75. $\log_3(x + 4) = \log_3 7$

76. $\log_2(x - 5) = \log_2 4$

77. $\log(x + 4) = \log x + \log 4$

78. $\log(5x + 1) = \log(2x + 3) + \log 2$

79. $\log(3x - 3) = \log(x + 1) + \log 4$

80. $\log(2x - 1) = \log(x + 3) + \log 3$

81. $2 \log x = \log 25$

82. $3 \log x = \log 125$

83. $\log(x + 4) - \log 2 = \log(5x + 1)$

84. $\log(x + 7) - \log 3 = \log(7x + 1)$

85. $2 \log x - \log 7 = \log 112$

86. $\log(x - 2) + \log 5 = \log 100$

87. $\log x + \log(x + 3) = \log 10$

88. $\log(x + 3) + \log(x - 2) = \log 14$

89. $\ln(x - 4) + \ln(x + 1) = \ln(x - 8)$

90. $\log_2(x - 1) - \log_2(x + 3) = \log_2\left(\dfrac{1}{x}\right)$

Practice PLUS

In Exercises 91–98, solve each equation.

91. $5^{2x} \cdot 5^{4x} = 125$

92. $3^{x+2} \cdot 3^x = 81$

93. $3^{x^2} = 45$

94. $5^{x^2} = 50$

95. $\log_2(x - 6) + \log_2(x - 4) - \log_2 x = 2$

96. $\log_2(x - 3) + \log_2 x - \log_2(x + 2) = 2$

97. $5^{x^2-12} = 25^{2x}$

98. $3^{x^2-12} = 9^{2x}$

Application Exercises

99. The formula $A = 37.3e^{0.0095t}$ models the population of California, A, in millions, t years after 2010.

   a. What was the population of California in 2010?

   b. When will the population of California reach 40 million?

100. The formula $A = 25.1e^{0.0187t}$ models the population of Texas, A, in millions, t years after 2010.

   a. What was the population of Texas in 2010?

   b. When will the population of Texas reach 28 million?

The function $f(x) = 20(0.975)^x$ models the percentage of surface sunlight, $f(x)$, that reaches a depth of x feet beneath the surface of the ocean. The figure shows the graph of this function. Use this information to solve Exercises 101–102.

Intensity of Sunlight Beneath the Ocean's Surface

$f(x) = 20(0.975)^x$

Percentage of Surface Sunlight Intensity

Ocean Depth (feet)

101. Use the function to determine at what depth, to the nearest foot, there is 1% of surface sunlight. How is this shown on the graph of f?

102. Use the function to determine at what depth, to the nearest foot, there is 3% of surface sunlight. How is this shown on the graph of f?

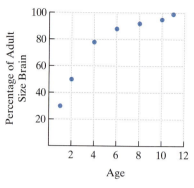

EXAMPLE 3   Choosing a Model for Data

The data in **Table 9.3** indicate that between the ages of 1 and 11, the human brain does not grow linearly, or steadily. A scatter plot for the data is shown in **Figure 9.26**. What type of function would be a good choice for modeling the data?

Table 9.3	Growth of the Human Brain
Age	**Percentage of Adult Size Brain**
1	30%
2	50%
4	78%
6	88%
8	92%
10	95%
11	99%

Figure 9.26

Source: Gerrig and Zimbardo, *Psychology and Life*, 18th Edition, Allyn and Bacon, 2008.

Solution   Because the data in the scatter plot increase rapidly at first and then begin to level off a bit, the shape suggests that a logarithmic function is a good choice for modeling the data.   ∎

✓ **CHECK POINT 3**   **Table 9.4** shows the populations of various cities, in thousands, and the average walking speed, in feet per second, of a person living in the city. Create a scatter plot for the data. Based on the scatter plot, what type of function would be a good choice for modeling the data?

Table 9.4	Population and Walking Speed
Population (thousands)	**Walking Speed (feet per second)**
5.5	0.6
14	1.0
71	1.6
138	1.9
342	2.2

Source: Mark H. Bornstein and Helen G. Bornstein, "The Pace of Life." *Nature*, 259, Feb. 19, 1976, pp. 557–559

Table 9.5	
x, Age	**y, Percentage of Adult Size Brain**
1	30
2	50
4	78
6	88
8	92
10	95
11	99

```
        LnReg
y=a+blnx
a=31.95404756
b=28.94733911
r²=.9806647799
r=.9902852013
```

Figure 9.27   A logarithmic model for the data in **Table 9.5**

How can we obtain a logarithmic function that models the data for the growth of the human brain? A graphing utility can be used to obtain a logarithmic model of the form $y = a + b \ln x$. **Because the domain of the logarithmic function is the set of positive numbers, zero must not be a value for x.** This is not a problem for the data giving the percentage of an adult size brain because the data begin at age 1. We will assign x to represent age and y to represent the percentage of an adult size brain. This gives us the data shown in **Table 9.5**. Using the logarithmic regression option, we obtain the equation in **Figure 9.27**.

From **Figure 9.27**, we see that the logarithmic model of the data, with numbers rounded to three decimal places, is

$$y = 31.954 + 28.947 \ln x.$$

The number r that appears in **Figure 9.27** is called the **correlation coefficient** and is a measure of how well the model fits the data. The value of r is such that $-1 \leq r \leq 1$. A positive r means that as the x-values increase, so do the y-values. A negative r means that as the x-values increase, the y-values decrease. **The closer that r is to -1 or 1, the better the model fits the data.** Because r is approximately 0.99, the model fits the data very well.

EXAMPLE 4   Choosing a Model for Data

Figure 9.28(a) shows world population, in billions, for seven selected years from 1950 through 2010. A scatter plot is shown in **Figure 9.28(b)**. Suggest two types of functions that would be good choices for modeling the data.

World Population, 1950–2010

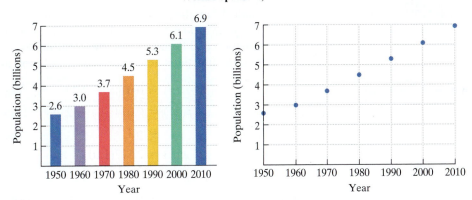

Figure 9.28(a)

Figure 9.28(b)

Source: U.S. Census Bureau, International Database

Solution   Because the data in the scatter plot appear to increase more and more rapidly, the shape suggests that an exponential model might be a good choice. Furthermore, we can probably draw a line that passes through or near the seven points. Thus, a linear function would also be a good choice for modeling the data. ∎

☑ **CHECK POINT 4**   **Table 9.6** shows the percentage of U.S. men who are married or who have been married, by age. Create a scatter plot for the data. Based on the scatter plot, what type of function would be a good choice for modeling the data?

Table 9.6

Percentage of U.S. Men Who Are Married or Who Have Been Married, by Age

Age	Percent
18	2
20	7
25	36
30	61
35	75

Source: National Center for Health Statistics

EXAMPLE 5   Comparing Linear and Exponential Models

The data for world population are shown in **Table 9.7**. Using a graphing utility's linear regression feature and exponential regression feature, we enter the data and obtain the models shown in **Figure 9.29**.

Table 9.7

x, Number of Years after 1949	y, World Population (billions)
1 (1950)	2.6
11 (1960)	3.0
21 (1970)	3.7
31 (1980)	4.5
41 (1990)	5.3
51 (2000)	6.1
61 (2010)	6.9

Although the domain of $y = ab^x$ is the set of all real numbers, some graphing utilities only accept positive values for x. That's why we assigned x to represent the number of years after 1949.

Figure 9.29   A linear model and an exponential model for the data in **Table 9.7**

Because r, the correlation coefficient, is close to 1 in each screen in **Figure 9.29**, the models fit the data very well.

a. Use **Figure 9.29** to express each model in function notation, with numbers rounded to three decimal places.

b. How well do the functions model world population in 2000?

c. By one projection, world population is expected to reach 8 billion in the year 2026. Which function serves as a better model for this prediction?

Figure 9.29 (repeated)

Solution

a. Using **Figure 9.29** and rounding to three decimal places, the functions

$$f(x) = 0.074x + 2.294 \quad \text{and} \quad g(x) = 2.577(1.017)^x$$

model world population, in billions, x years after 1949. We named the linear function f and the exponential function g, although any letters can be used.

b. **Table 9.7** on the previous page shows that world population in 2000 was 6.1 billion. The year 2000 is 51 years after 1949. Thus, we substitute 51 for x in each function's equation and then evaluate the resulting expressions with a calculator to see how well the functions describe world population in 2000.

$f(x) = 0.074x + 2.294$	This is the linear model.
$f(51) = 0.074(51) + 2.294$	Substitute 51 for x.
≈ 6.1	Use a calculator.
$g(x) = 2.577(1.017)^x$	This is the exponential model.
$g(51) = 2.577(1.017)^{51}$	Substitute 51 for x.
≈ 6.1	Use a calculator:

2.577 $\times$ 1.017 y^x 51 $=$
or 2.577 $\times$ 1.017 $\wedge$ 51 ENTER.

Because 6.1 billion was the actual world population in 2000, both functions model world population in 2000 extremely well.

c. Let's see which model comes closer to projecting a world population of 8 billion in the year 2026. Because 2026 is 77 years after 1949 ($2026 - 1949 = 77$), we substitute 77 for x in each function's equation.

$f(x) = 0.074x + 2.294$	This is the linear model.
$f(77) = 0.074(77) + 2.294$	Substitute 77 for x.
≈ 8.0	Use a calculator.
$g(x) = 2.577(1.017)^x$	This is the exponential model.
$g(77) = 2.577(1.017)^{77}$	Substitute 77 for x.
≈ 9.4	Use a calculator:

2.577 $\times$ 1.017 y^x 77 $=$
or 2.577 $\times$ 1.017 $\wedge$ 77 ENTER.

The linear function $f(x) = 0.074x + 2.294$ serves as a better model for a projected world population of 8 billion by 2026. ∎

Great Question!

How can I use a graphing utility to see how well my models describe the data?

Once you have obtained one or more models for the data, you can use a graphing utility's TABLE feature to numerically see how well each model describes the data. Enter the models as y_1, y_2, and so on. Create a table, scroll through the table, and compare the table values given by the models to the actual data.

✓ **CHECK POINT 5**   Use the models in Example 5(a) to solve this problem.

a. World population in 1970 was 3.7 billion. Which function serves as a better model for this year?

b. By one projection, world population is expected to reach 9.3 billion by 2050. Which function serves as a better model for this projection?

When using a graphing utility to model data, begin with a scatter plot, drawn either by hand or with the graphing utility, to obtain a general picture for the shape of the data. It might be difficult to determine which model best fits the data—linear, logarithmic, exponential, quadratic, or something else. If necessary, use your graphing utility to fit several models to the data. The best model is the one that yields the value r, the correlation coefficient, closest to 1 or −1. Finding a proper fit for data can be almost as much art as it is mathematics. In this era of technology, the process of creating models that best fit data is one that involves more decision making than computation.

3 Express an exponential model in base e. ▶

Expressing $y = ab^x$ in Base e

Graphing utilities display exponential models in the form $y = ab^x$. However, our discussion of exponential growth involved base e. Because of the inverse property $b = e^{\ln b}$, we can rewrite any model in the form $y = ab^x$ in terms of base e.

> **Expressing an Exponential Model in Base e**
>
> $$y = ab^x \quad \text{is equivalent to} \quad y = ae^{(\ln b) \cdot x}.$$

Blitzer Bonus ▶

Global Population Increase

By the time you finish reading Example 6 and working Check Point 6, more than 1000 people will have been added to our planet. By this time tomorrow, world population will have increased by more than 220,000.

EXAMPLE 6 Rewriting the Model for World Population in Base e

We have seen that the function

$$g(x) = 2.577(1.017)^x$$

models world population, $g(x)$, in billions, x years after 1949. Rewrite the model in terms of base e.

Solution We use the two equivalent equations shown in the voice balloons to rewrite the model in terms of base e.

$$y = ab^x \qquad\qquad y = ae^{(\ln b) \cdot x}$$

$$g(x) = 2.577(1.017)^x \quad \text{is equivalent to} \quad g(x) = 2.577e^{(\ln 1.017)x}.$$

Using $\ln 1.017 \approx 0.017$, the exponential growth model for world population, $g(x)$, in billions, x years after 1949 is

$$g(x) = 2.577e^{0.017x}. \quad \blacksquare$$

In Example 6, we can replace $g(x)$ with A and x with t so that the model has the same letters as those in the exponential growth model $A = A_0e^{kt}$.

$$A = \boxed{A_0} \; \boxed{e^{kt}} \quad \text{This is the exponential growth model.}$$

$$A = 2.577e^{0.017t} \quad \text{This is the model for world population.}$$

The value of k, 0.017, indicates a growth rate of 1.7%. Although this is an excellent model for the data, we must be careful about making projections about world population using this growth function. Why? World population growth rate is now 1.2%, not 1.7%, so our model will overestimate future populations.

☑ **CHECK POINT 6** Rewrite $y = 4(7.8)^x$ in terms of base e. Express the answer in terms of a natural logarithm and then round to three decimal places.

CONCEPT AND VOCABULARY CHECK

Fill in each blank so that the resulting statement is true.

1. Consider the model for exponential growth or decay given by

$$A = A_0e^{kt}.$$

   If k _____, the function models the amount, or size, of a growing entity. If k _____, the function models the amount, or size, of a decaying entity.

2. In the model for exponential growth or decay, the amount, or size, at $t = 0$ is represented by _____. The amount, or size, at time t is represented by _____.

The exponential growth models describe the population of the indicated country, A, in millions, t years after 2006.

Canada ➔ $A = 33.1e^{0.009t}$

Uganda ➔ $A = 28.2e^{0.034t}$

In Exercises 66–69, use this information to determine whether each statement is true or false. If the statement is false, make the necessary change(s) to produce a true statement.

66. In 2006, Canada's population exceeded Uganda's by 4.9 million.

67. By 2009, the models indicate that Canada's population exceeded Uganda's by approximately 2.8 million.

68. The models indicate that in 2013, Uganda's population exceeded Canada's.

69. Uganda's growth rate is approximately 3.8 times that of Canada's.

70. Over a period of time, a hot object cools to the temperature of the surrounding air. This is described mathematically by Newton's Law of Cooling:

$$T = C + (T_0 - C)e^{-kt},$$

where t is the time it takes for an object to cool from temperature T_0 to temperature T, C is the surrounding air temperature, and k is a positive constant that is associated with the cooling object. A cake removed from the oven has a temperature of 210°F and is left to cool in a room that has a temperature of 70°F. After 30 minutes, the temperature of the cake is 140°F. What is the temperature of the cake after 40 minutes?

Review Exercises

71. Divide:

$$\frac{x^2 - 9}{2x^2 + 7x + 3} \div \frac{x^2 - 3x}{2x^2 + 11x + 5}.$$

(Section 6.1, Example 7)

72. Solve: $x^{\frac{2}{3}} + 2x^{\frac{1}{3}} - 3 = 0$.

(Section 8.4, Example 5)

73. Simplify: $6\sqrt{2} - 2\sqrt{50} + 3\sqrt{98}$.

(Section 7.4, Example 2)

Preview Exercises

Exercises 74–76 will help you prepare for the material covered in the first section of the next chapter.

In Exercises 74–75, let $(x_1, y_1) = (7, 2)$ and $(x_2, y_2) = (1, -1)$.

74. Find $\sqrt{(x_2 - x_1)^2 + (y_2 - y_1)^2}$. Express the answer in simplified radical form.

75. Find the point represented by $\left(\dfrac{x_1 + x_2}{2}, \dfrac{y_1 + y_2}{2}\right)$.

76. Use a rectangular coordinate system to graph the circle with center $(1, -1)$ and radius 1.

Chapter 9 Summary

| **Definitions and Concepts** | **Examples** |

Section 9.1 Exponential Functions

The exponential function with base b is defined by $f(x) = b^x$, where $b > 0$ and $b \neq 1$. The graph contains the point $(0, 1)$. When $b > 1$, the graph rises from left to right. When $0 < b < 1$, the graph falls from left to right. The x-axis is a horizontal asymptote. The domain is $(-\infty, \infty)$; the range is $(0, \infty)$. The natural exponential function is $f(x) = e^x$, where $e \approx 2.71828$.

Graph $f(x) = 2^x$ and $g(x) = 2^{x-1}$.

x	$f(x) = 2^x$	$g(x) = 2^{x-1}$
-2	$2^{-2} = \frac{1}{4}$	$2^{-3} = \frac{1}{8}$
-1	$2^{-1} = \frac{1}{2}$	$2^{-2} = \frac{1}{4}$
0	$2^0 = 1$	$2^{-1} = \frac{1}{2}$
1	$2^1 = 2$	$2^0 = 1$
2	$2^2 = 4$	$2^1 = 2$

The graph of g is the graph of f shifted one unit to the right.

Definitions and Concepts	**Examples**

Section 9.1 Exponential Functions (continued)

Formulas for Compound Interest

After t years, the balance, A, in an account with principal P and annual interest rate r is given by the following formulas:

1. For n compounding periods per year: $A = P\left(1 + \dfrac{r}{n}\right)^{nt}$

2. For continuous compounding: $A = Pe^{rt}$

Select the better investment for \$4000 over 6 years:

- 6% compounded semiannually

$$A = P\left(1 + \frac{r}{n}\right)^{nt}$$

$$= 4000\left(1 + \frac{0.06}{2}\right)^{2\cdot 6} \approx \$5703$$

- 5.9% compounded continuously

$$A = Pe^{rt} = 4000e^{0.059(6)} \approx \$5699$$

The first investment is better.

Section 9.2 Composite and Inverse Functions

Composite Functions

The composite function $f \circ g$ is defined by
$$(f \circ g)(x) = f(g(x)).$$
The composite function $g \circ f$ is defined by
$$(g \circ f)(x) = g(f(x)).$$

Let $f(x) = x^2 + x$ and $g(x) = 2x + 1$.

- $(f \circ g)(x) = f(g(x)) = (g(x))^2 + g(x)$

  Replace x with $g(x)$.

$$= (2x + 1)^2 + (2x + 1) = 4x^2 + 4x + 1 + 2x + 1$$

$$= 4x^2 + 6x + 2$$

- $(g \circ f)(x) = g(f(x)) = 2f(x) + 1$

  Replace x with $f(x)$.

$$= 2(x^2 + x) + 1 = 2x^2 + 2x + 1$$

Inverse Functions

If $f(g(x)) = x$ and $g(f(x)) = x$, function g is the inverse of function f, denoted f^{-1} and read "f inverse." The procedure for finding a function's inverse uses a switch-and-solve strategy. Switch x and y, then solve for y.

If $f(x) = 2x - 5$, find $f^{-1}(x)$.

$$y = 2x - 5 \qquad \text{Replace } f(x) \text{ with } y.$$

$$x = 2y - 5 \qquad \text{Exchange } x \text{ and } y.$$

$$x + 5 = 2y \qquad \text{Solve for } y.$$

$$\frac{x + 5}{2} = y$$

$$f^{-1}(x) = \frac{x + 5}{2} \qquad \text{Replace } y \text{ with } f^{-1}(x).$$

Definitions and Concepts	**Examples**

Section 9.2 Composite and Inverse Functions (continued)

The Horizontal Line Test for Inverse Functions

A function, f, has an inverse that is a function, f^{-1}, if there is no horizontal line that intersects the graph of f at more than one point. A one-to-one function is one in which no two different ordered pairs have the same second component. Only one-to-one functions have inverse functions. If the point (a, b) is on the graph of f, then the point (b, a) is on the graph of f^{-1}. The graph of f^{-1} is a reflection of the graph of f about the line $y = x$.

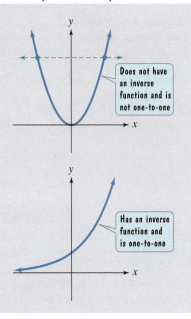

Does not have an inverse function and is not one-to-one

Has an inverse function and is one-to-one

Section 9.3 Logarithmic Functions

Definition of the logarithmic function: For $x > 0$ and $b > 0$, $b \neq 1$, $y = \log_b x$ is equivalent to $b^y = x$. The function $f(x) = \log_b x$ is the logarithmic function with base b. This function is the inverse function of the exponential function with base b.

- Write $\log_2 32 = 5$ in exponential form.
$$2^5 = 32 \quad y = \log_b x \text{ means } b^y = x.$$

- Write $\sqrt{49} = 7$, or $49^{\frac{1}{2}} = 7$, in logarithmic form.
$$\frac{1}{2} = \log_{49} 7 \quad b^y = x \text{ means } y = \log_b x.$$

The graph of $f(x) = \log_b x$ can be obtained from $f(x) = b^x$ by reversing coordinates. The graph of $f(x) = \log_b x$ contains the point $(1, 0)$. If $b > 1$, the graph rises from left to right. If $0 < b < 1$, the graph falls from left to right. The y-axis is a vertical asymptote. The domain is $(0, \infty)$; the range is $(-\infty, \infty)$. $f(x) = \log x$ means $f(x) = \log_{10} x$ and is the common logarithmic function. $f(x) = \ln x$ means $f(x) = \log_e x$ and is the natural logarithmic function. The domain of $f(x) = \log_b g(x)$ consists of all x for which $g(x) > 0$.

- Find the domain: $f(x) = \log_6(4 - x)$.
$$4 - x > 0$$
$$4 > x \quad (\text{or } x < 4)$$

The domain is $(-\infty, 4)$.

Basic Logarithmic Properties

Base b $(b > 0, b \neq 1)$	Base 10 (Common Logarithms)	Base e (Natural Logarithms)
$\log_b 1 = 0$	$\log 1 = 0$	$\ln 1 = 0$
$\log_b b = 1$	$\log 10 = 1$	$\ln e = 1$
$\log_b b^x = x$	$\log 10^x = x$	$\ln e^x = x$
$b^{\log_b x} = x$	$10^{\log x} = x$	$e^{\ln x} = x$

- $\log_8 1 = 0$ because $\log_b 1 = 0$.
- $\log_4 4 = 1$ because $\log_b b = 1$.
- $\ln e^{8x} = 8x$ because $\ln e^x = x$.
- $e^{\ln \sqrt[3]{x}} = \sqrt[3]{x}$ because $e^{\ln x} = x$.
- $\log_t t^{25} = 25$ because $\log_b b^x = x$.

Definitions and Concepts	**Examples**

Section 9.4 Properties of Logarithms

Properties of Logarithms

For $M > 0$ and $N > 0$:

1. *The Product Rule*: $\log_b(MN) = \log_b M + \log_b N$

2. *The Quotient Rule*: $\log_b\left(\dfrac{M}{N}\right) = \log_b M - \log_b N$

3. *The Power Rule*: $\log_b M^p = p \log_b M$

4. *The Change-of Base Property*:

The General Property	Introducing Common Logarithms	Introducing Natural Logarithms
$\log_b M = \dfrac{\log_a M}{\log_a b}$	$\log_b M = \dfrac{\log M}{\log b}$	$\log_b M = \dfrac{\ln M}{\ln b}$

• Expand: $\log_3(81x^7)$.
$$= \log_3 81 + \log_3 x^7$$
$$= 4 + 7 \log_3 x$$

• Write as a single logarithm: $7 \ln x - 4 \ln y$.
$$= \ln x^7 - \ln y^4 = \ln\left(\frac{x^7}{y^4}\right)$$

• Evaluate: $\log_6 92$.
$$\log_6 92 = \frac{\ln 92}{\ln 6} \approx 2.5237$$

Section 9.5 Exponential and Logarithmic Equations

An exponential equation is an equation containing a variable in an exponent. Some exponential equations can be solved by expressing both sides as a power of the same base. Then set the exponents equal to each other:

$$\text{If } b^M = b^N, \text{ then } M = N.$$

Solve: $4^{2x-1} = 64$.
$$4^{2x-1} = 4^3$$
$$2x - 1 = 3$$
$$2x = 4$$
$$x = 2$$
The solution is 2 and the solution set is $\{2\}$.

If both sides of an exponential equation cannot be expressed as a power of the same base, isolate the exponential expression. Take the natural logarithm on both sides for bases other than 10 and take the common logarithm on both sides for base 10. Simplify using

$$\ln b^x = x \ln b \quad \text{or} \quad \ln e^x = x \quad \text{or} \quad \log 10^x = x.$$

Solve: $7^x = 103$.
$$\ln 7^x = \ln 103$$
$$x \ln 7 = \ln 103$$
$$x = \frac{\ln 103}{\ln 7}$$
The solution is $\dfrac{\ln 103}{\ln 7}$ and the solution set is $\left\{\dfrac{\ln 103}{\ln 7}\right\}$.

A logarithmic equation is an equation containing a variable in a logarithmic expression. Logarithmic equations in the form $\log_b x = c$ can be solved by rewriting in exponential form as $b^c = x$. When checking logarithmic equations, reject proposed solutions that produce the logarithm of a negative number or the logarithm of zero in the original equation.

Solve: $\log_2(3x - 1) = 5$.
$$2^5 = 3x - 1$$
$$32 = 3x - 1$$
$$33 = 3x$$
$$11 = x$$
The solution is 11 and the solution set is $\{11\}$.

Solve: $3 \ln 2x = 15$.
$$\ln 2x = 5$$
$$\log_e 2x = 5$$
Exponential form $\rightarrow e^5 = 2x$
$$\frac{e^5}{2} = x$$
The solution is $\dfrac{e^5}{2}$ and the solution set is $\left\{\dfrac{e^5}{2}\right\}$.

17. $f(x) = \sqrt{x+2}$

18. $f(x) = 8x^3 + 1$

23. Use the graph of f in the figure shown to draw the graph of its inverse function.

Which graphs in Exercises 19–22 represent functions that have inverse functions?

19.

20.

21.

22.

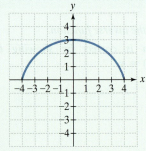

9.3 *In Exercises 24–26, write each equation in its equivalent exponential form.*

24. $\frac{1}{2} = \log_{49} 7$ **25.** $3 = \log_4 x$

26. $\log_3 81 = y$

In Exercises 27–29, write each equation in its equivalent logarithmic form.

27. $6^3 = 216$ **28.** $b^4 = 625$

29. $13^y = 874$

In Exercises 30–40, evaluate each expression without using a calculator. If evaluation is not possible, state the reason.

30. $\log_4 64$ **31.** $\log_5 \frac{1}{25}$

32. $\log_3(-9)$

33. $\log_{16} 4$ **34.** $\log_{17} 17$

35. $\log_3 3^8$ **36.** $\ln e^5$

37. $\log_3 \dfrac{1}{\sqrt{3}}$

38. $\ln \dfrac{1}{e^2}$

39. $\log \dfrac{1}{1000}$

40. $\log_3(\log_8 8)$

41. Graph $f(x) = 2^x$ and $g(x) = \log_2 x$ in the same rectangular coordinate system. Use the graphs to determine each function's domain and range.

42. Graph $f(x) = \left(\frac{1}{3}\right)^x$ and $g(x) = \log_{\frac{1}{3}} x$ in the same rectangular coordinate system. Use the graphs to determine each function's domain and range.

In Exercises 43–45, find the domain of each logarithmic function.

43. $f(x) = \log_8(x + 5)$

44. $f(x) = \log(3 - x)$

45. $f(x) = \ln(x - 1)^2$

In Exercises 46–48, simplify each expression.

46. $\ln e^{6x}$ **47.** $e^{\ln \sqrt{x}}$

48. $10^{\log 4x^2}$

49. On the Richter scale, the magnitude, R, of an earthquake of intensity I is given by $R = \log \dfrac{I}{I_0}$, where I_0 is the intensity of a barely felt zero-level earthquake. If the intensity of an earthquake is $1000I_0$, what is its magnitude on the Richter scale?

50. Students in a psychology class took a final examination. As part of an experiment to see how much of the course content they remembered over time, they took equivalent forms of the exam in monthly intervals thereafter. The average score, $f(t)$, for the group after t months is modeled by the function $f(t) = 76 - 18 \log(t + 1)$, where $0 \le t \le 12$.

   a. What was the average score when the exam was first given?

   b. What was the average score, to the nearest tenth, after 2 months? 4 months? 6 months? 8 months? one year?

   c. Use the results from parts (a) and (b) to graph f. Describe what the shape of the graph indicates in terms of the material retained by the students.

51. The formula

$$t = \frac{1}{c} \ln\left(\frac{A}{A - N}\right)$$

describes the time, t, in weeks, that it takes to achieve mastery of a portion of a task. In the formula, A represents maximum learning possible, N is the portion of the learning that is to be achieved, and c is a constant used to measure an individual's learning style. A 50-year-old man decides to start running as a way to maintain good health. He feels that the maximum rate he could ever hope to achieve is 12 miles per hour. How many weeks will it take before the man can run 5 miles per hour if $c = 0.06$ for this person?

9.4 *In Exercises 52–55, use properties of logarithms to expand each logarithmic expression as much as possible. Where possible, evaluate logarithmic expressions without using a calculator. Assume that all variables represent positive numbers.*

52. $\log_6(36x^3)$ **53.** $\log_4\left(\dfrac{\sqrt{x}}{64}\right)$

54. $\log_2\left(\dfrac{xy^2}{64}\right)$

55. $\ln\sqrt[3]{\dfrac{x}{e}}$

In Exercises 56–59, use properties of logarithms to condense each logarithmic expression. Write the expression as a single logarithm whose coefficient is 1.

56. $\log_b 7 + \log_b 3$

57. $\log 3 - 3 \log x$

58. $3 \ln x + 4 \ln y$

59. $\dfrac{1}{2}\ln x - \ln y$

In Exercises 60–61, use common logarithms or natural logarithms and a calculator to evaluate to four decimal places.

60. $\log_6 72{,}348$

61. $\log_4 0.863$

In Exercises 62–65, determine whether each equation is true or false. Where possible, show work to support your conclusion. If the statement is false, make the necessary change(s) to produce a true statement.

62. $(\ln x)(\ln 1) = 0$

63. $\log(x + 9) - \log(x + 1) = \dfrac{\log(x + 9)}{\log(x + 1)}$

64. $(\log_2 x)^4 = 4 \log_2 x$

65. $\ln e^x = x \ln e$

9.5 *In Exercises 66–71, solve each exponential equation. Where necessary, express the solution set in terms of natural logarithms and use a calculator to obtain a decimal approximation, correct to two decimal places, for the solution.*

66. $2^{4x-2} = 64$

67. $125^x = 25$

68. $9^x = \dfrac{1}{27}$

69. $8^x = 12{,}143$

70. $9e^{5x} = 1269$

71. $30e^{0.045x} = 90$

In Exercises 72–81, solve each logarithmic equation.

72. $\log_5 x = -3$

73. $\log x = 2$

74. $\log_4(3x - 5) = 3$

75. $\ln x = -1$

76. $3 + 4 \ln(2x) = 15$

77. $\log_2(x + 3) + \log_2(x - 3) = 4$

78. $\log_3(x - 1) - \log_3(x + 2) = 2$

79. $\log_4(3x - 5) = \log_4 3$

80. $\ln(x + 4) - \ln(x + 1) = \ln x$

81. $\log_6(2x + 1) = \log_6(x - 3) + \log_6(x + 5)$

82. The function $P(x) = 14.7e^{-0.21x}$ models the average atmospheric pressure, $P(x)$, in pounds per square inch, at an altitude of x miles above sea level. The atmospheric pressure at the peak of Mt. Everest, the world's highest mountain, is 4.6 pounds per square inch. How many miles above sea level, to the nearest tenth of a mile, is the peak of Mt. Everest?

83. **Newest Dinosaur: The PC-osaurus?** For the period from 2009 through 2012, worldwide PC sales stayed relatively flat, while tablet sales skyrocketed. The bar graph shows worldwide PC and tablet sales, in millions, from 2009 through 2012.

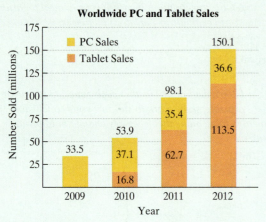

Worldwide PC and Tablet Sales

Source: Canalys

The function

$$f(t) = 33.4(1.66)^t$$

models worldwide PC and tablet sales combined, $f(t)$, in millions, t years after 2009. When does this model project that 421 million PC and tablets were sold? Round to the nearest year. Based on the relatively flat PC sales over the four years shown by the graph, estimate the number of tablet sales for that year.

84. **60 + Years after "Brown v. Board"** In 1954, the Supreme Court ended legal segregation in U.S. public schools. The bar graph shows the percentage of black and white U.S. adults ages 25 or older who completed high school since the end of segregation.

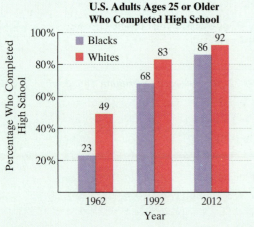

U.S. Adults Ages 25 or Older Who Completed High School

Source: Pew Research Center

The functions

$$B(x) = 16 \, lnx + 23 \text{ and } W(x) = 11 \, lnx + 49$$

model the percentage of black adults, $B(x)$, and white adults, $W(x)$, who had completed high school x years after 1961.

a. Use the appropriate model to determine the percentage of white adults who had completed high school in 2012. Round to the nearest percent. How well does this rounded value model the percents displayed by the graph?

b. Use the appropriate model to determine when 90% of black adults will have completed high school. Round to the nearest year.

85. Use the compound interest formula

$$A = P\left(1 + \frac{r}{n}\right)^{nt}$$

to solve this problem. How long, to the nearest tenth of a year, will it take $12,500 to grow to $20,000 at 6.5% annual interest compounded quarterly?

Use the compound interest formula

$$A = Pe^{rt}$$

to solve Exercises 86–87.

86. How long, to the nearest tenth of a year, will it take $50,000 to triple in value at 7.5% annual interest compounded continuously?

87. What interest rate is required for an investment subject to continuous compounding to triple in 5 years?

9.6

88. According to the U.S. Bureau of the Census, in 2000 there were 35.3 million residents of Hispanic origin living in the United States. By 2010, the number had increased to 50.5 million. The exponential growth function $A = 35.3e^{kt}$ describes the U.S. Hispanic population, A, in millions, t years after 2000.

a. Find k, correct to three decimal places.

b. Use the resulting model to project the Hispanic resident population in 2015.

c. In which year will the Hispanic resident population reach 70 million?

89. Use the exponential decay model, $A = A_0e^{kt}$, to solve this exercise. The half-life of polonium-210 is 140 days. How long will it take for a sample of this substance to decay to 20% of its original amount?

Exercises 90–92 present data in the form of tables. For each data set shown by the table,

a. *Create a scatter plot for the data.*

b. *Use the scatter plot to determine whether an exponential function, a logarithmic function, or a linear function is the best choice for modeling the data.*

90. Number of Seriously Mentally Ill Adults in the United States

Year	Number of Seriously Mentally Ill Adults
2006	9.0
2008	9.2
2010	9.4
2012	9.6

Source: U.S. Census Bureau

91. Percentage of Moderate Alcohol Users in the United States (Not Binge or Heavy Drinkers)

Age	Percentage of Moderate Drinkers
20	15%
25	24%
30	28%
35	32%
40	34%
45	35%

Source: Substance Abuse and Mental Health Services Administration

92. U.S. Electric Car Sales

Year	Number of Cars Sold
2010	5000
2011	19,000
2012	53,000
2013	98,000

Source: Electric Drive Transportation Association

In Exercises 93–94, rewrite the equation in terms of base e. Express the answer in terms of a natural logarithm and then round to three decimal places.

93. $y = 73(2.6)^x$

94. $y = 6.5(0.43)^x$

95. The figure shows world population projections through the year 2150. The data are from the United Nations Family Planning Program and are based on optimistic or pessimistic expectations for successful control of human population growth. Suppose that you are interested in modeling these data using exponential, logarithmic, linear, and quadratic functions. Which function would you use to model each of the projections? Explain your choices. For the choice corresponding to a quadratic model, would your formula involve one with a positive or negative leading coefficient? Explain.

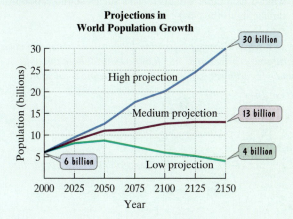

Projections in World Population Growth

CHAPTER 9 TEST

Step-by-step test solutions are found on the Chapter Test Prep Videos available in MyMathLab® or on YouTube (search "BlitzerInterAlg7e" and click on "Channels").

1. Graph $f(x) = 2^x$ and $g(x) = 2^{x+1}$ in the same rectangular coordinate system.

2. Use $A = P\left(1 + \dfrac{r}{n}\right)^{nt}$ and $A = Pe^{rt}$ to solve this problem.

Suppose you have $3000 to invest. Which investment yields the greater return over 10 years: 6.5% compounded semiannually or 6% compounded continuously? How much more (to the nearest dollar) is yielded by the better investment?

3. If $f(x) = x^2 + x$ and $g(x) = 3x - 1$, find $(f \circ g)(x)$ and $(g \circ f)(x)$.

4. If $f(x) = 5x - 7$, find $f^{-1}(x)$.

5. A function f models the amount given to charity as a function of income. The graph of f is shown in the figure.

a. Explain why f has an inverse that is a function.

b. Find $f(80)$.

c. Describe in practical terms the meaning of $f^{-1}(2000)$.

6. Write in exponential form: $\log_5 125 = 3$.

7. Write in logarithmic form: $\sqrt{36} = 6$.

8. Graph $f(x) = 3^x$ and $g(x) = \log_3 x$ in the same rectangular coordinate system. Use the graphs to determine each function's domain and range.

In Exercises 9–11, simplify each expression.

9. $\ln e^{5x}$ 10. $\log_b b$ 11. $\log_6 1$

12. Find the domain: $f(x) = \log_5 (x - 7)$.

13. On the decibel scale, the loudness of a sound, in decibels, is given by $D = 10 \log \dfrac{I}{I_0}$, where I is the intensity of the sound, in watts per meter^2, and I_0 is the intensity of a sound barely audible to the human ear. If the intensity of a sound is $10^{12} I_0$, what is its loudness in decibels? (Such a sound is potentially damaging to the ear.)

In Exercises 14–15, use properties of logarithms to expand each logarithmic expression as much as possible. Where possible, evaluate logarithmic expressions without using a calculator.

14. $\log_4 (64x^5)$

15. $\log_3 \left(\dfrac{\sqrt[3]{x}}{81} \right)$

In Exercises 16–17, write each expression as a single logarithm.

16. $6 \log x + 2 \log y$

17. $\ln 7 - 3 \ln x$

18. Use a calculator to evaluate $\log_{15} 71$ to four decimal places.

In Exercises 19–26, solve each equation.

19. $3^{x-2} = 81$

20. $5^x = 1.4$

21. $400e^{0.005x} = 1600$

22. $\log_{25} x = \dfrac{1}{2}$

23. $\log_6 (4x - 1) = 3$

24. $2 \ln(3x) = 8$

25. $\log x + \log(x + 15) = 2$

26. $\ln(x - 4) - \ln(x + 1) = \ln 6$

27. The function
$$A = 82.3e^{-0.004t}$$
models the population of Germany, A, in millions, t years after 2010.

a. What was the population of Germany in 2010?

b. Is the population of Germany increasing or decreasing? Explain.

c. In which year will the population of Germany be 79.1 million?

Use the formulas
$$A = P\left(1 + \frac{r}{n}\right)^{nt} \quad \text{and} \quad A = Pe^{rt}$$
to solve Exercises 28–29.

28. How long, to the nearest tenth of a year, will it take $4000 to grow to $8000 at 5% annual interest compounded quarterly?

29. What interest rate is required for an investment subject to continuous compounding to double in 10 years?

30. The 2010 population of Asia was 4121 million; in 2050, it is projected to be 5231 million. Write an exponential growth function that describes the population of Asia, in millions, t years after 2010.

31. Use the exponential decay model for carbon-14, $A = A_0 e^{-0.000121t}$, to solve this exercise. Bones of a prehistoric man were discovered and contained 5% of the original amount of carbon-14. How long ago did the man die?

In Exercises 32–35, determine whether the values in each table belong to an exponential function, a logarithmic function, a linear function, or a quadratic function.

32.

x	y
0	3
1	1
2	−1
3	−3
4	−5

33.

x	y
$\frac{1}{3}$	−1
1	0
3	1
9	2
27	3

34.

x	y
0	1
1	5
2	25
3	125
4	625

35.

x	y
0	12
1	3
2	0
3	3
4	12

36. Rewrite $y = 96(0.38)^x$ in terms of base e. Express the answer in terms of a natural logarithm and then round to three decimal places.

CUMULATIVE REVIEW EXERCISES   (CHAPTERS 1–9)

In Exercises 1–7, solve each equation, inequality, or system.

1. $8 - (4x - 5) = x - 7$

2. $\begin{cases} 5x + 4y = 22 \\ 3x - 8y = -18 \end{cases}$

3. $\begin{cases} -3x + 2y + 4z = 6 \\ 7x - y + 3z = 23 \\ 2x + 3y + z = 7 \end{cases}$

4. $|x - 1| > 3$

5. $\sqrt{x + 4} - \sqrt{x - 4} = 2$

6. $x - 4 \geq 0$ and $-3x \leq -6$

7. $2x^2 = 3x - 2$

In Exercises 8–12, graph each function, equation, or inequality in a rectangular coordinate system.

8. $3x = 15 + 5y$

9. $2x - 3y > 6$

10. $f(x) = -\dfrac{1}{2}x + 1$

11. $f(x) = x^2 + 6x + 8$

12. $f(x) = (x - 3)^2 - 4$

13. Evaluate:
$$\begin{vmatrix} 3 & 1 & 0 \\ 0 & 5 & -6 \\ -2 & -1 & 0 \end{vmatrix}.$$

14. Solve for c: $A = \dfrac{cd}{c + d}$.

In Exercises 15–17, let $f(x) = x^2 + 3x - 15$ and $g(x) = x - 2$. Find each indicated expression.

15. $f(g(x))$

16. $g(f(x))$

17. $g(a + h) - g(a)$

18. If $f(x) = 7x - 3$, find $f^{-1}(x)$.

In Exercises 19–20, find the domain of each function.

19. $f(x) = \dfrac{x - 2}{x^2 - 3x + 2}$

20. $f(x) = \ln(2x - 8)$

21. Write the equation of the linear function whose graph contains the point $(-2, 4)$ and is perpendicular to the line whose equation is $2x + y = 10$.

In Exercises 22–26, perform the indicated operations and simplify, if possible.

22. $\dfrac{-5x^3y^7}{15x^4y^{-2}}$

23. $(4x^2 - 5y)^2$

24. $(5x^3 - 24x^2 + 9) \div (5x + 1)$

25. $\dfrac{\sqrt[3]{32xy^{10}}}{\sqrt[3]{2xy^2}}$

26. $\dfrac{x + 2}{x^2 - 6x + 8} + \dfrac{3x - 8}{x^2 - 5x + 6}$

In Exercises 27–28, factor completely.

27. $x^4 - 4x^3 + 8x - 32$

28. $2x^2 + 12xy + 18y^2$

29. Write as a single logarithm whose coefficient is 1:
$$2 \ln x - \frac{1}{2}\ln y.$$

30. The length of a rectangular carpet is 4 feet greater than twice its width. If the area is 48 square feet, find the carpet's length and width.

31. Working alone, you can mow the lawn in 2 hours and your sister can do it in 3 hours. How long will it take you to do the job if you work together?

32. Your motorboat can travel 15 miles per hour in still water. Traveling with the river's current, the boat can cover 20 miles in the same time it takes to go 10 miles against the current. Find the rate of the current.

33. Use the formula for continuous compounding, $A = Pe^{rt}$, to solve this problem. What interest rate is required for an investment of $6000 subject to continuous compounding to grow to $18,000 in 10 years?

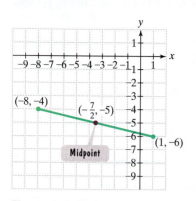

Figure 10.3 Finding a line segment's midpoint

EXAMPLE 2   Using the Midpoint Formula

Find the midpoint of the line segment with endpoints $(1, -6)$ and $(-8, -4)$.

Solution   To find the coordinates of the midpoint, we average the coordinates of the endpoints.

$$\text{Midpoint} = \left(\frac{1 + (-8)}{2}, \frac{-6 + (-4)}{2} \right) = \left(\frac{-7}{2}, \frac{-10}{2} \right) = \left(-\frac{7}{2}, -5 \right)$$

Average the x-coordinates.   Average the y-coordinates.

Figure 10.3 illustrates that the point $\left(-\frac{7}{2}, -5 \right)$ is midway between the points $(1, -6)$ and $(-8, -4)$. ■

✓ **CHECK POINT 2**   Find the midpoint of the line segment with endpoints $(1, 2)$ and $(7, -3)$.

Circles

Our goal is to translate a circle's geometric definition into an equation. We begin with this geometric definition.

Definition of a Circle

A **circle** is the set of all points in a plane that are equidistant from a fixed point, called the **center**. The fixed distance from the circle's center to any point on the circle is called the **radius**.

Figure 10.4 is our starting point for obtaining a circle's equation. We've placed the circle into a rectangular coordinate system. The circle's center is (h, k) and its radius is r. We let (x, y) represent the coordinates of any point on the circle.

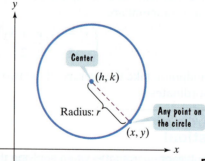

Figure 10.4 A circle centered at (h, k) with radius r

What does the geometric definition of a circle tell us about the point (x, y) in **Figure 10.4**? The point is on the circle if and only if its distance from the center is r. We can use the distance formula to express this idea algebraically:

The distance between (x, y) and (h, k) is always r.

$$\sqrt{(x - h)^2 + (y - k)^2} = r.$$

Squaring both sides of $\sqrt{(x - h)^2 + (y - k)^2} = r$ yields the *standard form of the equation of a circle*.

3 Write the standard form of a circle's equation.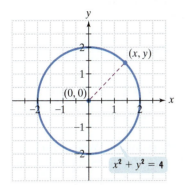

The Standard Form of the Equation of a Circle

The **standard form of the equation of a circle** with center (h, k) and radius r is

$$(x - h)^2 + (y - k)^2 = r^2.$$

EXAMPLE 3   Finding the Standard Form of a Circle's Equation

Write the standard form of the equation of the circle with center $(0, 0)$ and radius 2. Graph the circle.

Solution   The center is $(0, 0)$. Because the center is represented as (h, k) in the standard form of the equation, $h = 0$ and $k = 0$. The radius is 2, so we will let $r = 2$ in the equation.

$$(x - h)^2 + (y - k)^2 = r^2 \qquad \text{This is the standard form of a circle's equation.}$$
$$(x - 0)^2 + (y - 0)^2 = 2^2 \qquad \text{Substitute 0 for } h, \text{ 0 for } k, \text{ and 2 for } r.$$
$$x^2 + y^2 = 4 \qquad \text{Simplify.}$$

The standard form of the equation of the circle is $x^2 + y^2 = 4$. **Figure 10.5** shows the graph.   ■

✓ CHECK POINT 3   Write the standard form of the equation of the circle with center $(0, 0)$ and radius 4.

Figure 10.5   The graph of $x^2 + y^2 = 4$

Using Technology

To graph a circle with a graphing utility, first solve the equation for y.

$$x^2 + y^2 = 4$$
$$y^2 = 4 - x^2$$
$$y = \pm\sqrt{4 - x^2}$$

y is not a function of x.

Graph the two equations

$$y_1 = \sqrt{4 - x^2} \quad \text{and} \quad y_2 = -\sqrt{4 - x^2}$$

in the same viewing rectangle. The graph of $y_1 = \sqrt{4 - x^2}$ is the top semicircle because y is always positive. The graph of $y_2 = -\sqrt{4 - x^2}$ is the bottom semicircle because y is always negative. Use a ZOOM SQUARE setting so that the circle looks like a circle. (Many graphing utilities have problems connecting the two semicircles because the segments directly to the left and to the right of the center become nearly vertical.)

Example 3 and Check Point 3 involved circles centered at the origin. The standard form of the equation of all such circles is $x^2 + y^2 = r^2$, where r is the circle's radius. Now, let's consider a circle whose center is not at the origin.

EXAMPLE 4   Finding the Standard Form of a Circle's Equation

Write the standard form of the equation of the circle with center $(-2, 3)$ and radius 4.

Solution   The center is $(-2, 3)$. Because the center is represented as (h, k) in the standard form of the equation, $h = -2$ and $k = 3$. The radius is 4, so we will let $r = 4$ in the equation.

$$(x - h)^2 + (y - k)^2 = r^2 \qquad \text{This is the standard form of a circle's equation.}$$
$$[x - (-2)]^2 + (y - 3)^2 = 4^2 \qquad \text{Substitute } -2 \text{ for } h, \text{ 3 for } k, \text{ and 4 for } r.$$
$$(x + 2)^2 + (y - 3)^2 = 16 \qquad \text{Simplify.}$$

The standard form of the equation of the circle is $(x + 2)^2 + (y - 3)^2 = 16$.   ■

☑ **CHECK POINT 4**   Write the standard form of the equation of the circle with center $(5, -6)$ and radius 10.

4 Give the center and radius of a circle whose equation is in standard form. ▶

EXAMPLE 5   Using the Standard Form of a Circle's Equation to Graph the Circle

a. Find the center and radius of the circle whose equation is

$$(x - 2)^2 + (y + 4)^2 = 9.$$

b. Graph the equation.

Solution

a. We begin by finding the circle's center, (h, k), and its radius, r. We can find the values for h, k, and r by comparing the given equation to the standard form of the equation of a circle, $(x - h)^2 + (y - k)^2 = r^2$.

$$(x - 2)^2 + (y + 4)^2 = 9$$
$$(x - 2)^2 + (y - (-4))^2 = 3^2$$

This is $(x - h)^2$, with $h = 2$.   This is $(y - k)^2$, with $k = -4$.   This is r^2, with $r = 3$.

We see that $h = 2$, $k = -4$, and $r = 3$. Thus, the circle has center $(h, k) = (2, -4)$ and a radius of 3 units.

b. To graph this circle, first plot the center $(2, -4)$. Because the radius is 3, you can locate at least four points on the circle by going out three units to the right, to the left, up, and down from the center.

The points three units to the right and to the left of $(2, -4)$ are $(5, -4)$ and $(-1, -4)$, respectively. The points three units up and down from $(2, -4)$ are $(2, -1)$ and $(2, -7)$, respectively. Using these points, we obtain the graph in **Figure 10.6**. ■

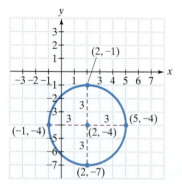

Figure 10.6 The graph of $(x - 2)^2 + (y + 4)^2 = 9$

Great Question!

What's the bottom line when identifying h and k in the standard form of a circle's equation?

It's easy to make sign errors when finding h and k, the coordinates of a circle's center, (h, k). Keep in mind that h and k are the numbers that *follow the subtraction signs* in a circle's equation:

$$(x - 2)^2 + (y + 4)^2 = 9$$
$$(x - 2)^2 + (y - (-4))^2 = 9.$$

The number *after* the subtraction is 2: $h = 2$.   The number *after* the subtraction is -4: $k = -4$.

☑ **CHECK POINT 5**   Find the center and radius of the circle whose equation is

$$(x + 3)^2 + (y - 1)^2 = 4$$

and graph the equation.

If we square $x - 2$ and $y + 4$ in the standard form of the equation of Example 5, we obtain another form for the circle's equation.

$$(x - 2)^2 + (y + 4)^2 = 9$$ This is the standard form of the equation from Example 5.

$$x^2 - 4x + 4 + y^2 + 8y + 16 = 9$$ Square $x - 2$ and $y + 4$.

$$x^2 + y^2 - 4x + 8y + 20 = 9 \qquad \text{Combine numerical terms and rearrange terms.}$$
$$x^2 + y^2 - 4x + 8y + 11 = 0 \qquad \text{Subtract 9 from both sides.}$$

This result suggests that an equation in the form $x^2 + y^2 + Dx + Ey + F = 0$ can represent a circle. This is called the *general form of the equation of a circle*.

> ### The General Form of the Equation of a Circle
>
> The **general form of the equation of a circle** is
>
> $$x^2 + y^2 + Dx + Ey + F = 0.$$

5 Convert the general form of a circle's equation to standard form. ▶

We can convert the general form of the equation of a circle to the standard form $(x - h)^2 + (y - k)^2 = r^2$. We do so by completing the square on x and y. Let's see how this is done.

Great Question!

Egad! I forgot how to complete the square. Any suggestions?

To complete the square on $x^2 + bx$, take half the coefficient of x. Then square this number. By adding the square of half the coefficient of x, a perfect square trinomial will result. For more detail, see pages 586–587 of Section 8.1.

EXAMPLE 6 Converting the General Form of a Circle's Equation to Standard Form and Graphing the Circle

Write in standard form and graph: $x^2 + y^2 + 4x - 6y - 23 = 0$.

Solution Because we plan to complete the square on both x and y, let's rearrange the terms so that x-terms are arranged in descending order, y-terms are arranged in descending order, and the constant term appears on the right.

$$x^2 + y^2 + 4x - 6y - 23 = 0 \qquad \text{This is the given equation.}$$

$$(x^2 + 4x \qquad) + (y^2 - 6y \qquad) = 23 \qquad \text{Rewrite in anticipation of completing the square.}$$

$$(x^2 + 4x + 4) + (y^2 - 6y + 9) = 23 + 4 + 9 \qquad$$

Complete the square on x: $\frac{1}{2} \cdot 4 = 2$ and $2^2 = 4$, so add 4 to both sides. Complete the square on y: $\frac{1}{2}(-6) = -3$ and $(-3)^2 = 9$, so add 9 to both sides.

> Remember that numbers added on the left side must also be added on the right side.

$$(x + 2)^2 + (y - 3)^2 = 36 \qquad \text{Factor on the left and add on the right.}$$

This last equation is in standard form. We can identify the circle's center and radius by comparing this equation to the standard form of the equation of a circle, $(x - h)^2 + (y - k)^2 = r^2$.

$$(x + 2)^2 + (y - 3)^2 = 36$$
$$(x - (-2))^2 + (y - 3)^2 = 6^2$$

This is $(x - h)^2$, with $h = -2$.   This is $(y - k)^2$, with $k = 3$.   This is r^2, with $r = 6$.

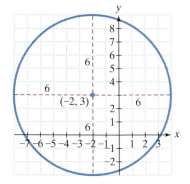

Figure 10.7 The graph of $(x + 2)^2 + (y - 3)^2 = 36$

We use the center, $(h, k) = (-2, 3)$, and the radius, $r = 6$, to graph the circle. The graph is shown in **Figure 10.7**. ■

We can sketch the ellipse by locating endpoints on the major and minor axes.

$$\frac{(x-1)^2}{2^2} + \frac{(y+2)^2}{3^2} = 1$$

| Endpoints of the minor axis are 2 units to the right and left of the center. | Endpoints of the major axis (the vertices) are 3 units up and down from the center. |

We categorize the observations in the voice balloons as follows:

For a Vertical Major Axis with Center (1, −2)

	Vertices	Endpoints of Minor Axis	
3 units above and below center	(1, −2 + 3) = (1, 1)	(1 + 2, −2) = (3, −2)	2 units right and left of center
	(1, −2 − 3) = (1, −5)	(1 − 2, −2) = (−1, −2)	

Using the center and these four points, we can sketch the ellipse, shown in **Figure 10.17.** ∎

$$\frac{(x-1)^2}{4} + \frac{(y+2)^2}{9} = 1$$

Figure 10.17 The graph of an ellipse centered at (1, −2)

✓ **CHECK POINT 3**   Graph the ellipse: $\dfrac{(x+1)^2}{9} + \dfrac{(y-2)^2}{4} = 1.$

3 Solve applied problems involving ellipses. ▶

Applications

Ellipses have many applications. German scientist Johannes Kepler (1571–1630) showed that the planets in our solar system move in elliptical orbits, with the sun at a focus. Earth satellites also travel in elliptical orbits, with Earth at a focus.

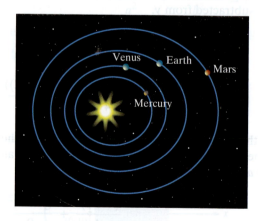

Planets move in elliptical orbits.

One intriguing aspect of the ellipse is that a ray of light or a sound wave emanating from one focus will be reflected from the ellipse to exactly the other focus. A whispering gallery is an elliptical room with an elliptical, dome-shaped ceiling. People standing at the foci can whisper and hear each other quite clearly, while persons in other locations in the room cannot hear them. Statuary Hall in the U.S. Capitol Building is elliptical. President John Quincy Adams, while a member of the House of Representatives, was aware of this acoustical phenomenon. He situated his desk at a focal point of the elliptical ceiling, easily eavesdropping on the private conversations of other House members located near the other focus.

Whispering in an elliptical dome

The elliptical reflection principle is used in a procedure for disintegrating kidney stones. The patient is placed within a device that is elliptical in shape. The patient is placed so the kidney is centered at one focus, while ultrasound waves from the other focus hit the walls and are reflected to the kidney stone. The convergence of the ultrasound waves at the kidney stone causes vibrations that shatter it into fragments. The small pieces can then be passed painlessly through the patient's system. The patient recovers in days, as opposed to up to six weeks if surgery is used instead.

Disintegrating kidney stones

Ellipses are often used for supporting arches of bridges and in tunnel construction. This application forms the basis of our next example.

EXAMPLE 4   An Application Involving an Ellipse

A semielliptical archway over a one-way road has a height of 10 feet and a width of 40 feet (see **Figure 10.18**). Your truck has a width of 10 feet and a height of 9 feet. Will your truck clear the opening of the archway?

Solution   Because your truck's width is 10 feet, to determine the clearance, we must find the height of the archway 5 feet from the center. If that height is 9 feet or less, the truck will not clear the opening.

Figure 10.18 A semielliptical archway

In **Figure 10.19**, we've constructed a coordinate system with the *x*-axis on the ground and the origin at the center of the archway. Also shown is the truck, whose height is 9 feet.

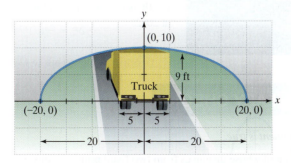

Figure 10.19

Using the equation $\frac{x^2}{a^2} + \frac{y^2}{b^2} = 1$, we can express the equation of the blue archway in **Figure 10.19** as $\frac{x^2}{20^2} + \frac{y^2}{10^2} = 1$, or $\frac{x^2}{400} + \frac{y^2}{100} = 1$.

Standard Forms of the Equations of a Hyperbola

The **standard form of the equation of a hyperbola** with center at the origin is

$$\frac{x^2}{a^2} - \frac{y^2}{b^2} = 1 \quad \text{or} \quad \frac{y^2}{a^2} - \frac{x^2}{b^2} = 1.$$

Figure 10.23(a) illustrates that for the equation on the left, the transverse axis lies on the *x*-axis. **Figure 10.23(b)** illustrates that for the equation on the right, the transverse axis lies on the *y*-axis. The vertices are *a* units from the center and the foci are *c* units from the center.

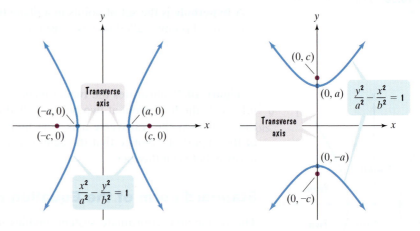

Figure 10.23(a) Transverse axis lies on the *x*-axis.

Figure 10.23(b) Transverse axis lies on the *y*-axis.

Great Question!

How can I tell from a hyperbola's equation whether the transverse axis is horizontal or vertical?

When the x^2-term is preceded by a plus sign, the transverse axis is horizontal. When the y^2-term is preceded by a plus sign, the transverse axis is vertical.

① Locate a hyperbola's vertices. ▶

Using the Standard Form of the Equation of a Hyperbola

We can use the standard form of the equation of a hyperbola to find its vertices. Because the vertices are *a* units from the center, begin by identifying a^2 in the equation. In the standard form of a hyperbola's equation, **a^2 is the number under the variable whose term is preceded by a plus sign (+)**. If the x^2-term is preceded by a plus sign, the transverse axis lies along the *x*-axis. Thus, the vertices are *a* units to the left and right of the origin. If the y^2-term is preceded by a plus sign, the transverse axis lies along the *y*-axis. Thus, the vertices are *a* units above and below the origin.

EXAMPLE 1   Finding Vertices from a Hyperbola's Equation

Find the vertices for each of the following hyperbolas with the given equation:

a. $\dfrac{x^2}{16} - \dfrac{y^2}{9} = 1$

b. $\dfrac{y^2}{9} - \dfrac{x^2}{16} = 1.$

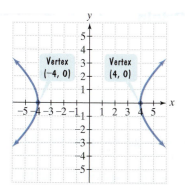

Figure 10.24 The graph of
$$\frac{x^2}{16} - \frac{y^2}{9} = 1$$

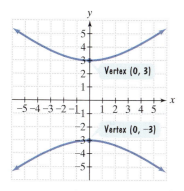

Figure 10.25 The graph of
$$\frac{y^2}{9} - \frac{x^2}{16} = 1$$

Solution Both equations are in standard form. We begin by identifying a^2 and b^2 in each equation.

a. The first equation is in the form $\dfrac{x^2}{a^2} - \dfrac{y^2}{b^2} = 1$.

$$\frac{x^2}{16} - \frac{y^2}{9} = 1$$

> $a^2 = 16$. This is the denominator of the term preceded by a plus sign.

> $b^2 = 9$. This is the denominator of the term preceded by a minus sign.

Because the x^2-term is preceded by a plus sign, the transverse axis lies along the x-axis. Thus, the vertices are a units to the *left* and *right* of the origin. Based on the standard form of the equation, we know that the vertices are $(-a, 0)$ and $(a, 0)$. Because $a^2 = 16$, $a = 4$. Thus, the vertices are $(-4, 0)$ and $(4, 0)$, shown in **Figure 10.24**.

b. The second given equation is in the form $\dfrac{y^2}{a^2} - \dfrac{x^2}{b^2} = 1$.

$$\frac{y^2}{9} - \frac{x^2}{16} = 1$$

> $a^2 = 9$. This is the denominator of the term preceded by a plus sign.

> $b^2 = 16$. This is the denominator of the term preceded by a minus sign.

Because the y^2-term is preceded by a plus sign, the transverse axis lies along the y-axis. Thus, the vertices are a units *above* and *below* the origin. Based on the standard form of the equation, we know that the vertices are $(0, -a)$ and $(0, a)$. Because $a^2 = 9$, $a = 3$. Thus, the vertices are $(0, -3)$ and $(0, 3)$, shown in **Figure 10.25**. ∎

✓ **CHECK POINT 1**   Find the vertices for each of the following hyperbolas with the given equation:

a. $\dfrac{x^2}{25} - \dfrac{y^2}{16} = 1$ **b.** $\dfrac{y^2}{25} - \dfrac{x^2}{16} = 1$.

The Asymptotes of a Hyperbola

 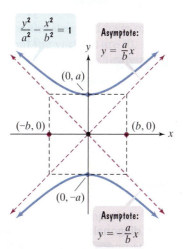

As x and y get larger, the two branches of the graph of a hyperbola approach a pair of intersecting lines, called **asymptotes**. The asymptotes pass through the center of the hyperbola and are helpful in graphing hyperbolas.

Figure 10.26 shows the asymptotes for the graphs of hyperbolas centered at the origin. The asymptotes pass through the corners of a rectangle. Note that the dimensions of this rectangle are $2a$ by $2b$.

Why are $y = \pm\dfrac{b}{a}x$ the asymptotes for a hyperbola whose transverse axis is horizontal? The proof can be found in the appendix.

Figure 10.26 Asymptotes of a hyperbola

The equation $x = y^2$ interchanges the variables in the equation $y = x^2$. By interchanging x and y in the two forms of a parabola's equation,

$$y = a(x - h)^2 + k \quad \text{and} \quad y = ax^2 + bx + c,$$

we can obtain the forms of the equations of parabolas that open to the right or to the left.

Parabolas Opening to the Left or to the Right

The graphs of

$$x = a(y - k)^2 + h \quad \text{and} \quad x = ay^2 + by + c$$

are parabolas opening to the left or to the right.

1. If $a > 0$, the graph opens to the right. If $a < 0$, the graph opens to the left.
2. The vertex of $x = a(y - k)^2 + h$ is (h, k).

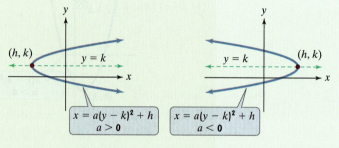

3. The y-coordinate of the vertex of $x = ay^2 + by + c$ is $y = -\dfrac{b}{2a}$.

---

1 Graph horizontal parabolas.

Graphing Parabolas Opening to the Left or the Right

Here is a procedure for graphing horizontal parabolas that are not functions. Notice how this procedure is similar to the one that we used in Chapter 8 for graphing vertical parabolas.

Graphing Horizontal Parabolas

To graph $x = a(y - k)^2 + h$ or $x = ay^2 + by + c$,

1. Determine whether the parabola opens to the left or to the right. If $a > 0$, it opens to the right. If $a < 0$, it opens to the left.
2. Determine the vertex of the parabola. The vertex of $x = a(y - k)^2 + h$ is (h, k). The y-coordinate of the vertex of $x = ay^2 + by + c$ is $y = -\dfrac{b}{2a}$.

   Substitute this value of y into the equation to find the x-coordinate.
3. Find the x-intercept by replacing y with 0.
4. Find any y-intercepts by replacing x with 0. Solve the resulting quadratic equation for y.
5. Plot the intercepts and the vertex. Connect these points with a smooth curve. If additional points are needed to obtain a more accurate graph, select values for y on each side of the axis of symmetry and compute values for x.

EXAMPLE 1 Graphing a Horizontal Parabola in the Form $x = a(y - k)^2 + h$

Graph: $x = -3(y - 1)^2 + 2$.

Solution We can graph this equation by following the steps in the preceding box. We begin by identifying values for a, k, and h.

$$x = a(y - k)^2 + h$$

$$\boxed{a = -3} \quad \boxed{k = 1} \quad \boxed{h = 2}$$

$$x = -3(y - 1)^2 + 2$$

Step 1. Determine how the parabola opens. Note that a, the coefficient of the squared expression in y, is -3. Thus, $a < 0$; this negative value tells us that the parabola opens to the left.

Step 2. Find the vertex. The vertex of the parabola is at (h, k). Because $k = 1$ and $h = 2$, the parabola has its vertex at $(2, 1)$.

Step 3. Find the x-intercept. Replace y with 0 in $x = -3(y - 1)^2 + 2$.

$$x = -3(0 - 1)^2 + 2 = -3(-1)^2 + 2 = -3(1) + 2 = -1$$

The x-intercept is -1. The parabola passes through $(-1, 0)$.

Step 4. Find the y-intercepts. Replace x with 0 in the given equation.

$x = -3(y - 1)^2 + 2$	This is the given equation.
$0 = -3(y - 1)^2 + 2$	Replace x with O.
$3(y - 1)^2 = 2$	Solve for y. Add $3(y - 1)^2$ to both sides of the equation.
$(y - 1)^2 = \dfrac{2}{3}$	Divide both sides by 3.
$y - 1 = \sqrt{\dfrac{2}{3}}$ or $y - 1 = -\sqrt{\dfrac{2}{3}}$	Apply the square root property.
$y = 1 + \sqrt{\dfrac{2}{3}} \qquad y = 1 - \sqrt{\dfrac{2}{3}}$	Add 1 to both sides in each equation.
$y \approx 1.8 \qquad\qquad y \approx 0.2$	Use a calculator.

The y-intercepts are approximately 1.8 and 0.2. The parabola passes through approximately $(0, 1.8)$ and $(0, 0.2)$.

Step 5. Graph the parabola. With a vertex at $(2, 1)$, an x-intercept at -1, and y-intercepts at approximately 1.8 and 0.2, the graph of the parabola is shown in **Figure 10.33**. The axis of symmetry is the horizontal line whose equation is $y = 1$.

Figure 10.33 The graph of $x = -3(y - 1)^2 + 2$

x	y
−1	2
−10	−1

Choose values for y.

Compute values for x using
$x = -3(y − 1)^2 + 2$:
$x = -3(2 − 1)^2 + 2 = -3 \cdot 1^2 + 2 = -1$
$x = -3(-1 − 1)^2 + 2 = -3(-2)^2 + 2 = -10.$

We can possibly improve our graph in **Figure 10.33** by finding additional points on the parabola. Choose values of y on each side of the axis of symmetry, $y = 1$. We use $y = 2$ and $y = -1$. Then we compute values for x. The values in the table of coordinates show that $(-1, 2)$ and $(-10, -1)$ are points on the parabola. Locate each point in **Figure 10.33** on the previous page. ∎

✅ **CHECK POINT 1** Graph: $x = -(y − 2)^2 + 1.$

EXAMPLE 2 Graphing a Horizontal Parabola in the Form $x = ay^2 + by + c$

Graph: $x = y^2 + 4y − 5.$

Solution

Step 1. Determine how the parabola opens. Note that a, the coefficient of y^2, is 1. Thus $a > 0$; this positive value tells us that the parabola opens to the right.

Step 2. Find the vertex. We know that the y-coordinate of the vertex is $y = -\dfrac{b}{2a}$. We identify a, b, and c in $x = ay^2 + by + c$.

$$x = y^2 + 4y − 5$$

$$a = 1 \qquad b = 4 \qquad c = -5$$

Substitute the values of a and b into the equation for the y-coordinate:

$$y = -\frac{b}{2a} = -\frac{4}{2 \cdot 1} = -2.$$

The y-coordinate of the vertex is -2. We substitute -2 for y into the parabola's equation, $x = y^2 + 4y − 5$, to find the x-coordinate:

$$x = (-2)^2 + 4(-2) − 5 = 4 − 8 − 5 = -9.$$

The vertex is at $(-9, -2)$.

Step 3. Find the x-intercept. Replace y with 0 in $x = y^2 + 4y − 5$.

$$x = 0^2 + 4 \cdot 0 − 5 = -5$$

The x-intercept is -5. The parabola passes through $(-5, 0)$.

Step 4. Find the y-intercepts. Replace x with 0 in the given equation.

$x = y^2 + 4y − 5$	This is the given equation.
$0 = y^2 + 4y − 5$	Replace x with O.
$0 = (y − 1)(y + 5)$	Use factoring to solve the quadratic equation.
$y − 1 = 0 \quad$ or $\quad y + 5 = 0$	Set each factor equal to O.
$y = 1 \qquad\qquad y = -5$	Solve.

The y-intercepts are 1 and -5. The parabola passes through $(0, 1)$ and $(0, -5)$.

Using Technology

To graph $x = y^2 + 4y − 5$ using a graphing utility, rewrite the equation as a quadratic equation in y.

$$y^2 + 4y + (-x − 5) = 0$$

$a = 1 \qquad b = 4 \qquad c = -x − 5$

Use the quadratic formula to solve for y and enter the resulting equations.

$$y_1 = \frac{-4 + \sqrt{16 − 4(-x − 5)}}{2}$$

$$y_2 = \frac{-4 − \sqrt{16 − 4(-x − 5)}}{2}$$

$[-10, 6, 1]$ by $[-8, 4, 1]$

Step 5. Graph the parabola. With a vertex at $(-9, -2)$, an x-intercept at -5, and y-intercepts at 1 and -5, the graph of the parabola is shown in **Figure 10.34**. The axis of symmetry is the horizontal line whose equation is $y = -2$. ∎

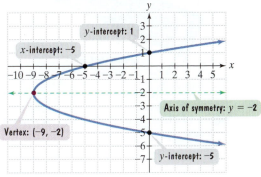

Figure 10.34   The graph of $x = y^2 + 4y - 5$

✓ **CHECK POINT 2**   Graph: $x = y^2 + 8y + 7$.

Applications

Parabolas have many applications. Cables hung between structures to form suspension bridges form parabolas. Arches constructed of steel and concrete, whose main purpose is strength, are usually parabolic in shape.

We have seen that comets in our solar system travel in orbits that are ellipses and hyperbolas. Some comets also follow parabolic paths. Only comets with elliptical orbits, such as Halley's Comet, return to our part of the galaxy.

If a parabola is rotated about its axis of symmetry, a parabolic surface is formed. **Figure 10.35(a)** shows how a parabolic surface can be used to reflect light. Light originates at the focus. Note how the light is reflected by the parabolic surface, so that the outgoing light is parallel to the axis of symmetry. The reflective properties of parabolic surfaces are used in the design of searchlights [see **Figure 10.35(b)**], automobile headlights, and parabolic microphones.

Figure 10.35(a) Parabolic surface reflecting light

Figure 10.35(b) Light from the focus is reflected parallel to the axis of symmetry.

Figure 10.36(a) Parabolic surface reflecting incoming light

Figure 10.36(b) Incoming light rays are reflected to the focus.

Figure 10.36(a) shows how a parabolic surface can be used to reflect *incoming* light. Note that light rays strike the surface and are reflected *to the focus*. This principle is used in the design of reflecting telescopes, radar, and television satellite dishes. Reflecting telescopes magnify the light from distant stars by reflecting the light from these bodies to the focus of a parabolic mirror [see **Figure 10.36(b)**].

2 Identify conic sections by their equations. ▶

Identifying Conic Sections by Their Equations

What do the equations of the conic sections in this chapter have in common? They contain x^2-terms, y^2-terms, or both. Furthermore, they do not contain terms with exponents greater than 2. **Table 10.2** shows how to identify conic sections by their equations.

Table 10.2 Recognizing Conic Sections from Equations

Conic Section	How to Identify the Equation	Example
Circle	When x^2- and y^2-terms are on the same side, they have the same coefficient.	$x^2 + y^2 = 16$
Ellipse	When x^2- and y^2-terms are on the same side, they have different coefficients of the same sign.	$4x^2 + 16y^2 = 64$ or (dividing by 64) $\dfrac{x^2}{16} + \dfrac{y^2}{4} = 1$
Hyperbola	When x^2- and y^2-terms are on the same side, they have coefficients with opposite signs.	$9y^2 - 4x^2 = 36$ or (dividing by 36) $\dfrac{y^2}{4} - \dfrac{x^2}{9} = 1$
Parabola	Only one of the variables is squared.	$x = y^2 + 4y - 5$

EXAMPLE 3 Recognizing Equations of Conic Sections

Indicate whether the graph of each equation is a circle, an ellipse, a hyperbola, or a parabola:

a. $4y^2 = 16 - 4x^2$ **b.** $x^2 = y^2 + 9$

c. $x + 7 - 6y = y^2$ **d.** $x^2 = 16 - 16y^2$.

Solution (Throughout the solution, in addition to identifying each equation's graph, we'll also discuss the graph's important features.) If both variables are squared, the graph of the equation is not a parabola. In this case, we collect the x^2- and y^2-terms on the same side of the equation.

a. $4y^2 = 16 - 4x^2$

> Both variables, x and y, are squared.

The graph is a circle, an ellipse, or a hyperbola. To see which one it is, we collect the x^2- and y^2-terms on the same side. Add $4x^2$ to both sides. We obtain

$$4x^2 + 4y^2 = 16.$$

The Cone of Apollonius (Detail), © 2011 Richard E. Prince.

Because the coefficients of x^2 and y^2 are the same, namely 4, the equation's graph is a circle. This becomes more obvious if we divide both sides by 4.

$$\frac{4x^2}{4} + \frac{4y^2}{4} = \frac{16}{4} \qquad \text{Divide both sides by 4.}$$

$$x^2 + y^2 = 4 \qquad \text{Simplify.}$$

$$(x - 0)^2 + (y - 0)^2 = 2^2 \qquad \text{Write in the form } (x - h)^2 + (y - k)^2 = r^2 \text{ with center } (h, k) \text{ and radius } r.$$

The graph is a circle with center at the origin and radius 2.

b. $x^2 = y^2 + 9$

Both variables, x and y, are squared.

The graph cannot be a parabola. To see if it is a circle, an ellipse, or a hyperbola, we collect the x^2- and y^2-terms on the same side. Subtract y^2 from both sides. We obtain

$$x^2 - y^2 = 9.$$

Because the x^2- and y^2-terms have coefficients with opposite signs, the equation's graph is a hyperbola. This becomes more obvious if we divide both sides by 9 to obtain 1 on the right.

$$\frac{x^2}{9} - \frac{y^2}{9} = 1$$

$a^2 = 9 \quad b^2 = 9$ The hyperbola's vertices are $(a, 0)$ and $(-a, 0)$, namely $(3, 0)$ and $(-3, 0)$.

c. $x + 7 - 6y = y^2$

Only one variable, y, is squared.

Because only one variable is squared, the graph of the equation is a parabola. We can express the equation of the horizontal parabola in the form $x = ay^2 + by + c$ by isolating x on the left. We obtain

$$x = y^2 + 6y - 7. \qquad \text{Add } 6y - 7 \text{ to both sides.}$$

Because the coefficient of the y^2-term, 1, is positive, the graph of the horizontal parabola opens to the right.

d. $x^2 = 16 - 16y^2$

Both variables, x and y, are squared.

The graph cannot be a parabola. To see if it is a circle, an ellipse, or a hyperbola, we collect the x^2- and y^2-terms on the same side. Add $16y^2$ to both sides. We obtain

$$x^2 + 16y^2 = 16.$$

Because the x^2- and y^2-terms have different coefficients of the same sign, namely 1 and 16, the equation's graph is an ellipse. This becomes more obvious if we divide both sides by 16 to obtain 1 on the right.

$$\frac{x^2}{16} + \frac{y^2}{1} = 1$$

$a^2 = 16 \quad b^2 = 1$ An equation in the form $\frac{x^2}{a^2} + \frac{y^2}{b^2} = 1$ is an ellipse. The vertices are $(4, 0)$ and $(-4, 0)$. ∎

✓ **CHECK POINT 3**   Identify whether the graph of each equation is a circle, an ellipse, a hyperbola, or a parabola:

a. $x^2 = 4y^2 + 16$

b. $x^2 = 16 - 4y^2$

c. $4x^2 = 16 - 4y^2$

d. $x = -4y^2 + 16y.$

CONCEPT AND VOCABULARY CHECK

Fill in each blank so that the resulting statement is true.

1. A/an _____ is the set of points in a plane that are equidistant from a fixed line, called the directrix, and a fixed point, called the _____.

2. The graph of $x = a(y - k)^2 + h$ or $x = ay^2 + by + c$ opens to the right if a _____.

3. The graph of $x = a(y - k)^2 + h$ or $x = ay^2 + by + c$ opens to the left if a _____.

Determine whether the graph of each equation opens upward, downward, to the left, or to the right.

4. $x = -3(y - 1)^2 + 2$ _____

5. $x = 3(y - 1)^2 + 2$ _____

6. $y = 3(x - 1)^2 + 2$ _____

7. $y = -3(x - 1)^2 + 2$ _____

8. $y = -x^2 + 2x + 1$ _____

9. $x = y^2 + 2y + 1$ _____

10. $x = -y^2 + 2y + 1$ _____

11. $y = -x^2 + 2x + 1$ _____

12. $y = x^2 - 2x - 1$ _____

13. The vertex of $x = a(y - k)^2 + h$ is _____.

14. The y-coordinate of the vertex of $x = ay^2 + by + c$ is $y =$ _____.

15. The circle, ellipse, hyperbola, and parabola are examples of _____ sections. Their equations contain x^2-terms, y^2-terms, or both.

   a. When these terms both appear, are on the same side, and have coefficients with opposite signs, the equation is that of a/an _____.

   b. When these terms both appear, are on the same side, and have different coefficients with the same sign, the equation is that of a/an _____.

   c. When these terms both appear, are on the same side, and have the same coefficients, the equation is that of a/an _____.

   d. When one of these terms does not appear, the equation is that of a/an _____.

10.4 EXERCISE SET   ▶ MyMathLab®

Practice Exercises

In Exercises 1–6, the equation of a horizontal parabola is given. For each equation: Determine how the parabola opens. Find the parabola's vertex. Use your results to identify the equation's graph. [The graphs, on the next page, are labeled (a) through (f).]

1. $x = (y - 2)^2 - 1$

2. $x = (y + 2)^2 - 1$

3. $x = (y + 2)^2 + 1$

4. $x = (y - 2)^2 + 1$

5. $x = -(y - 2)^2 + 1$

6. $x = -(y + 2)^2 + 1$

a.

b.

c.

d.

e.

f.

31. $x = \dfrac{1}{2}(y + 2)^2 + 1$ **32.** $x = \dfrac{1}{2}(y + 1)^2 + 2$

33. $x = y^2 + 2y - 3$ **34.** $x = y^2 - 6y + 8$

35. $x = -y^2 - 4y + 5$ **36.** $x = -y^2 - 6y + 7$

37. $x = y^2 + 6y$ **38.** $x = y^2 + 4y$

39. $x = -2y^2 - 4y$ **40.** $x = -3y^2 - 6y$

41. $x = -2y^2 - 4y + 1$ **42.** $x = -2y^2 + 4y - 3$

In Exercises 43–54, the equation of a parabola is given. Determine:

    a. *if the parabola is horizontal or vertical.*
    b. *the way the parabola opens.*
    c. *the vertex.*

43. $x = 2(y - 1)^2 + 2$

44. $x = 2(y - 3)^2 + 1$

45. $y = 2(x - 1)^2 + 2$

46. $y = 2(x - 3)^2 + 1$

47. $y = -(x + 3)^2 + 4$

48. $y = -(x + 1)^2 + 4$

49. $x = -(y + 3)^2 + 4$

50. $x = -(y + 1)^2 + 4$

51. $y = x^2 - 4x - 1$

52. $y = x^2 + 6x + 10$

53. $x = -y^2 + 4y + 1$

54. $x = -y^2 - 6y - 10$

In Exercises 55–64, indicate whether the graph of each equation is a circle, an ellipse, a hyperbola, or a parabola.

55. $x - 7 - 8y = y^2$

56. $x - 3 - 4y = 6y^2$

57. $4x^2 = 36 - y^2$

58. $4x^2 = 36 + y^2$

59. $x^2 = 36 + 4y^2$

60. $x^2 = 36 - 4y^2$

61. $3x^2 = 12 - 3y^2$

62. $3x^2 = 27 - 3y^2$

63. $3x^2 = 12 + 3y^2$

64. $3x^2 = 27 + 3y^2$

In Exercises 7–18, find the coordinates of the vertex for the horizontal parabola defined by the given equation.

7. $x = 2y^2$

8. $x = 4y^2$

9. $x = (y - 2)^2 + 3$

10. $x = (y - 3)^2 + 4$

11. $x = -4(y + 2)^2 - 1$

12. $x = -2(y + 5)^2 - 1$

13. $x = 2(y - 6)^2$

14. $x = 3(y - 7)^2$

15. $x = y^2 - 6y + 6$

16. $x = y^2 + 6y + 8$

17. $x = 3y^2 + 6y + 7$

18. $x = -2y^2 + 4y + 6$

In Exercises 19–42, use the vertex and intercepts to sketch the graph of each equation. If needed, find additional points on the parabola by choosing values of y on each side of the axis of symmetry.

19. $x = (y - 2)^2 - 4$ **20.** $x = (y - 3)^2 - 4$

21. $x = (y - 3)^2 - 5$ **22.** $x = (y + 2)^2 - 3$

23. $x = -(y - 5)^2 + 4$ **24.** $x = -(y - 3)^2 + 4$

25. $x = (y - 4)^2 + 1$ **26.** $x = (y - 2)^2 + 3$

27. $x = -3(y - 5)^2 + 3$ **28.** $x = -2(y + 6)^2 + 2$

29. $x = -2(y + 3)^2 - 1$ **30.** $x = -3(y + 1)^2 - 2$

In Exercises 65–74, indicate whether the graph of each equation is a circle, an ellipse, a hyperbola, or a parabola. Then graph the conic section.

65. $x^2 - 4y^2 = 16$

66. $7x^2 - 7y^2 = 28$

67. $4x^2 + 4y^2 = 16$

68. $7x^2 + 7y^2 = 28$

69. $x^2 + 4y^2 = 16$

70. $4x^2 + y^2 = 16$

71. $x = (y - 1)^2 - 4$

72. $x = (y - 4)^2 - 1$

73. $(x - 2)^2 + (y + 1)^2 = 16$

74. $(x - 1)^2 + (y + 2)^2 = 16$

Practice PLUS

In Exercises 75–80, use the vertex and the direction in which the parabola opens to determine the relation's domain and range. Is the relation a function?

75. $x = y^2 + 6y + 5$

76. $x = y^2 - 2y - 5$

77. $y = -x^2 + 4x - 3$

78. $y = -x^2 - 4x + 4$

79. $x = -4(y - 1)^2 + 3$

80. $x = -3(y - 1)^2 - 2$

In Exercises 81–86, find the solution set for each system by graphing both of the system's equations in the same rectangular coordinate system and finding points of intersection. Check all solutions in both equations.

81. $\begin{cases} x = (y - 2)^2 - 4 \\ y = -\dfrac{1}{2}x \end{cases}$

82. $\begin{cases} x = (y - 3)^2 + 2 \\ x + y = 5 \end{cases}$

83. $\begin{cases} x = y^2 - 3 \\ x = y^2 - 3y \end{cases}$

84. $\begin{cases} x = y^2 - 5 \\ x^2 + y^2 = 25 \end{cases}$

85. $\begin{cases} x = (y + 2)^2 - 1 \\ (x - 2)^2 + (y + 2)^2 = 1 \end{cases}$

86. $\begin{cases} x = 2y^2 + 4y + 5 \\ (x + 1)^2 + (y - 2)^2 = 1 \end{cases}$

Application Exercises

87. The George Washington Bridge spans the Hudson River from New York to New Jersey. Its two towers are 3500 feet apart and rise 316 feet above the road. The cable between the towers has the shape of a parabola and the cable just touches the sides of the road midway between the towers. The parabola is positioned in a rectangular coordinate system with its vertex at the origin. The point (1750, 316) lies on the parabola, as shown.

a. Write an equation in the form $y = ax^2$ for the parabolic cable. Do this by substituting 1750 for x and 316 for y and determining the value of a.

b. Use the equation in part (a) to find the height of the cable 1000 feet from a tower. Round to the nearest foot.

88. The towers of the Golden Gate Bridge connecting San Francisco to Marin County are 1280 meters apart and rise 140 meters above the road. The cable between the towers has the shape of a parabola and the cable just touches the sides of the road midway between the towers. The parabola is positioned in a rectangular coordinate system with its vertex at the origin. The point (640, 140) lies on the parabola, as shown.

a. Write an equation in the form $y = ax^2$ for the parabolic cable. Do this by substituting 640 for x and 140 for y and determining the value of a.

b. Use the equation in part (a) to find the height of the cable 200 meters from a tower. Round to the nearest meter.

89. A satellite dish is in the shape of a parabolic surface. Signals coming from a satellite strike the surface of the dish and are reflected to the focus, where the receiver is located. The satellite dish shown has a diameter of 12 feet and a depth of 2 feet. The parabola is positioned in a rectangular coordinate system with its vertex at the origin.

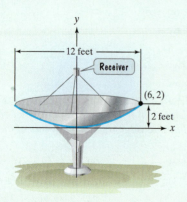

a. Write an equation in the form $y = ax^2$ for the parabola used to shape the dish.

b. The receiver should be placed at the focus $(0, p)$. The value of p is given by the equation $a = \dfrac{1}{4p}$. How far from the base of the dish should the receiver be placed?

90. An engineer is designing a flashlight using a parabolic reflecting mirror and a light source, as shown. The casting has a diameter of 4 inches and a depth of 2 inches. The parabola is positioned in a rectangular coordinate system with its vertex at the origin.

a. Write an equation in the form $y = ax^2$ for the parabola used to shape the mirror.

b. The light source should be placed at the focus $(0, p)$. The value of p is given by the equation $a = \dfrac{1}{4p}$. Where should the light source be placed relative to the mirror's vertex?

Moiré patterns, such as those shown in Exercises 91–92, can appear when two repetitive patterns overlap to produce a third, sometimes unintended, pattern.

    a. *In each exercise, use the name of a conic section to describe the moiré pattern.*

    b. *Select one of the following equations that can possibly describe a conic section within the moiré pattern:*

$$x^2 + y^2 = 1; \quad x^2 - y^2 = 1; \quad x^2 + 4y^2 = 4.$$

91.

92.

Explaining the Concepts

93. What is a parabola?

94. If you are given an equation of a parabola, explain how to determine if the parabola opens to the right, to the left, upward, or downward.

95. Explain how to use $x = 2(y + 3)^2 - 5$ to find the parabola's vertex.

96. Explain how to use $x = y^2 + 8y + 9$ to find the parabola's vertex.

97. Describe one similarity and one difference between the graphs of $x = 4y^2$ and $x = 4(y - 1)^2 + 2$.

98. How can you distinguish parabolas from other conic sections by looking at their equations?

99. How can you distinguish ellipses from hyperbolas by looking at their equations?

100. How can you distinguish ellipses from circles by looking at their equations?

Technology Exercises

Use a graphing utility to graph the parabolas in Exercises 101–102. Write the given equation as a quadratic equation in y and use the quadratic formula to solve for y. Enter each of the equations to produce the complete graph.

101. $y^2 + 2y - 6x + 13 = 0$

102. $y^2 + 10y - x + 25 = 0$

103. Use a graphing utility to graph any three of the parabolas that you graphed by hand in Exercises 19–42. First solve the given equation for y, possibly using the square root property. Enter each of the two resulting equations to produce the complete graph.

Critical Thinking Exercises

Make Sense? *In Exercises 104–107, determine whether each statement makes sense or does not make sense, and explain your reasoning.*

104. I graphed a parabola that opened to the right and contained a maximum point.

105. I'm graphing an equation that contains neither an x^2-term nor a y^2-term, so the graph cannot be a conic section.

106. Knowing that a parabola opening to the right has a vertex at $(-1, 1)$ gives me enough information to determine its graph.

107. I noticed that depending on the values for A and B, assuming that they are both not zero, the graph of $Ax^2 + By^2 = C$ can represent any of the conic sections other than a parabola.

In Exercises 108–111, determine whether each statement is true or false. If the statement is false, make the necessary change(s) to produce a true statement.

108. The parabola whose equation is $x = 2y - y^2 + 5$ opens to the right.

109. If the parabola whose equation is $x = ay^2 + by + c$ has its vertex at $(3, 2)$ and $a > 0$, then it has no y-intercepts.

110. Some parabolas that open to the right have equations that define y as a function of x.

111. The graph of $x = a(y - k) + h$ is a parabola with vertex at (h, k).

112. Look at the satellite dish shown in Exercise 89. Why must the receiver for a shallow dish be farther from the base of the dish than for a deeper dish of the same diameter?

113. The parabolic arch shown in the figure is 50 feet above the water at the center and 200 feet wide at the base. Will a boat that is 30 feet tall clear the arch 30 feet from the center?

Review Exercises

114. Graph: $f(x) = 2^{1-x}$. (Section 9.1, Example 4)

115. If $f(x) = \dfrac{1}{3}x - 5$, find $f^{-1}(x)$. (Section 9.2, Example 4)

116. Solve: $(x + 1)^2 + (x + 3)^2 = 4$. (Section 5.7, Example 2)

Preview Exercises

Exercises 117–119 will help you prepare for the material covered in the next section.

117. Solve by the substitution method:
$$\begin{cases} 4x + 3y = 4 \\ y = 2x - 7. \end{cases}$$

118. Solve by the addition method:
$$\begin{cases} 2x + 4y = -4 \\ 3x + 5y = -3. \end{cases}$$

119. Solve: $x^2 = 2(3x - 9) + 10$.

SECTION

10.5 Systems of Nonlinear Equations in Two Variables ▶

What am I supposed to learn?

After studying this section, you should be able to:

1 Recognize systems of nonlinear equations in two variables. ▶

2 Solve systems of nonlinear equations by substitution. ▶

3 Solve systems of nonlinear equations by addition. ▶

4 Solve problems using systems of nonlinear equations. ▶

Scientists debate the probability that a "doomsday rock" will collide with Earth. It has been estimated that an asteroid, a tiny planet that revolves around the sun, crashes into Earth about once every 250,000 years, and that such a collision would have disastrous results. In 1908, a small fragment struck Siberia, leveling thousands of acres of trees. One theory about the extinction of dinosaurs 65 million years ago involves Earth's collision with a large asteroid and the resulting drastic changes in Earth's climate.

   Understanding the path of Earth and the path of a comet is essential to detecting threatening space debris. Orbits about the sun are not described by linear equations in the form $Ax + By = C$. The ability to solve systems that contain equations that are not linear provides NASA scientists watching for troublesome asteroids with a way to locate possible collision points with Earth's orbit.

Systems of Nonlinear Equations and Their Solutions

① Recognize systems of nonlinear equations in two variables. ▶

A **system of** two **nonlinear equations** in two variables, also called a **nonlinear system**, contains at least one equation that cannot be expressed in the form $Ax + By = C$. Here are two examples:

$$\begin{cases} x^2 = 2y + 10 \\ 3x - y = 9 \end{cases}$$ Not in the form $Ax + By = C$. The term x^2 is not linear.

$$\begin{cases} y = x^2 + 3 \\ x^2 + y^2 = 9. \end{cases}$$ Neither equation is in the form $Ax + By = C$. The terms x^2 and y^2 are not linear.

A **solution** of a nonlinear system in two variables is an ordered pair of real numbers that satisfies all equations in the system. The **solution set** of the system is the set of all such ordered pairs. As with linear systems in two variables, the solution of a nonlinear system (if there is one) corresponds to the intersection point(s) of the graphs of the equations in the system. Unlike linear systems, the graphs can be circles, ellipses, hyperbolas, parabolas, or anything other than two lines. We will solve nonlinear systems using the substitution method and the addition method.

Eliminating a Variable Using the Substitution Method

② Solve systems of nonlinear equations by substitution. ▶

The substitution method involves converting a nonlinear system into one equation in one variable by an appropriate substitution. The steps in the solution process are exactly the same as those used to solve a linear system by substitution. However, when you obtain an equation in one variable, this equation may not be linear. In our first example, this equation is quadratic.

EXAMPLE 1   Solving a Nonlinear System by the Substitution Method

Solve by the substitution method:

$$\begin{cases} x^2 = 2y + 10 & \text{(The graph is a parabola.)} \\ 3x - y = 9. & \text{(The graph is a line.)} \end{cases}$$

Solution

Step 1. Solve one of the equations for one variable in terms of the other. We begin by isolating one of the variables raised to the first power in either of the equations. By solving for y in the second equation, which has a coefficient of -1, we can avoid fractions.

$$3x - y = 9 \qquad \text{This is the second equation in the given system.}$$
$$3x = y + 9 \qquad \text{Add } y \text{ to both sides.}$$
$$3x - 9 = y \qquad \text{Subtract 9 from both sides.}$$

Step 2. Substitute the expression from step 1 into the other equation. We substitute $3x - 9$ for y in the first equation.

$$y = \boxed{3x - 9} \qquad x^2 = 2\boxed{y} + 10.$$

This gives us an equation in one variable, namely

$$x^2 = 2(3x - 9) + 10.$$

The variable y has been eliminated.

Step 3. Solve the resulting equation containing one variable.

$$x^2 = 2(3x - 9) + 10 \qquad \text{This is the equation containing one variable.}$$
$$x^2 = 6x - 18 + 10 \qquad \text{Use the distributive property.}$$
$$x^2 = 6x - 8 \qquad \text{Combine numerical terms on the right.}$$
$$x^2 - 6x + 8 = 0 \qquad \text{Move all terms to one side and set the quadratic equation equal to 0.}$$
$$(x - 4)(x - 2) = 0 \qquad \text{Factor.}$$
$$x - 4 = 0 \quad \text{or} \quad x - 2 = 0 \qquad \text{Set each factor equal to 0.}$$
$$x = 4 \qquad\qquad x = 2 \qquad \text{Solve for } x.$$

Step 4. Back-substitute the obtained values into the equation from step 1. Now that we have the x-coordinates of the solutions, we back-substitute 4 for x and 2 for x into the equation $y = 3x - 9$.

If x is 4, $\quad y = 3(4) - 9 = 3$, $\quad$ so $(4, 3)$ is a solution.

If x is 2, $\quad y = 3(2) - 9 = -3$, $\quad$ so $(2, -3)$ is a solution.

Step 5. Check the proposed solutions in both of the system's given equations. We begin by checking $(4, 3)$. Replace x with 4 and y with 3.

$x^2 = 2y + 10$	$3x - y = 9$	These are the given equations.
$4^2 \stackrel{?}{=} 2(3) + 10$	$3(4) - 3 \stackrel{?}{=} 9$	Let $x = 4$ and $y = 3$.
$16 \stackrel{?}{=} 6 + 10$	$12 - 3 \stackrel{?}{=} 9$	Simplify.
$16 = 16$, true	$9 = 9$, true	True statements result.

The ordered pair $(4, 3)$ satisfies both equations. Thus, $(4, 3)$ is a solution of the system. Now let's check $(2, -3)$. Replace x with 2 and y with -3 in both given equations.

$x^2 = 2y + 10$	$3x - y = 9$	These are the given equations.
$2^2 \stackrel{?}{=} 2(-3) + 10$	$3(2) - (-3) \stackrel{?}{=} 9$	Let $x = 2$ and $y = -3$.
$4 \stackrel{?}{=} -6 + 10$	$6 + 3 \stackrel{?}{=} 9$	Simplify.
$4 = 4$, true	$9 = 9$, true	True statements result.

The ordered pair $(2, -3)$ also satisfies both equations and is a solution of the system. The solutions are $(4, 3)$ and $(2, -3)$, and the solution set is $\{(4, 3), (2, -3)\}$.

Figure 10.37 shows the graphs of the equations in the system and the solutions as intersection points. ■

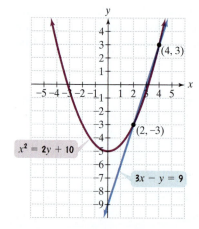

Figure 10.37 Points of intersection illustrate the nonlinear system's solutions.

✓ **CHECK POINT 1** Solve by the substitution method:
$$\begin{cases} x^2 = y - 1 \\ 4x - y = -1. \end{cases}$$

EXAMPLE 2 Solving a Nonlinear System by the Substitution Method

Solve by the substitution method:
$$\begin{cases} x - y = 3 & \text{(The graph is a line.)} \\ (x - 2)^2 + (y + 3)^2 = 4. & \text{(The graph is a circle.)} \end{cases}$$

Solution Graphically, we are finding the intersection of a line and a circle with center $(2, -3)$ and radius 2.

Great Question!

Why is the graph of $(x - 2)^2 + (y + 3)^2 = 4$ a circle?

Recall that $(x - h)^2 + (y - k)^2 = r^2$ describes a circle with center (h, k) and radius r.

Step 1. Solve one of the equations for one variable in terms of the other. We will solve for x in the linear equation—that is, the first equation. (We could also solve for y.)

$$x - y = 3 \qquad \text{This is the first equation in the given system.}$$
$$x = y + 3 \quad \text{Add } y \text{ to both sides.}$$

Step 2. Substitute the expression from step 1 into the other equation. We substitute $y + 3$ for x in the second equation.

$$x = \boxed{y + 3} \qquad (\boxed{x} - 2)^2 + (y + 3)^2 = 4$$

This gives an equation in one variable, namely

$$(y + 3 - 2)^2 + (y + 3)^2 = 4.$$

The variable x has been eliminated.

Step 3. Solve the resulting equation containing one variable.

$(y + 3 - 2)^2 + (y + 3)^2 = 4$	This is the equation containing one variable.
$(y + 1)^2 + (y + 3)^2 = 4$	Combine numerical terms in the first parentheses.
$y^2 + 2y + 1 + y^2 + 6y + 9 = 4$	Use the formula $(A + B)^2 = A^2 + 2AB + B^2$ to square $y + 1$ and $y + 3$.
$2y^2 + 8y + 10 = 4$	Combine like terms on the left.
$2y^2 + 8y + 6 = 0$	Subtract 4 from both sides and set the quadratic equation equal to O.
$2(y^2 + 4y + 3) = 0$	Factor out 2.
$2(y + 3)(y + 1) = 0$	Factor $2(y^2 + 4y + 3)$ completely.
$y + 3 = 0 \quad \text{or} \quad y + 1 = 0$	Set each variable factor equal to O.
$y = -3 \qquad\qquad y = -1$	Solve for y.

Step 4. Back-substitute the obtained values into the equation from step 1. Now that we have $y = -3$ and $y = -1$, or the y-coordinates of the solutions, we back-substitute -3 for y and -1 for y in the equation $x = y + 3$.

If $y = -3$: $x = -3 + 3 = 0,$ so $(0, -3)$ is a solution.
If $y = -1$: $x = -1 + 3 = 2,$ so $(2, -1)$ is a solution.

Step 5. Check the proposed solutions in both of the system's given equations. Take a moment to show that each ordered pair satisfies both equations. The solutions are $(0, -3)$ and $(2, -1)$, and the solution set of the given system is $\{(0, -3), (2, -1)\}$.

Figure 10.38 shows the graphs of the equations in the system and the solutions as intersection points. ■

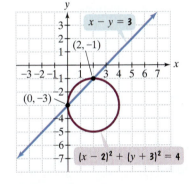

Figure 10.38 Points of intersection illustrate the nonlinear system's solutions.

✓ **CHECK POINT 2**   Solve by the substitution method:

$$\begin{cases} x + 2y = 0 \\ (x - 1)^2 + (y - 1)^2 = 5. \end{cases}$$

---

3 Solve systems of nonlinear equations by addition. ▶

Eliminating a Variable Using the Addition Method

In solving linear systems with two variables, we learned that the addition method works well when each equation is in the form $Ax + By = C$. For nonlinear systems, the addition method can be used when each equation is in the form $Ax^2 + By^2 = C$. If necessary, we will multiply either equation or both equations by appropriate numbers so that the coefficients of x^2 or y^2 will have a sum of 0. We then add equations. The sum will be an equation in one variable.

EXAMPLE 3 Solving a Nonlinear System by the Addition Method

Solve the system:

$$\begin{cases} 4x^2 + y^2 = 13 & \text{Equation 1 (The graph is an ellipse.)} \\ x^2 + y^2 = 10. & \text{Equation 2 (The graph is a circle.)} \end{cases}$$

Solution We can use the same steps that we did when we solved linear systems by the addition method.

Step 1. Write both equations in the form $Ax^2 + By^2 = C$. Both equations are already in this form, so we can skip this step.

Step 2. If necessary, multiply either equation or both equations by appropriate numbers so that the sum of the x^2-coefficients or the sum of the y^2-coefficients is 0. We can eliminate y^2 by multiplying Equation 2 by -1.

$$\begin{cases} 4x^2 + y^2 = 13 \xrightarrow{\text{No change}} \\ x^2 + y^2 = 10 \xrightarrow{\text{Multiply by } -1.} \end{cases} \begin{cases} 4x^2 + y^2 = 13 \\ -x^2 - y^2 = -10 \end{cases}$$

Steps 3 and 4. Add equations and solve for the remaining variable.

$$\begin{cases} 4x^2 + y^2 = 13 \\ \underline{-x^2 - y^2 = -10} \\ 3x^2 = 3 \end{cases}$$ Add equations.

$$x^2 = 1$$ Divide both sides by 3.

$$x = \pm 1$$ Use the square root property: If $x^2 = c$, then $x = \pm\sqrt{c}$.

Step 5. Back-substitute and find the values for the other variable. We must back-substitute each value of x into either one of the original equations. Let's use $x^2 + y^2 = 10$, Equation 2. If $x = 1$,

$$1^2 + y^2 = 10$$ Replace x with 1 in Equation 2.

$$y^2 = 9$$ Subtract 1 from both sides.

$$y = \pm 3.$$ Apply the square root property.

$(1, 3)$ and $(1, -3)$ are solutions. If $x = -1$,

$$(-1)^2 + y^2 = 10$$ Replace x with -1 in Equation 2.

$$y^2 = 9$$ The steps are the same as before.

$$y = \pm 3.$$

$(-1, 3)$ and $(-1, -3)$ are solutions.

Step 6. Check. Take a moment to show that each of the four ordered pairs satisfies the given equations, $4x^2 + y^2 = 13$ and $x^2 + y^2 = 10$. The solutions are $(1, 3)$, $(1, -3)$, $(-1, 3)$, and $(-1, -3)$, and the solution set of the given system is $\{(1, 3), (1, -3), (-1, 3), (-1, -3)\}$.

Figure 10.39 shows the graphs of the equations in the system, an ellipse and a circle, and the solutions as intersection points. ■

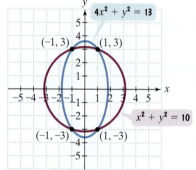

Figure 10.39 A system with four solutions

✓ **CHECK POINT 3** Solve the system:

$$\begin{cases} 3x^2 + 2y^2 = 35 \\ 4x^2 + 3y^2 = 48. \end{cases}$$

Great Question!

When solving nonlinear systems, do I really have to go to the hassle of checking my solutions?

Yes. Extra solutions may be introduced that do not satisfy both equations in the system. Therefore, you should get into the habit of checking all proposed pairs in each of the system's two equations.

In solving nonlinear systems, we will include only ordered pairs with real numbers in the solution set. We have seen that each of these ordered pairs corresponds to a point of intersection of the system's graphs.

EXAMPLE 4   Solving a Nonlinear System by the Addition Method

Solve the system:

$$\begin{cases} y = x^2 + 3 & \text{Equation 1 (The graph is a parabola.)} \\ x^2 + y^2 = 9. & \text{Equation 2 (The graph is a circle.)} \end{cases}$$

Solution   We could use substitution because Equation 1, $y = x^2 + 3$, has y expressed in terms of x, but substituting $x^2 + 3$ for y in $x^2 + y^2 = 9$ would result in a fourth-degree equation. However, we can rewrite Equation 1 by subtracting x^2 from both sides and adding the equations to eliminate the x^2-terms.

Notice how like terms are arranged in columns.

$$\begin{cases} -x^2 + y = 3 & \text{Subtract } x^2 \text{ from both sides of Equation 1.} \\ \underline{x^2 + y^2 = 9} & \text{This is Equation 2.} \\ y + y^2 = 12 & \text{Add the equations.} \end{cases}$$

We now solve this quadratic equation.

$$y + y^2 = 12 \qquad \text{This is the equation containing one variable.}$$

$$y^2 + y - 12 = 0 \qquad \text{Subtract 12 from both sides and write the quadratic equation in standard form.}$$

$$(y + 4)(y - 3) = 0 \qquad \text{Factor.}$$

$$y + 4 = 0 \quad \text{or} \quad y - 3 = 0 \qquad \text{Set each factor equal to 0.}$$

$$y = -4 \qquad\qquad y = 3 \qquad \text{Solve for } y.$$

To complete the solution, we must back-substitute each value of y into either one of the original equations. We will use $y = x^2 + 3$, Equation 1. First, we substitute -4 for y.

$$-4 = x^2 + 3$$

$$-7 = x^2 \qquad \text{Subtract 3 from both sides.}$$

Because the square of a real number cannot be negative, the equation $x^2 = -7$ does not have real-number solutions. We will not include the imaginary solutions, $x = \pm\sqrt{-7}$, or $i\sqrt{7}$ and $-i\sqrt{7}$, in the ordered pairs that make up the solution set. Thus, we move on to our other value for y, 3, and substitute this value into Equation 1.

$$y = x^2 + 3 \qquad \text{This is Equation 1.}$$

$$3 = x^2 + 3 \qquad \text{Back-substitute 3 for } y.$$

$$0 = x^2 \qquad \text{Subtract 3 from both sides.}$$

$$0 = x \qquad \text{Solve for } x.$$

We showed that if $y = 3$, then $x = 0$. Thus, $(0, 3)$ is the solution with a real ordered pair. Take a moment to show that $(0, 3)$ satisfies the given equations, $y = x^2 + 3$ and $x^2 + y^2 = 9$. The solution set of the system is $\{(0, 3)\}$. **Figure 10.40** shows the system's graphs and the solution as an intersection point. ■

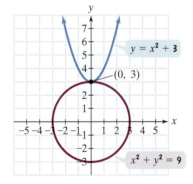

Figure 10.40 A system with one real solution

✓ **CHECK POINT 4**   Solve the system:

$$\begin{cases} y = x^2 + 5 \\ x^2 + y^2 = 25. \end{cases}$$

4 Solve problems using systems of nonlinear equations. ▶

Applications

Many geometric problems can be modeled and solved by the use of systems of nonlinear equations. We will use our step-by-step strategy for solving problems using mathematical models that are created from verbal conditions.

Figure 10.41 Building an enclosure

EXAMPLE 5   An Application of a Nonlinear System

You have 36 yards of fencing to build the enclosure in **Figure 10.41**. Some of this fencing is to be used to build an internal divider. If you'd like to enclose 54 square yards, what are the dimensions of the enclosure?

Solution

Step 1. Use variables to represent unknown quantities.   Let x = the enclosure's length and y = the enclosure's width. These variables are shown in **Figure 10.41**.

Step 2. Write a system of equations that models the problem's conditions.   The first condition is that you have 36 yards of fencing.

Fencing along both lengths	plus	fencing along both widths	plus	fencing for the internal divider	equals	36 yards.
$2x$	$+$	$2y$	$+$	y	$=$	36

Adding like terms, we can express the equation that models the verbal conditions for the fencing as $2x + 3y = 36$.

The second condition is that you'd like to enclose 54 square yards. The rectangle's area, the product of its length and its width, must be 54 square yards.

Length	times	width	is	54 square yards.
x	$\cdot$	y	$=$	54

Step 3. Solve the system and answer the problem's question.   We must solve the system

$$\begin{cases} 2x + 3y = 36 & \text{Equation 1} \\ xy = 54. & \text{Equation 2} \end{cases}$$

We will use substitution. Because Equation 1 has no coefficients of 1 or -1, we will work with Equation 2 and solve for y. Dividing both sides of $xy = 54$ by x, we obtain

$$y = \frac{54}{x}.$$

Now we substitute $\frac{54}{x}$ for y in Equation 1 and solve for x.

$2x + 3y = 36$	This is Equation 1.
$2x + 3 \cdot \dfrac{54}{x} = 36$	Substitute $\dfrac{54}{x}$ for y.
$2x + \dfrac{162}{x} = 36$	Multiply.
$x\left(2x + \dfrac{162}{x}\right) = 36 \cdot x$	Clear fractions by multiplying both sides by x.
$2x^2 + 162 = 36x$	Use the distributive property on the left side.
$2x^2 - 36x + 162 = 0$	Subtract $36x$ from both sides and write the quadratic equation in standard form.
$2(x^2 - 18x + 81) = 0$	Factor out 2.
$2(x - 9)^2 = 0$	Factor completely using $A^2 - 2AB + B^2 = (A - B)^2$.
$x - 9 = 0$	Set the repeated factor equal to zero.
$x = 9$	Solve for x.

We back-substitute this value of x into $y = \dfrac{54}{x}$.

$$\text{If } x = 9, \quad y = \frac{54}{9} = 6.$$

This means that the dimensions of the enclosure in **Figure 10.41** are 9 yards by 6 yards.

Step 4. Check the proposed solution in the original wording of the problem. Take a moment to check that a length of 9 yards and a width of 6 yards results in 36 yards of fencing and an area of 54 square yards. ■

☑ **CHECK POINT 5** Find the length and width of a rectangle whose perimeter is 20 feet and whose area is 21 square feet.

Achieving Success

Try to relax during a math test. Many of us have had the experience of feeling anxious and panicky during a test. If this happens to you, close your eyes and breathe deeply. Take a short break and think about something pleasant. This may help you to calm down and put the test in perspective.

CONCEPT AND VOCABULARY CHECK

Fill in each blank so that the resulting statement is true.

1. A system of two equations in two variables that contains at least one equation that cannot be expressed in the form $Ax + By = C$ is called a system of _____ equations.

2. When solving

$$\begin{cases} x^2 - 4y = 4 \\ x + y = -1 \end{cases}$$

by the substitution method, we obtain $x = -4$ or $x = 0$, so the solution set is _____.

3. When solving

$$\begin{cases} 3x^2 + 2y^2 = 35 \\ 4x^2 + 3y^2 = 48 \end{cases}$$

by the addition method, we can eliminate x^2 by multiplying the first equation by -4 and the second equation by _____, and then adding the equations.

4. When solving

$$\begin{cases} x^2 + 4y^2 = 16 \\ x^2 - y^2 = 1 \end{cases}$$

by the addition method, we obtain $y^2 = 3$, so the solution set is _____.

5. When solving

$$\begin{cases} x^2 + y^2 = 13 \\ x^2 - y = 7 \end{cases}$$

by the addition method, we can eliminate x^2 by multiplying the second equation by _____ and then adding the equations. We obtain _____, a quadratic equation in y.

6. When solving

$$\begin{cases} x^2 + 4y^2 = 20 \\ xy = 4 \end{cases}$$

by the substitution method, we can eliminate y by solving the second equation for y. We obtain $y =$ _____. Then we substitute _____ for _____ in the first equation.

Definitions and Concepts	**Examples**

Section 10.2 The Ellipse

An ellipse is the set of all points in a plane the sum of whose distances from two fixed points, the foci, is constant. The midpoint of the segment connecting the foci is the center of the ellipse.

Standard Forms of the Equations of an Ellipse Centered at the Origin ($a^2 > b^2$)

Horizontal with vertices $(a, 0)$ and $(-a, 0)$

$$\frac{x^2}{a^2} + \frac{y^2}{b^2} = 1$$

Endpoints of major axis a units left and right of center; minor axis b units up and down from center.

Vertical with vertices $(0, a)$ and $(0, -a)$

$$\frac{x^2}{b^2} + \frac{y^2}{a^2} = 1$$

Endpoints of major axis a units up and down from center; minor axis b units left and right of center.

The equations $\dfrac{(x - h)^2}{a^2} + \dfrac{(y - k)^2}{b^2} = 1$ and

$\dfrac{(x - h)^2}{b^2} + \dfrac{(y - k)^2}{a^2} = 1$ $(a^2 > b^2)$ represent ellipses centered at (h, k).

Graph: $\dfrac{(x + 2)^2}{9} + \dfrac{(y + 4)^2}{25} = 1.$

$$\frac{(x - (-2))^2}{9} + \frac{(y - (-4))^2}{25} = 1$$

$b^2 = 9$ $a^2 = 25$

Center, (h, k), is $(-2, -4)$. With $a^2 = 25$, vertices are 5 units above and below the center. With $b^2 = 9$, endpoints of minor axis are 3 units to the left and right of the center.

Section 10.3 The Hyperbola

A hyperbola is the set of all points in a plane the difference of whose distances from two fixed points, the foci, is constant.

Standard Forms of the Equations of a Hyperbola Centered at the Origin

$$\frac{x^2}{a^2} - \frac{y^2}{b^2} = 1$$

Vertices are $(a, 0)$ and $(-a, 0)$.

$$\frac{y^2}{a^2} - \frac{x^2}{b^2} = 1$$

Vertices are $(0, a)$ and $(0, -a)$.

As x and y get larger, the two branches of a hyperbola approach a pair of intersecting lines, called asymptotes. Draw the rectangle centered at the origin with sides parallel to the axes, crossing one axis at $\pm a$ and the other at $\pm b$. Draw the diagonals of this rectangle and extend them to obtain the asymptotes. Draw the two branches of the hyperbola by starting at each vertex and approaching the asymptotes.

Graph: $4x^2 - 9y^2 = 36.$

$$\frac{4x^2}{36} - \frac{9y^2}{36} = \frac{36}{36}$$

$$\frac{x^2}{9} - \frac{y^2}{4} = 1$$

$a^2 = 9$ $b^2 = 4$

Vertices are $(3, 0)$ and $(-3, 0)$. Draw a rectangle using -3 and 3 on the x-axis and -2 and 2 on the y-axis. Its extended diagonals are the asymptotes.

Definitions and Concepts	**Examples**

Section 10.4 The Parabola; Identifying Conic Sections

A parabola is the set of all points in a plane that are equidistant from a fixed line, the directrix, and a fixed point, the focus, that is not on the line. The line passing through the focus and perpendicular to the directrix is the axis of symmetry. The point of intersection of the parabola with its axis of symmetry is the vertex.

Equations of Horizontal Parabolas

$x = a(y - k)^2 + h$ $x = ay^2 + by + c$

Vertex is (h, k). y-coordinate of

vertex is $y = -\dfrac{b}{2a}$.

If $a > 0$, the parabola opens to the right. If $a < 0$, the parabola opens to the left.

Equations of Vertical Parabolas

$y = a(x - h)^2 + k$ $y = ax^2 + bx + c$

Vertex is (h, k). x-coordinate of

vertex is $x = -\dfrac{b}{2a}$.

If $a > 0$, the parabola opens upward. If $a < 0$, the parabola opens downward.

Find the vertex and graph:
$$x = -(y + 3)^2 + 4.$$

$x = -1(y - (-3))^2 + 4$

$\boxed{a = -1}$ $\boxed{k = -3}$ $\boxed{h = 4}$

Parabola opens to the left. Vertex, (h, k), is $(4, -3)$.
x-intercept: Let $y = 0$.
$$x = -(0 + 3)^2 + 4 = -9 + 4 = -5$$
y-intercepts: Let $x = 0$.
$$0 = -(y + 3)^2 + 4$$
$$(y + 3)^2 = 4$$
$$y + 3 = \pm\sqrt{4}$$
$$y = -1 \quad \text{or} \quad y = -5$$

Recognizing Conic Sections from Equations
Conic sections result from the intersection of a cone and a plane. Their equations contain x^2-terms, y^2-terms, or both. For parabolas, only one variable is squared. For circles, ellipses, and hyperbolas, both variables are squared. Collect x^2- and y^2-terms on the same side of the equation to identify the graph:

- Circle: x^2 and y^2 have the same coefficient of the same sign.
- Ellipse: x^2 and y^2 have different coefficients of the same sign.
- Hyperbola: x^2 and y^2 have coefficients with opposite signs.

- $9x^2 + 4y^2 = 36$

  $\boxed{\text{Different coefficients of the same sign; ellipse}}$

- $9y^2 = 25x^2 + 225$
  $$9y^2 - 25x^2 = 225$$

  $\boxed{\text{Coefficients with opposite signs; hyperbola}}$

- $x = -y^2 + 4y$

  $\boxed{\text{Only one variable, } y, \text{ is squared; parabola}}$

- $\dfrac{x^2}{9} + \dfrac{y^2}{9} = 1$
  $$x^2 + y^2 = 9$$

  $\boxed{\text{Same coefficient; circle}}$

Definitions and Concepts	**Examples**

Section 10.5 Systems of Nonlinear Equations in Two Variables

A system of two nonlinear equations in two variables, x and y, contains at least one equation that cannot be expressed in the form $Ax + By = C$. Systems can be solved by the substitution method or the addition method. Each real solution corresponds to a point of intersection of the system's graphs.

Solve: $\begin{cases} x^2 + y^2 = 25 \\ x - 3y = -5. \end{cases}$

Using substitution: $x = 3y - 5$

$$(3y - 5)^2 + y^2 = 25$$
$$9y^2 - 30y + 25 + y^2 = 25$$
$$10y^2 - 30y = 0$$
$$10y(y - 3) = 0$$
$$y = 0 \quad \text{or} \quad y = 3$$

If $y = 0$: $x = 3y - 5 = 3 \cdot 0 - 5 = -5$.
If $y = 3$: $x = 3y - 5 = 3 \cdot 3 - 5 = 4$.

The solutions are $(-5, 0)$ and $(4, 3)$, and the solution set is $\{(-5, 0), (4, 3)\}$.

CHAPTER 10 REVIEW EXERCISES

10.1   *In Exercises 1–2, find the distance between each pair of points. If necessary, express answers in simplified radical form and then round to two decimal places.*

1. $(-2, -3)$ and $(3, 9)$

2. $(-4, 3)$ and $(-2, 5)$

In Exercises 3–4, find the midpoint of the line segment with the given endpoints.

3. $(2, 6)$ and $(-12, 4)$

4. $(4, -6)$ and $(-15, 2)$

In Exercises 5–6, write the standard form of the equation of the circle with the given center and radius.

5. Center $(0, 0)$, $r = 3$

6. Center $(-2, 4)$, $r = 6$

In Exercises 7–10, give the center and radius of each circle and graph its equation.

7. $x^2 + y^2 = 1$

8. $(x + 2)^2 + (y - 3)^2 = 9$

9. $x^2 + y^2 - 4x + 2y - 4 = 0$

10. $x^2 + y^2 - 4y = 0$

10.2   *In Exercises 11–16, graph each ellipse.*

11. $\dfrac{x^2}{36} + \dfrac{y^2}{25} = 1$

12. $\dfrac{x^2}{25} + \dfrac{y^2}{16} = 1$

13. $4x^2 + y^2 = 16$

14. $4x^2 + 9y^2 = 36$

15. $\dfrac{(x - 1)^2}{16} + \dfrac{(y + 2)^2}{9} = 1$ **16.** $\dfrac{(x + 1)^2}{9} + \dfrac{(y - 2)^2}{16} = 1$

17. A semielliptic archway has a height of 15 feet at the center and a width of 50 feet, as shown in the figure. The 50-foot width consists of a two-lane road. Can a truck that is 12 feet high and 14 feet wide drive under the archway without going into the other lane?

10.3   *In Exercises 18–21, use vertices and asymptotes to graph each hyperbola.*

18. $\dfrac{x^2}{16} - y^2 = 1$ **19.** $\dfrac{y^2}{16} - x^2 = 1$

20. $9x^2 - 16y^2 = 144$ **21.** $4y^2 - x^2 = 16$

10.4   *In Exercises 22–25, use the vertex and intercepts to sketch the graph of each equation. If needed, find additional points on the parabola by choosing values of y on each side of the axis of symmetry.*

22. $x = (y - 3)^2 - 4$ **23.** $x = -2(y + 3)^2 + 2$

24. $x = y^2 - 8y + 12$ **25.** $x = -y^2 - 4y + 6$

In Exercises 26–32, indicate whether the graph of each equation is a circle, an ellipse, a hyperbola, or a parabola.

26. $x + 8y = y^2 + 10$

27. $16x^2 = 32 - y^2$

28. $x^2 = 25 + 25y^2$

29. $x^2 = 4 - y^2$

30. $36y^2 = 576 + 16x^2$

31. $\dfrac{(x + 3)^2}{9} + \dfrac{(y - 4)^2}{25} = 1$

32. $y = x^2 + 6x + 9$

In Exercises 33–41, indicate whether the graph of each equation is a circle, an ellipse, a hyperbola, or a parabola. Then graph the conic section.

33. $5x^2 + 5y^2 = 180$

34. $4x^2 + 9y^2 = 36$

35. $4x^2 - 9y^2 = 36$

36. $\dfrac{x^2}{25} + \dfrac{y^2}{1} = 1$

37. $x + 3 = -y^2 + 2y$

38. $y - 3 = x^2 - 2x$

39. $\dfrac{(x + 2)^2}{16} + \dfrac{(y - 5)^2}{4} = 1$

40. $(x - 3)^2 + (y + 2)^2 = 4$

41. $x^2 + y^2 + 6x - 2y + 6 = 0$

42. An engineer is designing headlight units for automobiles. The unit has a parabolic surface with a diameter of 12 inches and a depth of 3 inches. The situation is illustrated in the figure, where a coordinate system has been superimposed.

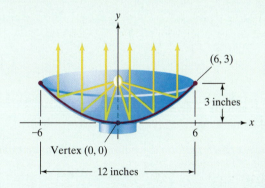

a. Use the point $(6, 3)$ to write an equation in the form $y = ax^2$ for the parabola used to design the headlight.

b. The light source should be placed at the focus $(0, p)$. The value of p is given by the equation $a = \dfrac{1}{4p}$. Where should the light source be placed? Describe this placement relative to the vertex.

10.5 In Exercises 43–53, solve each system by the method of your choice.

43. $\begin{cases} 5y = x^2 - 1 \\ x - y = 1 \end{cases}$

44. $\begin{cases} y = x^2 + 2x + 1 \\ x + y = 1 \end{cases}$

45. $\begin{cases} x^2 + y^2 = 2 \\ x + y = 0 \end{cases}$

46. $\begin{cases} 2x^2 + y^2 = 24 \\ x^2 + y^2 = 15 \end{cases}$

47. $\begin{cases} xy - 4 = 0 \\ y - x = 0 \end{cases}$

48. $\begin{cases} y^2 = 4x \\ x - 2y + 3 = 0 \end{cases}$

49. $\begin{cases} x^2 + y^2 = 10 \\ y = x + 2 \end{cases}$

50. $\begin{cases} xy = 1 \\ y = 2x + 1 \end{cases}$

51. $\begin{cases} x + y + 1 = 0 \\ x^2 + y^2 + 6y - x = -5 \end{cases}$

52. $\begin{cases} x^2 + y^2 = 13 \\ x^2 - y = 7 \end{cases}$

53. $\begin{cases} 2x^2 + 3y^2 = 21 \\ 3x^2 - 4y^2 = 23 \end{cases}$

54. The perimeter of a rectangle is 26 meters and its area is 40 square meters. Find its dimensions.

55. Find the coordinates of all points (x, y) that lie on the line whose equation is $2x + y = 8$, so that the area of the rectangle shown in the figure is 6 square units.

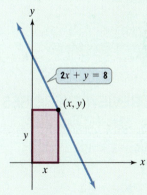

56. Two adjoining square fields with an area of 2900 square feet are to be enclosed with 240 feet of fencing. The situation is represented in the figure. Find the length of each side where a variable appears.

1. Find the distance between $(-1, 5)$ and $(2, -3)$. Express the answer in radical form and then round to two decimal places.

2. Find the midpoint of the line segment whose endpoints are $(-5, -2)$ and $(12, -6)$.

3. Write the standard form of the equation of the circle with center $(3, -2)$ and radius 5.

In Exercises 4–5, give the center and radius of each circle.

4. $(x - 5)^2 + (y + 3)^2 = 49$

5. $x^2 + y^2 + 4x - 6y - 3 = 0$

In Exercises 6–7, give the coordinates of the vertex for each parabola.

6. $x = -2(y + 3)^2 + 7$

7. $x = y^2 + 10y + 23$

In Exercises 8–16, indicate whether the graph of each equation is a circle, an ellipse, a hyperbola, or a parabola. Then graph the conic section.

8. $\dfrac{x^2}{4} - \dfrac{y^2}{9} = 1$

9. $4x^2 + 9y^2 = 36$

10. $x = (y + 1)^2 - 4$

11. $16x^2 + y^2 = 16$

12. $25y^2 = 9x^2 + 225$

13. $x = -y^2 + 6y$

14. $\dfrac{(x - 2)^2}{16} + \dfrac{(y + 3)^2}{9} = 1$

15. $(x + 1)^2 + (y + 2)^2 = 9$

16. $\dfrac{x^2}{4} + \dfrac{y^2}{4} = 1$

In Exercises 17–18, solve each system.

17. $\begin{cases} x^2 + y^2 = 25 \\ x + y = 1 \end{cases}$

18. $\begin{cases} 2x^2 - 5y^2 = -2 \\ 3x^2 + 2y^2 = 35 \end{cases}$

19. The rectangular plot of land shown in the figure is to be fenced along three sides using 39 feet of fencing. No fencing is to be placed along the river's edge. The area of the plot is 180 square feet. What are its dimensions?

20. A rectangle has a diagonal of 5 feet and a perimeter of 14 feet. Find the rectangle's dimensions.

CUMULATIVE REVIEW EXERCISES   (CHAPTERS 1–10)

In Exercises 1–7, solve each equation, inequality, or system.

1. $3x + 7 > 4$ or $6 - x < 1$

2. $x(2x - 7) = 4$

3. $\dfrac{5}{x - 3} = 1 + \dfrac{30}{x^2 - 9}$

4. $3x^2 + 8x + 5 < 0$

5. $3^{2x-1} = 81$

6. $30e^{0.7x} = 240$

7. $\begin{cases} 3x^2 + 4y^2 = 39 \\ 5x^2 - 2y^2 = -13 \end{cases}$

In Exercises 8–11, graph each function, equation, or inequality in a rectangular coordinate system.

8. $f(x) = -\dfrac{2}{3}x + 4$

9. $3x - y > 6$

10. $x^2 + y^2 + 4x - 6y + 9 = 0$

11. $9x^2 - 4y^2 = 36$

In Exercises 12–15, perform the indicated operations and simplify, if possible.

12. $-2(3^2 - 12)^3 - 45 \div 9 - 3$

13. $(3x^3 - 19x^2 + 17x + 4) \div (3x - 4)$

14. $\sqrt[3]{4x^2y^5} \cdot \sqrt[3]{4xy^2}$

15. $(2 + 3i)(4 - i)$

In Exercises 16–17, factor completely.

16. $12x^3 - 36x^2 + 27x$

17. $x^3 - 2x^2 - 9x + 18$

18. Find the domain: $f(x) = \sqrt{6 - 3x}$.

19. Rationalize the denominator: $\dfrac{1 - \sqrt{x}}{1 + \sqrt{x}}$.

20. Write as a single logarithm: $\dfrac{1}{3}\ln x + 7\ln y$.

21. Divide using synthetic division:

$(3x^3 - 5x^2 + 2x - 1) \div (x - 2)$.

22. Write a quadratic equation whose solution set is $\{-2\sqrt{3}, 2\sqrt{3}\}$.

23. Two cars leave from the same place at the same time, traveling in opposite directions. The rate of the faster car exceeds that of the slower car by 10 miles per hour. After 2 hours, the cars are 180 miles apart. Find the rate of each car.

24. Rent-a-Truck charges a daily rental rate of $39 plus $0.16 per mile. A competing agency, Ace Truck Rentals, charges $25 a day plus $0.24 per mile for the same truck. How many miles must be driven in a day to make the daily cost of both agencies the same? What will be the cost?

25. Three apples and two bananas provide 354 calories, and two apples and three bananas provide 381 calories. Find the number of calories in one apple and one banana.

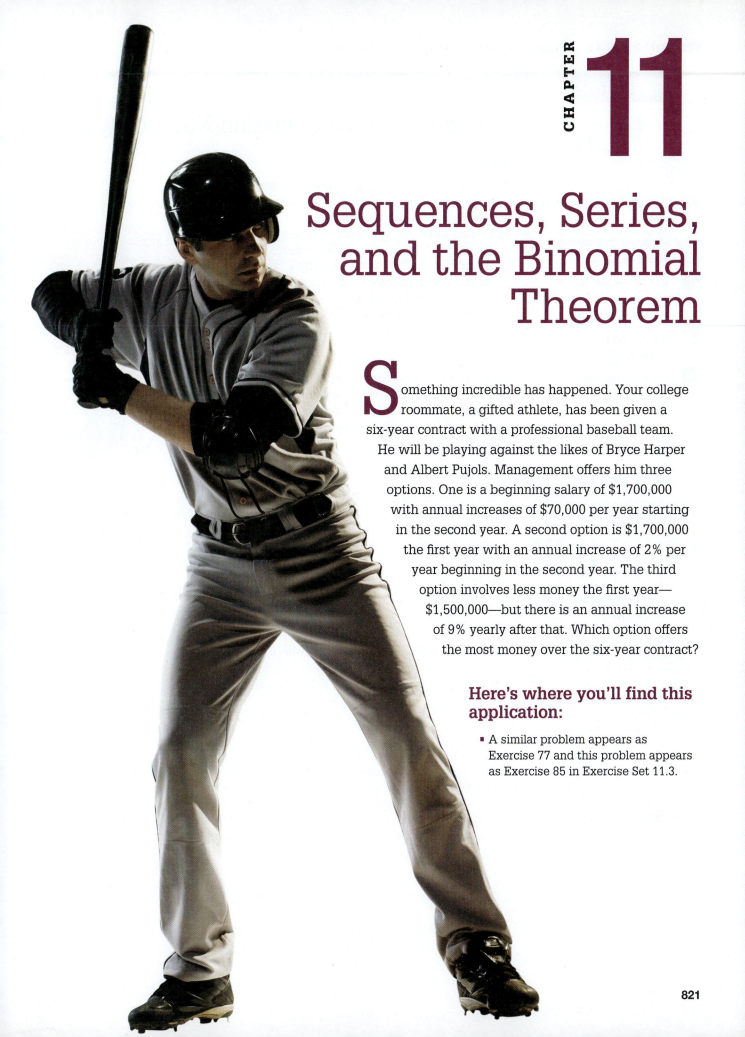

Sequences, Series, and the Binomial Theorem

Something incredible has happened. Your college roommate, a gifted athlete, has been given a six-year contract with a professional baseball team. He will be playing against the likes of Bryce Harper and Albert Pujols. Management offers him three options. One is a beginning salary of $1,700,000 with annual increases of $70,000 per year starting in the second year. A second option is $1,700,000 the first year with an annual increase of 2% per year beginning in the second year. The third option involves less money the first year—$1,500,000—but there is an annual increase of 9% yearly after that. Which option offers the most money over the six-year contract?

Here's where you'll find this application:

- A similar problem appears as Exercise 77 and this problem appears as Exercise 85 in Exercise Set 11.3.

SECTION

11.1

Sequences and Summation Notation

Sequences

Many creations in nature involve intricate mathematical designs, including a variety of spirals. For example, the arrangement of the individual florets in the head of a sunflower forms spirals. In some species, there are 21 spirals in the clockwise direction and 34 in the counterclockwise direction. The precise numbers depend on the species of sunflower: 21 and 34, or 34 and 55, or 55 and 89, or even 89 and 144.

This observation becomes more interesting when we consider a sequence of numbers investigated by Leonardo of Pisa, also known as Fibonacci, an Italian mathematician of the thirteenth century. The **Fibonacci sequence** of numbers is an infinite sequence that begins as follows:

$$1, 1, 2, 3, 5, 8, 13, 21, 34, 55, 89, 144, 233, \ldots.$$

The first two terms are 1. Every term thereafter is the sum of the two preceding terms. For example, the third term, 2, is the sum of the first and second terms: $1 + 1 = 2$. The fourth term, 3, is the sum of the second and third terms: $1 + 2 = 3$, and so on. Did you know that the number of spirals in a daisy or a sunflower, 21 and 34, are two Fibonacci numbers? The number of spirals in a pinecone, 8 and 13, and a pineapple, 8 and 13, are also Fibonacci numbers.

We can think of the Fibonacci sequence as a function. The terms of the sequence

$$1, 1, 2, 3, 5, 8, 13, 21, 34, 55, 89, 144, 233, \ldots$$

are the range values for a function f whose domain is the set of positive integers.

Domain:	1,	2,	3,	4,	5,	6,	7,	...
	↓	↓	↓	↓	↓	↓	↓	
Range:	1,	1,	2,	3,	5,	8,	13,	...

Thus, $f(1) = 1, f(2) = 1, f(3) = 2, f(4) = 3, f(5) = 5, f(6) = 8, f(7) = 13$, and so on.

The letter a with a subscript is used to represent function values of a sequence, rather than the usual function notation. The subscripts make up the domain of the sequence and they identify the location of a term. Thus, a_1 represents the first term of the sequence, a_2 represents the second term, a_3 the third term, and so on. This notation is shown for the first six terms of the Fibonacci sequence:

1,   1,   2,   3,   5,   8.

$a_1 = 1$ $a_2 = 1$ $a_3 = 2$ $a_4 = 3$ $a_5 = 5$ $a_6 = 8$

The notation a_n represents the nth term, or **general term**, of a sequence. The entire sequence is represented by $\{a_n\}$.

What am I supposed to learn?

After studying this section, you should be able to:

1. Find particular terms of a sequence from the general term.

2. Use factorial notation.

3. Use summation notation.

Blitzer Bonus

Fibonacci Numbers on the Piano Keyboard

One Octave

Numbers in the Fibonacci sequence can be found in an octave on the piano keyboard. The octave contains 2 black keys in one cluster and 3 black keys in another cluster, for a total of 5 black keys. It also has 8 white keys, for a total of 13 keys. The numbers 2, 3, 5, 8, and 13 are the third through seventh terms of the Fibonacci sequence.

Definition of a Sequence

An **infinite sequence** $\{a_n\}$ is a function whose domain is the set of positive integers. The function values, or **terms**, of the sequence are represented by

$$a_1, a_2, a_3, a_4, \ldots, a_n, \ldots.$$

Sequences whose domains consist only of the first n positive integers are called **finite sequences**.

---

① Find particular terms of a sequence from the general term. ▶

EXAMPLE 1   Writing Terms of a Sequence from the General Term

Write the first four terms of the sequence whose nth term, or general term, is given:

a. $a_n = 3n + 4$ **b.** $a_n = \dfrac{(-1)^n}{3^n - 1}$.

Solution

a. We need to find the first four terms of the sequence whose general term is $a_n = 3n + 4$. To do so, we replace n in the formula with $1, 2, 3$, and 4.

| a_1, 1st term | $3 \cdot 1 + 4 = 3 + 4 = 7$ | a_2, 2nd term | $3 \cdot 2 + 4 = 6 + 4 = 10$ |
| a_3, 3rd term | $3 \cdot 3 + 4 = 9 + 4 = 13$ | a_4, 4th term | $3 \cdot 4 + 4 = 12 + 4 = 16$ |

The first four terms are 7, 10, 13, and 16. The sequence defined by $a_n = 3n + 4$ can be written as

$$7, 10, 13, 16, \ldots, 3n + 4, \ldots.$$

Great Question!

What effect does $(-1)^n$ have on the terms of a sequence?

The factor $(-1)^n$ in the general term of a sequence causes the signs of the terms to alternate between positive and negative, depending on whether n is even or odd.

b. We need to find the first four terms of the sequence whose general term is $a_n = \dfrac{(-1)^n}{3^n - 1}$. To do so, we replace each occurrence of n in the formula with $1, 2, 3$, and 4.

| a_1, 1st term | $\dfrac{(-1)^1}{3^1 - 1} = \dfrac{-1}{3 - 1} = -\dfrac{1}{2}$ | a_2, 2nd term | $\dfrac{(-1)^2}{3^2 - 1} = \dfrac{1}{9 - 1} = \dfrac{1}{8}$ |
| a_3, 3rd term | $\dfrac{(-1)^3}{3^3 - 1} = \dfrac{-1}{27 - 1} = -\dfrac{1}{26}$ | a_4, 4th term | $\dfrac{(-1)^4}{3^4 - 1} = \dfrac{1}{81 - 1} = \dfrac{1}{80}$ |

The first four terms are $-\dfrac{1}{2}, \dfrac{1}{8}, -\dfrac{1}{26}$, and $\dfrac{1}{80}$. The sequence defined by $\dfrac{(-1)^n}{3^n - 1}$ can be written as

$$-\frac{1}{2}, \frac{1}{8}, -\frac{1}{26}, \frac{1}{80}, \ldots, \frac{(-1)^n}{3^n - 1}, \ldots. \quad \blacksquare$$

✓ **CHECK POINT 1**   Write the first four terms of the sequence whose nth term, or general term, is given:

a. $a_n = 2n + 5$ **b.** $a_n = \dfrac{(-1)^n}{2^n + 1}$.

Although sequences are usually named with the letter a, any lowercase letter can be used. For example, the first four terms of the sequence $\{b_n\} = \left\{ \left(\frac{1}{2}\right)^n \right\}$ are $b_1 = \frac{1}{2}, b_2 = \frac{1}{4}, b_3 = \frac{1}{8}$, and $b_4 = \frac{1}{16}$.

Using Technology

Graphing utilities can write the terms of a sequence and graph them. For example, to find the first six terms of

$$\{a_n\} = \left\{\frac{1}{n}\right\}, \text{ enter}$$

SEQ (1 ÷ x, x, 1, 6, 1).

The first few terms of the sequence in decimal form are shown in the viewing rectangle. By pressing the right arrow key to scroll right, you can see the remaining terms. The first six terms are also shown in fraction form.

Because a sequence is a function whose domain is the set of positive integers, the **graph of a sequence** is a set of discrete points. For example, consider the sequence whose general term is $a_n = \frac{1}{n}$. How does the graph of this sequence differ from the graph of the rational function $f(x) = \frac{1}{x}$? The graph of $f(x) = \frac{1}{x}$ is shown in **Figure 11.1(a)** for positive values of x. To obtain the graph of the sequence $\{a_n\} = \left\{\frac{1}{n}\right\}$, remove all the points from the graph of f except those whose x-coordinates are positive integers. Thus, we remove all points except $(1, 1), \left(2, \frac{1}{2}\right), \left(3, \frac{1}{3}\right), \left(4, \frac{1}{4}\right)$, and so on. The remaining points are the graph of the sequence $\{a_n\} = \left\{\frac{1}{n}\right\}$, shown in **Figure 11.1(b)**. Notice that the horizontal axis is labeled n and the vertical axis is labeled a_n.

Figure 11.1(a) The graph of $f(x) = \frac{1}{x}, x > 0$

Figure 11.1(b) The graph of $\{a_n\} = \{\frac{1}{n}\}$

Comparing a continuous graph to the graph of a sequence

Factorial Notation

Products of consecutive positive integers occur quite often in sequences. These products can be expressed in a special notation, called **factorial notation**.

> ### Factorial Notation
>
> If n is a positive integer, the notation $n!$ (read "n factorial") is the product of all positive integers from n down through 1.
>
> $$n! = n(n - 1)(n - 2) \ldots (3)(2)(1)$$
>
> 0! (zero factorial), by definition, is 1.
>
> $$0! = 1$$

2 Use factorial notation.

Using Technology

Calculators have factorial keys. To find 5 factorial, most calculators use one of the following:

Many Scientific Calculators:

5 $\boxed{x!}$

Many Graphing Calculators:

5 $\boxed{!}$ $\boxed{\text{ENTER}}$

Because $n!$ becomes quite large as n increases, your calculator will display these larger values in scientific notation.

The values of $n!$ for the first six positive integers are

$$1! = 1$$
$$2! = 2 \cdot 1 = 2$$
$$3! = 3 \cdot 2 \cdot 1 = 6$$
$$4! = 4 \cdot 3 \cdot 2 \cdot 1 = 24$$
$$5! = 5 \cdot 4 \cdot 3 \cdot 2 \cdot 1 = 120$$
$$6! = 6 \cdot 5 \cdot 4 \cdot 3 \cdot 2 \cdot 1 = 720.$$

Factorials affect only the number or variable that they follow unless grouping symbols appear. For example,

$$2 \cdot 3! = 2(3 \cdot 2 \cdot 1) = 2 \cdot 6 = 12$$

whereas

$$(2 \cdot 3)! = 6! = 6 \cdot 5 \cdot 4 \cdot 3 \cdot 2 \cdot 1 = 720.$$

In this sense, factorials are treated similarly to exponents in the order of operations.

EXAMPLE 2 Finding Terms of a Sequence Involving Factorials

Write the first four terms of the sequence whose nth term is

$$a_n = \frac{2^n}{(n - 1)!}.$$

Solution We need to find the first four terms of the sequence. To do so, we replace each n in the formula with $1, 2, 3,$ and 4.

a_1, 1st term $\quad \dfrac{2^1}{(1 - 1)!} = \dfrac{2}{0!} = \dfrac{2}{1} = 2$

a_2, 2nd term $\quad \dfrac{2^2}{(2 - 1)!} = \dfrac{4}{1!} = \dfrac{4}{1} = 4$

a_3, 3rd term $\quad \dfrac{2^3}{(3 - 1)!} = \dfrac{8}{2!} = \dfrac{8}{2 \cdot 1} = 4$

a_4, 4th term $\quad \dfrac{2^4}{(4 - 1)!} = \dfrac{16}{3!} = \dfrac{16}{3 \cdot 2 \cdot 1} = \dfrac{16}{6} = \dfrac{8}{3}$

The first four terms are $2, 4, 4,$ and $\frac{8}{3}$. ∎

☑ **CHECK POINT 2** Write the first four terms of the sequence whose nth term is

$$a_n = \frac{20}{(n + 1)!}.$$

③ Use summation notation. ▶

Summation Notation

It is sometimes useful to find the sum of the first n terms of a sequence. For example, consider the cost of raising a child born in the United States to a middle-income family. **Table 11.1** shows the cost from year 1 (child is under 1) to year 18 (child is 17). (Costs include housing, food, health care, and clothing, but exclude college costs.)

Table 11.1 The Cost of Raising a Child Born in the U.S. to a Middle-Income Family

Year	1	2	3	4	5	6	7	8	9
Average Cost	$10,600	$10,930	$11,270	$11,960	$12,330	$12,710	$12,950	$13,350	$13,760
	Child is under 1.	Child is 1.	Child is 2.	Child is 3.	Child is 4.	Child is 5.	Child is 6.	Child is 7.	Child is 8.

Year	10	11	12	13	14	15	16	17	18
Average Cost	$13,970	$14,400	$14,840	$16,360	$16,860	$17,390	$18,430	$19,000	$19,590
	Child is 9.	Child is 10.	Child is 11.	Child is 12.	Child is 13.	Child is 14.	Child is 15.	Child is 16.	Child is 17.

Source: U.S. Department of Agriculture

We can let a_n represent the cost of raising a child in year n. The terms of the finite sequence in **Table 11.1** are given as follows.

10,600, 10,930, 11,270, 11,960, 12,330, 12,710, 12,950, 13,350, 13,760,

a_1 a_2 a_3 a_4 a_5 a_6 a_7 a_8 a_9

13,970, 14,400, 14,840, 16,360, 16,860, 17,390, 18,430, 19,000, 19,590.

a_{10} a_{11} a_{12} a_{13} a_{14} a_{15} a_{16} a_{17} a_{18}

Why might we want to add the terms of this sequence? We do this to find the total cost of raising a child from birth through age 17. Thus,

$$a_1 + a_2 + a_3 + a_4 + a_5 + a_6 + a_7 + a_8 + a_9 + a_{10} + a_{11} + a_{12} + a_{13} + a_{14} + a_{15} + a_{16} + a_{17} + a_{18}$$
$$= 10{,}600 + 10{,}930 + 11{,}270 + 11{,}960 + 12{,}330 + 12{,}710 + 12{,}950 + 13{,}350 + 13{,}760$$
$$+ 13{,}970 + 14{,}400 + 14{,}840 + 16{,}360 + 16{,}860 + 17{,}390 + 18{,}430 + 19{,}000 + 19{,}590$$
$$= 260{,}700.$$

We see that the total cost of raising a child from birth through age 17 is $260,700.

There is a compact notation for expressing the sum of the first n terms of a sequence. For example, rather than write

$$a_1 + a_2 + a_3 + a_4 + a_5 + a_6 + a_7 + a_8 + a_9 + a_{10} + a_{11}$$
$$+ a_{12} + a_{13} + a_{14} + a_{15} + a_{16} + a_{17} + a_{18},$$

we can use *summation notation* to express the sum as

$$\sum_{i=1}^{18} a_i.$$

We read this expression as "the sum as i goes from 1 to 18 of a_i." The letter i is called the *index of summation* and is not related to the use of i to represent $\sqrt{-1}$.

You can think of the symbol Σ (the uppercase Greek letter sigma) as an instruction to add up the terms of a sequence.

Summation Notation

The sum of the first n terms of a sequence is represented by the **summation notation**

$$\sum_{i=1}^{n} a_i = a_1 + a_2 + a_3 + a_4 + \cdots + a_n,$$

where i is the **index of summation**, n is the **upper limit of summation**, and 1 is the **lower limit of summation**.

Any letter can be used for the index of summation. The letters i, j, and k are used commonly. Furthermore, the lower limit of summation can be an integer other than 1.

When we write out a sum that is given in summation notation, we are **expanding the summation**. Example 3 shows how to do this.

EXAMPLE 3   Using Summation Notation

Expand and evaluate the sum:

a. $\displaystyle\sum_{i=1}^{6}(i^2 + 1)$ **b.** $\displaystyle\sum_{k=4}^{7}\left[(-2)^k - 5\right]$ **c.** $\displaystyle\sum_{i=1}^{5} 3.$

Using Technology

Graphing utilities can calculate the sum of a sequence. For example, to find the sum of the sequence in Example 3(a), enter

| SUM | SEQ | $(x^2 + 1, x, 1, 6, 1)$.

Then press | ENTER |; 97 should be displayed. Use this capability to verify Example 3(b).

```
sum(seq(X²+1,X,1,6,1))
                    97
```

Some graphing calculators require that you fill in the information following | SEQ | on a separate screen and then paste it to the home screen.

Solution

a. To find $\sum_{i=1}^{6}(i^2 + 1)$, we must replace i in the expression $i^2 + 1$ with all consecutive integers from 1 to 6, inclusive. Then we add.

$$\sum_{i=1}^{6}(i^2 + 1) = (1^2 + 1) + (2^2 + 1) + (3^2 + 1) + (4^2 + 1)$$
$$+ (5^2 + 1) + (6^2 + 1)$$
$$= 2 + 5 + 10 + 17 + 26 + 37$$
$$= 97$$

b. The index of summation in $\sum_{k=4}^{7}[(-2)^k - 5]$ is k. First we evaluate $(-2)^k - 5$ for all consecutive integers from 4 through 7, inclusive. Then we add.

$$\sum_{k=4}^{7}[(-2)^k - 5] = [(-2)^4 - 5] + [(-2)^5 - 5]$$
$$+ [(-2)^6 - 5] + [(-2)^7 - 5]$$
$$= (16 - 5) + (-32 - 5) + (64 - 5) + (-128 - 5)$$
$$= 11 + (-37) + 59 + (-133)$$
$$= -100$$

c. To find $\sum_{i=1}^{5}3$, we observe that every term of the sum is 3. The notation $i = 1$ through 5 indicates that we must add the first five terms of a sequence in which every term is 3.

$$\sum_{i=1}^{5}3 = 3 + 3 + 3 + 3 + 3 = 15 \quad \blacksquare$$

✓ **CHECK POINT 3**   Expand and evaluate the sum:

a. $\displaystyle\sum_{i=1}^{6}2i^2$

b. $\displaystyle\sum_{k=3}^{5}(2^k - 3)$

c. $\displaystyle\sum_{i=1}^{5}4.$

Although the domain of a sequence is the set of positive integers, any integers can be used for the limits of summation. For a given sum, we can vary the upper and lower limits of summation, as well as the letter used for the index of summation. By doing so, we can produce different-looking summation notations for the same sum. For example, the sum of the squares of the first four positive integers, $1^2 + 2^2 + 3^2 + 4^2$, can be expressed in a number of equivalent ways:

$$\sum_{i=1}^{4}i^2 = 1^2 + 2^2 + 3^2 + 4^2 = 30$$

$$\sum_{i=0}^{3}(i + 1)^2 = (0 + 1)^2 + (1 + 1)^2 + (2 + 1)^2 + (3 + 1)^2$$
$$= 1^2 + 2^2 + 3^2 + 4^2 = 30$$

$$\sum_{k=2}^{5}(k - 1)^2 = (2 - 1)^2 + (3 - 1)^2 + (4 - 1)^2 + (5 - 1)^2$$
$$= 1^2 + 2^2 + 3^2 + 4^2 = 30.$$

EXAMPLE 4 Writing Sums in Summation Notation

Express each sum using summation notation:

a. $1^3 + 2^3 + 3^3 + \cdots + 7^3$ **b.** $1 + \dfrac{1}{3} + \dfrac{1}{9} + \dfrac{1}{27} + \cdots + \dfrac{1}{3^{n-1}}.$

Solution In each case, we will use 1 as the lower limit of summation and i for the index of summation.

a. The sum $1^3 + 2^3 + 3^3 + \cdots + 7^3$ has seven terms, each of the form i^3, starting at $i = 1$ and ending at $i = 7$. Thus,

$$1^3 + 2^3 + 3^3 + \cdots + 7^3 = \sum_{i=1}^{7} i^3.$$

b. The sum

$$1 + \frac{1}{3} + \frac{1}{9} + \frac{1}{27} + \cdots + \frac{1}{3^{n-1}}$$

has n terms, each of the form $\dfrac{1}{3^{i-1}}$, starting at $i = 1$ and ending at $i = n$. Thus,

$$1 + \frac{1}{3} + \frac{1}{9} + \frac{1}{27} + \cdots + \frac{1}{3^{n-1}} = \sum_{i=1}^{n} \frac{1}{3^{i-1}}. \quad \blacksquare$$

✓ **CHECK POINT 4** Express each sum using summation notation:

a. $1^2 + 2^2 + 3^2 + \cdots + 9^2$ **b.** $1 + \dfrac{1}{2} + \dfrac{1}{4} + \dfrac{1}{8} + \cdots + \dfrac{1}{2^{n-1}}.$

Achieving Success

Don't wait too long after class to review your notes. Reading notes while the classroom experience is fresh in your mind will help you to remember what was covered during lecture.

CONCEPT AND VOCABULARY CHECK

Fill in each blank so that the resulting statement is true.

1. $\{a_n\} = a_1, a_2, a_3, a_4, \ldots, a_n, \ldots$ represents an infinite _____, a function whose domain is the set of positive _____. The function values $a_1, a_2, a_3, \ldots$ are called the _____.

2. The nth term of a sequence, represented by a_n, is called the _____ term.

Write the first term of each sequence.

3. $a_n = 5n - 6$ _____

4. $a_n = \dfrac{(-1)^n}{4^n - 1}$ _____

5. 5!, called 5 _____, is the product of all positive integers from _____ down through _____. By definition, 0! = _____.

6. $\displaystyle\sum_{i=1}^{n} a_i =$ _____ + _____ + _____ + $\cdots$ + _____. In this summation notation, i is called the _____ of summation, n is the _____ of summation, and 1 is the _____ of summation.

11.1 EXERCISE SET   ▶ MyMathLab®

Practice Exercises

In Exercises 1–16, write the first four terms of each sequence whose general term is given.

1. $a_n = 3n + 2$

2. $a_n = 4n - 1$

3. $a_n = 3^n$

4. $a_n = \left(\dfrac{1}{3}\right)^n$

5. $a_n = (-3)^n$

6. $a_n = \left(-\dfrac{1}{3}\right)^n$

7. $a_n = (-1)^n(n + 3)$

8. $a_n = (-1)^{n+1}(n + 4)$

9. $a_n = \dfrac{2n}{n + 4}$

10. $a_n = \dfrac{3n}{n + 5}$

11. $a_n = \dfrac{(-1)^{n+1}}{2^n - 1}$

12. $a_n = \dfrac{(-1)^{n+1}}{2^n + 1}$

13. $a_n = \dfrac{n^2}{n!}$

14. $a_n = \dfrac{(n + 1)!}{n^2}$

15. $a_n = 2(n + 1)!$

16. $a_n = -2(n - 1)!$

In Exercises 17–30, find each indicated sum.

17. $\displaystyle\sum_{i=1}^{6} 5i$

18. $\displaystyle\sum_{i=1}^{6} 7i$

19. $\displaystyle\sum_{i=1}^{4} 2i^2$

20. $\displaystyle\sum_{i=1}^{5} i^3$

21. $\displaystyle\sum_{k=1}^{5} k(k + 4)$

22. $\displaystyle\sum_{k=1}^{4} (k - 3)(k + 2)$

23. $\displaystyle\sum_{i=1}^{4} \left(-\dfrac{1}{2}\right)^i$

24. $\displaystyle\sum_{i=2}^{4} \left(-\dfrac{1}{3}\right)^i$

25. $\displaystyle\sum_{i=5}^{9} 11$

26. $\displaystyle\sum_{i=3}^{7} 12$

27. $\displaystyle\sum_{i=0}^{4} \dfrac{(-1)^i}{i!}$

28. $\displaystyle\sum_{i=0}^{4} \dfrac{(-1)^{i+1}}{(i + 1)!}$

29. $\displaystyle\sum_{i=1}^{5} \dfrac{i!}{(i - 1)!}$

30. $\displaystyle\sum_{i=1}^{5} \dfrac{(i + 2)!}{i!}$

In Exercises 31–42, express each sum using summation notation. Use 1 as the lower limit of summation and i for the index of summation.

31. $1^2 + 2^2 + 3^2 + \cdots + 15^2$

32. $1^4 + 2^4 + 3^4 + \cdots + 12^4$

33. $2 + 2^2 + 2^3 + \cdots + 2^{11}$

34. $5 + 5^2 + 5^3 + \cdots + 5^{12}$

35. $1 + 2 + 3 + \cdots + 30$

36. $1 + 2 + 3 + \cdots + 40$

37. $\dfrac{1}{2} + \dfrac{2}{3} + \dfrac{3}{4} + \cdots + \dfrac{14}{14 + 1}$

38. $\dfrac{1}{3} + \dfrac{2}{4} + \dfrac{3}{5} + \cdots + \dfrac{16}{16 + 2}$

39. $4 + \dfrac{4^2}{2} + \dfrac{4^3}{3} + \cdots + \dfrac{4^n}{n}$

40. $\dfrac{1}{9} + \dfrac{2}{9^2} + \dfrac{3}{9^3} + \cdots + \dfrac{n}{9^n}$

41. $1 + 3 + 5 + \cdots + (2n - 1)$

42. $a + ar + ar^2 + \cdots + ar^{n-1}$

In Exercises 43–48, express each sum using summation notation. Use a lower limit of summation of your choice and k for the index of summation.

43. $5 + 7 + 9 + 11 + \cdots + 31$

44. $6 + 8 + 10 + 12 + \cdots + 32$

45. $a + ar + ar^2 + \cdots + ar^{12}$

46. $a + ar + ar^2 + \cdots + ar^{14}$

47. $a + (a + d) + (a + 2d) + \cdots + (a + nd)$

48. $(a + d) + (a + d^2) + \cdots + (a + d^n)$

Practice PLUS

In Exercises 49–56, use the graphs of $\{a_n\}$ and $\{b_n\}$ to find each indicated sum.

The Graph of $\{a_n\}$

The Graph of $\{b_n\}$

49. $\displaystyle\sum_{i=1}^{5} (a_i^2 + 1)$

50. $\displaystyle\sum_{i=1}^{5} (b_i^2 - 1)$

51. $\displaystyle\sum_{i=1}^{5} (2a_i + b_i)$

52. $\displaystyle\sum_{i=1}^{5} (a_i + 3b_i)$

53. $\displaystyle\sum_{i=4}^{5} \left(\dfrac{a_i}{b_i}\right)^2$

54. $\displaystyle\sum_{i=4}^{5} \left(\dfrac{a_i}{b_i}\right)^3$

55. $\displaystyle\sum_{i=1}^{5} a_i^2 + \sum_{i=1}^{5} b_i^2$

56. $\displaystyle\sum_{i=1}^{5} a_i^2 - \sum_{i=3}^{5} b_i^2$

Application Exercises

57. The bar graph shows the average number of hours per day that U.S. adult users spent on digital media (desktop/laptop, mobile, and other devices) from 2008 through 2014.

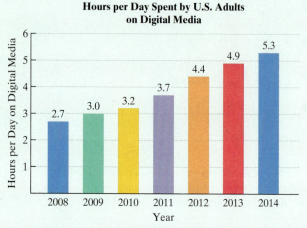

Hours per Day Spent by U.S. Adults on Digital Media

Source: 2015 Internet Trends Report

Let a_n represent the average number of hours per day that U.S. adult users spent on digital media n years after 2007.

a. Use the numbers given in the graph to find and interpret $\frac{1}{7}\sum_{i=1}^{7} a_i$. Round to the nearest tenth of an hour.

b. The finite sequence whose general term is

$$a_n = 0.5n + 2,$$

where $n = 1, 2, 3, 4, 5, 6, 7$, models the averge number of hours per day that U.S. adult users spent on digital media n years after 2007. Use this model to find $\frac{1}{7}\sum_{i=1}^{7} a_i$. Does this underestimate or overestimate the rounded value you obtained in part (a)? By how much?

58. The bar graph shows the number of Americans who renounced their U.S. citizenship, many over tax laws, from 2010 through 2014.

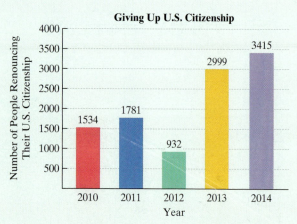

Giving Up U.S. Citizenship

Source: Internal Revenue Service

Let a_n represent the number of Americans who gave up their U.S. citizenship n years after 2009.

a. Use the numbers given in the graph to find and interpret $\sum_{i=1}^{5} a_i$.

b. The finite sequence whose general term is

$$a_n = 498n + 638,$$

where $n = 1, 2, 3, 4, 5$, models the number of Americans who gave up their U.S. citizenship n years after 2009. Use this model to find $\sum_{i=1}^{5} a_i$. Does this underestimate or overestimate the actual sum in part (a)? By how much?

59. A deposit of $6000 is made in an account that earns 6% interest compounded quarterly. The balance in the account after n quarters is given by the sequence

$$a_n = 6000\left(1 + \frac{0.06}{4}\right)^n, \quad n = 1, 2, 3, \ldots.$$

Find the balance in the account after five years. Round to the nearest cent.

60. A deposit of $10,000 is made in an account that earns 8% interest compounded quarterly. The balance in the account after n quarters is given by the sequence

$$a_n = 10,000\left(1 + \frac{0.08}{4}\right)^n, \quad n = 1, 2, 3, \ldots.$$

Find the balance in the account after six years. Round to the nearest cent.

Explaining the Concepts

61. What is a sequence? Give an example with your description.

62. Explain how to write terms of a sequence if the formula for the general term is given.

63. What does the graph of a sequence look like? How is it obtained?

64. Explain how to find $n!$ if n is a positive integer.

65. What is the meaning of the symbol Σ? Give an example with your description.

66. You buy a new car for $24,000. At the end of n years, the value of your car is given by the sequence

$$a_n = 24,000\left(\frac{3}{4}\right)^n, \quad n = 1, 2, 3, \ldots.$$

Find a_5 and write a sentence explaining what this value represents. Describe the nth term of the sequence in terms of the value of your car at the end of each year.

Technology Exercises

67. Use the ⟨SEQ⟩ (sequence) capability of a graphing utility to verify the terms of the sequences you obtained for any five sequences from Exercises 1–16.

68. Use the ⟨SUM⟩ ⟨SEQ⟩ (sum of the sequence) capability of a graphing utility to verify any five of the sums you obtained in Exercises 17–30.

Many graphing utilities have a sequence-graphing mode that plots the terms of a sequence as points on a rectangular coordinate system. Consult your manual; if your graphing utility has this capability, use it to graph each of the sequences in Exercises 69–72. What appears to be happening to the terms of each sequence as n gets larger?

69. $a_n = \dfrac{n}{n+1}$; *n*: [0, 10, 1] by a_n: [0, 1, 0.1]

70. $a_n = \dfrac{100}{n}$; *n*: [0, 1000, 100] by a_n: [0, 1, 0.1]

71. $a_n = \dfrac{2n^2 + 5n - 7}{n^3}$; *n*: [0, 10, 1] by a_n: [0, 2, 0.2]

72. $a_n = \dfrac{3n^4 + n - 1}{5n^4 + 2n^2 + 1}$; *n*: [0, 10, 1] by a_n: [0, 1, 0.1]

Critical Thinking Exercises

Make Sense?   *In Exercises 73–76, determine whether each statement makes sense or does not make sense, and explain your reasoning.*

73. Now that I've studied sequences, I realize that the joke in this cartoon is based on the fact that you can't have a negative number of sheep.

WHEN MATHEMATICIANS CAN'T SLEEP

74. By writing $a_1, a_2, a_3, a_4, \ldots, a_n, \ldots$, I can see that the range of a sequence is the set of positive integers.

75. It makes a difference whether or not I use parentheses around the expression following the summation symbol, because the value of $\displaystyle\sum_{i=1}^{8}(i+7)$ is 92, but the value of $\displaystyle\sum_{i=1}^{8} i + 7$ is 43.

76. Without writing out the terms, I can see that $(-1)^{2n}$ in $a_n = \dfrac{(-1)^{2n}}{3n}$ causes the terms to alternate in sign.

In Exercises 77–80, determine whether each statement is true or false. If the statement is false, make the necessary change(s) to produce a true statement.

77. $\displaystyle\sum_{i=1}^{2}(-1)^i 2^i = 0$

78. $\displaystyle\sum_{i=1}^{2} a_i b_i = \sum_{i=1}^{2} a_i \sum_{i=1}^{2} b_i$

79. $\displaystyle\sum_{i=1}^{4} 3i + \sum_{i=1}^{4} 4i = \sum_{i=1}^{4} 7i$

80. $\displaystyle\sum_{i=0}^{6}(-1)^i(i+1)^2 = \sum_{j=1}^{7}(-1)^j j^2$

In Exercises 81–88, find a general term, a_n, for each sequence. More than one answer may be possible.

81. $1, \dfrac{1}{2}, \dfrac{1}{3}, \dfrac{1}{4}, \ldots$ **82.** $1, 4, 9, 16, \ldots$

83. $-1, 1, -1, 1, \ldots$

84. $1 \cdot 3, 2 \cdot 4, 3 \cdot 5, 4 \cdot 6, \ldots$

85. $\dfrac{3}{2}, \dfrac{4}{3}, \dfrac{5}{4}, \dfrac{6}{5}, \ldots$

86. $5, 7, 9, 11, \ldots$

87. $\dfrac{4}{1}, \dfrac{9}{2}, \dfrac{16}{3}, \dfrac{25}{4}, \ldots$

88. $4, -8, 16, -32, \ldots$

89. Evaluate without using a calculator: $\dfrac{600!}{599!}$.

In Exercises 90–91, rewrite each expression as a polynomial in standard form.

90. $\dfrac{(n+4)!}{(n+2)!}$ **91.** $\dfrac{n!}{(n-3)!}$

In Exercises 92–93, expand and write the answer as a single logarithm with a coefficient of 1.

92. $\displaystyle\sum_{i=1}^{4} \log(2i)$

93. $\displaystyle\sum_{i=2}^{4} 2i \log x$

94. If $a_1 = 7$ and $a_n = a_{n-1} + 5$ for $n \geq 2$, write the first four terms of the sequence.

Review Exercises

95. Simplify: $\sqrt[3]{40x^4 y^7}$.

(Section 7.3, Example 5)

96. Factor: $27x^3 - 8$.

(Section 5.5, Example 9)

97. Solve: $\dfrac{6}{x} + \dfrac{6}{x+2} = \dfrac{5}{2}$.

(Section 6.6, Example 4)

Preview Exercises

Exercises 98–100 will help you prepare for the material covered in the next section.

98. Consider the sequence $8, 3, -2, -7, -12, \ldots$. Find $a_2 - a_1$, $a_3 - a_2, a_4 - a_3$, and $a_5 - a_4$. What do you observe?

99. Consider the sequence whose *n*th term is $a_n = 4n - 3$. Find $a_2 - a_1, a_3 - a_2, a_4 - a_3$, and $a_5 - a_4$. What do you observe?

100. Use the formula $a_n = 4 + (n-1)(-7)$ to find the eighth term of the sequence $4, -3, -10, \ldots$.

11.2

Arithmetic Sequences

What am I supposed to learn?

After studying this section, you should be able to:

1 Find the common difference for an arithmetic sequence. ▶

2 Write terms of an arithmetic sequence. ▶

3 Use the formula for the general term of an arithmetic sequence. ▶

4 Use the formula for the sum of the first n terms of an arithmetic sequence. ▶

1 Find the common difference for an arithmetic sequence. ▶

Your grandmother and her financial counselor are looking at options in case an adult residential facility is needed in the future. The good news is that your grandmother's total assets are $500,000. The bad news is that yearly adult residential community costs average $64,130, increasing by $1800 each year. In this section, we will see how sequences can be used to describe your grandmother's situation and help her to identify realistic options.

Arithmetic Sequences

The bar graph in **Figure 11.2** shows how much Americans spent on their pets, rounded to the nearest billion dollars, each year from 2001 through 2012.

The graph illustrates that each year spending increased by $2 billion. The sequence of annual spending

$$29, 31, 33, 35, 37, 39, 41, \ldots$$

shows that each term after the first, 29, differs from the preceding term by a constant amount, namely 2. This sequence is an example of an *arithmetic sequence.*

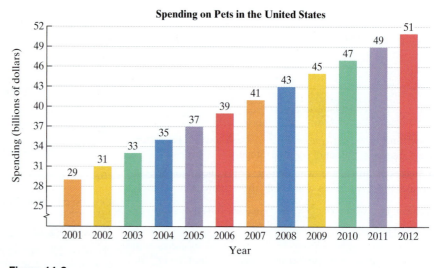

Spending on Pets in the United States

Figure 11.2

Source: American Pet Products Manufacturers Association

Definition of an Arithmetic Sequence

An **arithmetic sequence** is a sequence in which each term after the first differs from the preceding term by a constant amount. The difference between consecutive terms is called the **common difference** of the sequence.

The common difference, d, is found by subtracting any term from the term that directly follows it. In the following examples, the common difference is found by subtracting the first term from the second term, $a_2 - a_1$.

Arithmetic Sequence	Common Difference
$142, 146, 150, 154, 158, \ldots$	$d = 146 - 142 = 4$
$-5, -2, 1, 4, 7, \ldots$	$d = -2 - (-5) = -2 + 5 = 3$
$8, 3, -2, -7, -12, \ldots$	$d = 3 - 8 = -5$

Figure 11.3 shows the graphs of the last two arithmetic sequences in our list. The common difference for the increasing sequence in **Figure 11.3(a)** is 3. The common difference for the decreasing sequence in **Figure 11.3(b)** is -5.

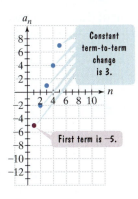

Figure 11.3(a) The graph of $\{a_n\} = -5, -2, 1, 4, 7, \ldots$

Figure 11.3(b) The graph of $\{b_n\} = 8, 3, -2, -7, -12, \ldots$

The graph of each arithmetic sequence in **Figure 11.3** forms a set of discrete points lying on a straight line. This illustrates that **an arithmetic sequence is a linear function whose domain is the set of positive integers**.

If the first term of an arithmetic sequence is a_1, each term after the first is obtained by adding d, the common difference, to the previous term.

2 Write terms of an arithmetic sequence.

EXAMPLE 1 Writing the Terms of an Arithmetic Sequence Using the First Term and the Common Difference

Write the first six terms of the arithmetic sequence with first term 6 and common difference -2.

Solution To find the second term, we add -2 to the first term, 6, giving 4. For the next term, we add -2 to 4, and so on.

$$a_1 \text{ (first term)} \quad = 6$$
$$a_2 \text{ (second term)} = 6 + (-2) = 4$$
$$a_3 \text{ (third term)} \quad = 4 + (-2) = 2$$
$$a_4 \text{ (fourth term)} \quad = 2 + (-2) = 0$$
$$a_5 \text{ (fifth term)} \quad = 0 + (-2) = -2$$
$$a_6 \text{ (sixth term)} \quad = -2 + (-2) = -4$$

The first six terms are

$$6, 4, 2, 0, -2, \text{ and } -4. \quad \blacksquare$$

✓ **CHECK POINT 1** Write the first six terms of the arithmetic sequence with first term 100 and common difference -30.

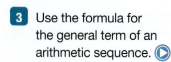

3 Use the formula for the general term of an arithmetic sequence.

The General Term of an Arithmetic Sequence

Consider an arithmetic sequence whose first term is a_1 and whose common difference is d. We are looking for a formula for the general term, a_n. Let's begin by writing the first six terms. The first term is a_1. The second term is $a_1 + d$. The third term is $a_1 + d + d$, or $a_1 + 2d$. Thus, we start with a_1 and add d to each successive term. The first six terms are

$$a_1, \quad a_1 + d, \quad a_1 + 2d, \quad a_1 + 3d, \quad a_1 + 4d, \quad a_1 + 5d.$$

| a_1, first term | a_2, second term | a_3, third term | a_4, fourth term | a_5, fifth term | a_6, sixth term |

Compare the coefficient of d and the subscript of a denoting the term number. Can you see that the coefficient of d is 1 less than the subscript of a denoting the term number?

a_3: third term $= a_1 + 2d$ a_4: fourth term $= a_1 + 3d$

One less than 3, or 2, is the coefficient of d.   One less than 4, or 3, is the coefficient of d.

Thus, the formula for the nth term is

$$a_n: n\text{th term} = a_1 + (n - 1)d.$$

One less than n, or $n - 1$, is the coefficient of d.

> ### General Term of an Arithmetic Sequence
>
> The nth term (the general term) of an arithmetic sequence with first term a_1 and common difference d is
>
> $$a_n = a_1 + (n - 1)d.$$

EXAMPLE 2   Using the Formula for the General Term of an Arithmetic Sequence

Find the eighth term of the arithmetic sequence whose first term is 4 and whose common difference is -7.

Solution   To find the eighth term, a_8, we replace n in the formula with 8, a_1 with 4, and d with -7.

$$a_n = a_1 + (n - 1)d$$
$$a_8 = 4 + (8 - 1)(-7) = 4 + 7(-7) = 4 + (-49) = -45$$

The eighth term is -45. We can check this result by writing the first eight terms of the sequence:

$$4, -3, -10, -17, -24, -31, -38, -45. \quad \blacksquare$$

☑ **CHECK POINT 2**   Find the ninth term of the arithmetic sequence whose first term is 6 and whose common difference is -5.

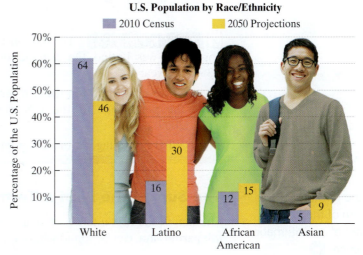

EXAMPLE 3 Using an Arithmetic Sequence to Model Changes in the U.S. Population

The graph in **Figure 11.4** shows the percentage of the U.S. population by race/ethnicity for 2010, with projections by the U.S. Census Bureau for 2050.

U.S. Population by Race/Ethnicity

Figure 11.4
Source: U.S. Census Bureau

The data show that in 2010, 64% of the U.S. population was white. On average, this is projected to decrease by approximately 0.45% per year.

a. Write a formula for the nth term of the arithmetic sequence that describes the percentage of the U.S. population that will be white n years after 2009.

b. What percentage of the U.S. population is projected to be white in 2030?

Solution

a. With a yearly decrease of 0.45%, we can express the percentage of the white population by the following arithmetic sequence:

$$64, \qquad 64 - 0.45 = 63.55, \qquad 63.55 - 0.45 = 63.10, \ \ldots .$$

a_1: percentage of whites in the population in 2010, 1 year after 2009

a_2: percentage of whites in the population in 2011, 2 years after 2009

a_3: percentage of whites in the population in 2012, 3 years after 2009

In the sequence 64, 63.55, 63.10, ... , the first term, a_1, represents the percentage of the population that was white in 2010. Each subsequent year this amount decreases by 0.45%, so $d = -0.45$. We use the formula for the general term of an arithmetic sequence to write the nth term of the sequence that describes the percentage of whites in the population n years after 2009.

$a_n = a_1 + (n-1)d$ This is the formula for the general term of an arithmetic sequence.

$a_n = 64 + (n-1)(-0.45)$ $a_1 = 64$ and $d = -0.45$.

$a_n = 64 - 0.45n + 0.45$ Distribute -0.45 to each term in parentheses.

$a_n = -0.45n + 64.45$ Simplify.

Thus, the percentage of the U.S. population that will be white n years after 2009 can be described by

$$a_n = -0.45n + 64.45.$$

b. Now we need to project the percentage of the population that will be white in 2030. The year 2030 is 21 years after 2009: Thus, $n = 21$. We substitute 21 for n in $a_n = -0.45n + 64.45$.

$$a_{21} = -0.45(21) + 64.45 = 55$$

The 21st term of the sequence is 55. Thus, 55% of the population is projected to be white in 2030. ∎

✓ **CHECK POINT 3** The data in **Figure 11.4** show that in 2010, 16% of the U.S. population was Latino. On average, this is projected to increase by approximately 0.35% per year.

    a. Write a formula for the nth term of the arithmetic sequence that describes the percentage of the U.S. population that will be Latino n years after 2009.

    b. What percentage of the U.S. population is projected to be Latino in 2030?

4 Use the formula for the sum of the first n terms of an arithmetic sequence. ▶

The Sum of the First n Terms of an Arithmetic Sequence

The sum of the first n terms of an arithmetic sequence, denoted by S_n, and called the **nth partial sum**, can be found without having to add up all the terms. Let

$$S_n = a_1 + a_2 + a_3 + \cdots + a_n$$

be the sum of the first n terms of an arithmetic sequence. Because d is the common difference between terms, S_n can be written forward and backward, as shown below.

Forward: Start with the first term, a_1. Keep adding d.

Backward: Start with the last term, a_n. Keep subtracting d.

$$S_n = a_1 \qquad + (a_1 + d) \quad + (a_1 + 2d) \quad + \cdots + a_n$$
$$S_n = a_n \qquad + (a_n - d) \quad + (a_n - 2d) \quad + \cdots + a_1$$
$$2S_n = (a_1 + a_n) + (a_1 + a_n) + (a_1 + a_n) + \cdots + (a_1 + a_n).$$ Add the two equations.

Because there are n sums of $(a_1 + a_n)$ on the right side, we can express this side as $n(a_1 + a_n)$. Thus, the last equation can be written as follows:

$$2S_n = n(a_1 + a_n).$$

$$S_n = \frac{n}{2}(a_1 + a_n) \qquad \text{Solve for } S_n, \text{ dividing both sides by 2.}$$

We have proved the following result:

> ### The Sum of the First n Terms of an Arithmetic Sequence
> The sum, S_n, of the first n terms of an arithmetic sequence is given by
> $$S_n = \frac{n}{2}(a_1 + a_n),$$
> in which a_1 is the first term and a_n is the nth term.

To find the sum of the terms of an arithmetic sequence using $S_n = \frac{n}{2}(a_1 + a_n)$, we need to know the first term, a_1, the last term, a_n, and the number of terms, n. The following examples illustrate how to use this formula.

EXAMPLE 4 Finding the Sum of *n* Terms of an Arithmetic Sequence

Find the sum of the first 100 terms of the arithmetic sequence: $1, 3, 5, 7, \ldots$.

Solution By finding the sum of the first 100 terms of $1, 3, 5, 7, \ldots$, we are finding the sum of the first 100 odd numbers. To find the sum of the first 100 terms, S_{100}, we replace *n* in the formula with 100.

$$S_n = \frac{n}{2}(a_1 + a_n)$$

$$S_{100} = \frac{100}{2}(a_1 + a_{100})$$

The first term, a_1, is 1.   We must find a_{100}, the 100th term.

We use the formula for the general term of an arithmetic sequence to find a_{100}. The common difference, *d*, of $1, 3, 5, 7, \ldots$, is 2.

$a_n = a_1 + (n - 1)d$ This is the formula for the *n*th term of an arithmetic sequence. Use it to find the 100th term.

$a_{100} = 1 + (100 - 1) \cdot 2$ Substitute 100 for *n*, 2 for *d*, and 1 (the first term) for a_1.

$\quad\quad = 1 + 99 \cdot 2$ Perform the subtraction in parentheses.

$\quad\quad = 1 + 198 = 199$ Multiply $(99 \cdot 2 = 198)$ and then add.

Now we are ready to find the sum of the 100 terms $1, 3, 5, 7, \ldots, 199$.

$S_n = \frac{n}{2}(a_1 + a_n)$ Use the formula for the sum of the first *n* terms of an arithmetic sequence. Let $n = 100$, $a_1 = 1$, and $a_{100} = 199$.

$S_{100} = \frac{100}{2}(1 + 199) = 50(200) = 10{,}000$

The sum of the first 100 odd numbers is 10,000. Equivalently, the 100th partial sum of the sequence $1, 3, 5, 7, \ldots$ is 10,000. ∎

☑ **CHECK POINT 4** Find the sum of the first 15 terms of the arithmetic sequence: $3, 6, 9, 12, \ldots$.

EXAMPLE 5 Using S_n to Evaluate a Summation

Find the following sum: $\displaystyle\sum_{i=1}^{25}(5i - 9)$.

Solution

$$\sum_{i=1}^{25}(5i - 9) = (5 \cdot 1 - 9) + (5 \cdot 2 - 9) + (5 \cdot 3 - 9) + \cdots + (5 \cdot 25 - 9)$$

$$= -4 \quad\quad + 1 \quad\quad + 6 \quad\quad + \cdots + 116$$

By evaluating the first three terms and the last term, we see that $a_1 = -4$, *d*, the common difference, is $1 - (-4)$ or 5, and a_{25}, the last term, is 116.

Using Technology

To find

$$\sum_{i=1}^{25}(5i-9)$$

on a graphing utility, enter:

SUM SEQ $(5x-9, x, 1, 25, 1)$.

Then press ENTER.

sum(seq(5X-9,X,1,25,1))
1400

$$S_n = \frac{n}{2}(a_1 + a_n)$$ Use the formula for the sum of the first n terms of an arithmetic sequence. Let $n = 25$, $a_1 = -4$, and $a_{25} = 116$.

$$S_{25} = \frac{25}{2}(-4 + 116) = \frac{25}{2}(112) = 1400$$

Thus,

$$\sum_{i=1}^{25}(5i-9) = 1400. \quad\blacksquare$$

✓ **CHECK POINT 5** Find the following sum: $\sum_{i=1}^{30}(6i-11)$.

EXAMPLE 6 Modeling Total Nursing Home Costs over a Six-Year Period

Your grandmother has assets of $500,000. One option that she is considering involves an adult residential community for a six-year period beginning in 2018. The model

$$a_n = 1800n + 64{,}130$$

describes yearly adult residential community costs n years after 2017. Does your grandmother have enough to pay for the facility?

Solution We must find the sum of an arithmetic sequence whose general term is $a_n = 1800n + 64{,}130$. The first term of the sequence corresponds to the facility's costs in the year 2018. The last term corresponds to costs in the year 2023. Because the model describes costs n years after 2017, $n = 1$ describes the year 2018 and $n = 6$ describes the year 2023.

$$a_n = 1800n + 64{,}130$$ This is the given formula for the general term of the sequence.

$$a_1 = 1800 \cdot 1 + 64{,}130 = 65{,}930$$ Find a_1 by replacing n with 1.

$$a_6 = 1800 \cdot 6 + 64{,}130 = 74{,}930$$ Find a_6 by replacing n with 6.

The first year the facility will cost $65,930. By year six, the facility will cost $74,930. Now we must find the sum of the costs for all six years. We focus on the sum of the first six terms of the arithmetic sequence

$$\underset{a_1}{65{,}930,} \; \underset{a_2}{67{,}730,} \; \ldots, \; \underset{a_6}{74{,}930.}$$

We find this sum using the formula for the sum of the first n terms of an arithmetic sequence. We are adding 6 terms: $n = 6$. The first term is 65,930: $a_1 = 65{,}930$. The last term—that is, the sixth term—is 74,930: $a_6 = 74{,}930$.

$$S_n = \frac{n}{2}(a_1 + a_n)$$

$$S_6 = \frac{6}{2}(65{,}930 + 74{,}930) = 3(140{,}860) = 422{,}580$$

Total adult residential community costs for your grandmother are predicted to be $422,580. Because your grandmother's assets are $500,000, she has enough to pay for the facility for the six-year period. ■

✓ **CHECK POINT 6** In Example 6, how much would it cost for the adult residential community for a ten-year period beginning in 2018?

CONCEPT AND VOCABULARY CHECK

Fill in each blank so that the resulting statement is true.

1. A sequence in which each term after the first differs from the preceding term by a constant amount is called a/an _____ sequence. The difference between consecutive terms is called the _____ of the sequence.

2. The nth term of the sequence described in Concept Check 1 is given by the formula $a_n = $ _____, where a_1 is the _____ and d is the _____ of the sequence.

3. The sum, S_n, of the first n terms of the sequence described in Concept Check 1 is given by the formula $S_n = $ _____, where a_1 is the _____ and a_n is the _____.

4. The first term of $\sum_{i=1}^{20}(6i-4)$ is _____ and the last term is _____.

5. The first three terms of $\sum_{i=1}^{17}(5i+3)$ are _____, _____, and _____. The common difference is _____.

11.2 EXERCISE SET ▶ MyMathLab®

Practice Exercises

In Exercises 1–6, find the common difference for each arithmetic sequence.

1. $2, 6, 10, 14, \ldots$
2. $3, 8, 13, 18, \ldots$
3. $-7, -2, 3, 8, \ldots$
4. $-10, -4, 2, 8, \ldots$
5. $714, 711, 708, 705, \ldots$
6. $611, 606, 601, 596, \ldots$

In Exercises 7–16, write the first six terms of each arithmetic sequence with the given first term, a_1, and common difference, d.

7. $a_1 = 200, d = 20$
8. $a_1 = 300, d = 50$
9. $a_1 = -7, d = 4$
10. $a_1 = -8, d = 5$
11. $a_1 = 300, d = -90$
12. $a_1 = 200, d = -60$
13. $a_1 = \dfrac{5}{2}, d = -\dfrac{1}{2}$
14. $a_1 = \dfrac{3}{4}, d = -\dfrac{1}{4}$
15. $a_1 = -0.4, d = -1.6$
16. $a_1 = -0.3, d = -1.7$

In Exercises 17–24, use the formula for the general term (the nth term) of an arithmetic sequence to find the indicated term of each sequence with the given first term, a_1, and common difference, d.

17. Find a_6 when $a_1 = 13, d = 4$.
18. Find a_{16} when $a_1 = 9, d = 2$.
19. Find a_{50} when $a_1 = 7, d = 5$.
20. Find a_{60} when $a_1 = 8, d = 6$.
21. Find a_{200} when $a_1 = -40, d = 5$.
22. Find a_{150} when $a_1 = -60, d = 5$.

23. Find a_{60} when $a_1 = 35, d = -3$.
24. Find a_{70} when $a_1 = -32, d = 4$.

In Exercises 25–34, write a formula for the general term (the nth term) of each arithmetic sequence. Then use the formula for a_n to find a_{20}, the 20th term of the sequence.

25. $1, 5, 9, 13, \ldots$
26. $2, 7, 12, 17, \ldots$
27. $7, 3, -1, -5, \ldots$
28. $6, 1, -4, -9, \ldots$
29. $-20, -24, -28, -32, \ldots$
30. $-70, -75, -80, -85, \ldots$
31. $a_1 = -\dfrac{1}{3}, d = \dfrac{1}{3}$
32. $a_1 = -\dfrac{1}{4}, d = \dfrac{1}{4}$
33. $a_1 = 4, d = -0.3$
34. $a_1 = 5, d = -0.2$
35. Find the sum of the first 20 terms of the arithmetic sequence: $4, 10, 16, 22, \ldots$.
36. Find the sum of the first 25 terms of the arithmetic sequence: $7, 19, 31, 43, \ldots$.
37. Find the sum of the first 50 terms of the arithmetic sequence: $-10, -6, -2, 2, \ldots$.
38. Find the sum of the first 50 terms of the arithmetic sequence: $-15, -9, -3, 3, \ldots$.
39. Find $1 + 2 + 3 + 4 + \cdots + 100$, the sum of the first 100 natural numbers.
40. Find $2 + 4 + 6 + 8 + \cdots + 200$, the sum of the first 100 positive even integers.
41. Find the sum of the first 60 positive even integers.
42. Find the sum of the first 80 positive even integers.
43. Find the sum of the even integers between 21 and 45.
44. Find the sum of the odd integers between 30 and 54.

For Exercises 45–50, write out the first three terms and the last term. Then use the formula for the sum of the first n terms of an arithmetic sequence to find the indicated sum.

45. $\displaystyle\sum_{i=1}^{17}(5i + 3)$

46. $\displaystyle\sum_{i=1}^{20}(6i - 4)$

47. $\displaystyle\sum_{i=1}^{30}(-3i + 5)$

48. $\displaystyle\sum_{i=1}^{40}(-2i + 6)$

49. $\displaystyle\sum_{i=1}^{100}4i$

50. $\displaystyle\sum_{i=1}^{50}(-4i)$

Practice PLUS

Use the graphs of the arithmetic sequences $\{a_n\}$ and $\{b_n\}$ to solve Exercises 51–58.

51. Find $a_{14} + b_{12}$.

52. Find $a_{16} + b_{18}$.

53. If $\{a_n\}$ is a finite sequence whose last term is -83, how many terms does $\{a_n\}$ contain?

54. If $\{b_n\}$ is a finite sequence whose last term is 93, how many terms does $\{b_n\}$ contain?

55. Find the difference between the sum of the first 14 terms of $\{b_n\}$ and the sum of the first 14 terms of $\{a_n\}$.

56. Find the difference between the sum of the first 15 terms of $\{b_n\}$ and the sum of the first 15 terms of $\{a_n\}$.

57. Write a linear function $f(x) = mx + b$, whose domain is the set of positive integers, that represents $\{a_n\}$.

58. Write a linear function $g(x) = mx + b$, whose domain is the set of positive integers, that represents $\{b_n\}$.

Use a system of two equations in two variables, a_1 and d, to solve Exercises 59–60.

59. Write a formula for the general term (the nth term) of the arithmetic sequence whose second term, a_2, is 4 and whose sixth term, a_6, is 16.

60. Write a formula for the general term (the nth term) of the arithmetic sequence whose third term, a_3, is 7 and whose eighth term, a_8, is 17.

Application Exercises

The bar graph shows changes in the percentage of college graduates for Americans ages 25 and older from 1990 to 2010. Exercises 61–62 involve developing arithmetic sequences that model the data.

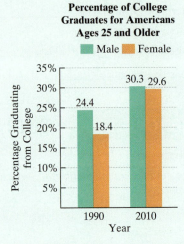

Source: U.S. Census Bureau

61. In 1990, 18.4% of American women ages 25 and older had graduated from college. On average, this percentage has increased by approximately 0.6 each year.

a. Write a formula for the nth term of the arithmetic sequence that models the percentage of American women ages 25 and older who had graduated from college n years after 1989.

b. Use the model from part (a) to project the percentage of American women ages 25 and older who will be college graduates by 2019.

62. In 1990, 24.4% of American men ages 25 and older had graduated from college. On average, this percentage has increased by approximately 0.3 each year.

a. Write a formula for the nth term of the arithmetic sequence that models the percentage of American men ages 25 and older who had graduated from college n years after 1989.

b. Use the model from part (a) to project the percentage of American men ages 25 and older who will be college graduates by 2019.

63. Company A pays $24,000 yearly with raises of $1600 per year. Company B pays $28,000 yearly with raises of $1000 per year. Which company will pay more in year 10? How much more?

64. Company A pays $23,000 yearly with raises of $1200 per year. Company B pays $26,000 yearly with raises of $800 per year. Which company will pay more in year 10? How much more?

The bar graph shows the average dormitory charges at public and private four-year U.S. colleges. Use the information displayed by the graph to solve Exercises 65–66.

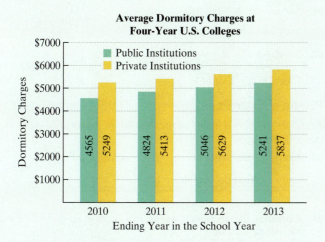

Average Dormitory Charges at Four-Year U.S. Colleges

Source: U.S. Department of Education

65. a. Use the numbers shown by the bar graph to find the total dormitory charges at public colleges for a four-year period from the school year ending in 2010 through the school year ending in 2013.

   b. The model

   $$a_n = 225n + 4357$$

   describes dormitory charges at public colleges in academic year n, where $n = 1$ corresponds to the school year ending in 2010, $n = 2$ to the school year ending in 2011, and so on. Use this model and the formula for S_n to find the total dormitory charges at public colleges for a four-year period from the school year ending in 2010 through the school year ending in 2013. Does this underestimate or overestimate the actual sum that you obtained in part (a)? By how much?

66. a. Use the numbers shown by the bar graph to find the total dormitory charges at private colleges for a four-year period from the school year ending in 2010 through the school year ending in 2013.

   b. The model

   $$a_n = 198n + 5037$$

   describes the dormitory charges at private colleges in academic year n, where $n = 1$ corresponds to the school year ending in 2010, $n = 2$ to the school year ending in 2011, and so on. Use this model and the formula for S_n to find the total dormitory charges at private colleges for a four-year period from the school year ending in 2010 through the school year ending in 2013. How does this compare with the actual sum you obtained in part (a)?

67. Use one of the models in Exercises 65–66 and the formula for S_n to find the total dormitory charges for your undergraduate education. How well does the model describe your anticipated costs?

68. A company offers a starting yearly salary of $33,000 with raises of $2500 per year. Find the total salary over a ten-year period.

69. You are considering two job offers. Company A will start you at $19,000 a year and guarantee a raise of $2600 per year. Company B will start you at a higher salary, $27,000 a year, but will only guarantee a raise of $1200 per year. Find the total salary that each company will pay over a ten-year period. Which company pays the greater total amount?

70. A theater has 30 seats in the first row, 32 seats in the second row, increasing by 2 seats each row for a total of 26 rows. How many seats are there in the theater?

71. A section in a stadium has 20 seats in the first row, 23 seats in the second row, increasing by 3 seats each row for a total of 38 rows. How many seats are in this section of the stadium?

Explaining the Concepts

72. What is an arithmetic sequence? Give an example with your explanation.

73. What is the common difference in an arithmetic sequence?

74. Explain how to find the general term of an arithmetic sequence.

75. Explain how to find the sum of the first n terms of an arithmetic sequence without having to add up all the terms.

Technology Exercises

76. Use the ⎣ SEQ ⎦ (sequence) capability of a graphing utility and the formula you obtained for a_n to verify the value you found for a_{20} in any five exercises from Exercises 25–34.

77. Use the capability of a graphing utility to calculate the sum of a sequence to verify any five of your answers to Exercises 45–50.

Critical Thinking Exercises

Make Sense?   In Exercises 78–81, determine whether each statement makes sense or does not make sense, and explain your reasoning.

78. Rather than performing the addition, I used the formula $S_n = \frac{n}{2}(a_1 + a_n)$ to find the sum of the first thirty terms of the sequence 2, 4, 8, 16, 32, ...

79. I was able to find the sum of the first fifty terms of an arithmetic sequence even though I did not identify every term.

80. The sequence for the number of seats per row in our movie theater as the rows move toward the back is arithmetic with $d = 1$ so people don't block the view of those in the row behind them.

81. Beginning at 6:45 A.M., a bus stops on my block every 23 minutes, so I used the formula for the nth term of an arithmetic sequence to describe the stopping time for the nth bus of the day.

In Exercises 82–85, determine whether each statement is true or false. If the statement is false, make the necessary change(s) to produce a true statement.

82. The sum of an arithmetic sequence cannot be negative.

83. The common difference for the arithmetic sequence 1, −1, −3, −5, . . . is 2.

84. An arithmetic sequence is a linear function whose domain is the set of natural numbers.

85. The sequence $\log_2 2, \log_2 4, \log_2 8, \log_2 16, \log_2 32, \ldots$ is arithmetic.

86. Give examples of two different arithmetic sequences whose fourth term, a_4, is 10.

87. In the sequence $21{,}700, 23{,}172, 24{,}644, 26{,}116, \ldots$, which term is $314{,}628$?

88. A *degree-day* is a unit used to measure the fuel requirements of buildings. By definition, each degree that the average daily temperature is below $65°F$ is 1 degree-day. For example, an average daily temperature of $42°F$ constitutes 23 degree-days. If the average temperature on January 1 was $42°F$ and fell $2°F$ for each subsequent day up to and including January 10, how many degree-days are included from January 1 to January 10?

89. Show that the sum of the first n positive odd integers,

$$1 + 3 + 5 + \cdots + (2n - 1),$$

is n^2.

Review Exercises

90. Solve: $\log(x^2 - 25) - \log(x + 5) = 3$.
(Section 9.5, Example 5)

91. Solve: $x^2 + 3x \le 10$.
(Section 8.5, Example 1)

92. Solve for P: $A = \dfrac{Pt}{P + t}$.
(Section 6.7, Example 1)

Preview Exercises

Exercises 93–95 will help you prepare for the material covered in the next section.

93. Consider the sequence $1, -2, 4, -8, 16, \ldots$. Find $\dfrac{a_2}{a_1}, \dfrac{a_3}{a_2}, \dfrac{a_4}{a_3},$ and $\dfrac{a_5}{a_4}$. What do you observe?

94. Consider the sequence whose nth term is $a_n = 3 \cdot 5^n$. Find $\dfrac{a_2}{a_1}, \dfrac{a_3}{a_2}, \dfrac{a_4}{a_3},$ and $\dfrac{a_5}{a_4}$. What do you observe?

95. Use the formula $a_n = a_1 3^{n-1}$ to find the 7th term of the sequence $11, 33, 99, 297, \ldots$.

SECTION

11.3

Geometric Sequences and Series

What am I supposed to learn?

After studying this section, you should be able to:

1. Find the common ratio of a geometric sequence.

2. Write terms of a geometric sequence.

3. Use the formula for the general term of a geometric sequence.

4. Use the formula for the sum of the first n terms of a geometric sequence.

5. Find the value of an annuity.

6. Use the formula for the sum of an infinite geometric series.

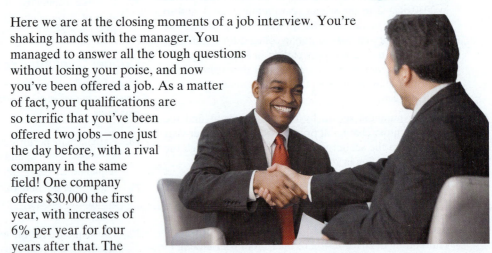

Here we are at the closing moments of a job interview. You're shaking hands with the manager. You managed to answer all the tough questions without losing your poise, and now you've been offered a job. As a matter of fact, your qualifications are so terrific that you've been offered two jobs—one just the day before, with a rival company in the same field! One company offers $30,000 the first year, with increases of 6% per year for four years after that. The other offers $32,000 the first year, with annual increases of 3% per year after that. Over a five-year period, which is the better offer?

If salary raises amount to a certain percent each year, the yearly salaries over time form a *geometric sequence*. In this section, we investigate geometric sequences and their properties. After studying the section, you will be in a position to decide which job offer to accept: You will know which company will pay you more over five years.

Geometric Sequences

Figure 11.5 shows a sequence in which the number of squares is increasing. From left to right, the number of squares is 1, 5, 25, 125, and 625. In this sequence, each term after the first, 1, is obtained by multiplying the preceding term by a constant amount, namely 5. This sequence of increasing numbers of squares is an example of a *geometric sequence*.

Figure 11.5 A geometric sequence of squares

Definition of a Geometric Sequence

A **geometric sequence** is a sequence in which each term after the first is obtained by multiplying the preceding term by a fixed nonzero constant. The amount by which we multiply each time is called the **common ratio** of the sequence.

1 Find the common ratio of a geometric sequence. ▶

The common ratio, r, is found by dividing any term after the first term by the term that directly precedes it. In the following examples, the common ratio is found by dividing the second term by the first term, $\dfrac{a_2}{a_1}$.

Geometric Sequence	Common Ratio
$1, 5, 25, 125, 625, \ldots$	$r = \dfrac{5}{1} = 5$
$4, 8, 16, 32, 64, \ldots$	$r = \dfrac{8}{4} = 2$
$6, -12, 24, -48, 96, \ldots$	$r = \dfrac{-12}{6} = -2$
$9, -3, 1, -\dfrac{1}{3}, \dfrac{1}{9}, \ldots$	$r = \dfrac{-3}{9} = -\dfrac{1}{3}$

Great Question!

What happens to the terms of a geometric sequence when the common ratio is negative?

When the common ratio is negative, the signs of the terms alternate.

Figure 11.6 shows a partial graph of the first geometric sequence in our list, $1, 5, 25, 125\ldots$. The graph forms a set of discrete points lying on the exponential function $f(x) = 5^{x-1}$. This illustrates that **a geometric sequence with a positive common ratio other than 1 is an exponential function whose domain is the set of positive integers**.

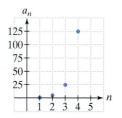

Figure 11.6 The graph of $\{a_n\} = 1, 5, 25, 125, \ldots$

2 Write terms of a geometric sequence. ▶

How do we write out the terms of a geometric sequence when the first term and the common ratio are known? We multiply the first term by the common ratio to get the second term, multiply the second term by the common ratio to get the third term, and so on.

EXAMPLE 1 Writing the Terms of a Geometric Sequence

Write the first six terms of the geometric sequence with first term 6 and common ratio $\frac{1}{3}$.

Solution The first term is 6. The second term is $6 \cdot \frac{1}{3}$, or 2. The third term is $2 \cdot \frac{1}{3}$, or $\frac{2}{3}$. The fourth term is $\frac{2}{3} \cdot \frac{1}{3}$, or $\frac{2}{9}$, and so on. The first six terms are

$$6, 2, \frac{2}{3}, \frac{2}{9}, \frac{2}{27}, \text{ and } \frac{2}{81}. \quad \blacksquare$$

✓ **CHECK POINT 1** Write the first six terms of the geometric sequence with first term 12 and common ratio $\frac{1}{2}$.

---

3 Use the formula for the general term of a geometric sequence. ▶

The General Term of a Geometric Sequence

Consider a geometric sequence whose first term is a_1 and whose common ratio is r. We are looking for a formula for the general term, a_n. Let's begin by writing the first six terms. The first term is a_1. The second term is $a_1 r$. The third term is $a_1 r \cdot r$, or $a_1 r^2$. The fourth term is $a_1 r^2 \cdot r$, or $a_1 r^3$, and so on. Starting with a_1 and multiplying each successive term by r, the first six terms are a_1, $a_1 r$, $a_1 r^2$, $a_1 r^3$, $a_1 r^4$, and $a_1 r^5$.

Compare the exponent on r and the subscript of a denoting the term number:

$$a_1, \qquad a_1 r, \qquad a_1 r^2, \qquad a_1 r^3, \qquad a_1 r^4, \qquad a_1 r^5.$$

| a_1, first term | a_2, second term | a_3, third term | a_4, fourth term | a_5, fifth term | a_6, sixth term |

Can you see that the exponent on r is 1 less than the subscript of a denoting the term number?

a_3: third term $= a_1 r^2$ a_4: fourth term $= a_1 r^3$

One less than 3, or 2, is the exponent on r.   One less than 4, or 3, is the exponent on r.

Thus, the formula for the nth term is

$$a_n = a_1 r^{n-1}.$$

One less than n, or $n-1$, is the exponent on r.

General Term of a Geometric Sequence

The nth term (the general term) of a geometric sequence with first term a_1 and common ratio r is

$$a_n = a_1 r^{n-1}.$$

Great Question!

When using a_1r^{n-1} to find the nth term of a geometric sequence, what should I do first?

Be careful with the order of operations when evaluating

$$a_1r^{n-1}.$$

First subtract 1 in the exponent and then raise r to that power. Finally, multiply the result by a_1.

EXAMPLE 2   Using the Formula for the General Term of a Geometric Sequence

Find the eighth term of the geometric sequence whose first term is -4 and whose common ratio is -2.

Solution   To find the eighth term, a_8, we replace n in the formula with 8, a_1 with -4, and r with -2.

$$a_n = a_1r^{n-1}$$
$$a_8 = -4(-2)^{8-1} = -4(-2)^7 = -4(-128) = 512$$

The eighth term is 512. We can check this result by writing the first eight terms of the sequence:

$$-4, 8, -16, 32, -64, 128, -256, 512. \quad \blacksquare$$

✓ **CHECK POINT 2**   Find the seventh term of the geometric sequence whose first term is 5 and whose common ratio is -3.

In Chapter 9, we studied exponential functions of the form $f(x) = b^x$ and used an exponential function to model the growth of the U.S. population from 1970 through 2010 (Example 1 on page 732). From 2000 through 2010, the nation's population grew at the slowest rate since the 1930s. Consequently, in our next example, we consider the country's population growth from 2000 through 2010. Because a geometric sequence is an exponential function whose domain is the set of positive integers, geometric and exponential growth mean the same thing.

EXAMPLE 3   Geometric Population Growth

The table shows the population of the United States in 2000 and 2010, with estimates given by the Census Bureau for 2001 through 2009.

Year	2000	2001	2002	2003	2004	2005	2006	2007	2008	2009	2010
Population (millions)	281.4	284.0	286.6	289.3	292.0	294.7	297.4	300.2	303.0	305.8	308.7

a. Show that the population is increasing geometrically.

b. Write the general term for the geometric sequence modeling the population of the United States, in millions, n years after 1999.

c. Project the U.S. population, in millions, for the year 2020.

Solution

a. First, we use the sequence of population growth, 281.4, 284.0, 286.6, 289.3, and so on, to divide the population for each year by the population in the preceding year.

$$\frac{284.0}{281.4} \approx 1.009, \quad \frac{286.6}{284.0} \approx 1.009, \quad \frac{289.3}{286.6} \approx 1.009$$

Continuing in this manner, we will keep getting approximately 1.009. This means that the population is increasing geometrically with $r \approx 1.009$. The population of the United States in any year shown in the sequence is approximately 1.009 times the population the year before.

b. The sequence of the U.S. population growth is

$$281.4, \ 284.0, \ 286.6, \ 289.3, \ 292.0, \ 294.7, \ldots$$

Because the population is increasing geometrically, we can find the general term of this sequence using

$$a_n = a_1r^{n-1}.$$

Blitzer Bonus ⓞ

Geometric Population Growth

Economist Thomas Malthus (1766–1834) predicted that population would increase as a geometric sequence and food production would increase as an arithmetic sequence. He concluded that eventually population would exceed food production. If two sequences, one geometric and one arithmetic, are increasing, the geometric sequence will eventually overtake the arithmetic sequence, regardless of any head start that the arithmetic sequence might initially have.

In the sequence 281.4, 284.0, 286.6, 289.3, ... , $a_1 = 281.4$ and [from part (a)] $r \approx 1.009$. We substitute these values into the formula $a_n = a_1 r^{n-1}$ for the general term. This gives the general term for the geometric sequence modeling the U.S. population, in millions, n years after 1999.

$$a_n = 281.4(1.009)^{n-1}$$

c. We can use the formula for the general term, a_n, in part (b) to project the U.S. population for the year 2020. The year 2020 is 21 years after 1999—that is, $2020 - 1999 = 21$. Thus, $n = 21$. We substitute 21 for n in $a_n = 281.4(1.009)^{n-1}$.

$$a_{21} = 281.4(1.009)^{21-1} = 281.4(1.009)^{20} \approx 336.6$$

The model projects that the United States will have a population of approximately 336.6 million in the year 2020. ∎

☑ **CHECK POINT 3**   Write the general term for the geometric sequence

$$3, 6, 12, 24, 48, \ldots.$$

Then use the formula for the general term to find the eighth term.

---

4 Use the formula for the sum of the first n terms of a geometric sequence. ⊙

The Sum of the First n Terms of a Geometric Sequence

The sum of the first n terms of a geometric sequence, denoted by S_n, and called the **nth partial sum**, can be found without having to add up all the terms. Recall that the first n terms of a geometric sequence are

$$a_1, a_1 r, a_1 r^2, \ldots, a_1 r^{n-2}, a_1 r^{n-1}.$$

We find the sum of $a_1, a_1 r, a_1 r^2, \ldots, a_1 r^{n-2}, a_1 r^{n-1}$ as follows:

$$S_n = a_1 + a_1 r + a_1 r^2 + \cdots + a_1 r^{n-2} + a_1 r^{n-1}$$ S_n is the sum of the first n terms of the sequence.

$$r S_n = a_1 r + a_1 r^2 + a_1 r^3 + \cdots + a_1 r^{n-1} + a_1 r^n$$ Multiply both sides of the equation by r.

$$S_n - r S_n = a_1 - a_1 r^n$$ Subtract the second equation from the first equation.

$$S_n(1 - r) = a_1(1 - r^n)$$ Factor out S_n on the left and a_1 on the right.

$$S_n = \frac{a_1(1 - r^n)}{1 - r}.$$ Solve for S_n by dividing both sides by $1 - r$ (assuming that $r \neq 1$).

We have proved the following result:

Great Question!

What is the sum of the first n terms of a geometric sequence if the common ratio is 1?

If the common ratio is 1, the geometric sequence is

$$a_1, a_1, a_1, a_1, \ldots.$$

The sum of the first n terms of this sequence is na_1:

$$S_n = \underbrace{a_1 + a_1 + a_1 + \cdots + a_1}_{\text{There are } n \text{ terms.}}$$

$$= na_1.$$

The Sum of the First n Terms of a Geometric Sequence

The sum, S_n, of the first n terms of a geometric sequence is given by

$$S_n = \frac{a_1(1 - r^n)}{1 - r}$$

in which a_1 is the first term and r is the common ratio ($r \neq 1$).

To find the sum of the terms of a geometric sequence, we need to know the first term, a_1, the common ratio, r, and the number of terms, n. The following examples illustrate how to use this formula.

EXAMPLE 4   Finding the Sum of the First n Terms of a Geometric Sequence

Find the sum of the first 18 terms of the geometric sequence: $2, -8, 32, -128, \ldots.$

Solution   To find the sum of the first 18 terms, S_{18}, we replace n in the formula with 18.

$$S_n = \frac{a_1(1 - r^n)}{1 - r}$$

$$S_{18} = \frac{a_1(1 - r^{18})}{1 - r}$$

The first term, a_1, is **2**.   We must find r, the common ratio.

We can find the common ratio by dividing the second term of $2, -8, 32, -128, \ldots$ by the first term.

$$r = \frac{a_2}{a_1} = \frac{-8}{2} = -4$$

Now we are ready to find the sum of the first 18 terms of $2, -8, 32, -128, \ldots$.

$$S_n = \frac{a_1(1 - r^n)}{1 - r}$$ Use the formula for the sum of the first n terms of a geometric sequence.

$$S_{18} = \frac{2[1 - (-4)^{18}]}{1 - (-4)}$$ a_1 (the first term) $= 2$, $r = -4$, and $n = 18$ because we want the sum of the first 18 terms.

$$= -27{,}487{,}790{,}694$$ Use a calculator.

The sum of the first 18 terms is $-27{,}487{,}790{,}694$. Equivalently, this number is the 18th partial sum of the sequence $2, -8, 32, -128, \ldots$.   ■

☑ **CHECK POINT 4**   Find the sum of the first nine terms of the geometric sequence: $2, -6, 18, -54, \ldots$.

EXAMPLE 5   Using S_n to Evaluate a Summation

Find the following sum: $\displaystyle\sum_{i=1}^{10} 6 \cdot 2^i$.

Solution   Let's write out a few terms in the sum.

$$\sum_{i=1}^{10} 6 \cdot 2^i = 6 \cdot 2 + 6 \cdot 2^2 + 6 \cdot 2^3 + \cdots + 6 \cdot 2^{10}$$

Do you see that each term after the first is obtained by multiplying the preceding term by 2? To find the sum of the 10 terms ($n = 10$), we need to know the first term, a_1, and the common ratio, r. The first term is $6 \cdot 2$ or 12: $a_1 = 12$. The common ratio is 2.

$$S_n = \frac{a_1(1 - r^n)}{1 - r}$$ Use the formula for the sum of the first n terms of a geometric sequence.

$$S_{10} = \frac{12(1 - 2^{10})}{1 - 2}$$ a_1 (the first term) $= 12$, $r = 2$, and $n = 10$ because we are adding ten terms.

$$= 12{,}276$$ Use a calculator.

Thus,

$$\sum_{i=1}^{10} 6 \cdot 2^i = 12{,}276.$$ ■

☑ **CHECK POINT 5**   Find the following sum: $\displaystyle\sum_{i=1}^{8} 2 \cdot 3^i$.

Some of the exercises in the previous Exercise Set involved situations in which salaries increased by a fixed amount each year. A more realistic situation is one in which salaries increase by a certain percent each year. Example 6 shows how such a situation can be described using a geometric sequence.

EXAMPLE 6 Computing a Lifetime Salary

A union contract specifies that each worker will receive a 5% pay increase each year for the next 30 years. One worker is paid $35,000 the first year. What is this person's total lifetime salary over a 30-year period?

Solution The salary for the first year is $35,000. With a 5% raise, the second-year salary is computed as follows:

Salary for year 2 = $35,000 + 35,000(0.05) = 35,000(1 + 0.05) = 35,000(1.05)$.

Each year, the salary is 1.05 times what it was in the previous year. Thus, the salary for year 3 is 1.05 times $35,000(1.05)$, or $35,000(1.05)^2$. The salaries for the first five years are given in the table.

Yearly Salaries					
Year 1	**Year 2**	**Year 3**	**Year 4**	**Year 5**	...
35,000	35,000(1.05)	$35,000(1.05)^2$	$35,000(1.05)^3$	$35,000(1.05)^4$	...

The numbers in the bottom row form a geometric sequence with $a_1 = 35,000$ and $r = 1.05$. To find the total salary over 30 years, we use the formula for the sum of the first n terms of a geometric sequence, with $n = 30$.

$$S_n = \frac{a_1(1 - r^n)}{1 - r}$$

$$S_{30} = \frac{35,000[1 - (1.05)^{30}]}{1 - 1.05}$$

Total salary over 30 years

Use $S_n = \dfrac{a_1(1 - r^n)}{1 - r}$ with $a_1 = 35,000$, $r = 1.05$, and $n = 30$.

$$\approx 2,325,360 \qquad \text{Use a calculator.}$$

The total salary over the 30-year period is approximately $2,325,360. ∎

✓ **CHECK POINT 6** A job pays a salary of $30,000 the first year. During the next 29 years, the salary increases by 6% each year. What is the total lifetime salary over the 30-year period?

5 Find the value of an annuity. ▶

Annuities

The compound interest formula

$$A = P(1 + r)^t$$

gives the future value, A, after t years, when a fixed amount of money, P, the principal, is deposited in an account that pays an annual interest rate r (in decimal form) compounded once a year. However, money is often invested in small amounts at periodic intervals. For example, to save for retirement, you might decide to place $1000 into an Individual Retirement Account (IRA) at the end of each year until you retire. An **annuity** is a sequence of equal payments made at equal time periods. An IRA is an example of an annuity.

Suppose P dollars is deposited into an account at the end of each year. The account pays an annual interest rate, r, compounded annually. At the end of the first year, the account contains P dollars. At the end of the second year, P dollars is deposited again. At the time of this deposit, the first deposit has received interest earned during the second year. The **value of the annuity** is the sum of all deposits made plus all interest paid. Thus, the value of the annuity after two years is

$$P + P(1 + r).$$

Deposit of P dollars at end of second year

First-year deposit of P dollars with interest earned for a year

The value of the annuity after three years is

$$P \quad + \quad P(1 + r) \quad + \quad P(1 + r)^2.$$

Deposit of P dollars at end of third year

Second-year deposit of P dollars with interest earned for a year

First-year deposit of P dollars with interest earned over two years

The value of the annuity after t years is

$$P + P(1 + r) + P(1 + r)^2 + P(1 + r)^3 + \cdots + P(1 + r)^{t-1}.$$

Deposit of P dollars at end of year t

First-year deposit of P dollars with interest earned over $t - 1$ years

This is the sum of the terms of a geometric sequence with first term P and common ratio $1 + r$. We use the formula

$$S_n = \frac{a_1(1 - r^n)}{1 - r}$$

to find the sum of the terms:

$$S_t = \frac{P[1 - (1 + r)^t]}{1 - (1 + r)} = \frac{P[1 - (1 + r)^t]}{-r} = P\frac{(1 + r)^t - 1}{r}.$$

This formula gives the value of an annuity after t years if interest is compounded once a year. We can adjust the formula to find the value of an annuity if equal payments are made at the end of each of n yearly compounding periods.

> ### Value of an Annuity: Interest Compounded n Times Per Year
>
> If P is the deposit made at the end of each compounding period for an annuity at r percent annual interest compounded n times per year, the value, A, of the annuity after t years is
>
> $$A = \frac{P\left[\left(1 + \dfrac{r}{n}\right)^{nt} - 1\right]}{\dfrac{r}{n}}.$$

EXAMPLE 7   Determining the Value of an Annuity

At age 25, to save for retirement, you decide to deposit $200 at the end of each month into an IRA that pays 7.5% compounded monthly.

a. How much will you have from the IRA when you retire at age 65?

b. Find the interest.

Solution

a. Because you are 25, the amount that you will have from the IRA when you retire at 65 is its value after 40 years.

$$A = \dfrac{P\left[\left(1 + \dfrac{r}{n}\right)^{nt} - 1\right]}{\dfrac{r}{n}}$$ Use the formula for the value of an annuity.

$$A = \dfrac{200\left[\left(1 + \dfrac{0.075}{12}\right)^{12 \cdot 40} - 1\right]}{\dfrac{0.075}{12}}$$ The annuity involves month-end deposits of $200: $P = 200$. The interest rate is 7.5%: $r = 0.075$. The interest is compounded monthly: $n = 12$. The number of years is 40: $t = 40$.

$$= \dfrac{200\left[(1 + 0.00625)^{480} - 1\right]}{0.00625}$$ Using parentheses keys, this can be performed in a single step on a graphing calculator.

$$= \dfrac{200\left[(1.00625)^{480} - 1\right]}{0.00625}$$

$$\approx \dfrac{200(19.8989 - 1)}{0.00625}$$ Use a calculator to find $(1.00625)^{480}$: 1.00625 $\boxed{y^x}$ 480 $\boxed{=}$.

$$\approx 604{,}765$$

After 40 years, you will have approximately $604,765 when retiring at age 65.

b. Interest = Value of the IRA − Total deposits

$$\approx \$604{,}765 - \$200 \cdot 12 \cdot 40$$

$200 per month × 12 months per year × 40 years

$$= \$604{,}765 - \$96{,}000 = \$508{,}765$$

The interest is approximately $508,765, more than five times the amount of your contributions to the IRA. ∎

☑ **CHECK POINT 7**   At age 30, to save for retirement, you decide to deposit $100 at the end of each month into an IRA that pays 9.5% compounded monthly.

a. How much will you have from the IRA when you retire at age 65?

b. Find the interest.

6 Use the formula for the sum of an infinite geometric series.

Geometric Series

An infinite sum of the form

$$a_1 + a_1 r + a_1 r^2 + a_1 r^3 + \cdots + a_1 r^{n-1} + \cdots$$

with first term a_1 and common ratio r is called an **infinite geometric series**. How can we determine which infinite geometric series have sums and which do not? We look at what happens to r^n as n gets larger in the formula for the sum of the first n terms of this series, namely

$$S_n = \dfrac{a_1(1 - r^n)}{1 - r}.$$

If r is any number between -1 and 1, that is, $-1 < r < 1$, the term r^n approaches 0 as n gets larger. For example, consider what happens to r^n for $r = \frac{1}{2}$:

$$\left(\frac{1}{2}\right)^1 = \frac{1}{2} \quad \left(\frac{1}{2}\right)^2 = \frac{1}{4} \quad \left(\frac{1}{2}\right)^3 = \frac{1}{8} \quad \left(\frac{1}{2}\right)^4 = \frac{1}{16} \quad \left(\frac{1}{2}\right)^5 = \frac{1}{32} \quad \left(\frac{1}{2}\right)^6 = \frac{1}{64}.$$

These numbers are approaching 0 as n gets larger.

Take another look at the formula for the sum of the first n terms of a geometric sequence.

$$S_n = \frac{a_1(1 - r^n)}{1 - r}$$

If $-1 < r < 1$, r^n approaches 0 as n gets larger.

Let us replace r^n with 0 in the formula for S_n. This change gives us a formula for the sum of an infinite geometric series with a common ratio between -1 and 1.

The Sum of an Infinite Geometric Series

If $-1 < r < 1$ (equivalently, $|r| < 1$), then the sum of the infinite geometric series

$$a_1 + a_1r + a_1r^2 + a_1r^3 + \cdots$$

in which a_1 is the first term and r is the common ratio is given by

$$S = \frac{a_1}{1 - r}.$$

If $|r| \geq 1$, the infinite series does not have a sum.

To use the formula for the sum of an infinite geometric series, we need to know the first term and the common ratio. For example, consider

First term, a_1, is $\frac{1}{2}$.

$$\frac{1}{2} + \frac{1}{4} + \frac{1}{8} + \frac{1}{16} + \frac{1}{32} + \cdots.$$

Common ratio, r, is $\frac{a_2}{a_1}$.

$$r = \frac{1}{4} \div \frac{1}{2} = \frac{1}{4} \cdot 2 = \frac{1}{2}$$

With $r = \frac{1}{2}$, the condition that $|r| < 1$ is met, so the infinite geometric series has a sum given by $S = \frac{a_1}{1 - r}$. The sum of the series is found as follows:

$$\frac{1}{2} + \frac{1}{4} + \frac{1}{8} + \frac{1}{16} + \frac{1}{32} + \cdots = \frac{a_1}{1 - r} = \frac{\frac{1}{2}}{1 - \frac{1}{2}} = \frac{\frac{1}{2}}{\frac{1}{2}} = 1.$$

Figure 11.7 The sum $\frac{1}{2} + \frac{1}{4} + \frac{1}{8} + \frac{1}{16} + \frac{1}{32} + \cdots$ is approaching 1.

Thus, the sum of the infinite geometric series is 1. Notice how this is illustrated in **Figure 11.7**. As more terms are included, the sum is approaching the area of one complete circle.

EXAMPLE 8 Finding the Sum of an Infinite Geometric Series

Find the sum of the infinite geometric series: $\frac{3}{8} - \frac{3}{16} + \frac{3}{32} - \frac{3}{64} + \cdots$.

Solution Before finding the sum, $\frac{3}{8} - \frac{3}{16} + \frac{3}{32} - \frac{3}{64} + \ldots$, we must find the common ratio.

$$r = \frac{a_2}{a_1} = \frac{-\frac{3}{16}}{\frac{3}{8}} = -\frac{3}{16} \cdot \frac{8}{3} = -\frac{1}{2}$$

Because $r = -\frac{1}{2}$, the condition that $|r| < 1$ is met. Thus, the infinite geometric series has a sum.

$$S = \frac{a_1}{1 - r}$$

This is the formula for the sum of an infinite geometric series. Let $a_1 = \frac{3}{8}$ and $r = -\frac{1}{2}$.

$$= \frac{\frac{3}{8}}{1 - \left(-\frac{1}{2}\right)} = \frac{\frac{3}{8}}{\frac{3}{2}} = \frac{3}{8} \cdot \frac{2}{3} = \frac{1}{4}$$

Thus, the sum of $\frac{3}{8} - \frac{3}{16} + \frac{3}{32} - \frac{3}{64} + \cdots$ is $\frac{1}{4}$. Put in an informal way, as we continue to add more and more terms, the sum is approximately $\frac{1}{4}$. ∎

☑ **CHECK POINT 8** Find the sum of the infinite geometric series: $3 + 2 + \frac{4}{3} + \frac{8}{9} + \cdots$.

We can use the formula for the sum of an infinite geometric series to express a repeating decimal as a fraction in lowest terms.

EXAMPLE 9 Writing a Repeating Decimal as a Fraction

Express $0.\overline{7}$ as a fraction in lowest terms.

Solution

$$0.\overline{7} = 0.7777\ldots = \frac{7}{10} + \frac{7}{100} + \frac{7}{1000} + \frac{7}{10,000} + \cdots$$

Observe that $0.\overline{7}$ is an infinite geometric series with first term $\frac{7}{10}$ and common ratio $\frac{1}{10}$. Because $r = \frac{1}{10}$, the condition that $|r| < 1$ is met. Thus, we can use our formula to find the sum. Therefore,

$$0.\overline{7} = \frac{a_1}{1 - r} = \frac{\frac{7}{10}}{1 - \frac{1}{10}} = \frac{\frac{7}{10}}{\frac{9}{10}} = \frac{7}{10} \cdot \frac{10}{9} = \frac{7}{9}.$$

An equivalent fraction for $0.\overline{7}$ is $\frac{7}{9}$. ∎

☑ **CHECK POINT 9** Express $0.\overline{9}$ as a fraction in lowest terms.

Infinite geometric series have many applications, as illustrated in Example 10.

EXAMPLE 10 Tax Rebates and the Multiplier Effect

A tax rebate that returns a certain amount of money to taxpayers can have a total effect on the economy that is many times this amount. In economics, this phenomenon is called the **multiplier effect**. Suppose, for example, that the government reduces taxes so that each consumer has $2000 more income. The government assumes that each person will spend 70% of this (=$1400). The individuals and businesses receiving this $1400 in turn spend 70% of it (=$980), creating extra income for other people to spend, and so on. Determine the total amount spent on consumer goods from the initial $2000 tax rebate.

Solution   The total amount spent is given by the infinite geometric series

$$1400 + 980 + 686 + \cdots .$$

70% of 1400   70% of 980

$1400

70% is spent.

$980

70% is spent.

$686

The first term is 1400: $a_1 = 1400$. The common ratio is 70%, or 0.7: $r = 0.7$. Because $r = 0.7$, the condition that $|r| < 1$ is met. Thus, we can use our formula to find the sum. Therefore,

$$1400 + 980 + 686 + \cdots = \frac{a_1}{1-r} = \frac{1400}{1-0.7} \approx 4667.$$

This means that the total amount spent on consumer goods from the initial $2000 rebate is approximately $4667. ■

✓ **CHECK POINT 10**   Rework Example 10 and determine the total amount spent on consumer goods with a $1000 tax rebate and 80% spending down the line.

CONCEPT AND VOCABULARY CHECK

Fill in each blank so that the resulting statement is true.

1. A sequence in which each term after the first is obtained by multiplying the preceding term by a fixed nonzero constant is called a/an _____ sequence. The amount by which we multiply each time is called the _____ of the sequence.

2. The nth term of the sequence described in Concept Check 1 is given by the formula $a = $ _____, where a_1 is the _____ and r is the _____ of the sequence.

3. The sum, S_n, of the first n terms of the sequence described in Concept Check 1 is given by the formula $S_n = $ _____, where a_1 is the _____ and r is the _____, $r \neq 1$.

4. A sequence of equal payments made at equal time periods is called a/an _____. Its value, A, after t years is given by the formula

$$A = \frac{P\left[\left(1 + \frac{r}{n}\right)^{nt} - 1\right]}{\frac{r}{n}},$$

where _____ is the deposit made at the end of each compounding period at _____ percent annual interest compounded _____ times per year.

5. An infinite sum of the form

$$a_1 + a_1 r + a_1 r^2 + a_1 r^3 + \cdots$$

is called a/an _____. If $-1 < r < $ _____, its sum, S, is given by the formula $S = $ _____. The series does not have a sum if _____.

6. The first four terms of $\displaystyle\sum_{i=1}^{6} 2^i$ are _____, _____, _____, and _____. The common ratio is _____.

Determine whether each sequence is arithmetic or geometric.

7. $4, 8, 12, 16, 20, \ldots$ _____

8. $4, 8, 16, 32, 64, \ldots$ _____

9. $1, -3, 9, -27, 81, \ldots$ _____

10. $-1, 1, 3, 5, 7, \ldots$ _____

11.3 EXERCISE SET ▶ MyMathLab®

Practice Exercises

In Exercises 1–8, find the common ratio for each geometric sequence.

1. $5, 15, 45, 135, \ldots$ 2. $5, 10, 20, 40, \ldots$

3. $-15, 30, -60, 120, \ldots$

4. $-2, 6, -18, 54, \ldots$

5. $3, \dfrac{9}{2}, \dfrac{27}{4}, \dfrac{81}{8}, \ldots$ 6. $4, \dfrac{8}{3}, \dfrac{16}{9}, \dfrac{32}{27}, \ldots$

7. $4, -0.4, 0.04, -0.004, \ldots$

8. $7, -0.7, 0.07, -0.007, \ldots$

In Exercises 9–16, write the first five terms of each geometric sequence with the given first term, a_1, and common ratio, r.

9. $a_1 = 2, r = 3$

10. $a_1 = 2, r = 4$

11. $a_1 = 20, r = \dfrac{1}{2}$

12. $a_1 = 24, r = \dfrac{1}{3}$

13. $a_1 = -4, r = -10$

14. $a_1 = -3, r = -10$

15. $a_1 = -\dfrac{1}{4}, r = -2$

16. $a_1 = -\dfrac{1}{16}, r = -4$

In Exercises 17–24, use the formula for the general term (the nth term) of a geometric sequence to find the indicated term of each sequence with the given first term, a_1, and common ratio, r.

17. Find a_8 when $a_1 = 6, r = 2$.

18. Find a_8 when $a_1 = 5, r = 3$.

19. Find a_{12} when $a_1 = 5, r = -2$.

20. Find a_{12} when $a_1 = 4, r = -2$.

21. Find a_6 when $a_1 = 6400, r = -\dfrac{1}{2}$.

22. Find a_6 when $a_1 = 8000, r = -\dfrac{1}{2}$.

23. Find a_8 when $a_1 = 1,000,000, r = 0.1$.

24. Find a_8 when $a_1 = 40,000, r = 0.1$.

In Exercises 25–32, write a formula for the general term (the nth term) of each geometric sequence. Then use the formula for a_n to find a_7, the seventh term of the sequence.

25. $3, 12, 48, 192, \ldots$

26. $3, 15, 75, 375, \ldots$

27. $18, 6, 2, \dfrac{2}{3}, \ldots$

28. $12, 6, 3, \dfrac{3}{2}, \ldots$

29. $1.5, -3, 6, -12, \ldots$

30. $5, -1, \dfrac{1}{5}, -\dfrac{1}{25}, \ldots$

31. $0.0004, -0.004, 0.04, -0.4 \ldots$

32. $0.0007, -0.007, 0.07, -0.7, \ldots$

Use the formula for the sum of the first n terms of a geometric sequence to solve Exercises 33–38.

33. Find the sum of the first 12 terms of the geometric sequence: $2, 6, 18, 54 \ldots$.

34. Find the sum of the first 12 terms of the geometric sequence: $3, 6, 12, 24, \ldots$.

35. Find the sum of the first 11 terms of the geometric sequence: $3, -6, 12, -24, \ldots$.

36. Find the sum of the first 11 terms of the geometric sequence: $4, -12, 36, -108, \ldots$.

37. Find the sum of the first 14 terms of the geometric sequence: $-\dfrac{3}{2}, 3, -6, 12, \ldots$.

38. Find the sum of the first 14 terms of the geometric sequence: $-\dfrac{1}{24}, \dfrac{1}{12}, -\dfrac{1}{6}, \dfrac{1}{3}, \ldots$.

In Exercises 39–44, find the indicated sum. Use the formula for the sum of the first n terms of a geometric sequence.

39. $\displaystyle\sum_{i=1}^{8} 3^i$ 40. $\displaystyle\sum_{i=1}^{6} 4^i$

41. $\displaystyle\sum_{i=1}^{10} 5 \cdot 2^i$ 42. $\displaystyle\sum_{i=1}^{7} 4(-3)^i$

43. $\displaystyle\sum_{i=1}^{6} \left(\dfrac{1}{2}\right)^{i+1}$ 44. $\displaystyle\sum_{i=1}^{6} \left(\dfrac{1}{3}\right)^{i+1}$

In Exercises 45–52, find the sum of each infinite geometric series.

45. $1 + \dfrac{1}{3} + \dfrac{1}{9} + \dfrac{1}{27} + \cdots$

46. $1 + \dfrac{1}{4} + \dfrac{1}{16} + \dfrac{1}{64} + \cdots$

47. $3 + \dfrac{3}{4} + \dfrac{3}{4^2} + \dfrac{3}{4^3} + \cdots$

48. $5 + \dfrac{5}{6} + \dfrac{5}{6^2} + \dfrac{5}{6^3} + \cdots$

49. $1 - \dfrac{1}{2} + \dfrac{1}{4} - \dfrac{1}{8} + \cdots$

50. $3 - 1 + \dfrac{1}{3} - \dfrac{1}{9} + \cdots$

51. $\displaystyle\sum_{i=1}^{\infty} 26(-0.3)^{i-1}$ 52. $\displaystyle\sum_{i=1}^{\infty} 51(-0.7)^{i-1}$

In Exercises 53–58, express each repeating decimal as a fraction in lowest terms.

53. $0.\overline{5} = \dfrac{5}{10} + \dfrac{5}{100} + \dfrac{5}{1000} + \dfrac{5}{10,000} + \cdots$

54. $0.\overline{1} = \dfrac{1}{10} + \dfrac{1}{100} + \dfrac{1}{1000} + \dfrac{1}{10,000} + \cdots$

55. $0.\overline{47} = \dfrac{47}{100} + \dfrac{47}{10,000} + \dfrac{47}{1,000,000} + \cdots$

56. $0.\overline{83} = \dfrac{83}{100} + \dfrac{83}{10,000} + \dfrac{83}{1,000,000} + \cdots$

57. $0.\overline{257}$ 58. $0.\overline{529}$

In Exercises 59–64, the general term of a sequence is given. Determine whether the sequence is arithmetic, geometric, or neither. If the sequence is arithmetic, find the common difference; if it is geometric, find the common ratio.

59. $a_n = n + 5$

60. $a_n = n - 3$

61. $a_n = 2^n$

62. $a_n = \left(\dfrac{1}{2}\right)^n$

63. $a_n = n^2 + 5$

64. $a_n = n^2 - 3$

Practice PLUS

In Exercises 65–70, let

$$\{a_n\} = -5, 10, -20, 40, \ldots,$$
$$\{b_n\} = 10, -5, -20, -35, \ldots,$$
and $\qquad \{c_n\} = -2, 1, -\tfrac{1}{2}, \tfrac{1}{4}, \ldots.$

65. Find $a_{10} + b_{10}$.

66. Find $a_{11} + b_{11}$.

67. Find the difference between the sum of the first 10 terms of $\{a_n\}$ and the sum of the first 10 terms of $\{b_n\}$.

68. Find the difference between the sum of the first 11 terms of $\{a_n\}$ and the sum of the first 11 terms of $\{b_n\}$.

69. Find the product of the sum of the first 6 terms of $\{a_n\}$ and the sum of the infinite series containing all the terms of $\{c_n\}$.

70. Find the product of the sum of the first 9 terms of $\{a_n\}$ and the sum of the infinite series containing all the terms of $\{c_n\}$.

In Exercises 71–72, find a_2 and a_3 for each geometric sequence.

71. $8, a_2, a_3, 27$

72. $2, a_2, a_3, -54$

In Exercises 73–74, round all answers to the nearest dollar.

73. Here are two ways of investing $30,000 for 20 years.

Lump-Sum Deposit	Rate	Time
$30,000	5% compounded annually	20 years

Periodic Deposits	Rate	Time
$1500 at the end of each year	5% compounded annually	20 years

After 20 years, how much more will you have from the lump-sum investment than from the annuity?

74. Here are two ways of investing $40,000 for 25 years.

Lump-Sum Deposit	Rate	Time
$40,000	6.5% compounded annually	25 years

Periodic Deposits	Rate	Time
$1600 at the end of each year	6.5% compounded annually	25 years

After 25 years, how much more will you have from the lump-sum investment than from the annuity?

Application Exercises

Use the formula for the general term (the nth term) of a geometric sequence to solve Exercises 75–78.

In Exercises 75–76 suppose you save $1 the first day of a month, $2 the second day, $4 the third day, and so on. That is, each day you save twice as much as you did the day before.

75. What will you put aside for savings on the fifteenth day of the month?

76. What will you put aside for savings on the thirtieth day of the month?

77. A professional baseball player signs a contract with a beginning salary of $3,000,000 for the first year and an annual increase of 4% per year beginning in the second year. That is, beginning in year 2, the athlete's salary will be 1.04 times what it was in the previous year. What is the athlete's salary for year 7 of the contract? Round to the nearest dollar.

78. You are offered a job that pays $30,000 for the first year with an annual increase of 5% per year beginning in the second year. That is, beginning in year 2, your salary will be 1.05 times what it was in the previous year. What can you expect to earn in your sixth year on the job? Round to the nearest dollar.

In Exercises 79–80, you will develop geometric sequences that model the population growth for California and Texas, the two most-populated U.S. states.

79. The table shows the population of California for 2000 and 2010, with estimates given by the U.S. Census Bureau for 2001 through 2009.

Year	2000	2001	2002	2003	2004	2005
Population in millions	33.87	34.21	34.55	34.90	35.25	35.60

Year	2006	2007	2008	2009	2010
Population in millions	36.00	36.36	36.72	37.09	37.25

a. Divide the population for each year by the population in the preceding year. Round to two decimal places and show that California has a population increase that is approximately geometric.

b. Write the general term of the geometric sequence modeling California's population, in millions, n years after 1999.

c. Use your model from part (b) to project California's population, in millions, for the year 2020. Round to two decimal places.

80. The table shows the population of Texas for 2000 and 2010, with estimates given by the U.S. Census Bureau for 2001 through 2009.

Year	2000	2001	2002	2003	2004	2005
Population in millions	20.85	21.27	21.70	22.13	22.57	23.02

Year	2006	2007	2008	2009	2010
Population in millions	23.48	23.95	24.43	24.92	25.15

a. Divide the population for each year by the population in the preceding year. Round to two decimal places and show that Texas has a population increase that is approximately geometric.

b. Write the general term of the geometric sequence modeling Texas's population, in millions, n years after 1999.

c. Use your model from part (b) to project Texas's population, in millions, for the year 2020. Round to two decimal places.

Use the formula for the sum of the first n terms of a geometric sequence to solve Exercises 81–85.

In Exercises 81–82, you save $1 the first day of a month, $2 the second day, $4 the third day, continuing to double your savings each day.

81. What will your total savings be for the first 15 days?

82. What will your total savings be for the first 30 days?

83. A job pays a salary of $24,000 the first year. During the next 19 years, the salary increases by 5% each year. What is the total lifetime salary over the 20-year period? Round to the nearest dollar.

84. You are investigating two employment opportunities. Company A offers $30,000 the first year. During the next four years, the salary is guaranteed to increase by 6% per year. Company B offers $32,000 the first year, with guaranteed annual increases of 3% per year after that. Which company offers the better total salary for a five-year contract? By how much? Round to the nearest dollar.

In Exercises 85–86, suppose you are serving on a financial team, analyzing the three options given to the professional baseball player described in the chapter opener on page 821.

85. Reread the chapter opener. Determine which option provides the most amount of money over the six-year contract and which provides the least.

86. Describe one advantage and one disadvantage to each option.

87. A pendulum swings through an arc of 20 inches. On each successive swing, the length of the arc is 90% of the previous length.

$$20, \quad 0.9(20), \quad 0.9^2(20), \quad 0.9^3(20), \ldots$$

After 10 swings, what is the total length of the distance the pendulum has swung? Round to the nearest hundredth of an inch.

88. A pendulum swings through an arc of 16 inches. On each successive swing, the length of the arc is 96% of the previous length.

$$16, \quad 0.96(16), \quad (0.96)^2(16), \quad (0.96)^3(16), \ldots$$

After 10 swings, what is the total length of the distance the pendulum has swung? Round to the nearest hundredth of an inch.

Use the formula for the value of an annuity to solve Exercises 89–94. Round answers to the nearest dollar.

89. To save money for a sabbatical to earn a master's degree, you deposit $2000 at the end of each year in an annuity that pays 7.5% compounded annually.

a. How much will you have saved at the end of five years?

b. Find the interest.

90. To save money for a sabbatical to earn a master's degree, you deposit $2500 at the end of each year in an annuity that pays 6.25% compounded annually.

a. How much will you have saved at the end of five years?

b. Find the interest.

91. At age 25, to save for retirement, you decide to deposit $50 at the end of each month in an IRA that pays 5.5% compounded monthly.

a. How much will you have from the IRA when you retire at age 65?

b. Find the interest.

92. At age 25, to save for retirement, you decide to deposit $75 at the end of each month in an IRA that pays 6.5% compounded monthly.

a. How much will you have from the IRA when you retire at age 65?

b. Find the interest.

93. To offer scholarship funds to children of employees, a company invests $10,000 at the end of every three months in an annuity that pays 10.5% compounded quarterly.

a. How much will the company have in scholarship funds at the end of ten years?

b. Find the interest.

94. To offer scholarship funds to children of employees, a company invests $15,000 at the end of every three months in an annuity that pays 9% compounded quarterly.

a. How much will the company have in scholarship funds at the end of ten years?

b. Find the interest.

Use the formula for the sum of an infinite geometric series to solve Exercises 95–97.

95. A new factory in a small town has an annual payroll of $6 million. It is expected that 60% of this money will be spent in the town by factory personnel. The people in the town who receive this money are expected to spend 60% of what they receive in the town, and so on. What is the total of all this spending, called the *total economic impact* of the factory, on the town each year?

96. How much additional spending will be generated by a $10 billion tax rebate if 60% of all income is spent?

97. If the shading process shown in the figure is continued indefinitely, what fractional part of the largest square will eventually be shaded?

Explaining the Concepts

98. What is a geometric sequence? Give an example with your explanation.

99. What is the common ratio in a geometric sequence?

100. Explain how to find the general term of a geometric sequence.

101. Explain how to find the sum of the first *n* terms of a geometric sequence without having to add up all the terms.

102. What is an annuity?

103. What is the difference between a geometric sequence and an infinite geometric series?

104. How do you determine if an infinite geometric series has a sum? Explain how to find the sum of such an infinite geometric series.

105. Would you rather have $10,000,000 and a brand new BMW or 1¢ today, 2¢ tomorrow, 4¢ on day 3, 8¢ on day 4, 16¢ on day 5, and so on, for 30 days? Explain.

106. For the first 30 days of a flu outbreak, the number of students on your campus who become ill is increasing. Which is worse: the number of students with the flu is increasing arithmetically or is increasing geometrically? Explain your answer.

Technology Exercises

107. Use the $\boxed{\text{SEQ}}$ (sequence) capability of a graphing utility and the formula you obtained for a_n to verify the value you found for a_7 in any three exercises from Exercises 25–32.

108. Use the capability of a graphing utility to calculate the sum of a sequence to verify any three of your answers to Exercises 39–44.

In Exercises 109–110, use a graphing utility to graph the function. Determine the horizontal asymptote for the graph of f and discuss its relationship to the sum of the given series.

109.   Function

$$f(x) = \dfrac{2\left[1 - \left(\dfrac{1}{3}\right)^x\right]}{1 - \dfrac{1}{3}}$$

Series

$$2 + 2\left(\dfrac{1}{3}\right) + 2\left(\dfrac{1}{3}\right)^2 + 2\left(\dfrac{1}{3}\right)^3 + \cdots$$

110.   Function

$$f(x) = \dfrac{4[1 - (0.6)^x]}{1 - 0.6}$$

Series

$$4 + 4(0.6) + 4(0.6)^2 + 4(0.6)^3 + \cdots$$

Critical Thinking Exercises

Make Sense? *In Exercises 111–114, determine whether each statement makes sense or does not make sense, and explain your reasoning.*

111. There's no end to the number of geometric sequences that I can generate whose first term is 5 if I pick nonzero numbers *r* and multiply 5 by each value of *r* repeatedly.

112. I've noticed that the big difference between arithmetic and geometric sequences is that arithmetic sequences are based on addition and geometric sequences are based on multiplication.

113. I modeled California's population growth with a geometric sequence, so my model is an exponential function whose domain is the set of natural numbers.

114. I used a formula to find the sum of the infinite geometric series $3 + 1 + \frac{1}{3} + \frac{1}{9} + \cdots$ and then checked my answer by actually adding all the terms.

In Exercises 115–118, determine whether each statement is true or false. If the statement is false, make the necessary change(s) to produce a true statement.

115. The sequence 2, 6, 24, 120, . . . is an example of a geometric sequence.

116. The sum of the geometric series $\frac{1}{2} + \frac{1}{4} + \frac{1}{8} + \cdots + \frac{1}{512}$ can only be estimated without knowing precisely which terms occur between $\frac{1}{8}$ and $\frac{1}{512}$.

117. $10 - 5 + \dfrac{5}{2} - \dfrac{5}{4} + \cdots = \dfrac{10}{1 - \dfrac{1}{2}}$

118. If the *n*th term of a geometric sequence is $a_n = 3(0.5)^{n-1}$, the common ratio is $\frac{1}{2}$.

119. In a pest-eradication program, sterilized male flies are released into the general population each day. Ninety percent of those flies will survive a given day. How many flies should be released each day if the long-range goal of the program is to keep 20,000 sterilized flies in the population?

120. You are now 25 years old and would like to retire at age 55 with a retirement fund of $1,000,000. How much should you deposit at the end of each month for the next 30 years in an IRA paying 10% annual interest compounded monthly to achieve your goal? Round to the nearest dollar.

Review Exercises

121. Simplify: $\sqrt{28} - 3\sqrt{7} + \sqrt{63}$.

(Section 7.4, Example 2)

122. Solve: $2x^2 = 4 - x$.

(Section 8.2, Example 2)

123. Rationalize the denominator: $\dfrac{6}{\sqrt{3} - \sqrt{5}}$.

(Section 7.5, Example 5)

Preview Exercises

Exercises 124–126 will help you prepare for the material covered in the next section.

Each exercise involves observing a pattern in the expanded form of the binomial expression $(a + b)^n$.

$(a + b)^1 = a + b$

$(a + b)^2 = a^2 + 2ab + b^2$

$(a + b)^3 = a^3 + 3a^2b + 3ab^2 + b^3$

$(a + b)^4 = a^4 + 4a^3b + 6a^2b^2 + 4ab^3 + b^4$

$(a + b)^5 = a^5 + 5a^4b + 10a^3b^2 + 10a^2b^3 + 5ab^4 + b^5$

124. Describe the pattern for the exponents on a.

125. Describe the pattern for the exponents on b.

126. Describe the pattern for the sum of the exponents on the variables in each term.

MID-CHAPTER CHECK POINT   Section 11.1–Section 11.3

 What You Know: We learned that a sequence is a function whose domain is the set of positive integers. In an arithmetic sequence, each term after the first differs from the preceding term by a constant, the common difference, d. In a geometric sequence, each term after the first is obtained by multiplying the preceding term by a nonzero constant, the common ratio, r. We found the general term of arithmetic sequences $[a_n = a_1 + (n - 1)d]$ and geometric sequences $[a_n = a_1 r^{n-1}]$ and used these formulas to find particular terms. We determined the sum of the first n terms of arithmetic sequences $\left[S_n = \dfrac{n}{2}(a_1 + a_n) \right]$ and geometric sequences $\left[S_n = \dfrac{a_1(1 - r^n)}{1 - r} \right]$. Finally, we determined the sum of an infinite geometric series, $a_1 + a_1 r + a_1 r^2 + a_1 r^3 + \cdots$, if $-1 < r < 1 \left(S = \dfrac{a_1}{1 - r} \right)$.

In Exercises 1–3, write the first five terms of each sequence. Assume that d represents the common difference of an arithmetic sequence and r represents the common ratio of a geometric sequence.

1. $a_n = (-1)^{n+1} \dfrac{n}{(n - 1)!}$

2. $a_1 = 5, d = -3$

3. $a_1 = 5, r = -3$

In Exercises 4–6, write a formula for the general term (the nth term) of each sequence. Then use the formula to find the indicated term.

4. $2, 6, 10, 14, \ldots ; a_{20}$

5. $3, 6, 12, 24, \ldots ; a_{10}$

6. $\dfrac{3}{2}, 1, \dfrac{1}{2}, 0, \ldots ; a_{30}$

7. Find the sum of the first ten terms of the sequence: $5, 10, 20, 40, \ldots$.

8. Find the sum of the first 50 terms of the sequence: $-2, 0, 2, 4, \ldots$.

9. Find the sum of the first ten terms of the sequence: $-20, 40, -80, 160, \ldots$.

10. Find the sum of the first 100 terms of the sequence: $4, -2, -8, -14, \ldots$.

In Exercises 11–14, find each indicated sum.

11. $\displaystyle\sum_{i=1}^{4} (i + 4)(i - 1)$

12. $\displaystyle\sum_{i=1}^{50} (3i - 2)$

13. $\displaystyle\sum_{i=1}^{6} \left(\dfrac{3}{2}\right)^i$

14. $\displaystyle\sum_{i=1}^{\infty} \left(-\dfrac{2}{5}\right)^{i-1}$

15. Express $0.\overline{45}$ as a fraction in lowest terms.

16. Express the sum using summation notation. Use i for the index of summation.

$$\frac{1}{3} + \frac{2}{4} + \frac{3}{5} + \cdots + \frac{18}{20}$$

17. A skydiver falls 16 feet during the first second of a dive, 48 feet during the second second, 80 feet during the third second, 112 feet during the fourth second, and so on. Find the distance that the skydiver falls during the 15th second and the total distance the skydiver falls in 15 seconds.

18. If the average value of a house increases 10% per year, how much will a house costing $500,000 in year 1 be worth in year 9? Round to the nearest dollar.

SECTION

11.4

The Binomial Theorem

Galaxies are groupings of billions of stars bound together by gravity. Some galaxies, such as the Centaurus galaxy shown here, are elliptical in shape.

What am I supposed to learn?

After studying this section, you should be able to:

1 Evaluate a binomial coefficient. ▶

2 Expand a binomial raised to a power. ▶

3 Find a particular term in a binomial expansion. ▶

Is mathematics discovered or invented? For example, planets revolve in elliptical orbits. Does that mean that the ellipse is out there, waiting for the mind to discover it? Or do people create the definition of an ellipse just as they compose a song? And is it possible for the same mathematics to be discovered/invented by independent researchers separated by time, place, and culture? This is precisely what occurred when mathematicians attempted to find efficient methods for raising binomials to higher and higher powers, such as

$$(x + 2)^3, (x + 2)^4, (x + 2)^5, (x + 2)^6,$$

and so on. In this section, we study higher powers of binomials and a method first discovered/invented by great minds in Eastern and in Western cultures working independently.

1 Evaluate a binomial coefficient. ▶

Binomial Coefficients

Before turning to powers of binomials, we introduce a special notation that uses factorials.

Definition of a Binomial Coefficient $\dbinom{n}{r}$

For nonnegative integers n and r, with $n \geq r$, the expression $\dbinom{n}{r}$ (read "*n* above *r*") is called a **binomial coefficient** and is defined by

$$\binom{n}{r} = \frac{n!}{r!(n-r)!}.$$

The symbol $_nC_r$ is often used in place of $\dbinom{n}{r}$ to denote binomial coefficients.

Can you see that the definition of a binomial coefficient involves a fraction with factorials in the numerator and the denominator? When evaluating such an expression, try to reduce the fraction before performing the multiplications. For example, consider $\frac{26!}{21!}$. Rather than writing out 26! as the product of all integers from 26 down to 1, we can express 26! as

$$26! = 26 \cdot 25 \cdot 24 \cdot 23 \cdot 22 \cdot 21!.$$

In this way, we can divide both the numerator and the denominator by the common factor, 21!.

$$\frac{26!}{21!} = \frac{26 \cdot 25 \cdot 24 \cdot 23 \cdot 22 \cdot 21!}{21!} = 26 \cdot 25 \cdot 24 \cdot 23 \cdot 22 = 7{,}893{,}600$$

EXAMPLE 1 Evaluating Binomial Coefficients

Using Technology

Graphing utilities can compute binomial coefficients. For example, to find $\binom{6}{2}$, many utilities require the sequence

6 [nCr] 2 [ENTER].

The graphing utility will display 15. Consult your manual and verify the other evaluations in Example 1.

Evaluate:

a. $\binom{6}{2}$ **b.** $\binom{3}{0}$ **c.** $\binom{9}{3}$ **d.** $\binom{4}{4}$.

Solution In each case, we apply the definition of the binomial coefficient.

a. $\binom{6}{2} = \frac{6!}{2!\,(6-2)!} = \frac{6!}{2!\,4!} = \frac{6 \cdot 5 \cdot 4!}{2 \cdot 1 \cdot 4!} = 15$

b. $\binom{3}{0} = \frac{3!}{0!(3-0)!} = \frac{3!}{0!\,3!} = \frac{1}{1} = 1$

Remember that $0! = 1$.

c. $\binom{9}{3} = \frac{9!}{3!\,(9-3)!} = \frac{9!}{3!\,6!} = \frac{9 \cdot 8 \cdot 7 \cdot 6!}{3 \cdot 2 \cdot 1 \cdot 6!} = 84$

d. $\binom{4}{4} = \frac{4!}{4!\,(4-4)!} = \frac{4!}{4!\,0!} = \frac{1}{1} = 1$ ∎

✓ **CHECK POINT 1** Evaluate:

a. $\binom{6}{3}$ **b.** $\binom{6}{0}$ **c.** $\binom{8}{2}$ **d.** $\binom{3}{3}$.

2 Expand a binomial raised to a power. ▶

The Binomial Theorem

When we write out the *binomial expression* $(a + b)^n$, where n is a positive integer, a number of patterns begin to appear.

$$(a + b)^1 = a + b$$
$$(a + b)^2 = a^2 + 2ab + b^2$$
$$(a + b)^3 = a^3 + 3a^2b + 3ab^2 + b^3$$
$$(a + b)^4 = a^4 + 4a^3b + 6a^2b^2 + 4ab^3 + b^4$$
$$(a + b)^5 = a^5 + 5a^4b + 10a^3b^2 + 10a^2b^3 + 5ab^4 + b^5$$

Each expanded form of the binomial expression is a polynomial. Observe the following patterns:

1. The first term in the expansion of $(a + b)^n$ is a^n. The exponents on a decrease by 1 in each successive term.

2. The exponents on b in the expansion of $(a + b)^n$ increase by 1 in each successive term. In the first term, the exponent on b is 0. (Because $b^0 = 1$, b is not shown in the first term.) The last term is b^n.

3. The sum of the exponents on the variables in any term in the expansion of $(a + b)^n$ is equal to n.

4. The number of terms in the polynomial expansion is one greater than the power of the binomial, n. There are $n + 1$ terms in the expanded form of $(a + b)^n$.

Using these observations, the variable parts of the expansion of $(a + b)^6$ are

$$a^6, \quad a^5b, \quad a^4b^2, \quad a^3b^3, \quad a^2b^4, \quad ab^5, \quad b^6.$$

The first term is a^6, with the exponents on a decreasing by 1 in each successive term. The exponents on b increase from 0 to 6, with the last term being b^6. The sum of the exponents in each term is equal to 6.

We can generalize from these observations to obtain the variable parts of the expansion of $(a + b)^n$. They are

$$a^n, \quad a^{n-1}b, \quad a^{n-2}b^2, \quad a^{n-3}b^3, \dots, \quad ab^{n-1}, \quad b^n.$$

> Exponents on a are decreasing by 1.
> Exponents on b are increasing by 1.

> Sum of exponents: $n - 1 + 1 = n$
> Sum of exponents: $n - 3 + 3 = n$
> Sum of exponents: $1 + n - 1 = n$

If we use binomial coefficients and the pattern for the variable part of each term, a formula called the **Binomial Theorem** can be used to expand any positive integral power of a binomial.

A Formula for Expanding Binomials: The Binomial Theorem

For any positive integer n,

$$(a + b)^n = \binom{n}{0}a^n + \binom{n}{1}a^{n-1}b + \binom{n}{2}a^{n-2}b^2 + \binom{n}{3}a^{n-3}b^3 + \cdots + \binom{n}{n}b^n$$

$$= \sum_{r=0}^{n}\binom{n}{r}a^{n-r}b^r.$$

EXAMPLE 2 Using the Binomial Theorem

Expand: $(x + 2)^4$.

Solution We use the Binomial Theorem

$$(a + b)^n = \binom{n}{0}a^n + \binom{n}{1}a^{n-1}b + \binom{n}{2}a^{n-2}b^2 + \binom{n}{3}a^{n-3}b^3 + \cdots + \binom{n}{n}b^n$$

to expand $(x + 2)^4$. In $(x + 2)^4$, $a = x$, $b = 2$, and $n = 4$. In the expansion, powers of x are in descending order, starting with x^4. Powers of 2 are in ascending order, starting with 2^0. (Because $2^0 = 1$, a 2 is not shown in the first term.) The sum of the exponents on x and 2 in each term is equal to 4, the exponent in the expression $(x + 2)^4$.

$$(x + 2)^4 = \binom{4}{0}x^4 + \binom{4}{1}x^3 \cdot 2 + \binom{4}{2}x^2 \cdot 2^2 + \binom{4}{3}x \cdot 2^3 + \binom{4}{4}2^4$$

> These binomial coefficients are evaluated using $\binom{n}{r} = \dfrac{n!}{r!(n-r)!}$.

$$= \frac{4!}{0!4!}x^4 + \frac{4!}{1!3!}x^3 \cdot 2 + \frac{4!}{2!2!}x^2 \cdot 4 + \frac{4!}{3!1!}x \cdot 8 + \frac{4!}{4!0!} \cdot 16$$

> $\dfrac{4!}{2!2!} = \dfrac{4 \cdot 3 \cdot 2!}{2! \cdot 2 \cdot 1} = \dfrac{12}{2} = 6$
> Take a few minutes to verify the other factorial evaluations.

$$= 1 \cdot x^4 + 4x^3 \cdot 2 + 6x^2 \cdot 4 + 4x \cdot 8 + 1 \cdot 16$$
$$= x^4 + 8x^3 + 24x^2 + 32x + 16 \quad \blacksquare$$

☑ **CHECK POINT 2** Expand: $(x + 1)^4$.

Using Technology

You can use a graphing utility's table feature to find the five binomial coefficients in Example 2.

Enter $y_1 = 4$ [nCr] x.

EXAMPLE 3 Using the Binomial Theorem

Expand: $(2x - y)^5$.

Solution Because the Binomial Theorem involves the addition of two terms raised to a power, we rewrite $(2x - y)^5$ as $[2x + (-y)]^5$. We use the Binomial Theorem

$$(a + b)^n = \binom{n}{0}a^n + \binom{n}{1}a^{n-1}b + \binom{n}{2}a^{n-2}b^2 + \binom{n}{3}a^{n-3}b^3 + \cdots + \binom{n}{n}b^n$$

to expand $[2x + (-y)]^5$. In $[2x + (-y)]^5$, $a = 2x$, $b = -y$, and $n = 5$. In the expansion, powers of $2x$ are in descending order, starting with $(2x)^5$. Powers of $-y$ are in ascending order, starting with $(-y)^0$. [Because $(-y)^0 = 1$, a $-y$ is not shown in the first term.] The sum of the exponents on $2x$ and $-y$ in each term is equal to 5, the exponent in the expression $(2x - y)^5$.

$$(2x - y)^5 = [2x + (-y)]^5$$

$$= \binom{5}{0}(2x)^5 + \binom{5}{1}(2x)^4(-y) + \binom{5}{2}(2x)^3(-y)^2 + \binom{5}{3}(2x)^2(-y)^3 + \binom{5}{4}(2x)(-y)^4 + \binom{5}{5}(-y)^5$$

> Evaluate binomial coefficients using $\binom{n}{r} = \dfrac{n!}{r!(n-r)!}$.

$$= \frac{5!}{0!5!}(2x)^5 + \frac{5!}{1!4!}(2x)^4(-y) + \frac{5!}{2!3!}(2x)^3(-y)^2 + \frac{5!}{3!2!}(2x)^2(-y)^3 + \frac{5!}{4!1!}(2x)(-y)^4 + \frac{5!}{5!0!}(-y)^5$$

> $\dfrac{5!}{2!3!} = \dfrac{5 \cdot 4 \cdot 3!}{2 \cdot 1 \cdot 3!} = 10$
> Take a few minutes to verify the other factorial evaluations.

$$= 1(2x)^5 + 5(2x)^4(-y) + 10(2x)^3(-y)^2 + 10(2x)^2(-y)^3 + 5(2x)(-y)^4 + 1(-y)^5$$

> Raise both factors in these parentheses to the indicated powers.

$$= 1(32x^5) + 5(16x^4)(-y) + 10(8x^3)(-y)^2 + 10(4x^2)(-y)^3 + 5(2x)(-y)^4 + 1(-y)^5$$

> Now raise $-y$ to the indicated powers.

$$= 1(32x^5) + 5(16x^4)(-y) + 10(8x^3)y^2 + 10(4x^2)(-y^3) + 5(2x)y^4 + 1(-y^5)$$

Multiplying factors in each of the six terms gives us the desired expansion:

$$(2x - y)^5 = 32x^5 - 80x^4y + 80x^3y^2 - 40x^2y^3 + 10xy^4 - y^5. \blacksquare$$

✓ **CHECK POINT 3** Expand: $(x - 2y)^5$.

---

3 Find a particular term in a binomial expansion. ▶

Finding a Particular Term in a Binomial Expansion

By observing the terms in the formula for expanding binomials, we can find a formula for finding a particular term without writing the entire expansion.

1st term   2nd term   3rd term

$$\binom{n}{0}a^n b^0 \qquad \binom{n}{1}a^{n-1}b^1 \qquad \binom{n}{2}a^{n-2}b^2$$

> The exponent on b is 1 less than the term number.

Based on the observation in the bottom voice balloon, the $(r + 1)$st term of the expansion of $(a + b)^n$ is the term that contains b^r.

Finding a Particular Term in a Binomial Expansion

The $(r + 1)$st term of the expansion of $(a + b)^n$ is

$$\binom{n}{r}a^{n-r}b^r.$$

EXAMPLE 4   Finding a Single Term of a Binomial Expansion

Find the fourth term in the expansion of $(3x + 2y)^7$.

Solution   The fourth term in the expansion of $(3x + 2y)^7$ contains $(2y)^3$. To find the fourth term, first note that $4 = 3 + 1$. Equivalently, the fourth term of $(3x + 2y)^7$ is the $(3 + 1)$st term. Thus, $r = 3$, $a = 3x$, $b = 2y$, and $n = 7$. The fourth term is

$$\binom{7}{3}(3x)^{7-3}(2y)^3 = \binom{7}{3}(3x)^4(2y)^3 = \frac{7!}{3!(7-3)!}(3x)^4(2y)^3.$$

> Use the formula for the $(r + 1)$st term of $(a + b)^n$: $\binom{n}{r}a^{n-r}b^r$.

> We use $\binom{n}{r} = \frac{n!}{r!(n-r)!}$ to evaluate $\binom{7}{3}$.

Now we need to evaluate the factorial expression and raise $3x$ and $2y$ to the indicated powers. We obtain

$$\frac{7!}{3!\,4!}(81x^4)(8y^3) = \frac{7 \cdot 6 \cdot 5 \cdot 4!}{3 \cdot 2 \cdot 1 \cdot 4!}(81x^4)(8y^3) = 35(81x^4)(8y^3) = 22{,}680x^4y^3.$$

The fourth term of $(3x + 2y)^7$ is $22{,}680x^4y^3$.   ■

✓ **CHECK POINT 4**   Find the fifth term in the expansion of $(2x + y)^9$.

Blitzer Bonus ◉
The Universality of Mathematics

Pascal's triangle is an array of numbers showing coefficients of the terms in the expansions of $(a + b)^n$. Although credited to French mathematician Blaise Pascal (1623–1662), the triangular array of numbers appeared in a Chinese document printed in 1303. The Binomial Theorem was known in Eastern cultures prior to its discovery in Europe. The same mathematics is often discovered/invented by independent researchers separated by time, place, and culture.

Binomial Expansions

$(a + b)^0 = 1$
$(a + b)^1 = a + b$
$(a + b)^2 = a^2 + 2ab + b^2$
$(a + b)^3 = a^3 + 3a^2b + 3ab^2 + b^3$
$(a + b)^4 = a^4 + 4a^3b + 6a^2b^2 + 4ab^3 + b^4$
$(a + b)^5 = a^5 + 5a^4b + 10a^3b^2 + 10a^2b^3 + 5ab^4 + b^5$

Pascal's Triangle

Coefficients in the Expansions

```
                1
              1   1
            1   2   1
          1   3   3   1
        1   4   6   4   1
      1   5  10  10   5   1
    1   6  15  20  15   6   1
  1   7  21  35  35  21   7   1
1   8  28  56  70  56  28   8   1
```

Chinese Document: 1303

Achieving Success

As you continue taking college math courses, remember that you are expected to study and do homework for at least two hours for each hour of class time. In order to make education a top priority, be consistent in the time and location for studying. Arrange your schedule so that you can work in a quiet location without distractions and at a time when you are alert and best able to study. Do this on a regular basis until the routine of studying and doing homework becomes familiar and comfortable.

CONCEPT AND VOCABULARY CHECK

Fill in each blank so that the resulting statement is true.

1. $\binom{n}{r}$ is called a/an _____ coefficient.

2. $\binom{8}{2} = \dfrac{\underline{}!}{\underline{}!\,\underline{}!}$

3. $\binom{n}{r} = $ _____

4. $(x + 2)^5 = \binom{5}{0}x^5 + $ _____ $x^4 \cdot 2 + $ _____ $x^3 \cdot 2^2 + $ _____ $x^2 \cdot 2^3 + $ _____ $x \cdot 2^4 + $ _____ $\cdot 2^5$

5. $(a + b)^n = \binom{n}{0}a^n + $ _____ $a^{n-1}b + $ _____ $a^{n-2}b^2 + $ _____ $a^{n-3}b^3 + \ldots + $ _____ b^n

   The sum of the exponents on a and b in each term is _____.

6. The formula in Concept Check 5 is called the _____ Theorem.

7. The $(r + 1)$st term of the expansion of $(a + b)^n$ is $\binom{n}{r}$ _____.

11.4 EXERCISE SET ▶ MyMathLab®

Practice Exercises

In Exercises 1–8, evaluate the given binomial coefficient.

1. $\binom{8}{3}$
2. $\binom{7}{2}$
3. $\binom{12}{1}$

4. $\binom{11}{1}$
5. $\binom{6}{6}$
6. $\binom{15}{2}$

7. $\binom{100}{2}$
8. $\binom{100}{98}$

In Exercises 9–30, use the Binomial Theorem to expand each binomial and express the result in simplified form.

9. $(x + 2)^3$
10. $(x + 4)^3$
11. $(3x + y)^3$
12. $(x + 3y)^3$
13. $(5x - 1)^3$
14. $(4x - 1)^3$

15. $(2x + 1)^4$
16. $(3x + 1)^4$
17. $(x^2 + 2y)^4$
18. $(x^2 + y)^4$
19. $(y - 3)^4$
20. $(y - 4)^4$
21. $(2x^3 - 1)^4$
22. $(2x^5 - 1)^4$
23. $(c + 2)^5$
24. $(c + 3)^5$
25. $(x - 1)^5$
26. $(x - 2)^5$
27. $(3x - y)^5$
28. $(x - 3y)^5$
29. $(2a + b)^6$
30. $(a + 2b)^6$

In Exercises 31–38, write the first three terms in each binomial expansion, expressing the result in simplified form.

31. $(x + 2)^8$

32. $(x + 3)^8$

33. $(x - 2y)^{10}$

34. $(x - 2y)^9$

35. $(x^2 + 1)^{16}$

36. $(x^2 + 1)^{17}$

37. $(y^3 - 1)^{20}$

38. $(y^3 - 1)^{21}$

In Exercises 39–48, find the indicated term in each expansion.

39. $(2x + y)^6$; third term

40. $(x + 2y)^6$; third term

41. $(x - 1)^9$; fifth term

42. $(x - 1)^{10}$; fifth term

43. $(x^2 + y^3)^8$; sixth term

44. $(x^3 + y^2)^8$; sixth term

45. $\left(x - \frac{1}{2}\right)^9$; fourth term

46. $\left(x + \frac{1}{2}\right)^8$; fourth term

47. $(x^2 + y)^{22}$; the term containing y^{14}

48. $(x + 2y)^{10}$; the term containing y^6

Practice PLUS

In Exercises 49–52, use the Binomial Theorem to expand each expression and write the result in simplified form.

49. $(x^3 + x^{-2})^4$

50. $(x^2 + x^{-3})^4$

51. $\left(x^{\frac{1}{3}} - x^{-\frac{1}{3}}\right)^3$

52. $\left(x^{\frac{2}{3}} - \frac{1}{\sqrt[3]{x}}\right)^3$

Exercises 53–54 involve expressions containing i, where $i = \sqrt{-1}$. Expand each expression and use powers of i to simplify the result.

53. $\left(-1 + i\sqrt{3}\right)^3$

54. $\left(-1 - i\sqrt{3}\right)^3$

In Exercises 55–56, find $\dfrac{f(x + h) - f(x)}{h}$ and simplify.

55. $f(x) = x^4 + 7$

56. $f(x) = x^5 + 8$

57. Find the middle term in the expansion of $\left(\dfrac{3}{x} + \dfrac{x}{3}\right)^{10}$.

58. Find the middle term in the expansion of $\left(\dfrac{1}{x} - x^2\right)^{12}$.

Application Exercises

The graph shows that U.S. smokers have a greater probability of suffering from some ailments than the general adult population. Exercises 59–60 are based on some of the probabilities, expressed as decimals, shown to the right of the bars. In each exercise, use a calculator to determine the probability, correct to four decimal places.

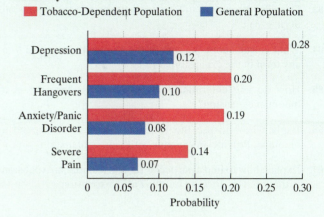

Probability That United States Adults Suffer from Various Ailments

Source: MARS 2005 OTC/DTC Pharmaceutical Study.

If the probability an event will occur is p and the probability it will not occur is q, then each term in the expansion of $(p + q)^n$ represents a probability.

59. The probability that a smoker suffers from depression is 0.28. If five smokers are randomly selected, the probability that three of them will suffer from depression is the third term of the binomial expansion of

$(0.28 + 0.72)^5.$

What is this probability?

60. The probability that a person in the general population suffers from depression is 0.12. If five people from the general population are randomly selected, the probability that three of them will suffer from depression is the third term of the binomial expansion of

$(0.12 + 0.88)^5$

What is this probability?

Explaining the Concepts

61. Explain how to evaluate $\binom{n}{r}$. Provide an example with your explanation.

62. Describe the pattern in the exponents on a in the expansion of $(a + b)^n$.

63. Describe the pattern in the exponents on b in the expansion of $(a + b)^n$.

64. What is true about the sum of the exponents on a and b in any term in the expansion of $(a + b)^n$?

65. How do you determine how many terms there are in a binomial expansion?

66. Explain how to use the Binomial Theorem to expand a binomial. Provide an example with your explanation.

67. Explain how to find a particular term in a binomial expansion without having to write out the entire expansion.

Technology Exercises

68. Use the $\boxed{\text{nCr}}$ key on a graphing utility to verify your answers in Exercises 1–8.

In Exercises 69–70, graph each of the functions in the same viewing rectangle. Describe how the graphs illustrate the Binomial Theorem.

69. $f_1(x) = (x + 2)^3$ $f_2(x) = x^3$
$f_3(x) = x^3 + 6x^2$ $f_4(x) = x^3 + 6x^2 + 12x$
$f_5(x) = x^3 + 6x^2 + 12x + 8$
Use a $[-10, 10, 1]$ by $[-30, 30, 10]$ viewing rectangle.

70. $f_1(x) = (x + 1)^4$ $f_2(x) = x^4$
$f_3(x) = x^4 + 4x^3$ $f_4(x) = x^4 + 4x^3 + 6x^2$
$f_5(x) = x^4 + 4x^3 + 6x^2 + 4x$
$f_6(x) = x^4 + 4x^3 + 6x^2 + 4x + 1$
Use a $[-5, 5, 1]$ by $[-30, 30, 10]$ viewing rectangle.

In Exercises 71–73, use the Binomial Theorem to find a polynomial expansion for each function. Then use a graphing utility and an approach similar to the one in Exercises 69 and 70 to verify the expansion.

71. $f_1(x) = (x - 1)^3$

72. $f_1(x) = (x - 2)^4$

73. $f_1(x) = (x + 2)^6$

Critical Thinking Exercises

Make Sense?   *In Exercises 74–77, determine whether each statement makes sense or does not make sense, and explain your reasoning.*

74. In order to expand $(x^3 - y^4)^5$, I find it helpful to rewrite the expression inside the parentheses as $x^3 + (-y^4)$.

75. Without writing the expansion of $(x - 1)^6$, I can see that the terms have alternating positive and negative signs.

76. I use binomial coefficients to expand $(a + b)^n$, where $\binom{n}{1}$ is the coefficient of the first term, $\binom{n}{2}$ is the coefficient of the second term, and so on.

77. One of the terms in my binomial expansion is $\binom{7}{5} x^2 y^4$.

In Exercises 78–81, determine whether each statement is true or false. If the statement is false, make the necessary change(s) to produce a true statement.

78. The binomial expansion for $(a + b)^n$ contains n terms.

79. The Binomial Theorem can be written in condensed form as
$$(a + b)^n = \sum_{r=0}^{n} \binom{n}{r} a^{n-r} b^r.$$

80. The sum of the binomial coefficients in $(a + b)^n$ cannot be 2^n.

81. There are no values of a and b such that
$(a + b)^4 = a^4 + b^4$.

82. Use the Binomial Theorem to expand and then simplify the result: $(x^2 + x + 1)^3$. [Hint: Write $x^2 + x + 1$ as $x^2 + (x + 1)$.]

83. Find the term in the expansion of $(x^2 + y^2)^5$ containing x^4 as a factor.

Review Exercises

84. If $f(x) = x^2 + 2x + 3$, find $f(a + 1)$.
(Section 5.2, Example 11)

85. If $f(x) = x^2 + 5x$ and $g(x) = 2x - 3$, find $f(g(x))$ and $g(f(x))$.
(Section 9.2, Example 1)

86. Subtract: $\dfrac{x}{x + 3} - \dfrac{x + 1}{2x^2 - 2x - 24}$.
(Section 6.2, Example 7)

Chapter 11 Summary

Definitions and Concepts	**Examples**

Section 11.1 Sequences and Summation Notation

An infinite sequence $\{a_n\}$ is a function whose domain is the set of positive integers. The function values, or terms, are represented by

$$a_1, a_2, a_3, a_4, \ldots, a_n, \ldots.$$

General term: $a_n = \dfrac{(-1)^n}{n^3}$.

$$a_1 = \frac{(-1)^1}{1^3} = -1, \quad a_2 = \frac{(-1)^2}{2^3} = \frac{1}{8},$$

$$a_3 = \frac{(-1)^3}{3^3} = -\frac{1}{27}, \quad a_4 = \frac{(-1)^4}{4^3} = \frac{1}{64}$$

First four terms are $-1, \frac{1}{8}, -\frac{1}{27}$, and $\frac{1}{64}$.

Factorial Notation

$$n! = n(n-1)(n-2) \cdots (3)(2)(1) \quad \text{and} \quad 0! = 1$$

$$6! = 6 \cdot 5 \cdot 4 \cdot 3 \cdot 2 \cdot 1 = 720$$
$$3! = 3 \cdot 2 \cdot 1 = 6$$

Summation Notation

$$\sum_{i=1}^{n} a_i = a_1 + a_2 + a_3 + a_4 + \cdots + a_n$$

In the summation shown here, i is the index of summation, n is the upper limit of summation, and 1 is the lower limit of summation.

$$\sum_{i=3}^{7}(i^2 - 4)$$
$$= (3^2 - 4) + (4^2 - 4) + (5^2 - 4) + (6^2 - 4) + (7^2 - 4)$$
$$= (9 - 4) + (16 - 4) + (25 - 4) + (36 - 4) + (49 - 4)$$
$$= 5 + 12 + 21 + 32 + 45$$
$$= 115$$

Section 11.2 Arithmetic Sequences

In an arithmetic sequence, each term after the first differs from the preceding term by a constant, the common difference. Subtract any term from the term that directly follows it to find the common difference.

Find the general term and the tenth term:

$$3, 7, 11, 15, \ldots.$$

$$\boxed{a_1 = 3} \quad \boxed{d = 7 - 3 = 4}$$

General Term of an Arithmetic Sequence

The nth term (the general term) of an arithmetic sequence with first term a_1 and common difference d is

$$a_n = a_1 + (n-1)d.$$

Using $a_n = a_1 + (n-1)d$,
$$a_n = 3 + (n-1)4 = 3 + 4n - 4 = 4n - 1.$$
The general term is $a_n = 4n - 1$.
The tenth term is $a_{10} = 4 \cdot 10 - 1 = 39$.

The Sum of the First n Terms of an Arithmetic Sequence

The sum, S_n, of the first n terms of an arithmetic sequence is given by

$$S_n = \frac{n}{2}(a_1 + a_n)$$

in which a_1 is the first term and a_n is the nth term.

Find the sum of the first ten terms:

$$2, 5, 8, 11, \ldots.$$

$$\boxed{a_1 = 2} \quad \boxed{d = 5 - 2 = 3}$$

First find a_{10}, the 10th term. Using $a_n = a_1 + (n-1)d$,
$$a_{10} = 2 + (10 - 1) \cdot 3 = 2 + 9 \cdot 3 = 29.$$
Find the sum of the first ten terms using

$$S_n = \frac{n}{2}(a_1 + a_n).$$

$$S_{10} = \frac{10}{2}(a_1 + a_{10}) = 5(2 + 29) = 5(31) = 155$$

Definitions and Concepts	**Examples**

Section 11.3 Geometric Sequences and Series

In a geometric sequence, each term after the first is obtained by multiplying the preceding term by a nonzero constant, the common ratio. Divide any term after the first by the term that directly precedes it to find the common ratio.

General Term of a Geometric Sequence

The nth term (the general term) of a geometric sequence with first term a_1 and common ratio r is

$$a_n = a_1 r^{n-1}.$$

Find the general term and the ninth term:

$$12, -6, 3, -\frac{3}{2}, \dots .$$

$a_1 = 12$ $r = \frac{-6}{12} = -\frac{1}{2}$

Using $a_n = a_1 r^{n-1}$,

$$a_n = 12\left(-\frac{1}{2}\right)^{n-1} \text{ is the general term.}$$

The ninth term is

$$a_9 = 12\left(-\frac{1}{2}\right)^{9-1} = 12\left(-\frac{1}{2}\right)^8 = \frac{12}{256} = \frac{3}{64}.$$

The Sum of the First n Terms of a Geometric Sequence

The sum, S_n, of the first n terms of a geometric sequence is given by

$$S_n = \frac{a_1(1 - r^n)}{1 - r}$$

in which a_1 is the first term and r is the common ratio ($r \neq 1$).

Find $\displaystyle\sum_{i=1}^{8} 4 \cdot 3^i$

$$= 4 \cdot 3 + 4 \cdot 3^2 + 4 \cdot 3^3 + \cdots + 4 \cdot 3^8.$$

$a_1 = 12$ $r = \frac{4 \cdot 3^2}{4 \cdot 3} = 3$

Using $S_n = \dfrac{a_1(1 - r^n)}{1 - r}$,

$$S_8 = \frac{12(1 - 3^8)}{1 - 3} = 39{,}360.$$

The Sum of an Infinite Geometric Series

If $-1 < r < 1$ (equivalently, $|r| < 1$), then the sum of the infinite geometric series

$$a_1 + a_1 r + a_1 r^2 + a_1 r^3 + \cdots$$

in which a_1 is the first term and r is the common ratio is given by

$$S = \frac{a_1}{1 - r}.$$

If $|r| \geq 1$, the infinite series does not have a sum.

Find the sum:

$$6 + \frac{6}{3} + \frac{6}{3^2} + \frac{6}{3^3} + \cdots .$$

$a_1 = 6$ $r = \frac{1}{3}$

Using $S = \dfrac{a_1}{1 - r}$, the sum is

$$S = \frac{6}{1 - \frac{1}{3}} = \frac{6}{\frac{2}{3}} = 6 \cdot \frac{3}{2} = 9.$$

Section 11.4 The Binomial Theorem

Definition of a Binomial Coefficient

$$\binom{n}{r} = \frac{n!}{r!\,(n-r)!}$$

$$\binom{8}{3} = \frac{8!}{3!\,(8-3)!} = \frac{8!}{3!\,5!}$$

$$= \frac{8 \cdot 7 \cdot 6 \cdot 5!}{3 \cdot 2 \cdot 1 \cdot 5!} = 56$$

Definitions and Concepts	Examples

Section 11.4 The Binomial Theorem (continued)

A Formula for Expanding Binomials: The Binomial Theorem

For any positive integer n,

$$(a + b)^n = \binom{n}{0}a^n + \binom{n}{1}a^{n-1}b$$
$$+ \binom{n}{2}a^{n-2}b^2 + \binom{n}{3}a^{n-3}b^3 + \cdots + \binom{n}{n}b^n.$$

Expand: $(3x - y)^4 = [3x + (-y)]^4$.

$$= \binom{4}{0}(3x)^4 + \binom{4}{1}(3x)^3(-y)$$
$$+ \binom{4}{2}(3x)^2(-y)^2 + \binom{4}{3}(3x)^1(-y)^3 + \binom{4}{4}(-y)^4$$
$$= 1 \cdot 81x^4 + 4 \cdot 27x^3(-y) + 6 \cdot 9x^2y^2 + 4 \cdot 3x(-y^3) + 1 \cdot y^4$$
$$= 81x^4 - 108x^3y + 54x^2y^2 - 12xy^3 + y^4$$

Finding a Particular Term in a Binomial Expansion

The $(r + 1)$st term in the expansion of $(a + b)^n$ is

$$\binom{n}{r}a^{n-r}b^r.$$

The 8th term, or $(7 + 1)$st term $(r = 7)$, of $(x + 2y)^{10}$ is

$$\binom{10}{7}x^{10-7}(2y)^7$$
$$= \binom{10}{7}x^3(2y)^7 = 120x^3 \cdot 128y^7$$
$$= 15,360x^3y^7.$$

CHAPTER 11 REVIEW EXERCISES

11.1 *In Exercises 1–4, write the first four terms of each sequence whose general term is given.*

1. $a_n = 7n - 4$

2. $a_n = (-1)^n \dfrac{n + 2}{n + 1}$

3. $a_n = \dfrac{1}{(n - 1)!}$

4. $a_n = \dfrac{(-1)^{n+1}}{2^n}$

In Exercises 5–6, find each indicated sum.

5. $\displaystyle\sum_{i=1}^{5}(2i^2 - 3)$

6. $\displaystyle\sum_{i=0}^{4}(-1)^{i+1}i!$

In Exercises 7–8, express each sum using summation notation. Use i for the index of summation.

7. $\dfrac{1}{3} + \dfrac{2}{4} + \dfrac{3}{5} + \cdots + \dfrac{15}{17}$

8. $4^3 + 5^3 + 6^3 + \cdots + 13^3$

11.2 *In Exercises 9–11, write the first six terms of each arithmetic sequence.*

9. $a_1 = 7, d = 4$

10. $a_1 = -4, d = -5$

11. $a_1 = \dfrac{3}{2}, d = -\dfrac{1}{2}$

In Exercises 12–14, use the formula for the general term (the nth term) of an arithmetic sequence to find the indicated term of each sequence.

12. Find a_6 when $a_1 = 5, d = 3$.

13. Find a_{12} when $a_1 = -8, d = -2$.

14. Find a_{14} when $a_1 = 14, d = -4$.

In Exercises 15–18, write a formula for the general term (the nth term) of each arithmetic sequence. Then use the formula for a_n to find a_{20}, the 20th term of the sequence.

15. $-7, -3, 1, 5, \ldots$

16. $a_1 = 200, d = -20$

17. $a_1 = -12, d = -\dfrac{1}{2}$

18. $15, 8, 1, -6, \ldots$

19. Find the sum of the first 22 terms of the arithmetic sequence: $5, 12, 19, 26, \ldots$.

20. Find the sum of the first 15 terms of the arithmetic sequence: $-6, -3, 0, 3, \ldots$.

21. Find $3 + 6 + 9 + \cdots + 300$, the sum of the first 100 positive multiples of 3.

In Exercises 22–24, use the formula for the sum of the first n terms of an arithmetic sequence to find the indicated sum.

22. $\displaystyle\sum_{i=1}^{16}(3i + 2)$

23. $\displaystyle\sum_{i=1}^{25}(-2i + 6)$

24. $\displaystyle\sum_{i=1}^{30}(-5i)$

25. In 2014, the average ticket price for top rock concerts, adjusted for inflation, had increased by 77% since 1995. This was greater than the percent increase in the cost of tuition at private four-year colleges during the same time period. The bar graph shows the average ticket price, in 2015 dollars, for rock concerts in 1995 and 2014.

Average Ticket Price for Rock Concerts

Source: Rolling Stone

In 1995, the average ticket price, in 2015 dollars, for a rock concert was $40.36. On average, this has increased by approximately $1.63 per year since then.

a. Write a formula for the *n*th term of the arithmetic sequence that describes the average ticket price, in 2015 dollars, for rock concerts *n* years after 1994.

b. Use the model to project the average ticket price, in 2015 dollars, for rock concerts in 2020.

26. A company offers a starting salary of $31,500 with raises of $2300 per year. Find the total salary over a ten-year period.

27. A theater has 25 seats in the first row and 35 rows in all. Each successive row contains one additional seat. How many seats are in the theater?

11.3 *In Exercises 28–31, write the first five terms of each geometric sequence.*

28. $a_1 = 3, r = 2$

29. $a_1 = \dfrac{1}{2}, r = \dfrac{1}{2}$

30. $a_1 = 16, r = -\dfrac{1}{4}$

31. $a_1 = -5, r = -1$

In Exercises 32–34, use the formula for the general term (the nth term) of a geometric sequence to find the indicated term of each sequence.

32. Find a_7 when $a_1 = 2, r = 3$.

33. Find a_6 when $a_1 = 16, r = \frac{1}{2}$.

34. Find a_5 when $a_1 = -3, r = 2$.

In Exercises 35–37, write a formula for the general term (the nth term) of each geometric sequence. Then use the formula for a_n to find a_8, the eighth term of the sequence.

35. $1, 2, 4, 8, \ldots$

36. $100, 10, 1, \frac{1}{10}, \ldots$

37. $12, -4, \frac{4}{3}, -\frac{4}{9}, \ldots$

38. Find the sum of the first 15 terms of the geometric sequence: $5, -15, 45, -135, \ldots$.

39. Find the sum of the first 7 terms of the geometric sequence: $8, 4, 2, 1, \ldots$.

In Exercises 40–42, use the formula for the sum of the first n terms of a geometric sequence to find the indicated sum.

40. $\displaystyle\sum_{i=1}^{6} 5^i$

41. $\displaystyle\sum_{i=1}^{7} 3(-2)^i$

42. $\displaystyle\sum_{i=1}^{5} 2\left(\dfrac{1}{4}\right)^{i-1}$

In Exercises 43–46, find the sum of each infinite geometric series.

43. $9 + 3 + 1 + \dfrac{1}{3} + \cdots$

44. $2 - 1 + \dfrac{1}{2} - \dfrac{1}{4} + \cdots$

45. $-6 + 4 - \dfrac{8}{3} + \dfrac{16}{9} - \cdots$

46. $\displaystyle\sum_{i=1}^{\infty} 5(0.8)^i$

In Exercises 47–48, express each repeating decimal as a fraction in lowest terms.

47. $0.\overline{6}$

48. $0.\overline{47}$

49. The table shows the population of Florida for 2000 and 2010, with estimates given by the U.S. Census Bureau for 2001 through 2009.

Year	2000	2001	2002	2003	2004	2005
Population in millions	15.98	16.24	16.50	16.76	17.03	17.30

Year	2006	2007	2008	2009	2010
Population in millions	17.58	17.86	18.15	18.44	18.80

a. Divide the population for each year by the population in the preceding year. Round to two decimal places and show that Florida has a population increase that is approximately geometric.

b. Write the general term of the geometric sequence modeling Florida's population, in millions, *n* years after 1999.

c. Use your the model from part (b) to project Florida's population, in millions, for the year 2030. Round to two decimal places.

50. A job pays $32,000 for the first year with an annual increase of 6% per year beginning in the second year. What is the salary in the sixth year? What is the total salary paid over this six-year period? Round answers to the nearest dollar.

In Exercises 51–52, use the formula for the value of an annuity and round to the nearest dollar.

51. You spend \$10 per week on lottery tickets, averaging \$520 per year. Instead of buying tickets, if you deposited the \$520 at the end of each year in an annuity paying 6% compounded annually,

  a. How much would you have after 20 years?

  b. Find the interest.

52. To save for retirement, you decide to deposit \$100 at the end of each month in an IRA that pays 5.5% compounded monthly.

  a. How much will you have from the IRA after 30 years?

  b. Find the interest.

53. A factory in an isolated town has an annual payroll of \$4 million. It is estimated that 70% of this money is spent within the town, that people in the town receiving this money will again spend 70% of what they receive in the town, and so on. What is the total of all this spending in the town each year?

11.4 *In Exercises 54–55, evaluate the given binomial coefficient.*

54. $\dbinom{11}{8}$

55. $\dbinom{90}{2}$

In Exercises 56–59, use the Binomial Theorem to expand each binomial and express the result in simplified form.

56. $(2x + 1)^3$

57. $(x^2 - 1)^4$

58. $(x + 2y)^5$

59. $(x - 2)^6$

In Exercises 60–61, write the first three terms in each binomial expansion, expressing the result in simplified form.

60. $(x^2 + 3)^8$

61. $(x - 3)^9$

In Exercises 62–63, find the indicated term in each expansion.

62. $(x + 2)^5$; fourth term

63. $(2x - 3)^6$; fifth term

CHAPTER 11 TEST

Step-by-step test solutions are found on the Chapter Test Prep Videos available in MyMathLab® or on YouTube (search "BlitzerInterAlg7e" and click on "Channels").

1. Write the first five terms of the sequence whose general term is $a_n = \dfrac{(-1)^{n+1}}{n^2}$.

2. Find the indicated sum: $\displaystyle\sum_{i=1}^{5}(i^2 + 10)$.

3. Express the sum using summation notation. Use i for the index of summation.

$$\frac{2}{3} + \frac{3}{4} + \frac{4}{5} + \cdots + \frac{21}{22}$$

In Exercises 4–5, write a formula for the general term (the nth term) of each sequence. Then use the formula to find the twelfth term of the sequence.

4. $4, 9, 14, 19, \ldots$

5. $16, 4, 1, \frac{1}{4}, \ldots$

In Exercises 6–7, use the formula for the sum of the first n terms of an arithmetic sequence.

6. Find the sum of the first ten terms of the arithmetic sequence: $-7, -14, -21, -28, \ldots$.

7. Find $\displaystyle\sum_{i=1}^{20}(3i - 4)$.

In Exercises 8–9, use the formula for the sum of the first n terms of a geometric sequence.

8. Find the sum of the first ten terms of the geometric sequence: $7, -14, 28, -56, \ldots$.

9. Find $\displaystyle\sum_{i=1}^{15}(-2)^i$.

10. Find the sum of the infinite geometric series:

$$4 + \frac{4}{2} + \frac{4}{2^2} + \frac{4}{2^3} + \cdots.$$

11. Express $0.\overline{73}$ in fractional notation.

12. A job pays \$30,000 for the first year with an annual increase of 4% per year beginning in the second year. What is the total salary paid over an eight-year period? Round to the nearest dollar.

13. Evaluate: $\dbinom{9}{2}$.

14. Use the Binomial Theorem to expand and simplify: $(x^2 - 1)^5$.

15. Use the Binomial Theorem to write the first three terms in the expansion and simplify: $(x + y^2)^8$.

CUMULATIVE REVIEW EXERCISES (CHAPTERS 1–11)

In Exercises 1–10, solve each equation, inequality, or system.

1. $\sqrt{2x + 5} - \sqrt{x + 3} = 2$

2. $(x - 5)^2 = -49$

3. $x^2 + x > 6$

4. $6x - 3(5x + 2) = 4(1 - x)$

5. $\dfrac{2}{x - 3} - \dfrac{3}{x + 3} = \dfrac{12}{x^2 - 9}$

6. $3x + 2 < 4$ and $4 - x > 1$

7. $\begin{cases} 3x - 2y + z = 7 \\ 2x + 3y - z = 13 \\ x - y + 2z = -6 \end{cases}$

8. $\log_9 x + \log_9(x - 8) = 1$

9. $\begin{cases} 2x^2 - 3y^2 = 5 \\ 3x^2 + 4y^2 = 16 \end{cases}$

10. $\begin{cases} 2x^2 - y^2 = -8 \\ x - y = 6 \end{cases}$

In Exercises 11–15, graph each function, equation, or inequality in a rectangular coordinate system.

11. $f(x) = (x + 2)^2 - 4$

12. $y < -3x + 5$

13. $f(x) = 3^{x-2}$

14. $\dfrac{x^2}{16} + \dfrac{y^2}{4} = 1$

15. $x^2 - y^2 = 9$

In Exercises 16–19, perform the indicated operations and simplify, if possible.

16. $\dfrac{2x + 1}{x - 5} - \dfrac{4}{x^2 - 3x - 10}$

17. $\dfrac{\dfrac{1}{x - 1} + 1}{\dfrac{1}{x + 1} - 1}$

18. $\dfrac{6}{\sqrt{5} - \sqrt{2}}$

19. $8\sqrt{45} + 2\sqrt{5} - 7\sqrt{20}$

20. Rationalize the denominator: $\dfrac{5}{\sqrt[3]{2x^2y}}$.

21. Factor completely: $5ax + 5ay - 4bx - 4by$.

22. Write as a single logarithm: $5 \log x - \dfrac{1}{2} \log y$.

23. Solve for p: $\dfrac{1}{p} + \dfrac{1}{q} = \dfrac{1}{f}$.

24. Find the distance between $(6, -1)$ and $(-3, -4)$. Round to two decimal places.

25. Find the indicated sum: $\displaystyle\sum_{i=2}^{5}(i^3 - 4)$.

26. Find the sum of the first 30 terms of the arithmetic sequence: $2, 6, 10, 14, \ldots$.

27. Express $0.\overline{3}$ as a fraction in lowest terms.

28. Use the Binomial Theorem to expand and simplify: $(2x - y^3)^4$.

In Exercises 29–31, find the domain of each function.

29. $f(x) = \dfrac{2}{x^2 + 2x - 15}$

30. $f(x) = \sqrt{2x - 6}$

31. $f(x) = \ln(1 - x)$

32. The length of a rectangular garden is 2 feet more than twice its width. If 22 feet of fencing is needed to enclose the garden, what are its dimensions?

33. With a 6% raise, you will earn $19,610 annually. What is your salary before this raise?

34. The function $F(t) = 1 - k \ln(t + 1)$ models the fraction of people, $F(t)$, who remember all the words in a list of nonsense words t hours after memorizing them. After 3 hours, only half the people could remember all the words. Determine the value of k and then predict the fraction of people in the group who will remember all the words after 6 hours. Round to three decimal places and then express the fraction with a denominator of 1000.

APPENDIX

Where Did That Come From? Selected Proofs

Section 9.4   Properties of Logarithms

The Product Rule

Let b, M, and N be positive real numbers with $b \neq 1$.

$$\log_b(MN) = \log_b M + \log_b N$$

Proof. We begin by letting $\log_b M = R$ and $\log_b N = S$.
Now we write each logarithm in exponential form.

$$\log_b M = R \quad \text{means} \quad b^R = M.$$
$$\log_b N = S \quad \text{means} \quad b^S = N.$$

By substituting and using a property of exponents, we see that

$$MN = b^R b^S = b^{R+S}$$

Now we change $MN = b^{R+S}$ to logarithmic form.

$$MN = b^{R+S} \quad \text{means} \quad \log_b(MN) = R + S.$$

Finally, substituting $\log_b M$ for R and $\log_b N$ for S gives us

$$\log_b(MN) = \log_b M + \log_b N,$$

the property that we wanted to prove.
The quotient and power rules for logarithms are proved using similar procedures.

The Change-of-Base Property

For any logarithmic bases a and b, and any positive number M,

$$\log_b M = \frac{\log_a M}{\log_a b}.$$

Proof. To prove the change-of-base property, we let x equal the logarithm on the left side:

$$\log_b M = x.$$

Now we rewrite this logarithm in exponential form.

$$\log_b M = x \quad \text{means} \quad b^x = M.$$

Because b^x and M are equal, the logarithms with base a for each of these expressions must be equal. This means that

$$\log_a b^x = \log_a M$$

$$x \log_a b = \log_a M \quad \text{Apply the power rule for logarithms on the left side.}$$

$$x = \frac{\log_a M}{\log_a b} \quad \text{Solve for } x \text{ by dividing both sides by } \log_a b.$$

In our first step we let x equal $\log_b M$. Replacing x on the left side by $\log_b M$ gives us

$$\log_b M = \frac{\log_a M}{\log_a b},$$

which is the change-of-base property.

Section 10.2   The Ellipse

The Standard Form of the Equation of an Ellipse with a Horizontal Major Axis Centered at the Origin

Proof. Refer to **Figure A.1**.

$$d_1 + d_2 = 2a \qquad \text{The sum of the distances from } P \text{ to the foci equals a constant, } 2a.$$

$$\sqrt{(x + c)^2 + y^2} + \sqrt{(x - c)^2 + y^2} = 2a \qquad \text{Use the distance formula.}$$

$$\sqrt{(x + c)^2 + y^2} = 2a - \sqrt{(x - c)^2 + y^2} \qquad \text{Isolate a radical.}$$

$$(x + c)^2 + y^2 = 4a^2 - 4a\sqrt{(x - c)^2 + y^2} \qquad \text{Square both sides.}$$
$$+ (x - c)^2 + y^2$$

$$x^2 + 2cx + c^2 + y^2 = 4a^2 - 4a\sqrt{(x - c)^2 + y^2} \qquad \text{Square } x + c \text{ and } x - c.$$
$$+ x^2 - 2cx + c^2 + y^2$$

$$4cx - 4a^2 = -4a\sqrt{(x - c)^2 + y^2} \qquad \text{Simplify and isolate the radical.}$$

$$cx - a^2 = -a\sqrt{(x - c)^2 + y^2} \qquad \text{Divide both sides by 4.}$$

$$(cx - a^2)^2 = a^2[(x - c)^2 + y^2] \qquad \text{Square both sides.}$$

$$c^2x^2 - 2a^2cx + a^4 = a^2(x^2 - 2cx + c^2 + y^2) \qquad \text{Square } cx - a^2 \text{ and } x - c.$$

$$c^2x^2 - 2a^2cx + a^4 = a^2x^2 - 2a^2cx + a^2c^2 + a^2y^2 \qquad \text{Use the distributive property.}$$

$$c^2x^2 + a^4 = a^2x^2 + a^2c^2 + a^2y^2 \qquad \text{Add } 2a^2cx \text{ to both sides.}$$

$$c^2x^2 - a^2x^2 - a^2y^2 = a^2c^2 - a^4 \qquad \text{Rearrange the terms.}$$

$$(c^2 - a^2)x^2 - a^2y^2 = a^2(c^2 - a^2) \qquad \text{Factor out } x^2 \text{ and } a^2, \text{ respectively.}$$

$$(a^2 - c^2)x^2 + a^2y^2 = a^2(a^2 - c^2) \qquad \text{Multiply both sides by } -1.$$

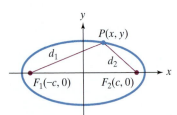

Figure A.1

Refer to the discussion on pages 769–770 and let $b^2 = a^2 - c^2$ in the preceding equation.

$$b^2 x^2 + a^2 y^2 = a^2 b^2$$

$$\frac{x^2}{a^2} + \frac{y^2}{b^2} = 1 \qquad \text{Divide both sides by } a^2 b^2.$$

Section 10.3   The Hyperbola

The Asymptotes of a Hyperbola Centered at the Origin

The hyperbola

$$\frac{x^2}{a^2} - \frac{y^2}{b^2} = 1$$

with a horizontal transverse axis has the two asymptotes

$$y = \frac{b}{a}x \quad \text{and} \quad y = -\frac{b}{a}x.$$

Proof. Begin by solving the hyperbola's equation for y.

$$\frac{x^2}{a^2} - \frac{y^2}{b^2} = 1 \qquad \text{This is the standard form of the equation of a hyperbola.}$$

$$\frac{y^2}{b^2} = \frac{x^2}{a^2} - 1 \qquad \text{We isolate the term involving } y^2 \text{ to solve for } y.$$

$$y^2 = \frac{b^2 x^2}{a^2} - b^2 \qquad \text{Multiply both sides by } b^2.$$

$$y^2 = \frac{b^2 x^2}{a^2}\left(1 - \frac{a^2}{x^2}\right) \qquad \text{Factor out } \frac{b^2 x^2}{a^2} \text{ on the right. Verify that this result is correct by multiplying using the distributive property and obtaining the previous step.}$$

$$y = \pm \sqrt{\frac{b^2 x^2}{a^2}\left(1 - \frac{a^2}{x^2}\right)} \qquad \text{Solve for } y \text{ using the square root property: If } u^2 = d, \text{ then } u = \pm\sqrt{d}.$$

$$y = \pm \frac{b}{a}x\sqrt{1 - \frac{a^2}{x^2}} \qquad \text{Simplify.}$$

As x increases or decreases without bound, the value of $\frac{a^2}{x^2}$ approaches 0. Consequently, the value of y can be approximated by

$$y = \pm\frac{b}{a}x.$$

This means that the lines whose equations are $y = \frac{b}{a}x$ and $y = -\frac{b}{a}x$ are asymptotes for the graph of the hyperbola.

Answers to Selected Exercises

CHAPTER 1

Section 1.1 Check Point Exercises

1. a. $8x + 5$ **b.** $\dfrac{x}{7} - 2x$ **2.** 21.8; At age 10, the average neurotic level is 21.8. **3.** 608 **4. a.** \$27,660 **b.** overestimates by \$978
5. a. true **b.** true **6. a.** -8 is less than -2.; true **b.** 7 is greater than -3.; true **c.** -1 is less than or equal to -4.; false
d. 5 is greater than or equal to 5.; true **e.** 2 is greater than or equal to -14.; true
7. a. $\{x | -2 \le x < 5\}$ **b.** $\{x | 1 \le x \le 3.5\}$ **c.** $\{x | x < -1\}$

Concept and Vocabulary Check

1. variable **2.** expression **3.** b to the nth power; base; exponent **4.** formula; modeling; models **5.** natural **6.** whole **7.** integers
8. rational **9.** irrational **10.** rational; irrational **11.** left **12.** 2; 5; 2; 5 **13.** greater than **14.** less than or equal to

Exercise Set 1.1

1. $x + 5$ **3.** $x - 4$ **5.** $4x$ **7.** $2x + 10$ **9.** $6 - \dfrac{1}{2}x$ **11.** $\dfrac{4}{x} - 2$ **13.** $\dfrac{3}{5 - x}$ **15.** 57 **17.** 10 **19.** $1\dfrac{1}{9}$ or $\dfrac{10}{9}$ **21.** 10 **23.** 44
25. 46 **27.** $\{1, 2, 3, 4\}$ **29.** $\{-7, -6, -5, -4\}$ **31.** $\{8, 9, 10, \ldots\}$ **33.** $\{1, 3, 5, 7, 9\}$ **35.** true **37.** true **39.** false **41.** true
43. false **45.** true **47.** false **49.** -6 is less than -2.; true **51.** 5 is greater than -7.; true **53.** 0 is less than -4.; false
55. -4 is less than or equal to 1.; true **57.** -2 is less than or equal to -6.; false **59.** -2 is less than or equal to -2.; true
61. -2 is greater than or equal to -2.; true **63.** 2 is less than or equal to $-\dfrac{1}{2}$.; false
65. $\{x | 1 < x \le 6\}$ **67.** $\{x | -5 \le x < 2\}$ **69.** $\{x | -3 \le x \le 1\}$
71. $\{x | x > 2\}$ **73.** $\{x | x \ge -3\}$ **75.** $\{x | x < 3\}$
77. $\{x | x < 5.5\}$ **79.** true **81.** false **83.** true **85.** false **87.** false **89.** 4.2 **91.** 0.4
93. 62%; underestimates by 1% **95.** 10°C **97.** 60 ft **117.** does not make sense
119. makes sense **121.** false **123.** true **125.** $2 \cdot 4 + 20 = 28$ and $\dfrac{1}{7}$ of 28 is 4, but $2(4 + 20) = 48$
and $\dfrac{1}{7}$ of 48 is about 6.86. You can't purchase 6.86 birds; the correct translation is $2 \cdot 4 + 20$. **127.** $8 + 2 \cdot (4 - 3) = 10$ **129.** -5 and 5
130. 8 **131.** 34; 34

Section 1.2 Check Point Exercises

1. a. 6 **b.** 4.5 **c.** 0 **2. a.** -28 **b.** 0.7 **c.** $-\dfrac{1}{10}$ **3. a.** 8 **b.** $-\dfrac{1}{3}$ **4. a.** -3 **b.** 10.5 **c.** $-\dfrac{3}{5}$
5. a. 25 **b.** -25 **c.** -64 **d.** $\dfrac{81}{625}$ **6. a.** -8 **b.** $\dfrac{8}{15}$ **7.** 74 **8.** -4 **9.** addition: $9 + 4x$; multiplication: $x \cdot 4 + 9$
10. a. $(6 + 12) + x = 18 + x$ **b.** $(-7 \cdot 4)x = -28x$ **11.** $-28x - 8$ **12.** $14x + 15x^2$ or $15x^2 + 14x$ **13.** $12x - 40$ **14.** $42 - 4x$

Concept and Vocabulary Check

1. negative number **2.** 0 **3.** positive number **4.** positive number **5.** positive number **6.** negative number **7.** positive number
8. divide **9.** subtract **10.** absolute value; 0; a **11.** a; $-a$ **12.** 0; inverse; 0; identity **13.** $b + a$ **14.** $(ab)c$ **15.** $ab + ac$
16. simplified

Exercise Set 1.2

1. 7 **3.** 4 **5.** 7.6 **7.** $\dfrac{\pi}{2}$ **9.** $\sqrt{2}$ **11.** $-\dfrac{2}{5}$ **13.** -11 **15.** -4 **17.** -4.5 **19.** $\dfrac{2}{15}$ **21.** $-\dfrac{35}{36}$ **23.** -8.2 **25.** -12.4
27. 0 **29.** -11 **31.** 5 **33.** 0 **35.** -12 **37.** 18 **39.** -15 **41.** $-\dfrac{1}{4}$ **43.** 5.5 **45.** $\sqrt{2}$ **47.** -90 **49.** 33 **51.** $-\dfrac{15}{13}$
53. 0 **55.** -8 **57.** 48 **59.** 100 **61.** -100 **63.** -8 **65.** 1 **67.** -1 **69.** $\dfrac{1}{8}$ **71.** -3 **73.** 45 **75.** 0 **77.** undefined

79. $\dfrac{9}{14}$ **81.** -15 **83.** -2 **85.** -24 **87.** 45 **89.** $\dfrac{1}{121}$ **91.** 14 **93.** $-\dfrac{8}{3}$ **95.** $-\dfrac{1}{2}$ **97.** 31 **99.** 37

101. addition: $10 + 4x$; multiplication: $x \cdot 4 + 10$ **103.** addition: $-5 + 7x$; multiplication: $x \cdot 7 - 5$ **105.** $(4 + 6) + x = 10 + x$

107. $(-7 \cdot 3)x = -21x$ **109.** $\left(-\dfrac{1}{3} \cdot -3\right)y = y$ **111.** $6x + 15$ **113.** $-14x - 21$ **115.** $-3x + 6$ **117.** $12x$ **119.** $5x^2$

121. $10x + 12x^2$ **123.** $18x - 40$ **125.** $8y - 12$ **127.** $16y - 25$ **129.** $12x^2 + 11$ **131.** $x - (x + 4)$; -4 **133.** $6(-5x)$; $-30x$

135. $5x - 2x$; $3x$ **137.** $8x - (3x + 6)$; $5x - 6$ **139.** -8 **141.** 50 **143.** by 7 **145.** -3 **147.** \$116.8 billion; underestimates by

\$0.2 billion   **149. a.** $1200 - 0.07x$ **b.** \$780   **169.** makes sense   **171.** does not make sense   **173.** false   **175.** false

177. $(8 - 2) \cdot 3 - 4 = 14$ **179.** -7 **180.** $\dfrac{10}{x} - 4x$ **181.** 42 **182.**   **183.** $-5; 0; 3; 4; 3; 0; -5$

184. $-8; -3; 0; 1; 0; -3; -8$ **185.** $3; 2; 1; 0; 1; 2; 3$

Section 1.3 Check Point Exercises

1.   **2.**   **3.**

4. a. 0 to 3 hr   **b.** 3 to 13 hr   **c.** 0.05 mg per 100 ml; after 3 hr   **d.** None of the drug is left in the body.   **5.** minimum x-value: -100; maximum x-value: 100; distance between tick marks on x-axis: 50; minimum y-value: -100; maximum y-value: 100; distance between tick marks on y-axis: 10

Concept and Vocabulary Check

1. x-axis   **2.** y-axis   **3.** origin   **4.** quadrants; four   **5.** x-coordinate; y-coordinate   **6.** solution; satisfies

Exercise Set 1.3

1–9.   **11.**   **13.**   **15.**

17.   **19.**   **21.**   **23.**

25.   **27.** c   **29.** b   **31.** c   **33.** no   **35.** $(2, 0)$ **37.** $(-2, 4)$ and $(1, 1)$ **39.** $y = 2x + 4$

41. $y = 3 - x^2$ **43.**   **45.**

47. 2013; 84%   **49.** 2009 and 2010; 71%   **51.** 2008 and 2009; 3%   **53.** 8 yr old; 1 awakening   **55.** about 1.9 awakenings   **57.** a
59. b   **61.** b   **63.** c   **73.** makes sense   **75.** makes sense   **77.** false   **79.** true   **81.** $15   **83.** 14.3   **84.** 3   **85.** $-14x - 25$
86. true   **87.** $7 - 3x$ **88.** $15x + 5$

Section 1.4 Check Point Exercises

1. 6 or {6}   **2.** -1 or $\{-1\}$ **3.** 6 or {6}   **4.** 1 or {1}   **5.** no solution or $\varnothing$; inconsistent equation
6. $\{x \mid x$ is a real number} or $(-\infty, \infty)$ or $\mathbb{R}$; identity   **7.** the school year ending 2021

Concept and Vocabulary Check

1. linear   **2.** equivalent   **3.** $b + c$ **4.** bc **5.** apply the distributive property   **6.** least common denominator; 12
7. inconsistent; $\varnothing$ **8.** identity; $(-\infty, \infty)$

Exercise Set 1.4

1. 3 or {3}   **3.** 11 or {11}   **5.** 11 or {11}   **7.** 7 or {7}   **9.** 13 or {13}   **11.** 2 or {2}   **13.** -4 or $\{-4\}$ **15.** 9 or {9}

17. -5 or $\{-5\}$ **19.** 6 or {6}   **21.** 19 or {19}   **23.** $\dfrac{5}{2}$ or $\left\{\dfrac{5}{2}\right\}$ **25.** 12 or {12}   **27.** 24 or {24}   **29.** -15 or $\{-15\}$ **31.** 5 or {5}

33. 13 or {13}   **35.** -12 or $\{-12\}$ **37.** $\dfrac{46}{5}$ or $\left\{\dfrac{46}{5}\right\}$ **39.** $\{x \mid x$ is a real number} or $(-\infty, \infty)$ or $\mathbb{R}$; identity   **41.** no solution or $\varnothing$; inconsistent
equation   **43.** 0 or {0}; conditional equation   **45.** -10 or $\{-10\}$; conditional equation   **47.** no solution or $\varnothing$; inconsistent equation
49. 0 or {0}; conditional equation   **51.** $3(x - 4) = 3(2 - 2x)$; 2 or {2}   **53.** $-3(x - 3) = 5(2 - x)$; 0.5 or {0.5}   **55.** 2   **57.** -7 **59.** -2 or $\{-2\}$
61. no solution or $\varnothing$ **63.** 10 or {10}   **65.** -2 or $\{-2\}$ **67. a.** model 1: $31,159; model 2: $31,736; Model 1 underestimates the cost by $542, and
model 2 overestimates the cost by $35.   **b.** the school year ending 2019   **69. a.** $22,000   **b.** $21,809; reasonably well   **c.** $21,726; reasonably well
71. model 2; overestimates by $26   **73.** 2020
87. 5 or {5}   **89.** -5 or $\{-5\}$ **91.** makes sense   **93.** does not make sense   **95.** false
97. false   **101.** 2   **102.** $\dfrac{3}{10}$ **103.** -60 **104.** $y = x^2 - 4$
105. a. $3x - 4 = 32$ **b.** 12
106. $x + 44$
107. $20,000 - 2500x$

Mid-Chapter Check Point Exercises

1. $3x + 10$ **2.** 6 or {6}   **3.** -15 **4.** -7 or $\{-7\}$ **5.** 3   **6.** $13x - 23$ **7.** 0 or {0}   **8.** $-7x - 34$ **9.** 7
10. no solution or $\varnothing$; inconsistent equation   **11.** 3 or {3}   **12.** -4 **13.** $\{x \mid x$ is a real number} or $(-\infty, \infty)$ or $\mathbb{R}$; identity   **14.** 2
15. $\{x \mid -2 \le x < 0\}$ **16.** $\{x \mid x \le 0\}$ **17.** $y = 2x - 1$ **18.** $y = 1 - |x|$ **19.** $y = x^2 + 2$

20. true   **21.** false   **22.** false   **23.** true

Section 1.5 Check Point Exercises

1. associate's degree: $33 thousand; bachelor's degree: $47 thousand; master's degree: $59 thousand   **2.** by 67 years after 1969, or in 2036
3. 20 bridge crossings per month   **4.** $1200   **5.** 50 ft by 94 ft   **6.** $w = \dfrac{P - 2l}{2}$ **7.** $h = \dfrac{V}{lw}$ **8.** $W = 106 + 6H$ or $W = 6H + 106$
9. $C = \dfrac{P}{1 + M}$

Concept and Vocabulary Check

1. $x + 658.6$ **2.** $31 + 2.4x$ **3.** $4 + 0.15x$ **4.** $x - 0.15x$ or $0.85x$ **5.** isolated on one side   **6.** distributive

Exercise Set 1.5

1. 6   **3.** 25   **5.** 120   **7.** 320   **9.** 19 and 45   **11.** 2   **13.** 8   **15.** all real numbers   **17.** goldfish: 43 years; horse: 64 years; human: 122 years
19. 94°, 47°, 39°   **21.** 59°, 60°, 61°   **23.** by 2015   **25.** by 2042   **27. a.** births: 384,000; deaths: 156,000   **b.** 83 million
c. approximately 4 years   **29.** 30 times   **31.** 20 times   **33. a.** 2017; 22,300 students   **b.** $y_1 = 13,300 + 1000x$; $y_2 = 26,800 - 500x$
35. $420   **37.** $150   **39.** $467.20   **41.** 50 yd by 100 yd   **43.** 36 ft by 78 ft   **45.** 2 in.   **47.** 11 min

49. $l = \dfrac{A}{w}$ **51.** $b = \dfrac{2A}{h}$ **53.** $P = \dfrac{I}{rt}$ **55.** $p = \dfrac{T - D}{m}$ **57.** $a = \dfrac{2A}{h} - b$ or $a = \dfrac{2A - hb}{h}$ **59.** $h = \dfrac{3V}{\pi r^2}$ **61.** $m = \dfrac{y - y_1}{x - x_1}$

63. $d_1 = Vt + d_2$ **65.** $x = \dfrac{C - By}{A}$ **67.** $v = \dfrac{2s - at^2}{2t}$ **69.** $n = \dfrac{L - a}{d} + 1$ or $n = \dfrac{L - a + d}{d}$ **71.** $l = \dfrac{A - 2wh}{2w + 2h}$ **73.** $I = \dfrac{E}{R + r}$

81.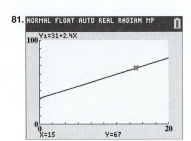

83. does not make sense　　**85.** makes sense　　**87.** false　　**89.** false　　**91.** $200

93. Mrs. Ricardo received $4000; boy received $8000; girl received $2000.　　**95.** $C = \dfrac{LV - SN}{L - N}$

96. $\{x \mid -4 < x \le 0\}$

97. $\dfrac{7}{3}$　　**98.** -8 or $\{-8\}$　　**99. a.** b^7　　**b.** b^{10}　　**c.** Add the exponents.　　**100. a.** b^4

b. b^6　　**c.** Subtract the exponents.　　**101.** -8

Section 1.6 Check Point Exercises

1. a. b^{11}　　**b.** $40x^5 y^{10}$　　**2. a.** $(-3)^3$ or -27　　**b.** $9x^{11}y^3$　　**3. a.** 1　　**b.** 1　　**c.** -1　　**d.** 10　　**e.** 1　　**4. a.** $\dfrac{1}{25}$　　**b.** $-\dfrac{1}{27}$　　**c.** 16

d. $\dfrac{3y^4}{x^6}$　　**5. a.** $\dfrac{4^3}{7^2} = \dfrac{64}{49}$　　**b.** $\dfrac{x^2}{5}$　　**6. a.** x^{15}　　**b.** $\dfrac{1}{y^{14}}$　　**c.** b^{12}　　**7. a.** $16x^4$　　**b.** $-27y^6$　　**c.** $\dfrac{y^2}{16x^{10}}$　　**8. a.** $\dfrac{x^{15}}{64}$　　**b.** $\dfrac{16}{x^{12}y^8}$　　**c.** $x^{15}y^{20}$

9. a. $-12y^9$　　**b.** $\dfrac{4y^{14}}{x^6}$　　**c.** $\dfrac{64}{x^9 y^{15}}$

Concept and Vocabulary Check

1. b^{m+n}; add　　**2.** b^{m-n}; subtract　　**3.** 1　　**4.** $\dfrac{1}{b^n}$　　**5.** false　　**6.** b^n　　**7.** true

Exercise Set 1.6

1. b^{11}　　**3.** x^4　　**5.** 32　　**7.** $6x^6$　　**9.** $20y^{12}$　　**11.** $100x^{10}y^{12}$　　**13.** $21x^5 yz^4$　　**15.** b^9　　**17.** $5x^5$　　**19.** $x^5 y^5$　　**21.** $10xy^3$　　**23.** $-8a^{11}b^8 c^4$

25. 1　　**27.** 1　　**29.** -1　　**31.** 13　　**33.** 1　　**35.** $\dfrac{1}{3^2} = \dfrac{1}{9}$　　**37.** $\dfrac{1}{(-5)^2} = \dfrac{1}{25}$　　**39.** $-\dfrac{1}{5^2} = -\dfrac{1}{25}$　　**41.** $\dfrac{x^2}{y^3}$　　**43.** $\dfrac{8y^3}{x^7}$　　**45.** $5^3 = 125$

47. $(-3)^4 = 81$　　**49.** $\dfrac{y^5}{x^2}$　　**51.** $\dfrac{b^7 c^3}{a^4}$　　**53.** x^{60}　　**55.** $\dfrac{1}{b^{12}}$　　**57.** 7^{20}　　**59.** $64x^3$　　**61.** $9x^{14}$　　**63.** $8x^3 y^6$　　**65.** $9x^4 y^{10}$　　**67.** $-\dfrac{x^6}{27}$

69. $\dfrac{y^8}{25x^6}$　　**71.** $\dfrac{x^{20}}{16y^{16}z^8}$　　**73.** $\dfrac{16}{x^4}$　　**75.** $\dfrac{x^6}{25}$　　**77.** $\dfrac{81x^4}{y^4}$　　**79.** $\dfrac{x^{24}}{y^{12}}$　　**81.** $x^9 y^{12}$　　**83.** $a^8 b^{12}$　　**85.** $\dfrac{1}{x^6}$　　**87.** $-\dfrac{4}{x}$　　**89.** $\dfrac{2}{x^7}$　　**91.** $\dfrac{10a^2}{b^5}$

93. $\dfrac{1}{x^9}$　　**95.** $-\dfrac{6}{a^2}$　　**97.** $\dfrac{3y^4 z^3}{x^7}$　　**99.** $3x^{10}$　　**101.** $\dfrac{1}{x^{10}}$　　**103.** $-\dfrac{5y^8}{x^6}$　　**105.** $\dfrac{8a^9 c^{12}}{b}$　　**107.** x^{16}　　**109.** $-\dfrac{27b^{15}}{a^{18}}$　　**111.** 1　　**113.** $\dfrac{81x^{20}}{y^{32}}$

115. $\dfrac{1}{100a^4 b^{12}c^8}$　　**117.** $10x^2 y^4$　　**119.** $\dfrac{8}{3xy^{10}}$　　**121.** $\dfrac{y^5}{8x^{14}}$　　**123.** $\dfrac{1}{128x^7 y^{16}}$　　**125. a.** 1000 aphids　　**b.** 16,000 aphids　　**c.** 125 aphids

127. a. one person　　**b.** 10 people　　**129. a.** $(0, 1)$　　**b.** $(4, 10)$　　**131.** d　　**133.** 0.55 astronomical unit　　**135.** 1.8 astronomical units

147. makes sense　　**149.** does not make sense　　**151.** false　　**153.** false　　**155.** false　　**157.** true　　**159.** x^{9n}　　**161.** $x^{3n}y^{6n+3}$

162.

$y = 2x - 1$

(2, 3)

(0, −1)　(3, 5)　(1, 1)

(−1, −3)

(−2, −5)

(−3, −7)

163. $y = \dfrac{C - Ax}{B}$　　**164.** 40 m by 75 m　　**165.** It moves the decimal point three places to the right.

166. It moves the decimal point two places to the left.　　**167. a.** 10^5; 100,000　　**b.** 10^6; 1,000,000

Section 1.7 Check Point Exercises

1. a. $-2{,}600{,}000{,}000$　　**b.** 0.000003017　　**2. a.** 5.21×10^9　　**b.** -6.893×10^{-8}　　**3.** 1.8×10^7　　**4. a.** 3.55×10^{-1}　　**b.** 4×10^8

5. $7756　　**6.** 3.1×10^7 mi or 31 million mi

Concept and Vocabulary Check

1. a number greater than or equal to 1 and less than 10; integer　　**2.** true　　**3.** false

Exercise Set 1.7

1. 380　　**3.** 0.0006　　**5.** $-7{,}160{,}000$　　**7.** 1.4　　**9.** 0.79　　**11.** -0.00415　　**13.** $-60{,}000{,}100{,}000$　　**15.** 3.2×10^4　　**17.** 6.38×10^{17}
19. -3.17×10^2　　**21.** -5.716×10^3　　**23.** 2.7×10^{-3}　　**25.** -5.04×10^{-9}　　**27.** 7×10^{-3}　　**29.** 3.14159×10^0　　**31.** 6.3×10^7
33. 6.4×10^4　　**35.** 1.22×10^{-11}　　**37.** 2.67×10^{13}　　**39.** 2.1×10^3　　**41.** 4×10^5　　**43.** 2×10^{-8}　　**45.** 5×10^3　　**47.** 4×10^{15}
49. 9×10^{-3}　　**51.** 6×10^{13} or $\{6 \times 10^{13}\}$　　**53.** -6.2×10^3 or $\{-6.2 \times 10^3\}$　　**55.** 1.63×10^{19} or $\{1.63 \times 10^{19}\}$　　**57.** -3.6×10^5 or $\{-3.6 \times 10^5\}$
59. 7.81×10^{10}　　**61.** 1.07×10^{10}　　**63.** approximately 67 hot dogs per person　　**65.** $2.5 \times 10^2 = 250$ chickens　　**67. a.** 1.0813×10^4; $10,813

b. $901　　**69.** Medicare; $3242　　**71.** 1.06×10^{-18} g　　**73.** 3.1536×10^7　　**81.** does not make sense　　**83.** makes sense　　**85.** false

87. false　　**89.** true　　**91.** 1.25×10^{-15}　　**94.** $85x - 26$　　**95.** 4 or $\{4\}$　　**96.** $\dfrac{y^6}{64x^8}$　　**97.** set 1　　**98.** -170　　**99.** $5a + 5h + 7$

Review Exercises

1. $2x - 10$　　**2.** $6x + 4$　　**3.** $\dfrac{9}{x} + \dfrac{1}{2}x$　　**4.** 34　　**5.** 60　　**6.** 15　　**7.** $\{1, 2\}$　　**8.** $\{-3, -2, -1, 0, 1\}$　　**9.** false　　**10.** true　　**11.** true

12. -5 is less than 2.; true　　**13.** -7 is greater than or equal to -3.; false　　**14.** -7 is less than or equal to -7.; true　　**15.** overestimates by 3.2 million

16. $\{x \mid -2 < x \le 3\}$

17. $\{x \mid -1.5 \le x \le 2\}$

18. $\{x \mid x > -1\}$

19. 9.7 **20.** 5.003 **21.** 0 **22.** -7.6 **23.** -4.4 **24.** 13 **25.** 60 **26.** $-\dfrac{1}{10}$ **27.** $-\dfrac{3}{35}$ **28.** -240 **29.** 16 **30.** -32

31. $-\dfrac{5}{12}$ **32.** 7 **33.** -9.1 **34.** 7 **35.** 9 **36.** -2 **37.** -18 **38.** 55 **39.** 1 **40.** -4 **41.** -13 **42.** $17x - 15$

43. $9x^2 + x$ **44.** $5y - 17$ **45.** $10x$ **46.** $-3x - 8$

47–49.

50. $y = 2x - 2$

51. $y = x^2 - 3$

52. $y = x$

53. $y = |x| - 2$

54. minimum x-value: -20; maximum x-value: 40; distance between tick marks on x-axis: 10; minimum y-value: -5; maximum y-value: 5; distance between tick marks on y-axis: 1

55. 20% **56.** 85 years old **58.** c **59.** 6 or $\{6\}$ **60.** -10 or $\{-10\}$ **61.** 5 or $\{5\}$

62. -13 or $\{-13\}$ **63.** -3 or $\{-3\}$ **64.** -1 or $\{-1\}$ **65.** 2 or $\{2\}$ **66.** 2 or $\{2\}$ **67.** $\dfrac{72}{11}$ or $\left\{\dfrac{72}{11}\right\}$

68. -12 or $\{-12\}$ **69.** $\dfrac{77}{15}$ or $\left\{\dfrac{77}{15}\right\}$ **70.** no solution or $\varnothing$; inconsistent equation

71. $\{x \mid x$ is a real number$\}$ or $(-\infty, \infty)$ or $\mathbb{R}$; identity **72.** 0 or $\{0\}$; conditional equation **73.** $\dfrac{3}{2}$ or $\left\{\dfrac{3}{2}\right\}$; conditional equation

74. no solution or $\varnothing$; inconsistent equation **75. a.** underestimates by 1% **b.** 2020 **76.** engineering: $\$76$ thousand; accounting: $\$63$ thousand; marketing: $\$57$ thousand **77.** $25°, 35°, 120°$ **78. a.** 2018 **b.** $\$1117$ billion **c.** They are shown by the intersection of the graphs at approximately $(2018, 1117)$. **79.** 5 GB **80.** $\$60$ **81.** $\$10,000$ in sales **82.** 44 yd by 126 yd

83. a. $14{,}100 + 1500x = 41{,}700 - 800x$ **b.** 2027; $32{,}100$ **84.** $h = \dfrac{3V}{B}$ **85.** $x = \dfrac{y - y_1}{m} + x_1$ or $x = \dfrac{y - y_1 + mx_1}{m}$

86. $R = \dfrac{E}{I} - r$ or $R = \dfrac{E - Ir}{I}$ **87.** $F = \dfrac{9C + 160}{5}$ or $F = \dfrac{9}{5}C + 32$ **88.** $g = \dfrac{s - vt}{t^2}$ **89.** $g = \dfrac{T}{r + vt}$ **90.** $15x^{13}$ **91.** $\dfrac{x^2}{y^5}$

92. $\dfrac{x^4 y^7}{9}$ **93.** $\dfrac{1}{x^{18}}$ **94.** $49x^6 y^2$ **95.** $-\dfrac{8}{y^7}$ **96.** $-\dfrac{12}{x^7}$ **97.** $3x^{10}$ **98.** $-\dfrac{a^8}{2b^5}$ **99.** $-24x^7 y^4$ **100.** $\dfrac{3}{4}$ **101.** $\dfrac{y^{12}}{125x^6}$

102. $-\dfrac{6x^9}{y^5}$ **103.** $\dfrac{9x^8 y^{14}}{25}$ **104.** $-\dfrac{x^{21}}{8y^{27}}$ **105.** $7{,}160{,}000$ **106.** 0.000107 **107.** -4.1×10^{13} **108.** 8.09×10^{-3} **109.** 1.26×10^8

110. 2.5×10^1 **111.** 2.88×10^{13}

Chapter Test

1. $4x - 5$ **2.** 170 **3.** $\{-4, -3, -2, -1\}$ **4.** true **5.** -3 is greater than -1.; false

6. $\{x \mid -3 \le x < 2\}$

7. $\{x \mid x \le -1\}$

8. underestimates by $\$13$ **9.** 17.9 **10.** -7.6 **11.** $\dfrac{1}{4}$

12. -60 **13.** $\dfrac{1}{8}$ **14.** 3.1 **15.** -3 **16.** 6 **17.** -4

18. $-5x - 18$ **19.** $6y - 27$ **20.** $17x - 22$

21.

22. $y = x^2 - 4$

23. 2 or $\{2\}$ **24.** -6 or $\{-6\}$ **25.** no solution or $\varnothing$; inconsistent equation **26.** 23 and 49

27. 5 yr **28.** 15 bridge crossings per month **29.** $\$50$ **30.** 120 yd by 380 yd

31. $h = \dfrac{3V}{lw}$ **32.** $y = \dfrac{C - Ax}{B}$ **33.** $-\dfrac{14}{x^5}$ **34.** $\dfrac{40}{x^3 y^8}$ **35.** $\dfrac{x^6}{4y^3}$ **36.** $\dfrac{x^{15}}{64y^6}$ **37.** $\dfrac{x^{16}}{9y^{10}}$

38. 0.0000038 **39.** 4.07×10^{11} **40.** 5×10^3 **41.** 1.38×10^{10}

CHAPTER 2

Section 2.1 Check Point Exercises

1. domain: $\{0, 10, 20, 30, 40\}$; range: $\{9.1, 6.7, 10.7, 13.2, 21.2\}$ **2. a.** not a function **b.** function **3. a.** 29 **b.** 65 **c.** 46 **d.** $6a + 6h + 9$
4. a. Every element in the domain corresponds to exactly one element in the range. **b.** domain: $\{0, 1, 2, 3, 4\}$; range: $\{3, 0, 1, 2\}$ **c.** 0
d. 2 **e.** $x = 0$ and $x = 4$

Concept and Vocabulary Check

1. relation; domain; range **2.** function **3.** f; x **4.** r; -2

Exercise Set 2.1

1. function; domain: $\{1, 3, 5\}$; range: $\{2, 4, 5\}$ **3.** not a function; domain: $\{3, 4\}$; range: $\{4, 5\}$ **5.** function; domain: $\{-3, -2, -1, 0\}$;

range: $\{-3, -2, -1, 0\}$ **7.** not a function; domain: {1}; range: {4, 5, 6}   **9. a.** 1   **b.** 6   **c.** -7 **d.** $2a + 1$ **e.** $a + 3$ **11. a.** -2
b. -17 **c.** 0   **d.** $12b - 2$ **e.** $3b + 10$ **13. a.** 5   **b.** 8   **c.** 53   **d.** 32   **e.** $48b^2 + 5$ **15. a.** -1 **b.** 26   **c.** 19
d. $2b^2 + 3b - 1$ **e.** $50a^2 + 15a - 1$ **17. a.** 7   **b.** -7 **c.** 13   **d.** 12   **19. a.** $\dfrac{3}{4}$ **b.** -3 **c.** $\dfrac{11}{8}$ **d.** $\dfrac{13}{9}$ **e.** $\dfrac{2a + 2h - 3}{a + h - 4}$
f. Denominator would be zero.   **21. a.** 6   **b.** 12   **c.** 0   **23. a.** 2   **b.** 1   **c.** -1 and 1   **25.** -2; 10   **27.** -38 **29.** $-2x^3 - 2x$
31. a. -1 **b.** 7   **c.** 19   **d.** 112   **33. a.** {(Iceland, 9.7), (Finland, 9.6), (New Zealand, 9.6), (Denmark, 9.5)}   **b.** Yes; each country
corresponds to exactly one corruption rating.   **c.** {(9.7, Iceland), (9.6, Finland), (9.6, New Zealand), (9.5, Denmark)}   **d.** No; 9.6 in the domain
corresponds to two countries in the range, Finland and New Zealand.   **39.** makes sense   **41.** makes sense   **43.** false   **45.** true   **47.** true

49. 3   **51.** $f(2) = 6; f(3) = 9; f(4) = 12$; no   **52.** 0   **53.** $\dfrac{y^{10}}{9x^4}$ **54.** $\{-15\}$

55.   **56.**

57. a. 3   **b.** -3 and 3   **c.** $(-\infty, \infty)$ **d.** $[1, \infty)$

Section 2.2 Check Point Exercises

1.   ; The graph of g is the graph of f shifted down by 3 units.

2. a. function   **b.** function   **c.** not a function   **3. a.** 400   **b.** 9   **c.** approximately 425
4. a. domain: $[-2, 1]$; range $[0, 3]$ **b.** domain: $(-2, 1]$; range: $[-1, 2)$ **c.** domain: $[-3, 0)$; range: $\{-3, -2, -1\}$

Concept and Vocabulary Check
1. ordered pairs   **2.** more than once; function   **3.** $[1, 3)$; domain   **4.** $[1, \infty)$; range

Exercise Set 2.2

1.

The graph of g is the graph of f
shifted up by 3 units.

3.

The graph of g is the graph of f
shifted down by 1 unit.

5.

The graph of g is the graph of f
shifted up by 1 unit.

7.

The graph of g is the graph of f
shifted down by 2 units.

9.

The graph of g is the graph of f
shifted up by 2 units.

11. function   **13.** not a function   **15.** function   **17.** not a function   **19.** -4 **21.** 4   **23.** 0   **25.** 2   **27.** 2   **29.** -2
31. domain: $[0, 5)$; range: $[-1, 5)$ **33.** domain: $[0, \infty)$; range: $[1, \infty)$ **35.** domain: $[-2, 6]$; range: $[-2, 6]$ **37.** domain: $(-\infty, \infty)$; range: $(-\infty, 2]$
39. domain: $\{-5, -2, 0, 1, 3\}$; range: {2}   **41. a.** $(-\infty, \infty)$ **b.** $[-4, \infty)$ **c.** 4   **d.** 2 and 6   **e.** $(1, 0), (7, 0)$ **f.** $(0, 4)$ **g.** $(1, 7)$
h. positive   **43. a.** 81; In 2010, the wage gap was 81%; (30, 81)   **b.** underestimates by 2%   **45.** 440; For 20-year-old drivers, there are 440
accidents per 50 million miles driven.; (20, 440)   **47.** $x = 45; y = 190$; The minimum number of accidents is 190 per 50 million miles driven and is
attributed to 45-year-old drivers.   **49.** 0.91; It costs \$0.91 to mail a 3-ounce first-class letter.   **51.** \$0.70

57.

The number of annual visits decreases until about age 20 and then increases; (20.3, 4.0): The minimum
is at about age 20 when there are about 4 annual physician visits.   **59.** makes sense   **61.** does not
make sense   **63.** true   **65.** false   **67.** false   **69.** -3 **70.** yes   **71.** $\dfrac{3}{2}$ or $\left\{\dfrac{3}{2}\right\}$

72. 76 yd by 236 yd   **73.** Division by 0 is undefined.   **74.** 19   **75.** $-2x^2 + 42x + 1582$

Section 2.3 Check Point Exercises

1. a. $(-\infty, \infty)$ **b.** $(-\infty, -5)$ or $(-5, \infty)$ **2. a.** $3x^2 + 6x + 6$ **b.** 78   **3. a.** $\dfrac{5}{x} - \dfrac{7}{x-8}$ **b.** $(-\infty, 0)$ or $(0, 8)$ or $(8, \infty)$ **4. a.** 23
b. $x^2 - 3x - 3; 1$ **c.** $\dfrac{x^2 - 2x}{x+3}; \dfrac{7}{2}$ **d.** -24 **5. a.** $(B + D)(x) = -3.2x^2 + 56x + 6406$ **b.** 6545.2 thousand   **c.** overestimates by 7.2 thousand

Concept and Vocabulary Check
1. zero   **2.** negative   **3.** $f(x) + g(x)$ **4.** $f(x) - g(x)$ **5.** $f(x) \cdot g(x)$ **6.** $\dfrac{f(x)}{g(x)}; g(x)$ **7.** $(-\infty, \infty)$ **8.** $(2, \infty)$ **9.** $(0, 3); (3, \infty)$

Exercise Set 2.3
1. $(-\infty, \infty)$ **3.** $(-\infty, -4)$ or $(-4, \infty)$ **5.** $(-\infty, 3)$ or $(3, \infty)$ **7.** $(-\infty, 5)$ or $(5, \infty)$ **9.** $(-\infty, -7)$ or $(-7, 9)$ or $(9, \infty)$
11. a. $5x - 5$ **b.** 20   **13. a.** $3x^2 + x - 5$ **b.** 75   **15. a.** $2x^2 - 2$ **b.** 48
17. $(f + g)(x) = 3x - 3; (f - g)(x) = 7x + 3; (fg)(x) = -10x^2 - 15x; \left(\dfrac{f}{g}\right)(x) = \dfrac{5x}{-2x - 3}$ **19.** $(-\infty, \infty)$ **21.** $(-\infty, 5)$ or $(5, \infty)$
23. $(-\infty, 0)$ or $(0, 5)$ or $(5, \infty)$ **25.** $(-\infty, -3)$ or $(-3, 2)$ or $(2, \infty)$ **27.** $(-\infty, 2)$ or $(2, \infty)$ **29.** $(-\infty, \infty)$ **31.** $x^2 + 3x + 2; 20$ **33.** 0
35. $x^2 + 5x - 2; 48$ **37.** -8 **39.** -16 **41.** -135 **43.** $\dfrac{x^2 + 4x}{2 - x}; 5$ **45.** -1 **47.** $(-\infty, \infty)$ **49.** $(-\infty, 2)$ or $(2, \infty)$ **51.** 5
53. -1 **55.** $[-4, 3]$
57.

59. -4 **61.** -4 **63. a.** $(M + F)(x) = 2.9x + 236$ **b.** 308.5 million   **c.** underestimates by 0.5 million
65. a. $\left(\dfrac{M}{F}\right)(x) = \dfrac{1.5x + 115}{1.4x + 121}$ **b.** 0.968   **c.** overestimates by approximately 0.003

71.

73.

75. No y-value is displayed; y_3 is undefined at $x = 0$.
77. makes sense   **79.** makes sense   **81.** true
83. false   **84.** $b = \dfrac{R - 3a}{3}$ or $b = \dfrac{R}{3} - a$
85. 7 or $\{7\}$ **86.** $6b + 8$ **87. a.** $\dfrac{3}{2}$ **b.** -2

88. a.

b. $(-2, 0)$ **c.** $(0, 4)$ **89.** $y = -\dfrac{5}{3}x - 4$

Mid-Chapter Check Point Exercises
1. not a function; domain: $\{1, 2\}$; range $\{-6, 4, 6\}$ **2.** function; domain: $\{0, 2, 3\}$; range: $\{1, 4\}$ **3.** function; domain: $[-2, 2)$; range: $[0, 3]$
4. not a function; domain: $(-3, 4]$; range: $[-1, 2]$ **5.** not a function; domain: $\{-2, -1, 0, 1, 2\}$; range: $\{-2, -1, 1, 3\}$ **6.** function; domain: $(-\infty, 1]$;
range: $[-1, \infty)$ **7.** No vertical line intersects the graph of f more than once.   **8.** 3   **9.** -2 **10.** -6 and 2   **11.** $(-\infty, \infty)$ **12.** $(-\infty, 4]$
13. $(-\infty, \infty)$ **14.** $(-\infty, -2)$ or $(-2, 2)$ or $(2, \infty)$ **15.** 23   **16.** 23   **17.** $a^2 - 5a - 3$ **18.** $x^2 - 5x + 3; 17$ **19.** $x^2 - x + 13; 33$
20. -36 **21.** $\dfrac{x^2 - 3x + 8}{-2x - 5}; 12$ **22.** $\left(-\infty, -\dfrac{5}{2}\right)$ or $\left(-\dfrac{5}{2}, \infty\right)$

Section 2.4 Check Point Exercises
1. $3x - 2y = 6$

2. a. 6   **b.** $-\dfrac{7}{5}$ **3.** $m = 2; b = -5$ **4.** $y = 4x - 3$

5. $f(x) = -\dfrac{2}{3}x$

6. $y = 3$

7. $x = -3$
8. -0.05; From 2005 to 2010, waste production for each American decreased by 0.05 pound per day each year. The rate of change per is -0.05 pound per day per year.
9. 0.01 mg per 100 ml per hr   **10. a.** $C(x) = 1.33x + 310$ **b.** 443 parts per million

Concept and Vocabulary Check
1. scatter plot; regression   **2.** standard   **3.** x-intercept; zero   **4.** y-intercept; zero   **5.** $\dfrac{y_2 - y_1}{x_2 - x_1}$ **6.** positive   **7.** negative   **8.** zero
9. undefined   **10.** $y = mx + b$; slope; y-intercept   **11.** $(0, 3); 2; 5$ **12.** horizontal   **13.** vertical   **14.** $y; x$

Exercise Set 2.4

1. $x + y = 4$

3. $x + 3y = 6$

5. $6x - 2y = 12$

7. $3x - y = 6$

9. $x - 3y = 9$

11. $2x = 3y + 6$

13. $6x - 3y = 15$

15. $m = 4$; rises   **17.** $m = \dfrac{1}{3}$; rises

19. $m = 0$; horizontal   **21.** $m = -\dfrac{4}{3}$; falls   **23.** $m = \dfrac{5}{2}$; rises   **25.** undefined slope; vertical   **27.** L_1: $\dfrac{2}{3}$; L_2: -2; L_3: $-\dfrac{1}{2}$

29. $m = 2$; $b = 1$;

31. $m = -2$; $b = 1$;

33. $m = \dfrac{3}{4}$; $b = -2$;

35. $m = -\dfrac{3}{5}$; $b = 7$;

37. $m = -\dfrac{1}{2}$; $b = 0$;

39. $m = 0$; $b = -\dfrac{1}{2}$;

41. a. $y = -2x$
b. $m = -2$; $b = 0$
c. $y = -2x$

43. a. $y = \dfrac{4}{5}x$
b. $m = \dfrac{4}{5}$; $b = 0$
c. $y = \dfrac{4}{5}x$

45. a. $y = -3x + 2$
b. $m = -3$; $b = 2$
c. $y = -3x + 2$

47. a. $y = -\dfrac{5}{3}x + 5$
b. $m = -\dfrac{5}{3}$; $b = 5$
c. $y = -\dfrac{5}{3}x + 5$

49. $y = 3$

51. $f(x) = -2$

53. $3y = 18$

55. $f(x) = 2$

57. $x = 5$

59. $3x = -12$

61. $x = 0$

63. $m = -\dfrac{a}{b}$; falls   **65.** undefined slope; vertical   **67.** $m = -\dfrac{A}{B}$; $b = \dfrac{C}{B}$ **69.** -2

71. $3x - 4f(x) = 6$

73. 5   **75.** m_1, m_3, m_2, m_4 **77.** $m = 55.7$; Smartphone sales are increasing by 55.7 million each year.
79. $m = -0.52$; The percentage of U.S. adults who smoke cigarettes is decreasing by 0.52% each year.
81. a. 30%   **b.** 50%   **c.** $m = 4$; average increase of 4% of marriages ending in divorce per year
83. a. 254; If no women in a country are literate, the mortality rate of children under 5 is 254 per thousand.
b. -2.4; For each 1% of adult females who are literate, the mortality rate decreases by 2.4 per thousand.
c. $f(x) = -2.4x + 254$ **d.** 134 per thousand   **85.** $P(x) = 0.24x + 29$

105. $m = 2$;

107. $m = -\dfrac{1}{2}$;

109. does not make sense **111.** does not make sense **113.** false **115.** true **117.** coefficient of x: -6; coefficient of y: 3
119. a. $mx_1 + mx_2 + b$ **b.** $mx_1 + mx_2 + 2b$ **c.** no **120.** $16x^4y^6$ **121.** 3.2×10^{-3} **122.** $3x + 17$ **123.** $y = 7x + 33$
124. $y = -\dfrac{7}{3}x - \dfrac{2}{3}$ **125. a.** $y = -\dfrac{1}{4}x + 2; -\dfrac{1}{4}$ **b.** 4

Section 2.5 Check Point Exercises

1. $y + 3 = -2(x - 4)$; $y = -2x + 5$ or $f(x) = -2x + 5$ **2. a.** $y + 3 = -2(x - 6)$ or $y - 5 = -2(x - 2)$ **b.** $y = -2x + 9$ or $f(x) = -2x + 9$
3. Answers will vary due to rounding.; $f(x) = 0.17x + 72.9$ or $f(x) = 0.17x + 73$; 83.1 yr or 83.2 yr
4. $y - 5 = 3(x + 2)$; $y = 3x + 11$ or $f(x) = 3x + 11$ **5. a.** $m = 3$ **b.** $y + 6 = 3(x + 2)$; $y = 3x$ or $f(x) = 3x$

Concept and Vocabulary Check

1. $y - y_1 = m(x - x_1)$ **2.** equal/the same **3.** -1 **4.** $-\dfrac{1}{5}$ **5.** $\dfrac{5}{3}$ **6.** $-4; -4$ **7.** $\dfrac{1}{2}; -2$

Exercise Set 2.5

1. $y - 5 = 3(x - 2)$; $f(x) = 3x - 1$ **3.** $y - 6 = 5(x + 2)$; $f(x) = 5x + 16$ **5.** $y + 2 = -4(x + 3)$; $f(x) = -4x - 14$
7. $y - 0 = -5(x + 2)$; $f(x) = -5x - 10$ **9.** $y + \dfrac{1}{2} = -1(x + 2)$; $f(x) = -x - \dfrac{5}{2}$ **11.** $y - 0 = \dfrac{1}{4}(x - 0)$; $f(x) = \dfrac{1}{4}x$
13. $y + 4 = -\dfrac{2}{3}(x - 6)$; $f(x) = -\dfrac{2}{3}x$ **15.** $y - 3 = 1(x - 6)$ or $y - 2 = 1(x - 5)$; $f(x) = x - 3$
17. $y - 0 = 2(x + 2)$ or $y - 4 = 2(x - 0)$; $f(x) = 2x + 4$ **19.** $y - 13 = -2(x + 6)$ or $y - 5 = -2(x + 2)$; $f(x) = -2x + 1$
21. $y - 9 = -\dfrac{11}{3}(x - 1)$ or $y + 2 = -\dfrac{11}{3}(x - 4)$; $f(x) = -\dfrac{11}{3}x + \dfrac{38}{3}$ **23.** $y + 5 = 0(x + 2)$ or $y + 5 = 0(x - 3)$; $f(x) = -5$
25. $y - 8 = 2(x - 7)$ or $y - 0 = 2(x - 3)$; $f(x) = 2x - 6$ **27.** $y - 0 = \dfrac{1}{2}(x - 2)$ or $y + 1 = \dfrac{1}{2}(x - 0)$; $f(x) = \dfrac{1}{2}x - 1$
29. a. 5 **b.** $-\dfrac{1}{5}$ **31. a.** -7 **b.** $\dfrac{1}{7}$ **33. a.** $\dfrac{1}{2}$ **b.** -2 **35. a.** $-\dfrac{2}{5}$ **b.** $\dfrac{5}{2}$ **37. a.** -4 **b.** $\dfrac{1}{4}$ **39. a.** $-\dfrac{1}{2}$ **b.** 2
41. a. $\dfrac{2}{3}$ **b.** $-\dfrac{3}{2}$ **43. a.** undefined **b.** 0 **45.** $y - 2 = 2(x - 4)$; $y = 2x - 6$ or $f(x) = 2x - 6$
47. $y - 4 = -\dfrac{1}{2}(x - 2)$; $y = -\dfrac{1}{2}x + 5$ or $f(x) = -\dfrac{1}{2}x + 5$ **49.** $y + 10 = -4(x + 8)$; $y = -4x - 42$ or $f(x) = -4x - 42$
51. $y + 3 = -5(x - 2)$; $y = -5x + 7$ or $f(x) = -5x + 7$ **53.** $y - 2 = \dfrac{2}{3}(x + 2)$; $y = \dfrac{2}{3}x + \dfrac{10}{3}$ or $f(x) = \dfrac{2}{3}x + \dfrac{10}{3}$
55. $y + 7 = -2(x - 4)$; $y = -2x + 1$ or $f(x) = -2x + 1$ **57.** $f(x) = 5$ **59.** $f(x) = -\dfrac{1}{2}x + 1$ **61.** $f(x) = -\dfrac{2}{3}x - 2$
63. $f(x) = 4x - 5$ **65.** $-\dfrac{A}{B}$ **67. a.** $y - 38.9 = 0.89(x - 20)$ or $y - 47.8 = 0.89(x - 30)$ **b.** $f(x) = 0.89x + 21.1$ **c.** 56.7%

69. a. & b.
Number of Smartphones Sold
in the United States

b. $y - 40.8 = 51.16(x - 1)$ or $y - 296.6 = 51.16(x - 6)$: $f(x) = 51.16x - 10.36$ **c.** 552.4 million

71. a. $m \approx 43.1$; The cost of Social Security is projected to increase at a rate of approximately \$43.1 billion per year. **b.** $m \approx 51.4$; The cost of Medicare is projected to increase at a rate of approximately \$51.4 billion per year. **c.** no; The cost of Medicare is projected to increase at a faster rate than the cost of Social Security.

81. a. **b. & d.** **c.**

 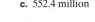

83. makes sense   **85.** makes sense   **87.** true   **89.** true   **91.** -4 **93.** $(-40, 74)$ and $(97, -200)$ **95.** 33   **96.** -56
97. $40°, 60°$, and $80°$ **98. a.** yes   **b.** yes   **99.** $(3, -4)$;   **100.** 1 or $\{1\}$

Review Exercises

1. function; domain: $\{3, 4, 5\}$; range: $\{10\}$ **2.** function; domain: $\{1, 2, 3, 4\}$; range: $\{12, 100, \pi, -6\}$ **3.** not a function; domain: $\{13, 15\}$;
range: $\{14, 16, 17\}$ **4. a.** -5 **b.** 16   **c.** -75 **d.** $14a - 5$ **e.** $7a + 9$ **5. a.** 2   **b.** 52   **c.** 70   **d.** $3b^2 - 5b + 2$
e. $48a^2 - 20a + 2$

6. $f(x) = x^2$
$g(x) = x^2 - 1$

The graph of g is the graph of
f shifted down by 1 unit.

7.

$f(x) = |x|$
$g(x) = |x| + 2$

The graph of g is the graph of
f shifted up by 2 units.

8. not a function   **9.** function   **10.** function   **11.** not a function   **12.** not a function   **13.** function   **14.** -3 **15.** -2
16. 3   **17.** $[-3, 5)$ **18.** $[-5, 0]$
19. a. For each time, there is only one height.
   b. 0; The eagle was on the ground after 15 seconds.
   c. 45 m
   d. 7 and 22; After 7 seconds and after 22 seconds, the eagle's height is 20 meters.
   e. Answers will vary.
20. $(-\infty, \infty)$ **21.** $(-\infty, -8)$ or $(-8, \infty)$ **22.** $(-\infty, 5)$ or $(5, \infty)$ **23. a.** $6x - 4$ **b.** 14
24. a. $5x^2 + 1$ **b.** 46   **25.** $(-\infty, 4)$ or $(4, \infty)$ **26.** $(-\infty, -6)$ or $(-6, -1)$ or $(-1, \infty)$ **27.** $x^2 - x - 5; 1$

28. 1   **29.** $x^2 - 3x + 5; 3$ **30.** 9   **31.** -120 **32.** $\dfrac{x^2 - 2x}{x - 5}; -8$ **33.** $(-\infty, \infty)$ **34.** $(-\infty, 5)$ or $(5, \infty)$

35. $x + 2y = 4$

36. $2x - 3y = 12$

37. $4x = 8 - 2y$

38. 2; rises   **39.** $-\dfrac{2}{3}$; falls

40. undefined; vertical
41. 0; horizontal

42. $m = 2; b = -1$;

$y = 2x - 1$

43. $m = -\dfrac{1}{2}; b = 4$;

$f(x) = -\dfrac{1}{2}x + 4$

44. $m = \dfrac{2}{3}; b = 0$;

$y = \dfrac{2}{3}x$

45. $y = -2x + 4; m = -2; b = 4$ **46.** $y = -\dfrac{5}{3}x; m = -\dfrac{5}{3}; b = 0$ **47.** $y = -\dfrac{5}{3}x + 2; m = -\dfrac{5}{3}; b = 2$

48. $y = 2$ **49.** $7y = -21$ **50.** $f(x) = -4$ **51.** $x = 3$ **52.** $2x = -10$

53. -0.27; Record time has been decreasing at a rate of 0.27 second per year since 1900.   **54. a.** 137; There was an average increase of
approximately 137 discharges per year.   **b.** -130; There was an average decrease of approximately 130 discharges per year.
55. a. $F = \dfrac{9}{5}C + 32$ **b.** $86°$ **56.** $y - 2 = -6(x + 3); y = -6x - 16$ or $f(x) = -6x - 16$ **57.** $y - 6 = 2(x - 1)$ or $y - 2 = 2(x + 1)$;
$y = 2x + 4$ or $f(x) = 2x + 4$ **58.** $y + 7 = -3(x - 4); y = -3x + 5$ or $f(x) = -3x + 5$ **59.** $y - 6 = -3(x + 2); y = -3x$ or $f(x) = -3x$
60. a. $y - 28.2 = 0.2(x - 2)$ or $y - 28.6 = 0.2(x - 4)$ **b.** $f(x) = 0.2x + 27.8$ **c.** 30.2 years

Chapter Test

1. function; domain: $\{1, 3, 5, 6\}$; range: $\{2, 4, 6\}$　**2.** not a function; domain: $\{2, 4, 6\}$; range: $\{1, 3, 5, 6\}$　**3.** $3a + 10$　**4.** 28

5. $g(x) = x^2 + 1$
$f(x) = x^2 - 1$

The graph of g is the graph of f shifted up by 2 units.

6. function　**7.** not a function　**8.** -3　**9.** -2 and 3　**10.** $(-\infty, \infty)$　**11.** $(-\infty, 3]$

12. $(-\infty, 10)$ or $(10, \infty)$　**13.** $x^2 + 5x + 2; 26$　**14.** $x^2 + 3x - 2; -4$　**15.** -15　**16.** $\dfrac{x^2 + 4x}{x + 2}; 3$

17. $(-\infty, -2)$ or $(-2, \infty)$

18. $4x - 3y = 12$

19. $f(x) = -\dfrac{1}{3}x + 2$

20. $f(x) = 4$

21. $-\dfrac{1}{2}$; falls　**22.** undefined; vertical　**23.** 176; In 2005, the number of Super Bowl viewers was 176 million.　**24.** 3.6; The number of Super Bowl viewers is increasing at a rate of 3.6 million per year.　**25.** $y + 3 = 1(x + 1)$ or $y - 2 = 1(x - 4)$; $y = x - 2$ or $f(x) = x - 2$

26. $y - 3 = 2(x + 2)$; $y = 2x + 7$ or $f(x) = 2x + 7$　**27.** $y + 4 = -\dfrac{1}{2}(x - 6)$; $y = -\dfrac{1}{2}x - 1$ or $f(x) = -\dfrac{1}{2}x - 1$

28. a. $y - 0.053 = 0.017(x - 3)$ or $y - 0.121 = 0.017(x - 7)$　**b.** $f(x) = 0.017x + 0.002$　**c.** 0.138

Cumulative Review Exercises

1. $\{0, 1, 2, 3\}$　**2.** false　**3.** 7　**4.** 15　**5.** $7 + 3x$ or $3x + 7$　**6.** -4 or $\{-4\}$　**7.** $\{x \mid x \text{ is a real number}\}$ or $(-\infty, \infty)$ or $\mathbb{R}$; identity　**8.** -6 or $\{-6\}$

9. \$2250　**10.** $t = \dfrac{A - p}{pr}$　**11.** $\dfrac{y^{10}}{9x^8}$　**12.** $\dfrac{9x^{10}}{y^{12}}$　**13.** 2.1×10^{-5}　**14.** function; domain: $\{1, 2, 3, 4, 6\}$; range: $\{5\}$

15.
$g(x) = |x| + 2$
$f(x) = |x| - 1$

The graph of g is the graph of f shifted up by 3 units.

16. $(-\infty, 15)$ or $(15, \infty)$　**17.** $2x^2 + x + 5; 6$

18. $f(x) = -2x + 4$

19. $x - 2y = 6$

20. $y + 5 = 4(x - 3)$; $y = 4x - 17$ or $f(x) = 4x - 17$

CHAPTER 3

Section 3.1 Check Point Exercises

1. a. not a solution　**b.** solution　**2.** $(1, 4)$ or $\{(1, 4)\}$　**3.** $(6, 11)$ or $\{(6, 11)\}$　**4.** $(-2, 5)$ or $\{(-2, 5)\}$　**5.** $\left(-\dfrac{1}{2}, 2\right)$ or $\left\{\left(-\dfrac{1}{2}, 2\right)\right\}$

6. $(2, -1)$ or $\{(2, -1)\}$　**7.** $\left(\dfrac{37}{7}, \dfrac{19}{7}\right)$ or $\left\{\left(\dfrac{37}{7}, \dfrac{19}{7}\right)\right\}$　**8.** no solution or $\varnothing$　**9.** $\{(x, y) \mid x = 4y - 8\}$ or $\{(x, y) \mid 5x - 20y = -40\}$

Concept and Vocabulary Check

1. satisfies both equations in the system　**2.** the intersection point　**3.** $\left\{\left(\dfrac{1}{3}, -2\right)\right\}$　**4.** -2　**5.** -3

6. $\varnothing$; inconsistent; parallel　**7.** $\{(x, y) \mid x = 3y + 2\}$ or $\{(x, y) \mid 5x - 15y = 10\}$; dependent; are identical or coincide

Exercise Set 3.1

1. solution　**3.** not a solution　**5.** solution　**7.** $\{(3, 1)\}$　**9.** $\left\{\left(\dfrac{1}{2}, 3\right)\right\}$　**11.** $\{(4, 3)\}$　**13.** $\{(x, y) \mid 2x + 3y = 6\}$ or $\{(x, y) \mid 4x = -6y + 12\}$

15. $\{(1, 0)\}$　**17.** $\varnothing$　**19.** $\{(1, 2)\}$　**21.** $\{(3, 1)\}$　**23.** $\varnothing$　**25.** $\{(2, 4)\}$　**27.** $\{(3, 1)\}$　**29.** $\{(2, 1)\}$　**31.** $\{(2, -3)\}$　**33.** $\varnothing$

35. $\{(3, -2)\}$　**37.** $\varnothing$　**39.** $\{(-5, -1)\}$　**41.** $\left\{(x, y) \mid y = \dfrac{2}{5}x - 2\right\}$ or $\{(x, y) \mid 2x - 5y = 10\}$　**43.** $\{(5, 2)\}$　**45.** $\{(2, -3)\}$

47. $\{(-1, 1)\}$　**49.** $\{(-2, -7)\}$　**51.** $\{(7, 2)\}$　**53.** $\{(4, -1)\}$　**55.** $\{(x, y) \mid 2x + 6y = 8\}$ or $\{(x, y) \mid 3x + 9y = 12\}$　**57.** $\left\{\left(\dfrac{29}{22}, -\dfrac{5}{11}\right)\right\}$

59. $\{(-2, -1)\}$　**61.** $\{(1, -3)\}$　**63.** $\{(1, -3)\}$　**65.** $\{(4, 3)\}$　**67.** $\{(x, y) \mid x = 3y - 1\}$ or $\{(x, y) \mid 2x - 6y = -2\}$　**69.** $\varnothing$　**71.** $\{(5, 1)\}$

73. $\left\{\left(\dfrac{32}{7}, -\dfrac{20}{7}\right)\right\}$　**75.** $\{(-5, 7)\}$　**77.** $\varnothing$　**79.** $\{(x, y) \mid x + 2y - 3 = 0\}$ or $\{(x, y) \mid 12 = 8y + 4x\}$　**81.** $\{(0, 0)\}$　**83.** $\{(6, -1)\}$

85. $\left\{\left(\dfrac{1}{a}, 3\right)\right\}$　**87.** $m = -4, b = 3$　**89.** $y = x - 4$; $y = -\dfrac{1}{3}x + 4$　**91. a.** 2039; 37%　**b.** by the intersection point (69, 37)

93. a. $y = 0.04x + 5.48$ **b.** $y = 0.17x + 1.84$ **c.** 2028; 6.6%; Medicare   **95. a.** $y = -0.54x + 38$ **b.** $y = -0.79x + 40$
c. 1993; 33.68%   **97. a.** 150 sold; 300 supplied   **b.** $100; 250   **109.** makes sense   **111.** makes sense   **113.** false   **115.** false
117. $a = 3, b = 2$ **119.** $\left\{ \left(\dfrac{b_2c_1 - b_1c_2}{a_1b_2 - a_2b_1}, \dfrac{a_1c_2 - a_2c_1}{a_1b_2 - a_2b_1} \right) \right\}$ **120.** $\left\{ \dfrac{10}{9} \right\}$ **121.** $-128x^{19}y^8$ **122.** 11   **123.** $0.15x + 0.07y$ **124.** 15 mL
125. $80x$

Section 3.2 Check Point Exercises

1. hamburger and fries: 1240; fettuccine Alfredo: 1500   **2.** $3150 at 9%; $1850 at 11%   **3.** 12% solution: 100 oz; 20% solution: 60 oz
4. boat: 35 mph; current: 7 mph   **5. a.** $C(x) = 300,000 + 30x$ **b.** $R(x) = 80x$ **c.** $(6000, 480,000)$; The company will break even when it
produces and sells 6000 pairs of shoes. At this level, both revenue and cost are $480,000.   **6.** $P(x) = 50x - 300,000$

Concept and Vocabulary Check

1. $1180x + 125y$ **2.** $0.12x + 0.09y$ **3.** $0.09x + 0.6y$ **4.** $x + y; x - y$ **5.** $4(x + y)$ **6.** revenue; profit   **7.** break-even point

Exercise Set 3.2

1. 3 and 4   **3.** first number: 2; second number: 5   **5. a.** 1500 units; $48,000   **b.** $P(x) = 17x - 25,500$ **7. a.** 500 units; $122,500
b. $P(x) = 140x - 70,000$ **9.** multiple times per day: 24%; once per day: 17%   **11.** 22 computers and 14 hard drives
13. $2000 at 6% and $5000 at 8%   **15.** first fund: $8000; second fund: $6000   **17.** $17,000 at 12%; $3000 at a 5% loss
19. California: 100 gal; French: 100 gal   **21.** 18-karat gold: 96 g; 12-karat gold: 204 g   **23.** cheaper candy: 30 lb; more expensive candy: 45 lb
25. 8 nickels and 7 dimes   **27.** plane: 130 mph; wind: 30 mph   **29.** crew: 6 km/hr; current: 2 km/hr   **31.** in still water: 4.5 mph; current: 1.5 mph
33. 86 and 74   **35.** $80°, 50°, 50°$ **37.** 70 ft by 40 ft   **39.** two-seat tables: 6; four-seat tables: 11   **41.** 500 radios
43. −6000; When the company produces and sells 200 radios, the loss is $6000.   **45. a.** $P(x) = 20x - 10,000$ **b.** $190,000
47. a. $C(x) = 18,000 + 20x$ **b.** $R(x) = 80x$ **c.** $(300, 24,000)$; When 300 canoes are produced and sold, both revenue and cost are $24,000.
49. a. $C(x) = 30,000 + 2500x$ **b.** $R(x) = 3125x$ **c.** $(48, 150,000)$; For 48 sold-out performances, both cost and revenue are $150,000.

59. $(6, 300)$;

63. does not make sense   **65.** makes sense   **67.** yes, 8 hexagons and 4 squares
69. 95   **71.** $y - 5 = -2(x + 2)$ or $y - 13 = -2(x + 6)$; $y = -2x + 1$ or $f(x) = -2x + 1$
72. $y - 0 = 1(x + 3)$; $y = x + 3$ or $f(x) = x + 3$ **73.** $(-\infty, 3)$ or $(3, \infty)$ **74.** yes
75. $11x + 4y = -3$ **76.** $16a + 4b + c = 1682$

Section 3.3 Check Point Exercises

1. $(-1) - 2(-4) + 3(5) = 22; 2(-1) - 3(-4) - 5 = 5; 3(-1) + (-4) - 5(5) = -32$ **2.** $(1, 4, -3)$ or $\{(1, 4, -3)\}$
3. $(4, 5, 3)$ or $\{(4, 5, 3)\}$ **4.** $y = 3x^2 - 12x + 13$ or $f(x) = 3x^2 - 12x + 13$

Concept and Vocabulary Check

1. triple; all   **2.** $-2; -4$ **3.** z; add Equations 1 and 3   **4.** quadratic   **5.** curve fitting

Exercise Set 3.3

1. not a solution   **3.** solution   **5.** $\{(2, 3, 3)\}$ **7.** $\{(2, -1, 1)\}$ **9.** $\{(1, 2, 3)\}$ **11.** $\{(3, 1, 5)\}$ **13.** $\{(1, 0, -3)\}$ **15.** $\{(1, -5, -6)\}$

17. no solution or $\varnothing$ **19.** infinitely many solutions; dependent equations   **21.** $\left\{ \left(\dfrac{1}{2}, \dfrac{1}{3}, -1 \right) \right\}$ **23.** $y = 2x^2 - x + 3$

25. $y = 2x^2 + x - 5$ **27.** 7, 4, and 5   **29.** $\{(4, 8, 6)\}$ **31.** $y = -\dfrac{3}{4}x^2 + 6x - 11$ **33.** $\left\{ \left(\dfrac{8}{a}, -\dfrac{3}{b}, -\dfrac{5}{c} \right) \right\}$ **35. a.** $(0, 5), (50, 31), (100, 15)$

b. $\begin{cases} 0a + 0b + c = 5 \\ 2500a + 50b + c = 31 \\ 10,000a + 100b + c = 15 \end{cases}$ **37. a.** $y = -16x^2 + 40x + 200$ **b.** 0; After 5 seconds, the ball hits the ground.   **39.** chemical engineering:

22 hours; mathematics: 16 hours; psychology: 14 hours   **41.** $1200 at 8%; $2000 at 10%; $3500 at 12%   **43.** 200 $8 tickets; 150 $10 tickets;
50 $12 tickets   **45.** 4 oz of food A; 0.5 oz of food B; 1 oz of food C   **55.** does not make sense   **57.** makes sense   **59.** false   **61.** false
63. 13 triangles, 21 rectangles, and 6 pentagons

65. $f(x) = -\dfrac{3}{4}x + 3$ **66.** $-2x + y = 6$ **67.** $f(x) = -5$

68. $\{(-3, 1)\}$; The value for y is given and the value for x can be found by back-substitution.   **69.** $\{(6, 3, 5)\}$; The value for z is given and the values of
the other variables can be found by back-substitution.   **70.** $\begin{bmatrix} 1 & 2 & -1 \\ 0 & -11 & -11 \end{bmatrix}$

Mid-Chapter Check Point Exercises

1. $\{(-1, 2)\}$ **2.** $\{(1, -2)\}$ **3.** $\{(6, 10)\}$ **4.** $\{(x, y)|y = 4x - 5\}$ or $\{(x, y)|8x - 2y = 10\}$ **5.** $\left\{\left(\dfrac{11}{19}, \dfrac{7}{19}\right)\right\}$ **6.** $\varnothing$ **7.** $\{(-1, 2, -2)\}$

8. $\{(4, -2, 3)\}$ **9.** $\{(3, 2)\}$ **10.** $\{(-2, 1)\}$ **11. a.** $C(x) = 400{,}000 + 20x$ **b.** $R(x) = 100x$ **c.** $P(x) = 80x - 400{,}000$

d. $(5000, 500{,}000)$; The company will break even when it produces and sells 5000 PDAs. At this level, both revenue and cost are $500,000.

12. 6 roses and 14 carnations **13.** $6300 at 5% and $8700 at 6% **14.** 13% nitrogen: 20 gal; 18% nitrogen: 30 gal

15. rowing rate in still water: 3 mph; current: 1.5 mph **16.** $4500 at 2% and $3500 at 5% **17.** $y = -x^2 + 2x + 3$

18. 8 nickels, 6 dimes, 12 quarters

Section 3.4 Check Point Exercises

1. a. $\begin{bmatrix} 1 & 6 & -3 & 7 \\ 4 & 12 & -20 & 8 \\ -3 & -2 & 1 & -9 \end{bmatrix}$ **b.** $\begin{bmatrix} 1 & 3 & -5 & 2 \\ 1 & 6 & -3 & 7 \\ -3 & -2 & 1 & -9 \end{bmatrix}$ **c.** $\begin{bmatrix} 4 & 12 & -20 & 8 \\ 1 & 6 & -3 & 7 \\ 0 & 16 & -8 & 12 \end{bmatrix}$ **2.** $(-1, 2)$ or $\{(-1, 2)\}$ **3.** $(5, 2, 3)$ or $\{(5, 2, 3)\}$

Concept and Vocabulary Check

1. matrix; elements

2. $\begin{bmatrix} 3 & -2 & -6 \\ 4 & 5 & -8 \end{bmatrix}$ **3.** $\begin{bmatrix} 2 & 1 & 4 & -4 \\ 3 & 0 & 1 & 1 \\ 4 & 3 & 1 & 8 \end{bmatrix}$ **4.** first; $\dfrac{1}{2}$ **5.** 3; second; –2; third **6.** false **7.** true

Exercise Set 3.4

1. $\begin{bmatrix} 1 & -\frac{3}{2} & 5 \\ 2 & 2 & 5 \end{bmatrix}$ **3.** $\begin{bmatrix} 1 & -\frac{4}{3} & 2 \\ 3 & 5 & -2 \end{bmatrix}$ **5.** $\begin{bmatrix} 1 & -3 & 5 \\ 0 & 12 & -6 \end{bmatrix}$ **7.** $\begin{bmatrix} 1 & -\frac{3}{2} & \frac{7}{2} \\ 0 & \frac{17}{2} & -\frac{17}{2} \end{bmatrix}$ **9.** $\begin{bmatrix} 1 & -3 & 2 & 5 \\ 1 & 5 & -5 & 0 \\ 3 & 0 & 4 & 7 \end{bmatrix}$ **11.** $\begin{bmatrix} 1 & -3 & 2 & 0 \\ 0 & 10 & -7 & 7 \\ 2 & -2 & 1 & 3 \end{bmatrix}$

13. $\begin{bmatrix} 1 & 1 & -1 & 6 \\ 0 & -3 & 3 & -15 \\ 0 & -4 & 2 & -14 \end{bmatrix}$ **15.** $\{(4, 2)\}$ **17.** $\{(3, -3)\}$ **19.** $\{(2, -5)\}$ **21.** no solution or $\varnothing$ **23.** infinitely many solutions;

dependent equations **25.** $\{(1, -1, 2)\}$ **27.** $\{(3, -1, -1)\}$ **29.** $\{(1, 2, -1)\}$ **31.** $\{(1, 2, 3)\}$ **33.** no solution or $\varnothing$

35. infinitely many solutions; dependent equations **37.** $\{(-1, 2, -2)\}$ **39.** $\begin{cases} w - x + y + z = 3 \\ x - 2y - z = 0 \\ y + 6z = 17 \\ z = 3 \end{cases}$; $\{(2, 1, -1, 3)\}$ **41.** $\begin{bmatrix} 1 & -1 & 1 & 1 & 3 \\ 0 & 1 & -2 & -1 & 0 \\ 0 & 2 & 1 & 2 & 5 \\ 0 & 6 & -3 & -1 & -9 \end{bmatrix}$

43. $\{(1, 2, 3, -2)\}$ **45. a.** $s(t) = -16t^2 + 56t$ **b.** 0; The ball hits the ground 3.5 seconds after it is thrown.; $(3.5, 0)$

47. yes: 34%; no: 61%; not sure: 5% **59.** makes sense **61.** does not make sense **63.** false **65.** false **67.** $-6a + 13$ **68.** 15

69. $-\dfrac{x^{11}}{3y^{36}}$ **70.** 2 **71.** 6 **72.** –31

Section 3.5 Check Point Exercises

1. a. -4 **b.** -17 **2.** $(4, -2)$ or $\{(4, -2)\}$ **3.** 80 **4.** $(2, -3, 4)$ or $\{(2, -3, 4)\}$

Concept and Vocabulary Check

1. $5 \cdot 3 - 2 \cdot 4 = 15 - 8 = 7$; determinant; 7 **2.** $x = \dfrac{\begin{vmatrix} 8 & 1 \\ -2 & -1 \end{vmatrix}}{\begin{vmatrix} 1 & 1 \\ 1 & -1 \end{vmatrix}}; y = \dfrac{\begin{vmatrix} 1 & 8 \\ 1 & -2 \end{vmatrix}}{\begin{vmatrix} 1 & 1 \\ 1 & -1 \end{vmatrix}}$ **3.** $3\begin{vmatrix} 3 & 1 \\ 1 & 1 \end{vmatrix} - 4\begin{vmatrix} 2 & 1 \\ 1 & 1 \end{vmatrix} + 5\begin{vmatrix} 2 & 1 \\ 3 & 1 \end{vmatrix}$ **4.** $\dfrac{\begin{vmatrix} 3 & -8 & 4 \\ 2 & 11 & -2 \\ 1 & 4 & -2 \end{vmatrix}}{\begin{vmatrix} 3 & 1 & 4 \\ 2 & 3 & -2 \\ 1 & -3 & -2 \end{vmatrix}}$

Exercise Set 3.5

1. 1 **3.** -29 **5.** 0 **7.** 33 **9.** $-\dfrac{7}{16}$ **11.** $\{(5, 2)\}$ **13.** $\{(2, -3)\}$ **15.** $\{(3, -1)\}$ **17.** $\{(4, 0)\}$ **19.** $\{(4, 2)\}$ **21.** $\{(7, 4)\}$

23. inconsistent; no solution or $\varnothing$ **25.** dependent equations; infinitely many solutions **27.** 72 **29.** -75 **31.** 0 **33.** $\{(-5, -2, 7)\}$

35. $\{(2, -3, 4)\}$ **37.** $\{(3, -1, 2)\}$ **39.** $\{(2, 3, 1)\}$ **41.** -42 **43.** $\begin{cases} 2x - 4y = 8 \\ 3x + 5y = -10 \end{cases}$ **45.** $\{-11\}$ **47.** $\{4\}$ **49.** 28 sq units **51.** yes

53. $\begin{vmatrix} x & y & 1 \\ 3 & -5 & 1 \\ -2 & 6 & 1 \end{vmatrix} = 0; y = -\dfrac{11}{5}x + \dfrac{8}{5}$ **65.** does not make sense **67.** does not make sense **69.** true **71.** false

73. The value is multiplied by -1. **76.** $(-\infty, \infty)$ **77.** $y = \dfrac{2x + 7}{3}$ **78.** $\{0\}$ **79.** $\{14\}$ **80.** $\{-3\}$ **81.** $(-\infty, \infty)$

Review Exercises

1. not a solution **2.** solution **3.** $\{(2, 3)\}$ **4.** $\{(x, y)|3x - 2y = 6\}$ or $\{(x, y)|6x - 4y = 12\}$ **5.** $\{(-5, -6)\}$ **6.** $\varnothing$ **7.** $\{(3, 4)\}$

8. $\{(23, -43)\}$ **9.** $\{(-4, 2)\}$ **10.** $\left\{\left(3, \dfrac{1}{2}\right)\right\}$ **11.** $\{(x, y)|y = 4 - x\}$ or $\{(x, y)|3x + 3y = 12\}$ **12.** $\left\{\left(3, \dfrac{8}{3}\right)\right\}$ **13.** $\varnothing$

14. TV: $350; stereo: $370 **15.** $2500 at 4%; $6500 at 7% **16.** 10 mL of 34%; 90 mL of 4% **17.** plane: 630 mph; wind 90 mph
18. 12 ft by 5ft **19.** loss of $4500 **20.** (500, 42,500); When 500 calculators are produced and sold, both cost and revenue are $42,500.
21. $P(x) = 45x - 22,500$ **22. a.** $C(x) = 60,000 + 200x$ **b.** $R(x) = 450x$ **c.** (240, 108,000); When 240 desks are produced and both cost and revenue are $108,000. **23.** no **24.** $\{(0, 1, 2)\}$ **25.** $\{(2, 1, -1)\}$ **26.** infinitely many solutions; dependent equations

27. $y = 3x^2 - 4x + 5$ **28.** war: 124 million; famine: 111 million; tobacco: 71 million **29.** $\begin{bmatrix} 1 & -8 & | & 3 \\ 0 & 1 & | & -2 \end{bmatrix}$ **30.** $\begin{bmatrix} 1 & -3 & | & 1 \\ 0 & 7 & | & -7 \end{bmatrix}$

31. $\begin{bmatrix} 1 & -1 & \frac{1}{2} & | & -\frac{1}{2} \\ 1 & 2 & -1 & | & 2 \\ 6 & 4 & 3 & | & 5 \end{bmatrix}$ **32.** $\begin{bmatrix} 1 & 2 & 2 & | & 2 \\ 0 & 1 & -1 & | & 2 \\ 0 & 0 & 9 & | & -9 \end{bmatrix}$ **33.** $\{(-5, 3)\}$ **34.** no solution or $\varnothing$ **35.** $\{(1, 3, -4)\}$ **36.** $\{(-2, -1, 0)\}$

37. 17 **38.** 4 **39.** -86 **40.** -236 **41.** $\left\{\left(\frac{7}{4}, -\frac{25}{8}\right)\right\}$ **42.** $\{(2, -7)\}$ **43.** $\{(23, -12, 3)\}$ **44.** $\{(-3, 2, 1)\}$

45. $y = \frac{5}{8}x^2 - 50x + 1150$; 30-year-old drivers are involved in 212.5 accidents daily and 50-year-old drivers are involved in 212.5 accidents daily.

Chapter Test

1. $\{(2, 4)\}$ **2.** $\{(6, -5)\}$ **3.** $\{(1, -3)\}$ **4.** $\{(x, y) | 4x = 2y + 6\}$ or $\{(x, y) | y = 2x - 3\}$ **5.** one-bedroom: 15 units; two-bedroom: 35 units
6. $2000 at 6% and $7000 at 7% **7.** 6% solution: 12 oz; 9% solution: 24 oz **8.** boat: 14 mph; current: 2 mph **9.** $C(x) = 360,000 + 850x$
10. $R(x) = 1150x$ **11.** (1200, 1,380,000); When 1200 computers are produced and sold, both cost and revenue are $1,380,000.
12. $P(x) = 85x - 350,000$ **13.** $\{(1, 3, 2)\}$ **14.** $\begin{bmatrix} 1 & 0 & -4 & | & 5 \\ 0 & -1 & 26 & | & -20 \\ 2 & -1 & 4 & | & -3 \end{bmatrix}$ **15.** $\{(4, -2)\}$ **16.** $\{(-1, 2, 2)\}$ **17.** 17 **18.** -10
19. $\{(-1, -6)\}$ **20.** $\{(4, -3, 3)\}$

Cumulative Review Exercises

1. 1 **2.** $15x - 7$ **3.** $\{-6\}$ **4.** $\{-15\}$ **5.** no solution or $\varnothing$ **6.** $2000 **7.** $-\dfrac{x^8}{4y^{30}}$ **8.** $-4a - 3$ **9.** $(-\infty, -3)$ or $(-3, \infty)$

10. $x^2 - 3x - 1$; -1

11. $f(x) = -\frac{2}{3}x + 2$ **12.** $2x - y = 6$

13. $y - 4 = -3(x - 2)$ or $y + 2 = -3(x - 4)$; $y = -3x + 10$ or $f(x) = -3x + 10$
14. $y - 0 = -3(x + 1)$; $y = -3x - 3$ or $f(x) = -3x - 3$ **15.** $\left\{\left(7, \frac{1}{3}\right)\right\}$
16. $\{(3, 2, 4)\}$ **17.** pad: $0.80; pen: $0.20 **18.** 23 **19.** $\{(3, -2, 1)\}$
20. $\{(-3, 2)\}$

CHAPTER 4

Section 4.1 Check Point Exercises

1. $(-5, \infty)$ **2.** $(-\infty, 4)$ **3.** $[13, \infty)$

4. a. $(-\infty, \infty)$ **b.** $\varnothing$ **5.** more than 720 miles

Concept and Vocabulary Check

1. $< b + c$ **2.** $< bc$ **3.** $> bc$ **4.** adding 4; dividing; -3; direction; $>$; $<$ **5.** $\varnothing$ **6.** $(-\infty, \infty)$ **7.** $x \geq 7$ **8.** $x \leq 7$
9. $x \leq 7$ **10.** $x \geq 7$

Exercise Set 4.1

1. $(-\infty, 3)$ **3.** $[7, \infty)$ **5.** $(-\infty, -4]$

7. $\left(-\infty, -\dfrac{2}{5}\right]$ **9.** $[0, \infty)$ **11.** $(-\infty, 1)$

13. $[6, \infty)$ **15.** $[-6, \infty)$ **17.** $(-\infty, -6)$

19. $[13, \infty)$ **21.** $(-\infty, \infty)$ **23.** $(-\infty, \infty)$

25. $\varnothing$ **27.** $(-6, \infty)$ **29.** $[-1, \infty)$

31. $(-\infty, -2)$

33. $(-\infty, 5)$ **35.** $\left[-\dfrac{4}{7}, \infty\right)$ **37.** $[6, \infty)$ **39.** $(-\infty, 2)$

41. $x < \dfrac{c - b}{a}$ **43.** $(-\infty, -3]$ **45.** $(-1.4, \infty)$ **47.** $(0, 4)$ **49.** intimacy $\geq$ passion or passion $\leq$ intimacy

51. commitment > passion or passion < commitment **53.** 9; after 3 years **55. a.** It's exact. **b.** $x > 40$; years after 2020

57. a. -8; Per-capita consumption of carbonated soda is decreasing by eight 12-ounce servings per year on average. **b.** years after 2019

59. more than 100 miles per day **61.** greater than \$32,000 **63.** more than 6250 DVDs **65.** 40 bags or fewer

75.

$(-\infty, -3)$ **77.** $\varnothing$

79. a. plan A: $4 + 0.10x$; plan B: $2 + 0.15x$

b.

c–d. more than 40 checks per month

81. makes sense **83.** makes sense **85.** false **87.** true **89.** Since $x > y$, $y - x < 0$. Thus, when both sides were multiplied by $y - x$, the sense of the inequality should have been changed. **90.** 29 **91.** $\{(-1, -1, 2)\}$ **92.** $\dfrac{x^9}{8y^{15}}$ **93. a.** $\{3, 4\}$ **b.** $\{1, 2, 3, 4, 5, 6, 7\}$ **94. a.** $(-\infty, 8)$ **b.** $(-\infty, 5)$ **c.** any number less than 5 **d.** any number in $[5, 8)$ **95. a.** $[1, \infty)$ **b.** $[3, \infty)$ **c.** any number greater than or equal to 3 **d.** any number in $[1, 3)$

Section 4.2 Check Point Exercises

1. $\{3, 7\}$ **2.** $(-\infty, 1)$ **3.** $\varnothing$ **4.** $[-1, 4)$;

5. $\{3, 4, 5, 6, 7, 8, 9\}$ **6.** $(-\infty, 1] \cup (3, \infty)$ **7.** $(-\infty, \infty)$

Concept and Vocabulary Check

1. intersection; $A \cap B$ **2.** union; $A \cup B$ **3.** $(-\infty, 9)$ **4.** $(-\infty, 12)$ **5.** middle

Exercise Set 4.2

1. $\{2, 4\}$ **3.** $\varnothing$ **5.** $\varnothing$

7. $(6, \infty)$

9. $(-\infty, 1]$

11. $[-1, 2)$

13. $\varnothing$

15. $(-6, -4)$

17. $(-3, 6]$

19. $(2, 5)$

21. $\varnothing$

23. $[0, 2)$

25. $(3, 5)$

27. $[-1, 3)$

29. $(-5, -2]$

31. $[3, 6)$

33. $\{1, 2, 3, 4, 5\}$ **35.** $\{1, 2, 3, 4, 5, 6, 7, 8, 10\}$ **37.** $\{a, e, i, o, u\}$

39. $(3, \infty)$

41. $(-\infty, 5]$

43. $(-\infty, \infty)$

45. $(-\infty, -1) \cup [2, \infty)$

47. $(-\infty, -3) \cup (4, \infty)$

49. $(-\infty, 1] \cup [3, \infty)$

51. $(-\infty, \infty)$

53. $(-\infty, 2)$

55. $(4, \infty)$ **57.** $(-\infty, 0) \cup (6, \infty)$

59. $\dfrac{b - c}{a} < x < \dfrac{b + c}{a}$ **61.** $[-1, 3]$

63. $(-1, 3)$ **65.** $[-1, 2)$ **67.** $\{-3, -2, -1\}$

69. a. years after 2016 **b.** years after 2020 **c.** years after 2020 **d.** years after 2016 **71.** $[5°C, 10°C]$

73. $[76, 126)$; If the highest grade is 100, then $[76, 100]$. **75.** more than 3 and less than 15 crossings per 3-month period

83. $(-2, 6)$;

85. $\left[2, \dfrac{5}{2}\right]$;

87. Exercise 83:

Exercise 85:

89. makes sense **91.** makes sense **93.** false **95.** false **97.** $(-\infty, 4]$ **99.** $[-1, 4]$ **101.** least: 4 nickels; greatest: 7 nickels

102. $-x^2 + 5x - 9$; -15 **103.** $f(x) = -\dfrac{1}{2}x + 4$ **104.** $17 - 2x$ **105.** $-\dfrac{1}{2}$ and 1 **106.** -1 and 3

107. a. -5 satisfies the inequality. **b.** no

Section 4.3 Check Point Exercises

1. -2 and 3, or $\{-2, 3\}$ **2.** $-\dfrac{13}{3}$ and 5, or $\left\{-\dfrac{13}{3}, 5\right\}$ **3.** $\dfrac{4}{3}$ and 10, or $\left\{\dfrac{4}{3}, 10\right\}$

4. $(-3, 7)$

5. $\left[-\dfrac{11}{5}, 3\right]$;

6. $(-\infty, 1] \cup [4, \infty)$

7. $[37.8, 44.2]$; The percentage of U.S. adults in the population who dread going to the dentist is between a low of 37.8% and a high of 44.2%.

Concept and Vocabulary Check

1. c; $-c$ **2.** v; $-v$ **3.** $-c$; c **4.** $-c$; c **5.** $<$ **6.** $<$ **7.** C **8.** E **9.** A **10.** B **11.** D **12.** F

Exercise Set 4.3

1. $\{-8, 8\}$ **3.** $\{-5, 9\}$ **5.** $\{-3, 4\}$ **7.** $\{-1, 2\}$ **9.** $\varnothing$ **11.** $\{-3\}$ **13.** $\{-11, -1\}$ **15.** $\{-3, 4\}$ **17.** $\left\{-\dfrac{13}{3}, 5\right\}$ **19.** $\left\{-\dfrac{2}{5}, \dfrac{2}{5}\right\}$

21. $\varnothing$ **23.** $\varnothing$ **25.** $\left\{\dfrac{1}{2}\right\}$ **27.** $\left\{\dfrac{3}{4}, 5\right\}$ **29.** $\left\{\dfrac{5}{3}, 3\right\}$ **31.** $\{0\}$ **33.** $\{4\}$ **35.** $\{4\}$ **37.** $\{-1, 15\}$

39. $(-3, 3)$

41. $(1, 3)$

43. $[-3, -1]$

45. $(-1, 7)$

47. $(-\infty, -3) \cup (3, \infty)$

49. $(-\infty, -4) \cup (-2, \infty)$

51. $(-\infty, 2] \cup [6, \infty)$

53. $\left(-\infty, \dfrac{1}{3}\right) \cup (5, \infty)$

55. $[-5, 3]$

57. $(-6, 0)$

59. $(-\infty, -5] \cup [3, \infty)$

61. $(-\infty, -3) \cup (12, \infty)$

63. $\varnothing$

65. $(-\infty, \infty)$

67. $[-9, 5]$

69. $(-\infty, 1) \cup (2, \infty)$

71. $(-\infty, -3) \cup (5, \infty)$

73. $(-\infty, -1] \cup [2, \infty)$

75. $-\dfrac{3}{2}$ and 4 **77.** -7 and -2 **79.** $\left[-\dfrac{7}{3}, 1\right]$ **81.** $(-\infty, -1) \cup (4, \infty)$ **83.** $\left(-\infty, -\dfrac{1}{3}\right] \cup [3, \infty)$ **85.** $\left(\dfrac{-c-b}{a}, \dfrac{c-b}{a}\right)$ **87.** $\{3, 5\}$

89. $[-2, 1]$ **91.** $[18, 24]$; The percentage of interviewers in the population turned off by the job applicant being arrogant is between a low of 18% and a high of 24%. **93.** $[50, 64]$; The monthly average temperature for San Francisco, CA, is between a low of 50°F and a high of 64°F. **95.** $[8.59, 8.61]$; A machine part that is supposed to be 8.6 cm is acceptable between a low of 8.59 and a high of 8.61 cm. **97.** If the number of outcomes that result in heads is 41 or less or 59 or more, then the coin is unfair.

105. $\{-6, 4\}$;

107. $\{2, 3\}$;

109. $(-2, 3)$;

111. $(-\infty, -3) \cup (4, \infty)$;

113. $(-\infty, \infty)$;

115. does not make sense
117. does not make sense **119.** false
121. true **123. a.** $|x - 4| < 3$
b. $|x - 4| \geq 3$ **125.** $\{1\}$

126.

127. $f(x) = -\dfrac{2}{3}x$

128. $f(x) = -2$

Mid-Chapter Check Point Exercises

1. $(-\infty, -4]$ **2.** $[3, 5)$ **3.** $\left\{\dfrac{1}{2}, 3\right\}$ **4.** $(-\infty, -1)$ **5.** $(-\infty, -9) \cup (-5, \infty)$ **6.** $\left[-\dfrac{2}{3}, 2\right]$ **7.** $\left\{\dfrac{1}{2}, \dfrac{13}{4}\right\}$ **8.** $(-4, -2]$ **9.** $(-\infty, \infty)$

10. $(-\infty, -3]$ **11.** $(-\infty, -3)$ **12.** $(-\infty, -1) \cup \left(-\dfrac{1}{5}, \infty\right)$ **13.** $(-\infty, -10] \cup [2, \infty)$ **14.** $\varnothing$ **15.** $\left(-\infty, -\dfrac{5}{3}\right)$ **16.** $(4, \infty)$ **17.** $\{-2, 7\}$

18. $\varnothing$ **19.** no more than 80 miles per day **20.** $[49\%, 99\%)$ **21.** at least $120,000 **22.** at least 750,000 discs each month

Section 4.4 Check Point Exercises

1. $4x - 2y \geq 8$ **2.** $y > -\dfrac{3}{4}x$ **3. a.** $y > 1$ **b.** $x \leq -2$

4. $B = (60, 20)$; Using $T = 60$ and $P = 20$, each of the three inequalities for grasslands is true: $60 \geq 35$, true; $5(60) - 7(20) \geq 70$, true; $3(60) - 35(20) \leq -140$, true.

5. $x - 3y < 6$
$2x + 3y \geq -6$

6. $x + y < 2$
$-2 \leq x < 1$
$y > -3$

Concept and Vocabulary Check

1. solution; x; y; $5 > 1$ **2.** graph **3.** half-plane **4.** false **5.** true **6.** false **7.** $x - y < 1$; $2x + 3y \geq 12$ **8.** false

Exercise Set 4.4

1. $x + y \geq 3$

3. $x - y < 5$

...

3. $x - y < 5$

5. $x + 2y > 4$

...

5. $x + 2y > 4$

7. $3x - y \leq 6$

9. $\frac{x}{2} + \frac{y}{3} < 1$

11. $y > \frac{1}{3}x$

13. $y \leq 3x + 2$

15. $y < -\frac{1}{4}x$

17. $x \leq 2$

...

17. $x \leq 2$

19. $y > -4$

...

19. $y > -4$

21. $y \geq 0$

23. $3x + 6y \leq 6$
$2x + y \leq 8$

25. $2x - 5y \leq 10$
$3x - 2y > 6$

27. $y > 2x - 3$
$y < -x + 6$

29. $x + 2y \leq 4$
$y \geq x - 3$

31. $x \leq 2$
$y \geq -1$

33. $-2 \leq x < 5$

35. $x - y \leq 1$
$x \geq 2$

37. $\varnothing$

39. $x + y > 4$
$x + y > -1$

41. $x - y \leq 2$
$x \geq -2$
$y \leq 3$

43. $x \geq 0$
$y \geq 0$
$2x + 5y \leq 10$
$3x + 4y \leq 12$

45. $3x + y \leq 6$
$2x - y \leq -1$
$x \geq -2$
$y \leq 4$

47. $y \geq -2x + 4$

49. $\begin{cases} x + y \leq 4 \\ 3x + y \leq 6 \end{cases}$

$x + y \leq 4$
$3x + y \leq 6$

51. $-2 \leq x \leq 2$; $-3 \leq y \leq 3$

$-2 \leq x \leq 2$
$-3 \leq y \leq 3$

53. $y > \frac{3}{2}x - 2$ or $y < 4$

55. no solution   **57.** infinitely many solutions   **59. a.** $A = (20, 150)$; A 20-year-old with a heart rate of 150 beats per minute is within the target range.   **b.** $10 \leq 20 \leq 70$, true; $150 \geq 0.7(220 - 20)$, true; $150 \leq 0.8(220 - 20)$, true   **61.** $10 \leq a \leq 70$; $H \geq 0.6(220 - a)$; $H \leq 0.7(220 - a)$

63. a. $\begin{cases} y \geq 0 \\ x + y \geq 5 \\ x \geq 1 \\ 200x + 100y \leq 700 \end{cases}$ **b.** $y \geq 0$
$x + y \geq 5$
$x \geq 1$
$200x + 100y \leq 700$

c. 2 nights   **75.**

...

75.

77.

83. makes sense   **85.** makes sense   **87.** false   **89.** true   **91.** $\begin{cases} x \geq -2 \\ y > -1 \end{cases}$ **93.** $y \geq 2x - 2$

95. $y \geq nx + b$ $y \leq mx + b$ **96.** $\{(3, 1)\}$ **97.** $\{(2, 4)\}$ **98.** 165   **99. a.**   **b.** $(1, 5), (8, 5), (8, -2)$

c. at $(1, 5)$: 13; at $(8, 5)$: 34; at $(8, -2)$: 20

100. a.   **b.** $(0, 0), (2, 0), (4, 3), (0, 7)$ **c.** at $(0, 0)$: 0; at $(2, 0)$: 4; at $(4, 3)$: 23; at $(0, 7)$: 35   **101.** $20x + 10y \leq 80,000$

Section 4.5 Check Point Exercises

1. $z = 25x + 55y$ **2.** $x + y \leq 80$ **3.** $30 \leq x \leq 80, 10 \leq y \leq 30; z = 25x + 55y;$ $\begin{cases} x + y \leq 80 \\ 30 \leq x \leq 80 \\ 10 \leq y \leq 30 \end{cases}$

4. 50 bookshelves and 30 desks; $2900   **5.** 30

Concept and Vocabulary Check

1. linear programming   **2.** objective   **3.** constraints; corner

Exercise Set 4.5

1. $(1, 2)$: 17; $(2, 10)$: 70; $(7, 5)$: 65; $(8, 3)$: 58; maximum: 70; minimum: 17   **3.** $(0, 0)$: 0; $(0, 8)$: 400; $(4, 9)$: 610; $(8, 0)$: 320; maximum: 610; minimum: 0

5. a.   **7. a.**   **9. a.**

b. $(0, 4)$: 8; $(0, 8)$: 16; $(4, 0)$: 12   **b.** $(0, 3)$: 3; $(0, 4)$: 4; $(6, 0)$: 24; $(3, 0)$: 12   **b.** $(1, 2)$: -1; $(1, 4)$: -5; $(5, 8)$: -1; $(5, 2)$: 11
c. maximum: 16; at $(0, 8)$ **c.** maximum: 24; at $(6, 0)$ **c.** maximum: 11; at $(5, 2)$

11. a.   **13. a.**   **15. a.** $z = 125x + 200y$

b. $\begin{cases} x \leq 450 \\ y \leq 200 \\ 600x + 900y \leq 360,000 \end{cases}$

c.

b. $(0, 2)$: 4; $(0, 4)$: 8; $\left(\frac{12}{5}, \frac{12}{5}\right)$: $\frac{72}{5} = 14.4$;   **b.** $(0, 0)$: 0; $(0, 6)$: 72; $(3, 4)$: 78; $(5, 0)$: 50
$(4, 0)$: 16; $(2, 0)$: 8   **c.** maximum: 78; at $(3, 4)$
c. maximum: 16; at $(4, 0)$

d. 0; 40,000; 77,500; 76,250; 56,250
e. 300; 200; 77,500

17. 40 of model A and 0 of model B   **19.** 300 boxes of food and 200 boxes of clothing   **21.** 100 parents and 50 students
23. 10 Boeing 727s and 42 Falcon 20s   **29.** does not make sense   **31.** makes sense   **33.** $5000 in stocks and $5000 in bonds   **35.** $54x^7y^{15}$
36. $L = \dfrac{12P + W}{2}$ **37.** 10   **38.** $4x^3 + 9x^2 - 13x - 3$ **39.** $5x^3 - 2x^2 + 12x - 15$ **40. a.** g **b.** f

Review Exercises

1. $[-2, \infty)$

2. $\left[\dfrac{3}{5}, \infty\right)$

3. $\left(-\infty, -\dfrac{21}{2}\right)$

4. $(-3, \infty)$

5. $(-\infty, -2]$

6. $\varnothing$

7. more than 50 checks   **8.** more than \$13,500 in sales   **9.** $\{a, c\}$ **10.** $\{a\}$ **11.** $\{a, b, c, d, e\}$ **12.** $\{a, b, c, d, f, g\}$

13. $(-\infty, 3]$

14. $(-\infty, 6)$

15. $(6, 8)$

16. $(-\infty, 1]$

17. $\varnothing$

18. $(-\infty, 1) \cup (2, \infty)$

19. $(-\infty, -4] \cup (2, \infty)$

20. $(-\infty, -2)$

21. $(-\infty, \infty)$

22. $(-5, 2]$

23. $\left[-\dfrac{3}{4}, 1\right]$

24. $[49\%, 99\%)$

25. $\{-4, 3\}$ **26.** $\varnothing$ **27.** $\left\{-\dfrac{11}{2}, \dfrac{23}{2}\right\}$ **28.** $\left\{-4, -\dfrac{6}{11}\right\}$

29. $[-9, 6]$

30. $(-\infty, -6) \cup (0, \infty)$

31. $(-3, -2)$

32. $(-\infty, -5] \cup [1, \infty)$;   **33.** $\varnothing$ **34.** Approximately 90% of the population sleeps between 5.5 hours and 7.5 hours daily, inclusive.

35. $3x - 4y > 12$

36. $x - 3y \leq 6$

37. $y \leq -\dfrac{1}{2}x + 2$

38. $y > \dfrac{3}{5}x$

39. $x \leq 2$

40. $y > -3$

41. $2x - y \leq 4$
$x + y \geq 5$

42. $y < -x + 4$
$y > x - 4$

43. $-3 \leq x < 5$

44. $-2 < y \leq 6$

45. $x \geq 3$
$y \leq 0$

46. $2x - y > -4$
$x \geq 0$

47. $x + y \leq 6$
$y \geq 2x - 3$

48. $3x + 2y \geq 4$
$x - y \leq 3$
$x \geq 0, y \geq 0$

49. $\varnothing$

50. $\left(\dfrac{1}{2}, \dfrac{1}{2}\right): \dfrac{5}{2}$; $(2, 2): 10$; $(4, 0): 8$; $(1, 0): 2$; maximum: 10; minimum: 2

51.

maximum: 24

52.

maximum: 33

53.

maximum: 44

54. a. $z = 500x + 350y$ **b.** $\begin{cases} x + y \leq 200 \\ x \geq 10 \\ y \geq 80 \end{cases}$ **c.**

d. $(10, 80): 33{,}000$; $(10, 190): 71{,}500$; $(120, 80): 88{,}000$

e. 120; 80; $88{,}000$

55. 480 of model A and 240 of model B

Chapter Test

1. $(-\infty, 12]$

2. $\left[\dfrac{21}{8}, \infty\right)$

3. more than 200 minutes
4. $\{4, 6\}$
5. $\{2, 4, 6, 8, 10, 12, 14\}$

6. $(-2, -1)$

7. $[-2, \infty)$

8. $(-\infty, 4)$

9. $(-\infty, -4] \cup (2, \infty)$

10. $\left[-7, \dfrac{13}{2}\right)$

11. $\left\{-2, \dfrac{4}{5}\right\}$ **12.** $\left\{-\dfrac{8}{5}, 7\right\}$

13. $(-3, 4)$

14. $(-\infty, -1] \cup [4, \infty)$

15. $(-\infty, 90.6) \cup (106.6, \infty)$; Hypothermia: Body temperature below 90.6°F; Hyperthermia: Body temperature above 106.6°F

16. $3x - 2y < 6$

17. $y \geq \dfrac{1}{2}x - 1$

18. $y \leq -1$

19. $x + y \geq 2$
$x - y \geq 4$

20. $3x + y \leq 9$
$2x + 3y \geq 6$
$x \geq 0$
$y \geq 0$

21. $-2 < x \leq 4$

22. maximum: 26   **23.** 50 regular and 100 deluxe; $35,000

Cumulative Review Exercises

1. $\{-1\}$ **2.** $\{8\}$ **3.** $-\dfrac{2y^7}{3x^5}$ **4.** $22; 4a^2 - 6a + 4$ **5.** $2x^2 + x + 2; 12$ **6.** $f(x) = -\dfrac{1}{2}x + 4$

7. $f(x) = 2x + 1$

8. $y > 2x$

9. $2x - y \geq 6$

10. $f(x) = -1$

11. $\{(-4, 2, -1)\}$ **12.** $\{(-1, 2)\}$ **13.** -17 **14.** 46 rooms with kitchen facilities and 14 without kitchen facilities   **15.** a. and b. are functions.

16. $[-7, \infty)$

17. $(-\infty, -6)$

18. $(-\infty, 3] \cup [5, \infty)$

19. $[-10, 7]$

20. $\left(-\infty, \dfrac{1}{3}\right) \cup (5, \infty)$

CHAPTER 5

Section 5.1 Check Point Exercises

1.

Term	Coefficient	Degree
$8x^4y^5$	8	9
$-7x^3y^2$	-7	5
$-x^2y$	-1	3
$-5x$	-5	1
11	11	0

The degree of the polynomial is 9, the leading term is $8x^4y^5$, and the leading coefficient is 8.

2. 16   **3.** The graph rises to the left and to the right.   **4.** This would not be appropriate over long time periods. Since the graph falls to the right, at some point the ratio would be negative, which is not possible.   **5.** The graph does not show the end behavior of the function. The graph should fall to the left.
6. $-3x^3 + 10x^2 - 10$ **7.** $9xy^3 + 3xy^2 - 15y - 9$ **8.** $10x^3 - 2x^2 + 8x - 10$ **9.** $13x^2y^5 + 2xy^3 - 10$

Concept and Vocabulary Check

1. whole **2.** standard **3.** monomial **4.** binomial **5.** trinomial **6.** n **7.** $n + m$ **8.** greatest; leading; leading **9.** true
10. false **11.** end; leading **12.** falls; rises **13.** rises; falls **14.** rises; rises **15.** falls; falls **16.** true **17.** true **18.** like
19. $-3x^3$ **20.** $-3x^3y$ **21.** $2x^5$ **22.** $10x^5y^2$ **23.** $12xy^2 - 12y^2$

Exercise Set 5.1

1.

Term	Coefficient	Degree
$-x^4$	-1	4
x^2	1	2

The degree of the polynomial is 4, the leading term is $-x^4$, and the leading coefficient is -1.

3.

Term	Coefficient	Degree
$5x^3$	5	3
$7x^2$	7	2
$-x$	-1	1
9	9	0

The degree of the polynomial is 3, the leading term is $5x^3$, and the leading coefficient is 5.

5.

Term	Coefficient	Degree
$3x^2$	3	2
$-7x^4$	-7	4
$-x$	-1	1
6	6	0

The degree of the polynomial is 4, the leading term is $-7x^4$, and the leading coefficient is -7.

7.

Term	Coefficient	Degree
x^3y^2	1	5
$-5x^2y^7$	-5	9
$6y^2$	6	2
-3	-3	0

The degree of the polynomial is 9, the leading term is $-5x^2y^7$, and the leading coefficient is -5.

9.

Term	Coefficient	Degree
x^5	1	5
$3x^2y^4$	3	6
$7xy$	7	2
$9x$	9	1
-2	-2	0

The degree of the polynomial is 6, the leading term is $3x^2y^4$, and the leading coefficient is 3.

11. 0 **13.** 12 **15.** 56 **17.** -29 **19.** -1
21–23. Graph #23 is not that of a polynomial function.
25. falls to the left and falls to the right; graph (b)
27. rises to the left and rises to the right; graph (a)
29. $11x^3 + 7x^2 - 12x - 4$ **31.** $-\frac{2}{5}x^4 + x^3 + \frac{3}{8}x^2$
33. $9x^2y - 6xy$ **35.** $2x^2y + 15xy + 15$
37. $-9x^4y^2 - 6x^2y^2 - 5x^2y + 2xy$ **39.** $5x^{2n} - 2x^n - 6$
41. $12x^3 + 4x^2 + 12x - 14$ **43.** $22y^5 + 9y^4 + 7y^3 - 13y^2 + 3y - 5$
45. $-5x^3 + 8xy - 9y^2$ **47.** $x^4y^2 + 8x^3y + y - 6x$

49. $y^{2n} + 2y^n - 3$ **51.** $8a^2b^4 + 3ab^2 + 8ab$ **53.** $5x^3 + 3x^2y - xy^2 - 4y^3$ **55.** $5x^4 - x^3 + 5x^2 - 5x + 2$ **57.** $-10x^2y^2 + 4x^2 + 3$
59. $-4x^3 - x^2 + 4x + 8; 7$ **61.** $-8x^2 - 2x - 1; -29$ **63.** $-9x^3 - x^2 + 13x + 20$ **65. a.** 3167; The world tiger population in 2010 was approximately 3167.; $(40, 3167)$ **b.** underestimates by 33 **67.** rises to the right; no; The model indicates an increasing world tiger population but the tiger population will actually decrease without conservation efforts. **69.** falls to the right; The number of viral particles eventually decreases as the days increase. **71.** falls to the right; Eventually the elk population will be extinct. **85.** Answers will vary; an example is $f(x) = -x^2 - x + 1$.
87. Answers will vary; an example is $f(x) = x^3 + x + 1$.

89. **91.** **93.**

95. does not make sense **97.** makes sense **99.** false **101.** true **103.** $10y^{3n} - 6y^{2n} + 5$ **105.** $\left\{\frac{2}{3}\right\}$

106. $2x - 3y < -6$

107. $y - 5 = 3(x + 2); y = 3x + 11$ or $f(x) = 3x + 11$ **108.** $10x^7y^9$ **109.** $16x^8 + 6x^5$
110. $3x^3 + 19x^2 + 43x + 35$

Section 5.2 Check Point Exercises

1. a. $-18x^7y^{11}$ **b.** $30x^{10}y^6z^8$ **2. a.** $12x^9 - 18x^6 + 24x^4$ **b.** $10x^5y^9 - 8x^7y^7 - 10x^4y^3$ **3.** $6x^3 - 2x^2 - x + 2$
4. $12x^2y^6 - 8x^2y^4 + 6xy^5 + 2y^2$ **5. a.** $x^2 + 8x + 15$ **b.** $14x^2 + xy - 4y^2$ **c.** $4x^6 - 12x^4 - 5x^3 + 15x$ **6. a.** $x^2 + 16x + 64$
b. $16x^2 + 40xy + 25y^2$ **7. a.** $x^2 - 10x + 25$ **b.** $4x^2 - 24xy^4 + 36y^8$ **8. a.** $x^2 - 9$ **b.** $25x^2 - 49y^2$ **c.** $25a^2b^4 - 16a^2$
9. a. $9x^2 + 12x + 4 - 25y^2$ **b.** $4x^2 + 4xy + y^2 + 12x + 6y + 9$ **10. a.** $x^2 - 10x + 21$ **b.** 5 **11. a.** $a^2 + a - 2$ **b.** $2ah + h^2 - 5h$

Concept and Vocabulary Check

1. add   **2.** distributive; $4x^5 - 8x^2 + 6; 7x^3$ **3.** $5x; 3;$ like   **4.** $3x^2; 5x; 21x; 35$ **5.** $A^2 + 2AB + B^2;$ squared; product of the terms; squared
6. $A^2 - 2AB + B^2;$ minus; product of the terms; plus   **7.** $A^2 - B^2;$ minus   **8.** $x; a + h$

Exercise Set 5.2

1. $15x^6$ **3.** $15x^3y^{11}$ **5.** $-6x^2y^9z^9$ **7.** $2x^{3n}y^{n-2}$ **9.** $12x^3 + 8x^2$ **11.** $2y^3 - 10y^2$ **13.** $10x^8 - 20x^5 + 45x^3$ **15.** $28x^2y + 12xy^2$
17. $18a^3b^5 + 15a^2b^3$ **19.** $-12x^6y^3 + 28x^3y^4 - 24x^2y$ **21.** $-12x^{3n} + 20x^{2n} - 2x^{n+1}$ **23.** $x^3 - x^2 - x - 15$ **25.** $x^3 - 1$ **27.** $a^3 - b^3$
29. $x^4 + 5x^3 + x^2 - 11x + 4$ **31.** $x^3 - 4x^2y + 4xy^2 - y^3$ **33.** $x^3y^3 + 8$ **35.** $x^2 + 11x + 28$ **37.** $y^2 - y - 30$ **39.** $10x^2 + 11x + 3$
41. $6y^2 - 11y + 4$ **43.** $15x^2 - 22x + 8$ **45.** $2x^2 + xy - 21y^2$ **47.** $14x^2y^2 - 19xy - 3$ **49.** $x^3 - 4x^2 - 5x + 20$
51. $8x^5 - 40x^3 + 3x^2 - 15$ **53.** $3x^{2n} + 5x^ny^n - 2y^{2n}$ **55.** $x^2 + 6x + 9$ **57.** $y^2 - 10y + 25$ **59.** $4x^2 + 4xy + y^2$
61. $25x^2 - 30xy + 9y^2$ **63.** $4x^4 + 12x^2y + 9y^2$ **65.** $16x^2y^4 - 8x^2y^3 + x^2y^2$ **67.** $a^{2n} + 8a^nb^n + 16b^{2n}$ **69.** $x^2 - 16$ **71.** $25x^2 - 9$
73. $16x^2 - 49y^2$ **75.** $y^6 - 4$ **77.** $1 - y^{10}$ **79.** $49x^2y^4 - 100y^2$ **81.** $25a^{2n} - 49$ **83.** $4x^2 + 12x + 9 - 16y^2$
85. $x^2 + 2xy + y^2 - 9$ **87.** $25x^2 + 70xy + 49y^2 - 4$ **89.** $25y^2 - 4x^2 - 12x - 9$ **91.** $x^2 + 2xy + y^2 + 2x + 2y + 1$
93. $x^4 - 1$ **95. a.** $x^2 + 4x - 12$ **b.** -15 **c.** -12 **97. a.** $x^3 - 27$ **b.** -35 **c.** -27 **99. a.** $a^2 + a + 5$ **b.** $2ah + h^2 - 3h$
101. a. $3a^2 + 14a + 15$ **b.** $6ah + 3h^2 + 2h$ **103.** $48xy$ **105.** $-9x^2 + 3x + 9$ **107.** $16x^4 - 625$ **109.** $x^3 - 3x^2 + 3x - 1$
111. $(2x - 7)^2 = 4x^2 - 28x + 49$ **113. a.** $x^2 + 6x + 4x + 24$ **b.** $(x + 6)(x + 4) = x^2 + 10x + 24$ **115. a.** $x^2 + 12x + 27$
b. $x^2 + 6x + 5$ **c.** $6x + 22$ **117. a.** $4x^2 - 36x + 80$ **b.** $4x^3 - 36x^2 + 80x$ **119. a.** $V(x) = -2x^3 + 10x^2 + 300x$ **b.** rises to the left
and falls to the right   **c.** no; Because the graph falls to the right, volume will eventually be negative, which is not possible.   **d.** 2000; Carry-on
luggage with a depth of 10 inches has a volume of 2000 cubic inches.   **e.** (10, 2000)   **f.** (0, 15), although answers may vary.

131.   **133.**   **135.** makes sense   **137.** makes sense, although
answers may vary   **139.** false   **141.** false
143. $x^2 + 2x$ **145.** $2x^3 + 12x^2 + 12x + 10$
147. 9 and 11   **148.** $\left(-\infty, -\dfrac{14}{3}\right] \cup [2, \infty)$
149. $[-3, \infty)$ **150.** 8.034×10^9 **151. a.** $3x^2$
b. $6x^2y^2$ **152.** $x^3 - 5x^2 + 3x - 15$
153. $3x^3 - 2xy + 12x - 8y$

 conclusion: $y_1 = y_2$ conclusion: $y_1 = y_2$

Section 5.3 Check Point Exercises

1. $10x(2x + 3)$ **2. a.** $3x^2(3x^2 + 7)$ **b.** $5x^3y^2(3 - 5xy)$ **c.** $4x^2y^3(4x^2y^2 - 2xy + 1)$ **3.** $-2x(x^2 - 5x + 3)$
4. a. $(x - 4)(3 + 7a)$ **b.** $(a + b)(7x - 1)$ **5.** $(x - 4)(x^2 + 5)$ **6.** $(x + 5)(4x - 3y)$

Concept and Vocabulary Check

1. factoring   **2.** greatest common factor; smallest/least   **3.** false   **4.** $-2x$ **5.** false

Exercise Set 5.3

1. $2x(5x + 2)$ **3.** $y(y - 4)$ **5.** $x^2(x + 5)$ **7.** $4x^2(3x^2 - 2)$ **9.** $2x^2(16x^2 + x + 4)$ **11.** $2xy(2xy^2 + 3)$ **13.** $10xy^2(3xy - 1)$
15. $2x(6y - 3z + 2w)$ **17.** $3x^2y^4(5xy^2 - 3x^2 + 4y)$ **19.** $5x^2y^4z^2(5xy^2 - 3x^2z^2 + 5yz)$ **21.** $5x^n(3x^n - 5)$ **23.** $-4(x - 3)$ **25.** $-8(x + 6)$
27. $-2(x^2 - 3x + 7)$ **29.** $-5(y^2 - 8x)$ **31.** $-4x(x^2 - 8x + 5)$ **33.** $-1(x^2 + 7x - 5)$ **35.** $(x + 3)(4 + a)$ **37.** $(y - 6)(x - 7)$
39. $(x + y)(3x - 1)$ **41.** $(3x - 1)(4x^2 + 1)$ **43.** $(x + 3)(2x + 1)$ **45.** $(x + 3)(x + 5)$ **47.** $(x + 7)(x - 4)$ **49.** $(x - 3)(x^2 + 4)$
51. $(y - 6)(x + 2)$ **53.** $(y + 1)(x - 7)$ **55.** $(5x - 6y)(2x + 7y)$ **57.** $(4x - 1)(x^2 - 3)$ **59.** $(x - a)(x - b)$ **61.** $(x - 3)(x^2 + 4)$
63. $(a - b)(y - x)$ **65.** $(a + 2b)(y^2 - 3)$ **67.** $(x^n + 1)(y^n + 3)$ **69.** $(a + 1)(b - c)$ **71.** $(x^3 - 5)(1 + 4y)$
73. $y^6(3x - 1)^4(6xy - 2y - 7)$ **75.** $(x^2 + 5x - 2)(a + b)$ **77.** $(x + y + z)(a - b + c)$ **79. a.** 16; The ball is 16 feet above the ground
after 2 seconds.   **b.** 0; The ball is on the ground after 2.5 seconds.   **c.** $-8t(2t - 5); f(t) = -8t(2t - 5)$ **d.** 16; 0; yes; no; Answers will vary.
81. a. $(x - 0.4x)(1 - 0.4) = (0.6x)(0.6) = 0.36x$ **b.** no; 36%
83. $A = P + Pr + (P + Pr)r = P(1 + r) + Pr(1 + r) = (1 + r)(P + Pr) = P(1 + r)(1 + r) = P(1 + r)^2$ **85.** $A = r(\pi r + 2l)$

93.   Graphs coincide.;
factored correctly   **95.**   Graphs do not coincide.;
$x^2 + 2x + x + 2 = (x + 2)(x + 1)$

97. does not make sense   **99.** makes sense   **101.** false   **103.** true   **105.** $x^{2n}(x^{2n} + 1 + x^n)$ **107.** $4y^{2n}(2y^4 + 4y^3 - 3)$

109. Answers will vary; an example is $6x^2 - 4x + 9x - 6$.   **110.** $\left\{\left(\dfrac{20}{11}, -\dfrac{14}{11}\right)\right\}$ **111. a.** function   **b.** not a function

112. length: 8 ft; width: 3 ft   **113.** 4   **114.** 2   **115.** 7

Section 5.4 Check Point Exercises

1. $(x + 4)(x + 2)$ or $(x + 2)(x + 4)$ **2.** $(x - 5)(x - 4)$ **3.** $(y + 22)(y - 3)$ **4.** $(x - 3y)(x - 2y)$ **5.** $3x(x - 7)(x + 2)$

6. $(x^3 - 5)(x^3 - 2)$ **7.** $(3x - 14)(x - 2)$ **8.** $x^4(3x - 1)(2x + 7)$ **9.** $(2x - y)(x - 3y)$ **10.** $(3y^2 - 2)(y^2 + 4)$ **11.** $(2x - 5)(4x - 1)$

Concept and Vocabulary Check

1. completely   **2.** greatest common factor   **3.** $+ 5$ **4.** $- 4$ **5.** $+ 16$ **6.** $- 3y$ **7.** $2x; 2x; - 18$ **8.** $- 11$ **9.** $2x + 9$ **10.** $- 6y$

Exercise Set 5.4

1. $(x + 3)(x + 2)$ **3.** $(x + 6)(x + 2)$ **5.** $(x + 5)(x + 4)$ **7.** $(y + 8)(y + 2)$ **9.** $(x - 3)(x - 5)$ **11.** $(y - 2)(y - 10)$

13. $(a + 7)(a - 2)$ **15.** $(x + 6)(x - 5)$ **17.** $(x + 4)(x - 7)$ **19.** $(y + 4)(y - 9)$ **21.** prime   **23.** $(x - 2y)(x - 7y)$

25. $(x + 5y)(x - 6y)$ **27.** prime   **29.** $(a - 10b)(a - 8b)$ **31.** $3(x + 3)(x - 2)$ **33.** $2x(x - 3)(x - 4)$ **35.** $3y(y - 2)(y - 3)$

37. $2x^2(x + 3)(x + 16)$ **39.** $(x^3 + 2)(x^3 - 3)$ **41.** $(x^2 - 6)(x^2 + 1)$ **43.** $(x + 6)(x + 2)$ **45.** $(3x + 5)(x + 1)$ **47.** $(x + 11)(5x + 1)$

49. $(y + 8)(3y - 2)$ **51.** $(y + 2)(4y + 1)$ **53.** $(2x + 3)(5x + 2)$ **55.** $(4x - 3)(2x - 3)$ **57.** $(y - 3)(6y - 5)$ **59.** prime

61. $(x + y)(3x + y)$ **63.** $(2x + y)(3x - 5y)$ **65.** $(5x - 2y)(3x - 5y)$ **67.** $(3a - 7b)(a + 2b)$ **69.** $5x(3x - 2)(x - 1)$

71. $2x^2(3x + 2)(4x - 1)$ **73.** $y^3(5y + 1)(3y - 1)$ **75.** $3(8x + 9y)(x - y)$ **77.** $2b(a + 3)(3a - 10)$ **79.** $2y(2x - y)(3x - 7y)$

81. $13x^3y(y + 4)(y - 1)$ **83.** $(2x^2 - 3)(x^2 + 1)$ **85.** $(2x^3 + 5)(x^3 + 3)$ **87.** $(2y^5 + 1)(y^5 + 3)$ **89.** $(5x + 12)(x + 2)$

91. $(2x - 13)(x - 2)$ **93.** $(x - 0.3)(x - 0.2)$ **95.** $\left(x + \dfrac{3}{7}\right)\left(x - \dfrac{1}{7}\right)$ **97.** $(ax - b)(cx + d)$ **99.** $-x^3y^2(4x - 3y)(x - y)$

101. $f(x) = 3x - 13$ and $g(x) = x - 3$, or vice versa   **103.** $2x + 1$ by $x + 3$ **105. a.** 32; The diver is 32 feet above the water after 1 second.

b. 0; The diver hits the water after 2 seconds.   **c.** $-16(t - 2)(t + 1)$; $f(t) = -16(t - 2)(t + 1)$ **d.** 32; 0

107. a. $x^2 + x + x + x + 1 + 1 = x^2 + 3x + 2$ **b.** $(x + 2)(x + 1)$ **c.** Answers will vary.

117.

Graphs coincide.; factored correctly

119.

Graphs coincide.; factored correctly

123. makes sense   **125.** makes sense   **127.** true   **129.** false   **131.** $-16, -8, 8, 16$ **133.** $(4x^n - 5)(x^n - 1)$ **135.** $b^2(b^n - 2)(b^n + 5)$

137. $d^n(2d - 3)(d - 1)$ **138.** $(5, \infty)$ **139.** $\{(2, -1, 3)\}$ **140.** $(x + 2)(4x^2 - 5)$ **141.** $(x + 7)(x + 7)$ or $(x + 7)^2$

142. $(x - 4)(x - 4)$ or $(x - 4)^2$ **143.** $(x + 5)(x - 5)$

Mid-Chapter Check Point Exercises

1. $-x^3 + 4x^2 + 6x + 17$ **2.** $-2x^7y^3z^5$ **3.** $30x^5y^3 - 35x^3y^2 - 2x^2y$ **4.** $3x^3 + 4x^2 - 39x + 40$ **5.** $2x^4 - x^3 - 8x^2 + 11x - 4$

6. $-x^2 - 5x + 5$ **7.** $-4x^3y - 6x^2y - 7y - 5$ **8.** $8x^2 + 18x - 5$ **9.** $10x^2y^2 - 11xy - 6$ **10.** $9x^2 - 4y^2$ **11.** $6x^3y - 9xy^2 + 2x^2 - 3y$

12. $49x^6y^2 - 25x^2$ **13.** $6xh + 3h^2 - 2h$ **14.** $x^4 - 6x^2 + 9$ **15.** $x^5 + 2x^3 + 2x^2 - 15x - 6$ **16.** $4x^2 + 20xy + 25y^2$

17. $x^2 + 12x + 36 - 9y^2$ **18.** $x^2 + 2xy + y^2 + 10x + 10y + 25$ **19.** $(x - 8)(x + 3)$ **20.** $(5x + 2)(3y + 1)$

21. $(5x - 2)(x + 2)$ **22.** $5(7x^2 + 2x - 10)$ **23.** $9(x - 2)(x + 1)$ **24.** $5x^2y(2xy - 4y + 7)$ **25.** $(3x + 1)(6x + 5)$

26. $(4x - 3y)(3x - 4)$ **27.** $(3x - 4)(3x - 1)$ **28.** $(3x^3 + 5)(x^3 + 2)$ **29.** $x(5x - 2)(5x + 7)$ **30.** $(x^3 - 3)(2x - y)$

Section 5.5 Check Point Exercises

1. a. $(4x + 5)(4x - 5)$ **b.** $(10y^3 + 3x^2)(10y^3 - 3x^2)$ **2.** $6y(1 + xy^3)(1 - xy^3)$ **3.** $(4x^2 + 9)(2x + 3)(2x - 3)$

4. $(x + 7)(x + 2)(x - 2)$ **5. a.** $(x + 3)^2$ **b.** $(4x + 5y)^2$ **c.** $(2y^2 - 5)^2$ **6.** $(x + 5 + y)(x + 5 - y)$ **7.** $(a + b - 2)(a - b + 2)$

8. a. $(x + 3)(x^2 - 3x + 9)$ **b.** $(x^2 + 10y)(x^4 - 10x^2y + 100y^2)$ **9. a.** $(x - 2)(x^2 + 2x + 4)$ **b.** $(1 - 3xy)(1 + 3xy + 9x^2y^2)$

Concept and Vocabulary Check

1. $(A - B)(A + B)$ **2.** $(A + B)^2$ **3.** $(A - B)^2$ **4.** $(A + B)(A^2 - AB + B^2)$ **5.** $(A - B)(A^2 + AB + B^2)$ **6.** $4x; 4x$

$(b + 3); (b + 3)$ **8.** $- 7$ **9.** $4x$ **10.** $+ 3; - 3x$ **11.** $- 10; + 100$ **12.** false   **13.** true   **14.** false   **15.** true   **16.** false

Exercise Set 5.5

1. $(x + 2)(x - 2)$ **3.** $(3x + 5)(3x - 5)$ **5.** $(3 + 5y)(3 - 5y)$ **7.** $(6x + 7y)(6x - 7y)$ **9.** $(xy + 1)(xy - 1)$

11. $(3x^2 + 5y^3)(3x^2 - 5y^3)$ **13.** $(x^7 + y^2)(x^7 - y^2)$ **15.** $(x - 3 + y)(x - 3 - y)$ **17.** $(a + b - 2)(a - b + 2)$

19. $(x^n + 5)(x^n - 5)$ **21.** $(1 + a^n)(1 - a^n)$ **23.** $2x(x + 2)(x - 2)$ **25.** $2(5 + y)(5 - y)$ **27.** $8(x + y)(x - y)$

29. $2xy(x + 3)(x - 3)$ **31.** $a(ab + 7c)(ab - 7c)$ **33.** $5y(1 + xy^3)(1 - xy^3)$ **35.** $8(x^2 + y^2)$ **37.** prime

39. $(x^2 + 4)(x + 2)(x - 2)$ **41.** $(9x^2 + 1)(3x + 1)(3x - 1)$ **43.** $2x(x^2 + y^2)(x + y)(x - y)$ **45.** $(x + 3)(x + 2)(x - 2)$

47. $(x - 7)(x + 1)(x - 1)$ **49.** $(x + 2)^2$ **51.** $(x - 5)^2$ **53.** $(x^2 - 2)^2$ **55.** $(3y + 1)^2$ **57.** $(8y - 1)^2$ **59.** $(x - 6y)^2$

61. prime   **63.** $(3x + 8y)^2$ **65.** $(x - 3 + y)(x - 3 - y)$ **67.** $(x + 10 + x^2)(x + 10 - x^2)$ **69.** $(3x - 5 + 6y)(3x - 5 - 6y)$

71. $(x^2 + x + 1)(x^2 - x - 1)$ **73.** $(z + x - 2y)(z - x + 2y)$ **75.** $(x + 4)(x^2 - 4x + 16)$ **77.** $(x - 3)(x^2 + 3x + 9)$

79. $(2y + 1)(4y^2 - 2y + 1)$ **81.** $(5x - 2)(25x^2 + 10x + 4)$ **83.** $(xy + 3)(x^2y^2 - 3xy + 9)$ **85.** $x(4 - x)(16 + 4x + x^2)$

87. $(x^2 + 3y)(x^4 - 3x^2y + 9y^2)$ **89.** $(5x^2 - 4y^2)(25x^4 + 20x^2y^2 + 16y^4)$ **91.** $(x + 1)(x^2 - x + 1)(x^6 - x^3 + 1)$

93. $(x - 2y)(x^2 - xy + y^2)$ **95.** $(0.2x + 0.3)^2$ or $\dfrac{1}{100}(2x + 3)^2$ **97.** $x\left(2x - \dfrac{1}{2}\right)\left(4x^2 + x + \dfrac{1}{4}\right)$

99. $(x - 1)(x^2 + x + 1)(x - 2)(x^2 + 2x + 4)$ **101.** $(x^4 + 1)(x^2 + 4)(x + 2)(x - 2)$ **103.** $(x + 1)(x - 1)(x - 2)(x^2 + 2x + 4)$

105. a. $(A + B)^2$ **b.** $A^2; AB; AB; B^2$ **c.** $A^2 + 2AB + B^2$ **d.** $A^2 + 2AB + B^2 = (A + B)^2$; factoring a perfect square trinomial

107. $25x^2 - 9 = (5x + 3)(5x - 3)$ **109.** $49x^2 - 36 = (7x + 6)(7x - 6)$ **111.** $3a^3 - 3ab^2 = 3a(a + b)(a - b)$

117. ; Graphs do not conincide.; $x^2 + 4x + 4 = (x + 2)^2$

119. ; Graphs do not coincide; $25 - (x^2 + 4x + 4) = (7 + x)(3 - x)$

121. ; Graphs do not coincide.; $(x - 3)^2 + 8(x - 3) + 16 = (x + 1)^2$

123. ; Graphs do not coincide.; $(x + 1)^3 + 1 = (x + 2)(x^2 + x + 1)$

125. makes sense   **127.** does not make sense   **129.** false   **131.** false   **133.** $(y + x)(y^2 - xy + x^2 + 1)$ **135.** $(x^n + y^{4n})(x^{2n} - x^ny^{4n} + y^{8n})$

137. $x^6 - y^6 = (x^3 + y^3)(x^3 - y^3) = (x + y)(x^2 - xy + y^2)(x - y)(x^2 + xy + y^2)$;

$x^6 - y^6 = (x^2 - y^2)(x^4 + x^2y^2 + y^4) = (x + y)(x - y)(x^4 + x^2y^2 + y^4); x^4 + x^2y^2 + y^4 = (x^2 - xy + y^2)(x^2 + xy + y^2)$

139. 1   **140.** $[5, 11]$ **141.** $\{(-2, 1)\}$ **142.** $(x + 7)(3x - y)$ **143.** $2x(x + 2)^2$ **144.** $5x(x - y)(x - 7y)$

145. $(x + y)(3b + 4)(3b - 4)$

Section 5.6 Check Point Exercises

1. $3x(x - 5)^2$ **2.** $3y(x + 2)(x - 6)$ **3.** $(x + y)(4a + 5)(4a - 5)$ **4.** $(x + 10 + 6a)(x + 10 - 6a)$ **5.** $x(x + 2)(x^2 - 2x + 4)(x^6 - 8x^3 + 64)$

Concept and Vocabulary Check

1. b **2.** e **3.** h **4.** c **5.** d **6.** f **7.** a **8.** g

Exercise Set 5.6

1. $x(x + 4)(x - 4)$ **3.** $3(x + 3)^2$ **5.** $3(3x - 1)(9x^2 + 3x + 1)$ **7.** $(x + 4)(x - 4)(y - 2)$ **9.** $2b(2a + 5)(a - 3)$
11. $(y + 2)(y - 2)(a - 4)$ **13.** $11x(x^2 + y)(x^2 - y)$ **15.** $4x(x^2 + 4)(x + 2)(x - 2)$ **17.** $(x - 4)(x + 3)(x - 3)$
19. $2x^2(x + 3)(x^2 - 3x + 9)$ **21.** $3y(x^2 + 4y^2)(x + 2y)(x - 2y)$ **23.** $3x(2x + 3y)^2$ **25.** $(x - 6 + 7y)(x - 6 - 7y)$ **27.** prime
29. $12xy(x + y)(x - y)$ **31.** $6b(x^2 + y^2)$ **33.** $(x + y)(x - y)(x^2 + xy + y^2)$ **35.** $(x + 6 + 2a)(x + 6 - 2a)$ **37.** $(x^3 - 2)(x^3 + 7)$
39. $(2x - 7)(2 + x^2)$ **41.** $2(3x - 2y)(9x^2 + 6xy + 4y^2)$ **43.** $(x + 5 + y)(x + 5 - y)$ **45.** $(x^4 + y^4)(x^2 + y^2)(x + y)(x - y)$
47. $xy(x + 4y)(x - 4y)$ **49.** $x(1 + 2x)(1 - 2x + 4x^2)$ **51.** $2(4y + 1)(2y - 1)$ **53.** $y(14y^2 + 7y - 10)$ **55.** $3(3x + 2y)^2$
57. $3x(4x^2 + y^2)$ **59.** $x^3y^3(xy - 1)(x^2y^2 + xy + 1)$ **61.** $2(x + 5)(x - 5)$ **63.** $(x - y)(a + 2)(a - 2)$
65. $(c + d - 1)(c + d + 1)[(c + d)^2 + 1]$ **67.** $(p + q)^2(p - q)$ **69.** $(x + 2y)(x - 2y)(x + y)(x - y)$
71. $(x + y + 3)^2$ **73.** $(x - y)^2(x - y + 2)(x - y - 2)$ **75.** $(2x - y^2)(x - 3y^2)$ **77.** $(x - y)(x^2 + xy + y^2 - 1)$
79. $(xy + 1)(x^2y^2 - xy + 1)(x - 2)(x^2 + 2x + 4)$ **81. a.** $x(x + y) - y(x + y)$ **b.** $(x + y)(x - y)$
83. a. $xy + xy + xy + 3x(x) = 3xy + 3x^2$ **b.** $3x(y + x)$ **85. a.** $8x^2 - 2\pi x^2$ **b.** $2x^2(4 - \pi)$

89.

Graphs do not coincide.; $4x^2 - 12x + 9 = (2x - 3)^2$

91.

Graphs coincide.; factored correctly

95. makes sense
97. does not make sense **99.** true
101. false **103.** $x^n(3x - 1)(x - 4)$
105. $(x^2 + 2x + 2)(x^2 - 2x + 2)$
107. $\{-1\}$ **108.** $\dfrac{8x^9}{y^3}$ **109.** 52
110. 0 **111.** 0
112. $(x - 5)(x + 3) = 0$

Section 5.7 Check Point Exercises

1. $-\dfrac{1}{2}$ and 5, or $\left\{-\dfrac{1}{2}, 5\right\}$ **2. a.** 0 and $\dfrac{2}{3}$, or $\left\{0, \dfrac{2}{3}\right\}$ **b.** 5 or $\{5\}$ **c.** -4 and 3, or $\{-4, 3\}$ **3.** $-2, -\dfrac{3}{2}$, and 2, or $\left\{-2, -\dfrac{3}{2}, 2\right\}$

4. after 3 seconds; $(3, 336)$ **5.** 2 ft **6.** 5 ft

Concept and Vocabulary Check

1. quadratic **2.** $A = 0$ or $B = 0$ **3.** x-intercepts **4.** subtracting $20x$ **5.** subtracting $8x$; adding 12
6. polynomial; 0; descending; highest/greatest **7.** right; hypotenuse; legs **8.** right; legs; the square of the length of the hypotenuse

Exercise Set 5.7

1. $\{-4, 3\}$ **3.** $\{-7, 1\}$ **5.** $\left\{-4, \dfrac{2}{3}\right\}$ **7.** $\left\{\dfrac{3}{5}, 1\right\}$ **9.** $\left\{-2, \dfrac{1}{3}\right\}$ **11.** $\{0, 8\}$ **13.** $\left\{0, \dfrac{5}{3}\right\}$ **15.** $\{-2\}$ **17.** $\{7\}$ **19.** $\left\{\dfrac{5}{3}\right\}$

21. $\{-5, 5\}$ **23.** $\left\{-\dfrac{10}{3}, \dfrac{10}{3}\right\}$ **25.** $\{-3, 6\}$ **27.** $\{-3, -2\}$ **29.** $\{4\}$ **31.** $\{-2, 8\}$ **33.** $\left\{-1, \dfrac{2}{5}\right\}$ **35.** $\{-6, -3\}$

37. $\{-5, -4, 5\}$ **39.** $\{-5, 1, 5\}$ **41.** $\{-4, 0, 4\}$ **43.** $\{0, 2\}$ **45.** $\{-5, -3, 0\}$ **47.** 2 and 4; d **49.** -4 and -2; c **51.** $\{-7, -1, 6\}$

53. $\left\{-\dfrac{4}{3}, 0, \dfrac{4}{5}, 2\right\}$ **55.** -4 and 8 **57.** $-2, -\dfrac{1}{2}$, and 2 **59.** -7 and 4 **61.** 2 and 3 **63.** -9 and 5 **65.** 1 second; 0.25, 0.5, 0.75, 1, 1.25

67. 7 **69.** $(7, 21)$ **71.** length: 9 ft; width: 6 ft **73.** 5 in. **75.** 5 m **77. a.** $4x^2 + 44x$ **b.** 3 ft **79.** length: 10 in.; width: 10 in.

81. length: 12 ft; width: 5 ft **83.** $30\dfrac{1}{8}$ ft

97.

$\{-3, -1, 1\}$

99.

$\{0, 2\}$

101. makes sense **103.** makes sense
105. false **107.** true **108.** Answers will vary;
an example is $x^2 - 4x - 21 = 0$.
111. $\left\{-2, \dfrac{10}{3}\right\}$ **112.** 12
113. \$1700 at 5% and \$1300 at 8%
114. 30 **115.** $(-\infty, 2) \cup (2, \infty)$
116. $\dfrac{(x - 9)(x + 2)}{(2x - 1)(x + 2)}; \dfrac{x - 9}{2x - 1}$

Review Exercises

1.

Term	Coefficient	Degree
$-5x^3$	-5	3
$7x^2$	7	2
$-x$	-1	1
2	2	0

The degree of the polynomial is 3, the leading term is $-5x^3$, and the leading coefficient is -5.

2.

Term	Coefficient	Degree
$8x^4y^2$	8	6
$-7xy^6$	-7	7
$-x^3y$	-1	4

The degree of the polynomial is 7, the leading term is $-7xy^6$, and the leading coefficient is -7.

3. -31 **4. a.** 284; There were 284,000 record daily high temperatures in the United States in the 2000s. **b.** underestimates by 6000
5. rises to the left and falls to the right; c **6.** rises to the left and rises to the right; b **7.** falls to the left and rises to the right; a
8. falls to the left and falls to the right; d **9. a.** rises to the right **b.** no; The model indicates increasing deforestation despite a declining rate at which the forest is being cut down. **c.** falls to the right **d.** no; The model indicates that the amount of forest cleared, in square kilometers, will eventually be negative, which is not possible. **10.** $x^3 - 6x^2 - x - 9$ **11.** $12x^3y - 2x^2y - 14y - 17$ **12.** $15x^3 - 10x^2 + 11x - 8$
13. $-2x^3y^2 - 4x^3y - 8$ **14.** $3x^3 + 5x^2y - xy^2 - 8y^3$ **15.** $-12x^6y^2z^7$ **16.** $2x^8 - 24x^5 - 12x^3$ **17.** $21x^5y^4 - 35x^2y^3 - 7xy^2$
18. $6x^3 + 29x^2 + 27x - 20$ **19.** $x^4 + 4x^3 + 4x^2 - x - 2$ **20.** $12x^2 - 23x + 5$ **21.** $15x^2y^2 + 2xy - 8$ **22.** $9x^2 + 42xy + 49y^2$
23. $x^4 - 10x^2y + 25y^2$ **24.** $4x^2 - 49y^2$ **25.** $9x^2y^4 - 16x^2$ **26.** $x^2 + 6x + 9 - 25y^2$ **27.** $x^2 + 2xy + y^2 + 8x + 8y + 16$
28. $2x^2 - x - 15; 21$ **29. a.** $a^2 - 9a + 10$ **b.** $2ah + h^2 - 7h$ **30.** $8x^2(2x + 3)$ **31.** $2x(1 - 18x)$ **32.** $7xy(3xy - 2y + 1)$
33. $9x^2y(2xy - 3)$ **34.** $-4(3x^2 - 2x + 12)$ **35.** $-1(x^2 + 11x - 14)$ **36.** $(x - 1)(x^2 - 2)$ **37.** $(y - 3)(x - 5)$ **38.** $(x - 3y)(5a + 2b)$
39. $(x + 5)(x + 3)$ **40.** $(x + 20)(x - 4)$ **41.** $(x + 17y)(x - y)$ **42.** $3x(x - 1)(x - 11)$ **43.** $(x + 7)(3x + 1)$ **44.** $(3x - 2)(2x - 3)$
45. $(5x + 4y)(x - 2y)$ **46.** $x(3x + 4)(2x - 1)$ **47.** $(x + 3)(2x + 5)$ **48.** $(x^3 + 6)(x^3 - 5)$ **49.** $(x^2 + 3)(x^2 - 13)$ **50.** $(x + 11)(x + 9)$
51. $(5x^3 + 2)(x^3 + 3)$ **52.** $(2x + 5)(2x - 5)$ **53.** $(1 + 9xy)(1 - 9xy)$ **54.** $(x^4 + y^3)(x^4 - y^3)$ **55.** $(x - 1 + y)(x - 1 - y)$
56. $(x + 8)^2$ **57.** $(3x - 1)^2$ **58.** $(5x + 2y)^2$ **59.** prime **60.** $(5x - 4y)^2$ **61.** $(x + 9 + y)(x + 9 - y)$ **62.** $(z + 5x - 1)(z - 5x + 1)$
63. $(4x + 3)(16x^2 - 12x + 9)$ **64.** $(5x - 2)(25x^2 + 10x + 4)$ **65.** $(xy + 1)(x^2y^2 - xy + 1)$ **66.** $3x(5x + 1)$ **67.** $3x^2(2x + 1)(2x - 1)$
68. $4x(5x^3 - 6x^2 + 7x - 3)$ **69.** $x(x - 2)(x - 13)$ **70.** $-2y^2(y - 3)(y - 9)$ **71.** $(3x - 5)^2$ **72.** $5(x + 3)(x - 3)$
73. $(2x - 1)(x + 3)(x - 3)$ **74.** $(3x - y)(2x - 7y)$ **75.** $2y(y + 3)^2$ **76.** $(x + 3 + 2a)(x + 3 - 2a)$ **77.** $(2x - 3)(4x^2 + 6x + 9)$
78. $x(x^2 + 1)(x + 1)(x - 1)$ **79.** $(x^2 - 3)^2$ **80.** prime **81.** $4(a + 2)(a^2 - 2a + 4)$ **82.** $(x^2 + 9)(x + 3)(x - 3)$ **83.** $(a + 3b)(x - y)$
84. $(3x - 5y)(9x^2 + 15xy + 25y^2)$ **85.** $2xy(x + 3)(5x - 4)$ **86.** $(2x^3 + 5)(3x^3 - 1)$ **87.** $(x + 5)(2 + xy)$ **88.** $(y + 2)(y + 5)(y - 5)$
89. $(a^4 + 1)(a^2 + 1)(a + 1)(a - 1)$ **90.** $(x - 4)(3 + y)(3 - y)$ **91. a.** $2xy + 2y^2$ **b.** $2y(x + y)$ **92. a.** $x^2 - 4y^2$ **b.** $(x + 2y)(x - 2y)$
93. $\{-5, -1\}$ **94.** $\left\{\dfrac{1}{3}, 7\right\}$ **95.** $\{-8, 7\}$ **96.** $\{0, 4\}$ **97.** $\{-5, -3, 3\}$ **98.** 9 seconds **99. a.** 20 miles per hour **b.** (20, 40)
c. As a car's speed increases, its stopping distance gets longer at increasingly greater rates. **100.** length: 9 ft; width: 6 ft **101.** 2 in.
102. 50 yd, 120 yd, and 130 yd

Chapter Test

1. degree: 3; leading coefficient: -6 **2.** degree: 9; leading coefficient: 7 **3.** 6; 4 **4.** falls to the left and falls to the right **5.** falls to the left and rises to the right **6.** $7x^3y - 18x^2y - y - 9$ **7.** $11x^2 - 13x - 6$ **8.** $35x^7y^3$ **9.** $x^3 - 4x^2y + 2xy^2 + y^3$ **10.** $21x^2 - 20xy - 9y^2$
11. $4x^2 - 25y^2$ **12.** $16y^2 - 56y + 49$ **13.** $x^2 + 4x + 4 - 9y^2$ **14.** $3x^2 + x - 10; 60$ **15.** $2ah + h^2 - 5h$ **16.** $x^2(14x - 15)$
17. $(9y + 5)(9y - 5)$ **18.** $(x + 3)(x + 5)(x - 5)$ **19.** $(5x - 3)^2$ **20.** $(x + 5 + 3y)(x + 5 - 3y)$ **21.** prime **22.** $(y + 2)(y - 18)$
23. $(2x + 5)(7x + 3)$ **24.** $5(x - 1)(x^2 + x + 1)$ **25.** $3(2x + y)(2x - y)$ **26.** $2(3x - 1)(2x - 5)$ **27.** $3(x^2 + 1)(x + 1)(x - 1)$
28. $(x^4 + y^4)(x^2 + y^2)(x + y)(x - y)$ **29.** $4x^2y(3y^3 + 2xy - 9)$ **30.** $(x^3 + 2)(x^3 - 14)$ **31.** $(x^2 - 6)(x^2 + 4)$ **32.** $3y(4x - 1)(x - 2)$
33. $y(y - 3)(y^2 + 2)$ **34.** $\left\{-\dfrac{1}{3}, 2\right\}$ **35.** $\left\{-1, \dfrac{6}{5}\right\}$ **36.** $\left\{0, \dfrac{1}{3}\right\}$ **37.** $\{-1, 1, 4\}$ **38.** 7 seconds **39.** length: 5 yd; width: 3 yd
40. 12 units, 9 units, and 15 units

Cumulative Review Exercises

1. $\{4\}$ **2.** $\left\{\left(-4, \dfrac{1}{2}\right)\right\}$ **3.** $\{(2, 1, -1)\}$ **4.** (2, 3) **5.** $(-\infty, -2] \cup [7, \infty)$ **6.** $\left\{1, \dfrac{5}{2}\right\}$
7. $\{-5, 0, 2\}$ **8.** $x = \dfrac{b}{c - a}$ **9.** $f(x) = 2x + 1$ **10.** winner: 1480 votes; loser: 1320 votes

11. $f(x) = -\dfrac{1}{3}x + 1$ **12.** $4x - 5y < 20$ **13.** $y \le -1$ **14.** $-\dfrac{y^{10}}{2x^6}$ **15.** 7.06×10^{-5} **16.** $9x^4 - 6x^2y + y^2$
17. $9x^4 - y^2$ **18.** $(x + 3)(x - 3)^2$
19. $x^2(x^2 + 1)(x + 1)(x - 1)$ **20.** $14x^3y^2(1 - 2x)$

CHAPTER 6

Section 6.1 Check Point Exercises

1. a. 80; The cost to remove 40% of the lake's pollutants is $80 thousand.; (40, 80) **b.** 180; The cost to remove 60% of the lake's pollutants is $180 thousand.; (60, 180) **2.** $(-\infty, -3) \cup \left(-3, \frac{1}{2}\right) \cup \left(\frac{1}{2}, \infty\right)$ **3.** $x + 5$ **4. a.** $\dfrac{x - 5}{3x - 1}$ **b.** $\dfrac{x + 4y}{3x(x + y)}$ **5.** $\dfrac{x + 3}{x - 4}$

6. $\dfrac{-4(x + 2)}{3x(3x - 2)}$ or $-\dfrac{4(x + 2)}{3x(3x - 2)}$ **7. a.** $9(3x + 7)$ **b.** $\dfrac{x + 3}{5}$

Concept and Vocabulary Check

1. polynomial; polynomial **2.** zero **3.** asymptote **4.** asymptote **5.** factoring; common factors **6.** $x + 5$ **7.** false

8. -1 **9.** numerators; denominators **10.** multiplicative inverse/reciprocal; $\dfrac{S}{R}; \dfrac{PS}{QR}$ **11.** $\dfrac{x^2}{70}$ **12.** $\dfrac{10}{7}$

Exercise Set 6.1

1. $-5; -3; 2$ **3.** 0; does not exist; $-\dfrac{21}{2}$ **5.** $-\dfrac{7}{2}; -5; \dfrac{11}{5}$ **7.** $(-\infty, 5) \cup (5, \infty)$ **9.** $(-\infty, -3) \cup (-3, 1) \cup (1, \infty)$ **11.** $(-\infty, -5) \cup (-5, \infty)$

13. $(-\infty, 3) \cup (3, 5) \cup (5, \infty)$ **15.** $\left(-\infty, -\dfrac{4}{3}\right) \cup \left(-\dfrac{4}{3}, 2\right) \cup (2, \infty)$ **17.** 4 **19.** domain: $(-\infty, -2) \cup (-2, 2) \cup (2, \infty)$; range: $(-\infty, 0] \cup (3, \infty)$

21. As x decreases, the function values are approaching 3.; $y = 3$ **23.** There is no point on the graph with x-coordinate -2.

25. The graph is not continuous. Furthermore, it neither rises nor falls without bound to either the left or the right. **27.** $x + 2$ **29.** $\dfrac{1}{x - 3}$

31. $\dfrac{4}{x}$ **33.** $\dfrac{4}{y + 5}$ **35.** $-\dfrac{1}{3x + 5}$ **37.** $\dfrac{y + 7}{y - 7}$ **39.** $\dfrac{x + 9}{x - 1}$ **41.** cannot be simplified **43.** $-\dfrac{x + 3}{x + 4}$ **45.** $\dfrac{x + 5y}{3x - y}$ **47.** $\dfrac{x^2 + 2x + 4}{x + 2}$

49. $x^2 - 3$ **51.** $\dfrac{3}{2}$ **53.** $\dfrac{x + 7}{x}$ **55.** $\dfrac{x + 3}{x - 2}$ **57.** $\dfrac{x + 4}{x + 2}$ **59.** $-\dfrac{2}{y(y + 3)}$ **61.** $\dfrac{y^2 + 2y + 4}{2y}$ **63.** $\dfrac{x^2 + x + 1}{x - 2}$ **65.** 4

67. $\dfrac{4(x + y)}{3(x - y)}$ **69.** 1 **71.** 1 **73.** $\dfrac{9}{28}$ **75.** $\dfrac{7}{10}$ **77.** $\dfrac{x}{3}$ **79.** $\dfrac{y - 5}{2}$ **81.** $\dfrac{(x^2 + 4)(x - 4)}{x - 1}$ **83.** $\dfrac{y - 7}{y - 5}$ **85.** $(x - 1)(2x - 1)$

87. $\dfrac{1}{x - 2y}$ **89.** $3x^2$ **91.** $\dfrac{x + 1}{x(x - 1)}$ **93.** $\dfrac{-(x - y)}{(x + y)(b + c)}$ **95.** $\dfrac{a^2 + 1}{b(a + 2)}$ **97.** $-\dfrac{a - b}{4c^2}$ or $\dfrac{b - a}{4c^2}$ **99.** 7 **101.** $2a + h - 5$

103. $\left(\dfrac{f}{g}\right)(x) = -x - 2; (-\infty, -2) \cup \left(-2, \dfrac{1}{2}\right) \cup \left(\dfrac{1}{2}, \infty\right)$ **105.** 195; The cost to inoculate 60% of the population is $195 million.; (60, 195)

107. 100; This indicates that we cannot inoculate 100% of the population. **109.** after 6 minutes; about 4.8 **111.** 6.5; Over time the pH level rises back to normal. **113.** 90; Lung cancer has an incidence ratio of 10 between the ages of 55 and 64. 90% of deaths from lung cancer in this group are smoking related.; (10, 90) **115.** $y = 100$; As the incidence ratio increases, the percentage of smoking-related deaths is approaching 100%.

129.

131.

135. makes sense
137. does not make sense
139. false
141. false

Graphs coincide.; multiplied correctly   Graphs do not coincide; right side should be $x + 3$.

143. $f(x)' = \dfrac{x^2 - x - 2}{x - 2}$ **145.** $\dfrac{y^n - 1}{y^n + 4}$ **147.** $4x - 5y \ge 20$ **148.** $2x^3 - 11x^2 + 3x + 30$

149. $16a^8 b^{26} c^2$

150. $\dfrac{2}{5}$ **151.** $\dfrac{7}{6}$

152. $\dfrac{1}{6}$

Section 6.2 Check Point Exercises

1. $\dfrac{x - 5}{x + 3}$ **2.** $\dfrac{1}{x - y}$ **3.** $18x^2$ **4.** $5x(x + 3)(x + 3)$ or $5x(x + 3)^2$ **5.** $\dfrac{21 + 4x}{18x^2}$ or $\dfrac{4x + 21}{18x^2}$ **6.** $\dfrac{2x^2 - 2x + 8}{(x - 4)(x + 4)}$ or $\dfrac{2(x^2 - x + 4)}{(x - 4)(x + 4)}$

7. $\dfrac{x^2 - 3x + 5}{(x - 3)(x + 1)(x - 2)}$ **8.** $\dfrac{3}{y + 2}$ **9.** $\dfrac{3x - 5y}{x - 3y}$

Concept and Vocabulary Check

1. $\dfrac{P + Q}{R}$; numerators; common denominator **2.** $\dfrac{P - Q}{R}$; numerators; common denominator **3.** $\dfrac{x - 5 + y}{3}$ **4.** $\dfrac{-1}{-1}$

factor denominators **6.** $x + 3$ and $x - 2$; $x + 3$ and $x + 1$; $(x + 3)(x - 2)(x + 1)$ **7.** $2x$ **8.** $3y + 4$ **9.** -1

Exercise Set 6.2

1. $\dfrac{2}{3x}$ 3. $\dfrac{10x + 3}{x - 5}$ 5. $\dfrac{2x - 1}{x + 3}$ 7. $\dfrac{y - 3}{y + 3}$ 9. $\dfrac{x + 1}{4x - 3}$ 11. $\dfrac{x - 3}{x - 1}$ 13. $\dfrac{4y + 3}{2y - 1}$ 15. $\dfrac{x^2 + xy + y^2}{x + y}$ 17. $175x^2$

19. $(x - 5)(x + 5)$ 21. $y(y + 10)(y - 10)$ 23. $(x + 4)(x - 4)(x - 4)$ 25. $(y - 5)(y - 6)(y + 1)$ 27. $(y + 2)(2y + 3)(y - 2)(2y + 1)$

29. $\dfrac{3 + 50x}{5x^2}$ 31. $\dfrac{7x - 2}{(x - 2)(x + 1)}$ 33. $\dfrac{3x - 4}{(x + 2)(x - 3)}$ 35. $\dfrac{2x^2 - 2x + 61}{(x + 5)(x - 6)}$ 37. $\dfrac{20 - x}{(x - 5)(x + 5)}$ 39. $\dfrac{13}{(y - 3)(y - 2)}$ 41. $\dfrac{1}{x + 3}$

43. $\dfrac{6x^2 + 14x + 10}{(x + 1)(x + 4)(x + 3)}$ 45. $-\dfrac{x^2 - 3x - 13}{(x + 4)(x - 2)(x + 1)}$ 47. $\dfrac{4x - 11}{x - 3}$ 49. $\dfrac{2y - 14}{(y - 4)(y + 4)}$ 51. $\dfrac{x^2 + 2x - 14}{3(x + 2)(x - 2)}$ 53. $\dfrac{16}{x - 4}$

55. $\dfrac{12x + 7y}{(x - y)(x + y)}$ 57. $\dfrac{x^2 - 18x - 30}{(5x + 6)(x - 2)}$ 59. $\dfrac{3x^2 + 16xy - 13y^2}{(x + 5y)(x - 5y)(x - 4y)}$ 61. $\dfrac{8x^3 - 32x^2 + 23x + 6}{(x + 2)(x - 2)(x - 2)(x - 1)}$

63. $\dfrac{16a^2 - 12ab - 18b^2}{(3a + 4b)(3a - 4b)(2a - b)}$ 65. $\dfrac{m + 6}{(m - 1)(m + 2)(2m - 1)}$ 67. $\dfrac{x^2 + 5x + 8}{(x + 2)(x + 1)}$ 69. 2   71. $-\dfrac{1}{x(x + h)}$ 73. $\dfrac{2d}{a^2 + ab + b^2}$

75. $(f - g)(x) = \dfrac{x + 2}{x + 3}; (-\infty, -5) \cup (-5, -3) \cup (-3, \infty)$ 77. 11; If you average 0 miles per hour over the speed limits, the total driving time is

about·11 hours.; $(0, 11)$ 79. $\dfrac{720x + 48,050}{(x + 70)(x + 65)}$; 11   81. 12 mph; Answers will vary.

83. **a.** 307; There are 307 arrests for every 100,000 drivers who are 20 years old.; $(20, 307)$ **b.** $f(x) = \dfrac{-5x^3 + 27,680x - 388,150}{x^2 + 9}$

c. 25; 356   85. $\dfrac{2x(2x + 15)}{(x + 7)(x + 8)}$ 91. Answers will vary.; $\dfrac{b + a}{ab}$ 93. makes sense   95. does not make sense   97. false   99. false

101. $\dfrac{1}{x^{2n} - 1}$ 103. $\dfrac{x - y + 1}{(x - y)(x - y)}$ 104. $\dfrac{y^{10}}{9x^4}$ 105. $\left[-\dfrac{13}{3}, 5\right]$ 106. $2x(5x + 3)(5x - 3)$ 107. $xy^2 + y^3$ 108. $-h$ 109. $\dfrac{x + 1}{x - 3}$

Section 6.3 Check Point Exercises

1. $\dfrac{y}{x + y}$ 2. $-\dfrac{1}{x(x + 7)}$ 3. $\dfrac{2x}{x^2 + 1}$ 4. $\dfrac{x + 2}{x - 5}$

Concept and Vocabulary Check

1. complex; complex   2. $\dfrac{7x + 5}{5x + x^2}$ 3. $\dfrac{1}{x + 3}; \dfrac{1}{x}; x; x + 3; -3; -\dfrac{1}{x(x + 3)}$

Exercise Set 6.3

1. $\dfrac{4x + 2}{x - 3}$ 3. $\dfrac{x^2 + 9}{(x - 3)(x + 3)}$ 5. $\dfrac{y + x}{y - x}$ 7. $\dfrac{4 - x}{5x - 3}$ 9. $\dfrac{1}{x - 3}$ 11. $-\dfrac{1}{x(x + 5)}$ 13. $-x$ 15. $-\dfrac{x + 1}{x - 1}$ 17. $\dfrac{(x + y)(x + y)}{xy}$

19. $\dfrac{4x}{x^2 + 4}$ 21. $\dfrac{2y^3 + 5x^2}{y^3(5 - 3x^2)}$ 23. $-\dfrac{12}{5}$ 25. $\dfrac{3ab(b + a)}{(2b + 3a)(2b - 3a)}$ 27. $-\dfrac{x + 10}{5x - 2}$ 29. $\dfrac{2y}{3y + 7}$ 31. $\dfrac{2b + a}{b - 2a}$ 33. $\dfrac{4(7x + 5)}{69(x + 5)}$

35. $\dfrac{3x - 2y}{13y + 5x}$ 37. $\dfrac{m(m + 3)}{(m - 2)(m + 1)}$ 39. $\dfrac{3a(a - 3)}{(a + 4)(3a - 5)}$ 41. $-\dfrac{4}{(x + 2)(x - 2)}$ 43. 6   45. $x(x + 1)$ 47. $\dfrac{x + 4}{x + 2}$

49. $-\dfrac{3}{a(a + h)}$ 51. **a.** $\dfrac{Pi(1 + i)^n}{(1 + i)^n - 1}$ **b.** \$527 per month   53. $R = \dfrac{R_1 R_2 R_3}{R_2 R_3 + R_1 R_3 + R_1 R_2}$; about 2.18 ohms

59.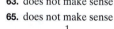

NORMAL FLOAT AUTO REAL RADIAN MP

Graphs coincide.; simplified correctly

61. NORMAL FLOAT AUTO REAL RADIAN MP

Graphs do not coincide.; $\dfrac{\dfrac{1}{x} + \dfrac{1}{3}}{\dfrac{1}{3x}} = x + 3$

63. does not make sense
65. does not make sense
67. $\dfrac{1}{(x + h + 1)(x + 1)}$
69. $\dfrac{a + 1}{a + 2}$
71. $\{3, 9\}$
72. $16x^4 - 8x^2y + y^2$
73. $(1, 5)$
74. $2xy^3$
75. $35 + \dfrac{2}{21}$
76. $7x$

Section 6.4 Check Point Exercises

1. $4x^2 - 8x + \dfrac{1}{2} + \dfrac{1}{x}$ 2. $3x^2y^2 - xy + \dfrac{2}{y}$ 3. $3x - 8$ 4. $2x^2 - x + 3 - \dfrac{6}{2x - 1}$ 5. $2x^2 + 7x + 14 + \dfrac{21x - 10}{x^2 - 2x}$

Concept and Vocabulary Check

1. $16x^3 - 32x^2 + 2x + 4; 4x$ 2. $2x^3 + 0x^2 + 6x - 4$ 3. $6x^3; 3x; 2x^2; 7x^2$ 4. $2x^2; 5x - 2; 10x^3 - 4x^2; 10x^3 + 6x^2$

5. $6x^2 - 10x; 6x^2 + 8x; 18x; -4; 18x - 4$ 6. $9; 3x - 5; 9; 3x - 5 + \dfrac{9}{2x + 1}$ 7. divisor; quotient; remainder; dividend

Exercise Set 6.4

1. $5x^4 - 3x^2 + 2$ **3.** $6x^2 + 2x - 3 - \dfrac{2}{x}$ **5.** $7x - \dfrac{7}{4} - \dfrac{4}{x}$ **7.** $-5x^3 + 10x^2 - \dfrac{3}{5}x + 8$ **9.** $2a^2b - a - 3b$ **11.** $6x - \dfrac{3}{y} - \dfrac{2}{xy^2}$

13. $x + 3$ **15.** $x^2 + x - 2$ **17.** $x - 2 + \dfrac{2}{x - 5}$ **19.** $x + 5 - \dfrac{10}{2x + 3}$ **21.** $x^2 + 2x + 3 + \dfrac{1}{x + 1}$ **23.** $2y^2 + 3y - 1$

25. $3x^2 - 3x + 1 + \dfrac{2}{3x + 2}$ **27.** $2x^2 + 4x + 5 + \dfrac{9}{2x - 4}$ **29.** $2y^2 + y - 2 - \dfrac{2}{2y - 1}$ **31.** $2y^3 + 3y^2 - 4y + 1$

33. $4x^2 + 3x - 8 + \dfrac{18}{x^2 + 3}$ **35.** $5x^2 + x + 3 + \dfrac{x + 7}{3x^2 - 1}$ **37.** $\left(\dfrac{f}{g}\right)(x) = 2x^2 - 9x + 10$ **39.** $\left(\dfrac{f}{g}\right)(x) = x^3 - x^2 + x - 2$

41. $x^3 - x^2y + xy^2 - y^3 + \dfrac{2y^4}{x + y}$ **43.** $3x^2 + 2x - 1$ **45.** $4x - 7 + \dfrac{4x + 8}{x^2 + x + 1}$ **47.** $x^3 + x^2 - x - 3 + \dfrac{-x + 5}{x^2 - x + 2}$

49. $4x^2 + 5xy - y^2$ **51.** $\left(\dfrac{f - g}{h}\right)(x) = 2x^2 - 5x + 8; \left(-\infty, -\dfrac{1}{4}\right) \cup \left(-\dfrac{1}{4}, \infty\right)$ **53.** $x = 3a^2 + 4a - 2$

55. 70; At the tax rate of 30%, the government tax revenue is \$70 tens of billions, or \$700 billion.; (30, 70)

57. $f(x) = 80 + \dfrac{800}{x - 110}$; 70; yes; Answers will vary.

65.

Graphs coincide.; division correct

67.

Graphs do not coincide; right side should be $3x^3 - 8x^2 - 5$.

69. makes sense **71.** makes sense **73.** false

75. false **77.** $x^{2n} + x^n + 3 + \dfrac{3}{x^n - 5}$ **79.** $x - 2$

81. $\left(-\infty, -\dfrac{1}{2}\right) \cup \left(\dfrac{7}{2}, \infty\right)$ **82.** 4.061×10^7

83. $22x + 12$ **84. a.** $5x^2 - 10x + 26 - \dfrac{44}{x + 2}$

b. $-10; 26; -44$; The numbers are the coefficients of the quotient and the remainder.

85. a. $3x^2 - 7x + 9 - \dfrac{10}{x + 1}$ **b.** $-7; 9; -10$; The numbers are the coefficients of the quotient and the remainder.

86. $2x^2 + 3x - 2; (x - 3)(2x - 1)(x + 2)$

Mid-Chapter Check Point Exercises

1. $\dfrac{x + 2}{x + 6}$ **2.** $\dfrac{3(x - 2)}{x - 1}$ **3.** $\dfrac{x^2 + 3x + 9}{x - 1}$ **4.** $\dfrac{5x - 3}{x - 2}$ **5.** $\dfrac{x^2 + 6x - 4}{x^2 + 6x + 1}$ **6.** $2x^3 - 5x^2 - 3x + 6$ **7.** $x + y$ **8.** $4x^6y^2 - 2x^4y + \dfrac{3}{7}y$

9. $\dfrac{x^2 - 14x - 16}{(x + 6)(x - 2)}$ **10.** $\dfrac{10}{(x + 2)(x - 2)}$ **11.** $\dfrac{2(x - 3)}{x - 1}$ **12.** $\dfrac{x - 5}{x - 7}$ **13.** $2x^2 - x - 3 + \dfrac{x + 1}{3x^2 - 1}$ **14.** $\dfrac{5x + 2}{3x - 1}$ **15.** $\dfrac{5x}{(x - 6)(x - 1)(x + 4)}$

16. $\dfrac{7x + 4}{4(x + 1)}$ **17.** $x + 2$ **18.** $16x^2 - 8x + 4 - \dfrac{2}{2x + 1}$ **19.** $\dfrac{3x + 2}{(x + 2)(x - 1)}$ **20.** Domain: $(-\infty, -7) \cup (-7, 2) \cup (2, \infty)$; $f(x) = \dfrac{5}{x + 7}$

Section 6.5 Check Point Exercises

1. $x^2 - 2x - 3$ **2.** -105 **3.** The remainder is zero.; $-1, -\dfrac{1}{3}$, and $\dfrac{2}{5}$, or $\left\{-1, -\dfrac{1}{3}, \dfrac{2}{5}\right\}$

Concept and Vocabulary Check

1. 4; 1; 5; -7; 1 **2.** -5; 4; 0, -8; -2 **3.** true **4.** $f(c)$

Exercise Set 6.5

1. $2x + 5$ **3.** $3x - 8 + \dfrac{20}{x + 5}$ **5.** $4x^2 + x + 4 + \dfrac{3}{x - 1}$ **7.** $6x^4 + 12x^3 + 22x^2 + 48x + 93 + \dfrac{187}{x - 2}$ **9.** $x^3 - 10x^2 + 51x - 260 + \dfrac{1300}{5 + x}$

11. $3x^2 + 3x - 3$ **13.** $x^4 + x^3 + 2x^2 + 2x + 2$ **15.** $x^3 + 4x^2 + 16x + 64$ **17.** $2x^4 - 7x^3 + 15x^2 - 31x + 64 - \dfrac{129}{x + 2}$ **19.** -25

21. -133 **23.** 240 **25.** 1 **27.** The remainder is 0.; $\{-1, 2, 3\}$ **29.** The remainder is 0.; $\left\{-\dfrac{1}{2}, 1, 2\right\}$ **31.** The remainder is 0.; $\left\{-5, \dfrac{1}{3}, \dfrac{1}{2}\right\}$

33. 2; The remainder is zero.; $-3, -1$, and 2, or $\{-3, -1, 2\}$ **35.** 1; The remainder is zero.; $\dfrac{1}{3}, \dfrac{1}{2}$, and 1, or $\left\{\dfrac{1}{3}, \dfrac{1}{2}, 1\right\}$ **37.** $4x^3 - 9x^2 + 7x - 6$

39. $0.5x^2 - 0.4x + 0.3$ **41. a.** The remainder is 0. **b.** 3 mm **49.** makes sense **51.** makes sense **53.** The remainder is 0.; $\{-2, -1, 2, 5\}$

54. $(-10, \infty)$ **55.** $\{-5, -1\}$ **56.** $\{(5, 3)\}$ **57.** $6x$ **58.** $3x$ **59.** $(x + 3)(x - 3)$

Section 6.6 Check Point Exercises

1. 6 or $\{6\}$ **2.** 4 or $\{4\}$ **3.** no solution or $\varnothing$ **4.** 4 and 6, or $\{4, 6\}$ **5.** 1 and 7, or $\{1, 7\}$ **6.** 50%

Concept and Vocabulary Check

1. least common denominator (LCD) **2.** 0 **3.** $2x$ **4.** $(x + 5)(x + 1)$ **5.** $x \neq 2; x \neq -4$ **6.** $5(x + 3) + 3(x + 4) = 12x + 9$

7. true

Exercise Set 6.6

1. $\{1\}$ **3.** $\{3\}$ **5.** $\{4\}$ **7.** $\{-4\}$ **9.** $\left\{\dfrac{9}{7}\right\}$ **11.** no solution or $\varnothing$ **13.** $\{2\}$ **15.** $\{2\}$ **17.** $\{5\}$ **19.** $\{-7, -1\}$ **21.** $\{-6, 3\}$

23. $\{-1\}$ **25.** no solution or $\varnothing$ **27.** $\{2\}$ **29.** no solution or $\varnothing$ **31.** $\left\{-\dfrac{4}{3}\right\}$ **33.** $\left\{-6, \dfrac{1}{2}\right\}$ **35.** $4, 10$ **37.** $-3, -\dfrac{3}{4}$ **39.** $\dfrac{x-2}{x(x+1)}$

41. $\{2\}$ **43.** $\{4\}$ **45.** $\dfrac{-2(x-4)}{(x-2)(x^2+2x+4)}$ **47.** 0 **49.** $1, 7$ **51.** 60% **53.** 10 days; $(10, 8)$ **55.** $y = 5$; On average, the students

remembered 5 words over an extended period of time. **57.** 11 learning trials; $(11, 0.95)$ **59.** As the number of learning trials increases, the proportion of correct responses increases.; Initially, the proportion of correct responses increases rapidly, but slows down as time increases.
61. 125 liters

71. ; $\{-5, 5\}$ **73.** 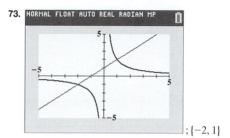 ; $\{-2, 1\}$

75. does not make sense **77.** makes sense **79.** false **81.** false **83.** no solution or $\varnothing$

87. $x + 2y \geq 2$
$x - y \geq -4$
88. $\{15\}$ **89.** $F = \dfrac{9C + 160}{5}$ **90.** $p = \dfrac{qf}{q - f}$ **91.** $\{-20, 30\}$ **92.** 40 miles per hour

Section 6.7 Check Point Exercises

1. $x = \dfrac{b - 2a}{a}$ **2.** $x = \dfrac{yz}{y - z}$ **3. a.** $C(x) = 500{,}000 + 400x$ **b.** $\overline{C}(x) = \dfrac{500{,}000 + 400x}{x}$ **c.** 10,000 wheelchairs **4.** 12 mph
5. 6 months **6.** experienced carpenter: 8 hr; apprentice: 24 hr

Concept and Vocabulary Check

1. xyz **2.** fixed; variable **3.** number of units produced **4.** distance traveled; rate of travel **5.** 1 **6.** $\dfrac{x}{19}$

Exercise Set 6.7

1. $P_1 = \dfrac{P_2 V_2}{V_1}$ **3.** $f = \dfrac{pq}{q + p}$ **5.** $r = \dfrac{A - P}{P}$ **7.** $m_1 = \dfrac{Fd^2}{Gm_2}$ **9.** $x = \bar{x} + zs$ **11.** $R = \dfrac{E - Ir}{I}$ **13.** $f_1 = \dfrac{ff_2}{f_2 - f}$
15. 50,000 wheelchairs **17.** $y = 400$; Average cost is nearing \$400 as production level increases. **19. a.** $C(x) = 100{,}000 + 100x$
b. $\overline{C}(x) = \dfrac{100{,}000 + 100x}{x}$ **c.** 500 bikes **21.** 5 mph **23.** The time increases as the running rate is close to zero miles per hour.
25. car: 50 mph; bus: 30 mph **27.** 6 mph **29.** 3 mph **31.** 500 mph **33.** 4.2 ft per sec **35.** 12 miles **37.** 18 min; yes **39.** 4 hr
41. 12 hr **43.** 10 hr **45.** 10 hr **47.** 10 min **49.** 3 **51.** $\dfrac{1}{3}$ and 3 **53.** 10 **55.** $x = \dfrac{ab}{a + b}$
67. makes sense **69.** makes sense **71.** false **73.** true **75.** $f = \dfrac{p - s}{ps - s}$ **77.** 8 hr **78.** $(x + 2 + 3y)(x + 2 - 3y)$
79. $\{(5, -3)\}$ **80.** $\{(3, -1, 2)\}$ **81. a.** 16 **b.** $y = 16x^2$ **c.** 400 **82. a.** 96 **b.** $y = \dfrac{96}{x}$ **c.** 32 **83.** 8

Section 6.8 Check Point Exercises

1. 66 gal **2.** about 556 ft **3.** 512 cycles per second **4.** 24 min **5.** 96π cubic feet

Concept and Vocabulary Check

1. $y = kx$; constant of variation **2.** $y = kx^n$ **3.** $y = \dfrac{k}{x}$ **4.** $y = \dfrac{kx}{z}$ **5.** $y = kxz$ **6.** directly; inversely **7.** jointly; inversely

Exercise Set 6.8

1. 156 **3.** 30 **5.** $\dfrac{5}{6}$ **7.** 240 **9.** 50 **11.** $x = kyz$; $y = \dfrac{x}{kz}$ **13.** $x = \dfrac{kz^3}{y}$; $y = \dfrac{kz^3}{x}$ **15.** $x = \dfrac{kyz}{\sqrt{w}}$; $y = \dfrac{x\sqrt{w}}{kz}$

17. $x = kz(y + w)$; $y = \dfrac{x - kzw}{kz}$ **19.** $x = \dfrac{kz}{y - w}$; $y = \dfrac{xw + kz}{x}$ **21.** 5.4 ft **23.** 80 in. **25.** about 607 lb **27.** 32°

29. a. $L = \dfrac{1890}{R}$ **b.** an approximate model **c.** 70 yr **31. a.** 90 beats per minute **b.** 95 beats per minute **c.** by the point $(63, 30)$

33. 90 milliroentgens per hour **35.** This person has a BMI of 24.4 and is not overweight. **37.** 1800 Btu **39.** $\dfrac{1}{4}$ of what it was originally

41. a. $C = \dfrac{kP_1P_2}{d^2}$ **b.** $k \approx 0.02$; $C = \dfrac{0.02P_1P_2}{d^2}$ **c.** approximately 39,813 daily phone calls

43. a.

49. z varies directly as the square root of x and inversely as the square of y. **53.** does not make sense **55.** makes sense **57.** The wind pressure is 4 times more destructive. **59.** Distance is increased by $\sqrt{50}$, or about 7.07, for the space telescope. **60.** -2 **61.** $(x + 3)(x - 3)(y - 3)$ **62.** 4 **63.** 3 **64.** 6 **65.** domain: $[-4, \infty)$; range: $[0, \infty)$

b. Current varies inversely as resistance. **c.** $R = \dfrac{6}{I}$

Review Exercises

1. a. $\dfrac{7}{4}$ **b.** $\dfrac{3}{4}$ **c.** does not exist **d.** 0 **2.** $(-\infty, -4) \cup (-4, 3) \cup (3, \infty)$ **3.** $(-\infty, -2) \cup (-2, 1) \cup (1, \infty)$ **4.** $\dfrac{x^2 - 7}{3x}$ **5.** $\dfrac{x - 1}{x - 7}$

6. $\dfrac{3x + 2}{x - 5}$ **7.** cannot be simplified **8.** $\dfrac{x^2 + 2x + 4}{x + 2}$ **9.** $\dfrac{5(x + 1)}{3}$ **10.** $\dfrac{2x - 1}{x + 1}$ **11.** $\dfrac{x - 7}{x^2}$ **12.** $\dfrac{1}{3(x + 3)}$ **13.** $\dfrac{1}{2}$

14. $\dfrac{(y + 4)(y + 2)}{(y - 6)(y^2 + 4y + 16)}$ **15.** $\dfrac{3x^2}{2x - 5}$ **16. a.** 50 deer **b.** 150 deer **c.** $y = 225$; The population will approach 225 deer over time.

17. 4 **18.** $\dfrac{1}{x + 3}$ **19.** $3x + 2$ **20.** $36x^3$ **21.** $(x + 7)(x - 5)(x + 2)$ **22.** $\dfrac{3x - 5}{x(x - 5)}$ **23.** $\dfrac{5x - 2}{(x - 3)(x + 2)(x - 2)}$

24. $\dfrac{2x - 8}{(x - 3)(x - 5)}$ or $\dfrac{2(x - 4)}{(x - 3)(x - 5)}$ **25.** $\dfrac{4x}{(3x + 4)(3x - 4)}$ **26.** $\dfrac{y - 3}{(y + 1)(y + 3)}$ **27.** $\dfrac{2x^2 - 9}{(x + 3)(x - 3)}$ **28.** $3(x + y)$ or $3x + 3y$

29. $\dfrac{3}{8}$ **30.** $\dfrac{x}{x - 5}$ **31.** $\dfrac{3x + 8}{3x + 10}$ **32.** $\dfrac{4(x - 2)}{2x + 5}$ **33.** $\dfrac{3x^2 + 9x}{x^2 + 8x - 33}$ **34.** $\dfrac{1 + x}{1 - x}$ **35.** $3x - 6 + \dfrac{2}{x} - \dfrac{2}{5x^2}$ **36.** $6x^3y + 2xy - 10x$

37. $3x - 7 + \dfrac{26}{2x + 3}$ **38.** $2x^2 - 4x + 1 - \dfrac{10}{5x - 3}$ **39.** $x^5 + 5x^4 + 8x^3 + 16x^2 + 33x + 63 + \dfrac{128}{x - 2}$ **40.** $2x^2 + 3x - 1$

41. $4x^2 - 7x + 5 - \dfrac{4}{x + 1}$ **42.** $3x^3 + 6x^2 + 10x + 10$ **43.** $x^3 - 4x^2 + 16x - 64 + \dfrac{272}{x + 4}$ **44.** 3 **45.** 4 **46.** solution

47. not a solution **48.** The remainder is 0.; $\left\{-1, \dfrac{1}{3}, \dfrac{1}{2}\right\}$ **49.** $\{6\}$ **50.** $\{52\}$ **51.** $\{7\}$ **52.** $\left\{-\dfrac{9}{2}\right\}$ **53.** $\left\{-\dfrac{1}{2}, 3\right\}$ **54.** $\{-23, 2\}$

55. $\{-3, 2\}$ **56.** 80% **57.** $C = R - nP$ **58.** $T_1 = \dfrac{P_1V_1T_2}{P_2V_2}$ **59.** $P = \dfrac{A}{rT + 1}$ **60.** $R = \dfrac{R_1R_2}{R_2 + R_1}$ **61.** $n = \dfrac{IR}{E - Ir}$

62. a. $C(x) = 50{,}000 + 25x$ **b.** $\overline{C}(x) = \dfrac{50{,}000 + 25x}{x}$ **c.** 5000 graphing calculators **63.** 12 mph **64.** 9 mph **65.** 2 hr; no

66. faster crew: 36 hr; slower crew: 45 hr **67.** 240 min or 4 hr **68.** \$4935 **69.** 1600 ft **70.** 440 vibrations per second **71.** 112 decibels **72.** 16 hr **73.** 800 cubic feet

Chapter Test

1. $(-\infty, 2) \cup (2, 5) \cup (5, \infty)$; **2.** $\dfrac{x}{x - 4}$ **3.** $\dfrac{(x + 3)(x - 1)}{x + 1}$ **4.** $\dfrac{x - 2}{x + 5}$ **5.** $\dfrac{3x - 24}{x - 3}$ **6.** $\dfrac{1}{3x + 5}$ **7.** $\dfrac{x^2 + 2x + 15}{(x + 3)(x - 3)}$
$\dfrac{x}{x - 5}$

8. $\dfrac{3x - 4}{(x + 2)(x - 3)}$ **9.** $\dfrac{5x^2 - 7x + 4}{(x + 2)(x - 1)(x - 2)}$ **10.** $\dfrac{x + 3}{x + 5}$ **11.** $\dfrac{x - 2}{x - 10}$ **12.** $\dfrac{x - 2}{8}$ **13.** $1 - x$ **14.** $3x^2y^2 + 4y^2 - \dfrac{5}{2}y$

15. $3x^2 - 3x + 1 + \dfrac{2}{3x + 2}$ **16.** $3x^2 + 2x + 3 + \dfrac{9 - 6x}{x^2 - 1}$ **17.** $3x^3 - 4x^2 + 7$ **18.** 12 **19.** solution **20.** $\{-7, 3\}$ **21.** $\{-1\}$

22. 3 years **23.** $a = \dfrac{Rs}{s - R}$ or $a = -\dfrac{Rs}{R - s}$ **24. a.** $C(x) = 300{,}000 + 10x$ **b.** $\overline{C}(x) = \dfrac{300{,}000 + 10x}{x}$ **c.** 20,000 players **25.** 12 hr **26.** 4 mph **27.** 45 foot-candles

Cumulative Review Exercises

1. $(-\infty, -6)$ **2.** $\left\{-\dfrac{1}{2}, 4\right\}$ **3.** $\{(-2, 0, 4)\}$ **4.** $\left[-2, \dfrac{14}{3}\right]$ **5.** $\{12\}$ **6.** $s = \dfrac{2R - Iw}{2I}$ **7.** $\{(3, 2)\}$ **8.** $f(x) = -3x - 2$

9.

$y = |x| + 2$

10. $y \geq 2x - 1$
$x \geq 1$

11. $2x - y < 4$

12. $x^2 + 4x + 4 - 9y^2$ **13.** $\dfrac{x(x + 1)}{2x + 5}$ **14.** $\dfrac{x + 25}{2(x - 4)(x - 5)}$

15. $3x + 4 + \dfrac{2}{x + 2}$ **16.** $(y - 6)(x + 2)$

17. $2xy(2x + 3)(6x - 5)$ **18.** 4 sec **19.** basic cable service: \$25; each movie channel: \$10 **20.** 1 ft

CHAPTER 7

Section 7.1 Check Point Exercises

1. a. 8   **b.** −7   **c.** $\dfrac{4}{5}$ **d.** 0.09   **e.** 5   **f.** 7   **2. a.** 4   **b.** $-\sqrt{24} \approx -4.90$ **3.** $[3, \infty)$ **4.** approximately 15.7 minutes
5. a. 7   **b.** $|x + 8|$ **c.** $|7x^5|$ or $7|x^5|$ **d.** $|x - 3|$ **6. a.** 3   **b.** −2   **7.** $-3x$ **8. a.** 2   **b.** −2   **c.** not a real number
d. −1   **9. a.** $|x + 6|$ **b.** $3x - 2$ **c.** 8

Concept and Vocabulary Check

1. principal   **2.** 8^2 **3.** $[0, \infty)$ **4.** $5x - 20 \geq 0$ **5.** $|a|$ **6.** 10^3 **7.** $(-5)^3$ **8.** a **9.** $(-\infty, \infty)$ **10.** nth; index
11. $|a|$; a **12.** true   **13.** false   **14.** true   **15.** false

Exercise Set 7.1

1. 6   **3.** −6   **5.** not a real number   **7.** $\dfrac{1}{5}$ **9.** $-\dfrac{3}{4}$ **11.** 0.9   **13.** −0.2   **15.** 3   **17.** 1   **19.** not a real number
21. 4; 1; 0; not a real number   **23.** $-5; -\sqrt{5} \approx -2.24; -1$; not a real number   **25.** 4; 2; 1; 6   **27.** $[3, \infty)$; c   **29.** $[-5, \infty)$; d   **31.** $(-\infty, 3]$; e
33. 5   **35.** 4   **37.** $|x - 1|$ **39.** $|6x^2|$ or $6x^2$ **41.** $-|10x^3|$ or $-10|x^3|$ **43.** $|x + 6|$ **45.** $-|x - 4|$ **47.** 3   **49.** −3
51. $\dfrac{1}{5}$ **53.** $-\dfrac{3}{10}$ **55.** 3; 2; −1; −4   **57.** −2; 0; 2   **59.** 1   **61.** 2   **63.** −2   **65.** not a real number   **67.** −1   **69.** not a real number
71. −4   **73.** 2   **75.** −2   **77.** x **79.** $|y|$ **81.** $-2x$ **83.** −5   **85.** 5   **87.** $|x + 3|$ **89.** $-2(x - 1)$

91.

domain: $[0, \infty)$; range: $[3, \infty)$

93.
$f(x) = \sqrt{x - 3}$
domain: $[0, \infty)$; range: $[3, \infty)$

95. $(-\infty, 15)$ **97.** $[1, 3)$ **99.** 3
101. a. 40.2 in.; underestimates by 0.6 in.   **b.** 0.9 in. per month
c. 0.2 in. per month; This is a much smaller rate of change.; The graph is not as steep between 50 and 60 as it is between 0 and 10.
103. 70 mph; The officer should not believe the motorist.;
Answers will vary.

115.

; Answers will vary   **117.**
domain of f: $[0, \infty)$; range of f: $[0, \infty)$
domain of g: $[0, \infty)$; range of g: $(-\infty, 0]$
domain of h: $(-\infty, 0]$; range of h: $[0, \infty)$
domain of k: $(-\infty, 0]$; range of k: $(-\infty, 0]$

119. does not make sense   **121.** makes sense   **123.** false   **125.** false   **127.** Answers will vary; an example is $f(x) = \sqrt{15 - 3x}$.
129. $|(2x + 3)^5|$ **131.** The graph of h is the graph of f shifted left 3 units.

132. $7x + 30$ **133.** $\dfrac{x^8}{9y^6}$ **134.** $\left(-\infty, -\dfrac{7}{3}\right) \cup (5, \infty)$ **135.** $\dfrac{2^7}{x}$ or $\dfrac{128}{x}$ **136.** $\dfrac{2}{x^3}$ **137.** $\dfrac{y^{12}}{x^8}$

Section 7.2 Check Point Exercises

1. a. $\sqrt{25} = 5$ **b.** $\sqrt[3]{-8} = -2$ **c.** $\sqrt[4]{5xy^2}$ **2. a.** $(5xy)^{1/4}$ **b.** $\left(\dfrac{a^3b}{2}\right)^{1/5}$ **3. a.** $(\sqrt[3]{8})^4 = 16$ **b.** $(\sqrt{25})^3 = 125$

c. $-(\sqrt[4]{81})^3 = -27$ **4. a.** $6^{4/3}$ **b.** $(2xy)^{7/5}$ **5. a.** $\dfrac{1}{100^{1/2}} = \dfrac{1}{10}$ **b.** $\dfrac{1}{8^{1/3}} = \dfrac{1}{2}$ **c.** $\dfrac{1}{32^{3/5}} = \dfrac{1}{8}$ **d.** $\dfrac{1}{(3xy)^{5/9}}$ **6. a.** $7^{5/6}$ **b.** $\dfrac{5}{x}$
c. $9.1^{3/10}$ **d.** $\dfrac{y^{1/12}}{x^{1/5}}$ **7. a.** $\sqrt{x}$ **b.** $2a^4$ **c.** $\sqrt[4]{x^2y}$ **d.** $\sqrt[6]{x}$ **e.** $\sqrt[6]{x}$

Concept and Vocabulary Check

1. $\sqrt{36} = 6$ **2.** $\sqrt[3]{8} = 2$ **3.** $\sqrt[n]{a}$ **4.** $(\sqrt[4]{16})^3 = (2)^3 = 8$ **5.** $(\sqrt[n]{a})^m$ or $\sqrt[n]{a^m}$ **6.** $\dfrac{5}{3}$ **7.** $\dfrac{1}{16^{3/2}} = \dfrac{1}{(\sqrt{16})^3} = \dfrac{1}{(4)^3} = \dfrac{1}{64}$

Exercise Set 7.2

1. $\sqrt{49} = 7$ **3.** $\sqrt[3]{-27} = -3$ **5.** $-\sqrt[4]{16} = -2$ **7.** $\sqrt[3]{xy}$ **9.** $\sqrt[5]{2xy^3}$ **11.** $(\sqrt{81})^3 = 729$ **13.** $(\sqrt[3]{125})^2 = 25$

15. $(\sqrt[5]{-32})^3 = -8$ **17.** $(\sqrt[3]{27})^2 + (\sqrt[4]{16})^3 = 17$ **19.** $\sqrt[7]{(xy)^4}$ **21.** $7^{1/2}$ **23.** $5^{1/3}$ **25.** $(11x)^{1/5}$ **27.** $x^{3/2}$ **29.** $x^{3/5}$

31. $(x^2y)^{1/5}$ **33.** $(19xy)^{3/2}$ **35.** $(7xy^2)^{5/6}$ **37.** $2xy^{2/3}$ **39.** $\dfrac{1}{49^{1/2}} = \dfrac{1}{7}$ **41.** $\dfrac{1}{27^{1/3}} = \dfrac{1}{3}$ **43.** $\dfrac{1}{16^{3/4}} = \dfrac{1}{8}$ **45.** $\dfrac{1}{8^{2/3}} = \dfrac{1}{4}$

47. $\left(\dfrac{27}{8}\right)^{1/3} = \dfrac{3}{2}$ **49.** $\dfrac{1}{(-64)^{2/3}} = \dfrac{1}{16}$ **51.** $\dfrac{1}{(2xy)^{7/10}}$ **53.** $\dfrac{5x}{z^{1/3}}$ **55.** 3 **57.** 4 **59.** $x^{5/6}$ **61.** $x^{3/5}$ **63.** $\dfrac{1}{x^{5/12}}$ **65.** 25

67. $\dfrac{1}{y^{1/6}}$ **69.** $32x$ **71.** $5x^2y^3$ **73.** $\dfrac{x^{1/4}}{y^{3/10}}$ **75.** 3 **77.** $27y^{2/3}$ **79.** $\sqrt[4]{x}$ **81.** $2a^2$ **83.** x^2y^3 **85.** x^6y^6 **87.** $\sqrt[5]{3y}$ **89.** $\sqrt[3]{4a^2}$

91. $\sqrt[3]{x^2y}$ **93.** $\sqrt[6]{2^5}$ or $6\sqrt{32}$ **95.** $\sqrt[10]{x^9}$ **97.** $\sqrt[12]{a^{10}b^7}$ **99.** $\sqrt[20]{x}$ **101.** $\sqrt{y}$ **103.** $\sqrt[8]{x}$ **105.** $\sqrt[4]{x^2y}$ **107.** $\sqrt[12]{2x}$

109. x^9y^{15} **111.** $\sqrt[4]{a^3b^3}$ **113.** $x^{2/3} - x$ **115.** $x + 2x^{1/2} - 15$ **117.** $2x^{1/2}(3 + x)$ **119.** $15x^{1/3}(1 - 4x^{2/3})$ **121.** $\dfrac{x^2}{7y^{3/2}}$ **123.** $\dfrac{x^3}{y^2}$

125. 58 species of plants **127.** about 1872 calories per day **129. a.** $C = 35.74 + 0.6215t - 35.74v^{4/25} + 0.4275tv^{4/25}$ **b.** 8°F

131. a. $C(v) = 35.74 - 35.74v^{4/25}$ **b.** $C(25) \approx -24$; When the air temperature is 0°F and the wind speed is 25 miles per hour, the windchill

temperature is $-24°$. **c.** the point $(25, -24)$ **133. a.** $L + 1.25S^{1/2} - 9.8D^{1/3} \le 16.296$ **b.** eligible **145.** simplified correctly

147. Right side should be $x^{1/2}$. **149.** does not make sense **151.** does not make sense **153.** false **155.** true **157.** $\dfrac{1}{4}$ of the cake

159. $[3, \infty)$ **160.** $y = -2x + 11$ or $f(x) = -2x + 11$ **161.** $y \le -\dfrac{3}{2}x + 3$ **162.** $\{(3, 4)\}$

163. a. 8 **b.** 8 **c.** $\sqrt{16} \cdot \sqrt{4} = \sqrt{16 \cdot 4}$ **164. a.** 17.32 **b.** 17.32 **c.** $\sqrt{300} = 10\sqrt{3}$ **165. a.** x^7 **b.** y^4

Section 7.3 Check Point Exercises

1. a. $\sqrt{55}$ **b.** $\sqrt{x^2 - 16}$ **c.** $\sqrt[3]{60}$ **d.** $\sqrt[5]{12x^4}$ **2. a.** $4\sqrt{5}$ **b.** $2\sqrt[3]{5}$ **c.** $2\sqrt[4]{2}$ **d.** $10|x|\sqrt{2y}$ **3.** $f(x) = \sqrt{3}|x - 2|$

4. $x^4y^5z\sqrt{xyz}$ **5.** $2x^3y^4\sqrt[3]{5xy^2}$ **6.** $2x^2z\sqrt[5]{x^2y^2z^3}$ **7. a.** $2\sqrt{3}$ **b.** $100\sqrt[3]{4}$ **c.** $2x^2y\sqrt[4]{2}$

Concept and Vocabulary Check

1. $\sqrt[n]{ab}$ **2.** 77 **3.** $\sqrt[3]{8} \cdot \sqrt[3]{5} = 2\sqrt[3]{5}$ **4.** $\sqrt{5}|x + 1|$ **5.** $x^5; x^4; x^5$

Exercise Set 7.3

1. $\sqrt{15}$ **3.** $\sqrt[3]{18}$ **5.** $\sqrt[4]{33}$ **7.** $\sqrt{33xy}$ **9.** $\sqrt[5]{24x^4}$ **11.** $\sqrt{x^2 - 9}$ **13.** $\sqrt[6]{(x - 4)^5}$ **15.** $\sqrt{x}$ **17.** $\sqrt[4]{\dfrac{3x}{7y}}$ **19.** $\sqrt[7]{77x^5y^3}$

21. $5\sqrt{2}$ **23.** $3\sqrt{5}$ **25.** $5\sqrt{3x}$ **27.** $2\sqrt[3]{2}$ **29.** $3x$ **31.** $-2y\sqrt[3]{2x^2}$ **33.** $6|x + 2|$ **35.** $2(x + 2)\sqrt[3]{4}$ **37.** $|x - 1|\sqrt{3}$

39. $x^3\sqrt{x}$ **41.** $x^4y^4\sqrt{y}$ **43.** $4x\sqrt{3x}$ **45.** $y^2\sqrt[3]{y^2}$ **47.** $x^4y\sqrt[5]{x^2z}$ **49.** $3x^2y^2\sqrt[3]{3x^2}$ **51.** $(x + y)\sqrt[3]{(x + y)^2}$ **53.** $y^3\sqrt[5]{y^2}$

55. $2xy^3\sqrt[5]{2xy^2}$ **57.** $2x^2\sqrt[4]{5x^2}$ **59.** $(x - 3)^2\sqrt[4]{(x - 3)^2}$ or $(x - 3)^2\sqrt{x - 3}$ **61.** $2\sqrt{6}$ **63.** $5\sqrt{2xy}$ **65.** $6x$ **67.** $10xy\sqrt{2y}$

69. $60\sqrt{2}$ **71.** $2\sqrt[3]{6}$ **73.** $2x^2\sqrt{10x}$ **75.** $5xy^4\sqrt[3]{x^2y^2}$ **77.** $2xyz\sqrt[4]{x^2z^3}$ **79.** $2xy^2z\sqrt[5]{2y^3z}$ **81.** $(x - y)^2\sqrt[3]{(x - y)^2}$

83. $-6x^3y^3\sqrt[3]{2yz^2}$ **85.** $-6y^2\sqrt[5]{2x^3y}$ **87.** $-6x^3y^3\sqrt{2}$ **89.** $-12x^3y^4\sqrt[4]{x^3}$ **91.** $6\sqrt{3}$ miles; 10.4 miles

93. $8\sqrt{3}$ ft per sec; 14 ft per sec **95. a.** $\dfrac{7.644}{2\sqrt[4]{2}} = \dfrac{3.822}{\sqrt[4]{2}}$ **b.** 3.21 liters of blood per minute per square meter; (32, 3.21)

103. Graphs are the same; simplification is correct. **105.** Graphs are not the same.; $\sqrt{3x^2 - 6x + 3} = |x - 1|\sqrt{3}$

107. makes sense **109.** makes sense **111.** false **113.** false **115.** Its square root is multiplied by $\sqrt{3}$. **117.** $g(x) = \sqrt[3]{4x^2}$

119. $[5, 11]$ **120.** $\{(2, -4)\}$ **121.** $(4x - 3)(16x^2 + 12x + 9)$ **122. a.** $31x$ **b.** $31\sqrt{2}$ **123. a.** $-8x$ **b.** $-8\sqrt[3]{2}$ **124.** $\dfrac{y\sqrt[4]{7y}}{x^3}$

Section 7.4 Check Point Exercises

1. a. $10\sqrt{13}$ **b.** $(21 - 6x)\sqrt[3]{7}$ **c.** $5\sqrt[4]{3x} + 2\sqrt[3]{3x}$ **2. a.** $21\sqrt{5}$ **b.** $-12\sqrt{3x}$ **c.** cannot be simplified

3. a. $-9\sqrt[3]{3}$ **b.** $(5 + 3xy)\sqrt[3]{x^2y}$ **4. a.** $\dfrac{2\sqrt[3]{3}}{5}$ **b.** $\dfrac{3x\sqrt{x}}{y^5}$ **c.** $\dfrac{2y^2\sqrt[3]{y}}{x^4}$ **5. a.** $2x^2\sqrt{5}$ **b.** $\dfrac{5\sqrt{xy}}{2}$ **c.** $2x^2y$

Concept and Vocabulary Check

1. $(5 + 8)\sqrt{3} = 13\sqrt{3}$ **2.** $\sqrt{9 \cdot 3} - \sqrt{4 \cdot 3} = 3\sqrt{3} - 2\sqrt{3} = \sqrt{3}$ **3.** $\sqrt[3]{27 \cdot 2} + \sqrt[3]{8 \cdot 2} = 3\sqrt[3]{2} + 2\sqrt[3]{2} = 5\sqrt[3]{2}$

4. $\dfrac{\sqrt[n]{a}}{\sqrt[n]{b}}$ **5.** $\dfrac{\sqrt[3]{8}}{\sqrt[3]{27}} = \dfrac{2}{3}$ **6.** $\sqrt{\dfrac{72x^3}{2x}} = \sqrt{36x^2} = 6x$

Exercise Set 7.4

1. $11\sqrt{5}$ **3.** $7\sqrt[3]{6}$ **5.** $2\sqrt[5]{2}$ **7.** $\sqrt{13} + 2\sqrt{5}$ **9.** $7\sqrt{5} + 2\sqrt[3]{x}$ **11.** $4\sqrt{3}$ **13.** $19\sqrt{3}$ **15.** $6\sqrt{2x}$ **17.** $13\sqrt[3]{2}$

19. $(9x + 1)\sqrt{5x}$ **21.** $7y\sqrt[3]{2x}$ **23.** $(3x - 2)\sqrt[3]{2x}$ **25.** $4\sqrt{x - 2}$ **27.** $5x\sqrt[3]{xy^2}$ **29.** $\dfrac{\sqrt{11}}{2}$ **31.** $\dfrac{\sqrt[3]{19}}{3}$ **33.** $\dfrac{x}{6y^4}$ **35.** $\dfrac{2x\sqrt{2x}}{5y^3}$

37. $\dfrac{x\sqrt[3]{x}}{2y}$ **39.** $\dfrac{x^2\sqrt[3]{50x^2}}{3y^4}$ **41.** $\dfrac{y\sqrt[4]{9y^2}}{x^2}$ **43.** $\dfrac{2x^2\sqrt[5]{2x^3}}{y^4}$ **45.** $2\sqrt{2}$ **47.** 2 **49.** $3x$ **51.** x^2y **53.** $2x^2\sqrt{5}$ **55.** $4a^5b^5$

57. $3\sqrt{xy}$ **59.** $2xy$ **61.** $2x^2y^2\sqrt[4]{y^2}$ or $2x^2y^2\sqrt{y}$ **63.** $\sqrt[3]{x + 3}$ **65.** $\sqrt[3]{a^2 - ab + b^2}$ **67.** $\dfrac{43\sqrt{2}}{35}$ **69.** $-11xy\sqrt{2x}$ **71.** $25x\sqrt{2x}$

73. $7x\sqrt{3xy}$ **75.** $3x\sqrt[3]{5xy}$ **77.** $\left(\dfrac{f}{g}\right)(x) = 4x\sqrt{x}$; domain: $(0, \infty)$ **79.** $\left(\dfrac{f}{g}\right)(x) = 2x\sqrt[3]{2x}$; domain: $(-\infty, 0) \cup (0, \infty)$

81. $P = 18\sqrt{5}$ ft; $A = 100$ sq ft   **83.** $12\sqrt{5}$ m   **85.** **a.** $5\sqrt{10}$; the projected increase in the number of Americans ages 65–84, in millions, from 2020 to 2050   **b.** 15.8; underestimates by 2.7 million

95. Graphs are not the same.; $\sqrt{16x} - \sqrt{9x} = \sqrt{x}$

97. makes sense   **99.** does not make sense   **101.** false   **103.** false   **105.** $2\sqrt{2}$

107. $x^2y^3a^5b\sqrt{b}$ **108.** $\{0\}$ **109.** $(x - 2y)(x - 6y)$ **110.** $\dfrac{3x^2 + 8x + 6}{(x + 3)^2(x + 2)}$

111. a. $7x + 35$ **b.** $x\sqrt{7} + \sqrt{35}$ **112. a.** $6x^2 + 33x + 15$ **b.** $27 + 33\sqrt{2}$

113. $\dfrac{5\sqrt[5]{8x^2y^4}}{x}$

Mid-Chapter Check Point Exercises

1. 13   **2.** $2x^2y^3\sqrt{2xy}$ **3.** $5\sqrt[3]{4x^2}$ **4.** $12x\sqrt[3]{2x}$ **5.** 1   **6.** $4xy^{1/12}$ **7.** $-\sqrt{3}$ **8.** $\dfrac{5x\sqrt{5x}}{y^2}$ **9.** $\sqrt[4]{x^3}$ **10.** $3x\sqrt[3]{2x^2}$ **11.** $2\sqrt[3]{10}$

12. $\dfrac{x^2}{y^4}$ **13.** $x^{7/12}$ **14.** $x\sqrt[3]{y^2}$ **15.** $(x - 2)\sqrt[7]{(x - 2)^2}$ **16.** $2x^2y^4\sqrt[4]{2x^3y}$ **17.** $14\sqrt[3]{2}$ **18.** $x\sqrt[3]{x^2y^2}$ **19.** $\dfrac{1}{25}$ **20.** $\sqrt[6]{32}$

21. $\dfrac{4x}{y^2}$ **22.** $8x^2y^3\sqrt{y}$ **23.** $-10x^3y\sqrt{6y}$ **24.** $(-\infty, 6]$ **25.** $(-\infty, \infty)$

Section 7.5 Check Point Exercises

1. a. $x\sqrt{6} + 2\sqrt{15}$ **b.** $y - \sqrt[3]{7y}$ **c.** $36 - 18\sqrt{10}$ **2. a.** $11 + 2\sqrt{30}$ **b.** 1   **c.** $a - 7$ **3. a.** $\dfrac{\sqrt{21}}{7}$ **b.** $\dfrac{\sqrt[3]{6}}{3}$

4. a. $\dfrac{\sqrt{14xy}}{7y}$ **b.** $\dfrac{\sqrt[3]{3xy^2}}{3y}$ **c.** $\dfrac{3\sqrt[5]{4x^3y}}{y}$ **5.** $12\sqrt{3} - 18$ **6.** $\dfrac{3\sqrt{5} + 3\sqrt{2} + \sqrt{35} + \sqrt{14}}{3}$ **7.** $\dfrac{1}{\sqrt{x + 3} + \sqrt{x}}$

Concept and Vocabulary Check

1. $350; -42\sqrt{10}; 30\sqrt{10}; -36$ **2.** $(\sqrt{10})^2 - (\sqrt{5})^2 = 10 - 5 = 5$ **3.** rationalizing the denominator   **4.** $\sqrt{5}$ **5.** $\sqrt[3]{9}$
6. $7\sqrt{2} - 5$ **7.** $3\sqrt{6} + \sqrt{5}$

Exercise Set 7.5

1. $x\sqrt{2} + \sqrt{14}$ **3.** $7\sqrt{6} - 6$ **5.** $12\sqrt{2} - 6$ **7.** $\sqrt[3]{12} + 4\sqrt[3]{10}$ **9.** $2x\sqrt[3]{2} - \sqrt[3]{x^2}$ **11.** $32 + 11\sqrt{2}$ **13.** $34 - 15\sqrt{5}$
15. $117 - 36\sqrt{7}$ **17.** $\sqrt{6} + \sqrt{10} + \sqrt{21} + \sqrt{35}$ **19.** $\sqrt{6} - \sqrt{10} - \sqrt{21} + \sqrt{35}$ **21.** $-48 + 7\sqrt{6}$ **23.** $8 + 2\sqrt{15}$

25. $3x - 2\sqrt{3xy} + y$ **27.** -44 **29.** -71 **31.** 6   **33.** $6 - 5\sqrt{x} + x$ **35.** $\sqrt[3]{x^2} + \sqrt[3]{x} - 20$ **37.** $2x^2 + x\sqrt[3]{y^2} - y\sqrt[3]{y}$ **39.** $\dfrac{\sqrt{10}}{5}$

41. $\dfrac{\sqrt{11x}}{x}$ **43.** $\dfrac{3\sqrt{3y}}{y}$ **45.** $\dfrac{\sqrt[3]{4}}{2}$ **47.** $3\sqrt[3]{2}$ **49.** $\dfrac{\sqrt[3]{18}}{3}$ **51.** $\dfrac{4\sqrt[3]{x^2}}{x}$ **53.** $\dfrac{\sqrt[3]{2y}}{y}$ **55.** $\dfrac{7\sqrt[4]{4x}}{2x}$ **57.** $\dfrac{\sqrt[3]{2x^2y}}{xy}$ **59.** $\dfrac{3\sqrt[4]{x^3}}{x}$

61. $\dfrac{3\sqrt[5]{4x^2}}{x}$ **63.** $x\sqrt[5]{8x^3y}$ **65.** $\dfrac{3\sqrt[3]{3y}}{xy}$ **67.** $-\dfrac{5a^2\sqrt{3ab}}{b^2}$ **69.** $\dfrac{\sqrt{2mn}}{2m}$ **71.** $\dfrac{3\sqrt[4]{x^3y}}{x^2y}$ **73.** $-\dfrac{6\sqrt[3]{xy}}{x^2y^3}$ **75.** $8\sqrt{5} - 16$

77. $\dfrac{13\sqrt{11} + 39}{2}$ **79.** $3\sqrt{5} - 3\sqrt{3}$ **81.** $\dfrac{a + \sqrt{ab}}{a - b}$ **83.** $25\sqrt{2} + 15\sqrt{5}$ **85.** $4 + \sqrt{15}$ **87.** $\dfrac{x - 2\sqrt{x} - 3}{x - 9}$ **89.** $\dfrac{3\sqrt{6} + 4}{2}$

91. $\dfrac{4\sqrt{xy} + 4x + y}{y - 4x}$ **93.** $\dfrac{3}{\sqrt{6}}$ **95.** $\dfrac{2x}{\sqrt[3]{2x^2y}}$ **97.** $\dfrac{x - 9}{x - 3\sqrt{x}}$ **99.** $\dfrac{a - b}{a - 2\sqrt{ab} + b}$ **101.** $\dfrac{1}{\sqrt{x + 5} + \sqrt{x}}$ **103.** $\dfrac{1}{(x + y)(\sqrt{x} - \sqrt{y})}$

105. $\dfrac{3\sqrt{2}}{2}$ **107.** $-2\sqrt[3]{25}$ **109.** $\dfrac{7\sqrt{6}}{6}$ **111.** $7\sqrt{3} - 7\sqrt{2}$ **113.** 0   **115.** 6   **117.** $\dfrac{\sqrt{5} + 1}{2}$; 1.62 to 1   **119.** $P = 8\sqrt{2}$ in.; $A = 7$ sq in.

121. $2\sqrt{6}$ in.

131. ; Graphs are not the same.;
$(\sqrt{x} - 1)(\sqrt{x} - 1) = x - 2\sqrt{x} + 1$

133. ; Graphs are not the same.;
$(\sqrt{x} + 1)^2 = x + 2\sqrt{x} + 1$

135. makes sense   **137.** does not make sense   **139.** false   **141.** true   **143.** $\left\{\dfrac{45}{7}\right\}$ **145.** $\dfrac{5\sqrt{2} + 3\sqrt{3} - 4\sqrt{6} + 2}{23}$ **146.** $\dfrac{2x + 7}{x^2 - 4}$

147. $\left\{-\dfrac{7}{2}\right\}$ **148.** 13   **149.** $x + 5 + 2\sqrt{x + 4}$ **150.** $\{0, 5\}$ **151.** $\{-5, 2\}$

Section 7.6 Check Point Exercises

1. 20 or $\{20\}$ **2.** no solution or $\varnothing$ **3.** -1 and 3, or $\{-1, 3\}$ **4.** 4 or $\{4\}$ **5.** -12 or $\{-12\}$ **6.** 2060

Concept and Vocabulary Check

1. radical   **2.** extraneous   **3.** $2x + 1; x^2 - 14x + 49$ **4.** $x + 2; x + 8 - 6\sqrt{x - 1}$ **5.** $2x + 3; 8$ **6.** true   **7.** false

Exercise Set 7.6

1. $\{6\}$ **3.** $\{17\}$ **5.** no solution or $\varnothing$ **7.** $\{8\}$ **9.** $\{0, 3\}$ **11.** $\{1, 3\}$ **13.** $\{3, 7\}$ **15.** $\{9\}$ **17.** $\{8\}$ **19.** $\{35\}$ **21.** $\{16\}$

23. $\{2\}$ **25.** $\{2, 6\}$ **27.** $\left\{\dfrac{5}{2}\right\}$ **29.** $\{5\}$ **31.** no solution or $\varnothing$ **33.** $\{2, 6\}$ **35.** $\{0, 10\}$ **37.** $\{8\}$ **39.** 4   **41.** -7

43. $V = \dfrac{\pi r^2 h}{3}$ or $V = \dfrac{1}{3}\pi r^2 h$ **45.** $l = \dfrac{8t^2}{\pi^2}$ **47.** 8   **49.** 9   **51.** 4 ft   **53.** by the point $(5.4, 1.16)$ **55. a.** 16; 16% of Americans earning

$25 thousand annually report fair or poor health.; overestimates by 1%   **b.** approximately $30 thousand   **57.** 27 sq mi   **59.** 149 million km

69. $\{4\}$ **71.** $\{1, 9\}$

73. does not make sense   **75.** does not make sense   **77.** false   **79.** true   **81.** $\sqrt{x - 7} = 3; \sqrt{x} = 4; 1 + \sqrt{x} = 5$ **83.** $\{16\}$

85. $4x^3 - 15x^2 + 47x - 142 + \dfrac{425}{x + 3}$ **86.** $\dfrac{x + 2}{2(x + 3)}$ **87.** $(y - 3 + 5x)(y - 3 - 5x)$ **88.** $6 + 13x$ **89.** $15x^2 - 29x - 14$

90. $\dfrac{54 + 43\sqrt{2}}{-46}$

Section 7.7 Check Point Exercises

1. a. $8i$ **b.** $i\sqrt{11}$ **c.** $4i\sqrt{3}$ **2. a.** $8 + i$ **b.** $-10 + 10i$ **3. a.** $63 + 14i$ **b.** $58 - 11i$ **4.** $-\sqrt{35}$ **5.** $\dfrac{18}{25} + \dfrac{26}{25}i$

6. $-\dfrac{1}{2} - \dfrac{3}{4}i$ **7. a.** 1   **b.** i **c.** $-i$

Concept and Vocabulary Check

1. $\sqrt{-1}; -1$ **2.** $4i$ **3.** complex; imaginary; real   **4.** $-6i$ **5.** $14i$ **6.** $18; -15i; 12i; -10i^2; 10$ **7.** $2 + 9i$ **8.** $2 + 5i$
9. $-4i$ **10.** $-1; 1$ **11.** $-1; -1; -i$

Exercise Set 7.7

1. $10i$ **3.** $i\sqrt{23}$ **5.** $3i\sqrt{2}$ **7.** $3i\sqrt{7}$ **9.** $-6i\sqrt{3}$ **11.** $5 + 6i$ **13.** $15 + i\sqrt{3}$ **15.** $-2 - 3i\sqrt{2}$ **17.** $8 + 3i$ **19.** $8 - 2i$
21. $5 + 3i$ **23.** $-1 - 7i$ **25.** $-2 + 9i$ **27.** $8 + 15i$ **29.** $-14 + 17i$ **31.** $9 + 5i\sqrt{3}$ **33.** $-6 + 10i$ **35.** $-21 - 15i$
37. $-35 - 14i$ **39.** $7 + 19i$ **41.** $-1 - 31i$ **43.** $3 + 36i$ **45.** 34   **47.** 34   **49.** 11   **51.** $-5 + 12i$ **53.** $21 - 20i$ **55.** $-\sqrt{14}$
57. -6 **59.** $-5\sqrt{7}$ **61.** $-2\sqrt{6}$ **63.** $\dfrac{3}{5} - \dfrac{1}{5}i$ **65.** $1 + i$ **67.** $\dfrac{28}{25} + \dfrac{21}{25}i$ **69.** $-\dfrac{12}{13} + \dfrac{18}{13}i$ **71.** $0 + i$ or i **73.** $\dfrac{3}{10} - \dfrac{11}{10}i$
75. $\dfrac{11}{13} - \dfrac{16}{13}i$ **77.** $-\dfrac{23}{58} + \dfrac{43}{58}i$ **79.** $0 - \dfrac{7}{3}i$ or $-\dfrac{7}{3}i$ **81.** $-\dfrac{5}{2} - 4i$ **83.** $-\dfrac{7}{3} + \dfrac{4}{3}i$ **85.** -1 **87.** $-i$ **89.** -1 **91.** 1
93. i **95.** 1   **97.** $-i$ **99.** 0   **101.** $-11 - 5i$ **103.** $-5 + 10i$ **105.** $0 + 47i$ or $47i$ **107.** $1 - i$ **109.** 0   **111.** $10 + 10i$
113. $\dfrac{20}{13} + \dfrac{30}{13}i$ **115.** $(47 + 13i)$ volts   **117.** $(5 + i\sqrt{15}) + (5 - i\sqrt{15}) = 10; (5 + i\sqrt{15})(5 - i\sqrt{15}) = 25 - 15i^2 = 25 + 15 = 40$

131. $\sqrt{-9} + \sqrt{-16} = 3i + 4i = 7i$ **133.** makes sense **135.** does not make sense **137.** false **139.** false **141.** $\dfrac{14}{25} - \dfrac{2}{25}i$

143. $\dfrac{8}{5} + \dfrac{16}{5}i$ **144.** $\dfrac{x^2}{y^2}$ **145.** $x = \dfrac{yz}{y-z}$ **146.** $\left\{\dfrac{21}{11}\right\}$ **147.** $\left\{-4, \dfrac{1}{2}\right\}$ **148.** $\{-3, 3\}$ **149.** $-\sqrt{6}$ is a solution.

Review Exercises

1. 9 **2.** $-\dfrac{1}{10}$ **3.** -3 **4.** not a real number **5.** -2 **6.** $5; 1.73; 0;$ not a real number **7.** $2; -2; -4$ **8.** $[2, \infty)$

9. $(-\infty, 25]$ **10.** $5|x|$ **11.** $|x + 14|$ **12.** $|x - 4|$ **13.** $4x$ **14.** $2|x|$ **15.** $-2(x + 7)$ **16.** $\sqrt[3]{5xy}$

17. $(\sqrt{16})^3 = 64$ **18.** $(\sqrt[5]{32})^4 = 16$ **19.** $(7x)^{1/2}$ **20.** $(19xy)^{5/3}$ **21.** $\dfrac{1}{8^{2/3}} = \dfrac{1}{4}$ **22.** $\dfrac{3x}{a^{4/5}b^{4/5}} = \dfrac{3x}{\sqrt[5]{a^4 b^4}}$ **23.** $x^{7/12}$ **24.** $5^{1/6}$

25. $2x^2 y$ **26.** $\dfrac{y^{1/8}}{x^{1/3}}$ **27.** $x^3 y^4$ **28.** $y\sqrt[3]{x}$ **29.** $\sqrt[6]{x^5}$ **30.** $\sqrt[6]{x}$ **31.** $\sqrt[15]{x}$ **32.** \$3150 million **33.** $\sqrt{21xy}$ **34.** $\sqrt[5]{77x^3}$

35. $\sqrt[6]{(x-5)^5}$ **36.** $f(x) = \sqrt{7}|x - 1|$ **37.** $2x\sqrt{5x}$ **38.** $3x^2 y^2 \sqrt[3]{2x^2}$ **39.** $2y^2 z\sqrt[4]{2x^3 y^3 z}$ **40.** $2x^2\sqrt{6x}$ **41.** $2xy\sqrt[3]{2y^2}$

42. $xyz^2\sqrt[5]{16y^4 z}$ **43.** $\sqrt{x^2 - 1}$ **44.** $8\sqrt[3]{3}$ **45.** $9\sqrt{2}$ **46.** $(3x + y^2)\sqrt[3]{x}$ **47.** $-8\sqrt[3]{6}$ **48.** $\dfrac{2}{5}\sqrt[3]{2}$ **49.** $\dfrac{x\sqrt{x}}{10y^2}$ **50.** $\dfrac{y\sqrt[3]{3y}}{2x^5}$

51. $2\sqrt{6}$ **52.** $2\sqrt[3]{2}$ **53.** $2x\sqrt[4]{2x}$ **54.** $10x^2\sqrt{xy}$ **55.** $6\sqrt{2} + 12\sqrt{5}$ **56.** $5\sqrt[3]{2} - \sqrt[3]{10}$ **57.** $-83 + 3\sqrt{35}$

58. $\sqrt{xy} - \sqrt{11x} - \sqrt{11y} + 11$ **59.** $13 + 4\sqrt{10}$ **60.** $22 - 4\sqrt{30}$ **61.** -6 **62.** 4 **63.** $\dfrac{2\sqrt{6}}{3}$ **64.** $\dfrac{\sqrt{14}}{7}$ **65.** $4\sqrt[3]{3}$

66. $\dfrac{\sqrt{10xy}}{5y}$ **67.** $\dfrac{7\sqrt[3]{4x}}{x}$ **68.** $\dfrac{\sqrt[4]{189x^3}}{3x}$ **69.** $\dfrac{5\sqrt[3]{xy^4}}{2xy}$ **70.** $3\sqrt{3} + 3$ **71.** $\dfrac{\sqrt{35} - \sqrt{21}}{2}$ **72.** $10\sqrt{5} + 15\sqrt{2}$ **73.** $\dfrac{x + 8\sqrt{x} + 15}{x - 9}$

74. $\dfrac{5 + \sqrt{21}}{2}$ **75.** $\dfrac{3\sqrt{2} + 2}{7}$ **76.** $\dfrac{2}{\sqrt{14}}$ **77.** $\dfrac{3x}{\sqrt[3]{9x^2 y}}$ **78.** $\dfrac{7}{\sqrt{35} + \sqrt{21}}$ **79.** $\dfrac{2}{5 - \sqrt{21}}$ **80.** $\{16\}$ **81.** no solution or $\varnothing$

82. $\{2\}$ **83.** $\{8\}$ **84.** $\{-4, -2\}$ **85. a.** $f(20) \approx 46.8$; In 2005 (20 years after 1985), approximately 46.8% of freshmen women described their health as above average.; The rounded value is the same as the value displayed by the graph. **b.** 36 years after 1985, or in 2021

86. 84 years old **87.** $9i$ **88.** $3i\sqrt{7}$ **89.** $-2i\sqrt{2}$ **90.** $12 + 2i$ **91.** $-9 + 4i$ **92.** $-12 - 8i$ **93.** $29 + 11i$ **94.** $-7 - 24i$

95. $113 + 0i$ or 113 **96.** $-2\sqrt{6} + 0i$ or $-2\sqrt{6}$ **97.** $\dfrac{15}{13} - \dfrac{3}{13}i$ **98.** $\dfrac{1}{5} + \dfrac{11}{10}i$ **99.** $\dfrac{1}{3} - \dfrac{5}{3}i$ **100.** 1 **101.** $-i$

Chapter Test

1. a. 6 **b.** $(-\infty, 4]$ **2.** $\dfrac{1}{81}$ **3.** $\dfrac{5y^{1/8}}{x^{1/4}}$ **4.** $\sqrt{x}$ **5.** $\sqrt[20]{x^9}$ **6.** $5|x|\sqrt{3}$ **7.** $|x - 5|$ **8.** $2xy^2\sqrt[3]{xy^2}$ **9.** $-\dfrac{2}{x^2}$

10. $\sqrt[3]{50x^2 y}$ **11.** $2x\sqrt[4]{2y^3}$ **12.** $-7\sqrt{2}$ **13.** $(2x + y^2)\sqrt[3]{x}$ **14.** $2x\sqrt[3]{x}$ **15.** $12\sqrt{2} - \sqrt{15}$ **16.** $26 + 6\sqrt{3}$ **17.** $52 - 14\sqrt{3}$

18. $\dfrac{\sqrt{5x}}{x}$ **19.** $\dfrac{\sqrt[3]{25x}}{x}$ **20.** $-5 + 2\sqrt{6}$ **21.** $\{6\}$ **22.** $\{16\}$ **23.** $\{-3\}$ **24.** 49 months **25.** $5i\sqrt{3}$ **26.** $-1 + 6i$

27. $26 + 7i$ **28.** $-6 + 0i$ or -6 **29.** $\dfrac{1}{5} + \dfrac{7}{5}i$ **30.** $-i$

Cumulative Review Exercises

1. $\{(-2, -1, -2)\}$ **2.** $\left\{-\dfrac{1}{3}, 4\right\}$ **3.** $\left(\dfrac{1}{3}, \infty\right)$ **4.** $\left\{\dfrac{3}{4}\right\}$ **5.** $\{-1\}$

6. $x + 2y < 2$
$2y - x > 4$

7. $\dfrac{x^2}{15(x+2)}$ **8.** $\dfrac{x}{y}$ **9.** $8x^3 - 22x^2 + 11x + 6$ **10.** $\dfrac{5x - 6}{(x - 5)(x + 3)}$ **11.** -64 **12.** 0 **13.** $-\dfrac{16 - 9\sqrt{3}}{13}$

14. $2x^2 + x + 5 + \dfrac{6}{x - 2}$ **15.** $-34 - 3\sqrt{6}$ **16.** $2(3x + 2)(4x - 1)$ **17.** $(4x^2 + 1)(2x + 1)(2x - 1)$

18. about 53 lumens **19.** \$1500 at 7% and \$4500 at 9% **20.** 2650 students

CHAPTER 8

Section 8.1 Check Point Exercises

1. $\pm\sqrt{7}$ or $\{\pm\sqrt{7}\}$ **2.** $\pm\dfrac{\sqrt{33}}{3}$ or $\left\{\pm\dfrac{\sqrt{33}}{3}\right\}$ **3.** $\pm\dfrac{3}{2}i$ or $\left\{\pm\dfrac{3}{2}i\right\}$ **4.** $3 \pm \sqrt{10}$ or $\{3 \pm \sqrt{10}\}$ **5. a.** $25; x^2 + 10x + 25 = (x + 5)^2$

b. $\dfrac{9}{4}; x^2 - 3x + \dfrac{9}{4} = \left(x - \dfrac{3}{2}\right)^2$ **c.** $\dfrac{9}{64}; x^2 + \dfrac{3}{4}x + \dfrac{9}{64} = \left(x + \dfrac{3}{8}\right)^2$ **6.** $-2 \pm \sqrt{5}$ or $\{-2 \pm \sqrt{5}\}$ **7.** $\dfrac{-3 \pm \sqrt{41}}{4}$ or $\left\{\dfrac{-3 \pm \sqrt{41}}{4}\right\}$

8. $\dfrac{3}{2} \pm i\dfrac{\sqrt{15}}{6}$ or $\left\{\dfrac{3}{2} \pm i\dfrac{\sqrt{15}}{6}\right\}$ **9.** 20% **10.** $10\sqrt{21}$ ft; 45.8 ft

Concept and Vocabulary Check

1. $\pm\sqrt{d}$ **2.** $\pm\sqrt{7}$ **3.** $\pm\sqrt{\dfrac{11}{2}}; \pm\dfrac{\sqrt{22}}{2}$ **4.** $\pm 3i$ **5.** 25 **6.** $\dfrac{9}{4}$ **7.** $\dfrac{4}{25}$ **8.** 9 **9.** $\dfrac{1}{9}$

Exercise Set 8.1

1. $\{\pm 5\}$ **3.** $\{\pm \sqrt{6}\}$ **5.** $\left\{\pm \dfrac{5}{4}\right\}$ **7.** $\left\{\pm \dfrac{\sqrt{6}}{3}\right\}$ **9.** $\left\{\pm \dfrac{4}{5}i\right\}$ **11.** $\{-10, -4\}$ **13.** $\{3 \pm \sqrt{5}\}$ **15.** $\{-2 \pm 2\sqrt{2}\}$ **17.** $\{5 \pm 3i\}$

19. $\left\{\dfrac{-3 \pm \sqrt{11}}{4}\right\}$ **21.** $\{-3, 9\}$ **23.** $1; x^2 + 2x + 1 = (x + 1)^2$ **25.** $49; x^2 - 14x + 49 = (x - 7)^2$ **27.** $\dfrac{49}{4}; x^2 + 7x + \dfrac{49}{4} = \left(x + \dfrac{7}{2}\right)^2$

29. $\dfrac{1}{16}; x^2 - \dfrac{1}{2}x + \dfrac{1}{16} = \left(x - \dfrac{1}{4}\right)^2$ **31.** $\dfrac{4}{9}; x^2 + \dfrac{4}{3}x + \dfrac{4}{9} = \left(x + \dfrac{2}{3}\right)^2$ **33.** $\dfrac{81}{64}; x^2 - \dfrac{9}{4}x + \dfrac{81}{64} = \left(x - \dfrac{9}{8}\right)^2$ **35.** $\{-8, 4\}$ **37.** $\{-3 \pm \sqrt{7}\}$

39. $\{4 \pm \sqrt{15}\}$ **41.** $\{-1 \pm i\}$ **43.** $\left\{\dfrac{-3 \pm \sqrt{13}}{2}\right\}$ **45.** $\left\{-\dfrac{3}{7}, -\dfrac{1}{7}\right\}$ **47.** $\left\{\dfrac{-1 \pm \sqrt{5}}{2}\right\}$ **49.** $\left\{-\dfrac{5}{2}, 1\right\}$ **51.** $\left\{\dfrac{-3 \pm \sqrt{6}}{3}\right\}$

53. $\left\{\dfrac{4 \pm \sqrt{13}}{3}\right\}$ **55.** $\left\{\dfrac{1}{4} \pm \dfrac{1}{4}i\right\}$ **57.** $\left\{\dfrac{5}{4} \pm i\dfrac{\sqrt{31}}{4}\right\}$ **59.** $-\dfrac{1}{5}, 1$ **61.** $-2 \pm 5i$ **63.** $2 \pm 2i$ **65.** $v = \sqrt{2gh}$ **67.** $r = \dfrac{\sqrt{AP}}{P} - 1$

69. $\left\{\dfrac{-1 \pm \sqrt{19}}{6}\right\}$ **71.** $\{-b, 2b\}$ **73. a.** $x^2 + 8x$ **b.** 16 **c.** $x^2 + 8x + 16$ **d.** $(x + 4)^2$ **75.** 20% **77.** 6.25%

79. a. 940 billionaires; underestimates by 6 billionaires **b.** 30 years after 1987, or 2017 **81.** $10\sqrt{3}$ sec; 17.3 sec **83.** $3\sqrt{5}$ mi; 6.7 mi

85. $20\sqrt{2}$ ft; 28.3 ft **87.** $50\sqrt{2}$ ft; 70.7 ft **89.** 10 m **97.** $\{-3, 1\}$ **101.** makes sense **103.** does not make sense

105. false **107.** false **109.** $\left\{\dfrac{-1 \pm \sqrt{1 - 4c}}{2}\right\}$ **111.** $\{\pm \sqrt{5}, \pm \sqrt{3}\}$ **112.** $4 - 2x$ **113.** $(1 - 2x)(1 + 2x + 4x^2)$

114. $x^3 - 2x^2 - 4x - 12 - \dfrac{42}{x - 3}$ **115. a.** $\left\{-\dfrac{1}{2}, \dfrac{1}{4}\right\}$ **b.** 36; yes **116. a.** $\left\{\dfrac{1}{3}\right\}$ **b.** 0 **117. a.** $3x^2 + 4x + 2 = 0$ **b.** -8

Section 8.2 Check Point Exercises

1. -5 and $\dfrac{1}{2}$, or $\left\{-5, \dfrac{1}{2}\right\}$ **2.** $\dfrac{3 \pm \sqrt{7}}{2}$ or $\left\{\dfrac{3 \pm \sqrt{7}}{2}\right\}$ **3.** $-1 \pm i\dfrac{\sqrt{6}}{3}$ or $\left\{-1 \pm i\dfrac{\sqrt{6}}{3}\right\}$ **4. a.** 0; one real rational solution

b. 81; two real rational solutions **c.** -44; two imaginary solutions that are complex conjugates **5. a.** $20x^2 + 7x - 3 = 0$

b. $x^2 - 50 = 0$ **c.** $x^2 + 49 = 0$ **6.** 26 years old; The point $(26, 115)$ lies approximately on the blue graph.

Concept and Vocabulary Check

1. $\dfrac{-b \pm \sqrt{b^2 - 4ac}}{2a}$ **2.** $2; 9; -5$ **3.** $1; -4; -1$ **4.** $2 \pm \sqrt{2}$ **5.** $-1 \pm i\dfrac{\sqrt{6}}{2}$ **6.** $b^2 - 4ac$ **7.** no **8.** two

9. the square root property **10.** the quadratic formula **11.** factoring and the zero-product principle **12.** false

Exercise Set 8.2

1. $\{-6, -2\}$ **3.** $\left\{1, \dfrac{5}{2}\right\}$ **5.** $\left\{\dfrac{-3 \pm \sqrt{89}}{2}\right\}$ **7.** $\left\{\dfrac{7 \pm \sqrt{85}}{6}\right\}$ **9.** $\left\{\dfrac{1 \pm \sqrt{7}}{6}\right\}$ **11.** $\left\{\dfrac{3}{8} \pm i\dfrac{\sqrt{87}}{8}\right\}$ **13.** $\{2 \pm 2i\}$ **15.** $\left\{\dfrac{4}{3} \pm i\dfrac{\sqrt{5}}{3}\right\}$

17. $\left\{-\dfrac{3}{2}, 4\right\}$ **19.** 52; two real irrational solutions **21.** 4; two real rational solutions **23.** -23; two imaginary solutions

25. 36; two real rational solutions **27.** -60; two imaginary solutions **29.** 0; one (repeated) real rational solution **31.** $\left\{-\dfrac{2}{3}, 2\right\}$

33. $\{1 \pm \sqrt{2}\}$ **35.** $\left\{\dfrac{1}{6} \pm i\dfrac{\sqrt{107}}{6}\right\}$ **37.** $\left\{\dfrac{3 \pm \sqrt{65}}{4}\right\}$ **39.** $\left\{0, \dfrac{8}{3}\right\}$ **41.** $\left\{\dfrac{-6 \pm 2\sqrt{6}}{3}\right\}$ **43.** $\left\{\dfrac{2 \pm \sqrt{10}}{3}\right\}$ **45.** $\{2 \pm \sqrt{10}\}$

47. $\left\{1, \dfrac{5}{2}\right\}$ **49.** $\{2 \pm 2i\sqrt{2}\}$ **51.** $x^2 - 2x - 15 = 0$ **53.** $12x^2 + 5x - 2 = 0$ **55.** $x^2 - 2 = 0$ **57.** $x^2 - 20 = 0$

59. $x^2 + 36 = 0$ **61.** $x^2 - 2x + 2 = 0$ **63.** $x^2 - 2x - 1 = 0$ **65.** b **67.** a **69.** $1 + \sqrt{7}$ **71.** $\left\{\dfrac{-1 \pm \sqrt{21}}{2}\right\}$

73. $\left\{-2\sqrt{2}, \dfrac{\sqrt{2}}{2}\right\}$ **75.** $\{-3, 1, -1 \pm i\sqrt{2}\}$ **77.** 33-year-olds and 58-year-olds; The function models the actual data well. **79.** 77.8 ft; (b)

81. 5.5 m by 1.5 m **83.** 17.6 in. and 18.6 in. **85.** 9.3 in. and 0.7 in. **87.** 7.5 hr and 8.5 hr

97. ; depth: 5 in.; maximum area: 50 sq in. **99.** does not make sense **101.** does not make sense

103. false **105.** true **107.** 2.4 m; yes

109. $\left\{-3, \dfrac{1}{4}\right\}$

106. $\{3, 7\}$

111. $\dfrac{5\sqrt{3} - 5x}{3 - x^2}$ **112.** 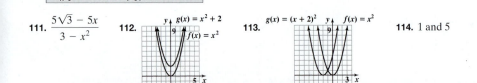 **113.** **114.** 1 and 5

Section 8.3 Check Point Exercises

1.

2. *f*(*x*) = (*x* − 2)² + 1

3. (−2, −9)   **4.** *f*(*x*) = −*x*² + 4*x* + 1   domain: (−∞, ∞); range: (−∞, 5]

5. a. minimum   **b.** Minimum is 984 at *x* = 2.   **c.** domain: (−∞, ∞); range: [984, ∞)   **6.** 33.7 ft   **7.** 4, −4; −16   **8.** 30 ft by 30 ft; 900 sq ft

Concept and Vocabulary Check

1. parabola; upward; downward   **2.** lowest/minimum   **3.** highest/maximum   **4.** (*h*, *k*)   **5.** $-\dfrac{b}{2a}; -\dfrac{b}{2a}$ **6.** 0; solutions   **7.** *f*(0)

Exercise Set 8.3

1. $h(x) = (x - 1)^2 + 1$ **3.** $j(x) = (x - 1)^2 - 1$ **5.** $h(x) = x^2 - 1$ **7.** $g(x) = x^2 - 2x + 1$ **9.** (3, 1)   **11.** (−1, 5)
13. (2, −5)   **15.** (−1, 9)

17. *f*(*x*) = (*x* − 4)² − 1   [−1, ∞)

19. *f*(*x*) = (*x* − 1)² + 2   [2, ∞)

21. *y* − 1 = (*x* − 3)²   [1, ∞)

23. *f*(*x*) = 2(*x* + 2)² − 1   [−1, ∞)

25. *f*(*x*) = 4 − (*x* − 1)²   (−∞, 4]

27. *f*(*x*) = *x*² − 2*x* − 3   [−4, ∞)

29. *f*(*x*) = *x*² + 3*x* − 10 $\left[-\dfrac{49}{4}, \infty\right)$

31. *f*(*x*) = 2*x* − *x*² + 3   (−∞, 4]

33. *f*(*x*) = *x*² + 6*x* + 3   [−6, ∞)

35. *f*(*x*) = 2*x*² + 4*x* − 3   [−5, ∞)

37. *f*(*x*) = 2*x* − *x*² − 2   (−∞, −1]

39. a. minimum   **b.** Minimum is −13 at *x* = 2.   **c.** domain: (−∞, ∞); range: [−13, ∞)   **41. a.** maximum   **b.** Maximum is 1 at *x* = 1.
c. domain: (−∞, ∞); range: (−∞, 1]   **43. a.** minimum   **b.** Minimum is $-\dfrac{5}{4}$ at $x = \dfrac{1}{2}$.   **c.** domain: (−∞, ∞); range: $\left[-\dfrac{5}{4}, \infty\right)$
45. domain: (−∞, ∞); range: [−2, ∞)   **47.** domain: (−∞, ∞); range: (−∞, −6]   **49.** $f(x) = 2(x - 5)^2 + 3$ **51.** $f(x) = 2(x + 10)^2 - 5$
53. $f(x) = -3(x + 2)^2 + 4$ **55.** $f(x) = 3(x - 11)^2$ **57. a.** 2 sec; 224 ft   **b.** 5.7 sec   **c.** 160; 160 feet is the height of the building.

d.

59. 8 and 8; 64   **61.** 8, −8; −64   **63.** length: 300 ft; width: 150 ft; maximum area: 45,000 sq ft
65. 12.5 yd by 12.5 yd; 156.25 sq yd   **67.** 5 in.; 50 sq in.   **69. a.** $C(x) = 525 + 0.55x$
b. $P(x) = -0.001x^2 + 2.45x - 525$
c. 1225 sandwiches; $975.63

77. a.   **b.** (20.5, −120.5)   **c.**   **d.** Answers will vary.

79. $(2.5, 185)$ **81.** $(-30, 91)$

83. does not make sense   **85.** does not make sense   **87.** false   **89.** false   **91.** $x = 3; (0, 11)$ **93.** $f(x) = -2(x + 3)^2 - 1$

95. 65 trees; 16,900 pounds   **96.** $\{7\}$ **97.** $\dfrac{x}{x - 2}$ **98.** $\{(6, -2)\}$ **99.** $\{-1, 9\}$ **100.** $\left\{-2, \dfrac{5}{2}\right\}$ **101.** $5u^2 + 11u + 2 = 0$

Mid-Chapter Check Point Exercises

1. $\left\{-\dfrac{1}{3}, \dfrac{11}{3}\right\}$ **2.** $\left\{-1, \dfrac{7}{5}\right\}$ **3.** $\left\{\dfrac{3 \pm \sqrt{15}}{3}\right\}$ **4.** $\{-3 \pm \sqrt{7}\}$ **5.** $\left\{\pm\dfrac{6\sqrt{5}}{5}\right\}$ **6.** $\left\{\dfrac{5}{2} \pm i\dfrac{\sqrt{7}}{2}\right\}$ **7.** $\{\pm i\sqrt{13}\}$

8. $\left\{-4, \dfrac{1}{2}\right\}$ **9.** $\{-3 \pm 2\sqrt{6}\}$ **10.** $\{2 \pm \sqrt{3}\}$ **11.** $\left\{\dfrac{3}{4} \pm i\dfrac{\sqrt{23}}{4}\right\}$ **12.** $\left\{\dfrac{-3 \pm \sqrt{41}}{4}\right\}$ **13.** $\{4\}$ **14.** $\{-5 \pm 2\sqrt{7}\}$

15. $f(x) = (x - 3)^2 - 4$ domain: $(-\infty, \infty)$; range: $[-4, \infty)$ **16.** $g(x) = 5 - (x + 2)^2$ domain: $(-\infty, \infty)$; range: $(-\infty, 5]$

17. $h(x) = -x^2 - 4x + 5$ domain: $(-\infty, \infty)$; range: $(-\infty, 9]$ **18.** $f(x) = 3x^2 - 6x + 1$ domain: $(-\infty, \infty)$; range: $[-2, \infty)$

19. two imaginary solutions   **20.** two real rational solutions   **21.** $8x^2 - 2x - 3 = 0$ **22.** $x^2 - 12 = 0$ **23.** 75 cabinets per day; $1200
24. $-9, -9; 81$ **25.** 10 in.; 100 sq in.

Section 8.4 Check Point Exercises

1. $-\sqrt{3}, -\sqrt{2}, \sqrt{2},$ and $\sqrt{3}$ or $\{\pm\sqrt{2}, \pm\sqrt{3}\}$ **2.** 16 or $\{16\}$ **3.** $-\sqrt{6}, -1, 1,$ and $\sqrt{6}$, or $\{-\sqrt{6}, -1, 1, \sqrt{6}\}$ **4.** -1 and 2, or $\{-1, 2\}$

5. $-\dfrac{1}{27}$ and 64, or $\left\{-\dfrac{1}{27}, 64\right\}$

Concept and Vocabulary Check

1. $x^2; u^2 - 13u + 36 = 0$ **2.** $x^{1/2}$ or $\sqrt{x}; u^2 - 2u - 8 = 0$ **3.** $x + 3; u^2 + 7u - 18 = 0$ **4.** $x^{-1}; 2u^2 - 7u + 3 = 0$
5. $x^{1/3}; u^2 + 2u - 3 = 0$

Exercise Set 8.4

1. $\{-2, 2, -1, 1\}$ **3.** $\{-3, 3, -\sqrt{2}, \sqrt{2}\}$ **5.** $\{-2i, 2i, -\sqrt{2}, \sqrt{2}\}$ **7.** $\{1\}$ **9.** $\{49\}$ **11.** $\{25, 64\}$ **13.** $\{2, 12\}$ **15.** $\{-\sqrt{3}, 0, \sqrt{3}\}$

17. $\{-5, -2, -1, 2\}$ **19.** $\left\{-\dfrac{1}{4}, \dfrac{1}{5}\right\}$ **21.** $\left\{\dfrac{1}{3}, 2\right\}$ **23.** $\left\{\dfrac{-2 \pm \sqrt{7}}{3}\right\}$ **25.** $\{-8, 27\}$ **27.** $\{-243, 32\}$ **29.** $\{1\}$ **31.** $\{-8, -2, 1, 4\}$

33. $-2, 2, -1,$ and $1; c$ **35.** $1; e$ **37.** 2 and 3; f   **39.** $-5, -4, 1,$ and 2   **41.** $-\dfrac{3}{2}$ and $-\dfrac{1}{3}$ **43.** $\dfrac{64}{15}$ and $\dfrac{81}{20}$ **45.** $\dfrac{5}{2}$ and $\dfrac{25}{6}$

47. ages 20 and 55; The function models the data well.   **53.** $\{3, 5\}$ **55.** $\{1\}$ **57.** $\{-1, 4\}$ **59.** $\{1, 8\}$ **61.** makes sense

63. does not make sense   **65.** true   **67.** false   **69.** $\left\{\sqrt[3]{-2}, \dfrac{\sqrt[3]{225}}{5}\right\}$ **71.** $\dfrac{1}{5x - 1}$ **72.** $\dfrac{1}{2} + \dfrac{3}{2}i$ **73.** $\{(4, -2)\}$ **74.** $\left\{-3, \dfrac{5}{2}\right\}$

75. $\{-2, -1, 2\}$ **76.** $\dfrac{-x - 5}{x + 3}$

Section 8.5 Check Point Exercises

1. $(-\infty, -4) \cup (5, \infty)$

2. $\left[\dfrac{-3 - \sqrt{7}}{2}, \dfrac{-3 + \sqrt{7}}{2}\right]$

3. $(-\infty, -3] \cup [-1, 1]$

4. $(-2, 5)$

5. $(-\infty, -1) \cup [1, \infty)$

6. between 1 and 4 seconds, excluding $t = 1$ and $t = 4$

Concept and Vocabulary Check

1. $x^2 + 8x + 15 = 0$; boundary   **2.** $(-\infty, -5), (-5, -3), (-3, \infty)$ **3.** true   **4.** true   **5.** $(-\infty, -2) \cup [1, \infty)$

Exercise Set 8.5

1. $(-\infty, -2) \cup (4, \infty)$

3. $[-3, 7]$

5. $(-\infty, 1) \cup (4, \infty)$

7. $(-\infty, -4) \cup (-1, \infty)$

9. $[2, 4]$

11. $\left[-4, \dfrac{2}{3}\right]$

13. $\left(-3, \dfrac{5}{2}\right)$

15. $\left(-1, -\dfrac{3}{4}\right)$

17. $(-\infty, 0] \cup [4, \infty)$

19. $\left(-\infty, -\dfrac{3}{2}\right) \cup (0, \infty)$

21. $[0, 1]$

23. $[2 - \sqrt{2}, 2 + \sqrt{2}]$

25. $\left(-\infty, \dfrac{2 - \sqrt{10}}{3}\right) \cup \left(\dfrac{2 + \sqrt{10}}{3}, \infty\right)$;

27. $\left(-\infty, \dfrac{5 - \sqrt{33}}{4}\right] \cup \left[\dfrac{5 + \sqrt{33}}{4}, \infty\right)$;

29. no solution or $\varnothing$

31. $[1, 2] \cup [3, \infty)$

33. $[-2, -1] \cup [1, \infty)$

35. $(-\infty, -3)$

37. $(-1, \infty)$

39. $\{0\} \cup [9, \infty)$

41. $(-\infty, -3) \cup (4, \infty)$

43. $(-4, -3)$

45. $[2, 4)$

47. $\left(-\infty, -\dfrac{4}{3}\right) \cup [2, \infty)$

49. $(-\infty, 0) \cup (3, \infty)$

51. $(-\infty, -5) \cup (-3, \infty)$

53. $\left(-\infty, \dfrac{1}{2}\right) \cup \left[\dfrac{7}{5}, \infty\right)$

55. $(-\infty, -6] \cup (-2, \infty)$

57. $\left(-\infty, \dfrac{1}{2}\right] \cup [2, \infty)$

59. $(-1, 1)$

61. $(-\infty, -8) \cup (-6, 4) \cup (6, \infty)$

63. $(-3, 2)$

65. $(-\infty, -1) \cup (1, 2) \cup (3, \infty)$

67. $\left[-6, -\dfrac{1}{2}\right] \cup [1, \infty)$ **69.** $(-\infty, -2) \cup [-1, 2)$ **71.** between 0 and 3 seconds, excluding $t = 0$ and $t = 3$ **73. a.** dry: 160 ft; wet: 185 ft

b. dry pavement: graph (b); wet pavement: graph (a)   **c.** extremely well; Function values and data are identical.   **d.** speeds exceeding 76 miles per hour; points on graph (b) to the right of $(76, 540)$ **75.** The company's production level must be at least 20,000 wheelchairs per month. For values of x greater than or equal to 20,000, the graph lies on or below the line $y = 425$.   **77.** The length of the shorter side cannot exceed 6 ft.

83. $\left[-3, \dfrac{1}{2}\right]$ **85.** $(-\infty, 3) \cup [8, \infty)$ **87.** $(-3, -1) \cup (2, \infty)$ **89. a.** $f(x) = 0.1375x^2 + 0.7x + 37.8$ **b.** speeds exceeding 52 mph

91. does not make sense   **93.** does not make sense   **95.** false   **97.** true   **99.** Answers will vary.; example: $\dfrac{x - 3}{x + 4} \geq 0$ **101.** $\{2\}$

103. $(-\infty, 2) \cup (2, \infty)$ **105.** $27 - 3x^2 \geq 0; [-3, 3]$ **106.** $(-19, 29)$ **107.** $\dfrac{2(x - 1)}{(x + 4)(x - 3)}$ **108.** $(x^2 + 4y^2)(x + 2y)(x - 2y)$

109.

110.

111.

Review Exercises

1. $\{\pm 8\}$ **2.** $\{\pm 5\sqrt{2}\}$ **3.** $\left\{\pm \dfrac{\sqrt{6}}{3}\right\}$ **4.** $\{4 \pm 3\sqrt{2}\}$ **5.** $\{-7 \pm 6i\}$ **6.** $100; x^2 + 20x + 100 = (x + 10)^2$

7. $\dfrac{9}{4}; x^2 - 3x + \dfrac{9}{4} = \left(x - \dfrac{3}{2}\right)^2$ **8.** $\{3, 9\}$ **9.** $\left\{\dfrac{7 \pm \sqrt{53}}{2}\right\}$ **10.** $\left\{\dfrac{-3 \pm \sqrt{41}}{4}\right\}$ **11.** 8% **12.** 14 weeks **13.** $60\sqrt{5}$ m; 134.2 m

14. $\{1 \pm \sqrt{5}\}$ **15.** $\{1 \pm 3i\sqrt{2}\}$ **16.** $\left\{\dfrac{-2 \pm \sqrt{10}}{2}\right\}$ **17.** two imaginary solutions **18.** two real rational solutions

19. two real irrational solutions **20.** $\left\{-\dfrac{2}{3}, 4\right\}$ **21.** $\{-5, 2\}$ **22.** $\left\{\dfrac{1 \pm \sqrt{21}}{10}\right\}$ **23.** $\{-4, 4\}$ **24.** $\{3 \pm 2\sqrt{2}\}$ **25.** $\left\{\dfrac{1}{6} \pm i\dfrac{\sqrt{23}}{6}\right\}$

26. $\{4 \pm \sqrt{5}\}$ **27.** $15x^2 - 4x - 3 = 0$ **28.** $x^2 + 81 = 0$ **29.** $x^2 - 48 = 0$ **30. a.** 261 ft; overestimates by 1 ft **b.** 40 mph
31. a. by the point $(35, 261)$ **b.** by the point $(40, 267)$ **8.** 8.8 sec

33. **34.** **35.** **36.**

37. 25 in. of rainfall per year; 13.5 in. of growth **38.** 12.5 sec; 2540 feet **39.** 7.2 h; 622 per 100,000 males **40.** 250 yd by 500 yd; 125,000 sq yard

41. -7 and $7; -49$ **42.** $\{-\sqrt{2}, \sqrt{2}, -2, 2\}$ **43.** $\{1\}$ **44.** $\{-5, -1, 3\}$ **45.** $\left\{-\dfrac{1}{8}, \dfrac{1}{7}\right\}$ **46.** $\{-27, 64\}$ **47.** $\{16\}$

48. $\left(-3, \dfrac{1}{2}\right)$; **49.** $\left(-\infty, -4\right] \cup \left[-\dfrac{1}{2}, \infty\right)$; **50.** $(-3, 0) \cup (1, \infty)$;

51. $(-\infty, -2) \cup (6, \infty)$; **52.** $(-\infty, 4) \cup \left[\dfrac{23}{4}, \infty\right)$; **53.** between 1 and 2 seconds, excluding $t = 1$ and $t = 2$
54. a. 200 beats per minute
b. between 0 and 4 minutes and more than 12 minutes after the workout; between 0 and 4 minutes; Answers will vary.

Chapter Test

1. $\left\{\pm \dfrac{\sqrt{10}}{2}\right\}$ **2.** $\{3 \pm 2\sqrt{5}\}$ **3.** $64; x^2 - 16x + 64 = (x - 8)^2$ **4.** $\dfrac{1}{25}; x^2 + \dfrac{2}{5}x + \dfrac{1}{25} = \left(x + \dfrac{1}{5}\right)^2$ **5.** $\{3 \pm \sqrt{2}\}$ **6.** $50\sqrt{2}$ ft

7. two real irrational solutions **8.** two imaginary solutions **9.** $\left\{-5, \dfrac{1}{2}\right\}$ **10.** $\{-4 \pm \sqrt{11}\}$ **11.** $\{-2 \pm 5i\}$ **12.** $\left\{\dfrac{3}{2} \pm \dfrac{1}{2}i\right\}$

13. $x^2 - 4x - 21 = 0$ **14.** $x^2 + 100 = 0$ **15. a.** $327.1 \approx 327$; underestimates by 5 **b.** 22 years after 2003, or 2025

16. **17.** **18.** after 2 sec; 69 ft **19.** 4.1 sec **20.** 23 computers; $169 hundreds or $16,900
21. $\{1, 2\}$ **22.** $\{-3, 3, -2, 2\}$ **23.** $\{1, 512\}$
24. $(-3, 4)$; **25.** $(-\infty, 3) \cup [10, \infty)$;

Cumulative Review Exercises

1. $\left\{\dfrac{2}{3}\right\}$ **2.** $\{(-5, 2)\}$ **3.** $\{(1, 4, -2)\}$ **4.** $[10, \infty)$ **5.** $(-\infty, -4]$ **6.** $(2, \infty)$ **7.** $(-2, 3)$ **8.** $\{3, 9\}$ **9.** no solution or $\varnothing$ **10.** $\{12\}$

11. $\left\{\dfrac{-2 \pm \sqrt{14}}{2}\right\}$ **12.** $\{8, 27\}$ **13.** $\left[-2, \dfrac{3}{2}\right]$

14. **15.** $f(x) = \dfrac{1}{2}x - 1$ **16.** $3x - 2y > -6$ **17.** $f(x) = -2(x - 3)^2 + 2$

18. $-16x + 24y$ **19.** $-\dfrac{20x^7}{y^4}$ **20.** $x^2 - 5xy - 16y^2$ **21.** $6x^2 + 13x - 5$ **22.** $9x^4 - 24x^2y + 16y^2$ **23.** $\dfrac{3x^2 + 6x - 2}{(x + 5)(x + 2)}$ **24.** $\dfrac{x - 3}{x}$

25. $\dfrac{x - 4}{3x + 6}$ **26.** $5xy\sqrt{2x}$ **27.** $9\sqrt{2}$ **28.** $44 + 6i$ **29.** $(9x^2 + 1)(3x + 1)(3x - 1)$ **30.** $2x(4x - 1)(3x - 2)$

31. $(x + 3y)(x^2 - 3xy + 9y^2)$ **32.** $x^2 + 2x - 13; 22$ **33.** $x + 5; (-\infty, 2) \cup (2, \infty)$ **34.** $2a + h + 3$

35. $3x^2 - 7x + 18 - \dfrac{28}{x + 2}$ **36.** $R = -\dfrac{Ir}{I - 1}$ or $R = \dfrac{Ir}{1 - I}$ **37.** $y = -3x - 1$ or $f(x) = -3x - 1$ **38.** 6 **39.** $\$620$

40. 13 yd by 4 yd   **41.** $\$2600$ at 12% and $\$1400$ at 14%   **42.** 11 amps

CHAPTER 9

Section 9.1 Check Point Exercises

1. approximately $\$160$; overestimates by $\$11$

2.   **3.**   **4.**   **5.**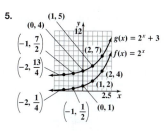

The graph of g is the graph of f shifted 1 unit to the right.

The graph of g is the graph of f shifted up 3 units.

6. approximately 4067   **7. a.** $\$14,859.47$ **b.** $\$14,918.25$

Concept and Vocabulary Check

1. $b^x; (-\infty, \infty); (0, \infty)$ **2.** $x; y = 0$; horizontal   **3.** e; natural; 2.72   **4.** $A; P; r; n$ **5.** semiannually; quarterly; continuous

Exercise Set 9.1

1. 10.556   **3.** 11.665   **5.** 0.125   **7.** 9.974   **9.** 0.387

11. ; d

x	f(x)
−2	$\frac{1}{9}$
−1	$\frac{1}{3}$
0	1
1	3
2	9

13. ; e

x	f(x)
−2	$-\frac{8}{9}$
−1	$-\frac{2}{3}$
0	0
1	2
2	8

15. ; f

x	f(x)
−2	9
−1	3
0	1
1	$\frac{1}{3}$
2	$\frac{1}{9}$

17.

x	f(x)
−2	$\frac{1}{16}$
−1	$\frac{3}{}$
0	1
1	4
2	16

19.

x	g(x)
−2	$\frac{4}{9}$
−1	$\frac{2}{3}$
0	1
1	$\frac{3}{2}$
2	$\frac{9}{4}$

21.

x	h(x)
−2	4
−1	2
0	1
1	$\frac{1}{2}$
2	$\frac{1}{4}$

23.

x	f(x)
−2	2.78
−1	1.67
0	1
1	0.6
2	0.36

25.
The graph of g is the graph of f shifted 1 unit to the left.

27.
The graph of g is the graph of f shifted 2 units to the right.

29.
The graph of g is the graph of f shifted up 1 unit.

31.
The graph of g is the graph of f shifted down 2 units.

33.
The graph of g is a reflection of the graph of f across the x-axis.

35.
The graph of g is the graph of f shifted 1 unit to the left and 1 unit down.

37.
The graph of g is the graph of f stretched vertically by a factor of $\frac{1}{3}$.

39. a. \$13,116.51   **b.** \$13,157.04
c. \$13,165.31
41. 7% compounded monthly
43. domain: $(-\infty, \infty)$; range: $(-2, \infty)$
45. domain: $(-\infty, \infty)$; range: $(1, \infty)$
47. domain: $(-\infty, \infty)$; range: $(0, \infty)$

49. $(0, 1)$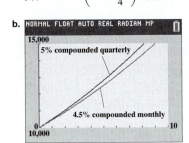

51.

53. a. 574 million   **b.** 1148 million   **c.** 2295 million   **d.** 4590 million
e. It appears to double.   **55.** \$832,744   **57. a.** 64%
b. 67%   **c.** the exponential model   **59. a.** 100%
b. about 68.5%   **c.** about 30.8%   **d.** about 20%
61. 11.3; About 11.3% of 30-year-olds have some coronary heart disease.

63. a. about 1429 people   **b.** about 24,546 people   **c.** The number of ill people cannot exceed the population.; The asymptote indicates that the number of ill people will not exceed 30,000, the population of the town.

69. a. $f(t) = 10,000\left(1 + \dfrac{0.05}{4}\right)^{4t}$; $f(t) = 10,000\left(1 + \dfrac{0.045}{12}\right)^{12t}$

b.

```
NORMAL FLOAT AUTO REAL RADIAN MP
15,000
  5% compounded quarterly



        4.5% compounded monthly
0                              10
10,000
```

the bank that pays 5% interest compounded quarterly

71. does not make sense   **73.** does not make sense   **75.** false   **77.** false   **79. a.** $y = \left(\dfrac{1}{3}\right)^x$ **b.** $y = \left(\dfrac{1}{5}\right)^x$ **c.** $y = 5^x$ **d.** $y = 3^x$

81. $b = \dfrac{Da}{a - D}$ **82.** -1 **83.** $\{-2, 5\}$ **84. a.** 11   **b.** 127   **85.** x **86.** $y = \dfrac{x + 5}{7}$

Section 9.2 Check Point Exercises

1. a. $5x^2 + 1$ **b.** $25x^2 + 60x + 35$ **2.** $f(g(x)) = 7\left(\dfrac{x}{7}\right) = x; g(f(x)) = \dfrac{7x}{7} = x$

3. $f(g(x)) = 4\left(\dfrac{x + 7}{4}\right) - 7 = (x + 7) - 7 = x; g(f(x)) = \dfrac{(4x - 7) + 7}{4} = \dfrac{4x}{4} = x$ **4.** $f^{-1}(x) = \dfrac{x - 7}{2}$ **5.** $f^{-1}(x) = \sqrt[3]{\dfrac{x + 1}{4}}$

6. (b) and (c)   **7.**

Concept and Vocabulary Check

1. composition; $f(g(x))$ **2.** $f; g(x)$ **3.** composition; $g(f(x))$ **4.** $g; f(x)$ **5.** false   **6.** false
7. inverse   **8.** $x; x$ **9.** horizontal; one-to-one   **10.** $y = x$

Exercise Set 9.2

1. a. $2x + 14$ **b.** $2x + 7$ **c.** 18   **3. a.** $2x + 5$ **b.** $2x + 9$ **c.** 9   **5. a.** $20x^2 - 11$ **b.** $80x^2 - 120x + 43$ **c.** 69
7. a. $x^4 - 4x^2 + 6$ **b.** $x^4 + 4x^2 + 2$ **c.** 6   **9. a.** $\sqrt{x - 1}$ **b.** $\sqrt{x} - 1$ **c.** 1   **11. a.** x **b.** x **c.** 2   **13. a.** x **b.** x **c.** 2

15. $f(g(x)) = x; g(f(x)) = x$; inverses   **17.** $f(g(x)) = x; g(f(x)) = x$; inverses   **19.** $f(g(x)) = \dfrac{5x - 56}{9}; g(f(x)) = \dfrac{5x - 4}{9}$; not inverses

21. $f(g(x)) = x; g(f(x)) = x$; inverses   **23.** $f(g(x)) = x; g(f(x)) = x$; inverses   **25. a.** $f^{-1}(x) = x - 3$ **b.** $f(f^{-1}(x)) = (x - 3) + 3 = x$ and

$f^{-1}(f(x)) = (x + 3) - 3 = x$ **27. a.** $f^{-1}(x) = \dfrac{x}{2}$ **b.** $f(f^{-1}(x)) = 2\left(\dfrac{x}{2}\right) = x$ and $f^{-1}(f(x)) = \dfrac{2x}{2} = x$ **29. a.** $f^{-1}(x) = \dfrac{x - 3}{2}$

b. $f(f^{-1}(x)) = 2\left(\dfrac{x - 3}{2}\right) + 3 = x$ and $f^{-1}(f(x)) = \dfrac{(2x + 3) - 3}{2} = x$ **31. a.** $f^{-1}(x) = \sqrt[3]{x - 2}$ **b.** $f(f^{-1}(x)) = (\sqrt[3]{x - 2})^3 + 2 = x$ and

$f^{-1}(f(x)) = \sqrt[3]{(x^3 + 2) - 2} = x$ **33. a.** $f^{-1}(x) = \sqrt[3]{x} - 2$ **b.** $f(f^{-1}(x)) = ((\sqrt[3]{x} - 2) + 2)^3 = x$ and $f^{-1}(f(x)) = \sqrt[3]{(x + 2)^3} - 2 = x$

35. a. $f^{-1}(x) = \dfrac{1}{x}$ **b.** $f(f^{-1}(x)) = \dfrac{1}{\frac{1}{x}} = x$ and $f^{-1}(f(x)) = \dfrac{1}{\frac{1}{x}} = x$ **37. a.** $f^{-1}(x) = x^2, x \geq 0$ **b.** $f(f^{-1}(x)) = \sqrt{x^2} = x$ and

$f^{-1}(f(x)) = (\sqrt{x})^2 = x$ **39. a.** $f^{-1}(x) = \sqrt{x - 1}$ **b.** $f(f^{-1}(x)) = (\sqrt{x - 1})^2 + 1 = x$ and $f^{-1}(f(x)) = \sqrt{(x^2 + 1) - 1} = x$

41. a. $f^{-1}(x) = \dfrac{3x + 1}{x - 2}$ **b.** $f(f^{-1}(x)) = \dfrac{2\left(\frac{3x + 1}{x - 2}\right) + 1}{\left(\frac{3x + 1}{x - 2}\right) - 3} = x$ and $f^{-1}(f(x)) = \dfrac{3\left(\frac{2x + 1}{x - 3}\right) + 1}{\left(\frac{2x + 1}{x - 3}\right) - 2} = x$ **43. a.** $f^{-1}(x) = (x - 3)^3 + 4$

b. $f(f^{-1}(x)) = \sqrt[3]{((x - 3)^3 + 4) - 4} + 3 = x$ and $f^{-1}(f(x)) = ((\sqrt[3]{x - 4} + 3) - 3)^3 + 4 = x$ **45.** no inverse   **47.** no inverse
49. inverse function   **51.**

53.

55. 5   **57.** 1   **59.** 2   **61.** 1
63. −6   **65.** −7   **67.** 3   **69.** 11

71. a. f represents the price after a \$400 discount, and g represents the price after a 25% discount (75% of the regular price).   **b.** $0.75x - 400; f \circ g$
represents an additional \$400 discount on a price that has already been reduced by 25%.   **c.** $0.75(x - 400) = 0.75x - 300; g \circ f$ represents
an additional 25% discount on a price that has already been reduced \$400.   **d.** $f \circ g; 0.75x - 400 < 0.75x - 300$, so $f \circ g$ represents the lower
price after the two discounts.   **e.** $f^{-1}(x) = x + 400; f^{-1}$ represents the regular price, since the value of x here is the price after a \$400 discount.
73. a. f: {(U.S., 1%), (U.K., 8%), (Italy, 5%), (France, 5%), (Holland, 30%)}   **b.** f^{-1}: {(1%, U.S.), (8%, U.K.), (5%, Italy), (5%, France),
(30%, Holland)}; No; The input 5% is associated with two outputs, Italy and France.   **75. a.** No horizontal line intersects the graph of f in more
than one point.   **b.** $f^{-1}(0.25)$, or approximately 15, represents the number of people who would have to be in the room so that the probability
of two sharing a birthday would be 0.25; $f^{-1}(0.5)$, or approximately 23, represents the number of people so that the probability would be 0.5;
$f^{-1}(0.7)$, or approximately 30, represents the number of people so that the probability would be 0.7.   **77.** $f(g(x)) = \dfrac{9}{5}\left[\dfrac{5}{9}(x - 32)\right] + 32 = x$ and

$g(f(x)) = \dfrac{5}{9}\left[\left(\dfrac{9}{5}x + 32\right) - 32\right] = x$

85.

; inverse function

87.

; no inverse function

89.

; inverse function

91.

; inverse function

93. ; inverses   **95.** does not make sense   **97.** makes sense   **99.** false   **101.** true

103. Answers will vary; Examples are $f(x) = \sqrt{x+5}$ and $g(x) = 3x^2$.

105. $f(f(x)) = \dfrac{3\left(\dfrac{3x-2}{5x-3}\right) - 2}{5\left(\dfrac{3x-2}{5x-3}\right) - 3} = \dfrac{3(3x-2) - 2(5x-3)}{5(3x-2) - 3(5x-3)} = \dfrac{9x - 6 - 10x + 6}{15x - 10 - 15x + 9} = \dfrac{-x}{-1} = x$

107. 5×10^8 **108.**   **109.** $\{5\}$

110. There is no method for solving $x = 2^y$ for y.   **111.** $\dfrac{1}{2}$ **112.** $(-\infty, 3) \cup (3, \infty)$

Section 9.3 Check Point Exercises

1. a. $7^3 = x$ **b.** $b^2 = 25$ **c.** $4^y = 26$ **2. a.** $5 = \log_2 x$ **b.** $3 = \log_b 27$ **c.** $y = \log_e 33$ **3. a.** 2   **b.** 1   **c.** $\dfrac{1}{2}$ **4. a.** 1   **b.** 0

5. a. 8   **b.** 17   **6.**   **7.** $(5, \infty)$ **8.** approximately 80%   **9.** 4

10. a. $(-\infty, 4)$ **b.** $(-\infty, 0) \cup (0, \infty)$ **11.** 34°; extremely well

Concept and Vocabulary Check

1. $b^y = x$ **2.** logarithmic; b **3.** 1   **4.** 0   **5.** x **6.** x **7.** $(0, \infty); (-\infty, \infty)$ **8.** $y; x = 0$; vertical
9. $5 - x > 0$ **10.** common; $\log x$ **11.** natural; $\ln x$

Exercise Set 9.3

1. $2^4 = 16$ **3.** $3^2 = x$ **5.** $b^5 = 32$ **7.** $6^y = 216$ **9.** $\log_2 8 = 3$ **11.** $\log_2 \dfrac{1}{16} = -4$ **13.** $\log_8 2 = \dfrac{1}{3}$ **15.** $\log_{13} x = 2$

17. $\log_b 1000 = 3$ **19.** $\log_7 200 = y$ **21.** 2   **23.** 6   **25.** -1   **27.** -3   **29.** $\dfrac{1}{2}$ **31.** $-\dfrac{1}{2}$ **33.** $\dfrac{1}{2}$ **35.** 1   **37.** 0   **39.** 7

41. 19   **43.**   **45.**   **47.** $(-4, \infty)$ **49.** $(-\infty, 2)$ **51.** $(-\infty, 2) \cup (2, \infty)$ **53.** 2   **55.** 7
57. 33   **59.** 0   **61.** 6   **63.** -6   **65.** 125   **67.** $9x$ **69.** $5x^2$
71. $\sqrt{x}$ **73.** $3^2 = x - 1; \{10\}$ **75.** $4^{-3} = x; \left\{\dfrac{1}{64}\right\}$ **77.** 0   **79.** 2
81. d   **83.** c   **85.** b   **87.** approximately 95.4%
89. a. 34%; It's the same.   **b.** 37%

91. approximately 188 decibels; yes

93. a. 88

b. 71.5; 63.9; 58.8; 55.0; 52.0; 49.5

c.

The students remembered less of the material over time.

103.

The graph of g is the graph of f shifted up 3 units.

105.

The graph of g is the graph of f shifted 2 units to the right and 1 unit up.

107. a.   **b.**   **c.**

d. In each case, the graphs of f and g are the same.; $\log_b(MN) = \log_b M + \log_b N$
e. the sum of the logarithms of the factors

109. makes sense   **111.** makes sense   **113.** false   **115.** false   **117.** $\dfrac{4}{.5}$ **119.** $\log_3 40$ **120.** $\{(-2, 3)\}$ **121.** $2(3x - y)(x - y)$

122. $(-\infty, -7] \cup [-2, \infty)$ **123. a.** 5   **b.** 5   **c.** $\log_2(8 \cdot 4) = \log_2 8 + \log_2 4$ **124. a.** 4   **b.** 4   **c.** $\log_2\left(\dfrac{32}{2}\right) = \log_2 32 - \log_2 2$

125. a. 4   **b.** 4   **c.** $\log_3 9^2 = 2\log_3 9$

Section 9.4 Check Point Exercises

1. a. $\log_6 7 + \log_6 11$ **b.** $2 + \log x$ **2. a.** $\log_8 23 - \log_8 x$ **b.** $5 - \ln 11$ **3. a.** $9 \log_6 8$ **b.** $\dfrac{1}{3}\ln x$ **c.** $2 \log(x + 4)$

4. a. $4 \log_b x + \dfrac{1}{3}\log_b y$ **b.** $\dfrac{1}{2}\log_5 x - 2 - 3\log_5 y$ **5. a.** $\log 100 = 2$ **b.** $\log\left(\dfrac{7x + 6}{x}\right)$ **6. a.** $\ln(x^2\sqrt[3]{x} + 5)$ **b.** $\log\left[\dfrac{(x - 3)^2}{x}\right]$

c. $\log_b\left(\dfrac{\sqrt[4]{x}}{25y^{10}}\right)$ **7.** $\dfrac{\log 2506}{\log 7} \approx 4.02$ **8.** $\dfrac{\ln 2506}{\ln 7} \approx 4.02$

Concept and Vocabulary Check

1. $\log_b M + \log_b N$; sum   **2.** $\log_b M - \log_b N$; difference   **3.** $p \log_b M$; product   **4.** $\dfrac{\log_a M}{\log_a b}$

Exercise Set 9.4

1. $\log_5 7 + \log_5 3$ **3.** $1 + \log_7 x$ **5.** $3 + \log x$ **7.** $1 - \log_7 x$ **9.** $\log x - 2$ **11.** $3 - \log_4 y$ **13.** $2 - \ln 5$ **15.** $3 \log_b x$

17. $-6 \log N$ **19.** $\dfrac{1}{5}\ln x$ **21.** $2 \log_b x + \log_b y$ **23.** $\dfrac{1}{2}\log_4 x - 3$ **25.** $2 - \dfrac{1}{2}\log_6(x + 1)$ **27.** $2 \log_b x + \log_b y - 2\log_b z$

29. $1 + \dfrac{1}{2}\log x$ **31.** $\dfrac{1}{3}\log x - \dfrac{1}{3}\log y$ **33.** $\dfrac{1}{2}\log_b x + 3\log_b y - 3\log_b z$ **35.** $\dfrac{2}{3}\log_5 x + \dfrac{1}{3}\log_5 y - \dfrac{2}{3}$ **37.** $\log 10 = 1$ **39.** $\ln(7x)$

41. $\log_2 32 = 5$ **43.** $\log\left(\dfrac{2x + 5}{x}\right)$ **45.** $\log(xy^3)$ **47.** $\ln(y\sqrt{x})$ **49.** $\log_b(x^2 y^3)$ **51.** $\ln\left(\dfrac{x^5}{y^2}\right)$ **53.** $\ln\left(\dfrac{x^3}{\sqrt[3]{y}}\right)$ **55.** $\ln\left[\dfrac{(x + 6)^4}{x^3}\right]$

57. $\ln\left(\dfrac{x^3 y^5}{z^6}\right)$ **59.** $\log_5\left[\dfrac{\sqrt{xy}}{(x + 1)^2}\right]$ **61.** 1.5937   **63.** 1.6944   **65.** -1.2304 **67.** 3.6193   **69.** $C - A$ **71.** $3A$ **73.** $\dfrac{1}{2}A - \dfrac{3}{2}C$

75. false; $\ln e = 1$ **77.** false; $\log_4(2x)^3 = 3\log_4(2x)$ **79.** true   **81.** true   **83.** false; $\log(x + 3) - \log(2x) = \log\left(\dfrac{x + 3}{2x}\right)$ **85.** true

87. true   **89. a.** 2   **b.** $\log_3\left(\dfrac{xy^4}{9}\right)$ **91. a.** $\dfrac{1}{2}$ **b.** $\log_{25}\left[\dfrac{x(x^2 - 1)}{5(x + 1)}\right] = \log_{25}\left[\dfrac{x(x - 1)}{5}\right]$ **93. a.** $D = 10\log\left(\dfrac{I}{I_0}\right)$ **b.** 20 decibels

103. a. & b. $y = \log_3 x$ is shifted up 2 units to obtain $y = 2 + \log_3 x$,

$y = \log_3 x$ is shifted to the left 2 units to obtain $y = \log_3(x + 2)$, and

$y = \log_3 x$ is reflected across the x-axis to obtain $y = -\log_3 x$.

37.

exponential function

39.

logarithmic function

41.

logarithmic function

43. $y = 100e^{(\ln 4.6)x}; y = 100e^{1.526x}$ **45.** $y = 2.5e^{(\ln 0.7)x}; y = 2.5e^{-0.357x}$

55. a. $y = 201.2(1.011)^x; r \approx 0.999$; Since r is close to 1, the model fits the data well.   **b.** $y = 201.2e^{\ln(1.011)x}; y = 201.2e^{0.0109x}$; by approximately 1%

57. $y = 2.657x + 197.923; r \approx 0.997$; Since r is close to 1, the model fits the data well.

59. $y = 201.2(1.011)^x; y = 2.657x + 197.923$; using exponential, by 2016; using linear, by 2021; Answers will vary.

61. Models will vary. Examples are given. Predictions will vary. For Exercise 37: $y = 1.402(1.078)^x$; For Exercise 39: $y = 120 + 4.343 \ln x$; For Exercise 41: $y = -175.582 + 56.808 \ln x$ **63.** does not make sense   **65.** makes sense   **67.** true   **69.** true

71. $\dfrac{x + 5}{x}$ **72.** $\{-27, 1\}$ **73.** $17\sqrt{2}$ **74.** $3\sqrt{5}$ **75.** $\left(4, \dfrac{1}{2}\right)$ **76.**

Review Exercises

1. ; d

x	f(x)
−2	$\frac{1}{16}$
−1	$\frac{1}{4}$
0	1
1	4
2	16

2. ; a

x	f(x)
−2	16
−1	4
0	1
1	$\frac{1}{4}$
2	$\frac{1}{16}$

3. ; b

x	f(x)
−2	−16
−1	−4
0	−1
1	$-\frac{1}{4}$
2	$-\frac{1}{16}$

4. ; c

x	f(x)
−2	−13
−1	−1
0	2
1	$\frac{11}{4}$
2	$\frac{47}{16}$

5.

The graph of g is the graph of f shifted to the right 1 unit.

6.

The graph of g is the reflection of f across the y-axis.

7.

The graph of g is the graph of f shifted down 1 unit.

8.

The graph of g is the reflection of f across the x-axis.

9. 5.5% compounded semiannually   **10.** 7% compounded monthly   **11. a.** 200°F   **b.** about 119°F   **c.** 70°F; The temperature of the room is 70°F.   **12. a.** $16x^2 - 8x + 4$ **b.** $4x^2 + 11$ **c.** 124   **13. a.** $\sqrt{x + 1}$ **b.** $\sqrt{x} + 1$ **c.** 2

14. $f(g(x)) = x - \dfrac{7}{10}; g(f(x)) = x - \dfrac{7}{6}$; not inverses   **15.** $f(g(x)) = x; g(f(x)) = x$; inverses

16. a. $f^{-1}(x) = \dfrac{x + 3}{4}$ **b.** $f(f^{-1}(x)) = 4\left(\dfrac{x + 3}{4}\right) - 3 = x$ and $f^{-1}(f(x)) = \dfrac{(4x - 3) + 3}{4} = x$

17. a. $f^{-1}(x) = x^2 - 2, x \geq 0$ **b.** $f(f^{-1}(x)) = \sqrt{(x^2 - 2) + 2} = x$ and $f^{-1}(f(x)) = (\sqrt{x + 2})^2 - 2 = x$

18. a. $f^{-1}(x) = \dfrac{\sqrt[3]{x - 1}}{2}$ **b.** $f(f^{-1}(x)) = 8\left(\dfrac{\sqrt[3]{x - 1}}{2}\right)^3 + 1 = x$ and $f^{-1}(f(x)) = \dfrac{\sqrt[3]{(8x^3 + 1) - 1}}{2} = x$

19. inverse function   **20.** no inverse function   **21.** inverse function   **22.** no inverse function

23.

The graph passes through $(0,0)$, $(4, 2)$, and $(-1, -3)$.

24. $49^{1/2} = 7$ **25.** $4^3 = x$ **26.** $3^y = 81$ **27.** $\log_6 216 = 3$ **28.** $\log_b 625 = 4$

29. $\log_{13} 874 = y$ **30.** 3   **31.** −2   **32.** −9 is not in the domain of $y = \log_3 x$.

33. $\dfrac{1}{2}$ **34.** 1   **35.** 8   **36.** 5   **37.** $-\dfrac{1}{2}$ **38.** −2   **39.** −3   **40.** 0

41. domain of f: $(-\infty, \infty)$; range of f: $(0, \infty)$; domain of g: $(0, \infty)$; range of g: $(-\infty, \infty)$

42. domain of f: $(-\infty, \infty)$; range of f: $(0, \infty)$; domain of g: $(0, \infty)$; range of g: $(-\infty, \infty)$

43. $(-5, \infty)$ **44.** $(-\infty, 3)$ **45.** $(-\infty, 1) \cup (1, \infty)$ **46.** $6x$ **47.** $\sqrt{x}$ **48.** $4x^2$ **49.** 3

50. a. 76

b. 67.4 after 2 months; 63.4 after 4 months; 60.8 after 6 months; 58.8 after 8 months; 55.9 after one year

c.

As time increases, the amount of material retained by the students decreases.

51. about 9 weeks **52.** $2 + 3\log_6 x$ **53.** $\frac{1}{2}\log_4 x - 3$ **54.** $\log_2 x + 2\log_2 y - 6$

55. $\frac{1}{3}\ln x - \frac{1}{3}$ **56.** $\log_b 21$ **57.** $\log\left(\frac{3}{x^3}\right)$ **58.** $\ln(x^3 y^4)$ **59.** $\ln\left(\frac{\sqrt{x}}{y}\right)$

60. 6.2448 **61.** -0.1063 **62.** true **63.** false; $\log(x + 9) - \log(x + 1) = \log\left(\frac{x + 9}{x + 1}\right)$

64. false; $\log_2 x^4 = 4\log_2 x$ **65.** true **66.** $\{2\}$ **67.** $\left\{\frac{2}{3}\right\}$ **68.** $\left\{-\frac{3}{2}\right\}$

69. $\left\{\frac{\ln 12{,}143}{\ln 8} \approx 4.52\right\}$ **70.** $\left\{\frac{\ln 141}{5} \approx 0.99\right\}$ **71.** $\left\{\frac{\ln 3}{0.045} \approx 24.41\right\}$ **72.** $\left\{\frac{1}{125}\right\}$

73. $\{100\}$ **74.** $\{23\}$ **75.** $\left\{\frac{1}{e}\right\}$ **76.** $\left\{\frac{e^3}{2}\right\}$ **77.** $\{5\}$ **78.** no solution or $\varnothing$ **79.** $\left\{\frac{8}{3}\right\}$ **80.** $\{2\}$ **81.** $\{4\}$ **82.** 5.5 mi

83. 5 years after 2009, or 2014; 385 million **84. a.** 92%; It's the same. **b.** 66 years after 1961, or 2027 **85.** 7.3 years **86.** 14.6 years

87. about 22% **88. a.** 0.036 **b.** 60.6 million **c.** 2019 **89.** 325 days

90. Mentally Ill Adults in ; linear function
United States

91. Moderate Alcohol Users in ; logarithmic function
the United States

92. U.S. Electric Car Sales ; exponential function

93. $y = 73e^{(\ln 2.6)x}$; $y = 73e^{0.956x}$ **94.** $y = 6.5e^{(\ln 0.43)x}$; $y = 6.5e^{-0.844x}$ **95.** Answers will vary.

Exercise Set 10.2

1.
$$\frac{x^2}{16} + \frac{y^2}{4} = 1$$

3.
$$\frac{x^2}{9} + \frac{y^2}{36} = 1$$

5.
$$\frac{x^2}{25} + \frac{y^2}{64} = 1$$

7.
$$\frac{x^2}{49} + \frac{y^2}{81} = 1$$

9.
$$25x^2 + 4y^2 = 100$$

11.
$$4x^2 + 16y^2 = 64$$

13.
$$25x^2 + 9y^2 = 225$$

15.
$$x^2 + 2y^2 = 8$$

17. $\dfrac{x^2}{4} + \dfrac{y^2}{1} = 1$ or $\dfrac{x^2}{4} + y^2 = 1$

19. $\dfrac{x^2}{1} + \dfrac{y^2}{4} = 1$ or $x^2 + \dfrac{y^2}{4} = 1$

21.
$$\frac{(x-2)^2}{9} + \frac{(y-1)^2}{4} = 1$$

23.
$$(x+3)^2 + 4(y-2)^2 = 16$$

25.
$$\frac{(x-4)^2}{9} + \frac{(y+2)^2}{25} = 1$$

27.
$$\frac{x^2}{25} + \frac{(y-2)^2}{36} = 1$$

29.
$$\frac{(x+3)^2}{9} + (y-2)^2 = 1$$

31.
$$9(x-1)^2 + 4(y+3)^2 = 36$$

33. $\dfrac{(x+1)^2}{4} + \dfrac{(y-1)^2}{1} = 1$ **35.** $\{(0,-1),(0,1)\}$ **37.** $\{(0,3)\}$ **39.** $\{(0,-2),(1,0)\}$

41.
$$y = -\sqrt{16-4x^2}$$

43. Yes, the height of the archway 4 feet from the center is approximately 9.64 feet.

45. a. $\dfrac{x^2}{48^2} + \dfrac{y^2}{23^2} = 1$ or $\dfrac{x^2}{2304} + \dfrac{y^2}{529} = 1$ **b.** approximately 42 feet

55. does not make sense   **57.** makes sense   **59.** false   **61.** true

63.
$$\frac{(x-2)^2}{25} + \frac{(y+1)^2}{9} = 1$$

65. a. 984 mi   **b.** 1016 mi   **67.** $(x+2)(x+2)(x-2)$ or $(x+2)^2(x-2)$

68. $2xy^2\sqrt[3]{5xy}$ **69.** $\{-1\}$ **70.** $\dfrac{x^2}{9} - \dfrac{y^2}{4} = 1$; Terms are separated by subtraction rather than by addition.   **71. a.** -4 and 4 **b.** The equation $y^2 = -9$ has no real solutions.   **72. a.** -3 and 3

b. The equation $x^2 = -16$ has no real solutions.

Section 10.3 Check Point Exercises

1. a. $(-5,0)$ and $(5,0)$ **b.** $(0,-5)$ and $(0,5)$ **2.**
$$\frac{x^2}{36} - \frac{y^2}{9} = 1$$
   3.
$$y^2 - 4x^2 = 4$$

Concept and Vocabulary Check

1. hyperbola; vertices; transverse   **2.** $(-5,0)$; $(5,0)$ **3.** $(0,-5)$; $(0,5)$ **4.** asymptotes; center   **5.** dividing; 36   **6.** ellipse
7. hyperbola   **8.** neither   **9.** hyperbola   **10.** ellipse

Exercise Set 10.3

1. $(-2, 0)$ and $(2, 0)$; b   **3.** $(0, -2)$ and $(0, 2)$; a

5.
$$\frac{x^2}{9} - \frac{y^2}{25} = 1$$

7.
$$\frac{x^2}{100} - \frac{y^2}{64} = 1$$

9.
$$\frac{y^2}{16} - \frac{x^2}{36} = 1$$

11.
$$\frac{y^2}{36} - \frac{x^2}{25} = 1$$

13.
$$9x^2 - 4y^2 = 36$$

15.
$$9y^2 - 25x^2 = 225$$

17.
$$4x^2 = 4 + y^2$$

19. $\dfrac{x^2}{9} - \dfrac{y^2}{25} = 1$

21. $\dfrac{y^2}{4} - \dfrac{x^2}{9} = 1$

23.
$$\frac{x^2}{9} - \frac{y^2}{16} = 1$$
domain: $(-\infty, -3] \cup [3, \infty)$; range: $(-\infty, \infty)$

25.
$$\frac{x^2}{9} + \frac{y^2}{16} = 1$$
domain: $[-3, 3]$; range: $[-4, 4]$

27.
$$\frac{y^2}{16} - \frac{x^2}{9} = 1$$
domain: $(-\infty, \infty)$; range: $(-\infty, -4] \cup [4, \infty)$

29. $\{(-2, 0), (2, 0)\}$ **31.** $\{(0, -3), (0, 3)\}$ **33.** 40 yd

41.   no; two lines; $y = \dfrac{b}{a}x$ and $y = -\dfrac{b}{a}x$ **43.** does not make sense   **45.** makes sense   **47.** false   **49.** true

51.
$$\frac{(x-2)^2}{16} - \frac{(y-3)^2}{9} = 1$$

53.
$$(x-3)^2 - 4(y+3)^2 = 4$$

55. $\dfrac{x^2}{36} - \dfrac{y^2}{576} = 1$

57.
$(-2, 9)$, $(0, 5)$, $(-5, 0)$, $(1, 0)$
$$y = -x^2 - 4x + 5$$

58. $\left(-\infty, -\dfrac{1}{3}\right] \cup [4, \infty)$ **59.** $\{21\}$ **60.**
$y = x^2 + 4x - 5$
$(1, 0)$, $(-5, 0)$, $(0, -5)$, $(-2, -9)$, $x = -2$

61.
$\left(1 - \dfrac{\sqrt{6}}{3}, 0\right)$, $(1, 2)$, $\left(1 + \dfrac{\sqrt{6}}{3}, 0\right)$, $(0, -1)$, $y = -3(x-1)^2 + 2$, $x = 1$

62. 0.2 and 1.8

Mid-Chapter Check Point Exercises

1.
$$x^2 + y^2 = 9$$

2.
$$(x-3)^2 + (y+2)^2 = 25$$

3.
$$x^2 + (y-1)^2 = 4$$

4.
$$x^2 + y^2 - 4x - 2y - 4 = 0$$

5.
$$\frac{x^2}{25} + \frac{y^2}{4} = 1$$

6.
$$9x^2 + 4y^2 = 36$$

73. $x^6 + 12x^5 + 60x^4 + 160x^3 + 240x^2 + 192x + 64$ **75.** makes sense   **77.** does not make sense   **79.** true

81. false   **83.** $10x^4y^6$ **84.** $a^2 + 4a + 6$ **85.** $f(g(x)) = 4x^2 - 2x - 6; g(f(x)) = 2x^2 + 10x - 3$ **86.** $\dfrac{2x^2 - 9x - 1}{2(x-4)(x+3)}$

Review Exercises

1. $3, 10, 17, 24$ **2.** $-\dfrac{3}{2}, \dfrac{4}{3}, -\dfrac{5}{4}, \dfrac{6}{5}$ **3.** $1, 1, \dfrac{1}{2}, \dfrac{1}{6}$ **4.** $\dfrac{1}{2}, -\dfrac{1}{4}, \dfrac{1}{8}, -\dfrac{1}{16}$ **5.** 95 **6.** -20 **7.** Answers will vary; an example is $\sum\limits_{i=1}^{15} \dfrac{i}{i+2}$.

8. Answers will vary; examples are $\sum\limits_{i=4}^{13} i^3$ or $\sum\limits_{i=1}^{10} (i+3)^3$.   **9.** $7, 11, 15, 19, 23, 27$ **10.** $-4, -9, -14, -19, -24, -29$ **11.** $\dfrac{3}{2}, 1, \dfrac{1}{2}, 0, -\dfrac{1}{2}, -1$

12. 20 **13.** -30 **14.** -38 **15.** $a_n = 4n - 11; a_{20} = 69$ **16.** $a_n = 220 - 20n; a_{20} = -180$ **17.** $a_n = -\dfrac{23}{2} - \dfrac{1}{2}n; a_{20} = -\dfrac{43}{2}$

18. $a_n = 22 - 7n; a_{20} = -118$ **19.** 1727 **20.** 225 **21.** $15,150$ **22.** 440 **23.** -500 **24.** -2325

25. a. $a_n = 1.63n + 38.73$ **b.** $\$81.11$ **26.** $\$418,500$ **27.** 1470 seats   **28.** $3, 6, 12, 24, 48$ **29.** $\dfrac{1}{2}, \dfrac{1}{4}, \dfrac{1}{8}, \dfrac{1}{16}, \dfrac{1}{32}$ **30.** $16, -4, 1, -\dfrac{1}{4}, \dfrac{1}{16}$

31. $-5, 5, -5, 5, -5$ **32.** $a_7 = 1458$ **33.** $a_6 = \dfrac{1}{2}$ **34.** $a_5 = -48$ **35.** $a_n = 1(2)^{n-1}$ or $a_n = 2^{n-1}; a_8 = 128$

36. $a_n = 100\left(\dfrac{1}{10}\right)^{n-1}; a_8 = \dfrac{1}{100,000} = 0.00001$ **37.** $a_n = 12\left(-\dfrac{1}{3}\right)^{n-1}; a_8 = -\dfrac{4}{729}$ **38.** $17,936,135$ **39.** $\dfrac{127}{8}$ or 15.875 **40.** $19,530$

41. -258 **42.** $\dfrac{341}{128}$ **43.** $\dfrac{27}{2}$ **44.** $\dfrac{4}{3}$ **45.** $-\dfrac{18}{5}$ **46.** 20 **47.** $\dfrac{2}{3}$ **48.** $\dfrac{47}{99}$ **49. a.** approximately 1.02 for each division

b. $a_n = 15.98(1.02)^{n-1}$ **c.** 28.95 million   **50.** approximately $\$42,823$; approximately $\$223,210$ **51. a.** $\$19,129$ **b.** $\$8729$ **52. a.** $\$91,361$

b. $\$55,361$ **53.** $\$9\dfrac{1}{3}$ million   **54.** 165 **55.** 4005 **56.** $8x^3 + 12x^2 + 6x + 1$ **57.** $x^8 - 4x^6 + 6x^4 - 4x^2 + 1$

58. $x^5 + 10x^4y + 40x^3y^2 + 80x^2y^3 + 80xy^4 + 32y^5$ **59.** $x^6 - 12x^5 + 60x^4 - 160x^3 + 240x^2 - 192x + 64$ **60.** $x^{16} + 24x^{14} + 252x^{12}$

61. $x^9 - 27x^8 + 324x^7$ **62.** $80x^2$ **63.** $4860x^2$

Chapter Test

1. $1, -\dfrac{1}{4}, \dfrac{1}{9}, -\dfrac{1}{16}, \dfrac{1}{25}$ **2.** 105 **3.** Answers will vary; examples are $\sum\limits_{i=2}^{21} \dfrac{i}{i+1}$ or $\sum\limits_{i=1}^{20} \dfrac{i+1}{i+2}$.   **4.** $a_n = 5n - 1; a_{12} = 59$

5. $a_n = 16\left(\dfrac{1}{4}\right)^{n-1}; a_{12} = \dfrac{1}{262,144}$ **6.** -385 **7.** 550 **8.** -2387 **9.** $-21,846$ **10.** 8 **11.** $\dfrac{73}{99}$ **12.** approximately $\$276,427$

13. 36 **14.** $x^{10} - 5x^8 + 10x^6 - 10x^4 + 5x^2 - 1$ **15.** $x^8 + 8x^7y^2 + 28x^6y^4$

Cumulative Review Exercises

1. $\{22\}$ **2.** $\{5 \pm 7i\}$ **3.** $(-\infty, -3) \cup (2, \infty)$ **4.** $\{-2\}$ **5.** no solution or $\varnothing$ **6.** $\left(-\infty, \dfrac{2}{3}\right)$ **7.** $\{(4, 0, -5)\}$ **8.** $\{9\}$

9. $\{(2, 1), (-2, 1), (2, -1), (-2, -1)\}$ **10.** $\{(-14, -20), (2, -4)\}$

11.   **12.**   **13.**   **14.**   **15.**

$y = (x+2)^2 - 4$ $y < -3x + 5$ $y = 3^{x-2}$ $\dfrac{x^2}{16} + \dfrac{y^2}{4} = 1$ $x^2 - y^2 = 9$

16. $\dfrac{2x^2 + 5x - 2}{(x-5)(x+2)}$

17. $-\dfrac{x+1}{x-1}$

18. $2\sqrt{5} + 2\sqrt{2}$

19. $12\sqrt{5}$ **20.** $\dfrac{5\sqrt[3]{4xy^2}}{2xy}$ **21.** $(x+y)(5a - 4b)$ **22.** $\log\left(\dfrac{x^5}{\sqrt{y}}\right)$ **23.** $p = \dfrac{qf}{q - f}$ **24.** $3\sqrt{10} \approx 9.49$ units   **25.** 208 **26.** 1800

27. $\dfrac{1}{3}$ **28.** $16x^4 - 32x^3y^3 + 24x^2y^6 - 8xy^9 + y^{12}$ **29.** $(-\infty, -5) \cup (-5, 3) \cup (3, \infty)$ **30.** $[3, \infty)$ **31.** $(-\infty, 1)$ **32.** 8 ft by 3 ft

33. $\$18,500$ **34.** $k = \dfrac{1}{2 \ln 4} \approx 0.3607$; about 0.298 or $\dfrac{298}{1000}$

Applications Index

Subject Index

Credits

Photo Credits

Front Matter

Discpicture/Shutterstock, **p. iii**(t) Jose Luis Pelaez Inc/Blend Images/Alamy, **p. iii**(b) Shizuo Kambayashi/AP Images, **p. iv** Mega Pixel/Shutterstock, **p. v** 103tnn/Fotolia, **p. vi** Bob Blitzer, **p. xvii**

Chapter 1

Discpicture/Shutterstock, **p. 1**(t) Jose Luis Pelaez Inc/Blend Images/Alamy, **p. 1**(b) ASK-Fotografie/Fotolia, **p. 2**(r) Raisa Kanareva/Fotolia, **p. 2**(l) Vectorskills/Fotolia, **p. 15** Ron Chapple/Dreamstime LLC, **p. 30** Iofoto/Shutterstock, **p. 40**(t) Stephen Coburn/Shutterstock, **p. 40**(b) Pictorial Press Ltd/Alamy, **p. 54**(l) Pictorial Press Ltd/Alamy, **p. 54**(r) Blitzer, Robert F., **p. 56** Scott Camazine/Alamy, **p. 65** Susumu Nishinaga/Science Source, **p. 70**(l) Dhoxax/Shutterstock, **p. 70**(r) Slanted Roof Studio/Alamy, **p. 82** Alexey Boldin/Shutterstock, **p. 98**

Chapter 2

Jan Martin Will/Shutterstock, **p. 103** Nancy Kaszerman/Zuma Press, Inc./Alamy, **p. 104**(a) Peter Brooker/Rex Features/Presselect/Alamy, **p. 104**(b) Creative Collection Tolbert Photo/Alamy, **p. 104**(c) Dee Cercone/Everett Collection Inc/Alamy, **p. 104**(d) Hyperstar/Alamy, **p. 104**(e) NIBSC/Science Source, **p. 114** Günay Mutlu/E+/Getty Images, **p. 125** LWA/Sharie Kennedy/Blend Images/Alamy, **p. 136** CP, Frank Gunn/AP Images, **p. 155**(l) Kippa/ANP/Newscom, **p. 155**(c) Chris Pizzello/AP Images, **p. 155**(r) French Government Tourist Office, **p. 167**

Chapter 3

Shizuo Kambayashi/AP Images, **p. 177** cypher0x/123RF, **p. 178** Hurst Photo/Shutterstock, **p. 194** Jules Selmes/Pearson Education, Inc., **p. 196**(l) FomaA/Fotolia, **p. 196**(r) Laura Rauch/AP Images, **p. 201** Universal Uclick, **p. 205** Silver-john/Fotolia, **p. 208** TANNEN MAURY/epa european pressphoto agency b.v./Alamy, **p. 221** SSPL/The Image Works, **p. 232**

Chapter 4

Jimmy Chin/National Geographic Magazines/Getty Images, **p. 253** Julián Maldonado/Fotolia, **p. 254** Gary Fabiano/Sipa/Newscom, **p. 266**(tl) UPI UPI Photo Service/Newscom, **p. 266**(tc) Pamela Price/Zumapress/

Newscom, **p. 266**(tr) Corbis/SuperStock, **p. 275** Jim Goldstein/Danita Delimont Photography/Newscom, **p. 287**(l) limitedqstock/Alamy, **p. 287**(r) AP Images, **p. 298** US Navy Photo/Alamy, **p. 299**

Chapter 5

Fuse/Getty Images, **p. 313** Zuma Press, Inc./Alamy, **p. 314** Fancy Collection/SuperStock, **p. 336** Monkey Business/Fotolia, **p. 341** Stock Connection/SuperStock, **p. 350** StudioSmart/Shutterstock, **p. 364** Rido/Fotolia, **p. 374** Warren Photographic Ltd., **p. 381**

Chapter 6

M. Spencer Green/AP Images, **p. 403** Claudia Otte/Shutterstock, **p. 404** Steve Skjold/Alamy, **p. 418** Mauritius/SuperStock, **p. 430** ginasanders/123RF, **p. 438** Nature's Images/Science Source, **p. 448** Songquan Deng/Shutterstock, **p. 455** Charlie Neuman/U-T San Diego/Zuma Press, Inc/Alamy, **p. 466** Holger Burmeister/Alamy, **p. 478** Gary718/Shutterstock, **p. 479**(tc) Anna Azimi/Shutterstock, **p. 479**(tr) Ty Allison/Photographer's Choice/Getty Images, **p. xxx** David Madison/Getty Images, **p. xxx** Jana Lumley/Fotolia, **p. 480** Exactostock-1527/SuperStock, **p. 482** Ullstein Bild/The Image Works, **p. 488**

Chapter 7

Ian Evans/Alamy, **p. 501** Kike Calvo/V&W/The Image Works, **p. 502** Ryan M. Bolton/Shutterstock, **p. 515** George Tooker "Mirror II," 1963. Egg tempera on gesso panel, 20 × 20 in. 1968.4. Gift of R.H. Donnelley Erdman (PA 1956). Addison Gallery of American Art, Phillips Academy, Andover, Massachusetts. © The Estate of George Tooker. Courtesy of DC Moore Gallery, New York. Photo courtesy of Art Resource. **p. 525** Paul Wootton/Science Source, **p. 533** Universal Uclick, **p. 542** John G. Ross/Science Source, **p. 551** Bikeriderlondon/Shutterstock, **p. 552** Justin Kase Conder/Icon SMI 745/Justin Kase Conder/Icon SMI/Newscom, **p. 560** Roz Chast/The New Yorker Collection/The Cartoon Bank, **p. 562** Richard F. Voss, **p. 569** Gordon Caulkins/Cartoon Stock, **p. 571**

Chapter 8

Photo courtesy of Professor Andrew Davidhazy/Rochester Institute of Technology (RIT), **p. 581** Alamy, **p. 582** Kazoka303030/Fotolia, **p. 596** Simon Price/Alamy, **p. 611** Plpchirawong/Fotolia, **p. 630** Copyright © 2011 by Warren Miller/The New Yorker Collection/The Cartoon Bank, **p. 638**

Chapter 9

Mega Pixel/Shutterstock, **p. 661** Javiindy/Fotolia, **p. 662** Seth Wenig/AP Images, **p. 668**(tl) Rolffimages/Dreamstime, **p. 668**(tr) Tim Davis/Corbis, **p. 668**(b) Archives du 7eme Art/Photos 12/Alamy, **p. 676**(r) Everett Collection, **p. 676**(l) John Trotter/Sacramento Bee/Newscom, **p. 691** Wieslaw Smetek/Science Source, **p. 704** Mitchel Gray/Superstock, **p. 716** Jamaway/Alamy, **p. 730** Franck Boston/Shutterstock, **p. 731** Courtesy of the French Ministry of Culture and Communication, **p. 734** Science Photo Library/SuperStock, **p. 736**

Chapter 10

Goddard/SDO/NASA, **p. 757** Millard H. Sharp/Science Source, **p. 758** Kevin Fleming/Corbis, **p. 768** Ronald Royer/Science Source, **p. 776** Pixtal Images/Photolibrary New York/Getty Images, **p. 780** NASA, **p. 790** NASA, **p. 795** Richard E. Prince "The Cone of Apollonius" (detail), fiberglass, steel, paint, graphite, 51 × 18 × 14 in. Collection: Vancouver Art Gallery, Vancouver, Canada. Photo courtesy of Equinox Gallery, Vancouver, Canada., **p. 796** Kolestamas/Fotolia, **p. 802**

Chapter 11

103tnn/Fotolia, **p. 821** Sergey Galushko/iStock/Getty Images, **p. 822** Xixinxing/Fotolia, **p. 832** Ariel Skelley/Blend Images/Alamy, **p. 842** Jokerpro/Shutterstock, **p. 845** Courtesy U.S. Bureau of Engraving and Printing., **p. 848**(a) Courtesy U.S. Bureau of Engraving and Printing., **p. 848**(b) Courtesy U.S. Bureau of Engraving and Printing., **p. 848**(c) Courtesy U.S. Bureau of Engraving and Printing., **p. 848**(d) Courtesy U.S. Bureau of Engraving and Printing., **p. 848**(e) Dr. Rudolph Schild/Science Source, **p. 859** Anna Omelchenko/Shutterstock, **p. 870**

Text Credits

Chapter 1

Screenshots from Texas Instruments. Courtesy of Texas Instruments, **p. 37**

Chapter 2

Screenshots from Texas Instruments. Courtesy of Texas Instruments, **p. 108**

Chapter 3

Quote by Henny Youngman, **p. 178**
Quote by Jeff Foxworthy, **p. 178**

Chapter 4

Quote by Cha Sa-Soon from Newsweek, **p. 261**

Chapter 7

Quote by Alan Bass, **p. 510**

Chapter 8

Quote from With Amusement for All: A History of American Popular Culture Since 1830 by LeRoy Ashby. Published by University Press of Kentucky, © 2006, **p. 630**